Project 233

IUGS UNESCO IGCP

Terranes In The Circum-Atlantic Paleozoic Orogens

R. D. DALLMEYER, Athens, USA
J. D. KEPPIE, Halifax, Canada
(Directors)

Volumes already published in the series:

R. D. DALLMEYER and E. MARTINEZ GARCIA (Eds.):
Pre-Mesozoic Geology of Iberia (1990)

R. D. DALLMEYER and P. P. LECORCHE (Eds.):
The West African Orogens and Circum-Atlantic Correlatives (1991)

J. D. KEPPIE (Ed.):
Pre-Mesozoic Geology in France and Related Areas (1994)

J. D. Keppie (Ed.)

Pre-Mesozoic Geology in France

and Related Areas

Associate Editors:
J. Chantraine, J. Rolet, D. S. Santallier
and A. Piqué

With 250 Figures

Springer-Verlag
Berlin Heidelberg New York
London Paris Tokyo
Hong Kong Barcelona
Budapest

J. D. Keppie
Department of Mines and Energy
P. O. Box 1087
Halifax, Nova Scotia B3J 2X1
Canada

ISBN-13:978-3-642-84917-6 e-ISBN-13:978-3-642-84915-2
DOI: 10.1007/978-3-642-84915-2

Library of Congress Cataloging-in-Publication Data. Pre-Mesozoic geology in France and related areas / J. D. Keppie (ed.). p. cm. Developed under IGCP Project 233–Terranes in the Circum-Atlantic Paleozoic Orogens. Includes bibliographical references and index. 1. Geology, Stratigraphic–Precambrian. 2. Geology, Stratigraphic–Paleozoic. 3. Geology–France. I. Keppie, J. Duncan. II. IGCP Project 233–Terranes in the Circum-Atlantic Paleozoic Orogens. QE653.P7437 1994 551.7'2'0944–dc20 93-44603 CIP

Softcover reprint of the hardcover 1st edition 1994

Production editor: Renate Münzenmayer
32/3130-5 4 3 2 1 0 – Printed on acid-free paper

Preface

This book represents the third in the series on the Precambrian and Palaeozoic geology of Circum-Atlantic regions developed under the IGCP (International Geological Correlation Programme) Project # 233 on Terranes in the Circum-Atlantic Palaeozoic Orogens. The book is an up-to-date and comprehensive account of the Pre-Mesozoic Geology of France (Armorican and Central Massifs) and related areas (Aquitaine Basin, Vosges, Crystalline Provence, Pyrenees, western Alps, Corsica and Sardinia) that systematically reviews the geophysics, stratigraphy, palaeontology, structure, metamorphism, igneous activity and metallogeny. The need for such a book became apparent at the international conference held in Montpellier, France, in August, 1988, where a vast amount of new information was presented. The collection of papers presented at the conference and published in Tectonophysics, however, does not give the comprehensive treatment of the geology required for terrane analysis that is provided by this book.

We sincerely appreciate the work of the reviewers whose comments have helped to ensure that this book has an international appeal. We also thank the authors whose diligent efforts have provided an excellent synthesis.

Halifax, Nova Scotia, Canada 1993 J. Duncan Keppie

Contents

Part III The Massif Central

Part IV Correlative Areas

Part V

List of Contributors

A. Autran
Bureau de Recherches
Géologiques
et Minières (BRGM)
BP 6009
Service Géologique National
45060 Orléans Cedex
France

Bernard Auvray
Centre Armoricain d'Etude Structurale
des Socles
Université de Rennes
Campus Beaulieu
35042 Rennes Cedex
France

Michel Ballèvre
Université de Rennes I
Géosciences Rennes
Laboratoire de Tectonique
Avenue du Général Leclerc
35042 Rennes Cedex
France

Serge Bogdanoff
Laboratoire de Géologie
Structurale & Appliquée
Université de Paris-Sud
Batiment 504
91405 Orsay Cedex
France

M. C. Boiron
Laboratoire de Minéralogie
Université de Toulouse
39 allée J. Guesde
31400 Toulouse
France

C. Bois
Institut Français du Pétrole
1–4, Avenue de Bois-Préau
92506 Rueil-Malmaison Cedex
France

Jean-Jacques Bonjour
Laboratoire de Pétrologie Sédimentaire
Institut de Géologie
Université de Rennes I
UPR 4661 du CNRS
35042 Rennes Cedex
France

B. Briand
Centre des Sciences de la Terre
Laboratoire de Pétrographie
Université Claude Bernard Lyon I
27–43 Bd du 11 novembre
69622 Villeurbanne Cedex
France

H. Bril
Laboratoire de Géologie
123 avenue A. Thomas
87060 Limoges Cedex
France

Jean-Pierre Brun
Centre Armoricain d'Etude Structurale
des Socles
Université de Rennes
Campus Beaulieu
35042 Rennes Cedex
France

Bruno Cabanis
Laboratoire de Géochimie comparée
et systématique
T. 16–26, 3° étage
Université P. et M. Curie
75252 Paris Cedex 05
France

Jean-Paul Carron
Département de Géologie
Faculté des Sciences
Université de Bretagne Occidentale
29287 Brest
France

M. Cazes
Elf Qauitaine
Tour Elf
2, place de la Coupole
92708 Paris-La Défense Cedex 45
France

Pierre Cavet
35 Avenue Maisdon Pajot
44100 Nantes
France

J. Chantraine
Service Géologique National
Bureau de Recherches
Géologiques et Minières
rue Picherit
44300 Nantes
France

L. Chauris
Directeur de Recherches au
CNRS
Département des Sciences de la Terre
Université de Bretagne Occidentale
29287 Brest
France

Jean-Jacques Chauvel
Centre Armoricain d'Etude Structurale
des Socles
Université de Rennes
Campus Beaulieu
35042 Rennes Cedex
France

P. Choukroune
Université de Rennes
Avenue du General Leclerc
35402 Rennes Cedex
France

Michel Colchen
Laboratoire de Tectonique
et Géodynamique
Université de Poitiers
40, Avenue du Recteur Pineau
86022 Poitiers Cedex
France

Sylvie Costa
Université de Montpellier
Pl. E. Bataillon
34095 Montpellier Cedex 5
France

Gilbert Crevola
Institut de Géodynamique
Université de Bordeaux III
Avenue des Facultés
33405 Talence Cedex
France

M. Cuney
Laboratoire de Minéralogie
Université de Toulouse
39 allée J. Guesde
31400 Toulouse
France

Pierre Debat
Dept. Géologie-Pétrologie
Géol. Struct. & Tectonophysique
38, rue des 36-Ponts
31400 Toulouse
France

N. Debeglia
Bureau de Recherches Géologiques
et Minières (BRGM)
BP 6009
Service Géologique National
45060 Orléans Cedex
France

Francis Doré
4 allée des Tilleuls
94220 Charenton-le-Pont
France

G. Dubuisson
Université de Paris VII
Laboratoire de Pétrologie Physique
et Institut de Physique du Globe
2, Place Jussieu
Tour 25/24, 1^0 étage
75251 Paris Cedex 05
France

Helmut Echtler
Geologisches Institut
Regionale Geologie
Universität Karlsruhe (TH)
Kaiserstr. 12
76131 Karlsruhe
FRG

J. B. Edel
Institut de Physique du Globe
de Strasbourg
Université Louis Pasteur
5, rue René Descartes
67084 Strasbourg Cedex
France

Raimund Feist
Laboratoire de Paléontologie
Université des Sciences et
Techniques du Languedoc
34095 Montpellier
France

Pierre Fluck
Institut de Géologie
Université de Strasbourg
1, rue Blessig
67084 Nice
France

O. Gariel
Institut Français du Pétrole
1–4 Avenue de Bois Preau
92506 Rueil-Malmaison
France

Jean Galtier
Laboratoire de Paléobotanique
Université des Sciences
et Techniques du Languedoc
34095 Montpellier
France

J. Girardeau
Université de Nantes
Département des Sciences
de la Terre 2, rue de la Houssinière
44072 Nantes Cedex 03
France

Gaston Godard
Université de Paris VII
Laboratoire de Pétrologie
4, Place Jussieu
55252 Paris Cedex 05
France

Jean-Claude Goujou
Université de Montpellier II
Centre Géologique et Géophysique
Place E. Bataillon
34095 Montpellier Cedex 05
France

Frédéric Gresselin
Caen University
Geology Department
Esplanade de la Paix
14–32 Caen Cedex
France

A. Hirn
Institut de Physique du Globe
Place Jussieu
75005 Paris
France

Pierre Jegouzo
Rennes University
Geology Institute
Tectonic Laboratory
Campus de Beaulieu
Avenue du General Leclerc
35042 Rennes Cedex
France

J. M. Lardeaux
Laboratoire de Sciences
de la Terre
U. C. B. Lyon I
U. R. A. CNRS 726
46, Allée d'Italie
69364 Lyon Cedex 07
France

HURBERT LARDEUX
Laboratoire de Stratigraphie
et Paléontologie
Université de Rennes I
CNRS URAD 1364
35042 Rennes Cedex
France

CLAUDE LE CORRE
Université de Rennes I
Campus de Beulieu
35042 Rennes Cedex
France

PATRICK LEDRU
Bureau de Recherches Géologiques
et Minières
Service Géologique National
BP 6009
45060 Orléans Cedex
France

J. P. LEFORT
Université de Rennes
Avenue du General Leclerc
35402 Rennes Cedex
France

L. B. LE GALL
Institut Français du Pétrole
1–4, Avenue de Bois Preau
92506 Rueil-Malmaison
France

MARTINE LE GUEN DE KERNEIZON
Département de Géologie
Faculté des Sciences
Université de Bretagne Occidental
29287 Brest
France

GILBERT LE POCHAT
Bureau de Recherches Géologiques
et Minières
SGR Aquitaine
Avenue du Docteur Albert Schweitzer
36600 Pessac
France

G. LIBOUREL
Département de Géologie
et Minéralogie
Centre de Recherches Volcanologiques
U. A. CNRS 10
5, rue Kessler
63038 Clermont Ferrand Cedex
France

JACQUES MARCHAND
Université de Nantes
Laboratoire de Pétrologie
et Minéralogie
2, rue de la Houssinière
44072 Nantes
France

CH. MARIGNAC
Laboratoire de Géologie
Ecole Nationale Supérieur
des Mines de Nancy
Parc de Saurupt
54042 Nancy Cedex
France

E. MARCOUX
BRGM, Service Géologique National
Département de Géologie
BP 6009
45060 Orléans Cedex 2
France

P. MATTE
Université des Sciences
et Techniques de Languedoc
Place Bataillon
34060 Montpellier
France

C. MAJESTÉ-MENJOULAS
Dept. Géologie-Pétrologie
Géol. Struct. & Tectonophysique
38, rue des 36 Ponts
31400 Toulouse
France

RENÉ-P. MÉNOT
Laboratoire de Pétrologie
Université Jean Monnet
CNRS UA 10
42023 Saint Etienne Cedex 02
France

J. C. MERCIER
Université de Paris VII
Laboratoire de Pétrologie Physique
et Institut de Physique du Globe
2, Place Jussieu
Tour 25/24, 1^0 étage
75251 Paris Cedex 05
France

PIERRE MORZADEC
Laboratoire de Micropaléontologie
et Paléontologie Marines
Institut de Géologie Université
de Rennes I
35042 Rennes Cedex
France

BERNARD MOUTHIER
COGEMA, BP 4
78141 Velizy-Villacoublay
France

HASSANE NACHIT
Laboratoire de Géologie
Faculté des Sciences
Université Ibnou Zohr
Agador
Morocco

G. NAUD
Laboratoire de Géologie
Dynamique
Université de Marseille
Centre Saint Jérome
13097 Marseille
France

M. OHNENSTETTER
Centre de Recherches sur la Synthèse
et la Chimie des Minéraux
1 A, rue de l Ferollerie
45071 Orléans Cedex 2
France

J. B. ORSINI
Laboratoire de Géologie-Pétrologie
U. A. CNRS 10
Université des Sciences
et Techniques de St. Etienne
42023 Saint Etienne Cedex 02
France

Dr. FLORENTIN PARIS
Laboratoire de Paléontologie
et Stratigraphie
Université de Rennes I
URAD 1364 du CNRS
35042 Rennes Cedex
France

ANNIK PELHATE
45, Rue Adolphe Leray
35000 Rennes Cedex
France

B. PINET
Institut Français du Pétrole
1–4, Avenue de Bois Preau
92506 Rueil-Malmaison
France

A. PIQUÉ
Département des Sciences
de la Terre
Université de Bretagne Occidentale
Avenue Le Gorgeu
29287 Brest Cedex
France

ALAIN PLOQUIN
Centre de Recherches Pétrographiques
et Géochimiques
B. P. 20
54501 Vandoeuvre-les-Nancy Cedex
France

DIDIER PONCET
Laboratoire de Géologie Strati-
graphique et Structurale
Université de Poitiers
40, Avenue du Recteur Pineau
86022 Poitiers Cedex
France

JEAN-PIERRE PUPIN
Laboratoire de Pétrologie-
Minéralogie
Université de Nice
06034 Nice
France

Jean-Michel Quénardel
Laboratoire de Géologie
Structurale & Appliquée
Université de Paris-Sud
Batiment 504
91405 Orsay Cedex
France

Dominique Rabu
Service Géologique National
Bureau de Recherches
Géologiques et Minières
rue Picherit
44300 Nantes
France

Patrick Racheboeuf
Centre des Sciences de la Terre
Université de Lyon
URA 11 du CNRS
43 Boulevard du 11 Novembre
69622 Villeurbanne Cedex
France

Michel Robardet
Laboratoire de Paléontologie
et Stratigraphie
Université de Rennes I
URA 1364 du CNRS
35042 Rennes Cedex
France

Joel Rolet
Bretagne Occidentale University (UBO)
Structural Geology
GDR (CNRS) "GEDO"
6, avenue Le Gorgeu
29287 Brest Cedex
France

Ph. Rossi
GIS BRGM/CNRS
1 A, rue de la Ferollerie
45071 Orléans Cedex
France

D. Santallier
Centre des Sciences de la Terre
Laboratoire de Pétrographie
Université Claude Bernard
27–43 Bd du 11 novembre
69622 Villeurbanne Cedex
France

Jean-Luc Schneider
Institut de Géologie
Université de Strasbourg
1, rue Blessig
67084 Strasbourg
France

Jean-Marc Stussi
Centre de Recherches Pétrographiques
et Géochimiques
BP 20
54501 Vandoevre-les-Nancy
France

Denis Thieblemont
SCOP 6
Laboratoire de Télédétection
T. 16–26, 1 α étage
Université P. et M. Curie
75252 Paris Cedex 05
France

F. Tollon
Laboratoire de Minéralogie
Université de Toulouse
39 allée J. Guesde
31400 Toulouse
France

J. L. Vigneresse
Laboratoire de Tectonophysique
Université de Nantes
77072 Nantes Cedex 03
France

G. Vivier
Institut Dolomieu
Université Joseph Fourier
U. R., CNRS 69
38031 Grenoble Cedex
France

Jürgen von Raumer
Institut de Minéralogie
et de Pétrographie
Université
Pérolles
1700 Fribourg
Switzerland

HUBERT WHITECHURCH
Institut de Géologie
Université de Strasbourg
1, rue Blessig
67084 Strasbourg
France

ROBERT WYNS
B. R. G. M.
Service Géologique National Bretagne
Pays de la Loire 10, rue H. Picherit
44071 Nantes Cedex 3
France

Part I Geophysics

Seismic Reflection Images of the Pre-Mesozoic Crust in France and Adjacent Areas

C. Bois, M. Cazes, P. Choukroune, O. Gariel, A. Hirn, B. Le Gall, J. P. Lefort, P. Matte and B. Pinet

1 Introduction

The contribution of deep seismic profiling to our understanding of the pre-Mesozoic crust in France and adjacent areas is of the utmost importance. Despite limitations which will be discussed in Section 1.2, the method can obtain images of the crust down to and deeper than 50 km. It is the only way to observe the deep structure of the crust in some detail, to identify features such as thrusts, decollements or detachments, decoupling levels, deep faults, etc., to trace the Mohorovicic Discontinuity and image upper mantle discontinuities in favorable cases.

However, interpretation of deep seismic sections in France and adjacent areas is made difficult by the fact that the crust underwent several tectonic deformations during its geological history. A schematic assignment of the features observed in the sections to a given framework, which was mainly acquired during the Variscan Orogeny (400 to 300 Ma), is not easy.

1. The Northern Variscan Foreland. Beneath the Mesozoic and Tertiary sedimentary cover, a number of pre-Variscan units have been identified north of the Variscan front. The Brabant Massif, deformed by the Caledonian Orogeny, forms the foreland in northern France, Belgium, and southwestern Germany (Michot 1980; Robaszynski and Dupuis 1983). It extends beyond the Strait of Dover in East Anglia and is fringed to the north by the Precambrian Midlands Block. Both units have been identified in this area in boreholes (Smith 1987). In Wales and southern Ireland, the Variscan Foreland corresponds to the southern segment of the Caledonian range bounded in the north by the Iapetus Suture. However, outcrops also indicate a Proterozoic subduction in the South Irish Sea near Anglesey Island (Cogné and Wright 1980; Harris et al. 1980).

2. The Variscan Frontal Thrust. This feature corresponds to gently dipping thrusts which bring folded and generally cleaved Ordovician to early Carboniferous formations to rest on top of the late Carboniferous paralic foredeep. This front has been identified clearly in France as the Midi Fault (Bouroz et al. 1961; Raoult and Meilliez 1986 a, b), and extends in Germany as the Aachen Fault (Teichmüller and Teichmüller 1979). The Variscan Front has also been noted in southern Wales (Hancock et al. 1981) and southern Ireland (Cooper et al. 1984), where its location is less obvious and is subject to debate (Gardiner and Sheridan 1981; Le Gall 1990).

3. The Rheic Suture. This is assumed to occur in southern Cornwall, marked by the middle Devonian Lizard ophiolitic slab (Badham and Kirby 1976; Davies 1984), which originated farther south in the presentday English Channel. The extension of this feature to the east, in the English Channel, is difficult to identify. In France it is often placed along the Bray-Vittel Fault, which is lined by a major gravity anomaly (Guillen and Debeglia 1984) and is assumed to cross the Vosges Mountains on the Lubine Suture (Burg 1988).

4. The Cadomian Block. A major magnetic anomaly occurs in the axial part of the English Channel. It is regarded as the trace of a Proterozoic suture which marks the northwestern boundary of a crustal block, whose structure is principally related to the Cadomian Orogeny (Lefort 1975; Graviou and Auvray 1985). Although this block largely escaped the Variscan shortening (Cogné and Wright 1980; Chantraine et al. 1988), its other boundaries correspond to Variscan features. To the east, its boundary is the Seine Fault, where a major magnetic anomaly occurs, and the southern boundary is located along the North Armorican Shear Zone.

5. The Variscan Axial Shear Zones. Major strike-slip ductile shear zones associated with Carboniferous syntectonic granite occur in the axial part of the Variscan Orogen. The main ones are the North- and South-Armorican shear zones which display a right-lateral strike-slip component (Jegouzo 1980) and extend toward the Massif Central, where they are asso-

J. D. Keppie (Ed.): Pre-Mesozoic Geology in France and Related Areas

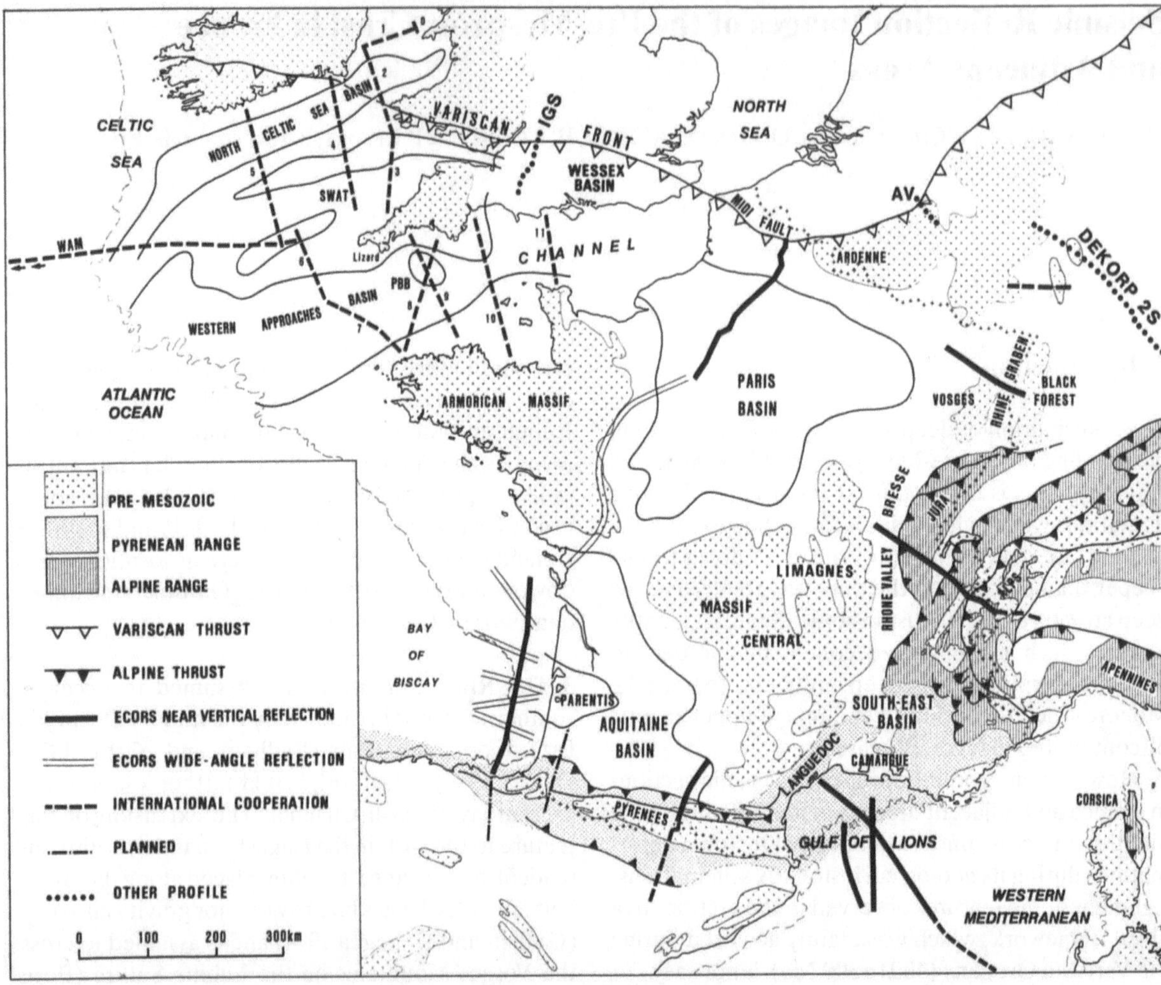

Fig. 1. Location map of the ECORS deep seismic profiles and main structural features in France and adjacent areas. (After Bois et al. 1989). *AV* Aachen-Venn profile; *IGS* Institute of Geological Science profile

ciated with strong gravity and magnetic anomalies (Matte 1983, 1986).

6. The South-Armorican Suture. A major suture zone should correspond to the oceanic domain which extended south of Brittany and in the Massif Central before the Variscan collision. This suture is located by most authors near the South-Armorican Shear Zone, a Carboniferous feature which obliterated the Silurian-Devonian suture, which is only evidenced by a few ultrabasic and high-pressure rocks (Matte 1983, 1986). This suture is believed to extend southeastwards in the northern edge of the Massif Central. Lefort (1975), however, places the suture farther south, along an offshore magnetic anomaly of major amplitude associated with a gravity high and an active seismic zone and extending southeastward at the southern edge of the Massif Central. A third alternative assumes the existence of two sutures enclosing an intervening domain (Girardeau et al. 1986).

7. The Aquitaine Domain. South of the above suture, the Aquitaine domain is sometimes regarded as a terrane of the African realm (Autran 1980), affected by the external thrusts of the Variscan Orogen (Paris and Le Pochat 1987). Its relationship to the Ebro Block, from which it is separated by a late Variscan and Mesozoic North-Pyrenean strike-slip fault, is conjectural.

The extension of the above features into southeastern France is difficult to ascertain because this area was strongly reworked by late Variscan strike-slip and Mesozoic and Tertiary tectonic phases, including the opening of the West Mediterranean and the rotation of the Corsica-Sardinia Block. The restoration of the pre-Mesozoic situation of these blocks (Fig. 2) should be regarded as approximate (Arthaud and Matte, 1977; Burrus 1984).

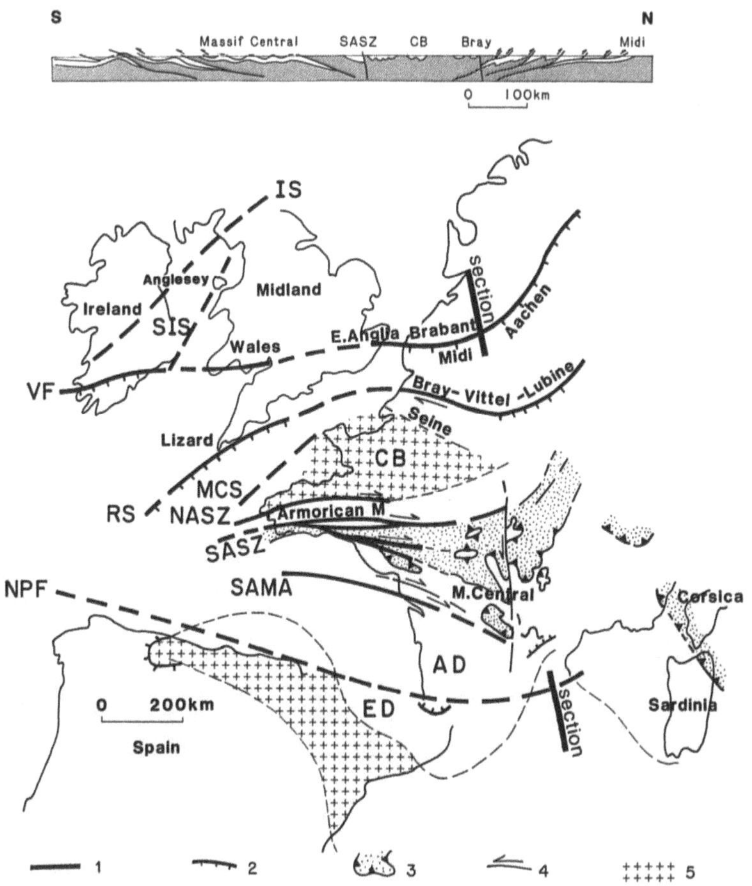

Fig. 2. Structural framework of France and adjacent areas. (After Matte 1986). *1* Major structure; *2* thrust; *3* innermost crystalline nappes; *4* ductile strike-slip fault; *5* block with moderate Variscan deformation. *AD* Aquitaine domains; *CB* Cadomian Block; *ED* Ebro domain; *IS* Caledonian Iapetus Suture; *MCS* Cadomian Mid-Channel Suture; *NASZ* North-Armorican Shear Zone; *NPF* North Pyrenean Fault; *RS* Rheic Suture; *SAMA* South-Armorican magnetic anomaly; *SASZ* South-Armorican Shear Zone; *SIS* Cadomian South Irish Sea Suture; *VF* Variscan frontal thrust

1.1 Deep Seismic Data

The ECORS (Etude Continentale et Océanique par Réflexion et Réfraction Sismiques) project of deep seismic reflection profiling in France was launched in 1983 (Bois and Allegre 1983). Similar projects were developed in neighboring countries: BIRPS (British Institutions Reflection Profiling Syndicate) in the United Kingdom (1981), DEKORP (Deutsches kontinentales Reflexions-seismisches Programm) in West Germany (1983), and other projects in Switzerland (1986), Italy (1986), Spain (1986), and Belgium (1986). The ECORS project has co-operated in one way or another with all these groups in the study of geological features straddling international borders.

Deep seismic data were recorded in France and adjacent areas before the beginning of the ECORS project. They consist of seismic refraction and wide-angle seismic reflection data coming from experiments carried out mainly in the eastern and southern parts of the country, Brittany, the English Channel, and the Celtic Sea (Fig. 3) (Avedik 1975; Blundell 1979; Weber et al. 1980). These experiments provided valuable information on the geometry of the Moho on a regional scale as well as on the overall distribution of P-wave velocity in the crust and the underlying upper mantle. They showed, for the first time, major Moho deepening in the Pyrenees and the Alps and the presence of a high-velocity layer in the lower third of the crust, which is regarded as a granulite layer. These data were very useful for the preparation of later deep seismic profiles. Wide-angle seismic reflection and seismic refraction techniques were also used during the course of the ECORS project to complement the data of the vertical seismic reflection.

The following ECORS profiles have been completed (Fig. 1): (1) the northern France profile, 230 km long, across the northern Variscan zone, complemented by a 400-km-wide angle reflection profile across most of the Variscan Orogen (Bois et al. 1986; Cazes et al. 1986; Cazes and Torreilles 1988; Matte and Hirn 1988); (2) the SWAT (South Western Approaches Traverse) survey, consisting of 1600 km of profiles in the English Channel and the Celtic Sea operated by BIRPS with the co-operation of the ECORS project; this survey aimed to study the northern Variscan area and the overlying Permian-Mesozoic basins (BIRPS and ECORS 1986; Bois et al.

Fig. 3. Seismic refraction profiles carried out in France and adjacent areas prior to the ECORS project. (Hirn, in Weber et al. 1980; Blundell 1979)

1990, 1991); (3) The WAM (Western Approaches Margin) profile, 645 km long, connecting the previous survey to the Atlantic margin and carried out in the same conditions (Peddy et al. 1989); (4) the Bay of Biscay profile, 300 km long, cutting across the Aquitaine basin shelf, complemented by six expanding spread profiles (Pinet et al. 1987 a, b); (5) the Pyrénées profile, 250 km long, running from the Aquitaine to the Ebro basins and carried out in co-operation with the Spanish (ECORS Pyrénées team 1988; Choukroune et al. 1990); (6) the Alps profile, 210 km long, from the Po Plain to the Jura Mountains, carried out in co-operation with the CROP project and complemented by the 100-km Jura-Bresse profile which reaches the Massif Central (Bayer et al. 1987; Nicolas et al. 1990); (7) the Rhine Graben survey consisting of two traverses across the graben amounting to 200 km, carried out in co-operation with the DEKORP project; and (8) the Gulf of Lions survey, consisting of 580 km of profiles aimed at studying the West Mediterranean rifting and the underlying Pyrenean crust (this survey was extended by a 300-km profile up to the Sardinia shelf by the CROP project).

A few industrial seismic surveys in the Ardenne and Aquitaine areas should also be mentioned even though they were not primarily aimed at studying the basement. The Aquitaine surveys were recorded to 4–6 s TWT (second Two-Way Travel Time) and occasionally to 12–15 s TWT, and provided interesting deep data on this part of the region.

The backbone of every ECORS experiment is a vertical seismic reflection profile carried out with industrial equipment adapted for recording seismic events coming from depths of 30 km and more. The sources are especially powerful, the geophone spreads very long, and the recording time is at least 15 s TWT (Bois et al 1987). In order to help the geological interpretation, other geophysical experiments, seismic refraction and wide-angle seismic reflection, gravity, magnetism, and magneto-telluric were carried out together with vertical seismic reflection. As much as possible, surface geological surveys were performed to link up the deep seismic data with the surface geology data.

Although these profiles studied geological features as different as the Paleozoic basement, Permo-Mesozoic basins, and Tertiary ranges, all of them have imaged the pre-Mesozoic crust which forms the ubiquitous basement of the study region, and deserve attention. However, the Jura-Bresse, Rhine Graben, and Gulf of Lions surveys have yet been scarcely published and this chapter must be limited to the northern, western, and southwestern parts of the country.

1.2 Limitations of the Seismic Reflection Technique

The deep seismic surveys presented above are generally long regional traverses without tie or cross-lines. Such a profile setting makes the interpretation of dipping seismic events observed in the sections difficult. These events can be not only real reflections but also diffractions, refracted reflections, or out-of-plane events. Diffractions have a characteristic hyperbola shape and can be fairly well identified in most cases. The presence of side events has been minimized in shooting the profiles perpendicular to the geological structures. Whenever cross-lines are available, e. g., in the South Irish Sea (WINCH 4 and SWAT 1), the Iapetus area (Beamish and Smythe 1986) or the Western Approaches (Hillis and Day 1987), the dipping events observed proved to be real reflections located in the plane of the sections.

Seismic sections, however, provide a distorted image of the dipping crustal features, which can be restored only by migration. Without entering into the difficult problems raised by the migration of very deep sections (see Warner 1987), let us say that the uncertainty as to the velocity distribution makes the migrated sections questionable anyway. Nevertheless, only the general features of the structures are studied in this chapter and they are not significantly modified by migration. This point can be checked by comparing the line drawings of migrated with unmigrated sections show in Fig. 5, 14, and 15.

One of the most difficult questions is the physical origin of the reflections observed in the crust. Most of them do not extend to the surface and we do not have any way to check their geological nature. It is generally assumed that dipping reflections with high amplitude correspond to faults, thrusts, or shear zones, either compressional or extensional (Smithson 1978; Fountain and Salisbury 1981; Fountain et al. 1984; Jones and Nur 1984; Hurich et al. 1985; Smithson et al. 1986). But more work has to be done, especially in the study area, to ascertain this interpretation. The horizontal reflections which characterize the lower crust in most of Western Europe are even more puzzling, and many different explanations have been proposed (see compilation in Bois et al. 1988). This point will be further discussed in the next sections.

Whatever the geological interpretation of the crustal reflections, they correspond to impedance contrasts related to changes in the rocks and/or their fluid content. However, the energy absorption which affects the seismic waves to an unknown extent can alter the reflectivity distribution in the crust. The absorption is probably moderate in the crust itself where the rocks are dense, while it is greater in the sediments, more especially in the superficial zone which may largely vary laterally. Hence, lateral variations in the crustal reflectivity may actually reflect variations in the superficial zone.

The width of the central Fresnel zone increases with the reflector depth and the predominant wavelength of the signal, and can reach 4 km in the lower crust. This width probably represents an upper limit of the lateral resolution of seismic reflection. In the seismic sections, deep geological features closer than 4 km should appear as a single feature, and the geological features could appear to be several kilometers too long.

Whereas the above limitations of the seismic reflection technique must be kept in mind, the seismic images, however, provide most valuable information on the crust at depth, information which can be surpassed only by well data.

2 Northern France

2.1 Regional Geological Setting

In northern France, the northern part of the Variscan belt is almost entirely hidden beneath the Mesozoic and Tertiary sediments of the Paris Basin (Fig. 4).

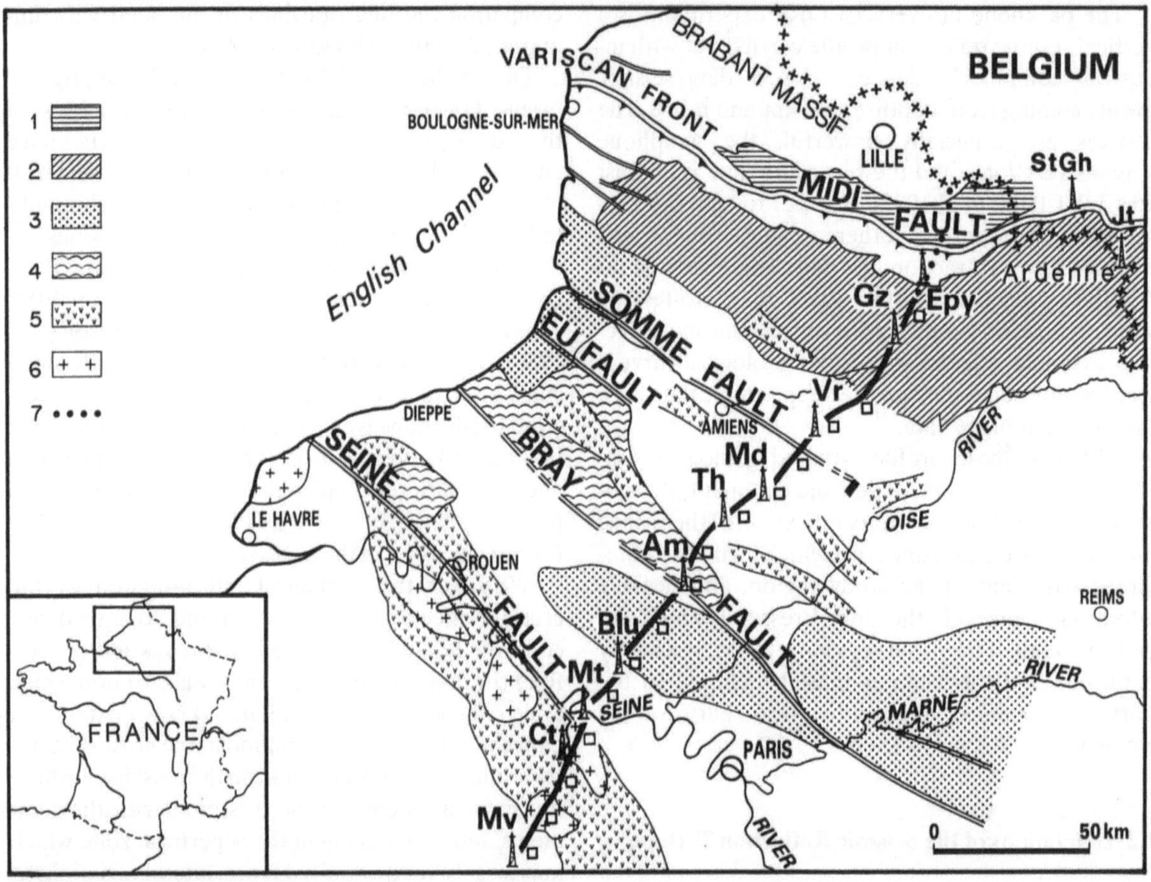

Fig. 4. Pre-Jurassic subcrop in northern France and location of the ECORS profile. (Cazes et al. 1986). *1* Namur Synclinorium (molassic coal basin); *2* Dinant Synclinorium (Devonian); *3* gravity low (Permo-Triassic?); *4* gravity high (metamorphic rocks); *5* magnetic anomaly; *6* gravity low (granite). *Epy*, *Gz* etc. are boreholes (see text). *Solid line* ECORS northern France profile. *Dotted line* CO 83 and 99 industrial profiles

Three structures are of particular interest in this area: the Variscan Front, the Bray Fault, and the Paris Basin magnetic anomaly. Many seismic lines and wells have been completed by oil companies (Heritier and Villemin 1971), and ten boreholes are located on the northern France profile or very near to it. Knowledge of the Paleozoic, however, remains scanty.

The Variscan orogenic belt is bounded in the north by the Midi Fault, a major thrust which extends laterally from Germany (Meissner et al. 1981, 1983) to England (Kenolty et al. 1981). North of this fault, the Variscan Foreland corresponds to the southern limb of the Brabant Massif overlain by a thin middle-late Devonian and early Carboniferous platform series on top of which a large molassic foredeep basin, the Namur Synclinorium, was deposited. This basin, which contains coal reserves, is well documented in subsurface studies (Bouroz et al. 1961). The Midi Fault is a low-angle thrust which flattens at depth and roots into a detachment underlying the thick terrigenous sequence of the Variscan frontal units. The Epinoy (Epy, Fig. 4), Jeumont and Saint-Ghislain wells cut the Midi and subsidiary faults which extend farther north into the Namur Synclinorium. This structure was regarded as a thin-skinned tectonic pattern from field mapping, mining operations, former seismic surveys, and wells [CFP(M) et al. 1969; Delmer 1977; Bless et al. 1980; Raoult and Meilliez 1986 a, b; Raoult 1988]. Before the completion of the northern France profile, the length of the detachment was believed to be a few tens of kilometers.

The Variscan frontal units (Rheno-Hercynian Zone) form the Dinant Nappe which outcrops to the west (Boulogne Inlier) and east (Ardenne) of the profile (Raoult and Meilliez 1986 b) (Fig. 4). It consists of a more basinal, thick Devonian to Carboniferous series including small amounts of Silurian rocks. The rocks were folded into a broad synclinorium, affected by vertical cleavage and thrust along the Midi Fault over the foredeep basin.

South of the Variscan frontal units, several east-trending zones are displayed, the boundaries of which have been determined approximately by gravity and magnetic data where the Variscan belt is overlain by the Paris Basin sediments (Fig. 4): the zones include: (1) the Ardenne-Picardie Anticlinorium of highly sheared Cambro-Ordovician rocks (Vr well), (2) micashist, gneiss, and granite terrains which are correlated with the Mid-German Crystalline Ridge and (3) the Precambrian Cadomian terrane, mainly deformed before the Paleozoic (700–650 Ma) and intruded by 650–550-Ma-old granite. Farther south, the South-Armorican domain shows evidence of early Variscan subduction, collision and deformation, obliterated by late Variscan ductile strike slipping (Fig. 7).

The Variscan belt is thought to result from continental collision following the closure of several Paleozoic oceanic domains that existed until the Devonian or Carboniferous (Matte 1986). The existence of the northern Rheic Ocean in the Bray area is inferred from the presence of magnetic and gravity anomalies assumed to reflect deep-seated mafic bodies similar to the Lizard Ophiolitic Complex in southern Cornwall (Figs. 4 and 5, see Sect. 3).

The whole region is dessected by several northwest-trending faults which controll the pattern of the rivers (Seine, Eu, Somme) and are clearly visible on the satellite photographs (Fig. 4). They are interpreted as late Variscan strike-slip and thrust faults (Arthaud and Matte 1977; Matte 1986) which were reactivated as normal faults in the Mesozoic and as strike-slip and thrust faults during the early Tertiary Pyrenean compression. The main Bray Fault is associated with a large surface anticline involving Jurassic and Cretaceous formations.

A large magnetic anomaly, coincident with the Seine Fault, extends over 400 km southward across the whole Paris Basin. Its origin is debated (Gerard and Weber 1971) (Fig. 4): it might be related to major Cadomian or Variscan crustal faulting. North of the Bray Fault, positive gravity and magnetic anomalies are regarded as mafic and ultra-mafic rocks associated with an inferred suture zone between the Cadomian terrane and the Brabant Massif.

Most of the gravity lows are interpreted as granite batholiths (Mégnien 1980). However, the negative anomaly around the Bray fault may be alternatively interpreted as a late or post-Variscan (late Carboniferous or Permo-Triassic) basin.

2.2 Seismic and Other Experiments

A first test of seismic refraction and wide-angle reflection (1983) was carried out in the area of the northern France profile. It consisted of two large dynamite blasts 95 km apart with self-recording three-phone stations positioned every 2 km between the shot points. This test provided the regional velocity layering pattern and the critical distances.

Continuous profiling was then carried out over a length of 230 km with industrial equipment (1983–1984). The same geophone spread (192 traces, 80 m spacing) and recording equipment were used for vertical and wide-angle seismic reflection and seismic refraction. The source was composed of five heavy vibrators thumping the ground 12 times every 6.6 m and explosive shots every 15 km. In addition, shots were fired at northward offsets of 90 and 105 km to obtain wide-angle reflection data. In the north, this profile was tied to the CO 83 and CO 99 industrial seismic profiles kindly provided by the Total-BP-AGIP-Elf Aquitaine Association.

Three heavy dynamite blasts in the north of the profile and two in the south were recorded by fan spreads of low-frequency geophones to study the geometry of the Moho and the deepest reflectors (1984). Similar profiling of the deep crust was achieved by five more shots to the southwest, up to the Atlantic coast (1985). This survey extended the vertical seismic reflection profile by a 400-km wide-angle reflection profile, almost connecting it to the offshore Bay of Biscay profile (Fig. 7).

In addition, 15 magneto-telluric stations were recorded along the vertical seismic reflection profile (1984) and three 400-km aeromagnetic profiles were flown in the same area at an elevation of 600 m. Details of the field work, data processing, and interpretation can be found in the book devoted to the northern France profile by Cazes and Torreilles (1988).

2.3 Geological Interpretation of the Near-Vertical Seismic Reflection Profile

On the line drawing in Fig. 5, three shallow horizons make up the sedimentary cover of the Paris Basin. According to wells and former seismic lines, they are, from top to bottom, the top Kimmeridgian, the top Dogger and the bottom Jurassic. They are continuous and interrupted only by steep faults with the greatest offset (0.3 s TWT) being observed in the Bray Fault. Beneath this sedimentary cover, the section displays three crustal segments showing different seismic im-

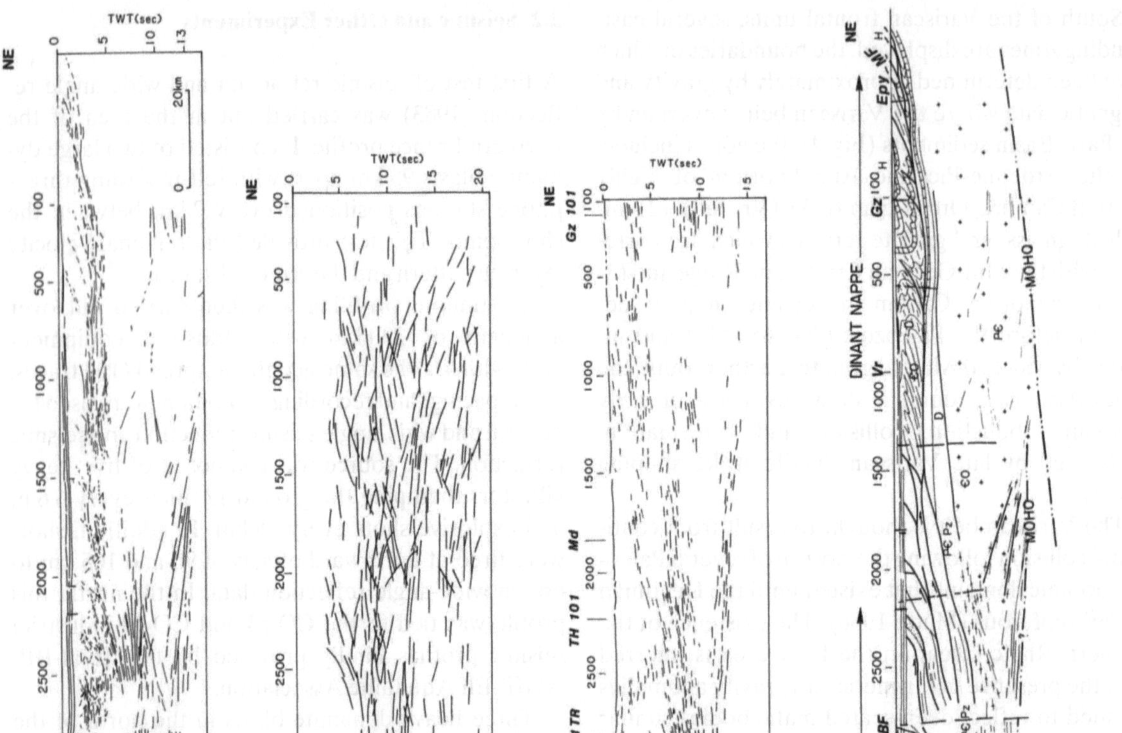

Fig. 5a–d

ages (Cazes et al. 1985, 1986; Bois et al. 1986; Cazes and Torreilles 1988).

The northern segment (Pts 100 to 1500) is characterized by a band of reflections between 2.5 and 5 s TWT with flat to gently south-dipping attitudes. Above this band, the comparatively transparent seismic section shows only a few south-dipping reflections. The flat reflective band has been followed northwards through the CO 83 and CO 99 industrial profiles, where it splits into two main markers and intervening subordinate ones (Fig. 9 a, b). The tectonic significance of the two main markers has been determined by correlation with the Epinoy and Jeumont wells (Figs. 4 and 9). The shallower south-dipping marker has been interpreted as the Midi Fault which carries the thick early Devonian terrigeneous unit of the Dinant Nappe northward over the middle Devonian to late Carboniferous formations of the Namur Synclinorium. The internal deformation of the Dinant Nappe, whose average thickness is estimated to be about 6 to 9 km, has been interpreted in the light of the Meuse River cross-section (Raoult 1988; Raoult

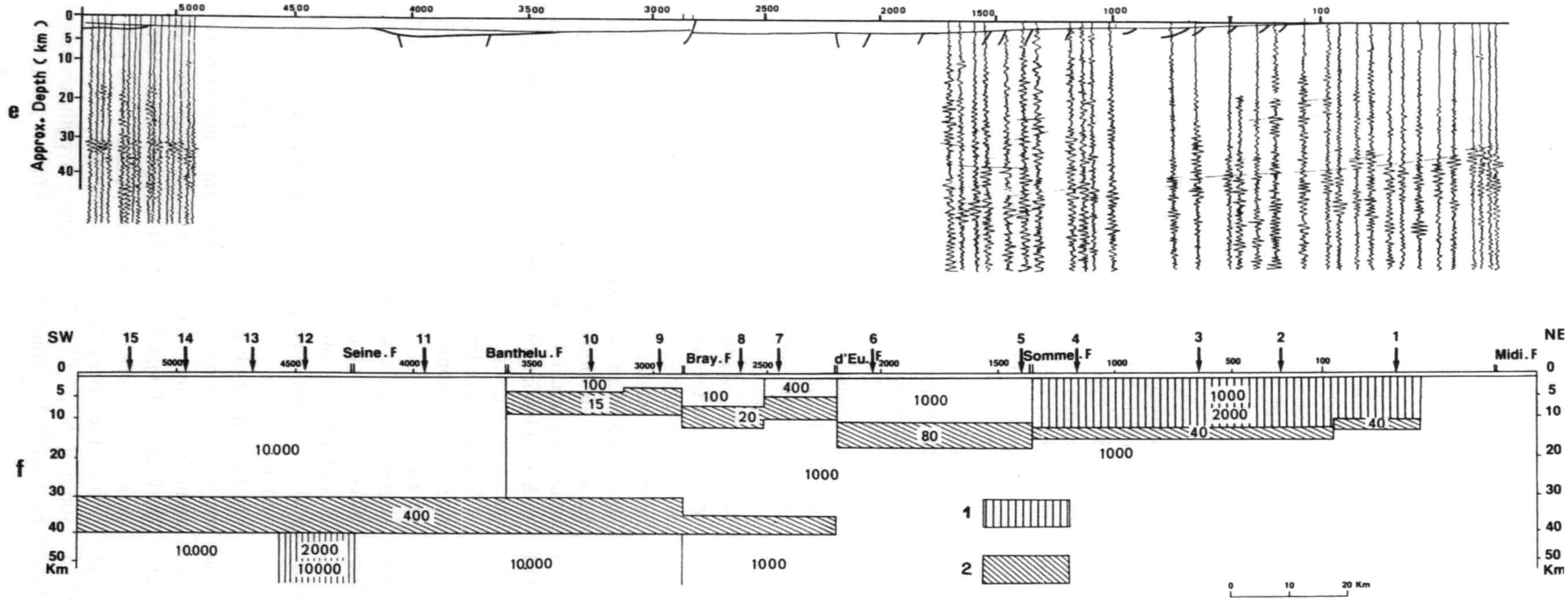

Fig. 5a–f. Northern France profile. (Cazes and Torreilles 1988). See location in Fig. 4. **a** Line drawing of the Vibroseis near-vertical seismic reflection profile (unmigrated), complemented by industrial profiles north of Pt 100. **b** Line drawing of the explosive seismic reflection profile (unmigrated). **c** Line drawing of the Vibroseis profile (migrated). **d** Geological interpretation. *CO* Cambro-Ordovician; *D* Devonian; *H* Carboniferous; *PC* Pre-Cambrian; *Pz* Paleozoic; *BF* Bray Fault; *LL C* layered lower crust; *MF* Midi Fault; *Epy, Gz, etc.* boreholes (see text). **e** Wide-angle seismic reflection data in the northern and southern ends of the near-vertical seismic reflection profile. **f** Magneto-telluric interpretation. Resistivity in ohm.m. *1* anisotropic zone; *2* conductive layer; *1* to *15* recording stations

and Meilliez 1986 a, b) located 100 km to the east (Fig. 9).

In the footwall of the Midi Fault, the Namur para-autochthonous sedimentary cover overlies the Brabant Caledonian basement along a flat-lying contact corresponding to the second deeper seismic marker which is seen to extend as far south as the Montdidier well (Fig. 9, Md). While this marker forms the stratigraphic contact between the middle Devonian and the Brabant basement in the northern limb of the Namur Synclinorium, it may become a decollement of greater and greater importance farther south. The southern extension of the para-autochthonous sediments is supported somewhat by the results of the magneto-telluric experiment which found a conductive layer located in this area at a depth of between 10 and 15 km (Pham 1988) (Fig. 5f). Underneath, only a few scattered reflections can be observed. This comparatively transparent unit is considered to be the Precambrian Brabant basement with a few traces of the Caledonian deformation at the northern end of the profile. However, south-dipping reflections are visible in the explosive source seismic section (Fig. 5b) beneath the top basement, and suggest that the upper part of the Brabant crust was involved in the shortening. There are also faint indications of the lower crustal layering deeper than 11 s TWT and of the Moho at 13 s TWT. The wide-angle experiment showed that the Moho dips southwards up to Pt 1500 (Fig. 5e), where it is connected to a flat reflection located at 18 s TWT in the explosive seismic section (Damotte 1988; Hirn 1988) (Fig. 5b).

The central segment (Pts 1500 to 3000) displays flat reflections with high amplitude, between 8 and 12.5 s TWT, which form the lower crust bottomed by the Moho. This characteristic layered seismic image of the lower crust (Figs. 5a and 6) was found in most of the profiles shot in Western Europe (McGeary 1987; Bois et al. 1988). The layered reflections fade northwards near a major set of north-dipping reflections clearly imaged by explosive seismics down to 18 s TWT (Damotte 1988). Farther north, the latter section shows a few indications of lower crustal reflections (Fig. 5b).

Above the layered lower crust, the profile displays a great number of south-dipping reflections, between 2 and 8 s TWT, which are interpreted as imbricated thrusts involving early Paleozoic and Precambrian rocks in which the Dinant Nappe is rooted. Between Pts 2900 and 3000, the dipping reflections extend into the lower crustal layering. The short horizontal reflections with very high amplitude situated at a depth of 2.5 s TWT near Pt 2600 (Fig. 5a) correspond to gravity and magnetic anomalies and might show the trace of mafic and ultra-mafic flakes. The low resistivity bodies might also represent these flakes (Pham 1988) (Fig. 5f).

The Bray Fault (Pt 2820), a steeply south-dipping feature in the Mesozoic cover, appears as a roughly vertical discontinuity in the lower crust where layering is interrupted over 2–3 km with a small offset. An explosive shot, recorded by a geophone spread straddling the surface fault trace, shows that this interruption could be an artifact and that the lower crust probably was not greatly affected by the fault (Damotte 1988) (Fig. 6). Accordingly, the fault is either rooted above the lower crust or crosses the crust with a negligible offset due to a principally transcurrent displacement in agreement with the vertical foliation of a mylonite sampled in a nearby well (Matte et al. 1986). An alternative hypothesis is that the layering was emplaced in the lower crust after the main displacement of the Bray Fault (Bois et al. 1988).

The southern segment (Pts 3000 to 5400) displays one of the most striking images of the lower crustal layering between 8 and 12 s TWT. Gently south-dipping reflections cross-cut the horizontal layering and some of them even extend beneath the Moho down to 13.5 s TWT. The upper crust is characterized by north- and south-dipping reflections which outline a broad antiform regarded as a large nappe called the Bray Nappe by Matte and Hirn (1988). A transparent body, the apex of which is located at a depth of 5 km beneath the Mantes well (Mt), coincides approximately with the model of the Paris basin magnetic anomaly (Galdeano and Guillon 1988). This faint seismic indication suggests that the magnetic body was involved in northward Variscan thrusting but leaves the question of its origin unanswered: Precambrian magnetic sediments, basic intrusions associated with a Cadomian rift, or a mafic body of oceanic origin associated with a Variscan suture?

The northern France profile shows the crustal image of a collisional orogenic belt. The suture which emerges north of the Bray Fault should be rooted somewhere to the south. But the large strike-slip displacement along the Bray Fault (Matte 1986; Matte et al. 1986) makes interpretation difficult. The dipping reflections observed across the layered lower crust and in the upper mantle are probably the trace of a crustal unit which was stacked over the foreland. It may have belonged to the Brabant, Cadomian, or some intervening terrane. The mafic flakes north of the Bray Fault may be related to the Paris Basin magnetic anomaly body, and the suture placed at the bottom of the "Bray Nappe" should cross the Moho south of the profile (Fig. 5d). Other hypotheses may be proposed in which the suture is farther north and

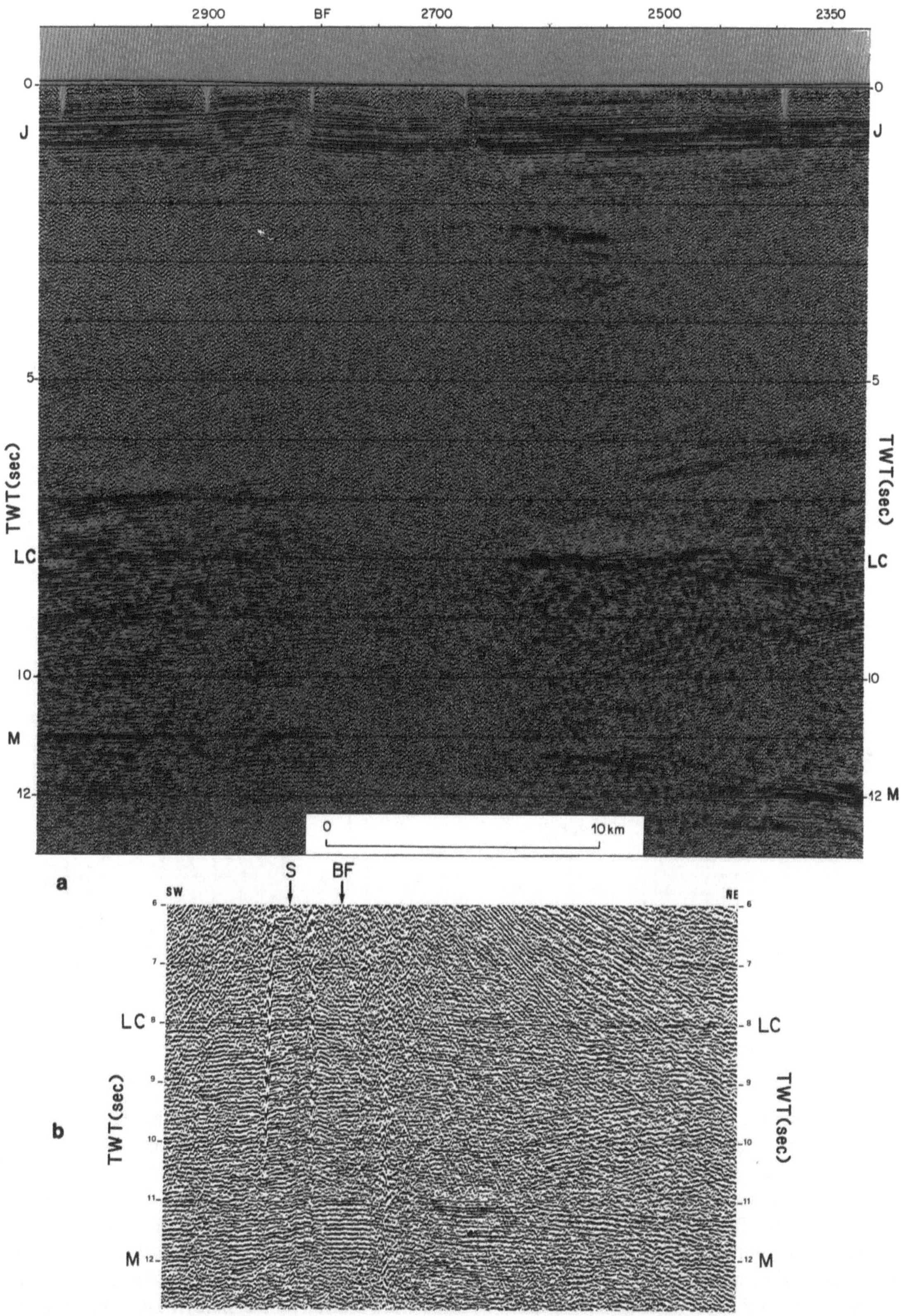

Fig. 6 a, b. Bray Fault area. (Bois et al. 1989). **a** Vibroseis seismic reflection section. **b** Explosive seismic reflection section. *BF* Bray Fault; *S* explosive source

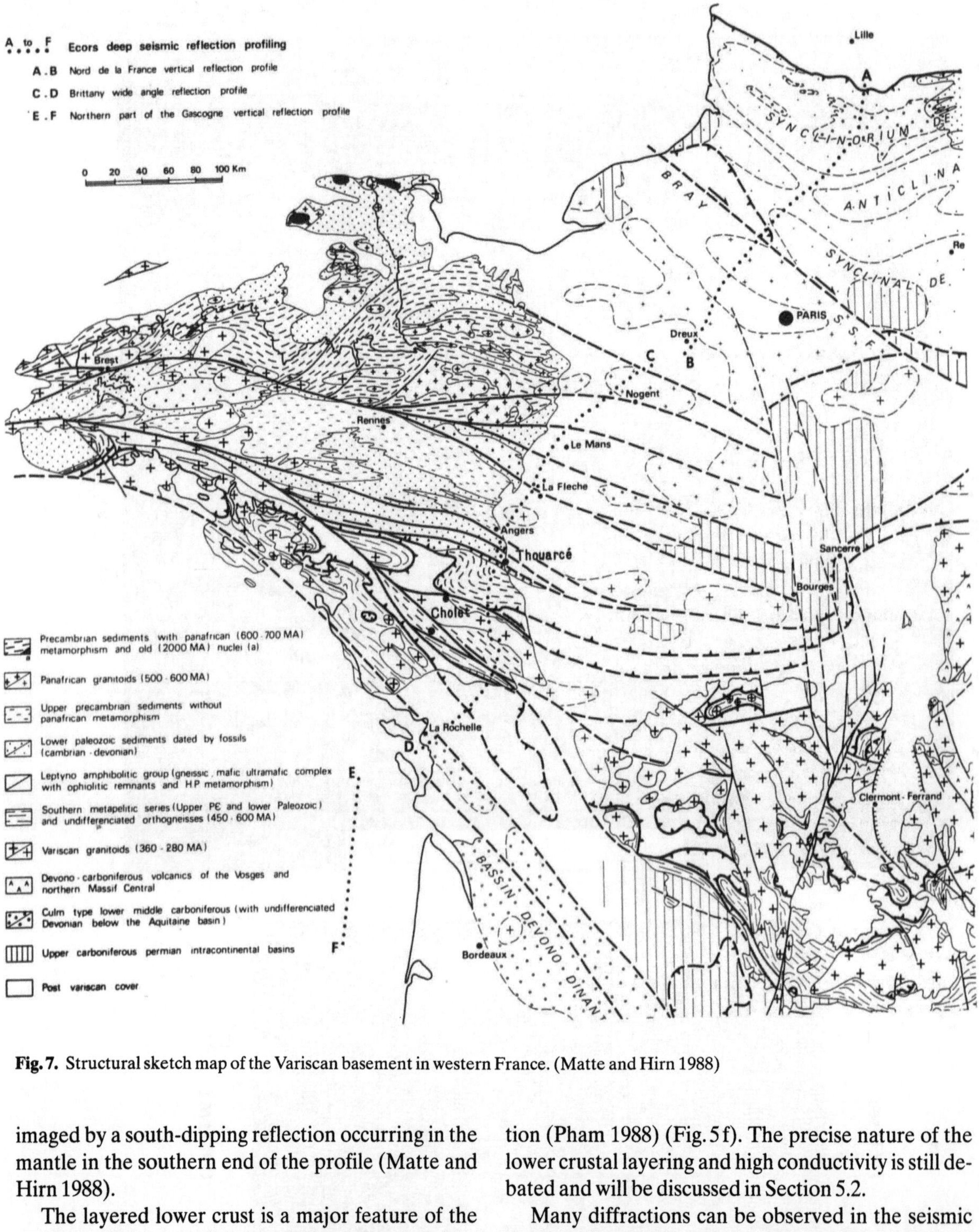

Fig. 7. Structural sketch map of the Variscan basement in western France. (Matte and Hirn 1988)

imaged by a south-dipping reflection occurring in the mantle in the southern end of the profile (Matte and Hirn 1988).

The layered lower crust is a major feature of the northern France profile. It is generally bounded by two bands of higher amplitude reflections. Explosive seismic and wide-angle experiments found that this unit is 8 km thick with a P-wave velocity of 6.6 km/s. It is overlain by an upper crustal unit, 27 km thick, with a P-wave velocity of 5.9 km/s. Magneto-telluric surveys indicated a conductive layer situated at a similar depth range and showing the same areal distribution (Pham 1988) (Fig. 5f). The precise nature of the lower crustal layering and high conductivity is still debated and will be discussed in Section 5.2.

Many diffractions can be observed in the seismic section. They often occur in the vicinity of the late Variscan faults, which were re-activated several times and controlled the Paris Basin subsidence and then underwent a Tertiary inversion (Mascle and Cazes 1988). However, the main displacement along these faults should have been strike-slip (320 Ma) (Matte et al. 1986), occurring before the slaty cleavage development in the Ardenne (Piqué et al. 1984).

The refraction profiling has shown the possible existence of a Carboniferous or Permo-Triassic basin between the basement and the Jurassic bottom around the Banthelu (Blu) and Marville (Mv) wells (Torreilles 1988). The thickness of such a sedimentary wedge does not exceed a few hundred meters, as confirmed by modeling of the gravity low situated south of the Bray Fault (Galdeano and Guillon 1988).

2.4 The Dreux-La Rochelle Wide-Angle Seismic Reflection Profile

The Dreux-La Rochelle wide-angle seismic reflection experiment extends over the northern France Vibroseis profile to the south by about 400 km and crosses the whole Cadomian Block up to the southern limb of the Variscan orogenic belt (Matte and Hirn 1988) (Fig. 7). In this region, the fold geometry and the associated steep cleavage that affect the late Proterozoic and Paleozoic formations may be related to a large-scale dextral shearing which is especially clear along the North- and South-Armorican shear zones.

Although the recording equipment was different from what was used for the Vibroseis profile and led to different deep crustal images, a satisfactory correlation has been established between the two techniques in the places where they were both used (Fig. 8). The layered lower crust does not appear to be 8 km thick in the wide-angle section, but it is generally reduced to a high-amplitude seismic event corresponding to the Moho and situated around 32–37 km. The profile can be subdivided into three segments showing different seismic images (Matte and Hirn 1988).

The northern segment (Dreux-Le Mans) displays a very clear Moho progressively shallowing southward from 37 to 35 km. In the crust, several seismic events can be seen, similar to the south-dipping reflections that occur at the southern end of the Vibroseis profile. Each one is interrupted sharply to the north, suggesting a fault similar to the Bray Fault. Such strike-slip faults are known in the Armorican Massif, which outcrops farther west (Fig. 7).

In the central segment (Le Mans-Thouarcé), the location of the Moho is not clear, and the crust-mantle

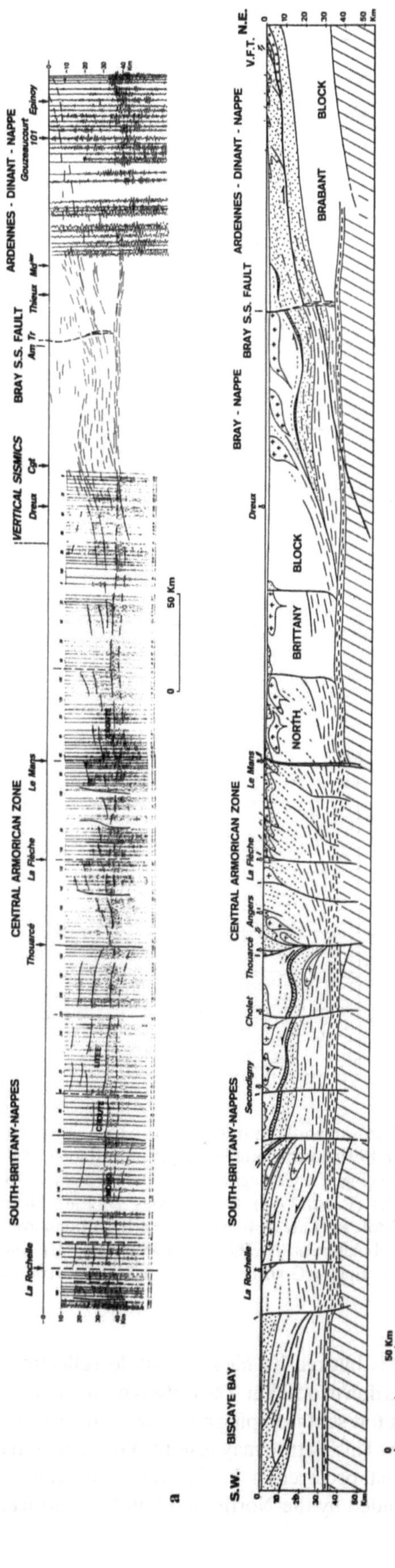

Fig. 8. **a** Wide-angle seismic reflection data in northern France and from Dreux to La Rochelle superimposed on the Vibroseis line-drawing. See location in Fig. 7. **b** Geological interpretation of a 800-km profile from northern France to the northern Bay of Biscay. *1* Mantle; *2* Moho; *3* lowermost layered crust; *4* foliation and cleavage; *5* Precambrian granite; *6* Paleozoic; *7* Hercynian granite; *8* Carboniferous; *9* Mesozoic and Tertiary. (Matte and Hirn 1988)

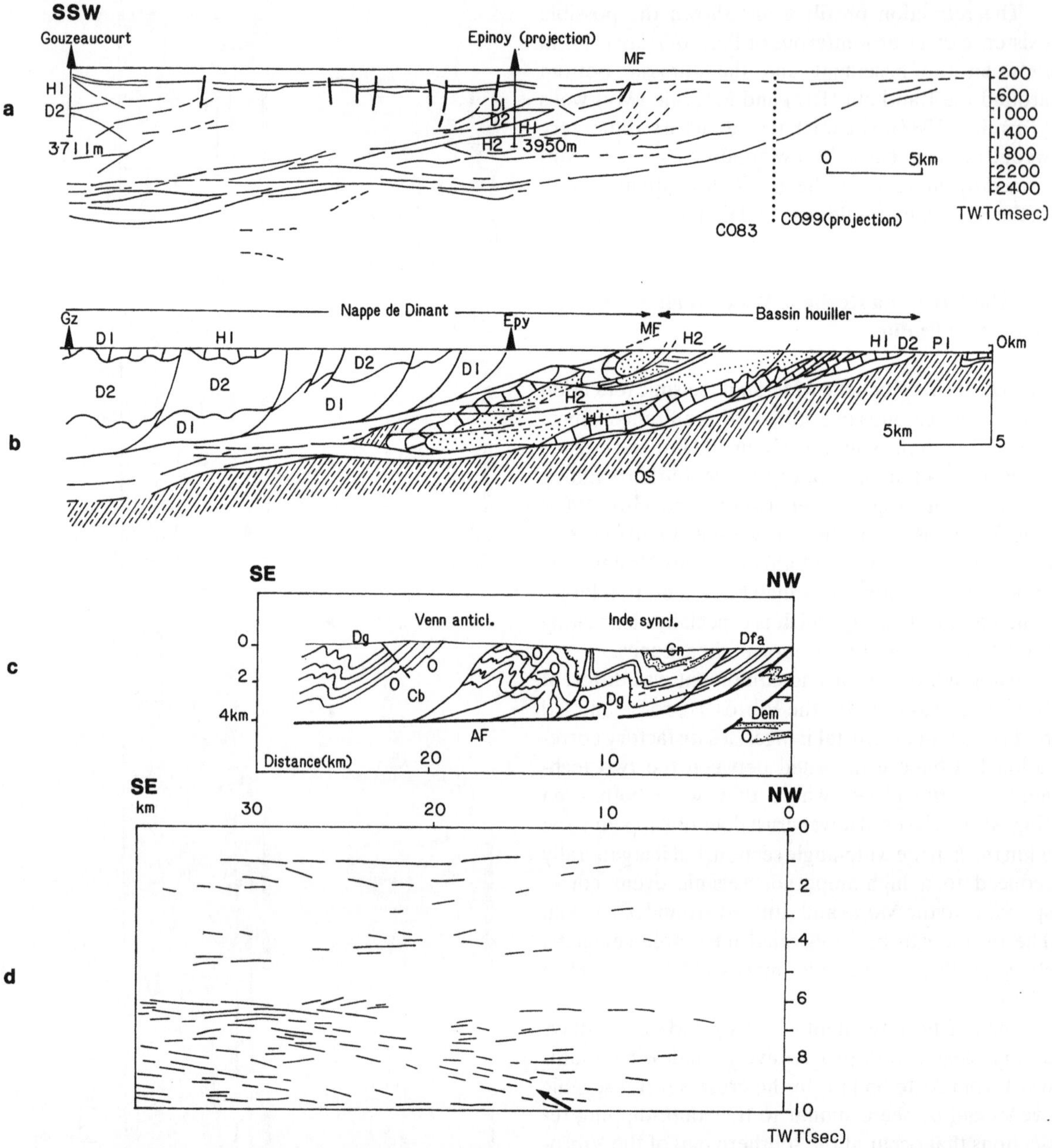

Fig. 9 a–d. Variscan northern frontal area. **a** Line drawing of the CO 83 and 99 industrial profiles (migrated). (Raoult 1988). See location in Fig. 4. **b** Geological interpretation of the CO 83 and 99 profiles. (Raoult 1988). *C* Cretaceous; *D1* Early Devonian; *D2* Middle and Late Devonian; *H1* Early Carboniferous; *H2* Late Carboniferous; *OS* Ordovician-Silurian; *MF* Midi fault; *Epy*, *Gz* Epinoy, Gouzaucourt wells. **c** Surface section of the Aachen-Venn area. (Meissner et al. 1983). See location in Fig. 1. *Cb* Cambrian; *Cn* Carboniferous; *Dem* Early Middle Devonian; *Dfa* Famennien; *Dg* Gedinian; *O* Ordovician; *AF* Aachen Fault. **d** Line drawing of the Aachen-Venn near-vertical seismic reflection. (Meissner et al. 1983). The *arrow* shows the major set of north dipping reflections

transition is blurred by high amplitude reflections, sometimes down to 50 km. The crust is poorly imaged and only a few south-dipping reflectors can be drawn tentatively. This picture may result from the regional shearing that occurred in the Central Armorican domain bounded by the North- and South-Armorican shear zones. These lineaments may have been re-activated later since they are still slightly active to-day.

The southern segment (Thouarcé-La Rochelle) mirrors the northern one, with the crustal reflections dipping northward. The Moho can be observed at a depth of 32–33 km, and a horizontal reflection around

25 km might correspond to the top of the lower crust which is clearly imaged in the northern end of the Bay of Biscay profile (Pinet et al. 1987 a, b). The north-dipping reflection, south of Thouarcé, may be related to the Champtoceaux Nappe which includes eclogite and blue schist within a mesocatazonal unit (Marchand 1981; Ballèvre et al. 1989). These relics of an oceanic crust (Bernard-Griffith and Cornichet 1985) are regarded as the possible trace of a southern Armorican suture (Matte 1986; Matte and Hirn 1988). The area was also later affected by the South-Armorican Shear Zone, and strike-slip faults are present as far south as La Rochelle. In that area, the crust-mantle transition is again blurred, as seen in the Bay of Biscay profile. This seismic image, together with the strong gravity gradient and the particular distribution of the Carboniferous, suggests major strike-slip movement. Farther south, the Bay of Biscay profile shows reflecting material in the upper crust, which outlines large south-vergent thrust sheets above the layered lower crust and a Moho located at a depth of 35 km (see Sect. 4.3.2).

The double vergent shape of the Variscan orogenic belt, which was already known from surface data (Behr et al. 1984; Matte 1986) (Fig. 2), is clearly present at depth as evidences by seismic data. The overall architecture of the belt is, however, obliterated by a number of vertical faults along which blocks of different characteristics have been brought together. These large transcurrent displacements may have led to a dramatic shortening of the original crustal section (Fig. 8 b) and to the disappearance of large fragments of the crust.

2.5 Comparison with Seismic Profiles in Germany

The northern France profile can be compared to deep seismic profiles carried out in Germany, especially the DEKORP 1 and 2 profiles. The external part of the Variscides is partly covered by the short Aachen-Venn (Meissner et al. 1981, 1983) (Fig. 1, AV) and Stavelot-Venn profiles (Durst 1985) which show very strong similaritites with the DEKORP 1 and 2 profiles. Beneath the Aachen Fault decollement, which is the eastern continuation of the Midi Fault, there are a number of flat and gently south-dipping reflections down to 4 s TWT. They may be related to the deformed foreland basin and the underlying Paleozoic platform sediments more or less detached from the basement (right side of Fig. 9 c). Beneath 6 s TWT, the lower part of the section shows a number of horizontal reflections interfering with a prominent set of north-dipping ones (Fig. 9 d) which can be correlated to the north-dipping reflection north of which the lower crust layering fades out in the northern France Vibroseis profile. This setting was interpreted as a southward pinch-out of the Brabant foreland (Fig. 5 a), the upper side of which was shaped by south-dipping imbricated thrusts and the lower side by deep north-dipping antithetic thrusts (Cazes and Torreilles 1988). However, in the northern France explosive seismic profile (Fig. 5 c) as well as in Germany (Fig. 9), indications of lower-crust layering are present north of the north-dipping reflection, suggesting that the northward lower crustal fading is not related directly to the north-dipping event. Moreover, there are a few north-dipping reflections in the crust farther north (Figs. 5 and 9 d) and therefore, the Brabant foreland may extend southwards beyond the major north-dipping reflection. This north-dipping reflection, which is located in the foreland beneath the main frontal detachment, could represent the trace of some earlier tectonic event, probably related to the Caledonian deformation which affected the Brabant Massif (Michot 1980). It may have been re-activated as an antithetic thrust by the Variscan compression.

The internal zone of the Variscan orogenic belt was investigated by the DEKORP 2S profile (DEKORP Research Group 1985; Behr and Heinrichs 1987) (Fig. 1). The structural control of its interpretation is better than that provided by the northern France profile because the Variscan basement outcrops in critical areas such as the Taunus, Spessart, and southern Moldanubian ones (Fig. 10 a). The Taunus area (Mid-German Crystalline Ridge) marks the boundary between the Rheno-Hercynian and Saxo-Thuringian zones and has its French equivalent in the imbricated thrust zone (Figs. 5 d and 10 b) which is located just north of the Bray Fault and is intercalated between the Dinant and Bray nappes. The main difference of the DEKORP section compared to the northern France section is the much lower reflectivity of the lower crust and the shallower Moho. The Taunus area is underlain by south-dipping thrusts which clearly cross the middle crust and may extend down to the Moho. In northern France, the corresponding features have been largely obliterated by the lower crust lavering, but indications of south-dipping reflections are visible in the section (Cazes and Torreilles 1988). The Spessart Dome, shaped by late Variscan events, is much steeper than the Bray Nappe antiform, but both units are rooted farther south in the Moho. The DEKORP 2 profile ends to the south with a high-amplitude south-dipping reflection cutting straight across the whole crust, but the northern France wide-angle profile could not image such a thrust properly. In both Germany and France, the

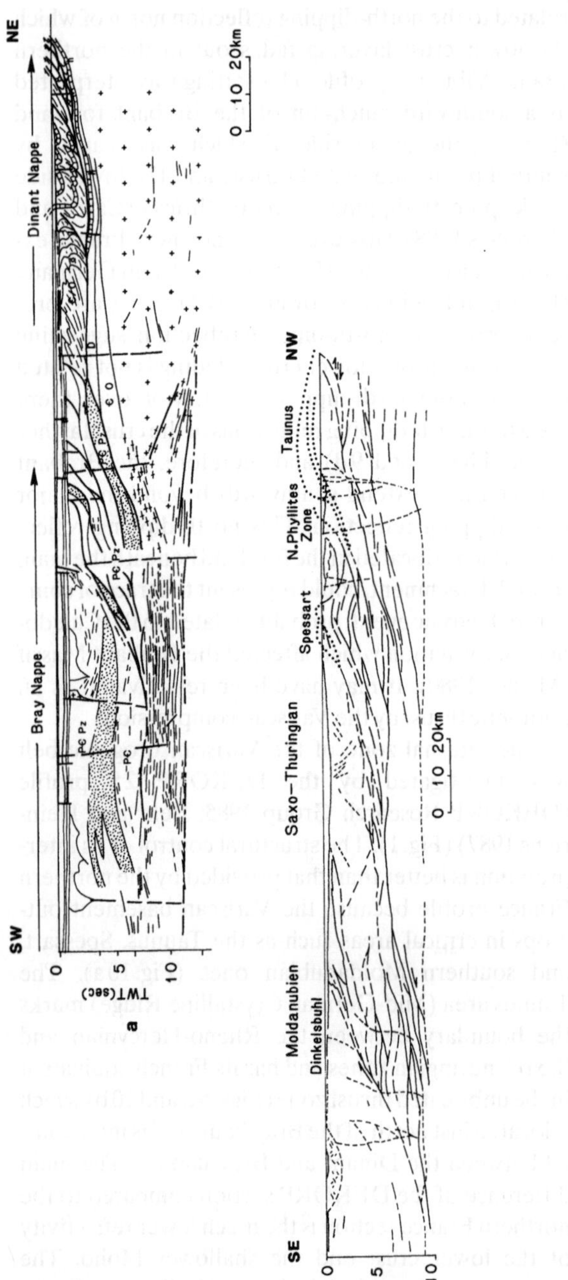

Fig. 10 a, b. Comparison between the northern France and DEKORP 2S profiles. See location in Fig. 1. **a** Geological interpretation of the northern France profile. (Cazes and Torreilles 1988). **b** Geological interpretation of the DEKORP 2S profile. (After Behr and Heinrichs 1987)

seismic sections show a great number of diffractions, which suggest that the crust has been cut by vertical faults. Whereas these faults are clearly visible in the overlying sediments, the crustal displacements remain problematic because of a lack of reliable markers. In both countries the most important and long-lived faults are located near a major boundary: the Taunus-Hunsrück Fault which controlled the late Carboniferous and Permian sedimentation of the Saar-Nahe Basin and the Bray Fault. In spite of all the above geometric similarities, there is a major geological difference between the DEKORP and northern France profiles. Whereas a suture is inferred north of the Bray Fault in the northern France profile, the DEKORP 2 profile has crossed two sutures. There is a suture in the Dinkelsbuhl area (Fig. 10) and another suture may be located between the Rheno-Hercynian and Saxo-Thuringian zones (Engel et al. 1983). This indicates that the Rheic domain was there composed of two narrow basins with a transitional to oceanic crust set up in complex pattern and more or less connected by zones of crustal weakness (pull apart?). The Dinkelsbuhl suture is underlined by a Moho step without an equivalent in the northern France profile.

3 Celtic Sea and English Channel

3.1 Regional Geological Setting

In the continental shelf of the Celtic Sea and the English Channel up to the Atlantic margin, deep seismic surveys have covered a large area where the crust was consolidated by at least three major orogenies (Cadomian 650–600 Ma, Caledonian 450–380 Ma, and Variscan 380–300 Ma) and was subsequently deformed by basin subsidence and the North Atlantic Ocean opening (Fig. 11). Although the traces of these different tectonic events are often superimposed, their specific structures can be identified. From north to south they are: (1) the north-dipping Caledonian Iapetus Suture crossing northern England and central Ireland (Dewey 1974); (2) the South Irish Sea Cadomian Suture (Cogné and Wright 1980); (3) the Church Stretton Fault in Wales, often regarded as the southern Caledonian Front (Coward and Siddans 1979); (4) the northern Variscan Front located by most authors in southern Ireland and Britain (Hancock et al. 1983; Cooper et al. 1984; Smallwood 1985); (5) several granite batholith aligments in Cornwall and the Celtic Sea; (6) the north-vergent Variscan Lizard suture (Badham and Kirby 1976); and (7) the English Channel magnetic anomaly regarded as the trace of a Cadomian suture (Lefort 1975, 1989; Graviou and Auvray 1985). Farther south, this picture is completed by the North- and South-Armorican shear zones associated with widespread syntectonic granite and by the large South-Armorican magnetic anomaly which affects the southern limb of the Variscan orogenic belt.

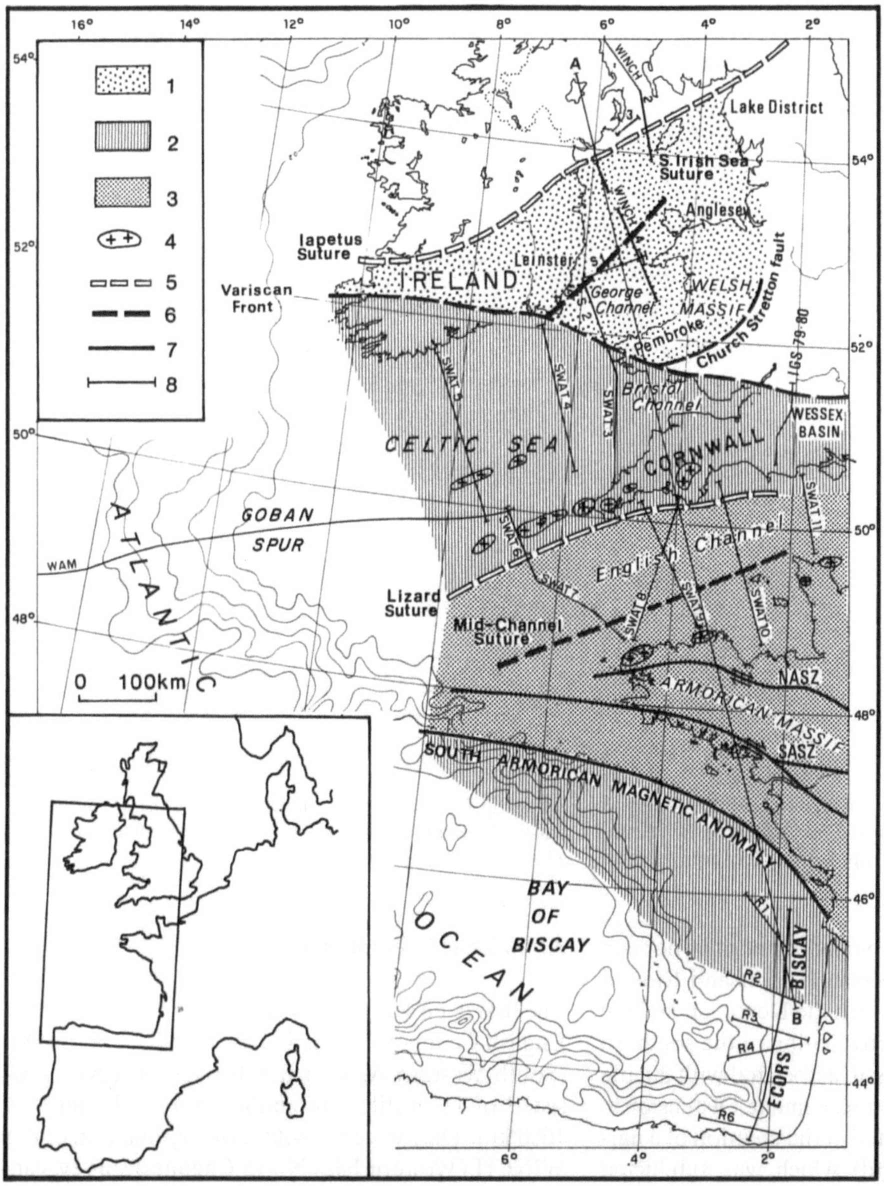

Fig. 11. Location map and main structural features in southern Ireland, southwestern Britain and western France. (Bois et al. 1990). *1* Caledonides; *2* northern and southern Variscides; *3* Central Variscides; *4* Late Carboniferous granite; *5* Caledonian or Variscan suture; *6* Proterozoic suture; *7* main fault; *8* deep seismic profile; *AB* regional cross-section

Figure 12 summarizes the regional evolution of the crust as it is known from the above references as well as from others such as Philipps et al. 1976; Moseley 1977, Johnson et al. 1979; Watson and Dunning 1979; Soper and Hutton 1984; Rolet et al. 1986 and Chantraine et al. 1988. In the Proterozoic three crustal blocks, probably belonging to the African realm (Perroud et al. 1984), were separated by two oceanic arms bounded by south-dipping subductions associated with volcanic arcs in the South Irish Sea and the English Channel. At the end of the Proterozoic, the Cadomian-Monian Orogeny welded these blocks together and resulted in intense deformation and widespread granite intrusion. The Iapetus Ocean opened up in the Late Proterozoic-Early Paleozoic between this continent and Laurentia and was bounded by two divergent subductions associated with accretion prisms and volcanism. A first deformation in the Late Ordovician probably marked the end of the southern subduction but the main collision occurred in the late Silurian, whereas Wales and southern Ireland were deformed and intruded by granite.

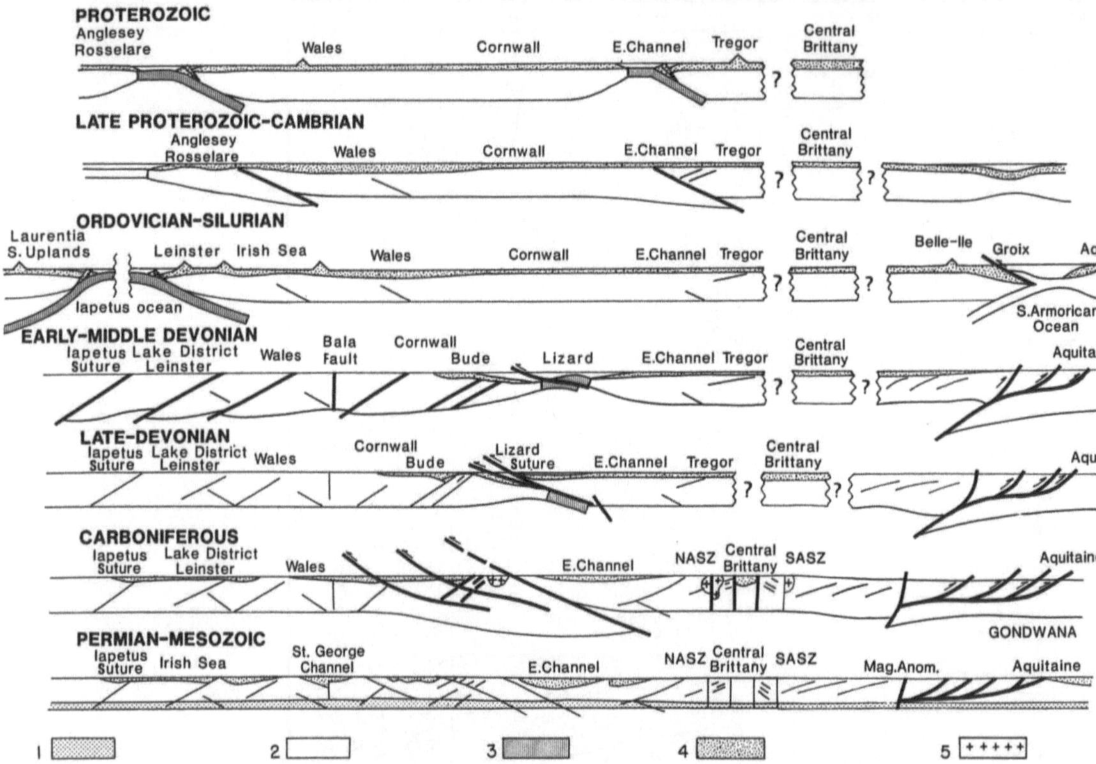

Fig. 12. Schematic evolution of southern Ireland, southwestern Britain, and western France from the Proterozoic to the Permo-Mesozoic. (Bois et al. 1990). Approximate location: traverse *AB* in Fig. 11. Not to scale. *1* layered lower crust; *2* continental crust; *3* oceanic crust; *4* volcanic and sedimentary series; *5* Late Carboniferous granite

Aquitaine was cut off from the Armorican domain by an ocean in the Ordovician (Lefort and Ribeiro 1980). The ocean was closed in the Devonian and the collision resulted in the deformation of the southern side of the Armorican Massif, associated with ductile strike-slip and migmatite emplacement. Distension in the English Channel resulted in the creation of a narrow oceanic basin (Lizard) which was subducted without any volcanic arc. Farther north, platform basins were infilled by sediments coming from the erosion of the Caledonian mountains.

The impingement of Africa against the northern terranes resulted in the overall Carboniferous shortening which led to the Lizard Ocean suturing, basin inversion, and large-scale thrusting associated with granite intrusion in southern Britain. South-vergent thrusting also affected the southern limb of the Variscan belt, and ductile strike-slip, lined by syntectonic granite, affected a large part of the area between the Lizard and Aquitaine. Over this multiphase crust, basins have developed since the Permo-Triassic with sedimentary piles that may locally exceed 10 km in thickness.

3.2 Seismic Experiments

The first deep seismic survey in the Celtic Sea and the English Channel was carried out in 1983. The SWAT (South Western Approaches Traverses) survey consisted of 11 profiles amounting to a total length of 1600 km. They were the southern continuation of the WINCH (Western Isles North Channel) survey, carried out in the Irish Sea by BIRPS (Brewer et al. 1983). Roughly oriented north-south, the SWAT profiles are perpendicular to the Variscan and Caledonian structures outcroppping in southern Ireland, southwestern Britain, and the northern Armorican Massif. Forming several parallel traverses, they provide an idea about the lateral evolution of the structural units. Unfortunately, the distance between the profiles is too great to achieve any precise correlation.

The WAM (Western Approaches Margin) profile was carried out in 1985 to extend the SWAT survey westward and to study the Atlantic margin in the Goban Spur and particularly the influence of the Atlantic opening on the lower crustal layering. This 645-km profile, tied to the SWAT 5 and SWAT 6 profiles, was shot perpendicular to the margin and

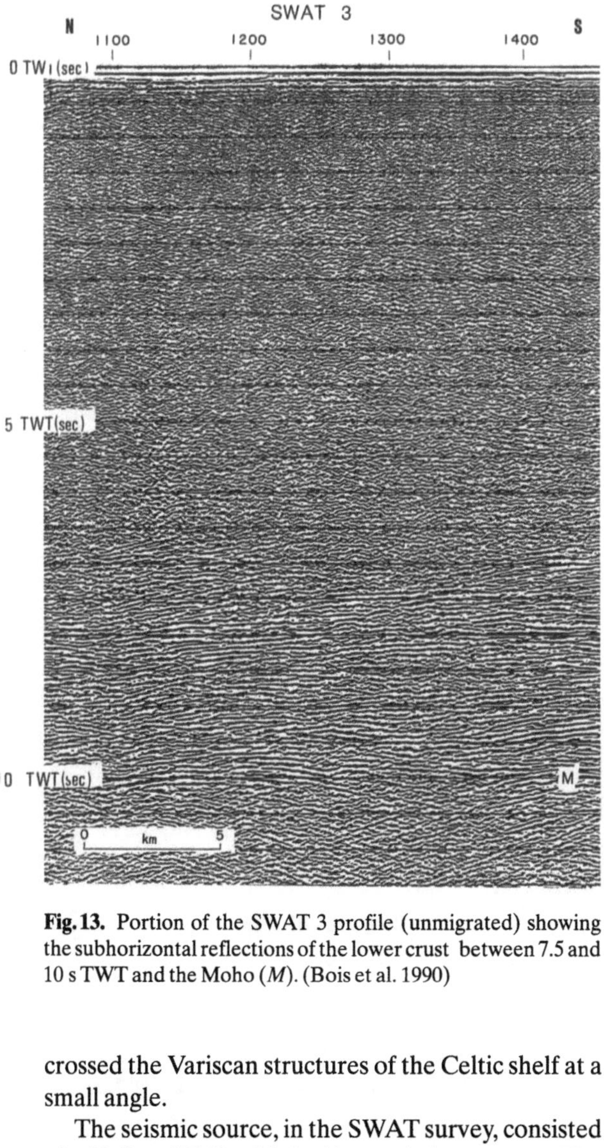

Fig. 13. Portion of the SWAT 3 profile (unmigrated) showing the subhorizontal reflections of the lower crust between 7.5 and 10 s TWT and the Moho (*M*). (Bois et al. 1990)

crossed the Variscan structures of the Celtic shelf at a small angle.

The seismic source, in the SWAT survey, consisted of a 91-litre (5560 cubic inch) tuned air gun array. The WAM source was more powerful: 118 litre (7235 cubic inch). In both surveys, the source was recorded by a 3000-m streamer with 60 traces, achieving a 30-fold coverage with a 25-m spacing between CDPs.

The results of the SWAT and WAM surveys are summarized in preliminary papers (BIRPS and ECORS 1986; Peddy et al. 1989) and presented in more detail in Bois et al. 1990 and particularly in the book devoted to this work (Bois et al. 1991).

3.3 Geological Interpretation of the Seismic Sections

3.3.1 Subhorizontal Features

The most striking feature observed in the SWAT and WAM profiles is certainly the flat lower crustal layering which forms a more or less continuous reflecting band, 2–4 s TWT in thickness (Fig. 13). Whereas the distribution of these reflections varies laterally (McGeary 1987) as in northern France, the bottom of the lower crust generally shows higher amplitude reflections and corresponds to the Moho, as demonstrated by seismic refraction and wide-angle reflection (Barton et al. 1984; Mooney and Brocher 1987). The Moho is located at about 10 s TWT, with the crust being about 30 km in thickness except in the easternmost SWAT profile in the English Channel (SWAT 11) where the Moho depth is 35 km, as it is beneath the Paris Basin. However, the Moho that looks flat in the time section should rise beneath the deep sedimentary basins if the lower velocity of the sediments is taken into account. Unfortunately, there are no precise velocity data available, but a reasonable estimate suggests that the Moho rise should not exceed a few kilometers, as confirmed by gravity modeling (Sibuet et al. 1989). As in northern France, the Moho and the lower crustal layering intersect dipping reflections which are regarded as Variscan features (e.g., SWAT 8 in the English Channel, Fig. 15). The geological nature and origin of the layered lower crust and the Moho will be discussed further in Section 5.2.

In all the SWAT and WINCH sections (Figs. 14, 15, and 17), subhorizontal reflections also occur at shallower depths in scattered packets, more or less separated from the lower crustal layering, and showing less continuity and amplitude. The nature of these reflections is uncertain but possible explanations include horizontal shearing, nappes, or flakes of reflecting material, petrologic differentiation related to granite emplacement or basic intrusions.

3.3.2 Variscan Features

The dipping reflections commonly display a criss-crossing pattern which illustrates the structural complexity of the crust (see SWAT 3 in Fig. 14). The last orogenic event which shaped this crust was the Variscan Orogeny. Bounded in the north by the Variscan Front and in the south by the Lizard Suture, the northern Variscan Unit, called the Celtic Unit, is characterized by south-dipping reflections (Figs. 14 and 15). The Variscan frontal detachment was identified in the SWAT 2, 3, 4 and 5 profiles by comparison

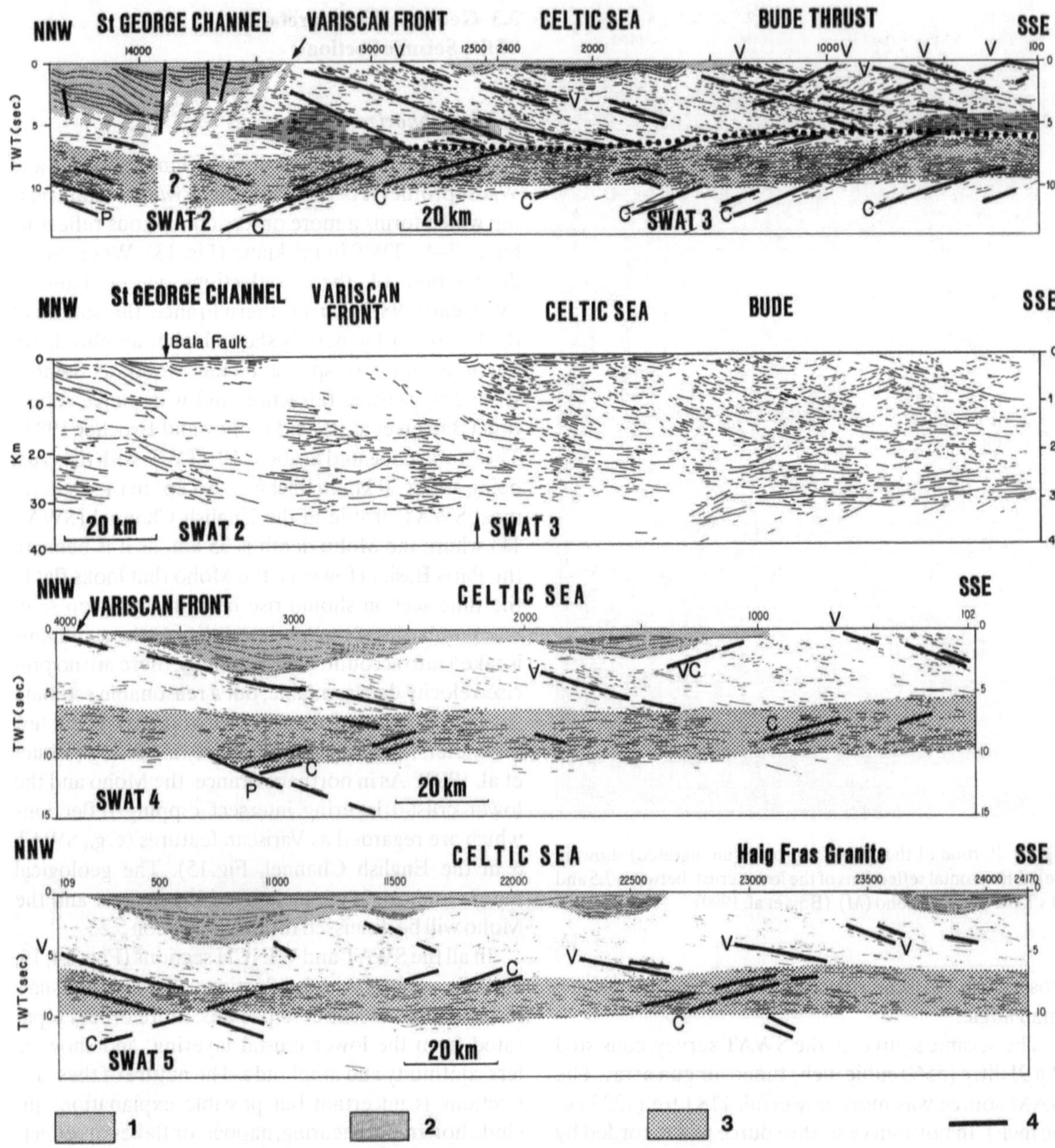

Fig. 14. Line drawing of the SWAT 2, 3, 4, and 5 profiles (unmigrated) and SWAT 3 profile (migrated) in the Celtic Sea and the Saint George Channel. See location in Fig. 11. (Bois et al. 1990) The main features are interpreted as follows: *V* Variscan; *C* Caledonian; *P* Cadomian; *VC* Caledonian rejuvenated as Variscan thrust. The boundary between the deep north-dipping and the shallow south-dipping features (*dotted line*) may correspond to the bottom of the Variscan frontal unit. *1* Sedimentary cover (*plain* Permo-Mesozoic; *shaded* Paleozoic); *2* shallow crustal layering; *3* lower crustal layering; *4* dipping reflection with high amplitude

with field data in southern Wales and Ireland (Le Gall 1990). Its seismic image in the Celtic Sea is a single reflection cutting straight across the upper and middle crust and rooted near the top of the layered lower crust (Fig. 14). This detachment does not show any broad flat as in northern France (Cazes et al. 1985, 1986; Cazes and Torreilles 1988) and Germany (Meissner et al. 1981, 1983). Only a narrow flat might have affected the detachment before the basin subsidence shown by the SWAT 4 profile. In the SWAT 2 profile, beyond the Variscan Front and beneath the Permo-Mesozoic basin, the presence of flat Paleozoic sediments is inferred from faint seismic indications and gravity modeling (Sibuet et al. 1989).

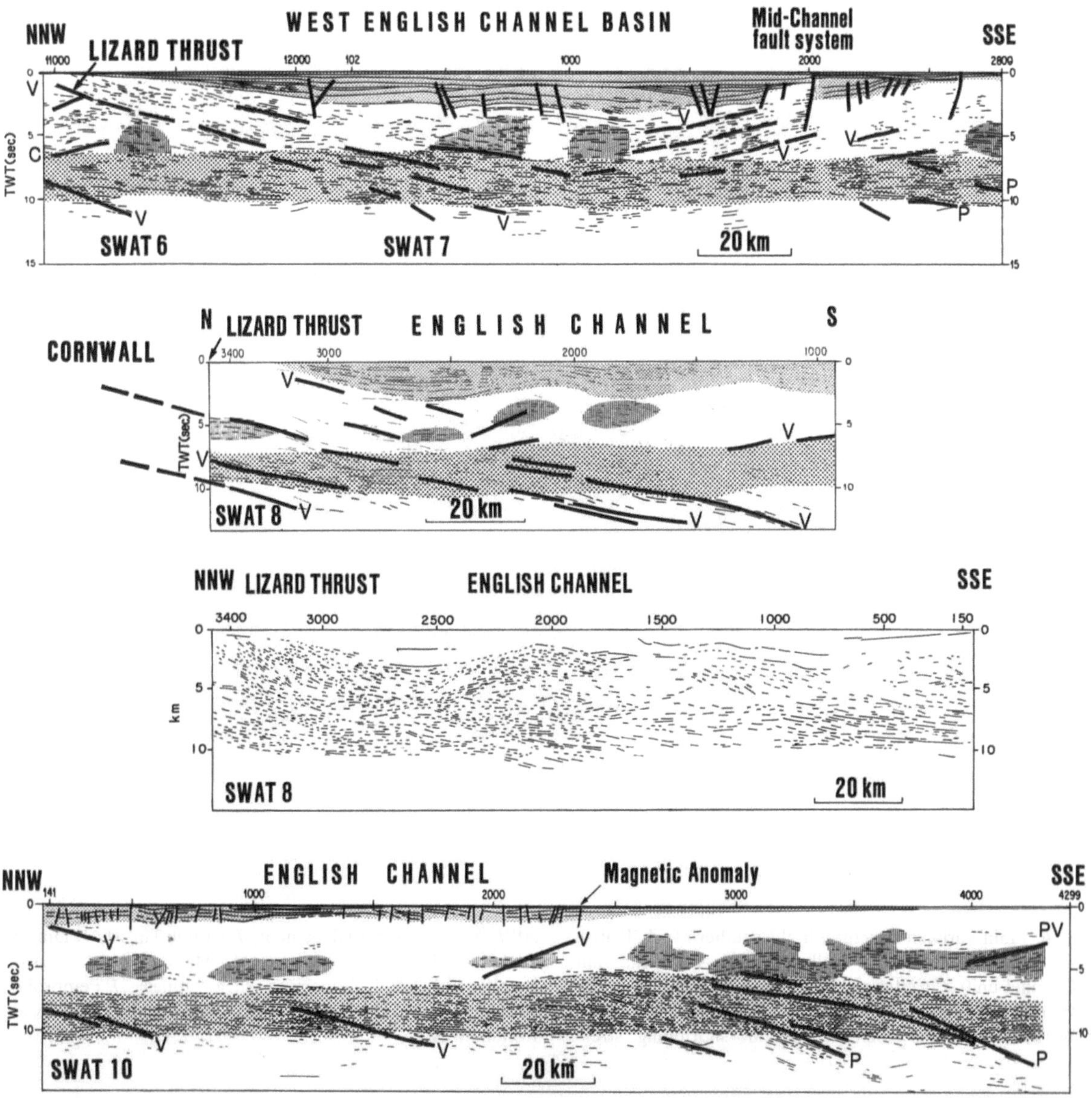

Fig. 15. Line drawing of the SWAT 6, 7, 8, and 10 profiles (unmigrated) and SWAT 8 profile (migrated) in the English Channel. (Bois et al. 1990). See location in Fig. 11. See legend to Fig. 14 for explanation

The frontal ramp emerges at the surface along a discontinuous thrust trending roughly 110° and corresponding to the Dingle-Mallow Thrust in Ireland and the Pembroke Thrust in Wales (Fig. 16). It carries a thick Devonian terrigenous series over a shallow Caledonian basement overlain by thin Devonian and Namurian series. A slight deformation of the Namurian series, however, occurs up to 80 km north of the frontal thrust in Ireland (Le Gall 1990). The external shallow detachment is not imaged in the SWAT 2 and 4 profiles because of their southern location. In this area, the Variscan shortening was also caused by major wrenching which led to a 20-km dextral offset of the frontal thrust between Ireland and Wales (Figs. 14 and 16).

South of the front, the crust is cut by straight south-dipping reflections interpreted as thrust structures because their vergence is consistent with the regional compression observed in Devon and Cornwall (Coward and Smallwood 1984). In the northern part of the Celtic Sea, these thrusts disappear near the top of the layered lower crust where a decoupling level is inferred above an autochthonous basement affected by north-dipping reflections that will be discussed in the next section.

Farther south, the thrusts seem to be rooted deep-

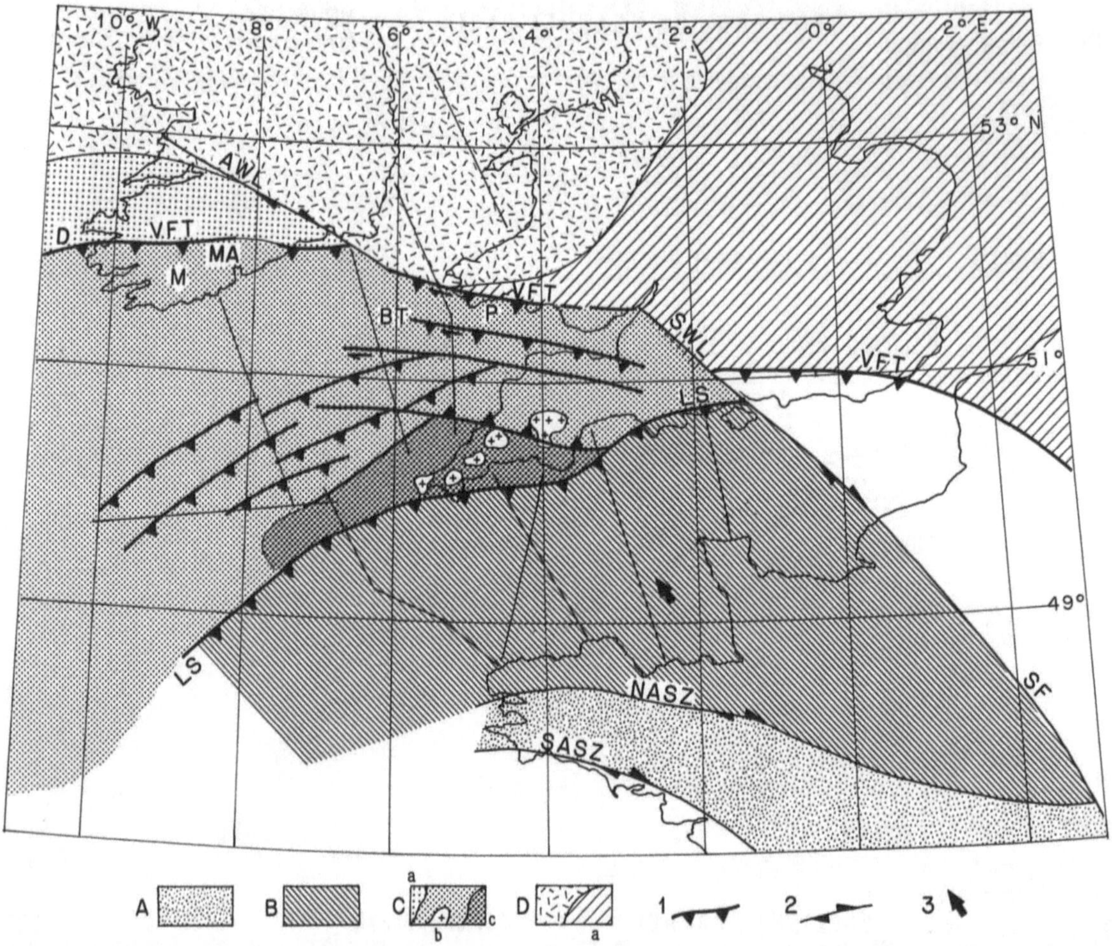

Fig. 16. Main Variscan structures in the southern Irish, British, and Armorican areas. (Le Gall 1990). *A* Armorican unit; *B* Channel unit; *C* Celtic unit (*a* thin-skinned unit; *b* granite batholith; *c* high-strain zone). *D* foreland (*a* Midland craton). *1* thrust; *2* strike-slip fault; *3* Variscan shortening stress. *AWL* Aran-Waterford lineament; *BT* Bude Thrust; *D* Dingle; *LS* Lizard Suture; *M* Munster; *MA* Mallow; *NASZ*, *SASZ* North- and South-Armorican shear zones; *P* Pembroke; *SF* Seine Fault; *SWL* Severn-Wight Lineament; *VFT* Variscan frontal thrust

er, beneath the present Moho (Fig. 15; SWAT 6, 8, and 10), and they may determine a deep crustal duplex beneath the Bude Carboniferous Basin (Le Gall 1990). Most of the deep crustal ramps do not reach the surface. In Devon and Cornwall they are overlain by a high-strain shear zone developed in the thick Devonian terrigenous series. The decoupling level which should be the bottom of this shear zone (Shackelton et al. 1982) and form the roof of the crustal duplex is not imaged by the seismic profiles, most likely because of subsequent deformation. Shallow north-dipping reflectors occur beneath the southern rim of the Bude Basin and can be related to Variscan south-vergent thrusts (Sanderson 1979) (Fig. 14, SWAT 3). The southern boundary of the Celtic Unit corresponds to south-dipping reflections, of very high amplitude, cutting across the whole crust down to the upper mantle in the English Channel (Fig. 15) and interpreted as the Lizard Suture. Correlation between profiles and field data shows that these reflections have a 070° trend which is parallel to the Cornwall Batholith alignment, and that they emerge near the Devonian Lizard Ophiolite Complex (Hillis and Day 1987). However, the straight gently dipping shape of such reflections is a rather unexpected geometry for a suture.

In the Celtic Unit, the areal distribution of the Variscan thrusts cannot be determined with confidence because of the great distance between the profiles. Between the 110° Variscan Front and the 070° Lizard Suture, the thrusts display a west-opening fan shape (Fig. 16), although gravity indicates that the 070° trend occurs in the Celtic Sea up to southern Ireland. This trend bends eastwards to 110° between Cornwall and Devon, while the deformation rate in-

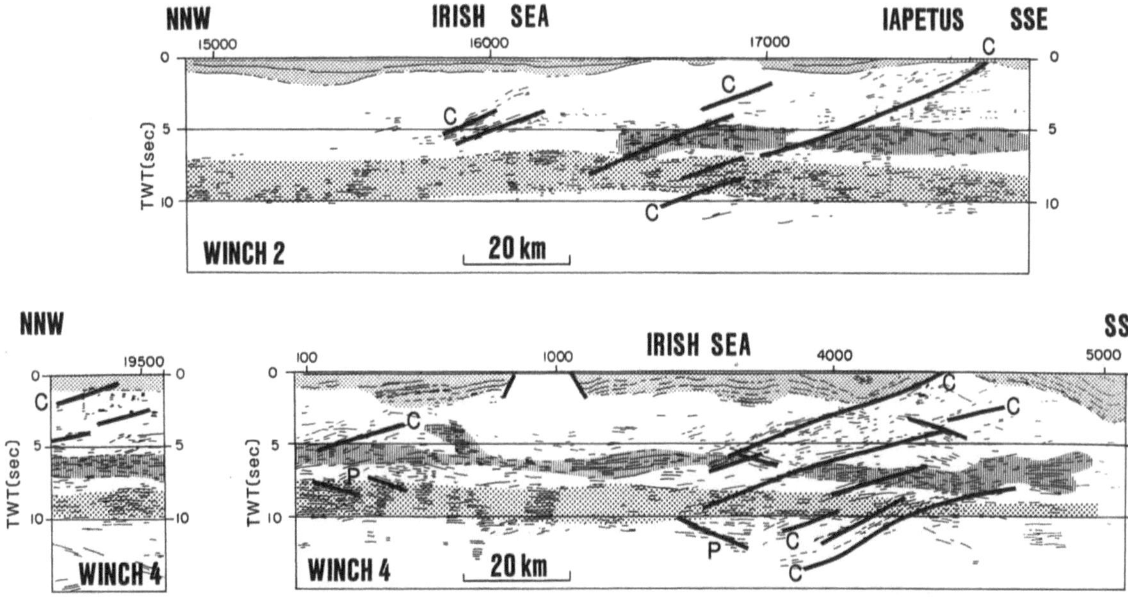

Fig. 17. Line drawing (unmigrated) of the WINCH 2 and 4 profiles in the South Irish Sea. (Bois et al. 1990) See location in Fig. 11. See legend to Fig. 14 for explanation

creases southwards and eastwards as illustrated by the greater number of south-dipping reflections in the SWAT 3 profile (Le Gall 1990) (Fig. 14).

The granite batholiths reflected by prominent gravity lows in the Celtic Sea (Sibuet et al. 1989) are not imaged by the SWAT sections. Their parallelism with major Variscan crustal thrusts indicates a crustal anatectic origin in agreement with geochemical isotopic and radiometric studies (Dodson and Rex 1971; Darbyshire and Shepherd 1985). By contrast, most of the Armorican granite is closely related to major ductile wrench shearing developed during several Carboniferous shortening stages (Berthé et al. 1979).

The Channel Block which was thrust over the Celtic Unit along the Lizard Suture was involved slightly in the Variscan deformation, as illustrated by a comparatively transparent upper crust. It shows a few north-dipping reflections correlated to south-vergent Variscan thrusts in northern Brittany and Normandy (Lefort 1991). Between the north- and south-dipping reflections occurring on both sides of the English Channel, the seismic sections show few reflections, with interference between dips having opposite directions. The detailed structure of this feature cannot be properly deciphered without back-up from boreholes.

During the Carboniferous shortening oriented 170°, the rigid Cadomian Channel Block was translated northward along its eastern and southern boundaries and punched into the Irish and Welsh foreland oriented 110°. The angle displayed by the colliding continents led to structural virgation, inhomogeneous strain, and late strike-slipping in the intervening Celtic Unit (Le Gall 1990) (Fig. 16). This setting might also be related to the formation of straight crustal thrusts in the whole Celtic area during the last stage of crustal shortening.

3.3.3 *Caledonian Features*

All the SWAT and WINCH sections, recorded in the Celtic and Irish seas, show northdipping reflections which are generally distributed in high-amplitude reflecting bands in the lower crust (Figs. 14 and 17). They may be regarded as thrusts related either to the Variscan antithetic thrusts in the Bude Basin or to the 060–070° Caledonian structures exposed in Wales and southeastern Ireland (Soper and Hutton 1984; Beamish and Smythe 1986) and to the Iapetus Suture (Brewer et al. 1983; Hall et al. 1984). Whereas some of the thrusts were active during the Variscan deformation, a Caledonian cause must be contemplated for those located north of the Variscan domain up to the Iapetus suture (Fig. 18). Beneath the Bude Basin, probable shallow Caledonian thrusts were rejuvenated by the Variscan compression.

The Caledonian reflections reach the surface and their trend can be appraised in the Irish Sea but they are generally deep seated in the Celtic Sea (Figs. 14, 17, and 18). Correlation of the Caledonian features across the Variscan Front cannot be established because the Variscan deformation should have removed

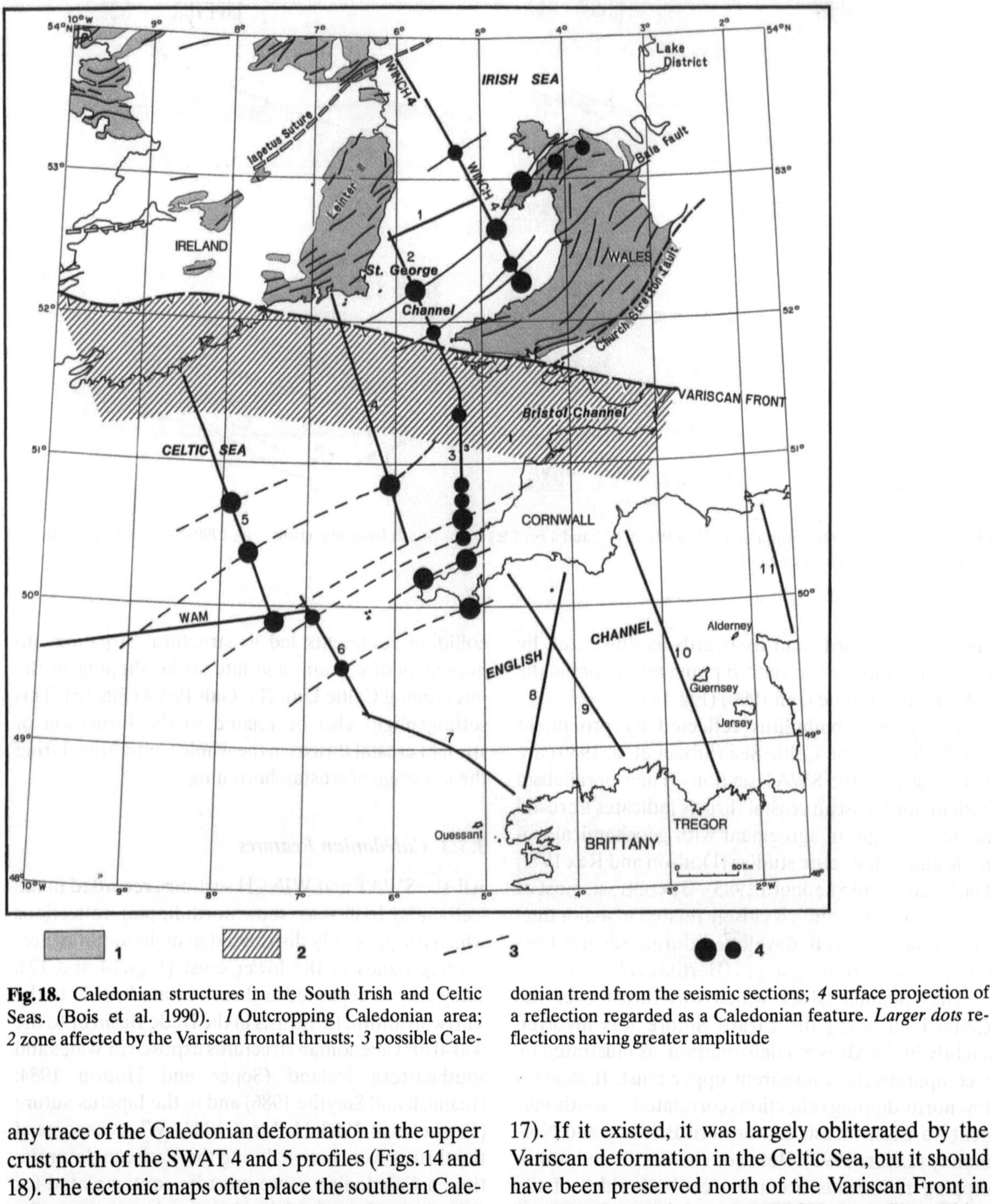

Fig. 18. Caledonian structures in the South Irish and Celtic Seas. (Bois et al. 1990). *1* Outcropping Caledonian area; *2* zone affected by the Variscan frontal thrusts; *3* possible Caledonian trend from the seismic sections; *4* surface projection of a reflection regarded as a Caledonian feature. *Larger dots* reflections having greater amplitude

any trace of the Caledonian deformation in the upper crust north of the SWAT 4 and 5 profiles (Figs. 14 and 18). The tectonic maps often place the southern Caledonian Front east of the Welsh Massif along the Church Stretton Fault (Dunning 1977; Harris 1985). This front was obliterated by the Variscan frontal units. North-dipping reflections, regarded as Caledonian features, cover the whole Celtic Sea to the Variscan granite batholith trend (Fig. 18). Therefore the location of the Caledonian Front in this area remains conjectural. Caledonian thin-skinned tectonics was assumed in Wales (Coward and Siddans 1979), but it is not visible in the seismic sections (Figs. 14 and 17). If it existed, it was largely obliterated by the Variscan deformation in the Celtic Sea, but it should have been preserved north of the Variscan Front in the SWAT 1 and 2 and WINCH 4 profiles. Thin-skinned tectonics was imaged no better in a seismic profile across the Church Stretton Fault (Smith 1987) where northwest-dipping reflections down to 4 s TWT are very similar to the SWAT and WINCH ones. A recent survey across the Norwegian Caledonides (Hurich et al. 1988) has revealed the same kind of image in an area where the existence of thin-skinned tectonics has been demonstrated in the field. The presence of straight north-dipping reflections regarded as

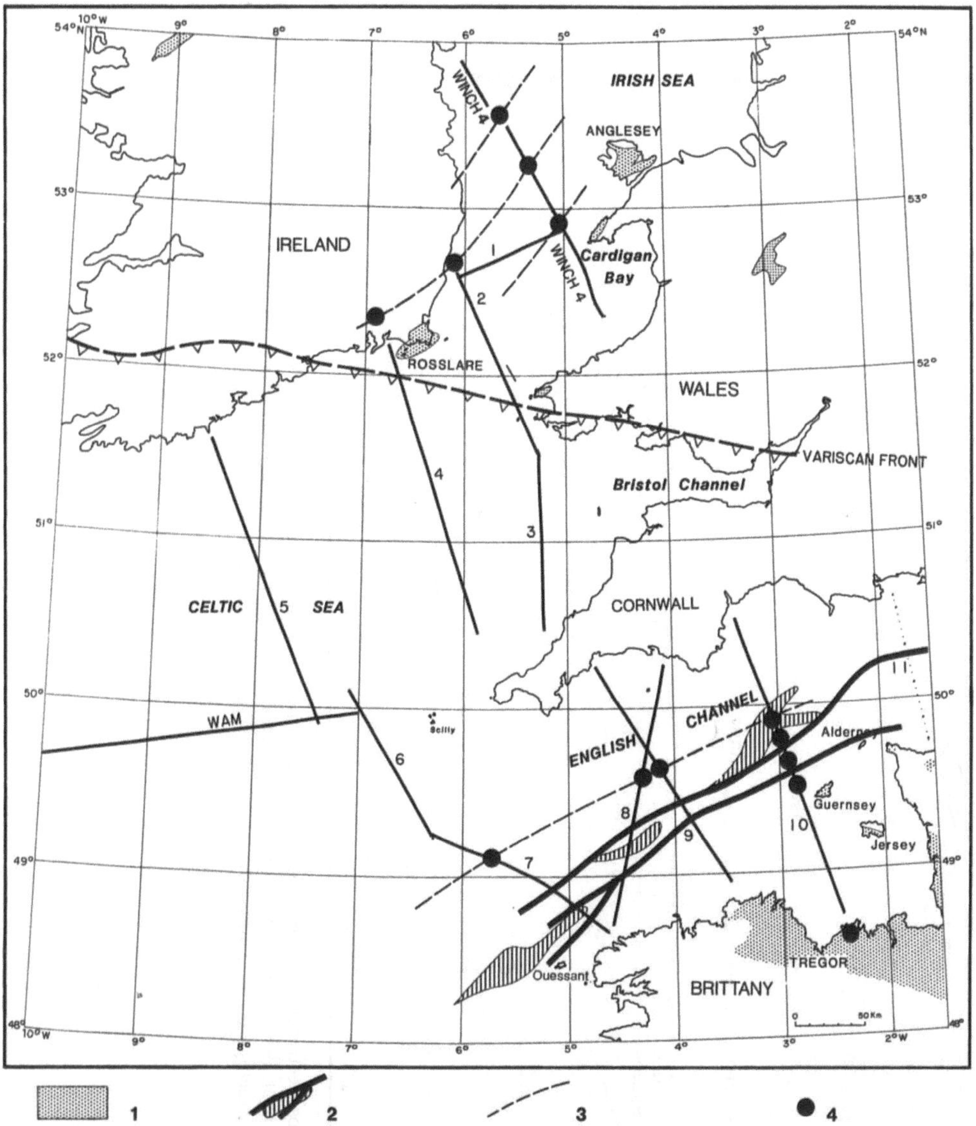

Fig. 19. Cadomian structures in the South Irish Sea and the English Channel. (Bois et al. 1990). *1* Outcropping Cadomian area; *2* mid-Channel magnetic anomaly and fault system; *3* possible Cadomian trend from the seismic sections; *4* surface projection of a reflection regarded as a Cadomian thrust

Caledonian thrusts across the whole crust, from the Celtic Sea to the Iapetus Suture, suggests a late shortening which deformed the thin-skinned detachments, making its seismic imaging impossible. Great obliquity between the colliding blocks (Philipps et al. 1976; Freeman et al. 1988) may have contributed to this overall crustal breaking which occurred when the colliding blocks were considerably compressed.

3.3.4 Cadomian Features

Most of the SWAT and WINCH profiles show south-dipping reflections that are parallel to the Variscan reflections but occur only in the crust-mantle transition. They have been tentatively regarded as Cadomian features. The most striking of these reflections can be observed south of the English Channel (Fig. 15), just southeast of a major magnetic anomaly. Together with the anomaly, they are located where a Cadomian suture was inferred on the basis of surface data (volcanic arc), shallow seismic and magnetic survey interpretation (Lefort 1975; Graviou and Auvray 1985). The magnetic anomaly is assumed to be the expression of mafic and ultramafic bodies associated with this suture. Figure 19 shows a clear offset of the anomaly with respect to the projection of the Cadomian reflections on the surface. This discrepancy may result from the Variscan reworking of the magnetic

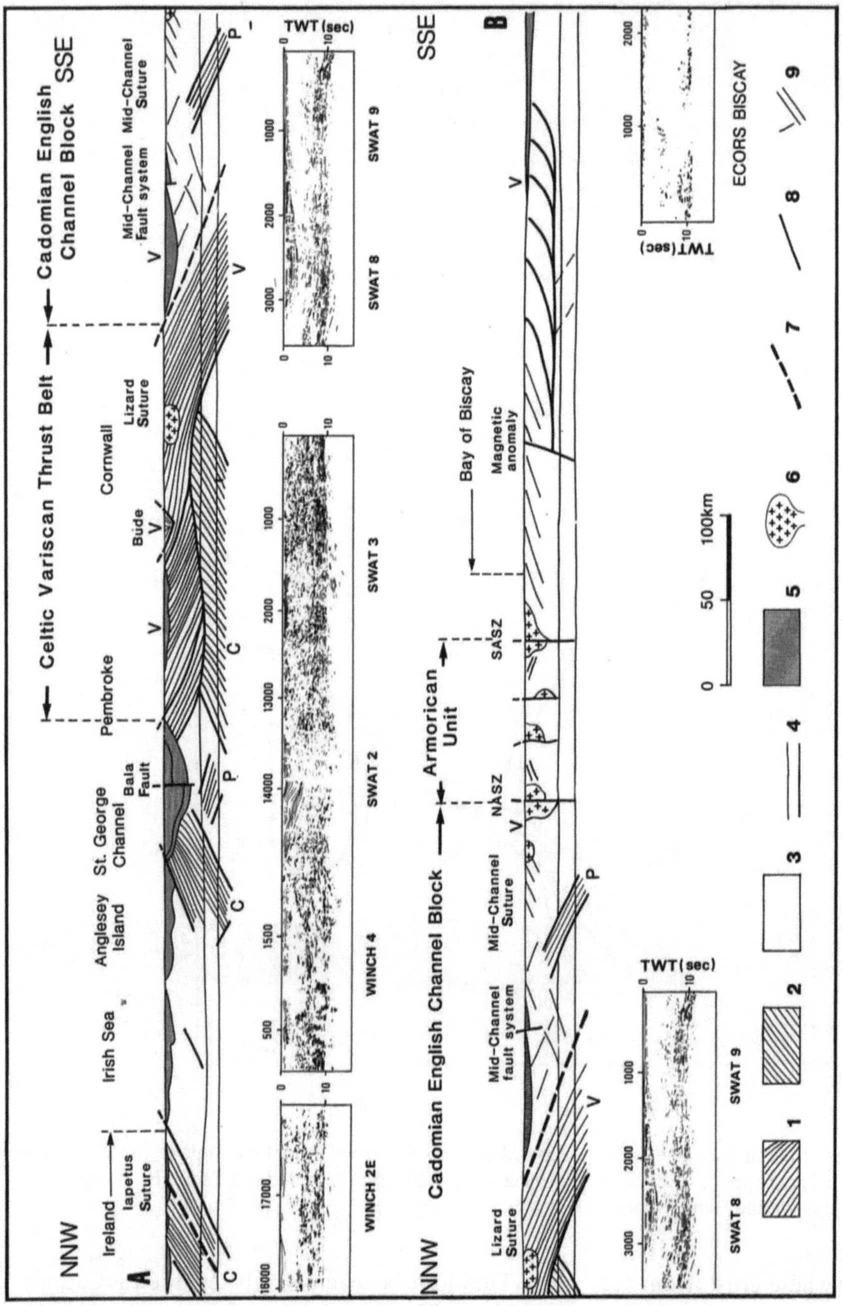

Fig. 20. Regional cross-section from southern Ireland to the northern Bay of Biscay (Bois et al. 1990). See location in Fig. 11. *1* Unit showing a north-vergent structural grain; *2* unit showing a south-vergent structural grain; *3* upper crust; *4* layered lower crust; *5* late- and post-tectonic sedimentary basin; *6* Late Carboniferous granite; *7* suture; *8* main thrust or fault; *9* subordinate thrust or fault

material by south-vergent thrusts (Lefort 1991; Lefort et al. 1991).

In the northern Celtic Sea and the southern Irish Sea, there are south-dipping reflections deeper than the Variscan frontal detachment, with some of them being beneath the Moho (Figs. 14 and 17). A rough correlation between the SWAT 2, 4, and 5 profiles leads to a 070° orientation that is oblique to the Variscan Front (Fig. 19). Similar reflections are also present in the WINCH 2 and 4 profiles (Fig. 17). All these reflections may be related to a south-dipping Cadomian-Monian subduction, which was active in the Proterozoic beneath Anglesey Island in the South Irish Sea (Cogné and Wright 1980) (Fig. 19).

The Cadomian thrusts were probably preserved in the crust because they were deeper than any later thin-skinned detachment and located within blocks that were little deformed by the Caledonian or Variscan orogenies. Another explanation is that the Proterozoic features themselves were rejuvenated by the subsequent tectonic events.

3.3.5 *Vertical Faults*

The subsidence affecting the crust during the Permo-Mesozoic was controlled in places by major vertical faults that probably reactivated former Proterozoic or Paleozoic faulted areas. Whereas the seismic sections clearly display such faults in the sedimentary cover, they are not visible in the upper crust and do not show any offset of the Moho and the layered lower crust. In the English Channel, a vertical fault system developed during the Mesozoic above the Cadomian Suture and was reactivated several times until the Oligocene (Boillot and Lefort 1974) (Figs. 15 and 19). The Moho and the layered lower crust are more or less continuous beneath it. The Bala Fault, which was active in the Jurassic and the Tertiary in the Saint George Channel, corresponds to a Caledonian feature observed in the field (Figs. 14 and 18). Beneath it, there is a blank in the layered lower crust but no obvious Moho offset can be observed in the seismic sections. Both faults may have been late tectonic vertical features, which could not be imaged in the upper crust and whose interpretation is very similar to that of the Bray Fault one (see Sect. 2.3).

3.3.6 *Regional Cross-Section*

Based on the SWAT data complemented by the WINCH and ECORS Bay of Biscay profiles, a 1000-km crustal section has been drawn from southern Ireland to northern Aquitaine (Figs. 11 and 20). This necessarily sketchy section covers a large portion of the crust from the Variscan foreland, affected by the Cadomian and Caledonian orogenies, to the Variscan internal zone. Seismic information has been completed in Cornwall and Brittany by previous deep seismic investigations (Hirn et al. 1975; Brooks et al. 1984). From north to south, Fig. 20 shows the Iapetus suture area, the south-dipping Cadomian features in the South Irish Sea, the north-dipping Caledonian thrusts in the South Irish Sea and the Celtic Sea, the south- and north-dipping Variscan thrusts in the Celtic Sea and the English Channel, and the south-dipping Cadomian features in the English Channel. The section ends with north-dipping Variscan thrusts and nappes in the northern Bay of Biscay (Pinet et al. 1987 a, b).

The dipping features may be divided into two categories. According to the geological context, the north-dipping features are either Caledonian or Variscan and the south-dipping ones either Cadomian or Variscan. More or less parallel dips in structures assigned to very different deformation periods suggest a tectonic heritage. The region between the two Proterozoic sutures was probably strongly affected by south-dipping thrusts during the Cadomian and early Caledonian orogenies. Some of them may have been reactivated during the Variscan deformation because of a stress in a similar direction. Variscan deformation may also have reactivated a number of north-dipping features regarded as Caledonian thrusts. An example has been observed beneath the Bude Basin where north-dipping Variscan thrusts may have been controlled by Caledonian ones. Together with the south-dipping ones, they may have guided the formation and inversion of the basin (Le Gall 1990). The upper crust in the English Channel was affected by Cadomian south-vergent antithetic thrusts. Their reactivation by the Variscan compression led to the double-vergent setting observed in the English Channel (Rolet et al. 1986; Lefort 1991).

3.4 Comparison with Seismic Profiles in England and Northern France

Deep seismic reflection data from southern England were published by Chadwick et al. (1983) (Fig. 11). The Variscan belt is covered by the Permo-Mesozoic series of the Wessex Basin, the subsidence of which was controlled by Paleozoic basement faults, some of them subsequently having been inverted (Chadwick et al. 1983). The Paleozoic series outcrops to the west, in the Mendip and Bristol area, where it shows asymmetric folds, overturned northward.

The line drawing of the IGS (Institute of Geological Science) 1979–1980 profiles (Figs. 11 and 21 b) shows comparatively few reflections, that have been interpreted in different ways, due to the uncertainty of their geological meaning (Chadwick et al. 1983; Williams and Brooks 1985). On the basis of a reasonable correlation with the Mendip area, all the authors agree that the northern set of south-dipping reflections is the Variscan Front. Despite the lack of any indication in the section, Williams and Brooks (1985) inferred a gentle subhorizontal detachment in which the frontal and other thrusts are rooted between 10 and 15 km. Two thrust zones appear in the seismic section as straight reflections with 25° southward dips. By means of cross-lines, they were both contoured down to 8–10 km (Chadwick et al. 1983) and found to be roughly planar features without any indication of decreasing dip with depth, in contrast to Williams and Brooks's (1985) interpretation. There is a striking similarity between the IGS line drawing and the SWAT sections in the Celtic Sea (Fig. 21 b, c, d), i. e., straight dipping reflections, flat layering of the foredeep sediments, shallow layering around 10–

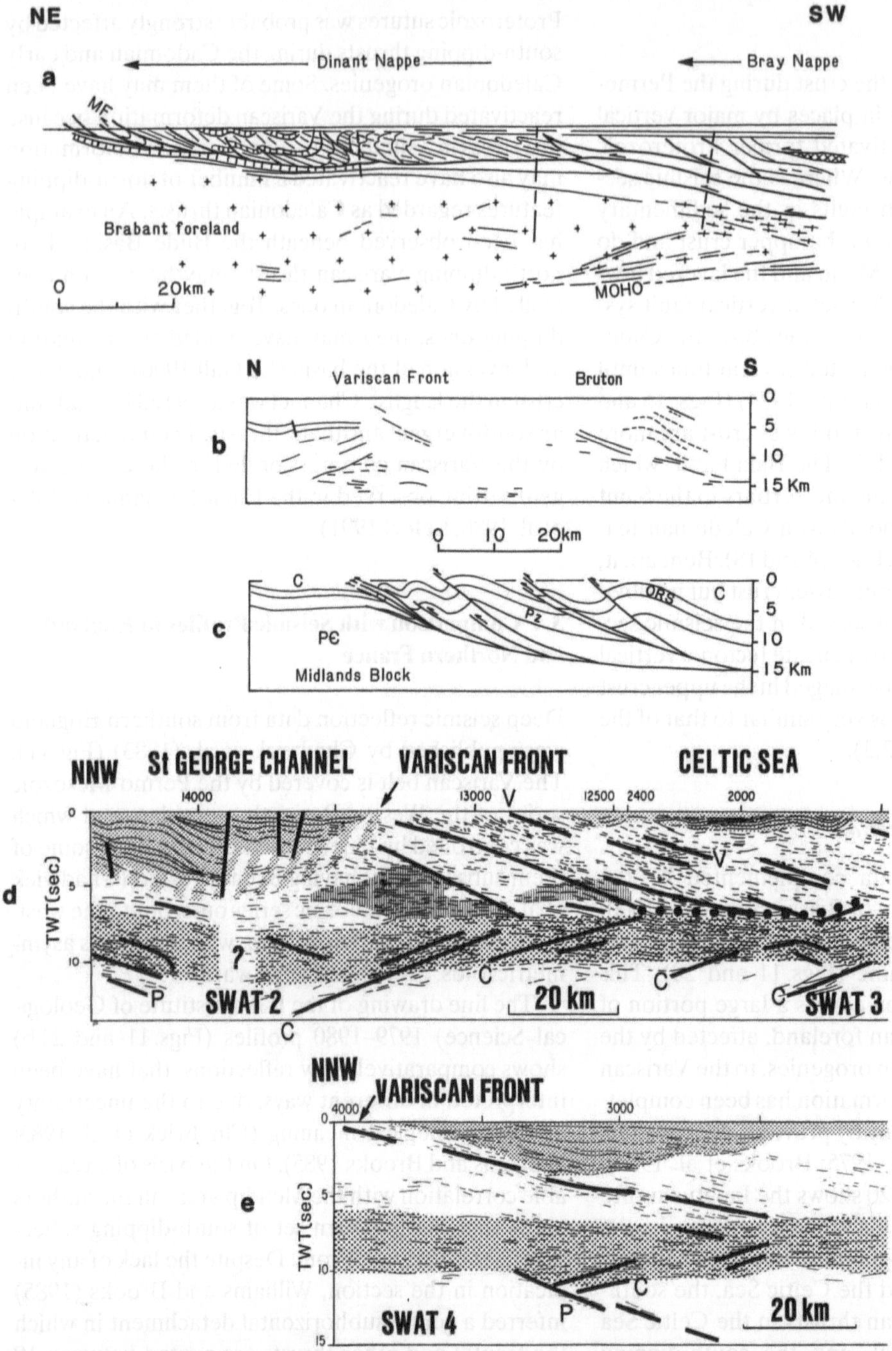

Fig. 21 a–e. Comparison between deep seismic profiles in the Celtic Sea, southern England, and northern France. **a** Geological interpretation of the northern France profile (see explanation in Fig. 5). Cazes and Torreilles 1988). **b** Line drawing (migrated and depth converted) of the IGS 1979–1980 profiles. (Chadwick et al. 1983). See location in Fig. 11. **c** Geological interpretation of the IGS 1979–1980 profiles. (Williams and Brooks 1985) (*C* Carboniferous; *ORS* Old Red Sandstone (Devonian); *Pz* Early Paleozoic; *PC* Pre-Cambrian). **d** Portion of the SWAT 2 and 3 profiles in the Saint George Channel and the Celtic Sea. See location in Fig. 11. **e** Portion of the SWAT 4 profile in the Celtic Sea. See location in Fig. 11. **c** and **d** Same as legend to Fig. 14

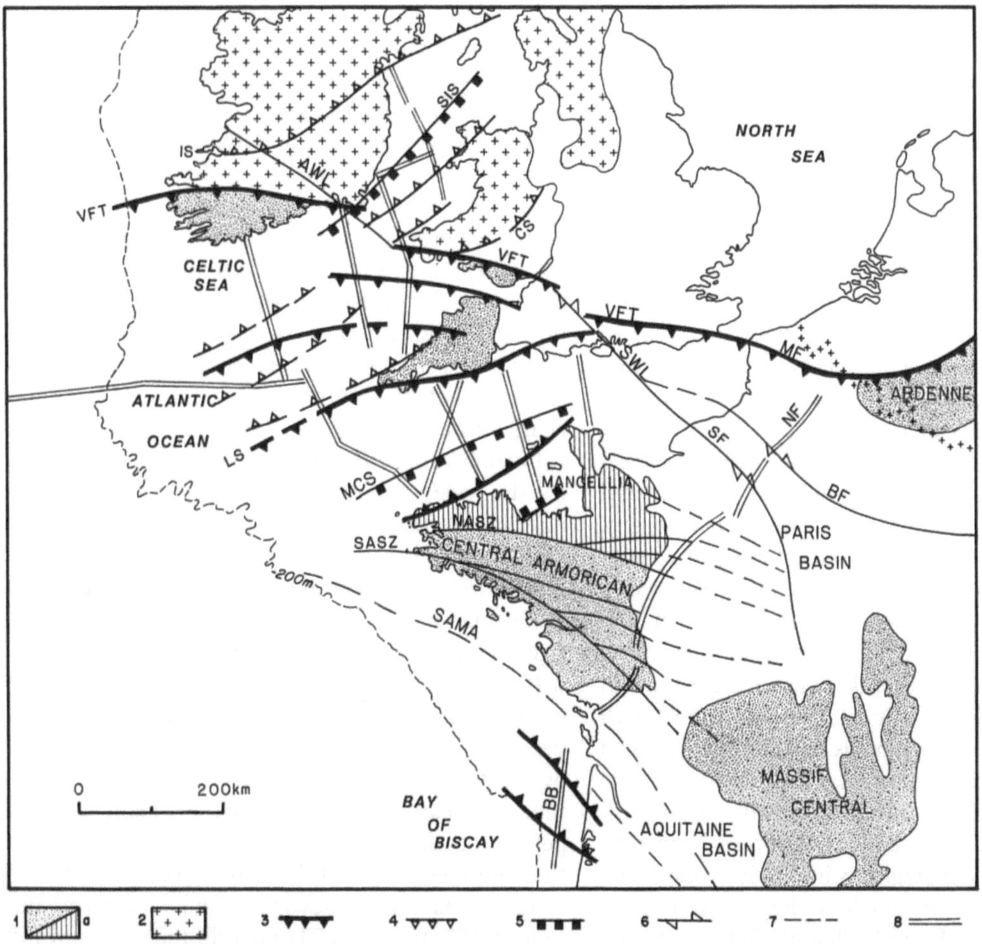

Fig. 22. Main structural features of the pre-Mesozoic crust in western France, southern Britain, and Ireland. *1* Variscan belt (*a* Cadomian block); *2* Caledonian belt; *3* Variscan thrust; *4* Caledonian thrust (projection to the surface); *5* Cadomian thrust (projection to the surface); *6* strike-slip fault; *7* major magnetic anomaly; *8* deep seismic profile. *AWL* Aran-Waterford Lineament; *BB* Bay of Biscay profile; *BF* Bray Fault; *LS* Lizard Suture; *MCS* mid-Channel magnetic anomaly; *MF* Midi Fault; *NASZ* North-Armorican Shear Zone; *NF* Northern France profile; *SAMA* South-Armorican magnetic anomaly; *SASZ* South-Armorican Shear Zone; *SF* Seine Fault; *SIS* South Irish Sea Suture; *SWL* Severn-Wight Lineament; *VFT* Variscan frontal thrust

15 km (equivalent to 3.5–5 s TWT), and north-dipping reflections in the foreland crust. Whatever the interpretation may be, the detachment is much deeper than in northern France (Fig. 21 a) and not associated with a thick band of subhorizontal reflections. The first interpretation (Chadwick et al. 1983) suggested basement involvement. The authors are inclined to believe that southern England, like the Celtic Sea, underwent Variscan oblique shortening leading to both an overall dextral shearing and a major strike-slip along the Severn-Wight Lineament which may extend southeastward into the Seine Fault (Le Gall 1990) (Fig. 16).

Seismic reflections that are deeper than 15 km are very scarce in the IGS 1979–1980 sections (Chadwick et al. 1983). Very short events situated near 10 s TWT were interpreted as indications of the Moho. Nowhere can the lower crustal layering be observed, in contrast to the SWAT sections, where it is present everywhere, whatever the age of the orogenic deformations may be, and to northern France, where it is observed in the southern and central parts of the profile but fades northward.

All the profiles in Fig. 21 show north-dipping reflections in what is regarded as the foreland. Caledonian deformation has been inferred in the Celtic Sea and the Brabant foreland (see Sect. 2.5). In southern England, however, the foreland is the Precambrian Midlands Block, and the north-dipping reflections are probably lithological boundaries within the Late Proterozoic-Early Paleozoic Series.

The Lizard Suture is associated with prominent

south-dipping reflections which cut across the lower crust and the upper mantle (Fig. 15). In northern France, there are only a few indications of such reflections in the lower crust, beneath the Bray Suture (Fig. 5). In the more internal zones of the Variscan belt, the SWAT profiles shot in the English Channel indicate a complex double-vergent structure (Figs. 15 and 20). In the corresponding segment of the northern France profile, only north-vergent features have been imaged (Figs. 5 and 8). However, the wide-angle reflection technique could not provide a resolution comparable to the near-vertical reflection one, and the structural relations between the two areas deserve further deep seismic investigation.

Figure 22 shows how complex the regional structure is because of the interference between several major orogenies. The last one, the Variscan Orogeny, had the greatest effect, with a structural grain trending from 120°–130° in the southeast of the study area to 040°–060° in the northwest. In the north, the Variscan external units obliterated the southeast-vergent Caledonian structures that appear clearly, north of the Variscan frontal thrust, between the Church Stretton Fault and the Iapetus Suture. The Variscan external units are themselves superimposed on the Cadomian deformation, illustrated by the South Irish Sea Suture. The core of the Variscan belt is formed by the Cadomian Block, which is intensely cross-cut by dextral faults. The northern boundary of this terrane corresponds to the English Channel Suture and the Bray Fault area, where a Variscan suture is believed to emerge. If the Bray and Lizard sutures were once correlated, they now show an offset resulting from some late Variscan strike-slip. The same can be said of the Variscan Front, which also shows dextral offset. South of the Cadomian Block, the Armorican fold-belt displays an overall ductile shear evidenced by dextral strike-slip faults (Gapais and Le Corre 1980), which may extend southward down to the South-Armorican magnetic anomaly and perhaps farther south.

4 Aquitaine and Pyrénées

4.1 Regional Geological Setting

The main feature of the Aquitaine region is a large sedimentary basin of triangular shape, which is bounded in the north by the Massif Central, in the south by the Pyrénées range, and in the west by the Atlantic margin of the Bay of Biscay (Fig. 23). The basin formation was initiated in the Triassic. The basin was subsequently overprinted by the early Cretaceous local depressions of the Parentis and Arzacq basins and later its southern edge was overlain by the late Cretaceous and Tertiary foredeep basin related to the Pyrénées (BRGM et al. 1974). Beneath the Mesozoic sedimentary cover, the nature and structure of the Paleozoic basement are scarcely known. A number of oil exploration wells have reached this basement, but nowhere have they investigated more than a few hundred meters of Paleozoic rocks, leading to a very schematic map of the pre-Mesozoic subcrop. Recent studies of the material coming from some of these wells have shown that most of the ages from the Ordovician to the late Carboniferous are represented in a series with a Gondwanian affinity (Paris et al. 1988). These Paleozeoic rocks are distributed in three different tectonometamorphic units: (1) the southwestern unit represented by early Ordovician terrigenous deposits and Silurian black shales, with a gently dipping cleavage and low-grade metamorphism which increases northwestward, (2) the median unit made up of mid-late Ordovician terrigenous and tuffite series and Devonian carbonates, without any cleavage or metamorphism, and (3) the eastern unit characterized by Carboniferous deposits overlying Silurian or Ordovician rocks, and affected by slight cleavage and low grade metamorphism. Farther northeast, in the edge of the Massif Central, pelites with dolomite lenses overlain by basic rocks of tholeiite composition (Genis unit) contain Ordovician, Silurian, and Devonian fauna (Guillot and Lefèvre 1975) (Fig. 23). They overlie the acidic tuff and ignimbrite of the Thiviers-Payzac unit, also present in the Terrasson Inlier and regarded as Proterozoic. Both these units show low-grade metamorphism and intense strain illustrated by tight folding and steep foliation. They overlie the high-grade Upper Gneiss Unit through a major shear zone dipping towards the south and southwest (Floc'h et al. 1984). This contact corresponds to Carboniferous refolded south-verging thrusts or back-thrusting and strike-slip on the southwestern limb of the Tulle Antiform, in the core of which the Lower Gneiss Unit outcrops with flat foliation (Ledru and Autran 1987).

The dips observed in most of the wells and the Ordovician rocks overlying Carboniferous rocks in the Avensac well (Fig. 23) suggest large-scale thrusting. Decollement may be present at several levels and particularly in the Silurian black shales which seem to be absent in the median unit. The overall strain and metamorphism of the Paleozoic rocks show that the Aquitaine area belongs to the external part of the Variscan belt. In the Montagne Noire-Mouthoumet area to the east (Fig. 23), there are also major south-verging thrusts which were active in the early Carboniferous.

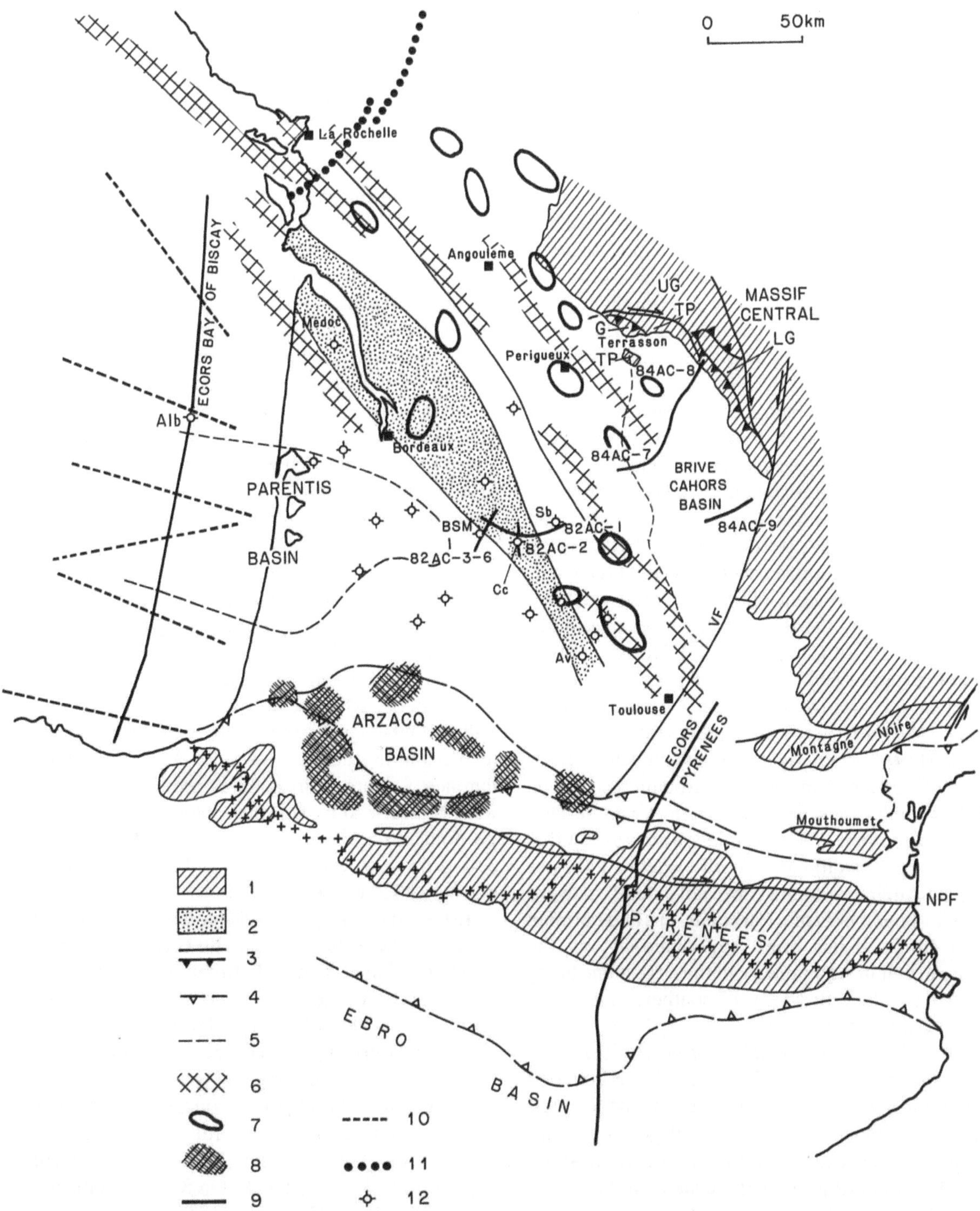

Fig. 23. Main features of the Aquitaine and Pyrénées region and location of the seismic profiles. *1* Pre-Permian outcrop; *2* median unit. (Paris et al. 1988); *3* Variscan fault; *4* Pyrenean thrust and fault; *5* Early Cretaceous basin boundary; *6* magnetic anomaly; *7* gravity low; *8* Elf Aquitaine deep seismic survey; *9* vertical seismic profile; *10* expanding spread profile; *11* wide-angle seismic reflection survey; *12* oil exploration well (*Alb* Albatros; *Av* Avensac; *BSM* Bouglon Sainte Marthe; *Cc* Clairac; *Sb* Soubirous); *G* Genis Unit; *LG* Lower Gneiss Unit; *NPF* North Pyrenean Fault; *SAMA* South Armorican magnetic anomaly; *TP* Thiviers-Payzac Unit; *UG* Upper Gneiss Unit; *VF* Villefranche fault

There are Variscan high-grade metamorphic rocks in the Pyrénées, and such a difference in metamorphism within so short a distance can be explained by the large disruption that affected the original Variscan setting in the late Carboniferous and Permian and brought together terranes from different tectonic zones along the north Pyrénées strike-slip fault (Arthaud and Matte 1977). Likewise, the setting of the Aquitaine median unit, which is less strained and metamorphosed than the adjacent units, suggests transcurrent displacements similar to those observed in the Armorican area. Gravity and boreholes indicate that the eastern unit has the present shape of a graben bounded by steep faults and infilled by Carboniferous rocks (Fig. 23).

The Variscan basement of Aquitaine is bounded by major faults. The north Pyrenean and Villefranche faults form the southern and eastern boundaries, respectively. The northern boundary may correspond to the main South-Armorican magnetic anomaly along which Lefort (1975) placed the suture between the Aquitaine and Armorican micro-plates. This anomaly, especially prominent offshore, can be extended southeastward by a series of smaller anomalies southwest of the Massif Central (Fig. 23). However, other authors (Matte 1986) have located the suture farther north in the area of the South-Armorican Shear Zone. Whatever its location, this feature was reworked by the overall right-lateral shear which affected all of western France in the Carboniferous. Some of these faults are seismically active (Veinante-Delhaye and Santoire 1980). Correlation across the Villefranche Fault is difficult, but all evidence indicates that the Aquitaine domain does not extend eastward from it. A number of late Carboniferous and Permian late-tectonic basins were superimposed on the Variscan belt in Aquitaine, with the largest one being the Brive-Cahors Basin on the southern side of the Massif Central (Fig. 23).

Aquitaine shows a number of gravity (Marchal 1988) and magnetic anomalies (Fig. 23). It is bounded to the north and northeast by steep magnetic gradients and anomalies associated with gravity highs, which indicate a north-northwest structural grain. The abrupt interruption of these anomalies between Angoulême and Périgueux suggests the presence of a northeast strike-slip fault. A west-northwest magnetic trend in the southern half of the area is related to basic intrusions in the Mesozoic cover. A number of gravity lows with round shapes are present in the area of the median and eastern units and, comparing them with outcrop and drillhole data, they are interpreted as late-Variscan granite batholiths.

4.2 Seismic Experiments

The 300-km ECORS Bay of Biscay profile parallel to the Aquitaine shore line was carried out in 1984 with the co-operation of REPSOL, the Spanish State Oil Company (Fig. 23). It was performed by two vessels: one towing a 140-l (8536 cu.in) airgun source was followed by a second vessel at a constant 7.5 km distance (Pinet et al. 1987 a). The shots were recorded to 15 s by the first vessel with a 3000-m streamer and to 16 s by the second one with a 2400-m streamer. Coverage was respectively 30-fold for the first vessel and 24-fold for the second, with the CDP spacing being 25 m. These wide aperture data provided a spectacular enhancement in the energy and continuity of the deep reflections. At intervals, six expanding-spread profiles (ESP), with a maximum offset of 100 km, were carried out perpendicular to the seismic reflection profile (Fig. 23). Their interpretation by two different methods provided information on velocity distribution (Marillier et al. 1987).

The 250-km ECORS-Pyrénées traverse was implemented in 1985 and 1986 in co-operation with ESSOREP (Société ESSO de Recherche et d'Exploitation Pétrolières), REPSOL, and other Spanish institutions (ECORS Pyrénées team 1988) (Fig. 23). Running from the Toulouse area to Balaguer in the Ebro Valley, it was mainly carried out by explosives except for 23 km in the narrow Salat Valley (northern Pyrénées) where Vibroseis was used. The coverage was 20-, 30-, or 60-fold for the explosive segments and 120-fold for the Vibroseis one. The sections were processed down to 15–20 s TWT and 25 s TWT in the northern 80 km.

In Central Aquitaine, a seismic test was carried out in 1982 by the Institut Français du Pétrole for a group of oil companies (British Petroleum, Elf Aquitaine, and ESSOREP) (Fig. 23). It consisted of a 50-km profile (82-AC 1) with two small traverses (AC 2 and 3). Five vibrators were recorded by a 10-km spread (192 traces, 50 m spacing) down to 4–6 s TWT, achieving a 96-fold coverage. Comparative explosive tests were carried out with a 24-fold coverage and one of them (AC 6) recorded down to 12 s TWT. This survey was completed by the 40-km 82-AC 7 Vibroseis profile in the Brive-Cahors Basin. In 1984, the seismic tests were resumed for Elf Aquitaine and ESSOREP and provided 80 km of Vibroseis profiles (84-AC 8 and 9) using one or two parallel 96-trace spreads (60 m spacing), providing 48- or 96-fold CDP. Three explosive shots carried out on the 84-AC 8 profile were recorded by a fixed 144-trace spread down to 30 s TWT.

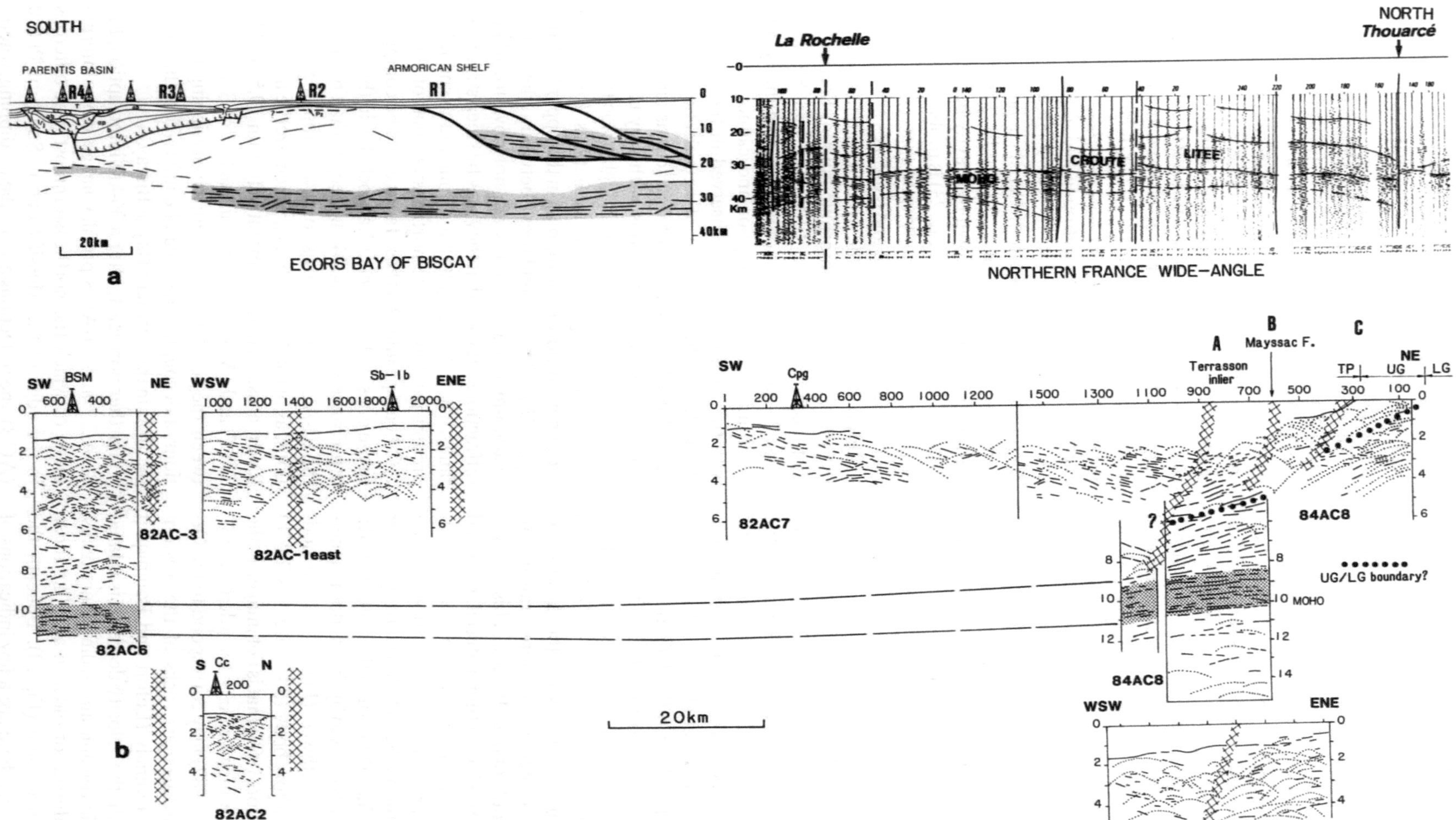

Fig. 24a, b. Seismic profiles in northern Aquitaine (see location in Fig. 23). **a** Geological interpretation of the ECORS Bay of Biscay profile (Pinet et al. 1987b) and southern end of the northern France wide-angle survey. (Matte and Hirn 1988). **b** Central Aquitaine industrial profile. *Heavy lines* high-amplitude reflections; *Dotted line* diffractions; *shaded* layered lower crust; *cross-hatched areas* strike-slip faults. *BSM* Bouglon-Sainte-Marthe; *Cc* Clairac; *Cpg* Campagne; *Sb* Soubirous)

4.3 Geological Interpretation of Seismic Data

The principal aim of the ECORS Bay of Biscay profile was to study the crust beneath the Parentis Basin whose formation is linked directly to the early Cretaceous opening of the Bay of Biscay (Marillier et al. 1987; Pinet et al. 1987a, b). Beneath the cover sediments which correspond to velocities of less than 5.7 km/s, the upper crust velocity ranges from 5.9 to 6.2 km/s and the lower crust velocity from 6.5 to 7.1 km/s. Major crustal thinning is imaged beneath the 15 km thick Mesozoic and Tertiary, accumulated in the Parentis Trough (Figs. 24 and 25). The Moho, which corresponds to a strong wide-angle reflection on each ESP and to the deepest high amplitude reflection on the vertical seismic data, is 35 km deep beneath the rims of the basin and only 20 km deep beneath its central part. This crustal thinning, associated with a strong fading of the lower crustal reflections, subdivides the crustal image into northern and southern segments.

4.3.1 Northern Aquitaine

Beneath the thin veneer of Mesozoic-Tertiary sediments, the northern part of the Bay of Biscay profile shows the usual lower crustal layering between 8 and 11 s TWT, with the Moho being about 33 km in depth (Marillier et al. 1987; Pinet et al. 1987a, b) (Fig. 24). At the northern end of this profile, flat reflections are also visible beneath the Moho down to 12 s TWT. The most striking feature is the shallower layering that occurs in three packets between 3 and 6 s TWT. Formed of short discontinuous reflections, the layering is bounded above and below by more transparent zones, and it fades out south of Pt 1300. This layering may be thrust sheets of reflecting material, bounded by listric north-dipping thrusts rooted in a sole detachment, located around 6 s TWT and associated with the nappes of the Variscan belt occurring farther north in Brittany. The northern end of the ECORS Bay of Biscay profile is crossed by the Medoc magnetic anomaly trending northwest (Fig. 23), whose modeling indicates a shallow body (3 km, i.e., 1 s TWT) (Poulpiquet and Lefort 1989) that is not imaged in the seismic section. The high velocity recorded by ESP R 1 (Marillier et al. 1987) may correspond to the Ordovician-Silurian terrigenous rocks of the western unit, south of the anomaly (Lefort and Poulpiquet 1990). Farther south, the velocity decrease is related to less metamorphic volcanic and sedimentary rocks, which are Carboniferous or older in age, like those found in the Albatros well (Fig. 23). Beneath the Variscan thrust sheets, there are a few indications of deep south-dipping reflections, some of which are in the upper mantle. They might be the trace of a foreland deformation older than the Variscan orogeny (Cadomian?).

The relationships between the ECORS Bay of Biscay and northern France profiles, briefly presented in Section 2.4, are shown in more detail in Fig. 24a. The northern France wide-angle profile displays deep south-dipping reflections south of La Rochelle (Matte and Hirn 1988) like those in the northern part of the Bay of Biscay profile. North of La Rochelle, north-dipping reflections can be correlated with South-Armorican south-vergent thrusts. A major late-Variscan strike-slip known on the surface near La Rochelle is expressed by reflection scattering and a blurred image of the crust-mantle transition associated with very deep flat reflections south of La Rochelle as in the northern end of the Bay of Biscay profile. The shallow Carboniferous graben shown by gravity between La Rochelle and the Sb well (Fig. 23) is probably bounded by vertical strike-slip faults (pull apart?).

The 82-AC 1 east, 2, 3, and 6 profiles are approximately perpendicular to the structural grain and cross the three tectono-metamorphic units of Section 4.1 that occur between the BSM, Cc, and Sb wells (Figs. 23 and 24). Reflection density, continuity, and amplitude are generally poor and cannot compare with the shallow layering observed in the northern Bay of Biscay. Identification of the reflections is hampered by a large number of diffractions and by the many flat multiple reflections, with very high amplitude, which come from Jurassic high-velocity flat carbonates overlying Paleozoic formations. Reflections are generally northeast-dipping in the upper 5 s TWT (Fig. 24b). The boundaries between the three tectono-metamorphic units are not obvious. Major diffractions indicate that they may be steep strike-slip faults. They also suggest that a similar fault should be present just beyond the northeastern end of the 82-AC 1 east profile. All these faults could be compared to the ones that occur near La Rochelle, in the northern France profile (Fig. 24a). Deeper than 6 s TWT, the 82-AC 6 profile shows southwest-dipping reflections, as in the Bay of Biscay, which overlie a layered lower crust that is comparatively thin (9.5 to 11 s TWT) and a Moho as deep as in the Bay of Biscay. Between the southwest- and northeast-dipping reflections, a decoupling level can be inferred at the same depth as in the northern Bay of Biscay.

Farther northeast, after a 36-km interruption, the 82-AC 7 and 84-AC 8 profiles image more internal units, of which the northeasternmost part outcrops in the edge of the Massif Central. South of Pt 850 (AC 8), the reflections are northeast-dipping, where-

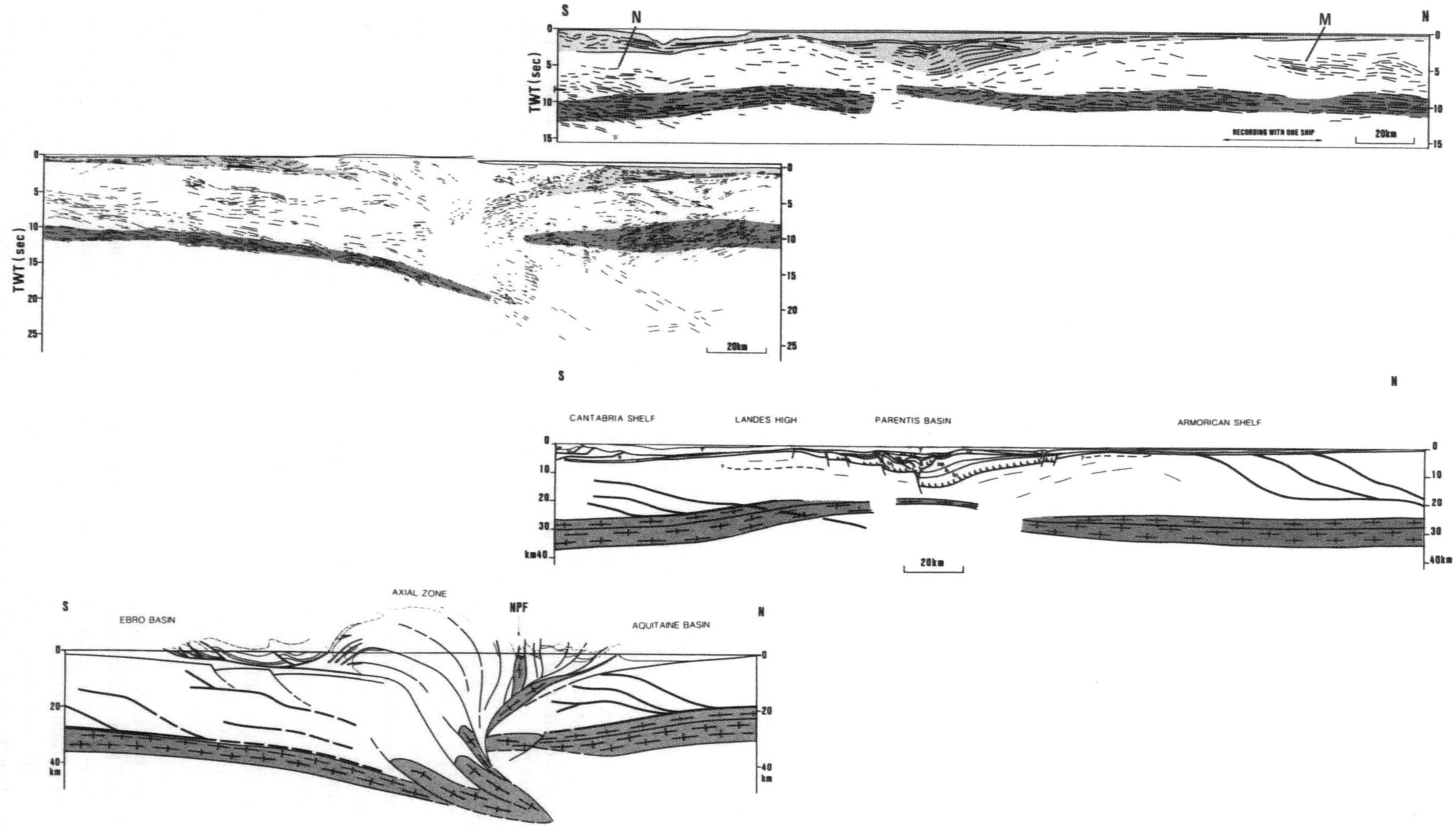

Fig. 25 a–d. Seismic profiles in Aquitaine and Pyrénées. (Choukroune et al. 1990). **a** Line drawing (unmigrated) of the ECORS Bay of Biscay profile. **b** Line drawing (unmigrated) of the ECORS Pyrénées profile. **c** Geological interpretation of the Bay of Biscay profile. **d** Geological interpretation of the Pyrénées profile. *NPF* North Pyrenean Fault; *stippled* layered lower crust as they are southwest-dipping farther north (Fig. 24b). A large domal structure, which is deeper than 3 s TWT (8 km) beneath Pt 850 (AC 7), can be related to a granite batholith that produces a gravity low, just northwest of the profile. Around Pt 1100 (AC-7), a fading of the reflections, opposite dips, and diffractions suggest a fault zone. But the main faults are farther north, near the Massif Central border. The first occurs near Pts 900–1100 (AC 8) between divergent dips and should correspond to the southern boundary of the Terrasson Inlier (Fig. 23). The northern boundary of the inlier is the Tertiary Meyssac Fault which does not appear in deep seismic data, because either it improbably dips northward or is rooted around 3–4 s TWT (± 8 km). The highly strained Genis and Thiviers-Payzac units produce only criss-crossing diffractions. In the northern end of the profile, another set of high-amplitude diffractions corresponds to the Upper Gneiss and underlying Lower Gneiss units outcropping in the southern limb of the Tulle Antiform. A few southwest-dipping reflections have been distinguished from the diffractions and may indicate the upper part of the Lower Gneiss Unit with basic and ultrabasic rocks. A strike-slip fault, which marks the boundary between the Thiviers-Payzac and Upper Gneiss units, has been tentatively placed at the southern end of these southwest-dipping reflections. It probably belongs to the same strike-slip and back-thrust fault system as the two Terrasson faults, which placed the Thiviers-Payzac Unit on top of the Upper Gneiss Unit in the Carboniferous. High-amplitude southwest-dipping reflections occur between 4 and 8 s TWT (8–20 km). Depending on the deep geometry of the above three faults, these reflections represent either the southern continuation of the Upper and Lower Gneiss units or more external units beneath the Thiviers-Payzac. The high amplitude of the reflections is suggestive of deeply metamorphosed formations that look similar to the shallow layering in the northern Bay of Biscay. Mafic lenses and eclogite relics in the Upper Gneiss, basic and ultrabasic rocks and micaschist intercalations in the Lower Gneiss, together with tectonic repetition, may explain the seismic image. This unit is underlain unconformably by the layered lower crust which is about 2 s TWT thick. The Moho shows a definite southwest dip from 10.2 to 10.8 s TWT. A few deeper seismic events are also observed in the upper mantle.

The 84-AC 9 profile provides an image that is roughly similar to the AC 8 one, although reflections appear poorer and more discontinuous among the many diffractions. A steep fault, inferred between reflections with opposite dips, may form the boundary between the Upper Gneiss in the east and the Thiviers-Payzac Unit in the west. In both the 84-AC 8 and 9 profiles, the sharp changes in reflection attitude can be compared with the image of the northern France profile, north of La Rochelle (Fig. 24), and regarded as the traces of late Variscan strike-slip faults. However, no equivalent of the south-dipping reflections which mark the southern limb of the Tulle Antiform have been observed in the northern France profile (Figs. 8 and 24).

4.3.2 South Aquitaine and the Pyrénées

In both the ECORS Bay of Biscay and Pyrénées profiles, north-dipping reflections are observed in the middle crust beneath the southern part of the Aquitaine Basin (Choukroune et al. 1990) (Fig. 26). In the Bay of Biscay profile, these reflections are readily truncated by the pre-Mesozoic unconformity and can be regarded as Paleozoic thrusts not reactivated by the Tertiary compression. They generally die out on top of the lower crust layering, which occurs between 9 and 12 s TWT with a geometry reflecting the flexure of the European plate beneath the Pyrénées range. Deeper north-dipping reflections, however, are imaged in the upper mantle beneath the southern limb of the Parentis Basin. They can be compared with the south-dipping reflections occurring in the upper mantle in the northern English Channel (SWAT 8, Fig. 15).

The ECORS Pyrénées profile was aimed at obtaining a complete seismic traverse across the Tertiary belt from the Aquitaine to the Ebro foreland (ECORS Pyrénées team 1988; Choukroune et al. 1990). Whereas the crust has been intensely deformed and thickened beneath the belt, the pre-Pyrenean structures have been preserved in both forelands. The section shows contrasting images of the crust on both sides of the Pyrénées Axial Zone (Fig. 25). The crust is 8 s TWT (25 km) thick and the layered lower crust 3 s TWT thick beneath Aquitaine, whereas these units are respectively 11 s TWT (35 km) and 1.5 s TWT thick beneath the Ebro Basin. Despite these differences, which are at least in part a late acquisition, both forelands display north-dipping and flat reflections in the middle crust. They appear from 3 to 5 s TWT beneath Aquitaine and die out near the top of the layered lower crust. These structures may be correlated to the Montagne Noire-Mouthoumet Variscan thrusts outcropping 100 km farther east. In the Ebro foreland, the deep reflections are displayed from the southern end of the section to the edge of the Axial Zone. All these reflections outline the classical flat-ramp geometry.

Since the late 1970's, Elf Aquitaine has recorded oil exploration seismic surveys in southern Aquitaine

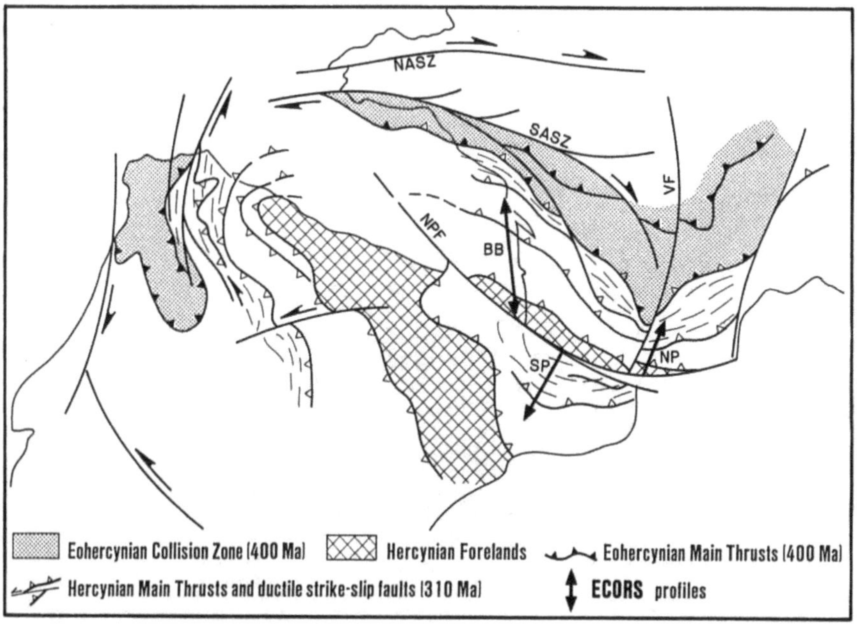

Fig. 26. Restoration of the Variscan central and southern belt setting in the Permian. (After Choukroune et al. 1990). *BB* Bay of Biscay profile; *NASZ* North-Armorican Shear Zone; *NP* Pyrénées profile (North); *NPF* North Pyrénées Fault; *SASZ* South-Armorican Shear Zone; *SP* Pyrénées profile (South); *VF* Villefranche Fault

and the northern Pyrénées down to 12 s TWT (Bois et al. 1986) (Fig. 23). The seismic sections have revealed the presence of north-dipping reflections between 5 and 8 s TWT beneath the pre-Mesozoic unconformity (Torreilles pers. commun.) which seem to form a link between the ECORS Pyrénées and Bay of Biscay profiles and show the continuity of these Paleozoic structures beneath the southern boundary of the Mesozoic platform overlain by the Pyrenean frontal thrust.

Any restoration of the Iberian Plate's position during the Variscan Orogeny is hampered by the late-Variscan break-up, mainly controlled by the North Pyrenean and Villefranche strike-slip faults (Arthaud and Matte 1977) and the early Cretaceous sinistral drift of the Iberian Plate (Choukroune 1976). The reconstruction in Fig. 26 suggests that the Variscan material outcropping in the Pyrénées and the Ebro Valley belongs to a more internal zone than the south Aquitaine structures (Choukroune et al. 1990). These latter are regarded as the roots of the southern Variscan Front, the more external part of which would have been destroyed by subsequent Mesozoic extension and Tertiary compression (Fig. 26).

5 Main Seismic Features of the Crust in France and Adjacent Areas

5.1 The Moho's Geometry

Both the near-vertical and wide-angle seismic reflection profiles, carried out since 1983 in France and adjacent areas, have brought valuable information on the Moho's geometry. In the near-vertical profiles, the Moho is placed at the bottom of the layered lower crust, above the more transparent upper mantle. The precision of its picking is no better than 0.5 s TWT, i. e., 1.5–1.8 km. There are few refraction experiments providing Pn waves along deep seismic reflection profiles in the study area. The coincidence between the refraction Moho and the bottom of the layered lower crust has, however, been demonstrated in the North Sea (Barton et al. 1984). The reflection Moho has been found to coincide with the wide-angle Moho in many places of the study area, e. g., northern France (Cazes et al. 1985, 1986; Hirn 1988), the Bay of Biscay (Marillier et al. 1987), the Pyrenees (Choukroune et al. 1990) and the Alps (Bayer et al. 1987; Nicolas et al. 1990). In the wide-angle profiles, the Moho corresponds to the first high-amplitude wave. The Moho contour map in Fig. 27 is largely based on Hirn's map in Weber et al. (1980), complemented by new data from Whittaker and Chadwick (1983), Meissner et al. (1986); Pinet et al. (1987 a, b), Lueschen et al. (1987), Matte and Hirn (1988), Choukroune et al. (1990) and Sibuet et al. (1989).

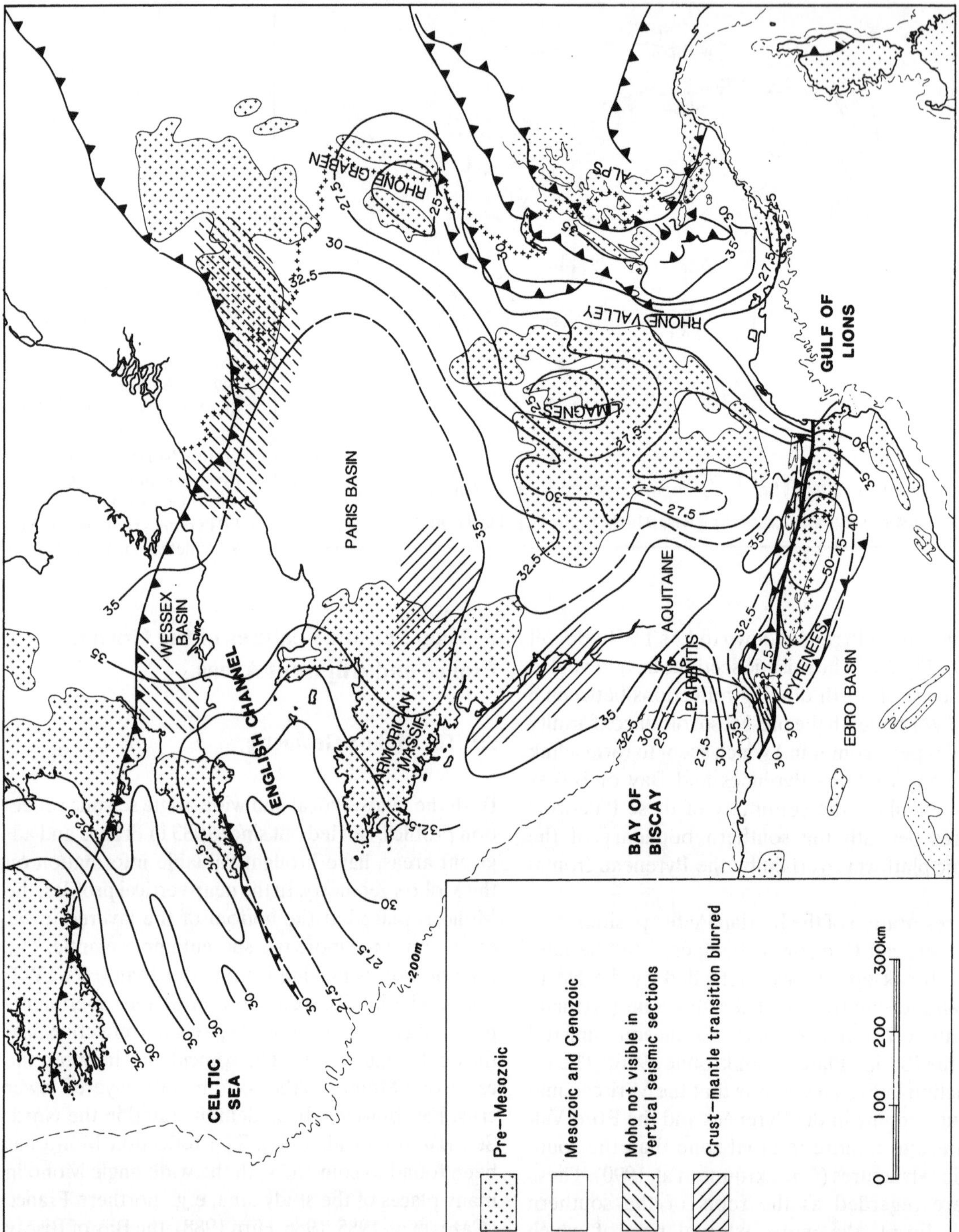

Fig. 27. Contour map of the Moho in France and adjacent areas. *Contour interval* 2.5 km. Sources: Hirn, in Weber et al. (1980); Whittaker and Chadwick (1983); Meissner et al. (1986); Lueschen et al. (1987); Pinet et al. (1987b); Matte and Hirn (1988); Choukroune et al. (1990); Sibuet et al. (1989)

Assuming that 30–35 km corresponds to the average thickness of the equilibrated crust in western Europe, the Moho is shallower in the southeastern half of France, within a number of anomalies more or less related to the early Cretaceous and/or early Tertiary rifting, namely Parentis, north-Pyrenean, Lower Rhone Valley, Gulf of Lions, and Limagne basins. Another region with a thin crust is found in the English Channel and the Celtic Sea, although the contours there should be regarded only as tentative because of the lack of velocity data leading to controversial interpretations (Meissner et al. 1986; Sibuet et al. 1989). Whatever the interpretation may be, the crust is probably thinner than 35 km in the whole area and thinner than 30 km beneath the thick Permo-Mesozoic basins.

By contrast, the Moho is deeper than 35 km in 2 types of areas: (1) cratonic areas beneath the Paris and Wessex basins, the South-Armorican and North-Aquitaine regions and the Ebro Valley, and (2) the axial zones of the Pyrénées and the Alps. In the orogenic belts, crustal thickening clearly results from Tertiary crustal stacking. The greater thickness of cratonic areas is not very well understood even though it is well documented in the northern France, northern Bay of Biscay, and southern Pyrénées profiles, despite crust-mantle blurring and Moho reflection fading in some places, as indicated in Fig. 27. The cause of this thickening might be underplating associated with a thick lower crustal layering and/or some crustal down-flexing related to the Tertiary compression. We can observe, however, a vague coincidence between the thick areas and the core of the pre-Variscan Cadomian, Aquitaine and Ebro terranes.

5.2 The Layered Lower Crust

Lower crustal layering has been observed in most of the deep seismic profiles shot in France and adjacent areas. It is characterized by flat reflections closely spaced in the lower part of the crust, between the comparatively transparent upper crust and upper mantle (Figs. 6 and 13). Such layering has also been found in the North Sea (Beach 1986), some places in southern England (Whittaker and Chadwick 1983, 1984), the Rhine Graben, and the Black Forest (Dohr and Meissner 1975, Lueschen et al. 1987), southern Germany (Meissner and Wever 1986), and Switzerland (Finck et al. 1984; Schweizerische Arbeitsgruppe Reflektionsseismik 1988). However, it fades in the northern part of the northern France profile and in Germany (Sect. 2.3 and 2.5) and is absent in Belgium (Boukhaert et al. 1988), in some parts of southern England (Sect. 2.4) and in the DEKORP 4 profile near the German-Czechoslovakian border (DEKORP Research Group 1988).

Wherever the layered lower crust is present, it corresponds to a P-wave velocity of 6.5–7 km/s (Pinet et al. 1987 a, b; Hirn 1988) while the upper crust velocity ranges between 5.9 and 6.2 km/s. These figures agree with a granulitic lower crust underlying an upper crust composed of lower grade metamorphic rocks with granite intrusions. Seismic refraction and wide-angle experiments carried out before the beginning of the ECORS project also found a 6.4–6.8 km/s lower crust lying deeper than 15–22 km in most of France (Hirn, in Weber et al. 1980). There is little overlap between these experiments and the near-vertical seismic reflection profiles of the ECORS project (Figs. 1 and 3). Wherever both kinds of data are available (Pyrénées, Alps, Jura, Bresse, Rhine Graben) the lower crust corresponds to a layered unit in seismic reflection sections. We can then assume that the layered lower crust is present over almost the whole French territory.

Several geological causes have been proposed to explain the nature and emplacement of lower crustal layering (see references in Bois et al. 1988). Whereas deep seismic data cannot thoroughly answer such a controversial question, Bois et al. (1988, 1989) have shown that seismic images can make a significant contribution to the timing of layering emplacement in the lower crust and the results are summarized here. In the Pyrénées (Fig. 25) and the Alps (Bayer et al. 1987), the Moho and the layered lower crust were deformed and offset by the Tertiary Orogeny. The upper mantle together with the lower crust are involved in imbricated thrusts. Although the Variscan range followed a structural model similar to the Alpine one (Matte 1986), the northern France and SWAT profiles show that the Moho and the layered lower crust are flat and not offset by dipping reflections related to the Variscan deformation, such as the ones occurring in the Lizard Suture area and extending into the upper mantle (Fig. 15). The emplacement of the Moho and crustal layering was then completed after the last Variscan tectonic event. Besides, it certainly occurred before the early Cretaceous because the layered lower crust was strongly attenuated by the Parentis Basin rifting (Fig. 25). By contrast, beneath the Celtic Sea and the English Channel basins, the layered lower crust does not change in thickness and seismic fabrics, with the whole crustal attenuation being concentrated within the upper crust (Figs. 14 and 15). These basins were initiated in the Permian-Triassic as the result of stress relaxation and thermal subsidence following Variscan crustal thickening and granite intrusion. A large share of the tectonic subsidence of the crust occurred during this period, even though subsidence went on in the Jurassic together with the rifting of the Tethys Ocean and the North Sea Graben System. The emplacement of the Moho and layering in the lower crust may then have resulted either from a late Variscan crust-mantle equilibration (Carboniferous, Permian, or Triassic) or from a thermal event associated with metamorphism, intrusion, moderate extension, and subsidence during the Jurassic.

Whatever its actual cause, the Moho and layering emplacement appears to have been largely a late process that obliterated the traces of the Variscan or older deformations in the crust and made the Paleozoic mountain roots disappear through deep metamorphism. This conclusion, however, does not preclude that part of the layering was acquired earlier, from shearing and delamination within the lower part of the crust (Matte and Hirn 1988), because its mechanical strength is weaker than that of the overlying upper crust and underlying upper mantle (Meissner and Wever 1986). In several instances, the imbricated thrusts related to the frontal Variscan unit seem to die out on top of this layering (northern France, Fig. 5 d; Celtic Sea, Fig. 14; Pyrenean crust, Fig. 25) and decoupling has been inferred at this level. Spectacular mylonites can be observed in an uplifted piece of Variscan lower crust which outcrops in the northern Pyrénées (Delay and Paquet 1988). However, no incipient Tertiary layering has been observed beneath the Pyrénées and the Alps, leading to the questioning of the formation of lower crustal layering from compressional processes. Besides, the poor reflectivity of the lower crust in Germany (Fig. 10) and the presence of well-developed layering in the Variscan foreland of the Celtic and South Irish Seas, beyond the Variscan Front (Figs. 14 and 17), indicate that the bulk of the layering is not related to the Variscan deformation.

In most of the seismic profiles, several reflections can cut across the lower crust and go down below the present Moho. They may be regarded as the traces of dipping Moho, thrusts, or sutures. Deep metamorphism may have transformed some of the crustal rocks that were deeper than 30–35 km into high velocity rocks (eclogites) which are not discernible from mantle rocks by seismic methods (Bois et al. 1988, 1989). The rocks that could not form high velocity eclogites (e.g., acid granulites) may also be present beneath the Moho reflection, where they image former crustal structures.

5.3 The Deep Crustal Faults and Shear Zones

Gentle ramps and shallow thin-skinned flats have been observed in front of the Variscan belt in northern France and Germany (Fig. 9). They are rooted within a set of dipping reflections regarded as imbricated thrusts and die out near the top of the layered lower crust which forms a major decoupling level at the bottom of the external orogenic unit. The same geometry and deep decoupling is present beneath Aquitaine and the Pyrénées (Fig. 25), but the corresponding shallow external flat was not imaged due to pre-Pyrenean and Pyrenean deformation overprinting.

Lower crustal decoupling is also inferred in the Celtic Sea, but the image of the crust is different (Fig. 14): straight features, with an apparent dip of 15–35 °, cut across large parts of the crust. They are interpreted as thrusts, possibly associated with strike-slip shear zones. Although the almost constant dip of these reflections makes them mimic seismic artifacts, their reality has been demonstrated by a few line crossings. The inferred thrusts probably correspond to shear fractures in a crust which should be roughly homogeneous. Their dip is related to the angle of internal friction of rocks whose overall properties should be about constant. The lack of shallow thin-skinned detachment images in the Celtic Sea seismic sections may be related to a late crustal shortening, largely controlled by these thrusts and subject to strike-slip in this area. Similar straight dipping reflections are found in more internal zones where they cut across the whole crust, their root being not visible in the upper mantle. They have also been regarded as thrusts, some of them being related to the Lizard Suture (Fig. 15). The unexpected straight thrust geometry in a suture zone suggests that the dipping features occurred late in the development of the Variscan deformation and overprinted the crustal weak zone related to the suture. A similar dipping feature has been observed in the Dinkelsbuhl Suture area in Germany (Fig. 10), but it has not been clearly observed in northern France, probably because the angle between the profile and the Variscan direction of transport significantly flattens the apparent dip of the reflections and makes it hard for them to be distinguished from the lower crustal layering. Moreover, the deep crust was disturbed in this area by the Bray Fault's strike-slip.

The area of Caledonian deformation shows a very similar picture, with the seismic reflections dipping northward in agreement with the overall vergence of the Caledonian belt and the Iapetus Suture (Figs. 14 and 17). Due to Variscan reworking, these reflections were mainly developed in the middle and lower parts of the crust and extend into the upper mantle where their roots are not visible. Their dip value is comparable to that of the Variscan features, probably due to the same mechanical cause. The Caledonides are well known to have resulted from an oblique collision (Phillips et al. 1976) and to have undergone intense late tectonic strike-slip. In both the Variscan and Caledonian deformations, the straight dipping features seem to have been superimposed on the previously deformed crust. They may have deformed the thin-skinned tectonic setting which is no longer visible in the seismic profiles.

That the dipping crustal reflections are post-Variscan extensional structures is very improbable. Their relationships with the Permo-Mesozoic basins are questionable, since sedimentary wedges are not associated with most of the dipping reflections. Paleozoic thrusts may, however, have been reactivated slightly by the Permo-Mesozoic and Tertiary extensional and compressional phases, with such a reactivation enhancing the acoustic contrasts, hence their seismic image.

5.4 Terrane Boundaries

Several sutures of various ages were crossed by deep seismic profiles, including the Iapetus, South Irish Sea, Lizard, English Channel, and Bray sutures. Unfortunately, the Variscan South-Armorican Suture was not readily covered. None of the sutures that have been imaged show any specific signature. The Iapetus suture is poorly imaged, with the main change across it being in the shallow crustal layering (Fig. 17). The dipping reflections associated with the other sutures are sometimes of high amplitude (e.g., Lizard, English Channel), but they may be later features. Suture zones in the Tertiary ranges generally correspond to very steep and/or geometrically complex features. The Pyrénées and Alps suture zones, for instance, do not provide any seismic image (Bayer et al. 1987; ECORS Pyrénées team 1988; Choukroune et al. 1990). Therefore the authors believe that the crustal reflections observed in the Proterozoic and Paleozoic suture zones come from thrusts associated with a late stage of crustal shortening. Very similar dipping reflectors are observed beneath the Penninic front, a possible suture in the Alps which may have been reworked by later strike-slip and thrust processes in the same way as the Variscan and Caledonian belts (Nicolas et al. 1990).

The different terranes postulated in the region under study hardly show any significant signature. Their crustal seismic image mainly results from the latest processes in the geological history, namely strike-slip faulting, Moho and lower crustal layering emplacement, and Mesozoic basin subsidence. All these processes affected the crust of the region under study whatever its original terrane and removed any difference in the terrane's seismic fabrics.

6 Conclusions

The seismic profiles, recorded since 1983 in France and adjacent areas, have provided images of the crust down to and even deeper than 30 km. Major tectonic features such as orogenic fronts, thrust sheets, suture zones, and strike-slip faults have been investigated, and their crustal geometry has been revealed to some extent. Traces of the Cadomian, Caledonian, and Variscan orogenic deformations, superimposed in the study region, seem to appear in the deep seismic images. Indications of north-verging Cadomian features may be present in the South Irish Sea and the southern English Channel sections. North-dipping seismic features interpreted as south-verging Caledonian thrusts are present in the deep crust in the South Irish Sea and the whole Celtic Sea despite the subsequent Variscan deformation. This latter deformation affected a shallower part of the crust in the Celtic Sea and the whole crust beneath the English Channel, the Paris and Aquitaine basins and the Pyrénées. The northern Variscan front investigated in northern France and the Celtic Sea shows great lateral variations in structural style. The southern Variscan front seems to be hidden beneath the southern part of the Aquitaine Basin. In all the regions investigated, the deep seismic sections suggest that Proterozoic and Paleozoic tectonic features have been preserved in the western European crust despite tectonic reworking including the Permo-Mesozoic subsidence and the Tertiary Alpine orogeny. The old tectonic features probably induced crustal heterogeneities which guided subsequent extensional and compressional tectonic events, causing the same features to be reactivated several times in the geological history, hence their seismic imaging.

Besides deep seismic data obtained before 1983, the deep seismic reflection profiles discussed in this article have filled the gap in our knowledge of the Pre-Mesozoic basement beneath the sedimentary basins of Paris and Aquitaine and the offshore areas of the South Irish Sea, Celtic Sea, English Channel, Western Approaches, and Bay of Biscay. This review of the seismic images of the Pre-Mesozoic crust is not complete. Recent profiles shot in eastern France (e.g., Rhine Graben, Alps, and Jura-Bresse) reveal prominent crustal features and should certainly help elucidate the Variscan Belt setting in this area where it is especially poorly understood. These profiles have not been discussed here because their Pre-Mesozoic features have not yet been thoroughly studied. It is also hoped that seismic profiles will be shot across the Massif Central and Armorican Massif as well as in the major sedimentary basins of Paris, Aquitaine, and southeastern France. They should complete the deep seismic coverage of the Pre-Mesozoic crust in France and provide at least one thorough cross-section of the Variscan belt from its northern to southern orogenic fronts.

The study of deep seismic sections has also raised many questions that still remain unanswered. The interpretation of particular features such as Moho deepening beneath sedimentary basins, lower crustal layering, and straight dipping reflections cutting across the whole crust has major scientific implications. In the present state of knowledge, none of these features can be regarded as definitely understood, and the authors are fully aware that their suggested interpretations must be taken with caution and regarded only as options. The main drawback to any seismic interpretation in the basement is certainly the lack of well-documented references where coincidental geological data and seismic features are available in the same area. Therefore a wide research field is open ahead, which requires not only cross-lines, velocity shooting, and S-wave studies near the profiles already completed, but entirely new methodological studies based on combined surface geology, shallow seismic and borehole investigation of characteristic basement features such as faults, thrusts, shear zones, mylonites, suture areas, intrusions, etc. We believe that the geology of the French territory offers a number of favorable areas to develop this kind of study, especially in the Massif Central, Armorican Massif, and northern Pyrenees.

Acknowledgments. The ECORS project was launched in 1983 by the Institut Français du Pétrole, the Institut National des Sciences de l'Univers (CNRS) and Elf Aquitaine. The Institut Français de Recherches pour l'Exploitation de la Mer joined the project in 1984 for the offshore operations.

The authors thank the Elf Aquitaine, ESSOREP, and BP oil companies for having kindly authorized publication of the line drawings of the AC-82 1 to 7 and AC-84 8 and 9 seismic profiles from central and northern Aquitaine.

A. Autran provided information and ideas that were especially useful for the interpretation of the northern Aquitaine profiles. The authors are also indebted to W. Franke for his fine review of the manuscript and to D. Keppie and D. Santallier for their useful remarks.

References

Arthaud F, Matte P (1977) Late Paleozoic strike-slip faulting in Southern Europe and Northern Africa; results of a right lateral shear zone between the Appalachians and the Urals. Geol Soc Am Bull 88: 1305–1320

Autran A (1980) Evolution structurale du Protérozoïque aux distensions post-hercyniennes. In: Autran A, Dercourt J (eds) Géologie de la France. 26 th IGC, Paris. Mem Bur Rech Geol Min 107, pp 10–17 (in French)

Avedik F (1975) The seismic structure of the Western approaches and the South Armorican continental shelf and its geological interpretation. In: Woodland AW (ed) Petroleum and the continental shelf of North West Europe. Applied Science, Barking, Essex, England, pp 29–43

Badham JNP, Kirby GA (1976) Ophiolites and the generation of oceanic crust, data from the Lizard complex, Cornwall. Bull Soc Geol Fr (7) XVIII, 4: 885–888

Ballèvre M, Pinardon JL, Kienast JR, Vuchard JP (1989) Reversal of Fe-Mg partitioning between garnet and staurolite in eclogite facies metapelite from the Champtoceaux nappe, Britanny, France. J Petrol 30 (6): 1321–1349

Barton P, Matthews D, Hall J, Warner M (1984) Moho beneath the North Sea compared on normal incidence and wide-angle seismic records. Nature 308: 5954, 55–56

Bayer R, Cazes M, Dal Piaz GV, Damotte B, Elter J, Gosso G, Hirn A, Lanza R, Lombardo B, Mugnier J-L, Nicolas A, Nicolich R, Polino R, Roure F, Sacchi R, Scarascia S, Tabacco I, Tapponnier P, Tardy M, Taylor M, Thouvenot F, Torreilles G, Villien A (1987) Premiers résultats de la traversée des Alpes occidentales par sismique réflexion verticale (Programme ECORS-CROP). C R Acad Sci Paris, 305, II: 1461–1470 (in French)

Beach A (1986) A deep seismic reflection profile across the northern North Sea. Nature 323: 53–55

Beamish D, Smythe DK (1986) Geophysical image of the deep crust: the Iapetus suture. J Geol Soc (Lond) 143: 489–497

Behr HJ, Heinrichs T (1987) Geological interpretation of DEKORP 2S: A deep seismic reflection profile across the Saxothuringian and possible implications for the Late Variscan structural evolution of Central Europe. Tectonophysics 142: 173–202

Behr HJ, Engel M, Franke W, Giese P, Weber K (1984) The Variscan Belt in: Central Europe; main structures, geodynamic implications, open questions. Tectonophysics 109: 15–40

Bernard-Griffith J, Cornichet J (1985) Origin of eclogites from south Brittany, France: a Sm-Nd isotopic and REE study. Chem Geol 52: 185–201

Berthé P, Choukroune P, Jegouzo P (1979) Orthogneiss, mylonites and non-coaxial deformation of granites; the example of the South Armorican Shear Zone. J Struct Geol 1: 31–42

BIRPS and ECORS (1986) Deep seismic reflection profiling between England, France and Ireland. J Geol Soc (Lond) 143: 45–62

Bless MJM, Bouckaert J, Paproth E (1980) Pre-Permian around the Brabant Massif in Belgium, the Netherlands and Germany. Meded Rijks Geol Dienst 32: 1–4

Blundell DJ (1979) The geology and structure of the Celtic Sea. In: Banner FT, Collins MB, Massie KS (eds) The north-west European Shelf seas: the sea bed and the sea in motion, I. Geology and sedimentology. Elsevier Oceanography Series 24 A, Amsterdam, pp 43–60

Boillot G, Lefort J-P (1974) Geological map of the English Channel, scale 1:1,000,000, Notice. Bureau de Recherches Géologiques et Minières, Orléans, France, 17 pp

Bois C, Allègre C (1983) The structure and dynamics of continental crust: the French ECORS Programme. First Break 5: 21–25

Bois C, Cazes M, Damotte B, Galdeano A, Hirn A, Mascle A, Matte P, Raoult J-F, Torreilles G (1986) Deep seismic reflection profiling of the crust in northern France: the ECORS project. In: Barazangi M, Brown L (eds) Reflection seismology: a global perspective. AGU Geodynamic Series 13. Am Geophys Union, Washington, DC, pp 21–29

Bois C, Biju-Duval B, Damotte B, Mascle A, Cazes M, Hirn A (1987) Operations and main results of the ECORS project in France. Geophys JR Astron Soc 89: 279–286

Bois C, Cazes M, Hirn A, Mascle A, Matte P, Montadert L, Pinet B (1988) Contribution of deep seismic profiling to the knowledge of the lower crust in France and neighbouring areas. Tectonophysics 145: 253–275

Bois C, Pinet B, Roure F (1989) Dating lower crustal features in France and adjacent areas from deep seismic profiles. In:

Mereu RF, Mueller S, Fountain DM (eds) Properties and processes of Earth's lower crust. AGU Geophysical Monograph 51, IUGG, vol 6. Am Geophys Union, Washington, DC, pp 17–31
Bois C, Lefort J-P, Le Gall B, Sibuet J-C, Gariel O, Pinet B, Cazes M (1990) Superimposed Variscan Caledonian and Proterozoic features inferred from deep seismic profiles recorded between southern Ireland, southwestern Britain and western France. Tectonophysics 177: 15–37
Bois C, Sibuet J-C, Gariel O (eds) (1991) Etude de la croûte terrestre par sismique profonde. Mer Celtique, Manche et ses Approches Occidentales: Profils SWAT et WAM. Mémoire 159, Société Geol France, Paris, vol 1: 220 p, vol 2: 22 pl
Boukhaert J, Fock W, Vandenberghe N (1988) First results of the Belgium Geotraverse 1986 (BELCORP). Ann Soc Geol Belg 111, 2: 279–290
Bouroz A, Chalard J, Dalinval A, Stievenard M (1961) La structure du bassin houiller du Nord de la région de Douai à la frontière belge, Ann Soc Geol Nord, Lille, France LXXXI: 173–220 (in French)
Brewer JA, Matthews DH, Warner MR, Hall J, Smythe DK, Whittington RJ (1983) BIRPS deep seismic reflection studies of the British Caledonides. Nature 305: 206–210
Brooks M, Doody J-J, Al-Rawi FRJ (1984) Major crustal reflectors beneath SW England. J Geol Soc (Lond) 141: 97–103
Bureau de Recherches Géologiques et Minières, Société Elf Recherche, Société ESSO de Recherche et Exploitation du Pétrole, Société Nationale des Pétroles d'Aquitaine (1974) Géologie du Bassin d'Aquitaine, 29 encl. Bureau de Rech Geol Min, Orléans, France (in French)
Burg P (1988) L'environnement du graben: les Vosges cristallophylliennes, histoire Varisque, in: Rapport d'implantation du profil ECORS-DEKORP Graben du Rhin, Programme ECORS, pp 65–75 (in-house rept) Institut Français du Pétrole, Rueil-Malmaison, France (in French)
Burrus J (1984) Contribution to a geodynamic synthesis of the Provençal Basin (north-western Mediterranean). Mar Geol 55: 247–269
Cazes M, Torreilles G (eds) (1988) Etude de la croûte terrestre par sismique profonde, Profil Nord de la France, structure hercynienne. Technip, Paris, 260 pp, 15 encl. (in French)
Cazes M, Torreilles G, Bois C, Damotte B, Galdeano A, Hirn A, Mascle A, Matte P, Pham VN, Raoult J-F (1985) Structure de la croûte hercynienne du Nord de la France: premiers résultats du profil ECORS. Bull Soc Geol Fr (8) I, 6: 925–941 (in French)
Cazes M, Mascle A, Torreilles G, Bois C, Damotte X, Matte P, Raoult X, Pham VN, Hirn A, Galdeano X (1986) Large Variscan overthrusts beneath the Paris basin. Nature London 323: 144–147
Centre National de la Recherche Scientifique, CFP(M), COPESEP, RAP and SNPA (1965) Contribution à la connaissance des bassins paléozoïques du Nord de la France. Ann Soc Géol Nord, Lille, France LXXXXV: 273–281 (in French)
CFP(M), COPESEP, RAP, SNPA (1969) Contribution à la connaissance des bassins paléozoïques du Nord de la France. Ann Soc Géol Nord, Lille 85: 273–281 (in French)
Chadwick RA, Kenolty N, Whittaker A (1983) Crustal structure beneath southern England from deep seismic reflection profiles. J Geol Soc (Lond) 140: 893–911
Chantraine J, Chauvel J-J, Balé P, Denis E, Rabu D (1988) Le Briovérien (Protérozoïque supérieur à terminal) et l'orogenèse cadomienne en Bretagne (France), 1988. Bull Soc Geol Fr (8) IV, 4: 815–829 (in French)
Choukroune P (1976) Structure et évolution tectonique de la zone nord-pyrénéenne, Mem Soc Geol Fr 55, 127: 1–116 (in French)
Choukroune P, Pinet B, Roure F, Cazes M (1990) Major Hercynian structures along the ECORS Pyrénées and Biscay lines. Bull Soc Geol Fr (8) VI, 2: 313–320
Cogné J, Wright AE (1980) L'orogène cadomien. In: Cogné J, Slansky M (eds) Géologie de l'Europe. 26th IGC, Coll. C. 6. Soc Geol du Nord and Bur Rech Geol Min France, Villeneuve d'Ascq, France, pp 28–55 (in French)
Cooper MA, Collins D, Ford M, Murphy FX, Trayner PM (1984) Structural style, shortening estimate and the thrust front of the Irish Variscides. In: Hutton DHW, Sanderson DJ (eds) Variscan tectonics of the North Atlantic region. Geol Soc London, Spec Public 14, pp 167–175
Coward MP, Siddans AWB (1979) The tectonic evolution of the Welsh Caledonides. In: Harris AL, Holland CJ, Leake BE (eds) The Caledonides of the British Isles – reviewed. Geol Soc London, pp 187–198
Coward MP, Smallwood S (1984) An interpretation of the Variscan tectonics of SW Britain. In: Hutton DHW, Sanderson DJ (eds) Variscan tectonics of the North Atlantic region. Geol Soc London, Soc Publ 14, pp 80–102
Damotte B (1988) Traitement et interprétation de la sismique réflexion à l'explosif. In: Cazes M, Torreilles G (eds) Etude de la croûte terrestre par sismique profonde. Profil Nord de la France. Structure hercynienne. Technip, Paris, pp 93–108 (in French)
Darbyshire DPF, Shepherd TJ (1985) Chronology of granite magmatism and associated mineralisation, SW England. J Geol Soc (Lond) 142: 1159–1177
Davies GR (1984) Isotopic evolution of the Lizard complex. J Geol Soc (Lond) 141: 3–14
DEKORP Research Group (1985) First results and preliminary interpretation of deep reflection seismic recordings along profile DEKORP 2 South. J Geophys 57: 137–163
DEKORP Research Group (1988) Results of the DEKORP 4/KBT Oberpfalz deep seismic reflection investigations. J Geophys 62: 69–101
Delay F, Paquet J (1988) Le Massif Nord Pyrénéen de l'Agly (Pyrénées Orientales). Un exemple de diapirisme crustal hercynien. 12ème Réunion Annuelle des Sciences de la Terre, Lille, p 43 (in French)
Delmer A (1977) Le Bassin du Hainaut et le sondage de St-Ghislain, Prof Paper, Serv Geol, Belgique, 143, 12 pp (in French)
Dewey JF (1974) The geology of the southern termination of the Caledonides. In: Nairn AEM, Stehli FG (eds) The ocean basins and margins, vol 2. The North Atlantic. Plenum Press, New-York, pp 205–231
Dodson MH, Rex DC (1971) Potassium-argon ages of slates and phyllites from southwest England. J Geol Soc (Lond) 126: 465–499
Dohr G, Meissner R (1975) Deep crustal reflections in Europe. Geophysics 40: 25–39
Dunning FW (1977) Caledonian-Variscan Relations in North-West Europe. In: La chaîne varisque d'Europe moyenne et occidentale. Coll Int CNRS, Rennes, France, 243: 165–180
Durst H (1985) Interpretation of reflection-seismic profile across the northeastern Stavelot-Venn Massif and its northern foreland. N Jh Geol Palaeont Abh 171: 441–446
ECORS Pyrénées team (1988) The ECORS deep seismic survey across the Pyrénées. Nature 331: 508–511
Engel W, Franke W, Grote C, Weber K, Ahrendt H, Eder FW (1983) Nappe tectonics in the southeastern part of the Rheinisches Schiefergebirge. In: Martin H, Eder RW (eds) Intracontinental Fold Belts; case studies in the Variscan Belt of Europe and the Damara Belt in Namibia. Springer, Berlin Heidelberg New York, pp 267–287

Finck P, Ansorge J, Mueller St, Sprecher C (1984) Deep crustal reflections from a Vibroseis survey in Northern Switzerland. Tectonophysics 109: 1–14

Floc'h JP, Santallier D, Henri B (1984) La série limousine: réflexion axée sur une synthèse géologique régionale et son utilisation pour la prospection minière (or). Chron Rech Min 474: 5–10 (in French)

Fountain DM, Salisbury MH (1981) Exposed cross-sections through the continental crust: implications for crustal structures, petrology and evolution. Earth Planet Sci Lett 56: 263–277

Fountain DM, Hurich CA, Smithson SB (1984) Seismic reflectivity of mylonitic zones in the crust. Geology 12: 195–198

Freeman B, Klemperer SL, Hobbs RW (1988) The deep structure of Northern England and the Iapetus suture zone, from BIRPS deep seismic reflection profiles. J Geol Soc (Lond) 145: 727–740

Galdeano A, Guillon J-C (1988) Interprétation des données magnétiques et gravimétriques. In: Cazes M, Torreilles G (eds) Etude de la croûte terrestre par sismique profonde. Profil Nord de la France. Structure Hercynienne. Technip, Paris, pp 127–140 (in French)

Gapais D, Le Corre C (1980) Is the Hercynian belt of Brittany a major shear zone? Nature 288: 574–576

Gardiner PRR, Sheridan DJR (1981) Tectonic framework of the Celtic Sea and adjacent areas with special reference to the location of the Variscan Front. J Struct Geol 3: 317–331

Gérard A, Weber C (1971) L'anomalie du Bassin de Paris interprétée comme un élément structural majeur dans l'histoire géologique de la France. C R Acad Sci Paris 272: 921–923 (in French)

Girardeau J, Dubuisson G, Mercier J-C (1986) Cinématique de mise en place des ophiolites et nappes cristallophylliennes du Limousin, Ouest du Massif Central français. Bull Soc Géol Fr (8) II, 5: 849–860 (in French)

Graviou P, Auvray B (1985) Caractérisation pétrographique et géochimique des granitoïdes cadomiens du domaine nord-armoricain: implications géodynamiques. C R Acad Sci Paris 301, II, 5: 315–318 (in French)

Guillen A, Debeglia N (1984) Programme Géologie Profonde de la France. Doc Bur Rech Géol Min (Orléans, France) 81: 2 (in French)

Guillot PL, Lefèvre J (1975) Découverte de conodontes dans le calcaire à entroques de Genis en Dordogne (série métamorphique du Bas-Limousin). C R Acad Sci Paris D 280: 1529–1530 (in French)

Hall J, Brewer JA, Matthews DH, Warner MR (1984) Crustal structure across the Caledonides from the "WINCH" seismic reflection profiles: influences on the evolution of the Midland Valley of Scotland. Trans R Soc Edinb Earth Sci 75: 97–105

Hancock PL, Dunne WM, Tringham ME (1981) Variscan structure in Southwest Wales. Geol Mijnbouw 60: 81–88

Hancock PL, Dunne WM, Trigham ME (1983) Variscan deformation in southwest Wales. In: Hancock PL (ed) The Variscan Fold Belt in the British Isles. Adam Hilger, Bristol, pp 47–73

Harris AL (1985) The nature and timing of orogenic activity in the Caledonian rocks of the British Isles. Geol Soc Lond Mem 9, 53 pp

Harris AL, Rathbone PA, Watson J (1980) The Pre-Caledonian evolution of the British Isles. In: Cogné J, Slansky M (eds) Géologie de l'Europe, Coll C6, 26th IGC, Paris, Soc Geol du Nord and Bur Rech Géol Min, France, pp 22–28

Héritier F, Villemin J (1971) Mise en évidence de la tectonique profonde du Bassin de Paris par l'exploration pétrolière. Bull Bur Rech Géol Min 1, 2: 1–30 (in French)

Hillis RR, Day GA (1987) Deep events in U.K. south western approaches. Geophys JR Astr Soc 89: 243–250

Hirn A (1988) Les opérations de sismique grand angle. In: Cazes M, Torreilles G (eds) Etude de la croûte terrestre par sismique profonde. Profil Nord de la France. Structure Hercynienne. Technip, Paris, pp 109–126 (in French)

Hirn A, Prodehl C, Steinmetz L (1975) An experimental test of models of the lower lithosphere in Bretagne (France). Ann Geophys 31, 4: 517–530

Hurich CA, Smithson SB, Fountain DM, Humphreys MC (1985) Mylonite reflectivity: evidence from Kettle Dome, Washington. Geology 13: 577–580

Hurich CA, Deemer SJ, Kristoffersen Y (1988) Deep structure of the southern Norwegian Caledonides: a possible link between Scandinavian and British Caledonides. In: Dooly JC (ed) International Conference on Seismic Probing of Continents and their Margins. Bur Min Res, Canberra, Australia

Jegouzo P (1980) The South-Armorican shear zone. J Struct Geol 2: 39–47

Johnson MRW, Sanderson DJ, Soper NJ (1979) Deformation in the Caledonides of England, Ireland and Scotland. In: Harris AL, Holland CH, Leake BE (eds) The Caledonides of the British Isles – reviewed. Geol Soc London, pp 165–186

Jones TD, Nur A (1984) Nature of seismic reflections from deep crustal faults zones. J Geophys Res 89, B 5: 3153–3171

Kenolty N, Chadwick RA, Blundell DJ, Bacon M (1981) Deep seismic reflection survey across the Variscan Front of southern England. Nature 293: 451–453

Ledru P, Autran A (1987) L'édification de la chaîne varisque dans le Limousin. Rôle de la faille d'Argentat à la limite Limousin-Millevaches. In: Programme Géologie Profonde de la France. Doc Bur Rech Géol Min (Orléans) 140: 51–81 (in French)

Lefort J-P (1975) Le socle péri-armoricain, étude géologique et géophysique du socle submergé à l'Ouest de la France. PhD Thesis, Univ. Rennes, 250 pp, 5 encl. (in French)

Lefort J-P (1991) L'organisation structurale du socle profond de la Manche occidentale d'après l'interprétation des profils sismiques SWAT: un exemple d'intersection entre une suture cadomienne et des structures hercyniennes divergentes. In: Bois C, Sibuet J-C, Gariel O (eds) Etude de la croûte terrestre par sismique profonde. Mer Celtique, Manche et ses Approches Occidentales: Profils SWAT et WAM. Mémoire 159, Soc Geol France, Paris, vol 1, pp 123–132 (in French)

Lefort JP, Poulpiquet de J (1990) Corrélations entre les profils ECORS-Gascogne et ECORS-Massif Armoricain. Bull Soc Geol Fr (8) VI, 2: 321-330 (in French)

Lefort J-P, Ribeiro A (1980) La faille Porto-Badajoz-Cordoue a-t-elle contrôlé l'évolution de l'océan paléozoïque sud-armoricain. Bull Soc Geol Fr 7, 22, 3: 455–462 (in French)

Lefort J-P, Bardy Ph, Poulpiquet de J (1991) Interprétation des anomalies magnétiques reconnues en Manche occidentale et en Mer Celtique le long des profils SWAT. In: Bois C, Sibuet J-C, Gariel O (eds) Etude de la croûte terrestre par sismique profonde. Mer Celtique, Manche et ses Approches Occidentales: Profils SWAT et WAM. Mémoire 159, Soc Geol France, Paris, vol 1, pp 133–143 (in French)

Le Gall B (1990) Evidence of an imbricate crustal thrust belt in the southern British Variscides. Contribution of SWAT deep seismic reflection profiling recorded in the English Channel and the Celtic Sea. Tectonics 9 (2): 283–302

Lueschen E, Wenzel F, Sandmeier K-J, Menges D, Ruhl T, Stiller M, Jaboth W, Keller F, Sollner W, Thomas R, Krohe A, Stenger R, Fuchs K, Wilhelm H, Eisbacher G (1987) Near-vertical and wide-angle seismic surveys in the Black Forest, SW Germany. J Geophys 62: 1–30

Marchal F (1988) Apport de la gravimétrie à la connaissance des structures profondes sous le Bassin Aquitain: extension de la chaîne hercynienne. PhD Thesis, Univ. Montpellier, France, 393 pp, 12 encl. (in French)
Marchand J (1981) Ecaillage d'un "mélange tectonique" profond: le complexe cristalophyllien de Champtoceaux (Bretagne méridionale). C R Acad Sci Paris 293: 223–228 (in French)
Marillier F, Tomassino A, Patriat P, Pinet B (1987) Deep structure of the Aquitaine shelf: constraints from expanding spread profiles on the ECORS Bay of Biscay transect. Mar Petr Geol 5, 1: 65–74
Mascle A, Cazes M (1988) La couverture sédimentaire du Bassin Parisien le long du profil ECORS. Nord de la France. Rev Inst Fr Pétr 42, 3: 303–316 (in French)
Matte P (1983) Two geotraverses across the Ibero-Armorican Variscan arc of western France. In: Rast N, Delany FM (eds) Profiles of Orogenic Belts. AGU Geodynamic Series 10. Am Geophys Union, Washington, DC, pp 53–81
Matte P (1986) Tectonics and plate tectonics model for the Variscan Belt of Europe. Tectonophysics 126: 329–374
Matte P, Hirn A (1988) Seismic signature and tectonic cross section of the Variscan crust in western France. Tectonics 7, 2: 141–155
Matte P, Respaut J-P, Maluski J, Lancelot J, Brunel M (1986) La faille NW-SE du Pays de Bray: un décrochement ductile dexte hercynien; déformation à 320 Ma d'un granite à 570 Ma dans le sondage pays de Bray 201. Bull Soc Geol Fr 8, II, 1: 55–69 (in French)
McGeary S (1987) Non typical BIRPS on the margin of the northern North Sea: the SHET survey. Geophys JR Astr Soc 89: 231–237
Mégnien C (ed) (1980) Synthèse géologique du Bassin de Paris, vol 3. Bull Bur Rech Géol Min, Orléans, France (in French)
Meissner R, Wever T (1986) Nature and development of the crust according to deep reflection data from the German Variscides. In: Barazangi M, Brown L (eds) Reflection seismology: a global perspective. AGU Geodyn Series 13. Am Geophys Union, Washington, DC, pp 31–42
Meissner R, Barthelsen J, Murawski J (1981) Thin-skinned tectonics in the northern Rhenish Massif, Germany. Nature 290: 399–401
Meissner R, Springer M, Murawski A, Barthelsen H, Fluh E, Duerschner H (1983) Combined seismic reflection-refraction investigations in the Rhenish Massif and their relation to recent tectonic movement. In: Fuchs K, Gehlen von K, Maelzer H, Murawski H, Semmel A (eds) Plateau uplift, a case history. Springer, Berlin Heidelberg New York, pp 276–287
Meissner R, Matthews D, Wever T (1986) The "Moho" in and around Great Britain. Ann Geophys 4, B, 6: 659–664
Michot P (1980) Belgique. In: Auboin J (ed) Géologie des Pays Européens: France, Belgique, Luxembourg. Dunod, Paris, pp 485–576 (in French)
Mooney WD, Brocher TM (1987) Coincident seismic reflection/refraction studies of the continental lithosphere: a global review. Geophys JR Astr Soc 89: 1–6
Moseley F (1977) Caledonian plate tectonics and the place of the English Lake District. Geol Soc Am Bull 88: 764–768
Nicolas A, Hirn A, Nicolich R, Polino R (1990) Lithospheric wedging in the Western Alps inferred from the ECORS-CROP traverse. Geology 18: 587–590
Paris F, Le Pochat G (1987) Synthèse géologique de l'Aquitaine nord. In: Bassins paléozoïques cachés dans l'Aquitaine. Doc Bur Rech Geol Min (Orléans) 144: 43–53 (in French)
Paris F, Le Pochat G, Pelhate A (1988) Le socle paléozoïque nord-Aquitain: caractéristiques principales et implications géodynamiques. CR Acad Sci Paris 306, II: 597–603 (in French)
Peddy C, Pinet B, Masson D, Sowtton R, Sibuet J-C, Warner MR, Lefort J-P, Schroeder IJ (BIRPS and ECORS) (1989) Crustal structure of the Goban Spur continental margin, Northeast Atlantic, from deep seismic reflection profiling. J Geol Soc (Lond) 146: 427–437
Perroud HJ, Van der Voo R, Bonhommet N (1984) Paleozoic evolution of the Armorican Plate on the basis of paleomagnetic data. Geology 12: 579–582
Pham NV (1988) Propriétés électriques et structure de la croûte sous le Bassin parisien par sondage magnéto-tellurique profond. In: Cazes M, Torreilles G (eds) Etude de la crûte terrestre par sismique profonde. Profil Nord de la France. Structure hercynienne. Technip, Paris, pp 141–156 (in French)
Phillips WEA, Stillman CJ, Murphy J (1976) A Caledonian plate tectonic model. J Geol Soc (Lond) 132: 579–609
Pinet B, Montadert L, Curnelle R, Cazes M, Marillier F, Rolet J, Tomassino A, Galdeano A, Patriat P, Brunet M-F, Olivet J-L, Schaming M, Lefort J-P, Arrieta A, Riaza C (1987 a) Crustal thinning on the Aquitaine Shelf, Bay of Biscay, from deep seismic data. Nature 325: 513–516
Pinet B, Montadert L, ECORS Scientific Party (1987 b) Deep seismic reflection and refraction profiling along the Aquitaine Shelf (Bay of Biscay). Geophys JR Astr Soc 89: 305–312
Piqué A, Huon S, Clauer N (1984) La schistosité hercynienne et le métamorphisme associé dans la vallée de la Meuse entre Charleville Mézières et Namur, Ardennes franco-belges. Bull Soc Belge Géol 93: 55–70 (in French)
Poulpiquet de J, Lefort J-P (1989) Modelling of structures representing the south Armorican suture. Tectonophysics 165: 93–103
Raoult J-F (1988) Le front varisque du Nord de la France: interprétation des principales coupes d'après les profils sismiques, la géologie de surface et les sondages. In: Cazes M, Torreilles G (eds) Etude de la croûte terrestre par sismique profonde. Profil Nord de la France. Structure hercynienne. Technip, Paris, pp 171–196 (in French)
Raoult J-F, Meilliez F (1986 a) Commentaires sur une coupe structurale de l'Ardenne selon le méridien de Dinant. Ann Soc Geol Nord, Lille, France CV: 97–109 (in French)
Raoult J-F, Meilliez F (1986 b) The Variscan Front and the Midi Fault between the Channel and the Meuse River. J Struct Geol 9: 473–479
Robaszynski F, Dupuis C (1983) Belgique. In: Pomerol C (ed) Guides géologiques régionaux. Masson, Paris, 204 pp (in French)
Rolet J, Le Gall B, Darboux J-R, Thonon R, Gravelle M (1986) L'évolution géodynamique de l'extrémité occidentale de la chaîne hercynienne d'Europe sur le transect Armorique-Cornwall. Bull Soc Géol Fr (8) II, 1: 43–54 (in French)
Sanderson DJ (1979) The transition from upright to recumbent folding in the Variscan fold belt of southwest England: a model based on the kinematics of simple shear. J Struct Geol 1: 171–180
Schweizerische Arbeitsgruppe für Reflektionsseismik (1988) Vorläufige Ergebnisse der Alpentraversen des NFP-20 "Geologische Tiefenstruktur der Schweiz". Bull Ver Schweiz Petrol Geol 54, 126: 1–30 (in German)
Shackleton RM, Ries AC, Coward MP (1982) An interpretation of the Variscan structure in SW England. J Geol Soc (Lond) 139: 533–541
Sibuet J-C, Dyment J, Bois C, Pinet B, Ondreas H (1989) Crustal structure of the Celtic Sea and Western Approaches from gravity data and deep seismic profiles: constraints on

the formation of continental basins. J Geophys Res 95 (B 7): 10999–11020

Smallwood S (1985) A thin-skinned thrust model for Variscan Pembrokeshire, Wales. J Struct Geol 7: 683–687

Smith NJP (1987) The deep geology of central England: the prospectivity of the Palaeozoic rocks. In: Brooks J, Glennie K (eds) Petroleum geology of North West Europe. Graham and Trotman, London, pp 217–224

Smithson SB (1978) Modeling continental crust: structural and chemical constraints. Geophys Res Lett 5, 9: 749–752

Smithson SB, Johnson RA, Hurich CA (1986) Crustal reflections and crustal structure. In: Barazangi M, Brown L (eds) Reflection seismology: the continental crust. AGU Geodynamic Series 14. Am Geophys Union, Washington, DC, pp 21–32

Soper MJ, Hutton DHW (1984) Late Caledonian sinistral displacement in Britain: implications for a three-plate collision model. Tectonics 3, 7: 781–794

Teichmüller M, Teichmüller R (1979) Ein Inkohlungsprofil entlang der linksrheinischen Geotraverse von Schleider nach Aachen und die Inkohlung in der Nord-Süd Zone der Eifel. Fortschr Geol Rheinl Westfalen 27: 323–355 (in German)

Torreilles G (1988) Traitement et interprétation de la sismique réfraction. In: Cazes M, Torreilles G (eds) Etude de la croûte terrestre par sismique profonde. Profil Nord de la France. Structure hercynienne. Technip, Paris, pp 79–92 (in French)

Veinante-Delhaye A, Santoire J-P (1980) Sismicité récente de l'arc sud-armoricain et du nord-ouest du Massif Central, mécanisme au foyer et néotectonique. Bull Soc Geol Fr 7, XXII, 1: 93–102 (in French)

Warner M (1987) Migration – why doesn't it work for deep continental data? Geophys JR Astr Soc 89: 21–26

Watson J, Dunning FW (1979) Basement-cover relations in the British Isles Caledonides. In: Harris AL, Holland CH, Leake BE (eds) The Caledonides of the British Isles – reviewed. Graham and Trotman, London, pp 67–91

Weber C, Hirn A, Philip H, Roche A (1980) Image geophysique de la France. In: Autran A, Dercourt J (eds) Géologie de la France. 26th IGC, Coll C7, Mem Bur Rech Géol Min, Orléans, France, 107, pp 25–50 (in French)

Whittaker A, Chadwick RA (1983) Deep seismic reflection profiling onshore United Kingdom. First Break 1, 9: 9–13

Whittaker A, Chadwick RA (1984) The large-scale structure of the Earth's crust beneath southern Britain. Geol Mag 121, 6: 621–624

Williams GD, Brooks M (1985) A reinterpretation of the concealed Variscan structure beneath southern England by section balancing. J Geol Soc (Lond) 142: 689–695

Gravity and Magnetic Expression of Terranes in France and Their Correlation Beneath Overstep Sequences

A. Autran, J. P. Lefort, N. Debeglia, J. B. Edel, and J. L. Vigneresse

1 Introduction

The geological basement of France appears to be composed of different terranes having various compositions. The presence of these bodies, even if concealed beneath the cover, is reflected by the pattern of gravity or magnetic data. These geophysical techniques are excellent criteria, that allow a real discrimination of granite-rich zones from mafic-rich areas in plan view. Sutures and plate boundaries are also easily characterized by studies of potential field character; their dip is best defined by gravity and magnetics, since mafic rocks incorporated in the major scars are never imaged on seismic reflection records.

In the following compilation and reinterpretation, France is divided into several provinces; each of them constitutes a different geophysical zone, which, in many cases, but not always, corresponds to a geological terrane. Figure 1 shows a structural subdivision of the French territory and the adjacent areas; it is based on both geological and geophysical data (Bur. Rech. Geol. Min. 1975, 1980). Although this figure will be used as a guide in the following sections, it has been considered useful to show how this part of the Variscan chain can be divided if geophysical data (magnetics and gravity) are used alone. The map of Debeglia and Autran (Fig. 2), based on a magnetic map reduced to the actual pole and on a gravity map stripped of the sediment effect in basinal areas, shows the prominent geophysical features recognized at depth in France.

One may recognize four relatively flat zones bounded by strong geophysical linears. At first glance, northern Brittany can easily be isolated in the west from the Aquitaine Basin. In the east a more complicated northern zone corresponds with a classical subdivision of this area into three main Hercynian belts that outcrop in the eastern Paris Basin and the Ardennes and Vosges mountains. It can be distinguished from a southern zone showing only a weak geophysical expression. This zone incorporates the Massif Central, the Alps, and the Provence basement. Most of the geophysical disruptions represented here have been recognized by their gradients (vertical or horizontal) across geophysical lineaments. The dip of the boundaries is also shown. Major strike-slip faults do not necessarily represent terrane boundaries because post- or late-orogenic features often do not correspond with major structural features.

In other respects, the strong imprint of the Alpine Orogeny in southern and eastern France can be held responsible for strong linear anomalies. Some of the clearest boundaries shown on potential field maps (Fig. 5), such as the South Aquitaine EW high or the NE-SW horizontal gradient surveyed in the eastern Massif Central, represent in fact very late transcurrent Hercynian faults later rejuvenated. Their weak original expression has been enhanced by the intense thinning of the crust which developed during the formation of the southern Mesozoic basins. The crust in France is complex, however, because it has been formed through a succession of orogenic events.

The definition of the geophysical characteristics and limits of a terrane, when it is composed of previously thickened crust (resulting, for instance, from the superposition of several allochtonous crustal units above a deep craton) can be of ambiguous interpretation when using only the potential fields. Because of this we have used also seismic profiling and surface structural mapping rather than gravity and magnetism only. Such is the case for the external Rhenohercynian zone and for the southern Brittany-Massif Central belt, which are both allochtonous.

2 Terranes of the Armorican Province

(Fig. 1: 1, 2, 3)

The Armorican massif is buried for about 60% of its surface by Mesozoic and Cenozoic sediments, and none of its boundaries crop out at the surface. So far as the boundaries are concerned, they correspond with major geophysical disruptions which separate the inner zone from the British Midland block in the North (Fig. 1), Brabant block in the north east (Fig. 1:

J. D. Keppie (Ed.): Pre-Mesozoic Geology in France and Related Areas

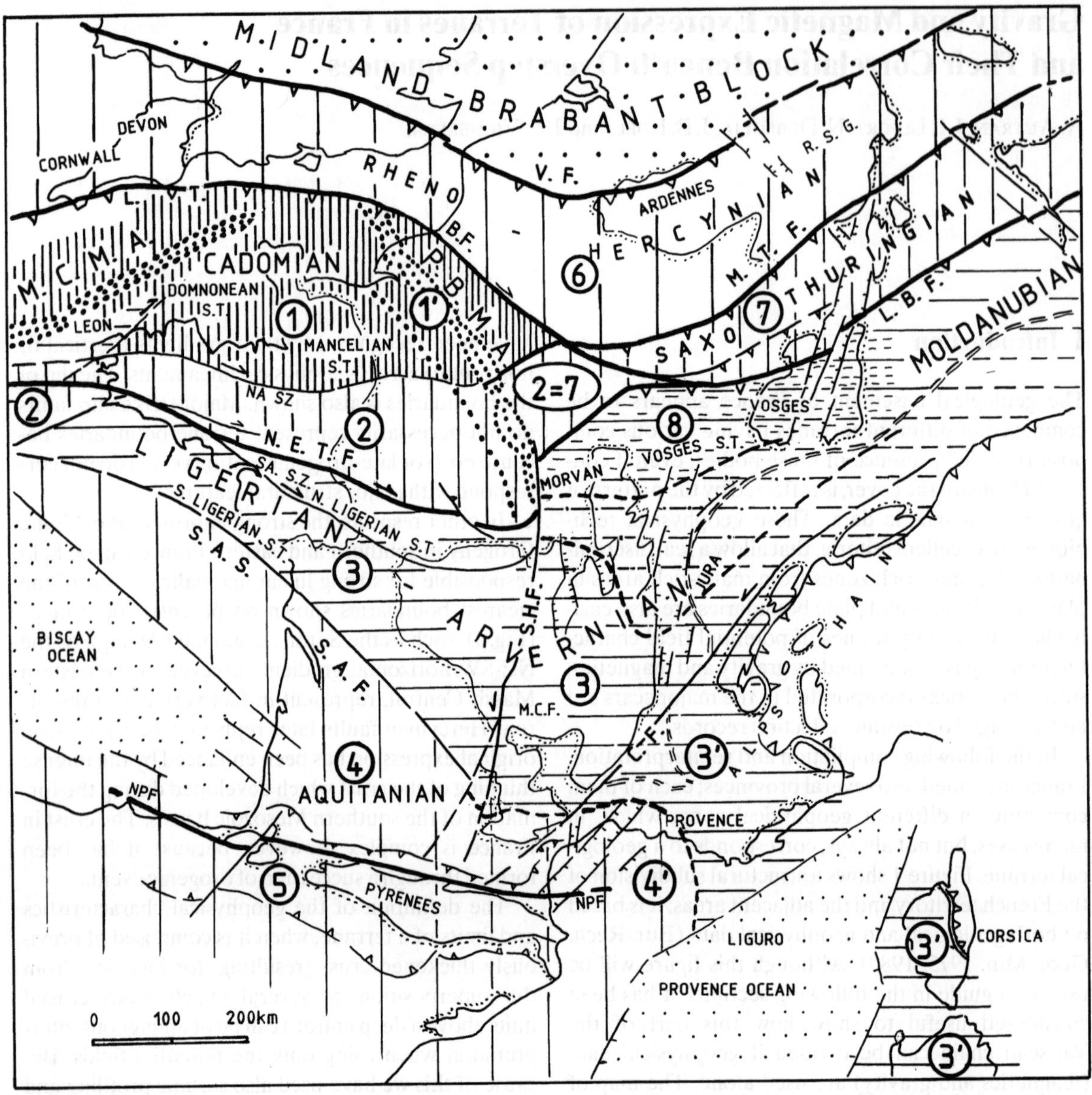

Fig. 1. Structural sketch map of France, showing the terranes identified by geophysics. *M. C. M. A.* Mid-Channel magnetic anomaly; *P. B. M. A.* Paris Basin magnetic anomaly; *M. C. F.* Massif Central of France; *V. F.* Variscan front of the Rhenohercynian nappes on the foreland bassin; *L. T.* Lizard thrust; *M. T. F.* Metz-Taunus fault; *B. F.* Bray fault; *N. A. S. Z.* North-Armorican shear zone; *S. A. S. Z.* South-Armorican shear zone; *N. E. T. F.* Nort-sur-Erdre Thouarcé fault; *L. B. F.* Lubine-Baden Baden fault; *S. A. S.-S. A. F.* South-Armorican suture or South Armorican front of internal nappes; *C. F.* Cévennes fault zone; *S. H. F.* Sillon Houiller fault; *N. P. F.* North Pyrenean Cretaceous faults

Alpinized area: Oceanized zone: *dots;* rifts: *symbol* of normal fault; limits of Alps and Pyrénées: *double line* or thrust symbol. Geophysical terranes of the Variscan belt in France and neighboring countries: *1* Cadomian terrane; *1'* Possible allochtonous part of Cadomian terrane (Bray nappe) with its rooting at the base of the crust shown by *dotted line; 2* Mid-Armorican terrane; *2'* W. part of Mid-Armorican terrane; *3* Ligerian Arvernian terrane; *3'* Southern extension of Ligerian Arvernian terrane; *4* Aquitanian terrane; *4'* Eastern extension of A. T.; *5* Pyrenees terrane; *6* Rhenohercynian terrane; *7* Saxothuringian terrane; *8* Morvano-Vosgian suspected terrane

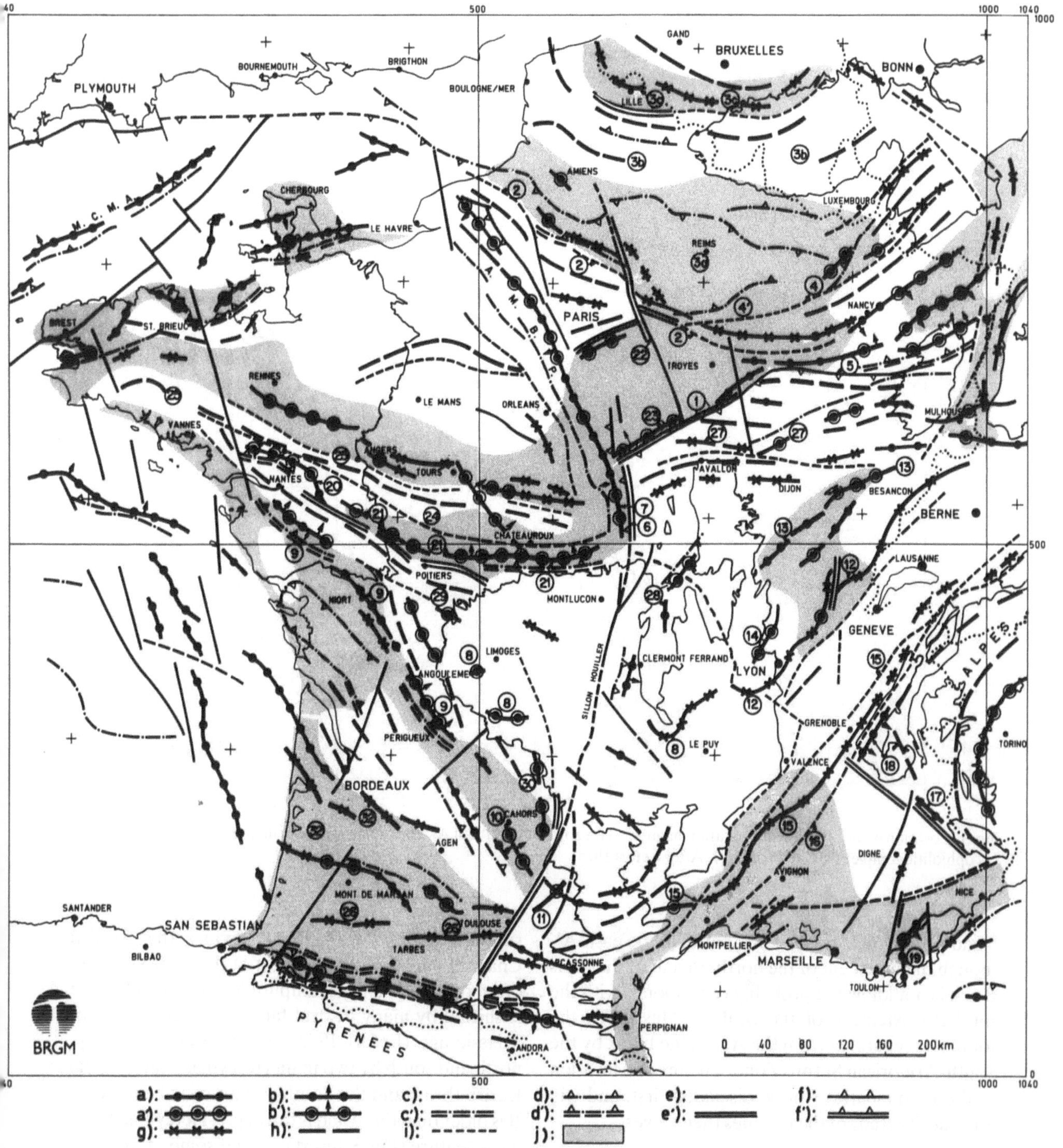

Fig. 2. Sketch map of the main magnetic and gravity features of France (A. Autran and N. Debeglia, original). This diagram results from the structural interpretation of the gravity map of France (BRGM 1975) and of the aeromagnetic survey of France (BRGM 1980; Le Mouel 1969). Magnetic and gravity data have been interpolated on a 10×10 km grid. For magnetism, the magnetic field reduced to the pole has been computed according to the hypothesis of induced magnetization. For gravity, the Bouguer anomaly has been corrected for the effect of the post permian sedimentary areas. Discontinuities and tops have been shown by using an automated procedure for the local analysis of horizontal and vertical gradients. *a* Magnetic axis; *b* direction of the magnetic dip (in the hypothesis of induced magnetization); *c* subvertical magnetic contact; *d* magnetic contact with low dip (the direction of dip is pointed out by *triangles*); *e* transverse discontinuity; *f* cross-cutting discontinuity with thrusting component shown by geology or seismics; *a'* to *f'* same structures or discontinuities confirmed by gravity interpretation; *g* positive gravity axis; *h* negative gravity axis; *i* linear horizontal gravity gradient; *j* area characterized by a positive gravity trend

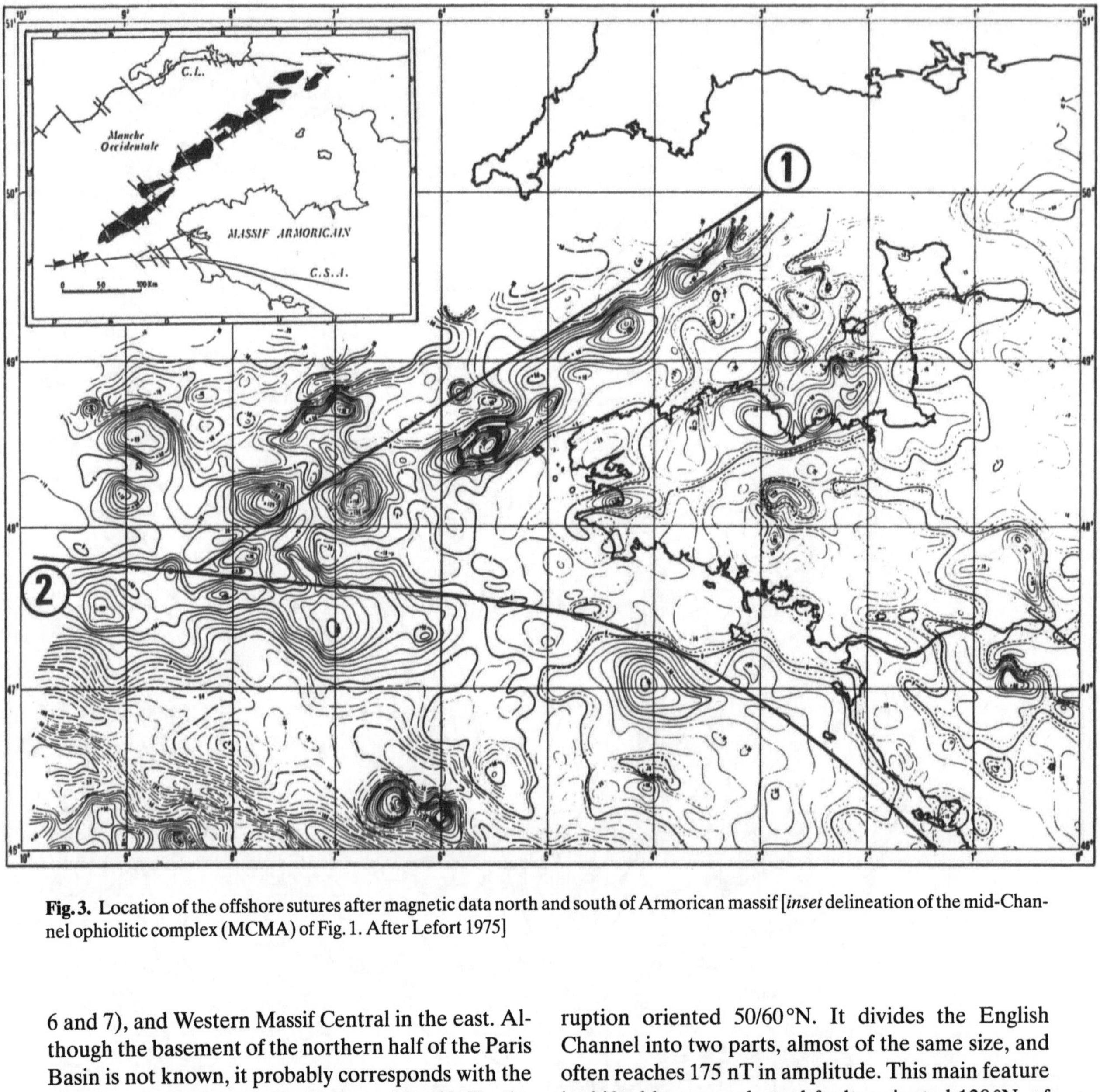

Fig. 3. Location of the offshore sutures after magnetic data north and south of Armorican massif [*inset* delineation of the mid-Channel ophiolitic complex (MCMA) of Fig. 1. After Lefort 1975]

6 and 7), and Western Massif Central in the east. Although the basement of the northern half of the Paris Basin is not known, it probably corresponds with the southern extension of the Brabant Massif. To the south it is separated from the Aquitaine block by the South-Armorican Suture Zone. The precise location of these boundaries will be discussed first, and will precede the study of the terranes themselves.

2.1 The Boundaries of the Armorican Province (J. P. Lefort)

2.1.1 The Mid-Channel Magnetic Anomaly (MCMA: Fig. 1; Fig. 3: 1 and Fig. 4: 2) J. P. Lefort

The Mid-Channel magnetic anomaly (MCMA) is a magnetic high that is superimposed above a linear disruption oriented 50/60 °N. It divides the English Channel into two parts, almost of the same size, and often reaches 175 nT in amplitude. This main feature is shifted by many dextral faults oriented 130 °N, of Triassic age (Lefort 1973), and is truncated to the north and south by two main Hercynian features, the Lizard thrust and the South-Armorican Shear Zone. It is older than these and is thus a pre-Hercynian basement feature (Fig. 3, insert), even if some of the shallow mafic bodies recognized in the middle of the Channel represent later igneous intrusions related to Triassic rifting in the English Channel (Lefort et al. 1991 a). The mean depth of this body is between 3100 and 4000 m. We consider (Lefort 1975; Lefort and Segoufin 1978) that the Mid-Channel magnetic anomaly corresponds with a Late Proterozoic suture, and represents the remnants of an ocean which subducted towards the south, beneath Brittany. This suture is not well evidenced on gravity maps, because of the thick sedimentary cover (Boillot and Lefort 1974). After reduction to the pole, the modeling of MCMA indicated (Gérard 1975) a northward dip (if due to induced magnetization) (see Fig. 2). However, new data (500 m elevation oil industry survey) and a new calculation show that the mafic bodies dip south (Bardy 1988). The southern part of the suture displays a southward offset (Fig. 3), which is related, as seen on the SWAT seismic records (Lefort, 1991 b), to a truncation of the Precambrian mafic body by a southward-directed Hercynian thrust with a displacement of approximately 20 km.

2.1.2 The South Armorican Suture *(SAS-SAF Fig. 1; Fig. 3: 2; Fig. 4: 1)*

In Southern Brittany there are many geological data supporting the existence of an Acadian Orogeny (Autran and Cogné 1980) and of a Siluro-Devonian volcanic-arc and back-arc basin (Jegouzo et al. 1986; Thieblemont and Cabanis 1986; Thieblemont et al. 1987 a, b; Cabanis and Thieblemont 1988; Wyns et al. 1989). This Siluro-Devonian orogenic evolution developed close to a north-dipping subduction zone that was directed towards Brittany (Cogné 1976; Cogné and Lefort 1985).

The location, orientation, and timing of this suture has been a matter of some discussion:

1. For some authors (Matte 1986 a; Matte and Burg 1981), the suture is located onshore, along the South-Armorican Shear Zone (SASZ, Fig. 1), where mafic and ultramafic rocks are found, closely associated with the plurifacial metamorphic crustal slabs of the leptyno-amphibolic complex (LAC) of Champtoceaux (Fig. 2: 20) and Audierne (Fig. 5: 21) that contain high pressure relics (eclogite) of Silurian age, retrogressed in medium pressure-high temperature metamorphic facies during Devonian nappe emplacement.

2. For Lefort, the mafic-ultramafic rocks found onshore only represent remnants of back-basins, and he regards the true suture as being located offshore, 60 km south of the Brittany coastline. There, a major crustal disruption coincident with gravity, magnetic and seismologic anomalies (Fig. 6) has been recognized (Lefort and de Poulpiquet 1990), trending over 800 km and showing characteristics of sutures known elsewhere (Lefort 1989 a). In Fig. 6, the three geophysical indications are not exactly superimposed because they do not display the same kind of information and are not all at to the same depth:

a) Magnetic modeling shows that the mafic bodies at depth can be considered as slabs dipping north at 70° (de Poulpiquet and Lefort 1989). These possible ophiolitic remnants have been calculated to lie between 750 and 2100 m depth.
b) Offshore (zone 1, Fig. 6 B) free air gravity anomalies are not accurate enough for modeling (Sibuet 1972). Zone II in the same figure corresponds with a break in the gravity high, associated with known granites. Beneath Aquitaine Basin (zone III, Fig. 6 B), filtering of the short wavelength anomalies (cut off at 0.14 cycles/km) (Marchal 1986) reveals the existence of a deeply rooted body lying just beneath the magnetic belt. This gravity high can be followed down to Toulouse (Fig. 5, Fig. 2: 10). Onshore, the magnetic body shows a northerly magnetic dip. (Fig. 2: 9–10).
c) The location of the earthquakes epicenters for 1977 (Laboratoire de Détection Géophysique 1977) (Fig. 6 C) shows an alignment which borders the magnetic and gravity highs. The poor overlap that can be seen between Fig. 6 A and 6 B on one hand and Fig. 6 C on the other hand might be explained by the bad location of epicenters offshore. The location of epicenters along sutures is explained by the preferential rejuvenation of fault planes in serpentinites which are often caught up in sutures (Kane 1977). Furthermore, the different calculations made about these earthquakes show that the dip of seismic fault planes (Veinante-Delhaye and Santoire 1980) is the same as the dip calculated from the magnetic data.

The three geophysical markers displayed in Fig. 6 also correspond locally with a zone where the Moho is 5 km deeper than elsewhere in NW France (Bois et al. 1989). All these structures cut the northern branch of the Ibero-Armorican Arc, in the north, which suggests that a chip of the Paleozoic Iberian plate is now docked along the Armorican basement. Seismic reflection surveys recorded over Médoc (Garonne River mouth in the Bordeaux area), which also revealed a different thickness of the crust on both sides of the suture (Lefort and de Poulpiquet 1990), reinforce this interpretation.

3. For Autran, these offshore and northern Aquitaine gravity and magnetic highs can be interpreted in two different ways:

a) The SAS, as described by Lefort, is the southermost border of a transported suture which is rooted onshore, beneath the Devonian crustal nappes of southern Brittany and Massif Central. These

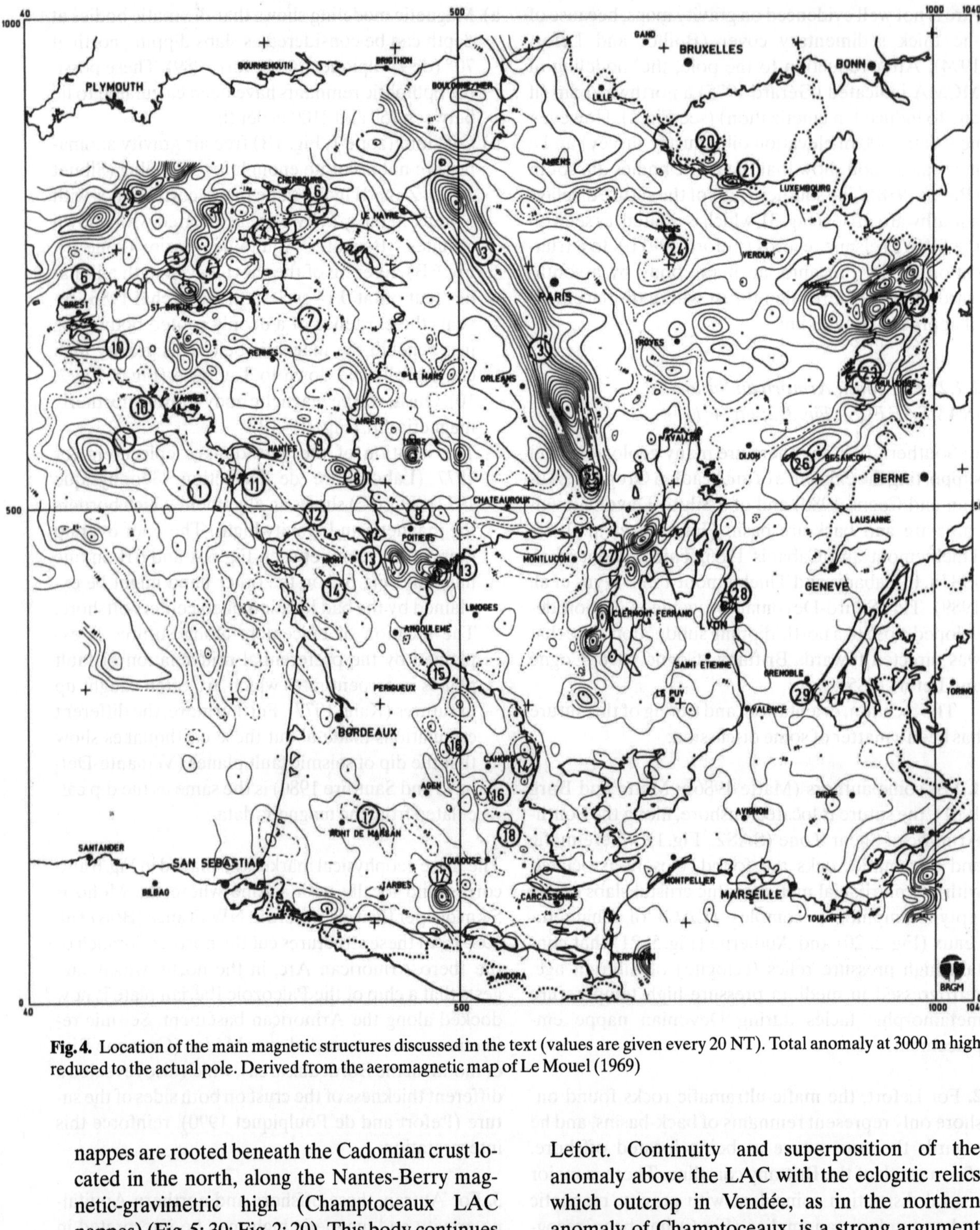

Fig. 4. Location of the main magnetic structures discussed in the text (values are given every 20 NT). Total anomaly at 3000 m high reduced to the actual pole. Derived from the aeromagnetic map of Le Mouel (1969)

nappes are rooted beneath the Cadomian crust located in the north, along the Nantes-Berry magnetic-gravimetric high (Champtoceaux LAC nappes) (Fig. 5: 30; Fig. 2: 20). This body continues eastward along the northern Massif Central following the Berry anomaly (Weber 1972) (Fig. 5: 29-28; Fig. 2: 21). The body crops out along the gravity-magnetic high of Loire mouth-Niort-Angoulême (Fig. 2: 9; Fig. 5: 32) South of Angoulême, this anomaly (Fig. 2: 10) coincides with the SAS of Lefort. Continuity and superposition of the anomaly above the LAC with the eclogitic relics which outcrop in Vendée, as in the northern anomaly of Champtoceaux, is a strong argument for this allochtonous suture hypothesis. Unfortunately, the wide angle seismic profile ECORS (Matte and Hirn 1988) is not accurate enough to support this interpretation unequivocally.

b) The SAS, offshore and south of Niort, may also represent a new feature. However, the seismic pro-

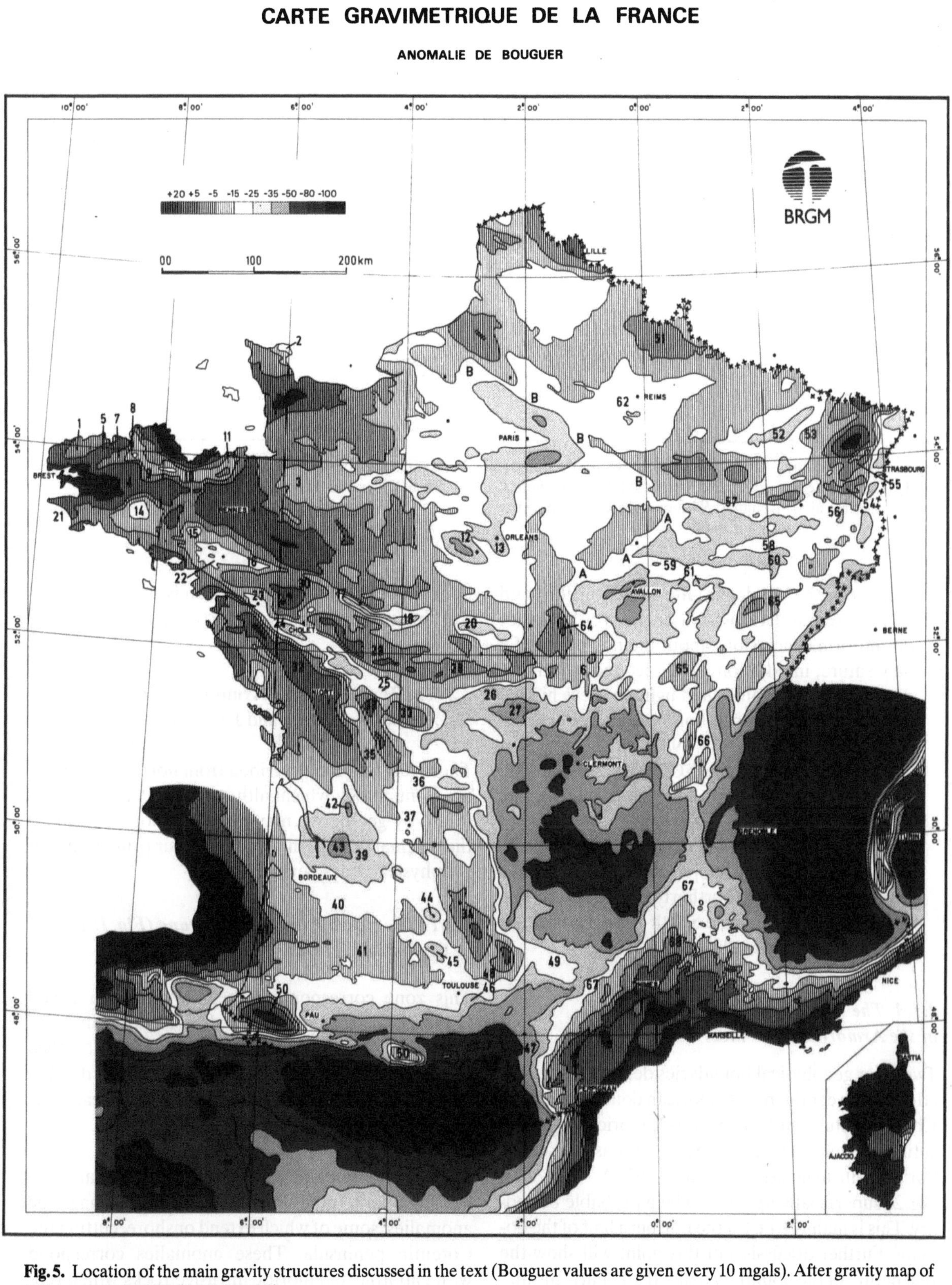

Fig. 5. Location of the main gravity structures discussed in the text (Bouguer values are given every 10 mgals). After gravity map of France at 1/1 000 000° (BRGM 1975)

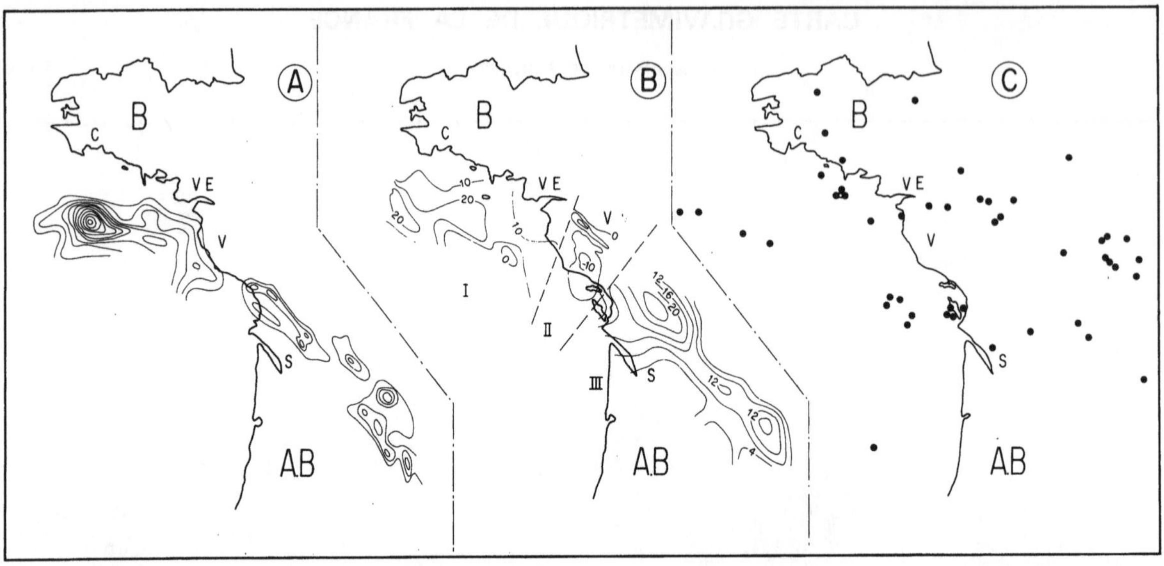

Fig. 6. Location of the South Armorican Suture after geophysical data. (After Lefort and de Poulpiquet 1989). *A* Magnetics; *B* gravity (*I* free air; *II* Bouguer; *III* anomaly after stripping the Aquitanian sediment effect); *C* Seismicity. (Taken from Laboratoire de Détection Geophysique report for 1977)

files across the Vendée (Matte and Hirn 1988) and along the coast (ECORS Gascogne profile) do not permit firm conclusions about the possibility of a new suture; in addition, no direct geological evidence exists that could shed light on these hidden bodies. The thick allochthonous slab (>10 km) of Variscan crust (Pinet et al. 1987) which appears in the northern part of the ECORS Gascogne line seems to include the bodies responsible for the magnetic anomalies that were taken by Lefort as the suture (Lefort and de Poulpiquet 1990). So this huge Carboniferous nappe could have further transported the same northerly rooted suture-like rocks. In Fig. 1, the name South-Armorican Front (SAF) is proposed in place of SAS.

2.1.3 The Western and Eastern Limits of the Armorican Province

Two clear geophysical boundaries define the western edges of the continental Armorican domain, the Mid-Channel suture and the South-Armorican offshore suture lineament (Fig. 3). Towards the east, the major Paris Basin magnetic anomaly (PBMA – Fig. 4: 3; Fig. 2) superficially appears to be a probable boundary. This is mainly true for the northern half of the domain. Further discussion of this point will show the real nature of the boundary beneath the Paris Basin.

The southern part of the Armorican domain extends into the Massif Central, at least up to the Sillon Houiller fault (Figs. 1 and 2), which is not the eastern boundary of the domain because it is a very late sinistral fault that cuts simply across the terrane.

2.2 The Terranes of the Armorican Province

(J. P. Lefort, A. Autran and J. L. Vigneresse)[1]

These terranes are described from north to south using terminology which, although sometimes already in use by geologists, may not correspond exactly to the original definition, because of our primary use of geophysical characteristics.

2.2.1 The North Armorican Terrane (Fig. 1: 1) or Cadomian Terrane

This zone corresponds largely to the area called Domnonea. It is characterized by an Icartian basement (2000 Ma) rejuvenated by the Cadomian (660 to 560 Ma) orogeny in the south; it can be divided into two subzones defined by their geophysical character.

1. The Domnonean Subterrane Proper is characterized by a large number of almost circular magnetic anomalies, some of which extend onshore, cutting the Cotentin peninsula. These anomalies correspond with gabbroic and dioritic intrusive rocks which represent the position of a Cadomian volcanic arc

[1] For the gravimetric effects of granitic pluton.

(Lefort 1975) (Fig. 4: 4). This subterrane corresponds roughly to the Normano-Breton Gulf, but extends also onshore in a N 60° direction towards the western tip of Brittany. Its general direction, which is parallel to the Mid-Channel Suture, traces the approximate extension of the Cadomian volcanic arc. The geophysical expression of this arc is blurred in three places by the Paleozoic Tregor Graben (Lefort 1975) (Fig. 4: 5) and by two Hercynian granitic batholiths in Léon (Fig. 4: 6; Fig. 5: 1) and North Cotentin (Fig. 4: 6 and Fig. 5: 2).

A recent interpretation of the Léon area north of Brest (Fig. 4: 6) (Bale and Brun 1986) suggests that this zone represents a fragment of the southern Brittany Devonian orogenic belt which has been transported towards the NE, along a N 40 dextral strike slip fault (Fig. 1). This would imply another pattern for the Armorican terranes; in this interpretation the MCMA would correspond with the direct extension of the onshore Cadomian belt. This displacement is difficult to support because the five deep seismic lines that pass across the Channel do not show any major faults cutting Precambrian thrusts (Lefort 1991 b).

2. The Mancellian Subterrane only occurs onshore. It includes many undeformed Cadomian granitic batholiths, usually responsible for clear gravity lows (Fig. 5: 3). Beneath the Paris Basin cover, the extension of the Cadomian granites towards the Paris Basin magnetic anomaly (PBMA) is very clearly shown by a vertical gradient map of the Bouguer anomaly. West of Paris, the ECORS deep seismic line shows the very probable superposition of the Mancellian subterrane extension above the PBMA itself, up to the Bray fault (Fig. 1: 1′) (Matte and Hirn 1988). Cadomian granodiorites are not responsible for deep gravity lows, probably because of the weak density contrast between these rocks and the surrounding Brioverian series. On the magnetic map the batholith is responsible for a flat low (Fig. 4: 7); its mean thickness is around 15 km.

2.2.2 The Mid-Armorican terrane (*Fig. 1: 2 and 2′*).

1. The Mid-Armorican Terrane Proper is characterized by a series of low frequency gravity highs which result from the contrast between the high densities Brioverian and Paleozoic series located in the middle of the zone, and the low density of the granitic belts which bound it on both sides. The deepest Paleozoic basins are located close to the highest gravity anomalies, as in Chateaulin (Fig. 5: 4).

The strong gravity high located south and east of Brest (Fig. 5) is probably not only the result of a continuation of the Devonian and Visean basalt-gabbro known south of Brest and in the Chateaulin basin, but it can also be interpreted as a western extension of the Cadomian magmatic arc (as in the Domnonean subterrane) or a slight uplift of the mantle in the west (Vigneresse 1978).

The Mid-Armorican terrane is intruded in the north by a series of granites that are responsible for gravity lows, such as in Huelgoat (Fig. 5: 7), Commana (Fig. 5: 8), Quintin (Fig. 5: 9), Moncontour (Fig. 5: 10), Dinan (Fig. 5: 11), Marchenoir (Fig. 5: 12), and Sennely (Fig. 5: 13). The first-mentioned granites reach a depth of 8 km (Vigneresse 1978) while the others are rooted at 10 km (Weber 1972).

In the south, a leucogranitic igneous intrusive belt that is about parallel with the northern one is responsible for deep lows that reach −25 milligals in places (Fig. 2: 25). This belt incorporates the following batholiths: Pontivy (Fig. 5: 14), Gehenno (Fig. 5: 15), La Gacilly (Fig. 5: 16), Angers (Fig. 5: 17), Saumur (Fig. 5: 18), Chinon (Fig. 5: 19), and Issoudun (Fig. 5: 20). Gravity modeling shows that the southern granites have deeper roots in the east than in the west (ranging from 14 to 1 km) (Weber 1972; Vigneresse 1978). In both cases the granitic belts are composed of Hercynian granitoids.

As in the North, where the Mid-Armorican terrane is separated from the Cadomian terrane by a shear zone (NASZ, Fig. 1), this terrane is also separated from the Ligerian terrane in the south by the South-Armorican Shear Zone (SASZ, Fig. 1), which was rejuvenated, and acted as a dextral Hercynian transcurrent fault in Carboniferous time. One can observe on the gravity and magnetic maps (Figs. 4 and 5) that this southern boundary is better defined by the granitic low ridge than by the SASZ itself, which is nonetheless a seismic and deep-seated fault.

The southern boundary of the Mid-Armorican terrane is well identified on the ECORS wide-angle seismic line, as the Nort-sur-Erdre/Thouarcé transcurrent fault, a northern branch of the SASZ (NETF, Fig. 1; Fig. 2: 24). This fault follows the southern border of the low gravity trend related to granites 14 to 20 (Fig. 5; Fig. 2: 25). On each side of this major fault one can observe in the field and on the seismic profile a divergent dip of the structures dips are to the north in the southern part and to the south in the northern side.

2. The Audierne Bay Zone (Fig. 1: 2′) corresponds with the westward extension of the Mid-Armorican terrane, which veers towards the southwest of Brittany. This curvature shows that the Mid-Armorican terrane is a part of the Ibero-Armorican arc. This zone, which is now separated from its parent terrane by the South-Armorican Shear Zone, is also characterized by a gravity high related with a thick Brioverian and Paleozoic clastic sedimentary succession (Fig. 5: 21). It is bounded in the south by an ophiolitic complex that was sheared and metamorphosed in Mid-Devonian (Peucat 1983), and is responsible for a narrow linear gravity and magnetic belt that is only seen on detailed marine surveys.

2.2.3 The Ligerian Terrane (Fig. 1: 3)

This terrane is bounded to the north by the northern branch of the South-Armorican Shear Zone (NETF Fig. 1; Fig. 2: 24) and to the south by the South-Armorican suture (Fig. 4: 1). It is characterized by strong gravity and magnetic lineations trending N 120 or N 130°, which are typical of the Ligerian belt (Autran and Cogné 1980). On maps 4 and 5 there is a clear link between the structures of southern Brittany and the western part of that of Massif Central. This terrane is cut in two by the southern branch of the South-Armorican Shear Zone, which has only a weak magnetic and gravity expression. This entire zone is characterized by the superimposition of Carboniferous deformation on pre-existing Devonian folding and metamorphism (these Eo- and Mid-Variscan phases are subsynchronous with the Acadian orogeny in the Appalachians).

1. The Northern Ligerian Subterrane (Fig. 1). The terrane occurs in a fan shape that is open to the east; it is separated from its neighboring northern and southern terranes by two strong gravity lows that are associated with granitic batholiths. The northern belt, characterized by flattened granites, has already been described. In the south, a series of granites showing an intense deformation extend in a narrow belt. The most important plutons are: Questembert (Fig. 5: 22), Vigneux (Fig. 5: 23), and Mortagne (Fig. 5: 24); they are all rooted between 2 and 6 km (Vigneresse 1978) and are leucocratic peraluminous. The Bressuire-Montluçon axis extends along this ridge beneath the cover (Fig. 5: 25) up to the La Marche (Fig. 5: 26) and Guéret granites (Fig. 5: 27), prolonging the Armorican batholith towards the Massif Central.

Between the two negative ridges runs a major gravity and magnetic high. This prominent geophysical marker (Fig. 2: 21; Fig. 4: 8; Fig. 5: 28–29) can be correlated with a magmatic unit that can be traced easterly from Cholet to the Loire fault (Fig. 2: 6) under the sedimentary cover of the southern Paris Basin (Weber 1972). This Berry gravity high is about 20 mgal in amplitude and 20 km in width, whereas the magnetic anomaly is still 20 nT at 3000 m flight elevation. Both anomalies can be correlated over 300 km. In the west, in northern Vendée, the anomalies coincide with a nonmetamorphosed Upper Ordovician plutonic gabbro cutting across a synchronous a granite-rhyolite complex. Further east, north of the Massif Central, the magnetic complex apparently dips north. To the west, north, and east of Nantes, similar magnetic and gravity highs (Fig. 2: 20; Fig. 5: 30; Fig. 4: 9) seem to extend westward along the Berry anomaly. There, this anomaly is superposed on a deep crustal slab, named the Champtoçeaux LAC nappes, which also contains an Ordovician plutonic complex (gabbro-ultrabasic-granites). It was highly metamorphosed and deformed (first in high pressure-medium temperature and then medium pressure-high temperature, between 440 and 380 Ma). It is almost certain that the Cholet section of the geophysical belt corresponds to the undeformed and unmetamorphosed steeply dipping Ordovician intrusive across the well-preserved Cadomian basement of the Mauges area. This basement shows the same high density as the Brioverian of the Central Brittany terrane. It constitutes the northern part of the eastern Ligerian terrane, just between the Nort-sur-Erdre fault (NEFT, Fig. 1; Fig. 2: 24) and the Berry anomaly. It is from beneath this weakly deformed Cadomian remnant, that the LAC, a highly metamorphosed and sheared crustal slab, was expelled toward the west in Devonian time (Lagarde 1978). This continuous Champtoceaux-Berry anomaly can be thus considered as the root-zone of the Devonian crustal nappes of northern Massif Central and Vendée (Bayer and Hirn 1987; Matte and Hirn 1988). We suggest that the Cholet magmatic component of the Berry anomaly represents the supracrustal expression of the same magmatic extensional belt (Thièblemont et al. 1987b), which also partly constitutes the LAC complex nappe.

2. The Southern Ligerian Subterrane (Fig. 2). This zone incorporates a large part of the volcanic arc related to the Eovariscan subduction (Audren and Lefort 1977; Autran and Cogné 1980). The main part of this arc occurs offshore, where "porphyroid" volcanism are responsible for a large magnetic low (Fig. 3: 3; Fig. 4: 10). This low can be extended down to Vendée (Fig. 4: 11) where "porphyroids" are also known onshore and have been dated as Silurian. The strong gravity high (Fig. 5: 32) and the magnetic high (Fig. 4: 12) which lie between Vendée and Niort correspond, near Les Essarts to gabbros and orthogneisses (Thièblemont et al. 1987a) that have been affected only by Barovian metamorphism. They may represent a deep structural equivalent of the volcanic arc recognized offshore. The same anomaly also incorporates a LAC-type slab of Champtoceaux affinity with high pressure-low temperature metamorphic relicts of the Eovariscan subduction.

The strongest magnetic high known in France, which is located south of Poitiers (Aubert 1978), is also related to diorites and gabbros, of Devonian age. These rocks show a laccolitic shape that roofs at 3 km depth (Fig. 2: 29; Fig. 4: 13; Fig. 5: 33). This calc alkaline basic plutonism shows the same type of geophysical signature from the Loire river mouth to the east of Cahors (SW of the Massif Central) (Fig. 2: 30). It forms a 500-km-long belt of Upper Devonian calc alkaline magmatism parallel to the SAS-SAF (Autran and Lameyre 1980).

South of the outcrops of the Massif Central, a roughly N-S gravity high (Fig. 5: 34) has been correlated with Stephano-Permian overstepping furrows between Brive and Castres. The thickness and density of these sediments, known from drilling and seismic reflection data, are not sufficient to produce the measured anomaly. On the northern and eastern rim of this structure, gabbro-diorite plutons of Devonian age (near Figeac) (Fig. 2: 30; Fig. 4: 14) are apparently responsible for the highest part of the magnetic anomaly. The major part of this Cahors (Fig. 2: 10; Fig. 4: 16; Fig. 5: 34) and Albi anomalies (Fig. 2: 11; Fig. 5: 35) represents the southern part of the Angoulême-Périgueux linear (SAS-SAF anomaly) (Fig. 2: 9). It can be interpreted either as the suture itself (Lefort and de Poulpiquet 1989) or as a LAC-type body setting in a klippe which is suggested by the metamorphosed gabbros, amphibolites, and leptynites cored beneath the Stephanian of Carmaux basin (Fig. 2: 11) (Autran et Cogné 1980; Guérangé-Lozes 1988).

The southern Ligerian terrane, generally dense, is locally spotted by small gravity lows which correspond to granites (Fig. 5: 35, 36, 37). Some of these bodies have been drilled south of Poitiers.

3 The Aquitanian Terrane (Fig. 1: 4)

(J. P. Lefort and A. Autran)

This triangular zone is bounded in the north by the South-Armorican Suture (SAS-SAF Fig. 1) and in the south by the Pyrénées (Fig. 1: 5). It is generally a positive gravity area with only a weak magnetic relief due to the thickness of the sediments. Many drillholes show that the Paleozoic sediments at the top of the basement are different from those of Brittany (Paris and Robardet 1985); fossils encountered there are more typical of the Cantabrian mountains than of the southern Massif Central (south Montagne Noire) (Paris and Le Pochat 1987).

3.1 The Médoc Subterrane (Fig. 1)

The gravity low between Agen and Bordeaux (Fig. 5: 39) represents the Medoc subterrane. It corresponds with an unmetamorphosed nappe of Westphalien age (Lefort et al. 1989b) that is mainly composed of Devonian sediments (Paris et al. 1988), shown to be no thicker than 3 km by magnetic modeling and deep seismic reflection data (Lefort and de Poulpiquet 1990). South of it, other nappes, but with older Paleozoic rocks (Ordovician) and more metamorphosed (Paris and Robardet 1985; Lefort and de Poulpiquet 1990) increase the gravity field (Fig. 5: 40). On a Bouguer anomaly map stripped of the Mesozoic and Cenozoic basinal effect, the NW-SE orientation of the front of nappe is well shown (Fig. 2).

3.2 South of the Garonne River

From Bordeaux to Toulouse, the geophysical expression of the Variscan imprint is progressively obliterated by the gravimetric and magnetic trends associated with the phase of extensional tectonics which occurred during Mesozoic time:

1. The Parentis Basin with its strong thinning of the crust and a consequent upraising of the mantle is responsible for the long wavelength gravity high seen between the coast and Toulouse (Fig. 5: 41; Fig. 2: 32).
2. Many basic intrusions that were emplaced during the Triassic, Liassic, and Cretaceous time are responsible for strong magnetic highs (Autran and Weber 1971; Fig. 4: 17).
3. In the southern part of the Aquitaine Basin, where the sediments are very thick (≥ 8000 m), all the crust is thinned throughout. Between Bayonne and Toulouse, deep vertical seismic reflection profiles have been interpreted as indicating north-dipping thrusts in the crust, probably of Variscan age, because they are overstepped by Mesozoic basinal rocks. North of those southerly directed nappes, a continuous gravimetric high, visible on the sediments stripped map (Marchal 1986) could represent a denser crust involved in these nappes (Fig. 2: 26).

Granites are scarce throughout the whole Aquitaine terrane. They are restricted to the middle part of the basin. Most of them display a sub-circular pattern and correspond with deep gravity lows (Fig. 5: 42, 43, 44, 45, 46 and 47). Some of the granites, for instance in Montauban (Fig. 5: 45) or south of Toulouse (Fig. 5: 47), have been cored.

3.3 The Eastern Part of the Aquitanian Terrane

This terrane is isolated from the main Aquitaine zone by the late Carboniferous Toulouse-Villefranche fault, which clearly controls north-south disruptions in gravity (Fig. 5: 48) and magnetic (Fig. 4: 18) patterns. This zone, which corresponds roughly to the Montagne Noire-Mouthoumet area, is part of Aquitaine because of a common gravity character (Fig. 5: 49), and because the nappes that are recognized in this region can be correlated with nappes delineated in the north Aquitaine subterrane (Lefort and de Poulpiquet 1990).

4 The Pyrenean Terrane (Fig. 1: 5)

Although the Pyrenees incorporates a Paleozoic tectonothermal belt that was rejuvenated during Cenozoic times, it is difficult to discriminate between the anomalies linked to the basement and the anomalies related to younger geophysical bodies or deformations. The two strong gravity highs located along the axis of this belt (Fig. 5: 50), however, are clearly associated with the Paleozoic North Pyrenean belt since it is recognized that a high density basement overrides the younger cover. These peculiar features are probably also increased by the effect of Mesozoic extensional tectonics and associated alkaline magmatism. Most of the other geophysical features, and particularly the huge gravity lows above the high chain (–100 mgals) are related to the Cenozoic foldbelt itself and its strongly thickened crust ($\simeq$50 km), as shown by ECORS seismic profiling (Choukroune 1989).

5 Terranes in Northeast France (Fig. 7)

(J. B. Edel)

In northeast France, the terranes delimited by the interpretation of geophysical maps coincide with the terrane classification given by Kossmat (1927): these are the Rhenohercynian, Saxothuringian and Moldanubian. We subdivide the Moldanubian into two subterranes because it contains two distinct zones having different structural orientations: the EW Morvanian-Vosgian subterrane occurs in the north and the Ligerian-Arvernian occurs in the south (Fig. 1). The western part of the Ligerian Arvernian terrane has been already described with the Armorican province (Sect. 2.2).

5.1 The Rhenohercynian Terrane (Fig. 1: 6)

The Ardennes. The Devonian-Dinantian serie, which is mainly composed of limestones, is characterized by intermediate densities and negligible magnetization; overall there are no significant anomalies. On the contrary, the Cambrian of the Rocroi and Givonne massifs is correlated with a large gravity high (Fig. 5: 51) and two EW magnetic anomalies (Fig. 4: 20, 21). The northernmost anomaly is caused by magmatic sills of uncertain age, intrusive into the Cambrian; susceptibilities around 10^{-2} SI and remanent magnetizations up to 10^{-1} Am^{-1} were measured in diabase sills (Edel and Coulon 1987).

The northern Lorraine-Hunsrück zone represents the western continuation of the Taunus-Hunsrück massifs and shows the same geophysical character; Devonian sequences with intermediate densities and low magnetization are expected under the Mesozoic cover. The boundary with the Saxothuringian zone is marked by a positive gravity linear of uncertain origin (Fig. 7).

5.2 The Saxothuringian Terrane (Fig. 1: 7)

This terrane comprises two subterranes:

1. **The External Subterrane.** In the Lorraine-Saar coal basin, thick sequences of late Carboniferous-Permian sediments are responsible for a 200-km-long gravity low (Fig. 5: 52; Fig. 7). The Saar-1 borehole, which is located on the negative axis, passed down through 4666 m of Westphalian-Stephanian and about 1000 m of Devonian-Dinantian sediments before reaching a crystalline ridge composed of granite dated at 400 Ma.
2. **The Internal Subterrane.** This zone is characterized by its intense gravity and magnetic positive anomalies. The dense basement is transected, however, by a series of lighter crystalline ridges that are characterized by negative gravity axes. On the Lunéville-Adamswiller ridge (Fig. 5: 53), the Adamswiller borehole cored gneiss and quartzites (Lienhardt 1961). In northern Vosges, the gravity highs are related to Cambrian and Ordovician-Silurian schist

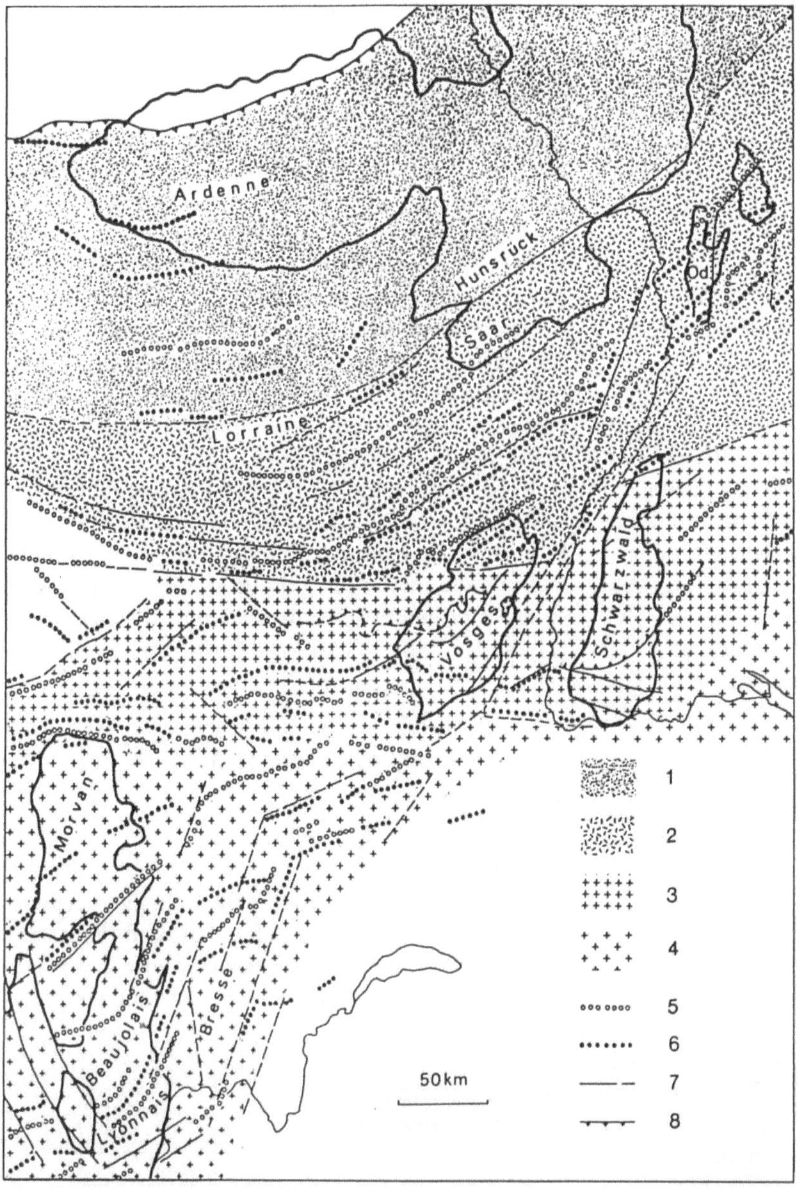

Fig. 7. The terranes in northeast France. (After Edel 1982). *1* Rhenohercynian; *2* Saxothuringian; *3* Morvanian-Vosgian; *4* Arvernian (ligerian); *5* "light" belts: granites and/or late Variscan basins; *6* "magnetic" belts: mafic intrusions and volcanics (metamorphic or not); *7–8* discontinuities: faults, thrusts

and hornfels, characterized by densities around 2.8 (Edel and Fluck 1989) (Fig. 5: 54), and to Devonian-Dinantian units, including thick volcanic series with a calc alkaline affinity, that display densities up to 2.9 (Gançarski 1977). Farther north, the prominent Saverne-Sarrebourg anomaly (Fig. 5: 55) is interpreted as being caused by several kilometers of lightly metamorphosed Early Paleozoic rocks which are covered by Devonian-Dinantian volcanic and sedimentary sequences.

The magnetic anomalies of this zone are amongst the most important in western Europe, even when considering the anomaly of the Paris Basin. The N 50–60 °-trending magnetic bodies which were identified in the Vosges (Fig. 4: 22), are mostly mafic intrusions (diorites, microdiorites, granodiorites) and flows (spilites) emplaced in the latest Devonian-Visean (Lauer and Taktak 1971; Lauer 1975; Gançarski 1977; Edel et al. 1982, 1986). Diorites and granodiorites have K-Ar ages in the range of 350-338 Ma (Edel et al. 1986; Montigny and Thuizat 1989). In the Odenwald (FRG), farther to the northeast, gabbros dated at 362 Ma (Kober et al. 1988) and diorites and granodiorites cooled in the 342-330 Ma range (Kreuzer and Harre 1975). All these Variscan intrusive bodies and their hidden counterparts are responsible for the intense magnetic anomalies (Fig. 7).

As a whole, the Saxothuringian basement is composed of "heavy" more or less metamorphic Paleozoic units that were intruded in the latest Devonian-Visean by "heavy" and highly magnetic gabbros and diorites, by granodiorites, and then finally by granites. In Late Carboniferous-Early Permian times, the zone underwent crustal extension (Lorenz and Nicholls 1984). Intramontane basins were filled and rhyolitic volcanics were emplaced, both resulting in gravity lows, as the Saar-Nahe trough.

5.3 The Vittel, Lalaye-Lubine, Baden-Baden Zone (Fig. 1, L. B. F.)

From the geophysical point of view, this zone corresponds to a gravity gradient separating the positive anomalies of the Saxothuringian terrane from the negative anomalies characteristic of the Moldanubian terrane (Fig. 5: 57). The EW-striking discontinuity is shifted to the NE by several N 35°-trending sinistral Variscan wrench-faults, the most important of these being located in the present Rhinegraben (Edel and Fluck 1989). In northern Vosges, along the Lalaye-Lubine zone (Fig. 5: 56), the geophysical discontinuity is the result of a polyphase deformation consisting of dextral wrench-faulting and northward thrusting (Fluck et al. 1987; Wickert and Eisbacher 1988). The right-lateral offset is estimated to be of at least 100 km (Edel 1982).[1]

5.4 The Morvanian-Vosgian Subterrane (Fig. 1: 8)

In the mainly NE-SW-striking Variscan belt, the geophysically EW-trending Morvanian-Vosgian area is in a particular position. Negative gravity anomalies dominate the northern part of the zone. Magnetic anomalies are rare and weak. In the middle Vosges, the basement consists of gneiss thrusted southwestward over Upper-Devonian to Visean sedimentary and volcanic sequences, and of Visean granites emplaced along thrusts and wrench faults (Fluck et al. 1987; Wickert and Eisbacher 1988). Small outcrops permit the almost horizontal thrust to be traced westward. The predominance of gravity lows is attributed to the presence of small granitic plutons.

In the south, two elongated EW gravity and magnetic anomaly belts of mostly intermediate amplitude (Fig. 5: 58–59; Fig. 4: 23) occur. The northern belt crosses the southern Vosges, where outcroping Devonian-Dinantian volcanic and sedimentary rocks are intruded by mafic stocks and Visean granites. The magnetic bodies have been identified as diorites, spilites, and andesites emplaced between 367-337 Ma (Edel and Lossy 1975; Edel 1982; Montigny and Thuizat 1989).

The EW belts continue toward the east in the southern Schwarzwald. The northern magnetic lineation seen in the Vosges disappears in the Badensweiler-Lenzkirch zone, which is a narrow belt of Devonian-Dinantian units that has been overthrust by the southeast-verging central Schwarzwald gneiss complex (Eisbacher et al. 1989).

From northern Morvan to the southern Vosges, the magnetic and gravity lineations are separated by negative axes caused by the presence of granites and particularly to leucogranites emplaced along wrench-faults (Edel 1982) (Fig. 5: 60–61).

In conclusion, the Morvanian-Vosgian subterrane is characterized by south-verging crustal thrusts, an E-W structural grain resulting from EW wrench-faulting and numerous granitic intrusions emplaced along Carboniferous thrusts and wrench-faults (Fluck et al. 1987; Wickert and Eisbacher 1989).

[1] See also a recent reinterpretation of these transcurent movement in order of 300 km.

6 Terranes in the Center and North of the Paris Basin

(A. Autran and N. Debeglia)

This region is a key area for the relationships between the four types of terranes defined around the Basin; their geophysical characteristics and boundaries are apparently different both in gravity and magnetics.

6.1 The Eastern Extension of the Armorican Terranes

In the west, the Cadomian and Central Armorican terranes appear to be truncated in the east at the Paris Basin magnetic anomaly (PBMA). However, the gravity maps suggest that the eastern boundary of those terranes is formed by two very strong, almost orthogonal horizontal gradients, a NE-SW one near Tonnerre in the southeast (Fig. 2: 1) and the similar WNW-ESE gradients of Bray/Vittel in the north (Fig. 2: 2).

6.2 The Western Extension of the Rhenohercynian and Saxothuringian Terranes

The area north of the well-known Bray/Vittel gravity gradient can best be described in gravity terms after removal of the negative effect of the Paris Basin sedi-

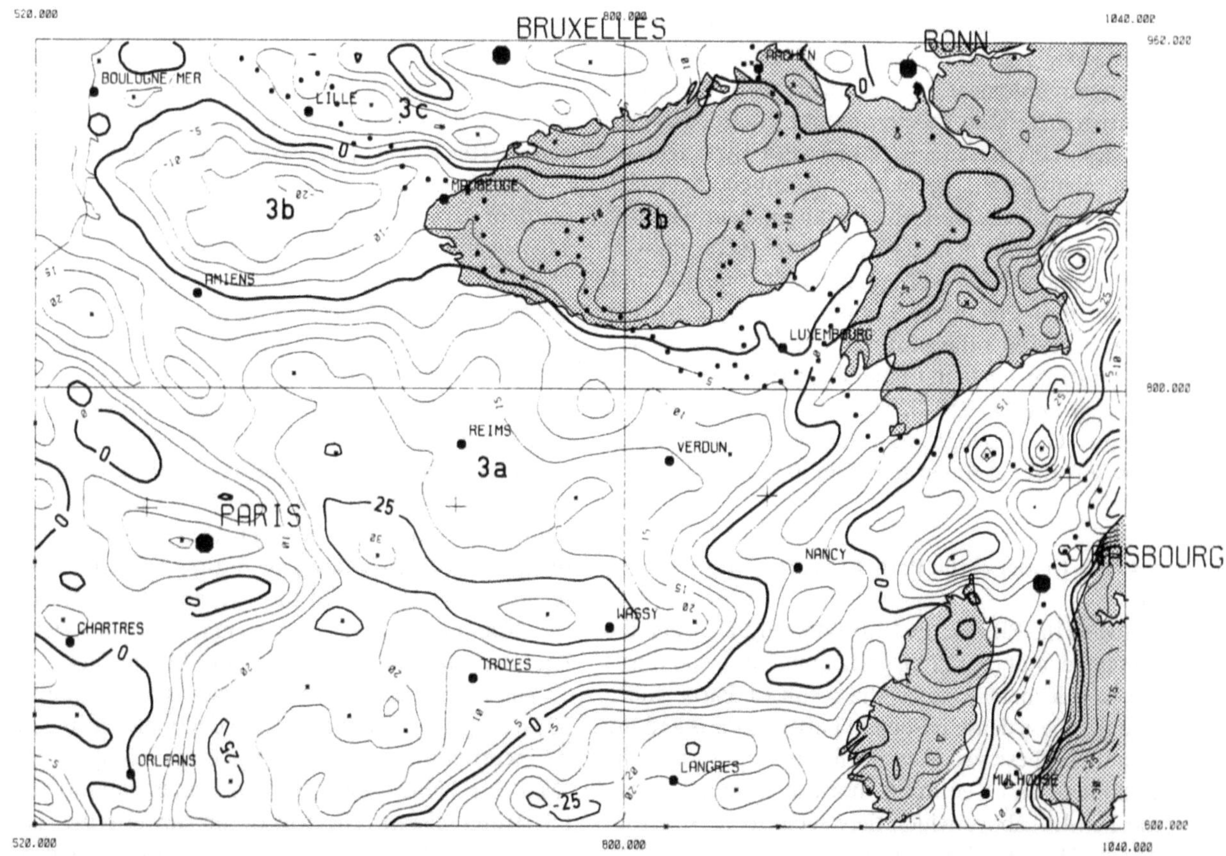

Fig. 8. Bouguer anomaly map after removal of the effect of Paris Basin sediments (with a 10×10 km grid). (N. Debeglia, original)

ments (Fig. 8). It is this sedimentary effect which causes the huge low anomaly of Reims (Fig. 5: 62). After stripping, this anomaly is replaced by a positive anomaly which coincides also with a probably deep-seated magnetic body (Fig. 2: 3 and Fig. 4: 8, 24). This magnetic body display a long wave length but its amplitude is almost ten times smaller than that of the PBMA.

South of the Metz/Hunsrück fault (Fig. 2: 4), which continues to the SW up to the Bray fault through a N-S gradient (Fig. 2: 4′), there is an area where NE-SW magnetic and dense stripes alternate with nonmagnetic and gravity lows. These stripes about the Vittel/Lubine/Baden-Baden fault (Fig. 2: 5; Fig. 5: 56–57). This feature which is characteristic of the Saxo-Thuringian terrane in Germany and the northern Vosges mountains, was described in Section 2.5. Northwest of the Metz-Hunsrück fault in the Paris Basin, the western extension of the Rhenohercynian terrane occurs. Its dense and moderately magnetic southern part corresponds with thick lower Devonian schists and their overthrusted upper sheets consisting mostly of basic volcanic rock and flysch, known on outcrops only in RFA. The latter originated in the Metz/Hunsrück "suture" (Franke et al. 1989). The heavy and magnetic Reims anomaly (Figs. 2: 3 a and 8: 3 a) probably corresponds with the same type of schists and volcanics. The even more dense metamorphic terranes that lie as a continuous strip north of the Bray fault could be considered as klippen of the Saxo-Thuringian area (Matte et al. 1986).

The low density and nonmagnetic area to the north (Figs. 2: 3 b and 8: 3 b) is the western extension of the Ardennes. It incorporates the Neufchateaux and Dinant synclines which contain upper Devonian carbonates. In the far north, the syncline of the marginal Namur basin forms a separate dense basement below the front of the major Rhenohercynian thrust sheets (Figs. 2: 3 c and 8: 3 c).

6.3 North-Western Extension of the Moldanubian-Arvernian Terrane

The Vittel/Lubine fault (Fig. 2: 5; Fig. 5: 57) (Sect. 2.5) forms a particularly spectacular terrane boundary because of its truncation of the Saxo-Thuringian NE-SW structures. South of this geophysical and geological fault, the Vosges/Morvan/Massif Central region is defined by a generally low density and only poorly magnetic basement. In contrast with the Saxo-Thuringian terrane, this area is poorly defined by geophysics. This terrane is linked with the Ligerian-Arvernian terrane previously described (Sect. 2.2, 2.3). Towards the west, this terrane is clearly limited by the truncation of its E-W structures by the entirely hidden major NE-SW Tonnerre fault (Fig. 1: 2) and, further south, by the N-S Loire fault (Fig. 2: 6). The latter represents the northern extension of the left lateral strike-slip fault of the Sillon Houiller, a major, but very late ($\simeq$ 310–320 Ma) fault of the Ligerian terrane. Good geophysical and geological markers show an offset of 80 km across this fault.

6.4 The Center of the Paris Basin at Depth

Between the PBMA, the Bray fault (Fig. 2: 2), and the Tonnerre fault (Fig. 2: 1), a triangular area contains a dense basement zone to the south of Paris, and a wedge of less dense basement between Paris and the Channel. These two areas form the extension of the Cadomian and Central Armorican terranes, across the PBMA, above which they seem to be superimposed. Thus, the less dense zone south of the Bray fault is composed largely of Cadomian granites which were metamorphosed during the Carboniferous (Matte et al. 1986), as shown by some cores of orthogneiss.

6.5 Discussion and Small-Scale Correlations Beneath the Paris Basin

The new data provided by the seismic reflection profiles such as ECORS (Bois et al. 1989) in Northern France and DEKORP (1985) parallel to the Rhine-Graben show that the deep structures are composed of large deep crustal nappes, which are typical of the northern branch of the Variscan chain. This indicates that the Bray/Vittel fault can no longer be regarded as a permanent terrane boundary. This fault, in fact, superimposed rather late in the Hercynian history (about 320 Ma) as a dextral strike-slip fault that cuts across the stacked Cadomian, Saxo-Thuringian, and Rheno-Hercynian terranes. These terranes must therefore be seen as allochthonous units that were thrust, one above the other, toward the north. For example, the dense crustal area north of the Bray fault probably joins the positive anomaly located in the middle of the Saxo-Thuringian terrane towards the east, in the direction of Nancy. This positive anomaly probably belongs to an allochthonous terrane, which, north of Paris, would have been sandwiched between the Rhenohercynian terrane (thrust over the Brabant Massif and its cover) and the overriding Cadomian terrane (the Bray nappe of Matte and Hirn 1988).

The dense and weakly magnetic area, located south of Paris and north of the Tonnerre fault (Fig. 2: 1), probably represents the eastern extension of the Mid-Armorican terrane. This area shows a strong geophysical resemblance to the southern half of the Saxo-Thuringian area in Germany with its large and dense Tauber-Neckar zone which terminates to the west below the Lubine/Baden-Baden thrust. Here, too, a Cadomian basement has been moderately deformed during the Carboniferous northward thrusting of the Moldanubian terrane. South of Paris, this dense block would thus belong to the southern border of either the Saxo-Thuringian or the Mid-Armorican terranes (Fig. 1: 2–7). In this hypothesis we can use this "Cadomian" marker to evaluate a 300-km right-lateral displacement along the Bray-Lubine fault.

The prominent PBMA magnetic anomaly (150 to 300 nT) appears, on the ECORS deep seismic line, to be incorporated within the Cadomian Bray nappe (Matte and Hirn 1988). The continuous, subhorizontal, strong seismic reflectors of the "lower crust" beneath the PBMA seem to exclude a post-tectonic intrusion origin for the body that is responsible of the anomaly (Galdeano and Guillon 1988). This body is very probably located above the reflectors used to define the sole of the Bray nappe (subterranes 1' and 2' in Fig. 2). Thus, being associated with the base of those allochtonous Armorican terranes, the PBMA cannot be used directly as a geophysical boundary. It is necessary to know the nature and the significance of the body in the basement; but it is completely hidden and, to date, its geological origin has not yet been elucidated. Its interpretation as a transported Paleozoic suture is preferred by Autran et al. (1986), Autran and Chantraine (1988), and by Matte and Hirn (1988).

Toward the southern end of the PBMA, where it becomes parallel to the major sinistral Sillon Houiller fault, the GPF (Géologie Profonde de la France program) borehole of Sancerre-Couy (Fig. 2: 7; Fig. 5: 64; Fig. 4: 25) has been drilled to 3500 m depth, but did not reach the PBMA. It passed through 2600 m of rocks whose Eo-variscan-suture significance is clear because of their MORB-type basic protolith which has been affected by high pressure/high temperature metamorphism ending about 380 Ma with an overprint of middle pressure/high temperature metamorphism (Burg et al. 1988). These rocks have the same age and metamorphic characteristics as the Leptynite-Amphibolite complex (LAC) rocks known to the south in the Massif Central, where they form the higher allochthonous unit in the structural succession. They also partly constitute the petrographic association of the LAC rocks found in the Champtoceaux nappe of the Loire-Atlantique at the northern border of the Ligerian terrane (Fig. 2: 20; Fig. 5: 29). Between the southern end of the PBMA and the Loire-Atlantique zone, the Berry anomaly (Fig. 2: 21; Fig. 5: 28) links these two areas beneath the sediments of the Paris Basin.

In the Sancerre borehole, the LAC-type body appears superimposed as a later nappe above the more magnetic PBMA body. Its primary kinematics of emplacement during early Devonian collision is not known and cannot be compared with the westward or southwestward vergence of the same type rocks in the Champtoceaux (Lagarde 1978) or in the Massif Central upper allochthonous unit. The gravity and magnetic properties of the LAC rocks in the Sancerre borehole can be compared with those of the Berry anomaly, which represents an extensional Cambro-Ordovician magmatic belt that was involved in the northern suture of the Ligerian terrane (Sect. 2.2.3). If the PBMA body is genetically related to these distensive suture events, it could represent, (with its dramatically higher magnetic signature), an oceanized-rift area in Mid-Paleozoic time, which closed with a general northward vergence.

In spite of the large uncertainty about the linkage of the PBMA body with LAC rocks of the Berry-Sancerre belt, the extension of this belt to the north or east can be discussed with potential field maps.

The extension of this "suture" zone to the east of the Loire fault (Fig. 2: 6) into the Morvanian/Vosgian terrane, to the NE of the Massif Central, is not clear. Dense and magnetic signatures are tenuous, not comparable with those of the Berry anomaly. Moreover, in the Vosges and the Black Forest, the highly metamorphic gneisses with high pressure metamorphic relicts, which form a vast overlying allochthonous unit, have until now revealed only their Carboniferous tectonic history (Blumenfeld 1986; Fluck et al. 1987; Eisbacher et al. 1989). This history consisted of thrusting around 330 Ma that was both northerly directed, in the Lubine/Baden-Baden fault area, and southerly directed, as is shown by the Badenweiler-Lenzkirch zone in the Black Forest and the Gerbépal area in the Vosges. The initial protoliths and ages of the early metamorphism are Cambrian to middle Devonian ($\simeq$ 380 Ma) which makes them older or comparable to northern units of the Massif Central. The Morvanian-Vosgian terrane, with its pre-Devonian history that is in part independent of that of the Massif Central, may be a new terrane, as shown in Section 2.5.

The extension of the Berry belt, north of the Massif Central, could be located along a series of weak positive E-W anomalies surveyed north of the Morvan (Fig. 5: 59; Fig. 2: 27). These anomalies coalesce to the west with the Charité-sur-Loire anomaly (north of the Sancerre GPF borehole), and to the east with the "autochthonous" gneisses of the southern Vosges and southern Black Forest which suffered only a single metamorphic event. This interpretation was implicitly proposed by Matte and Hirn (1988) in his map of the Variscan basement in France. Edel (1982), however, prefers an interpretation in which all the dense and magnetic anomalies are caused by basic and acidic Devonian/Carboniferous volcanic rocks that overlie the pre-Late Devonian basement with a marked unconformity. Support for this could come from detailed geophysical work on the extensive outcrops of the Morvan and Vosges Devonian-Carboniferous volcanic basins.

Another possibility, more compatible with the characteristics of the Berry anomaly, might be found along the Tonnerre fault (Fig. 2: 1). There from SW to NE a heavy and magnetic body similar to that of the Berry anomaly appears north of Sancerre and east of the PBMA. This Auxerre anomaly (Fig. 2: 23) is superimposed on the western part of the Tonnerre gravity gradient (Fig. 2: 1). It extends North of Auxerre, where the magnetic high ends at the NNW-SSE dextral Provins fault (>200 km long) (Fig. 2: 22). North of this fault the results of a high-precision aeromagnetic survey, in the area of Bar-sur-Seine, (BRGM 1980) suggests an extension of the Auxerre anomaly along the Tonnerre fault, up to the Lubine fault (Fig. 2: 5).

Thus, the possible continuity of a Berry-like anomaly trending from the southern border of the Mid-Armorican terrane to the south of the Saxothuringian terrane eastward seems to us a better hypothesis than the E-W extension inside the Morvano-Vosgian terrane. The northeastward extension is probably truncated by the eastern part of the NETF strike-slip fault (Fig. 1) and partially hidden under the late Westphalian northward-thrusting of the western part of the Morvano-Vosgian terrane.

These two hypotheses imply a very different structural framework during Devonian collision, especially the crucial problem of the location of the root zone of the Eovariscan nappes of the Massif Central.

7 Terranes in the Massif Central, Southeast and Southern France

(A. Autran, N. Debeglia)

This area, which consists of about one-third of the surface area of France, has been strongly influenced by the Alpine Orogeny. Its effects on the thickness of the continental crust and the modifications of the upper mantle caused very intense regional disturbances of the gravity field and local disturbance of the magnetic fields because of the associated magmatism. It is therefore difficult to recognize geophysical terranes within this area which can be considered unequivocally as Variscan.

7.1 The "Alpine" Effects in Eastern, Central, and Southern France

A number of regional geophysical anomalies are clearly associated with the Mesozoic and Tertiary reworking of the lithosphere (Weber et al. 1980; see also the Moho map of Fig. 10). The general Bouguer and magnetic maps (Figs. 4 and 5) show:

1. Areas of crustal thickening in the Alps (Choudhury et al. 1971) and Pyrenees (Gallart et al. 1981), where seismic refraction and reflection studies show that the crustal thickness is greater than 40 km; they are responsible for prominent negative gravity anomalies (–50 to –120 mgals).
2. Areas where thinned crust is known, such as in the Bay of Biscay (Ginzburg et al. 1985) or the periphery of the Mediterranean (Hirn et al. 1977); they coincide with the highest gravity anomalies (up to 25 mgals).
3. In the Massif Central, below Tertiary volcanics, the thinning of the crust (26 km) is not responsible for any positive anomaly because of the thickness of the low velocity and low density anomalous mantle beneath this area (Perrier et Ruegg 1973). The magnetic field is particularly disturbed by a large number of high frequency anomalies related with the Auvergne, Cantal and Velay Cenozoïc volcanics.
4. In the peri-Alpine Tertiary grabens (Rhône, Saône, and Rhine), the moderate crustal thinning indicated by seismics (Rhinegraben Research Group 1974; Edel et al. 1975) is marked by a slight gravity high (Fig. 5: 65). This effect is particularly clear in the NE-SW-oriented zone that links the south of the Rhinegraben with the thinned areas of the Massif Central (Perrier and Ruegg 1973; Michel 1978). This zone, which links Mulhouse, Belfort, Besançon, and Chalons-sur-Saône, is characterized by a long-wavelength magnetic high that passes into the Auvergne anomaly further west (Fig. 2: 13; Fig. 4: 26). A Variscan emplacement of basic magnetic bodies of upper Devonian age, analogous to that of the La Brevenne anomaly which spread north of Lyon (Fig. 2: 14) is also possible for this anomaly (Edel 1982).

In this "Alpinized" area, crustal interpretations based on good-quality seismic refraction data (Menard 1979; Hirn 1980), have been confirmed by the recent ECORS vertical deep seismic profiles. It is possible to define, both the thickness of the Mesozoic sedimentary basins, and the Moho topography (Fig. 10). Information concerning the sediment density provided by industrial deep wells, and the upper-mantle density, estimated from seismic velocities, have been used for the correction of the Bouguer anomalies for part of the post-Hercynian effects; this have resulted in a gravity anomaly map that is more representative of the Variscan crust (Fig. 9) than the classical Bouguer map.

7.2 The Variscan Geophysical Structures and Terranes

The corrected Bouguer map (Fig. 9) shows several major structures that suggest continuations of Variscan terranes eastward of the Massif Central.

7.2.1 The Arvernian Terrane

In the Massif Central, the widespread granite massifs are coincident with numerous negative anomalies and for an overall gravity low and weak magnetic regional field. A few localized, very dense and magnetically positive anomalies correspond to gabbro and calcalkaline intrusions of Devonian age (Limousin, SW Morvan) or to extensional rift-type gabbros such as in the Brévenne, north of Lyon (Fig. 2: 14; Fig. 4: 28; Fig. 5: 66) or between Riom and Moulins (Fig. 2: 28 and Fig. 4: 27) (Aubert et al. 1970).

Detailed studies show a correlation of leptynite-amphibolite complex (LAC) rocks with high pressure metamorphic relicts, outcropping in the heart of the Massif Central (Fig. 2: 8): the LAC rocks are more dense and have a more magnetic signature than the paragneiss of their substratum. Owing to their contrasting geophysical properties, the LAC rocks appear to occur entirely in a thrust sheet (Van de Meulebrouck et al. 1984; Bayer and Hirn 1987). By extrapolation from this key area, where the coher-

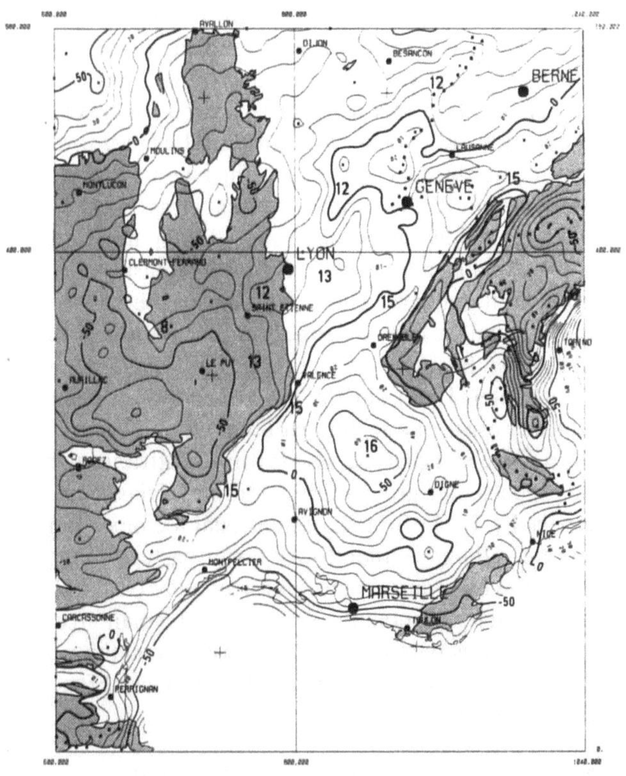

Fig. 9. Map of the Alpine part of France showing the Bouguer anomaly corrected for the effect of the Mesozoic-Cenozoic sedimentary basins and of the thinning or thickening of the crust (with a 10×10 km grid). (N. Debeglia, original)

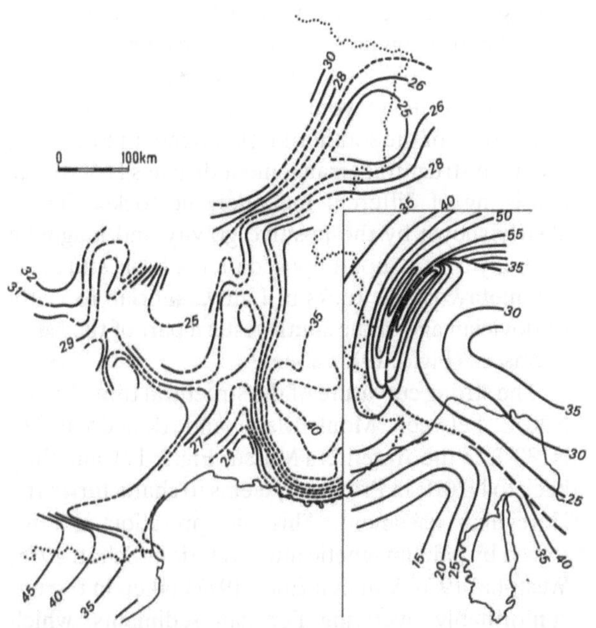

Fig. 10. Isobaths of the crust-mantle boundary (km) (Hirn 1980)

ence of the structural model and geophysical data have been established, it is generally possible to identify (through their gravity and magnetic expression) the LAC rocks in the Limousin (Autran et al. 1976; Dubuisson et al. 1988). The same lithological types can be followed below the sediments of the northern Aquitaine basin (Autran 1978; Autran and Cogné 1980; Marchal 1986), and perhaps between the Vendée to Périgueux (Fig. 2: 9), where they seem to plunge to the NE beneath the Massif Central (de Poulpiquet and Lefort 1989) or appear like klippes. A similar interpretation suggests that LAC klippes occur between Agen and Cahors (Fig. 2: 10; Fig. 5: 34), and even southwest of Rouergue (Fig. 2: 11) where the klippe structure can be seen in outcrop (Bodinier and Burg 1980; Guérangé-Lozes 1988). There, in this NW-SE alignment of high gravity and variably magnetic anomalies, run the southwestern boundary of the allochthonous upper gneiss, or LAC terrane of the Massif Central-Vendée. The same anomalies are interpreted by de Poulpiquet and Lefort (1989) as the South-Armorican Suture (Fig. 1, SAS), rather than a southerly transported suture. This is mainly based on a magnetic modeling which suggests deep roots for the mafic bodies (Sect. 2.2).

As described in other parts of this volume devoted to the lithostratigraphy, metamorphic, and tectonic characteristics of the Massif Central, the internal zone of the Variscan chain is regarded as entirely allochthonous and constituted by units of varying geological histories that were tectonically stacked on a southern foreland. However, in terms of geophysics the area can only be described as a composite (allochthonous) terrane. Its triangular shape, with a sharp corner towards the NW near the Loire mouth, and its increasing width towards the east over a length of more than 400 km, is a general characteristic of the **Ligerian terrane**, which opens transitionally eastward into the Arvernian terrane, without a clear margin.

Towards the east, the LAC rocks (allochthonous upper gneiss unit) outcropping west of Lyon (Fig. 2: 12 and Fig. 9) can be connected with the Auvergne outcrops (Fig. 2: 8). This belt continues northeast of Lyon, below the Jura Mountains, in the direction of Basle, (Switzerland). This long, dense, and highly magnetic strip lies south of the linear "Alpine" gravity high, and magnetic anomaly that connects Mulhouse with Chalons-sur-Saône (Fig. 2: 13; Fig. 5: 65). Even though continuity with the LAC rocks in the Lyon area seems certain, a possible connection with the late Devonian Brevenne gravity high and very magnetic rift zone (Fig. 2: 14) cannot be excluded. In any case, this rift zone has an entirely different geostructural significance.

This Jura-Lyon anomaly, whose direction is the same as that of the Devonian extensional lineation in the Lyonnais basement rocks, evidently corresponds to a geophysical feature residing in the Variscan basement, since it provides a strong contrast in an otherwise relatively stable Mesozoic platform area. This area is bordered in the south by a light and nonmagnetic zone that seems to be the NE extension of the low anomaly caused by the huge Carboniferous anatectic dome of Velay (Fig. 9: 13).

In the northern part of the terrane, there is no clear geophysical boundary between the Morvano-Vosgian terrane in the north and the Arvernian to the south. On the gravity map, a clear horizontal gradient in a zone oriented NW-SE is superposed along the northern border of the Cenozoic basins in northeastern MCF, from Moulins to the Lyonnais area, where it connects with the LAC upper unit. This NW-SE zone separates the Arvernian and Morvanian terranes characterized by their low gravity level.

Figure 1 defines in the NE of MCF two possible terranes by two limits: one at the southern border of the NW-SE gravity gradient discussed above and a second, more northerly, based on a slight difference of orientation of geophysical linears. The terrane between these two limits could be the root zone of the LAC upper nappe of the Arvernian terrane, as indicated by its linkage with the Lyonnais LAC, some occurrence of eclogites, and high metamorphism throughout (Goddard 1990), and the overstepping marine platform deposits of Giveto-Frasnian age.

7.2.2 The Southeastern Expression of the Ligerian-Arvernian Terrane

1. The Hidden Basement. The spectacular 600-km-long zone from the Languedoc to Switzerland having a NW-SE horizontal gravimetric gradient (Fig. 2: 15, Fig. 9: 15) probably cannot be considered as the eastern Hercynian boundary of the low density Ligerian-Arvernian terrane. This zone corresponds to the western margin of the southeastern sub-Alpine basin; from the Languedoc to Grenoble it appears to be a dense crustal segment that continues southeastward below the entire Southeast Basin (Figs. 2: 16 and 9: 16). None of the realistic density hypothesis taken for the Bouguer correction of these very thick (over 10 km) Mesozoic sediments, or for that of the mantle, has been able to compensate this gravity anomaly. This basin is moderately magnetic along its margins where sediments are thinner. Progressive attenuation of the magnetic signature towards the center of the basin reveals the deepening of the top of the magnetic basement.

Two seismic refraction profiles, which are perpendicular to this gravity horizontal gradient (Sapin and Hirn 1974), show a rapid thinning of the Hercynian crust from 29 to 15 km associated with an increasing proportion of the dense lower crust. The latter forms about one third of the crustal thickness below the Massif Central, but increases to as much as one half of the crust below the Rhône valley. Taken alone, this value explains about one half of the intensity of the horizontal gradient. The strict coincidence between the original western paleogeographic margin of the Southeast Basin of France (Debrand-Passard and Courbouleix 1984), and this huge horizontal gradient, which consists of a rectilinear attenuation of crust of the passive margin type, also favors the attribution of this spectacular gravimetric feature to Triassic and, especially, Early Jurassic, extensional tectonics. The margin of the basin therefore is not an eastern boundary of the Ligerian terrane. Unfortunately, east of the passive margin anomaly, no clear Hercynian structural features persist in the crust which was thinned during the Early Jurassic below the Southeast French Basin (Fig. 9).

2. The Outcropping Basement. The southern and eastern margins of the Southeast French Basin are partly incorporated in basements of the external crystalline massifs of the Alps and of Provence. They show clear Variscan geophysical signatures which consist of a generally low density basement that is only slightly magnetized. A magnetic high is superimposed on the ophiolites of Chamrousse (Fig. 4: 29; Fig. 2: 18) and on the Devonian-Early Carboniferous volcanic rocks of the southern Belledonne. In Provence, the deep structural arrangement displays N-S stripes consisting of different metamorphic rocks. This is clearly shown by the positive gravity and magnetic anomaly that is superimposed on the basic and alkaline meta-volcanic rocks of Late Cambrian to Early Ordovician age of the southwestern part of the Maures massif (Fig. 2: 19).

The strong curvature of the structural trends, from N30°E between Mont Blanc and Belledonne to N130°E in the Argentèra Massif (Fig. 2: 17), and then back to N20°E in Provence, seems to characterize the Hercynian basement. This interpretation is reinforced by paleomagnetic interpretation (Schott 1976; Westphal 1976; Van den Ende 1977) taken in the unconformably overlying Permian sediments, which clearly indicate that no important Alpine rotation occurred here. This is true for the entire basement, except for a part of the Haut Dauphiné Massif that is located south of Belledonne. Furthermore, major NNE-SSW sinistral shear zones that developed blastomylonites cut the Belledonne and Mont Blanc massifs (Vivier et al. 1987) before the Stephanian deposits (K/Ar age of 320 Ma). It was probably along these major transcurrent faults that the western passive margin of the French Southeast Basin developed during Liassic times.

Convincing analogies of tectono-metamorphic evolution and lithology, can be made between the basement of the external crystalline massifs of the Alps, Provence, Corsica, and Sardinia, and the basement of the Ligerian terrane. The N-S to N20°E direction (if we take into account the movement of the Argentèra Massif) of the belt, which can be followed over 1000 km (after restoration of the early Cenozoic pre-Neogene-oceanization state), seems to be compatible with the southern extension of the NW-SE structure that form the western boundary of the Ligerian terrane. It is thus possible to suggest that before the NNE-SSW strike-slip sinistral faulting of Westphalian age, along the Sillon Houiller (Fig. 1, SHF) and Belledonne-Cévennes faults (Fig. 1, CF), the entire block that now includes the external Alps as far as Sardinia in the south, had a different position; a 300-km displacement is probable. Following this interpretation, the Pyrenees were located farther to the west, which is compatible with Paleozoic paleogeography. (Engel 1984; Raymond 1987; Chap. III.1.1, this Vol.). The Westphalian/Stephanian fault system between Languedoc and Valais (in Switzerland) could have been the boundary of a western subterrane of the Ligerian terrane, which, before Westphalian time, during the Devonian and Early Carboniferous shortening, extended largely towards the south to Sardinia and towards the East in the Moldanubian of southern Bavaria and Bohemia. Its true eastern boundary should probably be sought in the basement of the internal Alps (proto Apulian promontory of Gondwana) (Autran and Cogné 1980).

8 Conclusions

France can be divided into eight geophysical terranes which can be followed beneath cover and offshore.

The interpretation of boreholes and geophysical data beneath the Paris Basin allows us to make longitudinal correlations between Brittany and the northeastern part of France. In the north, the Midland-Brabant Precambrian craton forms a major block against which the Hercynian belt was thrust towards the north. Although it is not possible to follow in detail the Hercynian zoning recognized in the east, the classical subdivision of Kossmat (1927) can be partly extended at least to Brittany. These belts are only disturbed west of the Paris Basin magnetic anomaly by the large Cadomian massif. Deep seismic reflection data show that between the Cadomian and the Brabant Precambrian blocks, the Paleozoïc terranes were intensively caught up during the Hercynian Orogeny; they now appear sandwiched and thrust between these two blocks.

To the south, a major boundary lies between southern Brittany and the Paris Basin magnetic anomaly; it can be extended towards the east up to the Northern Vosges. This boundary represents either an oceanic suture or some other significant geological feature. All the terranes located south of this suture are considered to be equivalent to the Moldanubian belt; these were later disturbed by the Alpine Orogeny in the southeast.

In the southwest, the Aquitaine terrane is isolated from Brittany and the Massif Central by a prominent geophysical marker, which can be followed over 800 km. This feature could represent either an "Acadian" suture or the southern extension of mafic sheets that were rooted in a linear disruption located to the North.

Acknowledgments. This work was possible with the support of the BRGM research program. Thanks to the reviewers M.D. Max and R.T. Haworth, whose constructive remarks greatly helped to improve the text.

References

Aubert M (1978) Interprétation de la plus intense anomalie magnétique mesurée en France à 3 kilomètres d'altitude. Bull Bur Rech Géol Min II 2: 134–147

Aubert M, Cornet J, Naudy H, Roques M, Sanselme H, Weber C (1970) Apports de l'aeromagnetisme à lá connaissance du socle du Morvan. Bull Bur Rech Géol Min 52-2: 13–24 and cartes

Audren C, Lefort JP (1977) Géologie du plateau continental sud-armoricain entre les îles de Glénan et de Noirmoutier. Bull Soc Géol Fr XIX: 395–404

Autran A (1978) Synthèse provisoire des événements orogéniques calédoniens en France. PICG n° 27. Geol Surv Can Pap 78–13: 159–175

Autran A, Chantraine J (1988) L'anomalie du Bassin de Paris et le sondage GPF de Sancerre-Couy. Discussion de quelques hypothèses. Doc Bur Rech Géol Min 137: 253–270

Autran A, Cogné J (1980) La zone interne de l'orogène varisque dans l'Ouest de la France et sa place dans le développement de l'orogenèse. In: 26th IGC, Paris Coll C 6 Mém Bur Rech Géol Min 108: 90–111

Autran A, Lameyre J (1980) Les granitoïdes de France. In: Autran A et Dercourt J. Evolutions géologiques de la France. Mem Bur Rech Géol Min (Orleans) 107: 51–97

Autran A, Weber C (1971) Données sur les anomalies magnétiques du bassin aquitain et de sa marge continentale. In: Histoire structurale du golfe de Gascogne. Série IFP-CNEXO Technip (Paris) 22, 1, IV 10: 1–8

Autran A, Gerard A, Weber C (1976) La carte gravimétrique de la France, exemples d'utilisation géologique. Bull Soc Géol Fr (7), XVIII, 5: 1119–1132

Autran A, Castaing C, Debeglia N, Guillen A, Weber C (1986) Nouvelles contraintes géophysiques et géodynamiques pour l'interprétation de l'Anomalie magnétique du Bassin de Paris. Hypothèse d'un rift paléozoïque refermé au Carbonifère. Bull Soc Géol Fr Paris (8) II: 125–141

Balé J, Brun JP (1986) Les complexes métamorphiques du Léon (NW Bretagne): un segment du domaine eo-hercynien sud armoricain translaté au Dévonien. Bull Soc Géol Fr II, 3: 471–478

Bardy Ph (1988) L'orogène cadomien dans le nord-est du Massif Armoricain et en Manche Occidentale. Thesis, University of Rennes, 328 pp

Bayer R, Hirn A (1987) Données géophysiques sur la structure profonde de la croûte hercynienne dans l'arc ibéro-armoricain et le Massif Central français. Bull Soc Géol Fr (8) III, 3: 561–574

Blumenfeld P (1986) Déformation et fusion partielle dans la croûte continentale: l'unité occidentale des Vosges moyennes. Thesis, University of Nancy, 156 pp

Bodinier JL, Burg JP (1980) Evolution métamorphique et tectonique des séries cristallophylliennes du Rouergue occidental: mise en évidence d'un chevauchement dans la région de Najac. Bull Bur Rech Géol Min, série Géol Fr 4: 315–339

Boillot G, Lefort JP (1974) La Manche, scale 1/1.000.000. Bur Rech Géol Min, Orléans

Bois Ch et al. (1989) Seismic reflection images of the pre-Mesozoic crust in France and adjacent areas. In: Keppie JD, Santallier DS, Piqué A (eds) Springer, Berlin Heidelberg New York

Bur Rech Géol Min (1975) Carte gravimétrique de la France au 1/1 000 000, 1st edn. Orleans (Two sheets)

Bur Rech Géol Min (1980) Carte magnétique de la France à 1/1 000 000 anomalies détaillées du champ total, 1st edn. Orléans (Two sheets)

Burg JP, Castaing Ch, Chantraine J, Hottin AM, Kienast JR, Megnien C, Turland M, Vezat R, Weber C (1988) Les formations métamorphiques traversées par le sondage Sancerre-Couy (GPF). Nouveau jalon de la chaine varisque. CR Acad Sci Paris 307, 5, II: 1819–1824

Cabanis B, Thieblemont D (1988) La discrimination des tholeïtes continentales et des basaltes arrière-arc. Proposition d'un nouveau diagramme. Bull Soc Géol Fr (8), IV, 6: 927–935

Choudhury M, Giese P, de Visintini G (1971) Crustal structure of the Alps: some general features from explosion seismology. Boll Geof teor ed appl, vol. XIII, 51–52: 211–240

Choukroune P (1989) The ECORS Pyrenean deep seismic profile reflection data and the overall structure of an orogenic belt. Tectonics 8, 1: 23–31

Cogne J (1976) La chaîne hercynienne ouest européenne correspond-elle à un orogène par collision? Propositions pour une interprétation géodynamique globale. In: Ecologie et géologie de l'Himalaya. Coll Int CNRS 268: 111–129

Cogne J, Lefort JP (1985) The Ligerian orogeny: a Proto-Variscan event, related to the Siluro-Devonian evolution of the Thetys I ocean. In: Gee D, Sturts B (eds) The Caledonide orogen, Scandinavia and related areas. Wiley, London, pp 1185–1193

Debrand-Passard S, Courbouleix S (1984) Synthèse géologique du sud-Est de la France. Mém Bur Rech Géol Min (Orleans) 125, 126, 620 pp, 6O pl

DEKORP Research Group (1985) First results and preliminary interpretation of deep-reflection seismic recordings along profile Dekorp-2-South. J Geophys 57: 137–163

Dubuisson G, Hirn A, Girardeau J, Mercier JC, Veinante JL (1988) Multiple variscan nappes in Limousin, Western Massif Central, France: geophysical constraints to the geological model and geodynamic implications. Tectonophysics 147: 19–31

Edel JB (1982) Le socle varisque de l'Europe moyenne. Apports du magnétisme et de la gravimétrie. Sci Géol Bull 35, 4: 207–224

Edel JB, Coulon M (1987) A paleomagnetic cross-section through the Ardennes and the Brabant Massifs (France-Belgium). J Geophys 61: 21–29

Edel JB, Fluck P (1989) The Upper Rheinish shield basement (Vosges, Upper Rheingraben, Swartzwald): Main structural features deduced from magnetics, gravimetrics and geological data. Tectonophysics 169: 303–316

Edel JB, Lossy D (1975) Magnétisme et géologie dans les Vosges méridionales. Sci Géol Bull 28, 2: 149–159

Edel JB, Fuchs K, Gelbke C, Prodehl C (1975) Deep structure of the Southern Rhinegraben area from seismic refraction investigations. J Geophys 41: 333–356

Edel JB, Campos-Enriquez O, Goupillot M, Kiro KN (1982) Levé magnétique du fossé rhénan supérieur. Interprétation géologique. Bull Bur Rech Géol Min 2, 1/3: 179–192

Edel JB, Montigny R, Royer JJ, Thuizat R, Trolard F (1986) Paleomagnetic investigations and K-Ar dating on the Variscan plutonic massif of the Champ du Feu and its volcanic-sedimentary environment, northern Vosges, France. Tectonophysics, 122, pp 165–185

Eisbacher GH, Luschen E, Wickert F (1989) Crustal-socle thrusting and extension in the hercynian Schwarzwald and Vosges. Tectonics, vol 8, 1: 1–21

Engel W (1984) Migration of folding and flysch sedimentation on the southern flank of the Variscan belt (Montagne Noire – Monthoumet, Pyrénées) Z dtsch Géol Ges 135: 279–292

Fluck P, Edel JB, Gagny C, Montigny R, Piquet A, Schneider JL, Whitechurch M (1987) Le socle vosgien, segment de la chaîne varisque de l'Europe. Doc Bur Rech Géol Min (Orleans) 146, 97 pp

Franke W, Floyd PA, Holder M, Leveridge B (1989) The rhenohercynian ocean revisited. EUG V – Abstract Terra Cognita, Strasbourg, p 365

Galdeano A, Guillon JC (1988) Interprétation des données magnétiques et gravimétriques. In: Programme ECORS profil Nord de la France. Technip, Paris, pp 127–140

Gallart J, Banda E, Daignieres M (1981) Crustal structure of the Paleozoic axial zone of the Pyrénées and transition to the North pyrenean zone. Ann Geophys 37, 3: 457–480

Gançarski S (1977) Contribution à l'étude des propriétés magnétiques du socle rhénan. Ing., Dipl, University of Strasbourg, 153 pp

Gerard A (1975) La tectonique du socle sous la Manche occidentale d'après les données du magnétisme aéroporté. Philos Trans R Soc Lond A 279: 55–68

Ginzburg A, Whitmarsh RB, Roberts DG, Montadert L, Camus A, Avedik F (1985) The deep seismic structure of the northern continental margin of the bay of Biscay. Ann Geophys 3, 4: 499–510

Godard G (1990) Decouverte d'eclogites et d'amphibolites a anorthite, spinelle et corindon dans le Morvan. CR Acad Sciences, Paris 310, 2: 227–232

Guérangé-Lozes J (1988) Les nappes varisques de l'Albigeois cristallin. Thèsis, University of Toulouse. Doc Bur Rech Géol Min 135, 260 pp

Hirn A (1980) Le cadre structural profond d'après les profils sismiques. In: Evolutions géologiques de la France. Mém Bur Rech Géol Min (Orleans) 107: 34–39

Hirn A, Steinmetz L, Sapin M (1977) A long-range seismic profile in the Western Mediterranean basin: structure of the upper mantle. Ann Geophys 3: 373–384

Jegouzo P, Peucat JJ, Audren C (1986) Caractérisation et signification géodynamiques des orthogneiss calco-alcalins d'âge ordovicien de Bretagne méridionale. Bull Soc Géol Fr 5: 839–848

Kane M (1977) Correlation of major eastern earthquarke centres with mafic/ultramafic basement masses. Geol Surv Prof Pap 1028: 199–204

Kober B, Lippolt HJ, Kirsch H (1988) Intrusionsalter und Abkuhlgeschichte des varistischen Gabbros Frankenstein (Odenwald) mit der Ar-Ar Technik und mit Radiogenbleianalysen an einzelnen Zirkonen. 4. Rundgespräch Geodynamik des europäischen Varistikums Saxothuringikum. Heidelberg, p 14 (Abstr)

Kossmat F (1927) Gliederung des Varistisches Gebirgebaues. Abh Sächs Geol Landesamtes 1: 1–39

Kreuzer H, Harre W (1975) K-Ar Altersbestimmungen an Hornblenden und Biotiten des Kristallinen Odenwaldes. Aufschluß, Sonderbd 27: 71–77

Laboratoire de Détection Géophysique (1977) Séismes proches, localisés par le réseau LDG – France. LDG du CEA, Int Repl, Montrouge

Lagarde JL (1978) La déformation des roches dans les domaines à schistosité subhorizontale: application à la nappe du Canigou et de Champtoceaux. Thèse, Université of Rennes, 154 pp

Lauer JP (1975) Contribution à l'étude de l'influence du métamorphisme thermique sur les propriétés magnétiques des roches: exemples pris dans les Vosges cristallines du Nord. Thèsis, University of Strasbourg I, 217 pp

Lauer JP, Taktak AG (1971) Propriétés magnétiques des roches au voisinage du contact métamorphique des schistes de Steije et des granites d'Andlau et du Hohwald (Vosges cristallines du Nord). Cr Acad Sci Paris 272: 924–927

Lefort JP (1973) La zonale Biscaye-Labrador: mise en évidence de cisaillements dextres antérieurs à l'ouverture de l'Atlantique Nord. Mar Géol 14: 33–38

Lefort JP (1975) Le socle Peri-Armoricain: étude géologique et géophysique du socle submergé à l'Ouest de la France. Thesis, University of Rennes, 150 pp

Lefort JP (1989 a) Basement correlation across the North Atlantic. Springer, Berlin Heidelberg New York, pp 1–148

Lefort JP (1991 b) L'organisation structurale du socle profond de la Manche Occidentale, d'après l'interprétation des profils sismiques SWAT. Un exemple d'intersection entre suture cadomienne et structures hercyniennes divergentes. Mem Soc Géol Fr 159: 123–132

Lefort JP, de Poulpiquet J (1990) Correlations entre les profils de sismique réflexion profonde ECORS-GASCOGNE et ECORS-Nord de la France au travers de la suture sud Armoricaine. Bull Soc Géol Fr VI (2): 321–330

Lefort JP, Segoufin J (1978) Etude comparée des structures profondes et des anomalies magnétiques allongées reconnues en Manche Occidentale et en Baie d'Audierne: existence possible d'une suture cryptique au nord-ouest du Massif Armoricain (France). Tectonophysics 46: 65–75

Lefort JP, Bardy Ph, de Poulpiquet J (1991 a) Interprétation des anomalies magnétiques reconnues en Manche Occidentale et en Mer Celtique le long des profils SWAT. Mem Soc Géol Fr 159: 133–143

Lefort JP, Liewig N, Peucat JJ, Cazes M, Pinet E (1989 b) Apport de la géochronologie à la connaissance des structures du socle sous l'Aquitaine. CR Acad Sci Paris (in press)

Le Mouel (1969) Carte magnétique de la France à 1/1 000 000. CNRS-I.P.G., Paris

Lienhardt MJ (1961) Etude stratigraphique et structurale du socle anté-permien du Bassin de Paris. Ann Soc Géol Nord, Lille 81: 233–241

Lorenz V, Nicholls IA (1984) Plate and intraplate processes of Hercynian Europe during the late Paleozoic. Tectonophysics 107: 25–56

Marchal F (1986) Apport de la gravimétrie à la connaissance des structures profondes sous le Bassin Aquitain. Extension de la chaîne hercynienne. Thèsis, University of Montpellier, 195 pp

Matte Ph (1986 a) Tectonics and plate tectonics model for the Variscan belt of Europe. Tectonophysics 126: 329–374

Matte Ph (1986 b) La chaîne varisque parmi les chaînes paleozoïques peri-atlantiques, modèle d'évolution et position des grands blocs continentaux au Permo-Carbonifère. Bull Soc Géol Fr II: 9–24

Matte Ph, Hirn A (1988) Seismic signature and tectonic cross section of the Variscan crust in western France. Tectonics 7, 2: 141–155

Matte Ph, Burg JP (1981) Sutures, thrusts and nappes in the Variscan Arc of western Europe: plate tectonic implications. Geol Soc (Lond) 8: 353–358

Matte P, Respaut JP, Maluski H, Lancelot JR, Brunel M (1986) La faille NWSE du pays de Bray, un décrochement ductile dextre hercynien: déformation à 320 MA d'un granite à 570 MA dans le sondage Bray 201. Bull Soc Géol Fr (8) II, 1: 69–77

Menard G (1979) Relations entre structures profondes et superficielles dans le SE de la France. Thèsis, University of Grenoble, 175 pp

Menot RP (1987) Magmatismes et structuration orogénique paléozoïque de la chaîne de Belledonne. Le domaine sud occidental. Géol Alp 63: 55–93

Michel B (1978) La croûte entre la vallée du Rhin et la vallée du Rhône: interprétation de profils sismiques et résultats structuraux. Thèsis IIIrd cycle, University of Paris VII, 133 pp

Montigny R, Thuizat R (1989) K-Ar and Ar-Ar ages on cristalline rocks of the Vosges (France). Terra Cognita Strasbourg: 352

Paris F, Le Pochat G (1987) Synthèse géologique du Paléozoïque de l'Aquitaine nord. Doc Bur Rech Géol Min 144: 43–57

Paris F, Robardet M (1985) Evaluation des affinites entre le Paléozoïque caché sous l'Aquitaine et les formations armoricaines contemporaines. In: Geologie Profonde de la France. Doc Bur Rech Géol Min (Orleans) 95 (7): 11–36

Paris F, Le Pochat G, Pelhate A (1988) Le socle palézoïque nord aquitain: caractéristiques principales et implications géodynamiques. CR Acad Sci Paris 306, S II: 597–603

Perrier G, Ruegg JC (1973) Structure profonde du Massif Central français. Ann Geophys 29, 4: 425–502

Peucat JJ (1983) Geochronologie des roches métamorphiques (Rb-Sr et U-Pb). Thesis, Universiy of Rennes, 158 pp

Pinet B; Montadert L, ECORS party (1987) Deep seismic reflection and refraction profiling along the Aquitaine shelf (Bay of Biscay). JR Astr Soc Lond Geophys 89: 305–312

Poulpiquet de J, Lefort JP (1989) Modelling of magnetic structures representing the South Armorican suture; Tectonophysics 165: 93–103

Raymond D (1987) Le Dévonien et le Carbonifère inférieur du Sud Ouest de la France (Pyrénées, Montagne Noire, Mouthoumet): sédimentation dans un bassin flexural en bordure nord de la chaîne de collision varisque. Geol Rundsch 76: 795–803

Rhinegraben Research Group for explosion seismology (1974) The 1972 seismic refraction experiment in the Rhinegraben; first results. In: Approaches to taphrogenesis, Inter-Union

commission on geodynamics, scientific report 8, Stuttgart, pp 122–137
Sapin M, Hirn A (1974) Results of explosion seismology in the southern Rhône valley. Ann Geophysic 30: 181–201
Schott JJ (1976) Etude paléomagnétique des sédiments permiens de l'Argentèra Mercantour. Thèse doctorat 3ème cycle, Université Louis Pasteur, Strasbourg
Sibuet JC (1972) Contribution de la gravimétrie à l'étude de la Bretagne et du Plateau adjacent. CR Soc Géol Fr 3: 124–129
Thieblemont D, Cabanis B (1986) Découverte d'une association de volcanites d'arc et de basaltes de type MORB dans la série poleo volcanique dévonienne de la Meilleraie, Vendée (France). CR Acad Sci Paris II 302: 641–646
Thieblemont D, Cabanis B, Wyns R, Treuil M (1987a) Etude géochimique de la formation amphibolitique de Saint-Martin des Noyers (complexe cristallophyllien des Essarts). Mise en évidence d'un paléo-arc insulaire dans la zone interne de l'orogène varisque en Vendée; Bull Soc Géol Fr III 2: 371–378
Thieblemont D, Cabanis B, Le Metour J (1987b) Etude géochimique d'un magmatisme de distension intracontinentale: la série bimodale ordovicienne du Choletais (Vendée). Géol Fr Bur Rech Géol Min 1: 65–76
Van de Meulebrouck J, Bayer R, Burg JP (1984) Density and magnetic tomography of the upper continental crust: application to a thrust area of the French Massif Central. Ann Geophys 2–5: 579–592
Van den Ende C (1977) Paleomagnetism of Permian red beds of the dome de Barrot. Thèsis, Rijks University, Utrecht
Veinante-Delhaye A, Santoire JP (1980) Sismicité récente de l'arc sudarmoricain et du nord-ouest du Massif Central; mécanisme au foyer et néotectonique. Bull Soc Géol Fr (7) XXII 1: 93–102
Vigneresse JL (1978) Gravimétrie et granites armoricains. Thesis, University of Rennes, 95 pp
Vivier G, Menot RP, Giraud P (1987) Magmatismes et structuration orogénique paléozoïque de la chaine de Belledone (domaine nord oriental). Géol alp 63: 25–54
Weber C (1972) Le socle antetriasique sous la partie sud du Bassin de PARIS, d'après les données géophysiques. Thesis, University of Paris VI, 123 pp
Weber C, Hirn A, Philip H, Roche A (1980) Image géophysique de la France in "Evolutions géologiques de la France" ed. par A. Autran et J. Dercourt. Mém Bur Rech Géol Min 107: 25–50
Westphal M (1976) Contribution du paleomagnétisme à l'étude des déplacements continentaux autour de la Méditerranée occidentale. Thèsis, University Louis Pasteur, Strasbourg, 304 pp
Wickert F, Eisbacher GM (1989) Two-sided Variscan thrust tectonics in the Vosges Mountains. Geodyn Acta, Paris 8–3: 101–120
Wyns R, Lardeux H, Weyant M (1989) Présence de Dévonien dans le Groupe de Réaumur (synclinal de Chantonnay, Vendée); conséquences sur l'évolution géodynamique varisque de la Vendée. CR Acad Sci Paris 308, série II: 855–860

Part II A
The Cadomian Orogeny in the Armorican Massif

Introduction

J. Chantraine, B. Auvray, J. P. Brun, J. J. Chauvel, and D. Rabu (translated by M. S. N. Carpenter)

1 Definition and Location

Amongst the various massifs which constitute the West European Variscan fold belt, the Armorican Massif has a special place of interest. This is due to the importance of Cadomian remnants (cf. Pan-African) belonging to the northern border of Gondwana that was dislocated during Paleozoic orogenies (Fig. 1).

In North Brittany and the Cotentin area (northern part of the Armorican Massif), the tectonic pattern was acquired at the end of the Proterozoic and has been little affected by Variscan reactivation. The geological formations belong to the Brioverian (Barrois 1899), which should now be considered as a Supergroup, whose Upper Proterozoic age is well established. These rocks were involved in the Cadomian orogenic cycle before an Early Paleozoic transgression, which is diachronous in different parts of the Armorican Massif. In the Bocage Normand domain (E. part of the Massif), the beginning of the transgression is in the Lower Cambrian; Middle Cambrian sediments unconformably overlie the Cadomian basement, which is made up of an extensive granitic pluton (Mancellian Batholith) and its metamorphic envelope. In the Vendée area (S. E. part of the Massif), transgression occurred during the Middle Cambrian, and the underlying cratonized basement is thought to have recorded Cadomian effects. The Central Brittany domain is characterized by a Paleozoic transgression with basal red beds passing up into the Grès Armoricain Formation of Arenig age; recent radiometric results show that these red beds are also Arenig in age. In this domain, however, the Brioverian basement is devoid of any deformation pre-dating the Variscan orogeny. The South and Central Brittany domains are widely affected by Variscan deformation.

The Cadomian orogenic belt and outlying Mancellian domain has a particular style of tectonics, metamorphism, and magmatic activity, now defined by numerous radiometric data (Fig. 1). Following a belt running from the Trégor through the Channel Islands to the northern Cotentin, there are outcrops of orthogneiss dated at 2000 Ma which constitue the basement to the Brioverian in the Cadomian belt. They belong to Icartian massifs and are relicts of Lower Proterozoic age. Locally, conglomerates in the lower part of the Brioverian in the Baie de Saint-Brieuc contain clasts which yield ages of 655 and 665 Ma. Volcanic rocks associated with Brioverian sediments in the Baie de Lannion have been dated at about 640 Ma.

In the northern part of the Armorican Massif (Trégor and Cotentin), the emplacement of plutons associated with the Cadomian orogeny has been dated at several localities, yielding ages of between 615 and 585 Ma; a late orogenic magmatism, of crustal or hybrid origin, is widely developed and extends from ca. 560 to 480 Ma. Finally, the Paleozoic magmatic rocks that post-date Cadomian structures in this northern sector (Trégor-Channel Islands-Cotentin) have yielded ages of 520–480 Ma.

Thus, evidence for Cadomian orogenic activity in the Armorican Massif is now well established, especially in the north, where the effects are widely distributed. The N. W. part of the Massif (Pays du Léon) was long considered to be part of the Cadomian belt, but is now interpreted as a displaced segment of the Variscan South Armorican domain (Balé and Brun 1986). In contrast to previous hypotheses, there is no evidence to support the existence of Cadomian deformation in the South Armorican domain prior to the major Variscan tectonometamorphic event (Cogné 1976). However, in the Vendée area (Massif des Mauges), where Variscan reworking is only moderate, there is some indication of Cadomian effects. Furthermore, some granitic orthogneisses have been dated at 590 and 610 Ma in South Brittany and West Vendée.

Brioverian sediments contain microfossils which are commonly abundant, but of very low diversity (Chauvel and Schopf 1978; Chauvel and Mansuy 1981 a, b; Mansuy 1983). The genera *Palaeocryptidium* and *Bavlinella* are always well represented and correspond, respectively, to vegetative cells and to the endospores of a colonial form *(Sphaerocongregus)*

J. D. Keppie (Ed.): Pre-Mesozoic Geology in France and Related Areas

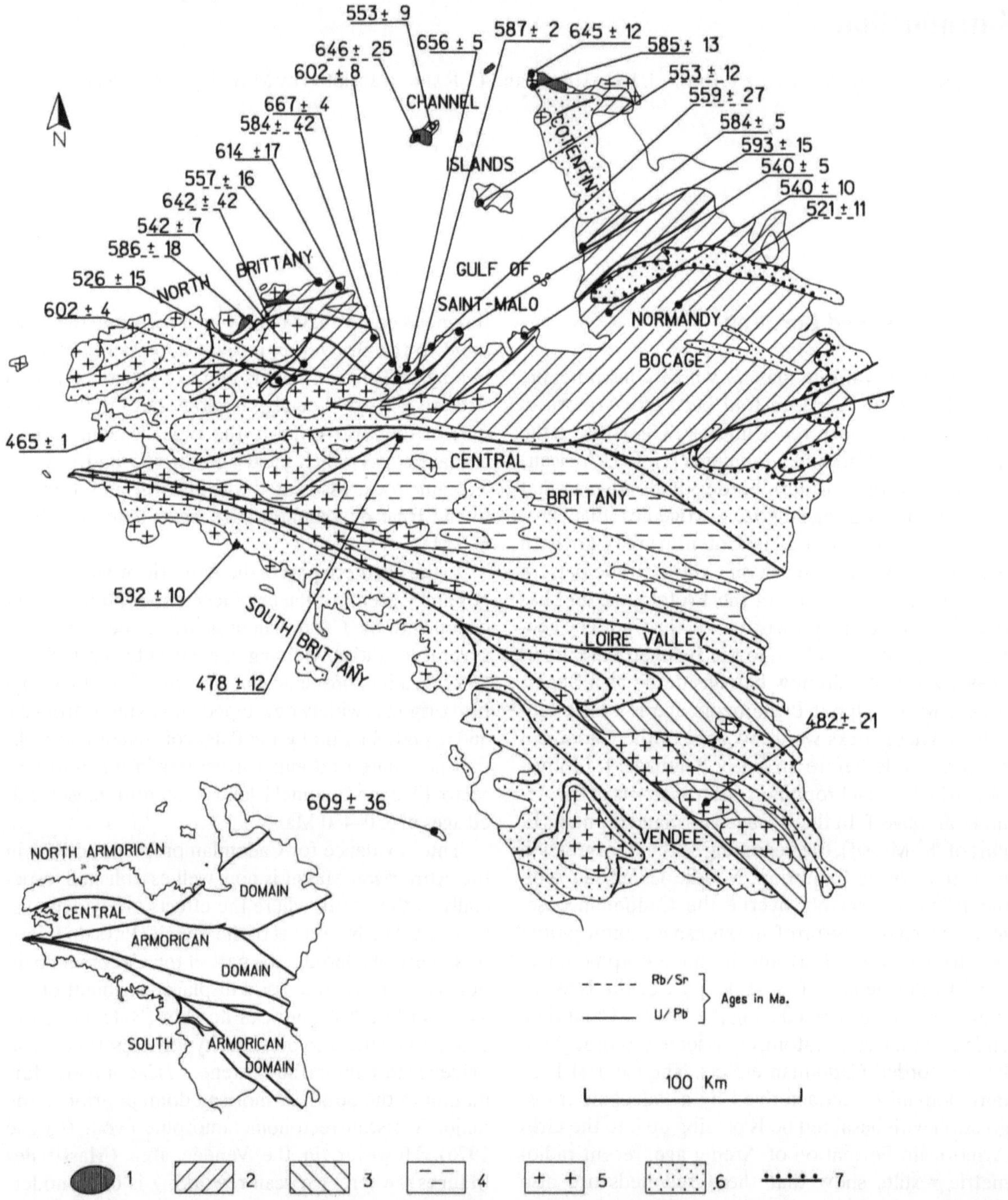

Fig. 1. Sketch map showing distribution of radiometric dates in parts of the Armorican Massif affected by the Cadomian orogeny. The compiled data include U/Pb zircon ages, and Rb/Sr whole-rock isochron ages (λ^{87}Rb=1.42×10^{-11} an^{-1}). References are given in the text. *1* Icartian basement; *2* North-Armorican Cadomian domain; *3* South-Armorican Cadomian domain; *4* Brioverian of Central Armorican domain; *5* Variscan granites; *6* Paleozoic cover; *7* Cambrian unconformity

which is very similar to certain presentday Cyanophyceans belonging to the Orders Pleurocapsalia and Chroococcalia. The two genera cited above are found in all Brioverian formations, except the calcareous facies. In the black chert-bearing Brioverian, there are some forms which have been assigned to the genus *Gloeodiniopsis*. In the Brioverian with chert clasts, apart from the genus *Sphaerocongregus*, there are abundant colonial forms of uncertain affinity which are assigned to the genus *Favososphaera*

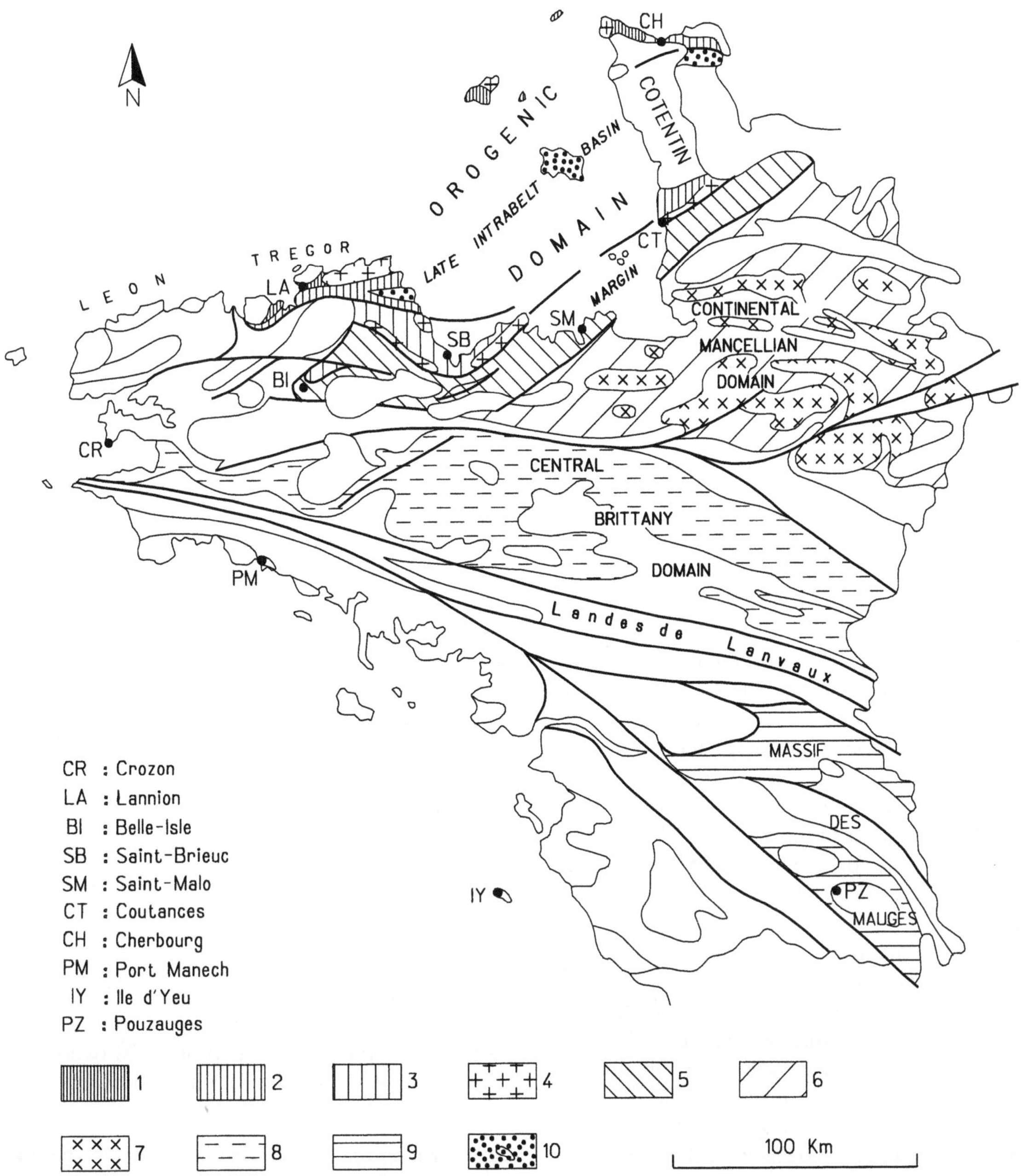

Fig. 2. Sketch map showing Cadomian structural domains in the Armorican Massif. *1* Icartian basement relicts; *2* volcanic arc domain; *3* back-arc basin; *4* syn-orogenic plutonism; *5* continental margin; *6* continental platform; *7* Mancellian Batholith; *8* Central Brittany domain; *9* inland Vendée block; *10* late intrabelt basin

(*F. sola, F. conglobata, F. grandifavosa,* and *F. aequabilis*). All these forms observed in the Brioverian have been recognized in various other parts of the world, but it appears impossible to use them to establish a precise biostratigraphy, since most of them range at least into the Ordovician.

2 Geodynamic Setting

During the Variscan orogeny, the Armorican Massif was the site of faulting linked to a general continental collision regime. These faults led to an important wrenching and thrusting of the Proterozoic tectonic pattern, although the displacements are hard to evaluate. One of the most important faults corresponds to the Landes de Lanvaux axis (Fig. 2), which continued to be active throughout the Paleozoic. On each side of this axis, it is very difficult to establish the relationship between different Cadomian relicts. Within the Cadomian belt itself, toward the north, Variscan faults have probably re-used ancient Cadomian structures, thus obscuring any reconstruction of Brioverian paleogeography. However, the rather short (250 × 100 km) segment of fold belt preserved today in the Armorican Massif has nevertheless enabled us to show that the Cadomian orogeny is the result of the development of an active continental margin at the end of the Proterozoic (Auvray 1979; Cogné 1984 a; Chantraine et al. 1988).

2.1 The Cadomian Belt in the North Armorican Domain

The Cadomian belt shows a general SW-NE orientation (Fig. 2); it can be recognized on land in the Trégor and Cotentin areas and can be traced offshore across the Golf de Saint-Malo. It also extends transversally into the Baie de Saint-Brieuc. The Cadomian belt is made up of a tectonic pile, of various lithologic units, separated by shear zones that have acted as thrust or wrench faults. A study of the juxtaposed structural units has revealed a zonation of the orogenic belt, with the internal zones in the N. W. and the external zones in the S. E. (Cogné and Wright 1980; Cogné 1984 a; Chantraine et al. 1988).

Along the N. W. edge of the belt, there are relicts of Icartian basement composed mainly of granitic gneisses dated at ca. 2000 Ma.

The orogenic belt itself (Domnonean domain, Fig. 3) is made up of a volcanic arc coupled with a back-arc basin. The sedimentation is essentially volcaniclastic in nature and the magmatism is mainly of the active margin type (Auvray and Maillet 1977). The internal arc basin recognized in the Trégor appears to split into two belts in the Cotentin, where they are separated by a late intrabelt basin.

The orogenic belt abuts against a continental margin in the S. E., separated from it by a major suture (Balé and Brun 1983). This continental margin is characterized on both sides of the Golfe de Saint-Malo by an abundance of bedded black cherts (the "phtanites" of French authors) which are locally associated with tholeiitic volcanics. In the Saint-Malo and Guingamp areas, important crustal remelting has occurred yielding migmatitic gneisses and anatectic granites. This has been interpreted as the result of overthrusting of the cordillera onto the continent (Graviou and Auvray 1985).

The continental platform (Mancellian domain, Fig. 3) recognized in the Bocage Normand shows a remarkable development of granitic plutonism of crustal origin (Mancellian Batholith). The sedimentation is mainly terrigenous and characterized by the presence of black chert debris derived from the erosion of bedded chert formations in the cordilleran terrain. In this way, a major stratigraphic discontinuity can be inferred within the Brioverian Supergroup between a bedded chert-bearing group and younger formations which contain chert clasts (Le Corre 1977).

2.2 The Central Armorican Domain: Cadomian Foreland

Even though it is isolated from the Cadomian domain belt by a main Variscan fault (the North-Armorican Shear Zone), the Brioverian of Central Brittany displays many similarities with the Brioverian succession of the Bocage Normand. Although these two domains cannot be directly correlated, they both contain Brioverian sediments with black chert fragments typical of the continental platform domain (Mancellia). The nature of the underlying basement in this area remains unknown. However, there are some important differences between the Bocage Normand and Central Brittany domain which reflect their geodynamic histories:

- in contrast with the Bocage Normand, where Cadomian plutonism and deformation are pronounced, the Brioverian of Central Brittany is devoid of granitic intrusions and shows only the effects of Variscan deformation;
- taking into account the Arenig age of the post-Brioverian cover and the absence of Cadomian intrusions, it is possible that part of the Brioverian in the Central Brittany domain could be Lower Paleozoic in age.

Southward, on the Landes de Lanvaux axis, there is an unusual sedimentary unit (Groupe de Bains-sur-Oust) which may correspond to continuous sedimentation extending from the Brioverian up into the Early Paleozoic (Arenig). This presents several problems concerning the development of sedimentary basins during this period in an area where there was probably no Cadomian deformation.

2.3 Cadomian Relicts in the South Armorican Domain

The main Variscan orogenic belt is developed south of the Landes de Lanvaux axis, but there are very

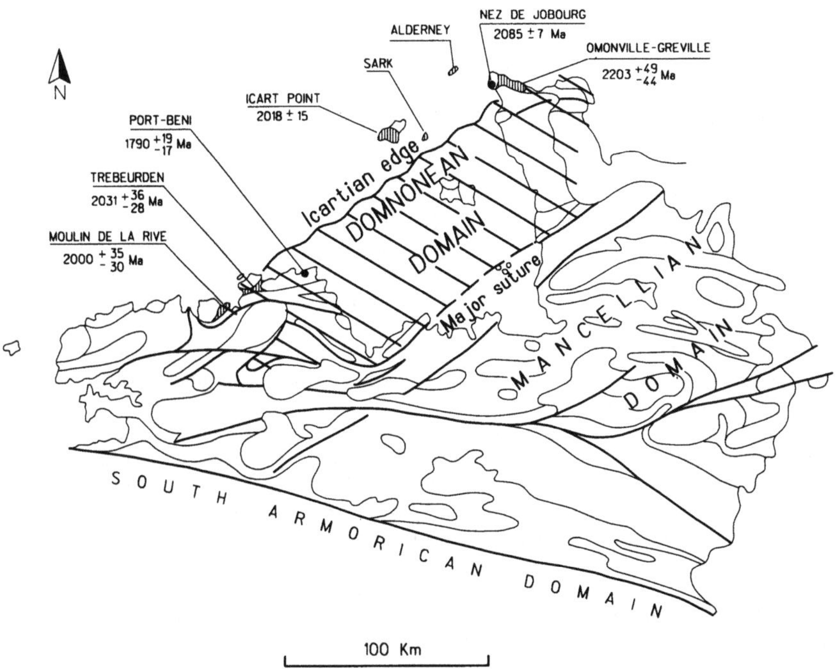

Fig. 3. Distribution of Icartian basement relicts in the North-Armorican domain. (After Auvray et al. 1980*a*; Vidal et al. 1981). All age data obtained by U/Pb zircon method (see text for references)

few indications of a Cadomian basement except in the Vendée, between the two branches of a major Variscan fault (the South Armorican Shear Zone).

In the Vendée area (Le Métour and Wyns 1984), a cratonic block known as the Massif des Mauges is characterized by multiphase deformation in a basement unconformably overlain by undeformed Middle Cambrian sediments (Cléré-sur-Layon unconformity; Cavet et al. 1966). On its southern flank, the Massif des Mauges is intruded by the Pouzauges granite, whose emplacement has been dated at 482 ± 21 Ma. (Calvez, unpubl.). In contrast to many other Lower Paleozoic plutons in the South Armorican domain, the Pouzauges granite has not been converted into an orthogneiss.

To the south of the South Armorican Shear Zone, orthogneisses contained within the Variscan terranes of South Brittany, Vendée, and South Limousin show protolith ages typical of Cadomian igneous activity, dated around 600 Ma. These units include the Port Manech orthogneiss (South Brittany), the Ile d'Yeu orthogneiss (West Vendée), and the Palanges orthogneiss (South Limousin). Even though the nature of their country rocks has not yet been elucidated, these orthogneisses demonstrate the existence of Cadomian relicts within the main Variscan domain.

All these Cadomian relicts are markers of a basement that has been re-worked in the South Armorican Variscan Belt. Their ages suggest some connection with the Cadomian orogenic domain of the northern Armorican Massif, but the two domains have been dislocated by a set of intra-crustal thrust and wrench faults, which have shown large transcurrent movements during the Variscan orogeny (Lefort 1975). This may explain why it is difficult to re-assemble the different structural domains, particularly the basement of Vendée, into a coherent pattern that integrates with the Cadomian orogenic belt of the Northern Armorican Massif.

Despite the Variscan "jig-saw puzzle", the identification of a Cadomian orogenic belt – at least in the Northern Armorican Massif – is nevertheless possible through the application of lithostratigraphic, tectonometamorphic, geochemical, and geophysical criteria. The recognition of such a belt enables us to point out structural zonations in time and space and, ultimately, to integrate all the features in a coherent geodynamic model.

References

Auvray B, Maillet P (1977) Volcanisme et subduction au Protérozoïque supérieur dans le Massif Armoricain (France). Bull Soc Géol Fr XIX-5: 953–957

Auvray B (1979) Génèse et évolution de la croûte continentale dans le Nord du Massif Armoricain. Thèse d'Etat, Rennes, 670 pp

Balé P, Brun JP (1983) Les chevauchements cadomiens de la Baie de Saint-Brieuc (Massif Armoricain). CR Acad Sci Paris 297: 359–362

Balé P, Brun JP (1986) Les complexes métamorphiques du Léon (NW Bretagne): un segment du domaine éo-hercynien sud-armoricain, translaté au Dévonien. Bull Soc Géol Fr II-3: 471–477

Barrois C (1899) Brioverian system in a sketch of the geology of Central Brittany. Proc Geol Assoc XVI: 101–132

Cavet P, Gruet M, Pillet J (1966) Sur la présence du Cambrien à Paradoxides à Cléré-sur-Layon (Maine-et Loire) dans le NE du Bocage vendéen. CR Acad Sci Paris 263: 1685–1688

Chantraine J, Chauvel JJ, Balé P, Denis E, Rabu D (1988) Le Briovérien (Protérozoïque supérieur à terminal) et l'orogenèse cadomienne en Bretagne (France). Bull Soc Géol Fr IV, 5: 815–829

Chauvel JJ, Schopf JW (1978) Late Precambrian microfossils from Brioverian cherts and limestones of Brittany and Normandy, France. Nature 275: 640–642

Chauvel JJ, Mansuy C (1981 a) Micropaléontologie du Protérozoïque du Massif Armoricain (France). Precambrian Res 15: 25–42

Chauvel JJ, Mansuy C (1981 b) Palaeocryptidium cayeuxi, une cyanophycée du Précambrien du Massif Armoricain. Bull Soc Géol Fr XXIII, 5: 449–458

Cogné J (1976) La chaîne hercynienne ouest-européenne correspond-elle à un orogène par collision? Propositions pour une interprétation géodynamique globale. Coll Int CNRS Paris 268: 111–129

Cogné J (1984 a) Geodynamic evolution of the Cadomian fold belt in the Armorican Massif; present data and new problems. In: Zoubek V (ed) Precambrian in younger fold belts. Wiley, New York, pp 470–483

Cogné J, Wright AE (1980) L'orogène cadomien. Vers un essai d'interprétation paléogéodynamique unitaire desphénomènes orogéniques fini-précambriens d'Europe moyenne et occidentale. In: Coll Géologie de l'Europe. 26ème CGI Paris; Mém Bur Rech Géol Min 108: 29–55

Graviou P, Auvray B (1985) Caractérisation pétrographique et géochimie des granitoïdes cadomiens du domaine nord-armoricain: implications géodynamiques. CR Acad Sci Paris 301: 315–318

Le Corre C (1977) Le Briovérien de Bretagne centrale: essai de synthèse lithologique et structurale. Bull Bur Rech Géol Min I-3: 219–254

Lefort JP (1975) Le socle péri-armoricain. Etude géologique et géophysique du socle submergé à l'Ouest de la France. Thèse d'Etat, Rennes. Mém Soc Géol Fr Paris 130, (1977), 68 pp

Le Métour J, Wyns R (1984) The Precambrian of the Vendée. In: Zoubek V (ed) Precambrian in younger fold belts. Wiley, New-York, pp 487–511

Mansuy C (1983) Les microsphères du Protérozoïque supérieur armoricain (Briovérien); nature, répartition stratigraphique, affinités biologiques. Thèse 3e cycle, Rennes, 108 pp

Lithostratigraphy

J. Chantraine, J. J. Chauvel, and D. Rabu (translated by M. S. N. Carpenter)

1 The Icartian Basement (Lower Proterozoic)

The Icartian basement is recognized at the extreme outer edge of the Cadomian belt, which is mostly offshore. It occurs as isolated pieces, several kilometers in size, which are often included as enclaves in Cadomian magmatic units. These Icartian relicts form a SW-NE-trending belt extending from the Trégor through the Channel Islands to the Cotentin, and have been dated by the U/Pb zircon method at several localities yielding ages of ca. 2000 Ma. (Auvray et al. 1980a; Vidal et al. 1981) (Fig. 3):

- Moulin-de-la-Rive: granodioritic orthogeneisses with metasedimentary enclaves (2000 + 35 / – 30 Ma);
- Trébeurden: granodioritic orthogeneisses with metasedimentary enclaves (2031 + 36 /– 29 Ma);
- Port Béni: banded gneisses included as enclaves in orthogneisses dated at 1790 + 19 /– 17 Ma;
- Guernsey: monzogranitic orthogneisses (Icart Gneiss) or granodioritic orthogneisses (Perelle Gneiss), intruded into a metamorphic envelope of sedimentary and volcanic origin showing multistage deformation. The gneisses at Icart Point yield an age of 2018 ± 15 Ma;
- Sark and Alderney: various gneisses of magmatic origin and probable Icartian age (Roach 1977);
- Cap de La Hague: migmatitic orthogneisses and mylonites, associated with a more ancient basement complex (Nez de Jobourg and Omonville-Gréville), yielding ages of 2083 ±7 Ma and 2203 + 49 /– 44 Ma, respectively (U/Pb zircon method; Piton 1985).

The Guernsey orthogneisses constitute the most important occurrence of ancient basement in the region, and were the first rocks to yield an age of 2000 Ma (Adams 1976; Calvez and Vidal 1978), hence the term Icartian has been extended to all basement relicts of this age. Previously, the Pentevrian gneiss complex of N. Brittany was thought to represent the basement to the Brioverian (Cogné 1959), but these rocks have never given ages other than typically Cadomian. Therefore, they should now be assigned to the Brioverian Supergroup.

Despite the very sporadic information available on the oldest basement material in the West European Variscan belt, now cropping out within the Armorican Massif, it is still possible to recognize a single major tectonic and metamorphic event in the Icartian gneisses. However, this does not enable a reconstruction of the geodynamic history of the basement segment dated at between 1.8 and 2.2 Ga. Such an age – and particularly the absence of any event between 1.7 and 1.0 Ga – in fact suggests an affinity with Gondwana rather than the Baltic Shield (Auvray 1979), which is further supported by the paleomagnetic data of Perroud et al. (1984). Finally, it should be mentioned that the dates obtained refer to the emplacement ages of granitic rocks, a period which covers 400 Ma. Thus, the history of this ancient basement is probably far more complex than is suggested by the partial record left today in the scattered Icartian relicts of the Armorican Massif.

2 Brioverian North of the Landes de Lanvaux

To the north of the Landes de Lanvaux axis, the Brioverian Supergroup appears to form a single sedimentary succession, but particular features are very variable between the different structural domains recognized within the Cadomian belt.

2.1 The Orogenic Cordilleran Domain

Here, the Brioverian is influenced by volcanic activity; there are interbedded acid and basic volcanics associated with volcaniclastic and hydrothermally affected sediments. In North Brittany, where Variscan reworking is slight, two units can be recognized (Fig. 2):

J. D. Keppie (Ed.): Pre-Mesozoic Geology in France and Related Areas

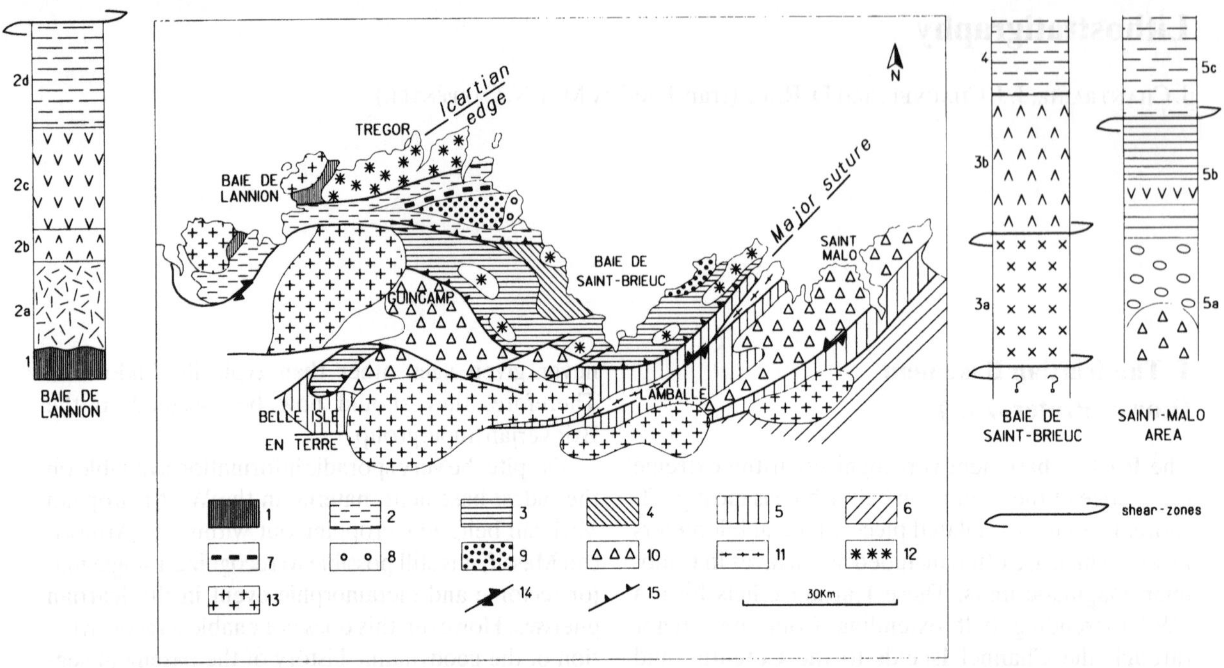

Fig. 4. The Cadomian orogenic belt of North Brittany. *1* Icartian basement; *2* volcanic arc (*2a* Locquirec-Tréguier Fm; *2b* Rugunay-Malabri Fm; *2c* Armorique-Trédrez Fm; *2d* Plestin-les-Grèves and Bois-de-la-Roche Fms); *3* back-arc basin igneous complex (*3a* Yffiniac Fm; *3b* Lanvollon Fm); *4* back-arc basin sediments (Binic-Le Légué Fm); *5* continental margin (*5a* Saint-Cast Fm; *5b* Lamballe and la Fresnaye Fms; *5c* Vallée de la Rance Fm); *6* continental platform; *7–9* late intrabelt basin (*7* Paimpol Spilites; *8* Le Minard Formation; *9* Plouézec-Plourivo and Erquy-Fréhel Formations); *10* Saint-Malo and Guingamp migmatites; *11* orthogeneisses; *12* orogenic plutonism; *13* Variscan granites; *14* wrench-fault; *15* thrust-fault

- an internal volcanic unit, seen in the Baie de Lannion and the Trégor area;
- an external volcanic unit, seen in the Baie de Saint-Brieuc. These two units can be distinguished mainly on the basis of the geochemistry of their volcanism; the internal unit contains volcanic arc-type rocks and the external unit is typical of a back-arc basin.

In the Cotentin area, where outcrops are discontinuous beneath Paleozoic and Mesozoic cover, the Cadomian orogenic belt is split into two: one branch in the north passes near Cherbourg and the other, to the south, near Coutances (Fig. 2). In contrast with the commonly proposed model (Dissler 1987), which tends to link the Coutances branch to the external volcanic unit of the Baie de Saint-Brieuc, the geochemical characteristics of the volcanism (Cabanis 1986) rather suggest that the two branches of the Cadomian belt in Normandy both belong to the internal volcanic unit as observed in the Trégor. The Cherbourg branch appears to be of the frontal arc type, whereas the Coutances branch seems to be of the internal arc type. These two arcs are separated by a late basin of Uppermost Proterozoic-Lower Paleozoic age, which is widely developed in the Cotentin but scarcely initiated in Brittany.

The imbricate pattern of these structural units produced overthrusting of the entire orogenic pile towards the south, which accounts for the following observations (Figs. 4 and 5):

- on the one hand, the back-arc basin recognized in Brittany does not appear to have any equivalent in the Cotentin, where it may be occluded or concealed beneath the thrusted volcanic arc;
- on the other hand, the isolated klippe of Belle-Isle en-Terre can be correlated with the back-arc basin and considered as evidence for the thrusting of this external volcanic domain onto the continental margin.

2.1.1 The Volcanic Arc Domain

This unit is juxtaposed, on its internal margin, with Icartian relicts. The normal contact of the Brioverian Supergroup with its basement is often intrusive or tectonic, being directly observed only in the Baie de Lannion (Fig. 4). Between the Icartian orthogneiss at Moulin-de-la-Rive and the basal unit of the volcanic arc (Locquirec Formation), the moderate degree of shear deformation has enabled recognition of the transgressive relationship of this cover to the basement (Autran et al. 1979). Above the fractured orthogneiss, occurring as lumps enveloped in an arkosic matrix, the basal member grades up, over a distance of about 100 m, from coarse arkosic facies (with potassic feldspar) and local and discrete conglomerates, into banded facies with albite-chlorite assemblages typical of the Locquirec Formation. With the exception of this limited border zone – and contrary to Lees et al. (1987), who proposed that the Locquirec Formation corresponds to the mylonitized border of the granitic orthogneisses – the volcaniclastic character of the basal unit of the volcanic arc is apparent through petrographic criteria (e. g., corroded quartz, presence of albite clasts, absence of potassic feldspar, preserved microlitic texture).

The lithostratigraphic succession of the Trégor volcanic arc is defined on its western side, in the Baie de Lannion, where it is more complete. The eastern side, on the Baie de Saint-Brieuc, is more disrupted by faulting. From north to south, and from base to top, the succession is as follows (Verdier 1968; Autran et al. 1979):

The Locquirec-Tréguier Formation is composed of a basal arkosic member, derived from the erosion of the Icartian basement, overlain by a thick unit (several 100 m) of volcaniclastic rocks (metatuffs) of soda-rich rhyolite composition; this formation contains some conglomeratic interbeds and becomes progressively more terrigenous towards the upper part.

The Rugunay-Malabri Formation is marked by the return of acid volcanism over a thickness of at least 100 m in the form of flows and massive sills which show rhyodacitic compositions. This volcanism is associated with hydrothermally affected (exhalative) sediments.

The Armorique-Trédrez Formation is a thick accumulation of basic volcanics (several 100 m) showing calc-alkaline characteristics; it is mainly composed of metabasalts (greenschists) with alternations of pillow-lavas and massive sills. At the base, sediments form some discrete interbeds which become more abundant towards the top, where metagraywackes predominate.

The Plestin-les-Grèves and Bois-de-la-Roche Formations are essentially sedimentary in character, but clastic supply is dominantly volcanic and some volcanic horizons are present. From base to top, this sequence changes from immature metagraywackes to alternations of more mature metasiltstone, metasandstone, meta-arkose, and black shale.

These Brioverian formations are only well developed (several 100 m) in the area of Le Bois-de-la-Roche, where their southerly limit is not clearly determined; they appear to pass upward into quartzites, but the continuity of outcrop is not established and the age of the quartzites remains undefined. Whatever the stratigraphic relations of these quartzites, they show a major tectonic contact with the Paleozoic of Variscan West Brittany.

In the Cotentin, the structural pattern of the internal volcanic terrain, in two volcanic arcs on both sides of the late intrabelt basin, prevents us from establishing a single lithostratigraphic succession:

The Internal Arc of the Coutances Area (Fig. 5) is composed of volcanic rocks of tholeiitic affinity, making up a thick spilite-keratophyre complex – the Montsurvent Formation (Dupret 1984; Dupret et al. 1985 a). The lithological succession of this formation is poorly defined and the formation appears to form an imbricated mass of lavas, volcanic breccias, volcaniclastic sediments, and tuffs of either basic (basaltic-andesitic) or acid composition (keratophyric and rhyodacitic). Towards the south, this complex grades into a more metamorphosed association of amphibolites and quartzo-feldspathic gneisses, which appear to envelope thrust-sheets and/or bodies of serpentinized ultrabasic cumulate (dunites and wehrlites) (Bardy 1988).

In the Frontal Arc of the Cherbourg Area (Fig. 5), the Brioverian is quite different in character, being composed mainly of a clastic sequence (metagraywackes and metashales/siltstones) where scarce volcanic horizons are interbedded with sediments. This is known as the Le Vast Formation (Le Gall et al. 1986; Dissler 1987), characterized by tholeiitic volcanics and subordinate acid rock types interbedded with tuffaceous sediments.

2.1.2 The Back-Arc Basin

This unit is best defined in the Baie de Saint-Brieuc where the succession of different formations has been established from the well-exposed section on the west side of the bay. From observations on this section, it was originally proposed that an important structural unconformity existed within the Brioverian succession (Jeannette and Cogné 1968). This hypothesis led to the subdivision of the Brioverian into two groups and the recognition of two orogenic phases (Cadomian I and II; Cogné 1972), but the discordance is now interpreted as a tectonic discontinuity (Balé and Brun 1983; Rabu et al. 1983). The Brioverian of this region is thus considered to form a single group. Around the southern end of the Baie de Saint-Brieuc, the back-arc basin formations are separated by a shear zone from the "Pentevrian Complex" defined by Cogné in 1959 as the basement to the Brioverian in this region (Fig. 4). This interpretation was held for many years (Cogné 1972; Jeannette 1972) and is still favored by some (Shufflebotham, 1989). However, recent studies in the Saint-Brieuc area suggest that the Pentevrian Complex makes up the basal part of the back-arc basin succession (Balé and Brun 1983); nevertheless, some slices of Icartian basement might have formed imbricate thrust-sheets during Cadomian deformation (Rabu et al. 1983).

The Saint-Brieuc Type Succession is composed of a volcanic and plutonic complex (Yffiniac and Lanvollon Formations), of tholeiitic affinity at the base, passing upward into a clastic sequence (Binic-Le Légué Formation) showing turbiditic facies at the top. The succession is affected by regional metamorphism with increasing grade towards the south. Generally, deformation becomes more intense in the same direction, but it is concentrated in several shear zones.

The Yffiniac Formation is made up of a complex of amphibolites, pyroxenites, and basic gneisses (often with garnet), probably derived from gabbros and diorites that remain locally as well-preserved relicts. Quartzo-feldspathic assemblages are rare, but there are some rafts of serpentinized ultrabasic rock (Bardy 1988) and the complex is intruded by several Cadomian plutons which appear to be syn-orogenic. The distribution and thickness of this crystalline complex is certainly very great but still poorly known.

The Basic Massif of Belle-Isle-en-Terre (Fig. 4) lies off to the SW in the midst of a migmatitic terrain near Guingamp. It has been interpreted as equivalent to the Yffiniac Formation, being composed of a klippe that has resulted from the thrusting of a volcanic-plutonic complex onto the passive continental margin (Hirbec 1979).

The Lanvollon Formation is at least 1000 m thick; it overlies the Yffiniac Formation but is separated from it by one of the shear zones affecting the succession. It is composed of medium-grade metavolcanic rocks (mostly amphibolites) of basaltic and andesitic composition. Along the coastal sections (Roselier in the west, Hillion in the east), some pre-metamorphic features are still observable: pillow lavas, massive sills, volcano-sedimentary intercalations, and some rare acid rock types.

The Binic-Le Légué Formation is composed of low-grade metasediments of volcaniclastic origin, making up sequences of wackes, banded siltstones, and laminated siltstones. The volcanic influence is everywhere present and is particularly apparent in the tuffaceous sandstones and tuffs. However, volcanic rocks are almost totally absent except at the base of the formation. Calc-silicate lenses appear typical of this succession. The Binic-Le Légué Formation is several hundred meters thick, with abundant way-up criteria that show it is everywhere stratigraphically above the Lanvollon Formation. Locally, there is a fault separating these two formations, which was previously interpreted as a major regional unconformity, but which is now considered to be a local discontinuity in the Brioverian succession.

In the classic type section of the Brioverian described in the Baie de Saint-Brieuc (Cogné 1984b), there are two units which have been left out of the type succession given here because they form discontinuous intercalations at the base and top of the Lanvollon Formation. They are, respectively:

- the Cesson conglomerate,
- the Port Martin member.

Both units are rather thin (ca. 100 m) and are very poorly defined. The Cesson conglomerate is characterized by the polygenetic nature of the clasts and has, up to now, been considered to represent the base of the Brioverian succession because of the presence of large blocks of orthogneiss which are taken as evidence for the proximity of basement. Recent dating of these clasts has yielded Cadomian ages, which suggest that they represent the erosion of early plutons associated with the opening of the Saint-Brieuc basin.

The Port Martin Member is composed of an alternation of banded volcano-clastic beds, containing acid and basic volcanic rocks, with polygenetic conglomerates and carbonaceous sediments. These latter have been incorrectly correlated with the black chert series of the continental margin domain. Recent studies have shown that the Port Martin member is a transition which marks the passage between the Lanvollon and Binic-Le Légué Formations, thus affirming the continuity of Brioverian sedimentation in the Baie de Saint-Brieuc.

2.1.3 Age of the Brioverian Succession

Several isotopic dating results enable us to bracket the ages of the Trégor volcanic arc and the Saint-Brieuc back-arc basin. Dating of the Tréguier tuffs has yielded an age of 642 ± 42 Ma. (Rb-Sr isochron of

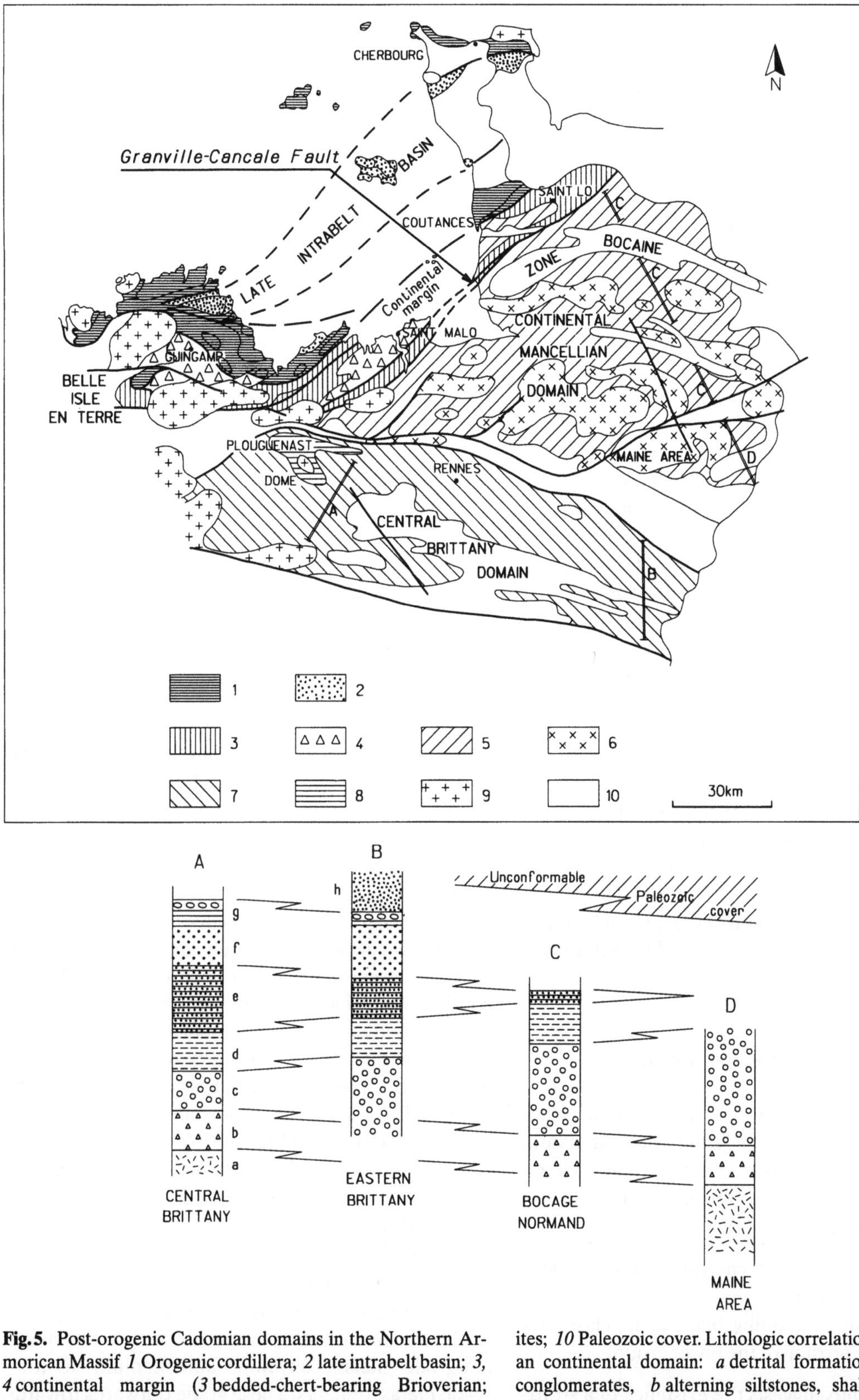

Fig. 5. Post-orogenic Cadomian domains in the Northern Armorican Massif *1* Orogenic cordillera; *2* late intrabelt basin; *3, 4* continental margin (*3* bedded-chert-bearing Brioverian; *4* Saint-Malo and Guingamp migmatites); *5, 6* continental platform (*5* post-bedded-chert Brioverian; *6* Mancellian Batholith); *7, 8* Central Brittany foreland (*7* post-bedded-chert Brioverian; *8* Plouguenast dome Complex); *9* Variscan granites; *10* Paleozoic cover. Lithologic correlations in the Cadomian continental domain: *a* detrital formation with polygenic conglomerates, *b* alterning siltstones, shales and arenites, *c* lithic wackes, *d* siltstones with volcaniclastic layers, *e* arenites, siltstones and carbonated sandstones, *f* massive siltstones, *g* slaty beds and quartz pebbles conglomerate, *h* top quartzites

Vidal 1976), whereas an isochron obtained from cleaved acid veins intruded into the schistose basic rocks of the Armorique Formation gives 586 ± 18 Ma (Charlot in: Autran et al. 1979). Therefore, these data indicate the period of activity of the Trégor volcanic arc, an episode which is associated with and terminated by the emplacement of syn-orogenic plutons at 614 $^{+17}_{-10}$ Ma (U/Pb zircon age; Graviou 1984) in the Trégor and 584 ± 5 Ma (U/Pb zircon age; Guerrot 1989) in the Cotentin. These data fit in well with previous ages obtained from the Saint-Brieuc quartz-diorite intruded into the Lanvollon Formation (604 ± 96 Ma; Rb-Sr isochron of Vidal 1976), as well as the Saint-Quay-Portrieux granodiorite intruded into the Binic-Le Légué Formation (584 ± 42 Ma; Rb-Sr isochron of Vidal et al. 1972). The age of the Fort-la-Latte pluton (593 ± 15 Ma; U/Pb zircon data of Vidal et al. 1974) also falls within the range of the Cadomian syn-orogenic suite. In the Baie de Saint-Brieuc, recently obtained results indicate older ages of 667 ± 4 and 656 ± 5 Ma (U/Pb zircon; Guerrot and Peucat, in press) for orthogneissic clasts in the Cesson conglomerate. Finally, the dating of the basic igneous complexes of Yffiniac (602 ± 8 Ma; U/Pb zircon age; Guerrot and Peucat 1990) and Belle-Isle-en-Terre (602 ± 6 Ma; U/Pb zircon age; Peucat et al. 1981) lends support to the hypothesis that the former complex belongs to the back-arc basin and not to the underlying basement.

All these data taken together show that the two successions – the Trégor volcanic arc and the Saint-Brieuc back-arc basin – can be considered as contemporaneous within experimental error. The active margin was initiated around 670 Ma ago and continued up to 600 Ma, when the subduction system was blocked.

2.2 The Mancellian Continental Domain

This contrasts very clearly with the previously described cordilleran terrain in having predominantly terrigenous sediments with common black cherts (phtanites) that enable us to distinguish two successive groups within the Brioverian:

- a lower group with black chert horizons interbedded in the succession;
- an upper group containing black chert clasts derived from the erosion of the lower group.

Such a relationship suggests the existence of a stratigraphic discontinuity within the Brioverian Supergroup. Although this unconformity is never directly observed, the contact between the two groups is marked by a major Cadomian shear fault, the Cancale-Granville Fault (Fig. 5). The two successive parts of the Brioverian are thus situated on either side of this fault and correspond to two different structural terranes:

- to the N.W. the Brioverian with bedded black cherts corresponds to the continental margin;
- to the S.E. the Brioverian with black chert clasts corresponds to the continental platform.

It is thus impossible to determine whether the two Brioverian groups – which follow each other in time – were deposited in the same basin in vertical succession or whether they were accumulated in two distinct basins with lateral facies variations. In both domains, the Brioverian succession contains some volcanic horizons whose continental affinity contrasts sharply with the orogenic cordillera.

Lastly, it should be noted that the Brioverian with black chert debris was also deposited in the late intra-belt basin (Le Minard Formation) which was opened within the orogenic domain after the rise of the Cadomian belt and then filled with material eroded from the cordillera.

2.2.1 The Continental Margin

Situated behind the cordillera and on the edge of the platform, the passive continental margin is represented by a group of formations making up a belt passing from Guingamp, through Saint-Malo to Saint-Lô (Fig. 5). These formations typically belong to the continental domain, having mainly terrigenous sedimentation and associated volcanic rocks with continental within-plate affinities. This belt corresponds thus to a continental margin that has been overthrust by the cordillera and incorporated into the orogen during the development of the Cadomian belt. In this zone of confrontation between cordillera and continent, geodynamic evolution resulted in partial melting of the crust that has led to the Guingamp and Saint-Malo migmatites (Brun and Martin 1978), as well as the development of major shear faults, which make it very difficult to establish a regional lithostratigraphic succession (Balé and Brun 1989).

In the Saint-Malo area, the formations are arranged around a migmatitic gneiss dome (Brun 1975, 1977) composed of a central core of migmatites (diatexites and metatexites) rimmed by a high- and medium-grade gneissic envelope and medium- to low-grade metasediments. These grade outwards towards less metamorphosed formations on the periphery.

The tectono-metamorphic event affecting this dome is now clearly assigned to the Cadomian (Peucat 1982), after previous attributions to the Cadomian (Jeannette 1972) and the "Pentevrian" (Brown et al. 1971). The protolith formations are considered to be part of Brioverian Supergroup, even though the hypothesis of an ancient basement within the core of the massif cannot be entirely excluded.

Authors are in agreement on the general sedimentary origin of the migmatite source-rocks, i.e., metagraywackes and meta-arkoses (Jeannette 1972; Brown 1978, 1979). These constitute a thick clastic sequence thought to correspond to the base of the Brioverian succession, known locally as the Saint-Cast Formation (Hameurt and Jeannette 1971). The metamorphic envelopes to the gneiss domes are also composed of metasedimentary rocks, i.e., shales/siltstones and graywackes with coarse clastic members towards the lower part of the succession. There are numerous black chert layers, apparently near the top of this group, in the Lamballe Formation. These bedded cherts (phtanites) are associated with basaltic volcanism of alkaline affinity (La Fresnaye Formation), which indicates a continental setting for the Brioverian of this domain (Cabanis et al. 1986).

The Vallée de la Rance Formation is in tectonic contact with the previously mentioned formations. It is composed of immature shale/sandstone deposits, showing volcaniclastic influence, which are interpreted locally as characterizing the upper part of the bedded chert-bearing Brioverian. However, an analysis of the structural relations with surrounding rocks and the particular lithological features of this formation suggest an alternative hypothesis (Balé 1986), which interprets this formation as a klippe resulting from the overthrusting of internal units onto the continental margin.

The westward extension of a stable margin domain into the Belle-Isle-en-Terre area (Fig. 5) is very likely, but the general development of crustal melting (migmatites and granites) in this region prevents any clear identification. Only the Callac Formation can be recognized as an equivalent of the Lamballe formation to the South of the North Armorican Shear Zone, overlying the migmatitic formations and overthrust by the Belle-Isle-en-Terre volcanic/plutonic complex.

The Cotentin is separated from Northern Brittany by only 50 km across the Golfe de Saint-Malo, but there are no gneiss domes and the coarse basal deposits of the Brioverian are missing. The Saint-Lô Formation (Fig. 5), which contains bedded black cherts, is very similar to the Lamballe Formation and thus provides evidence for the easterly extension of the continental margin. The intercalated volcanics (La Terrette Formation) display tholeiitic characteristics typical of a distensive setting, which is interpreted as indicating a marginal tectonic environment (Le Gall et al. 1986). This formation seems to pass upward into immature deposits (Rampan and Saint Pair Formations; Dupret 1984) and is unconformably overlain by the Granville Formation, characterized by the general occurrence of black chert debris – thus, it belongs to the upper part of the Brioverian.

2.2.2 The Continental Platform

All the Bocage Normand and the Maine area situated to the southeast of the Granville-Cancale Fault belong to the Continental Cadomian domain (see Fig. 5). Here, the Brioverian is terrigenous in character throughout showing a monotonous succession which belongs to the post-bedded-chert group. The sedimentation is represented by a homogeneous sequence of rhythmic beds, without any notable discontinuities or marker horizons; only vertical trends of a gradual nature can be observed. The sediments are very immature, with ill-sorted and poorly graded textures, being mainly composed of an association of wackes, siltstones, and mudstones containing abundant conglomeratic interbeds. Carbonates are almost entirely absent. The oolitic limestone lens at La Meauffe, although isolated at the extreme edge of this domain, appears, in view of its detrital content, to belong to the Brioverian group with chert clasts.

The tectonic disruption and metamorphic effects brought about by the emplacement of the Mancellian Batholith make it very difficult to recognize the original lithostratigraphic relations within the Brioverian. Several sedimentary facies may be described, but their relationships to each other remain unknown. They would appear to reflect both vertical and lateral variations across three main areas (Chantraine et al. 1982), viz:

- a northerly succession (on either side of the "Zone Bocaine" in the South Cotentin), which is characterized by the development of massive wackes and polygenetic conglomerates (Granville Formation) in a rhythmic sequence of proximal turbidites (La Laize Formation);
- a thick succession in South Normandy, forming the country rocks of the Mancellian Batholith, showing more homogeneous and deeper water facies with intercalations of distal turbidites which are probable lateral equivalents of the La Laize Formation;
- a southerly succession (Maine area), mainly composed of shale/siltstone alternations of greater regularity, characterized by conglomeratic interbeds of the pebbly-mudstone type (these formations

contain the important sulfide ore-body of Rouez; Chantraine et al. 1982).

This sedimentary pattern has been interpreted in the context of a marginal basin which opened after the rise of the Cadomian cordillera – and its overthrusting onto the continent – during a period of crustal thinning and distensive tectonics in the foreland (Garlan 1985).

Such an interpretation is corroborated by the distensive geochemical affinity of the few interbedded lava flows (Vassy Spilites; Dupret et al. 1985b). The difference between the northern flank of this basin, which corresponds to a steep slope near the cordillera in the Cotentin, and the more gently sloping southern flank (Maine) may explain the asymmetrical nature of the sedimentation.

The underlying basement to the Brioverian in the continental domain is unknown. Given the migration of sedimentation and deformation that is recognized from the internal to the more external zones of the Cadomian orogen, it is possible that the Brioverian group with black chert clasts may overlap directly onto an ancient basement. Under this condition, it would be only locally, at the continental margin, that the superposition of the two Brioverian groups could be observed. This hypothesis is supported by the description of a discordance in the expected area at Quibou (Roblot 1962).

After deformation and the intrusion of granites, the continental Cadomian domain was overlain by a Paleozoic cover of Lower-Mid-Cambrian age (depending on the region) which shows very few effects of the Variscan orogeny (Doré 1972).

2.2.3 *Age of the Successions*

The age of the formations deposited on the continental margin is not known with any precision. However, these formations have been involved in thrusting movements which have transposed the cordillera onto the continental margin and they have thus been incorporated into the Cadomian orogenic belt. The development of the fold belt can be dated by syn-orogenic plutons which place an upper limit on the age of deposition of the Brioverian. The Fort-la-Latte pluton has yielded an age of 593 ± 15 Ma (U/Pb zircon method; Vidal et al. 1974), whereas the Coutances Diorite has been dated at 584 ± 5 Ma (U/Pb zircon method; Guerrot et al. 1989). In this way, the bedded-chert-bearing Brioverian of the continental margin appears to be broadly contemporaneous with the volcanic Brioverian group of the orogenic cordillera. Given the present state of knowledge, it is not possible to go further with chronostratigraphic correlations between these two groups.

On the continental platform, the age of the main Cadomian orogenic phase (593–584 Ma) fixes a lower limit on the deposition of Brioverian sediments that contain eroded debris of black chert derived from the cordillera. An upper limit is placed by the ages of Mancellian granitoids which are intruded into the chert clast-bearing Brioverian. At present, only the ages of 540 ± 10 Ma (U/Pb zircon method; Pasteels and Doré 1982) and 521 ± 11 Ma (Rb/Sr isochron; Fouillac, et al. 1986) appears consistent with other data on Cadomian geodynamic evolution. Previously obtained results are based on pooled isochrons (ca. 570–600 Ma; Adams 1976; Vidal 1976) which yield imprecise and improbable old ages. To summarize, the post-bedded-chert Brioverian appears to have been deposited on a continental platform between about 590 and 540 Ma ago, i.e., in the time interval which corresponds to the various different assignments for the Precambrian-Cambrian boundary.

2.3 The Central Brittany Foreland

The Central Brittany domain is bounded by two major Variscan faults – the North-Armorican Shear Zone and the Landes de Lanvaux axis – so it is, strictly speaking, isolated from the rest of the Cadomian orogenic belt. Despite this, it has always been considered part of the Cadomian domain (Cogné 1972).

In fact, the general field relations of Central Brittany are very similar to those observed in the Bocage Normand. Terrigenous Brioverian formations with chert clasts, assigned to the uppermost Proterozoic, are overlain by red beds which have long been considered to be of Cambrian age. These pass upward into the fossiliferous Paleozoic succession. In this context, the infra-Paleozoic unconformity was considered to be the result of a Paleozoic transgression onto the external zones of the Cadomian orogen. However, this model must be revised in light of several fundamental observations:

- the Brioverian formations of Central Brittany are not intruded by late Cadomian intrusions (ca. 540–520 Ma); they bear no trace of any pre-Variscan deformation different in type to that affecting the Paleozoic cover; so the infra-Paleozoic unconformity is now interpreted as marking an episode of crustal distension (Ballard et al. 1986);
- the major transgression in Central Brittany seems to have taken place in the Early Ordovician, and not during Cambrian times; this is implied by the

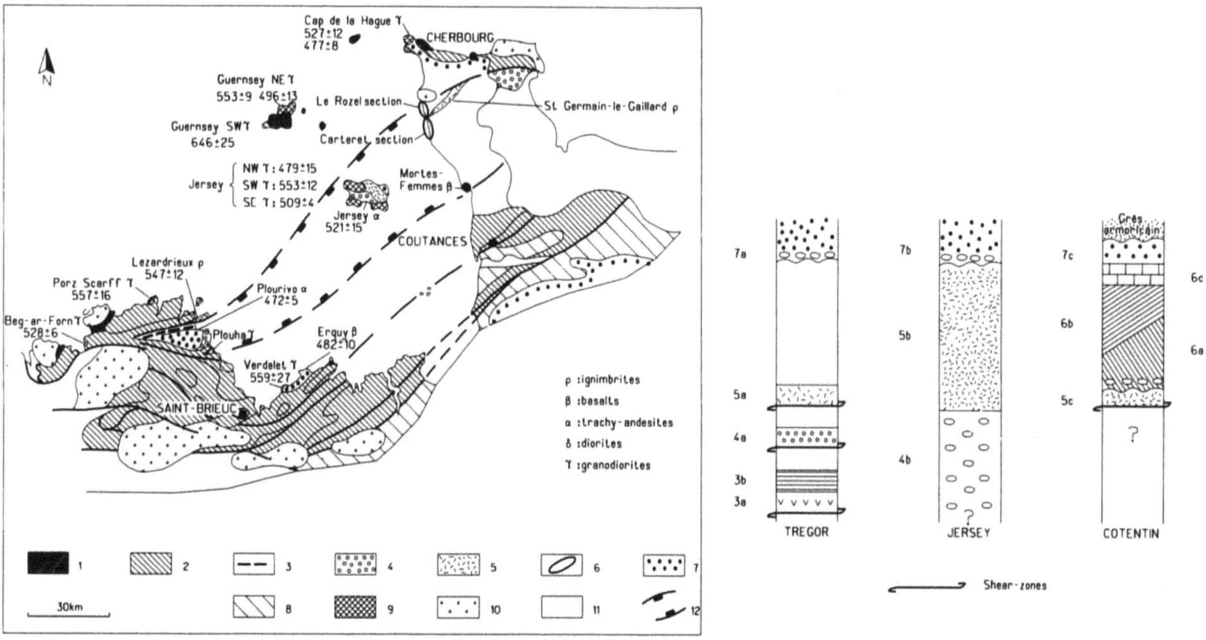

Fig. 6. Late intrabelt basin in the Northern Armorican Massif. *1* Icartian basement; *2* Cadomian cordillera; *3* Brioverian without chert clasts (*3a* Paimpol Spilites; *3b* La Roche-Derrien Formation); *4* Brioverian with chert clasts (*4a* Le Minard; *4b* Jersey Formations); *5* ignimbritic volcanics (*5a* Lézardrieux; *5b* Jersey; *5c* Saint-Germain-le-Gaillard); *6* marine Cambrian facies (*6a* Le Rozel; *6b* Carteret; *6c* Saint-Jean-de-la-Rivière); *7* red beds (*7a* Trégor; *7b* Jersey; *7c* Cotentin); *8* continental platform; *9* late-Cadomian granites; *10* Variscan granites; *11* post-Cambrian cover; *12* intrabelt basin limits. The indicated ages were determined by the Rb/Sr whole-rock isochron method, except at Beg-ar-Forn (U/Pb method)

date of 465 ± 1 Ma recorded from volcanic rocks intercalated in the basal red beds of the Paleozoic cover (U/Pb zircon method; Bonjour et al. 1988).

Thus, the entire Brioverian succession of Central Brittany should not necessarily be considered as Proterozoic. The results cited above tend to indicate that the uppermost parts of the Brioverian could have been deposited during the Early Paleozoic.

In Central Brittany, Brioverian formations make up an important terrigenous succession (Chauvel and Rabu 1984), at least 1000 m thick, which is composed of wackes, siltstones, and mudstones arranged in alternating sequences. Vertical stratigraphic changes as well as lateral facies variations have been demonstrated by Chantraine et al. (1982). The lower part of the succession is characterized by an important lithic component; clastic sediments with polygenetic conglomerates can be seen at the base, passing up into intercalated mudstones, siltstones, and sandstones overlain by immature sandstones with volcaniclastic interbeds. Above this lower part is developed a rythmic sequence made up of a series of sandy and silty beds, with local carbonate cement. In the upper part of the succession, there are massive siltstones passing up into slate beds at the top, which are associated with conglomerates containing quartz and chert pebbles (Gourin conglomerate). Towards the east of the domain, there are mature quartzites at the top of the exposed section.

Otherwise, it should be noted that the generally oolitic carbonate-bearing lenses of Saint-Thurial and Les Rochelles are very similar to the limestones at La Meauffe in the Bocage Normand. In both cases, the absence of normal contacts with the surrounding formations makes it very difficult to establish their true stratigraphic relations. Again, both these calcareous bodies are situated very near the contact between the Brioverian and the Paleozoic cover; in Central Brittany, the basal Paleozoic conglomerate contains clasts of limestone in the vicinity of the observed contact. As a result, two hypotheses may be envisaged: either the calcareous lenses form part of the uppermost Brioverian succession, despite their exotic character with respect to the normal sedimentation; or these lenses represent the remnants of a carbonate platform that was destroyed during the Arenig transgression; the period of limestone accumulation would thus be Cambrian, by analogy with carbonate-bearing deposits of this age in the North Cotentin.

Finally, we may remark that, in the Central Brittany domain, there are minor occurrences of interbedded volcanics throughout the Brioverian succession; some pillow lavas associated with massive sills (Telgruc spilites) are seen in areas of good outcrop on the coast in Western Brittany.

In summary, the Brioverian successions of Central Brittany and the Bocage Normand are fairly similar. Some attempts at correlation (Chantraine et al. 1982) suggest that the immature sediments of Lower Normandy form the basal part of the succession and that the more mature sediments of Central Brittany are younger. The folded Brioverian of Lower Normandy must be Proterozoic in age because it is unconformably overlain by Lower-Mid-Cambrian strata. By contrast, the Brioverian of Central Brittany is younger and may partly be of Cambrian age. In this way, the pre-Ordovician discordance may be merely the record of intra-basinal disturbances leading up to the Variscan orogeny.

2.4 The Late Intrabelt Basin

The Cadomian orogenic belt displays an internal volcanic arc which is divided in two by a late basin initiated in the Trégor and widely developed in the Cotentin. This basin opened due to post-orogenic crustal distension within the cordillera (Fig. 6) shows some very interesting features for the interpretation of Cadomian geodynamic evolution since it has recorded posthumous sedimentary and magmatic phenomena.

In the Trégor, the intrabelt basin is flanked by almost undeformed schistose formations classically assigned to the Brioverian, i.e.:

- the Minard Formation is composed of turbiditic facies and is characterized by ubiquitous black chert clasts. It is correlated with the post-bedded-chert Brioverian which is widespread over the continental platform;
- the Paimpol Spilites and La Roche Derrien Formation occur at the periphery of the basin and are made up of a continuous succession with calc-alkaline lavas and sills at the base overlain by turbiditic sediments.

These formations are often compared – sometimes correlated (Vidal 1976), sometimes separated (Auvray 1979) – with the volcanic arc domain. It seems now that the late intrabelt formations have not been affected by the main tangential tectonics observed in the volcanic arc domain (Balé 1986). Furthermore, their outcrop appears to be limited by fractures that are locally associated with volcanism (Lézardrieux ignimbrites, dated by the Rb/Sr method at 547 ± 12 Ma; Auvray 1979) or the injection of granitic intrusions (Plouha granite) that are undated but which may be contemporaneous with neighboring plutons (Loguivy granite) dated at 548 ± 24 Ma (Rb/Sr method; Auvray 1979).

The core of the intrabelt basin is bounded by normal faults of syn-sedimentary character, situated on major strike-slip fault zones (pull-apart basin tectonics; Dauteuil et al. 1987). It is filled by the Plouézec conglomeratic red beds, which are uncleaved and undeformed, passing up into the interbedded trachyandesites of Plourivo, dated at 472 ± 5 Ma (Rb/Sr method; Auvray et al. 1980 b), that show a hybrid continental affinity.

Further to the south, the Erquy Spilites have yielded an Rb/Sr isochron age of 482 ± 10 Ma (Vidal et al. 1971) and would appear to occupy a similar tectonic setting to the Plourivo lavas. However, the Erquy succession dips vertically and is overlain discordantly by the horizontal Fréhel red beds. The Erquy succession was formerly considered to be the stratotype of the Lower Brioverian (Graindor 1957; Cogné 1962), but geochronological (Vidal et al. 1971) and micropaleontological data (Deunff et al. 1973) have tended to favor a Lower Paleozoic age. Some authors continue to refute such an age (Roach et al. 1990).

Thus, the Trégor intrabelt basin is characterized by post-orogenic deposits which constitue a discontinuous succession laid down between about 600 and 470 Ma. Their sedimentation was probably controlled by periods of crustal distension.

A fairly similar pattern has been revealed in Jersey (Fig. 6), where the turbiditic Brioverian with black chert clasts is overlain in apparent concordance by about 3000 m of volcanic rocks (Bishop and Keen 1982). These include andesites at the base (St Saviour's Formation) with ignimbritic rhyolites at the top (St John's and Bouley's Formations). The andesites have been dated at 521 ± 15 Ma (Rb/Sr method; Duff 1978), thus fixing a Cambrian age for the Jersey Brioverian which is unconformably overlain by a conglomeratic red bed formation assigned to the Cambro-Ordovician.

In the Cotentin, despite the extent of Ordovician, Silurian and Devonian cover, the late intrabelt basin can still be recognized through the occurrence of Cambrian sedimentary facies that are different from the Cambrian facies of the continental Mancellian domain. Here, the Cambrian is locally unconformable onto Mancellian Batholith dated around 540 Ma.

Within the intrabelt basin, the oldest outcropping units are ignimbritic flows at Saint-Germain-le-Gaillard – undated (in prep.) but typically postorogenic in

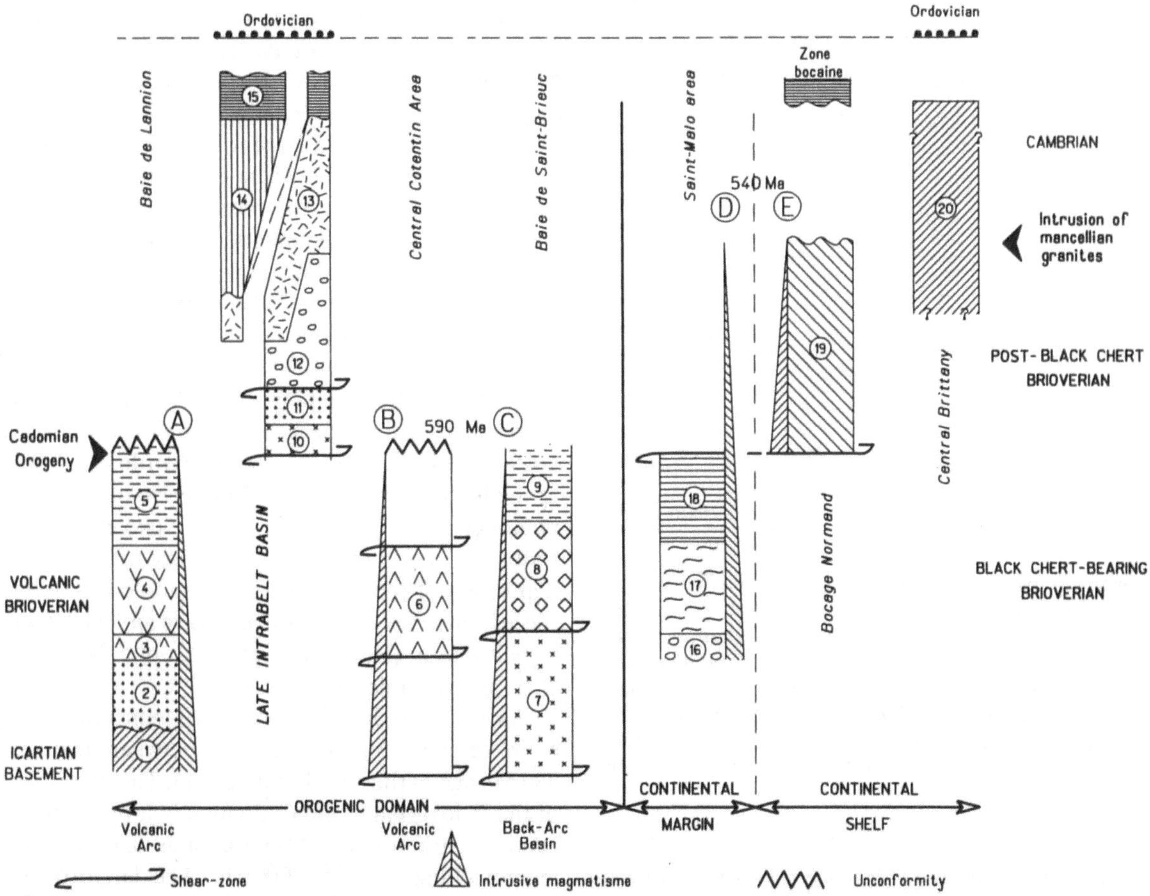

Fig. 7. Schematic cross-section of the Brioverian Supergroup across the Cadomian orogenic belt. (After Chantraine et al. 1988). *1* Moulin-de-la-Rive orthogneisses; *2* Locquirec-Tréguier Formation; *3* Rugunay-Malabri Formation; *4* Armorique-Trédez Formation; *5* Plestin-les-Grèves and Bois-de-la-Roche Formations; *6* Montsurvent Formation; *7* Yffiniac Formation; *8* Lanvollon Formation; *9* Binic-Le Legué Formation; *10* Paimpol spilites; *11* La Roche Derrien Formation; *12* Le Minard-Jersey late intrabelt deposits; *13* ignimbritic volcanics of Lézardrieux, Jersey and Saint-Germain-le-Gaillard; *14* marine Lower Cambrian; *15* Cambrian red beds; *16* Saint-Cast Formation; *17* Lamballe Formation; *18* Vallée de la Rance Formation; *19* Bocage Normand post-black chert formations; *20* Central Brittany post-black chert formations. Dated igneous activity: *A* Baie de Lannion acid veins; *B* syn-orogenic plutons in the Cotentin; *C* syn-orogenic plutons in the Baie de Saint-Brieuc; *D* anatectic granites in the Saint-Malo Massif; *E* Mancellian Batholith granites

character – overlain by a thick terrigenous succession (ca. 3000 m; Doré 1972). This is composed of conglomerates, shales, and sandstones (Le Rozel Formation), the middle part of which is correlated with the basal clastic parts of the Carteret Formation dated as Tommotian by the appearance of the first body-fossils (i. e., Precambrian/Cambrian boundary); the Carteret Formation is overlain by biogenic sediments (Saint-Jean-de-la-Rivière) limestones of Mid-Lower Cambrian age followed by a further marine clastic sequence (red sandstones and shales); these red bed units are overlain by Ordovician cover rocks (Grès Armoricain Formation). It is clear that the intrabelt basin deposits form an unusually thick succession compared with the Cambrian in other adjoining areas (Doré 1972). According to some authors (Choubert and Faure-Muret 1979), sedimentation may have continued here from uppermost Vendian through to Early Cambrian times.

In Val de Saire, the late intrabelt basin deposits are composed of immature sandstones (Montaigu Formation) and conglomerates, graywackes and siltstones making up turbiditic sequences (Saussemesnil Formation). The immature units contain numerous acidic and basic clasts, indicating the vicinity of a volcanic activity or the erosion of an ancient basement.

In this way, the existence of a late Cadomian intrabelt basin is well established. After the deposition of the post-bedded-chert Brioverian, observed towards the S. W. where distension is more restricted, the maximum development of the basin corresponds to a peri-

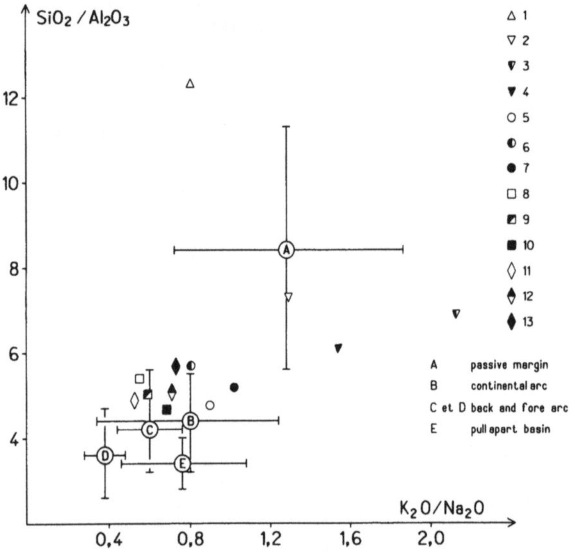

Fig. 8. Plot of Brioverian sedimentary formations in the tectonic setting discriminant diagram of Maynard et al. (1982). *1* W. Brittany (Douarnenez); *2* Central Brittany (Sérent); *3* Central Brittany (Janzé); *4* Mayenne area; *5* North of Zone Bocaine; *6* South of Zone Bocaine; *7* Maine area; *8* La Roche Derrien Formation; *9* Le Minard Formation; *10* Binic Formation; *11* Callac Formation; *12* Lamballe Formation; *13* Saint-Lô Formation. Si02/A1203 vs K20/Na20 diagram after Denis and Dabard (1988)

od of continental-type volcanism followed by the deposition of a thick Early Paleozoic succession accumulated in a syn-sedimentary tectonic setting. The succession is thickest in the N. E. where the basin was more open, and is characterized by marine clastic sequences that locally continue across the Precambrian/Cambrian boundary. The geodynamic history of this basin differs radically from the continental platform domain, which shows a major break at 540 Ma (plutonism, uplift) followed by the deposition of continental redbeds of Cambrian age.

3 Litho-Stratigraphic Correlations

In recent years, advances made in our knowledge of the Upper Proterozoic of the Armorican Massif have demonstrated that the classic idea of a Brioverian succession similar throughout the Cadomian domain is no longer valid. In each structural domain, there are particular successions which are more or less distinct from one another according to their exact tectonic setting. The main distinction is made between the orogenic domain and the continental domain.

Correlations are difficult across the boundary between the orogenic domain and the continental margin – this discontinuity corresponds to a major suture. However, some correlations can be proposed on the basis of the lithological, structural, and geochronological criteria discussed above (see Fig. 7).

From the lithological point of view, a difference can be recognized between the dominantly volcanogenic character of the orogenic domain and the essentially clastic sedimentation on the continental platform. This latter domain is distinguished by the general occurrence of black chert debris, which leads us to separate the bedded chert-bearing Brioverian from the Brioverian with black chert clasts.

From the structural point of view, the major break corresponds with the orogenic phase which affects the volcanic Brioverian in the orogenic cordillera as well as the bedded chert-bearing Brioverian of the continental margin. Deformation in these two groups was probably penecontemporaneous, occurring before deposition of the late- and post-tectonic Brioverian succession in the continental platform and the late intrabelt basin (Brioverian group containing black chert and volcanic fragments).

As far as radiometric dating is concerned, the main orogenic phase occurred between 600 and 580 Ma ago in the cordilleran terrain. The oldest part of the Brioverian in the Cadomian belt began to accumulate around 670 Ma on top of an Icartian basement yielding ages of 2000 Ma. The late-orogenic Brioverian succession was deposited from about 580 Ma through to a minimum age limit fixed by the 540 Ma date on Mancellian Batholith in the Bocage Normand. The age of the basic plutonic rocks of the Plouguenast Dome (Fig. 5; U/Pb zircon method; Calvez, pers. commun.) incorporated in the post-orogenic Brioverian succession of Central Brittany fixes a possible limit of 480 Ma. Chronostratigraphic correlations between these two domains are still hypothetical in nature.

3.1 Geodynamic Setting of Brioverian Sedimentation

The tectonic settings of the various Proterozoic units are based mainly on the geochemical characteristics of igneous rocks; this has led to the currently proposed model of geodynamic evolution for the region. The chemical compositions of sedimentary rocks in different basins and during various periods of deposition (Denis and Dabard 1988) would appear to be compatible with this model.

Clastic sequences containing interbedded cherts (Callac, Lamballe, and Saint-Lô Formations) display wacke-type homogeneous compositions which are characterized by K20/Na20 ratios generally less than unity and (Fe203 + Mg0) near 6% (see Fig. 8). This indicates that the depositional environment was of the active margin/continental arc type, a conclusion that is supported by the Ti/Zr and La/Sc ratios. The volcanic source area appears to have been mainly composed of acidic rocks.

The wackes of the Binic-Le Legué Formation are chemically very similar to the previously mentioned formations with, however, rather higher Ca0 and (Fe203 + Mg0) contents which may reflect some contribution from basic volcanic material. This difference is interpreted as resulting from a vertical sequence of deposition (Dabard 1989). In North Brittany, the Lamballe Formation may correspond to the initial stages of filling of a back-arc basin supplied by the erosion of acid volcanic rocks on the active margin (Tréguier and Locquirec tuffs). Later, the Binic-Le Legué Formation was deposited only after the accretion of basaltic material started during extension of the back-arc basin (Lanvollon Formation). The post-orogenic Roche-Derrien and Minard Formations have compositions close to the Binic-Le Legué Formation (Denis 1987), but the Minard Formation is clearly distinct in having a widely variable clast content composed of volcanic fragments.

In the Bocage Normand domain, the post-bedded-chert formations also show some volcanic influence, but less well pronounced compared with the North Brittany. The composition of Brioverian wackes generally appears compatible with an active margin type environment (Fig. 8), with REE patterns indicating a volcanic influence. However, these wackes also display a regional compositional trend from north to south reflected in an increase in the K20/Na20 ratio. In the north of the Cotentin, a volcanic influence is well marked – as in the Trégor area – but this falls off to the south where the sediments are much richer in basement rock debris (e.g., in the Maine area).

In the Central Brittany domain, the wackes show a quite different composition with K20/Na20 ratios greater than unity and Si02/A1203 ratios significantly higher than in Brioverian rocks from other areas. These wackes are chemically more mature, thus indicating a passive margin type source area. By contrast, the Brioverian sediments of the extreme western part of the Central Brittany (Douarnenez) show K20/Na20 ratios less than unity coupled with very high Si02/A1203 ratios; these features are quite different from all other Brioverian deposits of the Armorican Massif.

The mineralogical compositions of the Brioverian sediments of Central Brittany indicate a source area of the continental block and/or recycled orogen type, except in the east where the presence of volcanic fragments probably reflects contemporaneous volcanism in the sedimentary basin (Chantraine et al. 1982).

References

Adams CJ (1976) Geochronology of the Channel Islands and adjacent French mainland. J Geol Soc (Lond) 132: 233–250

Autran A, Chantraine J, Rabu D (1979) Lithostratigraphie et déformation du Briovérien de la baie de Lannion. Implications sur les relations entre les cycles cadomien et hercynien. Bull Bur Rech Géol Min 1–4: 277–292

Auvray B (1979) Génèse et évolution de la croûte continentale dans le Nord du Massif Armoricain. Thèse d'Etat, Rennes, 670 pp

Auvray B, Charlot R, Vidal P (1980 a) Données nouvelles sur le Protérozoïque inférieur du domaine nord-armoricain (France): âge et signification. Can J Earth Sci 17: 532–538

Auvray B, Macé J, Vidal P, Van Der Voo R (1980 b) Rb-Sr dating of the Plouézec volcanics, N Brittany: implications for the age of red beds ("séries rouges") in the northern Armorican massif. J Geol Soc (Lond) 137: 207–210

Balé P (1986) Tectonique cadomienne en Bretagne nord. Interaction décrochement, chevauchement: champs de déformation et modélisation expérimentale. Thèse d'Université, Rennes, 361 pp

Balé P, Brun JP (1983) Les chevauchements cadomiens de la Baie de Saint-Brieuc (Massif Armoricain). CR Acad Sci Paris 297: 359–362

Ballard JP, Brun JP, Durand J (1986) La discordance Briovérien-Paléozoïque inférieur en Bretagne centrale: signature d'un épisode de distension ordovicienne. CR Acad Sci Paris 303: 1327–1332

Bardy P (1988) L'orogène cadomien dans le Nord-Est du Massif armoricain et en Manche occidentale. Thèse d'Université. Rennes, 380 pp

Bishop AC, Keen DH (1982) Geological map of Jersey. Channel Islands Sheet 2 – 1/25000. Institute of Geological Sciences, London

Bonjour JL, Peucat JJ, Chauvel JJ, Paris F, Cornichet J (1988) U-Pb zircon dating of the early Paleozoic (Arenigian) transgression in Western Brittany (France): a new constraint for the lower Paleozoic time-scale. Chem Geol 72: 329–336

Brown M (1978) The tectonic evolution of the Precambrian rocks of the Saint-Malo Region, Armorican Massif, France. Precambrian Res 6: 1–21

Brown M (1979) The petrogenesis of the Saint-Malo Migmatite Belt, Armorican Massif, France, with particular reference to the diatexites. N Jb Miner Abh 1: 48–74

Brown M, Barber AJ, Roach RA (1971) Age of the Saint-Malo Migmatite Belt, Northern Brittany, Nat Phys Sci 236–66: 77–79

Brun JP (1977) La zonation structurale des dômes gneissiques. Un exemple: le massif de Saint-Malo (Massif Armoricain, France). Can J Earth Sci 14: 1697–1707

Brun JP, Martin J (1978) Relations métamorphisme – déformation au cours de l'évolution dynamique d'un dôme migmatitique: le massif de Saint-Malo (France). Bull Soc Geol Fr XX-1: 91–101

Cabanis B (1986) Identification des séries magmatiques dans les socles métamorphiques. Exemples d'application dans le Massif Armoricain et le Massif Central. Implications géodynamiques. Thèse d'Etat, Paris, 493 pp

Calvez JY, Vidal P (1978) Two billion years old relicts in the Hercynian Belt of Western Europe. Contrib Mineral Petrol 65: 395–399

Chantraine J, Chauvel JJ, Dupret L. Gatinot F, Icart JC, Le Corre C, Rabu D, Sauvan P, Villey M (1982) Inventaire lithologique et structural du Briovérien (Protérozoïque supérieur) de la Bretagne centrale et du Bocage normand. Bull Bur Rech Géol Min I(1–2): 3–17

Chauvel JJ, Rabu D (1984) Brioverian in Central Brittany. In: Zoubek V (ed) Precambrian in younger fold belts. Wiley, New York, pp 462–470

Choubert G, Faure-Muret A (1979) Au sujet de la couverture protérozoïque cambrienne en Normandie. Cr Acad Sci Paris 288: 795–798

Cogné J (1959) Données nouvelles sur l'Antécambrien dans l'Ouest de la France: Pentévrien et Briovérien en baie de Saint-Brieuc. Bull Soc Géol Fr I-1: 112–118

Cogné J (1972) Le Briovérien et le cycle orogénique cadomien dans le cadre des orogènes fini-précambriens. Coll Int CNRS Paris 192; Notes Mem Serv Geol (Rabat) 236: 193–218

Cogné J (1984b) Problem of the Lower Brioverian in the context of a two-phase Cadomian orogenic cycle: the Precambrian of the Penthièvre Massif and its adjoining Mancellian boundary. In: Zoubek V (ed) Precambrian in younger fold belts. Wiley, New York, pp 434–444

Dabard MP (1989) Les formations du Briovérien inférieur de Bretagne nord: apport de l'étude des fragments volcaniques. CR Acad Sci Paris 308: 1167–1170

Dauteuil O, Durand J, Brun JP (1987) Arguments en faveur de décrochements synchrones du dépôt des séries rouges d'Erquy-Fréhel. CR Acad Sci Paris 304: 83–88

Denis E, Dabard MP (1988) Sandstone petrography and geochemistry of Late Proterozoic sediments of the Armorican Massif (France). A key to basin development during the Cadomian orogeny. Precambrian Res 42: 189–206

Deunff J, Auvray B, Cogné J, Hameurt J, Jeanette D, Vidal P (1973) Confirmation micropaléontologique de l'âge radiométrique ordovicien inférieur du groupe spilitique d'Erquy (Côtes-du-Nord). CR Acad Sci Paris 276: 935–937

Dissler E (1987) Evolution géodynamique cadomienne du Nord-Cotentin (Massif Armoricain). Thèse d'Université, Caen, 255 pp

Doré F (1972) La transgression majeure du Paléozoïque inférieur dans le Nord-Est du Massif armoricain. Bull Soc Géol Fr 7, XIV: 79–83

Duff BA (1978) Rb-Sr Whole-rock age determination of the Jersey Andesite Formation, Jersey CI J Geol Soc (Lond) 135: 153–156

Dupret L (1984) The Proterozoïc of Northeastern Armorican massif. In: Zoubek V (ed) Precambrian in Younger Folds belts. Wiley, New-York, pp 444–461

Dupret L, Cabanis B, Le Gall J, Doré F (1985a) Les tholéiites d'arc briovériennes de Montsurvent (Manche). Leur place dans l'évolution orogénique cadomienne du Nord de l'Armorique. Hercynica I-1: 39–46

Dupret L, Le Gall J, Doré F, Gatinot F, Dissler E (1985b) Les spilites de Vassy (Calvados), témoin d'un volcanisme sous-marin tholéiitique et distensif, dans la sédimentation du Briovérien supérieur de Normandie (NE du Massif Armoricain). CR Acad Sci Paris 300: 687–692

Fouillac AM, Cocherie A, Rossi P, Calvez JY, Autran A, (1986) – Etude geochimique au batholite mancellien – Ref Bur Rech Géol Min (unpublished), 15 pp

Garlan T (1985) Sédimentologie du Briovérien supérieur de Normandie et du Maine. Thèse 3e cycle, Caen, 166 pp

Graindor MJ (1957) Le Briovérien dans le Nord-Est du Massif Armoricain. Mem Carte Géol Fr Bur Rech Géol Min 211 pp

Graviou P (1984) Pétrogenèse des magmas calco-alcalins: exemple des granites cadomiens de la région trégorroise (Massif armoricain). Thèse 3e cycle, Rennes, 236 pp

Guerrot C (1989) Archéen et Protérozoïque dans la chaîne hercynienne ouest-européenne: géochimie isotopique (Sr, Nd, Pb) et géochronologie U/Pb sur zircons. Thèse d'université, Rennes, 164 pp

Guerrot C, Peucat JJ (1990) U/Pb geochronology of the Late Proterozoic Cadomian orogeny in the Northern Armorican Massif. France. In: D'Lemos RS, Strachan RA, Topley CG (eds) The Cadomian Orogeny. Geol Soc Spec Publ 51: 13–26

Guerrot C, Peucat JJ, Dupret L (1989) Données nouvelles sur l'âge du système briovérien (Protérozoïque supérieur) dans le Nord du Massif Armoricain. CR Acad Sci Paris 308: 88–92

Hameurt J, Jeannette D (1971) Mise en évidence d'un terme litho-stratigraphique nouveau du Briovérien: le groupe détritique de Saint-Cast (Bretagne septentrionale). CR Acad Sci Paris 273: 1767–1770

Hirbec Y (1979) Le complexe basique de Belle-Isle-en-Terre (Côtes-du-Nord) Thèse 3e cycle, Rennes, 147 pp

Jeannette D (1972) Analyse tectonique de formations précambriennes. Etude du Nord-Est de la Bretagne. Thèse d'Etat, Strasbourg. Mem Soc Geol Strasbourg 36, 175 pp

Jeannette D, Cogné J (1968) Une discordance majeure au sein du Briovérien du flanc ouest de la baie de Saint-Brieuc. CR Acad Sci Paris 226: 2211–2214

Lees GS, Roach RA, Shufflebotham MM, Griffiths NH (1987) Upper Proterozoic basaltic volcanism in the northern Massif Armoricain, France. Geochemistry and mineralization of Proterozoic volcanism suites. Geol Soc Sp Publ London, pp 503–523

Le Gall J, Dissler J, Dupret L (1986) Signification géodynamique des volcanismes briovériens dans le nord-est du Massif armoricain. CR Acad Sci Paris 303: 1587–1592

Pasteels P, Doré F (1982) Age of the Vire-Carolles granite. In: Odin GS (ed) Numerical Dating in Stratigraphy. John Wiley, New-York, pp 784–790

Peucat JJ (1982) Géochronologie des roches métamorphiques (Rb-Sr et U-Pb). Exemples choises au Groenland, en Laponie et dans le Massif armoricain. Thèse d'Etat, Rennes. Mem Soc Géol Min Bretagne 28 (1983), Rennes, 158 pp

Peucat JJ, Hirbec Y, Auvray B, Cogné J, Cornichet J (1981) Late Proterozoic zircon ages from a basic-ultrabasic complex: a possible Cadomian orogenic complex in the Hercynian belt of western Europe. Geology 9: 169–173

Piton P (1985) Etude géochimique et géochronologique de la croûte continentale protérozoïque (1.8 à 2.2) au Nord du Massif armoricain. Diplôme, Rennes, 51 pp

Rabu D, Chauvel JJ, Chantraine J (1983) Nouvelles propositions pour la lithostratigraphie du Briovérien (Protérozoïque supérieur) et pour l'évolution géodynamique en baie de St Brieuc. Bull Soc Géol Fr XXV-4: 615–621

Roach RA (1977) A review of the Precambrian rocks of the British Variscides and their relationships with the Precambrian of NW France in: La chaîne varisque d'Europe moyenne et occidentale. Coll Int CNRS 243: 61–79

Roach RA, Lees GJ, Shufflebotham MM (1990) Brioverian volcanism and Cadomian tectonics, Baie de St Brieuc, Brittany. In: D'Lemos RS, Strachan RA, Topley CG (eds) The Cadomian Orogeny. Geol Soc Spec Publ 51: 41–67

Roblot MM (1962) Sur une coupe stratigraphique à l'intérieur du Briovérien. CR Acad Sci Paris 254: 3720–3722

Verdier (1968) Etude pétrographique et structurale du Trégor occidental (baie de Lannion). Thèse 3e cycle, Strasbourg, 51 pp

Vidal P (1976) L'évolution polyorogénique du Massif armoricain: apport de la géochronologie et de la géochimie isotopique du strontium. Thèse d'Etat, Rennes. Soc Géol Min Bretagne 21 (1980), Rennes, 162 pp

Vidal P, Auvray B, Chauvet JF, Cogné J (1972) L'âge radiométrique de la diorite de Saint-Quay-Portrieux – ses conséquences sur le Briovérien de la baie de Saint-Brieuc. CR Acad Sci Paris 279: 631–634

Vidal P, Deutsch S, Martineau F, Cogné J (1974) Nouvelles données radiométriques en Baie du Saint-Brieuc. Le problème d'un socle antécadomien nord armoricain. CR Acad Sci Paris 279: 631–634

Vidal P, Auvray B, Charlot R, Cogné J (1981) Precadomian relicts in the Armorican massif: their age and role in the evolution of the western and central cadomian-hercynian belt. Precambrian Res 14: 1–20

Structure and Metamorphism

D. Rabu, J. P. Brun, and J. Chantraine (translated by M. S. N. Carpenter)

1 Introduction

The Brioverian formations of the Armorican Massif have been involved in orogenic events generated by the plate tectonic regime during latest Proterozoic-Early Cambrian times; these gave rise to the Cadomian fold-belt which is contemporaneous with the Pan-African (sensu lato) and Avalonian orogenies.

The deformational and metamorphic features of the Cadomian orogen are well-exposed in the northern part of the Armorican Massif where Variscan reworking has not strongly affected the Brioverian formations. The only Variscan effects are apparent in the reactivation of some faults within the Cadomian domain. The area affected by Cadomian deformation is bounded by intracrustal shear zones, in the vicinity of which the influence of Variscan deformation becomes dominant. Locally, it is still difficult to clearly distinguish Cadomian from Variscan effects.

2 The Orogenic Cordillera

Cadomian orogenesis occurred during a single major tectono-metamorphic event, but several phases of deformation can be observed. The Brioverian formations display heterogeneous deformation and a metamorphic grade generally increasing from NW to SE. Deformation has produced well-developed foliation, generally steeply dipping, which is associated with isoclinal-type folds showing local shearing on their limbs. A stretching lineation is more or less well marked but is always present.

It shows a complex spatial distribution which is consistent with sinistral wrenching combined with southwestward thrusting (Balé and Brun 1983). The southern limit of the orogenic cordillera is marked by a major transverse thrust fault (Main Cadomian Thrust: MCT) which brings the volcanic arc and back-arc basin domains into contact with the continental margin autochthon.

Even within the orogenic cordillera, important combined thrust-wrench faults separate the various structural units (as, for example, the boundary between the volcanic arc and the back-arc basin; South Trégor Thrust: STT). Metamorphic parageneses vary from greenschist to amphibolite facies, but the pressure/temperature conditions during metamorphism are still poorly known.

The characteristics of deformation and metamorphism are described on the two type sections of Figs. 9 and 10:

- the Baie de Lannion section concerns the volcanic arc domain and its relations with the Icartian basement (Verdier 1968; Autran et al. 1979; Balé 1986);
- the Baie de Saint-Brieuc section illustrates the back-arc basin and its relations with the volcanic arc domain to the north (Jeannette 1972; Balé and Brun 1983; Rabu et al. 1983; Balé 1986).

2.1 The Baie de Lannion Section

2.1.1 Deformation

Formations belonging to the Icartian basement (Moulin-de-la-Rive orthogneiss) and the Brioverian (Locquirec-Tréguier, Rugunay-Malabri, Armorique-Trédrez, Plestin-les-Grèves, and Bois-de-la-Roche Formations) are all affected by the same deformation which gives rise to a strong foliation. This foliation is always well developed, but with variable aspects according to the lithological features. It is also remarkably well developed in the acid tuffs of Locquirec-Tréguier and rather less developed in the basic units of the Armorique-Trédrez Formation. Sub-parallel to the sedimentary banding, it strikes N50° E on the west side of the bay and N70° E on the east (Fig. 9 A); in both cases, the strata are steeply dipping (70° to the north). In thinly bedded units, the foliation is associated with generally isoclinal folds which become more open in the more thickly bedded rocks. In massive rocks (e. g., basement orthogneisses, massive

J. D. Keppie (Ed.): Pre-Mesozoic Geology in France and Related Areas

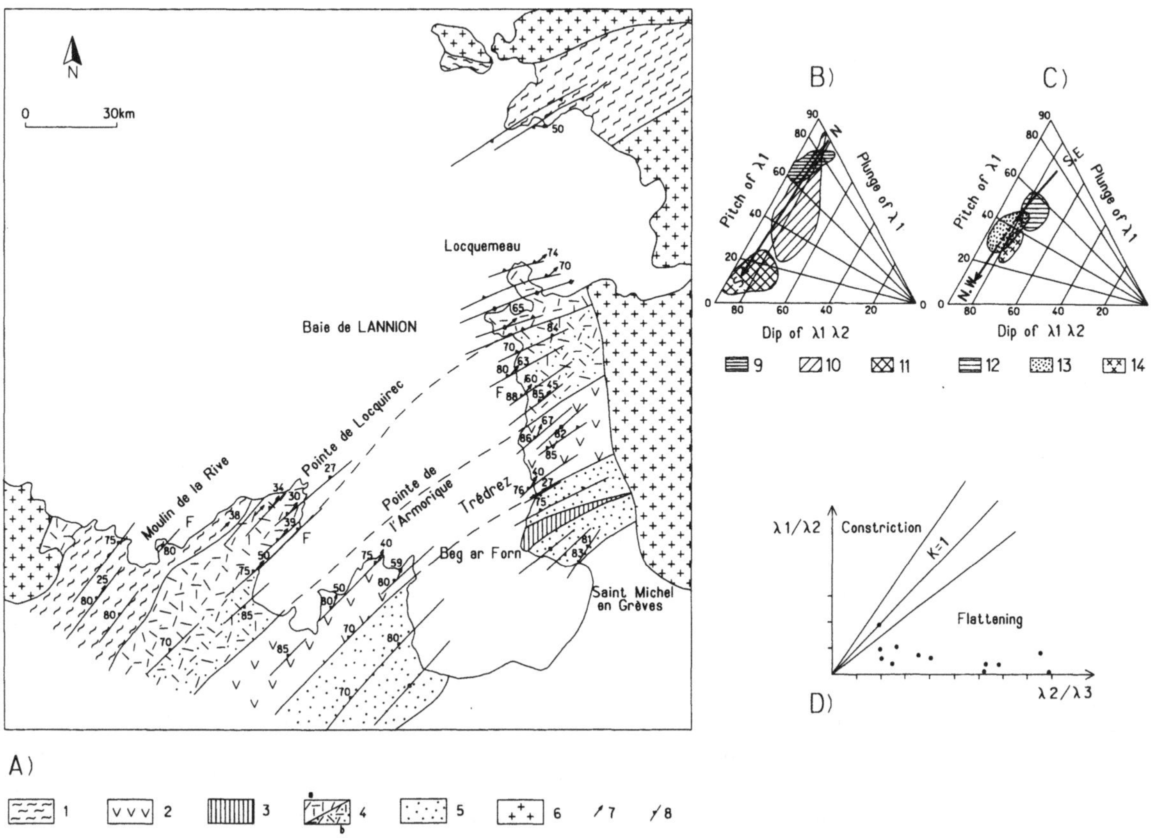

Fig. 9 A–C. Deformation of the volcanic arc domain (Baie de Lannion section). (After Autran et al. 1979; Balé 1986). **A** Geological sketch map. *1* Moulin-de-la-Rive orthogneiss (Icartian basement); *2* Armorique-Trédez Fm (metabasalts and metasediments); *3* Beg-Ar-Forn granodiorite; *4* Locquirec-Tréguir and Rugunay-Malabri Fms (*4a* keratophyric tuff; *4b* volcaniclastic units); *5* Plestin les-Grèves Fm (wackes); *6* Variscan granites; *7* stretching lineation, showing plunge; *8* foliation, showing dip. **B, C** Stretching lineation and foliation on the east and west sides of the bay (*arrows* show trends of increasing deformation). Locality key: *9* Locquémeau; *10* Trédez area; *11* Saint-Michel-en-Grèves; *12* Pointe de l'Armorique; *13* Pointe de Locquire; *14* Moulin-de-la-Rive. **D** Measurements of finite strain ellipsoids potted on a Flinn diagram

sills, and lavas) shear-bands envelop slightly deformed amygdaloidal structures containing preserved original textures.

Stretching lineations are present, although the analysis of various deformational markers (pressure-shadows, conglomerate pebbles, pillow-lavas) indicates a finite strain ellipsoid of the flattening type (i.e., $K < 1$ in the Flinn diagram; see Fig. 9D). The stretching lineations show a constant easterly plunge with, however, a fairly large scatter in the plane of schistosity.

On the eastern side of the bay (Trédrez coast), stretching lineation plunge (70° E) increases towards the north with the intensity of deformation (Fig. 9B). The shear component is not very strong, but shear sense can be estimated in many places. Various criteria indicate sinistral wrenching and/or thrusting shear towards the SW in the Brioverian pile. The southern end of the section on the east side of the bay (south of Beg ar Forn) contrasts with the rest of the Baie de Lannion by the occurrence of an almost horizontal stretching lineation, dextral shear sense indicators, and deformation intensity increasing towards the south. These unusual features are considered to be the result of Variscan influence at the boundary of the Cadomian domain.

On the western side of the bay, deformation intensity increases towards the NW, associated with a foliation dip increase and stretching lineations which become more horizontal (30° E) (Fig. 9C). The rotation of pre-tectonic markers and the existence of sinistral shear bands show that the deformation was noncoaxial. The overall pattern of deformation in this part of the bay can be interpreted in terms of a sinistral wrenching shear component in the vicinity of the boundary with Icartian basement. In the basement itself, deformation is mainly represented by conjugate shear bands, with a predominance of sinistral bands consistent with the tectonic style recorded in the Brioverian succession.

A second strain-slip cleavage (N30° E), without metamorphic recrystallization, is locally apparent on both eastern and western sides of the Baie de Lannion. Apparently not related to larger-scale structures disturbing the regional Cadomian tectonic pattern, this strain-slip cleavage could reflect the influence of Variscan effects at the boundary of the Cadomian domain.

2.1.2 Metamorphism

Metamorphic parageneses do not appreciably vary from one end of the Baie de Lannion section to the other. The increasing of deformation intensity towards the south is accompanied by only a slight change in metamorphic recrystallization phenomena. However, on the eastern side of the bay, the features of regional metamorphism are largely obliterated by mineral recrystallization brought about by the intrusion of the Variscan Trédrez granite (Fig. 9 A).

Apart from this contact zone, and mainly on the western side of the bay, Cadomian parageneses are medium grade (biotite zone), belonging for the most part to the greenschist facies with the following mineral assemblages: biotite, muscovite, chlorite, calcite, epidote, and albite in the meta-acid rocks; actinolite, chlorite, epidote, calcite, sphene, and albite in the metabasics.

Some recent interpretations (Lees et al. 1987) have proposed that the Locquirec-Tréguier Formation in the Baie de Lannion might represent tectonites derived at the expense of the Moulin-de-la-Rive orthogneiss. The lithostratigraphic and structural work carried out here over many years (Verdier 1968; Autran et al. 1979; Balé 1986) shows that this hypothesis is not valid. Firstly, original igneous textures are equally well preserved in the acid rocks (Locquirec-Tréguier Formation) as in the intermediate (Rugunay-Malabri Formation) or basic rocks (Armorique-Trédrez Formation). Secondly, the evolution of deformation intensity (Balé 1986) clearly shows that, on the west side of the bay, deformation increases in importance continuously right across the entire volcano-sedimentary pile; it is not restricted to a blastomylonitic zone at the edge of the Icartian basement as proposed previously.

2.1.3 Age of Deformation

The age of the Cadomian tectono-metamorphic event in the Baie de Lannion is bracketed by two dates: 586 ± 18 Ma (Charlot, in: Autran et al. 1979), obtained on acid veins intruded into the Amorique-Trédrez Formation, the veins being themselves metamorphosed and affected by the deformation; 528 ± 6 Ma (Guerrot 1989) obtained on zircons from the Beg-ar-Forn granodiorite (Fig. 9 A) which post-dates deformation and metamorphism in the Armorique-Trédrez Formation.

2.2 The Baie de Saint-Brieuc Section

On its western flank, the Baie de Saint-Brieuc presents a remarkable cross-section through the axial zone of the Cadomian orogen. It is made up from the superposition of two main thrust slices, the Trégor Unit in the north and the Saint-Brieuc Unit in the south. These imbricate thrust units are superimposed onto the paleogeographic pattern; in this way, the Saint-Brieuc section reveals the structural relationships between the volcanic arc domain, the back-arc basin and the passive continental margin (Figs. 10 and 11). The thrusting is post-dated by the late intrabelt basin developed within the Trégor Unit. The major tectonic contacts which border these two main units (i.e., South Trégor Thrust: STT; Main Cadomian Thrust: MCT) show a SW vergence. The thrust slice units are defined as follows:

The Trégor Unit (= volcanic arc domain) is bounded on its northern side by the North Trégor Batholith and on the south by the wrench-fault at Plage du Palus (Balé 1988) (Fig. 10 A). It is formed from the juxtaposition of several sub-units, each showing a particular style of deformation. The late intrabelt basin – occurring between Paimpol and Bréhec – is filled with volcanic slightly deformed products (Paimpol spilites) and sediments (Minard, La Roche-Derrien, and Plourivo-Bréhec Formations) that belong partly to the Brioverian Supergroup. This basin is bounded by faults which trend N50–70° and N140°.

The Saint-Brieuc Unit (= back-arc basin) crops out over most of the western part of the Baie de Saint-Brieuc from the wrench-fault at Plage du Palus to just south of Yffiniac, where it is thrust over the Lamballe and La Fresnaye Formations (Fig. 10 A) which likely represent the passive continental margin.

2.2.1 Deformation and Metamorphism in the Trégor Unit

On the northern side of the late intrabelt basin, the Tréguier tuffs show a foliation, steeply dipping to the south, which is of variable intensity (Auvray 1979) and which contains nearly horizontal stretching lineations. South of the late intrabelt basin, two struc-

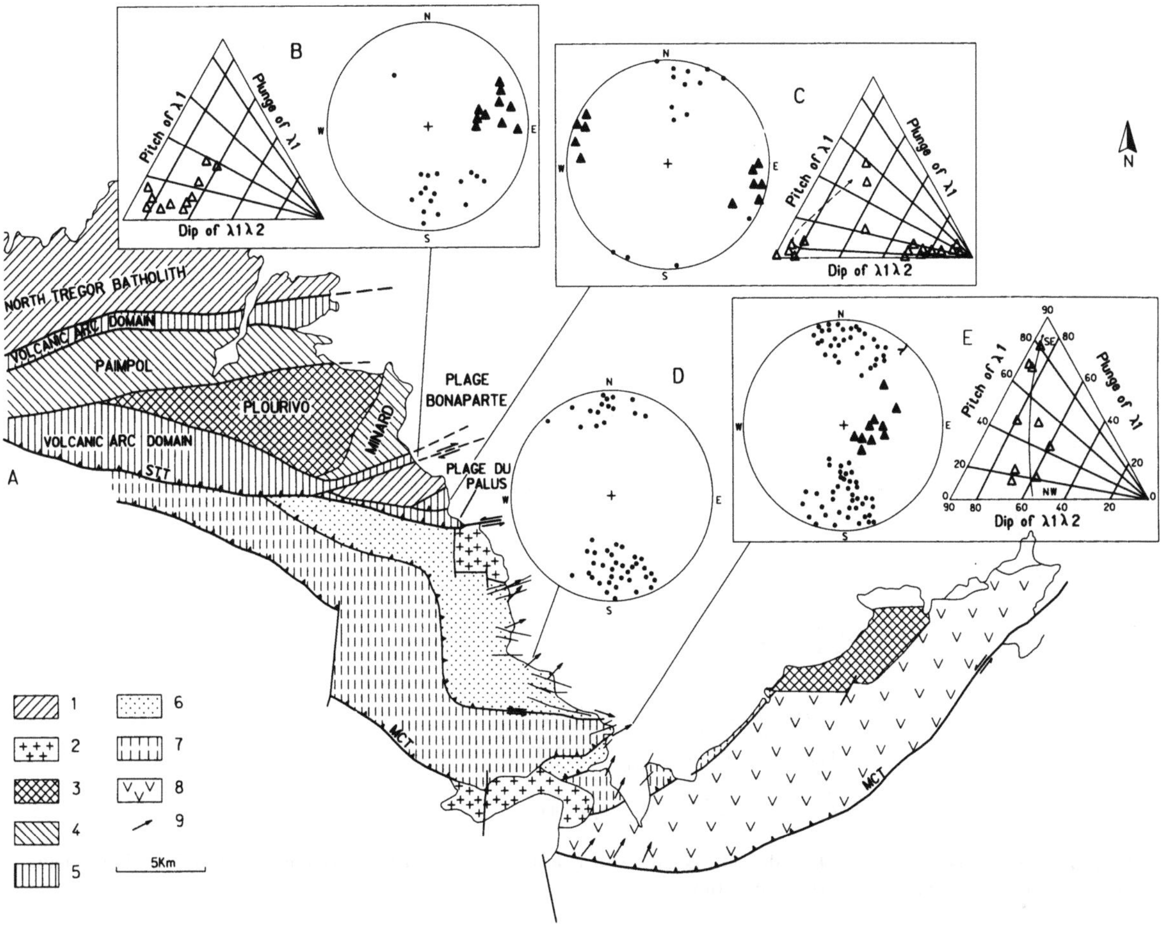

Fig. 10 A–E. Deformation in the back-arc basin (Baie de Saint Brieuc section). (After Balé and Brun 1983; Rabu et al. 1983; Balé 1986). **A** Geological sketch map. *1* North Trégor Batholith; *2* syn- to late-orogenic granodiorites and quartz-diorites; *3* Plourivo and Erquy Fms; *4* Paimpol-La Roche Derrien-Minard Fm; *5* Locquirec-Tréguier Fm; *6* Binic-Le Légué Fm; *7* Lanvollon Fm; *8* Yffiniac Fm; *9* stretching lineation **B, C, D, E** stretching lineation (▲) and foliation (●) and foliation-lineation (△): stereographic projection (lower hemisphere) and geometrical variation

tural sections are recognized (Balé 1986, 1988; Balé and Brun 1989) (Fig. 10 A):

- in the north, the Plage Bonaparte section bordering the southern limit of the late intrabelt basin and extending southward as far as the Plouha granite;
- in the south, the Plage du Palus section cropping out between the Plouha granite and the wrench-fault of Plage du Palus.

The Plage du Palus section is characterized by a main phase of thrusting directed southward. Two superposed units (a lower volcanic sequence and an upper volcano-sedimentary complex) are separated by a shallow dipping contact almost parallel to the main foliation (striking N80° E, dipping 30° N). The foliation is associated with sheath folds developed under medium-grade conditions and contains a sub-horizontal stretching lineation (Fig. 10 C). In both units, the shear criteria are consistent, indicating a relative translation towards the SW. The lower unit shows a single phase of deformation and is composed of acid and basic metavolcanics similar to those of the Armorique-Trédrez Formation in the Baie de Lannion. The uppermost unit shows multiphase deformation and may represent a slice of Icartian basement (Balé 1986); however, this correlation remains to be established.

Towards the south, the foliation planes steepen to vertical, but stretching lineations generally remain close to horizontal (Figs. 10 C). Shear criteria here indicate a sinistral wrenching (Plage du Palus wrench). The intensity of deformation measured in this zone (Balé 1988) is incompatible with the degree of shearing needed to generate syn-metamorphic sheath folds. So the sinistral wrenching of the Plage du Palus is probably superimposed on the thrust front devel-

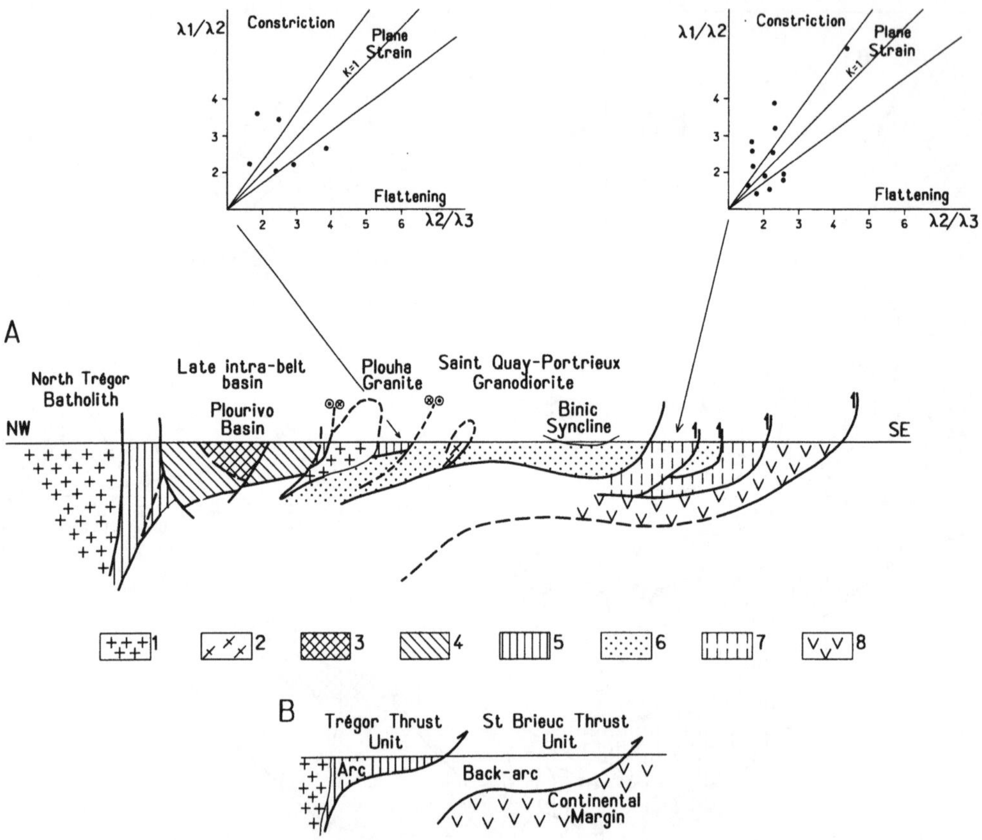

Fig. 11. A Geological cross-section of Baie de Saint-Brieuc, with measurements of finite strain ellipsoids plotted on a Flinn diagram (captions, see Fig. 10). **B** Schematic structural section

oped at an earlier stage between the volcanic arc and the back-arc basin. Measurements of the finite strain ellipsoid demonstrate that the K value is close to 1 (plane strain). The sinistral wrenching is synchronous with emplacement of the Saint-Quay-Portrieux pluton into the back-arc basin succession (Binic-Le Légué Formation) and also with the main phase of deformation that accompanied closure of this basin (see below). This ductile deformation which combines thrusting and wrenching in space and time, was developed at a medium metamorphic grade in the amphibolite facies (co-existence of biotite and hornblende). However, in the metamorphic contact aureola of Saint-Quay-Portrieux pluton, the lithology of the Binic-Le Légué Formation was profoundly modified (Rabu et al. 1983) and staurolite-garnet-cordierite gneisses occur.

The Plage Bonaparte section includes the Plouha Granite in the south and an acid/basic volcanic sequence in the north similar to the Armorique-Trédrez Formation. Both the granite and the volcanics are affected by a sub-vertical foliation containing a sub-horizontal stretching lineation resulting from dextral strike-slip shearing (Fig. 10B). The associated mineral assemblages indicate a lower grade of metamorphic conditions than that estimated for Plage du Palus. Biotite is stable at an early stage, but is replaced by chlorite at later stages of metamorphism. The dextral ductile shearing in the Plage Bonaparte terrain would thus appear to be a late Cadomian event possible related to the opening of the late intrabelt basin (i.e., "pull-apart" basin) and superimposed on the major early structures.

2.2.2 Deformation and Metamorphism in the Saint-Brieuc Unit

In the thrust slice of Saint-Brieuc, the main phase of Cadomian deformation has affected the Lanvollon amphibolites and the metasedimentary succession of Binic-Le Légué (Fig. 10 A). The Yffiniac complex constitutes the plutonic sole to the thrust sheet (belonging to the back-arc basin) during its translation onto the passive continental margin. These three lithostratigraphic units have been subjected to deformation that increases in intensity towards the south (Balé and Brun 1983; Rabu et al. 1983; Strachan and Roach 1990), with the exception of local effects developed around the syn-kinematic Saint-Quay-Portrieux intrusion in the northern part of the back-arc basin. The foliation is always steeply dipping and grades from fracture cleavage in low-grade zones of the north (Binic area, Fig. 10D) to metamorphic banding in the medium-grade zones of the south (Fig. 10E) near the contact with the Yffiniac complex. Here, the parageneses indicate intense tectono-metamorphic activity under upper amphibolite facies conditions.

The cleavage planes contain stretching lineations that are particularly well developed in areas of medium-grade metamorphism beneath the biotite isograd. The orientation of these lineations in the plane of foliation has enabled the recognition of areas with a sinistral wrenching component and areas showing a thrusting component towards the SW. The strain is of the planar ($K = 1$) or constrictive type ($K > 1$) according to intensity (Fig. 11 A). Apart from the general increase in deformational intensity from north to south, some local gradients are associated with internal contacts between the three lithological formations; these faulted contacts have thus acted as surfaces of preferential detachment during the closure of the basin.

A major discontinuity in the tectonic pile (Fig. 11 A) places the high-grade metamorphic complex of Yffiniac in contact with weakly metamorphosed and little deformed sediments and volcanic rocks (Lamballe and La Fresnaye Formations). This contact has the characteristics of a thrust at the south end of the Baie de Saint-Brieuc, but acquires a sinistral wrenching component towards the NW (Baie de la Fresnaye). Cadomian deformation in the back-arc basin has resulted mainly from the imbrication of thrusts directed towards the SW and their passage into ductile sinistral shears at the edge of the domain. The combinations of thrusting and wrenching, respectively along frontal and lateral ramps, are responsible for the arcuate pattern of the Cadomian belt at the regional scale (Balé and Brun 1983, 1989). On a larger scale it could indicate an oblique convergence environment (Brun and Balé 1990).

The basic/ultrabasic complex of Belle-Ile-en-Terre (Hirbec 1979) is situated inland SW of Saint-Brieuc; it is overthrust onto a sequence containing interbedded black cherts known locally as the Callac Formation (cf. Lamballe Formation). This basic/ultrabasic complex can be integrated into the regional geodynamic model and then would represent an isolated klippe of the orogenic cordillera lying on top of the continental margin domain. At the present day, the Guingamp migmatites lie structurally beneath this klippe and may be explained by crustal thickening caused by the piling of nappes.

2.3 Sections in the Cotentin

At present, the orogenic belt in the Cotentin appears to be divided into two branches by a late intrabelt basin; the southern branch corresponds to the Coutances-Montsurvent terrain and the northern to the Cherbourg-Val-de-Saire terrain (Fig. 5) (Dupret 1984).

2.3.1 Deformation and Metamorphism in the Coutances-Montsurvent Terrain

Here, Cadomian deformation has affected a volcanic sequence (Montsurvent Formation) as well as a plutonic complex (Coutances Diorite).

The Montsurvent Formation (basic/acid volcanics and volcanosedimentary units) has been affected by syn-metamorphic deformation which increases in intensity towards the south. In the weakly metamorphosed northern area, foliation is particularly well developed in the acid members. This foliation, steeply dipping and striking N70°–80° E, locally contains a steeply plunging (40–60°) stretching lineation (Bardy 1988), defined by amphibole and/or plagioclase patches. The associated folds are overturned towards the south: although shear criteria are rather uncommon, they indicate a top to the south shearing.

Farther south, in the vicinity of the Coutances diorite, the volcanics and volcaniclastic sediments of Montsurvent have been intensely metamorphosed to form banded amphibolites with pyroxene and garnet, micachists, and gneisses with hornblende and garnet and metaquartzites. These rocks show parageneses indicative of a multi-stage regional metamorphism. The first phase of prograde syn-tectonic metamorphism is characterized by the assemblage of hornblende, pyroxene, garnet, labrador-andesine, biotite, and epidote. This is followed by a late retrogressive phase which is characterized by the assemblage of chlorite, epidote, actinolite, and andesine-oligoclase. The evolution of metamorphic conditions is represented by a pressure-temperature-time path, consistent with a thrusting model, under amphibolite facies conditions (Bardy 1988). The metamorphic foliation strikes N30–40° E and dips southeastward at 30–40°. The stretching lineation defined by hornblende and pyroxene is oriented N130° E with variable plunges (40–70°). Shear criteria, particularly the asymmetry of pressure shadows, indicate a top to the south shearing. A late-stage rough cleavage defined by chlorite and actinolite is seen to cut the main foliation.

The Coutances diorite is bounded on its southern flank by a southward trending thrust fault. According to Dupret (1984), the syn-kinematic pluton was intruded into a Cadomian shear-zone trending ENE-WSW and developed a contact metamorphism in the Montsurvent Formation. According to Bardy (1988), the Coutances diorite is allochthonous with its root zone situated several kilometers to the north; the massif would have been emplaced during a phase of southward-directed tangential deformation that also affected the Montsurvent Formation.

The Coutances diorite displays planar-linear fabrics – foliation and stretching lineation on the northern and mylonites on the southern flank of the massif – which are nevertheless present throughout the core of the massif. Along the northern contact, the diorite lies above the Montsurvent Formation metamorphosed in the lower amphibolite facies. Along the southern contact, the diorite lies above weakly metamorphosed and little deformed sediments and volcanic rocks (Saint-Lô and La Terrette Formations). These units have been affected by a steeply dipping foliation striking N70° E which contains scattered stretching lineations trending N-S (Bardy 1988).

The foliation in the diorite along the northern and southern margins is concordant with major lithological boundaries, thus suggesting a tectonic control rather than simply intrusive (Bardy 1988). The N-S stretching lineation, defined by amphibole phenocrysts, indicates a relative movement from the north towards the south; this direction of transport is confirmed by shear criteria observed in pressure shadows and on C/S planes (or C' in more highly deformed zones). However, foliation trajectories and their obliquity with respect to major contacts suggest – at least for the last strain increments – a dextral wrench component. Even if the development of internal fabrics within the Coutances Diorite during a regional thrusting event is established, the allochtonous nature of the massif remains to be demonstrated.

Dated at 584 ± 5 Ma (Guerrot et al. 1989), the Coutances diorite can be placed in the same age bracket as the Coëtmieux-Fort-la-Latte tonalite of North Brittany (U/Pb zircon age of 593 ± 15 Ma; Vidal et al. 1974). Syn-kinematic character of their emplacement fix the age of the deformation for this domain of the Cotentin.

2.3.2 Deformation and Metamorphism in the Cherbourg-Val de Saire Terrane

According to Dissler (1987), the main deformation affecting the Brioverian rocks of the Val-de-Saire (Cherbourg area) can be assigned to the Variscan deformation. However, pre-existing Cadomian deformation is strongly marked in various places by a foliation which contains a stretching lineation. The attitude of the stretching lineation on the foliation planes is variable (e.g., from horizontal to vertical on the Plage de Breteville-en-Saire section) without any relation to Variscan tectonics (open upright folds). These elements are more compatible with a deformation combining thrusting and wrenching.

3 The Continental Margin

The Continental Margin domain constitutes a fairly narrow belt characterized by the presence of interbedded black chert members and the particular geochemistry of the associated volcanism. These units were deformed during the Cadomian orogeny and their style of deformation can best be described through the study of two type-sections (Figs. 12 and 13)

- the Saint-Malo-Saint-Cast Section (N. Brittany);
- the Coutances-Granville Section (S. Cotentin).

3.1 The Saint-Malo-Saint-Cast Section

From a study of the Saint-Malo-Saint-Cast section, the Brioverian rocks of this area have been interpreted as representing the passive continental margin external to the Cadomian orogenic belt. To the northwest, the margin is overthrust by the back-arc basin and towards the southeast it is in contact with the continental platform domain (Mancellia). The exact location and nature of this latter contact is currently being studied.

The main tectonic unit traversed in the segment is broadly composed of two associations (Fig. 12 A, B)

- a mainly sedimentary and volcanic sequence (Lamballe, La Fresnaye, and Saint-Cast Formations) which is generally weakly metamorphosed, showing a concentration of tectono-metamorphic effects into corridors oriented N50° E (Balé 1986; Balé and Brun 1989);
- a group of migmatites and mica schists (Saint-Malo Massif; Brun 1977; Brown 1979) formed during Cadomian deformation and post-dated by late-stage intrusions at 540 Ma (Peucat 1982).

On the whole, the sedimentary and volcanic sequence is little deformed and poorly metamorphosed. In contact with the gneisses, meta-gabbros, and amphibolites of Yffiniac (cf. Saint-Brieuc section), the volcanic La Fresnaye Formation shows low-grade assemblages of actinolite, chlorite, albite, and epidote (Jeannette 1972; Rabu et al. 1983). The sedimentary

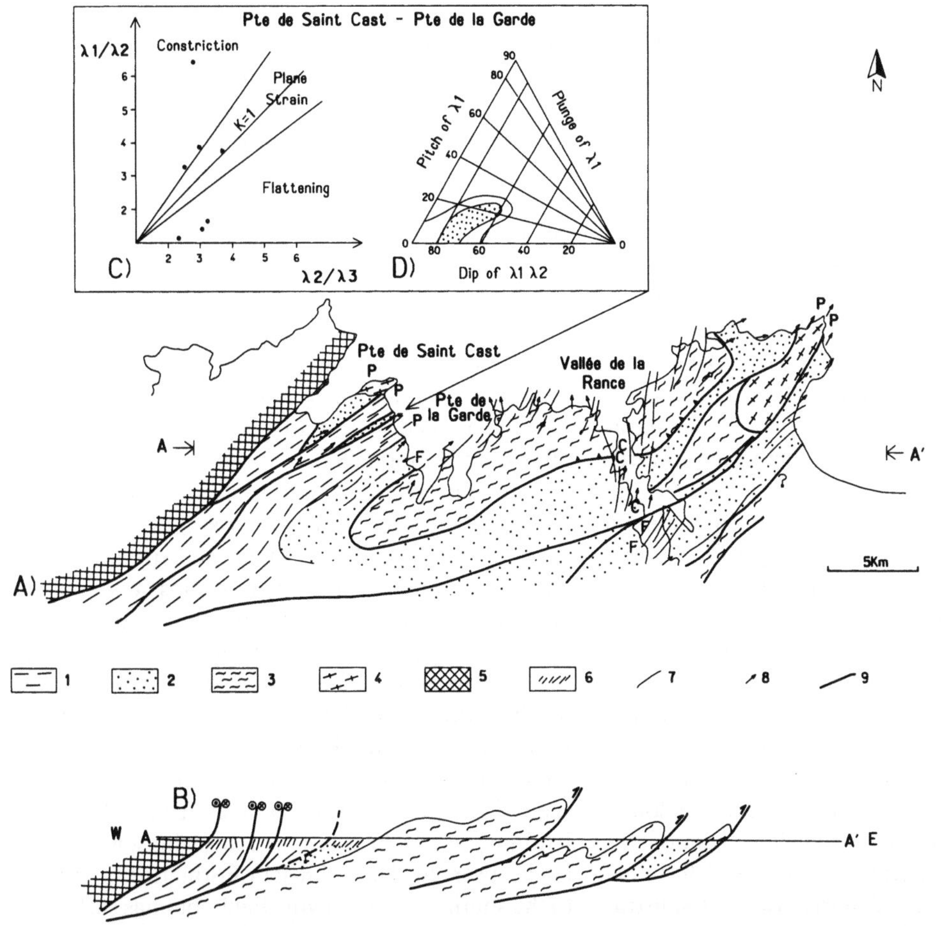

Fig. 12. Deformation in the passive continental margin (Saint-Cast-Saint-Malo section). (After Brun 1975; Balé 1986; Balé and Brun 1989). **A** Geological sketch map. **B** Structural section. **C** Measurements of finite deformation ellipsoid plotted on a Flinn diagram. **D** Stretching lineation and foliation – geometrical variation; *dotted area* shows greatest concentration of measurements. General key: *1* Lamballe Formation with interbedded black cherts and Baie de la Fresnaye volcanics; *2* gneisses and micaschists; *3* migmatites; *4* granites; *5* Yffiniac basic gneiss complex; *6* attitude of foliation; *7* trend of foliation; *8* stretching lineation; *9* thrust and wrench faults; *P* plane strain; *F* flattening; *C* constriction

Lamballe Formation is affected by low-grade metamorphism (parageneses including chlorite, albite, and epidote) associated with a steeply dipping foliation sometimes slightly overturned to the SE. Towards the southeast, the metamorphic grade increases and the schists and black cherts of the Lamballe Formation display lower amphibolite facies assemblages (biotite-muscovite micaschists). A break in metamorphic zonation is marked out by the presence of sheets of orthogneiss that were intensely deformed during the Cadomian orogeny. The conditions of outcrop on the coastal part of the section enable a more detailed appreciation of the structure. Finite strain in this area (Balé 1986; Treloar and Strachan 1990) results from a sinistral wrenching (e.g., Pointe de Saint-Cast and Pointe de la Garde, Fig. 12 A), associated with the up-doming of the Saint-Malo migmatite belt.

According to Balé (1986) and Balé and Brun (1989), this wrenching event was preceded by a previous thrusting towards the SW (evidence preserved in the Vallée de la Rance section, Fig. 12 A). This earlier stage of deformation was responsible for the thickening and the probably related partial melting. The associated penetrative deformation includes a N40–50° E-trending metamorphic foliation and stretching lineations plunging slightly towards the NE (Fig. 12 D). Within the shear zones, deformation is essentially of the constriction type ($K = 1.5 - 3$, $L > S$), whereas nearer the migmatites the strain is taken up by flattening ($K = 0.3$, $S > L$) (Fig. 12 C). Criteria of noncoaxial deformation are numerous, indicating a sinistral wrenching. However, there are some late-stage movements, also affecting the migmatites, which are developed in narrow dextral shear belts.

These latter are probably equivalent to shear zones of the same type recognized in the orogenic cordillera (e.g., Plage Bonaparte) which are linked to the opening of the late intrabelt basin. Bordered by the Saint-Cast and Cancale sinistral wrench zone, the Saint-Malo Massif, with its migmatitic core, represents a gneiss dome (Brun 1975, 1977) emplaced during multi-phase deformation events and post-dated by the intrusion of late-stage anatectic granites. The Vallée de la Rance section cuts across the metamorphic envelope to the gneiss dome and includes low-grade rocks at the periphery which may belong to the continental margin domain. Since there is no major structural discontinuity in this succession, a continuous metamorphic and deformational gradient increasing towards the core of the dome has been proposed (Brun and Martin 1978). The folds developed during the main phase of deformation (corresponding to maximum anatexis) are seen to change from overfolds within the internal zone to upright or slightly overturned (toward the SE) folds in the more external parts of the dome towards the SE. The evolution in structural style is accompanied by a modification in strain regime from constriction ($K > 1$) in the internal domain to flatttening in the external domain ($K = 0.3$); this is a consequence of the diapiric rise of the dome. The re-orientation of early stretching lineations in shearing structures is interpreted in terms of the interaction between thrusting/diapiric rise and sinistral wrenching (Brun and Balé 1990).

Even though the migmatite bodies are very often bounded by NE-SW-directed shear zones (Balé 1986; Balé and Brun 1989), no structural or metamorphic discontinuity appears to mark the passage between the continental margin and the platform. A recent study (Darlet et al. 1990) of formations assigned to the platform domain in the southeast part of Vallée de la Rance section (Fig. 12 A) has revealed a clearly volcaniclastic character and the absence of any black chert clasts. Therefore, these facies are very different from those usually described from the continental platform (Chantraine et al. 1988; Rabu et al. 1990). Because of this, the boundary between the two domains is not defined with any precision in North Brittany – perhaps it should be sought still farther to the SE.

3.2 The Coutances-Granville Section

In the South Cotentin area, the continental margin domain is represented by the bedded-chert-bearing formation of Saint-Lô, the volcanic formation of La Terrette (Le Gall et al. 1986), and the immature sediments of Rampan and Saint-Pair (Dupret 1984). This domain is bounded on the north by the southward-thrusted Coutances Diorite and on the south by a major fault zone which runs into the Granville Fault (Fig. 13). In fact, there are some grounds – as in Northern Brittany – to place the southern limit farther south since outcrops of the Saint-Pair Formation occur to the south of the Granville Fault. The latter fault was active during the Cadomian orogeny; on the one hand, it has thrusted the Saint-Pair Formation towards the south (Dupret 1974) and, on the other hand, it has operated as a ductile sinistral wrench fault which has deformed part of the Mancellian Batholith (Bardy 1988). Hercynian reactivation of this fault is also to be noted since it affects Paleozoic formations (Bardy and Lefort 1987).

The main phase of deformation is characterized near Coutances by a foliation striking N70–80° E and dipping steeply to the north; at the extreme eastern end of the terrain, the foliation dips to the south (Chantraine et al. 1982). The foliation planes display stretching lineations which vary across the domain; in the Coutances area, lineations plunge steeply to the north, indicating a relative displacement towards the south. Farther east, subhorizontal lineations are E-W indicating strike-slip movements. In this manner, the same duality of structural style as seen in Northern Brittany, i.e., the interaction of thrusting and wrenching is observed. The Saint-Pair Formation is situated to the south of the Granville Fault, but has suffered a comparable deformation to other parts of the continental margin domain. The foliation in these low-grade metamorphic rocks contains a streching lineation which is parallel to the axial planes of isoclinal folds (Dupret 1974).

3.2.1 *Age of Deformation*

No direct dating evidence exists for the South Cotentin part of the continental margin. The northern border of the Mancellian Domain was deformed at the same time as the Coutances Diorite and the Montsurvent Formation. It would appear that tectonic events in the South Cotentin were synchronous with deformation of the North Brittany marginal domain, where syn-kinematic intrusions are dated at 590 Ma and late-stage anatectic granites at 540 Ma.

4 The Continental Platform

The Continental Platform Domain is made up of a terrain which crops out extensively across the Bocage Normand. To the NW, it is bounded by the Granville

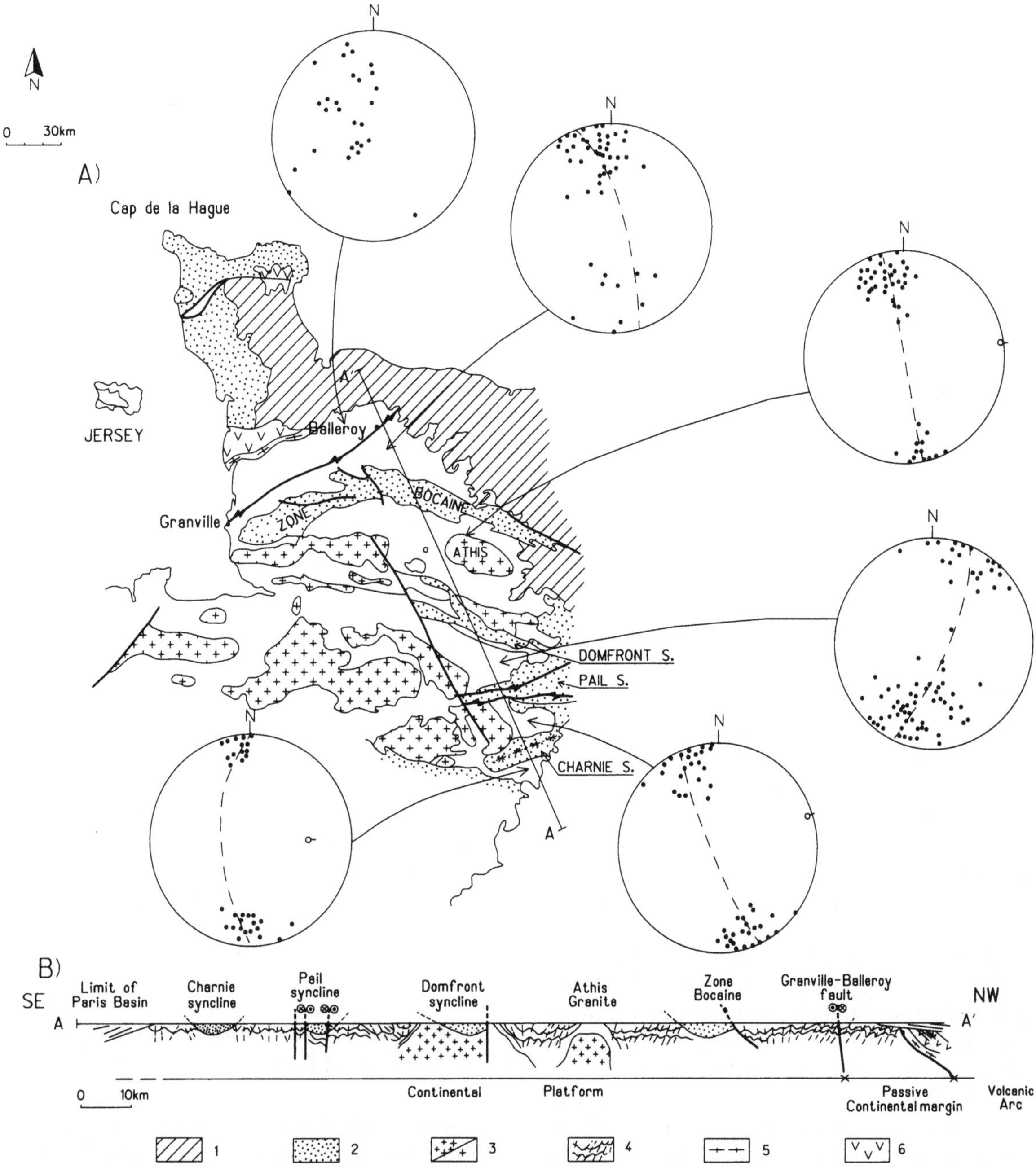

Fig. 13 A, B. Deformation in the continental platform (Mancellia section). (After Chantraine et al. 1982). **A** Geological sketch map and stereographic projection of poles of cleavage (●) – (○): measured of fold axis. **B** Structural section. *1* Post-Paleozoic cover; *2* Paleozoic; *3* late-tectonic granites; *4* Brioverian sediments, showing trend of bedding and attitude of cleavage; *5* Cadomian plutonic rocks; *6* Brioverian volcanics rocks; *A-A'* line of section

Fault in S Normandy and by the Saint-Malo Massif in N Brittany (Fig. 13 A). The North-Armorican Shear Zone (Variscan) form the southern limit of this domain. The continental platform is characterized by voluminous plutonism (Mancellian Batholith) which has developed a syn to late-tectonic contact metamorphism (Dissler et al. 1988). A NNW-SSE section across the eastern edge of the domain (Chantraine et al. 1982) (Fig. 13 B) is complemented by observations at various localities in the Granville Formation, which lies unconformably on formations of the continental margin domain.

On the type section shown as A-A' on Fig. 13 B, deformation has produced meso-scale folds with

N80–100° E-trending axes plunging very slightly to the east. They are associated with fairly widely spaced cleavage planes (fracture cleavage to slaty cleavage). No stretching lineations appear to be associated with this deformation. The intensity of deformation increases generally towards the north, accompanied by an increase in the degree of overturning of folds (transition to slaty cleavage and northward-facing overfolds to the north of the Domfront syncline). The boundary with the continental margin domain is marked by the Granville Fault; there is no major discontinuity in tectonic style, even though stretching lineations appear on cleavage planes north of the fault (Chantraine et al. 1982). South of the Domfront syncline the vertical axial planar cleavage is associated with upright open folds slightly overturned to the south. Throughout this domain, the metamorphic grade remains low or very low.

North of the boundary fault, near Granville, Dupret (1974) has described a formation (the Granville Formation) containing reworked black cherts and pebbles of plutonic rock, which is lithologically distinct from the underlying continental margin succession. The type of deformation in the Granville Formation is also quite different from the shales and bedded black cherts of the continental margin domain, a fact which is supported by the observation of a structural discordance at Quibou (Roblot 1962).

In fact, the Granville Formation is affected by open folding associated with a poorly developed cleavage, a tectonic style very different to that described for the continental margin formations. All these various observations can be integrated into a broad pattern in which the Granville Formation belongs to a late-orogenic succession rather like the deposits of the late intrabelt basin and younger in age than the sediments of the continental platform domain cropping out in the Bocage Normand.

In the Bocage Normand area, the age of main deformation is older than the intrusion of the Mancellian Batholith dated around 540 Ma. In Central Brittany, the major deformation is assigned to the Variscan orogeny (Le Corre 1977; Chantraine et al. 1982; Chauvel and Rabu 1984). The unconformity between the Brioverian formations and the Arenig red beds, as well as scattering of bedding-cleavage intersection lineations, is generally now interpreted (Ballard et al. 1986) as the result of Ordovician syn-sedimentary faulting.

5 The Late Intrabelt Basin

The late intrabelt basin is opened in the midst of the orogenic cordillera. It is bounded by shear zones acting as dextral strike-slip faults as described in several parts of the fold belt (e.g., Plage Bonaparte, Val-de-Saire). These shears were developed under very low-grade conditions (greenschist facies cataclasis). The sedimentary and volcanic rocks that fill this basin are weakly deformed. At the western end of the basin, the Paimpol volcanics and the La Roche-Derrien and Minard sediments are affected by meso-scale open folds and a fracture cleavage which is sometimes extremely rough (Auvray 1979; Denis 1987). On Jersey, the deformation is of very weak intensity; it has produced open folding almost without any associated cleavage (Helm 1983).

In the Val-de-Saire, the formations which characterize the late intrabelt basin are well separated from other Brioverian rocks assigned to the orogenic domain. These formations (Montaigu and Saussemesnil Formations) show a weak deformation and a lack of metamorphism; they display a widely spaced fracture cleavage without any neocrystallization or stretching lineation.

In the Trégor, deformation is bracketed between the eruption of the Lézardrieux ignimbrites (547 ± 12 Ma, Rb/Sr isochron; Auvray 1979) and the deposition of Plourivo volcanics and Frehel red beds of Ordovician age. On Jersey, this late-stage deformation occurred after the oldest phase of granitic plutonism dated at 553 ± 12 Ma (SW Granite, Rb/Sr isochron; Adams 1976).

6 Structural Zonation and Evolution of the Cadomian Belt

The segment of the Cadomian fold belt studied here is formed from the juxtaposition of structural zones, each characterized by its tectonic style, metamorphic facies, and associated igneous activity. The variations of these features within the different zones, and from one zone to another, enable us to propose a general geodynamic model for the evolution of the Cadomian belt in the Armorican Massif (Brun and Balé 1990; Rabu et al. 1990). (Table 1; Fig. 14).

The Zone of Thrusts and Sinistral Wrenches is characterized by the combination of thrusting with sinistral wrenching that re-utilizes some of the major thrusts initiated at earlier stages. Thrust displacements, which are also probably the earliest, are large-

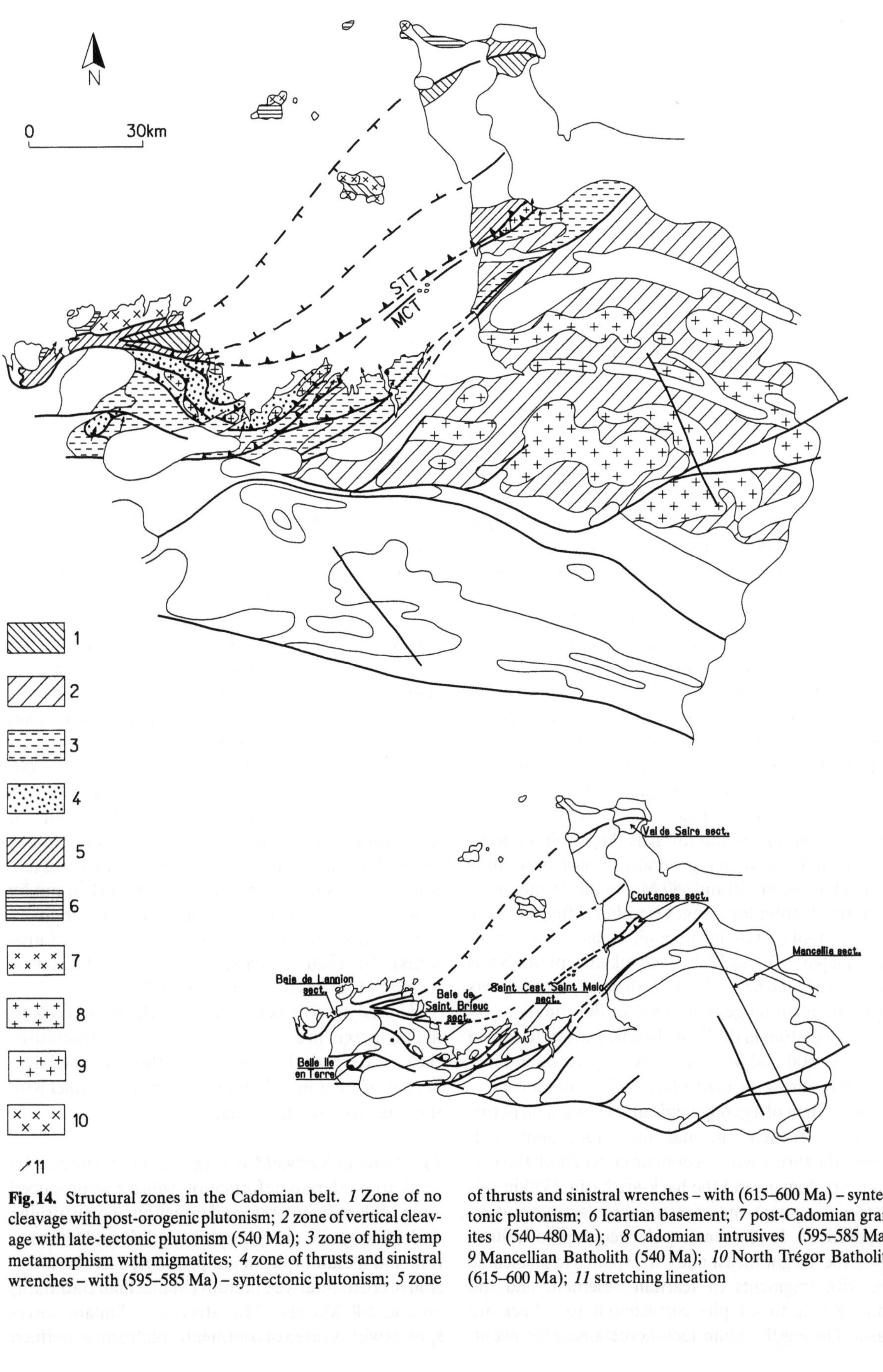

Fig. 14. Structural zones in the Cadomian belt. *1* Zone of no cleavage with post-orogenic plutonism; *2* zone of vertical cleavage with late-tectonic plutonism (540 Ma); *3* zone of high temp metamorphism with migmatites; *4* zone of thrusts and sinistral wrenches – with (595–585 Ma) – syntectonic plutonism; *5* zone of thrusts and sinistral wrenches – with (615–600 Ma) – syntectonic plutonism; *6* Icartian basement; *7* post-Cadomian granites (540–480 Ma); *8* Cadomian intrusives (595–585 Ma); *9* Mancellian Batholith (540 Ma); *10* North Trégor Batholith (615–600 Ma); *11* stretching lineation

Table 1. Structural and plutonic zonation of the Cadomian orogenic belt

Structural zones	Geodynamic setting	Tectonic style	Plutonic activity	Type and level of Intrusion	Age of Plutonism
Thrusts and sinistral wrench zones	volcanic arc active margin	thrusting towards the SW sinistral	North Tregor Batholith Guernsey Complex	syn-tectonic high level	615–600 Ma
	north: back-arc basin	shearing ductile	Saint-Quay-Portrieux Fort-la-Latte	shallow level syn-tectonic	595–585 Ma
High temp migmatite belt	south: passive continental margin	deformation medium to high grade metamorphism	Coutances Saint-Malo Guingamp migmatites	deep level migmatitic gneiss dome	540 Ma
vertical cleavage	continental platform	moderate flattening and tightening	Mancellian Batholith	late tectonic	540 Ma
No cleavage	late intrabelt basin	very weak flattening and tightening	Jersey granites	post tectonic "after-effects"	540–480 Ma

ly dominant over the entire zone while, wrenching is localized at the boundaries of the zone. This pattern is interpreted in terms of frontal and lateral ramps, where the interaction of two mechanisms has produced the acuate shape of the fold belt in Northern Brittany. Several structural units, separated by thrust contacts, have been piled up during the S- to SW-directed translation (i.e., Trégor and Saint-Brieuc thrust units). These major contacts usually correspond to important jumps in metamorphic grade and deformation intensity which normally increase progressively to the south or southwest. Metamorphism in the Saint-Brieuc Unit increases from low- to medium- or even high-grade towards the south. Plutonic activity in this domain is generally restricted to a time bracket between 620 and 580 Ma. In the details, however, the distribution of ages is rather different in the two different parts of he orogenic domain. In the most internal part (Trégor Unit), calc-alkaline plutonism is grouped around 620–600 Ma with no apparent effects from the main tangential phase of deformation. By contrast, plutons in the Saint-Brieuc Unit yield ages in the range 600–580 Ma; they were emplaced syn-kinematically during the main phase of deformation.

In the light of a geodynamic model based on petrographic, geochemical, and lithostratigraphic evidence, the thrust-wrench domain covers both the volcanic arc domain and the back-arc basin. Within this orogenic segment, major tectonic contacts are superimposed on fundamental paleogeographic boundaries: the Trégor thrust pile corresponds to a volcanic arc with fragments of Icartian basement and the Saint-Brieuc thrust pile corresponds to a back-arc basin. During the main tectonometamorphic event, the volcanic arc – which was probably already deformed at an early stage of the orogeny – was overthrust onto the back-arc basin, which itself came to be thrust onto the continental margin.

The Migmatitic Belt shows a deformation which also results from the interaction of tangential and transcurrent mechanisms. Nevertheless, the contribution of sinistral wrenching can locally appear predominant (Saint-Cast area), leading to an extensive re-orientation of previous structures. Partial melting in this domain can result from thickening brought about by early thrusting, although the displacement of migmatitic domes themselves along lateral ramps (Saint-Malo area) may be considered. The main tectonic contacts within this domain are marked out by outcrops of orthogneiss of indeterminate origin (? slices of Icartian basement or early Cadomian intrusions). The plutonism associated with the formation of migmatites has been dated at 540 Ma.

The migmatitic belt represents the passive continental margin in a paleogeographic reconstruction of the Celtic Ocean. It has been overthrust by the back-arc basin and is itself in possible tectonic contact with the external foreland domain.

The Zone of Vertical Cleavage has been affected by weak to moderate deformation which has produced open folds with no preferred direction of overturning under low-grade metamorphic conditions. Deformation and metamorphism are post-dated by the intrusion of crustal-derived plutons (Mancellian Batholith) around 540 Ma ago. This structural domain corresponds with an area of continental platform as defined in the paleogeographic reconstruction of the region. In comparison with the structural domains discussed above, the platform domain is clearly distinct, having a lower intensity of deformation and metamorphism as well as yielding younger ages. The development of voluminous plutonism in this area (Mancellian Batholith) is not actually understood from a structural point of view, but the crustal thickening in this domain could have been obtained by imbricate thrusting of the platform during the final stages of the main convergence which would have brought the orogenic cordillera onto the continental margin.

The Zone of No Cleavage is characterized by extremely weak deformation. A schistosity front is only observed in the older formations of the intrabelt basin; the metamorphism is very low grade. This zone corresponds to the outcrop of the late intrabelt basin, where the deposits are clearly post-orogenic with respect to the main Cadomian deformation. At the margins of the basin, the rocks are affected by a dextral shear deformation that is linked to a pull-apart opening mechanism. It should be noted that opening was initiated at a very early stage, since the dextral shears – recognized as late-stage structures in various parts of the transverse thrusting zone – at first acted under greenschist facies conditions which then changed to a cold brittle regime towards the end of faulting. The sediments accumulated in this basin record several tectonic episodes after the main Cadomian event. It is possible that the syn-schistosity deformation observed in the older formations of the basin may be contemporaneous with the cleavage in the vertical cleavage zone, since undeformed ignimbritic deposits are dated between 540 Ma (Lézardrieux) and 520 Ma (Jersey). Various plutonic rocks cutting the weakly deformed sediments of this basin show block tectonics and range in age from 540 to 480 Ma. Thus, the history of this basin belongs to a post-orogenic phase of Cadomian geodynamic evolution. However, this structural feature remained active for a long time because subsidence persisted at least into the Cambrian; the first body fossils appeared at a time when the rest of the Cadomian foldbelt was being cratonized (Brioverian/Cambrian unconformity and the presence of Cambrian red beds).

The structural zonation of the Cadomian orogen is intimately linked to the paleogeographic outline of the Icartian continent between 640 and 540 Ma ago (i. e., volcanic arc, back-arc basin, passive margin, and platform). Taking account of the particular features of tectogenesis in each of the paleogeographic domains (tectonic style, metamorphic grade, age of deformation, and magmatism), it would appear that the more internal units (i. e., the arc) were deformed at an early stage of the orogeny (ca. 615–600 Ma) when continental collision had just started. With time, deformation migrated across these units to affect the back-arc basin which, between 595 and 585 Ma ago, came to be thrust onto the continental margin. Although the interaction between thrusting and wrenching existed from the onset of closure of the plate system, it is very likely that wrench shearing played a dominant role as soon as the passive margin presented an obstacle to the advance of nappes during crustal shortening. The incorporation of this passive margin into the orogenic cordillera took place between 580 and 540 Ma ago. The structure of the continental platform was formed during the final stages of crustal shortening around 540 Ma ago, at the same time as the end of deformation on the passive margin.

The post-bedded chert sedimentation on the platform was thus contemporaneous with deformation in the orogenic cordillera, drawing its supply from the erosion of black chert-bearing formations on the margin. These were involved very early in tangential tectonics during their "welding" onto the cordillera. As the orogenic cordillera began to rise, a basin opened up at the tectonic boundary between the volcanic arc domain and the back-arc basin; the first deposits accumulated in this late intrabelt basin record the after-effects of shortening at the same time as sedimentation on the platform. Subsequently, an important phase of post-orogenic extension affected the basin, accompanied by widespread volcanic activity (e.g., Trégor, Jersey, and Cotentin ignimbrites).

References

Adams CJ (1976) Geochronology of the Channel Islands and adjacent French mainland. J Geol Soc (Lond) 132: 233–250

Autran A, Chantraine J, Rabu D (1979) Lithostratigraphie et déformation du Briovérien de la baie de Lannion. Implications sur les relations entre les cycles cadomien et hercynien. Bull Bur Rech Géol Min 1–4: 277–292

Auvray B (1979) Génèse et évolution de la croûte continentale dans le Nord du Massif Armoricain. Thèse d'Etat, Rennes, 670 pp

Balé P (1986) Tectonique cadomienne en Bretagne nord. Interaction décrochement, chevauchement: champs de déformation et modélisation expérimentale. Thèse d'Université, Rennes, 361 pp

Balé P (1988) La tectonique cadomienne dans l'Est du Trégor (Massif Armoricain). Géol Fr 1: 3–16

Balé P, Brun JP (1983) Les chevauchements cadomiens de la Baie de Saint-Brieuc (Massif Armoricain). CR Acad Sci Paris 297: 359–362

Balé P, Brun JP (1989) Late Precambrian thrusts and wrench zones in Northern Brittany (France). J Struct Geol 11-4: 391–405

Ballard JP, Brun JP, Durand J (1986) La discordance Briovérien-Paléozoïque inférieur en Bretagne centrale: signature d'un épisode de distension ordovicienne. CR Acad Sci Paris 303: 1327–1332
Bardy P (1988) L'orogène cadomien dans le Nord-Est du Massif armoricain et en Manche occidentale. Thèse d'Université, Rennes, 380 pp
Bardy P, Lefort JP (1987) Structure et stratigraphie des formations antémésozoïques du Golfe normano-breton d'après l'interprétation des données sismiques. CR Acad Sci Paris 304: 997–1000
Brown M (1979) The petrogenesis of the Saint-Malo Migmatite Belt, Armorican Massif, France, with particular reference to the diatexites. N Jb Miner Abh 1: 48–74
Brun JP (1977) La zonation structurale des dômes gneissiques. Un exemple: le massif de Saint-Malo (Massif armoricain, France). Can J Earth Sci 14: 1697–1707
Brun JP, Martin J (1978) Relations métamorphisme – déformation au cours de l'évolution dynamique d'un dôme migmatitique: le massif de Saint-Malo (France). Bull Soc Géol Fr XX-1: 91–101
Brun JP, Balé P (1990) Cadomian tectonics in Northern Brittany. In: D'Lemos RS, Strachan RA, Topley CG (eds) The Cadomian Orogeny. Geol Soc Spec Publ 51: 95–114
Chantraine J, Chauvel JJ, Dupret L, Gatinot F, Icart JC, Le Corre C, Rabu D, Sauvan P, Villey M (1982) Inventaire lithologique et structural du Briovérien (Protérozoïque supérieur) de la Bretagne centrale et du Bocage normand. Bull Bur Rech Géol Min I (1–2): 3–17
Chantraine J, Chauvel JJ, Balé P, Denis E, Rabu D (1988) Le Briovérien (Protérozoïque supérieur à terminal) et l'orogenèse cadomienne en Bretagne (France). Bull Soc Géol Fr IV, 5: 815–829
Chauvel JJ, Rabu D (1984) Brioverian in Central Brittany. In: Zoubek V (ed) Precambrian in younger fold belts. Wiley, New York, pp 462–470
Darlet PA, Dabard MP, Chauvel JJ (1989) Les métasédiments briovériens de la Rance (Bretagne du Nord) dans l'évolution cadomienne du Massif armoricain. CR Acad Sci Paris 310: 779–786
Denis E (1987) Les sédiments briovériens (Protérozoïque supérieur) de Bretagne septentrionale et occidentale: nature, mise en place et évolution. Thèse d'Université, Rennes, 202 pp
Dissler E (1987) Evolution géodynamique cadomienne du Nord-Cotentin (Massif armoricain). Thèse d'Université, Caen, 255 pp
Dissler E, Doré F, Dupret L, Gresselin F, Le Gall J (1988) L'évolution géodynamique cadomienne du Nord-Est du Massif armoricain. Bull Soc Géol Fr IV, 5: 801–814
Dupret L (1984) The Proterozoïc of Northeastern Armorican massif. In: Zoubek V (ed) Precambrian in Younger Folds belts. Wiley, New York, pp 444–461
Guerrot C (1989) Archéen et Protérozoïque dans la chaîne hercynienne ouest-européenne: géochimie isotopique (Sr, Nd, Pb) et géochronologie U/Pb sur zircons. Thèse d'université, Rennes, 164 pp
Helm DG (1983) The structure and tectonic evolution of the Jersey Shale Formation, St Ouen's Bay, Jersey, Channel Islands. Proc Geol Assoc 94: 201–216
Hirbec Y (1979) Le complexe basique de Belle-Isle-en-Terre (Côtes-du-Nord) Thèse 3e cycle, Rennes, 147 pp
Jeannette D (1972) Analyse tectonique de formations précambriennes. Etude du Nord-Est de la Bretagne. Thèse d'Etat, Strasbourg. Mém Soc Géol Strasbourg 36, 175 pp
Le Corre C (1977) Le Briovérien de Bretagne centrale: essai de synthèse lithologique et structurale. Bull Bur Rech Géol Min I-3: 219–254
Lees GS, Roach RA, Shufflebotham MM, Griffiths NH (1987) Upper Proterozoic basaltic volcanism in the northern Massif Armoricain, France. Geochemistry and mineralization of Proterozoic volcanism suites. Geol Soc Sp Publ London, pp 503–523
Le Gall J, Dissler J, Dupret L (1986) Signification géodynamique des volcanismes briovériens dans le nord-est du Massif armoricain. CR Acad Sci Paris 303: 1587–1592
Rabu D, Chauvel JJ, Chantraine J (1983) Nouvelles propositions pour la lithostratigraphie du Briovérien (Protérozoïque supérieur) et pour l'évolution géodynamique en baie du St Brieuc. Bull Soc Géol Fr XXV-4: 615–621
Rabu D, Chantraine J, Chauvel JJ, Denis E, Balé P, Bardy P (1990) The Brioverian (late Proterozoic) and the Cadomian Orogeny in the Armorican Massif. In: D'Lemos RS, Strachan RA, Topley CG (eds) The Cadomian Orogeny. Geol Soc Spec Publ 51: 91–94
Roblot MM (1962) Sur une coupe stratigraphique à l'intérieur du Briovérien. CR Acad Sci Paris 254: 3720–3722
Strachan RA, Roach RA (1990) Tectonic evolution of the Cadomian belt in north Brittany. In: D'Lemos RS, Strachan RA, Topley CG (eds) The Cadomian Orogeny. Geol Soc Spec Publ 51: 133–150
Verdier (1968) Etude pétrographique et structurale du Trégor occidental (baie de Lannion). Thèse 3e cycle, Strasbourg, 51 pp
Vidal P, Deutsch S, Martineau F, Cogné J (1974) Nouvelles données radiométriques en Baie de Saint-Brieuc. Le problème d'un socle antécadomien nord armoricain. CR Acad Sci Paris 279: 631–634

Igneous Activity

J. Chantraine, B. Auvray, and D. Rabu (translated by M. S. N. Carpenter)

1 Introduction

The Cadomian belt in the Armorican Massif is characterized by intense magmatic activity – plutonic and volcanic – associated with an active margin-type geodynamic setting. This developed into a post-orogenic stage terminated by the deposition of Paleozoic transgressive facies. This magmatism was at first generated in the context of subduction-collision processes, i. e., volcanic units intercalated within the different structural domains of the Brioverian and syn-orogenic plutons in the orogenic cordillera. Subsequently, igneous activity originated from, on the one hand, crustal remelting produced by collisional overthrusting of the cordillera onto the continent, and, on the other hand, the injection of juvenile magmas during post-orogenic crustal distension in the intrabelt basin or on the external continental platform.

2 The Orogenic Cordillera

This is the domain where igneous activity is most intense; the characteristics of the magmatic units are comparable with recent or present-day active margin rocks.

2.1 Active Margin-Type Volcanism

2.1.1 The Volcanic Arc Domain

The different volcanic units that are interbedded in the Brioverian formations allow us to envisage their tectonic environments through petrological and/or geochemical criteria.

In the Trégor arc, volcanism started with important pyroclastic activity (Locquirec-Tréguier Formation) passing upward into lava flows of intermediate composition (Rugunay-Malabri Formation). This more or less acidic sequence is overlain by a basic volcanic complex (Armorique-Trédrez Formation) which corresponds to the maximum and youngest phase of activity in the Trégor volcanic arc (Fig. 15).

In the Cotentin, the pyroclastic deposits corresponding to the onset of arc formation appear to be absent. Volcanic activity in the frontal arc (Cherbourg area) is of moderate intensity, characterized by basic lava flows associated with some acid members (Le Vast Formation); in the internal arc (Coutances area), volcanism is far more developed and is made up of a predominantly basic igneous complex (Montsurvent Formation) which, however, includes various acid lavas and tuffs whose relations with the basic rocks are unknown.

The pyroclastic flows of the Locquirec-Tréguier Formation constitute a fairly homogeneous unit, sometimes massive and tuffaceous, and sometimes banded or reworked. Even though the original structures are often obliterated, these rocks are evidently of pyroclastic origin (Verdier 1968; Autran et al. 1979; Auvray 1979). A dominant and constant feature of these rocks is the abundance of albitic phenoclasts up to a few millimeters in size, and the presence of large clasts of volcanic quartz. These clasts are "floating" in a siliceous sericitic matrix rich in small albite laths (An5–An10); the schistosity is defined by biotite and chlorite. Detrital K-feldspar is present only in the basal part of the formation, where it is derived from erosion of the underlying basement. Epidote is abundant, and amphibole is discrete. The preserved microlitic and microgranular textures are consistent with the presence of flow units and sills within the formation. The conglomeratic interbeds mentioned previously reflect a local detrital influence which increases towards the top of the formation.

The major element chemistry of these rocks shows a wide range of compositions, notably a variation in alkalis which is probably due to their mobility during magmatic and metamorphic processes. Nevertheless, these data confirm the soda-rhyolitic and quartz-rhyolitic composition of the flows (i. e., Na20 + K20 = 6–8 %, K20 < 3 %, Ca0 = 1–3 %, Si02 = 60–70 %, A1203 = 15–16 %). In the absence of trace element

J. D. Keppie (Ed.): Pre-Mesozoic Geology in France and Related Areas

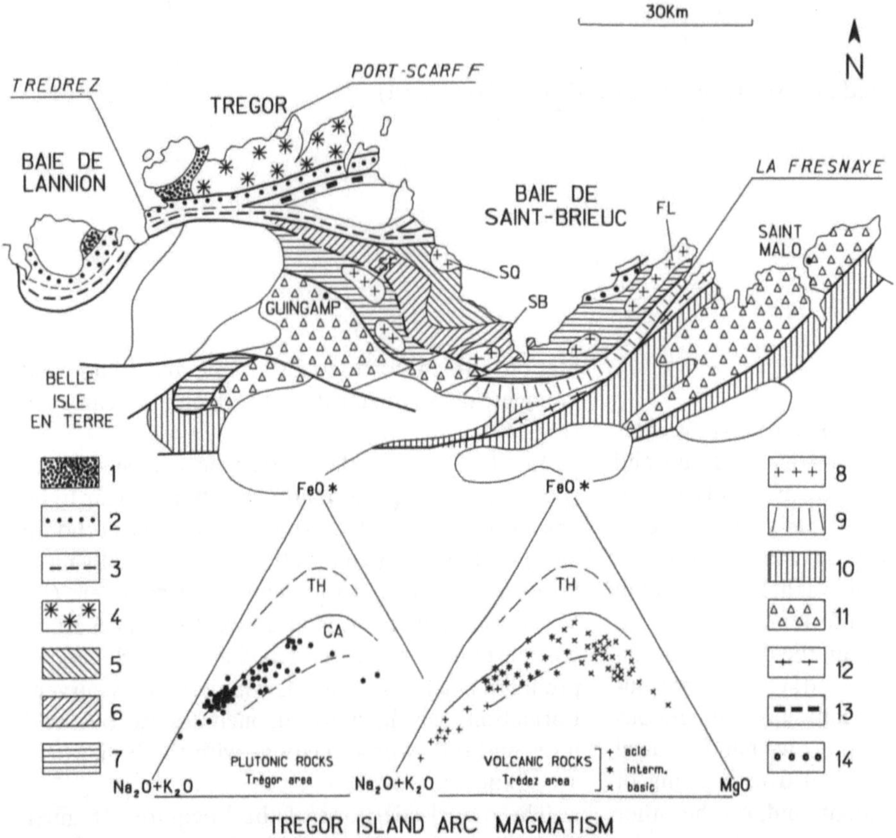

Fig. 15. Sketch map showing magmatism in the Cadomian orogenic belt of North Brittany. *1* Icartian basement; Trégor volcanic arc and back-arc basin: *2* acidic volcanics (Locquirec-Tréguier Formation); *3* intermediate (Rugunay-Malabri Formation) and basic volcanics (Armorique-Trédrez Formation); *4* North Trégor Batholith; *5* turbiditic sediments (Binic-Le Légué Formation); *6* basic volcanics (Lanvollon Formation); *7* basic plutonics (Yffiniac Formation); *8* syn-tectonic plutonic complexes, Squiffiec *(SF)*; Saint-Quay Portrieux *(SQ)*; Fort-la-Latte *(FL)*. Continental margin: *9* extensional domain volcanism (Baie de la Fresnaye); *10* bedded black cherts sediments (Lamballe Formation); *11* Saint-Malo and Guingamp migmatites; *12* Cadomian orthogneisses. Late intrabelt basin: *13* calc-alkaline spilites (Paimpol Formation); *14* tholeiitic spilites (Erquy Formation). (Chemical data after Auvray 1979 and Cabanis 1986)

data, the magmatic history and tectonic setting of this volcanism remains uncertain. The presently available data suggests either an origin by crustal melting at the beginning of subduction or magmatic differentiation following a calc-alkaline trend in an arc-type setting (Fig. 15). The volcanism has been dated at 642 ± 12 Ma (Rb/Sr method; Vidal 1976).

The lavas and sills of the Rugunay-Malabri Formation mark a return to volcanic activity in the Trégor arc; there are 10–100-m-thick massive members alternating with sedimentary interbeds. Deformation is moderate and the original textures are generally preserved; the lavas are characterized by microlitic and trachytic textures, whereas the sills are microgranular and more or less porphyritic. The intermediate composition of these rocks is reflected in their parageneses which include albite-oligoclase, quartz, chlorite, biotite, epidote, calcite, and actinolite. This is confirmed by the major element analyses, which show andesitic and dacitic compositions (SiO2 = 58–68 %). Various parameters, such as the variation of total iron as a function of FeO/MgO ratio, the TiO2 contents of the more basic members and total alkalis compared with silica all suggest a calc-alkaline affinity. Such a conclusion is supported by unpublished trace element data which tend to show that these volcanic rocks belong to the same calc-alkaline suite as the overlying basic formation (Fig. 15).

The basic volcanics of the Armorique-Trédrez Formation in the Trégor and the Montsurvent and Le Vast Formations of the Cotentin have a calc-alkaline character which enables us to locate the Cadomian active margin. In the Trégor, this phase of volcanism is represented by a pile of thick basaltic pillow lavas and massive sills with thin volcaniclastic interbeds at the base becoming more abundant towards the top of the

Table 2. Average major element geochemical compositions for the main basic volcanic formations in the Brioverian

Formations		Si02	A1203	Fe0+Fe203	Mn0	Mg0	Ca0	Na20	K20	Ti02	P205	P.F.
Armorique	M	48.79	14.88	9.34	0.18	9.67	8.46	2.39	0.18	0.64	0.12	3.87
N = 21	E	0.67	0.33	0.29	0.005	0.81	0.81	0.26	0.02	0.05	0.01	0.27
Paimpol	M	48.75	18.27	9.17	0.14	7.54	7.79	3.05	0.49	0.88	–	3.49
N = 17	E	0.47	0.42	0.14	0.004	0.49	0.32	0.17	0.05	0.02	–	0.14
Lanvollon	M	53.11	13.65	12.35	0.2	4.84	7.51	2.87	0.83	2.1	0.25	1
N = 24	E	0.55	0.2	0.2	0.004	0.21	0.32	0.12	0.06	0.08	0.02	0.08
La Fresnaye	M	48.61	14.47	11.82	0.17	5.31	7.66	3.62	0.51	3.3	0.44	2.75
N = 7	E	0.60	0.34	0.62	0.008	0.49	0.41	0.28	0.05	0.17	0.08	0.45
Erquy	M	56.49	13.62	11.42	0.17	3.94	3.51	4.51	0.68	2	0.31*	3.26
N = 16	E	1.03	0.3	0.28	0.01	0.22	0.46	0.24	0.08	0.1	0.02*	0.14
Le Vast	M	49.52	15.65	9.66	0.14	7.54	6.84	2.84	0.68	1.05	0.18	5.24
N = 12	E	0.94	0.36	0.54	0.01	0.92	0.43	0.36	0.14	0.11	0.02	0.53
Montsurvent	M	52.89	15.88	10.63	0.19	4.96	5.9	3.5	0.49	0.69	0.06**	3.58
N = 28	E	0.71	0.38	0.34	0.01	0.39	0.73	0.25	0.12	0.04	0.008**	0.33
Mortes Femmes	M	52.99	14.04	13.28	0.19	4	4.59	4.31	0.53	2.29	0.27	2.81
N = 6	E	1.4	0.42	1.19	0.03	0.47	1.21	0.35	0.12	0.15	0.03	0.43

N number of analyses; M mean values; E standard deviations; * average of 8 analyses; ** average of 18 analyses

succession. An effusive origin is clearly shown by the structures and the occasional microlitic or doleritic textures preserved in the cores of massive flows and sills. The mineral parageneses generally belong to the greenschist facies, i.e., oligoclase (An10–An20), actinolite, chlorite, epidote, calcite, and Fe-Ti oxides. Quartz and biotite are common accessories. There is a more or less dense actinolite-chlorite fabric, within the fine-grained sericitic/feldspathic matrix (rich in opaques), which picks out the foliation and contains epidote grains and plagioclase laths. In addition, small gabbro bodies are locally associated with the volcanism.

In the Cotentin, the volcanic arc domain splits into two branches. The northern branch is characterized by the occurrence of thin volcanic intercalations (Le Vast Formation) within the marine Brioverian deposits. This volcanic formation shows medium-grade metamorphism and moderate deformation; the magmatic nature and texture of the rocks are sparsely identifiable and the association is made up of acidic members with metabasalts and meta-andesites showing doleritic and microlitic textures (Dissler 1987). The southern branch is characterized by a thick volcanic complex (Montsurvent Formation), where acid or basic lavas and sills are associated with volcaniclastic sediments. This complex is affected by deformation and metamorphism which shows a marked gradient from north to south. In the north, the parageneses are low grade, i.e., albite, chlorite, epidote, calcite, Fe-Ti oxides, and pyroxene relicts in the metabasics and quartz, albite, K-feldspar, muscovite, and chlorite in the meta-acidic rocks. The igneous textures are often preserved in doleritic to microlitic metabasalts and meta-andesites, as well as in the trachytic to microlitic metadacites and metarhyolites (Dupret et al. 1985). Towards the south, parageneses become high-grade, with the appearance of actinolite in the metabasic rocks followed by green hornblende and clinopyroxene. Biotite and garnet are developed in the meta-acid rocks. Within a highly deformed zone near Coutances, the Montsurvent volcanics contain some slices of serpentinized ultra-basic rock with cumulate textures that have been identified as wehrlites belonging to the formation (Saunier, pers. commun.; Bardy 1988).

The geochemistry of these different basic formations in the Tregor and Cotentin is comparable with present-day or recent island-arc volcanics (see Table 2), but a distinction can be made between the various parts of the Cadomian arc. The Armorique-Trédrez Formation is composed of a very poorly differentiated suite of basalts whose calc-alkaline affinity is suggested by the absence of Fe-enrichment, even in the rare differentiated rocks, as well as the low Ti02 contents (< 1 %; see Fig. 16) (Cabanis et al. 1986). The basic rocks of the Montsurvent and Le Vast Formations constitute a more highly differentiated suite, containing basalts and andesites with low Ti02 contents (ca. 1 %) (Fig. 16) typical of calc-alkaline series. The Fe-enrichment observed in certain members of the Montsurvent Formation suggests an arc tholeiite-type trend (Dupret et al. 1985), whereas the magnesian character (Mg0 contents up to 12 %) of parts of the Le Vast Formation indicates perhaps the rather primitive nature of the frontal arc (Le Gall et al. 1986).

A study of trace element geochemistry, particularly the more incompatible elements, has confirmed the calc-alkaline character of the Trégor and Montsurvent lavas and has provided some additional information. The high Th/Ta (10–20) and La/Th (50–100) ratios are evidence of emplacement in an orogenic environment. However, the higher Hf/Ta and Hf/Th ratios for the Montsurvent Formation compared with the Armorique-Trédrez Formation are in favor of a more tholeiitic character for the volcanic arc in the Cotentin and a more calc-alkaline affinity in the Trégor (Cabanis 1986). These results are summarized in the Th-Hf-Ta diagram of Wood et al. (1979) and the Th-Tb-Ta diagram of Cabanis and Thièblemont (1988) (Fig. 16).

In this way, the volcanic rocks of the Trégor and Cotentin enable us to recognize (Auvray and Maillet 1977; Auvray 1979; Cabanis 1986; Cabanis et al. 1987) the existence of an active margin during the Upper Proterozoic linked to subduction processes beneath the Icartian edge. Within this volcanic arc domain, early activity appears to be mainly pyroclastic, giving way to voluminous basalts and andesites belonging to the tholeiitic and calc-alkaline differentiation series according to the local degree of maturity of the arc.

2.1.2 The Back-Arc Basin

Within the orogenic cordillera, the Saint-Brieuc coastal area is characterized by the development of a predominantly basic and extensive volcanic/plutonic complex composed of the Yffiniac and Lanvollon Formations (Fig. 15). Despite numerous internal shear zones, these formations appear to constitute a pile several thousands of meters thick affected by deformation and metamorphism increasing in intensity towards the south. The southern boundary of this complex corresponds to a major suture of the Cadomian orogen (Balé and Brun 1983).

Heterogeneous shearing deformation and metamorphic recrystallization have obliterated the original textures of these rocks, which have now been transformed into metagabbros and amphibolites. However, the general stratigraphic relations are recognizable with mainly plutonic rocks at the base (Yffiniac Formation) and volcanic rocks at the top (Lanvollon Formation).

The Yffiniac Formation is principally composed of medium to high-grade metagabbros with parageneses including hornblende, pyroxene, plagioclase, and garnet. This formation was previously considered (Cogné 1972) as part of the "Pentevrian" basement to the Brioverian succession in the Baie de Saint-Brieuc, but is now interpreted as the plutonic sole to the thrust pile. It could represent a crustal segment that was emplaced into the back-arc basin during the Cadomian orogeny. Such a hypothesis is supported by a recently obtained U/Pb zircon age (602 ± 8 Ma; Guerrot 1989) which, furthermore, is identical to the U/Pb zircon age of the Belle-Isle-en-Terre basic/ultrabasic complex (Peucat et al. 1981). This latter is situated south of the North Armorican Shear Zone and may be in tectonic contact with the Callac Formation (cf. Lamballe Formation), thus forming a klippe that has resulted from overthrusting of the cordillera onto the continental margin. Finally, it should be noted that the metagabbros of Yffiniac and Belle-Isle-en-Terre contain slices of serpentinized ultrabasic material; these ultrabasics mark a major suture of the Cadomian orogen that runs from the Trégor to the Coutances area of the Cotentin.

The Lanvollon Formation is composed of amphibolites that represent a fairly homogeneous basic volcanic pile of flows and sills. There are some thin sedimentary interbeds in the basal part which become more frequent towards the top, where acid volcanics are associated with clastic and volcaniclastic metasediments. An intense deformation has obliterated most of the igneous textures, with the exception of clearly identifiable pillow structures. The parageneses belong to amphibolite facies, i.e., blue-green amphibole, intermediate plagioclase, epidote, biotite, and Fe-Ti oxides.

Geochemical data on the metagabbros of Yffiniac and Belle-Isle-en-Terre are still rather sparse (Peucat et al. 1981; Rabu et al. 1983), but they tend to suggest an affinity with orogenic magmatic suites. However, this conclusion should be considered as incomplete and preliminary in nature. Studies of the Lanvollon Formation are relatively more comprehensive (Cabanis et al. 1987; Lees et al. 1987) and make it possible to characterize the volcanism. The amphibolites have the compositions of basaltic andesites (Si02 = 50–56 %) with moderate degrees of differentiation typical of tholeiitic suites (see Table 2). Their representative points in the Al-Fe-Mg and Al-Ti-Mg diagrams of Fig. 16 plot in the continental tholeiitic field; this affinity is confirmed by incompatible trace element data (Ti, V and Zr) as well as the La/Nb ratios (Lees et al. 1987).

The study of highly incompatible trace elements (Cabanis et al. 1987; Roach et al. 1990) has revealed an environment for these volcanics which is interme-

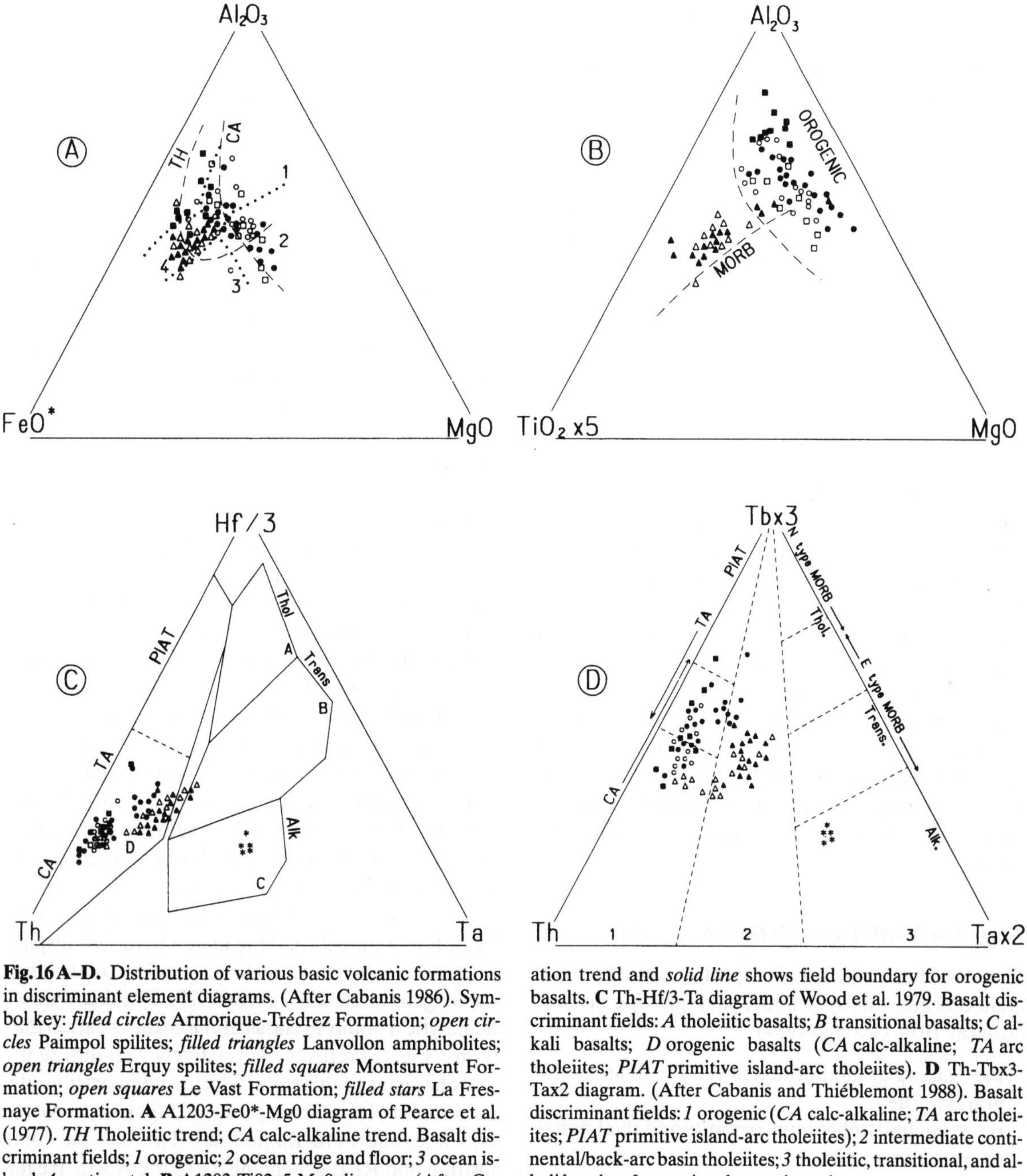

Fig. 16 A–D. Distribution of various basic volcanic formations in discriminant element diagrams. (After Cabanis 1986). Symbol key: *filled circles* Armorique-Trédrez Formation; *open circles* Paimpol spilites; *filled triangles* Lanvollon amphibolites; *open triangles* Erquy spilites; *filled squares* Montsurvent Formation; *open squares* Le Vast Formation; *filled stars* La Fresnaye Formation. **A** A1203-Fe0*-Mg0 diagram of Pearce et al. (1977). *TH* Tholeiitic trend; *CA* calc-alkaline trend. Basalt discriminant fields; *1* orogenic; *2* ocean ridge and floor; *3* ocean island; *4* continental. **B** A1203-Ti02x5-Mg0 diagram. (After Cabanis 1986). *Dashed line with arrow* indicates MORB differentiation trend and *solid line* shows field boundary for orogenic basalts. **C** Th-Hf/3-Ta diagram of Wood et al. 1979. Basalt discriminant fields: *A* tholeiitic basalts; *B* transitional basalts; *C* alkali basalts; *D* orogenic basalts (*CA* calc-alkaline; *TA* arc tholeiites; *PIAT* primitive island-arc tholeiites). **D** Th-Tbx3-Tax2 diagram. (After Cabanis and Thiéblemont 1988). Basalt discriminant fields: *1* orogenic (*CA* calc-alkaline; *TA* arc tholeiites; *PIAT* primitive island-arc tholeiites); *2* intermediate continental/back-arc basin tholeiites; *3* tholeiitic, transitional, and alkali basalts of extensional tectonic settings

diate between an orogenic domain and a rift, similar in character to continental tholeiites, but also the volcanic rocks of back-arc basins bordering the continents. Both these types of tectonic environment show medium Th/Ta (2–3) and La/Ta ratios, a feature which is reflected in the Th-Hf-Ta and Th-Tb-Ta diagrams of Fig. 16, where the points are scattered between the two corresponding fields. These intermediate characteristics are probably due to magmatic mixing phenomena which are often invoked for the petrogenesis of continental and back-are basin tholeiites.

Various structural arguments on the regional scale, such as the proximity of the active margin, would rather suggest – in our opinion (Cabanis et al. 1987; Chantraine et al. 1988) – a back-arc basin type setting for this volcanism. A detailed study of the plutonic rocks in the Yffiniac Formation might provide some answers to this problem.

2.2 Plutonic Activity in the Orogenic Domain

The emplacement of plutonic massifs, associated with the evolution of the Cadomian active margin and developed throughout the orogenic belt, is contemporaneous with the main phase of tectonic activity around 600 Ma. The hypothesis of an earlier magmatic episode associated with a more ancient tectonic event is based on unreliable radiometric data. Previously, Cadomian geodynamic evolution was interpreted in terms of multi-stage deformation (Cogné 1972), but nowadays a model of progressive deformation is favored (Balé and Brun 1983; Rabu et al. 1983).

Nevertheless, the existence of an early stage of magmatism suggested by the 646 ± 25 Ma age obtained on foliated metagranites at L'Erée, Guernsey, (Rb/Sr isochron of Adams 1976) is confirmed by the dating of orthogneiss clasts included in the Brioverian succession of the Baie de Saint-Brieuc (656 ± 5 and 667 ± 4 Ma U/Pb zircon ages; Guerrot 1989) and further work in progress on other formations. Although the significance of this magmatism is presently unknown, it would not appear to be linked to an early tectonic phase of the Cadomian orogeny.

Apart from a few isolated massifs, syn-orogenic Cadomian plutonism is represented by the North Trégor Batholith – whose origin is linked to the development of the volcanic arc – and by a series of intrusions which run along the border of the southern suture left by the overthrusting of the cordillera onto the continent.

2.2.1 The North Trégor Batholith (Fig. 15)

This complex body is made up of an association of granodioritic, monzogranitic, and granitic massifs (Auvray 1979; Graviou 1984):

- the Talberg granodiorite and the Pleubian microgranodiorite;
- the Pomelin-Bréhat monzogranite and the Launay micromonzogranite;
- the Port-Blanc granite.

These different massifs are co-genetic magmatic intrusions which are probably penecontemporaneous. There are numerous basic enclaves (gabbros and diorites) within the batholith, and a dense network of veins (Loguivy microgranites) is intruded into the eastern part. It is interesting to note that the North Trégor Batholith is only deformed in the west under the influence of Variscan tectonics – the relationship of the plutonic rocks to the volcanic arc domain is unknown due to the masking effect of a late-stage fault (Auvray 1979).

The mineral composition of this magmatic suite is characterized by the general occurrence of amphibole (magnesio- and ferro-hornblende) associated with essential biotite and some relict clinopyroxene. The Talberg granodiorite has a granular texture and the Pleubian Microgranodiorite is porphyritic – both facies are intimately associated, often showing mixing phenomena and a diffuse border between the two. In the microgranular facies, plagioclase, amphibole, and biotite are euhedral and form phenocrysts. The plagioclase is zoned (An50–An10), and the K-feldspar is composed of perthitic orthoclase restricted to the quartz-bearing groundmass. The Pomelin-Bréhat monzogranite and the Launay micromonzogranite show the same relations as the facies mentioned above. Perthitic K-feldspar is more abundant, often associated with graphic quartz texture in the groundmass of porphyritic microgranular varieties. Phenocrysts include amphibole, biotite, and euhedral plagioclase (generally zoned: An30–An10); exceptionally, there is some quartz or K-feldspar. The Port-Blanc granite is leucocratic and equigranular, being mainly composed of quartz and feldspar. The K-feldspar is perthitic microcline, the euhedral plagioclase is oligoclase (An20–An10) and globular quartz is very common. Although the amphibole is rather rare, it is always associated with biotite; even in the more leucocratic varieties, there is no primary muscovite.

The associated rock types of the North Trégor Batholith form a continuous magmatic differentiation series varying from granodiorites to leucogranites; the basic end-member is represented by enclaves of dioritic composition. Geochemical data show that the different massifs making up the batholith display a cogenetic suite which is closely similar to calc-alkaline magmatic trends (Auvray 1979). The distribution of analytical points in classical chemical variation diagrams (Fig. 15) reflects the relatively well-differentiated character of this suite, which is comparable with granitic complexes in orogenic settings (Sierra Nevada, Chile, and Peru Batholiths). The isotopic composition and REE geochemistry of the North Trégor Batholith are consistent with a calc-alkaline magma type derived from the mantle with some evident crustal contribution (Graviou 1984). One possible model involves partial melting of the upper mantle under the influence of "enriched" fluids coming from the subducted oceanic slab. Crystallization of the magmas took place at relatively low pressures (< 5 kb), and the emplacement of the batholith is dated at 614 + 17/− 10 Ma (U/Pb zircon method; Graviou 1984).

The Porz-Scarff Granite (Fig. 15), dated at 557 ± 15 Ma (Rb/Sr method; Auvray 1979), is one of the late-stage intrusions that cut the North Trégor Batholith. It is a leucocratic granite with a granular texture, composed of essential quartz, albite, and orthoclase. The parageneses includes accessory biotite and some rare amphibole. Chemical analyses show high silica contents (ca. 75 %) and an alkaline tendancy ($Na_2O + K_2O > 8\%$; $CaO < 1\%$). The relative proportions of quartz, albite, and orthoclase are close to the low pressure entectic ($P \approx 2$ kb), thus suggesting a very high level emplacement.

2.2.3 The Guernsey Plutonic Complex (Fig. 17)

This complex body occurs in a setting similar to the North Trégor Batholith, being intrusive into the Icartian basement on the outermost edge of the orogenic cordillera. It is composed of a set of several intrusions which are probably penecontemporaneous and which vary from gabbros to granites. The main units are as follows (French and Keen 1986; Topley et al. 1990):

- the Saint Peter Port gabbro, forming a small foliated massif with olivine-bearing rock types at the base passing upward into hornblende and pyroxene occurring together and finally hornblende on its own at the top;
- the Bordeaux diorite forms the major part of the complex and is composed of a mixed association of dioritic facies with tonalites and granodiorites containing biotite and hornblende; the Alderney diorite and the granodiorites of Herm and Sark probably belong to the same suite;
- the Cobo monzogranite, of adamellitic composition, is clearly intrusive, into the Guernsey Plutonic Complex, which has yielded ages of 558 ± 9 Ma. (Rb/Sr method; French and Keen 1986) and 553 ± 9 Ma (Rb/Sr method; de Brémond d'Ars 1986).

These results show that at least part of the Guernsey Plutonic Complex was intruded at a late stage of the Cadomian orogeny (i. e., post-580 Ma). Its supposed calc-alkaline character poses a problem similar to that of the basic plutons at the continental margin (Brée, Trégomar, Ernée; see below for discussion). It would seem that the geochemical characteristics of the Guernsey Plutonic Complex were acquired up to 50 Ma after the main Cadomian subduction ceased, due to hybridization of magmas derived from mantle sources combined with crustal re-melting during post-orogenic distension. The range of initial $^{87}Sr/^{86}Sr$ ratios obtained (0.706–0.708) also supports such a hypothesis.

2.2.4 Other Syn-Orogenic Plutons (Fig. 15)

Although other plutons in the Cadomian orogenic domain are less diversified than the North Trégor Batholith, they are nevertheless quite widely developed.

The Saint-Quay-Portrieux diorite is particularly interesting because it is intrusive into an intermediate structural domain. It has been the object of detailed studies (Arnaud 1980; Barrat 1986) which show that the noritic-tonalitic suite has an aluminous calc-alkaline trend derived from the partial melting of juvenile basaltic material, of mantle origin, slightly contaminated by continental crust. Crystallization occurred at intermediate depths (ca. 7 kb pressure) at around 584 ± 42 Ma (Rb/Sr isochron; Vidal et al. 1972).

A line of intrusions, which follows the major suture to the south and east of the orogenic domain, is less well characterized. It is made up of diorites at Squiffiec and Saint-Brieuc, on the western side of the Baie de Saint-Brieuc, as well as the Coetmieux-Fort-la-Latte tonalite on the eastern side; it continues into the Cotentin with the Coutances diorite.

Some recent studies (Shufflebotham 1986, 1989; Bardy 1988) suggest that all these plutons belong to a calc-alkaline magmatic suite of the orogenic type. The mineralogical and geochemical characteristics of these plutons are fairly homogeneous: i. e., tonalite with intermediate plagioclase (An35–An65), hornblende, biotite, as quartz essential minerals; K-feldspar is an accessory. They contain enclaves of more basic composition (gabbros and diorites) and may be associated with lighter-colored facies (leuco-quartz-diorites, leucogranodiorites) that appear to have been intruded at a late stage. Finally, the REE abundance patterns are very uniform, showing highly fractionated HREE.

These calc-alkaline plutons display tectonic fabrics, such as foliation and lineation, which are similar to the surrounding metamorphic formations. This suggests that their emplacement was contemporaneous with the tectono-metamorphic evolution of the region (Balé 1986), under amphibolite facies conditions; however, the exact relations between magmatic intrusions and deformation phases remain to be determined. Two of the plutons are well dated: i. e., the Fort-la-Latte Tonalite at 593 ± 15 Ma (U/Pb zircon method; Vidal et al. 1974) and the Coutances Diorite at 584 ± 5 Ma (U/Pb zircon method; Guerrot et al. 1989).

Altogether, this set of results shows the importance of volcanic and plutonic activity around 600 Ma ago linked to subduction processes beneath the Cadomian active margin. The volume of this mag-

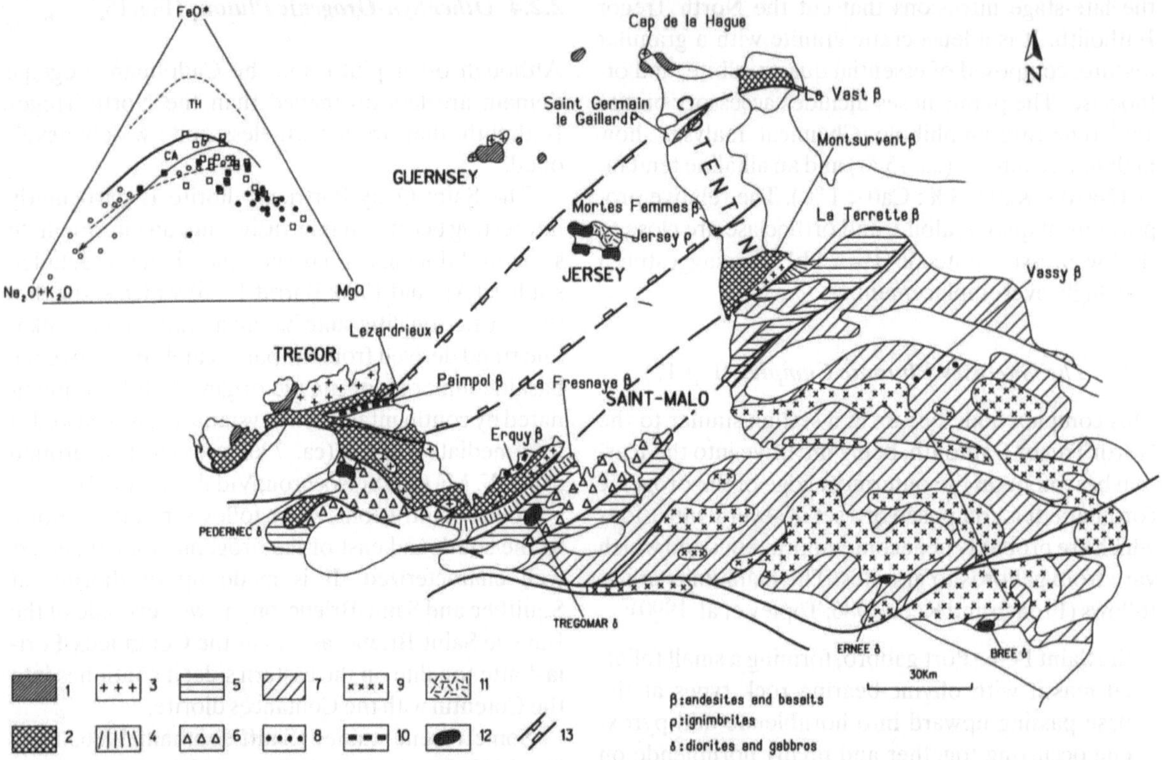

Fig. 17. Sketch map showing Cadomian magmatism in the Northern Armorican Massif. *1* Icartian basement; *2* orogenic cordillera volcanism; *3* orogenic cordillera plutonism; *4* extensional volcanism domain; *5* continental margin domain; *6* Saint-Malo migmatites; *7* continental platform domain; *8* rift-type volcanism; *9* Mancellian Batholith. Late-and post-stage Cadomian magmatism: *10* Paimpol and Erquy spilites; *11* late intrabelt basin ignimbrites; *12* post-orogenic plutonic suite (see AFM diagram); *13* late intrabelt basin limits. On the AFM triangular diagram (Le Gall and Barrat 1987), the *solid line with an arrow* indicates the differentiation trend of the Saint-Quay-Portrieux Diorite (Arnaud 1980). *Symbol key: filled circles* Trégomar; *open squares* Ernée; *filled squares* Pédernec; *open circles* Brée; *TH* tholeiitic suites; *CA* calc-alkaline suites

matism is large, as witnessed by one of the most intense gravimetric and magnetic anomalies of the Variscan belt.

3 The Continental Domain

Apart from the Mancellian Batholith, which is very extensive and of crustal origin, the magmatism associated with the tectonic evolution of this external Cadomian domain is less developed than that of the orogenic cordillera. In this continental domain, the magmatism is poorly characterized and still poses many problems, including the true significance of a calc-alkaline suite separated in time and place from the syn-orogenic plutons of the Cadomian belt. The petrogenesis of the Mancellian Batholith has not yet been satisfactorily interpreted in the context of Cadomian geodynamic evolution.

3.1 Calc-Alkaline Plutonism

Rocks of calc-alkaline affinity occur in both the continental margin and platform domains; they constitute small basic plutons that are clearly post-tectonic. These occurrences include the Pédernec diorite, intrusive into the Guingamp migmatites, the Trégomar gabbro, intruded into the bedded-chert-bearing Lamballe Formation, and also gabbros at Brée and Ernée in the post-bedded-chert Brioverian of the Bocage Normand (Fig. 17).

These plutons are highly variable and zoned on the small scale, but are broadly composed of a differentiation series from gabbros and norites (with or without olivine; including, notably, cumulate assemblages: troctolites and anorthosites) towards quartz-diorites and granodiorites (Le Gall and Mary 1983; Le Gall and Barrat 1987). The mineralogical assemblages of the different facies are analogous with the various zones observed in layered basic intrusions. The geochemical characteristics of the different massifs show close similarities which suggest that they are derived from a common mantle source. Their major element variation trends and REE abundance patterns are compatible with calc-alkaline magmatic suites (Le Gall and Barrat 1987; Fig. 17).

The absolute age of these plutons is unknown, but their field relations with the Brioverian formations leave no doubt as to their relative stratigraphic position. They intruded into the continental domain, itself affected by late Cadomian deformation which is no younger than 540 Ma. This compares with an age of 590 Ma for the main tectonic event in the orogenic domain. However, the calc-alkaline nature of the late-stage plutons presents a problem for the supposed distensive tectonic regime in this domain. Hybridization processes could have affected juvenile mantle-derived magmas as they rose through a thick continental crust during post-orogenic distension. This mechanism has already been invoked for the Guernsey Plutonic Complex, rather than supposing a second phase of subduction 50 Ma after the cessation of the main phase.

3.2 Extensional Domain Volcanism

The assumption that calc-alkaline magmatism is associated with a compressive tectonic regime in the continental domain is also in conflict with the persistance, throughout the history of the area, of volcanic activity that is characteristic of an extensional environment. This type of volcanism occurs as intercalations within the bedded-chert-Brioverian of the continental margin (La Fresnaye Formation in N. Brittany and La Terrette Formation in S. Cotentin) (Fig. 17), as well as in the post-bedded-chert group of the continental platform (Vassy Spilites). The volcanic horizons are rather thin and of limited extent (i.e., several 100s of meters thick over kilometer-scale outcrops), being composed of an association of basic sills and lavas with well-preserved textures (pillows breccias, microlitic, and quench textures). The parageneses are low grade, including albite, chlorite, actinolite, epidote, Fe-Ti oxides, and relict pyroxene. The geochemical characteristics are typical of volcanic rocks in distensive tectonic settings, showing an alkaline affinity for the la Fresnaye Formation (see Table 2) and a tholeiitic trend for the formations at La Terrette and Vassy (Cabanis et al. 1986; Le Gall et al. 1986) (Fig. 16).

It is interesting to note that the volcanism at la Fresnaye and la Terrette, considered as contemporaneous with magmatic activity in the Cadomian orogenic domain, shows no trace of an influence from subduction-related processes. The present-day proximity of these two types of volcanism – juxtaposed on either side of a major suture – underlines the magnitude of crustal shortening during the Cadomian orogeny.

On the continental foreland, there is a particular problem concerning the volcanics (Telgruc Spilites) included in the Brioverian succession in the Central Brittany domain (Baie de Douarnenez). Here, the geochemistry of the volcanics has been compared with calc-alkaline lavas in the volcanic arc domain of the Cadomian belt (Auvray and Maillet 1977). Recent geochemical studies (Cabanis 1986) suggest, rather, that the Brioverian volcanics of this area have an affinity with continental tholeiites. This new hypothesis fits better with the tectonic setting of this external part of the Cadomian orogen.

3.3 The Saint-Malo-Guingamp Migmatites and the Mancellian Batholith

The development of extensive crustal melting on the Cadomian continental margin (Saint-Malo and Guingamp migmatite belts; Fig. 17) can be integrated into a model of regional geodynamic evolution. Thrusting of the orogenic cordillera onto the continental margin during collision could have produced overthickening of the crust that led to re-melting. The partial melting of graywacke-type sediments (Brown 1979) can lead to the progressive development of a migmatite dome (Brun 1977), with the formation of metatexites, diatexites and, finally, anatectic granites. Associated with this anatexis, numerous granitic sills, veins, and stocks of granodioritic to leucogranitic composition were intruded into the country rock. This plutonic activity shows geochemical characteristics which confirm a crustal origin. The emplacement of the anatectic granites has been dated at 540 ± 5 Ma in the Saint-Malo region (U/Pb zircon method; Peucat 1982), and 542 ± 7 Ma near Guingamp (U/Pb zircon method; Andriamarofahatra and de La Boisse 1988). The previous idea of a pre-Brioverian age for these migmatites (Brown et al. 1971) has now been refuted by the radiometric age data.

The granite rocks of the Mancellian Batholith have approximately the same age as the Saint-Malo-Guingamp migmatites. The Mancellian Batholith (Fig. 17) has a very large surface extent (150 × 100 km) comprising several massifs showing the following representative facies types:

- cordierite-bearing granodiorites (Vire):
- granodiorites without cordierite (Louvigné-Lanhélin);
- quartz diorites (Bois-du-Gast);
- leucogranites (small isolated stocks).

The gray Vire-type granodiorite is the most widely represented; it is characterized by the coexistence of cordierite and biotite, the predominance of oligoclase-andesine over K-feldspar and an abundance of aluminous enclaves. The whitish and blue granodiorite varieties of Louvigné and Lanhélin are rather less widespread. They have compositions similar to the Vire type, but contain less enclave material and lack cordierite. The Bois-du-Gast quartz diorite contains only very rare K-feldspar and a little muscovite. Finally, the sporadic leucogranite bodies show a fairly homogeneous composition showing essential albite with perthite and accessory biotite and muscovite; aluminosilicates are scarce but always present.

The field relations between these different facies are not defined, but the leucogranites are seen to be intrusive into the other granitic massifs. The batholith as a whole is generally undeformed, apart from the occurrence of Variscan faulting; the SE part (Maine area) is largely cataclased, being situated between two branches of a fault system, the North Armorican Shear Zone.

Although the geochemical data are somewhat fragmentary, these granitoids appear to have a peraluminous composition. Georget (1986) has suggested that they are derived from a source region with a dominant crustal component; there results were confirned recently by Brown and D'Lemos (1991). The origin of the leucogranites remains uncertain, but their formation probably involved the partial melting of a similar crustal type source (Fouillac et al. 1986).

Numerous attempts have been made to date the granitoids of this batholith, whose intrusion appears to terminate the structural evolution of the region. The pooled Rb/Sr isochrons obtained from the Vire massif originally suggested an older age for the granodiorites (592 ± 15 Ma, Adams 1976; 617 ± 12 Ma, Vidal 1976) compared with the leucogranites (524 ± 6 Ma, Jonin and Vidal 1975). These results appear incompatible with more modern radiometric age data for Cadomian ignous activity. At present, many authors agree that the Mancellian Batholith is rather younger, favoring an age based on more recent dating results (i.e., 540 ± 10 Ma, U/Pb zircon method, Pasteels and Doré 1982; 521 ± 11 Ma, Rb/Sr isochron Fouillac et al. 1986). A recent critical review (Graviou et al. 1988) leads to the same result.

Despite the fact that the crustal origin and age of the Mancellian Batholith is now well established, the mechanisms which initiated melting are still problematic. It would seem that only the emplacement of juvenile magmas at the base of the crust could bring about such an anatexis (Graviou and Auvray 1985; Brown and D'Lemos 1991). Evidence of this mantle-derived magmatism can be seen in the continental domain (gabbros at Brée, Trégomar, and Ernée), where calc-alkaline basic compositions reflect the importance of crustal contamination in a post-tectonic environment. The andesitic volcanism which is intercalated in the Cambrian cover of this domain is associated with late-stage ignimbritic activity (Le Gall and Cabanis 1985) possibly related to Cadomian "after-effects".

4 The Late Intrabelt Basin

This basin, associated with rifting or pull-apart tectonics in a late intrabelt environment, is characterized by voluminous magmatism which spans the period between the main Cadomian tectonometamorphic event and the early Ordovician transgression. Formations belonging to this period include the calc-alkaline Paimpol Spilites of N. Brittany, the ignimbritic volcanics of the Trégor and Cotentin and the intracontinental basalts and trachy-andesites of Ordovician age in N. Brittany (Erquy-Plourivo). None of these formations has been affected by the tangential Cadomian tectonics that is characteristic of the entire orogenic cordillera; the intrabelt basin is almost undeformed and has been intruded by various late-stage plutons, particularly in Jersey.

4.1 Calc-Alkaline Brioverian Volcanism

The Paimpol Spilites of the Trégor (Fig. 15) have up to recently been considered as typical of the orogenic volcanic arc domain (Auvray and Maillet 1977), but structural studies demonstrate that this formation has not been affected by tangential Cadomian tectonics (Balé 1986). It is composed mainly of pyroclastic material (breccias and tuffs) associated with massive sills and pillow lavas. Although the strata are vertical, there is very little deformation, and igneous textures are well preserved (i.e., doleritic or microlitic sills, quench textures in the pillows, porphyritic microgranular to porphyroclastic breccias and tuffs). The paragenesis of the spilitic lavas includes albite, actinolite, epidote, chlorite, and Fe-Ti oxides, whereas the sills contain relicts of primary augite.

Geochemical studies (Auvray 1979) have shown the basaltic composition of these volcanics (see Table 2); they have an affinity with calc-alkaline magmatic suites which lack Fe-enrichment in the intermediate differentiates and which display low Ti02 contents (< 1%). This comparison is confirmed by the high Th/Ta (10–20) and La/Ta (50–100) ratios which are typical of orogenic basalts. Furthermore, the representative points in Th-Hf-Ta and Th-Tb-Ta diagrams (Fig. 16) show a distribution which is entirely comparable to other formations of the Trégor volcanic arc domain (Cabanis 1986).

The contradiction between the orogenic geochemical affinity of the Paimpol spilites and their post-tectonic setting is not clearly understood, especially since the absolute age of this formation is not well established. Currently available data lead us to stress the post-tectonic character of the formation, thus favoring an interpretation in terms of a late intrabelt basin. However, more detailed evidence should be sought, particularly as regards the true age of eruption.

4.2 Ignimbritic Volcanism

The island of Jersey (Fig. 17) is situated in the midst of the late intrabelt basin and is characterized by intense volcanic activity near the Proterozoic-Paleozoic boundary. Volcanic rocks make up a complex succession about 3000 m thick broadly composed of andesitic flows at the base with a main sequence of rhyolites and ignimbrites above. These rocks are undeformed and only slightly weathered (Bishop and Keen 1982). The andesites display doleritic to microlitic textures and contain quartz, oligoclase-andesine and altered ferro-magnesian minerals (amphibole and/or pyroxene). No published geochemical results exist to demonstrate the andesitic composition of these rocks. The Jersey Andesites have been dated at 521 ± 15 Ma, with an initial 87 Sr/86 Sr ratio of 0.704 (Rb/Sr isochron; Duff 1978).

The main volcanic sequence is composed of a suite of porphyritic and glassy lavas, volcaniclastic or pyroclastic rocks of rhyolitic composition and well-developed ignimbrites showing typical textures (i. e., vitroclastic-devitrified, spherolitic, or perlitic). The explosive nature of these flows is confirmed by numerous interbeds of agglomerate and volcanic breccia. These lavas are hololeucocratic rocks, oversaturated in silica, showing a monotonous paragenesis with essential quartz, K-feldspar, and albitic plagioclase. The few available chemical analyses indicate hypersiliceous compositions (Si02 ≈ 75 %) and an alkali-enrichment trend (K20 + Na20 = 7–9 %, CaO < 1 %). Their location on a quartz-albite-orthoclase ternary diagram is near the eutectic minimum, thus suggesting a probable crustal origin, as often assigned to these types of rock. However, the lack of isotope or trace element data is a handicap in arguing such an interpretation.

In the Trégor (Fig. 17), volcanics and shallow intrusives are found near the margin of the Plouézec-Plourivo basin (Auvray 1975). These include the ignimbritic rhyolites of Lézardrieux – injected into faults which bound the basin – and the Loguivy microgranites which form a dense vein network in the eastern part of the North Trégor Batholith. They are undeformed, but are quite weathered. The volcanic rocks have a fluidal, lava-like or tuffaceous texture with local fiamme structures and vitroclastic fragments typical of ignimbrites. The microgranites have porphyritic and/or granophyric textures and display leucocratic parageneses containing quartz, albite and perthite. Geochemical data show the silicification (Si02 up to 80 %) and decalcification (Ca0 not detected) of the ignimbrites and an alkali affinity for the microgranites (Na20 + K20 ≈ 9 %). These rocks, which have often been compared with the Jersey ignimbrites, provide evidence for the extensional character of the late intrabelt basin. The ignimbrites and microgranites are dated, respectively, at 547 ± 12 Ma (Rb/Sr isochron; Auvray 1979) and 584 ± 24 Ma (Rb/Sr isochron; Auvray 1979) with corresponding initial 87Sr/86Sr ratios of 0.706 and 0.709. It should also be noted that these rocks display very similar characteristics to the Porz-Scarff Granite mentioned previously, which is intrusive into the North Trégor Batholith (Auvray 1979).

In the Cotentin (Fig. 17), the rhyolites and ignimbrites of Saint-Germain-le-Gaillard occupy a stratigraphic position above the Rozel and Carteret formations of uppermost Precambrian-basal Cambrian age. However, this ignimbritic volcanism is undated (in progress) and poorly characterized. It is composed of an association of generally porphyritic rhyolites, pyroclastic or volcaniclastic sediments, and vitroclastic ignimbrites (Boyer et al. 1972). Parageneses include quartz, albite, perthite, and occasional ferro-magnesian minerals. The few chemical data that are available indicate a composition close to the ternary minimum in the quartz-albite-orthoclase system – this reflects a probable partial melting origin. These ignimbritic flows could possibly be correlated with the Cambrian volcanic activity recorded in the Maine region (Le Gall et Cabanis 1985).

4.3 Intra-Continental Early Paleozoic Volcanism

Situated on the eastern side of the Baie de Saint-Brieuc (Fig. 17), the Erquy Spilites Formation was long considered to belong to the Lower Brioverian type succession (Graindor 1957; Cogné 1962). Due to radiometric and palynological dating, a Lower Paleozoic age is now assigned. However, its geochemistry is very similar to that of the Lanvollon Formation which forms part of the Brioverian succession on the western side of the Baie de Saint-Brieuc – certain authors still maintain a correlation between the Erquy and Lanvollon Formations (Roach et al. 1990).

The Erquy Formation is composed of a basal member, mainly represented by basic sills intruded into volcaniclastic sediments, and an upper part containing an association of pyroclastics, pillow lavas, and hyaloclastites. The parageneses are apilitic with albite, augite, chlorite, epidote, Fe-Ti oxides, and quartz (Auvray 1968). Furthermore, it may be noted that the Erquy spilites are associated with coarse-grained igneous rocks that crop out nearby: i.e., the Caroual gabbros (Auvray 1969).

The geochemical data reveal a fairly differentiated tholeiitic-type suite at Erquy and a clear calc-alkaline affinity for the Caroual gabbros. This difference is well marked as far as Ti02 contents are concerned, with values typical of continental tholeiites (2–3%) at Erquy contrasting with lower values at Caroual (ca. 1%) typical of orogenic basalts. The La/Ta ratios are around 20 for Erquy and between 50 and 100 for Caroual, thus confirming the major element discrimination (Cabanis 1986).

The geochemical characteristics of the Erquy spilites (see Table 2) are very similar to the Lanvollon amphibolites, as illustrated by the Th-Hf-Ta and Th-Tb-Ta diagrams (Fig. 16), where representative points show comparable distributions between the orogenic and extensional discriminant fields. This kind of intermediate setting is typical both of continental tholeiites and back-arc basin magmatic suites. The structural geology and age relations of the Erquy Formation enable a distinction to be drawn with the Lanvollon Formation. The latter showing widely different metamorphic facies and tectonic style. Because of this, the Erquy Formation may be assigned to the late intrabelt basin. The age of the Erquy lavas, dated at 482 ± 10 Ma (Rb/Sr isochron; Vidal et al. 1971) suggests rather that the eruption is linked to the early Paleozoic volcanism represented by the Plourivo trachy-andesites dated at 472 ± 5 Ma (Rb/Sr isochron; Auvray et al. 1980 b). This is also the case with the undated Mortes-Femmes Formation (Le Gall et al. 1986; Fig. 17) which shows features similar to the Erquy Formation.

4.4 Late-Stage Cadomian Plutons

The late intrabelt basin, as well as the adjoining parts of the Cadomian orogenic cordillera, are intruded by numerous late-stage plutons whose ages range from 560 to 480 Ma. The Porz-Scarff granite, dated at 557 ± 15 Ma, probably belongs to this suite – it is discussed above in the section on the North Trégor Batholith. Other plutons situated in the Trégor possibly also belong to this suite:

- the post-orogenic granites at Le Paon and Plouha are located on either side of the late intrabelt basin;
- the Beg-ar-Forn granodiorite, at the western extremity of the Trégor, has recently yielded an age of 528 ± 5 Ma (U/Pb zircon method; Guerrot 1989).

However, these plutons are most widely developed in the Channel Islands. The Guernsey Plutonic Complex described above has been dated at around 550 Ma (French and Keen 1986; de Brémond d'Ars 1986), or even younger for the latest intrusion of Cobo, which even gives a Cambro-Ordovician age (496 ± 13 Ma; Rb/Sr method; D'Lemos 1987); these young ages suggest that the complex belongs to the post-orogenic plutonic suite. On Jersey, there are three granitic massifs (in the SW, NW, and SE parts of the island) of relatively young age that are sometimes associated with gabbros and diorites; these massifs are typically post-orogenic in character and have yielded ages of 553 ± 12, 479 ± 15 and 509 ± 4 Ma, respectively (Rb/Sr isochron; Adams 1976). As on Guernsey, the petrogenesis of these plutons – although still poorly understood – may probably be sought in complex exchange mechanisms between crust and mantle during post-orogenic extension.

Finally, in the Cotentin, several small intrusive bodies are observed in the Icartian basement of the Cap de la Hague – these plutons display similar characteristics and have ages ranging from 520 to 480 Ma (Rb/Sr isochron; Power et al. 1990), thus being comparable to the post-orogenic suite elsewhere.

5 Tectonic Setting of Magmatism in the Cadomian Belt

As described in the previous sections, the various phases of Cadomian orogenesis are characterized by important manifestations of diverse magmatic activity whose petrological and geochemical features enable us to infer the tectonic regime of emplacement. These results also provide constraints on the proposed geodynamic model and, moreover, reveal a regional pattern of magmatic developement which is linked to structural zonation in the Cadomian belt (see Fig. 18) (Brown et al. 1990; Graviou et al. 1988).

During a pre-orogenic phase – which began around 670 Ma according to available radiometric ages – Brioverian volcanism occurred in an active margin / back-arc basin system associated with the de-

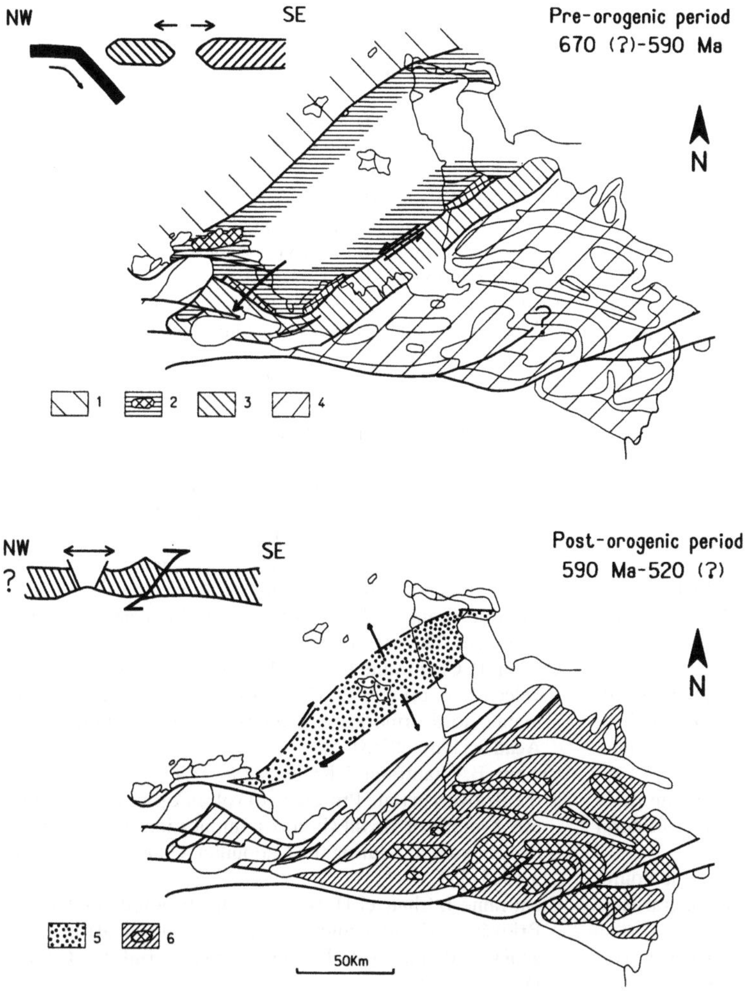

Fig. 18. Geodynamic evolution of magmatic activity in the Cadomian belt. *1* Icartian edge; *2* orogenic cordillera with main syntectonic plutons; *3* passive continental margin; *4* emergent continental platform (?); *5* late intrabelt basin; *6* molasse-type sedimentation on continental platform with Mancellian Batholith. The *arrows* indicate the sense of tectonic movements

velopement of subduction beneath the extreme outer edge of the Icartian block. Towards the inner part of the orogenic belt, volcanic activity typical of a distensive regime was developed on the borders of a passive continental margin. At present, these two domains are juxtaposed on either side of a major suture; they have been brought together by crustal shortening and thrusting of the orogenic cordillera onto the continent.

There is a complete lack of information on the geodynamic evolution of the external continental domain (Mancellia) during this early stage of the Cadomian orogeny.

Early plate tectonic events in the belt came to an end when the volcanic arc collided with the continental foreland, thus bringing about crustal shortening and closure of the back-arc basin. This was accompanied by a major phase of orogenic-type plutonism which is apparent at all levels in the tectono-metamorphic pile. Igneous activity is generally syn-tectonic in nature near the base of the thrust pile; here, intrusive rocks were deformed with their country rocks, and show high-grade parageneses. Towards the top of the pile, the North Trégor Batholith was intruded at a relatively high level in the crust. This orogenic phase of plutonism has been dated in several massifs at around 590–600 Ma.

A later stage of post-orogenic evolution corresponds to crustal overthickening due to the thrusting of tectonic slices within the orogenic Cordillera and, particularly, onto the adjoining continental margin. This has led to the hybridization of juvenile magmas injected into the cratonized cordillera during post-tectonic distensive movements. At the same time, there was important crustal remelting in the thrusted continental margin domain.

Post-orogenic magmatism is associated with the development of a late intrabelt basin within the cordillera, characterized by crustal-contaminated volcanism and, finally, the eruption of ignimbrites over a wide area. The genesis of late-stage basic calc-alkaline plutons, intrusive into the intrabelt basin or continental margin domain, has not yet been fully elucidated. However, it is thought that the associated granitoids, which are well developed on Jersey, have probably resulted from intra-continental hybridization.

The external structural domain (continental platform) is characterized by intense crustal anatexis near the margin – giving rise to migmatitic gneiss domes – but the emplacement of the Mancellian Batholith is still poorly understood. The age of post-tectonic magmatism in the Cadomian belt ranges from about 560 to 520 Ma, with some minor intrusive activity continuing to 480 Ma.

References

Adams CJ (1976) Geochronology of the Channel Islands and adjacent French mainland. J Geol Soc (Lond) 132: 233–250

Arnaud G (1980) L'intrusion gabbro-dioritique de Saint-Quay-Portrieux (Côtes-du-Nord). Thèse 3e cycle, Paris, 141 pp

Autran A, Chantraine J, Rabu D (1979) Lithostratigraphie et déformation du Briovérien de la baie de Lannion. Implications sur les relations entre les cycles cadomien et hercynien. Bull Bur Rech Géol Min 1–4: 277–292

Auvray B (1968) Le volcanisme spilitique de la Pointe de la Hussaye-~Erquy (Côtes-du-Nord). Bull Soc Géol Mineral Bretagne: 101–118

Auvray B (1969) Sur la nature des roches grenues de la Pointe de Caroual (Côtes-du-Nord). Bull Soc Géol Mineral Bretagne I-1: 1–9

Auvray B (1975) Relations entre plutonisme acide et volcanisme ignimbritique: exemple des manifestations magmatiques cambriennes du Nord de la Bretagne. Pétrologie I-2: 125–138

Auvray B (1979) Génèse et évolution de la croûte continentale dans le Nord du Massif armoricain. Thèse d'Etat, Rennes, 670 pp

Auvray B, Maillet P (1977) Volcanisme et subduction au Protérozoïque supérieur dans le Massif armoricain (France). Bull Soc Géol Fr XIX-5: 953–957

Auvray B, Macé J, Vidal P, Van Der Voo R (1980b) Rb-Sr dating of the Plouézec volcanics, N Brittany: implications for the age of red beds ("séries rouges") in the northern Armorican massif. J Geol Soc (Lond) 137: 207–210

Balé P (1986) Tectonique cadomienne en Bretagne nord. Interaction décrochement, chevauchement: champs de déformation et modélisation expérimentale. Thèse d'Université, Rennes, 361 pp

Balé P, Brun JP (1983) Les chevauchements cadomiens de la Baie de Saint-Brieuc (Massif armoricain). CR Acad Sci Paris 297: 359–362

Bardy P (1988) L'orogène cadomien dans le Nord-Est du Massif armoricain et en Manche occidentale. Thèse d'Université, Rennes, 380 pp

Barrat JA (1986) Pétrologie du magmatisme tardi-cadomien du domaine nord-armoricain: géochimie de l'intrusion de Saint-Quay. Hercynica II-2: 125–133

Boyer C, Roblot MM, Graindor MJ (1972) Les ignimbrites de Saint-Germain-le-Gaillard (Manche). CR Acad Sci Paris 274: 21–23

Brémond d'Ars J de (1986) Différenciation magmatique – approche théorique des phénomènes de second ordre: exemple du complexe plutonique de Guernesey. Diplôme, Rennes 67 pp (unpublished)

Brown M (1979) The petrogenesis of the Saint-Malo Migmatite Belt, Armorican Massif, France, with particular reference to the diatexites. N Jb Miner Abh 1: 48–74

Brown M, D'Lemos RS (1991) The Cadomian granites of Mancellia, northeast Armorican Massif of France: relationship to the St. Malo migmatite belt, petrogenesis and tectonic setting. Precambrian Res 51: 393–427

Brown M, Barber AJ, Roach RA (1971) Age of the Saint-Malo Migmatite Belt, Northern Brittany. Nat Phys Sci 236-66: 77–79

Brown M, Power GM, Topley CG, D'Lemos RS (1990) Cadomian magmatism in the North Armorican Massif. In: D'Lemos RS, Strachan RA, Topley CG (eds) The Cadomian Orogeny. Geol Soc Spec Publ 51: 181–213

Brun JP (1977) La zonation structurale des dômes gneissiques. Un exemple: le massif de Saint-Malo (Massif armoricain, France). Can Earth Sci 14: 1697–1707

Cabanis B (1986) Identification des séries magmatiques dans les socles métamorphiques. Exemples d'application dans le Massif armoricain et le Massif central. Implications géodynamiques. Thèse d'Etat, Paris, 493 pp

Cabanis B, Chantraine J, Rabu D (1986) Etude géochimique du volcanisme briovérien (Protérozoïque supérieur) du domaine nord armoricain; Implications géodynamiques. Cr Acad Sci Paris 303: 51–56

Cabanis B, Chantraine J, Rabu D (1987) Geochemical study of the Brioverian (late Proterozoic) volcanic rocks in the northern Armorican Massif (France). Implications for the geodynamic evolution during the Cadomian. Geol Soc Spec Publ 33: 525–539

Chantraine J, Chauvel JJ, Balé P, Denis E, Rabu D (1988) Le Briovérien (Protérozoïque supérieur à terminal) et l'orogenèse cadomienne en Bretagne (France). Bull Soc Géol Fr IV, 5: 815–829

Cogné J (1962) Le Briovérien: esquisse des caractères stratigraphiques, métamorphiques, structuraux et paléogéographiques de l'Antécambrien récent dans le Massif armoricain. Bull Soc Géol Fr IV-3: 413–430

Cogné J (1972) Le Briovérien et le cycle orogénique cadomien dans le cadre des orogènes fini-précambriens. Coll Int CNRS Paris 192; Notes Mém Serv Géol (Rabat) 236: 193–218

Dissler E (1987) Evolution géodynamique cadomienne du Nord-Cotentin (Massif armoricain). Thèse d'Université, Caen, 255 pp

D'Lemos RS (1987) The evolution of the Northern Igneous Complex of Guernsey, Channel Islands; some isotopic evidence. Proc Ussher Soc 6: 498–501

Duff BA (1978) Rb-Sr Whole-rock age determination of the Jersey Andesite Formation, Jersey CI J Geol Soc (Lond) 135: 153–156

Dupret L, Cabanis B, Le Gall J, Doré F (1985) Les tholéiites d'arc briovériennes des Montsurvent (Manche). Leur place dans l'évolution orogénique cadomienne du Nord de l'Armorique. Hercynica I-1: 39–46

Fouillac AM, Cocherie A, Rossi P, Calvez JY, Autran A (1986) Etude géochinique du Batholite mancellien. Rap Bur Rech Géol Min (unpublished), 15 pp

French WJ, Keen DH (1986) Geological map of Guernsey. Channel Islands Sheet 1–1/25000. British Geological Survey, London

Graindor MJ (1957) Le Briovérien dans le Nord-Est du Massif armoricain. Mém Carte Géol Fr Bur Rech Géol Min, 211 pp

Graviou P (1984) Pétrogenèse des magmas calco-alcalins: exemple des granites cadomiens de la région trégorroise (Massif armoricain). Thèse 3e cycle, Rennes, 236 pp

Graviou P, Auvray B (1985) Caractérisation pétrographique et géochimie des granitoïdes cadomiens du domaine nord-armoricain: implications géodynamiques. CR Acad Sci Paris 301: 315–318

Graviou P, Peucat JJ, Auvray B, Vidal P (1988) The Cadomian Orogeny in the Northern Armoican Massif: petrological and geochronological contraints on a geodynamic model. Hercynica IV-1: 1–13

Guerrot C (1989) Archéen et Protérozoïque dans la chaîne hercynienne ouest-européenne: géochimie isotopique (Sr, Nd, Pb) et géochronologie U/Pb sur zircons. Thèse d'université, Rennes, 164 pp

Guerrot C, Peucat JJ, Dupret L (1989) Données nouvelles sur l'âge du système briovérien (Protérozoïque supérieur) dans le Nord du Massif armoricain. CR Acad Sci Paris 308: 88–92

Lees GS, Roach RA, Shufflebotham MM, Griffiths NH (1987) Upper Proterozoic basaltic volcanism in the northern Massif armoricain, France. Geochemistry and mineralization of Proterozoic volcanism suites. Geol Soc Sp Publ London, pp 503–523

Le Gall J, Barrat JA (1987) Pétrologie du magmatisme tardicadomien du domaine nord-armoricain: l'exemple des complexes basiques d'Ernée et de Trégomar. Géologie de la France, Bur Rech Géol Min 1: 3–22

Le Gall J, Cabanis B (1985) Caractères minéralogiques et géochimiques des ignimbrites et des laves cambriennes de l'Est du Massif armoricain. Implications pétrogénétiques. Hercynica I-2: 99–109

Le Gall J, Mary G (1983) Place et signification du complexe basique de Brée dans l'histoire cadomo-varisque de l'Est du Massif armoricain. Bull Soc Géol Mineral Bretagne 15-2: 169–180

Le Gall J, Dissler J, Dupret L (1986) Signification géodynamique des volcanismes briovériens dans le nord-est du Massif armoricain. CR Acad Sci Paris 303: 1587–1592

Pasteels P, Doré F (1982) Age of the Vire-Carolles granite. In: Odin GS (ed) Numerical Dating in Stratigraphy. John Wiley, New-York, pp 784–790

Peucat JJ (1982) Géochronologie des roches métamorphiques (Rb-Sr et U-Pb). Exemples choisis au Groenland, en Laponie et dans le Massif armoricain. Thèse d'Etat, Rennes. Mém Soc Géol Minéral Bretagne 28 (1983), Rennes, 158 pp

Peucat JJ, Hirbec Y, Auvray B, Cogné J, Cornichet J (1981) Late Proterozoic zircon ages from a basic-ultrabasic complex: a possible Cadomian orogenic complex in the Hercynian belt of western Europe. Geology 9: 169–173

Power GM, Brewer TS, D'Lemos RS (1990) The Cadomian plutonic complex of La Hague, Manche, N. France. In: D'Lemos RS, Strachan RA, Topley CG (eds) The Cadomian Orogeny. Geol Soc Spec Publ 51: 215–229

Rabu D, Chauvel JJ, Chantraine J (1983) Nouvelles propositions pour la lithostratigraphie du Briovérien (Protérozoïque supérieur) et pour l'évolution géodynamique en baie de St Brieuc. Bull Soc Géol Fr XXV-4: 615–621

Roach RA, Lees GJ, Shufflebotham MM (1990) Brioverian volcanism and Cadomian tectonics, Baie de St Brieuc, Brittany. In: D'Lemos RS, Strachan RA, Topley CG (eds) The Cadomian Orogeny. Geol Soc Spec Publ51: 41–67

Topley CG, Brown M, Power GM, D'Lemos RS, Roach RA (1990) The northern igneous complex of Guernsey, Channel Islands. In: D'Lemos RS, Strachan RA, Topley CG (eds) The Cadomian Orogeny. Geol Soc Spec Publ 51: 245–259

Verdier (1968) Etude pétrographique et structurale du Trégor occidental (baie de Lannion). Thèse 3e cycle, Strasbourg, 51 pp

Vidal P (1976) L'évolution polyorogénique du Massif armoricain: apport de la géochronologie et de la géochimie isotopique du strontium. Thèse d'Etat, Rennes. Soc Géol Minéral Bretagne 21 (1980), Rennes, 162 pp

Vidal P, Auvray B, Cogné J, Hameurt J, Jeannette D (1971) Données géochronologiques sur la série spilitique d'Erquy: problèmes nouveaux à propos du Briovérien de Bretagne septentrionale. CR Acad Sci Paris 273: 132–135

Vidal P, Auvray B, Chauvet JF, Cogné J (1972) L'âge radiométrique de la diorite de Saint-Quay-Portrieux – ses conséquences sur le Briovérien de la baie de Saint-Brieuc. CR Acad Sci Paris 279: 631–634

Vidal P, Deutsch S, Martineau F, Cogné J (1974) Nouvelles données radiométriques en Baie de Saint-Brieuc. Le problème d'un socle antécadomien nord armoricain. CR Acad Sci Paris 279: 631–634

Wood DA, Joron JL, Treuil M (1979) A re-appraisal of the use of trace element to classify and discriminate between magma series erupted in different tectonic settings. Earth Planet Sci Lett 45: 326–336

Conclusions

J. CHANTRAINE, B. AUVRAY, J. P. BRUN, J. J. CHAUVEL, and D. RABU (translated by M. S. N. CARPENTER)

1 Introduction

A model for Cadomian orogenesis, built up in recent years by the various authors working on the preserved segment of this belt in the Armorican Massif, can be integrated into the geodynamic context of the evolution of an active margin on the northern edge of a continent that was affected by Pan-African tectonic events and dislocated by pre-Variscan rifting.

The development of Cadomian orogenesis resulted from subduction of the Celtic Ocean crust beneath the 2000-Ma-old Icartian continental edge during the Late Proterozoic. Since Lefort (1975) has interpreted the Mid-Channel magnetic anomaly as the trace of a buried ocean, most authors (Auvray and Maillet 1977; Cogné and Wright 1980; Rabu et al. 1983; Graviou and Auvray 1985; Chantraine et al. 1988; Rabu et al. 1990) have accepted this as the Celtic Ocean suture. Within such a scheme, the following synopsis (Fig. 19) can be proposed for the Cadomian orogeny in the Armorican Massif.

2 Subduction of the Celtic Oceanic Crust Beneath the Icartian Continent: Birth of an Arc-Back-Arc System

Associated with the subduction of the Celtic Ocean crust, an active continental margin became established on the NW edge of the Icartian continent, leading to the development of island-arc type magmatism. Volcanic activity began with substantial acidic volcanism, of at least partly crustal origin, dated at about 640 Ma. This activity continued with the eruption of basalts of calc-alkaline affinity, related to the North Trégor Batholith, which was emplaced during crustal shortening around 615 Ma ago. It should be noted, however, that early plutonic activity at about 660 Ma has been recognized in the form of reworked clasts in some conglomerates.

Behind the volcanic arc domain, extension of the continental margin provoked the opening of a marginal basin in which a thick succession of intermediate orogenic/continental andesites accumulated, associated with thick volcaniclastic turbidites (Chantraine et al. 1988). In the southeast, on the passive continental margin, terrigenous sediments were deposited containing, on the one hand, distinctive bedded black cherts and, on the other hand, volcanic intercalations of extensional tholeiitic character.

3 Other Subduction Hypotheses

Although subduction-related processes generated the Cadomian orogenic belt, controversy still remains concerning the location and dip of the subduction zone, since no direct evidence has been preserved. Whatever the exact geometry of subduction involved, the generally accepted model of geodynamic evolution described above is only slightly affected.

Classically, the Mid-Channel magnetic anomaly is interpreted as the scar left over from Celtic Ocean subduction (Lefort 1975; later modified by Auvray 1979). Southward-directed subduction would then account for the observed zonation in magmatic activity (Graviou et al. 1988), as well as the migration of deformation from north to south (Balé and Brun 1989). Now supported by further evidence, this hypothesis appears to be the most straightforward way of explaining Cadomian orogenic history (Chantraine et al. 1988). However, the real significance of the Mid-Channel magnetic anomaly is still poorly understood. Brun and Balé (1990) have proposed that this anomaly is the westward-displaced equivalent of similar anomalies found in the Saint-Brieuc area; these authors consider the later anomalies as the trace of Cadomian subduction which occurred along the main suture of the Cadomian orogen (MCT).

Such a hypothesis implies the previous opening of an ocean between the Icartian edge and the continental platform domain to the southeast. This ocean crust would have been totally consumed by northward subduction beneath the Trégor Arc before a final phase

J. D. Keppie (Ed.): Pre-Mesozoic Geology in France and Related Areas

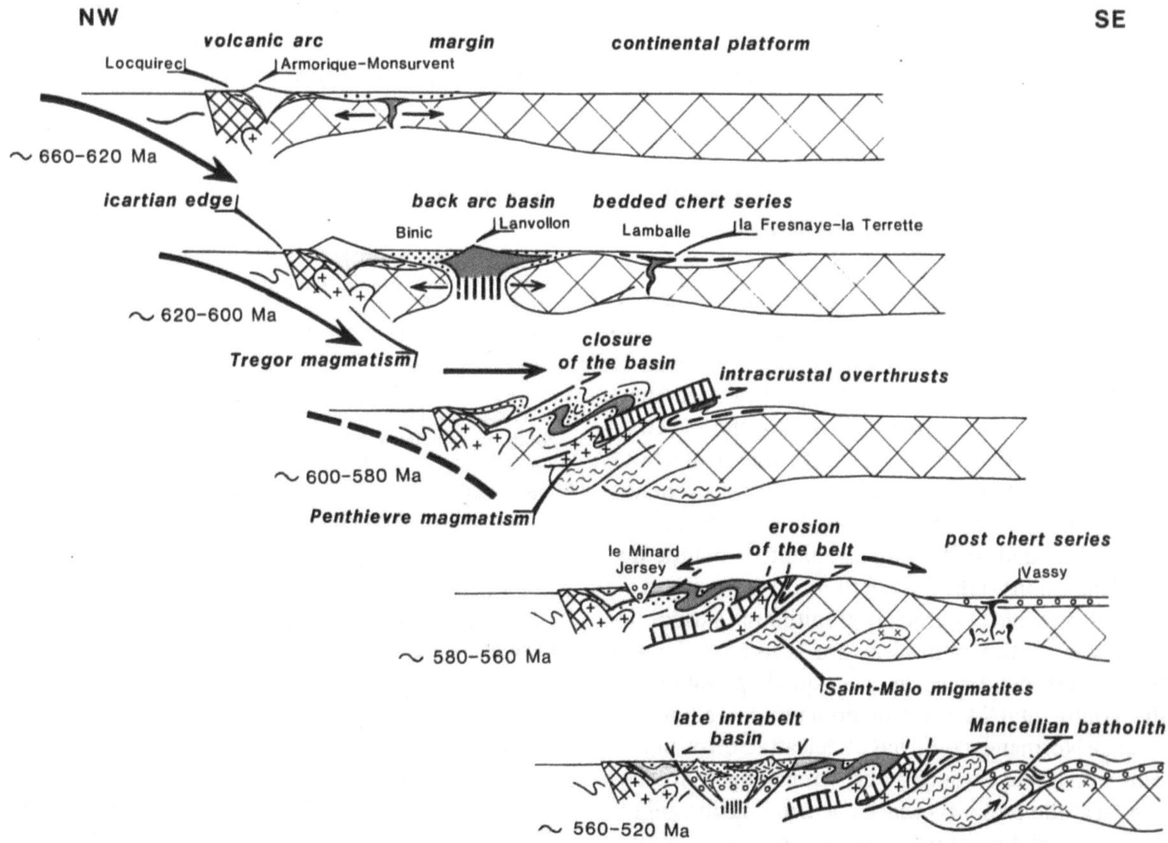

Fig. 19. Schematic interpretation of the Cadomian evolution in Armorican Massif. (After Chantraine et al. 1988)

of obduction onto the continental margin in the southeast. This continental margin would thus correspond to the boundary between oceanic and continental crust subjected to shearing deformation during thrusting of the tectonic pile. Under this condition, the Saint-Brieuc Basin can be interpreted as a fore-arc basin underlain by thinned Icartian basement. This, however, requires further investigation. In the absence of other arguments to support the hypothesis of Brun and Balé – which concerns only the geodynamic evolution of the Saint-Brieuc area – the classic interpretation of the Mid-Channel anomaly should be retained as a general framework for understanding the Cadomian orogenic belt in the Armorican Massif.

4 Collision and Obduction of the Back-Arc Basin: Rise of the Orogenic Cordillera

Blocking of the subduction system was brought about by the closure of the Celtic Ocean and collision of the Icartian continental margin with a northern continental plate. The effects of this collision appear to have been limited to the marginal domain, without major thrusting of the internal units of the orogen over the external ones. The tectonic shortening involved the deformation of the Cadomian orogenic belt, dated at between 615 and 585 Ma by the associated plutonism. Thrusting of the orogenic belt (cordillera) over the continental margin provoked the occlusion of the marginal basin and continued with the obduction of the segment of crust generated in this basin. The interaction of southeastward thrusting and sinistral wrenching, interpreted in a model of oblique convergence (Balé and Brun 1989), takes into account the arcuate shape of the belt and suggests continental accretion rather than frontal collision.

The orogenic overthrusting led to overthickening of the crust, which in turn led to crustal anatexis, with development of the diapiric Guingamp-St Malo migmatites and the incorporation of the continental margin into the orogenic domain during southward migration of the tangential tectonics. The anatectic rocks and those affected by the accompanying deformation were intruded by plutons dated at 540 Ma, just after the erosion of the orogenic belt supplied the late-intrabelt basin (in the Minard, Jersey and Cotentin areas) and the external continental domain (in the Bocage Normand and Central Brittany).

5 Cratonisation of the Orogenic Domain; Deposition of the Cadomian Molasse

After cratonization of the Cordillera, the effects of the Cadomian orogeny migrated southeastward into the external domains. Crustal thinning occurred in the Bocage Normand and the post-bedded-chert Brioverian was deformed only during the development of the Mancellian Batholith dated at about 540 Ma. The petrogenesis of these plutons remains unclear; it could have been initiated by crustal anatexis, itself related to the emplacement of mantle-derived intrusions at the base of the crust (Graviou et al. 1988; Brown and D'Lemos 1991) although surface occurrences of this plutonism are scarce (Brée and Ernée gabbros). However, it is equally possible that the continental Mancellian domain (including the Bocage Normand) could have been affected by imbricate transverse thrusting and wrenching – not observed at the present – which occurred during migration of the deformation and which produced partial melting by crustal thickening.

The late intrabelt basin is filled with a discontinuous accumulation of terrigenous sediments which were deposits near the Precambrian/Cambrian boundary. It is also characterized by voluminous ignimbritic and intermediate volcanism of possible crustal origin. The intrabelt succession is only weakly deformed, but is intruded by numerous plutons of hybrid type which were emplaced over a span of time extending up into the Ordovician.

In Central Brittany, the crustal thinning regime continued from the uppermost Proterozoic into the early Paleozoic, without interruption from the intrusion of granitic rocks. The Brioverian succession here was not deformed until the Variscan orogeny. Nevertheless, at Plouguenast, for example, it is still not known whether some gneiss domes could have been formed before the main phase of extension that resulted in the main transgression of the Ordovician unconformably over the Brioverian just before the opening of the proto-Tethys Ocean.

References

Auvray B (1979) Génèse et évolution de la croûte continentale dans le Nord du Massif Armoricain. Thèse d'Etat, Rennes, 670 pp

Auvray B, Maillet P (1977) Volcanisme et subduction au Protérozoïque supérieur dans le Massif armoricain (France). Bull Soc Géol Fr XIX-5: 953–957

Balé P, Brun JP (1989) Late Precambrian thrusts and wrench zones in Northern Brittany (France). Struct Geol 11-4: 391–405

Bronw M, D'Lemos RS (1991) The Cadomian granites of Mancellia, northeast Armorican Massif of France: relationship to the St. Malo migmatite belt, petrogenesis and tectonic setting. Precambrian Res 51: 393–427

Brun JP, Balé P (1990) Cadomian tectonics in Northern Brittany. In: D'Lemos RS, Strachan RA, Topley CG (eds) The Cadomian Orogeny. Geol Soc Spec Publ 51: 95–114

Chantraine J, Chauvel JJ, Balé P, Denis E, Rabu D (1988) Le Briovérien (Protérozoïque supérieur à terminal) et l'orogenèse cadomienne en Bretagne (France). Bull Soc Géol Fr IV, 5: 815–829

Cogné J, Wright AE (1980) L'orogène cadomien. Vers un essai d'interprétation paléogéodynamique unitaire des phénomènes orogéniques fini-précambriens d'Europe moyenne et occidentale. In: Coll Géologie de l'Europe. 26ème C.G.I., Paris; Mém Bur Rech Géol Min 108: 29–55

Graviou P, Auvray B (1985) Caractérisation pétrographique et géochimie des granitoïdes cadomiens du domaine nord-armoricain: implications géodynamiques. CR Acad Sci Paris 301: 315–318

Graviou P, Peucat JJ, Auvray B, Vidal P (1988) The Cadomian Orogeny in the Northern Armorican Massif: petrological and geochronological contraints on a geodynamic model. Hercynica, IV-1: 1–13

Lefort JP (1975) Le socle péri-armoricain. Etude géologique et géophysique du socle submergé à l'Ouest de la France. Thèse d'Etat, Rennes. Mém Soc Géol Fr Paris 130, (1977), 68 pp

Rabu D, Chauvel JJ, Chantraine J (1983) Nouvelles propositions pour la lithostratigraphie du Briovérien (Protérozoïque supérieur) et pour l'évolution géodynamique en baie de St Brieuc. Bull Soc Géol Fr XXV-4: 615–621

Rabu D, Chantraine J, Chauvel JJ, Denis E, Balé P, Bardy P (1990) The Brioverian (late Proterozoic) and the Cadomian Orogeny in the Armorican Massif. In: D'Lemos RS, Strachan RA, Topley CG (eds) The Cadomian Orogeny. Geol Soc Spec Publ 51: 91–94

Part II B
The Variscan Orogeny in the Armorican Massif

The Variscan

F. PARIS and M. ROBARDET

The major subdivisions in the Variscan of the Armorican Massif are based either upon the Paleozoic faunas and sedimentary successions or upon the tectono-metamorphic and geodynamic evolution of the different zones or terranes.

The sedimentary and faunal characteristics allow distinction of (Fig. 1):

1. a large and rather homogenous Medio-North Armorican Domain (Normandy, Trégor, northern and central Brittany);
2. South-Armorican Domains (southern Brittany, Ligeria, Vendée) comprising distinct allochtonous units with peculiar faunas and deposits.

The limit between these two domains runs along the northern branch of the South-Armorican Shear Zone. However, the true geotectonic boundary might correspond to the Nort-sur-Erdre fault, as the Paleozoic Series of the Saint-Georges-sur-Loire unit have probably been thrust toward the north.

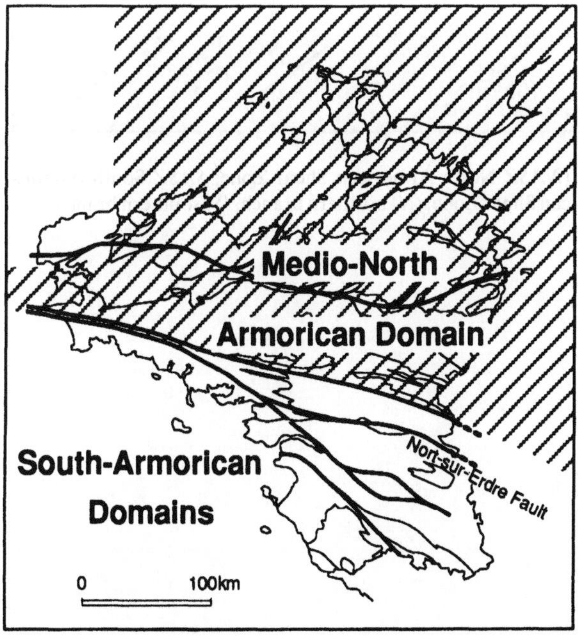

Fig. 1. The main paleogeographic domains of the Armorican Massif

The tectono-metamorphic evolution of the Armorican Massif leads to distinction of (Fig. 2):

1. an "internal Zone" (Cordillière ligérienne sensu Cogné 1976) including various high grade metamorphic units (ortho or para origin) with few oceanic and high pressure metamorphism relics. This zone comprises different domains, e. g., southern Brittany, western Vendée and Léon (displaced terrane);
2. an "external Zone" comprising the weakly or non-metamorphozed Paleozoic Series of Normandy, Trégor, central Brittany, Ligeria, and eastern Vendée.

The southern branch of the South-Armorican Shear Zone separates these two zones. The chronology of the main tectonometamorphic episodes shows obvious differences in the two regions with an early (Ordovician and Silurian) and strong development in the "internal zone" and a late (Carboniferous) and weak development in the "external zone".

Reference

Cogné J (1986) La chaîne hercynienne ouest-européenne correspond-elle à un orogène par collision? Propositions pour une interprétation géodynamique globale. Colloq Int CNRS 268: 111–129

J. D. Keppie (Ed.): Pre-Mesozoic Geology
in France and Related Areas

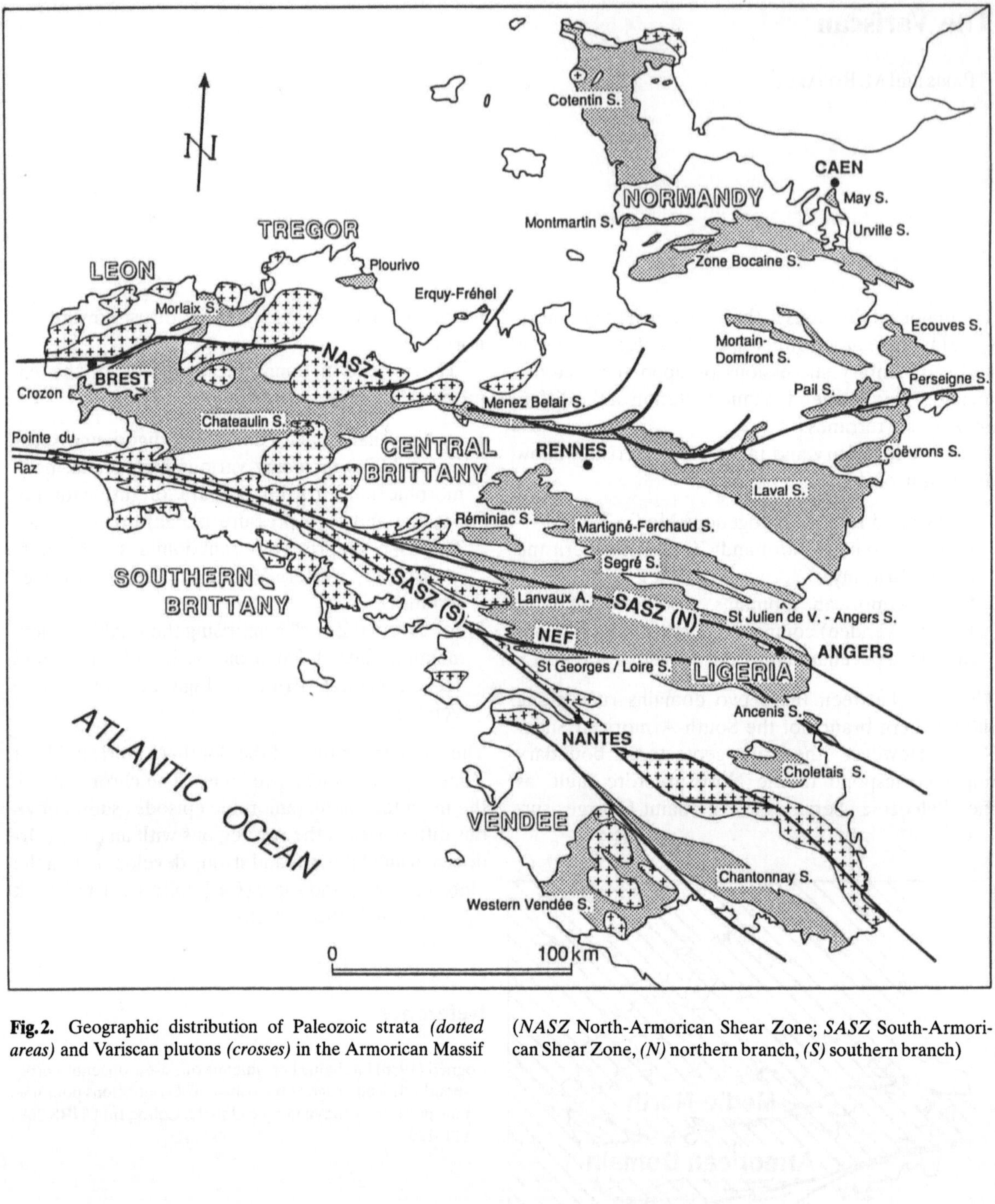

Fig. 2. Geographic distribution of Paleozoic strata *(dotted areas)* and Variscan plutons *(crosses)* in the Armorican Massif (*NASZ* North-Armorican Shear Zone; *SASZ* South-Armorican Shear Zone, *(N)* northern branch, *(S)* southern branch)

Introduction

F. PARIS and M. ROBARDET

The Armorican Paleozoic deposits extend from the lowermost Cambrian up to the lower Permian. The pre-orogenic and syn-orogenic sediments are preserved within E-W and SE-NW folded structures separated by Precambrian outcrops (Fig. 2). The Upper Carboniferous and Lower Permian post-orogenic rocks rest unconformably on older strata and constitute small subsident lacustrine basins.

During the Paleozoic the Armorican Massif was part of the North Gondwana margin, which also included other regions of southern Europe and North Africa. Both paleomagnetic and paleoclimatic data indicate that these areas drifted northward during Paleozoic times, from southern high latitudes in the Ordovician to a subequatorial position in the Carboniferous (Perroud 1985; Paris and Robardet 1990). Consequently, terrigenous deposits with poorly diversified faunas prevailed during the Early Paleozoic whereas carbonates and reefs developed only from the Devonian onwards, the whole Paleozoic succession being represented mainly by shelf deposits.

Generally, macro- and microfossils that occur in the Armorican Paleozoic deposits give precise stratigraphic age assignments. On the other hand, both lithologic and paleontologic data also help to define paleonvironments and to evaluate sedimentary and faunal affinities among the distinct terranes that now compose the Armorican Massif and, on a larger scale, southwestern Europe. Thus it appears that all the regions extending north of the northern branch of the South Armorican Shear Zone [SASZ(N) of Fig. 2], running from Pointe-du-Raz to Angers, contain Early Paleozoic sequences (up to the Devonian included) with rather uniform and coherent characteristics. These regions constitute the Medio-North Armorican Domain (Paris and Robardet 1977). The Early Paleozoic formations that occur to the south have different faunal and sedimentary features. These terranes (Ligeria, Vendée) most probably consist of several units tectonically juxtaposed during Variscan times and they will be studied separately. In addition, part of the Lower Paleozoic rocks of these terranes are now regarded as olistoliths within Early Carboniferous deposits (see Ligerian and Vendean Domains, this Vol.).

Paleontological and stratigraphic studies have been carried out of the Paleozoic of the Armorican Massif for more than 100 years and have resulted in numerous publications that cannot be listed completely here. A selected list of references must include:

1. The main paleontological monographs concerning different fossil groups, such as Ordovician cystoids (Chauvel 1941), Devonian brachiopods (Renaud 1942; Racheboeuf 1981; Gourvennec 1989), graptolites (Philippot 1950), acritarchs (Deunff 1958; Henry 1969), bivalves (Babin 1966), tentaculites (Lardeux 1969b), Ordovician trilobites (Henry 1980) and Devonian trilobites (Pillet 1972; Morzadec 1983a), chitinozoans (Paris 1981), Devonian crinoids (Le Menn 1985), Ordovician ostracods (Vannier 1986a, b), Carboniferous corals (Vuillemin 1990), and Silurian and Devonian spores (D'Erceville 1979; Le Herissé 1983).
2. Collective memoirs concerning Lower Devonian biostratigraphy (Lardeux 1976; Morzadec et al. 1981; Plusquellec 1980).
3. Regional monographs on the Cambrian (Doré 1969), the Ordovician and the Silurian (Chauvel 1968; Wyns 1980; Régnault 1981; Robardet 1981; Guillocheau 1983; Durand 1984; Bonjour 1988), the Devonian (Poncet 1968; Morzadec 1983b; Dubreuil 1986; Taillebois 1987) and the Carboniferous (Pelhate 1971); and the Paleozoic series of Vendée (Ters 1979; Poncet 1992).
4. Stratigraphic synthetic reviews (Babin et al. 1976; Lardeux et al. 1977; Paris et al. 1986; Morzadec et al. 1988).

In addition, the explanatory notes of recent geological maps, published by the Bureau de Recherches Géologiques et Minières (BRGM), contain numerous and detailed stratigraphic and paleontologic information.

The references listed above will not be referred to in subsequent chapters.

J. D. Keppie (Ed.): Pre-Mesozoic Geology
in France and Related Areas

References

Babin C (1966) Mollusques Bivalves et Céphalopodes du Paléozoïque armoricain. Thèse, Rennes University. Imp. commerciale et administrative Brest, pp 1–470

Babin C, Arnaud A, Blaise J, Cavet P, Chauvel JJ, Deunff J, Henry JL, Lardeux H, Mélou M, Nion J, Paris F, Plaine J, Quété Y, Robardet M (1976) The Ordovician of the Armorican Massif (France). In: Bassett MG (ed) The Ordovician System. University Wales Press and Nat Mus Wales, Cardiff, pp 359–385

Babin C, Deunff J, Mélou M, Paris F, Pelhate A, Plusquellec Y, Racheboeuf P (1979) La coupe de Porz Ar Vouden (Pridoli de la Presqu'ile de Crozon) Massif Armoricain, France. Lithologie et Biostratigraphie. Palaeontogr Abt A 164: 52–84

Bonjour JL (1988) Sédimentation paléozoïque initiale dans le Domaine Centre-armoricain (Massif Armoricain). Thèse, Rennes University, pp 1–257

Chauvel J (1941) Recherches sur les Cystoïdes et les Carpoïdes armoricains. Mém Soc Géol Minéral Bretagne 5: 1–286

Chauvel JJ (1968) Contribution à l'étude des minerais de fer de l'Ordovicien inférieur de Bretagne. Mém Soc Géol Minéral Bretagne 16: 1–243

D'Erceville MA (1979) Les spores des formations siluro dévoniennes de la coupe de Saint-Pierre-sur-Erve (Synclinorium médian armoricain). Palaeontogr Abt B 171: 79–121

Deunff J (1958) Microorganismes planctoniques du Primaire armoricain. Ordovicien du Veryhac'h (presquîle de Crozon). Bull Soc Géol Minéral Bretagne NS 2: 1–41

Doré F (1969) Les formations cambriennes de Normandie. Thèse, Caen University, pp 1–790

Dubreuil M (1986) Evolution géodynamique du Paléozoïque Ligérien (Massif Armoricain). Thèse, Nantes University, pp 1–258

Durand J (1984) Le Grès armoricain. Sédimentologie, traces fossiles, milieux de dépôt. Mém Doc CAESS Rennes 3: 1–150

Gourvennec R (1989) Brachiopodes Spiriferida du Dévonien inférieur du Massif Armoricain. Systématique, Paléoécologie, Evolution, Biostratigraphie. Biostratigraphie du Paléozoïque Mem 9: 1–281

Henry JL (1969) Micro-organismes incertae sedis (Acritarches et Chitinozoaires) de l'Ordovicien de la presqu'île de Crozon (Finistère): gisements de Mort-Anglaise et de Kerglintin. Bull Soc Géol Minéral Bretagne NS: 59–100

Henry JL (1980) Trilobites ordoviciens du Massif Armoricain. Mém Soc Géol Minéral Bretagne 22: 1–250

Lardeux H (1969b) Les Tentaculites d'Europe occidentale et d'Afrique du Nord. Cahiers Paléont CNRS, Paris, pp 1–238

Lardeux H (ed) (1976) Les Schistes et Calcaires eodévoniens de Saint-Céneré (Massif Armoricain, France); Sédimentologie, Paléontologie, Stratigraphie. Mém Soc Géol Minéral Bretagne 19: 1–328

Lardeux H, Chauvel JJ, Henry JL, Morzadec P, Paris F, Racheboeuf P, Robardet M (1977) Evolution géologique du Massif Armoricain au cours des temps ordoviciens, siluriens et dévoniens. Colloq Int CNRS 243: 181–192

Le Hérissé A (1983) Les spores du Dévonien inférieur du Synclinorium de Laval (Massif armoricain). Palaeontogr Abt B 188: 1–81

Le Menn J (1985) Les Crinoïdes du Dévonien inférieur et moyen du Massif Armoricain. Systématique, Paléobiologie, Evolution, Paléoécologie, Biostratigraphie. Mém Soc Géol Minéral Bretagne 30: 1–268

Morzadec P (1983a) Trilobites du Dévonien (Emsien-Famennien) de la Rade de Brest (Massif Armoricain). Palaeontogr Abt A 181: 103–184

Morzadec P (1983b) Le Dévonien (Emsien-Famennien) de la rade de Brest (Massif armoricain). Lithologie, cartographie, stratigraphie, paléogéographie. Géol Fr 4: 269–310

Morzadec P, Paris F, Racheboeuf P (eds) (1981) La tranchée de la Lézais, Emsien supérieur du Massif armoricain. Sédimentologie, Paléontologie, Stratigraphie. Mém Soc Géol Minéral Bretagne 24: 1–313

Morzadec P, Paris F, Plusquellec Y, Racheboeuf P, Weyant M (1988) Devonian stratigraphy and palaeogeography of the Armorican Massif (NW France). Second Intern Symposium Devonian System. Calgary 1987. Can Soc Petrol Geol Mem 14 1: 401–420

Paris F (1981) Les Chitinozoaires dans le Paléozoïque du Sud-Ouest de l'Europe (Cadre géologique – étude systématique – biostratigraphie). Mém Soc Géol Minéral Bretagne 26: 1–402

Paris F, Robardet M (1977) Paléogéographie et relations ibéro-armoricaines au Paléozoïque anté-Carbonifère. Bull Soc Géol Fr 7 19: 1121–1126

Paris F, Robardet M (1990) Early Palaeozoic palaeobiogeography of the Variscan regions. Tectonophysics 177: 193–213

Paris F, Morzadec P, Le Hérissé A, Pelhate A (1986) Late Devonian-Early Carboniferous events in the Armorican Massif (Western France): a review. Ann Soc Geol Belg 109: 187–195

Pelhate A (1971) Le Carbonifère inférieur du Bassin de Laval, Massif Armoricain. Mém Soc Géol Minéral Bretagne 15: 1–315

Perroud H (1985) Synthèse des résultats paléomagnétiques sur le Massif Armoricain. Hercynica 1 1: 65–71

Philippot A (1950) Les Graptolites du Massif Armoricain. Etude stratigraphique et paléontologique. Mém Soc Géol Minéral Bretagne 8: 1–295

Pillet J (1972) Les Trilobites du Dévonien inférieur et du Dévonien moyen du Sud-Est du Massif Armoricain. Mém Soc Etud Sci Anjou 1: 1–307

Plusquellec Y (ed) (1980) Les Schistes et Calcaires de l'Armorique (Dévonien inférieur, Massif Armoricain). Sédimentologie, Paléontologie, Stratigraphie. Mém Soc Géol Minéral Bretagne 23: 1–317

Poncet J (1968) Contribution à l'étude sédimentologique et stratigraphique du Dévonien de Basse Normandie. Thèse, Caen University, pp 1–367

Poncet D (1992) Analyse pétrostructurale et étude de la déformation dans les granitoïdes hercyniens du Haut Bocage Vendéen. Thèse, Poitiers University (in press)

Racheboeuf P (1981) Chonetacés (Brachiopodes) siluriens et dévoniens du Sud-Ouest de l'Europe (Systématique, Phylogénie, Biostratigraphie, Paléobiogéographie). Mém Soc Géol Minéral Bretagne 27: 1–294

Régnault S (1981) Stratigraphie et structure du Paléozoïque dans le Ménez-Bélair occidental (Synclinorium médian armoricain). Bull Soc Géol Minéral Bretagne 13 1: 1–105

Renaud A (1942) Le Dévonien du Synclinorium médian Brest-Laval. Mém Soc Géol Minéral Bretagne 7 1: 1–184, 2: 1–438

Robardet M (1981) Evolution géodynamique du Nord-Est du Massif Armoricain au Paléozoïque. Mém Soc Géol Minéral Bretagne 20: 1–342

Taillebois E (1987) Cadre géologique des indices sulfurés à Zn, Pb, Cu, Fe, du secteur de Gouezec-St Thois, Dévono-Carbonifère du flanc sud du Bassin de Châteaulin (Finistère). Mém Doc CAESS Rennes 16: 1–159

Ters M (1979) Les synclinoriums paléozoïques et le Précambrien sur la façade occidentale du Massif vendéen. Stratigraphie et structure. Bull Bur Rech Geol Min I 4: 293–301

Vannier J (1986a) Ostracodes Binodicopa de l'Ordovicien (Arenig-Caradoc) ibéro-armoricain. Palaeontogr Abt A 193: 77–143

Vannier J (1986b) Ostracodes Palaeocopa de l'Ordovicien (Arenig-Caradoc) ibéro-armoricain. Palaeontogr Abt A 193: 145–218

Vuillemin C (1990) Les Tetracoralliaires (Rugosa) du Carbonifère inférieur du Massif Armoricain. Cahiers Paléont CNRS Paris, pp 1–225

Wyns R (1980) Contribution à l'étude du Haut-Bocage vendéen: le Précambrien et le Paléozoïque dans la région de Chantonnay (Vendée). Thèse 3ème cycle, Paris VI University, pp 1–124

Cambrian of the Armorican Massif

F. Doré

1 Introduction

In the Armorican Massif, Cambrian rocks outcrop in the northeast (Normandy and Maine provinces) and to a lesser extent in the south (Vendée province). Brittany was mostly emerged and was only covered during the Arenig or pre-Arenig transgression. This partition roughly reflects the paleogeography (Fig. 3). The present geological setting results from a Cadomian structural heritage, and a Cambrian mobility caused by spreading and subsidence. Cambrian sediments unconformably overlie the Cadomian basement; however, the Variscan remobilization is observed in the far north and south. Deposits are mostly detrital with some carbonates in the Lower Cambrian; the sedimentary environments varied from fluvial-deltaic to shallow marine. Unfortunately these sediments contain only rare fossils. A typical Lower Cambrian fauna is known only in North Cotentin; Middle and Upper Cambrian outcrops are identified locally in the Vendée area. The upper parts of Cambrian sections generally contain some red-bed formations which might be Lower Ordovician in age, like those of Central Brittany. Under these conditions, reliable correlations are very difficult to establish between the different basins of the Armorican Massif and also with the other Cambrian rocks of the world. Post-orogenic volcanic activity occurred during the Cambrian, mostly in the eastern and southern parts of the Armorican Massif. Because of the Variscan evolution, the Armorican Massif must be divided into two domains, as was the case for the Cambrian sedimentary evolution.

2 Cambrian Sedimentation in the North-Armorican Domain (Normandy and Maine Provinces)

On both sides of the Mancellia land, which remained emerged, Cambrian basins were filled by red piedmont flats or foreshore conglomerates. The true epoch of marine transgression must be defined in the only area of Armorican Massif where radiometric and biostratigraphic data have been obtained.

2.1 Early Lower Cambrian Transgression (Fig. 4)

The unconformable Cambrian cover rocks post-date the Mancellia late-kinematic granodiorites which intrude the folded Upper Brioverian flysch. The flysch contains a microflora comparable with that of Late Proterozoic series in northern Europe and North America (Chauvel and Schopf 1978; Choubert and Faure-Muret 1980). No trace fossils have been found in the Brioverian of Normandy and Maine; a Vendian age may be suggested for the flysch. Several granodiorites of the Mancellia have yielded an isochron of 551 ± 28 Ma (Rb–Sr method) on whole rocks (Graviou et al. 1989). The age of 540 ± 10 Ma on monazite of the Vire granodiorite (Pasteels and Doré 1982), which unconformably lies just under the Cambrian red conglomerates, points to a maximum age for red conglomerates.

In the synclines of North Cotentin and south of Caen, the first silty strata above the basal conglomerate contain many trace fossils: *Taphrhelminthopsis, Monomorphichnus, Neonereites, Phycodes, Helminthoidea,* and, more rarely, *Rusophycus* and *Skolithos* (Doré et al. 1984; Dissler et al. 1988). This differentiated ichnofacies is compatible with a late Vendian as well as a Tommotian age. Higher in the successions, the first skeletonized microfossils are thought to belong to the Tommotian stage (the pre-trilobitic stage of the Lower Cambrian).

Thus in Normandy, the Vendian-Tommotian boundary should be within strata located between the Cadomian peneplain and the first skeletonized faunas, unless it coincides with the stratigraphic break.

J. D. Keppie (Ed.): Pre-Mesozoic Geology in France and Related Areas

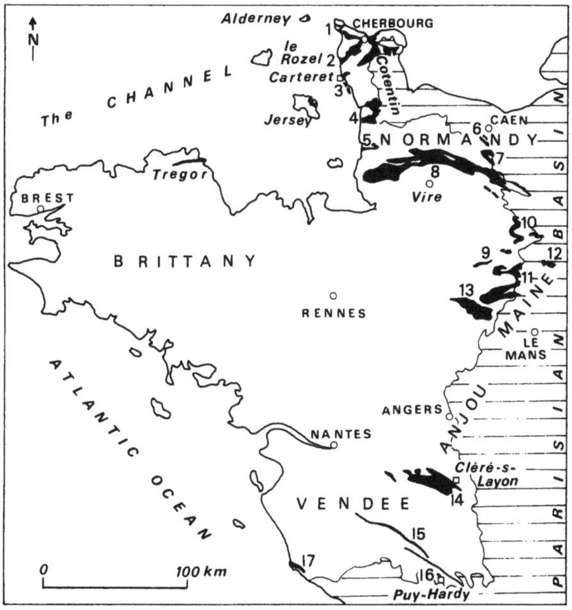

Fig. 3. Distribution of Cambrian rocks *(black)* in the Armorican Massif. *1* The Hague; *2* southern flank of the Siouville Syncline (Le Rozel); *3* Carteret area; *4* Lessay Syncline; *5* Montmartin Syncline; *6* May Syncline; *7* Urville Syncline; *8* Zone Bocaine Syncline; *9* Pail Syncline; *10* Ecouves Anticlinorium; *11* Alpes Mancelles; *12* Perseigne Horst; *13* Coëvrons and Charnie Synclines; *14* Choletais Syncline; *15* Chantonnay Synclinorium; *16* Puy-Hardy Slice; *17* Sables d'Olonne and Sauveterre Series. *Horizontal hatching* Mesozoic cover of the Parisian Basin

2.2 Typical Successions of the North Cotentin Synclines

The sections of the North Cotentin Synclines are the most representative of the Cambrian times, although only the Lower Cambrian seems to occur (Doré 1969, 1972). A SW-NE-trending trough was filled by a sequence ranging in thickness from 590 m (the Hague Peninsula) to 2100 m (Le Rozel). The substratum consists of crystalline Cadomian basement or conformable Saint-Germain-le-Gaillard ignimbrite. A southwestern laterally equivalent ignimbrite, in the Tregor Peninsula (North Brittany, cf. Fig. 3), yielded a Rb-Sr whole rock isochron date of 547 ± 8 Ma (Auvray 1979). The synthetic and most complete sedimentary succession in the North Cotentin displays 4 formations described below from bottom to top (Fig. 4 and 5).

1. The Couville Conglomerate and Arkose Formation, up to 400 m thick in the southern flank of the Siouville Syncline (Le Rozel section), contain acidic volcanics which are mostly reworked.
2. The Carteret Formation consists of irregular and thin alternating green or purplish mudstones, siltstones, and silty sandstones with a few gritty lenses. More or less parallel to the grain-size increase, there are syneresis cracks, wave and current ripples, ball and pillow structures, and hummocky cross-lamination. The littoral environment was characterized by medium to low energy with temporary local emergence. Trace fossils are numerous. A skeletonized fauna is mostly found inside calcareous nodules (Doré and Reid 1965). The hyolithids *(Eotheca)* and phosphatic spicules of heteractinid sponges *(Allonnia)* correlate with the Tommotian stage of the Lower Cambrian.
3. The Saint-Jean-de-la-Rivière Formation is made up of alternating dark limestones and siltstones. The latter contain some salt pseudomorphs. The individual carbonate sequences which range from a few decimetres to 1 m thick show subtidal to tidal environment conditions: oolitic limestones, algae limestones *(Epiphyton, Botomaella, Renalcis),* and digited and columnar stromatolites. Although rather scarce, the archaeocyathid bioherms *(Protopharetra, Aldanocyathus, Retecoscinus, Sibirecyathus)* belong to the oldest assemblages in western Europe (Rozanov and Debrenne 1974). The trilobite *Bigotina* (Bigot 1925) occurs higher in the strata of the succession. This genus has since been found in the Lower Cambrian of other countries: Siberian Platform, Yunnan Province, and the Cordoba Sierra in Spain (Linan et al. 1981). Thus archeocyathids and trilobites of the Saint-Jean-de-la-Rivière Formation indicate an early Atdabanian stage, the first trilobitic stage of the Lower Cambrian (Sdzuy 1959; Doré et al. 1984).
4. A regressive and azoic red bed formation is the last of the succession before the pre-Arenig break.

2.3 Successions in Middle Normandy (Fig. 5)

Two sedimentary cycles were deposited in Middle Normandy (Doré 1969, 1972). The first consists of three formations: (1) Rocreux purplish conglomerate, (2) shales and limestones, and (3) feldspathic sandstones. They constitute the Cambrian section of the May and Urville Synclines, and the lower part of the Zone Bocaine Syncline. The carbonate formation (2) shows three members from bottom to top: magnesian limestones with loferitic and bird's-eye layers, *Rosnaiella* detrital limestones, and stromatolites of the *Collenia* group interbedded with siltstones. A skeletonized fauna (*Halkieriidae* and *Lapworthellidae*), whose age may be Tommotian, is found at 250 m above the basal unconformity. In the western part of the Zone Bocaine Syncline the sedimentation is mostly terrigenous.

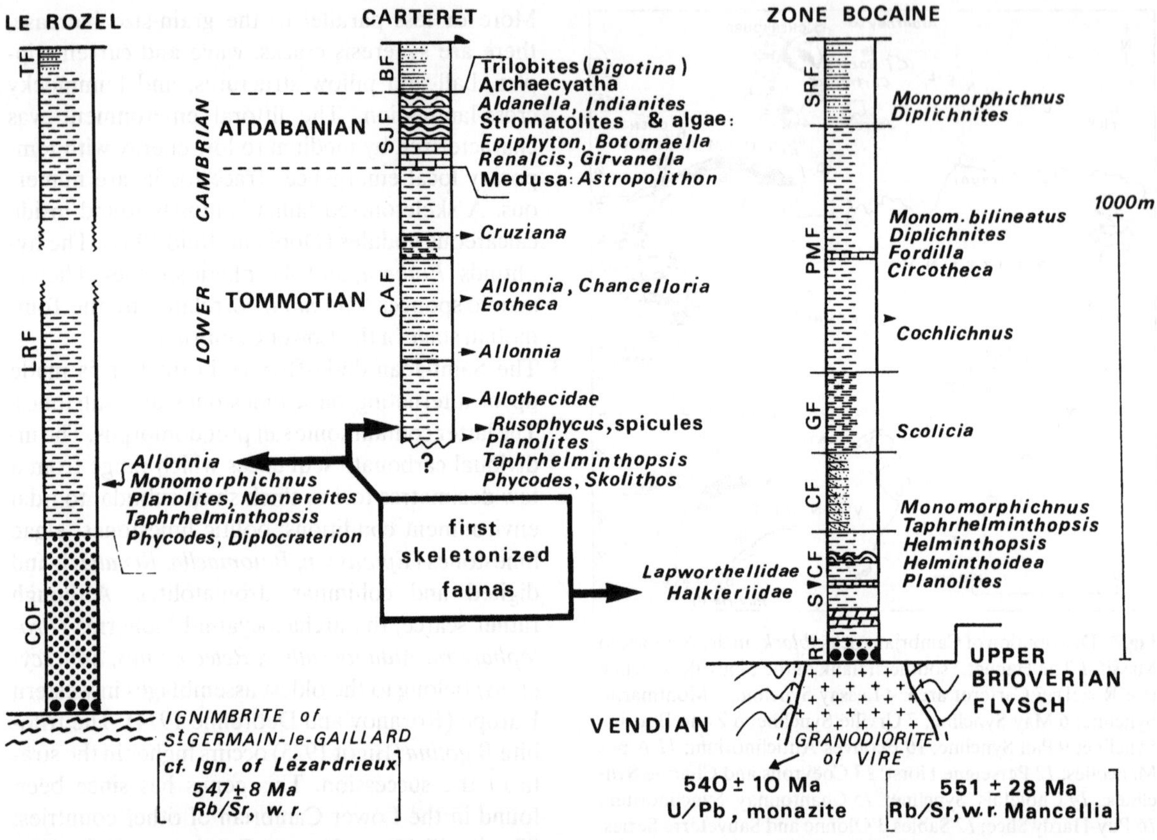

Fig. 4. The problem of the Precambrian-Cambrian boundary as documented by successions in the northern Armorican Massif. Le Rozel Section: *COF* Couville conglomerates and arkoses Formation; *LRF* Le Rozel siltstones and sandstones Formation; *TF* Teurtheville red siltstones Formation. Carteret Section: *CAF* Carteret siltstones and sandstones Formation; *SJF* Saint-Jean-de-la-Rivière limestones Formation; *BF* Bosquet red sandstones Formation. Zone Bocaine Section: *RF* Rocreux conglomerates Formation; *SCF* Siltstones and limestones Formation; *CF* Caumont feldspathic sandstones Formation; *GF* Gouvix purplish siltstones Formation; *PMF* Pont-de-la-Mousse green siltstones Formation; *SRF* Saint-Rémy red siltstones Formation

The second cycle is composed of green shales and siltstones whose thin and irregular alternation suggests a subtidal environment. The very scarce shelly fossils *(Circotheca, Fordilla)* in the middle part of the cycle are probably Lower Cambrian in age. The green shales pass upward into a red shale and sandstone formation which is barren except for trace fossils *(Diplichnites)* and therefore the question of its age remains open.

2.4 Successions in Southern Normandy and Maine

Southeast of the emerged Mancellia land, a SW-NE-trending flexure trough was filled by a Cambrian sequence which increases in thickness from a few meters along the Mancellia side to 2500–3000 m in the axis (Le Gall et al. 1975). In the latter area (Coëvrons and Charnie Synclines, section 13, Fig. 5), a thick carbonate formation overlying foreshore red conglomerates shows two members interbedded with green or gray siltstones: (1) light-colored dolostones or dolomitic limestones with stromatolites, and (2) dark and detrital limestones with a few oolitic horizons and Cyanophyta boundstones. Although they are devoid of any shelly fossils, these rocks can be correlated with those of the Lower Cambrian platform south of Caen (Fig. 6) (Doré 1969).

The following Sainte-Suzanne Sandstones are feldspathic and somewhat tuffaceous owing to volcanics which are more abundant upwards in the formation. The sedimentary features and the distribution of brachiopods *(Leptobolus, Dinobolus)* and trace fossils *(Skolithos, Rusophycus, Gyrichnites)* reflect more littoral conditions higher in the succession relating to an epirogenic episode.

The transgressive and overlying Sillé Formation consists mostly of green or purplish siltstones with brachiopods *Tomasina* and *Palaeoglossa* (Barrois and Pruvost 1931), whose age must be re-examined. Over-

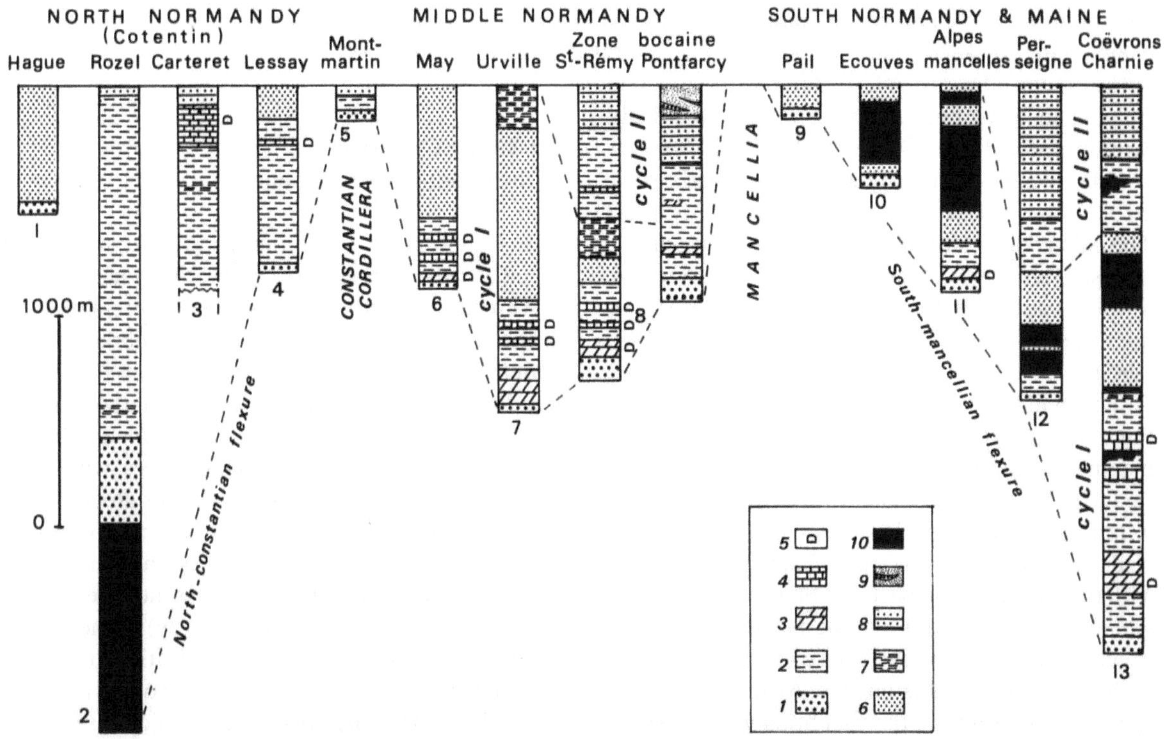

Fig. 5. Lithostratigraphic successions of Cambrian (sensu lato) strata in the northeastern Armorican Massif. See the location of sections *(numbers)* 1 to 13 in Fig. 3. *Inset: 1* Purplish conglomerates and arkoses; *2* thinly alternating green siltstones and sandstones; *3* magnesian limestones or dolostones; *4* detrital or oolitic limestones interbedded with siltstones; *5* stromatolites or algal mats; *6* feldspathic sandstones; *7* thin alternation of purplish siltstones and sandstones; *8* red sandstones and siltstones; *9* thin arkosic sandstones with gritty lenses; *10* rhyolites, ignimbrites, pyroclastic rocks with some andesitic lavas

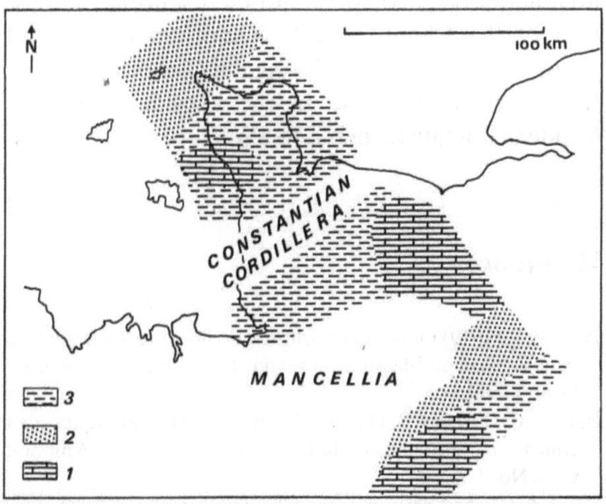

Fig. 6. Lower Cambrian paleogeography (Atdabanian Stage) in the northeastern Armorican Massif. The Mancellia and residual Constantian cordillera remain emerged. *Inset: 1* Dark-colored limestones (oolitic, algal, detrital) interbedded with siltstones; *2* feldspathic sandstones; *3* thin alternation of green siltstones and sandstones

lying the siltstones are yellow and red sandstones (the Blandouët Formation) bearing some mud-cracked surfaces and having a reddened top which lead up to a widespread emersion of the whole Normandy and Maine provinces. Given that these rocks contain brachiopods and trace fossils *(Cruziana, Didymaulichnus, Skolithos)* for which some Ordovician affinities are admitted, the stratigraphic break before the Armorican Sandstone deposition (Arenig) must have been brief. Morever, the Sillé and Blandouët Formations have some facies similar to the Initial Red Beds of Central Brittany. The Initial Red Beds rest unconformably on the Brioverian basement and they underlie the Armorican Sandstone with a gradual transition. In western Brittany, their radiometric age is 465 ± 1 Ma (Bonjour et al. 1988).

2.5 Post-Orogenic Volcanism During the Cambrian

A post-orogenic volcanism occurred along two volcano-tectonic structures: (1) in North Cotentin where ignimbrite sequences (Saint-Germain-le-Gaillard), up to 1000 m thick, underlie the transgressive Lower

Cambrian deposits; and (2) in south Normandy and Maine – the so-called Maine-graben – where several volcanic episodes are interbedded with sediments ranging in age from Lower Cambrian to possibly Arenig (Le Gall et al. 1975; Le Gall 1976, 1978).

The volcanics are mostly composed of acidic, welded or nonwelded, ash-flow tuffs and pyroclastic rocks. The calc-alkaline affinities of a minor component of andesitic lavas are marked by geochemical grounds: low TiO_2 and Y contents, enrichment in LREE, and high values of Th/Ta ratio (Le Gall and Cabanis 1985).

The acidic volcanic magmas are thought to have the same crustal origin as the Mancellia granodiorite magmas owing to similar trace elements compositions. In the structural framework of the plate tectonic model proposed by Dissler et al. (1988) for the Cadomian orogeny, crustal thickening associated with the closure of the back-arc Mancellian basin, together with the intrusion of basic magmas, might have triggered fusion of the crust. The composition and age of this crust remain questionable. At any rate, the volcanic eruptions during the Cambrian were linked to Late Cambrian stress relaxation and controlled by a SW-NE-trending normal fault system.

3 Cambrian Deposits of the South-Armorican Domain (Vendée Province)

Epimetamorphic Cambrian rocks with some fauna occur only in the Vendée area, where they unconformably overlie mesozonal Brioverian graywackes and slates known as the Mauves Group.

3.1 Middle Cambrian of the Northeastern Vendean Massif

The Cléré-sur-Layon Formation (Fig. 3, station 14), which locally can be 900 m thick, consists, from the base up, of siltstones and conglomerates; thick rhyolite and ignimbrite flows (Le Métour 1980). A gap in deposition is suspected before another volcanic series which has been dated as Silurian. Trilobites found 250 m above the basal disconformity (Cavet et al. 1966) include: *Paradoxides, Eccaparadoxides, Conocoryphe,* and *Couloumania.* This is the only identified fauna of Middle Cambrian age in the Armorican Massif. The species point to correlations either with the *P. paradoxissimus* zonal group of the middle part of the Middle Cambrian of Sweden (Cavet et al. 1967), or with *P. oelandicus* of the lower part of the Middle Cambrian (Sdzuy 1972).

3.2 Upper Cambrian of the Chantonnay Synclinorium

The Chantonnay Synclinorium (Fig. 3, station 15) belongs to the north side of the internal Variscan zone in Vendée. Beneath the Ordovician succession (Armorican sandstone), a putative Cambrian series about 1500 to 2500 m thick occurs. From bottom to top, it is composed of green shales, quartzitic sandstones and conglomerates with interbedded rhyolites (Montournais Formation), black shales with limy lenses (Gerbaudières Formation), and green shales and sandstones (Bourgneuf Formation). Although these detrital deposits are devoid of any fauna, they are related to the Cambrian. This is indicated by an isolated fossiliferous slice (Puy-Hardy) in the southern part of the synclinorium (Fig. 3, station 16) which is thought to be located toward the top of the regional succession. The badly preserved brachiopods of Puy-Hardy are assigned to *Billingsella,* an index genus of the Upper Cambrian (Wyns 1980, 1984; Bouton 1990).

Lastly it must be noted that the metamorphic Sables d'Olonne and Sauveterre Formations (2250 m) (Fig. 3, station 17) of the coastal Vendée area might be correlated with the Cambrian succession of the Chantonnay Synclinorium (Wyns and Chantraine 1984). If so, during the Cambrian period, there was an increase in the subsidence of the Vendean domain from NE to SW.

References

Auvray B (1979) Genèse et évolution de la croûte continentale dans le Nord du Massif armoricain. Thèse, Rennes University, pp 1–680

Barrois C, Pruvost P (1931) Relations stratigraphiques des couches cambriennes de la Bretagne et du Maine. Ann Soc Géol Nord 56: 80–125

Bigot A (1925) Sur la présence de Trilobites et d'Archéocyathidés dans les couches cambriennes des environs de Carteret (Manche). C R Acad Sci Paris 180: 1237–1239

Bonjour JL, Peucat JJ, Chauvel JJ, Paris F, Cornichet J (1988) U-Pb zircon dating of the Early Palaeozoic (Arenigian) transgression in Western Brittany (France); a new constraint for the Lower Palaeozoic time-scale. Chem Geol (Isotope Geoscience section) 72: 329–336

Bouton P (1990) Géologie du Haut Bocage Vendéen. Thèse, Poitiers University, pp 1–378

Cavet P, Gruet M, Pillet J (1966) Sur la présence du Cambrien à *Paradoxides* a Clére-sur-Layon (M.-et L.) dans le Nord-Est du bocage vendéen (Massif armoricain). C R Acad Sci Paris 263: 1685–1688

Cavet P, Gruet M, Pillet J (1967) Les trilobites mésocambriens de Clére-sur-Layon (Massif armoricain). Bull Soc Géol Fr 7 9 5: 750–757

Chauvel JJ, Schopf JW (1978) Late Precambrian microfossils from Brioverian cherts and limestones of Brittany and Normandy, France. Nature 275: 640–642

Choubert G, Faure-Muret A (1980) The Precambrian in North Peri-Atlantic and South Mediterranean mobile zones: general results. Earth Sci Rev 16: 85–219

Dissler E, Doré F, Dupret L, Gresselin F, Le Gall J (1988) L'évolution géodynamique cadomienne du Nord-Est du Massif armoricain. Bull Soc Géol Fr 8 4 5: 801–814

Doré F (1969) Les formations cambriennes de Normandie. Thèse, Caen University, pp 1–790

Doré F (1972) La transgression majeure du Paléozoïque inférieur dans le Nord-Est du Massif armoricain. Bull Soc Géol Fr 7 14 5: 801–814

Doré F, Reid R (1965) *Allonnia tripodophora* nov. gen. nov. sp., nouvelle éponge du Cambrien inférieur de Carteret (Manche). C R Somm Soc Géol Fr: 20–21

Doré F, Dupret L, Le Gall J (1984) Les premiers temps de la transgression cambrienne sur le socle dans le Cotentin. Bull Soc Géol Minéral Bretagne C 16 1: 49–59

Graviou P, Peucat JJ, Auvray B, Vidal P (1989) L'orogenèse cadomienne dans le Nord du Massif armoricain: contraintes pétrologiques et géochronologiques. Hercynica 4 1: 1–13

Le Gall J (1976) Les manifestations volcaniques acides dans le Cambrien des Coëvrons orientaux (E du Massif armoricain). Bull Soc géol minéral Bretagne C 8 1–2: 65–74

Le Gall J (1978) Les pyroclastites acides du Cambrien de la Charnie (E du Massif armoricain): pétrographie, encadrement lithostratigraphique. Bull Soc Linn Normandie 106: 27–34

Le Gall J, Cabanis B (1985) Caractères minéralogiques et géochimiques des ignimbrites et des laves cambriennes de l'Est du Massif armoricain. Hercynica I 2: 99–109

Le Gall J, Doré F, Giordano R, Pottier Y (1975) Position stratigraphique et cadre tectono-sédimentaire des manifestations volcaniques cambriennes dans le Nord-Est du Massif armoricain. Bull Soc Géol Fr 7 17 6: 1101–1109

Le Métour J (1980) La bordure nord de la zone orogénique éohercynienne dans le Nord-Est de la Vendée. In Réunion extraordinaire de la Société géologique de France en Limousin, Vendée, Bretagne Sud. Bull Soc Geól Fr 7 22 2: 163–166

Linan E, Moreno-Eiris E, Perejon A, Schmitt M (1981) Fossils from the basal levels of the Pedroche Formation, Lower Cambrian (Sierra Morena, Cordoba, Spain). Bol R Soc Esp Hist Nat 79: 277–286

Pasteels P, Doré F (1982) Age of the Vire-Carolles granite. In: Odin G (ed) Numerical dating in stratigraphy. John Wiley, New York, pp 784–790

Sdzuy K (1972) Das Kambrium der acadobaltischen Faunenprovinz. Zentralbl Geol Paläontol II 1/2: 1–91

Wyns R (1980) Contribution à l'étude du Haut-Bocage vendéen: le Précambrien et le Paléozoïque dans la région de Chantonnay (Vendée). Thèse 3ème cycle, Paris VI University, pp 1–124

Wyns R, Chantraine J (1984) Les séries métamorphiques de Sauveterre et des Sables d'Olonne (côte vendéenne) sontelles cambriennes? 10ème Réun Ann Sci Terre Bordeaux: 554

Ordovician, Silurian, and Devonian of the Medio-North-Armorican Domain

M. ROBARDET, J. L. BONJOUR, F. PARIS, P. MORZADEC and P. R. RACHEBOEUF

1 Ordovician

(M. Robardet, J. L. Bonjour, and F. Paris)

1.1 From the Initial Red Beds to the Grès Armorician Formation: the Arenigian Transgression

(J. L. Bonjour)

During the Arenig, a marine transgression flooded most of the Medio-North-Armorican Domain. To the west, on the southern flank of the Chateaulin Syncline and to the south, in the Martigné-Ferchaud, Segré, and Angers Synclines, the Initial Red Beds, respectively called the Cap de la Chèvre and the Pont-Réan Formations, rest unconformably on the late Proterozoic basement (Brioverian). These formations (Fig. 7) are mainly composed of conglomerates, sandstones, and siltstones which are typically red and were deposited in continental or coarse-grained delta environments. Petrological and geochemical data show that these sediments resulted from the reworking of the underlying Brioverian basement, after a weathering period of unknown duration (Bonjour 1989).

Rapid variations in thickness (0 to about 800 m), intense contemporaneous volcanic activity resulting in acidic effusions and numerous volcaniclastic levels, and also tectonic reconstructions suggest a tectonic setting which included rifting.

Although rare lingulids have been reported, biogenic activity is mainly represented by trace fossils (e.g., *Cruziana rugosa, C. furcifera,* and *C. goldfussi;* Fig. 8) whose concurrent-range biozone supports an Early Ordovician age.

In the Crozon Peninsula a volcaniclastic layer in the Cap de la Chèvre Formation has been dated at

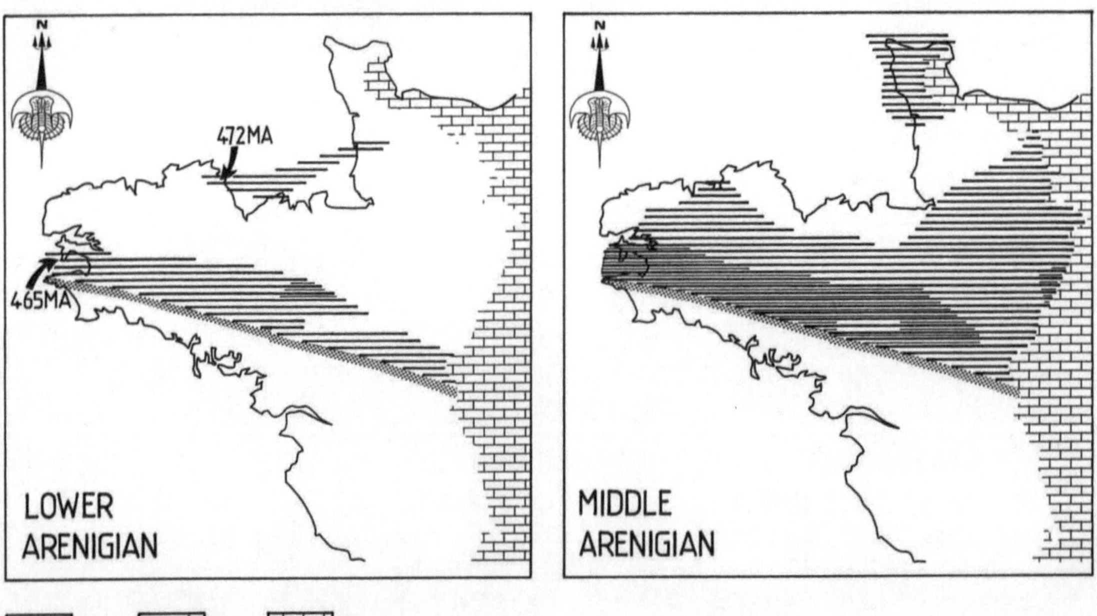

Fig. 7. Extension of Early and Middle Arenig deposits and radiometric data. *1* Boundary between the Medio-North and the South-Armorican Domains (major Variscan shear zone); *2* extension of the deposits and location of the more subsident areas *(doubled lines)*; *3* Mesozoic cover

J. D. Keppie (Ed.): Pre-Mesozoic Geology in France and Related Areas

	IRB	GAFm
Arenicolites		••••••••••
Arthrophycus		••••••••••
Cruziana	••••••••••	••••••••••
Daedalus	••••••••••	••••••••••
Didymaulichnus	••••••••••	••••••••••
Dimorphichnus	••••••••••	••••••••••
Diplichnites	••••••••••	
Diplocraterion		••••••••••
Imbrichnus		••••••••••
Lingulichnites		••••••••••
Monocraterion		••••••••••
Monomorphichnus	••••••••••	
Planolites	••••••••••	••••••••••
Phycodes		••••••••••
Skolithos	••••••••••	••••••••••
Teichichnus		••••••••••
Trichichnus		••••••••••

Fig. 8. Trace fossils of the Initial Red Beds *(IRB)* and of the Grès Armoricain Formation *(GAFm)*

Macrofauna	Microfauna
Bivalves	Acritarchs
Actinodonta	*Acanthodiacrodium*
Lyrodesma	*Baltisphaeridium*
Nuculites	*Micrhystridium*
Synek	*Priscogalea*
	Veryhachium
Brachiopods	Chitinozoans
Lingulobolus	*Conochitina*
Ectonoglossa	*Cyathochitina*
	Eremochitina
Trilobites	*Jenkinochitina*
Ogyginus	*Lagenochitina*
Platycoryphe	*Tanuchitina*

Fig. 9. Macrofauna and paleoplankton of the Grès Armoricain Formation

465±1 Ma (U-Pb zircon dating method; Bonjour et al. 1988). Stratigraphical constraints from the well-dated overlying Grès Armorician Formation suggest an Early Arenig age for the Initial Red Beds although the isotopic age is significantly younger than the usual radiochronological upper limit proposed for the Arenig (Odin 1985).

Continental red beds also occur in Normandy (Montmartin Syncline) and in northern Brittany (Plourivo and Erquy-Fréhel areas) (Fig. 2). In the Plourivo region, interbedded trachyandesitic lavas have been dated at 472±5 Ma (Rb/Sr whole rock; Auvray et al. 1980). These deposits could represent the northern counterpart of the Initial Red Beds of Central Brittany.

In the Medio-North-Armorican Domain, the large extent of the marine arenaceous deposits of the Grès Armoricain Formation indicates general transgression (Fig. 7). In areas where the Initial Red Beds had been deposited, there is a gradual transition to the Grès Armorician Formation. In other areas, this formation rests either disconformably on the Brioverian basement or conformably on Cambrian strata. The Grès Armorician Formation mainly consists of very mature massive sandstones, with accessory shaly horizons. The development of sandstone-shale alternations with a rather constant position in the sequence allows division of the formation into three members. Some ironstone levels are known, principally in the lower member in the Segré Syncline.

The sediments of the Grès Armoricain Formation were deposited in shallow marine environments on a large shelf under tidal, shore current, and storm influences (Durand and Noblet 1986). Significant variations of thickness (a few tens of meters to about 700 m) indicate the influences of residual topographies and differential subsidence (Noblet 1983). These deposits contain inarticulate brachiopods, bivalves, rare trilobites (Henry 1971), microfossils (Fig. 9), and numerous trace fossils (Durand 1984) similar to those of the Initial Red Beds (Fig. 8). Chitinozoans indicate that the whole Grès Armoricain Formation belongs to the *Eremochitina brevis* biozone which correlates with the early Middle Arenig and/or the upper part of the Lower Arenig (Paris 1990).

1.2 Outer Shelf Shaly Facies: Maximum Extent of the Transgression

From the Middle Arenig (Paris and Skevington 1979) or locally from the lowermost Llanvirn (e.g., Caen region), dark argillaceous deposits developed all over the Medio-North-Armorican Domain. These sediments correspond to the maximum of the Ordovician transgression which flooded the rare and restricted areas (May-sur-Orme, Saint-Rémy) that had not been submerged during the Lower Arenig. Locally (Cotentin), an arenaceous sedimentation (e.g., Moitiers d'Allonne Formation) persisted temporarily into the Early Llanvirn. The outer shelf shaly sequences, namely the Urville, Le Pissot, Andouillé, Postolonnec, Traveusot, and Angers Formations, extend into the Llandeilo or the lowermost Caradoc (Fig. 10). Oolitic ferruginous layers commonly occur in the lower part of these formations, especially in eastern Normandy (iron ore).

The poorly diversified benthic faunas comprise trilobites (e.g., *Neseuretus tristani, Crozonaspis morenensis mayensis, Placoparia cambriensis*), ostra-

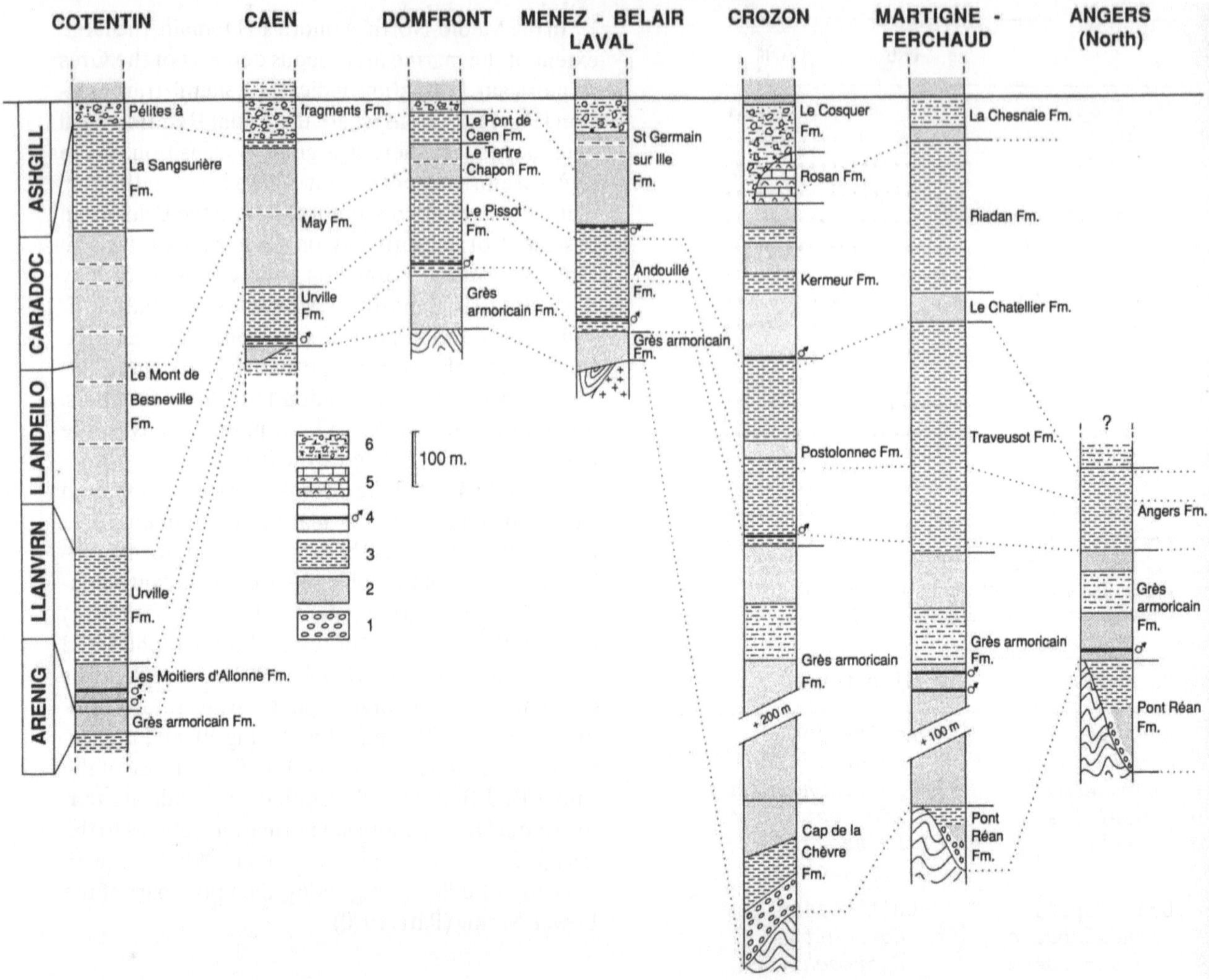

Fig. 10. The Ordovician succession in the Medio-North-Armorican Domain. *1* Conglomerate; *2* sandstone; *3* siltstone and shale; *4* ironstone (+phosphatic and ferruginous levels); *5* limestone and volcanics; *6* glacio-marine deposits

codes (e.g., *Gracquina hispanica, Rivellina henningsmoeni*), brachiopods (e.g., *Cacemia riberoi*), echinoderms, and bivalves (e.g., *Cardiolaria beirensis, Actinodonta naranjoana*). Age assignments are based on graptolites (e.g., *Corymbograptus v-similis* in the Middle Arenig, *Didymograptus artus* and *D. murchisoni* respectively in the Lower and Upper Llanvirn) and chitinozoans (e.g., *Pseudodesmochitina ornensis, Cyathochitina protocalix, C. calix,* and *Siphonochitina formosa*).

In western Brittany, the argillaceous sedimentation described above was temporarily interrupted by a minor sandy episode (the nearshore sandstones of the Kerarvail Member). Except in Normandy, where arenaceous sequences occur as early as the Llandeilo (see below), the argillaceous deposits of the Medio-North-Armorican Domain persisted to the lowermost Caradoc.

The geographic distribution of the Llandeilo shelf deposits indicates a north-south polarity from nearshore environments in the northern regions toward more distal and probably deeper environments in the south. This sedimentary pattern agrees with the postulated existence of a South-Armorican Ocean extending southward from the Medio-North-Armorican Domain.

The Llandeilo strata yield abundant but rather poorly diversified benthic faunas which show pronounced affinities with those of other Northern Gondwana regions, especially with those of Central Iberia (Henry et al. 1974, 1976; Paris and Robardet 1977; Hammann and Henry 1978). The nearshore arenaceous sequences of Normandy (e.g., May and Besneville Formations) contain shallow water Eohomalonotid trilobites (e.g., *Eohomalonotus vicaryi, E. brongniarti, Iberocoryphe bonissenti*) and brachiopods (e. g., *Tafilaltia valpyana*). The faunas of the inner shelf argillaceous deposits include abundant trilobites (e. g., *Neseuretus tristani, Placoparia tournemini, Phacopidina micheli, Crozonaspis struvei, Morgatia hupei*), ostracodes (e. g., *Lardeuxella bussacensis, Krutatia tromelini, Reuentalina riberiana*), brachiopods (e. g., *Appolonorthis bussacensis, Heterorthina kerfornei, Eorhipidomella musculosa;* cf. Mélou 1975, 1976), echinoderms and bivalves (e. g., *Cardiolaria bussacensis* and *Redonia deshayesi*).

The more distal environments in the southern part of Central Brittany, in addition to several species listed above, contain the trilobites *Guichenia dufouri* and *Zeliszkella lapeyrei* and the brachiopod *Aegiromena mariana.* In the uppermost levels of the Llandeilo, the trilobite *Marrolithus bureaui* is reported from several localities.

All these benthic faunas do not permit precise correlations with the standard succession of the British Isles. Thus the age assignments are based on rare graptolites (e. g., *Glyptograptus teretiusculus, Gymnograptus linnarsonni*) and principally on chitinozoans (e. g., *Linochitina pissotensis, Lagenochitina deunffi, L. dalbyensis*).

1.3 Inner Shelf Sandy Deposits: Regressive Episodes

The regressive trend noted in the lowermost Caradoc in Brittany occurred as early as the beginning of the Llandeilo in the May-sur-Orne and Cotentin areas and in the late Lower Llandeilo to the south, in the Domfront Syncline. It resulted in thick nearshore sandstone sequences, namely (from north to south) the May-sur-Orne, Tertre-Chapon, Saint-Germain-sur-Ille, Kermeur and Le Chatelier Formations (Fig. 10). In some areas (Ménez-Bélair and Laval Synclines), the arenaceous sedimentation persisted into the Ashgill whereas in other localities, it was succeeded by more distal and probably deeper silty and shaly deposits (e. g., La Sangsurière, Le Pont-de-Caen, and Riadan Formations). The Caradoc faunas mainly come from these argillaceous units and consist of trilobites (e. g., *Onnia grenieri, Crozonaspis dujardini*), ostracodes (e. g., *Hastatellina normandiensis, Satellina henningsmoeni*), brachiopods (e. g., *Svobodaina armoricana, Drabovia* cf. *redux*), bivalves (e. g., *Concavodonta ponderata*), and rare graptolites (e. g., *Orthograptus amplexicaulis, Diplograptus compactus*). Once again, precise stratigraphic assignments are provided by chitinozoans (e. g., *Jenkinochitina tanvillensis, Belonechitina robusta*). In the Saint-Germain-sur-Ille Formation, which extends into the Ashgill, additional faunas include trilobites (e. g., *Calymenella bayani*), brachiopods (e. g., *Drabovinella erratica*) and early Ashgill graptolite assemblages (e. g., *Orthograptus truncatus truncatus, O. truncatus abbreviatus,* and *O. pauperatus;* cf. Skevington and Paris 1975).

1.4 Late Ordovician Sedimentation: Limestones, Volcanics, and Glacio-Marine Deposits

In western Brittany (southern limb of the Chateaulin Syncline), the Rosan Formation includes volcaniclastic deposits with a more or less abundant calcareous matrix, volcanic breccias, and lava flows with pillow lavas. The top of the formation is a massive limestone bed a few meters thick. No other Late Ordovician volcanics are known in the rest of the Medio-North-Armorican Domain. On the other hand, limestones occur locally in southern Normandy (Les Vaux Formation). The Rosan Formation yields an abundant and fairly diversified fauna with a few trilobites (e. g., *Cekovia munieri*), numerous brachiopods (e. g., *Nicolella actoniae ramosa, Leptestiina aonensis;* cf. Mélou 1971; Babin et al. 1988), cystoids (e. g., *Heliocrinites* cf. *rouvillei*), crinoids, and bryozoans. Moreover, the limestones of Les Vaux Formation and of the upper part of the Rosan Formation contain similar conodont assemblages (e. g., *Amorphognathus ordovicicus, Icriodella superba*) consistent with an Ashgill age assignment (Weyant et al. 1977; Paris et al. 1981).

In Normandy and in other regions (e. g., Ménez-Bélair, Laval and Chateaulin Synclines), the uppermost part of the Ordovician succession includes terrigenous deposits which consist of fine-grained argillaceous sandstones, siltstones, or shales bearing clasts of various size and lithology, the so-called pélites à fragments (Dangeard and Doré 1971; Doré 1981). These diamictic formations generally lack distinctive fossils. Poorly preserved chitinozoan assemblages *(Armoricochitina nigerica, Calpichitina lenticularis)* recorded in a few localities (Babin et al. 1988) suggest a late Ashgill age for these peculiar strata (pélites à fragments, Upper Member of the Saint-Germain-sur-Ille Formation, Le Cosquer Formation). Such an age assignment is consistent with the Hirnantian age of the north African glacio-marine deposits (Paris 1990). Moreover, the uppermost Ordovician brachiopod *Hirnantia sagittifera* has been recently recorded from western Brittany (Mélou 1987). The precise stratigraphic position of the fossiliferous level in the Le Cosquer Formation remains obscure.

The undisputed glacial origin of the north African uppermost Ordovician deposits (Beuf et al. 1971)

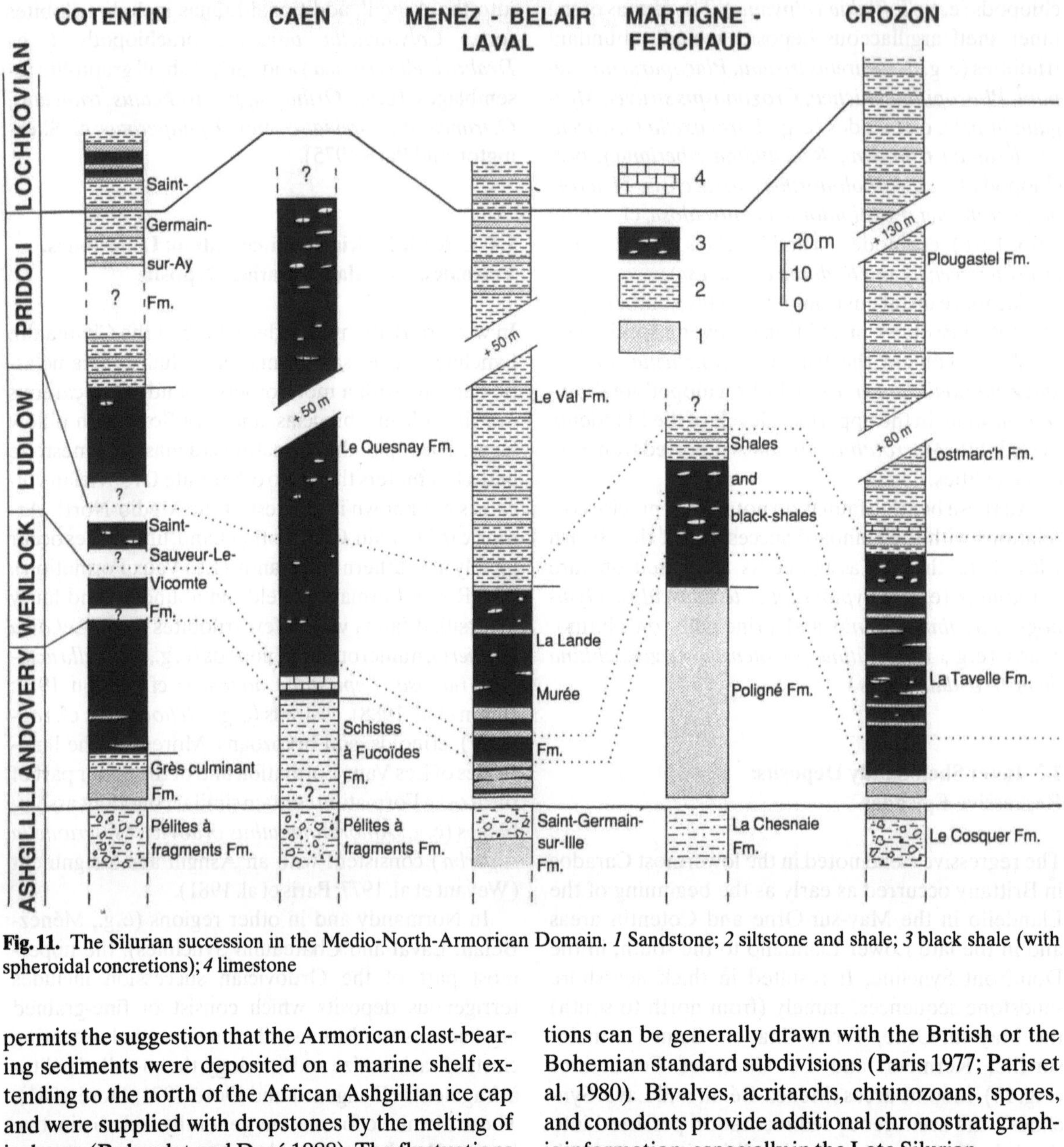

Fig. 11. The Silurian succession in the Medio-North-Armorican Domain. *1* Sandstone; *2* siltstone and shale; *3* black shale (with spheroidal concretions); *4* limestone

permits the suggestion that the Armorican clast-bearing sediments were deposited on a marine shelf extending to the north of the African Ashgillian ice cap and were supplied with dropstones by the melting of icebergs (Robardet and Doré 1988). The fluctuations of the ice sheet and the related eustatic variations explain the stratigraphic hiatus which occurs between the late Ashgill and the late Llandovery paleontologically dated strata (Babin et al. 1988).

2 Silurian

(F. Paris and M. Robardet)

The Silurian succession (Fig. 11) is fairly different from the other Paleozoic Systems because of the abundance of black shales (the "ampélites"). Based on graptolite evidence, precise stratigraphic correlations can be generally drawn with the British or the Bohemian standard subdivisions (Paris 1977; Paris et al. 1980). Bivalves, acritarchs, chitinozoans, spores, and conodonts provide additional chronostratigraphic information, especially in the Late Silurian.

2.1 The Anoxic Sedimentation

Neither Rhuddanian nor Aeronian graptolites have been recorded in the available material. Thus the Silurian transgression most likely flooded the region by Telychian times, as the first Silurian graptolites recorded belong to the *turriculatus* Zone. The resulting stratigraphic gap, as mentioned earlier, seems to be related to the eustatic effects of the Late Ordovician glaciation.

The Early Silurian sequence is mainly composed of black pyritic quartzites with subordinate black shale intercalations, with the exception of western Normandy, where sandy shales and siltstones ("schistes à fucoides" Formation) occur. During the Telychian and the Wenlock, a condensed sequence of graptolite-bearing black shales (a few meters thick) was deposited over the whole Medio-North-Armorican Domain. Locally, rare sandstone or calcareous layers and spheroidal concretions occur. The Silurian black shales are fine-grained pyritic argillaceous (chlorite, illite, and sometimes pyrophyllite) sediments. Their main feature is a high organic carbon content (from 5 to 15 %, with exceptional values up to 35 %) and an abnormally high concentration of trace elements, e.g., vanadium (up to 5000 ppm), strontium, and rubidium, which were most likely scavenged by the organic matter (Dabard and Paris 1986).

The faunas reported from the black shale formations (i.e., La Lande Murée, Poligné, La Tavelle, Saint-Sauveur-Le-Vicomte, and Le Quesnay Formations) are almost entirely pelagic or epipelagic. Among the most representative groups are graptolites (numerous species of *Monograptus, Pristiograptus, Cyrtograptus, Saetograptus*), myodocopid ostracodes *(Bolbozoe* cf. *bohemica),* crinoids *(Scyphocrinites elegans),* orthocone nautiloids, thin-shelled bivalves *(Mytilarca esuriens, Pterochaenia,* e.g., *tenuis),* and eurypterid remains. It is worth noting the total absence of benthic faunas, endo-faunas, and trace fossils (the anoxic conditions were too severe near the sea bottom). Among the organic microfossils, the occurrence of poorly preserved acritarchs, chitinozoans, spores, and abundant leiospheres should be stressed.

The depositional characteristics of the black shales have led to various interpretations of their depositional environment, including deep water and marine-restricted environments (more or less isolated areas or lagoons during regressive events). However, the peculiar lithology, the faunal content, and the huge geographic distribution of the black shales over the whole North Gondwana marine platform is better explained by stratification of the sea water column. The upper oxidic layer allowed the proliferation of planktonic and epiplanktonic organisms, while the redox and anoxic environment of the lower layer prevented both the development of benthic faunas and the destruction of the organic matter deposited on the sea bottom. The presumed stratification of the water column was most likely related to the melting of the Late Ordovician ice cap (cold stagnant water near the sea bottom).

2.2 Back to Normal Oxygenated Environment

Although anoxic conditions prevailed locally to the Late Pridoli (upper part of Le Quesnay Formation in Normandy), in the other regions an oxic sedimentation progressively settled down from the Late Ludlow-Early Pridoli (Fig. 11). Thus, overlying the black shales are gray-green siltstones with shale and sandstone interbeds (Le Val, Lostmarc'h, and Plougastel Formations). The environmental changes were accompanied by an acceleration of the subsidence (several hundred meters of terrigenous material during the Pridoli). The benthic fauna is still poorly represented although accumulations of brachiopods (e.g., *Clarkeia puilloni, Salopina missendenensis*) are known (Babin et al. 1979). Microfossils are much better represented with conodonts (e.g., *Spathognathodus steinhornensis eosteinhornensis*), spores, acritarchs, and especially typical Pridolian chitinozoans (e.g., *Urnochitina urna, Margachitina elegans, Pterochitina perivelata*). This sedimentary evolution announced the thick arenaceous deposits of the earliest Devonian which succeeded the Silurian succession with no hiatus (Jaeger and Robardet 1973).

3 Devonian

(P. Morzadec and P. R. Racheboeuf)

In the Medio-North-Armorican Domain, the marine Devonian platform-type sedimentation (for a recent and extensive bibliography, see Morzadec et al. 1988) was subjected to terrigenous influences and to global events which affected the northern Gondwana province. The stratigraphic sections show rapid horizontal and vertical lithological variations (Fig. 12). There is no evidence of deep sea deposits and the position of the continents remains hypothetical. Some acidic and basic volcanic events, indicating an extensional regime, occur within the succession in Central Brittany.

Above the Silurian-Devonian boundary, which has been dated by *Monograptus uniformis uniformis* and *Eisenackitina bohemica* in the upper part of a thick aternating sequence of siltstone and quartzite (Saint-Germain-sur-Ay Formation, Plougastel Formation, lower part of Gahard Formation) a gritty sedimentation begins. These deposits almost always contain ferrugineous layers. Apart from the Kerdeniel Member, which is restricted to the Rade de Brest area, the Eo-Devonian arenaceous formations are remarkably homogeneous in their lithology and faunas. This succession consists of siltstones and quartzarenites with cal-

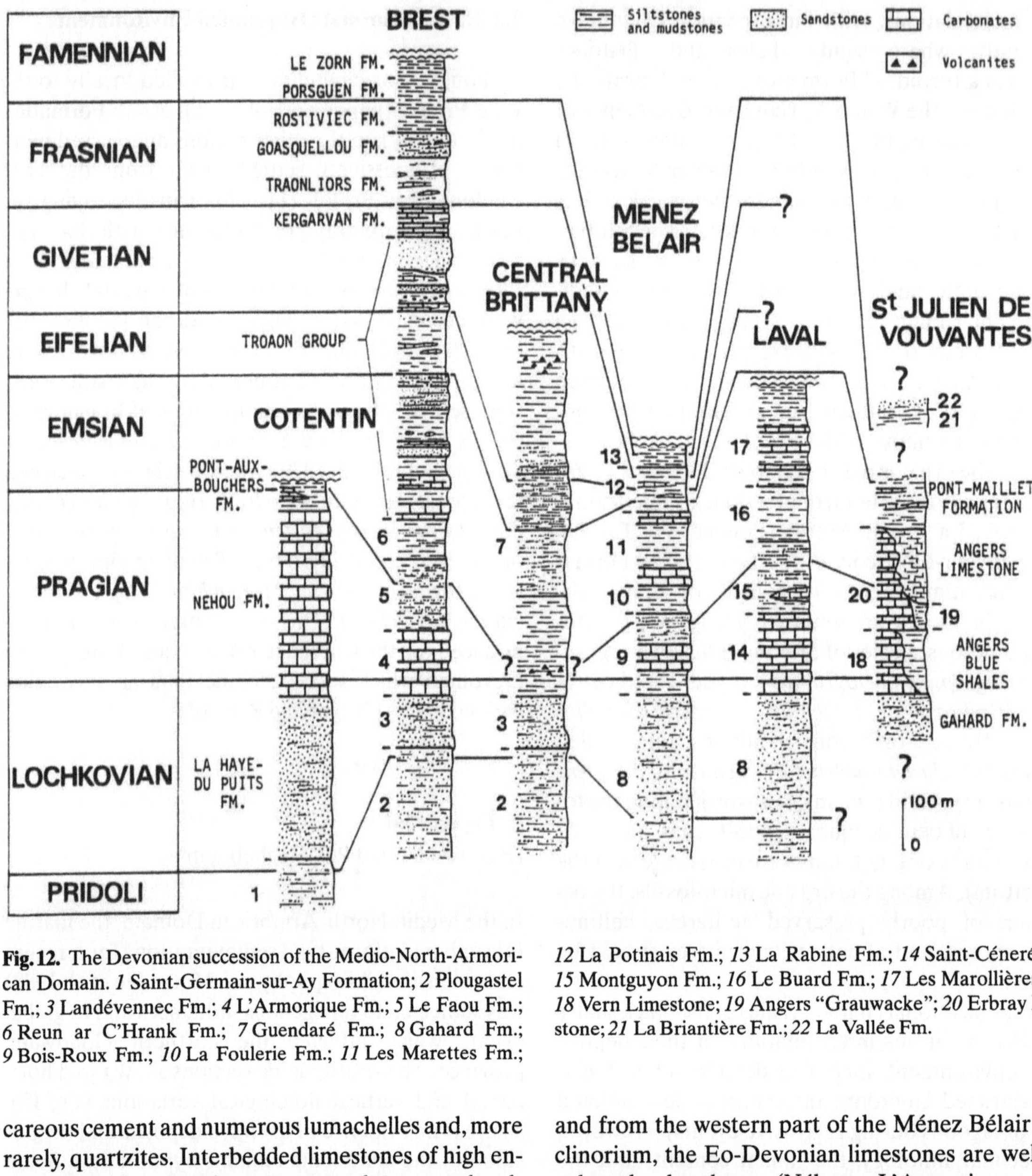

Fig. 12. The Devonian succession of the Medio-North-Armorican Domain. *1* Saint-Germain-sur-Ay Formation; *2* Plougastel Fm.; *3* Landévennec Fm.; *4* L'Armorique Fm.; *5* Le Faou Fm.; *6* Reun ar C'Hrank Fm.; *7* Guendaré Fm.; *8* Gahard Fm.; *9* Bois-Roux Fm.; *10* La Foulerie Fm.; *11* Les Marettes Fm.; *12* La Potinais Fm.; *13* La Rabine Fm.; *14* Saint-Céneré Fm.; *15* Montguyon Fm.; *16* Le Buard Fm.; *17* Les Marollières Fm.; *18* Vern Limestone; *19* Angers "Grauwacke"; *20* Erbray Limestone; *21* La Briantière Fm.; *22* La Vallée Fm.

careous cement and numerous lumachelles and, more rarely, quartzites. Interbedded limestones of high energy depositional environments may be present locally. Organic microplankton is abundant. The macrofauna mainly consists of benthic forms: e.g., *Grammysia armorica, Platyorthis monnieri, Howellella mercurii, Cleistopora geometrica, Acastella,* and *Digonus.*

The passage from the arenaceous formations to the overlying argillaceous-calcareous formations is marked by the more or less progressive disappearance of coarse detrital sedimentation, replaced by an argillaceous sedimentation with interbedded limestones. The end of the arenaceous sedimentation nearly coïncides with the Lower-Upper Lochkovian boundary. Although absent from the entire Chateaulin Synclinorium (except the Rade de Brest) and from the western part of the Ménez Bélair Synclinorium, the Eo-Devonian limestones are well developed elsewhere (Néhou, L'Armorique, Bois Roux, Saint-Céneré, and Vern d'Anjou Formations). Globally, they have been interpreted as resulting from platform sedimentation in shoals, which began during an extensional phase of the basin formation. The calcareous sedimentation begins in the Upper Lochkovian and persists through most of the Pragian. In the South of the Armorican Massif it continues into the Emsian. The lower part of the limestones is characterized by *Icriodus postwoschmidti, I. lotzei, Ctenochonetes jouannensis, Acastella crenulata,* and *Parahomalonotus diablintianus.* The Lochkovian-Pragian boundary is placed in the lower quarter of the formation because of *Icriodus steinachensis eta, Eognathodus sulcatus, Angochitina comosa, Eucharitina oehlerti, Vandercammenina, Nowakia acuaria, Pseudocryphaeus,* and *Protacanthina* (see Morzadec et al. 1991). Locally the limestones contain reefs, bioherms, and biostromes of the Pointe de l'Armorique.

A detrital sedimentation directly follows the argillaceous-calcareous deposits. The sediments are composed of alternating shales, calcareous sandstones, and lumachelles (Pont-aux-Bouchers, Le Faou, and Montguyon Formations) of upper Pragian to lower Emsian age. Locally, reefs occur in the succession: the biostromes and bioherms of Baubigny (Normandy) and the biostromes in the Laval Synclinorium. These formations contain a rich benthos in which brachiopods and crinoids predominate (e.g., *Ctenochonetes aremoricensis, Arduspirifer, Acrospirifer primaevus, Leptaenopyxis, Diamenocrinus, Seilloucrinus*). The upper part of the formations, attributed to the lower Emsian, contains an original fauna of uncommonly large size which is usually called the monster's fauna.

The Eo-Devonian succession is well established in the Rade de Brest but changes rapidly eastward. From the Chateaulin Synclinorium to the Ménez Bélair Synclinorium, the arenaceous deposits (Landévennec Formation) are overlain by a thick succession of silts which are interbedded with thin sandstones or quartzites (Guendaré Formation, Quenec'h Guen Group).

At the top of the Le Faou and Montguyon Formations, the incursion of a lower Zlichovian allochthonous fauna of Hercynian type (e.g., *Nowakia* cf. *praecursor, Odontochile seillouensis, Platyscutellum* aff. *viator*) corresponds to an important temporary decrease in the rhenan benthic fauna (mainly the brachiopods). It seems related to the setting of a more distal environment, as indicated by the shales of the Reun ar C'Hrank and le Buard Formations and to the final phase of the homogenization of the basin. In the Angers Synclinorium the Hercynian character is even more pronounced (Angers-Erbray Formations) and extends into the upper Emsian.

At the end of the upper Emsian (Troaon Group), abrupt vertical and lateral variations are observed in the lithology and the faunas. These changes are linked to the instability of the environment, which induces an alternation of proximal sandy formations and of more distal argillaceous-calcareous formations.

The sandy formations are composed of terrigenous detrital sediments with storm deposits and synsedimentary deformations; they also contain a rich benthic fauna (brachiopods, corals, trilobites, crinoids). The distal formations contain a more pelagic fauna (dacryoconarides, goniatites, trilobites).

In Normandy and in the Laval Synclinorium, the Middle and the Upper Devonian are not observed.

During the Eifelian the characteristics of each region become more pronounced. In the central synclinorium, the Devonian succession is continuous; but in the Angers Synclinorium, the Upper Eifelian, the Givetian, and the Frasnian are missing.

At the base of the Eifelian, shales with calcareous nodules (Saint-Fiacre Formation) contain goniatites (*Subanarcestes macrocephalus*), ostracodes, trilobites (e.g., *Heliopyge troaonensis, Kayserops cantarmoricus*), and bivalves (*Panenka* sp.). In the Angers Synclinorium, the upper part of the Pont-Maillet Formation, which contains *Pinacites jugleri,* corresponds to these levels.

From the top of the Eifelian to the base of the Frasnian more proximal facies, such as limestones and shales with nodules alternating sandstones (Tibidy Formation) which contain storm deposits are observed in the Rade de Brest. The benthic fauna is well diversified and includes *Receptaculites, Crassicyclus, Metrionaxon paucitabulatum, Atrypa squamifera, Haplocrinites, Ancyrocrinus armoricanus, Nuculana rostellata, Heliopyge renaudae,* and *Phacops.*

At the end of the Givetian, *hermanni-cristatus* zone, limestones with corals, stromatoporoids, bryozoans, and crinoids occur (Kergarvan Formation).

The succession is interrupted at the base of the Frasnian by sandstones and shales with rare beds of limestones (Traonliors Formation), lower *asymmetricus* zone. The Goasquellou sandstone shows storm deposits, convolute beddings, and slumps. This detrital sedimentation phase may be the first evidence of movements which developed at the end of the Devonian. These layers contain *Apousiella* gr. *bouchardi, Cyrtospirifer verneuili, Tornoceras* sp., *Productella* sp., and *Heliopyge porsguenensis.*

The end of the Frasnian and the lower Famennian (Rostiviec, Porsguen, and Le Zorn Formations) were calm times in tectono-sedimentary history. The sedimentation began with black limestones containing oolites and phosphatic nodules and a poor benthic fauna (middle *asymmetricus* zone). The black shales which contain large calcareous spheroids and later silico-pyritic nodules (*Ancyro. triangularis,* upper *gigas,* and *crepida* zones) were deposited. These deposits correspond to a restricted marine environment with Homoctenides, *Tornoceras, Cheiloceras, Buchiola, Posidonia,* and floated plant debris (*Protolepidodendron* sp.). The black shales with nodules and goniatites extend into the Menez Belair Synclinorium (La Rabine Formation).

In the Angers Synclinorium, shales with phosphatic nodules, sandstones, thin beds of lydites and limestones (La Briantière and La Vallée Formations) were deposited. The conodonts, the ostracods and the clymenids imply a "doII" to "doV" age.

In the Rade de Brest, the Le Zorn Formation, which consists of black shales and limestones with ostracodes and rare trilobites *(Trimerocephalus caecus)* (Fa2a), contains reworked spores from the Emsian. To the north of Rennes and in the Angers Synclinorium shales or sandstones with plant remains occur. These indicate emersions and erosions during the Upper Famennian.

In Central Brittany, the peculiar sedimentation established in the Lower Devonian continues up to the Givetian on the southern limb of the Chateaulin Synclinorium and to the Frasnian-Famennian (?) on the northern limb. These deposits form a sedimentary entity constituted by chloritoid shales and quartzites.

The history of the end of the Armorican Devonian (Strunian) is closely linked to that of the carboniferous basins.

References

Auvray B, Macé J, Vidal P, Van der Voo R (1980) Rb-Sr dating of the Plouézec volcanics, N Brittany: implications for the age of red beds ("Séries rouges") in the northern Armorican Massif. J Geol Soc (Lond) 137: 207–210

Babin C, Deunff J, Mélou M, Paris F, Pelhate A, Plusquellec Y, Racheboeuf P (1979) La coupe de Porz Ar Vouden (Pridoli de la Presqu'ile de Crozon) Massif Armoricain, France. Lithologie et Biostratigraphie. Palaeontogr Abt A 164: 52–84

Babin C, Feist R, Mélou M, Paris F (1988) La limite Ordovicien-Silurien en France. Bull Br Mus (Nat Hist) (Geol) 43: 73–79

Beuf S, Biju-Duval B, De Charpal O, Rognon P, Gariel O, Bennacef A (1971) Les grès du Paléozoïque inférieur du Sahara. Publ Inst Fr Pétrole 18: 1–464

Bonjour JL (1989) Paléoaltération du substratum briovérien (Protérozoïque supérieur) de Bretagne centrale: une étape vers les séries rouges de l'Arenig. Bull Soc Géol Fr 8 5: 717–721

Bonjour JL, Peucat JJ, Chauvel JJ, Paris F, Cornichet J (1988) U-Pb zircon dating of the Early Palaeozoic (Arenigian) transgression in Western Brittany (France); a new constraint for the Lower Palaeozoic time-scale. Chem Geol (Isotope Geoscience section) 72: 329–336

Dabard MP, Paris F (1986) Palaeontological and geochemical characteristics or Silurian black shale formations from the Central Brittany domain of the Armorican Massif (N. W. France). Chem Geol 55: 17–29

Dangeard L, Doré F (1971) Faciès glaciaires de l'Ordovicien supérieur en Normandie. Mém Bur Rech Géol Min 73: 119–125

Doré F (1981) The Late Ordovician tillite in Normandy (Armorican Massif). In: Hambrey M, Harland W (eds) Earth's pre-Pleistocene glacial record. Cambridge Univ. Press, Cambridge, pp 582–584

Durand J (1984) Le Grès armoricain. Sédimentologie, traces fossiles, milieux de dépot. Mem Doc CAESS Rennes 3: 1–150

Durand J, Noblet C (1986) Paléocourants dans la Formation du Grès armoricain: persistence des mécanismes de transport en domaine cratonique. Rev Géol Dyn Géogr Phys 27: 13–24

Hammann W, Henry JL (1978) Quelques espèces de Calymenella, Eohomalonotus et Kerfornella (Trilobita, Ptychopariida) de l'Ordovicien du Massif Armoricain et de la Péninsule ibérique. Senck Lethaea 59: 401–429

Henry JL (1971) Les Trilobites Asaphidae et Eohomalonotidae du grès armoricain supérieur (arénigien?) de l'Ouest de la France. Mém 73: 65–77

Henry JL, Nion J, Paris F, Thadeu D (1974) Chitinozoaires, Ostracodes et Trilobites de l'Ordovicien du Portugal (Serra de Buçaco) et du Massif Armoricain: essai de comparaison et signification paléogéographique. Com Serv Geol Port 57: 303–345

Henry JL, Mélou M, Nion J, Paris F, Robardet M, Skevington D, Thadeu D (1976) L'apport de Graptolites de la zone à G. teretiusculus dans la datation de faunes benthiques lusitano-armoricaines. Ann Soc Géol Nord 96: 275–281

Jaeger H, Robardet M (1973) Découverte de *Monograptus uniformis* Pribyl, Graptolite gedinnien dans les "schistes et quartzites siluro-dévoniens" du Cotentin (Normandie, France). C R Acad Sci Paris 277: 2129–2132

Mélou M (1971) Nouvelle espece de *Leptestiina* dans l'Ordovicien supérieur de l'Aulne (Finistère). Mém Bur Rech Géol Min 73: 93–101

Mélou M (1975) Le genre *Heterorthina* (Brachiopoda, Orthida) dans la Formation des Schistes de Postolonnec (Ordovicien), Finistère, France. Geobios 8: 191–208

Mélou M (1976) Orthida (Brachiopoda) de la Formation de Postolonnec (Ordovicien), Finistère, France. Geobios 9: 693–717

Mélou M (1987) Découverte de *Hirnantia sagittifera* (M'Coy 1851) (Orthida, Brachiopoda) dans l'Ordovicien supérieur (Ashgillien) de l'extrémité occidentale du Massif Armoricain. Geobios 20: 679–685

Morzadec P, Paris F, Plusquellec Y, Racheboeuf P, Weyant M (1988) Devonian stratigraphy and palaeogeography of the Armorican Massif (NW France). Second Intern Symposium Devonian System. Calgary 1987. Can Soc Petrol Geol Mem 14 1: 401–420

Noblet C (1983) Interprétation des cartes d'isopaques de la Formation du Grès armoricain levées en Bretagne et Normandie. Géol Fr 4: 311–320

Paris F (1977) Les formations siluriennes du Synclinorium du Ménez-Bélair: comparaison avec d'autres formations siluriennes du Massif armoricain. Bull Bur Rech Géol Min 1: 75–87

Paris F (1990) The Ordovician chitinozoan biozones of the Northern Gondwana Domain. Rev Palaeobot Palynol 66: 181–209

Paris F, Robardet M (1977) Paléogéographie et relations ibéro-armoricaines au Paléozoïque anté-Carbonifère. Bull Soc Géol Fr 7 19: 1121–1126

Paris F, Skevington D (1979) Présence de Graptolites de l'Arenig moyen à la base de la Formation de Postolonnec (Massif Armoricain): conséquences stratigraphiques et paléogéographiques. Geobios 12: 907–911

Paris F, Rickards B, Skevington D (1980) Les assemblages de

Graptolites du Llandovery dans le Synclinorium du Ménez-Bélair (Massif armoricain). Geobios 13: 153–171

Paris F, Pelhate A, Weyant M (1981) Présence de Conodontes de l'Ashgill dans la Formation de Rosan dans la coupe de Lostmarc'h (Finistère, Massif armoricain). Bull Soc Géol Minéral Bretagne 13 2: 15–35

Robardet M, Doré F (1988) The Late Ordovician diamictic formations from southwestern Europe: north-Gondwana glaciomarine deposits. Palaeogeogr Palaeoclimatol Palaeoecol 66: 19–31

Skevington D, Paris F (1975) Les Graptolites de la Formation de Saint-Germain-sur-Ille (Ordovicien supérieur du Massif Armoricain). Bull Soc Géol Fr 17: 260–266

Weyant M, Doré F, Le Gall J, Poncet J (1977) Un épisode calcaire ashgillien dans l'Est du Massif armoricain: incidence sur l'âge des dépôts glacio-marins fini-ordoviciens. C R Acad Sci Paris 284: 1147–1149

Paleozoic of the Ligerian Domain

H. LARDEUX and P. CAVET

1 Introduction

To the south of the faulted boundary of the Medio-North-Armorican Domain, three WNW-ESE-trending Paleozoic units have been tectonically juxtaposed during Variscan times in the northern part of the Ligerian Domain (Cavet et al. 1981; Dubreuil 1986): the Saint-Julien-de-Vouvantes, the Lanvaux-Les-Ponts-de-Cé, and the Saint-Georges-sur-Loire units. Beyond the Nort-sur-Erdre Fault, which is underlain by the Upper Carboniferous deposits of the Sillon Houiller de la Basse Loire, the substratum of the Ancenis basin consists of metamorphic rocks (Les Mauges Group, Champtoceaux Complex) which extend to the South Armorican Shear Zone. These four units (Fig. 13) show a very distinct development of facies and faunas during the Ordovician, Silurian, and Devonian (Fig. 14).

2 Saint-Julien-de-Vouvantes Unit

The Saint-Julien-de-Vouvantes Unit was first considered as the Silurian and Devonian core of a complex synclinorium (Péneau 1929). According to Dubreuil (1987), it should rather be regarded as a pull-apart basin, initiated in the Dinantian and composed of an azoic complex of siltstones and laminated sandstones containing numerous olistolites of older Paleozoic rocks. The rich faunas that have been described in this unit, the most recent being of Famennian age, mainly come from these olistolites, which originated partly from the north (Devonian faunas of Rhenish type) and partly from a southern domain which is at present occluded (Bohemian faunas).

Fossiliferous formations with Rhenish affinities have already been characterized (5, 6, 7, and 8, Fig. 14). Among the fossiliferous outcrops showing

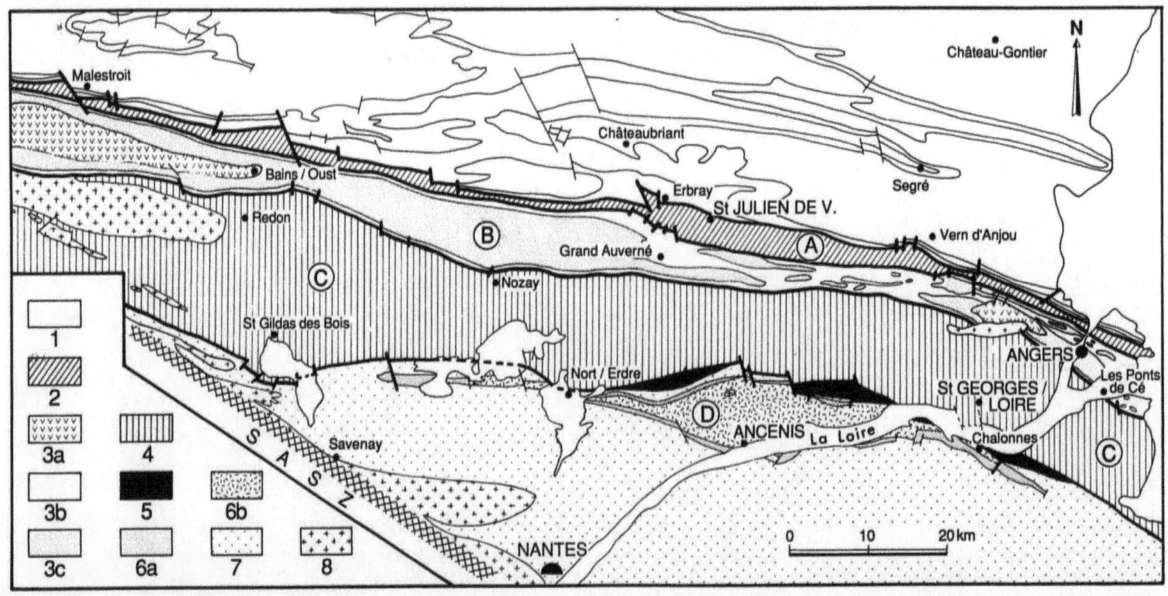

Fig. 13. The Ligerian Domain. *1* Medio-North-Armorican Domain; *2* Saint-Julien-de-Vouvantes Unit *(A)*; *3* Lanvaux-Les-Ponts-de-Cé Unit *(B)*; *3a* Landes de Lanvaux Orthogneiss; *3b* Bains-sur-Oust Group; *3c* Grand-Auverné Formation; *4* Saint-Georges-sur-Loire Unit *(C)*; *5* Sillon Houiller de la Basse-Loire; *6* Ancenis Basin *(D)*; *6a* "pre-Frasnian" Group; *6b* Ancenis Complex; *7* metamorphic rocks of Les Mauges Group and Champtoceaux Complex; *8* Variscan granites; *SASZ* South-Armorican Shear Zone

J. D. Keppie (Ed.): Pre-Mesozoic Geology in France and Related Areas

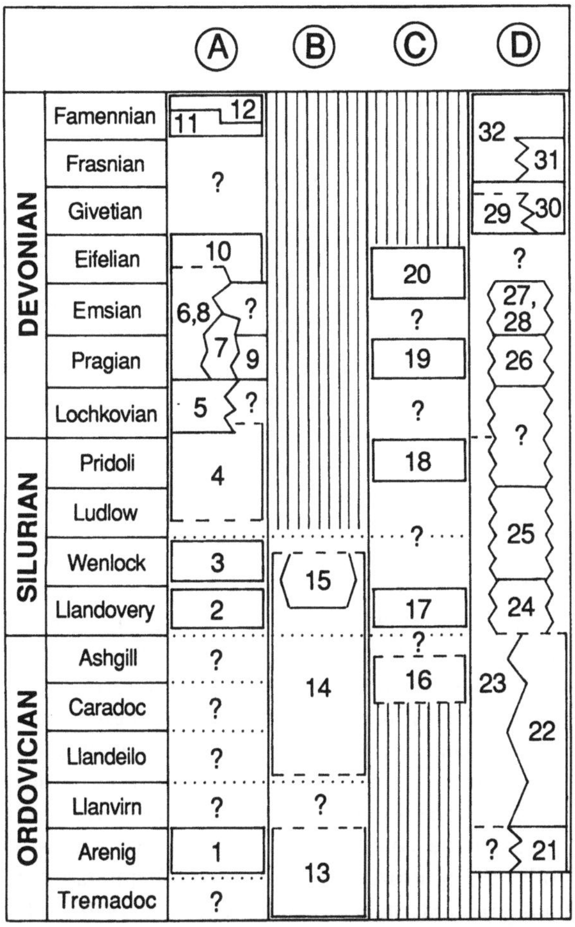

Fig. 14. Fossiliferous formations of the Ligerian Paleozoic. *A* Saint-Julien-de-Vouvantes Unit; *B* Lanvaux-Les-Ponts-de-Cé Unit; *C* Saint-Georges-sur-Loire Unit; *D* Ancenis Basin. *1* La Pile Formation; *2* black cherts with *Monograptus lobiferus; 3* spheroid-bearing shales; *4* La Meignanne black shales and limestones; *5* "Grès à *Platyorthis*" Formation; *6* Angers Limestone; *7* Vern d'Anjou Limestone; *8* Erbray Limestone; *9* siltstones and limestones with *Nowakia acuaria; 10* Pont-Maillet Formation; *11* La Briantière Limestone; *12* La Vallée Formation; *13* Bains-sur-Oust Group; *14* Grand-Auverné Formation; *15* Le Houx black shales; *16* Foret d'Ancenis Sandstone; *17* black cherts with *M. lobiferus; 18* Les Roches Limestone (Lower); *19* Les Roches Limestone (Upper); *20* Moulin Bachelot Limestone; *21–30* pre-Frasnian rocks; *21* Moulin de Chateaupanne Formation; *22* Le Fresne Complex; *23* Pierre-Melière Complex; *24–28* Tombeau Leclerc Complex; *24* black cherts with *M. lobiferus; 25* siltstones and limestones with graptolites; *26* siltstones and limestones with *N. acuaria; 27* La Grange Limestone; *28* Le Valet Limestone; *29* Chalonnes Limestone; *30* "Grès à Psilophytes" Formation; *31* Cope-Choux Limestone; *32* L'Ecochère Group

Bohemian affinities, the following deserve mention: Llandoverian black cherts (phtanites) with *Monograptus lobiferus;* Ludlovian, Pridolian, and Lochkovian black shales and ampelitic limestones with bivalves, graptolites *(Monograptus uniformis uniformis),* and chitinozoans *(Urnochitina urna)* at La Meignanne (Kříž and Paris 1982); Pragian siltsones and limestones with *Nowakia acuaria* and *Guerichina strangulata* in numerous locatities (Lardeux 1969 a); Famennian limestones with conodonts at La Briantière, and Famennian nodular shales with clymenids *(Wocklumeria sphaeroides),* trilobites *(Drevermannia schmidti),* and ostracodes at La Vallée (Péneau 1929).

The La Pile Formation must be considered separately: only two small outcrops are known at La Pile and La Boserie (Péneau 1929). In these localities, siltstones and sandstones that are not easily separable from the presumed Dinantian Complex have yielded Late Arenig Didymograptids.

3 Lanvaux-Les-Ponts-de-Cé Unit

The narrow, 200 km long Lanvaux-Les-Ponts-de-Cé unit (Chauvel 1960; Chauris and Lucas 1966) is bounded to the north by the Malestroit-Angers Fault and to the south by the Freigné Fault (Cavet et al. 1986). The core of the structure is occupied by the orthogneiss of the Landes de Lanvaux in the west and of Saint-Clément-de-la-Place in the east. These magmatic rocks are surrounded by the shales, siltstones, microconglomerates, and sandstones (locally with volcanic components) of the Bains-sur-Oust Group (Barrois and Pruvost 1938; Cavet et al. 1976; Berthé et al. 1977). The uppermost 20 m are composed of well-bedded sandstones with very rare *Cruziana* (Péneau 1947) and have been regarded as an equivalent of the Grès Armoricain Formation (Arenig pro parte). The age of the underlying beds is still debated and the lowermost siltstone outcrops are sometimes considered as Late Proterozoic in age (Chantraine et al. 1988). Upwards, the Bains-sur-Oust Group passes gradually into the dark shales of the Grand-Auverné Formation. These rocks, which show a pronounced slaty cleavage (Angers slates), contain a deformed trilobite fauna (e.g., *Placoparia tournemini, Zeliszkella lapeyrei*) of Llandeilo age (Pillet 1988) in the La Pouèze and Trélazé localities. The recent discovery (Cavet et al. 1986) of *Monograptus priodon* and other graptolites in black shales (the so-called Ampélites du Houx), which cannot be easily separated from the Angers slates, shows that the Grand-Auverné Formation reaches at least the Llandovery-Wenlock boundary.

4 Saint-Georges-sur-Loire Unit

East of Nozay (Fig. 13), the wide Saint-Georges-sur-Loire unit (Lucas 1960; Cavet et al. 1976) comprises a thick succession of slightly metamorphic and locally variegated shales and siltstones. From north to south, three more or less distinct areas can be distinguished. In the northern area, devoid of black cherts, the Forêt d'Ancenis Sandstone has yielded the Upper Ordovician trilobite *Calymenella bayani* (Bureau 1900). To the south, very numerous and thin intercalations of radiolarian-bearing black cherts contain Llandoverian graptolites (Philippot 1950). The southernmost area is characterized by an abundance of volcanic rocks (spilites, rhyolites, and tuffs: Maillet et al. 1977; Carpenter et al. 1982). Also in this area, two minute limestone outcrops (Les Roches) have yielded Pridolian conodonts and Pragian dacryoconarids (Lardeux 1978). Dubreuil and Vachard (1982) have reported Early to Middle Devonian algae from carbonate rocks (Moulin Bachelot Limestone) closely associated with spilites.

West of Nozay, the sedimentary and volcanic rocks mentioned above also occur, but they have a more irregular distribution (Barrois and Pruvost 1938; Barrois 1949). The assignment to the Grès Armoricain Formation of the sandstones which occur near Saint-Gildas-des-Bois (Pivette 1978; Fourniguet and Trautmann 1985) seems rather equivocal (Marchand and Bossière 1989).

In the present state of knowledge, the Saint-Georges-sur-Loire unit can only be regarded as composed of a thick monotonous and mostly azoic complex ranging from the Late Ordovician to the Early or Middle Devonian.

5 Ancenis Basin

The spindle-shaped, 90-km-long Ancenis Basin is generally separated from the Sillon Houiller de la Basse-Loire, located to the north, by the narrow block-faulted unit known as Horst de Pouillé, which is composed of epischists identical to those of the Mauges Group. The Ancenis complex (Rivière 1977) mainly consists of brackish shales, siltstones, and grits of Dinantian age (see Carboniferous chapter, this Vol.). Its lowermost part also includes lenses or olistoliths (Dubreuil 1986) of fossiliferous Late Devonian marine rocks. The most famous are the Cope-Choux Limestone, with *Hypothiridina cuboides* and other Frasnian brachiopods (Bureau 1900) and the L'Ecochère siltstones and limestones mainly of Frasnian age as indicated by the occurrence of *Manticoceras* sp. and of *Cyrtospirifer verneuili* (Péneau 1933), but also including at their top Lower Famennian levels with conodonts (Cavet et al. 1978).

The so-called pre-Frasnian shales, siltstones, sandstones, phtanites, and limestones regroup both clearly defined formations and sedimentary bodies without well-defined boundaries and will be described here from south to north.

South of the Loire River, the limit between the "pre-Frasnian" rocks and the Late Proterozoic of the Mauges Group is generally faulted. However, near Chalonnes, in small outcrops, gently tilted Arenigian rocks of the Moulin de Chateaupanne Formation unconformably overlie the folded epischists (Blaise et al. 1970). Above a thin basal coarse detrital level, the formation mainly consists of red-brown siltstones with conodonts (*Acantiodus franconicus*) and inarticulate brachiopods.

The Fresne Complex (Cavet et al. 1971; Babin et al. 1976) comprises dark micaceous siltstones with sandy levels regarded as succeeding conformably to the previous formation. They contain numerous trace fossils *(Chondrites)* and rare fragments of trinucleids, the latter suggesting a Middle or Late Ordovician age.

The Pierre-Melière Complex occurs on both flanks of the Ancenis Basin, but the contacts with both the Proterozoic series and the Dinantian Complex are systematically faulted (Diot 1980). The Pierre-Melière Complex consists of dark micaceous shales with arenaceous lenses (with rare *Scolithos*) and numerous siliceous nodules yielding Llanvirnian trilobites with typical Bohemian affinities (*Pricyclopyge synophtalma, Ormathops* cf. *atavus;* Cavet and Pillet 1964). Badly preserved monograptid fragments suggest that the Pierre-Melière Complex grades up into the Silurian (Cavet et al. 1978).

The Chalonnes Limestone Formation occurs in the easternmost part of the Ancenis Basin (Cavet et al. 1970; Blaise et al. 1986). Large lenses of reef limestones underlie, over a distance of approximately 18 km, the northern boundary of the Le Fresne Formation. The relations between the limestone lenses and the poorly fossiliferous surrounding shales have not yet been elucidated (Cavet et al. 1971). The age assignment close to the Emsian-Eifelian boundary was based upon the study of the rich coral fauna with *Favosites styriaca* and *Cyathophyllum elongatum* Le Maître 1934). More recently, a Givetian age was proposed on the basis of micropaleontological data (Dubreuil and Vachard 1979). Brachiopods *(Amphigenia bureaui, Pentamerus davyi)* are common, but trilobites are very rare and neither tentaculites nor conodonts have been found. These limestone bodies are at present interpreted as resulting from the disintegration of a reef barrier (Dubreuil 1986).

The "Grès à Psilophytes" Formation (Cavet et al. 1971) apparently conformably overlies the Chalonnes Limestone. These well-bedded gritty-pelitic terrigenous deposits mark the end of the calcareous sedimentation.

The Tombeau Leclerc Complex occupies a narrow area between the Chalonnes Limestone and the Dinantian Ancenis Complex near Chalonnes. Its peculiar characteristics have been emphasized for a long time (Davy 1906). It had first been regarded as an anticlinal structure or as a block-faulted unit (Cavet et al. 1971) but is at present interpreted as a large olistostrome included at the very base of the Dinantian Complex (Dubreuil 1986).

Several isolated localities have yielded rich and diversified faunas of Silurian or Lower Devonian age that are not known farther south. These include: the Llandoverian phtanites with *Monograptus lobiferus, Rastrites peregrinus,* and *Retiolites obesus* (Philippot 1950); the Eifelian limestone (Le Valet) with numerous brachiopods and trilobites (Le Maître 1934; Pillet 1972); and the Emsian limestone at La Grange (Lardeux et al. 1979) with conodonts *(Polygnathus dehiscens, P. laticostatus),* tentaculites *(Nowakia barrandei, N. elegans, N. cancellata),* and goniatites *(Anetoceras advolvens).*

References

Babin C, Arnaud A, Blaise J, Cavet P, Chauvel JJ, Deunff J, Henry JL, Lardeux H, Mélou M, Nion J, Paris F, Plaine J, Quété Y, Robardet M (1976) The Ordovician of the Armorican Massif (France). In: Bassett MG (ed) The Ordovician System. University Wales Press and Nat Mus Wales, Cardiff, pp 359–385

Barrois C (1949) Carte géologique détaillée de la France à 1/80000: Saint-Nazaire. Bur Rech Géol Min, Orléans

Barrois C, Pruvost P (1938) Carte géologique détaillée de la France à 1/80000: Redon. Bur Rech Géol Min, Orléans

Berthé D, Chauvel JJ, Jégouzo P, Plaine J (1977) Données nouvelles sur l'origine du "Complexe" des Schistes et arkoses de Bains (Cambro-Arenig de Bretagne centrale). 5ème Réun Ann Sci Terre Rennes: 66

Blaise J, Cavet P, Lardeux H (1970) Les relations entre Briovérien et Paléozoïque sur la rive gauche de la Loire entre Saint-Florent-le-Vieil et Chalonnes (M. et L., SE du Massif armoricain). Bull Soc Géol Fr 7 12: 5–14

Blaise J, Arnaud A, Brossé R, Cavet P, Gruet M, Lardeux H (1986) Carte géologique à 1/50000: Thouarcé. Bur Rech Géol Min, Orléans

Bureau L (1900) Notice sur la géologie de la Loire-Inférieure. In Nantes et la Loire-Inférieure. Grimaud impr.: 99–522

Carpenter MSN, Peucat JJ, Pivette B (1982) Geochemical and geochronological characteristics of paleozoic volcanism in the Saint-Georges-sur-Loire synclinorium (S of the armorican Massif); evidence for pre-hercynian tectonic evolution. Bull Bur Rech Géol Min 21: 63–79

Cavet P, Pillet J (1964) Contribution à l'étude stratigraphique et paléontologique du synclinal d'Ancenis (Massif armoricain). Bull Soc Géol Fr 7 5: 318–329

Cavet P, Arnaud A, Barbaroux L, Blaise J, Brossé R, Chauris L, Gruet M, Lardeux H, Rivière LM (1970) Carte géologique à 1/50000: Chalonnes-sur-Loire. Bur Rech Géol Min, Orléans

Cavet P, Lardeux H, Philippot A (1971) Ordovicien et Silurien aux environs de Montjean et Chalonnes (M. et L., SE du Massif armoricain). Mém Bur Rech Géol Min 73: 199–212

Cavet P, Arnaud A, Blaise J, Brossé R, Chauris L, Gruet M, Lardeux H (1976) Carte géologique à 1/50000: Angers. Bur Rech Géol Min, Orléans

Cavet P, Arnaud A, Blaise J, Gruet M, Lardeux H, Marchand J, Rivière LM (1978) Carte géologique à 1/50000: Ancenis. Bur Rech Géol Min, Orléans

Cavet P, Blaise J, Marchand J (1981) Le domaine ligérien entre Angers et Nantes; synthèse cartographique et précisions nouvelles. Mém Soc Et Sci Anjou 4: 23–24

Cavet P, Dubreuil M, Blaise J, Willefert S (1986) Les ampélites du Houx (Silurien); leurs relations avec les unités paléozoïques de Lanvaux-les-Ponts-de-Cé et de Saint-Georges-sur-Loire (Massif armoricain). Bull Soc Sci Nat Ouest Fr 8 3: 136–143

Chantraine J, Chauvel JJ, Balé P, Denis E, Rabu D (1988) Le Briovérien (Protérozoïque supérieur) et l'orogenése cadomienne en Bretagne (France). Bull Soc Géol Fr 8 4: 815–829

Chauris L, Lucas G (1966) Les environs de Bécon-les-Granits (feuille Ancenis au 1/80000). Bull Serv Carte Géol Fr 60: 25–33

Chauvel JJ (1960) Sur l'anticlinal de Lanvaux et ses enveloppes sédimentaires entre Peillac et Bains-sur-Oust. C R Somm Soc Géol Fr 1: 12

Davy L (1906) Ce que l'on croit savoir aujourd'hui sur la constitution géologique des environs de Chalonnes. Bull Soc Et Sci Angers 35 5: 91–125

Diot H (1980) Recherches structurales dans la partie orientale du Domaine ligérien (Massif armoricain). Thèse 3e cycle, Nantes University, pp 1–147

Dubreuil M (1986) Evolution géodynamique du Paléozoïque Ligérien (Massif Armoricain). Thèse, Nantes University, pp 1–258

Dubreuil M (1987) Le bassin en décrochement de Saint-Julien-de-Vouvantes-Angers (Carbonifère inférieur du Massif armoricain). Bull Soc Géol Fr 8 3: 215–221

Dubreuil M, Vachard D (1979) Sur l'âge givetien de la Formation de Chalonnes (Bassin d'Ancenis, Massif armoricain) et ses conséquences. C R Acad Sci Paris 289 D: 241–244

Dubreuil M, Vachard D (1982) La série de Saint-Georges-sur-Loire et les événements de la limite Silurien-Dévonien dans le domaine ligérien du Massif Armoricain. 9e Réun Ann Sci Terre, Paris: 213

Kříž J, Paris F (1982) Pridolian and Lochkovian in la Meignanne (armorican Massif) Biostratigraphy and correlations based on Bivalvia and Chitinozoa. Geobios 15 2: 391–421

Lardeux H (1967) Présence du Dévonien inférieur dans le synclinal de Saint-Georges-sur-Loire (Massif armoricain). C R Somm Soc Géol Fr 8: 341

Lardeux H (1969a) Le Dévonien du synclinal de Saint-Julien-de-Vouvantes (SE du Massif armoricain): état des recherches et précisions nouvelles. Bull Soc Et Sci Anjou 7: 3–40

Lardeux H, Morzadec P, Bultynck P, Walliser O (1979) Massif armoricain. Intern Subcommission Devonian Stratigraphy Field excursion guide, Rennes University, pp 1–27

Le Maître D (1934) Etudes sur la faune des calcaires dévoniens du Bassin d'Ancenis. Mém Soc Géol Nord 12: 1–267

Maillet P, Pivette B, Chauvel JJ (1977) Données nouvelles sur les volcanites du synclinorium de Saint-Georges-sur-Loire (Massif armoricain). 5e Réun Ann Sci Terre, Rennes: 326

Marchand J, Bossière G (1989) Carte géologique à 1/50000: Savenay, Bur Rech Géol Min, Orléans

Péneau J (1929) Recherches stratigraphiques et paléontologiques dans le Sud-Est du Massif armoricain. Bull Soc Sci Nat Ouest Fr 4 8: 1–300

Péneau J (1933) La faune du Dévonien supérieur de l'Ecochère, près Ancenis. C R Somm Soc Géol Fr: 33–34

Philippot A (1950) Les Graptolites du Massif Armoricain. Etude stratigraphique et paléontologique. Mém Soc Géol Minéral Bretagne 8: 1–295

Pillet J (1972) Les Trilobites du Dévonien inférieur et du Dévonien moyen du Sud-Est du Massif Armoricain. Mém Soc Etud Sci Anjou 1: 1–307

Pillet J (1988) Les Trilobites des "ardoises" d'Angers. III: Dalmanitidae, Cheiruridae, Trinucleina. Bull Soc Et Sci Anjou N S: 1–15

Pivette B (1978) Le synclinorium de Saint-Georges-sur-Loire (Massif armoricain): sa place dans l'évolution géodynamique de la Bretagne méridionale au Paléozoïque. Thèse 3e cycle, Rennes University, pp 1–108

Rivière LM (1977) Le Culm frasno-dinantien du synclinal d'Ancenis (SE du Massif armoricain) au N de la Loire. Bull Soc Géol Minéral Bretagne C 9 1: 19–57

Paleozoic of the Vendean Domain

M. Colchen and D. Poncet

1 Introduction

Two main Vendean domains are generally recognized and are separated by a narrow elongated structure known as Les Essarts-Mervent unit (Fig. 15): (1) The Bas-Bocage Domain consists of low grade metamorphic Paleozoic formations (from Ordovician to lower Carboniferous) locally intruded by Hercynian alkali granites (La Roche-sur-Yon and Bournezeau plutons), medium to high grade metamorphic rocks (Sauveterre and Les Sables d'Olonne series, and, to the northwest, of an allochthonous volcano-sedimentary series called the Porphyroids unit; (2) The Haut-Bocage Domain includes two distinct lithostructural units: the Chantonnay Synclinorium with both sedimentary and volcanic sequences (from Cambrian up to Devonian) and the Paleozoic granitoids suite which consists of intrusives in a monotonous medium grade metamorphic series, assigned to the Brioverian (Mathieu 1937).

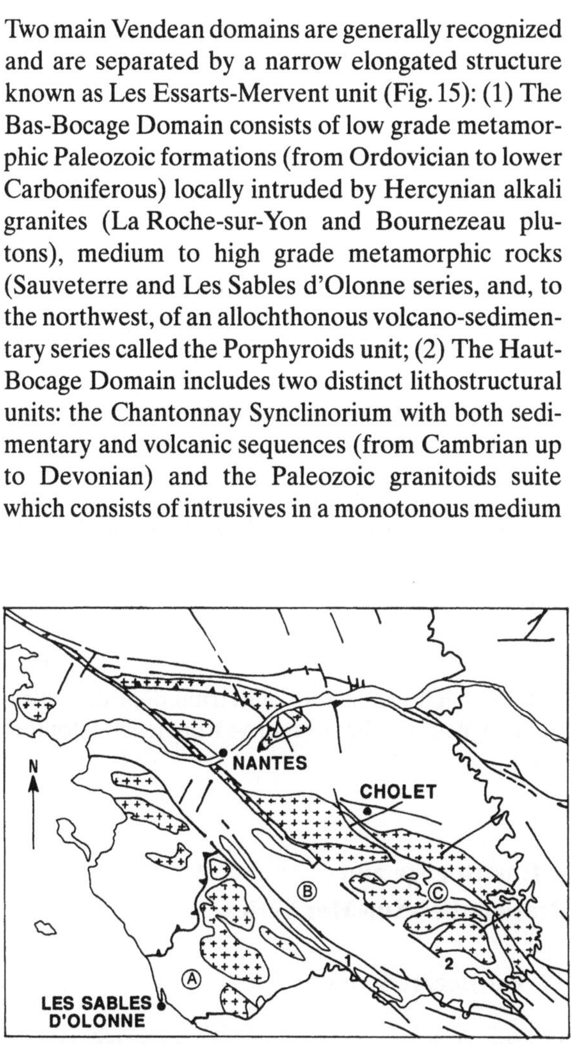

Fig. 15. Geological setting of Vendée. *A* Bas-Bocage: Sables d'Olonne metamorphites, Brétignolles sedimentary series (Silurian-Tournaisian), granitoids, porphyroids (Silurian-Devonian); *B* Chantonnay Synclinorium: sedimentary, volcanic, and volcano-sedimentary formations (Cambrian-Devonian); *C* Haut-Bocage: Hercynian calco-alkalic granitoids and leucogranites, Proterozoic and Paleozoic (?) metasedimentary formations. *1* Les Essarts-Mervent unit: orthogneiss and amphibolites, and coal-bearing of Faymoreau; *2* Ardelay Absie: NW-SE transcurrent dextral fault

2 Haut-Bocage Series

The Haut-Bocage Series rocks, the most complete Paleozoic succession known in Vendée, outcrop in a NW-SE-oriented structure, which corresponds to the Chantonnay Synclinorium (Wyns 1980). The stratigraphic sequence established to the NW (Wyns 1980, Wyns et al. 1989) and the SE (Bouton 1990) was deposited from the Cambrian up to Devonian. The lithostratigraphic series are about 3500 to 5500 m thick, and are made of (Fig. 16): (1) the Montournais Formation (500 to 900 m), graywackes, shales, pelites, and volcanics, the Les Gerbaudières Formation (600 to 1000 m), shales and carbonate lenses, and the Bourgneuf Formation (500 to 1000 m), greywackes and shales; despite the lack of any fauna, these three formations are tentatively referred to the Cambrian; (2) La Châtaigneraie Ignimbrites Formation (0 to 200 m), of continental tholeiitic affinity, indicates an episode of distension (rifting) of the continental crust (Thieblemont 1987). In the southwestern part of the Chantonnay Synclinorium, the Sigournais Formation (0 to 500 m), composed of polygenic conglomerates, sandstones, pelites with *Billingsella,* Orthidae from the Upper Cambrian, and volcanic rocks, is interpreted as a lateral facies of the La Châtaigneraie Ignimbrites Formation; (3) La Châtaigneraie Quartzite Formation (10 to 50 m), tentatively referred to the Ordovician (Arenig, cf. Armorican quartzite?); (4) Réaumur Formation, consisting of about 700 m of shales and gritty pelites with a phtanite bed (lower part of the Réaumur Group). New faunas have been discovered (Wyns et al. 1989): Caradoc or Lower Silurian (f_1) graptolites, Eifelian to Givetian tentaculites and conodonts (f_2); (5) La Meilleraie Formation (upper part of the Réaumur Group, 1000 to 1500 m of basaltic rocks with pillow lavas, is characterized by the association of

J. D. Keppie (Ed.): Pre-Mesozoic Geology
in France and Related Areas

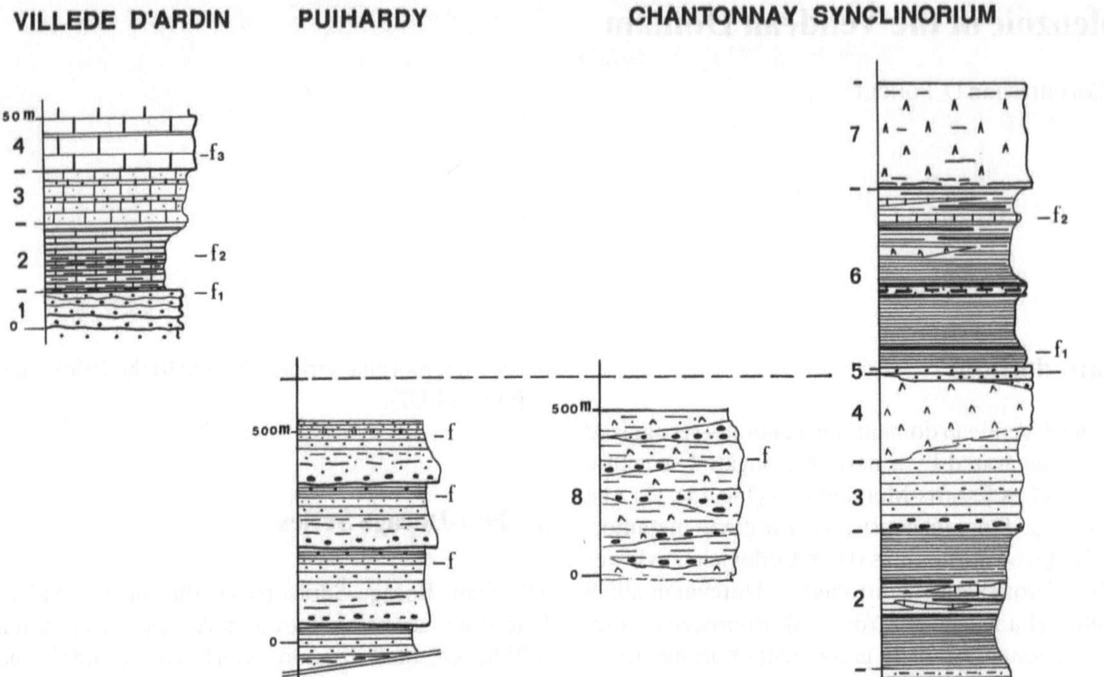

Fig. 16. Haut-Bocage Series. *Chantonnay Synclinorium: 1* Montournais Fm.; *2* Les Gerbaudières Fm.; *3* Bourgneuf Fm.; *4* La Châtaigneraie ignimbrites Fm.; *5* La Châtaigneraie quartzite Fm.; *6* Réaumur Fm.; *7* La Meilleraie Fm.; *8* Sigournais Fm.; f_1, f_2 fossils (see text). *Villedé d'Ardin: 1* quartzitic conglomerate and sandstones with f_1: *Dechenella* sp., cf. *setosa* from Givetian; *2* calcschists with f_2: corals and brachiopods; *3* arenic limestones; *4* limestones with f_3: *Stringocephalus burtini* (Givetian). *Puihardy:* detrital deposits with Upper Cambrian brachiopods *(f)*

oceanic and arc tholeiites, suggesting a back-arc basin (Thieblemont and Cabanis 1986; Thieblemont 1987).

Between the Chantonnay Synclinorium and the "Les Essarts-Mervent" unit at the juxtaposition of St-Martin-des-Noyers amphibolites and relict lenses of eclogites (Godard 1981) and the Mervent orthogneiss (446 ± 12 Ma; Vidal 1976), three other units have been recognized: (1) La Villedé-d'Ardin unit (Givetian). This is a nonmetamorphic formation (Camuzard et al. 1969) which was most likely formed during early Variscan events (Autran 1978). It is interpreted as an allochthonous unit, which is pinched in the Hercynian transcurrent shear zone and lies the Faymoreau Namuro-Stephanian coalfield; (2) Puihardy-Champdeniers unit. This is an allochthonous unit (Bouton 1989), consisting of differentiated detrital deposits with fossiliferous silty facies which contain a typical brachiopod fauna from the Upper Cambrian (Ters and Pillet 1987). (3) Roc-Cervelle unit. This unit is made up of detrital deposits which are assigned to the Cambro-Ordovician (Bouton 1990).

3 Bas-Bocage and Bretignolles-Sur-Mer Series

3.1 Bas-Bocage Series

The Bas-Bocage Series consists of mainly detrital formations which were first considered Brioverian (Mathieu 1937), but for which a Paleozoic age is now clearly demonstrated. Dated beds are middle Ordovician (Brillanceau 1962) and Silurian (Brillanceau and Mathieu 1965; Ters 1972). The Mareuil-sur-Lay tuffo-ignimbritic Formation, which has lithological, petrographical, and structural features like those of the "Porphyroids" unit (Maillet 1984), is interbedded

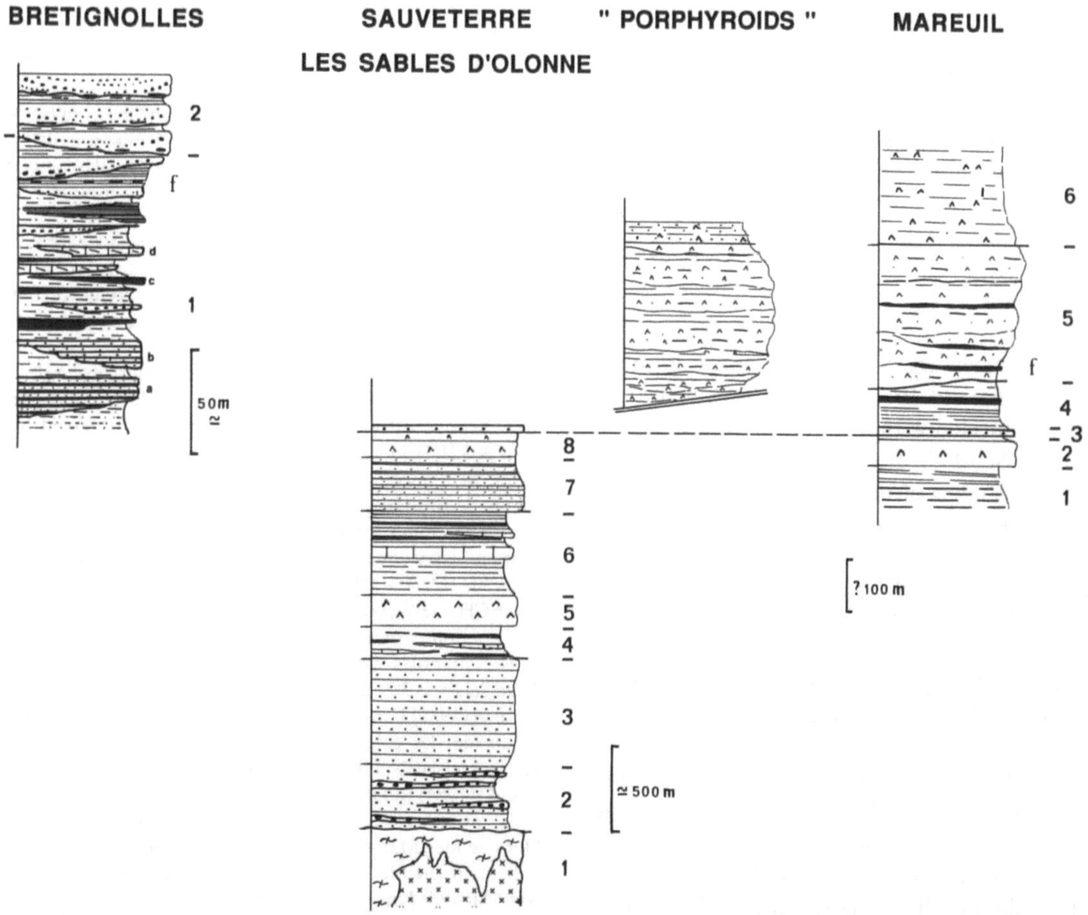

Fig. 17. Bas-Bocage and Vendean Coastline Series. Brétignolles: *1* Brétignolles Fm., olistostrome with pelitic matrix dated by *Siphonodella crenulata* (Middle Tournaisian) *(f)* and blocks and olistolites: *a* Sainte-Véronique phtanite (Llandovery?); *b* quartzites and red shales dated by *Monograptus priodon* (Lower Wenlock); *c* black shales (Upper Ordovician?); *d* limestone; *2* La"Parée Fm. Sauveterre-Les Sables d'Olonne: *1* orthogneiss and anatexis granite; *2* augen gneiss (cf. metaconglomerates); *3* micaschists and gneiss; *4* metashales and dolomitic limestones; *5* rhyolite; *6* metashales, graphitic shales and dolomitic limestones; *7* sericito-schists and sandstones; *8* rhyolite (Arenigian?). Mareuil: *1* black shales; *2* ignimbrites cf. La Châtaigneraie Fm.; *3* quartzites; *4* shales and phtanite layers; *5* Mareuil volcano-sedimentary Fm., with lens-shapped phtanite layers [*Ancyrochitina* (Silurian-Devonian)]; *6* tuffaceous-shales and sandstones

with a tuffaceous sandstone series with lens-shaped phtanitic layers (Chalet 1985) (Fig. 17). In the latter, Siluro-Devonian fauna was discovered (Chalet et al. 1983). U-Pb zircon ages of 404 ± 5 Ma, obtained from the tuff-ignimbritic formation (Peucat et al. 1986), give evidence for a Siluro-Devonian age of this acidic volcanic episode.

3.2 Brétignolles-sur-Mer Series (Fig. 17)

This series occurs in a synclinorium which trends NW-SE and is overfolded to the southwest. Its reversed northern limb is overthrust by the "Porphyroids" Nappe and is intensively sliced (Colchen and Poncet 1987). Two lithostratigraphic units can be recognized: (1) The Brétignolles Formation (120 to 150 m), is an olistolith-bearing unit composed of blocks of different ages wrapped up in a Tournaisian pelitic matrix (Colchen and Poncet 1989). Blocks (tens to hundreds of meters in size) include: black shales probably from Upper Ordovician (Comple, pers. comm.); dolomitic limestones interbedded with black shales; alternating quartzite and red shale beds (centimeter-thick) with phosphatic nodules dated by graptolites (e.g., *Monograptus priodon*) from the Lower Wenlock (Ters 1970); phtanites and red shale probably from the Llandovery (Ters 1972). The pelitic matrix, with arkosic graywacke and magmatic blue quartz, contains siliceous phosphatic nodules with a typical fauna of conodonts from the Middle to Upper Tournaisian (e.g., *Siphonodella crenula-*

ta) (Comble et al. 1985). This sequence is interpreted as a Dinantian olistostrome (Colchen and Poncet 1989). (2) The La Parée Formation (30 to 50 m), represents the most recent unit of the series and consists of fine-grained feldspathic sandstones with cross-bedding and black mud balls alternating with shaly layers.

3.3 Conclusions

Paleozoic sedimentary series recognized in both the Bas-Bocage and the Haut-Bocage can be respectively connected with the North-Aquitano-Vendean and the South-Armorican continental margins. Geochemical studies of interbedded volcanic series of the Chantonnay Synclinorium allow the interpretation of the geodynamic evolution of the Vendée area. It is characterized by: (1) an intracontinental distension (Cambrian-Ordovician) which occurs in an anorogenic context (Autran and Cogné 1980; Wyns and Chantraine 1984), and then (2) a convergence (Silurian-Devonian) (Wyns and Chantraine 1984).

The Haut-Bocage Series can be considered for this period as the active South-Armorican continental margin. The Bas-Bocage Series is attributed to the North Aquitano-Vendean margin which was passive during the Cambro-Ordovician period, and then perhaps active during the Siluro-Devonian acidic volcanic episode (e. g., Mareuil-sur-Lay and Porphyroids tuffo-ignimbrites). According to this interpretation, the Vendean massif was a collision zone between two continental margins and the Les Essarts-Mervent unit represents the suture zone.

References

Autran A (1978) Synthèse provisoire des événements orogéniques calédoniens en France. In: Tozer ET, Schenk PE (eds) Caledonian Appalachian orogen of the North Atlantic region. P.I.C.G., project 27. Geol Surv Can Pap 78–113: 159–175

Autran A, Cogné J (1980) La zone externe de l'orogène varisque dans l'Ouest de la France et sa place dans le développement de la chaîne hercynienne. Mém Bur Rech Géol Min 208: 7–17

Bouton P (1990) Géologie du Haut Bocage Vendéen. Thèse, Poitiers University, pp 1–378

Brillanceau A (1962) Sur la présence d'Ordovicien dans le Bas-Bocage vendéen. Région de la Roche-sur-Yon. C R Acad Sci Paris D 255: 2477–2478

Brillanceau A, Mathieu G (1965) Extension notable du Silurien dans le Bas-Bocage de Vendée. C R Acad Sci Paris D 261: 1879–1881

Camuzard JP, Lardeux H, Pillet J, Rivière LM (1969) Données nouvelles sur les calcaires givétiens de la Villedé d'Ardin au gisement de la fontaine de la Marbrière. Bull Soc Et Sci Anjou 7: 41–52

Chalet M (1985) Contribution à l'étude de la chaîne hercynienne d'Europe occidentale. Etude lithostratigraphique et structurale des "Porphyroïdes" et formations paléozoïques associées du Bas-Bocage vendéen (région de Mareuil-sur-Lay). Thèse, Poitiers University, pp 1–214

Chalet M, Colchen M, Paris F, Wyns R (1983) Caracteres lithologiques, petrographiques et structuraux de la formation volcano-sedimentaire "porphyroïde" de Mareuil-sur-Lay, Bas-Bocage vendéen. Bull Soc Géol Minéral Bretagne C 15 2: 181–195

Colchen M, Poncet D (1987) Sur l'âge post-Tournaisien de la tectonique hercynienne synschisteuse a Brétignolles-sur-Mer, Vendée littorale, Massif Armoricain. C R Acad Sci Paris 305 II: 1255–1258

Colchen M, Poncet D (1989) Présence, dans la série paléozoïque de Brétignolles-sur-Mer (Vendée, Sud du Massif Armoricain), d'une formation à blocs et olistolites d'âge Dinantien. Conséquences géodynamiques. C R Acad Sci Paris 309 II: 1503–1507

Comble CH, De Wever P, Ters M, Weyant M (1985) Découverte de Conodontes et de Radiolaires d'âge Tournaisien dans les schistes bariolés de Brétignolles-sur-Mer (Vendée). C R Acad Sci Paris 300 II 17: 899–904

Godard G (1981) Lambeaux probables d'une croûte océanique subductée: les éclogites de Vendée (Massif Armoricain), Thèse 3e cycle, Nantes University, pp 1–153

Maillet D (1984) Relation des Porphyroïdes et schistes de Saint-Gilles avec les formations siluriennes de Brétignolles-sur-Mer (Vendée Maritime). Une tectonique tangentielle par cisaillement ductile pendant l'orogenèse acadienne. Thèse, Aix-Marseille University, pp 1–160

Mathieu G (1937) Recherches géologiques sur les terrains paléozoïques de la région vendéenne. Thèse, Lille University, pp 1–416

Peucat JJ, Paris F, Chalet M (1986) U-Pb zircon dating of volcanic rocks close to the Silurian-Devonian boundary from Vendée (Western France). Chem Geol (Isotope Geoscience section) 59: 133–142

Ters M (1970) Découverte d'un gisement de graptolithes du Silurien (Wenlock) dans la série de Brétignolles, Vendée. C R Acad Sci Paris D 271: 1060–1062

Ters M (1972) Sur l'extension du Paléozoïque en Vendée littorale. Stratigraphie et structure. C R Somm Soc Géol Fr: 146

Ters M, Pillet J (1987) Le Cambrien supérieur de la vallée du Saumort et sa faune de Brachiopodes (Synclinorium de Chantonnay, Vendée, Massif Armoricain). Bull Soc Sci Nat Ouest Fr n s 9, 3: 105–116

Thieblemont D (1987) Le magmatisme paléozoïque en Vendée: apport de la géochimie des éléments traces et de la pétrologie du métamorphisme à la compréhension du développement orogénique varisque. Thèse, Paris VI University, pp 1–365

Thieblemont D, Cabanis B (1986) Découverte d'une association de volcanites d'arc et de basaltes de type MORB dans la série paléovolcanique de la Meilleraie-Vendée-France. C R Acad Sci Paris 302 II: 641–646

Vidal Ph (1976) Evolution polyorogenique du Massif Armoricain. Apport de la geochronologie et de la géochimie isotopique du strontium. Mém Soc Géol Minéral Bretagne 21: 1–140

Wyns R (1980) Contribution à l'étude du Haut-Bocage

vendéen: le Précambrien et le Paléozoïque dans la région de Chantonnay (Vendée). Thèse 3ème cycle, Paris VI University, pp 1–124

Wyns R, Chantraine J (1984) Les séries métamorphiques de Sauveterre et des Sables d'Olonne (côte vendéenne) sont-elles cambriennes ? 10ème Réun Ann Sci Terre Bordeaux: 554

Wyns R, Lardeux H, Weyant M (1989) Présence de Dévonien dans le Groupe de Réaumur (Synclinal de Chantonnay, Vendée); conséquences sur l'évolution géodynamique varisque de la Vendée. C R Acad Sci Paris 308: 855–860

Carboniferous of the Armorican Massif

A. PELHATE

1 Lower Carboniferous

Following the Bretonian (post-Famennian II) orogenic activity, sedimentary basins developed in response to dextral strike-slip faults (Villey 1978; Rolet 1984) in the Laval, Montmartin, Ménez Bélair, Morlaix, Châteaulin, Saint-Julien-de-Vouvantes and Ancenis areas (Fig. 18).

1.1 Paleontology

During the last 20 years, precise biostratigraphic correlations have been established in the Lower Carboniferous rocks of the Armorican Massif; in the terrigenous formations by using the spore zonation identified in Western Europe and in the Donetz Basin (Clayton et al. 1977) and in the carbonate formations by using the conodont (cc zones) and the foraminifera (cf zones) zonations established in Western Europe (Paproth et al. 1983). The conodont assemblages characterize shelf edge (Eotaphrid), foreslope (Gnathodid-Pseudopolygnathid) or high starved basin (Scaliognathid-Doliognathid) environments. The foraminiferal associations contain the Dinantian guide species and some endemic forms. The Lower Carboniferous formations also contain other taxa of lesser stratigraphic value which allow recognition of paleoenvironments. The carbonate formations comprise: algae (Paleoberesellae, Dasycladacean, Solenoporacean, filamentous Codiacean) characterizing the shallow platform facies of the Visean Tethys; in the Laval basin, ostracodes indicate inner infralittoral or intertidal environments with shallowing from the Tournaisian to the Visean (Crasquin 1986); rugose corals include nondissepimented species common in the other west European regions and endemic dissepimented forms (Vuillemin 1990; trilobites (Phillipsidae) and brachiopods are known in the uppermost Tournaisian and in the Visean with Orthida, Strophomenida (Chonetids, Productids), Rhynchonellida, Spiriferida, and Terebratulida. The lacustrine or brackish water terrigenous formations have yielded floral associations with Lycophytes *(Lepidodendropsis)* and Pteridophyllae *(Sphenopteridium);* these characteristic Dinantian forms indicate a hot and dry or momentarily moist climate. The uppermost levels of the succession comprise Namurian guide species: *Sphenopteris adiantoides* and *Sphenophyllum tenerrimum,* which are abundant in Europe and in Asia Minor. In these formations, crustacea phyllopods *(Leaia, Estheria)* and crossopterygian fish (Carpentier et al. 1935) are also recorded.

1.2 Lithostratigraphy

The characteristics of the sedimentary evolution (Figs. 19 and 20) allow distinction of types of basins or domains.

Type 1. *In the Laval, Montmartin, and Ménez-Bélair Basins,* three major tectonic phases can be recognized. (1) There is an extensional phase (Tournaisian) with acidic rhyolites and ignimbrites, (342 ± 6 Ma; Boyer 1974) and basic volcanic rocks (Le Hérissé and Plaine 1982) in the Laval Basin, rhyolites and tuffs in the Ereac Group (Régnault 1981) and in the eastern part of the Ménez-Bélair Syncline, but unknown in Montmartin Basin. During this phase, thick terrigenous and conglomeratic sequences were deposited at the foot of marginal reliefs which were reactivated by extensional growth faulting and overlain by coarse and fine wackes, siltstones, and mudstones. The sequences do not all contain the lithologic terms in a granulometric order. Coal beds developed in sheltered locations and their formation migrated from west to east in the laval Basin (Houlgatte et al. 1988). These formations (l'Huisserie Formation in the Laval and Ménez-Bélair Basins and the Robillard Sandstone Formation in the Montmartin Basin) contain Tournaisian miospore assemblages (Poncet 1974; Paris et al. 1982; Houlgatte et al. 1988). The depositional environment was paralic in the Laval Basin (marine levels with Productids), nonmarine or deltaic in the Ménez-Bélair

J. D. Keppie (Ed.): Pre-Mesozoic Geology
in France and Related Areas

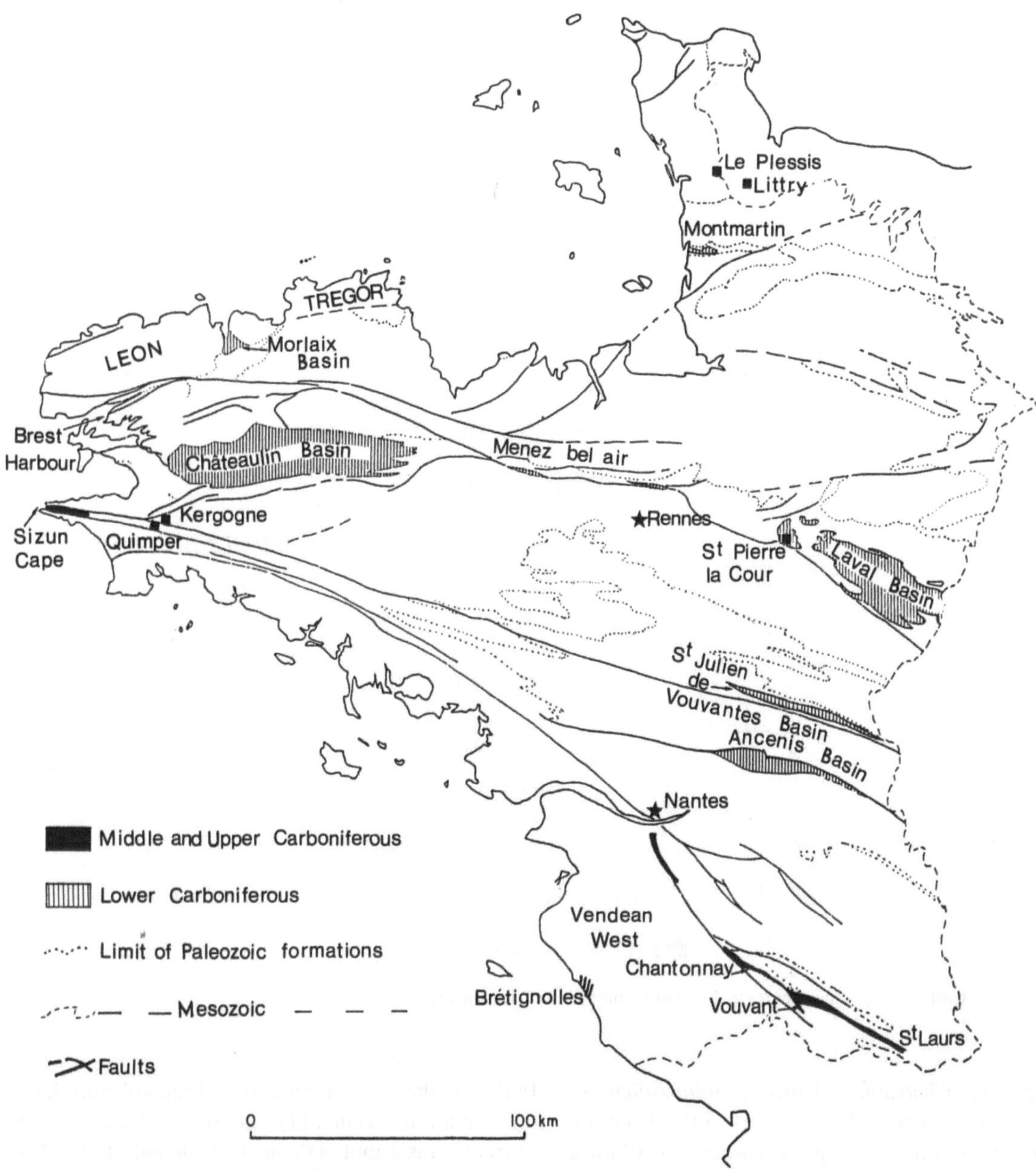

Fig. 18. Carboniferous rocks in the Armorican Massif

Basin and fluviatile in the Montmartin Basin. (2) The carbonate platform phase (uppermost Tournaisian-middle Visean: cf3–cf5 biozones and cc3–cc5 biozones) is characterized by platform limestone deposits; bioclastic, oolitic, or grapestone limestones of the Sablé Formation (Laval Basin), the Montmartin-Régneville Formation (Montmartin Basin) and the Waulsortian and periwaulsortian limestones of the Laval Formation (Laval Basin) and the Quenon Formation (Ménez-Bélair Basin). The facies sequence and the foraminiferal biozones are identical to those identified in the Dinant Basin in Belgium. The thin series in this eastern Armorican domain corresponds to a gulf, flooded by the large Visean transgression. (3) The pull-apart basin phase (upper Visean and Namurian) is only recognized in the Laval Basin (Laval Schists Formation) where Namurian plants occur (Roy 1979). The reason for its absence in Ménez-Bélair and Montmartin Basins is hypothetical: e.g., erosional gap or nondeposition by early filling. In the Laval Basin, during this period, coals formed again in sheltered areas and fine terrigenous deposits filled the basin. The environment was paralic.

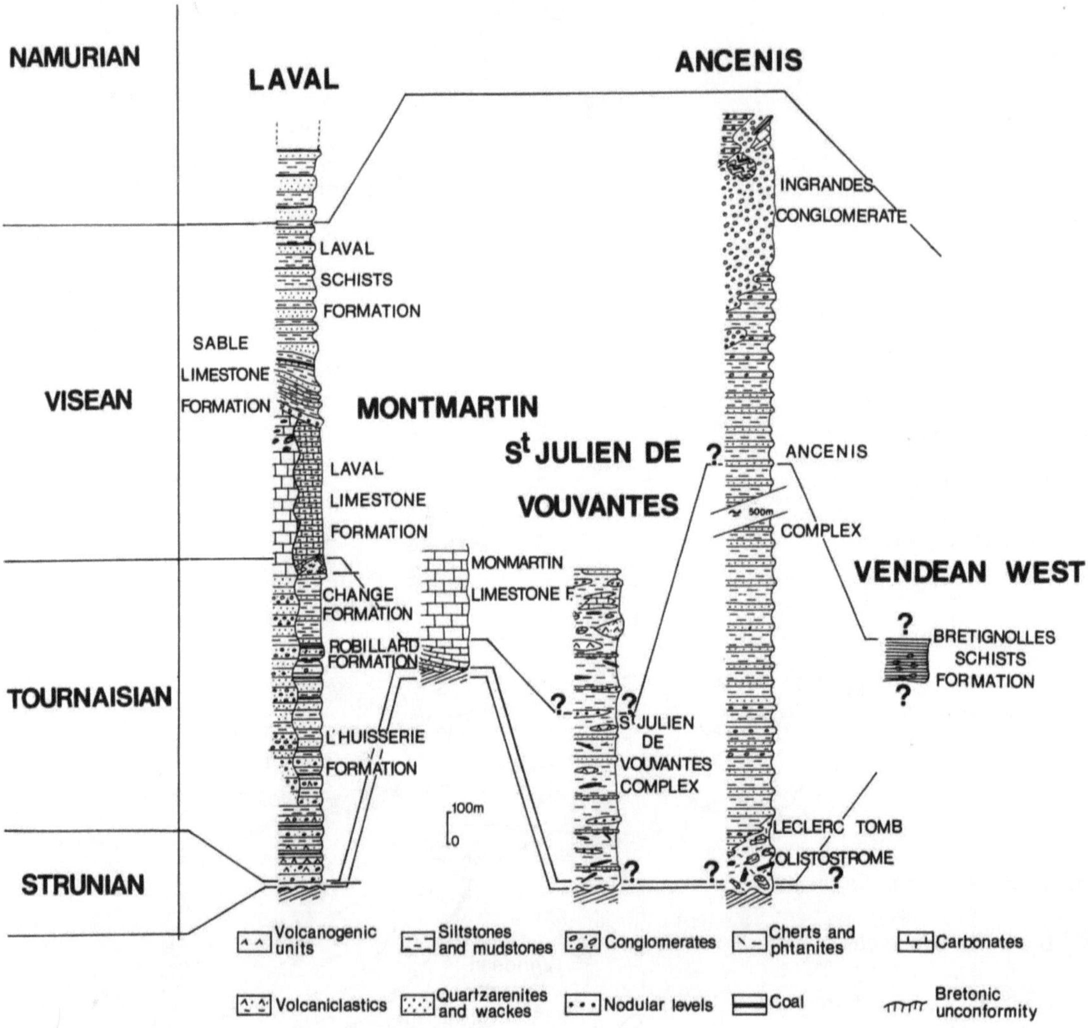

Fig. 19. Lower Carboniferous successions in the eastern units of the Armorican Massif

Type 2. *The Châteaulin, Morlaix, Saint-Julien-de-Vouvantes and Ancenis Basins.* These differ from the preceding domain by having a sedimentary evolution (Figs. 19 and 20) which was induced entirely by strike-slip fault activity during the Lower Carboniferous. The material of the basin, fine terrigenous deposits and volcanics, includes reworked elements, such as blocks or gravity flows from the reactivated basin margins. Each basin contains a specific sedimentary sequence related to the petrographic nature of the bordering areas and to the basin slope.

Châteaulin Basin. The opening phase of the Châteaulin Basin lasted from the Strunian to the Lower Visean. The lozengic sedimentation area, with gently sloping edges (Guillocheau and Rolet 1982), was induced by dextral wrench faults oriented N50° and N110° which were downwarping. The deposits differ according to the geographic position (Fig. 21). In the northwest a succession of three volcano-detrital formations occurs. (1) The Strunian Kermerrien Formation is about 1000 m thick (Rolet et al. 1986) and contains Siluro-Devonian exotic blocks and olistostromes including dispersed bodies issuing from the Ordovician to Devonian sedimentary heap. (2) The Kertanguy Formation has a minimum thickness of 500 m and is presumably Tournaisian. (3) The Kerroch Formation, minimum thickness of 500 m, is lower Visean (?) and at the base consists of mudflows with pebbles in a autoclastic tuffaceous matrix and overlain by a submarine or subaerial tholeiitic-volcanism, characterizing a geodynamic stable area (Maillet 1977; Rolet and Thonon 1979). In the west is the Quivit Limestone, with upper Tournaisian conodonts (*anchoralis* to *latus* conodont zones) which is considered to have been set down in a graben, isolated from the actual border of the Visean deposits (Plusquellec

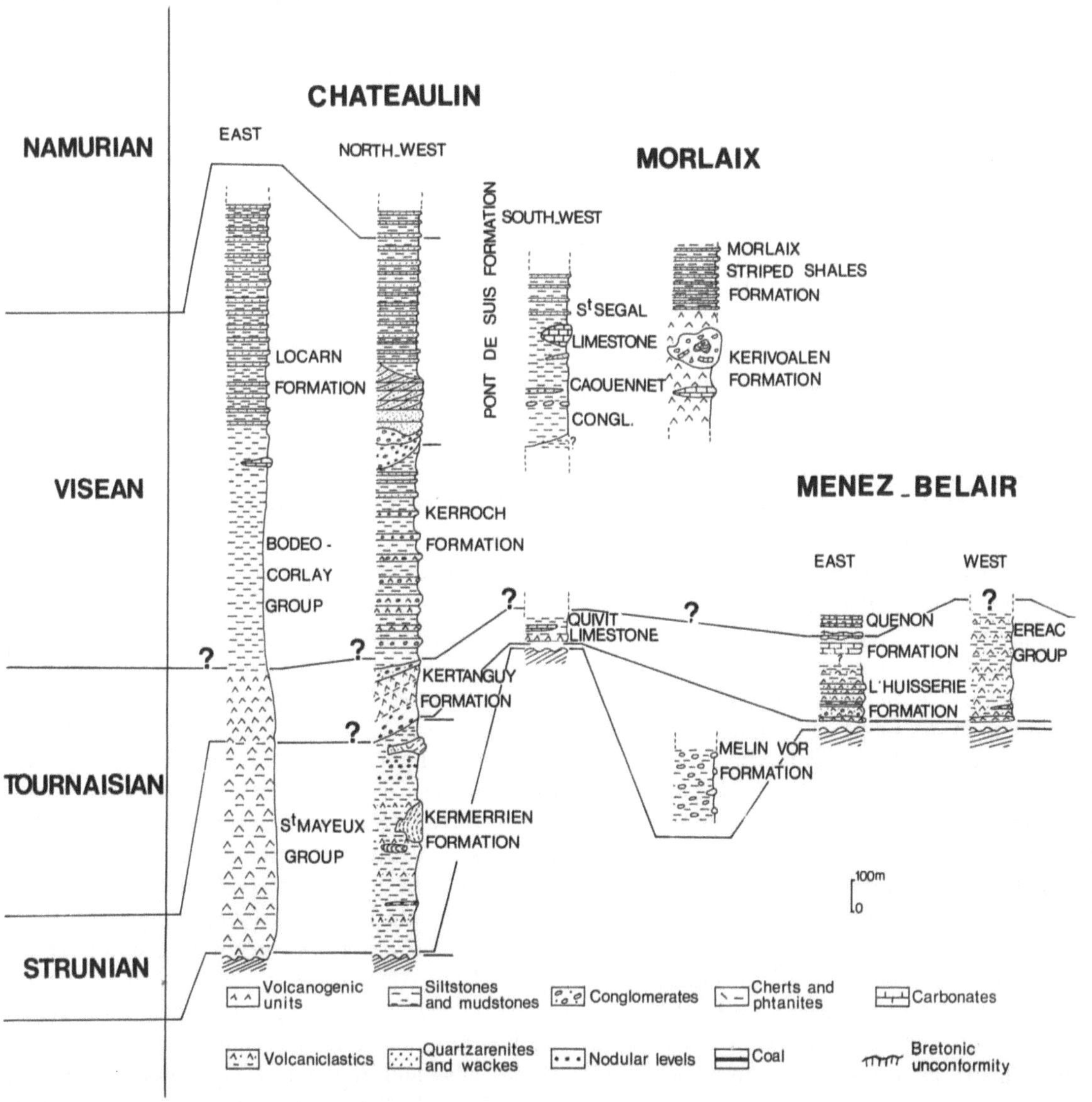

Fig. 20. Lower Carboniferous successions in the western and central units of the Armorican Massif

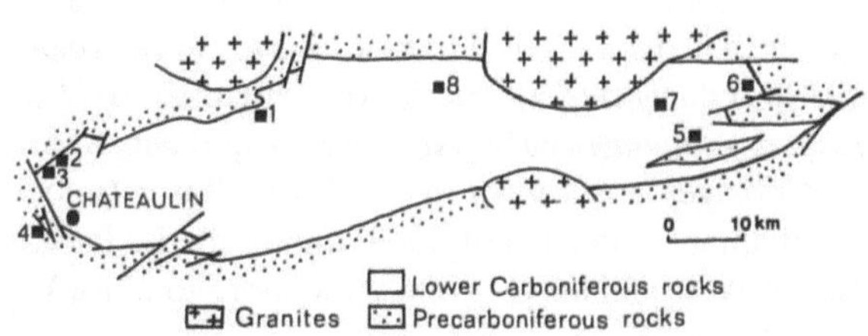

Fig. 21. Location of the type sections of the Lower Carboniferous lithological units in the Chateaulin Basin. *1* Kerroc'h Formation; *2* Pont-de-Buis Fm; *3* Douffine Bridge rhyolites, arkoses, and shales; *4* Quivit Limestone; *5* Saint-Mayeux Group; *6* and *7* Bodéo-Corlay Group; *8* Locarn Formation

et al. 1982). Also in the west are the Douffine Bridge Rhyolites, arkoses, and black shales of Tournaisian-Visean age (Morzadec et al. 1975). In the east are spilito-keratophyric units (Saint-Mayeux Group) and schists with limestone lenses (olistoliths ?) and the Bodéo Group of lower to middle Visean age (cf foraminiferal zones; Boulanger and Sagon 1970). The filling of the Châteaulin Basin took place during the upper Visean (V_3b–V_3c) and the Namurian A (palynologic associations; Doubinger and Pelhate 1976), with the deposits of the Pont-de-Buis and Locarn Formations. At the base are conglomeratic discharges overlain by sandy cones (Pelhate 1970) and shallow prodeltaic turbidites (Guillocheau and Rolet 1982) containing upper Visean (V_3b) carbonate platform-olistoliths (Pelhate 1973; Plusquellec et al. 1982).

Morlaix Basin. The stratigraphic column of the Morlaix Basin was established in spite of intense tectonization and scarce stratigraphic data by Cabanis et al (1979). It reveals a sedimentary evolution similar to those of the Châteaulin Basin. The Dourduff Breccia consists of two volcanosedimentary formations (Rolet et al. 1986): the Melin Vor Formation, a heterogranular conglomerate of Strunian age (palynologic assemblages; Coquel and Deunff 1977) and the Kerivoalen Formation, a gravity olistostrome with platform limestone blocks, some of which have an upper Tournaisian to lower Visean age (presence of *Gnathodus,* e.g., *pseudosemiglaber*), others presumably top of lower to middle Visean age (presence of algae *Koninckopora inflata*). The overlying azoïc Morlaix Striped Shales Formation is assigned to the upper Visean on geometrical grounds. In the Châteaulin and Morlaix Basins, upper Tournaisian to middle Viséan reworked limestones are evidence of a carbonate platform, located on the basin margins and similar to the autochthonous carbonates of the type 1 basins. The upper Visean (V_3b) reworked limestones do not outcrop in the Laval basin, but they may be concealed by the Laval Schists Formation.

Ancenis Basin. The basal chaotic sediments of the Ancenis Basin were deposited by gravity processes (Tombeau Leclerc unit; Dubreuil 1980) and the overlying strata (Ancenis Complex) represent gritty-pelitic violet-colored deposits with plant remains, deposited by gravity flows. The Ingrandes Conglomerate corresponds to proximal cones. In these sediments related to a scarp morphology, the Frasnian limestones are reworked (Dubreuil 1986) in a Dinantian matrix as attested by the flora (Beaupère 1973).

Saint-Julien-de-Vouvantes Basin. The individualization of the Saint-Julien-de-Vouvantes Basin began in the uppermost Devonian or the lowermost Carboniferous by a (dextral ?) strike slip fault north of the Lanvaux-Pont-de-Cé Axis. The basin fill consists of volcanic rocks, siltstones, and laminated sandstones, enclosing olistoliths with Bohemian faunas and Ordovician to Devonian reworked blocks possibly of Medio-North-Armorican origin. Many intraformationnal breccia have been observed. The Dinantian age is based on the existence of Famennian material among the reworked elements (Dubreuil 1986, 1987; see Ligerian Domain).

Type 3. *West Vendean Domain.* The Bretignolles Schists Formation, with scarce siliceous and partly phosphatic nodules, is assigned to middle or upper Tournaisian by conodont assemblages (Comble et al 1985). This lower Carboniferous domain has no lithologic affinity with synchronous sediments in the other Armorican basins (see Vendean Domain).

Type 4. *Namurian Basins.* The Sillon Houiller de la Basse Loire Basin, north of the Ancenis Basin (Cavet 1978), La Baconnière Basin (Mayenne), and the Vouvant Basin on the eastern extremity of the Vendean Sillon Houiller were formed during the lowermost Namurian. They comprise a terrigenous filling reworking the margins surrounding rocks and coal measures. The floral remains are Namurian (Bureau 1914; Carpentier 1920; Mathieu 1937).

2 Middle and Upper Carboniferous

2.1 Paleontology

The paleontological record is rather limited with limnic bivalves: (*Anthracomya,* Saint-Pierre-la-Cour), insects: (*Protoblattopteris,* Saint-Pierre-la-Cour) and plant remains of Westphalian C-D age (*Linopteris subbrongniarti, Eusphenopteris neuropteroides, Neuropteris tenuifolia;* Mathieu 1937) in Chantonnay Mine. In Stephanian basins (Picquenard 1920, 1921, 1924; Vetter 1957), some forms are common throughout the Stephanian (Doubinger and Vetter 1985): cannelate Sigillaria *(Annularia stellata), Sphenopsids (Calamites cisti, C. suckovi),* pecopteris *(P. cyathea, P. plumosa-dentata, P. polymorpha),* alethopterids (*Callipteridium pteridium,* restricted to Kergogne Basin), Odontopterids (*O. minorzeilleri,* Kergogne Basin only) and Sphenophyllae (*Sp. oblongifolium,* Saint-Pierre-la-Cour only). Nevertheless, in spite of the common assemblage, the floral associations are not similar in all basins. This fact reflects specific ecological communities in each intramontane basin, rather than distinct stratigraphic levels (Delénin et al. 1988). A characteristic species of Stephanian B, *Neuropteris cordata,* occurs in the Cap Sizun Basin. Spore assemblages are known in the Littry Basin (upper Stephanian; Coquel et al. 1969) and in the Le Plessis Basin (Westphalian, OT zone of the spore zonation).

2.2 Lithostratigraphy

Subsequent to the major orogenic events (late Namurian to early Westphalian), dextral strike-slip faults become active again and produced small limnic intramontagne pull-apart basins (Fig. 18). The Cap Sizun Basin and the Quimper Basin are linked to the graben bounded to the North by the Pointe-du-Van-

Quimper-Angers lineament and to the South by the Pointe-du-Raz-Quimper-Parthenay lineament. The Vendean Sillon Houiller is located along the South-Armorican Shear Zone extension and is divided into the Accident Houiller de Chantonnay and the Stephanian Vouvant Basin. The Accident Houiller comprises, from NW to SE, the Saint-Marc-de-Coutais, Vieille Vigne, and Malabrit scarps and the Chantonnay Basin, itself subdivided into amygdaloid subsidiary basins: the Chantonnay Mine of Westphalian Age and the La Marzelle Mine with Stephanian resting on the Namurian. Some basins, such as the north Quimper, NW-SE-directed Kergogne Basin, and the Saint-Pierre-la-Cour Basin were opened by submeridian fault action. Northeastward in the Carentan Depression, the upper Carboniferous formations overlain by the Autunian and the Trias are located in the Littry and Le Plessis basins, with alkaline lavas associated with sediments (see Permian). These volcanic rocks are linked to the post-Variscan distensional events (Pareyn 1954; Le Gall et al 1989). In all of the above-mentioned basins, the sedimentary accumulation comprises conglomerates, very coarse to fine clastics of which the reworked elements reflect the petrography of the basin borders. The material deposition resulted from cliff falls or torrential run off during active phases (climatic or tectonic), without mixing (Pelhate 1957, 1961) in the intramontane basin; the vegetal accumulation developed during calm phases or in sheltered locations.

References

Beaupère C (1973) Contribution à l'étude de la flore fossile du "Culm" du synclinal d'Ancenis. Thèse 3ème cycle, Paris VI University, pp 1–111

Boulanger D, Sagon J (1970) Découverte d'une microfaune d'âge viséen inférieur dans des calcaires silicifiés du bassin de Châteaulin (Massif armoricain). C R Acad Sci Paris 271: 1175–1177

Boyer C (1974) Volcanismes acides paléozoïques dans le Massif armoricain. Thèse, Paris Sud-Orsay University, pp 1–384

Bureau E (1914) Bassin houiller de la Basse-Loire, tome 2: description des flores fossiles. Etud. gites minéraux France, Paris, pp 1–417

Cabanis B, Chantraine J, Herrouin Y (1979) Le Bassin de Morlaix, unité circonscrite et indépendante dans le contexte geologique régional. Bull Bur Géol Min 4: 269–276

Carpentier A (1920) Sur des empreintes de fructifications recueillies dans plusieurs gisements carbonifères de l'Ouest de la France. Bull Soc Nat Ouest 3 6: 109–119

Carpentier A, Delépine G, Martel H (1935) Recherches paléontologiques sur quelques gisements carbonifères de la Sarthe. Bull Soc Géol Fr 5 5: 77–84

Cavet P (1978) Notice explicative de la carte géologique de la France à 1/50000: Ancenis. Bur Géol Min, Orléans

Clayton G, Coquel R, Doubinger J, Gueinn KJ, Loboziak S, Owens B, Streel M (1977) Carboniferous Miospores of western Europe: illustration and zonation. Meded Rijksgeol Dienst 29: 1–71

Comble CH, De Wever P, Ters M, Weyant M (1985) Découverte de Conodontes et de Radiolaires d'âge Tournaisien dans les schistes bariolés de Brétignolles-sur-Mer (Vendée). C R Acad Sci Paris 300 II 17: 899–904

Coquel R, Deunff J (1977) Sur la découverte de spores du passage Dévonien-Carbonifère (Strunien) dans le complexe schisteux de la "Brèche du Dourduff", région de Morlaix-Finistère et sa signification. C R Acad Sci Paris 285 D: 15–18

Coquel R, Loboziak S, Pareyn C (1969) Confirmation palynologique de l'âge stéphanien supérieur de la couche de houille du bassin de Littry (Calvados). Ann Soc Géol Nord 89 4: 309–315

Crasquin S (1986) Les Ostracodes dinantiens du Synclinal de Laval et du Synclinorium de Namur (Systématique-Biostratigraphie-Paléoécologie). Thèse, Lille University, t 1 pp 1–169, t 2 pp 1–71

Delénin P, Clermonté J, Courel L, Dumain M, Laversanne J (1988) Remise en cause des charriages dans le bassin houiller stéphanien des Cévennes (Gard, France). C R Acad Sci Paris 307 II: 1237–1243

Doubinger J, Pelhate A (1976) Nouvelles observations sur l'âge des schistes de Châteaulin (Massif armoricain). C R Acad Sci Paris 283 D: 467–470

Dubreuil M (1980) Hypothèse sur la mise en place, au Dinantien, du complexe du Tombeau Leclerc (Bassin d'Ancenis, Sud-Est du Massif armoricain) sous forme d'un olistostrome. Conséquences géodynamiques. C R Acad Sci Paris 290 D: 1455–1458

Dubreuil M (1986) Evolution géodynamique du Paléozoïque Ligérien (Massif Armoricain). Thèse, Nantes University, pp 1–258

Dubreuil M (1987) Le bassin en décrochement de Saint-Julien-de-Vouvantes-Angers (Carbonifère inférieur du Massif armoricain). Bull Soc Géol Fr 8 3: 215–221

Guillocheau F, Rolet J (1982) La sédimentation paléozoïque ouest-armoricaine. Histoire sédimentaire: relations tectonique-sédimentation. Bull Soc Géol Minéral Bretagne C 14 2: 45–62

Houlgatte E, Le Hérissé A, Pelhate A, Rolet J (1988) Evolution géodynamique du bassin carbonifère de Laval. Géol Fr 1: 27–46

Le Gall J, Doré F, Gresselin F, Pareyn C (1989) Le magmatisme alcalin de la distension post-varisque dans le Nord du Massif armoricain: exemples des volcanites carbonifères du bassin de Carentan et des lamprophyres du Nord-Cotentin. Ann Soc Géol Nord 108: 25–33

Le Hérissè A, Plaine J (1982) Volcanisme basique dans le Carbonifère inférieur du synclinorium de Laval (Massif armoricain, France). C R Acad Sci Paris 294 II: 1199–1202

Maillet P (1977) Etude géochimique de quelques séries spilitiques du Massif armoricain. Implications géotectoniques. Thèse 3ème cycle, Rennes University, pp 1–134

Mathieu G (1937) Recherches géologiques sur les terrains paléozoïques de la région vendéenne. Thèse, Lille University, pp 1–416

Morzadec P, Paris F, Pelhate A, Streel M (1975) Datation par les spores du début de la sédimentation carbonifère dans l'Ouest du synclinorium de Châteaulin, Massif armoricain. C R Acad Sci Paris 281 D: 1955–1957

Paproth E, Conil R, Bless MJM, Boonen P, Bouckaert J, Carpentier N, Coen M, Delcambre B, Deprijck Ch, Deuzon S, Dresses R, Groessens E, Hance L, Hennebert M, Hibo D,

Hahn GR, Hislaire O, Kasig W, Laloux M, Lauwers A, Lys M, Op De Beek K, Overlau P, Pirlet H, Poty E, Ramsbottom W, Streel M, Swennen R, Thorez J, Vanguestaine M, Van Steenwinkel M, Vieslet JL (1983) Bio- and lithostratigraphic subdivisions of the Dinantian in Belgium, a review. Ann Soc Géol Belg 106: 185–239

Pareyn C (1954) Le bassin houiller de Littry (Calvados). Publ Bur Géol Min 14: 1–132

Paris F, Le Hérissé A, Pelhate A, Weyant M (1982) Les formations carbonifères et la phase bretonne dans le synclinorium du Ménez-Bélair: essai de synthèse. Bull Soc Géol Minéral Bretagne 14 2: 19–33

Pelhate A (1957) Contribution à l'étude sédimentologique du Carbonifère de Quimper (Finistère). Bull Soc Géol Minéral Bretagne, n s 3 2: 35–47

Pelhate A (1961) Etude sédimentologique du Stéphanien de Kergogne (Nord de Quimper, Finistère). Bull Soc Géol Minéral Bretagne n s 6 1: 70–98

Pelhate A (1970) Caractéristiques sédimentologiques des "Schistes de Châteaulin" aux environs de Locmaria-Berrien et Poullaouen (Finistère). Bull Soc Géol Minéral Bretagne C2 1: 1–10

Pelhate A (1973) Précisions sur l'âge de la sédimentation carbonatée dans l'Ouest du bassin de Châteaulin et conséquences paléogeographiques. C R Acad Sci Paris 277 D: 2305–2307

Picquenard C (1920) Excursions géologiques à l'Ouest de Quimper. Bull Soc Géol Minéral Bretagne I 4: 240–250

Picquenard C (1921) Prodrome de la flore fossile du bassin houiller de Kergogne (Finistère). Bull Soc Géol Minéral Bretagne 2 3: 340–353

Picquenard C (1924) Contribution à l'étude de la flore fossile des bassins houillers du Finistère. Bull Soc Géol Minéral Bretagne 5 1: 97–98

Plusquellec Y, Pelhate A, Rolet J, Weyant M (1982) Découverte de calcaire tournaisien supérieur et de conglomérat (Viséen supérieur probable) près de la bordure occidentale du Bassin de Châteaulin (Massif armoricain, France). Intérêt stratigraphique et tectonique. Bull Soc Géol Minéral Bretagne C 14(2): 1–17

Poncet J (1984) La série stratigraphique dans le Synclinal de Montmartin-sur-Mer (Manche). Bull Soc Linn Normandie 104: 59–69

Régnault S (1981) Stratigraphie et structure du Paléozoïque dans le Ménez-Bélair occidental (Synclinorium médian armoricain). Bull Soc Géol Minéral Bretagne 13 1: 1–105

Rolet J (1984) Graben losangiques (pull-apart) en régime de décrochement. Le rôle des coulissements hercyniens dans l'individualisation des bassins carbonifères du Massif armoricain. Ann Soc Géol Nord 103: 209–220

Rolet J, Thonon P (1979) Mise en evidence de trois complexes volcano-detritiques d'âge devonien interieur a moyen, strunien et viséen inférieur sur la bordure nord du bassin de Châteaulin. Bull Bur Géol Min Orléans 2 1 4: 303–305

Rolet J, Plusquellec Y, Babin C, Deunff J (1986) Famennian regression and strunian grabens in the Armorican Massif, a key area: western Britanny. Ann Soc Géol Belg 109: 197–203

Roy C (1979) Contribution à la connaissance de la flore du Carbonifère du Bassin de Laval (Ouest de la France). Thèse 3ème cycle, Lyon University, pp 1–133

Vetter P (1957) Révision de la flore fossile du bassin houiller de Saint-Pierre-la-Cour (Mayenne). Ann Soc Géol Nord 67: 105–120

Villey M (1978) Décrochements hercyniens dans le Briovérien normand (Balleroy, feuille à 1/50000). Bull Bur Géol Min Orléans 2 1 4: 365–369

Permian of the Armorican Massif

F. DORÉ

1 Introduction

During the Permian most of the Armorican Massif was uplifted and block-faulted. The Permian rocks are located in a trough of the Cotentin Peninsula which is commonly known as the Carentan Basin (Fig. 22). Another prominent Permian-filled trough stretches eastward (Villequier Trough) beneath the Paris Basin succession. There are no Permian deposits between those two areas.

Only the Lower Permian (Autunian) is represented in the Armorican Massif. The Autunian overlies either the folded Paleozoic (Cambrian to Devonian) with marked unconformity resulting from Variscan movements, or the post-orogenic Carboniferous strata (Westphalian or Stephanian coal measures) with some disconformity. Triassic rocks overlying and overlapping the Permian deposition area with minor disconformity are Upper Norian or Rhaetian in age. In the field, owing to some similar red facies, the scarcity of fossils leads to serious difficulties in placing the Upper Stephanian-Autunian and Autunian-Norian boundaries.

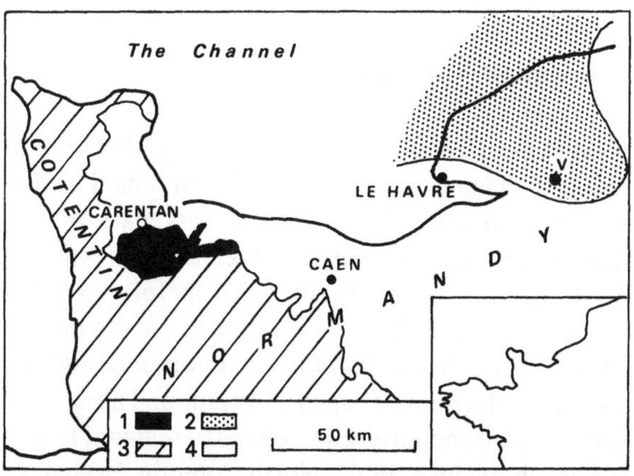

Fig. 22. Distribution of the Permian in the northeastern Armorican Massif. *1* Outcrops in the Carentan Basin and location of the St. Fromond borehole *(white triangle)*; *2* concealed Permian rocks in the Villequier Trough; *3* Cadomian and Variscan basement; *4* Mesozoic cover

2 Lithological Successions in the Carentan Basin (Fig. 23)

The thickest Permian sequences were discovered in coal research boreholes (cf. Bigot 1946; Pareyn 1954). Many variations of lithology and thickness are observed. Lower Permian deposits are composed of a red bed sequence with some conglomerates, sandstones, and limestones. The thickness increases and the succession is more complete from east to west. Volcanic rocks seem located along the bordering faults of the subsiding graben.

In the middle of the Carentan Basin, the Saint-Fromond borehole (831 m) yielded the most representative Permian sedimentation (Bigot and Pruvost 1925; Aubry 1982). Figure 23 depicts its evolution, gathering together data from other boreholes and outcrops. At the base, shale, sandstone, and conglomerate with coal measures are thought to belong to the Upper Stephanian (Coquel et al. 1969). The conformable and thick Lower Permian (550 m) consists of five formations.

1. Lower Red Siltstone. This detrital sequence comprises alternating sandstone and conglomerate which pass upward into red or brown siltstone. Cross-lamination and mudcracks have been observed in the siltstone.
2. Saint-Martin-de-Blagny Limestone. The repetition of gray and laminated lacustrine limestone with red siltstone and bituminous laminae characterize the formation. The lamination of small-scale limy cycles is disrupted upward by "bird's-eye" fabric, bioturbation, and rather scarce salt pseudomorphs.
3. Middle Red Siltstone. The thinly bedded siltstone with very fine lenses of sandstone shows shrinkage cracks and rain-drop prints.
4. Cartigny-l'Epinay Conglomerate and Sandstone. The coarse continental breccias were derived locally and deposited on the proximal parts of an alluvial fan.
5. Saint-Jean-de-Daye upper Red Mudstone. In this thick and monotonous formation of red or wine-

J. D. Keppie (Ed.): Pre-Mesozoic Geology in France and Related Areas

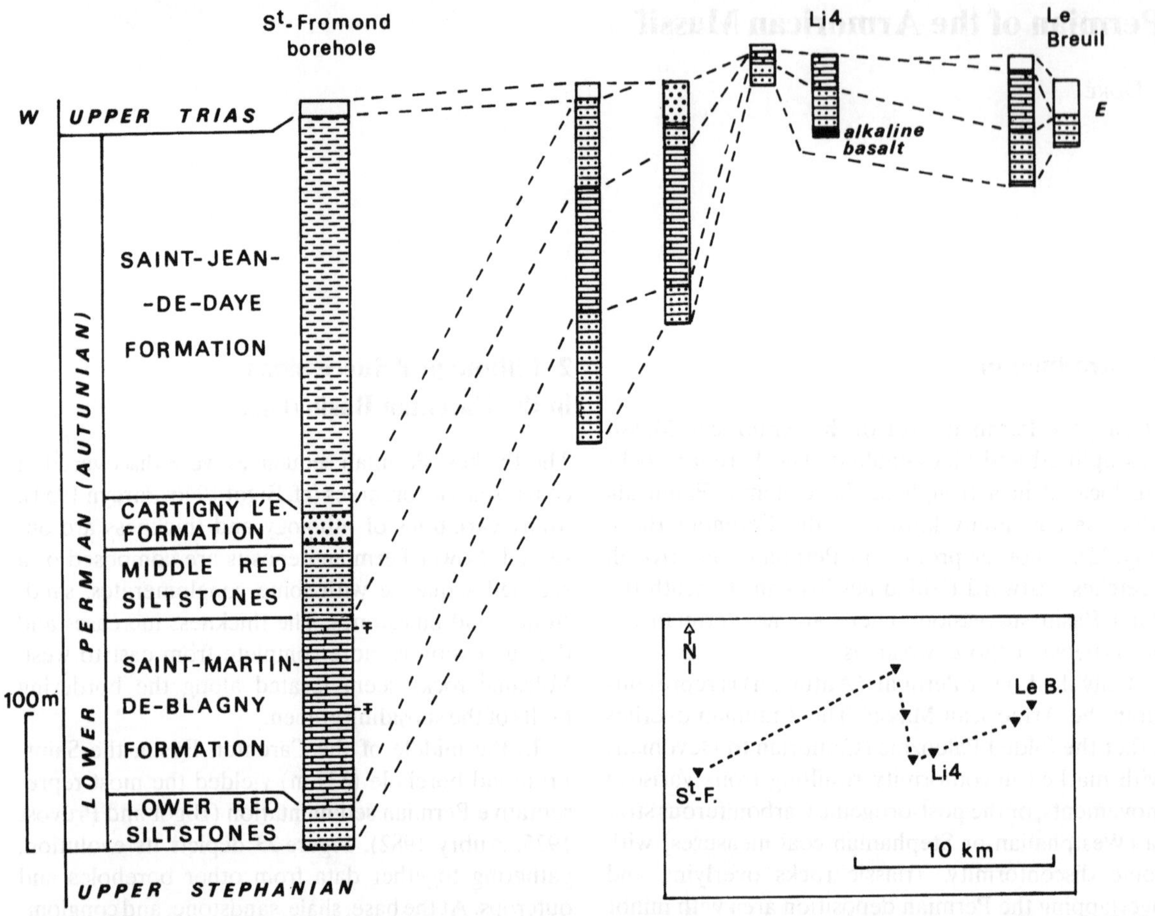

Fig. 23. Thickness variation of the Lower Permian rocks in the Carentan Basin, from the central part to the east as shown by the *inset.* (Aubry 1982)

colored mudstones, some scarce carbonate-rich mudstones occur. An ash level and a "caliche" with pisolitic concretions have been noticed toward the top of the sequence.

The clay minerals in the Permian succession include illite (30 to 100%), kaolinite (0 to 30%) and some expandable interstratified minerals (Aubry 1982).

The fossil content of Lower Permian sediments is sparse and is restricted to the Saint-Martin-de-Blagny Formation. The fossils include bivalves *(Anthracomya carbonaria),* phyllopods *(Estheria tenella),* teeth of *Diplodus,* and caudal spines of *Acanthodes,* all of which are indicative of nonmarine conditions. Despite the fossil record paucity, a stratigraphical relationship has been proposed with the Middle and Upper Autunian of the French Central Massif or the Lower Rotliegende of the Saar (West Germany).

Upper Triassic rocks overlie different Lower Permian formations in the Carentan Basin. An Upper Norian or Rhaetian age has been determined according to ongoing palynological studies (Pareyn, unpubl.) for the laterally interfingering sequences which include calcretes, red and plastic clays, and fan conglomerates.

3 Permian Volcanics

Lower Permian volcanism consisted of a rather limited eruptive episode which ended the Stephanian and Westphalian lava flows. In the east part of the Carentan Basin, the ancient and recent Saint-Martin-de-Blagny boreholes (Vieillard 1873; Termier 1918, Aubry 1982) intersected some sills of olivine basalts and biotite trachybasalts which are interbedded within the Saint-Martin-de-Blagny Formation. These alkaline lavas are chemically analogous to regional K-rich lamprophyre dykes of the North Cotentin. Both have high concentrations of transition elements (Cr, Ni) and are characterized by enrichment of incompatible elements (K, Ba, LREE). These data are interpreted in terms of derivation of the cogenetic magmas from partial melting of metasomatized mantle and without important crustal contamination (Le Gall et al. 1989).

In Western Brittany, several dykes of lamprophyres and microgranites were also emplaced toward the Stephanian-Permian time boundary as shown by the isotopic ages (Bellon et al. 1988).

References

Aubry J (1982) Formations permiennes et triasiques du bassin de Carentan. Thèse 3ème cycle, Caen University, pp 1–285

Bigot A (1946) Le terrain houiller de Basse-Normandie. Bull Soc Linn Normandie 9 5: 94–104

Coquel R, Loboziak S, Pareyn C (1969) Confirmation palynologique de l'âge stéphanien supérieur de la couche de houille du bassin de Littry (Calvados). Ann Soc Géol Nord 89 4: 309–315

Le Gall J, Doré F, Gresselin F, Pareyn C (1989) Le magmatisme alcalin de la distension post-varisque dans le Nord du Massif armoricain: exemples des volcanites carbonifères du bassin de Carentan et des lamprophyres du Nord-Cotentin. Ann Soc Géol Nord 108: 25–33

Pareyn C (1954) Le bassin houiller de Littry (Calvados). Publ Bur Rec Géol Min 14: 1–132

Termier P (1918) Roches éruptives interstratifiées dans le terrain houiller de Littry. C R Acad Sci Paris 167: 107–112

Vieillard E (1873) Le terrain houiller de Basse-Normandie, ses ressources, son avenir. Bull Soc Linn Normandie 2 7: 231–389

Paleogeographic Synthesis

F. Paris and M. Robardet

The paleogeographic reconstructions that have been proposed for the Paleozoic generally agree upon the existence of two major distinct paleogeographic units in Europe: Baltica and North Gondwana, separated by a mid-European Rheic Ocean. The evolution of this ocean, from its opening in the earliest Paleozoic until its closure during the Carboniferous, influenced the geological development of the entire Armorican Massif and more generally that of southwestern Europe. This evolution controlled: (1) the thick accumulation of sediments during the initial period of rifting and those which were possibly related to marginal accretionary prisms (Ardennes?); (2) the successive latitudinal positions of the distinct paleogeographic units and the resulting climatic conditions which influenced the distribution of facies and faunas; (3) the igneous activity, especially in relation with subduction zones; and (4) the tectonic and metamorphic processes when the ocean closed and the continents collided.

The precise paleogeographic pattern and the exact timing of the evolution of the Rheic Ocean are not yet unanimously accepted. Somes authors consider that this ocean succeeded a Tornquist's Sea and opened in the latest Ordovician (Cocks and Fortey 1982), whereas others propose that the opening occurred during the Cambrian (Paris and Robardet 1990). A similar controversy concerns the precise location of the ocean. In some reconstructions, the Ardennes and the Rhenish Schiefergebirge are included in the North Gondwana margin whereas, in our view, they composed the southern margin of the Baltica plate which also included the southern part of Great Britain, Poland, Podolia, and possibly eastern Newfoundland and Nova Scotia (Fig. 24). All these regions were separated by the Rheic Ocean from the North Gondwana Province comprising the Armorican Massif, the Iberian Peninsula, Aquitaine, the Montagne Noire, Bohemia, the Moesian Platform (?), and North Africa. Some reconstructions, generally based on paleomagnetic data (see references in Perroud 1985), propose the existence of intermediate blocks, which evolved independently between Baltica and Gondwana. In this manner an "Armorica Plate", mainly consisting of the Armorican Massif, the Iberian Peninsula, and Bohemia, has been postulated (Van der Voo 1979). This concept, which implies the existence of a Prototethys Ocean separating Gondwana from Armorica during the Silurian and the Devonian, is not accepted here: the original paleomagnetic data have proved to be wrong and, in addition, faunas clearly show the paleobiogeographic unity of North Gondwana and Southern Europe during these periods (Robardet et al. 1990). The paleogeographic reconstruction adopted here (Fig. 24) shows that the Armorican Massif was located in the northern part of North Gondwana and consequently was directly involved in the collision which gave rise to the Variscan Belt during the Carboniferous.

Another oceanic area, namely the South Armorican Ocean (=the Massif Central or the Ligerian Ocean), which would have closed during the Late Silurian-Early Devonian period, is regarded here only as a branch of the Rheic Ocean. Indeed, despite strong tectono-metamorphic events related to its closure and to the subsequent continental collision, the South-Armorican Ocean did not induce completely independent faunal evolutions on its opposite margins during the Early Paleozoic. The effects of the evolution of the Rheic and the South Armorican Oceans upon the Armorican sedimentation and faunas are exemplified below for different periods of the Paleozoic (Fig. 24).

The thick, mainly terrigenous Cambrian sequences in the northeast of the Armorican Massif (Normandy) indicate fairly high subsidence rates (Figs. 25 and 26) compatible with an extensional episode which would correspond to the earliest stages of the Rheic Ocean rifting. The evolution of the affinities of the North Gondwanan trilobite faunas during the Cambrian (Destombes and Feist 1987) supports the opening of a mid-European Rheic Ocean as early as the Middle Cambrian.

Up to the Lower Llanvirn, this ocean was not wide enough to prevent migrations of benthic organisms, and some trilobites are common to the Armorican Massif and Wales. These faunal affinities faded away

J. D. Keppie (Ed.): Pre-Mesozoic Geology
in France and Related Areas

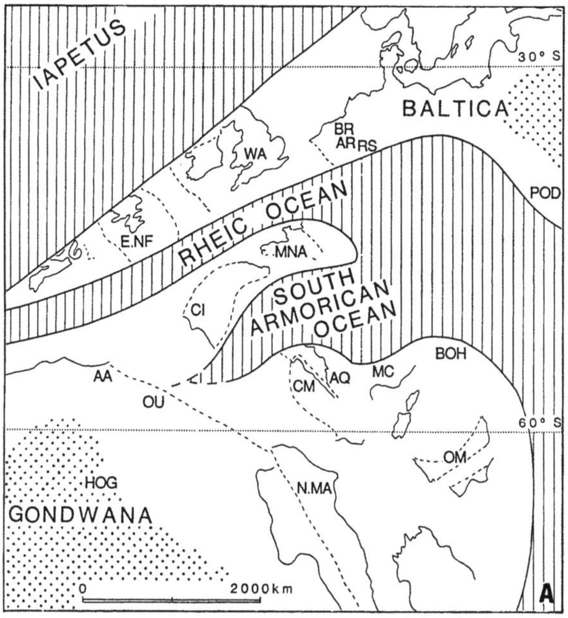

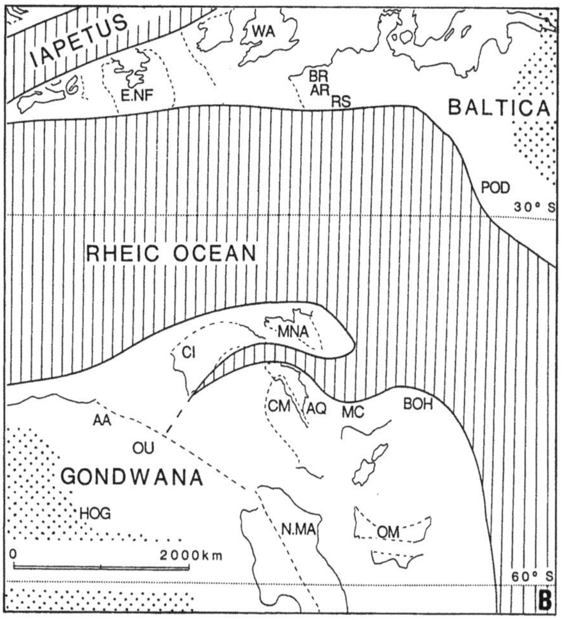

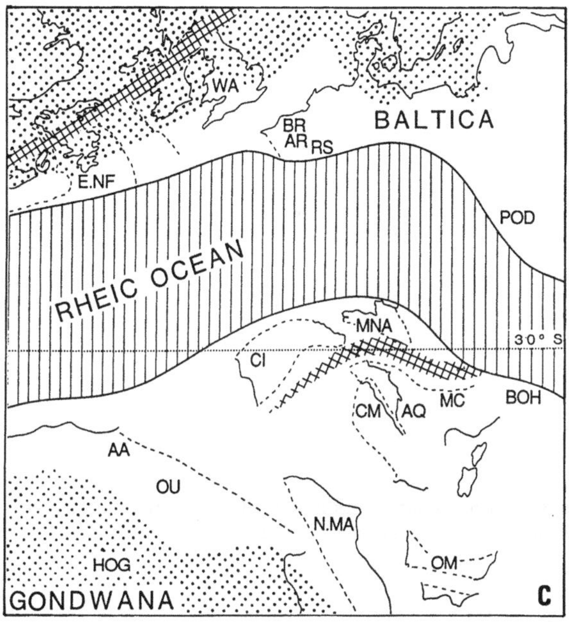

Fig. 24 A–C. Successive paleogeographic positions of the Armorican Massif during the Early Paleozoic. (After Paris and Robardet 1990.) **A** Early Ordovician. **B** Middle Silurian. **C** Early Devonian. *Vertical hatching* oceanic areas; *white* continental shelves and slopes; *stippled* mainland areas; *crossed areas* main oceanic sutures. *North Gondwana: AA* Anti-Atlas; *AQ* Aquitaine; *BOH* Bohemia; *CI* Central Iberia; *CM* Cantabrian Mountains; *HOG* Hoggar; *MC* French Massif Central; *MNA* Medio-North-Armorican Domain; *NMA* North Maghreb; *OM* Ossa Morena; *OU* Ougarta Chains; *PY* Pyrenées. *Baltica: AR* Ardennes; *BR* Brabant; *ENF* Eastern Newfoundland; *POD* Podolia; *RS* Rhenish Schiefergebirge; *WA* Wales

from the Upper Llanvirn and during the rest of the Ordovician only pelagic and epipelagic faunas (e.g., chitinozoans and graptolites) are common to both sides of the Rheic Ocean. The geographic position proposed for the Rheic oceanic barrier is well supported by the existence in the Brabant area of trilobites and brachiopods with Baltic affinities (Lespérance and Sheehan 1987) during the Ashgill at the very time when Bohemian and Armorican faunas maintained North Gondwana affinities. The geographic distribution of late Ashigillian glacio-marine deposits, whose northernmost occurrences are in Normandy and Thuringia (Robardet and Doré 1988), corroborates the rather high latitudinal position of North Gondwana regions and their separation from Baltica which was located in lower latitudes.

During the Silurian, the anoxic conditions prevailing all over the North Gondwana area prevented the development of the marine benthos and comparisons with the rich warm faunas of Baltica are therefore impossible. The distribution and the abundance of the paleoplankton indicate, however, that the Armorican Massif was located in a high phytoplankton productivity zone (cold to temperate waters of the *Neoveryachium carminae* belt of Cramer 1971) whereas Baltica and Laurentia were in tropical or equatorial latitudes (*Deunffia* and *Domasia* belts).

During the earliest Devonian, faunal differences were still perceptible between the Armorican Massif and the southern margin of Baltica (e.g., Artois and Ardennes). However, from the late Pragian-early Emsian there was a gradual uniformization of the benthic assemblages. The narrowing Rheic Ocean no longer constituted a faunal barrier and the same ben-

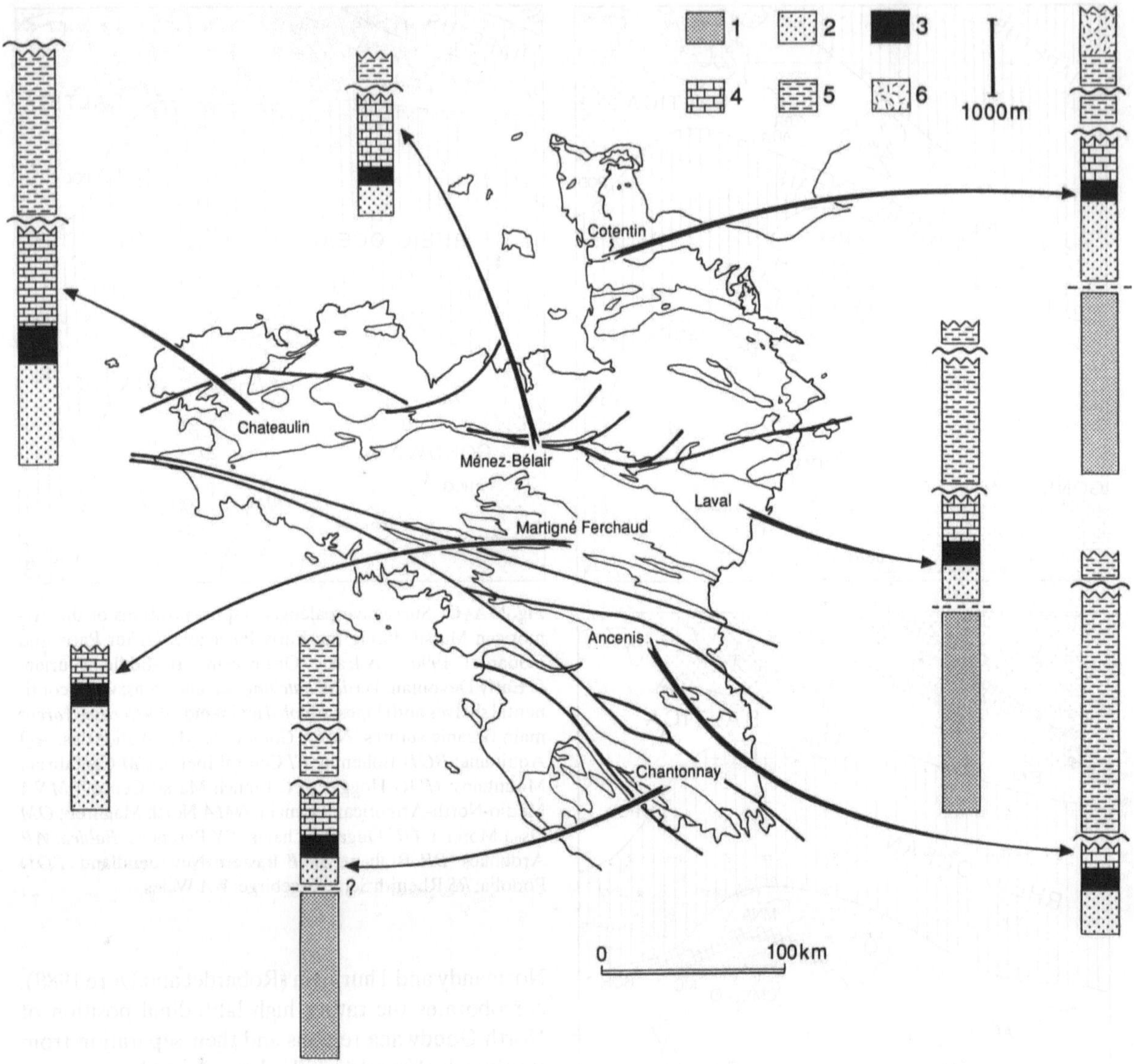

Fig. 25. Comparative thickness of the Paleozoic Systems in the Armorican Massif. *1* Cambrian; *2* Ordovician; *3* Silurian; *4* Devonian; *5* Carboniferous; *6* Permian; *wavy line* tectonic episode; *broken line* erosion; *dashed line* stratigraphic hiatus; *thicknesses* are regional average values

thic species developed in similar environments on both sides of the ocean. Moreover, the gradual increase of carbonate deposits (platform limestones and build-ups) shows that the North Gondwana regions had migrated into warmer temperate and tropical latitudes.

Paleomagnetic data (Perroud 1985) and paleoclimatic conclusions (Besly 1987) indicate an equatorial position of western Europe in the Carboniferous and Early Permian. The juxtaposition of northern and southern Europe, which had been separated by about 30° in latitude during the Silurian, agrees with the closure of the Rheic Ocean and the collision between Laurussia (Laurentia and Baltica) and North Gondwana.

The paleogeographic evolution which has been summarized above and the final continental collision from which the Variscan Belt originated, give an account of the general evolution of the entire Armorican Massif during the Paleozoic.

Distinct sedimentary and faunal characteristics allow, however, definition of distinct paleogeographic domains or units whose original positioning within the North Gondwana Province has most probably been considerably modified during the Variscan tectonic evolution (Robardet et al. 1990).

The Armorican Massif at present includes two main parts: the Medio-North-Armorican Domain and the South-Armorican Domains (cf. Fig. 1). The

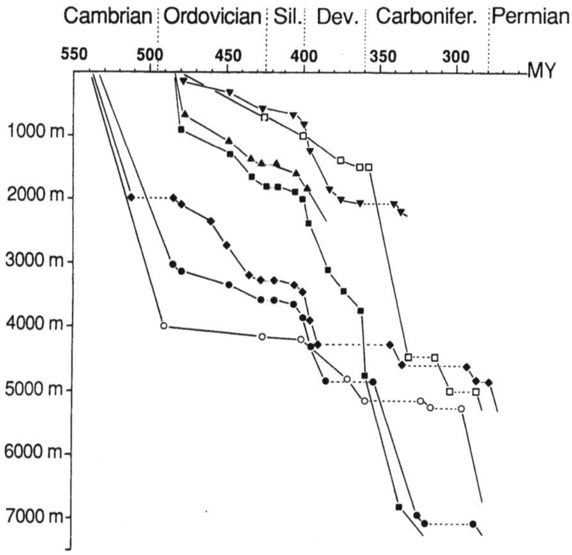

Fig. 26. Comparative sedimentation rates within the Armorican Massif during the Paleozoic (time scale after Odin 1985). The thickness of the Paleozoic formations (or in some cases, systems) corresponds to regional average values, without correction for compaction. *Black symbols* Medio-North Armorican Domain (*triangles* Cotentin; *circles* Laval Syncline (north); *reversed triangles* Ménez-Bélair Syncline; *squares* Chateaulin Syncline (west); *diamonds* Martigné-Ferchaud and northern Angers Synclines); *open symbols* South-Armorican Domains (*squares* Ancenis unit; *circles* Chantonnay Syncline)

Medio-North-Armorican Domain displays strong affinities with Central Iberia (Paris and Robardet 1977). In these regions a shelf sedimentation developed from the Ordovician to the Carboniferous. The Lower Paleozoic succession is broadly continuous, with minor hiatuses corresponding to global variations of sea level. Volcanic activity occurred only locally and during restricted periods in the Late Ordovician (Rosan Formation in the Chateaulin Syncline) and possibly in the Early Devonian (eastern Chateaulin Syncline). With these rare and local exceptions, the marine succession shows no obvious trace of tectonic activity even when the South Armorican Ocean is considered to have closed (Late Silurian to Lower Devonian). However, the sedimentation rate increased during this time interval (Fig. 26). The absence of thick marginal sedimentary sequences in the Medio-North-Armorican Domain, although this unit might represent the northermost part of the North Gondwana platform, should be stressed. This would mean that the true margin was located northward and is now missing or hidden.

The South Armorican Domains (Ligeria and Vendée) had clearly distinct characteristics with deeper or more distal environments and faunas with Bohemian affinities. They are more or less reminiscent of the Paleozoic sequences of the Aquitaine basement (see the Aquitaine Basin, this Vol.). This suggests that the South Armorican terranes were located in a more southeastern position within the North Gondwana Province (cf. Fig. 24) and that they were displaced during Variscan times, especially due to the action of the South-Armorican Shear Zone (SASZ). As the Paleozoic successions of the South-Armorican Domains are less well established than those of the Medio-North Armorican Domain, no clear indications of Silurian or Early Devonian tectonic activity can be documented in Ligeria. Possible echos of Eo-Hercynian events are only reported from Vendée with the volcaniclastic tuff-ignimbrites of the Mareuil-sur-Lay Formation (uppermost Silurian and/or lowermost Devonian) (Peucat et al. 1986) and the back-arc volcanic rocks of the La Meilleraie Formation which overlie fossiliferous Devonian strata. The first major tectonometamorphic event in this region would therefore have occurred in the Late Devonian or in the Early Carboniferous (Wyns et al. 1989). These events, which most probably resulted from the closure of the South-Armorican Ocean, are more or less contemporaneous with the Bretonian episode reported in the Medio-North-Armorican Domain.

Apart the initial Paleozoic deposits (e.g., Initial Red Beds) firm evidence of tectonically controlled sedimentation did not appear in the Armorican Massif before the uppermost Devonian and the Early Carboniferous (Figs. 25 and 26). Following the Bretonian episode, olistolith-bearing sequences and pull-apart basins developed in the uppermost Devonian and in the Early Carboniferous. Then, after the Early Carboniferous marine transgression, small post-orogenic lacustrine basins succeeded the late Namurian-early Westphalian major orogenic activity. During both periods, the thick terrigenous and volcaniclastic continental or shallow marine sequences correspond to high sedimentation rates (Fig. 26) and indicate a very active subsidence, probably related to the subduction of the Rheic oceanic floor and to the subsequent continental collision.

After these main Variscan orogenic events (Late Namurian to Early Westphalian), the Armorican Late Paleozoic sediments were restricted to fault-bounded limnic basins. These intramontane deposits were affected only by late movements of the northern and southern branches of the South-Armorican Shear Zone.

References

Besly BM (1987) Sedimentological evidence for Carboniferous and Early Permian palaeoclimates of Europe. Ann Soc Géol Nord 106: 131–143

Cocks LRM, Fortey RA (1982) Faunal evidence for oceanic separation in the Palaeozoic of Britain. J Geol Soc (Lond) 139: 465–471

Cramer FH (1971) A palynostratigraphic model for Atlantic pangea during Silurian time. Mém Bur Rech Geol Min 73: 229–235

Destombes J, Feist R (1987) Découverte du Cambrien supérieur en Afrique (Anti-Atlas central, Maroc). C R Acad Sci Paris 304: 719–724

Lespérance PJ, Sheehan PM (1987) Trilobites et Brachiopodes ashigilliens (Ordovicien supérieur) de "l'Assise" de Fosse, bande de Sambre et Meuse (Belgique). Bull Inst R Sci Nat Belg Sci Terre 57: 91–121

Paris F, Robardet M (1977) Paléogéographie et relations ibéro-armoricaines au Paléozoïque anté-Carbonifère. Bull Soc Géol Fr 7 19: 1121–1126

Paris F, Robardet M (1990) Early Palaeozoic palaeobiogeography of the Variscan regions. Tectonophysics 177: 193–213

Perroud H (1985) Synthèse des résultats paléomagnétiques sur le Massif Armoricain. Hercynica 1 1: 65–71

Peucat JJ, Paris F, Chalet M (1986) U-Pb zircon dating of volcanic rocks close to the Silurian-Devonian boundary from Vendée (Western France). Chem Geol (Isotope Geoscience section) 59: 133–142

Robardet M, Doré F (1988) The Late Ordovician diamictic formations from southwestern Europe: north-Gondwana glaciomarine deposits. Palaeogeogr Palaeoclimatol Palaeoecol 66: 19–31

Robardet M, Paris F, Racheboeuf P (1990) Palaeogeographic evolution of southwestern Europe during early Palaeozoic times. In: Mc Kerrow WS, Scotese CR (eds) Palaeozoic palaeogeography and biogeography. Geol Soc Mem 12, London, pp 409–417

Van der Voo R (1979) Paleozoic assembly of Pangea: a new plate tectonic model for the Taconic, Caledonian and Hercynian orogenies. EOS Trans Am Geophys Union 602: 241 (Abstr)

Wyns R, Lardeux H, Weyant M (1989) Présence de Dévonien dans le Groupe de Réaumur (Synclinal de Chantonnay, Vendée); conséquences sur l'évolution géodynamique varisque de la Vendée. C R Acad Sci Paris 308: 855–860

Introduction

J. Rolet

The Armorican Massif (Fig. 1) is cross-cut by several major late Carboniferous shear zones which subdivide it into seven geographically distinct zones. Each of these is characterized by its own lithotectonic pile. This explains the striking inhomogeneity of the Armorican Massif, although more or less well-preserved fossiliferous Paleozoic strata can be observed in nearly all zones as described by Robardet et al. this Vol. Generally speaking, four different orogenic events, still not too well dated in some areas, can be distinguished. Although they are never superimposed in one and the same area, their chronological relationships have been clearly established. These are: (1) the Cadomian orogeny, (2) the early Cambrian-Ordovician distensive events (Le Corre), (3) the Eohercynian (Silurian and Devonian) closure and collisional phases (Ballèvre et al.) and (4) the Hercynian (Carboniferous) intracontinental hypercollisional phases (Rolet et al.). Some of these Carboniferous shear zones are 100 km long; they were active for a long time, producing early ductile mylonites then later brittle breccias. Among the best known are the North-Armorican, South-Armorican, Cholet and Nort-sur-Erdre complex shear zones described by Jegouzo. These induced the development of Carboniferous pull-apart basins such as Chateaulin, Laval, Ancenis, and Loire-Vendée (Rolet et al.).

From north to south, the geographical scheme we use is the following (Fig. 2):

- Zone 1 is the Icartian craton described by Chantraine et al. (this Vol.).
- Zone 2 is the Cadomian Domain (Chantraine et al.).
- Zone 3 corresponds to the Cotentin and Mancellian block, where a weak Cadomian structure is slightly overprinted by the Paleozoic one, as discribed by Gresselin.
- Zone 4 comprises several different regions: South Brittany (Ballèvre et al.), coastal Vendée, Chantonnay synclinorium and Léon area (Wyns and Rolet). It is composed of several different piles of Paleozoic nappes stacked during Eohercynian and/or Carboniferous orogenic events. High pressure-bearing units have been described by Ballèvre in most of them. The coexistence of low temperature blueschists and high temperature eclogites is good evidence for the subduction of both oceanic and marginal series. Several successive and more or less widespread migmatizations affected all the metamorphic rocks between the lower Devonian and Stephanian.
- Zone 5 is the Champtoceaux-les Mauges area. The northern limit is the major Nort-sur-Erdre shear zone. It corresponds to a second pile of Paleozoic and Precambrian units with widespread Silurian eclogitic relicts emphasizing an early subduction stage. It was stacked during Eohercynian time and overprinted at least by several successive Carboniferous shear zones. It is described by Ballèvre for the Western part and by Wyns for the Eastern part.
- Zone 6 is the Saint-Georges-sur-Loire-Lanvaux area. It shows a pile of Paleozoic tectonic units stacked during Eohercynian time and overprinted by the Bretonian phase, as described by Ballèvre et al. for the northern part and by Wyns for the southern part.
- Zone 7 corresponds to Central Brittany. It is an external Hercynian zone, where the Paleozoic tectonic phases have been paleontologically dated. The main structural features are Upper Devonian to Lower Carboniferous (Bretonian phase) thin-skin and transcurrent tectonics as exposed by Rolet.

Zones 1, 2 and 3 correspond to the Proterozoic basement or North Armorican Domain, more or less involved in the Hercynian events.

Zones 4, 5 and 6 correspond to the Eohercynian (Silurian-Devonian) mobile belt, the inner zone of the Variscan orogeny or the South Armorican Domain.

Zones 7 and 3 are the Hercynian (upper Devonian, Carboniferous) foreland or Central Armorican Domain.

J. D. Keppie (Ed.): Pre-Mesozoic Geology in France and Related Areas

Fig. 1. Structural map of the armorican massif. *Cadomian basement: 1 a* Icartian relics (2 ba); *1 b* Cadomian complex (570 to 540 Ma); *1 c* undeformed Cambrian red beds (cratonic cover). *Hercynian foreland: 2 a* Hercynian folds on Paleozoic series; *2 b* Siluro-Devonian volcanic arc series; *2 c* Lower Carboniferous pull-apart basin; *2 d* Carboniferous granites and gabbro. *Eohercynian Inner Zone: 3 a* Ordovician orthogneiss; *3 b* MT/HT metamorphic series; *3 c* HT metamorphism on Paleozoic series; *3 d* porphyroid nappe; *3 e* blueschists and eclogitic klippes

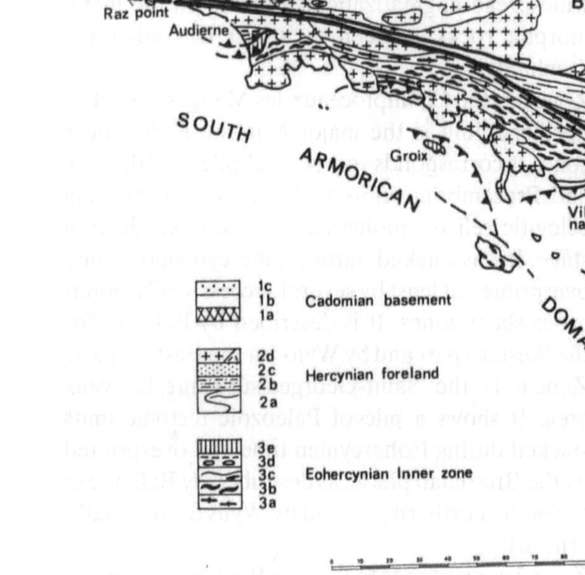

Fig. 2. Main Armorican Zones. *1* Icartian continent; *2* Cadomian craton; *3* Cadomian foreland; *4* South Armorican and Leon nappes; *5* Champtoceaux nappe and Mauges craton; *6* St Georges-sur-Loire synform and Lanvaux antiform; *7* Central Brittany. *LU* Leon underthrust; *STF* South Trégor Fault; *LT* La Hague thrust; *GCF* Granville Cancale fault; *NASZ* North Armorican shear zone; *SASZ* South Armorican shear zone; *(N)* Northern branch; *(S)* Southern branch; *NEF* Nort-sur-Erdre fault

Early Tectonic Events (Ordovician)

C. LE CORRE

Silurian, Devonian, or Carboniferous tectonic and metamorphic reworking make the early Paleozoic evolution in the Armorican Massif difficult to establish clearly. Nevertheless, two types of events should be emphasized which occurred especially during the Ordovician period (1) extensional tectonic events, and (2) emplacement of plutonic rocks.

1 Ordovician Extensional Events

The significance of the Proterozoic/Paleozoic main unconformity in Central Brittany has long been discussed. Some authors (Chauvel and Phillipot 1960; Cogné 1962) have interpreted the angular unconformity between Brioverian rocks and infra-Paleozoic red beds as the result of Cadomian folding, followed by erosion and deposition of a new Paleozoic series. However, it is now clearly established in Central Brittany that the main deformation in the Brioverian Series (folding and slaty cleavage) occured during the Carboniferous orogenic period (Le Corre 1977). Therefore, other authors consider the unconformity to be a result of only a moderate synsedimentary deformation of the Upper-Brioverian Series prior to Paleozoic deposition (Darboux et al. 1975; Le Corre 1977). A third interpretation has recently been proposed, which involves block faulting and tilting of Brioverian sediments during the deposition of Lower Ordovician red beds (Pont-Réan Formation) and before the deposition of the Grès Armoricain Formation (Ballard et al. 1986; Brun et al. 1991).

This interpretation has been deduced from map pattern analysis and field observations of the Brioverian and Lower Ordovician Formations in Central Brittany. Geological map contours and rapid thickness variations of the Pont-Réan Formation (0–700 m) suggest an interference between an early NE-SW trend (i.e. normal faulting) and the later E W Carboniferous refolding. Various interference models, assuming different obliquities between normal faulting and refolding, have been computerized and compared with the geological map of Central Brittany. With a 20° obliquity, the computerized pattern is in good agreement with the geological map (Fig. 1 a). For such a model the major early normal faulting has a N 70° trend, a dip toward the SE, and bound half-graben structures containing red-bed deposits (Fig. 1 b). A similar prefolding N 70° trend is also observed for bedding directions in many Brioverian localities, in agreement with block tilting. The model can also explain some sedimentological data of the Pont-Réan Formation, such as conglomerate facies, red beds, thickness variations, etc. (see Chap. II.B: Ordovician) suggesting continental rifting. The syn-rift red-bed series has been recently dated by radiometric measurements (U/Pb in zircon) on volcanic beds providing an age of 466±1 Ma (Bonjour and Chauvel 1988; Bonjour et al. 1988). During these times there was intense volcanic activity in various Armorican localities, compatible in some with an extensional environment (Thieblemont et al. 1987; see Chap. II.B: Igneous Activity). A possible generalization of the rifting model with a NW-SE extensional direction can be suggested for the North Armorican area, where some structures have the same N 70° trend implicated in the Cambrian and Lower Ordovician Series (Charnie, Coëvron, zone Bocaine, etc., see Chap. II.B: Ordovician).

2 Early Plutonic Rocks

The last igneous activities connected with Cadomian events provide radiometric data older than 540 Ma in the North Armorican area (Mancellia, Trégor, see Chap. A: Igneous Activity). However, during the Lower Paleozoic, many ages between 490 and 440 Ma have also been reported (Fig. 2 a) corresponding to Ordovician time. These Ordovician granitic plutons occur especially in the South-Armorican area, rarely in the Central Armorican Domain, and never in the North-Armorican area (Fig. 2 b).

The Ordovician plutons have been extensively transformed by tectonic and metamorphic processes

J. D. Keppie (Ed.): Pre-Mesozoic Geology in France and Related Areas

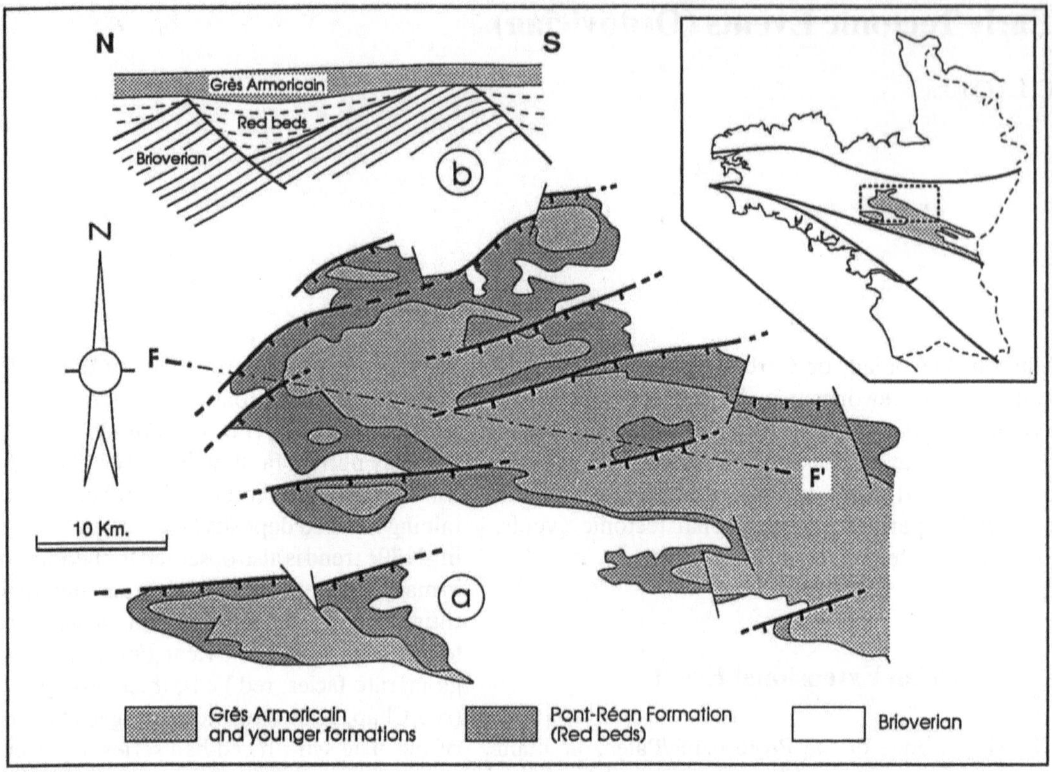

▲
Fig. 1 a, b. Ordovician block tilting model. **a** Geological map and location of deduced normal faults. **b** Geometric relations between the Brioverian Series (pre-rift), the Pont-Réan red beds (syn-rift) and the Grès Armoricain Formation (post-rift). *FF′* Carboniferous folding direction. (After Brun et al. 1991)

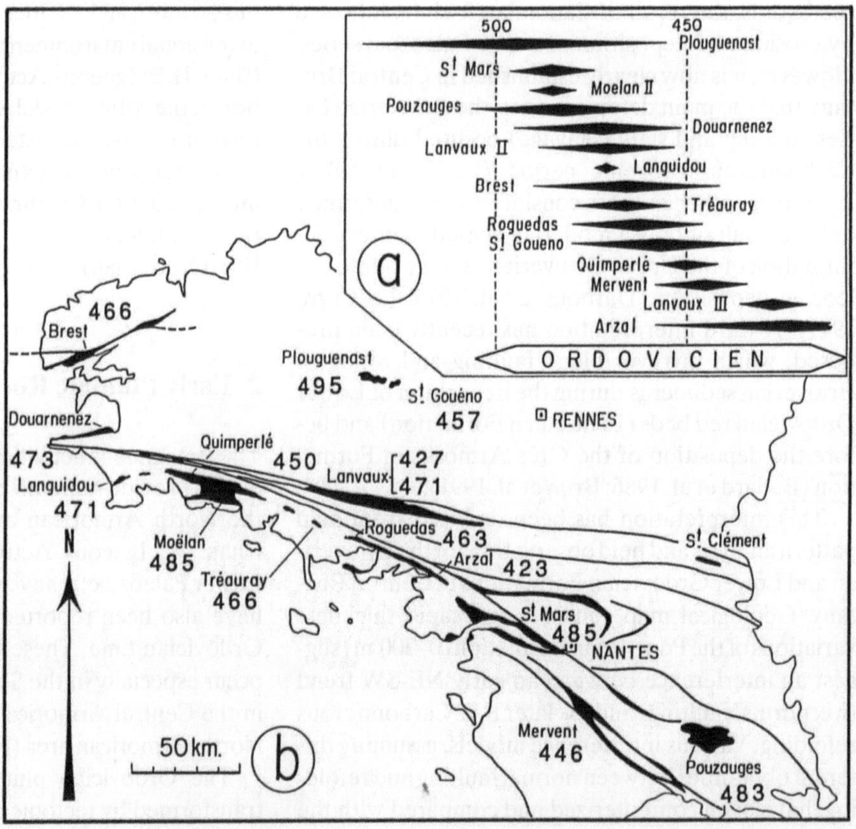

during the main Devonian and Carboniferous Variscan events. Hence, they generally occur as large orthogeneissic bodies (Brest, Lanvaux, Quimperlé, etc.) or as scattered outcrops (Tréauray, St Nazaire, etc.) characterized by intense metamorphic foliation.

The post-emplacement deformation of Ordovician plutons has been clearly demonstrated in South Brittany, for example (Peucat et al. 1978; Le Métour and Audren 1979). Some orthogneissic bodies are clearly involved in early nappe structures with other metamorphic terranes such as in the Léon area (Le Corre et al. 1999) or in the Champtoceaux nappe (Marchand 1981; Paquette et al. 1984), or are deformed along Carboniferous shear zones (Lanvaux; Cogné et al. 1983).

Strain and metamorphism, including local partial melting, often hide the original features of the Ordovician plutonic rocks, but available petrological data show a wide range of compositions including granites s. s., alkali-granites, granodiorites, quartz-diorites, trondhjemites, and gabbros. Sufficient chemical data are not yet available to establish an accurate geodynamic environment for the Ordovician plutons. Some of them (trondhjemite cf. Douarnenez) may be related to crustal stretching conditions, although some chemical data show a calc-alkaline trend, suggesting plate convergence conditions (Jegouzo et al. 1986).

Nevertheless it is now clear that Ordovician plutonic activity must be related to the beginning of Variscan plate tectonic movements (Autran and Cogné 1980). These data must be related to geodynamic events occurring at the same time in various Armorican domains such as Ordovician extensional tectonics in Central Brittany, the Ordovician extensional volcanic series in the Vendée (see Chap. II.B: Igneous Activity), and also Upper-Ordovician high pressure metamorphism in the Léon, Vendée, and Champtoceaux areas (see Chap. II.B: Eo-Hercynian Events).

◀ **Fig. 2 a, b.** Isotopic age data (**a**) and location map (**b**) of early plutonic rocks of the Armorican Belt (time scale from Odin and Gale 1982). Plouguenast: 495±40, Rb-Sr isochr. (Vidal 1980); St. Mars: 485±6, U-Pb zircon (Paquette et al. 1984); Moëlan II: 474±5, Rb-Sr isochr., MSWD = 0.8 (Vidal 1980); Pouzauges: 483±22, Rb-Sr isochr. (Wyns et al. 1988); Douarnenez: 473±23, Rb-Sr isochr., MSWD = 7 (Barrière et al. 1971); Lanvaux II: 471±18, Rb-Sr isochr., MSWD = 2.4 (Vidal 1972); Languidou: 471±3, U-Pb zircon (Peucat 1983); Brest: 466±25, U-Pb zircon (Cabanis et al. 1977; Cabanis 1986); Tréauray: 466±12, U-Pb zircon (Jégouzo et al. 1986); Roguédas: 463±17, Rb-Sr isochr., MSWD = 2 (Peucat et al. 1978); St. Gouéno: 457±23, Rb-Sr isochr. (Vidal 1980; Saunier 1986); Quimperlé: 450±8, Rb-Sr isochr., MSWD = 3.5 (Jégouzo et al. 1986); Mervent: 446±12, Rb-Sr isochr. (Vidal 1980; Wyns et al. 1988); Lanvaux III: 427±9, Rb-Sr isochr., MSWD = 1.6 (Vidal 1972; Jégouzo et al. 1986); Arzal: 423±35, Rb-Sr isochr., MSWD = 35 (Jégouzo et al. 1986)

References

Autran A, Cogné J (1980) La zone interne de l'orogène varisque dans l'Ouest de la France et sa place dans le développement de la chaîne hercynienne. Mem Bur Rech Geol Min 108: 90–111

Ballard JF, Brun JP, Durand J (1986) La discordance Briovérien-Paléozoîque inférieur en Bretagne Centrale: signature d'un épisode de distension ordovicienne. CR Acad Sci Paris 303: 1327–1332

Barrière M, Cogné J, Vidal Ph (1971) La trondhjémite de Douarnenez: une intrusion magmatique ordovicienne dans les schistes cristallins du Sud-Finistère. CR Acad Sci Paris 273: 1556–1559

Bonjour JL, Chauvel JJ (1988) Un example de sédimentation initiale dans un bassin paléozoîque: étude pétrographique et géochimique de l'Ordovicien inférieur de la presqu'île de Crozon (Finistère). Bull Soc Géol Fr IV: 81–91

Bonjour JL, Peucat JJ, Chauvel JJ, Paris F, Cornichet J (1988) U-Pb zircon dating of the early Paleozoic (Arenigian) transgression in Western Brittany (France): a new constraint for the lower paleozoic time-scale. Chem Geol 72: 329–336

Brun JP, Ballard JF, Le Corre Cl (1991) Identification of Ordovician block tilting in the Hercynian fold belt of Central Brittany (France): field evidence and computer models. J Struct Geol 13, 4: 419–429

Cabanis B (1986) Identification des séries magmatiques dans les socles métamorphiques sur la base de critères géologiques, pétrographiques et géochimiques. Exemples d'applications dans le Massif Armoricain. Implications géodynamiques. Contribution à une méthode d'étude basée sur les éléments traces. Thesis, Univ. Pet M. Curie Paris, pp 1–640 (unpublished)

Cabanis B, Michot J, Deutsch S (1977) Remise en question de la datation géochronologique des gneiss de Brest (Bretagne occidentale). CR Acad Sci Paris 284: 883–886

Chauvel JJ, Philippot A (1960) Sur la discordance de base du paléozoïque dans la région de Rennes: trois carrières démonstratives. Bull Soc Geol Mineral Bretagne 1: 1–7

Cogné J (1962) Le Briovérien. Bull Soc Geol Fr IV: 413–430

Cogné JP, Choukroune P, Cogné J (1983) Cisaillements varisques superposés dans le Massif de lanvaux (Bretagne Centrale). CR Acad Sci Paris 296: 773–776

Darboux JR, Le Corre Cl, Cogné J (1975) Tectoniques superposées cadomiennes et hercyniennes dans le Briovérien du Nord de la Baie de Douarnenez (Finistère). Bull Soc Géol Fr XVII: 680–685

Jégouzo P, Peucat JJ, Audren Cl (1986) Caractérisation et signification géodynamique des orthogneiss calco-alcalins d'âge ordovicien de Bretagne méridionale. Bull Soc Géol Fr II: 839–848

Le Corre Cl (1977) Le Briovérien de Bretagne centrale: essai de synthèse lithologique et structurale. Bull Bur Rech Geol Min I-3: 219–254

Le Corre Cl, Balé P, Georget Y (1999) The Leon area: an exotic domain in the variscan belt of Brittany (NW Armorican Massif, France). Geodin Acta 3, 2: 57–71

Le Métour J, Audren Cl (1979) Relations structurales entre l'orthogneiss ordovicien de Roguedas et son encaissant migmatitique. Conséquences sur l'âge des événements tectonométamorphiques en Bretagne méridionale. Bull Soc Geol Mineral Bretagne IX: 113–123

Marchand J (1981) Ecaillage d'un "mélange tectonique" profond: le complexe cristallophyllien de Champtoceaux (Bretagne méridionale). CR Acad Sci Paris 293: 223–228

Odin JS, Gale NH (1982) Mise à jour de l'échelle des temps calédoniens et hercyniens. CR Acad Sci Paris 294: 453–456

Paquette JL, Marchand J, Peucat JJ (1984) Absence de tectonique cadomienne dans le complexe de Champtoceaux (Bretagne méridionale)? Comparaison des système Rb-Sr et U-Pb d'un métagranite. Bull Soc Géol Fr XXVI: 907–912

Peucat JJ (1983) Géochronologie des roches métamorphiques (Rb-Sr et U-Pb). Exemples choisis au Groenland, en Laponie, dans le Massif Armoricain et en Grande Kabylie. Mem Soc Geol Mineral Bretagne 28: 1–158

Peucat JJ, Le Métour J, Audren Cl (1978) Arguments géochronologiques en faveur de l'existence d'une double ceinture métamorphique en Bretagne méridionale. Bull Soc Géol Fr XX: 163–167

Saunier JF (1986) Un domaine cristallophyllien dans le Protérozoïque supérieur (Briovérien) de Bretagne centrale: le dôme de Plouguenast (Côtes du Nord). Doc Bur Rech Geol Min 109: 1–119

Thiéblemont D, Cabanis B, Le Métour J (1987) Etude géochimique d'un magmatisme de distension intracontinentale: la série bimodale ordovicienne du Choletais (masif vendéen). Géol Fr Bur Rech Geol Min 1: 65–76

Vidal Ph (1972) L'axe granitique Moëlan-Lanvaux (sud du massif armoricain): mise en évidence par la méthode Rb-Sr de trois épisodes de plutonisme pendant le Paléozoïque inférieur. Bull Soc Géol Mineral Bretagne 4: 75–89

Vidal Ph (1980) L'évolution polyorogénique du massif armoricain. Apport de la géochronologie et de la géochimie isotopique du strontium. Mém Soc Geol Mineral Bretagne 21: 1–162

Wyns R, Lablanche G, Lefavrais-Raymond A (1988) Notice explicative de la feuille Chantonay à 1/50 000. n° 563. Bur Rech Geol Min, Orléans, pp 1–76

Eo-Hercynian Events in the Armorican Massif

M. BALLÈVRE, J. MARCHAND, G. GODARD, J.-C. GOUJOU, JEAN-CHRISTIAN and R. WYNS

1 Introduction

Eo-Hercynian events (i.e., significant deformation and metamorphism prior to strike-slip movements along transcurrent shear zones) are known only in the South-Armorican domain and in the Leon area (Fig. 1).

The South-Armorican domain can be conveniently divided into three major zones:

Zone 6 (between the northern branch of the South Armorican shear zone and the Nort sur Erdre fault) is classically divided into the Lanvaux-les-Ponts-de-Cé anticlinorium to the north and the Saint-Georges-sur-Loire synclinorium to the south. Although intensely deformed, the metamorphism is of greenschist-facies grade. The age of the main deformational event has not been established with certainty, but must be younger than the Silurian (according to paleontological data) or even the Devonian (i.e., the probable age of the metavolcanics from the St Georges/Loire synclinorium). Definite proof for the existence of true Eo-Hercynian events in this zone remains to be shown.

Zone 5 has the shape of a wedge, limited to the north by the Nort-sur-Erdre fault and to the south by the southern branch of the South-Armorican shear zone. Zone 5 mainly consists of the Champtoceaux nappe, a unit thrust over presumed Upper-Proterozoic metasediments, and the Mauges area, where undeformed or weakly deformed Paleozoic sediments rest unconformably over Upper Proterozoic sediments. However, the relationships between the Champtoceaux nappe and the Mauges area are not clearly understood. Evidence for Eo-Hercynian events is found in the Champtoceaux nappe.

Zone 4 is located to the south of the southern branch of the South-Armorican shear zone. In this domain, the Proterozoic basement has not been recognized, although rare, strongly deformed orthogneisses yield crystallization ages for the magmatic protoliths at about 600 Ma (U-Pb data on zircons: Port Manec'h: Vidal 1980). Fossiliferous Paleozoic sediments are recognized only in the Vendée area. Numerous Carboniferous granites intrude this zone. The largest exposure of zone 4 occurs in the Vendée area. Correlations with sections to the west are difficult to establish. This could be the result of:

- reworking of Eo-Hercynian structures by thrust and wrench faulting during the last stages of the Hercynian orogeny;
- brittle faulting associated with the opening of the Atlantic ocean.

In this section, we describe Eo-Hercynian events in the South-Armorican domain and in the Leon area. Our purpose is to provide a critical data base combining structural, petrological, geochemical, and geochronological studies concerning Eo-Hercynian events. Some constraints on the geodynamic significance of the Eo-Hercynian events in the Armorican Massif are briefly discussed. Detailed comments on the zone 6 are not made because our knowledge of Eo-Hercynian events in that zone is too imprecise.

2 Eo-Hercynian Events in Zone 5: the Champtoceaux Nappe

The Champtoceaux nappe (Fig. 2) is thrust over low-grade metasediments of presumed Upper Proterozoic age (the "Mauves-sur-Loire micaschists") and consists of several stacked units, as shown by detailed mapping (Marchand 1981; Marchand et al. 1989) and gravimetric modeling (Bayer and Hirn 1987).

2.1 Lithology

The footwall of the Champtoceaux nappe consists of albite-bearing micaschists, which contain greenschist-facies assemblages (quartz + white mica + biotite + chlorite + garnet).

The Champtoceaux nappe itself can be subdivided into two major units, which consists of different lithologies.

J. D. Keppie (Ed.): Pre-Mesozoic Geology in France and Related Areas

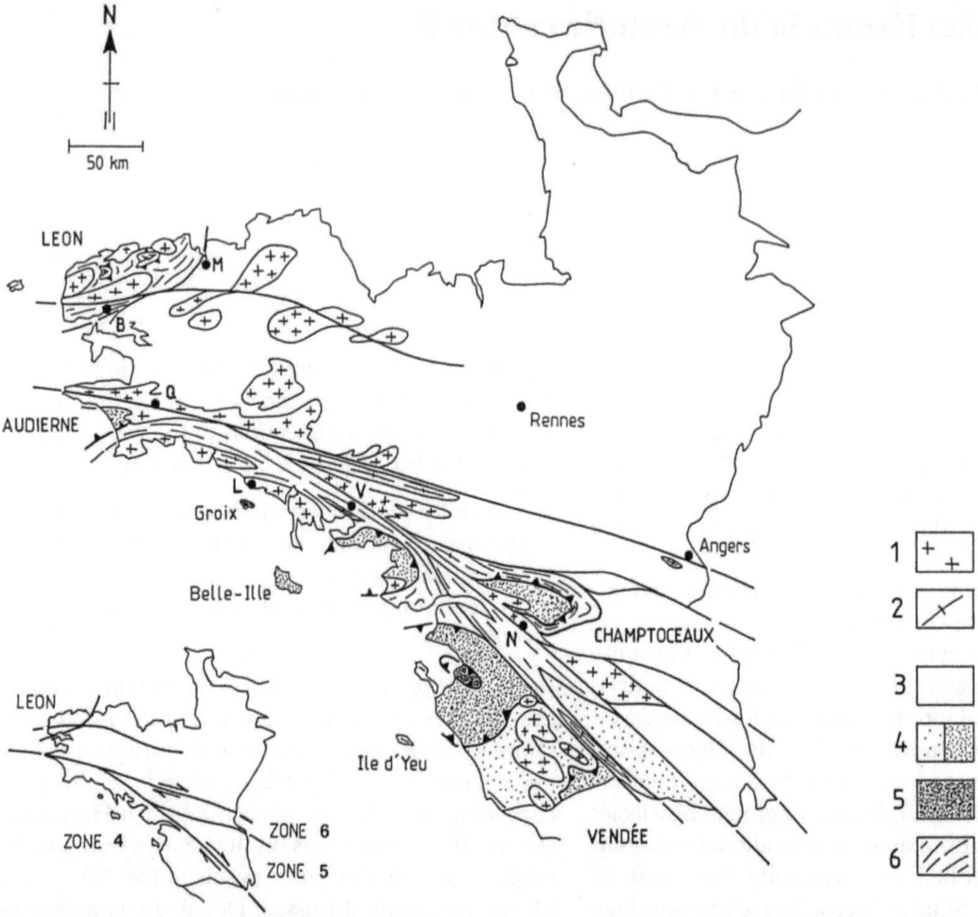

Fig. 1. Eo-Hercynian events in the Armorican massif: a synthetic map. *B* Brest; *L* Lorient; *M* Morlaix; *N* Nantes; *Q* Quimper; *V* Vannes. *1* Carboniferous granites; *2* Ordovician orthogneisses; *3* Unmetamorphosed sediments of Palaeozoic or Upper Proterozoic age; *4* Very low-grade to low-grade metasediments and metavolcanics; *5* Blueschists; *6* Medium- to high-grade metamorphic rocks

The lower unit is essentially made of leptynites, i.e., leucocratic orthogneisses, micaschists, and small but numerous eclogite lenses. Relicts of undeformed granites are locally observed within leptynites (La Picherais: Lasnier et al. 1973). Geochronological studies suggest that the magmatic protoliths are of Lower Palaeozoic age (La Picherais: Vidal et al. 1980) or of Upper Ordovician age (Saint Mars du Désert: Paquette et al. 1984). In addition, rare relicts of metahornfelses (Barbaroux et al. 1983; Marchand et al. 1989) show that these granites developed contact aureoles into their sedimentary country rocks. The basic rocks have a rather restricted range of compositions and REE patterns ranging from N-MORB to T-MORB types (Paquette 1987). In consequence, the lower unit of the Champtoceaux nappe is interpreted as part of the upper continental crust prior to the Eo-Hercynian events. Basic rocks are not remnants of an oceanic crust but are instead thought to represent dykes intruded in a thinning continental crust.

The upper unit consists of (1) migmatitic para- and orthogneisses, (2) fine-grained leptynites containing amphibolitized eclogites and (3) serpentinized peridotites and amphibolitized gabbros. The age of the protoliths, except for metagabbros, is unknown. A U-Pb study of zircons from the metagabbros gave lower intercept age of 371 ± 3 Ma, which is interpreted either as the age of the magmatic protolith or that of the metamorphic overprint (Paquette 1987). Some undeformed gabbros preserve evidence of LP granulite-facies parageneses (Lasnier 1974; Carlier 1978).

2.2 Structure

Two main deformational events can be distinguished in the Champtoceaux nappe (Lagarde 1978). The first

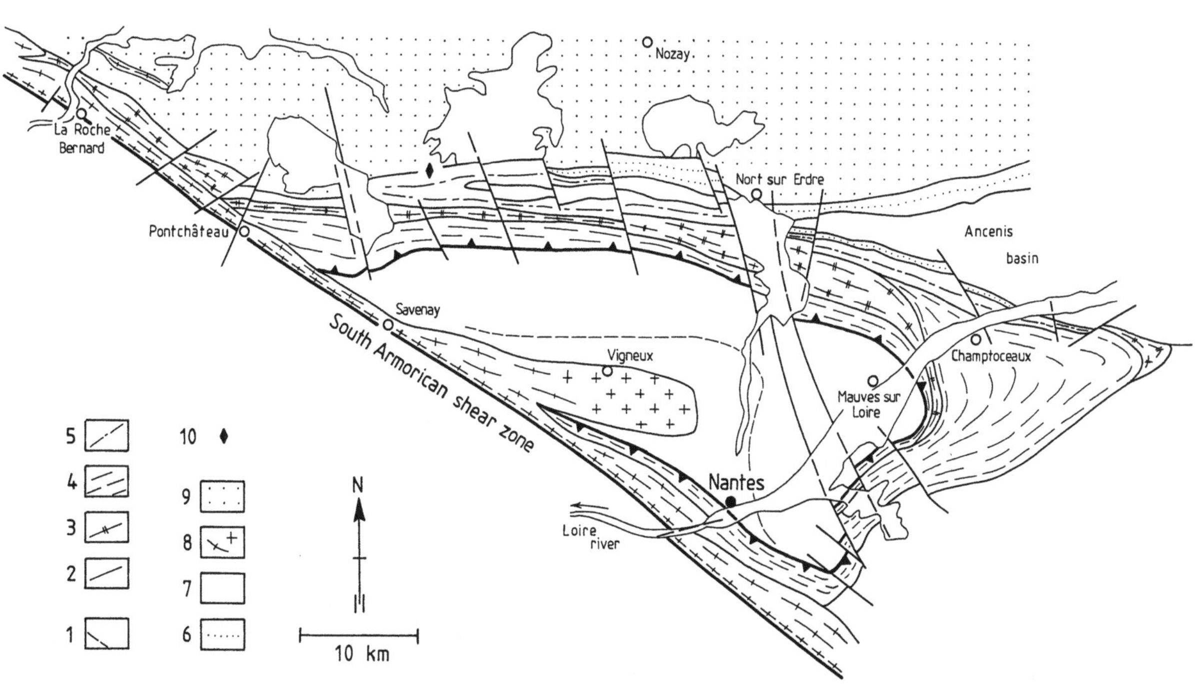

Fig. 2. A schematic structural map of the Champtoceaux nappe. (After Marchand et al. 1989). *1* Mauves-sur-Loire micaschists; *2* Leptynites, micaschists, and eclogite lenses; *3* Saint-Mars-du-Désert orthogneiss, deriving from an Ordovician granite, and La Mauvoisinière orthogneiss; *4* Medium- to high-grade micaschists and gneisses, including migmatites; *5* Le Hâvre micaschists; *6* Upper Proterozoic sediments; *7* Paleozoic sediments from the Ancenis unit; *8* Carboniferous granites; *9* Saint Georges sur Loire unit; *10* Garnet peridotite

event was the main ductile deformation, easily recognized on the field by a pervasive foliation and a prominent stretching lineation. This deformation is associated with the thrusting of the Champtoceaux nappe (from east to west) over the Upper Proterozoic sediments. The second event was the folding of the earlier structures (i. e., foliations, stretching lineations, and thrust surfaces between units within the nappe) around an east-west axis. This folding was probably synchronous with the dextral strike-slip motion along the South-Armorican shear zone and of the intrusion of the associated two-mica leucogranites (e. g., Vigneux). The southern part of the Champtoceaux nappe is strongly sheared along the South-Armorican shear zone.

2.3 Metamorphism

Recent detailed works have contributed to major advances in our understanding of the P-T history of the Champtoceaux nappe (Godard et al. 1981; Ballèvre et al. 1987, 1989; Godard, 1988; Ballèvre and Marchand 1991). Three main results have been obtained:

First, each unit of the Champtoceaux nappe preserves evidence of an earlier HP event. Eclogite-facies parageneses are found in both units. P-T conditions for the eclogite-facies event are estimated at 18–20 kbar, 700 °C for the lower unit. Pressures equal to or higher than 14 kbar at temperatures of the order of 700 °C have been estimated in the upper unit. The main consequences of these results are:

The continental crust was buried to depths of the order of 60 km at the beginning of the collision history; The thrusting of the Champtoceaux nappe occurred later than the eclogite-facies event.

Second, the specificity of the lower unit of the Champtoceaux nappe in the whole Hercynian belt is highlighted, this unit being comparable only to the Malpica-Tuy unit in northwestern Spain (Ibarguchi and Ortega Girones 1985). Indeed, synkinematic early eclogite-facies parageneses are observed in all lithologies from the lower unit. They are:

- garnet + omphacite + amphibole + quartz + rutile in the metabasites;
- quartz + garnet + jadeite + phengite + paragonite or quartz + omphacite + kyanite in the quartzites;
- quartz + phengite + garnet + chloritoid or kyanite + rutile in the metapelites.

Coronitic transformations (i. e., garnet growth at the contact between magmatic plagioclase and biotite) are also observed within undeformed metagranites.

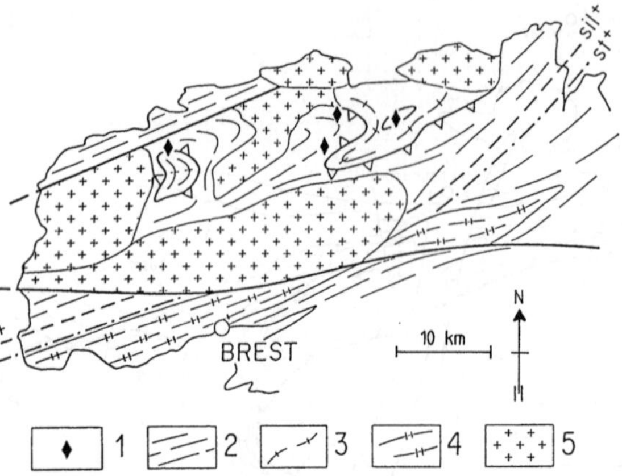

Fig. 3. A schematic structural map of the Léon area. *1* Eclogites; *2* medium- to high-grade micaschists and gneisses (upper unit); *3* orthogneisses (lower unit); *4* Brest orthogneisses, deriving from an Ordovician granite; *5* carboniferous granite. Isograds (*st* staurolite; *sil* sillimanite) after Cabanis et al. (1979)

Moreover, mineral assemblages within metapelites as well as metabasites from the lower unit of the Champtoceaux nappe reveal regionally increasing P-T conditions during the eclogite-facies event (Ballèvre and Marchand 1991). Estimated P-T conditions increase from around 15 kbar, 550 °C on the left bank of the River Loire to approximately 20 kbar, 650 °C near Pontchâteau. This records a variation in depth of the studied unit during the early collisional history.

Third, each unit of the Champtoceaux nappe is characterized by its own P-T path. In the lower unit, the temperature decreases immediately after the pressure peak. By contrast, the upper unit followed an isothermal decompressional P-T path during its uplift. This difference explains the good preservation of HP parageneses in the lower unit and the widespread occurrence of partial melting in the upper unit.

2.4 Timing

Geochronological data can be used to constrain the timing of the tectonic events.

1. The Lower Paleozoic age of the protoliths show that evidence for Proterozoic events are lacking in the Champtoceaux nappe. The entire history belongs to the Paleozoic: it must be later than ca. 480 Ma (the age of the Saint-Mars-du-Désert granite in the lower unit) or perhaps even 370 Ma (the possible age of the gabbros in the upper unit).
2. The age of the HP event is not well defined. U-Pb data on zircons have been obtained from two eclogites. Zircons from a kyanite-bearing eclogite, whose bulk composition and REE pattern is quite anomalous, give a lower intercept age of 413 ± 16 Ma. This was interpreted as the age of the HP event (Paquette et al. 1985) but could equally well be the age of the protolith. The upper intercept age of 2208 ± 39 Ma is now regarded as without geological significance (Paquette 1987). Zircons from a glaucophane-bearing eclogite give an upper intercept age of 358 ± 2 Ma (Paquette 1987). In this sample, whose chemistry and REE pattern are representative of the Champtoceaux eclogites, zircon grains are always localized in the glaucophane porphyroblasts. The HP event thus occurred at about 360 Ma or, if glaucophane growth took place at the beginning of the retrograde history (Godard et al. 1981), slightly earlier than 360 Ma.
3. A whole-rock-white mica Rb-Sr study from a glaucophane-bearing eclogite shows that temperatures of less than 450–500 °C were achieved 340 Ma ago (Peucat 1983).
4. The folding of the Champtoceaux nappe is coeval with the intrusion of the Vigneux two-mica granite, whose exact age is unknown. Nevertheless, similar synkinematic granites along the southern branch of the South-Armorican shear zone are dated at about 310–320 Ma (Vidal 1980; Guineberteau 1984).

3 Eo-Hercynian Events in Zone 4

3.1 Léon

Despite the fact that the Léon area was ascribed to the Proterozoic basement of the North-Armorican domain (e.g., Cogné 1974), an assertion which is still maintained by some authors (Cogné 1988; Matte and Hirn 1988), several lines of evidence show that its history took place during the Variscan orogeny and that Proterozoic events cannot be recognized.

The Leon area (Fig. 3) is made of medium-grade metamorphic rocks which are intruded by several generations of Carboniferous plutons (Le Corre et al. 1989). The metamorphic rocks belong to two units:

- the lower unit is essentially made of coarse-grained augen-gneisses, derived from a porphyritic granodiorite of presumed Paleozoic age (Cabanis et al. 1979);
- the upper unit is thrust over the previous one (Balé and Brun 1986) and consists of ortho- and paragneisses, containing numerous eclogite lenses close to the contact with the lower unit.

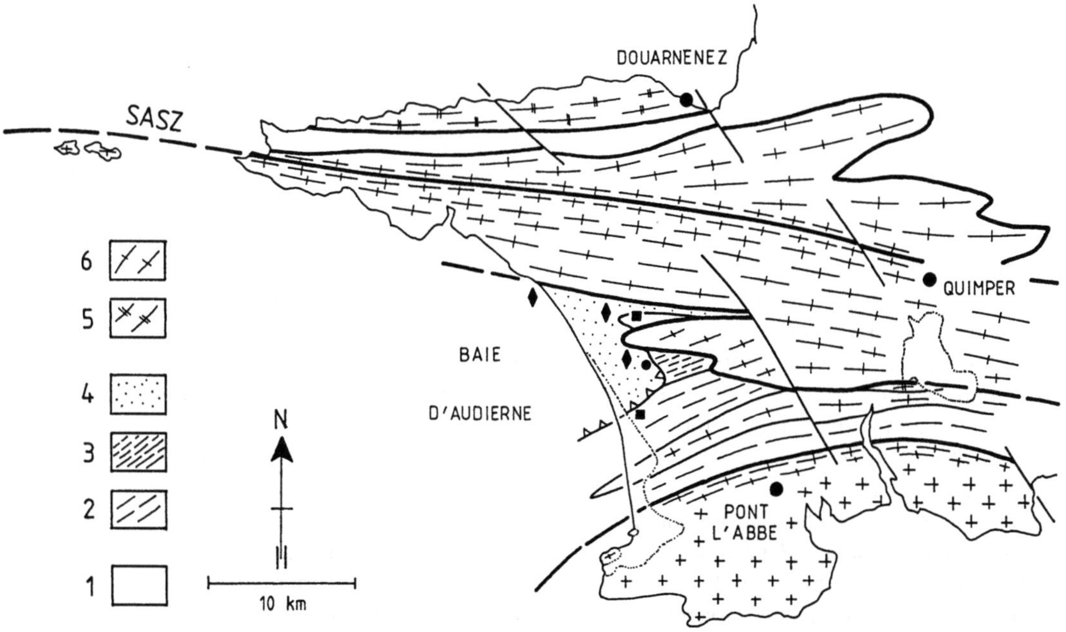

Fig. 4. A schematic structural map of the Baie d'Audierne area. *1* Paleozoic or Upper Proterozoic metasediments of the Medio-Armorican domain; *2* Trunvel and Tréogat formations; *3* Peumerit and Ty-Lan formations; *4* micaschists and amphibolites (upper unit of the Baie d'Audierne); *5* Ordovician orthogneisses; *6* carboniferous granites, more or less deformed along the South Armorican shear zone *(SASZ)*. The location of eclogites and high-pressure granulites are shown by *circles* and *squares*, respectively. Andalusite-bearing hornfelses *(lozenges)* develop near Carboniferous granites

High-pressure relicts are observed only in the eclogite lenses (Cabanis and Godard 1987; Paquette et al. 1987), where they consist of garnet + omphacite + kyanite + zoisite in magnesian bulk-rock compositions and garnet + omphacite + quartz + rutile in ferrous to ferrotitaniferrous bulk-rock compositions. These eclogites have geochemical affinities with N-MORBs and yield a lower intercept age of 442 + 14 −28 Ma by the U-Pb method on zircons. They are consequently thought to derive from oceanic material subducted at about 440 Ma ago.

Medium-grade metamorphism is observed in the whole area and is characterized by an increasing grade from southeast to northwest, as shown by the location of the staurolite and sillimanite isograds (Cabanis et al. 1979). Migmatites occur frequently within the sillimanite zone. This medium-grade metamorphism is not dated with accuracy, but must be earlier than the first generation of granitic intrusions (Saint Renan-Kersaint), which occurred at about 330–340 Ma (Deutsch and Chauris 1965; Leutwein et al. 1969) and is thus most probably Devonian in age.

3.2 Baie d'Audierne

The Baie d'Audierne area, at the western extremity of the South-Armorican domain, has a triangular shape bounded by two conjugate trancurrent shear zones – the dextral South-Armorican shear zone and a sinistral shear zone limiting the Pont l'Abbé granite. According to Peucat and Cogné (1974), two main units are recognized in the Baie d'Audierne (Fig. 4). Their extension offshore has been studied by geological (coring) and geophysical (magnetic anomalies) methods (Lefort and Peucat 1974; Poulpiquet 1988).

The lower unit consists of several lithologies which are briefly described here from south to north, i.e., in their order of superposition.

1. Albite-bearing micaschists (Trunvel Formation) are metamorphosed semi-pelitic rocks, which contain a garnet + chloritoid-bearing pelitic layer.
2. Prasinites (Tréogat Formation) are volcanic rocks strongly deformed in the albite-epidote amphibolite facies, as shown by synkinematic albite + blue-green amphibole + epidote + chlorite assemblages.
3. Amphibolites (Peumerit Formation) are basic rocks which were intensely deformed in the amphibolite facies: synkinematic assemblages are plagioclase + amphibole but relicts of earlier phases, notably garnet, zoisite, and rutile, are frequently ob-

served. Hence, a metamorphic gap occurs between prasinites and amphibolites. These latter rocks are perhaps strongly retrogressed basic granulites.
4. The largest outcrop of ultrabasic rocks in the South-Armorican domain occurs in the Baie d'Audierne, where serpentinites (Ty-Lan Formation) preserve rare olivine and clinopyroxene and locally contain chromite.

In addition, two further rock types occur in the lower group, although their relationships with other lithologies are not clearly established.
1. Metagabbros occur sparsely within amphibolites as well as serpentinites. While some of these metagabbros are undeformed and, more rarely, exhibit coronitic reactions (Lasnier 1970), most are strongly deformed and amphibolitized.
2. Two outcrops of basic granulites (Keramoine and Méot) preserve evidence of a HP granulite facies event (Velde 1972), shown by garnet + clinopyroxene + plagioclase + quartz + rutile assemblages. The basic granulites have REE patterns similar to T-MORBs (Bernard-Griffiths and Cornichet 1985). U-Pb zircon data give a lower intercept age of 384 ± 6 Ma (Peucat et al. 1982), interpreted as the age of the HP granulite-facies event.

The upper unit consists mainly of semipelites with some metabasic intercalations. Numerous aplitic and pegmatitic dykes are now parallel to the foliation of the micaschists, due to high strains during Hercynian deformation. Carboniferous granites develop contact aureoles in the micaschists from the upper unit.

At the boundary between the two main units there occur slices of:

- mylonitic to ultramylonitic orthogneisses with coarse recrystallized K-feldspar;
- garnet- and kyanite-bearing micaschists;
- eclogites (garnet + omphacite + kyanite + quartz + rutile) (Marchand 1982).

A U-Pb study of zircons from the orthogneisses and the micaschists gives similar lower intercept ages (470 ± 7 and 480 ± 8, respectively) (Peucat 1983).

3.3 Lorient-Vannes Section

This area occupies the central part of the South-Armorican domain. The geometry of the belt in this area is difficult to define accurately, because some units occur only on certain islands (e.g., Groix and Belle Ile), while others are only recognized onshore. Extensive sampling offshore has helped clarify the extent of these units (Audren and Lefort 1977; Lefort et al. 1982). Three main units are distinguished:

1. The lower unit consists of migmatitic paragneisses, orthogneisses and pyroxene-bearing amphibolites (Le Métour and Audren 1977; Le Métour 1978; Brown 1983; Jones and Brown 1989, 1990). In general, the age of the protoliths is not known, but a Rb-Sr and U-Pb study of the orthogneisses showed that their magmatic protoliths crystallized during the Lower Palaeozoic (Peucat et al. 1978; Peucat 1983). The late stages of the Eo-Hercynian event are defined by the intrusion of S-type granites, notably the cordierite-bearing granite of St. Anne d'Auray, dated at 376 ± 19 Ma by the Rb-Sr method (Vidal 1973) and at 372 ± 24 Ma by the U-Pb method (Peucat 1983).

The Eo-Hercynian event is characterized by widespread partial melting, which occurred mainly during the isothermal decompression part of a clockwise P-T path (Brown 1983; Jones and Brown 1989, 1990). The early P-T history of the migmatitic gneisses is recorded by successive assemblages preserved in inclusions within garnet, namely kyanite + ilmenite then kyanite + staurolite + rutile and finally sillimanite + staurolite + rutile. This sequence suggests that both P and T increased during garnet growth. Peak P-T conditions have been estimated at approximately 7–9 kbar, 750–800 °C. The late decompression history is marked by the growth of cordierite and/or biotite at the expense of garnet and occurred under nearly isothermal conditions.
2. The middle unit consits of three main groups (nappes?), which are here described from bottom to top. The first group occurs to the north along the Vilaine estuary and is composed of amphibolites and staurolite- or kyanite-bearing micaschists (Triboulet and Audren 1985, 1988). The second group mainly consists of albite-bearing micaschists (Audren 1974). Only greenschist-facies parageneses are recorded. The third group is made of very fine-grained metasediments with local intercalations of metavolcanics ("porphyroides") of Upper Silurian-Lower Devonian age (Peucat et al. 1986). This latter group outcrops extensively at Belle Ile.

The middle unit was thrust over the lower unit during the Carboniferous, as demonstrated by synkinematic S-C structures within two-mica leucogranites located along the thrust plane (Audren 1987) and dated at 300 ± 5 Ma by the Rb-Sr method (Vidal 1973).
3. The upper unit is famous since it is one of the few occurrences of blueschists in the Hercynian belt. It is well exposed in the Ile de Groix (Triboulet 1974; Carpenter 1976) and extends offshore 20 km towards the southeast (Audren and Lefort 1977). It is mainly characterized by the occurrence of oceanic metavolcanics and metasediments which underwent an intense duc-

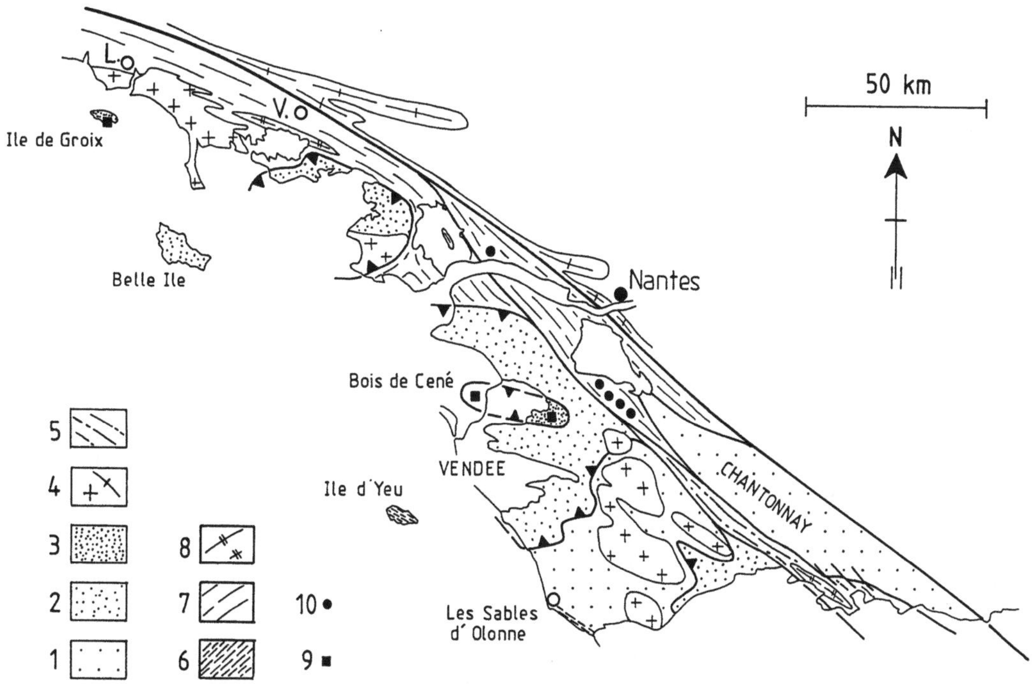

Fig. 5. A schematic structural map of the western part of Zone 4. *1* Low-grade Palaeozoic sediments from the Vendée littorale and Chantonnay areas; *2* metavolcanics ("porphyroids") and schists with greenschist-facies parageneses; *3* blueschists from Ile de Groix and Bois de Cené; *4* carboniferous granites; *5* medium-grade metamorphic rocks; *6* medium- to high-grade Barrovian series from Les Sables d'Olonne and Ile d'Yeu; *7* high-grade metamorphic rocks; *8* Ordovician orthogneisses; *9* blueschists; *10* eclogites

tile deformation during a blueschist-facies metamorphism. Structural studies (Quinquis et al. 1978; Quinquis and Choukroune 1981) show that the deformation results from intense north- to northwest-directed shearing, believed to be related to the obduction of oceanic crust towards the north. Lithological and petrological data distinguish two main zones (Carpenter 1976):

a) The higher-grade zone is defined by the presence of garnet in metabasites. Critical parageneses are garnet + omphacite/chloromelanite + glaucophane + epidote in the metabasites and garnet + chloritoid + glaucophane in the metapelites. Pink to red manganiferous quartzitic layers and nodules occur locally in the micaschists (Kiénast and Triboulet 1973). Deerite-bearing cherts are also known (Dudek and Kiénast 1989).

The P-T conditions of the blueschist-facies event are not well defined but values of the order of 10–12 kbar, 500 °C are generally quoted. The prograde P-T path is recorded by the occurrence of lawsonite pseudomorphs. The development of albite + barroisite + chlorite assemblages in the metabasites and of albite + chlorite + biotite or stilpnomelane associations in the metapelites suggest that the retrograde P-T path was a nearly isothermal decompression.

b) The lower-grade zone is defined by the lack of garnet within metabasites, which show glaucophane/crossite + epidote parageneses. Metapelites exhibit garnet + chloritoid + chlorite assemblages. Black quartzites are frequently interstratified within metapelites. Major uncertainties exist on P-T estimates within this zone (? 6–8 kbar, 400 °C).

Geochemical data on volcanics are inconclusive because most metabasites are not true lavas but result from mixing of sedimentary and volcanic material to various degrees. Nevertheless, Bernard-Griffiths et al. (1986) suggest that volcanics of tholeiitic and alkaline series are present.

Detailed geochronological studies have shown the existence of two events (Peucat 1986). The first one, at about 400–420 Ma, is recorded by three different isotopic systems, namely micaschist Rb-Sr whole rock isochron, Rb-Sr phengite, and U-Pb zircon ages. This event is thought to represent the age of the blueschist metamorphism. A second event, at about 320 Ma, partially reset the Rb-Sr and K-Ar phengite systems and is recorded by the K-Ar and Ar-Ar glaucophane systems. This latter event could be associated with the late thrusting of the nappe.

3.4 Vendée

The Vendée area (Fig. 5), in the southeastern part of the Armorican Massif, consists of several stacked units. The geometry of the stacking is not fully understood but one may distinguish two groups of units.

1. The first group is characterized by subhorizontal or moderately dipping foliations, east-west-stretching lineations, and a top to the west sense of shear (Burg 1981; Cannat and Bouchez 1986; Vauchez et al. 1987). Some north-south-trending stretching lineations are locally known (les Sables d'Olonne). Several units are observed in this group, which are described here from the base to the top.

a) Medium-pressure metamorphism is observed at the base of the sequence (les Sables d'Olonne and Ile d'Yeu). Increasing grade from North to South is documented by the appearance of garnet, biotite, staurolite, kyanite, and finally sillimanite (Ters 1972; Bossière 1988).

b) Fossiliferous Paleozoic sediments outcrop along the coast (Ters 1979; Comble et al. 1985), where they are structurally higher than the metamorphic series, and in the Chantonnay area (Wyns et al. 1988). The sediments are strongly deformed and have undergone very low-grade to low-grade metamorphism. Several granites intrude the Paleozoic sediments, where they develop a contact aureole (La Roche sur Yon) (Stussi 1976).

c) A first thrust nappe consists of strongly deformed ignimbritic volcanic rocks ("porphyroides") and very fine-grained metasediments ("schistes de Saint Gilles").

d) The second thrust nappe (Bois de Cené) comprises metabasic volcanic rocks, serpentinites and metapelitic sediments of unknown age (Lower Paleozoic?), which show an intense ductile deformation in the blueschist facies. This unit is comparable to the blueschists from the Ile de Groix (Anthonioz and Brillanceau 1969; Guiraud et al. 1987).

The age of the main deformation and metamorphic event is not well constrained. Conflicting interpretations result from consideration of paleontological data, which favor an Upper Devonian or Lower Carboniferous age, or from geochronological data, which suggest either a late Proterozoic (i.e., Cadomian) or Devonian age for the medium-grade metamorphism. The U-Pb age of the volcanics (405 ± 5 Ma, Peucat et al. 1986) shows the main deformation event to be Devonian or later.

2. The second group is characterized by steeply dipping foliations and N140-trending stretching lineations. Two superposed units have been defined in this group:

a) The lower unit is made of metabasic volcanic rocks and some granodioritic orthogneisses. Rb-Sr isotopic data on the latter rocks suggest a Lower Paleozoic age for the magmatic protoliths (Vidal 1980). This unit was strongly deformed and metamorphosed in the albite-epidote amphibolite facies. Synkinematic parageneses are albite + barroisite + epidote + chlorite. Estimated P-T conditions are 6–8 kbar, 500 °C (Thiéblemont et al. 1988).

b) The upper unit consists of micaschists, gneisses, serpentinites, and metabasites. These latter rocks were the subject of numerous studies because they occur as eclogite lenses of variable but sometimes large sizes (up to several kilometers long).

Geochemical data (e.g., major-element chemistry and REE patterns) suggest that the basic rocks are remnants of an oceanic crust because they belong to a tholeiitic suite (Godard 1988) of N-MORB type (Montigny and Allègre 1974; Bernard-Griffiths and Cornichet 1985).

Detailed analyses of the P-T history have been published recently (Godard 1988). The HP event developed garnet + omphacite + clinozoisite + kyanite associations in the magnesian bulk-rock compositions and garnet + omphacite + quartz + rutile in the ferrous and ferro-titaniferous bulk-rock compositions. P-T conditions are estimated at about 700 °C (± 100 °C) for pressures higher than 15 kbar. Retrograde transformations within eclogites (e.g., the replacement of omphacite by diopside-plagioclase symplectites, the breakdown of phengite, etc.) are related to near-isothermal decompression.

U-Pb data on zircons from two eclogite samples (Peucat et al. 1982) give an upper intercept age of 1297 ± 60 Ma and a lower intercept age of $436 + 15 - 12$ Ma. The upper intercept was originally thought to represent the age of the magmatism (i.e., of the oceanic crust) but this interpretation must now be abandoned. The upper intercept probably results from a contamination of the studied samples by crustal zircons (Paquette 1987). The lower intercept age is interpreted as the age of the eclogite-facies metamorphism.

In contrast to the eclogites, little information is available on the micaschists and gneisses. In the northern part of the upper unit, these rocks were strongly deformed in the upper amphibolite facies, as shown by the occurrence of sillimanite and migmatites (Lasnier 1968). Recent investigations in the southern part of the upper unit (Godard and Kiénast 1991) have uncovered relicts of the pre-eclogite facies history within the gneisses. The latter were migmatitic paragneisses and show incomplete transformations during the HP event (e.g., cordierite pseudomorphosed by garnet + kyanite + quartz and sillimanite replaced by kyanite).

Upper Carboniferous sediments nearby contain rare pebbles of eclogites. We thus have direct evidence in favor of a strong erosion of the Eo-Hercynian belt before the Upper Carboniferous (Mathieu 1937; Godard 1988).

4 Summary and Discussion

A synthetic view of the tectonic history of the South-Armorican domain is not easy to draw. This results from the fact that large offsets along late transcurrent shear zones (more than 200 km along the southern branch of the South-Armorican shear zone according to Jegouzo and Rossello 1988) have obliterated primary continuities between nappe stacks. Examples are (1) the Champtoceaux nappe, which has no equivalent to the south of the southern branch of the South-Armorican shear zone and (2) the Baie d'Audierne area, where peculiar lithologies are present which are not found in other sectors of the Armorican massif.

Early attempts to describe the tectonic history of the South-Armorican domain compared it to a paired belt (Carpenter and Civetta 1976; Cogné 1976; Audren and Lefort 1977; Peucat et al. 1978; Lefort 1979; Lefort et al. 1982). This idea must be abandoned now for several reasons. The South-Armorican domain clearly results from the progressive stacking of crustal units during a continous period of continent-continent or continent-arc collision (e.g. Jones 1991). This history is mainly recorded by the metamorphic history, i.e., an early HP event followed by thermal reequilibration during uplift which occurred before the transcurrent movements along major ductile shear zones. As a result, any interpretation of the Eo-Hercynian events in the South-Armorican domain is critically dependent on two points: the significance of the HP event and its age.

4.1 The Significance of the HP Event

The HP metamorphism results from the burial of oceanic or continental crust to great depths in subduction or collision zones. In this respect, two cases must be distinguished:

First, the eclogite-facies metamorphism in the Leon and Vendée areas and the blueschist-facies metamorphism in the Ile de Groix and Bois de Cené show that oceanic crust was buried to depths of the order of 40–50 km (Léon and Vendée) or around 30 km (Groix and Bois de Cené). Whether this occurred in a subduction zone or during a continental collision cannot be determined.

Second, the eclogite-facies metamorphism in the lower unit of the Champtoceaux nappe testify that a piece of continental crust was buried to depths of the order of 60 km during a continental collision.

In addition, the HP metamorphism can be used to define the spatial extent of the thickened domain during the Eo-Hercynian events. Significant thickening occurred in the Léon area and in the South-Armorican domain. That the Léon area was part of the South-Armorican domain has been argued by Balé and Brun (1986) and reconstructs large displacements along the transcurrent shear zones. In the South-Armorican domain sensu stricto, the northern limit of the thickened domain, i.e., the Nort-sur-Erdre fault defines the main suture zone of the Hercynian belt. This is in agreement with the occurrence of mantle-derived garnet peridotites along this fault (Marchand 1988).

4.2 The Age of the HP Event

The age of the main deformational event can be constrained in three ways, bearing in mind that it could be diachronous through the belt.

First, the deformation is clearly later than the age of the protolith. Geochronological data show that it is later than about 480 Ma in the higher-grade units of zone 5 and zone 4 or even 400 Ma in the lower-grade units of zone 4. Paleontological data are available only for the lower-grade units in zone 4. They suggest that the main deformational event is later than the Middle Devonian in the Chantonnay area (Wyns et al. 1988) and the Tournaisean in the Vendée littorale area (Colchen and Poncet 1987).

Second, the Eo-Hercynian events predate the granitic intrusions, which develop contact aureoles. When this contact metamorphism is observed in rocks which were previously submitted to a high-grade metamorphism, it shows that a major episode of cooling occurred before granite emplacement. Such an example is observed in the Baie d'Audierne (Fig. 3), where the Pors-Poulhan granite is dated at 345 ± 8 Ma (Peucat 1973).

Third, geochronological data can be used to define the age of the HP event itself. U-Pb zircon ages are the most reliable data but their interpretation is rarely straightforward. The main results of the isotopic studies are as follows.

- The eclogite-facies metamorphism in the Vendée and Léon area could have occurred at about

440 Ma (Peucat et al. 1982; Paquette et al. 1987) provided that lower intercept ages represent episodic loss of radiogenic Pb during the HP metamorphism.
- The blueschist-facies metamorphism in the Groix and Bois de Cené unit is dated at about 400–420 Ma (Peucat 1986).
- The eclogite-facies metamorphism in the lower unit of the Champtoceaux nappe occurred at or slightly before 360 Ma (Paquette 1987).
- The granulite-facies metamorphism in the Baie d'Audierne is dated at about 380 Ma (Peucat et al. 1982).

References

Anthonioz PM, Brillanceau A (1969) Introduction à la géologie de la région de Bois-de-Cené (Vendée): un nouveau jalon du métamorphisme de haute-pression dans le Massif armoricain. C R Acad Sci Paris D 269: 1050–1052

Audren C (1974) Les schistes cristallins de la Vilaine (Bretagne méridionale). Bull Soc Géol Minéral Bretagne C VI: 1–4

Audren C (1987) Evolution structurale de la Bretagne méridionale au Paléozoique. Mém Soc Géol Minéral Bretagne 31: 1–361

Audren C, Lefort JP (1977) Géologie du plateau continental Sud-armoricain entre les Iles Glénan et de Noirmoutier. Implications géodynamiques. Bull Soc Géol Fr (7) XIX: 395–404

Balé P, Brun JP (1986) Les complexes métamorphiques du Léon (NW Bretagne): un segment du domaine éohercynien sud-armoricain translaté au Dévonien. Bull Soc Géol Fr (8) II: 471–477

Ballèvre M, Marchand J (1991) Zonation du métamorphisme éclogitique dans la nappe de Champtoceaux (Massif armoricain, France). C R Acad Sci Paris (II) 312: 705–711

Ballèvre M, Kiénast JR, Paquette JL (1987) Le métamorphisme éclogitique dans la nappe hercynienne de Champtoceaux (Massif armoricain). CR Acad Sci Paris (II) 305: 127–131

Ballèvre M, Kiénast JR, Pinardon JL, Vuichard JP (1989) Reversal of Fe-Mg partitioning between garnet and staurolite in eclogite-facies metapelites from the Champtoceaux nappe (Brittany, France). J Petrol 30: 1321–1349

Barbaroux L, Barbaroux-Maurissen C, Bousquet B, Cavet P, Marchand J, Margerel JP, Sellier D (1983) Notice explicative de la feuille Nort-sur-Erdre à 1/50000. Bur Rech Geol Min, Orléans, 48 pp

Bayer R, Hirn A (1987) Données géophysiques sur la structure profonde de la croûte hercynienne dans l'arc ibéro-armoricain et le Massif central francais. Bull Soc Géol Fr (8) III: 561–574

Bernard-Griffiths J, Cornichet J (1985) Origin of eclogites from South Brittany, France: a Sm-Nd isotopic and REE study. Chem Geol 52: 185–201

Bernard-Griffiths J, Carpenter MSN, Peucat JJ, Jahn BM (1986) Geochemical and isotopic characteristics of blueschist facies rocks from the Ile de Groix, Armorican Massif (northwest France). Lithos 19: 235–253

Bossière G (1988) Evolutions chimico-minéralogiques du grenat et de la muscovite au voisinage de l'isograde biotite-staurotide dans un métamorphisme prograde de type barrovien: un exemple en Vendée littorale (Massif armoricain). C R Acad Sci Paris (II) 306: 135–140

Brown M (1983) The petrogenesis of some migmatites from the Presqu'Ile de Rhuys, Southern Brittany, France. In: Atherton MP, Gribble CD (eds) Migmatites, melting and metamorphism. Shira Publications, Nantwich, pp 174–200

Burg JP (1981) Tectonique tangentielle hercynienne en Vendée littorale: signification des linéations d'étirement E-W dans les porphyroides à foliation horizontale. C R Acad Sci Paris (II) 293: 849–854

Cabanis B, Godard G (1987) Les éclogites du Pays de Léon (Nord-Ouest du Massif armoricain): étude pétrologique et géochimique, implications géodynamiques. Bull Soc Géol Fr (8) III: 1133–1142

Cabanis B, Peucat JJ, Michot J, Deutsch S (1979) Remise en cause de l'existence d'un socle orthogneissique antécambrien dans le pays de Léon (domaine nord-armoricain): étude géochronologique par les méthodes Rb-Sr et U-Pb des orthogneiss de Tréglonou et Plounevez-Lochrist. Bull Bur Rech Geol Min (I) 1979: 357–364

Cannat M, Bouchez JL (1986) Linéations N-S et E-W en Vendée littorale (Massif armoricain). Episodes tangentiels successifs éo-hercyniens en France occidentale. Bull Soc Géol Fr (8) II: 299–310

Carlier G (1978) Etude des associations réationnelles à structure coronitique dans les métagabbros de la région nantaise. Thèse Doctorat 3ème cycle, Univ. Paris VI, 252 pp

Carpenter MSN (1976) A petrogenetic study of glaucophane schists and associated rocks from the Ile de Groix, Brittany, France. PhD Thesis, Oxford Univ., 271 pp

Carpenter MSN, Civetta L (1976) Hercynian high-pressure/low-temperature metamorphism in the Ile de Groix blueschists. Nature 262: 276–277

Cogné J (1974) Le Massif Armoricain. In: Debelmas J (ed) Géologie de la France, vol I. Doin, Paris, pp 105–161

Cogné J (1976) La chaine hercynienne ouest-européenne correspond-elle à un orogène par collision? Propositions pour une interprétation géodynamique globale. In: Géologie de l'Himalaya, Colloq Int CNRS 268: 111–129

Cogné J (1988) The Precambrian in the metamorphic terrains of the Ligerian domain and the problem of Precambrian geodynamic evolution in the mobile Variscan belt of Western France. In: Zoubek V (ed) Precambrian in younger fold belts: European variscides, the Carpathians and Balkans. John Wiley, Chichester, pp 529–552

Colchen M, Poncet D (1987) Sur l'âge post-Tournaisien de la tectonique hercynienne synschisteuse à Brétignolles-sur-Mer, Vendée littorale, Massif armoricain. C R Acad Sci Paris (II) 305: 1255–1258

Comble C, De Wewer P, Ters M, Weyant M (1985) Découverte de Conodontes et de Radiolaires d'age Tournaisien dans les schistes bariolés de Brétignolles sur Mer (Vendée). C R Acad Sci Paris (II) 300: 899–904

Deutsch S, Chauris L (1965) Age de quelques formations cristallophylliennes et granitiques du Pays de Léon (Finistère). C R Acad Sci Paris (D) 260: 615–617

Dudek K, Kiénast JR (1989) Deerite from Ile de Groix, Brittany, France. Mineral Mag 53: 603–612

Gil Ibarguchi JL, Ortega Girones E (1985) Petrology, structure and geotectonic implications of glaucophane-bearing eclogites and related rocks from the Malpica-Tuy unit, Galicia, northwest Spain. Chem Geol 50: 145–162

Godard G (1988) Petrology of some eclogites in the Hercynides: the eclogites from the Southern Armorican massif,

France. In: Smith DC (ed) Eclogites and eclogite-facies rocks. Elsevier, Amsterdam, pp 451–519

Godard G, Kiénast JR (1991) Réactions métamorphiques d'éclogitisation des gneiss encaissant les éclogites de Vendée. Bull Liaison Soc Fra Minéral Cristal 1991 (2): 37 (résumé)

Godard G, Kiénast JR, Lasnier B (1981) Retromorphic development of glaucophane in some eclogites, East of Nantes (Massif armoricain, France). Contrib Mineral Petrol 78: 126–135

Guineberteau B (1984) Le massif granitique de Mortagne sue Sèvre (Vendée). Structure, gravimétrie, mise en place, distribution de U-Th-K. Thèse Doctorat 3ème cycle, Nantes Univ., 183 pp

Guiraud M, Burg JP, Powell R (1987) Evidence for a Variscan suture zone in the Vendée, France: a petrological study of blueschist facies rocks from Bois de Cené. J Metamorphic Geol 5: 225–237

Jegouzo P, Rossello EA (1988) La Branche Nord du Cisaillement Sud-Armoricain (France): un essai d'évaluation du déplacement par l'analyse des mylonites. C R Acad Sci Paris (II) 307: 1825–1831

Jones KA (1991) Palaeozoic continental margin tectonics in southern Armorica. J Geol Soc Lond 148: 55–64

Jones KA, Brown M (1989) The metamorphic evolution of the Southern Brittany migmatite belt, France. In: Daly JS, Cliff RA, Yardley BWD (eds) Evolution of the metamorphic belts. Geol Soc London Spec Publ 43, pp 501–505

Jones KA, Brown M (1990) High-temperature "clockwise" P-T paths and melting in the development of regional migmatites: an example from southern Brittany, France. J Metamorph Geol 8: 551–578

Kiénast JR, Triboulet C (1973) Sur la piémontite de l'île de Groix (Morbihan, France). CR Acad Sci Paris (D) 276: 1377–1379

Lagarde JL (1978) La déformation des roches dans les domaines à schistosité subhorizontale. Applications à la nappe du Canigou-Roc de France (Pyrénées orientales) et au complexe cristallophyllien de Champtoceaux (Massif Armoricain). Thèse Doctorat 3ème cycle, Univ. Rennes, 164 pp

Lasnier B (1968) Le massif de péridotites serpentinisées à amas d'écologites de Sem-en-Donges et ses annexes (Loire-Atlantique). Bull Soc Sci Nat Ouest Fr 65: 3–17

Lasnier B (1970) Le métamorphisme régional des gabbros d'après la littérature internationale. Etude préliminaire des gabbros coronitiques du Massif Armoricain et du Massif des Maures (France). Thèse Doctorat 3ème cycle, Univ. Nantes, 295 pp

Lasnier B (1974) Origine du corindon secondaire dans les gabbros, norites et pyroxénolites à spinelle du Pônt de Louen, Bretagne méridionale, France. Bull Soc Géol Minéral Bretagne (C) VI: 109–130

Lasnier B, Leyreloup A, Marchand J (1973) Découverte d'un granite "charnockitique" au sein de gneiss oeillés. Perspectives nouvelles sur l'origine de certaines leptynites du Massif armoricain méridional (France). Contrib Mineral Petrol 41: 131–144

Le Corre C, Balé P, Georget Y (1989) Le Léon: un domaine exotique au NW de la chaîne varisque armoricaine (France). Geodin Acta 3: 57–71

Lefort JP (1979) Iberian-Armorican arc and Hercynian orogeny in Western Europe. Geology 7: 384–388

Lefort JP, Peucat JJ (1974) Le socle antémesozoique submergé à l'Ouest de la Baie d'Audierne (Finistère). CR Acad Sci Paris (D) 279: 635–637

Lefort JP, Audren C, Max MD (1982) The Southern part of the Armorican orogeny: a result of crustal shortening related to reactivation of a pre-hercynian mafic belt during Carboniferous time. Tectonophysics 89: 9–377

Le Métour J (1978) Petrogenesis of migmatites and associated granites in South Brittany. Neues Jahrb. Mineral Mh 1978: 364–376

Le Métour J, Audren C (1977) Relations structurales entre l'orthogneiss ordovicien de Roguédas et son encaissant migmatitique. Conséquences sur l'âge des evenements tectonométamorphiques en Bretagne méridionale. Bull Soc Géol Minéral Bretagne (C) IX: 113–123

Leutwein F, Chauris L, Sonet J, Zimmermann JL (1969) Etudes géochronologiques et géotectoniques dans le Nord-Finistère (Massif armoricain). Sci Terre 14: 331–358

Marchand J (1981) Ecaillage d'un "mélange tectonique" profond: le complexe cristallophyllien de Champtoceaux (Bretagne méridionale). C R Acad Sci Paris (II) 293: 223–228

Marchand J (1982) Une véritable éclogite en Bretagne occidentale (Baie d'Audierne, France). Terra Cognita 2: 312 (Abstr)

Marchand J (1988) Existence de péridotite à grenat jalonnant l'accident majeur de Nord/Erdre (Bretagne Sud). R S T XII: 92 (résumé)

Marchand J, Sellier D, Bossière G (1989) Notice explicative de la feuille Savenay à 1/50000. Bur Rech Geol Min, Orléans, 70 pp

Mathieu G (1937) Recherches géologiques sur les terrains paléozoiques de la région vendéenne. Thèse, Univ. Lille

Matte P, Hirn A (1988) Seismic signature and tectonic cross-section of the Variscan crust in Western France. Tectonics 7: 141–155

Montigny R, Allègre CJ (1974) A la recherche des océans perdus: les éclogites de Vendée, témoins métamorphisés d'une ancienne croûte océanique. CR Acad Sci Paris (D) 279: 534–545

Paquette JL (1987) comportement des systèmes isotopiques U-Pb et Sm-Nd dans le métamorphisme éclogitique. Chaîne hercynienne et chaine alpine. Mem Doc Centre Arm Et Struct Socles, Rennes, 14, 190 pp

Paquette JL, Marchand J, Peucat JJ (1984) Absence de tectonique cadomienne dans le complexe de Champtoceaux (Bretagne méridionale)? Comparaison des systèmes Rb-Sr et U-Pb d'un métagranite. Bull Soc Géol Fr (7) XXVI: 907–912

Paquette JL, Peucat JJ, Bernard-Griffiths J, Marchand J (1985) Evidence for old Precambrian relics shown by U-Pb zircon dating of eclogites and associated rocks in the Hercynian belt of South Brittany, France. Chem Geol 52: 203–216

Paquette JL, Balé P, Ballèvre M, Georget Y (1987) Géochronologie et géochimie des éclogites du Léon: nouvelles contraintes sur l'évolution géodynamique du Nord-Ouest du Massif armoricain. Bull Minér 110: 683–696

Peucat JJ (1973) Les schistes cristallins de la Baie d'Audierne (Massif armoricain, France). Etude pétrographique et structurale. Thèse Doctorat 3ème cycle, Univ. Rennes, 114 pp

Peucat JJ (1983) Géochronologie des roches métamorphiques (Rb-Sr et U-Pb). Exemples choisis au Groenland, en Laponie, dans le Massif armoricain et en Grande-Kabylie. Mém Soc Géol Minéral Bretagne 28, 158 pp

Peucat JJ (1986) Rb-Sr and U-Pb dating of the blueschists of the Ile de Groix. Geol Soc Am Mem 164: 229–238

Peucat JJ, Cogné J (1974) Les schistes cristallins de la Baie d'Audierne (Sud Finistère): un jalon intermédiaire entre la Meseta ibérique et les régions sud-armoricaines. C R Acad Sci Paris (D) 278: 1809–1812

Peucat JJ, Le Métour J, Audren C (1978) Arguments géochronologiques en faveur de l'existence d'une double ceinture métamorphique d'âge siluro-dévonien en Bretagne méridionale. Bull Soc Géol Fr (7) XX: 163–167

Peucat JJ, Vidal P, Godard G, Postaire B (1982) Precambrian U-Pb zircon ages in eclogites and garnet pyroxenites from South-Brittany (France): an old oceanic crust in the West European Hercynian belt? Earth Planet Sci Lett 60: 70–78
Peucat JJ, Paris F, Chalet M (1986) U-Pb zircon dating of volcanic rocks, close to the Silurian-Devonian boundary, from Vendée (Western France). Chem Geol. (Isotope Geo Sec) 59: 133–142
Poulpiquet J de (1988) Etude magnétique et sismique du massif basique et ultrabasique de la Baie d'Audierne (Massif armoricain). Géol Fr 1988: 11–22
Quinquis H, Choukroune P (1981) Les schistes bleus de l'Ile de Groix dans la chaîne hercynienne: implications cinématiques. Bull Soc Géol, Fr (7) XXIII: 409–418
Quinquis H, Audren C, Brun JP, Cobbold P (1978) Intense progressive shear in Ile de Groix blueschists and compatibility with subduction or obduction. Nature 273: 43–45
Stussi JM (1976) Le massif granitique de la Roche-sur-Yon (Vendée). Zonalité pétrographique et différentiation chimico-minéralogique. Sci Terre XX: 219–263
Ters M (1972) Notice de la carte géologique au 1/80 000, feuille Palluau - Ile d'Yeu Bur Rech Geol Min, Orléans, 60 pp
Ters M (1979) Les synclinoriums paléozoîques et le Précambrien sur la facade occidentale du Massif vendéen. stratigraphie et struture. Bull Bur Rech Geol Min (I) 1979: 293–301
Thiéblemont D, Triboulet C, Godard G (1988) Mineralogy, petrology and P-T-t paths of Ca-Na amphibole assemblages, Saint-Martin des Noyers formation, Vendée, France. J Metamorphic Geol 6: 697–715
Triboulet C (1974) Les glaucophanites et roches associées de l'Ile de Groix (France): étude minéralogique et pétrogénétique. Contrib Mineral Petrol 45: 65–90
Triboulet C, Audren C (1985) Continuous reactions between biotite, garnet, staurolite, kyanite-sillimanite-andalusite and P-T-time-deformation path in micaschists from the estuary of the river Vilaine, South Brittany, France. J Metamorphic Geol 3: 91–105
Triboulet C, Audren C (1988) Controls on P-T-t deformation path from amphibole zonation during progressive metamorphism of basic rocks (estuary of the River Vilaine, South Brittany, France). J Metamorphic Geol 6: 117–133
Vauchez A, Maillet D, Sougy J (1987) Strain and deformation mechanism in the Variscan nappes of Vendée, South Brittany, France. J Struct Geol 9: 31–40
Velde B (1972) The origin of some granulite facies rocks from the Baie d'Audierne, Finistère. Bull Soc Géol Minéral Bretagne (C) 4: 91–95
Vidal P (1973) Premières données géochronologiques sur les granites hercyniens du Sud du Massif armoricain. Bull Soc Géol Fr (7) XV: 239–245
Vidal P (1980) L'évolution polyorogénique du Massif armoricain: apport de la géochronologie et de la géochimie isotopique de strontium. Mém Soc Géol Minéral Bretagne 21, 162 pp
Vidal P, Peucat JJ, Lasnier B (1980) Dating of granulites involved in the Hercynian fold-belt of Europe: an example taken from the granulite facies orthogneisses at La Picherais, Southern Armorican Massif, France. Contrib Mineral Petrol 72: 283–289
Wyns R, Lablanche G, Lefavrais-Raymond A (1988) Notice explicative de la feuille Chantonnay à 1/50000. Bur Rech Geol Min, Orléans, 76 pp

Intracontinental Hercynian Events in the Armorican Massif

J. Rolet, F. Gresselin, P. Jegouzo, P. Ledru, R. Wyns

1 Introduction

In the Armorican Massif, at the end of the Devonian, all the oceanic spaces are closed and the collision process has already started.

At this stage (the Breton phase), the Cambro-Devonian epicontinental distensive basins are folded, particularly in the western part of the Armorican Massif.

The deformation occurs in Lower and Upper Carboniferous pull-apart basins and in several shear zones.

The seven zones (Chap. III, B.1) are more or less affected by the Hercynian events. The first (Zone 1) or Icartian craton is only cut, during the Devonian, by dolerite dykes. The Cadomian basement (Zone 2) edge is affected by some Hercynian shear zones.

The other zones are more involved in the Hercynian tectonic and metamorphic process. They are described below in more detail.

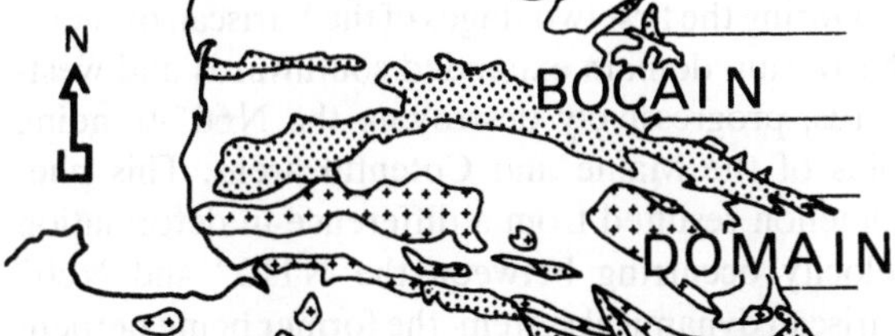

Fig. 1. Location map of the Hercynian domains of the NE Armorican Massif. *1* Post-Paleozoic cover; *2* Variscan granites; *3* Paleozoic cover; *4* Cadomian granites; *5* Brioverian

2 Hercynian Events in Normandy

F. Gresselin

As a result of detailed field analysis (Doré et al. 1977, 1982, 1986; Juignet et al. 1984; Dissler 1987; Gresselin 1990), the northeastern part of the Armorican Massif can now be regarded as composed of three main Variscan units: the Cotentin Peninsula, the Bocain domain, and the Maine area (Fig. 1). These were involved in polyphase Carboniferous deformation which, in most of the region, took place under very low-grade metamorphic conditions.

2.1 The Cotentin Peninsula

The Cotentin Peninsula provides evidence for major Variscan tectonogenesis during three principal events Dc 1, Dc 2, and Dc 3, respectively.

Dc 1 results from a Variscan remobilization of Cambrian normal faults which trend WSW-ENE and dip steeply northward (Gresselin 1990). This gave rise to a succession of regional open or SSW-facing folds (Fig. 2 A), more or less parallel to the Cambrian regional strike. These folds are locally associated with an intense cleavage (greenschist facies).

Dc 2. South of a line running from Eculleville to Octeville, the Fc 2 folds are upright or steeply inclined, verging SSW, and striking N 110° (Fig. 2 B). North of this zone, the Dc 2 deformation increases in intensity, controlling (1) eastward, the rotation of the Dc 1 structures and the emplacement of the Barfleur granite, and (2) westward, the development of a shear zone.

J. D. Keppie (Ed.): Pre-Mesozoic Geology in France and Related Areas

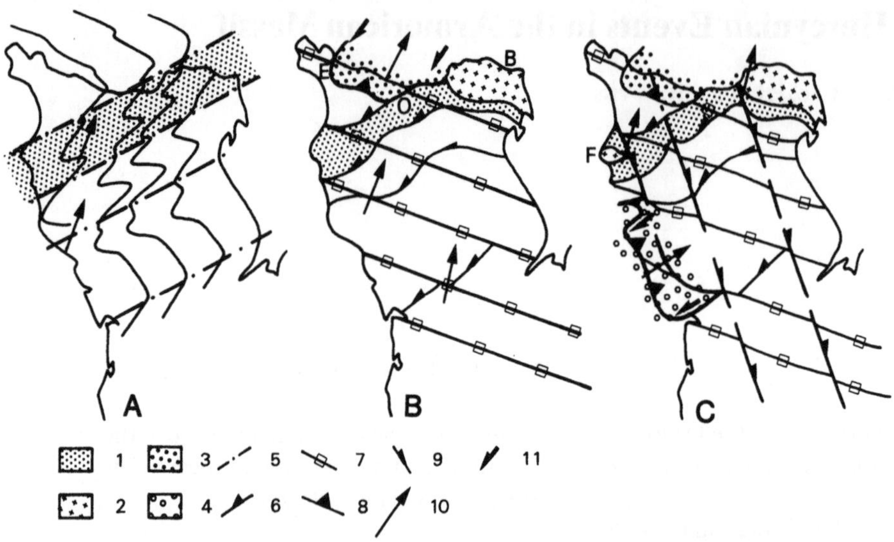

Fig. 2 A–C. A summary of the Variscan events in the Cotentin peninsula. **A** Dc1. **B** Dc2. **C** Dc3. *1* Area of intense Dc1 deformation (formation of granites); *2* Variscan granites; *3* area of intense Dc2 deformation (greenschist facies); *4* area of intense Dc3 deformation; *5* principal Cambrian faults; *6* sinistral thrust faults (Dc1); *7* folds or thrusts of Dc2 age; *8* principal thrust faults; *9* dextral thrust faults of Dc3 age; *10* stretching lineation; *11* sense of displacement. *B* Barfleur granite; *E* Eculleville; *F* Flamanville granite; *O* Octeville

The shear zone, composed of a thick multilayered Brioverian (Infracambrian, Vendian) and Cambrian pile, is limited southward by a WNW-ESE décollement surface (Fig. 2 B), westward by a dextral wrench fault (the N 30° Eculleville lateral ramp) and eastward by a N 60°-trending sinistral thrust (the Octeville lateral ramp). Within the thrust unit, the Sc2 foliation, oriented N 110° except close to the lateral ramps, undergoes an important typological variation from top to base of the sheet, leading us to distinguish three main structural units: (1) a zone of upright or steeply inclined folds, associated with relatively weak deformation; (2) a zone of steeply or gently inclined folds verging SSW; (3) a zone of isoclinal sheath-like folds with axes sub-parallel to the regional stretching lineation (N 20°). According to the analysis of the pyrite pressure fringes, the bulk of the deformation appears here to have been produced by simple shear (γ exceeding 4).

Dc3. In the Cotentin peninsula, the last stage of the Variscan evolution gave rise to the general development of N 160° striking faults which northward controlled the intrusion of the Flamanville granite and southward led to the development of a WSW-verging thrust cell (Fig. 2 C).

2.2 The Bocain Domain

Away from the Cambrian grabens of the Maine and the Cotentin, the Paleozoic cover was folded perpendicular to the Variscan strain field. The Bocain domain is also composed of a succession of regional upright to steeply inclined folds, which strike N 110° and are associated with a weak cleavage. This latter event is developed throughout the domain but intensifies in narrow shear zones affecting the north flank of the main regional synclines. Within these SSW-verging zones, deformed Cambrian conglomerates show a moderately intense stretching lineation (N 20°) and rotation about Y.

During the first two stages of the Variscan orogeny, the Bocain domain expanded southwards and westwards, progressively destroying the N 60°-trending folds of the Maine and Cotentin areas. This phenomenon resulted from a difference in deformation velocity occurring between the N 110° and N 60° Variscan dynamical system, the former being perpendicular to the strain field, in contrast to the latter.

2.3 The Maine Area

As in the Cotentin Peninsula, the first stage of the Variscan evolution of the Maine (Dm 1) corresponded to the mobilization of a succession of Cambrian normal faults, leading to the formation of open to asymmetrical SSW-facing folds and to the development of

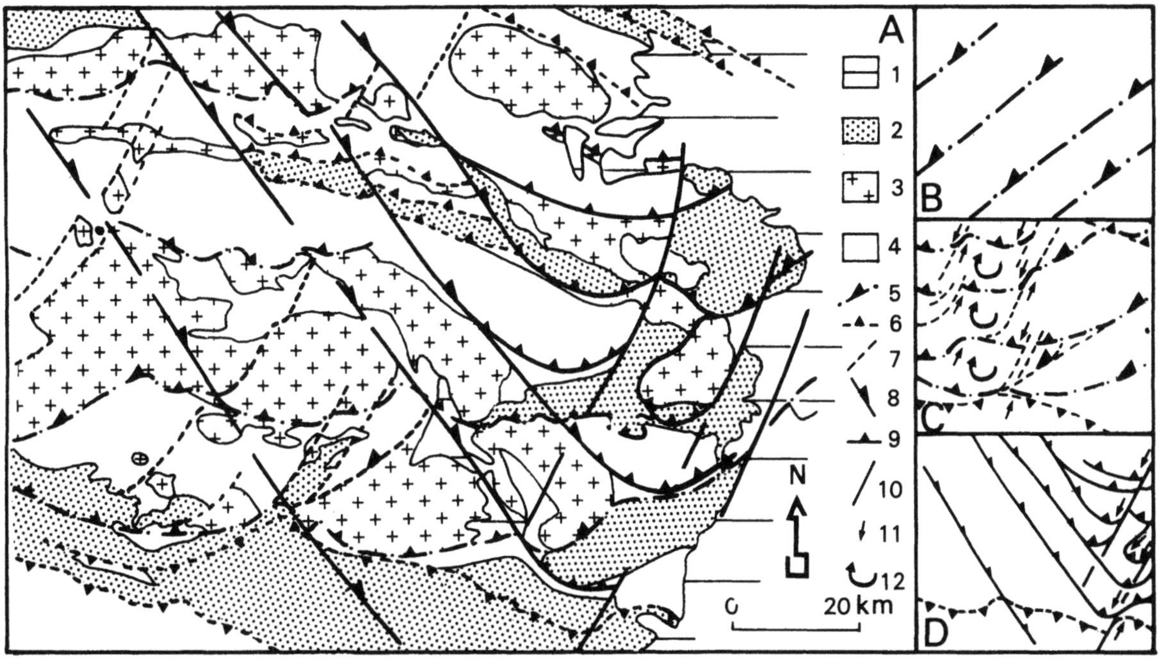

Fig. 3 A–D. Sketch map showing the Variscan evolution of the Maine area. **A** Structural map. **B** The Dm 1 evolution. **C** One stage of the Dm 2 evolution. **D** One stage of the Dm 3 evolution. *1* Post-Paleozoic cover; *2* Paleozoic cover; *3* Cadomian granites; *4* Brioverian turbidites; *5* sinistral thrust faults of Dm 1 age; *6* thrust faults of Dm 2 age; *7* strike-slip faults of Dm 2 age; *8* dextral thrust faults of Dm 3 age; *9* thrust faults of Dm 3 age; *10* strike-slip faults of Dm 3 age; *11* sense of displacement; *12* sense of rotation of the Dm 1 units during the Dm 2 events

sinistral thrust faults (Fig. 3 A, B). These N 60 °-trending faults were progressively folded (1) northward by the southern expression of the Bocain system, and (2) southward by the development of a succession of NNW-verging thrusts (Dm 2 events; Fig. 3 C). All these structures were then refolded by the expression of small SSW-verging thrust cells (Dm 3). As in the north, these cells are bounded by three main faults which are: (1) a basal décollement surface, generally orientated N 100 ° and dipping steeply or gently northward; (2) a western dextral lateral ramp, striking N 150 ° for the oldest cells and N 20 ° for the others; (3) an eastern sinistral lateral ramp, trending N 20 ° irrespective of the age of the cell (Fig. 3 A and D). Four successive generations of cells have been identified, all developed under chlorite or biotite facies metamorphic conditions (Gresselin 1990).

2.4 Conclusions

In the northeastern part of the Armorican Massif, the Variscan deformation was polyphase. It began in the Late Devonian (Gresselin 1990) when the remobilization of ancient Cambrian faults occurred. In the Maine area, these N 70 °-striking faults were folded in the Early Carboniferous by both the development of NNE verging thrusts (south Maine) and the expression of N 110 °-striking folds (north Maine). In the Cotentin peninsula, only the N 110 °-striking folds appeared. In this area, the age of the second Variscan phase is still unknown, although it was contemporaneous with the emplacement of the Barfleur granite. The third Variscan phase began in the Maine during the Lower Carboniferous, and its age is also unknown to the north. This stage corresponded to the development of thrust units, generally verging SSW. Thus, during the Variscan events, the NE Armorican Massif was only affected by thrust movements verging NNE or SSW.

3 Hercynian Events in Central Brittany

J. Rolet (translated by S. Bogdanoff
and M. C. Leroux)

This area comprises the regions which have undergone more or less intensive Hercynian tectonic events, and have accurately dated Paleozoic rocks, such as the Western Central Armorican zone, the Rennes block, whose evolution is strictly Hercynian, but also Mancellia, the Bocage Normand, and the Cotentin (Sect. 2, this Chap.) all with a Cadomian heritage (Fig. 4).

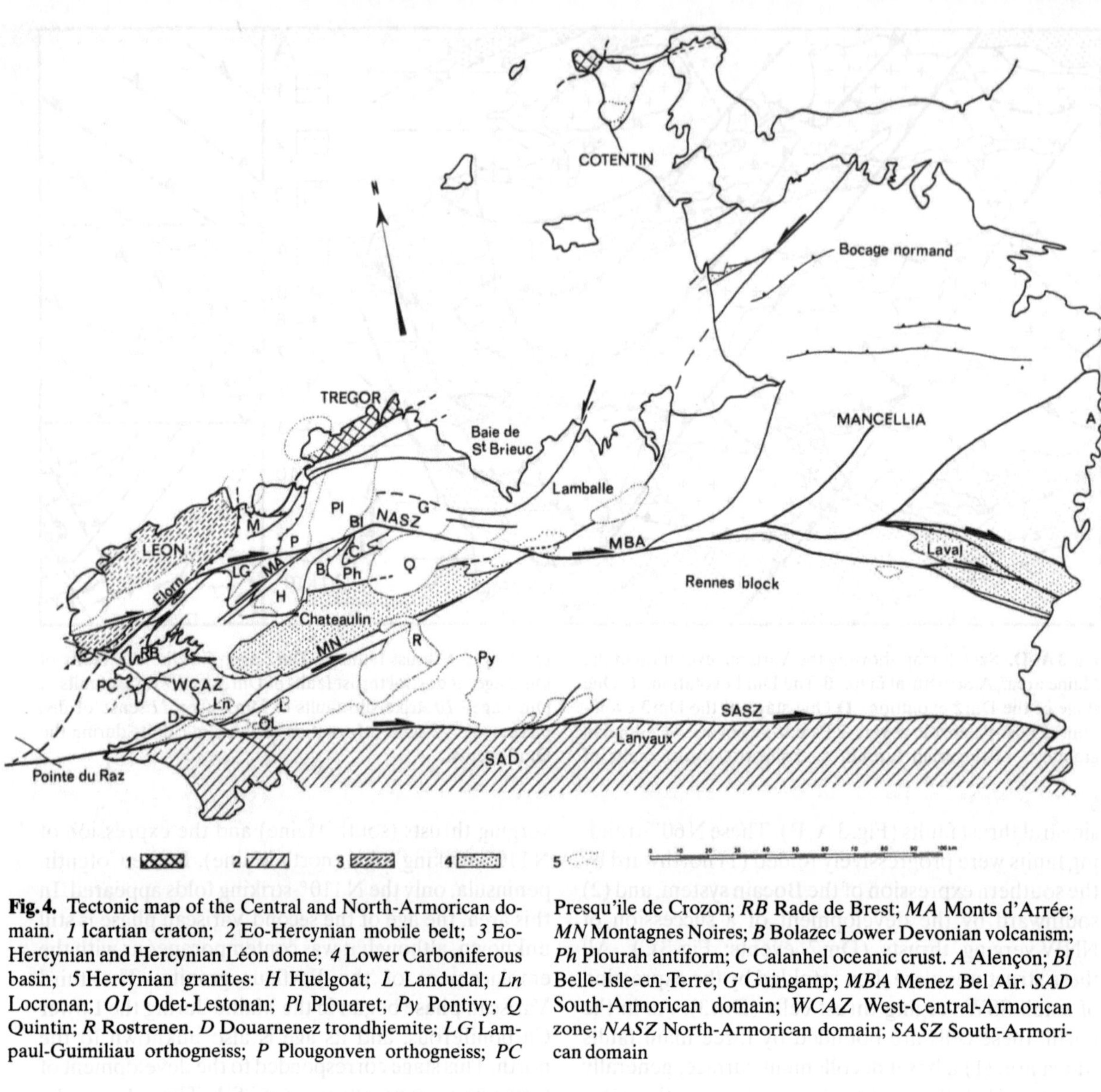

Fig. 4. Tectonic map of the Central and North-Armorican domain. *1* Icartian craton; *2* Eo-Hercynian mobile belt; *3* Eo-Hercynian and Hercynian Léon dome; *4* Lower Carboniferous basin; *5* Hercynian granites: *H* Huelgoat; *L* Landudal; *Ln* Locronan; *OL* Odet-Lestonan; *Pl* Plouaret; *Py* Pontivy; *Q* Quintin; *R* Rostrenen. *D* Douarnenez trondhjemite; *LG* Lampaul-Guimiliau orthogneiss; *P* Plougonven orthogneiss; *PC* Presqu'ile de Crozon; *RB* Rade de Brest; *MA* Monts d'Arrée; *MN* Montagnes Noires; *B* Bolazec Lower Devonian volcanism; *Ph* Plourah antiform; *C* Calanhel oceanic crust. *A* Alençon; *Bl* Belle-Isle-en-Terre; *G* Guingamp; *MBA* Menez Bel Air. *SAD* South-Armorican domain; *WCAZ* West-Central-Armorican zone; *NASZ* North-Armorican domain; *SASZ* South-Armorican domain

Fig. 5. Thin-skin tectonics (flat and ramps) in Ordovician Armorican sandstone of the Crozon Peninsula

The Medio-North-Armorican area is limited in the north by the Trégor and Baie de Saint-Brieuc thrusts, and by the Hercynian crystalline dome of Léon. These thrusts, sometimes of considerable extent (Calanhel, for example) can be attributed to the Breton episode (Upper Devonian). Towards the south, this area is separated from the southern area by a plethora of vertical ductile faults, tightening towards the west (SASZ), cutting the Eo-Hercynian thrusts again during the Carboniferous.

A strong contrast is apparent between the deformation intensities, metamorphism and granitic plutonism in the eastern and western parts.

Two large subdivisions can be recognized:

- an Eastern subarea overlain by thin and often incomplete Paleozoic formations, moderately deformed during the Hercynian, and slightly modified by the granitic intrusion with the exception of its borders;
- a Western area (Crozon, Rade de Brest, Morlaix basin, Monts d'Arrée, Montagnes Noires) extremely mobile during the whole of the Devonian and Carboniferous times deformed by tectonic thrusts leading to an overthickening of thousands of meters, accompanied by many granitic intrusions (Barrière et al. 1983).

3.1 Deformation Regimes and Structures

The two subareas defined above are also characterized by their deformation regimes. The whole of the eastern area is affected by simple deformation in dextral transcurrent shear while the western part shows complex deformation where thrusts and strike slip-faults are superposed.

3.1.1 Thrusts

The Ordovician to Devonian formations in the Western area are affected by early thrusts (phase Bretonne) characterizing thin-skin tectonics accompanied by epizonal metamorphism.

Numerous characteristic flats and ramps (Fig. 5) can be observed in the Grès Armoricain Formation, in the Presqu'île de Crozon. These thrusts accompany schistose folds that show very long limbs, steepening slowly to the NW. Their short limbs are rarely inverted and very often sheared to the SE. A slight stretching in the NW-SE to E-W direction sometimes accompanies these thrusts.

On a regional scale, Breton tectonics are most apparent:

- bringing together of the North Crozon and South Crozon blocks accompanying the rock scaling of the Devonian sediments of the Rade de Brest showing many intruded dykes (Rolet and Thonon 1978);
- the probable thrusting of Silurian formations of the Monts d'Arrée over the volcanic Lower Devonian of Bolazec (spilites), which itself is allochtonous on the phtanitic Brioverian sediments of Lamballe on the Plourah antiform nappe (Rolet and Deunff 1982) (Fig. 4);
- the thrusting of Siluro-Devonian formations (SE of the Elorn river and S of Morlaix) over the orthogneisses of Lampaul-Guimillau and Plougonven.

A part of the Trégor slices and the Saint-Brieuc Brioverian formations towards the S-SE may be linked to this episode.

Along narrow corridors (Monts d'Arrée, Montagnes Noires), this thrusting episode seems to turn into a transcurrent dextral shearing (Darboux 1981, Darboux and Le Gall 1988; Fig. 6). L1 lineations and P1 folds are rolled round subvertical planes striking ENE-WSW, on which P2 folds with nearly vertical axes develop.

During this pre-Carboniferous epizonal deformation, the intrusion of aluminous granites of crustal origin and Huelgoat type (342±6 Ma; Castaing et al. 1987) takes place.

The transcurrent shears do not necessarily mean that the stress field changes. On the contrary, these ductile and vertical corridors may well correspond to lateral or oblique ramps, formed during the last stages of the thrusting towards the west (Figs. 1 and 2; Gresselin Chap. III, B.4.2).

3.1.2 Transcurrent Shears

In the eastern area, the deformation is not so complex and is less pronounced. The major structures correspond to upright folds, the axes of which strike approximately E-W. A regional, generally vertical, cleavage is associated with them and a stretching lineation parallels the fold axes. This geometry and the close links that exist, particularly at the southern border of the Rennes block, between the finite deformation ellipsoid shape and the dextral movement of the South-Armorican shear zone induced numerous authors to propose a model of dextral deformational regime (Berthe et al. 1979; Jegouzo 1980; Yesou and Rolet 1990; Rolet and Yesou 1990) as a characteristic of the whole of Central Brittany situated between the North- and South-Armorican shear zones (Le Corre 1978; Barbotin 1987; Fig. 7).

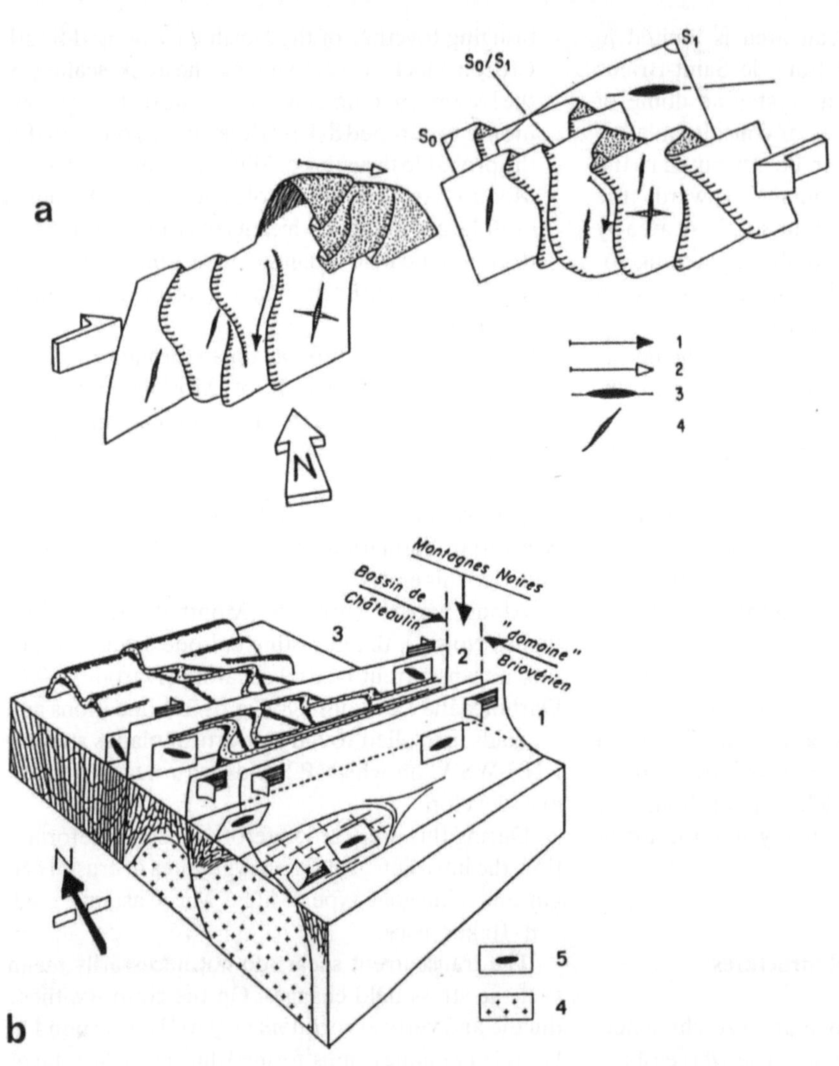

Fig. 6 a, b. Dextral transcurrent shear in the Montagnes Noires. (Darboux et Le Gall 1988). **a** Syn-cleavage curvilinear fold axes in the Siluro-Devonian rock of the Montagnes Noires. *1* First fold axes; *2* second fold axis; *3* extension direction; (X axis of the finite strain ellipsoid); *4* tension gash. **b** Regional structural model. *1* Late Proterozoic rocks (Brioverian); *2* Montagnes Noires: Paleozoic rocks (Lower Ordovician to Middle Devonian); *3* Chateaulin Basin: Lower Carboniferous rocks (Tournaisian to Upper Visean); *4* Carboniferous syntectonic Landudal leucogranite; *5* extension direction

Nevertheless, recall the fact that in the western area the transcurrent shear zones come before significant thrusts.

The influence of the transcurrent shear zones at the border of the South-Armorican shear zone and of the associated leucocratic granites decrease progressively towards the north as one passes from slaty cleavage to fracture cleavage (Le Corre 1978).

The deformation affecting the Carboniferous basins is often regionally parallel to the previous structures (Breton phase); this is particularly true of cleavage trajectories (Barbotin 1987; Fig. 7). This parallelism, a consequence of the accordance between structures, often led workers to infer only one post-Dinantian deformation episode.

Nevertheless, this parallelism is not always perfect. Changes in trend of the structures in the western part of the Chateaulin Basin and other facts, such as the sedimentology of the basins, reorganization paleogeography, geochronology, etc. indicate two separate periods.

3.2 Hercynian (Upper Devonian to Carboniferous) Metamorphic Episodes in Central Brittany

All the Paleozoic rocks of Central Brittany show weak metamorphism. This very low to low grade metamorphism appears to be Hercynian.

In the eastern part (Rennes block and Normandy) the temperature conditions increased very quickly near the northern (Cotentin) and the south (Lanvaux) boundaries. The eastern part of central Brittany appears to be cratonic block, reworked only along its borders (Fig. 8).

In contrast, the western part was largely involved in intense and complex tectono-metamorphic events, reaching catazonal conditions, and crustal melting in

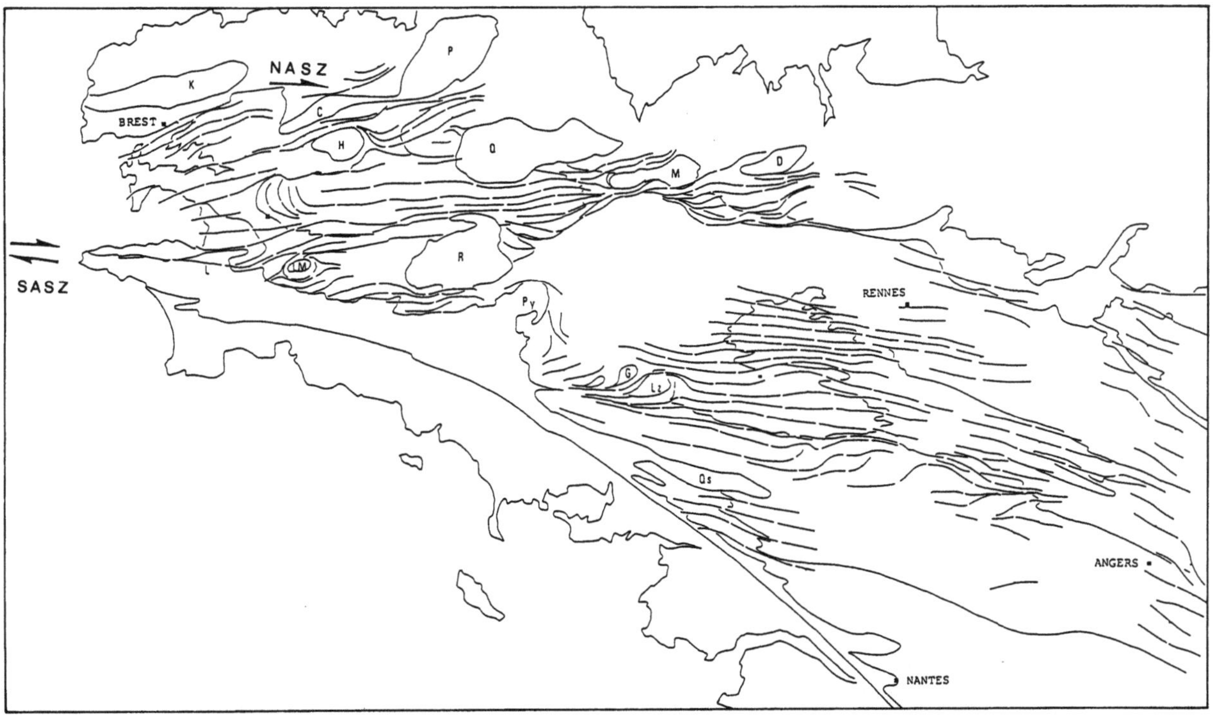

Fig. 7. Hercynian cleavage, regional trajectory (Barbotin 1987). *Hercynian granites:* C Commana; *D* Dinan; *H* Huelgoat; *G* Guéhenno; *K* Kersaint; *L* Locronan; *LM* Le Merdy; *Lz* Lizio; *M* Moncontour; *P* Plouaret; *Py* Pontivy; *Q* Quintin; *Qs* Questembert; *R* Rostrenen

the Léon domain and in the Pointe du Raz-Pontivy area.

Some examples will be described and their relationship with tectonic structures and chronology will be discussed.

3.2.1 Regional Data

1. Metamorphism along the South-Armorican Shear Zone (SASZ, Fig. 8). At the southern boundary of the Central Brittany domain, mineral assemblages show increasing P/T conditions from north to south and from east (South Rennes area) to west (Pointe du Raz-Pontivy area).

The southern Rennes area is characterized by chlorite + muscovite + albite + chloritoid assemblage (Le Corre 1978) that seems to be related to endo-plutonic granites associated with shear zone working.

The western region underwent stronger tectonometamorphic events than the eastern, but is still largely affected by a greenschist facies metamorphism characterized by a widespread development of chloritoid in aluminous Paleozoic metasediments (Martinez and Rolet 1988).

In the southern Rennes block, in contrast to the complex western zone, tectonics and metamorphism appear as a simple and single phase. The age of this metamorphism is the same as that of the associated granites, probably about 345 Ma (Bernard-Griffiths et al. 1985).

2. Pointe du Raz – Pontivy Area. This triangular zone, about 150 km long, is bounded to the north by the Chateaulin basin and to the south by the SASZ.

The metamorphism increases very quickly from north (Montagnes Noires) to south (Quimper-Baud) with very closely spaced isograds marking the progressive appearance of the following parageneses in Paleozoic and Brioverian pelitic rocks:

- quartz + muscovite + chlorite + albite + pyrophyllite + chloritoid;
- quartz + muscovite + biotite + andalusite;
- quartz + muscovite + biotite + staurolite + almandine.

Some very rich aluminous schists provide andalusite hornfels locally ("guerphalite" from the village of Guerphales near Rostrenen) which are quarried for aluminum production.

Index minerals do not always appear in the same order, testifying to lateral variations in pressure and temperature (Hanmer et al. 1982) and to superposed phases.

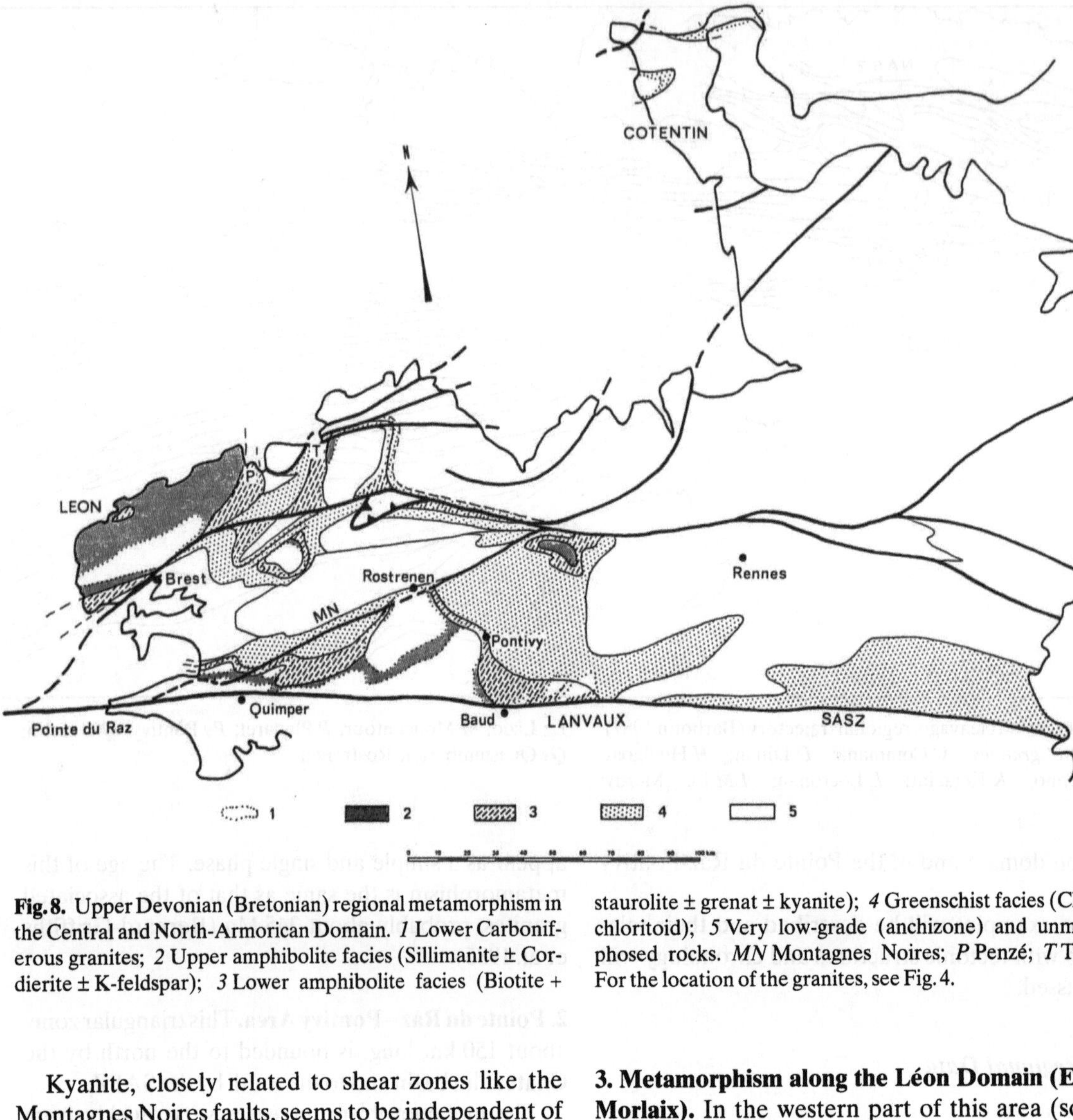

Fig. 8. Upper Devonian (Bretonian) regional metamorphism in the Central and North-Armorican Domain. *1* Lower Carboniferous granites; *2* Upper amphibolite facies (Sillimanite ± Cordierite ± K-feldspar); *3* Lower amphibolite facies (Biotite + staurolite ± grenat ± kyanite); *4* Greenschist facies (Chlorite ± chloritoid); *5* Very low-grade (anchizone) and unmetamorphosed rocks. *MN* Montagnes Noires; *P* Penzé; *T* Treduder. For the location of the granites, see Fig. 4.

Kyanite, closely related to shear zones like the Montagnes Noires faults, seems to be independent of the regional isograds. It can be found in quartz veins, associated with staurolite or andalusite (Barrière et al. 1973) and chloritoid. Kyanite appears later than chloritoid and staurolite, and probably indicates extreme local pressures.

Cordierite as a late high-temperature mineral is associated with biotite, sillimanite, and sometimes almandine in high-grade pelitic gneisses found as enclaves in the north of the Pontivy-Rostrenen granite pluton.

In the North-Quimper area, the earlier zones (chloritoid, biotite, etc.) are cut obliquely by a late sillimanite isograd and by the onset of melting, resulting in anatexis associated with subautochtonous leucogranitic plutons as in Locronan, Landudal, and Odet-Lestonan (Jouvin 1986).

3. Metamorphism along the Léon Domain (Elorn to Morlaix). In the western part of this area (south of Brest), weakly deformed Paleozoic sediments show a general metamorphism increasing from south to north (Paradis 1981; Paradis et al. 1983). Devonian aluminous metasediments are characterized by pyrophyllite development (Melou and Plusquellec 1967).

Three zones are defined:

- kaolinite + chlorite + pyrophyllite (1 or 2 kbar, 280 °C);
- kaolinite + chlorite + illite + pyrophyllite;
- illite + chlorite + chloritoid + albite (2 kbar, 400 °C).

In this area, chloritoid seems to be limited to the Elorn lineament, but in the western area (Sizun-Morlaix) chloritoid is common and probably characterizes a Paleozoic thrust unit on the Brest-Lampaul-Guimiliau-Plougonven orthogneisses.

The different isograds are concentrated in the Léon domain (Cabanis and Fabriès 1972) and show a prograde metamorphism which rises to high grade cordierite, and sillimanite development such as in the Pointe du Raz-Pontivy area.

Local reverse metamorphic isograds are found, for example, near Landivisiau (unpubl.) and in the Plourah antiform (Bril 1975; Rolet and Deunff 1982). In these few examples, weakly deformed sediments appear in a nappe inlier under Paleozoic thrusts.

More frequently, in eastern Léon, antiformal structures expose orthogneisses showing early melting cut by aluminous Carboniferous granites, e.g., near Treduder and Penze. These melting conditions in the sole of the thrusts favor the underthrusting of Léon under Central Brittany (Rolet et al. 1986b).

3.2.2 Conditions of Metamorphism

Apart from Kyanite, which indicates local high pressure, probably along shear zones, all of Central Brittany and especially the western part shows low to medium pressure metamorphic conditions. High grade paragenesis with low pressure minerals like cordierite mark the metamorphic peak, and indicates a progressive pressure decrease and temperature increase as metamorphism progressed (Martinez and Rolet 1988).

The high grade metamorphism and associated melting is related to a Hercynian crustal thickening (Rolet et al. 1986b) events occurring during a hypercollisional process.

3.2.3 Carboniferous Granites and Metamorphism

Isograds are very often narrow and closely, border the Carboniferous granitic plutons. Some geological works tend to relate these events to a thermal contact metamorphism.

The thermal effect of granitic pluton is obvious for the Huelgoat granite (Darboux and Garreau 1976), Quintin and Rostrenen granites (Sagon 1976), or hidden pluton as in Montagnes Noires (Darboux and Le Gall 1988).

Nevertheless, these aluminous granites such as Huelgoat, North Plouaret, Locronan, or Landudal granites are frequently preceded by high grade gneisses. These gneisses, often anatectic, indicate high temperature anomalies preceding granite emplacement.

The granites themselves cut and enclose these gneisses, appearing as the products of temperature anomalies and not the cause.

3.2.4 Tectonics and Metamorphism

In the whole of Central Brittany, anchizonal to epizonal metamorphism is contemporaneous with a regional slaty cleavage. In the western part (Crozon Peninsula and Rade de Brest) asymmetric folds and ramps developed during this epizonal metamorphism.

The thrust episode (probably Bretonian phase) reaches mesozonal, catazonal metamorphic conditions approaching the Léon.

Along transcurrent shear zones like the SASZ, ductile deformation is generally developed under epizonal conditions. Local retrograde metamorphism, as in the Brest gneisses along the Elorn lineament, could be related to such shear zones.

3.2.5 Chronology

Few radiometric data exist for metamorphic minerals. Some micas give an Upper Carboniferous age (about 300 Ma) in Odet-Lestonan leucogranite (Jouvin 1986).

The associated granites are separated in two groups:

- pre-Visean, e.g., Huelgoat and Pontivy;
- post-Visean, e.g., Quintin and Rostrenen.

Radiometric data and geological evidence, e.g., reworked granitic pebbles in the Visean conglomerate of the Chateaulin basin (Rolet 1984), confirm these two granitic episodes and the associated metamorphism.

Some orthogneisses, (Douarnenez tronhjemites, for instance) well known as enclaves in leucogranites (Jouvin 1986), testify to superposed metamorphism in the western part of Central Brittany.

4 The Saint-Georges-sur-Loire Unit

P. Ledru

The Saint-Georges-sur-Loire Synclinorium is located (Fig. 9) between the Central Armorican domain, comprising weakly deformed sediments of Paleozoic age, and the high grade metamorphic terranes of the Variscan belt (Champtoceaux complex). It is constituted from bottom to top of acid and basic volcanic rocks, pyro- and epiclastic rocks. The basic rocks are of Silurian age (400±24 Ma, whole rock Rb/Sr isochron; Carpenter et al. 1982). They have been interpreted as emplaced first in a rift zone, developing then in a volcanic arc system (Carpenter et al. 1982; Cabanis and Godard 1987).

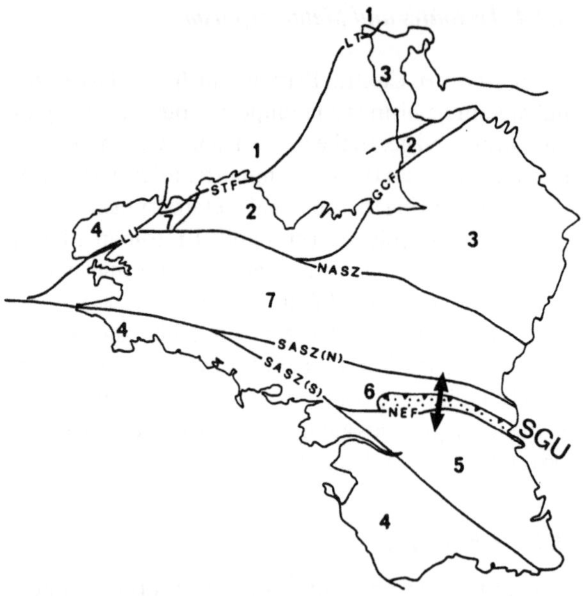

Fig. 9. St Georges-sur-Loire unit map location. Zones *1* to *7*, see legend to Fig. 2, p. 178

The northern margin of the synclinorium is a major thrust zone (Fig. 10) related to northward tangential tectonics (Ledru et al. 1986). The southern limb is affected by a major transcurrent fault, considered as sinistral in an early stage (prior to the Frasnian) corresponding to the opening of the Ancenis Basin (Diot 1980) and then dextral, connected with the South-Armorican shear zone (Fig. 1).

The formations are affected by a very penetrative foliation marked by the crystallization of phengitic micas and chlorite. This initially flat-lying foliation in the thrust zone shows a N-S-trending stretching lineation with northward kinematic criteria.

The analysis of phengitic micas (Table 1) indicates a pressure around 5 kbar for a temperature estimated around 300 °C. This medium-pressure metamorphism is consistent with the amphibole-stilpnomelane synfolial paragenesis discovered in metabasic rocks near the southern margin (Ledru et al. 1986). The amphibole has a crossite composition (Table 2) with a Na_2O content around 7%. These data correspond to the upper limit of the greenschist facies metamorphism, at the limit of the blueschist metamorphism. To summarize, the Saint-Georges-sur-Loire Synclinorium is affected by a LT-MP metamorphism, contemporaneous with the mylonitic foliation related to the thrusting of a rift and volcanic arc system over the Central Armorican domain. This event is dated between 360 and 350 Ma, as it occurred before the Frasnian, and Praguian sediments are involved in the thrusting. It may correspond to the obduction of an arc system during the Medio-Variscan collision.

5 Hercynian Events in the Vendée

R. Wyns (translated by J. Kemp)

5.1 Introduction

The Variscan belt in the southeast of the Armorican Massif is still insufficiently well known to enable a reliable overview of the succession of events to be drawn up. In particular, in the Vendée s.l., neither the ages of the tectonometamorphic events nor that of the final emplacement of the nappes have been determined with any certainty. The until recently accepted attribution of the major deformation in this area to the Devonian (Autran and Cogné 1980) needs to be more carefully considered in the light of the discovery of terranes of Late Paleozoic age affected by the regional metamorphism – Tournaisian in the coastal Vendée (Comble et al. 1985) and Middle Devonian in the Chantonnay Synclinorium (Wyns et al. 1989a).

5.2 Hercynian Intracontinental Deformation

Zones 4 to 6 described below correspond to those defined by Ballèvre et al. (Chap. II, B.12; see Figs. 11 and 12 and Table 3).

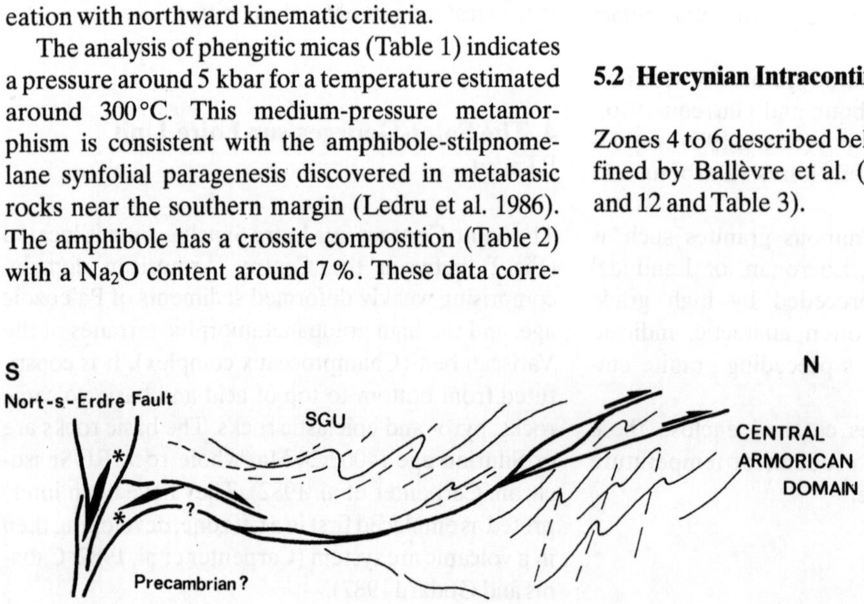

Fig. 10. Schematic cross-section of the St Georges-sur Loire unit. * Metabasic volcanic rocks in the blueschist facies; *SGU* St Georges-sur-Loire unit

Table 1. Micas (phengite): analytical data from metapelites (Retaud SC1 drill core) (BRGM microprobe analyses)

	95	96	97	98
SiO_2	0.5184	0.5193	0.5123	0.5181
FeO	0.0199	0.0176	0.0219	0.0205
Na_2O	0.0000	0.0000	0.0000	0.0000
CaO	0.0000	0.0000	0.0003	0.0000
Al_2O_3	0.2441	0.2371	0.2347	0.2443
MnO	0.0007	0.0006	0.0007	0.0001
MgO	0.0451	0.0445	0.0448	0.0429
K_2O	0.1070	0.1095	0.1069	0.1068
Cr_2O	0.0010	0.0000	0.0000	0.0000
TiO_2	0.0023	0.0029	0.0044	0.0025
H_2O	0.0443	0.0440	0.0436	0.0442
	0.9827	0.9754	0.9698	0.9795
Si	7.0165	7.0809	7.0420	7.0307
Fe	0.2248	0.2003	0.2519	0.2327
Na	0.0004	0.0000	0.0000	0.0000
Ca	0.0000	0.0002	0.0052	0.0000
Al	3.8931	3.8102	3.8022	3.9074
Mn	0.0075	0.0073	0.0084	0.0015
Mg	0.9098	0.9035	0.9183	0.8686
K	1.8478	1.9042	1.8751	1.8485
Cr	0.0102	0.0000	0.0000	0.0000
Ti	0.0229	0.0298	0.0457	0.0252
	13.9330	13.9364	13.9487	13.9146
OH	4.0000	4.0000	4.0000	4.0000
FM	0.2034	0.1868	0.2208	0.2124

5.2.1 *Zone 6 – Between the Northern Branch of the South-Armorican Shear Zone (SASZ) and the Nort-sur-Erdre Fault*

This zone is occupied by the allochthonous St-Georges-sur-Loire unit (SGLU), overthrust northward onto the Murs-Erigné unit (MEU), which itself is thrust onto the Lanvaux (Les-Ponts-de-Cé) axis. The deformation is considered to be post-Praguian and pre-Frasnian (Ledru et al. 1986).

The Bécon granite (BG) cuts structures associated with thrusting, but northward, towards the Lanvaux axis, it becomes affected by WNW-ESE dextral shearing with a southward thrust component (Janjou, work in progress, Segré 1:50000 geological map), where it has been given the name St-Clément-de-la-Place orthogneiss (SCPG). This deformation is concentrated along the Lanvaux axis, forming the northern branch of the South-Armorican Shear Zone (SASZ), and occurred during the major phase of activity along this zone between 345 and 320 Ma (Diot et al. 1983). Farther north, in the St-Julien-de-Vouvantes unit (SJVU), the same event is represented by a strain-slip cleavage with folds overturned south-westward, as in Central Brittany (Rolet, this Chap.). Southward, in the SGLU, this episode takes the form of a late strain-slip cleavage overturned southwestward (Ledru et al. 1986).

Table 2. Amphiboles: analytical data from south St-Georges-sur-Loire unit (BRGM) microprobe analyses). Ca and/or Na amphiboles calculation method from B. E. Leake (1978)

	111	112	
SiO_2	54.96	54.48	0.00
TiO_2	0.04	0.00	0.00
Al_2O_3	6.55	4.59	0.00
Cr_2O_3	0.01	0.03	0.00
FeO	21.17	22.76	0.00
ZnO	0.00	0.00	0.00
MnO	0.05	0.00	0.00
MgO	5.67	6.04	0.00
CaO	0.37	0.64	0.00
Na_2O	7.03	6.78	0.00
K_2O	0.01	0.00	0.00
F	0.00	0.00	
Cl	0.00	0.00	
FeOCalc	15.61	15.11	0.00
Fe_2O_3Calc	6.18	8.50	0.00
H_2OCalc	2.05	2.03	0.00
F=O	0.00	0.00	0.00
Cl=O	0.00	0.00	0.00
Total	98.53	98.20	0.00
	Ions/23 oxygen + 1 H_2O		
Si	8.035	8.053	
Al_{IV}	0.000	0.000	
Al_{VI}	1.129	0.800	0.000
Ti	0.004	0.000	
Cr	0.001	0.004	
Fe^{3+}	0.680	0.946	0.000
Fe^{2+}	1.909	1.868	
Zn	0.000	0.000	
Mn	0.006	0.000	
Mg	1.236	1.331	
Subtotal	13.000	13.000	0.000
Ca	0.058	0.101	
NaB	1.942	1.899	
NaA	0.051	0.044	0.000
K	0.002	0.000	
Total	15.053	15.044	0.000
OH	2.000	2.000	0.000
F	0.000	0.000	
Cl	0.000	0.000	
Fe/Mg + Fe	0.607	0.584	
Amphibole	Crossite	Crossite	

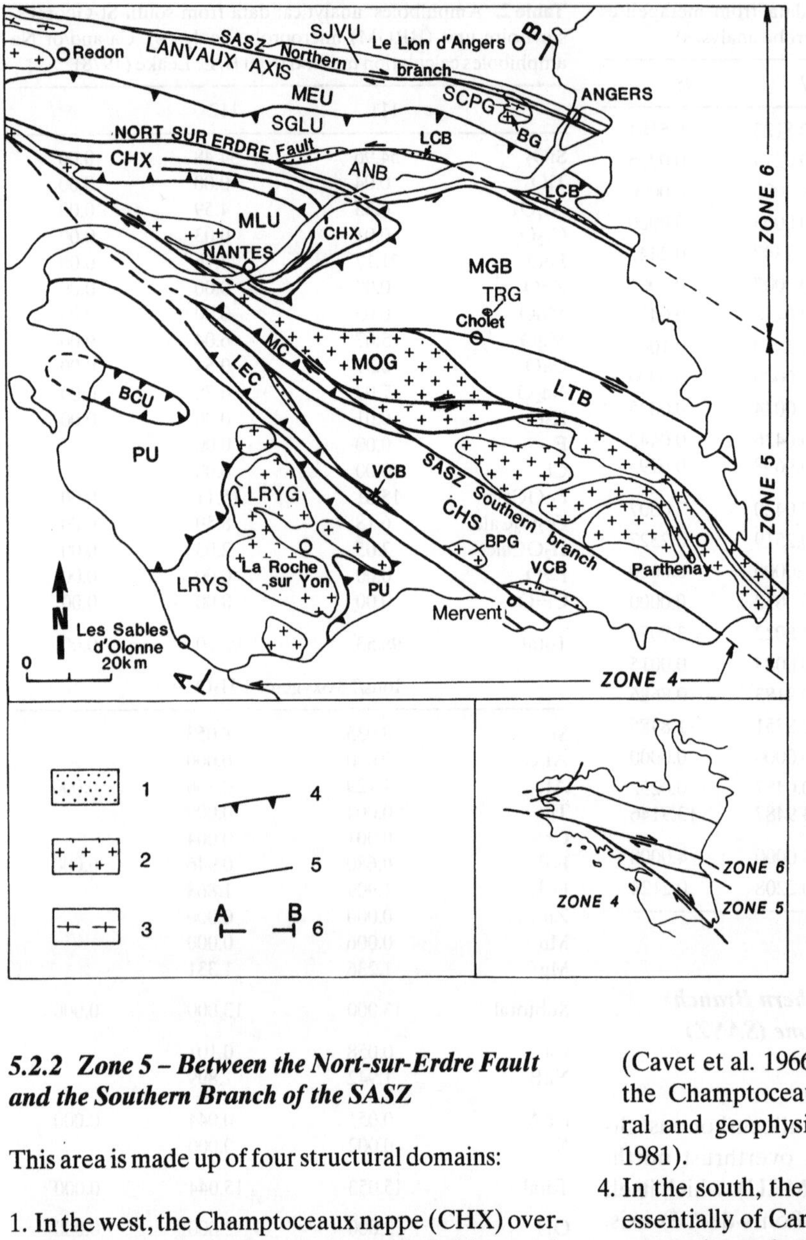

Fig. 11. Structural sketch of the Vendée. *1* Coal basin (Namurian to Stephanian); *2* granitoid; *3* orthogneiss; *4* thrust fault; *5* fault, strike-slip fault or possible thrust fault; *6: A–B* cross-section of Fig. 12. See Table 3 for unit names

5.2.2 Zone 5 – Between the Nort-sur-Erdre Fault and the Southern Branch of the SASZ

This area is made up of four structural domains:

1. In the west, the Champtoceaux nappe (CHX) overthrusts the Mauves-sur-Loire micaschists unit (MLU), its age of emplacement being dated as Devonian (Ballèvre et al. Chap. III, B.3).
2. In the north, the Ancenis Basin (ANB), interpreted by Diot (1980) and Dubreuil (1986) as a pull-apart basin opened during the Late Devonian-Early Carboniferous along a sinistral transcurrent fault (the Nort-sur-Erdre fault) separating the Mauges domain (MGB) and the Saint-Georges-sur-Loire unit (SGLU).
3. In the center, the Mauges domain (MGB), a microblock deformed and metamorphosed during the Cadomian (Le Métour and Wyns 1988) and unconformably overlain by a series of volcanogenic epicontinental deposits of Early Paleozoic age (Cavet et al. 1966). This block is emplaced above the Champtoceaux nappe, as shown by structural and geophysical data (Diot 1980; Marchand 1981).
4. In the south, the Nantes-Parthenay granitic zone, essentially of Carboniferous age, whose intrusions cut gneiss and schist possibly attributed either to the Cadomian or to the Variscan (the La Tessouale and Haut Bocage domains: LTB, Le Métour and Wyns 1988; Le Métour 1989).

Eo-Hercynian deformation in Zone 5 appears to be confined to the Champtoceaux arc and the Mauves-sur-Loire unit (MLU). The Mauges block, like the La Tessouale and Haut Bocage blocks, constitutes the highest unit of the nappe pile, and remained undeformed during emplacement.

Hercynian deformation is represented by a poorly developed WNW-ESE trending crenulation in the Cadomian metamorphic rocks (Le Métour and Wyns 1988), becoming more marked in the Paleozoic sedi-

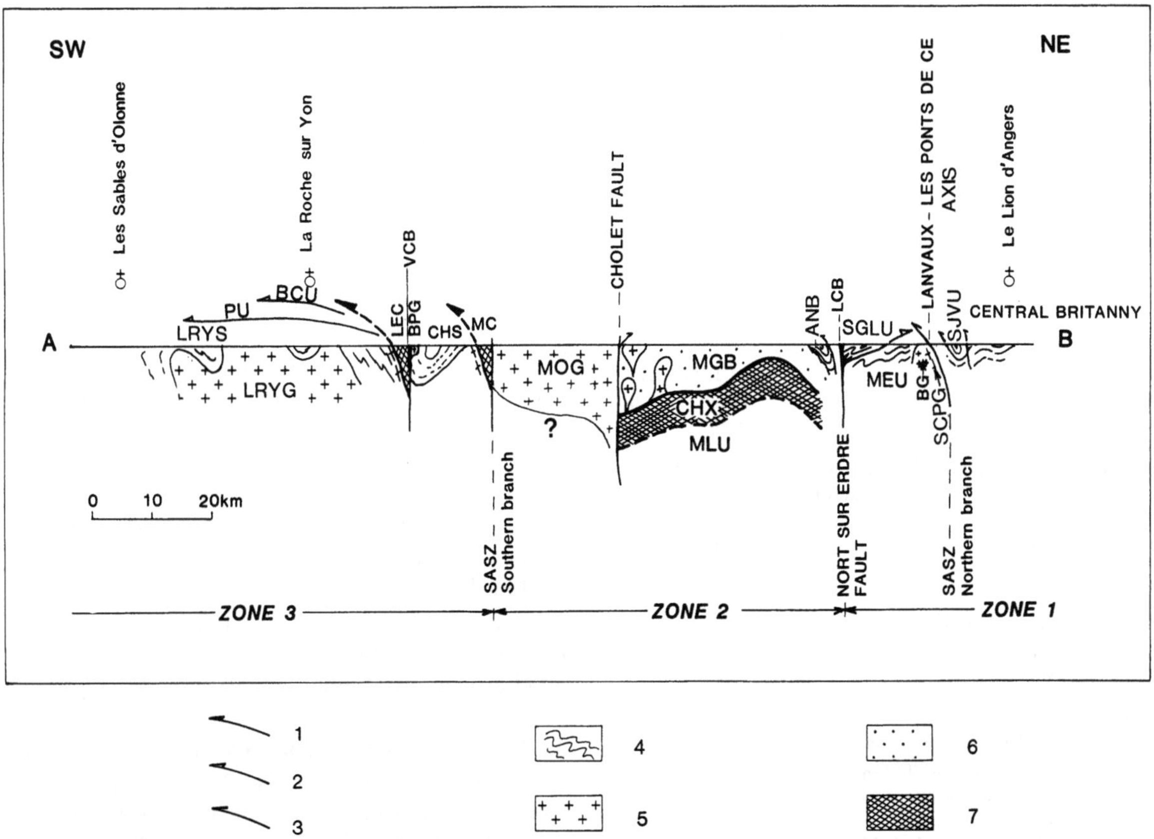

Fig. 12. Interpretative schematic cross-section through the Vendean Massif. *1* Major thrust fault, Pre-Dinantian; *2* other thrust fault, Pre-Dinantian; *3* minor thrust fault (late Hercynian), *4* folded Paleozoic formations; *5* granitoids; *6* Cadomian microblock; *7* high-pressure, high-temperature complexes. *SASZ* South-Armorican shear zone. See Table 3 for unit names

mentary rocks of the Ancenis Basin and in the Cholet area (Le Métour 1989), where a strain-slip cleavage develops. With increasing proximity to the SASZ, the deformation becomes penetrative and assumes a ductile character with dextral shear displacement (Cholet Fault, Le Métour 1989). The Trémentines granite (TRG), contemporaneous with this deformation, has been dated at 347±14 Ma (Rb-Sr, Le Métour 1989). The arching of the Champtoceaux Nappe, giving it its present arcuate shape, is also customarily attributed to this episode.

In the southern part of Zone 5, the SASZ forks into a number of branches and offshoots, delimiting blocks of various sizes. Movement on the various branches appears to have been reactivated at different times, which becomes particularly evident when the relations between the faults and the various stages of emplacement of the granitic intrusions of the Nantes-Parthenay zone are examined. Around the Mortagne batholith (MOG), the first-emplaced rocks have been affected (Le Métour 1989) by ductile NW-SE dextral movement, which could have been contemporaneous with the Cholet shearing (Visean) and the imposition of the regional cleavage/schistosity. A second ductile shearing event, which was sinistral and trends WNW-ESE, cut the dextral shears and could have been contemporaneous with the emplacement in a pull-apart type structure of the main part of the batholith during the Namurian-Westphalian (313± 15 Ma (Rb-Sr, Guineberteau 1986; Guineberteau et al. 1987). Lerouge and Quénardel (1988) propose the same relations for the Parthenay Granite: emplacement during a sinistral shearing after the dextral shearing of Cholet, which these authors correlate to that of the Marches-Combrailles shear zone in the north of the Massif Central. Finally, most of the faults were affected by brittle dextral movement during the Westphalian-Stephanian.

5.2.3 Zone 4 – South of the SASZ

This zone can be subdivided, from northeast to southwest, into three domains, the Chantonnay synclinorium (CHS), the Les Essarts complex (LEC) and the La-Roche-sur-Yon synclinorium (LRYS).

Table 3. Main units of Vendean basement

Structural units		
MGB	Mauges microblock	Upper units, undeformed
LTB	La Tessouale microblock	during Paleozoic thrustings
CHX	Champtoceaux complex	HP, HT
MC	Montaigu complex	allochthonous
LEC	Les Essarts complex	units
BCU	Bois-de-Cené unit	HP, LT allochthonous unit
PU	Porphyroids unit	M-LP, M-LT
MLU	Mauves-sur-Loire unit	allochthonous
SGLU	Saint-Georges-sur-Loire unit	units
SJVU	Saint-Julien-de-Vouvantes unit	Allochthonous
LRYS	La-Roche-sur-Yon synclinorium	or para-autochthonous
CHS	Chantonnay synclinorium	thonous
MEU	Murs-Erigné unit	units
Other units		
VCB	Vendean coal basin	
LCB	Ligerian coal basin	
ANB	Ancenis basin	
Hercynian granites		
LRYG	La-Roche-sur-Yon granite	
BPG	Bazoges-en-Pareds granite	
MOG	Mortagne granite	
BG	Bécon granite	
SCPG	Saint-Clément-de-la Place orthogneiss	

1. The Chantonnay Synclinorium. This is occupied by Cambrian to mid-Devonian rocks (Wyns et al. 1988, 1989 a) comprising metasediments overlain by a back-arc metavolcanic sequence (Thièblemont and Cabanis 1986). They were deformed and metamorphosed during two phases in the Late Devonian to Early Carboniferous. During the first phase, thrust tectonics imposed a schistosity under greenschist facies conditions. The second phase resulted in the formation of a steeply dipping NW-SE strain-slip cleavage, related to the regional dextral shearing (Bouton 1990). This cleavage passes progressively to dextral ductile shearing along the southwest margin of the synclinorium, the grade of metamorphism rising to amphibolite facies (Wyns et al. 1988). The Bazoges-en-Pareds granite (BPG) was converted to orthogneiss during this second phase. The rocks at this margin were later thrust southwestward (Wyns et al. 1989 b) over unmetamorphosed rocks of the Vendée coal basin (VCB), of Namurian to Stephanian age, extending from Grandlieu Lake in the northwest to Saint-Laurs in the southeast.

2. The Les Essarts Complex (LEC). Here, the St-Martin-des-Noyers amphibolites, medium pressure and temperature rocks (Tièblemont et al. 1987), are juxtaposed with high grade, high pressure metamorphic rocks containing relics of eclogite (Godard 1981). The amphibolites are in probable faulted contact with the Angle orthogneiss to the southwest that intrudes the Paleozoic rocks of the La-Roche-sur-Yon synclinorium. Pebbles of orthogneiss and eclogite in conglomerate in the coal basin to the northeast (Godard 1981) indicate that the deformation and metamorphism took place before the Namurian.

Although it has not been positively identified everywhere, the NW-SE dextral ductile shearing probably affected this unit before the deposition of the first (Namurian) coal beds, as in the Mervent area (Bouton 1990). Renault and Bouton (1986) and Bouton (1990) suggest that this deformation continued as brittle dextral transcurrent movement along the Vouvant coal basin (VCB) during Namurian to Stephanian, but they did not identify in the VCB the NW-SE sinistral movement related to Namurian-Westphalian by Guineberteau (1986), Guineberteau et al. (1987), and Lerouge and Quenardel (1988).

3. The La-Roche-sur-Yon Synclinorium (LRYS). This structure includes at least two nappes (Burg 1981; Brun and Burg 1982) – one comprising the Porphyroids and the Saint-Gilles schists (PU), overlain by another consisting of the Bois de Cené glaucophane schists (BCU). The Porphyroids nappe occupies extensive areas in coastal Vendée and probably in southern Brittany (the Guérande peninsula and Belle Ile island).

The nappes overlie, more or less horizontally, weakly metamorphosed Paleozoic rocks in which the Silurian has been identified at Brétignolles-sur-Mer (Ters 1979). Carboniferous sediments dated as Tournaisian (Comble et al. 1985) and affected by two cleavages are associated with this series, but their relationship with the Silurian remains unclear.

In the southwest, the Paleozoic rests on the metamorphic rocks of the Sauveterre-Les-Sables-d'Olonne zone, where the kyanite-zone Barrovian metamorphism has been dated at 388±8 Ma by anatectic granite (Rb-Sr, recalculated, Vidal 1976). The nature of the contact is as yet unknown.

The age of the major thrusting in this area (D 1), up to now regarded as Early Devonian by reference to the anatexis at Les Sables d'Olonne (Autran and Cogné 1980), could be as young as Visean were it to be shown that the Tournaisian at Brétignolles was deformed at the same time as the rest of the Paleozoic rocks (Colchen and Poncet 1987). The second deformation, with the formation of the steeply dipping NW-SE strain-slip cleavage, ended with the emplace-

Table 4. Principal Hercynian geodynamic events in the Vendée. (Time scale after Haq and Van Eysinga 1987)

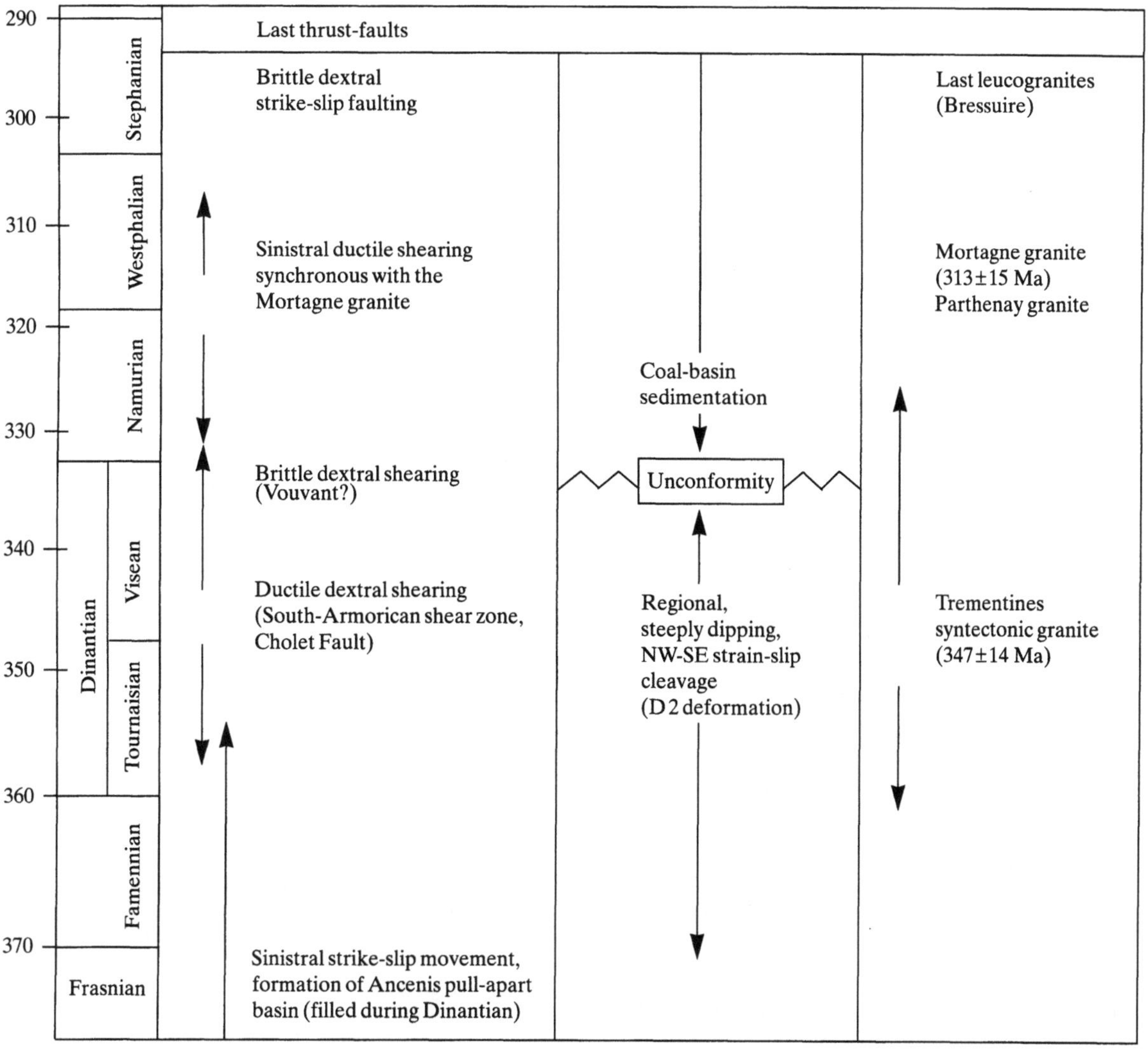

ment of the Bournezeau granite, dated at 307±19 Ma (Rb-Sr, Wyns et al. 1989b), and can therefore realistically be attributed to the Visean or earliest Namurian.

5.3 Conclusions: Evolution of Deformation in the Carboniferous (Table 4)

After the main thrust event, the end of which is not everywhere dated with certainty (Early Devonian or Late Devonian to Early Carboniferous), Hercynian intra-continental deformation appears to be coherent throughout the area dealt with here.

The first identifiable event was the opening of the Ancenis pull-apart Basin along a sinistral NW-SE strike-slip fault during the Frasnian-Dinantian (Diot 1980). No corresponding event has been identified elsewhere.

The major event during the Hercynian was dextral NW-SE shear deformation of variable intensity, related to the movement on the South-Armorican shear zone. A number of lines of evidence indicate a pre-Namurian, probably Visean, age for this deformation in the Vendée, as in South Brittany. Later, around 313±5 Ma (Namurian-Westphalian), a sinistral NW-SE shearing develops together with the emplacement of leucogranites of Mortagne and Parthenay; but this episode is as yet unknown at the scale of Armorican Massif.

After the formation of coal basins (Namurian to Stephanian), the "opening mechanisms of which are not yet clearly established" (Bouton 1990), a brittle dextral NW-SE strike-slip movement caused the closing of these basins. The small-scale thrust-faulting observed along the Vendée coal basin occurred during or after this episode (late to end-Stephanian).

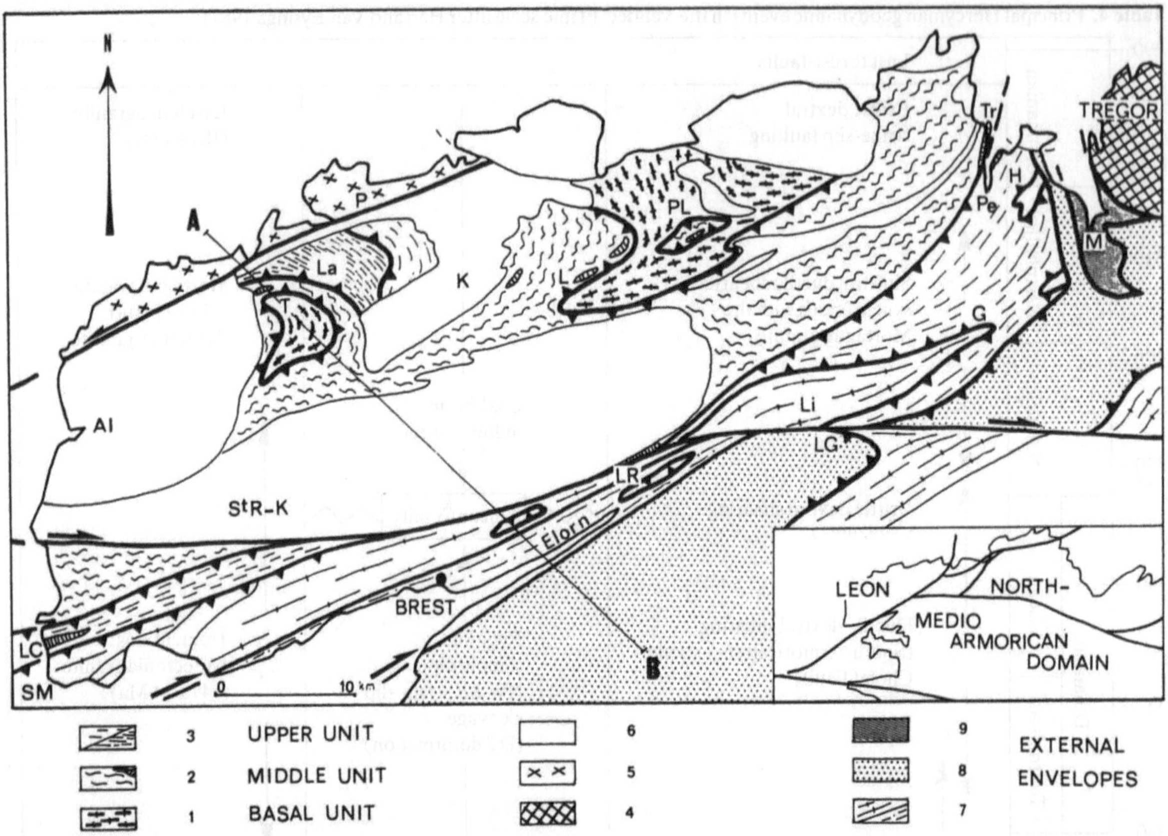

Fig. 13. Léon tectonic map. Internal nappe pile: *1* Basal unit, Treglonou and Plounevez-Lochrist orthogneiss; *2* middle unit, Lesneven paragneiss with mafic ultramafic remnants; *3* upper unit, Le Conquet-Penze micaschists and Lannilis amphibolites. External envelopes: *7* Brest-Lampaul Guimiliau orthogneiss and Elorn brioverian schists; *8* Paleozoic (Ordovician to Devonian) formations of Central Brittany; *9* Carboniferous Morlaix striped shales; *4* Trégor, pre-Cambrian microblock; *5* Plouguerneau migmatite; *6* Carboniferous granites. *A–B* Cross-section location; *Al* Aber Ildut; *G* Guiclan; *H* Henvic; *K* Kernilis; *L* Lesneven; *La* Lannilis; *LC* Le Conquet-Pors Liogan; *LG* Lampaul Guimiliau; *Li* Landivisiau; *LR* La Roche Maurice; *M* Morlaix; *P* Plouguerneau; *Pe* Penze; *PL* Plounevez Lochrist; *SM* Saint Mathieu; *StRK* St Renan Kersaint; *T* Treglonou; *Tr* Tregondern

6 Hercynian Events in Léon

J. Rolet (translated by S. Bogdanoff and M. C. Leroux)

Recent research on the Léon (Rolet et al. 1986 b); Cabanis and Godard 1987; Bale et al. 1989) highlights common features between the Léon and the two previously described domains (Medio-Northern and Southern Brittany), indicating that the former was once situated in between. Its correlation with the North-Armorican Cadomian no longer holds.

6.1 Structure

The Léon domain is a granitic and gneissic dome striking WSW-ENE for 70 km, with a width of 30 km. It is probably larger if one considers its continuation below sea level to the north and west. To the south, the limits are the faults of the Elorn and to the east the gneissic antiform disappears under the Penze schists and the striped shales of the Morlaix river (Fig. 13).

The hypothesis of a nappe pile, as in the southern domain, seems to explain the general structure of the Léon.

From some sections (Treglonou, Lesneven, Le Conquet), Bale et al. (1989) suggest a tectonic pattern characterizing the internal part of Léon. We shall describe the external structures, comparing them with the Northern Medio-Armorican Domain.

6.1.1 Internal Structures

Three main units may be distinguished:

1. a basal orthogneissic unit (porphyritic orthogneisses of Treglonou and Plounevez-Lochrist); intensely

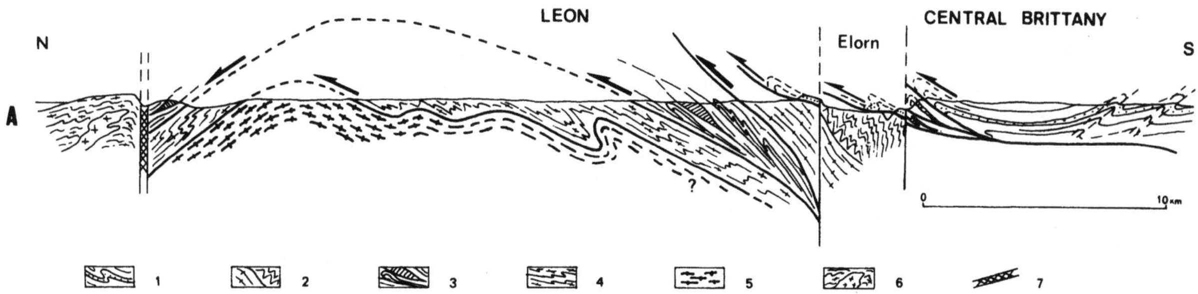

Fig. 14. Léon nappe antiform cross-section. *A–B* Location on Fig. 13. The St Renan-Kersaint leucogranite is intentionally not drawn. *1* Paleozoic formations; *2* Brest orthogneiss and Elorn Proterozoic shales; *3* Upper unit, Le Conquet micaschists and mafic ultramafic slices; *4* Middle unit, Lesneven high temperature metamorphic (sillimanite +) paragneiss; *5* Basal unit, Treglonou orthogneiss; *6* Plouguerneau migmatite; *7* Carboniferous shear zone (mylonite)

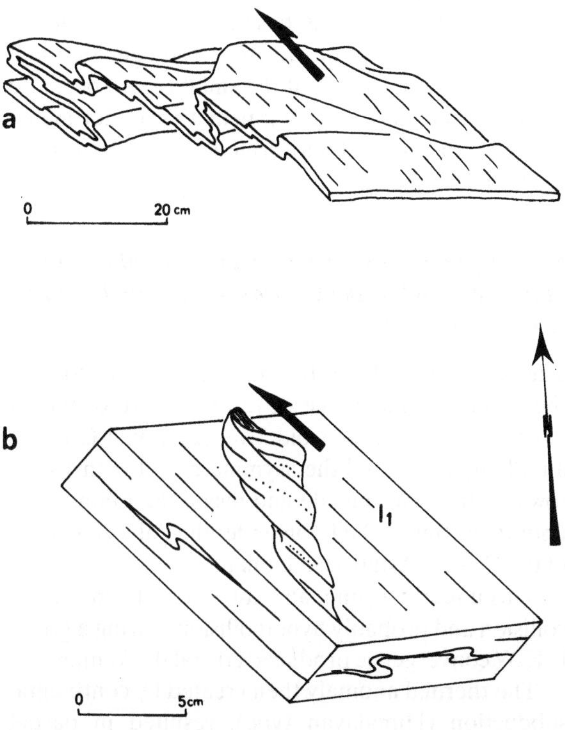

Fig. 15 a, b. Curved axis and synfolial sheath folds from Penze micaschists. *L* Stretching lineation

deformed, almost anatectic, and appearing as antiformal windows among other units (Fig. 14);

2. a middle unit, the anatectic paragneiss nappe of Lesneven lying on the orthogneisses via a mylonitic sole with slices of various mafic-ultramafic basic and ultra basic rocks, metagabbros, garnet amphibolites with eclogitic relics (Cabanis and Godard 1987);
3. an upper heterogeneous unit made up by Conquet-Pors-Liogan schists in the South and Lannilis amphibolites in the North.

The internal nappe pile is bordered to the south and east by a discontinuous belt of sometimes layered metagabbros (Pors-Liogan, La Roche Maurice, Landivisiau, Tregondern) which could represent the trace of a major contact with the external units.

Post-tectonically, the thrust pile was intruded by the granites of Kersaint-Saint-Renan (330–340 Ma), Aber Ildut (300 Ma) and Ploudalmezeau-Kernilis (292±15 Ma).

6.1.2 Deformation Events

The whole thrust pile shows a nearly flat-lying foliation, parallel to the basal contacts and contains a stretching lineation, pressure shadows around potassic feldspar phenocrysts, amphibolite boudins, sillimanite fibers, etc.

The shear criteria associated with this major deformation, although scattered, are concentrated around a SE-NW direction that could record a thrust transport (Bale et al. 1988) and shows an inverse asymmetry towards the NW (garnet rotation of Pors-Liogan-Le Conquet and sheath folds near Penze, Fig. 15).

A relict lineation is often observed trending from N 90–N 150° and reworked by a N 60–70°, pronounced lineation parallel to the axes of southernly overturned folds, developed during anatectic metamorphism. This reworking develops sillimanite fibers parallel to the lineation (N 60–70°) and could be contemporaneous with the earliest intruded leucocratic granites (Saint-Renan, Kersaint). The later granitic intrusions (Aber-Ildut, Kernilis) are synchronous with localized shear zones (Bale et al. 1988) but always in a partial melting context.

6.2 Structural Relationship with the Northern Medio-Armorican Domain

The Léon was considered for a long time as a metamorphic Precambrian basement beneath the Brioverian and Paleozoic envelopes which appeared to transgress over an old and eroded mountain belt (Chauris 1967).

In contrast recent structural and geochronological arguments (Cabanis et al. 1979; Paquette et al. 1987) suggest a Paleozoic orogenic evolution for the entire domain. The connection between the metamorphosed Léon, its metasedimentary envelopes, and the Northern Medio-Armorican Domain must now be viewed in tectonic terms rather than by stratigraphic relationships.

6.2.1 *Structures of the External Envelopes*

Three different lithological groups can be differentiated, from the deepest (to the north) to the shallowest (to the south) (Fig. 14).

1. Orthogneisses of the Brest-Saint-Mathieu Group. These orthogneisses are ancient granites and granodiorites which were foliated in mesozonal metamorphic conditions, sometimes retrogressed in low epizonal metamorphism.

The foliation dips systematically towards the south and it accompanies thrusting the internal superior unit of the Conquet (section Saint-Mathieu-Pors-Liogan) northwards.

2. Penze Schists and Brioverian Elorn Shales. At the eastern extremity of the Léon, the Brest orthogneisses dip (Guiclan region) under the Penze schists and then reappear in a window (Henvic region, Fig. 13). The tectonic nature of the contact and frequently shallow inclined attitude of the foliation suggest a general allochtony of the Penze schists thrust over the Brest orthogneisses.

3. Paleozoic Units. The earliest Paleozoic unit, La Roche Maurice quartzite formation (Chauris and Hallegouet 1973) appears to be made up of fragments of white metaquartzites, locally attributed to the Grès Armoricain Formation. The unit is horizontally thrust on various formations: Paleozoic of the South Elorn and Penze micaschists (Fig. 14).

The second, lower unit comprises the Siluro-Devonian formations of the Southern Elorn and Morlaix, which are thrust over the orthogneisses of Lampaul-Guimiliau and Plougonven, the eastern and south-eastern equivalents of the Brest orthogneisses.

6.2.2 *Deformation*

The entire external Léon units are allochthonous, and display a horizontal foliation or schistosity formed at a lower to middle metamorphic grade. The foliation carries a stretching lineation, often deformed by recumbent folds with curved axes (sometimes sheath folds), indicate of a NW thrust transport (Fig. 15).

To the south, along the North-Armorican Shear Zone, and the Elorn Shear Zone, these thrusts are reworked by a transcurrent dextral shear zone.

6.2.3 *Hypothesis About the Léon Translation*

In order to interpret the common Eo-Hercynian history of the Léon and the South-Armorican Domain, it has been suggested (Bale and Brun 1986) that at the end of the Devonian, lateral displacement of 150 km took place from the south along a transcurrent dextral fault. This hypothesis is based on the transcurrent dextral deformation of the Brest orthogneisses and may explain the closing of the small Lizard ocean to the north. However, these kinematics do not match well with the general antiformed nappe structure of the Léon.

6.2.4 *Hypothesis Concerning the Underthrusting of the Léon Below the Central Armorican Domain (Rolet et al. 1986b)*

The general dip of the structures towards the SE and the E, the reappearance of the Brest orthogneisses from beneath the Paleozoic cover (Lampaul-Guimiliau, Plougonven) and the asymmetry of the thrusting towards the NW, outside and inside the Léon, favor underthrusting of the Léon under the Paleozoic rocks of the Central Armorican Domain.

This underthrusting, an expression of continental collision and probably hypercollision, during a grossly E-W convergence produced crustal thickening.

The thermal anomaly then created by continental subduction (Himalayan type), resulted in partial melting of the crust, generating migmatites and aluminous granites during the entire Devono-Carboniferous period.

The aluminous granites (Huelgoat type) of the occidental part of the Central Armorican Domain could be the products of this crustal melting, which intrude the superstructures of the thrust pile 5 or 10 km above the source.

In such a model, the thrusts with an E and SE vergence, made up by the cover of the Central Armorican Domain, could be the antithetics of the underthrusting of the Léon, the transcurrent dextral shear zones being the lateral ramps of the thrusts.

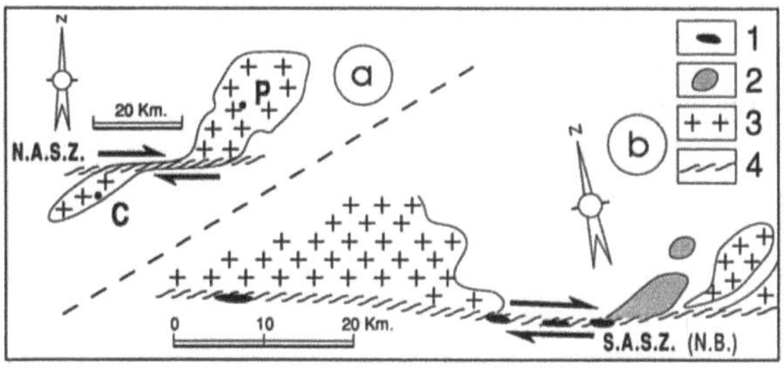

Fig. 16. a Late movements along the NASZ offsets the Plouaret *(P)* - Commana *(C)* complex by ca. 15 km after emplacement. (After Guillet et al. 1985). **b** Dark granodioritic ultramylonites *(1)* derived from Bignan granodiorite *(2)* imply some 40 km dextral offset along NB of SASZ; *3* ca. 340 Ma two-mica granites; *4* shear zone. (After Jégouzo and Rossello 1988)

6.3 Chronology

Whatever the mechanisms, all workers infer an Eo-Hercynian history, which is common to the South-Armorican Domain and the Léon, the high grade metamorphic relicts (eclogitic lenses of Treglonou and Lesneven) being the evidence of the collision stage accompanied by obduction during the Silurian (420–440 Ma).

The major thrusting responsible for the thrust stacking started as early as Eo-Hercynian times (380 Ma) and were probably continuous during the Devonian-Carboniferous period (345–290 Ma), during which all the aluminous granites were generated.

This final period clearly differentiates the Léon from the South-Armorican Domain and indicates its proximity to the Central Armorican Domain during the Hercynian period sensu stricto.

7 Main Shear Zones

P. Jegouzo

During Hercynian and Late Hercynian times, the whole of the Armorican Massif seems to have undergone deformation related to a major continental collision (Cogné 1977) and there is a characteristic development of large scale dextral shear zones: the North-Armorican (NASZ) and the South-Armorican Shear Zone (SASZ).

7.1 The North Armorican Shear Zone

The NASZ, recognized by Chauris (1969) as passing through the Molene-Moncontour lineament, separates the North-Armorican domain, characterized principally by Cadomian magmatism, metamorphism, and deformation from a Central Armorican domain where Cadomian effects are weak.

To the west, the shear zone is localized at the southern border of the St Renan-Kersaint granite (Goré and Le Corre 1987) passes through the Elorn-Morlaix metamorphic domain (Fig. 4) and cuts across the Plouaret-Commana plutonic complex (Guillet et al. 1985). Further east, it affects gneisses and migmatites of the Belle-Isle-en-Terre-Guingamp area (Hirbec 1979; Watts and Williams 1979) and to a lesser extent Quintin and Moncontour granites. Paleozoic formations of the Menez-Belair syncline, the western part of Laval syncline, and the adjacent Broverian metasediments and Cadomian granites are variably affected either by east-west transcurrent faults or by N 50 °E striking thrusts with southeastwards vergence (Paris and Jegouzo 1976). The eastern termination of the shear zone (Alençon area) also displays thrusting – with sinistral components – towards the south (Le Corre et al. 1990).

The deformation produced an early mylonitization – dominant in the western part of the shear zone – followed by late cataclasis and faulting. Limited occurrences of pseudotachylites, geometrically related to cataclasites, have been reported. The early development of the shear zone was broadly synchronous (Guillet et al. 1985) with the emplacement of the plutonic complex of Plouaret-Commana which has been dated at 329±9 Ma (Peucat et al. 1984). The Quintin (291±9 Ma, Peucat et al. 1979) and Moncontour granites, which display only protocataclastic textures, appear to have been only dextrally displaced during the late stage of shear movement. This movement does not affect micrograniitic veins of Bas-Léon (292±9 Ma, Chauris et al. 1977). A Hercynian offset from 10 to 15 km along the shear zone has been proposed by various authors (Fig. 16 a).

7.2 The South Armorican Shear Zone

The SASZ (Cogné 1960) lies between two domains within the Armorican Massif which have suffered contrasting metamorphic and structural histories. To the north, the Central Armorican domain is made up of the Late Proterozoic (Brioverian) sedimentary succession and its Paleozoic cover. To the south lies the South Brittany metamorphic complex, composed mainly of granites and medium to high grade schists and gneisses (Chap. III.B). The two main branches of the SASZ become difficult to distinguish from each other westwards from Quimper (Jegouzo 1980). External sinistral shear zones (strike N 50–60 °E) can be inter-

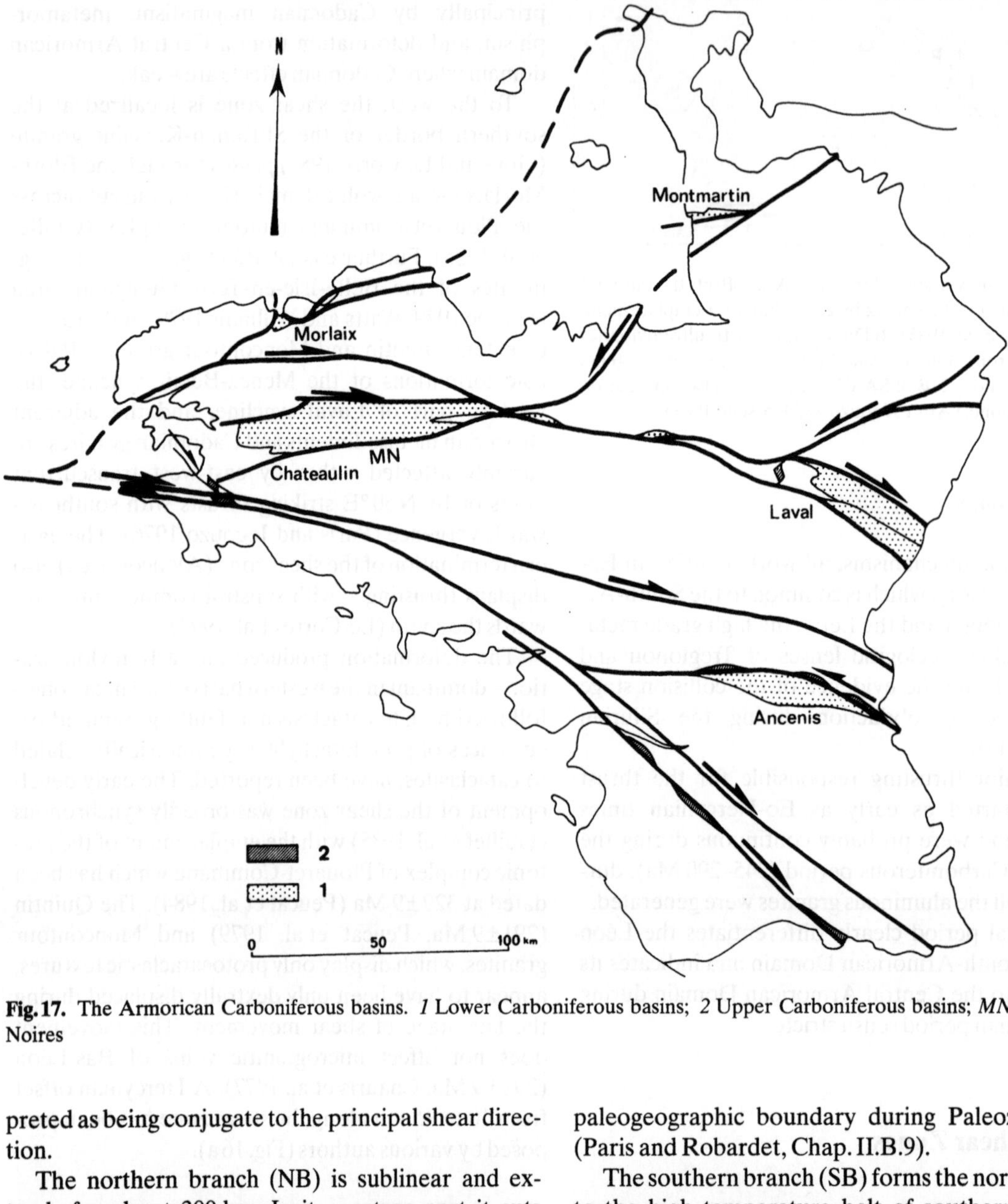

Fig. 17. The Armorican Carboniferous basins. *1* Lower Carboniferous basins; *2* Upper Carboniferous basins; *MN* Montagnes Noires

preted as being conjugate to the principal shear direction.

The northern branch (NB) is sublinear and extends for about 300 km. In its western part it cuts across two mica granites dated ca. 340 Ma (Peucat et al. 1979) and displays a subvertical mylonitic schistosity and a subhorizontal stretching lineation. Mesoscopic and microscopic structures provide evidence of dextral shearing (Berthé et al. 1979). To the east, a zone of faulted and unconformable contacts can be observed in the Paleozoic succession. Conjugate folds also demonstrate the operation of an important dextral component (Berthé and Brun 1980). On petrological and geochemical arguments – similarities between mylonites, ultramylonites, and Bignan granodiorite – a 40 km minimal dextral offset along this branch has been proposed (Fig. 16b; Jégouzo and Rossello 1988). The NB corresponds to an important paleogeographic boundary during Paleozoic times (Paris and Robardet, Chap. II.B.9).

The southern branch (SB) forms the northern limit to the high temperature belt of southern Brittany. Within this branch, the schistosity rotates gradually from N 100 °E near Quimper to N 130 °E in the region of Nantes. This major vertical shear zone is contained within 380 Ma anatectic granites (Vidal 1976) and 320 Ma two-mica granites (Bernard-Griffiths et al. 1985). A northward-dipping mylonite zone occurs to the west of Vannes. The strike of the schistosity in this mylonite zone is parallel or slightly discordant to that observed in adjacent vertical shear zones (Jegouzo 1980). The stretching lineation displays a slight plunge toward the northwest and a dominant shear component can easily be demonstrated. This northward-dipping shear zone is enclosed between vertical shear zones; it corresponds to an early shearing

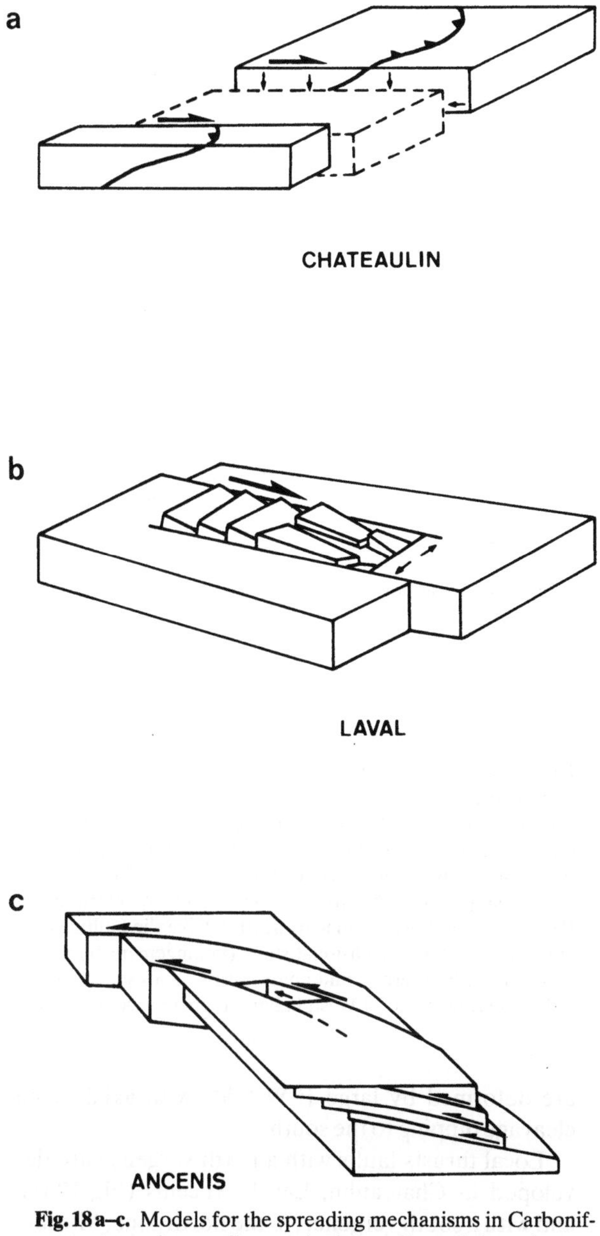

Fig. 18 a–c. Models for the spreading mechanisms in Carboniferous Armorican pull-apart basins. **a** Dextral lateral ramp graben in the Chateaulin basin. **b** Dextral tension relay in the Laval basin. (Houlgatte et al. 1988). **c** Sinistral shear above Vendean thrusts in the Ancenis basin. (Cannat and Bouchez 1986)

episode and would explain the absence to the west of the St Georges-sur-Loire and Champtoceaux complexes.

As observed in the NASZ, cataclasis overprints the mylonitic foliation and is itself overprinted by deformation which produced pseudotachylitic rocks. A displacement magnitude of the order of some 150 to 200 km has been proposed for this branch (Jégouzo and Rossello 1988).

8 Carboniferous Tectonics and Pull-Apart Basin Development

J. Rolet

Following the orogenic events of the Breton phase (Upper Devonian), the Lower Carboniferous paleogeography was completely transformed in the Armorican Massif. At that time, the marginal Devonian areas were almost completely emerged and the few "intramontane" Carboniferous basins were not generally open to the sea to the north (cf. Pelhate, Chap. II.B.7, this Vol).

Two orogenic periods can be distinguished showing different Carboniferous basin types (Rolet 1984).

- Lower Carboniferous (Tournaisian, Visean to Namurian).

This period is represented by the main Dinantian basin of Chateaulin-Laval, about 90 km long, and also Ancenis, Morlaix, and Montmartin.

These basins show long life (10 to 30 Ma) and a complex dynamic history.

- Middle and Upper Carboniferous (Westphalian to Stephanian) showing sporadic narrow coal-field graben closely controlled by the main shear zone (SASZ), like the french "Sillon Houiller".

All these activities were accompanied by fracturing, crustal deformation, and magmatism.

8.1 Lower Carboniferous Period

The main Dinantian basins are typically rhomboid with high subsidence (pull-apart) directly inherited from Upper Devonian transcurrent faulting.

The Chateaulin and Laval basins were opened during dextral motion (Fig. 17) along the North-Armorican Shear Zone (Rolet 1984; Houlgatte et al. 1988) and Ancenis basin during a sinistral motion (Diot 1980) along the South Armorican Shear Zone (Dubreuil 1986).

The opening mechanisms may be:

- tension relay for Laval basin (Fig. 18 b);
- lateral ramp graben for Chateaulin basin (Fig. 18 a);
- tear opening above thrust stacking for Ancenis basin (Cannat and Bouchez 1986; Fig. 18 c).

Some of them receive exotic blocks (olistoliths and olistostromes) suggesting gravity tectonics during the Tournaisian (Chateaulin and Ancenis basin) and the Visean (Morlaix basin).

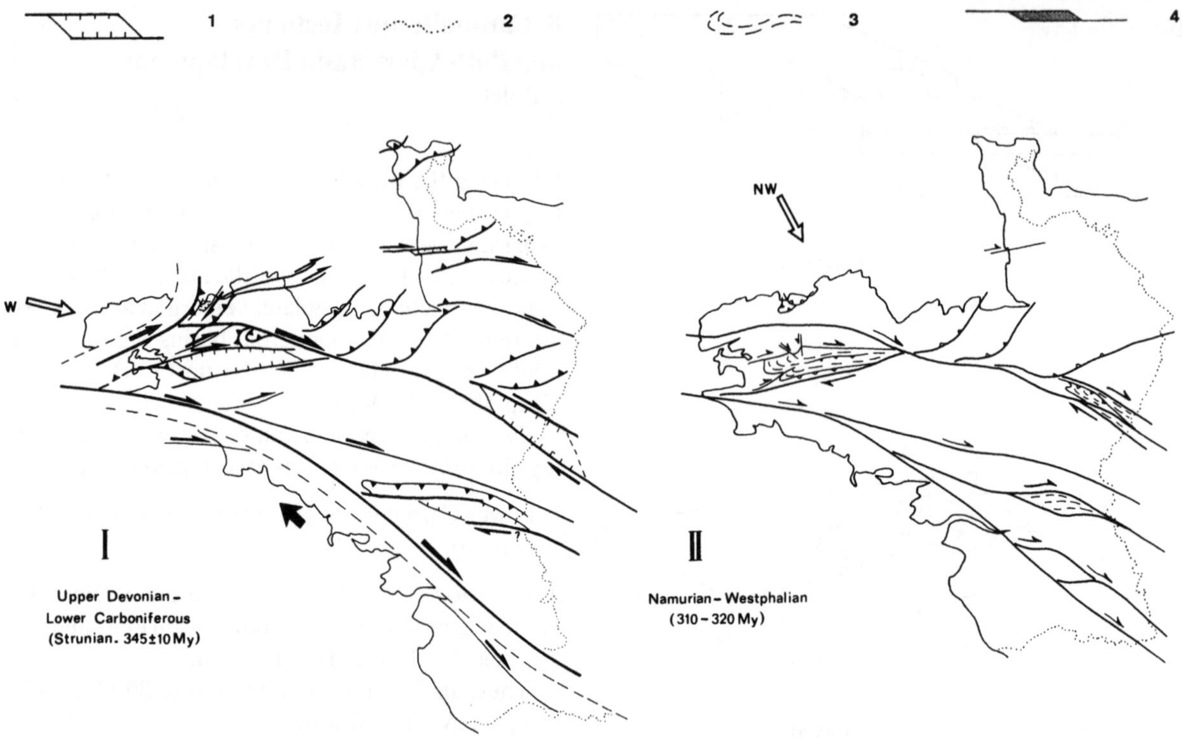

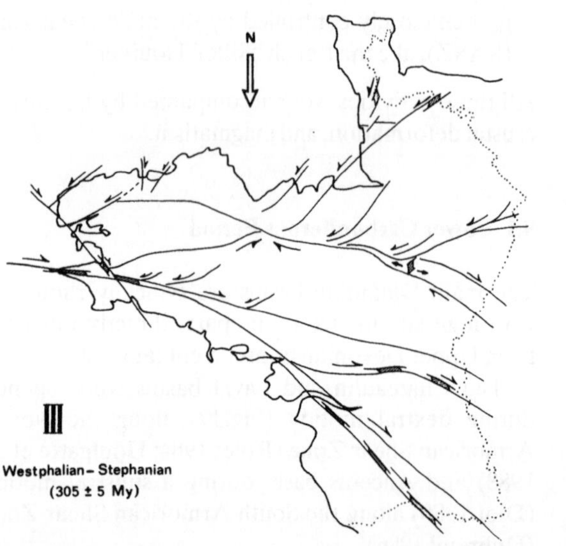

Fig. 19 I–III. Rotation of the regional compressive stress direction during Hercynian time, in the Armorican massif. (After Castaing et al. 1987). **I** Upper Devonian thrust tectonic ("phase Bretonne") and lower Carboniferous pull-apart spreading on right-lateral strike-slip movement. **II** Lower Carboniferous pull-apart basin closure and folding by dextral shear. **III** Narrow coal basin formed by dextral strike slip faulting during Late Hercynian collisional stage (impingement tectonic). *1* Lower Carboniferous pull-apart basins; *2* Mesozoic cover; *3* slaty cleavage trace; *4* Westphalian-Stephanian coal basins

They also show intense edge activities with new relief and fissural volcanics (Chateaulin and Laval).

During the Upper Visean to the Lower Namurian, the filling and closing stage appears after high erosion activity. This Lower Carboniferous basin history is concluded by a new orogenic period (Namurian to Westphalian), with closing and buckling of basins by dextral wrench fault reactivation and granitization in a N-S strain field. Dinantian sediments and volcanics are deformed by large E-W folds with axial plane cleavage dipping to the south.

Local thrusts faults with a north vergence are developed in Chateaulin, Laval, Ancenis (Fig. 19.II), and with a south vergence, in the north zone, Morlaix and Normandy (cf. Gresselin, this Chap. 4.2). All deformation is accompanied by low grade metamorphism and S-type granite intrusion.

Typical "flower structures" are developed along the regional lineament in Montagnes Noires (Darboux and Le Gall 1988).

The South-Armorican Carboniferous granitic belt is closely related to transcurrent shear along the SASZ (Jegouzo, this Chap. 4.7, Fig. 16). A pull-apart sinistral mechanism was proposed by Guineberteau et al. (1987) to explain shape and mode of emplacement for the Mortagne pluton (location Fig. 11).

8.2 Middle and Upper Carboniferous Period

During this period (Westphalian to Stephanian) the Armorican Massif emerged and a few small intracontinental graben basins remained. They are bounded by normal fault cliffs with typical breccia and coal deposits (Pelhate, Chap. II.B.7, this Vol.).

All these narrow pull-apart basins have different ages and a very short life, deformed only at the end of the filling up.

Deformation consists of local crushing with tectonic breccia, horizontal slickensides and reactivation of strike-slip faulting along the main lineaments, such as SASZ.

This brittle tectonics took place without any metamorphism. The shear direction if often dextral, but not exclusively so (Lerouge 1988). Later, closely spaced NW-SE fracturing cut through the continental Armorican Massif (Rolet and Yesou 1990; Yesou and Rolet 1990) (Fig. 19 III). The NS shortening induced tension fracture opening. These open fractures played the role of chemical exchange guides for the last metallogenic products (Chauris and Marcoux, Part II.C, Metallogeny).

During Carboniferous times in the Armorican Massif regional shortening direction rotated from EW in the beginning to NS in the end (Fig. 19). During this rotation, the temperature progressively decreased and less S-type granites were intruded. The last Upper Carboniferous stage appears as typical intracontinental cold tectonics (Fig. 19 III). The kinematics agree with the dextral rotation of Europe Hercynides during the Upper Carboniferous as indicated by paleomagnetic data (Edel 1987).

References

Autran A, Cogne J (1980) La zone interne de l'orogène varisque dans l'ouest de la France et sa place dans le développement de la chaîne hercynienne 26th IGC, Paris Coll C6 Bur Rech Géol Min Mém. 108 (Orléans): 233–246

Bale P, Brun JP (1986) Les complexes métamorphiques du Léon (NW Bretagne): un segment du domaine éohercynien sud-armoricain translaté au Dévonien. Bull Soc Géol Fr (8), II: 471–477

Bale P, Georget Y, Le Corre C (1989) Le Léon: un domaine exotique an Nord-Ouest de la chaine varisque armoricaine (France). Geodin Acta (Paris) 3 (2): 57–71

Barbotin E (1987) Trajectoires de déformation finie et interprétation cinématique. Modéles numériques et applications sur des exemples régionaux (Bretagne Centrale et Alpes Penniques). Thèse, Univ. Rennes (inédit)

Barriere M, Chauris L, Le Bail F (1973) Nodules de silicates d'aluminium autour des granites en Bretagne Occidentale. Bull Soc Mineral Cristallogr 96: 150–154

Barriere M, Rolet J, Thonon P (1983) Le magmatisme, marqueur de l'évolution orogénique, en domaine hercynien ouest-armoricain. CR Acad Sci Paris 296, II: 917–922

Bernard-Griffiths J, Peucat JJ, Sheppard S, Vidal Ph (1985) Petrogenesis of hercynian leucogranites from the southern Armorican Massif. Contribution of REE and isotopic (Sr, Nd, Pb and O) geochemical data to the study of source rock characteristics and ages. Earth Planet Sci Lett 74: 235–250

Berthe D, Brun JP (1980) Evolution of folds during progressive shear in South Armorican Shear Zone, France. J Struct Geol 2: 127–133

Berthe D, Choukroune P, Jegouzo P (1979) Orthogneiss, mylonite and noncoaxial deformation of granites: the example of the South Armorican Shear Zone. J Struct Geol 1: 31–42

Bouton P (1990) Analyse stratigraphique, pétrographique et structurale du segment vendéen de la chaîne hercynienne-partie orientale du synclinorium de Chantonnay-Massif Armoricain. Thèse d'Université, Poitiers, 378 pp, 77 fig

Bril H (1975) La minéralisation à Cu, Pb, Zn de Bodennec en Bolazec dans son cadre géologique. Thèse 3è cycle, Univ. Paris (inédit)

Brun JP, Burg JP (1982) Combined thrusting and wrenching in the Ibero-Armorican arc, a corner effect during continental collision. Earth Planet Sci Lett 61: 319–330

Burg JP (1981) Tectonique tangentielle hercynienne en Vendée littorale: signification des linéations d'étirement E-W dans les porphyroïdes à foliation horizontale. CR Acad Sci Paris 293 II: 849–854

Cabanis B (1985) Identification des séries magmatiques dans les socles métamorphiques, exemples d'application dans le Massif Armoricain et le Massif Central. Critères géologiques, pétrographiques et géochimiques. Implications géodynamiques. Thesis, Université Paris 6, 493 pp

Cabanis B, Fabries J (1972) Modalités du métamorphisme hercynien dans la région de Morlaix (Bretagne septentrionale). CR Acad Sci Paris 275 II: 2311–2314

Cabanis B, Godard G (1987) Les éclogites du pays de Léon (Nord-Ouest du Massif Armoricain): étude pétrologique et géochimique, implications géodynamiques. Bull Soc Géol Fr (8), III: 1133–1142

Cabanis B, Peucat JJ, Michot J, Deutsch S (1979) Remise en cause de l'existence d'un socle orthogneissique antécambrien dans le pays de Léon (domaine nord-armorican): étude géochronologique par les méthodes Rb-Sr et U-Pb des orthogneiss de Tréglonou et Plounevez-Lochrist. Bull Bur Rech Géol Min I: 357–364

Cannat M, Bouchez JL (1986) Linéations NS et NW en Vendée littorale (Massif Armoricain). Episodes tangentiels successifs éo-hercyniens en France occidentale. Bull Soc Géol Fr (8), II: 299–310

Carpenter MSN, Peucat JJ, Pivette B (1982) Geochemical and geochronogical characteristics of Palaeozoic volcanism in the Saint Georges sur Loire synclinorium (S. Armorican Massif). Evidence for pre-Hercynian tectonic evolution. Bull Bur Rech Geol Min (2) Orléans I, 1–2: 63–79

Castaing C, Rolet J, Chevremont P, Le Calvez JY, Thonon P, (1987) La région de Huelgoat (Finistère Central) dans le contexte géodynamique armoricain. Bull Bur Rech Géol Min Fr 1: 23–36

Cavet P, Gruet M, Pillet J (1966) Sur la présence de Cambrien à Paradoxidés à Cléré-sur Layon (Maine et Loire) dans le nord-est du Bocage vendéen (Massif armoricain) CR Acad Sci Paris 263 II: 1685–1688

Chauris L (1967) Les grandes lignes du pays de Léon (Finistère) et la gravimétrie. Mém Bur Rech Géol Min 52: 55–63

Chauris L (1969) Sur un important accident structural dans le

Nord-Ouest de l'Armorique. CR Acad Sci Paris 268 II: 2859–2861
Chauris L, Hallegouet B (1973) Les "quartzites" des gneiss de Brest (Finistère). Bull Soc Géol Minéral Bretagne (C) V 2: 93–99
Chauris L, Hallegouet B, Thonon P, Vidal Ph (1977) Le champ filonien microgranitique du Bas-Léon (Massif Armoricain). Bull Soc Géol Mineral Bretagne C IX: 5–17
Cogné J (1960) Schistes cristallins et granites en Bretagne méridionale: le domaine de l'Anticlinal de Cornouailles. Mem Serv Garte Géol Fr: 382
Cogné J (1977) La chaine hercynienne ouest-européenne correspond-elle un orogène par collision? Proposition pour une interprétation géodynamique globale. Colloq Int CNRS 268: 11–129
Colchen M, Poncet D (1987) Sur l'âge tournaisien de la tectonique synschisteuse à Brétignolles-sur-Mer, Vendée littorale, Massif armoricain. CR Acad Sci Paris, 307 II: 277–280
Comble C, De Wever P, Ters M, Weyant M (1985) Découverte de Conodontes et de Radiolaires d'âge Tournaisien dans les schistes bariolés de Brétignolles-sur-Mer (Vendée). CR Acad Sci Paris 300, II, 17: 899–904
Darboux JR (1981) Caractérisation du régime cisaillant de la déformation hercynienne dans les Monts d'Arrée (Massif Armoricain, France). CR Acad Sci Paris 292 II: 1497–1500
Darboux JR, Garreau J (1976) Précisions sur la structure de l'Arrée et des piémonts dans leur terminaison occidentale (Massif Armoricain, France). CR Acad Sci Paris 283 II: 1007–1010
Darboux JR, Le Gall B (1988) Les Montagnes Noires: cisaillement bordier méridional du bassin carbonifère de Chateaulin (Massif Armoricain, France). Caractéristiques structurales et métamorphiques. Geodin Acta 2, 3: 121–133
Deutsch S, Chauris L (1965) Age de quelques formations cristallophylliennes et granitiques du Pays de Léon (Finistère). CR Acad Sci Paris 260 II: 615–617
Diot H (1980) Recherches structurales dans la partie orientale du domaine ligérien (massif armoricain). Thèse 3è cycle, Université de Nantes, 147 pp
Diot H, Bouchez JL, Blaise J (1983) La bande ductile Lanvaux – Les-Ponts-de-Cé entre Redon et Angers. Une composante du cisaillement hercynien du Massif Armoricain (France). Bull Soc Géol Fr 7, XXV, 2: 155–167
Dissler E (1987) Evolution géodynamique cadomienne du Nord Cotentin. Thesis, Université Caen (unpublished)
Doré F, Dupret L, Le Gall J, Chalot-Prat F (1977) Carte géologique de la France à 1/50000: la Ferté-Macé. Serv Géol National, Bur Rech Géol Min Orléans
Doré F, Le Gall J, Rioult M, Kuntz G, Dassibat C, Verron-G (1982) Carte géologique de la France à 1/50000: Alençon. Serv Géol National, Bur Rech Géol Min, Orléans
Doré F, Le Gall J, Dupret L, Lebert A, Rioult M (1986) Carte géologique de la France à 1/50000: Villaines-la-Juhel. Serv Géol National, Bur Rech Géol Min, Orléans
Dubreuil M (1986) Evolution géodynamique du Paléozoïque ligérien (Massif Armoricain) Thèse, Univ Nantes (inédit)
Edel JB (1987) Paleopositions of the Western Europe Hercynides during the late Carboniferous deduced from paleomagnetic data: consequences for "Stable Europe". Tectonophysics 139: 31–41
Gapais D, Le Corre C (1980) Is the Hercynian belt of Brittany a major shear zone? Nature 288, 5791: 574–576
Godard G (1981) Lambeaux probables d'une croûte océanique subductée: les éclogites de Vendée (massif armoricain, France). Thèse 3è cycle, Université de Nantes
Gore B, Le Corre C (1987) Cinématique hercynienne du cisaillement nord-armoricain à la bordure du granite syntectonique de Saint Renan-Kersaint (Finistère). Bull Soc Géol Fr (8), III, 5: 811–819
Gresselin F (1990) Evolution structurale d'une transversale du socle cadomo-varisque du Nord-Est de l'Armorique. Insertion dans un profil ouest-européen. Thesis, Université Caen (unpublished)
Guillet Ph, Bouchez JL, Vigneresse JL (1985) Le complexe granitique de Plouaret (Bretagne): mise en évidence structurale et gravimétrique de diapirs emboités. Bull Soc Géol Fr (8), I: 503–513
Guineberteau B (1986) Le massif granitique de Mortagne sur Sèvre (Vendée). Structure, gravimétrie, mise en place; distribution de U-Th-K Thèse 3è cycle, Université de Nantes, mémoire CREGU 11
Guineberteau B, Bouchez JL, Vigneresse JL (1987) The Mortagne granite pluton (France) emplaced by pull-apart along a shear zone: Structural and gravimetric arguments and regional implication. Geol Soc Am Bull 99: 763–770
Hanmer SK, Le Corre C, Berthe D (1982) The role of hercynian granites in the deformation and metamorphism of Brioverian and Paleozoic rocks of Central Brittany. J Geol Soc (Lond) 139: 85–93
Haq BU, Van Eysinga FWB (1987) Geological time table (Fourth edition). Elsevier, Amsterdam
Hirbec Y (1979) Le complexe basique de Belle-Isle-en-Terre (Côtes-du-Nord): sa place dans l'évolution géodynamique du nord du Massif Armoricain. Thèse 3ème cycle, Université Rennes, 146 pp (unpubl)
Houlgatte E, Le Herisse A, Pelhate A, Rolet J (1988) Evolution géodynamique du bassin carbonifère de Laval. Bull Bur Rech Géol Min, (Orléans) Géol Fr 1: 27–46
Jegouzo P (1980) The South Armorican Shear Zone. J Struct Geol 2: 39–47
Jegouzo P, Rossello EA (1988) La Branche Nord du Cisaillement Sud-Armoricain (France): un essai d'évaluation du déplacement par l'analyse des mylonites. CR Acad Sci Paris 307, II: 1825–1831
Jouvin F (1986) Etude du contrôle structural et des conditions P.T. lors de la mise en place d'un leucogranite en contexte de collision: exemple du massif d'Odet-Lestonan (Finistère). Rapp. DEA, Univ Brest (inédit)
Juignet P, Lebert A, Le Gall J (1984) Carte géologique de la France à 1/50000: Fresnay-sur-Sarthe. Serv Géol National, Bur Rech Géol Min, Orléans
Leake BE (1978) Nomenclature of amphiboles. Am Mineral 63: 1023–1052
Le Corre C (1978) Approche quantitative des processus synschisteux. L'exemple du segment hercynien de Bretagne Centrale. Thèse, Univ Rennes (inédit)
Ledru P, Marot A, Herrouin Y (1986) Le synclinorium de Saint-Georges-sur Loire: une unité ligérienne charriée sur le domaine centre armoricain. Découverte de métabasite à glaucophane sur la bordure sud de cette unité. (The Saint-Georges-sur-Loire synclinorium: a ligerian unit thrust over the centro-armorican domain. Discovery of glaucophane occurence in a metabasite on the southern limb of this unit). CR Acad Sci Paris 303, II, 10: 963–968
Le Metour J (1989) Carte géologique à 1/50000 Cholet (n° 510) et notice. Bur Rech Géol Min, Orléans
Le Metour J, Wyns R (1988) The Precambrian of the Vendée. In: Zoubek V (ed) Precambrian in Younger Fold Belts. Wiley, New York, PICG, project 22, pp 487–511
Lerouge G (1988) Tectogenèse comparée de deux segments de la chaîne hercynienne: le Massif Central français septentrional et le Sud du Massif Armoricain. Mémoire Géodiffusion Paris, 2, 363 pp, 130 fig

Lerouge G, Quenardel JM (1988) Les zones de cisaillement carbonifères dans les plutons vendéens et leurs prolongations dans le nord-ouest au Massif central français). Bull Soc Géol Fr (8), IV, 5: 831–838
Marchand J (1981) Ecaillage d'un "mélange tectonique" profond: le complexe cristallophyllien de Champtoceaux CR Acad Sci Paris 293 II: 223–228
Martinez F, Rolet J (1988) Late palaeozoic metamorphism in the Northwestern Iberian Peninsula, Brittany and related areas in SW Europe. In: Harris AL, Fettes DJ (eds) The Caledonian-Appalachian orogen. Geol Soc Spec Publ 38: 611–620
Melou M, Plusquellec Y (1967) Répartition de la pyrophyllite dans quelques niveaux du Briovérien et du Primaire armoricain. CR Acad Sci Paris 265 II: 14–16
Paquette JL, Bale P, Ballevre M, Georget Y (1987) Géochronologie et géochimie des éclogites du Leon: nouvelles contraintes sur l'évolution géodynamique du Nord-Ouest du Massif Armoricain. Bull Minéral 110: 683–696
Paradis S (1981) Le métamorphisme hercynien dans le domaine Centre Armoricain Occidental. Thèse, Univ Brest (inédit)
Paradis S, Velde B, Nicot E (1983) Chloritoïd-Pyrophyllite-Rectorite facies rocks from Brittany, France. Contrib Mineral Petrol 83: 342–347
Paris F, Jegouzo P (1976) La bordure Mancellia-Synclinorium Médian Armoricain: une limite géotectonique majeure de l'édifice armoricain. 4ème Réun Ann Sci Terre Res: 317
Peucat JJ, Charlot R, Mifdal A, Chantraine J, Autran A (1979) Définition géochronologique de la phase bretonne en Bretagne centrale. Etude Rb-Sr de granites du domaine centre armoricain. Bull Bur Rech Géol Min (Orléan) (2), I, 4: 349–356
Peucat JJ, Auvray B, Hirbec Y, Calvez JY (1984) Granites et cisaillements hercyniens dans le Nord du Massif Armoricain: géochronologie Rb-Sr. Bull Soc Géol Fr (7), XXVI: 1365–1373
Renault JF, Bouton P (1986) Le bassin décrochant Carbonifère de Faymoreau (Vendée). 11è réunion des Sciences de la Terre, Clermont-Ferrand, p 157 (Abstr) (Livre en dépôt à la Société Géologique de France)
Rolet J (1984) Grabens losangiques (pull-apart) en régime de décrochement. Le rôle des coulissements hercyniens dans l'individualisation des bassins carbonifères du Massif Armoricain. Ann Soc Géol Nord C, III: 209–220
Rolet J, Deunff J (1982) Les formations paléozoïques à la bordure du dôme briovérien du Plourac'h (Côtes-du-Nord): arguments en faveur de l'allochtonie des terrains paléozoïques. Bull Bur Rech Géol Min, Géol Fr I, 1–2: 39–45
Rolet J, Thonon P (1978) La semelle d'un charriage hercynien majeur effondrée par un réseau de fractures en régime coulissant dextre; sa mise en place grâce aux marqueurs filoniens de la Rade de Brest (Massif Armoricain). CR Acad Sci Paris 287 II: 1099–1102
Rolet J, Plusquellec Y, Babin C, Deunff J (1986a) Famennian regression and strunian grabens in the Armorican Basin. A key-area: Western Brittany. Ann Soc Géol Belg Special volume "Aachen 1986", 109: 197–203
Rolet J, Le Gall B, Darboux JR, Thonon P, Gravelle M (1986b) L'évolution géodynamique dévono-carbonifère de l'extrémité occidentale de la chaîne hercynienne d'Europe sur le transect Armorique-Cornwall. Bull Soc Géol Fr (8) II, 1: 43–54
Rolet J. et Yesou H. (1990) Apport de l'imagerie SPOT à la reconnaissance en pays couvert de plans de cisaillement C et de schistosité S. Le cas du Cisaillement Sud Armoricain (France). CR Acad Sci Paris, 311, II: 1075–1080
Sagon JP (1976) Contribution à l'étude géologique de la partie orientale du Bassin de Chateaulin (Massif Armoricain): stratigraphie, volcanisme, tectonique. Thèse, Univ Paris (inédit)
Ters M (1979) Les synclinoriums paléozoïques et le Précambrien sur la façade occidentale du Massif vendéen. Bull Bur Rech Géol Min I, 4: 293–301
Thiéblemont D, Cabanis B (1986) Découverte d'une association de volcanites d'arc et de basaltes de type "MORB" dans la formation paléo-volcanique silurienne de la Meilleraie, Vendée, France. CR Acad Sci Paris 302, II, 9: 641–646
Thiéblemont D, Cabanis B, Wyns R, Treuil M (1987) Etude géochimique (majeurs et traces) de la formation amphibolique de Saint-Martin-des-Noyers (complexe cristallophyllien des Essarts, Vendée). Mise en évidence d'un paléoarc insulaire dans la partie interne de l'orogène varisque. Bull Soc Géol Fr (8), 2: 371–378
Vidal P (1976) L'évolution polyorogénique du Massif armoricain: apport de la géochronologie et de la géochimie isotopique du strontium. Thèse d'Etat, Université de Rennes, 142 pp
Watts MJ, Williams GD (1979) Fault rocks as indicators of progressive deformation in the Guingamp region, Brittany. J Struct Geol I: 323–332
Wyns R, Lablanche G, Lefavrais-Raymond A, Bresson G, Biron R (1988) Carte géologique à 1/50000 Chantonnay (n° 563) et notice. Bur Rech Géol Min, Orléans
Wyns R, Lardeux H, Weyant M (1989a) Présence de Dévonien dans le Groupe de Réaumur (synclinal de Chantonnay, Vendée); conséquences sur l'évolution géodynamique Varisque de la Vendée. CR Acad Sci Paris 308, II: 855–860
Wyns R, Godard G, Lablanche G, Biron R, Bresson G (1989b) Carte géologique à 1/50000 La Roche-sur-Yon (n° 562) et notice. Bur Rech Géol Min Orléans
Yesou H, Rolet J (1990) Regional mapping of the South Armorican shear zone (Brittany, France) using remotely sensed SPOT imagery. ISPRS T Photogrammetry Remote Sensing 45: 419–427

Magmatic Record of the Geodynamic Evolution of Brittany and Vendée During Paleozoic Times: Trace Element Constraints

D. THIÉBLEMONT and B. CABANIS

1 Introduction

The Paleozoic terranes of W. France are classically divided into two tectonic domains (Fig. 1): the "external" (epizonal) Middle Armorican Domain (Cogné 1974) and the epi-to catazonal Ligerian Domain, which is considered as the internal Variscan zone (Autran and Cogné 1980). The Ligerian Domain appears, in fact, as a complex puzzle. It is composed of geological formations with contrasting tectonic and metamorphic evolution. The geochronological, petrological, structural, and geochemical data are still insufficient to allow a detailed geodynamic model to be drawn. For instance, there is no consensus about the mode and timing of the pre-tectonic events such as rifting, oceanization, or subduction (or subductions?).

Geological, petrological, geochronological, and geochemical data are available for many Paleozoic volcanics of Brittany and Vendée. Only the geochronological and geochemical data will be summarized here. In particular, the geochemical investigation (Cabanis 1986; Thiéblemont 1988 and references therein) has shown that some trace elements (Th, Ta, Hf, Tb, Zr and, in most cases, La) are especially stable during alteration and metamorphism. We will show that geotectonic discrimination diagrams based on such trace elements provide constraints on the pre-tectonic evolution of W. France during Paleozoic times.

2 Geological Outline

This study includes igneous rocks from the Middle Armorican Domain and Ligerian Domain (South Brittany and Vendée). Nine areas have been investigated (Fig. 1): the Crozon, Bolazec, and Maine areas (Middle Armorican domain), the Ile de Groix, Moëlan-Concarneau and Saint-Georges-sur-Loire sections (South Brittany), the Saint-Martin des Noyers, Choletais and Chantonnay areas (Vendée). Geochronological data (ages and techniques) for the igneous rocks are summarized in Table 1 and Fig. 1. The Moëlan, Ile de Groix, and Saint-Martin des Noyers areas have suffered a meso- to catazonal metamorphism. On the other hand, the metamorphism is epizonal in the Choletais, Chantonnay, and Saint-Georges-sur-Loire areas and in the Middle Armorican Domain (see Fig. 1). No ages are available for the Saint-Martin des Noyers Formation. Because of the low variability of the Rb/Sr ratio and absence of zircon, attempts to date these rocks by Rb-Sr and U-Pb methods have been unsuccessful (Calvez and Thiéblemont, unpubl. data). The petrological study (Thiéblemont et al. 1988) shows, however, that the metabasites have suffered a major phase of Variscan recrystallization without evidence for an older regional event. This leads us to propose a Paleozoic age for the emplacement of the magmatic rocks (Thiéblemont 1988).

As is shown in Table 1, the unconformity of the Paleozoic terranes over the Cadomian basement is Cambrian to Ordovicivan in the Crozon, Choletais, and Maine areas and more than 100 Ma younger in the Bolazec area. Rhyolites and ignimbrites of Cambrian to Tremadocian age outcrop in the Maine, Choletais (Le Métour and Bernard-Griffiths 1979), and Chantonnay (Wyns 1980) areas. In the Crozon area, Bonjour and Chauvel (1988) have recognized "the influence of a syn-depositional acid(ic) volcanism" in the Ordovician initial sedimentation. Evidence for the occurrence of reworked felsic volcanics has also been described by Wyns (1980) in the Cambrian to Ordovician sediments of the Chantonnay area. The existence of a felsic-dominated magmatic event of Lower Paleozoic age is also indicated by the occurrence of Cambrian to Ordovician granitic plutons, both in Vendée (Pouzauges granite, Wyns and Le Métour 1983; Saint-Mars du Désert orthogneiss, Paquette et al. 1984) and South Brittany (Lanvaux Granite, Vidal 1980; Téauray and Languidou orthogneisses; Peucat 1983; Douarnenez Trondhjémite; Barrière et al. 1971; Vidal 1980). In the Moëlan-Concarneau area, the Moëlan orthogneiss has yielded consistent

J. D. Keppie (Ed.): Pre-Mesozoic Geology
in France and Related Areas

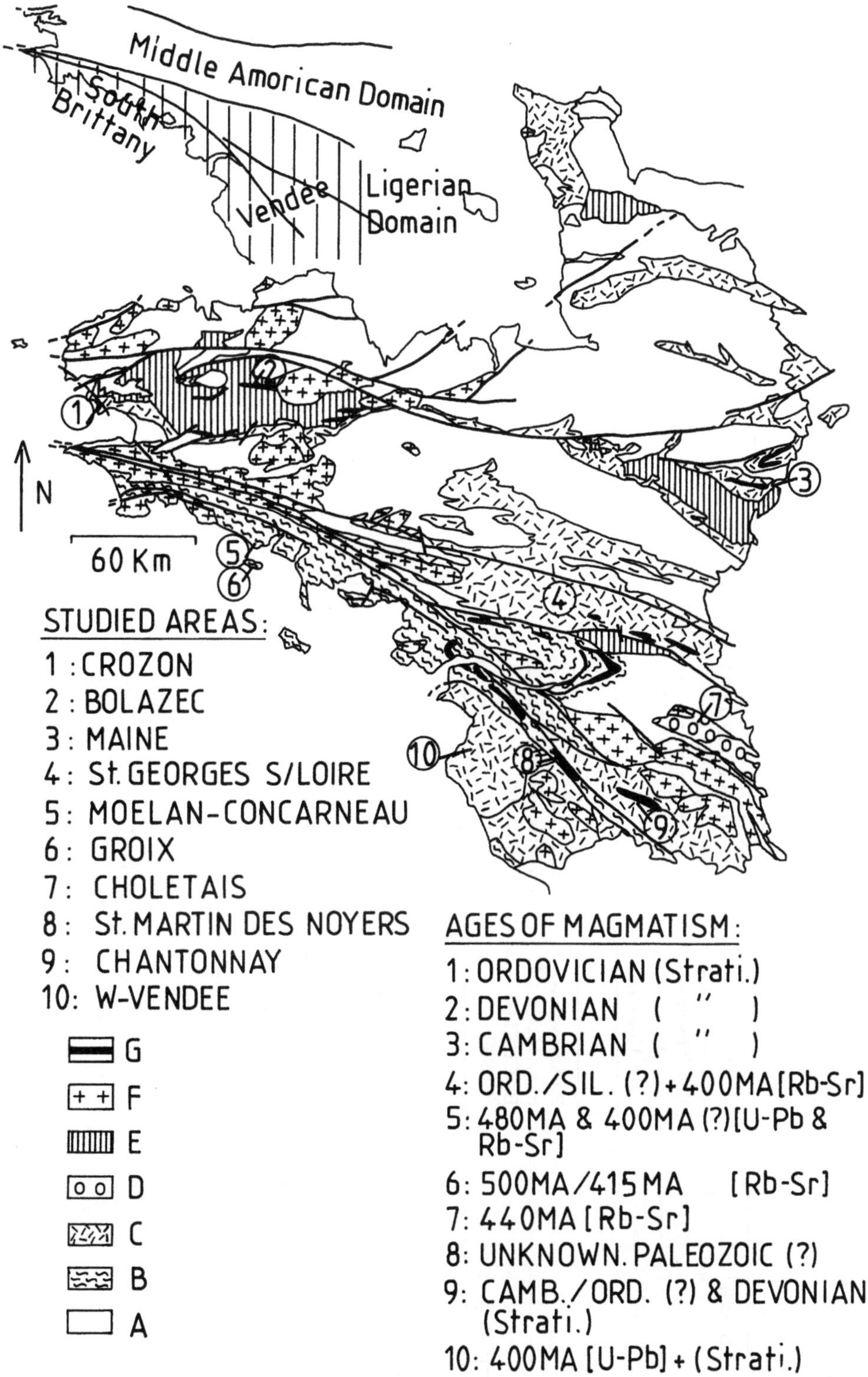

Fig. 1. Simplified geological map of the Armorican Massif (Bébien and Gagny 1980), location of the studied areas and summary of the available geochronological data for the igneous formations. References of the data are as follows: *1* Barrière et al. (1975); *2* Cabanis et al. (1982); *3* Le Gall and Cabanis (1985); *4* Carpenter et al. (1982); *5* Calvez (1976); Vidal (1980), and Béchennec et al. (in press); *6* Bernard-Griffiths et al. (1986); *7* Le Métour and Bernard-Griffiths (1979) and Thiéblemont (1988); *8* Thiéblemont (1988); *9* Wyns et al. (1988, 1989); *10* Peucat et al. (1986). *A* Precambrian basement; *B* strongly deformed and metamorphosed Precambrian to Devonian terranes of the Ligerian domain; *C* slightly deformed and metamorphosed pre-Variscan terranes; *D* volcanic and hypovolcanic Choletais complex; *E* Devonian and Carboniferous terranes (mainly Dinantian); *F* metabasalts and amphibolites; *G* granitoids

Table 1. Summary of the nature and age of some Paleozoic volcanic series from the Armorican Massif. The references of the data are given in the text and in the legend to Fig. 1. Lithostratigraphical correlations allow four periods of magmatic activity to be defined: the first around 500–470 Ma, the second at the Ordovician/Silurian boundary, the third at the Silurian/Devonian boundary and the fourth around 380–360 Ma. In the mesozonal terranes of the Ligerian Domain (South-Brittany and W. Vendée), a major tectono-metamorphic event (anatexis) is dated around 380 Ma (Vidal 1980; Peucat 1983). The (?) symbol indicates vaguely dated formations or events

	← MIDDLE ARMORICAN DOMAIN →		← LIGERIAN DOMAIN →						
	← EPIZONAL TERRANES →						← MESOZONAL TERRANES →		
	CROZON	BOLAZEC	MAINE	St GEORGES	CHOLETAIS	CHANTONNAY	ILE DE GROIX	MOELAN CONCARNEAU	W-VENDEE
350MA									
DEVONIAN		Basaltic volcanism (Gedinian /Givetian)				Basaltic volcanism (MEILLERAIE series)	---MAJOR TECTONO-METAMORPHIC EVENT---		
				Basaltic volcanism (INGRANDES series) (400Ma)				Orthogneisses (400Ma), minor metabasites (MERRIEN member)	Felsic volcanism (400Ma) ('PORPHYROIDES')
400MA SILURIAN	--UNCONFORMITY--								
	Basaltic volcanism (Caradoc)			Basaltic volcanism (MOULIN BACHELOT series) (Ord/Sil)	Rhyolites, microgranites (440Ma) and associated dolerites	? Rhyolites and dolerites ?			
450MA ORDOVICIAN	Reworked acidic magmatic rocks (473Ma)		?	? --UNCONFORMITY-- ?			? Metabasites (Blueschists) (Ordovician ?) ?	Granite (480Ma) minor associated basic rocks (amphibolites) (NERLY member)	
	--UNCONFORMITY--								
500MA			Felsic volcanism		Felsic volcanism	Felsic volcanism			
			? --UNCONFORMITY-- ?						
CAMBRIAN 550MA					? --UNCONFORMITY-- ?	? Felsic volcanism --UNCONFORMITY-- ?			

U-Pb and Rb-Sr ages of 470MA-480MA (Calvez 1976; Vidal 1980; Béchennec et al. in press). In this metagranite, we have studied metabasic units that we interpret as pene-contemporaneous dykes (Thiéblemont et al. 1989b).

Following this event, basalts and rhyolites were emplaced close to the Ordovician/Silurian boundary. Basalts are well dated in the Crozon area (Barrière et al. 1975), where pillow lavas are interbedded within the fossiliferous Ordovician sedimentary sequence. In the Saint-Georges-sur-Loire area, a similar stratigraphic position is suggested for the basic lavas of the Moulin-Bachelot series (Carpenter et al. 1982). In these two areas, the magmatism is strictly basic and forms only a minor part of the whole Ordovician sequence (Barrière et al. 1975; Carpenter et al. 1982). In Vendée, the Choletais massif is an old volcanic to hypovolcanic bimodal complex of Upper Ordovician age (Le Métour and Bernard-Griffiths 1979). The volcanic pile is very thick (more than 1000 m?) and strictly composed of rhyolites and ignimbrites (Boyer 1974; Le Métour and Bernard-Griffiths 1979; Thiéblemont 1988). To the east, the lavas show a gradual contact (Thiéblemont 1988) with a hypovolcanic complex composed of microgranites with subordinate gabbros, dolerites and hybrid rocks. Field and petrographical evidence (lobate contacts between felsic and basic magmas, occurrence of hybrid rocks consisting of finely mixed granitic and gabbroic "patches", etc.), shows that the felsic and basic magmas were emplaced contemporaneously (Autran in Le Métour and Bernard-Griffiths 1979; Thiéblemont 1988).

Magmatic rocks of Ordovician age have also been described in the Chantonnay area (rhyolites and associated dolerites of the Sigournais Formation, Wyns et al. 1988) and on the Ile de Groix, where a broad "Lower Paleozoic" age (500 to 400 Ma) is inferred from the Rb-Sr data obtained on micashists interbedded with the metabasites (Bernard-Griffiths et al. 1986). In these two areas, the magmatic rocks are mostly metabasites but form only a minor part of the whole Paleozoic sequence.

A Devonian basaltic volcanism is recognized (and well dated, Fig. 1) in the Bolazec (Cabanis et al. 1982), Saint-Georges-sur-Loire (Ingrandes series) (Carpenter et al. 1982) and Chantonnay (Meilleraie series) (Wyns et al. 1989) areas. In the Middle Armorican Domain the Saint-Jean du Doigt gabbro and Barnenès amphibolites are probable extensions of the Bolazec Formation (Cabanis 1986). In the Moëlan-Concarneau area, the Siluro-Devonian magmatism is represented by the orthogneisses and metabasites of the Merrien group (Thiéblemont et al. 1989b). The felsic rocks have yielded whole-rock Rb-Sr ages of 415 Ma (Béchennec et al. in press). This age is close to that of the Porphyroides Formation (Peucat et al. 1986), a thick volcanic and sedimentary sequence located in W-Vendée, at the SW part of the Ligerian domain (Fig. 1) and probably overthrust by the Saint-Martin des Noyers Amphibolites (Brun and Burg 1982).

In the Ligerian Domain, the Devonian basaltic volcanism appears, in some areas (Chantonnay area), to post-date the main tectono-metamorphic phase, which has been dated around 380 Ma by Rb-Sr (whole rocks) and U-Pb (zircon) methods (Peucat et al. 1978; Vidal 1980; Peucat 1983).

The Devonian volcanism is strictly mafic in the Chantonnay area, where the 1 to 1.5 km basaltic Meilleraie series conformably overlies the Paleozoic sedimentary sequence (Wyns 1980). In the Saint-Georges-sur-Loire area, the Ingrandes series is also mainly basaltic (Carpenter et al. 1982). Its thickness is not known. On the other hand, the Siluro-Devonian magmatic rocks of the Merrien Group (Moëlan-Concarneau area) are mainly orthogneisses (meta-rhyolites?), and metabasites outcrop as narrow and scarce units (Thiéblemont et al. 1989b). In W. Vendée, the voluminous Porphyroides Formation is strictly composed of ignimbrites and rhyolites (Boyer 1974; Ters 1979; Chalet 1985) and associated terrigeneous sediments.

Finally, the Saint-Martin-des-Noyers Formation is composed of mesozonal metabasites (amphibolit to greenschist facies) (Thiéblemont et al. 1988) with subordinate garnet-gneisses. The mafic and acidic units are narrow (10 cm to 10 m thick) and closely interlayered. They are interpreted as ancient flows, sills, or dykes (Thiéblemont et al. 1987). The whole formation appears as an old volcanic to hypovolcanic complex.

3 Geochemical Results and Geotectonic Discrimination

The major and trace element analyses of the igneous rocks listed in Table 1 and Fig. 1 are available in Cabanis (1986), Cabanis et al. (1982), Thiéblemont et al. (1987), Thiéblemont (1988), and Thiéblemont et al. (1989b), and discussed in detail therein. All trace elements have been analyzed by INA at the P. Sue Laboratory (Saclay, France), using the technique described by Chayla et al. (1973). The precision and accuracy of the method have been estimated by Jaffrezic et al. (1980).

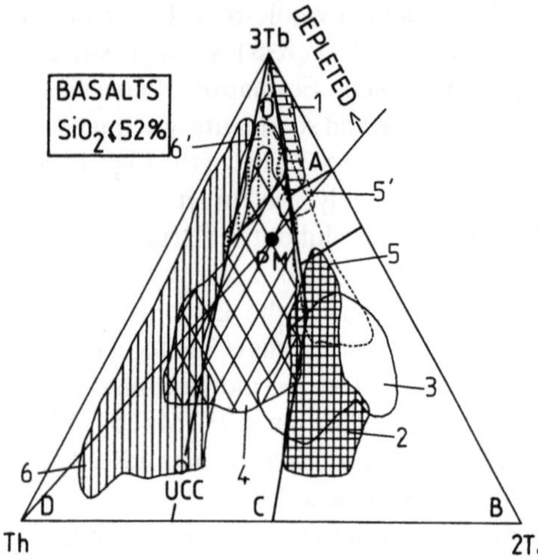

Fig. 2. Th-3Tb-2Ta diagram (Cabanis and Thiéblemont 1988) for the geotectonic discrimination of the basaltic rocks (Si02 < 52%). The fields noted *A, B, C, D,* and *D'* are defined as follows. *A* Passive margins (post-rift/pre-oceanic) basalts and N-MORB; *B* within-plate (oceanic + continental) and rift-related alkaline and transitional basalts; *C* tholeiitic basalts related to continental rifts ("continental tholeiites"); *D* subduction-related basalts; *D'* fore-arc and back-arc basin basalts. The values for the Primordial Mantle *(PM)* and upper continental crust *(UCC)* are from Wood et al. (1979b) and Taylor and Mac Lennan (1985), respectively. Depleted basalts are defined as having lower Ta/Tb ratios than the Primordial Mantle

3.1 Geotectonic Discrimination

In the following sections we will discuss the geodynamic implications of the trace element data by the mean of three discrimination diagrams: the Th-Tb-Ta (Cabanis and Thiéblemont, 1988), Yb vs. Ta (Pearce et al. 1984) and Rb-Tb-Ta (Thiéblemont and Cabanis 1990) plots. The Th-Tb-Ta and Rb-Tb-Ta diagrams are largely inspired from the Th-Hf-Ta triangle of Wood et al. (1979a) and Wood (1980), the Th/Yb vs. Ta/Yb diagram of Pearce (1982) and Ta+Yb vs. Rb diagram of Pearce et al. (1984), but include recent and sometimes unpublished data. They are based on geochemical parameters which are also accounted for by MORB- or chondrites-normalized "spidergrams", nevertheless they better fit a statistical approach to the problem of geotectonic discrimination.

As is shown in Fig. 2, the Th-Tb-Ta diagram almost perfectly discriminates between the basic magmas emplaced in areas of active subduction and those from anorogenic domains. The basalts from oceanic basins (fields 1, 2 and 5') plot close to the Ta-Tb side, along a trend which we will refer to as the Oceanic Mantle Trend. Considering that the partial melting of a given source tends to lower the Ta/Tb ratio in the residue, a basalt with a Ta/Tb ratio lower than that of chondrites (or Primordial Mantle) appears to derive from a depleted reservoir. Amongst the oceanic basalts (excluding the back-arc basin basalts) the N-type MORB are the only ones which show Ta/Tb ratios lower than that of the Primordial Mantle (Fig. 2). These depleted MORB are clearly discriminated from the Ocean Island Basalts (OIB) and Plume-related MORB (the E-type MORB of Wood et al. 1979a). On the other hand, the continental transitional and alkaline basalts (depicted as pre-rift and syn-rift, field 3, in Fig. 2) cannot be discriminated from OIB. In the Th-Tb-Ta diagram, the continental tholeiites define a separate field (field 4) lying to the left of the Oceanic Mantle Trend (fields 1+2) and slightly overlapping to the fields of subduction-related and arc-basin basalts (fields 6 and 6'). This shift might reflect the contamination of mantle-derived magmas by the upper continental crust (Dupuy and Dostal 1984), and/or the contamination of the mantle by a component with a high Th content and high Th/Ta ratio. Very few continental basalts show evidence for the involvement of a strongly depleted mantle such as the N-MORB reservoir. From the pre-rift and syn-rift continental, to the post-rift oceanic basalts, the Th-Tb-Ta diagram shows the involvement of an increasingly depleted mantle. The location of the subduction-related basalts close to the Th-Tb-side of the diagram reflects high values of their Th/Ta ratio (Joron and Treuil 1977; Wood et al. 1979a; Pearce 1982). In island-arc settings, the high Th/Ta ratio, and the low Ta/Tb ratio are generally interpreted in terms of mixing between a variably depleted oceanic mantle and Th-rich and Ta-poor fluids derived from the subducted oceanic plate (Gill 1981; Pearce 1982; Ryerson and Watson 1987). The basalts and andesites erupted at active continental margins show higher Ta/Tb ratios than island-arc basalts, which suggests that they derive from a slightly depleted to enriched part of the subcontinental mantle.

Pearce et al. (1984) have shown that it is possible to partly discriminate the geodynamic setting of granites using Yb vs Tb or Yb+Ta vs Rb diagrams. More recently, we have shown (Thiéblemont and Cabanis 1990) that the discrimination may be completed by means of a Rb-Tb-Ta triangle. As is shown in Fig. 3, this diagram shows a clear separation between the syn-collision, post-collision and anorogenic (or with-

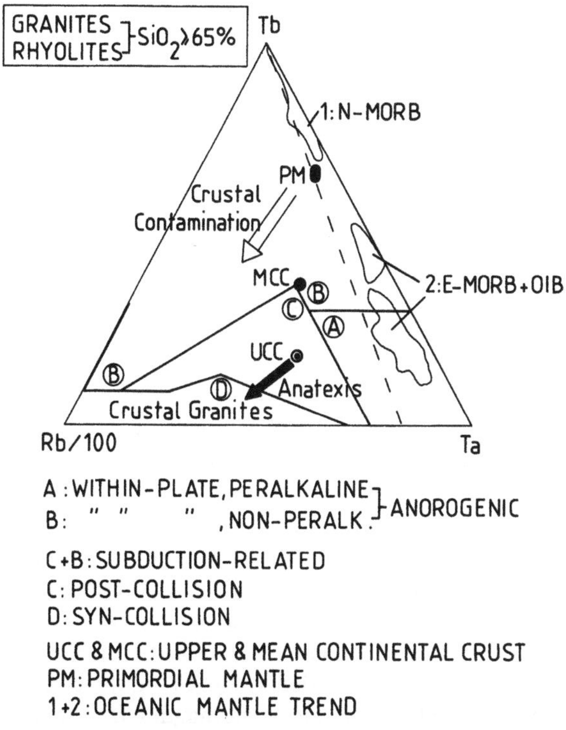

Fig. 3. Rb/100-Tb-Ta diagram (Thiéblemont and Cabanis 1990) for the discrimination of granites and rhyolites (Si02 > 65 %). Note that the subduction-related felsic rocks overlap fields *B* and *C*. The values for the upper and mean continental crust are from Taylor and MacLennan (1981) (assuming Ta = Nb/16). Note that the Oceanic Mantle Trend is situated close to the Tb-Ta side of the triangle. The *empty arrow* exemplifies the process of Crustal Contamination as it is defined in the British Tertiary Volcanic Province. (Walsh at al. 1979 and Thompson et al. 1982). The *full arrow* exemplifies the process of anatexis of the upper continental crust assuming that (D)Rb < (D)Ta < (D)Tb (where D is the solid/liquid partition coefficient during partial melting). This theoretical trend accounts for the high Ta/Tb and Rb/Ta ratios of the syn-collision "crustal" granites. (See Thiéblemont and Cabanis 1990 for further details)

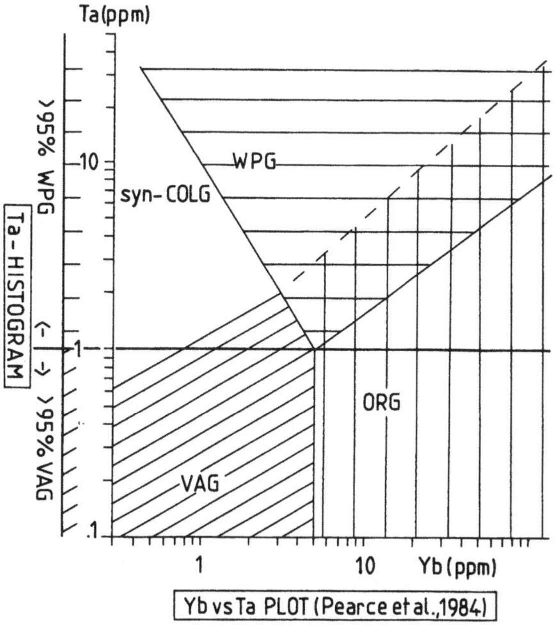

Fig. 4. Yb vs. Ta diagram of Pearce et al. (1984) for the geotectonic discrimination of granitic rocks. *VAG* volcanic arc granites ("subduction related"); *ORG* ocean ridges granites; *syn-COLG* syn-collision granites; *WPG* within-plate granites. From this plot and our own data (see data base in Thiéblemont and Cabanis 1990) it appears that the Ta content provides an almost perfect discriminant between the within-plate and subduction-related granites and rhyolites

in-plate) granites and rhyolites. On the other hand, the field of subduction-related acidic magmatic rocks overlap to those of anorogenic and post-collision granites and rhyolites. The Yb vs. Ta diagram of Pearce et al. (1984) allows the within-plate granites to be discriminated from the subduction-related granites (Fig. 4). We have not analyzed Yb; nervertheless, as is shown in Fig. 4, the Yb vs. Ta discrimination is principally based on the Ta concentration. Our own data base confirms that the Ta concentration is lower that 1 ppm in more than 95 % of the available analyses of subduction-related felsic rocks and higher than 1 ppm in more than 95 % of the within-plate granites and rhyolites.

Some petrogenetic implications of the Rb-Tb-Ta triangle are summarized in Fig. 3. It must be noted that because Rb is a mobile element, the use of this diagram must be restricted to unaltered magmatic rocks.

3.2 Lower Paleozoic Magmatism

Trace element analyses of the Cambrian (Maine, Chantonnay, and Moëlan areas) to Ordovician (Choletais and Chantonnay areas) acidic magmatic rocks have been plotted in the Rb-Tb-Ta diagram (Fig. 5). The Moëlan samples show high Ta/Tb and Ta/Rb ratios. Such values are characteristic of the peralkaline granites and rhyolites associated with anorogenic settings. As noted by Vidal (1980), the low Sr isotopic ratio (around 0.703; Calvez 1976; Vidal 1980) of these rocks is consistent with a within-plate setting. The rhyolites and microgranites from other areas display lower Ta/Rb and Ta/Tb ratios. The Ta/Tb ratio is generally lower than that of collision-related acidic magmas (< 1, see Fig. 3). Moreover, Fig. 6 shows that the Ta concentration is generally higher than 1 ppm, except for the Maine samples. As noted earlier, such a value is very uncommon in subduction-related acidic magmas, which suggests a within-plate setting for the Lower to Upper Ordovician felsic meta-volcanics of

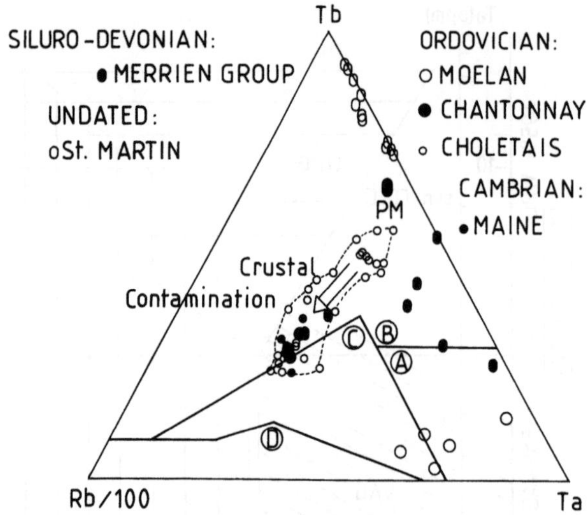

Fig. 5. Rb-Ta-Tb diagram for the Paleozoic acidic magmatic rocks from Brittany and Vendée. Fields *A, B, C,* and *D* are defined in Fig. 3. Only very few analyses plot in the field of collision granites and rhyolites and most of them are located in the field *B* which is common to subduction-relatd and anorogenic nonperalkaline felsic rocks. The trend for the rhyolites and microgranites from the Choletais Massif may be interpreted in terms of crustal contamination of basaltic magmas by the continental crust. This model has been developed in Thiéblemont (1988)

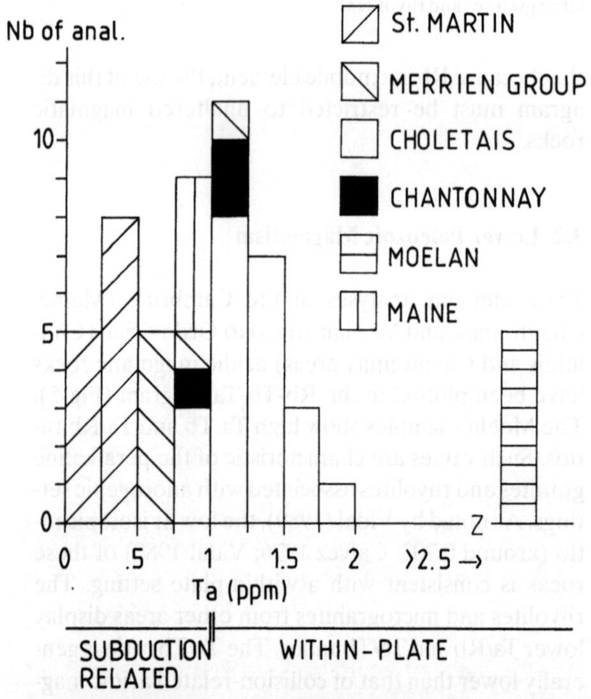

Fig. 6. Ta-histogram for the Paleozoic acidic magmatic rocks from Brittany and Vendée. From this plot and the Rb-Tb-Ta diagram (Fig. 5) the gneisses from the Saint-Martin-des-Noyers Formation and the Merrien Group may clearly be defined as subduction-related. On the other hand, the rhyolites from the Chantonnay and Choletais areas and the Moëlan orthogneiss appear as within-plate

the Ligerian Domain (Moëlan, Chantonay and Choletais areas). On the other hand, the Cambrian lavas from the Maine area show trace element characteristics which resemble those of recent subduction-related acidic magmatic rocks.

In the Th-Tb-Ta diagram (Fig. 7), the analyses of Ordovician basic lavas from Brittany and Vendée plot outside the fields of N-MORB, subduction-related basalts and arc-basin basalts. The Moëlan and Saint-Georges samples plot on the Oceanic Mantle Trend, which precludes a significant contamination by the continental crust. The Crozon, Choletais, and Chantonnay samples plot in the field of continental tholeiites, to the left of the Oceanic Mantle Trend. The Ta/Tb ratio in the Moëlan, Saint-Georges and Crozon metabasites is generally higher than that of the Primordial Mantle. This ratio is close to the Primordial Mantle value in the Choletais and Chantonnay metabasites. In the Th-Ta-Tb diagram, these analyses lie on two segments. Such trends are consistent with a process of mixing (Wood 1980). Considering the field and petrographical evidence suggesting the contemporaneity of the basic and acidic magmas (see Sect. 2), we interpret the trends of Fig. 7 as the result of mixing

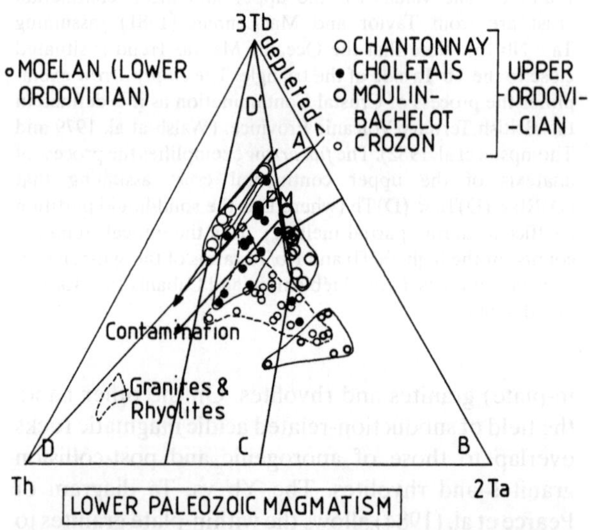

Fig. 7. Th-Tb-Ta diagram for the Lower Paleozoic metabasites from the Middle Armorican (Crozon area) and Ligerian (Chantonnay, Moëlan-Concarneau and Chantonnay areas and Moulin-Bachelot series) domains. Fields *A, B, C, D,* and *D'* are defined in the legend to Fig. 2. In the Moëlan-Concarneau area, the metabasites occur as pene-contemporaneous units (dykes) enclosed in the Moëlan orthogneiss. Note the location of the analyses in the within-plate fields *(B* and *C)* of the diagram and their undepleted to slightly depleted geochemical characteristics. Field, petrographical, and geochemical data detailed in Thiéblemont (1988) suggest that, in the Choletais and Chantonnay areas, the basaltic magmas have been variably contaminated by the contemporaneous acidic magmas

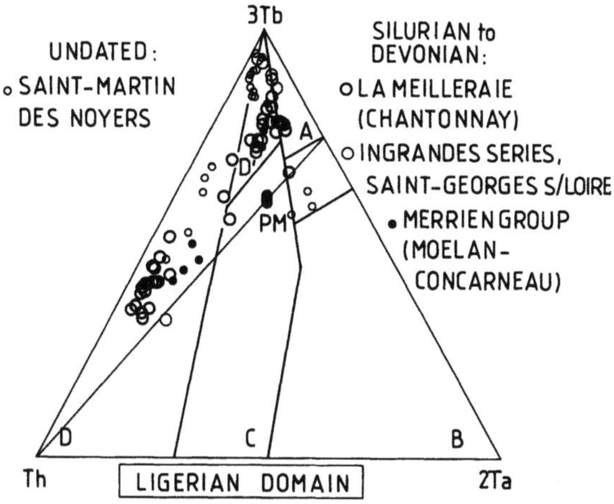

Fig. 8. Th-Tb-Ta diagram for the Siluro-Devonian to Devonian and undated metabasites of the Ligerian Domain. Fields *A, B, C, D,* and *D'* are defined in the legend to Fig. 2. Note the location of the analyses in the arc-basins and subduction-related fields *(D* and *D')* of the diagram and their depleted geochemical features

between the two magmas. This model has been developed in more detail in Thiéblemont (1988).

From rare-earth element patterns and Nd-Sr isotopic data, Bernard-Griffiths et al. (1986) have interpreted the metabasites from Ile de Groix to be related to ocean-island magmatic activity. We have recently noted (Thiéblemont et al. 1989; Triboulet et al. in press) that this conclusion is not consistent with the abundance of acidic metasedimentary rocks in the lithological sequence of Ile de Groix. Considering that the OIB cannot be discriminated from the continental alkaline basalts (Fig. 2), we argue (Triboulet et al. in press) that the geochemical and lithological features of the Ile de Groix sequence best fits an intracontinental rift setting.

3.3 Devonian Magmatism

The Rb-Th-Ta diagram (Fig. 5) shows that the Siluro-Devonian gneisses from the Concarneau area (Merrien group) display a Ta/Tb ratio generally lower than that of peralkaline and collision-related granites and rhyolites. Some analyses plot close to the Tb-Ta side of the diagram, which probably reflects a loss of Rb during alteration. Nevertheless, because Ta and Tb are immobile or slightly mobile elements, the Ta/Tb values may be considered as "primary". The Ta histogram (Fig. 6) shows that the Ta content of the Devonian gneisses from the Merrien group is lower than that of the Ordovician felsic metavolcanites (Fig. 6), but in the range of that of subduction-related rhyolites and granites. Figures 5 and 6 show that the Ta/Tb ratio and Ta content of the garnet-gneisses from the Saint-Martin des Noyers Formation are even lower and their Rb content generally close to O ppm. The low rubidium content may be attributed to the loss of Rb during alteration (Thiéblemont 1988).

In the Th-Tb-Ta diagram (Fig. 8), the Devonian metabasites from the Ligerian domain (Chantonnay, Saint-Georges and Moëlan-Concarneau areas) define a broad linear trend which extends from the field of N-MORB toward the Th apex. The analyses lie across the field of subduction-related and arc-basin basalts. In the Meilleraie series, the detailed field study (Thiéblemont and Cabanis 1986; Thiéblemont 1988) shows that lavas with trace element characteristics of island-arc tholeiites erupted together with lavas and sills geochemically similar to recent back-arc basin basalts. Shortly after, MORB-type dolerites emplaced as shallow level intrusive units (sills and dykes). This succession is thought to reflect the rifting of the arc. The evolution from arc-type to MORB-type basalts has been described in recent back-arc basins of the W-Pacific margin (Hussong and Uyeda 1981; Wood et al. 1981).

In the Th-Tb-Ta diagram (Fig. 8), the basalts from the Ingrandes series (Saint-Georges-sur-Loire area) show a gradual evolution from the field of N-MORB to that of primitive arc-tholeiites. This volcanism could reflect the initial stage of development of an island arc (Cabanis 1986). These metabasites have also been interpreted as ancient back-arc basin basalts (Carpenter et al. 1982).

Figure 8 shows that the metabasites from the Merrien member (Moëlan-Concarneau area) display trace element characteristics of both fore-arc basalts and island-arc tholeiites (Thiéblemont et al. 1989b). These results are consistent with the "subduction-related" affinity of the associated gneisses (Figs. 5 and 6). Such a conclusion may also be drawn for the Saint-Martin des Noyers Formation in which metabasites overlap to the fields of subduction-related and arc-basin basalts (Fig. 8).

In the Th-Tb-Ta triangle (Fig. 9), the Devonian basalts and dolerites of the Bolazec, Saint-Jean-du-Doigt, and Barnenes areas plot along a sub-linear trend close to the Oceanic Mantle Trend. As is noted on Figs. 2 and 9, such a trend is also displayed by the pre- to post-rift basalts from recent continental extensional domains, such as the Afar-Red Sea Rift. On the other hand, the low Th/Ta ratio suggests that neither the continental crust, nor a "subduction-zone component" played a significant role in the genesis of the Devonian basalts from Middle Brittany. Further evidence for the continental rift-setting of the metavolcanites has been reported by Cabanis et al. (1982) and Cabanis (1986).

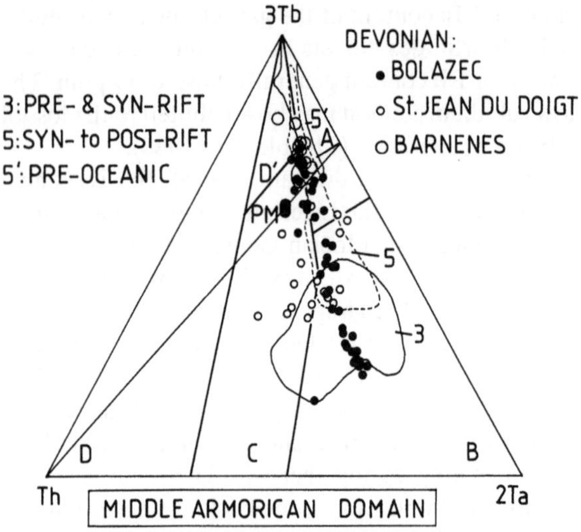

Fig. 9. Th-Tb-Ta diagram for the Devonian metabasites of the Middle Armorican Domain. Fields *A, B, C, D,* and *D′* are defined the legend to Fig. 2. Note the location of the analyses from the within-plate field *(B)* to the pre-oceanic + N–MORB field *(A)* of the diagram and their undepleted to depleted geochemical features

4 Geodynamic Implications

The trace element data support the definition of two main periods of magmatism in the Lower Ordovician to Devonian evolution of the Middle Armorican and Ligerian domains (Autran and Cogné 1980; Vidal 1980). During Lower Paleozoic times, the Armorican Massif evolves in a context of intracontinental rifting. The absence of metabasites with strongly depleted geochemical features (N-MORB type) suggests that oceanization has not occurred. A continental environment is also indicated by the occurrence of voluminous felsic magmatic rocks in the Choletais Massif and the wide distribution of terrigeneous sediments in the Lower Paleozoic sequences of the Ligerian and Middle Armorican domains. The variable thickness of the Upper Ordovician volcanic sequences suggests that the areas have suffered different rates of crustal thinning. This hypothesis is also supported by the geochemical data. In has been shown (Treuil and Varet 1973) that the Ta/Tb (and La/Yb) ratio decreases from the margins toward the axis of continental rifts. This results is illustrated by Fig. 2 (note the location of the pre-rift to post-rift basalts in this diagram). Likewise, the low Ta/Tb ratio of the Choletais and Chantonnay metabasites (Fig. 7) suggests that the original basalts erupted close to the axis of the Upper Ordovician rift.

The trace element data obtained from the Devonian ineous rocks show that, at this time, the Ligerian and Middle Armorican domains evolved in clearly distinct geodynamic contexts. The two main geochemical characteristics of the Devonian metabasites from South-Brittany and Vendée are their depleted (low Ta/Tb ratio) and arc- to arc-basin type (variable Th/Ta ratio) characteristics. Because all the sections studied show evidence for a Lower Paleozoic magmatism, we suggest that the depleted characteristics of the Devonian metabasites reflects the extraction of magmas during the Lower to Upper Ordovician. The arc-type features may be attributed to the occurrence of one or more subduction processes of Siluro-Devonan to Upper Devonian age. The available geochronological data show that at least two subductions have existed: one around 400 Ma (Moëlan and Saint-Georges areas) and the other around 360 Ma (Chantonnay area). In W. Vendée, the Porphyroides Formation has also been related to the existence of a subduction zone at the Silurian / Devonian boundary (Peucat et al. 1986). Finally, it may be noted that the 380 Ma age obtained from the migmatites of W. Vendée (Vidal 1980) and South-Brittany (Peucat 1983) suggests that subduction and collision have alternated.

In the Middle Armorican Domain, the Devonian magmatism is characteristic of a rifting event. The Bolazec series directly overlies the Cadomian basement. The results obtained from the paleomagmatism confirm the external character of the Middle Armorican Domain (Cogné 1974).

References

Autran A, Cogné J (1980) La zone interne de l'orogène varisque dans l'ouest de la France, et sa place dans le développement de la chaîne hercynienne. In: Cogné J, Slanski M (eds) Géologie de l'Europe, Colloque C6, 26° Congrès géologique international, Paris 1980. Mém Bur Rech Géol Min 108: 90–111

Barrière M, Cogné J, Vidal P (1971) La trondhjémite de Douarnenez: une intrusion magmatique ordoviciene dans les schistes cristallins du Sud-Finistère. CR Acad Sci Paris 273 (D): 1556–1559

Barrière M, Plusquellec Y, Garreau J, Chauvel JJ, Le Corre C, Mélou M, Thonon P, Darboux JR, Chauris L, Mullot B (1975) Geological map of Douarnenez and explanotary note. Bur Rech Géol Min, Orléans, France

Bébien J, Gagny C (1980) Volcanites du Précambrien au Crétacé et leur signification géostructurale. In: Autran A, Dercourt J (eds) Géologie de la France, Colloque C6, 26° Congrès géologique international, Paris 1980. Mém Bur Rech Géol Min 107: 99–135

Béchennec F, Thiéblemont D, Guenoc P, Delanoe Y, Lefort JP, Colleau A Geological map of Concarneau and explanatory note. Bur Rech Géol Min, Orléans, France (in press)

Bernard-Griffiths J, Carpenter MSN, Peucat JJ, Jahn BM (1986) Geochemical and isotopic characteristics of blueschists facies rocks from the Ile de Groix, Armorican Massif (northwest France). Lithos 19: 235–253

Bonjour JL, Chauvel JJ (1988) Un exemple de sédimentation initiale dans un bassin paléozoïque: étude pétrographique et géochimique de l'Ordovicien inférieur de la presqu'île de Crozon (Finistère). Bull Soc Géol Fr (8) IV 1: 81–91

Boyer C (1974) Volcanisme acide paléozoïque dans le Massif Armoricain. Thesis, University of Orsay

Brun JP, Burg JP (1982) Combined thrusting and wrenching in the Ibero-Armorican arc: a corner effect during continental collision. Earth Planet Sci Lett 61: 319–332

Cabanis B (1986) Identification des séries magmatiques dans les socles métamorphigues sur la base de critères géologiques, pétrologiques et géochimiques. Thesis, University P et M Curie, Paris

Cabanis B, Thiéblemont D (1988) La discrimination des tholéiites continentales et des basaltes arrière-arc: proposition d'un nouveau diagramme; le triangle Th-3xTb-2xTa. Bull Soc Géol Fr (8) IV 6: 927–935

Cabanis B, Chantraine J, Herrouin Y, Treuil M (1982) Etude géochimique (majeurs et traces) des spilites et dolérites de Bolazec. Mise en évidence d'un domaine en distension crustale au Dévonien inférieur en Bretagne Centre-Ouest. Bull Bur Rech Géol Min (2) 1–2: 47–61

Calvez JY (1976) Comportements des systèmes uranium-plomb et rubidium-strontium dans les orthogneiss d'Icart et de Moëlan (Massif armoricain). Theses, University of Rennes

Carpenter MSN, Peucat JJ, Pivette B (1982) Geochemical and geochronological characteristics of Paleozoic volcanism in the St Georges sur Loire synclinorium, South Armorican Massif, France. Evidence for prehercynian tectonic evolution. Bull Bur Rech Géol Min (2) I 1–2: 63–79

Chalet M (1985) Contribution à l'étude de la Chaîne hercynienne d'Europe occidentale. Etude stratigraphique et structurale des "porphyroïdes" et formations paléozoïques associées du Bas Bocage Vendéen (région de Mareuil-sur-Lay). Thesis, University of Poitiers

Chayla B, Jaffrezic H, Joron JL (1973) Analyse par activation dans les neutrons épithermiques. Application à la détermination d'éléments traces dans les roches. CR Acad Sci Paris 277 (D): 273–275

Cogné J (1974) Le Massif Armoricain. In: Debelmas J (ed) Géologie de la France. Doin, Paris, pp 105–161

Dupuy C, Dostal J (1984) Trace element of some continental tholeiites. Earth Planet Sci Lett 60: 207–225

Gill JB (1981) Orogenic andesites and plate tectonics. Springer, Berlin Heidelberg New York

Hussong DM, Uyeda S (1981) Tectonic processes and the history of the Mariana arc: a synthesis of the results of Deep Sea Drilling Project Leg 60. In: Hussong DM, Uyeda S et al. (eds) Init. Rep. of the Deep Sea Drilling Project, LX. US Govt Print Off, Washington, pp 909–929

Jaffrezic H, Joron JL, Treuil M, Joron JL (1980) A study of the precision attained by neutron activation analysis using international standard GS-N and BCR-l as examples. J Radioanal Chem 55 2: 417–425

Joron JL, Treuil M (1977) Utilisation des propriétés des éléments fortement hygromagmaphiles pour l'étude de la composition chimique et de l'hétérogénéité du manteau. Bull Soc Géol Fr XIX 6: 1197–1205

Le Gall J, Cabanis B (1985) Caractères minéralogiques et géochimiques des ignimbrites et des laves cambriennes de l'Est du Massif Armoricain. Implications pétrogénétiques. Hecynica 1 2: 99–110

Le Métour J, Bernard-Griffiths J (1979) Age (limite Ordovicien-Silurien) de mise en place du massif hypovolcanique de Thouars (Massif Vendéen). Implications géologiques. Bull Bur Rech Géol Min (2) I 4: 365–371

Paquette JL, Marchand J, Peucat JJ (1984) Absence de tectonique cadomienne dans le complexe de Champtoceaux (Bretagne méridionale)? Comparaison des systèmes Rb-Sr et U-Pb d'un métagranite. Bull Soc Géol Fr (7) XXVI 5: 907–912

Pearce JA (1982) Trace element characteristics of lavas from destructive plate boundaries. In: Thorpe RS (ed) Andesites. Orogenic Andesites and Related Rocks. John Willey and Sons, Chichester, pp 525–548

Pearce J, Harris BWH, Tindle AG (1984) Trace element discrimination diagrams for the tectonic interpretation of granitic rocks. J Petrol 25 4: 956–983

Peucat JJ (1983) Géochronologie des roches métamorphiques (Rb-Sr, U-Pb). Exemples choisis au Groënland, en Laponie, dans le Massif Armoricain et en Grande Kabylie. Mém Soc Géol Minéral Bretagne 28, 158 pp

Peucat JJ, Le Métour J, Audren C (1978) Arguments géochronologiques en faveur de l'existence d'une double ceinture métamorphique d'âge siluro-dévonien en Bretagne méridionale. Bull Soc Géol Fr (7) XX 2: 163–167

Peucat JJ, Paris F, Chalet M (1986) U-Pb zircon dating of volcanc rocks of the Silurian-Devonian boundary in Vendée (W. France). Chem Geol (isot Geosci Sect) 59: 133–142

Ryerson FJ, Watson EB (1987) Rutile saturation in magmas: implications for Ti-Nb-Ta depletion in island-arc basalts. Earth Planet Sci Lett 86: 225–239

Taylor SR, Mac Lennan M (1981) The composition and evolution of the continental crust: rare earth element evidence from sedimentary rocks. Philos Trans R Soc Lond A 301: 381–399

Taylor SR, Mac Lennan M (1985) The Continental Crust: its Composition and Evolution. Blackwell Sci Publ, Oxford, 312 pp

Ters M (1979) Les synclinoriums paléozoïques et la Précambrien sur la façade occidentale du Massif vendéen. Stratigraphie et structure. Bull Bur Rech Géol Min (2) I 4: 293–301

Thiéblemont D (1988) Le magmatisme paléozoïque en Vendée: apport de la géochimie des éléments traces et de la pétrologie du métamorphisme à la compréhension du développement orogénique varisque. Doc Bur Rech Géol Min, Orléans 157, 365 pp

Thiéblemont D, Cabanis B (1986) Découverte d'une association de tholéiites d'arc et de basaltes de type MORB dans la série paléovolcanique silurienne de la Meilleraie. Vendée, France. CR Acad Sci Paris II 302: 641–646

Thiéblemont D, Cabanis B (1990) Utilisation d'un diagramme Rb/100-Tb-Ta pour la discrimination géotectonique et l'étude pétrogénétique des roches magmatiques acides. Bull Soc Géol Fr (8) VI 1: 23–35

Thiéblemont D, Cabanis B, Wyns R, Treuil M (1987) Etude géochimique (majeurs et traces) de la formation amphibolitique de Saint-Martin des Noyers (complexe cristallophyllien des Essarts, Vendée). Mise en évidence d'un paléo-arc insulaire dans la partie interne de l'orogène varisque. Bull Soc Géol Fr (8) III 2: 371–378

Thiéblemont D, Triboulet C, Godard G (1988) Mineralogy, petrology and P-T-t path of Ca-Na amphibole assemblages, Saint-Martin des Noyers formation, Vendée, France. J Metamorphic Geol 6: 697–715

Thiéblemont D, Béchennec F, Triboulet C, Cabanis B, Chantraine (1989 a) Two geodynamic regimes in the Variscan orogen of South Brittany? Terra Cognita 1 1: 14

Thiéblemont D, Béchennec F, Cabanis B, Chantraîne J (1989 b) Lithostratigraphie et géochimie des formations paléomagmatiques dans le secteur Moëlan-Concarneau (Bretagne méridionale). Nouvelles contraintes sur l'évolution géodynamique du domaine sud-armoricain au cours du Paléozoïque. Geodin Acta 3 (2): 1–16

Thompson RN, Dickin AP, Gibson IL, Morisson MA (1982) Elemental fingerprints of isotopic contamination of Hebridean Paleocene mantle-derived magmas by archean sial. Contrib Mineral Petrol 79: 159–168

Treuil M, Varet J (1973) Critères volcanologiquels, pétrologiques et géochimiques de la genèse et la différenciation des magmas basaltiques: exemple de l'Afar. Bull Soc Géol Fr (7) XV: 5–6

Triboulet C, Audren C, Vigneresse A, Lefort JP, Thiéblemont D Geological map of Ile de Groix and explanatory. Bur Rech Géol Min Orléans (in press)

Vidal P (1980) L'évolution poly-orogénique du Massif armoricain. Apport de la géochronologie et de la géochimie isotopique du strontium. Mém Soc Géol Mineral Bretagne, Rennes, 21, 162 pp

Walsh JN, Beckinsale RD, Skelhorn RR, Thorpe RS (1979) Geochemistry and petrogenesis of Tertiary granitic rocks from the Island of Mull, Northwest Scotland. Contrib Mineral Petrol 71: 99–116

Wood DA (1980) The application of Th-Hf-Ta diagram to problems of tectonomagmatic classification, and to establishing the nature of crustal contamination of basaltic lavas of the British Tertiary Province. Earth Planet Sci Lett 50: 11–30

Wood DA, Joron JL, Treuil M (1979 a) A re-apparaisal of the use of trace element to classify and discriminate between magma series erupted in different tectonic settings. Earth Planet Sci Lett 45: 326–336

Wood DA, Joron JL, Treuil M, Norry CM, Tarney J (1979 b) Elemental and Sr isotope variations in basic lavas from Iceland and the surrounding ocean floor. Contrib Mineral Petrol 70: 319–339

Wood DA, Marsh NG, Tarney J, Joron JL, Fryer P, Treuil M (1981) Geochemistry of igneous rocks recovered from a transect across the Mariana Trough, arc, fore-arc, and trench, sites 453 through 461, DSDP Leg 60. In: Hussong DM, Uyeda S et al. (eds) Init Rep of the Deep Sea Drilling Project, LX. US Govt Print Off, Washington, pp 657–677

Wyns R (1980) Contribution à l'étude du Haut Bocage vendéen: le Précambrien et le Paléozoïque dans la région de Chantonnay (Vendée). Thesis, University P et M Curie, Paris

Wyns R, Le Métour J (1983) Le Précambrien du Massif Vendéen. Etude détaillée de deux coupes de référence (coupes de l'Evre et de la Divatte) et synthèse des données récentes. Doc Bur Rech Géol Min, Orléans 68, 60 pp

Wyns R, Lablanche G, Lefavrais-Raymond A (1988) Geological map of Chantonnay and explanotary note. Bur Rech Géol Min, Orléans, France

Wyns R, Lardeux H, Weyant M (1989) Présence de Dévonien dans le groupe de Réaumur (synclinal de Chantonnay, Vendée); conséquences sur l'évolution géodynamique varisque de la Vendée. CR Acad Sci Paris 308 (II): 855–860

Variscan Granites from Brittany

J.-P. Carron, M. Le Guen de Kerneizon, and H. Nachit

1 Introduction

From Barrois (1884) to Barrière (1980) it is has been classical to divide Armorican Granites into three main units (Fig. 1):

1. A leucogranite belt which extends from the Pointe du Raz to Bressuire and Parthenay, emplaced during activity on the South Armorican Shear Zone (SASZ, Cogné 1960) and of its subsidiary faults.
2. A red-granite belt which extends along the northern shore of Brittany from Aber Ildut to the Cotentin Peninsula over 300 km in length. In its western part, it is parallel to the Porspoder-Guisseny mylonitic zone (Chauris 1965).
3. Middle-Armorican Batholith, with large elliptic plutons of porphyritic gray granites, extending westwards from the Mancellia Cadomian formations, emplaced during activity on the North Armorican Shear Zone (NASZ, Hirbec 1979).

Although recent mineralogical and petrochemical studies, together with isotopic data, lead to a more complex interpretation, the above divisions are a satisfactory basis to describe the granitic rocks. For the nomenclature of these granites, we do not use the IUGS classification, which is concerned only with the quartzo-feldspathic rocks, but the major element classification of Debon and Le Fort (1983), which is also used in other sections of this book. The compositions of the biotites has proved to be a rapid and suitable tool for magmatic typology (Nachit et al. 1985), and has been used in many cases. Typical whole-rock and mineral compositions are given in Tables 1 and 2, to illustrate the most striking features of the different groups. The numerous isotopic data will not be cited for every massif: they are mainly from Vidal (1976), Peucat et al. (1979, 1984), and Bernard-Griffiths et al. (1985), and also include some unpublished data or personal communications. Most of these dates have been performed by the Rb-Sr method; the initial values of the $^{87}Sr/^{86}Sr$ ratios will be noted I_{Sr}.

2 The Southern Leucogranite Belt

These granites display very constant mineralogical and geochemical characteristics:

1. They are commonly fine-grained and grey-yellow colored by alteration; they contain perthitic K-feldspars, sodic plagioclases, and abundant quartz crystals.
2. The CIPW norm calculations indicate that the magmas are peraluminous (corundum > 3%), and the Qz-Or-Ab proportions are close to 30%, with An < 3%. (analyses 1, 2, 3, Table 1)
3. Muscovite is always present and biotite rarely absent. This mineral is characterized (Nachit 1986) by its high alumina content (Al_2O_3 > 18%; analyses 1, 2, 3, Table 2). Andalusite and/or fibrolite can be observed sometimes in small quantities.
4. Mafic xenoliths are absent, whereas metamorphic mesozonal enclaves are frequent.
5. U and Sn mineralization is frequent.

Isotopic and structural data (Bernard-Griffiths et al. 1985) indicate that the leucogranite belt may be divided into three main units:

1. The northern unit is composed of large plutons emplaced at about 330–340 Ma, such as those of Locronan, Pontivy (344 ± 8 Ma; I_{Sr} = 0.7054 ± 9) studied by Hameurt (1961) and Marcoux (1980), Lizio and Guéhenno (338 ± 13 Ma; I_{Sr} = 0.7053 ± 17). Geophysical data suggest that they are flattened "mushroom-shaped" diapirs. (Hanmer and Vigneresse 1980)
2. The middle unit is composed of two series of plutons on the northern and southern rims of the SASZ: Pointe du Raz (320 ± 25 Ma) Vigneux, Questembert (329 ± 21 Ma; I_{Sr} = 0.709 ± 4) with a geochemical and petrographic study by Charoy (1971), Nantes, Mortagne (Renard 1974; Guineberteau 1984), dated at 315 ± 15 Ma, with I_{Sr} = 0.7124 ± 17, Bressuire and Parthenay. (Dhoste 1980)
3. The southern unit is the youngest one (ca. 300 Ma). It includes the granites of Pont l'Abbé (305 ± 6 Ma;

J. D. Keppie (Ed.): Pre-Mesozoic Geology in France and Related Areas

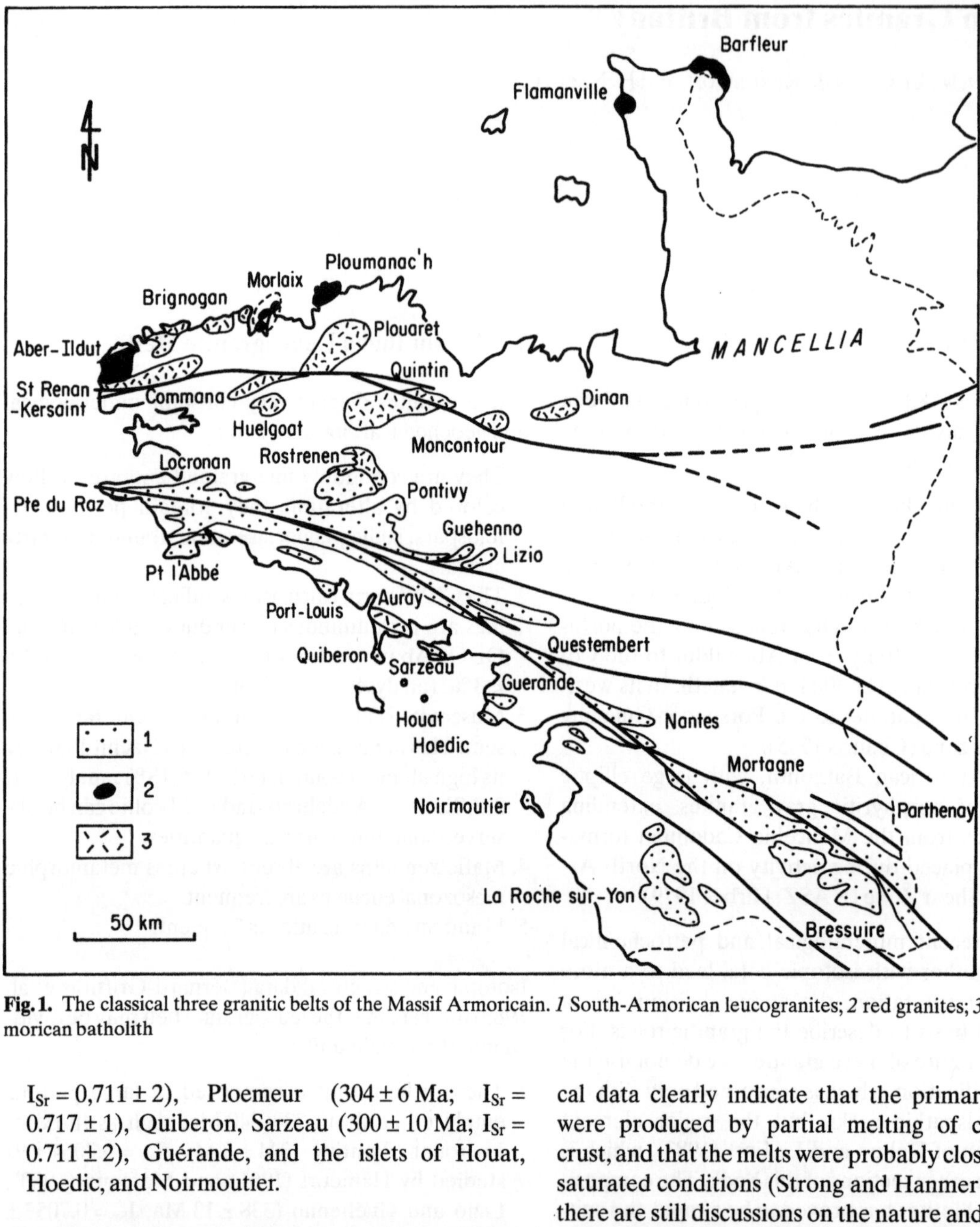

Fig. 1. The classical three granitic belts of the Massif Armoricain. *1* South-Armorican leucogranites; *2* red granites; *3* Middle-Armorican batholith

I_{Sr} = 0,711 ± 2), Ploemeur (304 ± 6 Ma; I_{Sr} = 0.717 ± 1), Quiberon, Sarzeau (300 ± 10 Ma; I_{Sr} = 0.711 ± 2), Guérande, and the islets of Houat, Hoedic, and Noirmoutier.

Deuteric alteration could lead to the growth of secondary muscovite, albite, and tourmaline or to complete kaolinization (Ploemeur area). Detailed geochemical studies and mapping have been done by de La Roche et al. (1980), using a data bank on uraniferous granites.

All these characteristics are typical of the S-type granites (Chappel and White 1974), or C-type granites (Didier et al. 1982). Pichavant and Stussi (1986) showed, however, that these leucogranites have many significant differences with the S-granites from Australia, and should be considered as a separate LG type (leucogranites). Experimental and mineralogical data clearly indicate that the primary magmas were produced by partial melting of continental crust, and that the melts were probably close to water-saturated conditions (Strong and Hanmer 1981), but there are still discussions on the nature and age of the metasedimentary materials. Bernard-Griffiths et al. (1985) provided numerous accurate geochemical data: I_{Sr} increases from 0.706 to 0.717 while the age decreases from 340 to 300 Ma; REE abundances are relatively low, with a strong enrichment of LREE relative to HREE (Fig. 2); $\delta^{18}O$ is high and uniform with typical "sedimentary" values close to 12. Nd and Pb isotopes data suggested that the model age of the source rocks was not very old (600–1200 Ma), but more recently, Peucat et al. (1988) explained the same data by the mixing of an ancient basement (1800 Ma) with a juvenile crust (late Precambrian or early Paleozoic).

Table 1. Rock analyses and CIPW norms

	1	2	3	4	5	6	7	8	9	10	11	12	13	14	15	16
SiO_2	73.0	72.6	71.8	66.9	69.3	74.8	55.8	48.3	71.3	71.6	55.3	70.5	57.7	54.3	69.1	67.1
TiO_2	0.1	0.1	0.2	0.5	0.5	0.2	1.2	1.2	0.3	0.2	1.4	0.4	1.0	1.4	0.5	0.6
Al_2O_3	15.0	15.1	15.2	15.0	14.5	12.6	15.8	13.5	15.2	15.3	15.9	15.3	16.8	16.8	14.5	15.0
Fe_2O_3	1.2	1.4	1.8	3.0	3.1	1.7	5.7	11.4	2.0	1.9	7.4	2.5	6.7	7.0	3.0	3.6
MgO	0.4	0.4	0.5	2.1	0.9	0.2	4.1	12.8	0.7	0.6	4.7	0.9	4.3	5.3	1.4	1.7
CaO	0.7	0.6	0.8	2.8	1.9	1.1	5.1	8.0	0.8	1.0	5.3	1.4	5.8	6.1	2.1	2.8
Na_2O	3.3	3.3	3.3	4.3	3.6	3.4	3.7	2.3	3.5	3.2	2.7	3.5	3.2	3.6	3.9	3.9
K_2O	4.8	4.8	4.9	4.1	5.2	5.2	6.0	1.6	4.7	4.9	5.6	4.4	3.0	3.0	4.4	4.0
P_2O_5	0.2	0.2	0.2	0.3	0.2	0.3	1.2	0.2	0.2	0.2	0.6	0.2	0.4	1.4	0.2	0.2
L.O.I	1.1	1.3	1.1	0.8	0.6	0.3	1.2	0.5	1.1	0.9	0.7	0.7	0.9	0.9	0.7	0.9
Qz	33.2	32.9	31.0	17.0	22.2	32.4			29.8	30.8		28.3	7.5	1.3	22.2	19.7
Or	28.4	28.4	28.9	24.2	30.8	30.7	35.5	9.5	27.8	28.9	33.1	26.0	17.7	17.7	26.0	23.7
Ab	27.9	27.9	27.9	36.6	30.4	28.7	31.3	19.4	29.6	27.0	22.8	29.6	27.0	31.3	32.9	33.0
An	2.3	1.8	2.8	9.5	8.0	3.7	8.7	21.8	2.8	3.8	14.7	5.8	22.6	20.3	9.0	11.6
Ol							8.1	23.6		6.6	3.2	1.0	0.2	0.9		
Di				2.2	0.2	0.1	7.7	13.5	3.8	3.6	15.7	4.8	16.1	19.4	6.4	7.4
Hy	2.4	2.6	3.2	7.2	5.3	2.3	0.7	5.5	0.4	0.4	1.6	0.5	1.5	1.5	0.6	0.8
Mt	0.3	0.3	0.4	0.7	0.7	0.4	1.2	2.5	0.6	0.4	2.7	0.8	1.9	2.7	0.9	1.1
Il	0.2	0.2	0.4	0.9	0.9	0.3	2.2	2.3	0.4	0.4	1.3	0.4	0.9	3.1	0.4	0.4
Ap	0.4	0.4	0.4	0.6	0.4	0.5	2.6	0.4	3.3	3.3		2.7				
Cor	3.5	3.8	3.4													

1: Pontivy leucogranite. 2: Lizio leucogranite. 3: Questembert leucogranite. 4: Flamanville biotite-hornblende subalkalic granodiorite. 5: Traouieros porphyric red green subalkalic granite (Ploumanac'h). 6: Clarté porphyric red subalkalic granite (Ploumanac'h). 7: Mafic xenolith, Flamanville. 8: Ste Anne olivine gabbro, (Ploumanac'h). 9: Huelgoat cordierite granite. 10: Brignogan cordierite granite. 11: Mafic xenolith, Brignogan. 12: Rostrenen porphyroïde cordierite granite. 13: Plélauff diorite (Rostrenen). 14: Plouaret diorite. 15: Quintin biotite subalkaline granite 16: Quintin hornblende biotite subalkaline granite.

Table 2. Microprobe mineral analyses

	1	2	3	4	5	6	7	8	9	10	11	12	13	14	15	16
SiO_2	36.10	35.95	34.70	38.25	35.70	36.70	35.85	48.60	43.05	47.25	35.53	47.30	36.20	35.00	36.80	50.20
TiO_2	3.70	2.80	2.90	3.05	4.20	2.55	3.40	1.10	1.70		2.85		0.75	3.55	4.45	0.80
Al_2O_3	18.45	18.95	19.75	13.65	14.65	14.20	13.55	6.05	9.15	32.40	19.63	31.90	20.30	19.80	13.85	4.40
FeO	18.80	19.85	21.45	16.60	21.95	23.80	28.90	13.25	22.70	8.85	19.00	10.00	18.65	19.85	21.50	17.15
MnO	0.25	0.25	0.25	0.20	0.20	0.35	0.60	0.40	0.50	0.25	0.27	0.50	0.25	0.20	0.25	0.55
MgO	9.60	8.30	6.75	14.80	8.65	9.50	5.65	15.00	7.30	6.50	8.08	6.25	8.85	7.10	10.50	12.35
CaO								11.90	10.80							11.50
Na_2O	0.15	0.10	0.15	0.15	0.10	0.05	0.05	1.60	1.90	1.20	0.17	1.05	0.15	0.25	0.10	1.00
K_2O	9.80	9.35	9.65	10.05	9.70	9.60	9.45	0.65	1.40		9.79	9.75	9.90	9.85	0.50	
Total	96.80	95.70	96.85	96.75	95.15	96.75	97.45	98.55	98.50	96.45	95.32	96.90	94.90	95.65	97.30	98.45

1: "Peraluminous" biotite from Locronan leucogranite. 2: "Peraluminous" Biotite from Pontivy leucogranite. 3: "Peraluminous" Biotite from Lizio leucogranite. 4: "Subalkaline" Biotite from Flamanville granodiorite. 5: "Subalkaline" Biotite from Aber Ildut monzogranite. 6: "Subalkaline" Biotite from Traouieros monzogranite. 7: "Subalkaline" Biotite from Clarté monzogranite. 8: Hornblende from Flamanville granodiorite. 9: Hornblende from Traouieros monzogranite. 10: Cordierite from Huelgoat porphyric granite. 11: Biotite from Huelgoat porphyric granite. 12: Cordierite from Brignogan granite. 13: Normal red biotite from Brignogan cordierite granite. 14: Secondary green biotite from Brignogan cordierite granite. 15: "subalkaline" Biotite from Quintin granite. 16: Hornblende from Quintin granite. All analyses are from Nachit (1986), except 10 and 11 from Georget (1986). Camebax automated Microprobe (Brest) 10 kV, 15 nA.

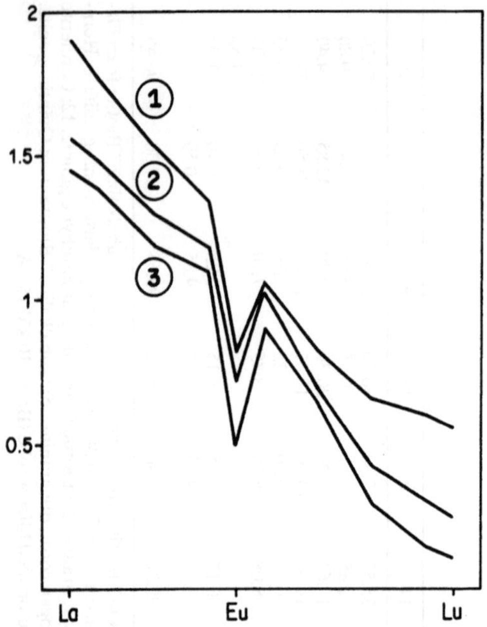

Fig. 2. Chondrite-normalized REE pattern for southern leucogranites. (Data from Bernard-Griffiths et al. 1985). Mass spectrometric isotope dilution method. *1* Questembert; *2* Pontivy; *3* Pont-l'-Abbé

3 The Northern Red Granites Belt

Five plutons of red granodiorites to syenogranites have been emplaced at ca. 300 Ma in a linear SW-NE arrangment. They contrast with the southern leucogranitic belt by the following characteristics:

1. lack of normative corundum and presence of minor diopside. (analyses 4, 5, 6, Table 1);
2. typical "subalkalic" biotites, with Al_2O_3 < 15% (analyses 4, 5, 6, 7, Table 2);
3. absence of magmatic muscovite in the main part of the plutons; possible presence of titanite and hornblende (analyses 8, 9, Table 2);
4. constant presence of mafic xenoliths, frequently modified by K enrichment (analysis 7, Table 1), which suggest the coexistence of partially molten granitic and mafic magmas;
5. frequent stocks of mafic plutonic rocks, mainly dioritic, but sometimes gabbroic. (analysis 8, Table 1);
6. absence of hydrothermal associated mineralization.

All these characteristics correspond to the I-type (Chappel and White 1974), or M-type, (Didier et al. 1982), and more precisely to the subalkalic-type (SAG) of Debon and Le Fort (1983).

1. The Aber Ildut Massif (Chauris 1966) is cut on its northern rim by an important shear zone, and is half hidden under the sea. It is a biotite-monzogranite with large perthitic pink orthoclase phenocrysts and zoned plagioclase. It has been dated by Deutsch and Chauris (1965) at ca. 300 Ma (310 ± 15 Ma by the Rb/Sr method on biotite, 280 ± 30 Ma by the U/Pb method on zircons). On its outer rim, it merges into a white granite, enriched in muscovite and tourmaline. Its interior is occupied by the younger two-mica granite of Ploudalmezeau, which has been dated by Leutwein et al. (1969) at 290 ± 10 Ma, with a puzzling I_{Sr} of 0.725. *The Morlaix Bay* pluton outcrops only in separate islets, but it appears on continental margin maps as a massif very similar in size and composition to that of Aber Ildut.

2. The Ploumanac'h Massif (302 ± 15 Ma) is well exposed on the seashore, and in a number of quarries. It has been carefully studied by Barrière (1977), who distinguished three successive intrusions (Fig. 3):

a) An outer rim of porphyritic granites, associated with olivine gabbros (St Anne). Hornblende is present in the "Clarté" red type, which grades inward into the more silica-rich Traouieros red-green type. Recently Chauris et al. (1989) described a small occurrence of orbicular facies.
b) A middle zone of medium-grained pink granites and aplites, which were forcefully injected in the porpyritic granites.
c) An inner zone of fine-grained gray granites (Ile Grande), which contain muscovite and has some characteristics of a ring-dyke intrusion.

From petrological and geochemical data, Barrière concluded that the red and pink granites were derived by fractionnal crystallization from a transitional basaltic magma, with the possibility of minor contamination by crustal material, whereas the inner gray granites were a product of crustal anatexis, induced by the ascent of the major molten mass.

Further studies confirmed this model, with some modifications and better accuracy. Albarède et al. (1980) made an oxygen isotope study and recalculated the I_{Sr} values for the three main types: they found $I_{Sr} = 0.705$ and $\delta^{18}O = 6$ for the mafic rocks, $I_{Sr} = 0.706$ and $\delta^{18}O = 9$ for the red granites, $I_{Sr} = 0.708$–0.712 and $\delta^{18}O = 11$–12 for the Ile Grande dyke. They therefore concluded a greater contribution of crustal material. Fourcade (1981) made a study of trace elements and REE (Fig. 4), and suggested that the red granites were probably the result of mixing between a crustal material and evolved magmas, derived from alkali-basalts. However,

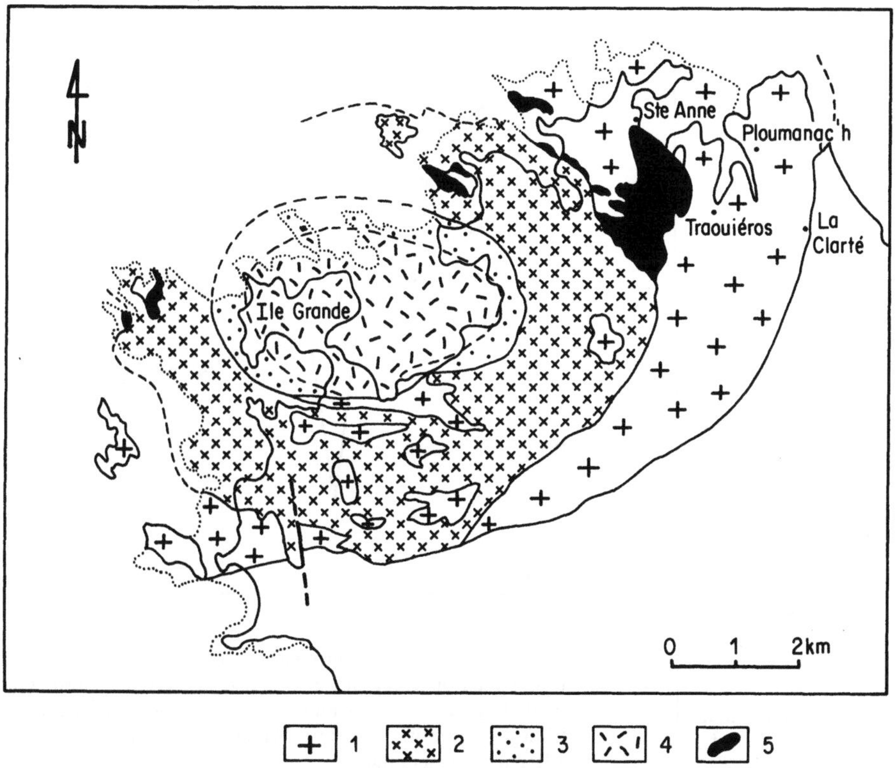

Fig. 3. Geological sketch map of the Ploumanac'h pluton. *1* Porphyroïd monzogranites (La Clarté, Traouièros); *2* Intermediate granites; *3* and *4* Ile Grande leucogranites; *5* Ste Anne gabbro. *Dotted line* is the lowest level of spring tides

he pointed out that the models are complicated by the role of minor minerals, such as allanite, sphene, or monazite, with very large crystal-liquid partition coefficients. It is interesting to note that all materials from this massif are rich in thorium (5 ppm in the mafic rocks, up to 100 ppm in the Clarté granite).

3. Flamanville is a pluton of historical importance in French petrology, since it was in the 1950s the subject of debate between the "solid-state" school, and the "magmatic" school. Martin (1952) clearly showed that the planar textures of micas and xenoliths correspond to an inward-dipping conic structure, instead of the outward-dipping structure proposed by Michel-Levy (1893), who considered that all granites extended downward to a continuous "primitive layer". This granodiorite contains medium grained K-feldspars and zoned plagioclase, together with Mg-rich biotite, hornblende, (analyses 4 and 8, Table 2), titanite, and allanite. Goni (1966) gave evidence that some chalcophile elements are presently not located within the minerals, but between intergranular spaces, and can be extracted by mild leaching. Carron et al. (1983) calculated that the mass budget of lithophile elements is, in contrast, entirely consistent with an intracrystalline distribution. Ledru and Brun (1977) showed that the relationships of the regional schistosity and of the internal structures of the granite indicate that this pluton is a typical syntectonic intrusion, whereas it has the appearance of a post-tectonic intrusion on large-scale maps.

4. Barfleur granite has been neither dated nor studied for mineralogy or geochemistry, because of poor exposure and evidence of deformation.

4 The Middle-Armorican Plutons

These include a number of intrusive bodies of various sizes, which range in age from 340 to 290 Ma. (Le Guen de Kerneizon and Barrière 1977). The main type is a gray-blue porphyroid monzogranite, with large perthitic megacrysts. Their structure is often complex, with concentric zoning, and complex associations with peraluminous or mafic bodies. This batholith can be divided, in fact, into two series, with quite distinctive chemical and mineralogical characteristics.

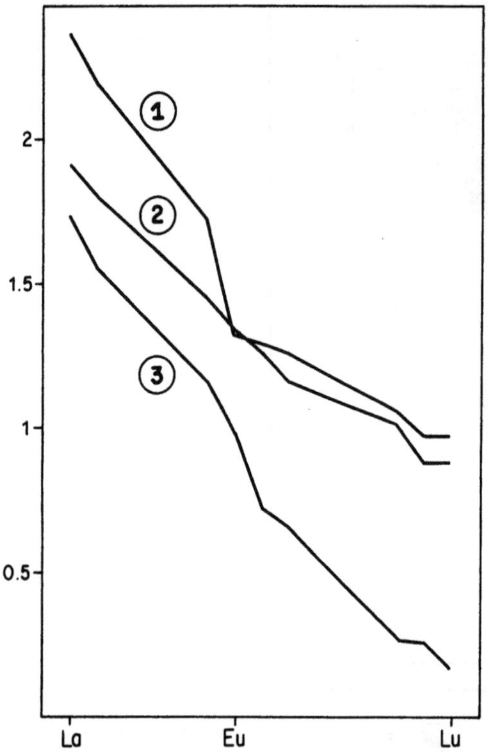

Fig. 4. Chondrite-normalized REE pattern for Ploumanac'h granites. (Data from Fourcade 1981). Mass spectrometric isotope dilution method. *1* Traouièros granite; *2* Ste Anne gabbro; *3* Ile grande external leucogranitic ring-dyke

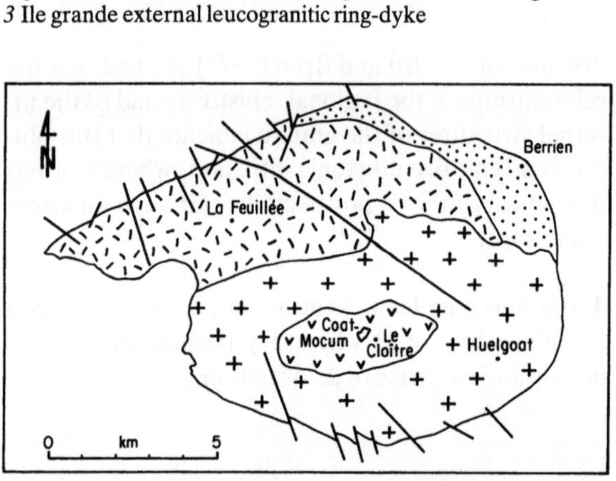

Fig. 5. Geological sketch map of the Huelgoat pluton. *1* Coat-Mocun aplitic plug; *2* Le Cloître cordierite granite, *3* Huelgoat cordierite granite; *4* La Feuillée biotite granite; *5* Berrien biotite muscovite granite

4.1 Cordierite-Bearing Granites

(analyses 12, 14, Table 1)

These are very similar in composition to the leucogranites, with corundum values at 2.5 to 3.5, and somewhat higher contents of Ca, Fe, Mg.

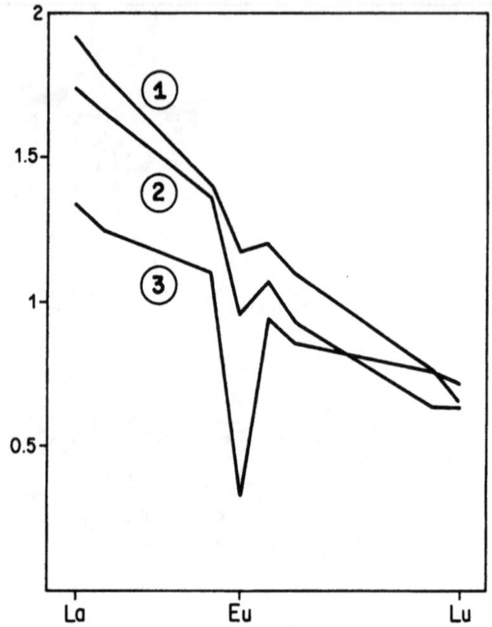

Fig. 6. Chrondrite-normalized REE pattern for Huelgoat granites. (Data from Georget 1986) Neutron activation analysis. Gd plotted by linear interpolation between Sm and Tb values. *1* Le Cloître cordierite granite; *2* Huelgoat cordierite granite; *3* Berrien leucogranite

1. The Huelgoat Massif (Fig. 5) has been studied by Conquéré (1969) and Georget (1986). It has been dated at 340 ± 9 Ma, with $I_{Sr} = 0.7066 \pm 7$. The central part includes the porphyric granites of Huelgoat ss. str. (analysis 9, Table 1) and Le Cloitre, which contain large idiomorphic crystals of cordierite (analysis 10, Table 2), together with aluminous biotites (analysis 11, Table 2) and rare secondary muscovite. On its northern margin this unit intrudes the cordierite-free granites of Berrien, where post-magmatic fluid circulation induced the formation of an important kaolin lode. Both units are of similar composition, and their different mineralogy needs some explanation. From comparison with the experimental data of Clemens and Wall (1981), Georget (1986) suggested that the cordierite granites were produced at greater depth and higher temperature than the muscovite granites, and had lower initial XH_2O and fO_2. In both cases, the crystallization conditions were close to 3–4 kbar and 700–750 °C. The REE pattern (Fig. 6) has a LREE enrichment similar to that of the leucogranites, but a lower La/Lu ratio.

2. The Brignogan-Plouescat Massif (Le Guen de Kerneizon 1975) has some analogies with Huelgoat (analysis 10, Table 1) and is surrounded by a muscovite granite probably produced by outward fluid circulation. The presence of rare mafic inclusions,

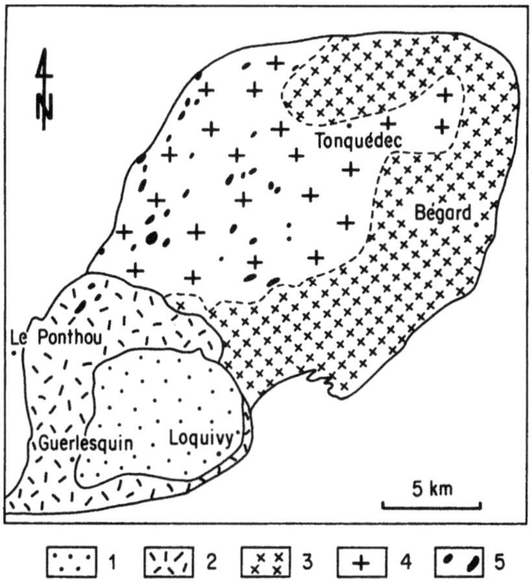

Fig. 7. Geological sketch map of the Plouaret complex *1* Loquivy leucogranite; *2* Guerlesquin-Le Ponthou leucogranites; *3* Bégard granodiorite; *4* Tonquédec granite; *5* Dioritic plugs

(analysis 11, Table 1), rather unusual in peraluminous granites, suggests the influence of basaltic liquids on the melting of the lower continental crust. It can be noted that cordierite (analysis 12, Table 2) frequently reacts with the later melts ("peritectic reaction"), producing andalusite and a green biotite distinguishable from the red one by its low titanium content (analyses 13, 14, Table 2). A recent date at 292 ± 15 Ma, with $I_{Sr} = 0.7074 \pm 11$ (Georget et al. 1986) indicates that it is slightly younger than Aber Ildut, and thus must be related to the red-granite generation, as well as the Kernilis leucogranite, which is intermediate in location and age.

3. The Rostrenen Massif, which is well known for its huge K feldspar megacrysts (up to 15 cm), also has a peraluminous composition (analysis 12, Table 1), and is associated with medium to small scale stocks of mafic rocks, such as the Plelauff diorite, (analysis 13, Table 1), which shows evidence of magma mixing. (Didier and Giret 1972). The main granite is intruded by a leucogranite, very similar to that of Pontivy.

All these magmas intruded in high crustal levels, where they induced large metamorphic aureoles with sillimanite, cordierite, or andalusite, which may form spectacular chiastolite prisms in the aluminous Calymene schists. They differ from the Australian S-type granites by the absence of garnet crystals, and Pichavant and Stussi (1986) suggested that they probably originated by noneutectic melting of a mixture of pelitic and volcanic material, such as graywackes. The name of CG (cordierite granites) seems to us rather convenient.

4.2 Cordierite-Free Monzogranites

Although they have the same location and tectonic relationships that the CG, their chemical and mineralogical features are completely distinct, and similar to the subalkalic granites.

1. In the Plouaret Complex (Chauris and Garreau 1983), the northern part is subalkaline, and contains dispersed small outcrops of gabbros and diorites (analysis 14, Table 2), associated with granodiorite and monzogranites (Fig. 7). They have been dated at 329 ± 5 Ma with $I_{Sr} = 0.7056 \pm 3$. The southern part consists of many leucogranitic concentric intrusions, and the Commana cordierite granite has been right-laterally displaced along the NASZ. The internal structures have been studied by Guillet et al. (1985): the northern unit is supposed to be a thin (2–3 km) laccolith, whereas the southern unit has the shape of a cone, with roots at about 7 km in depth.

2. The Quintin Pluton (291 ± 9 Ma; $I_{Sr} = 0.7068 \pm 4$) is the largest in size among Hercynian granites (650 km^2). Beurrier et al. (1979) showed that it is built by a concentric succession of porphyritic or fine grained granites. The innermost granites contain alumina-poor biotites (analysis 15, Table 2) and hornblende (analysis 16, Table 1), which, together with the whole-rock compositions (analyses 15, 16, Table 1), clearly indicate a subalkalic affinity. On its southern rim, it intrudes the Carboniferous rocks along a very low dipping surface.

3. Other Plutons. The Moncontour granite is rather similar to that of Quintin, whereas the Dinan pluton is probably peraluminous. The St. Renan-Kersaint granites have been dated at 350 ± 30 Ma by Deutsch and Chauris (1965) (U/Pb on zircons), with a greisen at 285 ± 5 Ma (Rb/Sr on muscovite). The western part is tourmaline-rich, whereas the eastern part contains cordierite. The Auray granite (376 ± 19 Ma, $I_{Sr} = 0.7062 \pm 5$) is described as anatectic, intruding migmatites (Le Metour 1976). A similar origin is proposed for the south-eastern pluton of La Roche sur Yon (Stussi 1976).

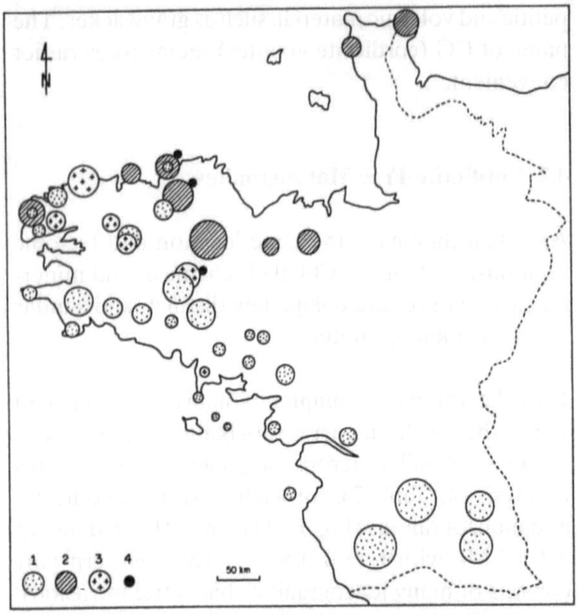

Fig. 8. Schematic map of the Variscan granites, with indication of their magmatic typology. *1* Leucogranites (LG); *2* subalkaline granites (SAG); *3* cordierite granites (CG); *4* mafic plugs associated with SAG or CG. The *radius of each circle* is approximately proportional to the area of the related massif and of its submarine extension

5 Conclusions

From this short synthesis, and although many more studies are needed, a number of points of interest already emerge:

1. Most of the leucogranites (LG) have been emplaced along the SASZ, and result from the regional anatexis of metasediments of various ages. Heat production and water supply are related to major tectonic events, such as overthrusting and subsequent vertical shearing. Out of this region, minor LG intrusions have been produced by local anatexis around SAG or CG ascending magmas.
2. The subalkaline granites or suites (SAG) probably resulted from fractional crystallization of basaltic magmas, with an important contribution of crustal materials. They are not restricted to the classical red-granite belt, since large plutons of the Middle-Armorican region exhibit the same geochemical and mineralogical patterns (except their feldspar coloration).
3. The cordierite granites or complexes (CG) have an important crustal component, but are also related mafic magmas, which probably supplied a part of the heat of melting. They do not constitute a single batholith, since they are scattered all over the peninsula, and range from 375 to 290 Ma.

A schematic map (Fig. 8) of the Variscan granites from Brittany indicates the present state of their magmatic typology.

References

Albarède F, Dupuis C, Taylor HP (1980) $^{18}O/^{16}O$ evidence for non-cogenetic magmas associated in a 300-Ma-old concentric pluton at Ploumanac'h (Brittany, France). J Geol Soc (Lond) 137: 641–647

Barrière M (1977) Le complexe de Ploumanac'h, Massif Armoricain. Essai sur la mise en place et l'évolution pétrologique d'une association plutonique subalcaline tardi-orogénique. Thèse, Univ. Brest, 291 pp

Barrière M (1980) Les granitoïdes paléozoïques armoricains. 26ᵉ Congr Géol Int, Paris, Coll C 7. Bur Rech Geol Min, Orléans, pp 56–63

Barrois C (1884) Observations sur la constitution géologique de la Bretagne. Ann Soc Géol Nord 11: 278–285

Bernard-Griffiths J, Peucat J-J, Sheppard S, Vidal P (1985) Petrogenesis of hercynian leucogranites from the southern Armorican Massif; contribution of REE and isotopic (Sr, Nd Pb and O) geochemical data to the study of source rock characteristics and ages. Earth Planet Sci Lett 74: 235–250

Beurrier M, Bles J-L, Gros Y, Lozes J, Marot A (1979) Etude de la partie occidentale du massif granitique de Quintin. Pétrographie, géochimie, fracturation. Bull Bur Rech Geol Min 4: 317–336

Carron J-P, Lagache M, Villemant B (1983) La granodiorite de Flamanville. Bull Mineral 106 (5): 109–110 (Suppl)

Chappel BW, White AJR (1974) Two contrasting granites types Pac Geol 8: 173–174

Charoy C (1971) Différenciations pétrographiques et géochimiques dans les granites à deux micas du massif de Questembert (Morbihan). Sci Terre (Nancy) 21: 5–44

Chauris L (1965) Sur la structure des massifs granitiques du Nord-Ouest du pays de Léon (Finistère). C R Acad Sci Paris 260: 4351–4353

Chauris L (1966) Le granite de l'Aber-Ildut (Finistère). Bull Carte Géol Fr 278: 9–30

Chauris L, Garreau J (1983) Le massif de Plouaret. C R Acad Sci Paris 269: 1591–1594

Chauris L, Hallegouet B, Riva N (1989) Données préliminaires sur un faciès orbiculaire dans le granite rouge de Ploumanac'h (Massif Armoricain). C R Acad Sci Paris 309: 59–62

Clemens JD, Wall VJ (1981) Origin and crystallisation of some peraluminous (S-type) granitic magmas. Can Minéral 19: 111–131

Cogné J (1960) Schistes cristallins et granites en Bretagne Méridionale: le Domaine de l'Anticlinal de Cornouille. Thèse, Univ Strasbourg

Conquéré F (1969) Le massif granitique de Huelgoat (Finistère). Mém Mus Nat Hist Nat 21: 42

Debon F, Le Fort P (1983) A chemical-mineralogical classification of common plutonic rocks and associations. Trans R Soc Edinb Earth Sci 73: 135–149

Deutsch S, Chauris L (1965) Age de quelques formations cristallophylliennes et granitiques du Pays de Léon (Finistère). C R Acad Sci Paris 260: 615–617

Dhoste M (1980) Les granitoïdes de la moitié Nord du Département des Deux-Sèvres: pétrographie, minéralogie, chimie de la moitié Est de l'axe Nantes-Parthenay, des massifs de Moncoutant et de Neuvy-Bouin. Thése, Univ. Poitiers, 493 pp

Didier J, Giret A (1972) Origine et rapports mutuels du granite de Rostrenen et de la granodiorite de Plélauff. C R Somm Soc Géol Fr 3: 143

Didier J, Duthou J-L, Lameyre J (1982) Mantle and crustal granites: genetic classification of orogenic granites and the nature of their enclaves. J Volc Geotherm Res 14: 169–186

Fourcade S (1981) Géochimie des granitoïdes. Thése, Univ. Paris 7, 211 pp

Georget Y (1986) Nature et origine des granites peralumineux à cordiérite et des roches associées. Thèse, Univ. Rennes, 250 pp

Georget Y, Martineau F, Capdevila R (1986) Age tardi-hercynien et origine crustale du granite de Brignogan (Finistère, France). Conséquences sur l'interprétation des granites Nord-Armoricains. C R Acad Sci Paris 302: 237–242

Goni J (1966) Contribution à l'étude de la localisation et de la distribution des éléments en traces dans les minéraux et les roches granitiques. Mém Bur Rech Geol Min (Orléans) 45, 66 pp

Guillet P, Bouchez J-L, Vigneresse J-L (1985) Le complexe granitique de Plouaret (Bretagne): mise en évidence structurale et gravimétrique de diapirs emboités. Bull Soc Géol Fr 4: 503–513

Guineberteau B (1984) Le Massif Granitique de Mortagne-sur-Sèvre (Vendée): structure, gravimétrie, mise en place, distribution de U-Th-K. Thèse, Univ. Nantes, 218 pp

Hameurt J (1961) Etude géologique et pétrographique du massif de Pontivy. Thèse, Univ. Nancy, 108 pp

Hanmer S, Vigneresse J-L (1980) Mise en place de diapirs syntectoniques dans la chaîne hercynienne: exemple des massifs leucogranitiques de Locronan et de Pontivy (Bretagne centrale). Bull Soc Géol Fr 22: 193–202

Hirbec Y (1979) Le complexe basique de Belle-Isle-en-Terre (Côtes-du-Nord): sa place dans l'évolution géodynamique du Nord du Massif Armoricain. Thèse, Univ. Rennes, 146 pp

La Roche H (de), Stussi J-M, Chauris L (1980) Les granites à deux micas hercyniens français. Essais de cartographie et de corrélations géochimiques appuyés sur une banque de données. Sci Terre (Nancy) 24: 5–121

Ledru P, Brun J-P (1977) Utilisation des fronts et des trajectoires de schistosité dans l'étude des relations entre tectonique et intrusion granitique: exemple du granite de Flamanville. C R Acad Sci Paris 285: 1199–1202

Le Guen de Kerneizon M (1975) Etude pétrographique du massif granitique de Brignogan-Plouescat. Thèse, Univ. Clermont-Ferrand, 83 pp

Le Guen de Kerneizon M, Barrière M (1977) Le massif granitique de Brignogan-Plouescat dans le cadre pétrographique du batholite hercynien médio-armoricain. C R Acad Sci Paris 284: 2317–2320

Le Metour J (1976) Contribution à l'étude pétrologique, géochimique et structurale du diapir migmatitique Hennebont-Vannes. (Bretagne méridionale). Thèse, Univ. Rennes, 80 pp

Leutwein F, Chauris L, Sonet J, Zimmermann J-L (1969) Etudes géochronologiques et géotectoniques dans le Nord-Finistère. (Massif Armoricain). Sci Terre (Nancy) 4: 329–358

Marcoux E (1980) Le district de Pontivy, sa place dans la métallogénie plombo-zincifère du Massif Armoricain. Thèse, Univ. Clermont-Ferrand, 252 pp

Martin NR (1952) The structure of the granite massif of Flamanville. Quart. Geol Soc (Lond) 108: 311–338

Michel-Levy A (1893) Contribution à l'étude du granite de Flamanville et des granites français en général. Bull Carte Géol Topog Souter 36, 41 pp

Nachit H (1986) Contribution à l'étude analytique et expérimentale des biotites des granitoïdes. Applications typologiques. Thése, Univ. Brest, 191 pp

Nachit H, Razafimahefa N, Stussi J-M, Carron J-P (1985) Composition chimique des biotites et typologie magmatique des granitoïdes. C R Acad Sci Paris 301: 813–818

Peucat J-J, Charlot R, Mifdal A, Chantraine J, Autran A (1979) Définition géochronologique de la phase bretonne en Bretagne centrale. Etude Rb/Sr de granites du domaine centre-armoricain. Bull Bur Rech Geol Min 4: 349–356

Peucat J-J, Auvray B, Hirbec Y, Calvez J-Y (1984) Granites et cisaillements hercyniens dans le Nord du Massif Armoricain: géochronologie Rb-Sr. Bull Soc Géol Fr 6: 1365–1373

Peucat J-J, Jegouzo P, Vidal P, Bernard-Griffiths J (1988) Continental crust formation seen through the Sr and Nd isotope systematics of S-type granites in the Hercynian belt of western France. Earth Planet Sci Lett 88: 60–68

Pichavant M, Stussi J-M (1986) Leucogranites à deux micas et granites "S": deux types de magmatisme crustal. 1 léme RST, Clermont-Ferrand. Societe Géologique de France, Paris, p 147

Renard J-P (1974) Etude pétrographique et géochimique des granites du district uranifère de Vendée. Sci Terre (Nancy) 30, 132 pp

Strong DF, Hanmer SK (1981) The Leucogranites of Southern Brittany: origin by faulting, frictional heating, fluid flux and fractional melting. Can Mineral 19: 163–176

Stussi J-M (1976) Le massif granitique de La Roche-sur-Yon (Vendée). Zonalité pétrographique et différenciation chimico-minéralogique. Sci Terre (Nancy) 20: 219–263

Vidal P (1976) L'évolution polyorogénique du Massif Armoricain. Apport de la géochronologie et de la géochimie isotopique du strontium. Thèse, Univ. Rennes, 142 pp

Part II C
Metallogeny

Metallogeny of the Armorican Massif

L. Chauris and E. Marcoux

1 Introduction

Successive metallogenic episodes, spread over several hundred million years in the Armorican Massif, can be related to the following major events (Chauris and Guigues 1969; Chauris 1980):

1. The Cadomian orogeny (southward subduction under the old Icartian continent) with volcanosedimentary and hydrothermal-sedimentary formations (Brioverian polymetallic sulfide deposits), followed by granitic intrusions (cassiterite-wolframite-bearing mineralization).
2. A shelf-forming period with deposition of sedimentary mineralization (Zr-Ti, Fe) followed by extension accompanied by volcanism (Silurian and Devonian polymetallic sulfide deposits).
3. The Hercynian orogeny (northward subduction of the north Aquitaine plate followed by collision) evidenced by the intrusion of thick granitic belts (cassiterite-wolframite-bearing deposits). This was followed by intense fracturing (kaolin deposits) which was gradually replaced by other major fracturing episodes that may be interpreted as the distal impacts of Atlantic rifting on the old peneplaned Hercynian basement (Au, U, Sb, Pb-Zn-Ag, Ba).

Within the framework of the Hercynian of western Europe, the Armorican Massif is of major interest, and is characterized by a Cadomian metallogeny (Chauris and Jonin 1978). In other parts of the Hercynian orogeny, the deformation is so intense that it is difficult, and even impossible, to determine metallogenic processes related to earlier orogenic events.

2 Background

The Armorican Massif, like other parts of the Hercynian range of central and western Europe, is an old mining region (see Bur Rech Géol Min 1960–64 and 1979) and the Breton Peninsula, along with Cornwall and Iberia on the Atlantic coastline, claims to belong to the Cassiterides (sources of tin-bearing ores in ancient times). In the 18th century, the Poullaouen-Huelgoat silver-lead mines were the most extensive in the kingdom of France, whilst in 1909 France was the leader of antimony-producing countries with its La Lucette mine. A great number of place names bear witness to the extent of the mining and metallurgical industry in the past.

Many mining operations in formerly prosperous districts have considerably slowed down or completely disappeared. Mining of the lead-zinc deposits has not occurred for a long time and, in Normandy, Soumont is the only producing mine among the major iron deposits. In Anjou, the Segré iron mines closed recently and, after a brief mining attempt, the Montbelleux tin and tungsten mine was closed again in 1983.

Mining activity has not ceased completely; it is more localized and different commodities are being mined (Chauris, 1989). The Vendée is one of the major uranium districts in France with more than 13000 t of U metal produced and will be the only producer of antimony in the country when mining of the Brouzils mine begins in 1990. Exploitation of Rouez, Sarthe, since summer 1989, (see later) provides the only gold production since the La Bellière mine closed in 1942. Brittany, is by far the main producer of kaolin in France (89%) with its Ploëmeur (150000 t/yr), Berrien (57000 t/yr) and Quessoy deposits. The Guerphalès andalusite deposit (50000 t/yr) ranks first in Europe and second in the world (after South Africa). Exploration by the Bureau de Recherches Géologiques et Minières (BRGM) and recently by the Société nationale ELF Aquitaine (Production) [SNEA(P)] has identified previously unknown polymetallic (Pb-Zn-Cu-Ag) sulfide deposits (Guigues 1978; Bernazeaud 1981).

J. D. Keppie (Ed.): Pre-Mesozoic Geology
in France and Related Areas

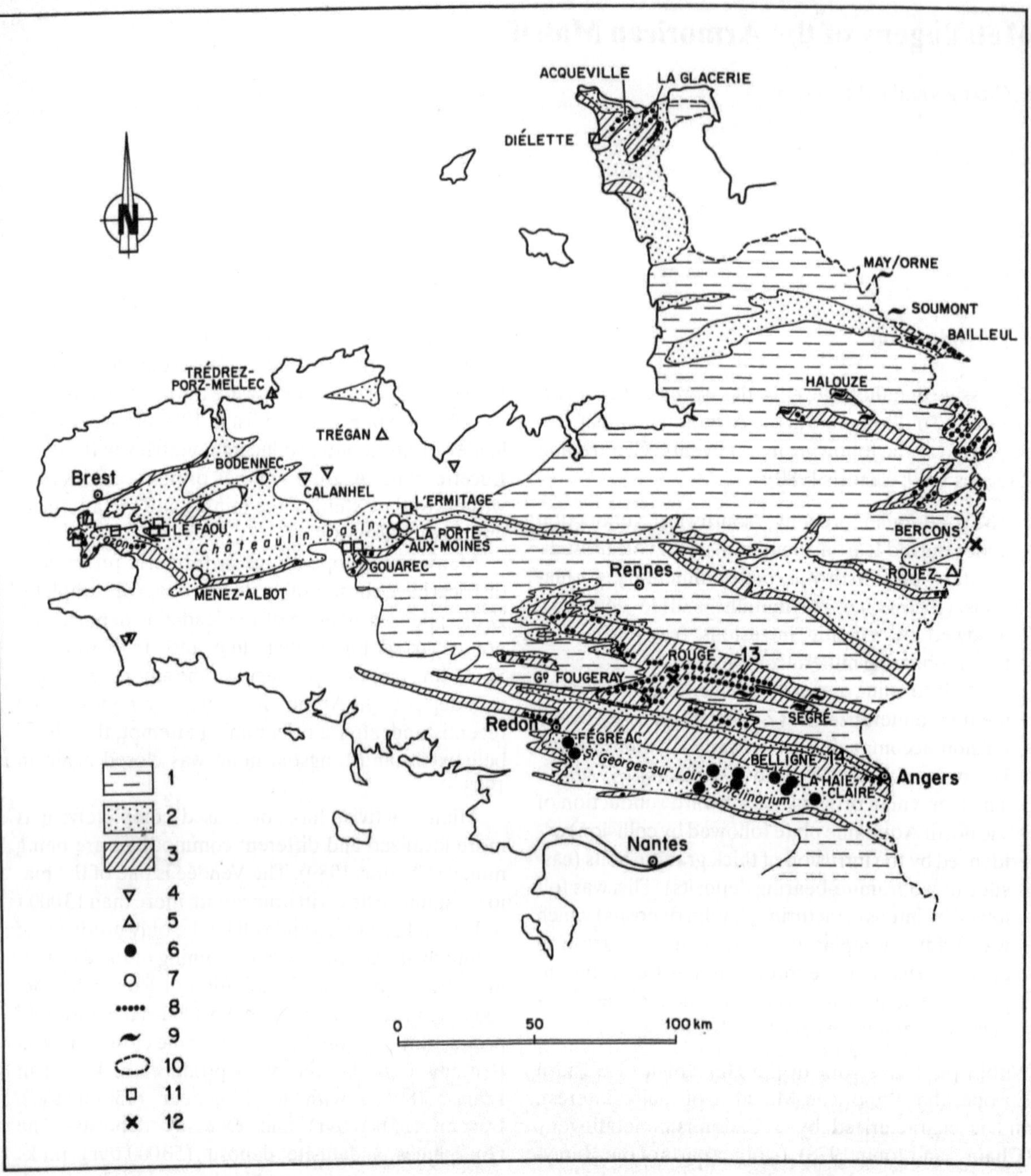

Fig. 1. Orthomagmatic, sedimentary, and volcanosedimentary mineralization of the Armorican Massif. Geological setting: *1* Brioverian epimetamorphic domain; *2* undifferentiated epimetamorphic Paleozoic (except the Ordovician); *3* epimetamorphic Ordovician. *Orthomagmatic mineralization (chromite):* 4 Brioverian. *Polymetallic volcano- and hydrothermal-sedimentary mineralization:* 5 Brioverian; *6* Silurian; *7* Devonian. *Sedimentary mineralization:* 8 Ordovician rutile and zircon sandstones; *9* Ordovician iron deposits; *10* Llandvirnian monazite schists; *11* Devonian iron deposits; *12* Post-Hercynian iron deposits; *13* Martigné-Ferchaud syncline; *14* Saint-Julien-de-Vouvantes syncline

3 The Brioverian and the Cadomian Orogeny

Only the Normano-Breton domain can be subjected to a detailed metallogenic study given the present state of knowledge (Chauris and Jonin 1978). The known mineralization is fairly significant, but is not economically valuable: chromite-bearing chrysotile associated with lenses of serpentinite (Calanhel), and disseminated sulfides (Pb, Zn, Fe, Cu) in cherts (Trégan (Le Fur 1978) or in volcanosedimentary complexes in Lannion Bay (Trédrez, Porz-Mellec) (Fig. 1). Exceptions are the Rouez and Montbelleux deposits.

3.1 Cadomian Massive Sulfide Deposit

The only extensive economically significant ore deposit (Bernazeaud 1981; Icart and Safa 1981) is the Rouez hydrothermal-sedimentary massive sulfide body (Fig. 1). It is hosted by Upper Brioverian detrital rocks in the vicinity of the secondary margin of the Paris basin. These turbiditic rocks consist of alternating wacke, pelite, and argillite with intercalations of conglomeratic lenses but no volcanics. They underwent schistose deformation and a very low-grade metamorphism of debated age.

The sulfide body, 800 m long and 60–70 m thick, is stratiform and has reserves estimated at 100 Mt of sulfides. The huge metal stocks (Table 1) are devalued by the low base-metal contents which would require complex upgrading processes of the pyrite – the high cost of these processes is why mining is being deferred, except for gold enriched in the surficial alteration zone, where it has been mined by the open-pit method since June 1989. The very fine-grained ore (20–30 μm) is massive, banded, or microbrecciated, and shows a simple mineralogical association (Table 1). In the vicinity of the sulfide body a double alteration halo of chlorite and sericite has been produced by hydrothermal fluids.

3.2 Cadomian Granitic Rocks

The Cadomian granitic rocks of the Mancellian batholith have a granodiorite-dominated composition. Near the margins of the batholith, recent, differentiated small leucogranite bodies are associated with cassiterite-wolframite mineralization (Fig. 2). The granite massif has been affected by pneumatolytic evolution, as evidenced by wall greisen and cassiterite-, scheelite- and arsenopyrite-bearing quartz veinlets, in the vicinity of Montaigu.

The only economically significant deposit is that of Monbelleux (Fig. 2), south of the Fougères Massif (production of 360 t WO_3 and 40 t SnO_2 between 1970–1957 and 1977–1982; Chauris et al. 1989). The Montbelleux stock, with 78.5% SiO_2, 4.75% Na_2O, 1.55% K_2O and 20 ppm Ba, forms a subvertical dike (20–25 m thick and 250 m long at level –130) of clearly differentiated albitic granite which intrudes the Brioverian schists. Although incomplete, greisenization is more developed towards the top of the stock. The contact with the schists is marked by a "stockscheider" or intense silicification. The Montbelleux deposit is partly included in the albitic granite where it consists of a system of quartz veinlets with cassiterite, topaz and wolframite (0.3% $SnO_2 + WO_3$ at level –30 m), and partly in the schists, where it is formed by a bundle of subparallel quartz veins rich in wolframite, topaz, fluorite and sulfides (arsenopyrite, pyrite, molybdenite) (Pillard et al. 1985; Chauris et al. 1989). The Montbelleux paragenesis is therefore distinctly fluor-bearing. K/Ar dating on muscovites associated with the cassiterite of the greisen yielded an age of 426 ± 15 Ma (Baubron, pers. commun.), which suggests that the intrusion developed much later than the Fougères Massif itself.

Late Cadomian gold mineralization is strongly suspected in the Mancellian domain: a suspicion supported by arsenopyrite-dominated quartz veins with native gold in the Mont-Dol and Dompierre Cadomian granites near Montbelleux gold occurrences, probably stockworks, located in the Brioverian host rocks (Loudéac, Craon, see Fig. 4); significant place names such as Laurière and Orière on the Pre-Paleozoic formations; alluvial maps etc. Sulfide mineralization produced by the evolution of Late Cadomian granites however is not very common (Dompierre, Channel Islands). Very recently, the large Saint-Jacut tourmaline and beryl potassic pegmatites (Fig. 2), were shown to be related to the Late Cadomian granites by their lead isotope composition ($^{206}Pb/^{204}Pb$ = 17.95), similar to that of the K-feldspar of the Mancellian granites ($^{206}Pb/^{204}Pb$ = 17.54–17.91) (Marcoux 1987).

No uranium-bearing mineralization has been discovered in the Mancellian Cadomian granites except for a few microscopic traces of uraninite at Montbelleux. This feature distinguishes the Cadomian from the Hercynian granites.

4 Paleozoic Sedimentation

4.1 The Ordovician Rutile and Zircon Sandstones

The Ordovician rutile and zircon sandstones were first discovered around 1955 as a result of anomalous radioactivity detected in the Arenigian Armorican sandstones (Fig. 1) in various localities of the Armorican Massif. Investigations by BRGM showed that these were actually zircon and rutile sandstones occurring in the upper part of the Armorican sandstones and that the radioactivity was caused by thorium contained in zircon and monazite (Faure 1978).

Table 1. Main characteristics of volcanosedimentary deposits. Lead isotope data are from Marcoux (1987) except those of Lower Silurian deposits. (Oh et al. 1989)

Age	Deposit	Mineable tonnage and grades	Main paragenesis	Accessory minerals	Lead isotopic ratios		
					206 Pb/204 Pb	207 Pb/204 Pb	208 Pb/204 Pb
Brioverian	Rouez	100 Mt of sulfides with 1.5 Mt Zn, 0.6 Mt Cu, 0.3 Mt Pb, 150 t Au. (Bernazeaud 1981)	Pyrite, pyrrhotite, siderite	Sphalerite, chalcopyrite, galena and gold are accessory minerals. Arsenopyrite, magnetite, and cassiterite (as inclusions in sphalerite) are rarer (Icart and Safa 1981)	17.700	15.564	37.689
Lower Silurian	Deposits of the St-Georges-sur-Loire synclinorium	No estimation	Pyrite, chalcopyrite, sphalerite and galena in varying proportions	Cassiterite as inclusions in sphalerite; arsenopyrite, tetrahedrite-tennantite, gold	17.58 to 17.64	15.52 to 15.56	37.66 to 37.78
Lower Devonian (?)	Bodennec	2.11 Mt with 1.16 % Cu, 0.7 % Pb, 2.89 % Zn, 55.8 g/t Ag (Aye 1983)	Sphalerite, galena, chalcopyrite and minor amounts of pyrite	Pyrrhotite, arsenopyrite tetrahedrite, Pb/Sb sulfosalts	(Average ± 1 σ of 24 analyses) 18.154 ± 0.013	15.638 ± 0.010	38.231 ± 0.032
	Porte-aux-Moines	1.86 Mt with 7.83 % Zn, 1.68 % Pb, 0.76 % Cu and 96.6 g/t Ag (Aye 1983)	Pyrite (dominant sulfide), sphalerite, galena, chalcopyrite	Tetrahedrite, bournonite, barite	(Average ± 1 σ of 12 analyses) 18.108 ± 0.006	15.623 ± 0.004	38.154 ± 0.010
	Menez-Albot	No estimation	Sphalerite-galena beds, pyrite beds	Arsenopyrite, chalcopyrite, pyrrhotite, cobaltite, smythite.	(Average ± 1 σ of 4 analyses) 18.104 ± 0.006	15.643 ± 0.007	38.236 ± 0.023

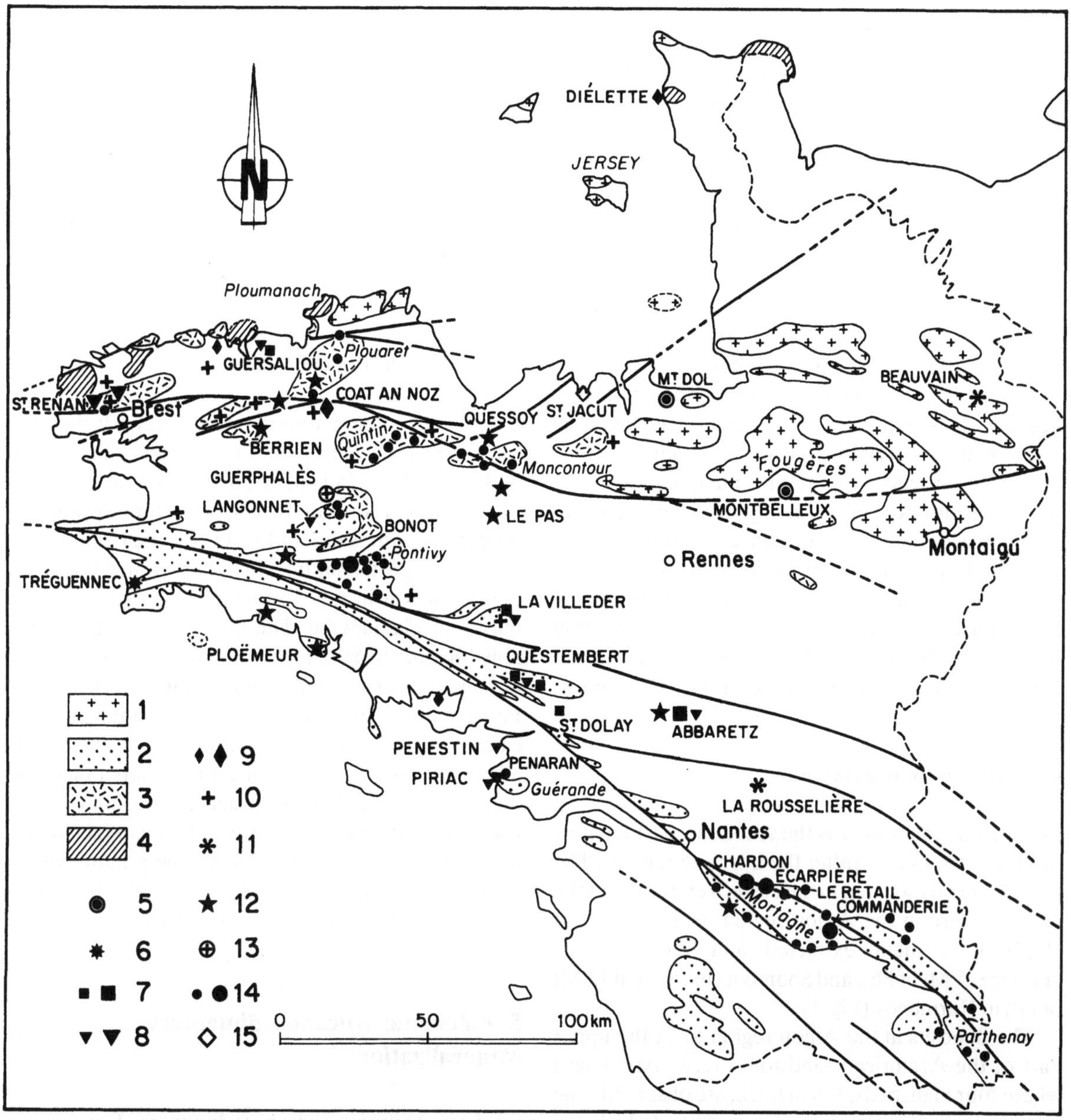

Fig. 2. Granite batholiths and tin, tungsten, kaolin, andalusite, and uranium deposits (or occurrences) in the Armorican Massif. *1* Cadomian and Late Cadomian granites; *2–4* Hercynian granites: *2* South leucogranite belt; *3* central belt; *4* north belt. *5* Late Cadomian Sn-W-As-Au deposits; *6* Sn-aplite; *7* cassiterite veins; *8* cassiterite alluvium; *9* skarns (scheelite); *10* various wolframite or scheelite occurrences; *11* Beauvain and La Rousselière polymetallic mineralization; *12* kaolin deposits; *13* andalusite deposit (Guerphalès); *14* uranium deposits (and occurrences); *15* Late-Cadomian beryl pegmatites

These mineralized sandstones are composed of an aggregate of fine-grained minerals such as quartz, zircon, rutile, anatase and monazite with bedding outlined by reddish rutile. They were generated by the erosion of Proterozoic formations, probably of distal origin, as indicated by the maturity of the deposits, and very likely were deposited on the upper parts of the Ordovician beaches. The possibility of mining these fossil placers has often been examined, but the fineness of the mineralization and the hardness of the sandstones require special grinding which constitutes a double handicap.

The mineralized formations are commonly extensive: 13 km long and 3–6 m thick in the Bailleul district, and about 20 km long and about 8 m thick in the Acqueville district. The heavy mineral concentrations are commonly very high: at Bailleul, the average grade is 15% Zr + Ti over 3 m (with grades of up to

35 % in places) and up to 3 % monazite; at La Glacerie, the mineralized formations are 4 m thick with an average grade of 15 % Zr + Ti and a maximum grade of 50 % (with up to 17 % zircon); at Acqueville, titaniferous minerals and zircon account for up to 42 %.

4.2 Ordovician Europium-Bearing Monazite Schists

Prospecting by BRGM (Guigues and Devismes 1969; Donnot et al. 1973) led to the discovery of several grey monazite occurrences in the Angers Llanvirnian schists. The monazite (values of about 50–200 g/t) occurs as highly flattened microcrystalline fibroradiate nodules, 0.1–2.0 mm in diameter, concentrated in 100-m scale lenses within schists. The nodules, which display a particular marine phosphate genesis, may have concentrated subsequently in placers that are being considered for mining since the monazite has very high europium grades (0.48 % Eu_2O_3). In 1967/1968, a mining test was performed in the alluvial grey monazite concentrations along the Aron river in the vicinity of Grand-Fougeray (Fig. 1); the monazite content of the alluvium ranges from 400 to 4000 g/m^3.

4.3 Ordovician Iron Deposits

The Armorican Massif is the second largest ferriferous region in France (after the Lorraine region). The main mines are located in the eastern extremity of the Massif: Segré in the Anjou region, with more than 10 Mt of Fe metal extracted; and May-sur-Orne, Halouze (5 Mt of Fe), and Soumont (more than 12 Mt of Fe) in Normandy (Fig. 1).

The deposits in the Anjou region lie in the upper part of the Armorican sandstones (e.g., Arenigian) where four main beds, 1–6 m thick, are observed. The mined ore consists mainly of magnetite (which appears to have been metamorphically derived) with lower and varying proportions of hematite, chlorite, and siderite. The deposits are located in highly elongate synclines such as Segré, Martigné-Ferchaud, and Saint-Julien-de-Vouvantes.

In Normandy, the iron-bearing synclines generally trend WNW-ESE. The deposits are located towards the base of the Calymènes schists, which are Middle Ordovician in age. Only one bed, 3–8 m thick, is mineable: it is divided into two parts – an upper chloritic-carbonaceous stratum and a lower hematite-rich stratum.

All the Ordovician ores are of the oolitic type, are semi-phosphorous (0.4 to 0.8 %), siliceous (6–18 %), and are depleted in lime with iron grades ranging from 35 to 48 %. The distribution and present shape of the iron deposits in the Hercynian synclinal keels (that is in the zones not affected by the post-Hercynian erosion) show no relationship with the ferriferous sedimentary basins that were formed under Ordovician paleogeographic conditions. In Normandy, the ferriferous areas lie in a particular position in relation to the Llanvirnian coastline; they occur at a distance offshore from the coast and schematically follow the exact shape of the coastline.

4.4 Devonian Iron Deposits

Ferriferous sedimentation re-occurred during the Devonian, especially in the Crozon Peninsula (Fig. 1), and also in the vicinity of Le Faou, Gouarec, and L'Ermitage, and in Normandy. The only economically significant deposit is that of Diélette in the Cotentin region. It is a particular type of deposit which is located below the sea in the contact aureole of the granite massif of Flamanville. The iron grades are remarkably high (48 to 58 %) with the ore consisting mainly of magnetite associated with hematite, and accessory Fe-Cu sulfides in a gangue rich in metamorphic silicates (pyroxene, garnet, and amphibole). The mineralization is interpreted as an oolitic sedimentary iron deposit of Lower Devonian age (according to fossil remnants), subsequently affected by contact metamorphism caused by the Flamanville granite, which explains the abundance of magnetite and the transformation of the limestones into skarns.

5 Paleozoic Volcanosedimentary Mineralization

5.1 Lower Silurian Volcanosedimentary Deposits

Lower Silurian volcanosedimentary deposits are located in the Saint-Georges-sur-Loire synclinorium, a 20-km-wide and 150-km-long trough which stretches from the vicinity of Redon to south of Angers (Fig. 1). The synclinorium was formed in the same sedimentary basin as the Central Brittany synclines. As early as the Upper Ordovician a back-arc basin developed and became filled with pyroclastic and detrital deposits with intercalations of sills and mafic flows. During a major thrust movement of probable Middle Devonian age, the basin was thrust over the northern Paleozoic formations.

The Zn-Pb-Cu-(Ag) mineralization is grouped in two clusters in two zones which show evidence of Silurian-Devonian volcanic activity. The most extensive deposits are grouped in the eastern portion of the synclinorium (e.g., La Haie-Claire) and are being investigated by core drilling. The meter-thick mineralization lies between the top of a coarse-grained unit of lava and acidic tuffs, Ashgillian to Middle Silurian in age, and the base of a unit of shales and silts associated with lavas and spilitized andesitic agglomerates, Upper Silurian to Lower Devonian in age (Oh et al. 1989). The ore is banded, pseudobrecciated, or disseminated, and the paragenesis is quite classical (Table 1). The gangue consists of siderite, quartz, chlorite and locally barite. The mineralization can be classified as proximal or distal on the basis of mineralogical data. The geothermometric study based on the chemical composition of chlorites which are contemporaneous with the deposition of sulfides yielded temperatures ranging from 280 to 300 °C (method of Cathelineau and Nieva 1985).

The lead isotope compositions of the galena (Oh et al. 1989) are highly clustered (Table 1). The anomalous unradiogenic lead of this Silurian mineralization ($\simeq$ 435 Ma) contrasts significantly with the more "crustal" lead of the Devonian volcanosedimentary deposits of the Châteaulin basin (cf. below) and suggests that this mineralization originated from the Paleozoic sediments formed from the erosion of a uranium-depleted basement which underwent high-grade metamorphism. Volcanism is very likely the thermal generator of the mineralization.

5.2 The Lower Devonian Volcanosedimentary Deposits[1]

The Lower Devonian polymetallic sulfide bodies are the first volcanosedimentary deposits identified in the Armorican Massif. Three main orebodies have been discovered: Bodennec, La Porte-aux-Moines and Menez-Albot (Fig. 1).

5.2.1 Geological Setting and Lithostratigraphic Data

The mineralization is interstratified in a horizon of black sediments (shales, silts and cherts) at the top of the Plougastel Gedinnian schist and quartzite formation (Bril 1975; Aye 1983). In the three main orebodies the chemically generated layers (chert, chlorite, and sericitite) invade the sedimentary basin. Terrigenous sediments are not common at Bodennec and La Porte-aux-Moines, and are limited to black shales and fine-grained silts. At Menez-Albot, they are extensive and include chloritoid black shales and light-colored sandstones.

5.2.2 Depositional Site of the Mineralization

The deposition of the mineralization during periods of intense seismic and tectonic activity was almost contemporaneous with ignimbritic, explosive, acidic volcanic activity (rhyolitic to rhyodacitic) which started before the Lower Siegenian (Aye 1983). However, the initial distribution of the orebodies has been disturbed deeply by various Hercynian tectonic events.

A reconstruction of the paleogeographic depositional site from the study of the volcanic material and sediments (Aye 1983) shows La Porte-aux-Moines orebody to have been rooted at the top of a volcanic vent very close to the outlet, and the Bodennec orebody to have been deposited on the flank of a vent probably less than 5 km from the outlet – both deposits having been formed in a marine environment far from the continent. The Menez-Albot orebody appears to have been deposited the farthest from the outlet and closer to the continent, in a lagoonal epicontinental environment. Intensely hydrothermally mafic volcanics of abyssal tholeiitic composition postdate the emplacement of the mineralization at various stratigraphic levels.

5.2.3 Mineralization

The mineral textures of the mineralization are varied and intensely imbricated (Table 1). The banded massive or brecciated ores and disseminated sulfides are dominant at La Porte-aux-Moines and Bodennec and display characteristics common to remobilization processes. The La Porte-aux-Moines orebody originally consisted of two massive sulfide bodies rooted in the hanging wall by a stockwork of sulfide veinlets; the upper lens must have been several hundred meters in extent. The Bodennec orebody was originally a horizon of sulfides, less than 1 km in extent and less than 1 m thick, enclosing brecciated fragments. The Menez-Albot orebody consists of finely disseminated sulfide beds, 50 m thick, that extend laterally for more than 2.5 km.

[1] Recent stratigraphic and structural re-examinations question this Lower Devonian age (Lescuyer, pers. comm.). In fact, these deposits seem more related to a Devono-Dinantian hydrothermal event, as are most of the stratiform polymetallic sulfides deposits in western Europe (Chessy, France; South Iberian pyrite belt).

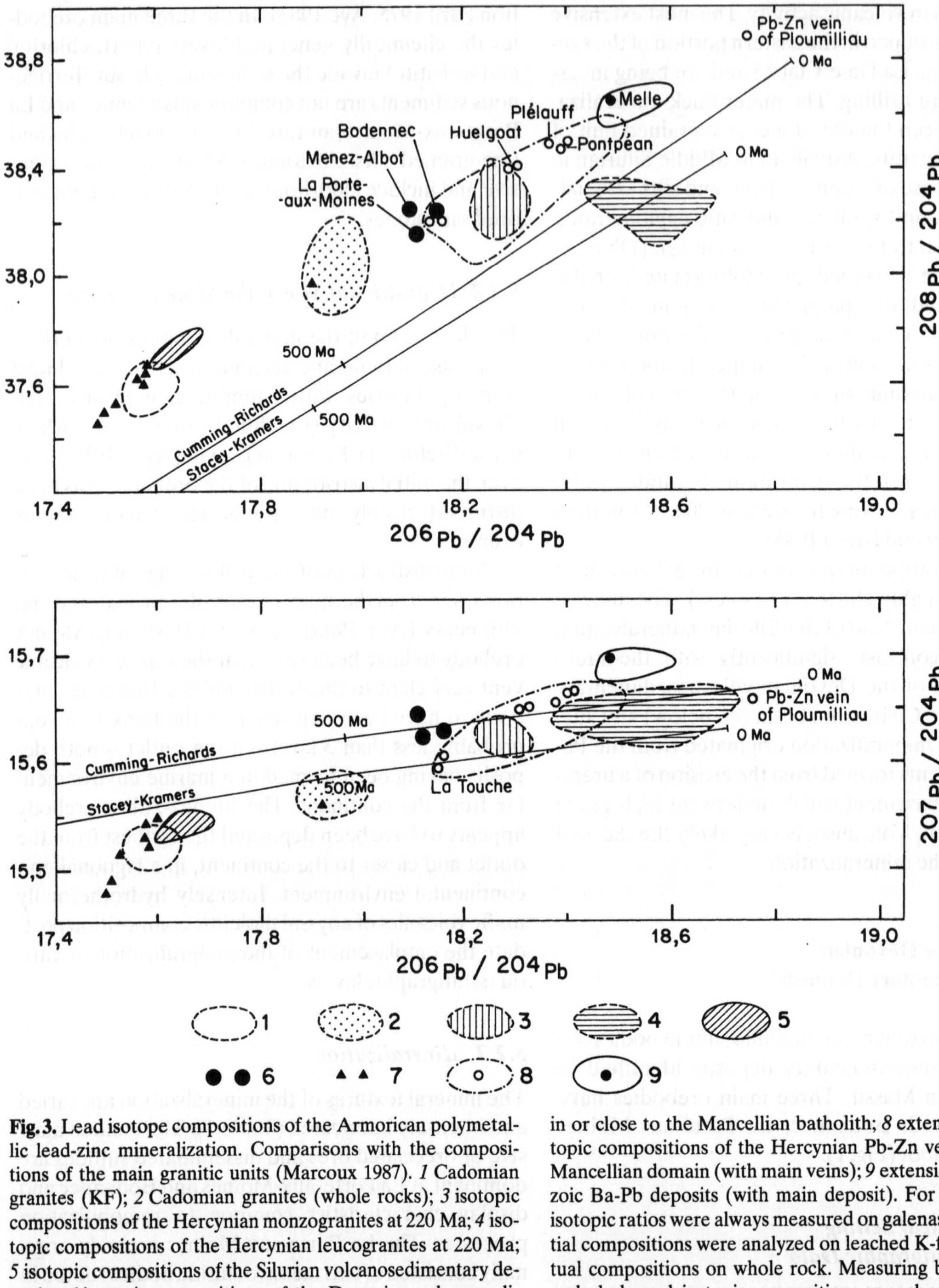

Fig. 3. Lead isotope compositions of the Armorican polymetallic lead-zinc mineralization. Comparisons with the compositions of the main granitic units (Marcoux 1987). *1* Cadomian granites (KF); *2* Cadomian granites (whole rocks); *3* isotopic compositions of the Hercynian monzogranites at 220 Ma; *4* isotopic compositions of the Hercynian leucogranites at 220 Ma; *5* isotopic compositions of the Silurian volcanosedimentary deposits; *6* isotopic compositions of the Devonian volcanosedimentary deposits; *7* Cadomian and Hercynian lead veins hosted in or close to the Mancellian batholith; *8* extension of the isotopic compositions of the Hercynian Pb-Zn veins outside the Mancellian domain (with main veins); *9* extension of the Mesozoic Ba-Pb deposits (with main deposit). For mineralization, isotopic ratios were always measured on galenas. For rocks, initial compositions were analyzed on leached K-feldspar and actual compositions on whole rock. Measuring both K-feldspar and whole-rock isotopic compositions on each sample of Hercynian granites, the isotopic composition is calculated at 220 Ma

5.2.4 Origin of the Mineralization

Lead isotope analysis on the three main polymetallic orebodies (Marcoux 1987) showed similar values (Table 1). Consequently, the depositional sites of the mineralization may have been nearly identical paleogeographic environments, and the orebodies may share a common metallic source. The analyses also showed a remarkable homogeneity of the isotope compositions within each deposit (Fig. 3); and so the slight variations observed from one orebody to the other are of great significance. The variations indicate gradual changes in the source of metals in relation to the respective depositional sites.

Most of the lead mineralization shows characteristics of a thorium-rich mean continental crust (Cumming and Richards 1975; Stacey and Kramers 1975) with $^{238}U/^{204}Pb$ (μ) and $^{232}Th/^{204}Pb$ (W) ratios of 9.9 and 39.0 respectively. This implies that the hydrothermal fluids coeval with volcanism acquired most of their metallic load by percolating through the sedimentary and plutonic rocks of the continental crust, and that the slight variations observed from one body to the other are caused by detrital sediments rather than by a lithological change within the crust. The influence of the mantle appears to have been negligible in the genesis of the mineralization.

The isotopic results are consistent with the model devoloped by Aye (1983): the subsynchronous deposition of the mineralization during a pre-oceanization period is located in an intracratonic distension zone characterized by an ignimbritic, then a mafic volcanism. These events are probably indicative of paroxysmal distensional episodes of the central Armorican trough.

6 Hercynian Granitization

Several major groups of mineralization are genetically related to Hercynian granitization.

6.1 Cassiterite (and Wolframite) Deposits

The Armorican Massif forms part of the well-known cassiterite province of central and western Europe which stretches from the Erzgebirge (in eastern Europe) to the Iberian Peninsula. In Armorica, except for the Montbelleux occurrences and those in the region of Mayenne (which are both related to the Late Cadomian granitic evolution of the Mancellian batholith), the Sn (W) deposits are associated mainly with three major Hercynian granitic belts (Chauris 1981). These can be distinguished from one another by their location in the orogenic structure, the shape of their massifs, their petrographic composition, their origin and age (between 345 and 290 Ma), and their significantly different stanniferous occurrences (Fig. 2) (see Chap. II.4, this Vol.).

In the south, the leucogranite belt consists of commonly mylonitized sheets and cupola-shaped lobes related to major linear faults. Farther north, the belt of biotite granites with early formed diorites and granodiorites and late leucogranitic injections (medio-Armorican batholith) consists of a string of nearly ovoid plutons; the emplacement of the plutons was controlled by the medio-Armorican lineament. In the northernmost part of the Armorican Massif, the rectilinear red granite belt is composed of small, nearly circular massifs locally displaying concentric structures; these are associated locally with the final member of leucogranitic intrusives. The granite belt appears to have been emplaced during the Hercynian reactivation of the Cadomian lineament. Each granitic belt has its specific Sn-W deposits of which main characteristics are summarized in Table 2. Only the most important formations are detailed below.

6.1.1 The Sodalithic Formation

The sodalithic formation is very uncommon and is represented by the Tréguennec aplitic-pegmatitic vein intrusive into micaschists. The predominant aplitic facies is hololeucocratic and albitic, and clearly sodic (5.3% Na_2O, 2.8% K_2O). It contains exceptional accessories such as lepidolite, amblygonite, and cassiterite, with locally high values of tantalum and niobium (~6% Ta_2O_5, 4.4% Nb_2O_5, Kosakévitch, unpubl.) and columbotantalite.

6.1.2 The Quartz-Cassiterite Formation

The classic quartz-cassiterite formation comprises the main Hercynian cassiterite occurrences identified in the Armorican Massif, preferentially associated with the southern leucogranitic belt.

The quartz veins at La Villeder are located in the eastern extremity of a leucogranitic lobe. Albitization, but not greisenization, developed at the contact of the veins. Muscovite crystallized as a rim in the wall rocks or in the quartz, whereas tourmaline is common only in the schistose wall rocks. Bluish crystals of apatite (several centimetres in size) and very common beryl are enclosed in the quartz. Locally large crystals of brownish cassiterite are commonly associated with sulfides (Table 2). Small Pliocene bodies of alluvial, fluviatile (or marine) cassiterite (La Hye) have been mined in the vicinity of the primary deposits.

The veins in the schists at Abbaretz consist mainly of quartz, commonly occurring as small lenses, with abundant muscovite. Tourmaline is extensively developed in the schistose wall rocks which have been transformed into tourmalite by boron metasomatism. Cassiterite molds beryl, and generally occurs as irregular bodies with fissures cemented by veinlets of quartz and muscovite.

Table 2. Main characteristics of the Sn-W occurrences related to the three Hercynian granitic belts. Most of the information is from Chauris (1980, 1981)

	Leucogranite belt	Biotite granite belt	Red granite belt
Cassiterite/tin	Abundant primary deposits (Abbaretz, La Villeder, Piriac), mainly quartz veins and small Sn pegmatite-aplite; and alluvial deposits (La Hye)	Numerous but small occurrences. No mined primary deposit. One large alluvial deposit worked (St-Renan)	Few to nil occurrences (except Guersaliou district)
Wolframite and scheelite/tungsten	Rare or absent	Common, associated with cassiterite	Sometimes wolframite; common scheelite in skarn
Beryl/beryllium	Relatively abundant	Rare	Rare or absent
Tourmalinization	Intense at the schistose walls of the quartz veins	Common	Slight
Greisenization	Absent	Extensive	Slight
Sulfides	Rare, with the exception of arsenopyrite	Relatively abundant	In places abundant molybdenite (pegmatite, skarn)
Main deposits with dominant minerals	La Villeder: beryl, cassiterite, arsenopyrite, sphalerite, chalcopyrite, bismuthinite, molybdenite. (Production: about 200 t Sn during 19th century)	St Renan alluvial deposit: cassiterite ilmenite, garnet, staurolite, tourmaline, rutile, zircon, monazite (more than 6000 t SnO_2 between 1960–1975)	Guersaliou: a) quartz veins with microscopic cassiterite, abundant arsenopyrite, chalcopyrite b) alluvial deposit (460000 m^3 with 1.09 kg/Sn/m^3). Partially exploited (1971–1973)
	Abbaretz: beryl, cassiterite, arsenopyrite, pyrite, chalcopyrite, rare molybdenite (4000 t SnO_2 between 1951–1957)	Coat-an-Noz: a) quartz veins with wolframite, scheelite, molybdenite and late sulfides (chalcopyrite, sphalerite Bi minerals) b) skarn formations with scheelite, chalcopyrite, arsenopyrite, grossular, vesuvianite. 1100000 metric tons at a grade of 1.3% WO_3 and 0.3% Cu (Mirabail and Terrée 1986) Locarn: quartz veins with wolframite, native gold, beryl, and sulfides (Lulzac 1968)	Diélette: skarn with scheelite and molybdenite

6.1.3 The Quartz-Cassiterite-Wolframite Formation

The quartz-cassiterite-wolframite formation is widespread in Armorica, with the better known examples being located in the biotite granite belt (Table 2). The association of tungsten (wolframite and commonly scheelite) with the tin distinguishes it from the quartz-cassiterite formation (see Sect. 6.1.2. above). The exposed occurrences in the Saint-Renan district are very numerous (Fig. 4) but are always of limited extent as at Kervenguy (wolframite in a hydrothermal tourmaline breccia), Penfeunteun (occurrence in the internal margin of granite bounded by a "stockscheider") and Kergonguy (small stock of quartz veinlets with cassiterite, wolframite and scheelite). The minerals are found in three commonly associated types of occurrence: tourmalite, greisen, and quartz veins.

No primary ore of economic value has been discovered in the Saint-Renan district to date, but the great number of cassiterite occurrences coupled with the morphological features of Saint-Renan valley induced the formations of Sn-rich fluviatile alluvium which has been recently mined (Moussu 1963) (Table 2). The cross-section through the flat enabled the distinction of old, typically fluviatile alluvium (1–2 m thick), and recent alluvium (2–3 m thick) corresponding to the spreading of Weichselian periglacial cones of coarse-grained mineralized sand in the valley.

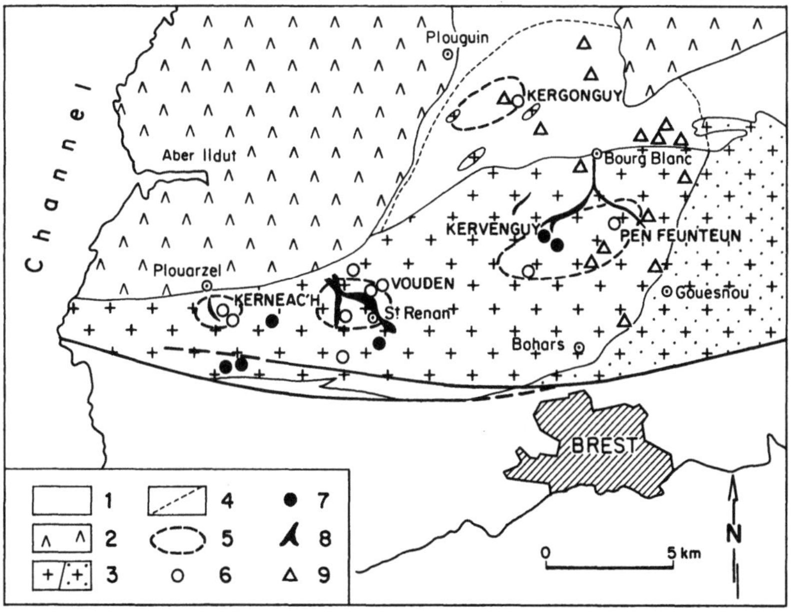

Fig. 4. Main Sn-W occurrences in the western part of the Saint-Renan Massif. *1* Metamorphic basement; *2* Hercynian red granites; *3* Saint-Renan granite (*with dots:* Kersaint facies); *4* probable northern limit of the Saint-Renan granite; *5* zones rich in greisen and quartz veinlets; *6* Sn-W occurrences; *7* wolframite (and/or scheelite) occurrences; *8* alluvial stanniferous placers; *9* alluvium with more than 5 kg/SnO_2/m^3. (After Chauris 1981)

6.1.4 Other Cassiterite or Wolframite Formations

Other cassiterite or wolframite formations show particular parageneses which cannot be related easily to one of the groups described in Sections 6.1.1–6.1.3 above (Table 2). Among these are relatively rare molybdenite and scheelite skarns (Diélette), although one with grossular, vesuvianite, and various sulfides of economic significance has been identified recently at Coat-an-Noz by SNEA (P) (Mirabail and Terrée 1986).

Finally, two original deposits, which are not clearly related to one of the three major Hercynian granitic belts are: (1) The Late Hercynian Beauvain Mo-porphyry (299 ± 4 Ma, U/Pb method on zircons), related to complex, peraluminous, subvolcanic intrusions (K_2O-rich calc-alkaline magmatism) (Callier 1989), in which the Mo(Sn-Cu) mineralization (0.1–0.2% MoS_2), linked to potassic metasomatism, would represent the roots of an eroded orebody (Callier 1989); and (2) the telescoping (Mo, W, Sn, Cu, Zn) mineralization of La Rousselière in the fissured apex of a Dinantian hypovolcanic intrusive granitic complex (Allon 1980).

6.2 Kaolin Deposits

Major kaolin deposits, such as in the leucogranite belt of the Ploëmeur district, and the Ménez-Molvé Berrien deposit in the medio-Armorican batholith (Fig. 2), were formed from the post-magmatic hydrothermal alteration of some Hercynian granite massifs (Nicolas 1956; Charoy 1975, 1979). The Ménez-Molvé deposit (Chauris 1988) located on the margin of the Berrien granite is taken here as an example. From initial transformation of the margin of the granite massif (development of a paragenesis of albite, quartz and muscovite) the extension of fracturing (mainly in the preferential direction N10–20 ° W of regional fracturing) was favorable for the development of limited greisenization, scattered tourmalinization, sericitization, and especially kaolinization, all located on the flank of the intrusion. The extensiveness of the kaolinization indicates that particularly persistent structural and thermal conditions occurred together: fissuring, high permeability, and high thermal gradient leading to efficient hydrothermal convection. In spite of the lack of isotopic data, several arguments support a deep post-magmatic hydrothermal origin of the kaolinization process: (1) a spatial relationship between kaolinization and early hydrothermal transformations along a regional tectonic fracturing zone; (2) location of the deepest portion of the kaolin deposit beneath a thick hanging wall of schist and quartzite; (3) an intense hydrothermal alteration halo (kaolinization) in the schists at the contact with the kaolinized granite; and (4) an absence of efficient supergene processes (absence of Tertiary lateritic alteration profile).

6.3 The Guerphales Andalusite Deposit

The contact metamorphism of the Rostrenen Hercynian granite (Fig. 2) is related to the formation of extremely andalusite-rich hornfels (10% andalusite) in the Lower Ordovician schists. These have been mined since 1970 by the open-pit method for the production of refractories. Chiastolite occurs as 5–8 cm long, pink, transparent and translucent, randomly oriented crystals.

7 Late Hercynian Mineralization

7.1 Gold Mineralization

Gold mineralization mainly occurs in the eastern part of the Armorican Massif along E-W to N60 ° W faults which lie parallel to major shear faults (Fig. 5). The mineralization may be of the shear zone type (Bonnemaison and Marcoux 1987), but this hypothesis has only been confirmed in the old La Bellière mine, where the mineralization is clearly related to a thick mylonitic zone probably one branch of the South Armorican fault system. La Lucette, the second largest gold deposit in the Armorican Massif, displays a mainly stibnite mineralization (see later).

Only the La Bellière-Bégrolles vein system has been mined recently (1905–1942). Hosted by Proterozoic schists and micaschists, it is oriented east-west and stretches for 15 km, its width being from 3 to 7 km. It is formed by fairly continuous quartz lenses, at most 20 m thick. The paragenesis (Table 3) comprises predominant pyrite, marcasite (formerly pyrrhotite) and numerous sulfides with native gold common as patches in the quartz and in the sulfides. The mineralized lenses developed within shear faults in extension zones during a sinistral displacement, probably related to a movement of the South-Armorican shear zone system.

The Beslé-Moisdon-La Pouëze axis is another gold district of the Armorican Massif. The mineralization is hosted by Upper Ordovician and Lower Silurian schists and sandstones, generally in N70–80 ° W oriented quartz lenses, and marks the axis irregularly over more than 100 km. Most lenses are marked at the surface by old surface workings, but only the Beslé deposit has been investigated recently. The paragenesis is poor, consisting of pyrite, arsenopyrite, chalcopyrite, and native gold in a quartz gangue (Table 3).

No geochronological data are available for the mineralization. However, the deposits are contemporaneous with the last shear movements of the faults and hence post-date the emplacement of the Hercynian granites but pre-date the Permian-Triassic N10–20 ° W fracturing episode. It can be assumed, therefore, that these gold deposits have ages of between approximately 300 and 240 Ma.

The location of gold in the eastern part of the Armorican Massif is being re-examined following the recent discovery of gold occurrences in the western regions (Finistère Department). Some of the occurrences are most likely related to shear zones developed in the Brioverian schists; others (Lopérec) are still being investigated by BRGM and have yet to be defined. The last magmatic events related to the mobile zones (~ 300 Ma?) may have been responsible for the emplacement of the mineralization, and the source of the gold may have been a reconcentration of gold occurrences disseminated in the Brioverian (Loudéac, Craon); these primary deposits generate small but valuable gold placers (Loudéac region).

7.2 Uranium

Uranium was until 1991 the principal metal extracted in the Armorican Massif, and its occurrences are obviously associated with the Hercynian peraluminous intrusive granites (La Roche de et al. 1980). The only major deposits investigated are related to the south Armorican leucogranitic belts: the districts of Mortagne (where the mines of La Commanderie, l'Ecarpière, Le Chardon have produced respectively 3300 t, 3120 t and 2745 t of U metal[1]), Parthenay, Guérande (where the Penaran mine at Piriac have produced 470 t of U metal[1]) and Pontivy (with the ancient mines of Le Bonot: 400 t of U metal produced, Kerlec'h and Quistiav) (Geffroy et al. 1960; Gerstner et al. 1962; Germain et al. 1964; Cathelineau 1981 a, b). The biotite granites of the medio-Armorican batholith are distinctly less mineralized (small occurrences north of Brest, at Plouaret, Quintin, and Moncontour) and the red granite belt seems to be barren.

An exceptional polarity, similar to that of tin, is displayed in the distribution of the uranium occurrences related to the Hercynian granites (Fig. 2). The main deposits are located generally within the granitic internal contact, with some extensions in the metamorphic rocks (Le Chardon, l'Ecarpière), or are strictly within the metamorphic basement (Retail, Penaran) but remaining always close to the granitic contact. This preferential distribution is particularly remarkable in the Mortagne Massif (Kosztolanyi 1971; Renard 1971; Cathelineau 1981 a) and in the Guérande Massif (Cathelineau 1981 b). In the Pontivy district the uranium deposits are mainly intragranitic; the

[1] Production at 01/01/1987.

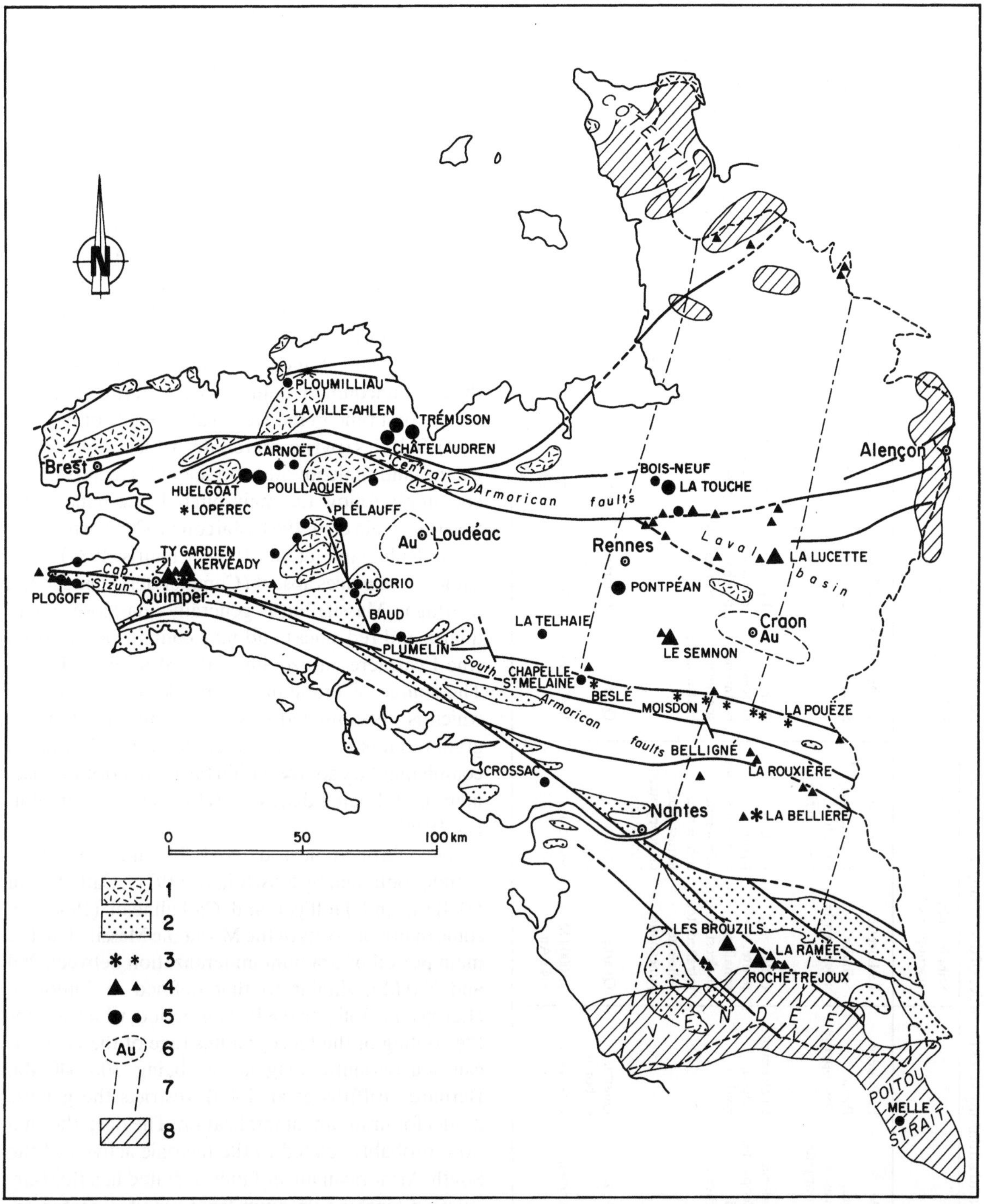

Fig. 5. Late Hercynian gold, antimony, and lead-zinc mineralization and post-Hercynian stratiform barite-lead mineralization. *1* Hercynian biotite granites; *2* Hercynian leucogranites; *3* gold deposits; *4* antimony veins; *5* lead-zinc veins; *6* zones of disseminated gold in Brioverian formations associated with small placers; *7* east Armorican Sb(Hg) strip; *8* extension of the post-Hercynian Ba(Pb) mineralization

Table 3. Mineralogy of the gold and antimony veins (in order of decreasing abundance). (Fouquet 1980; Marcoux et al. 1984; Pillard et al. 1985; pers. commun.)

	Gold Deposits			Stibnite-gold arsenopyrite type		Stibnite-quartz type					Stibnite cinnabar type
Deposit	La Bellière	Beslé	La Pouëze	La Lucette	Le Semnon	Rochetrejoux	La Ramée	Les Brouzils	Ty Gardien	Kerveady	La Rouxière
Main minerals	Arsenopyrite, pyrite, sphalerite, galena	Pyrite, marcasite, arsenopyrite	Pyrite, arsenopyrite	Stibnite, arsenopyrite	Berthierite, stibnite, arsenopyrite	Stibnite	Stibnite	Stibnite	Stibnite, berthierite	Berthierite, stibnite	Stibnite, cinnabar
Accessory minerals	Chalcopyrite, gold, tetradymite, cosalite, bismuthinite	Sphalerite, chalcopyrite, gold, pyrrhotite	Chalcopyrite, galena, gold	Gold, pyrite, sphalerite, chalcostibite, tetrahedrite, scheelite, jamesonite	Gold, pyrite, native antimony, ullmannite, sphalerite, aurostibite, chalcopyrite, bournonite	Pyrite, marcasite, berthierite, sphalerite, chalcostibite	Pyrite, marcasite, pyrrhotite, arsenopyrite	Arsenopyrite, berthierite, pyrite	Marcasite, arsenopyrite, pyrite, sphalerite, zinkenite, semseyite, plagionite, native antimony	Pyrite, native antimony, arsenopyrite, zinkenite, plagionite	Sphalerite, pyrite, marcasite, tetrahedrite-tennantite
Gangue	Quartz, siderite	Quartz	Quartz, carbonate	Quartz	Quartz calcite	Quartz	Quartz (barite)	Quartz	Quartz	Quartz	Quartz
Production	10.4 t Au	Unknown	Unknown	42000 t Sb 8.4 t Au	500 t Sb	16400 t Sb	800 t Sb	Opening in 1990 ~ 8000 t Sb	565 t Sb	–	–

granite is cut by a system of transverse mineralized fractures (N20 ° W and N60 ° W extension faults) preferentially located in the northern margin of the massif or along collapse corridors occurring within the granite, and everywhere spatially related to the surface portion of the granite massif.

Most of the vein deposits are polyphase with a classical layered texture. The main ore is pitchblende with spherulites of up to several centimetres in diameter (Germain et al. 1964; Pierrot et al. 1980). At depth, microscopic coffinite is commonly associated with pitchblende which it epigenizes or overlies. Near the surface, weathering transforms the pitchblende into "gummite" and abundant secondary minerals, one of the most uncommon being uranosphaerite identified for the first time in the Kerségalec vein (Pontivy district). Sulfides (pyrite, marcasite, in places sphalerite, galena, and chalcopyrite) and selenides (clausthalite) commonly follow the main pitchblende mineralization (Germain et al. 1964; Marcoux 1980; Pierrot et al. 1980; Ruhlmann et al. 1980; Chauris 1984). The gangue is in places absent (Commanderie mine in the Mortagne Massif) but is generally quartzose (comb quartz and red silica) and may contain rare fluorite (the Ecarpière antozonite in the Mortagne Massif). The source of uranium is very likely the uraninite which is disseminated in some granitoids ("fertile" granites,Gerstner et al. 1962; Renard 1971), but a complementary source within the metamorphic basement is still being discussed (Holliger and Cathelineau 1986).

The chronological data (U/Pb ages on pitchblende) obtained by Kosztolanyi (1971), Cathelineau (1981 a), and Holliger and Cathelineau (1986) on some major deposits of the Mortagne Massif, date the main period of uranium mineralization between 290 and 270 Ma, similar to that defined in Limousin (Leroy and Holliger 1984). This period, which is after the cooling of the leucogranites (age of the Armorican leucogranitic magmatism being 300–340 Ma; Bernard-Griffiths et al. 1985), restricts the genetic model for uranium mineralization. The deep thermal flow, probably related to the tectonic activity of the South-Armorican linear faults, resulted in a fluid circulation leaching uranium from granites, metamorphic basement rocks, and mylonites (Fehn et al. 1978; Darnley 1982) with the uranium deposited in late transverse fractures, in the linear faults or not far away from them. Such fractures thus appear to be one of the major metallotects in the metallogeny of uranium.

7.3 Antimony-Bearing Veins

The antimony deposits are as highly asymmetrically distributed as the gold deposits. Most of the deposits (major veins of La Lucette and Rochetrejoux, Fig. 5) are located in the eastern part of the Armorican Massif. The veins are less extensive in the west (Ty Gardien, Kerveady).

The La Lucette deposit, which was mined from 1899 to 1934, is located in the Laval basin. It consists of a swarm of meter-thick, roughly parallel, N40 ° E-oriented quartz veins that intersect the Ordovician to Upper Silurian schists and sandstones in an overturned anticlinal fold. The paragenesis is complex, with dominant stibnite and native gold (Table 3). These veins, with the nearby Semnon berthierite-stibnite-arsenopyrite and gold mine (Chauris et al. 1985), are a good illustration of the paragenetic stibnite-gold-arsenopyrite type defined by Chauris and Guigues (1969).

The Rochetrejoux deposit is part of a major antimony district where more than 20 veins have been mined or investigated since the 18th century. Mining of the Brouzils vein, recently discovered by BRGM, started in 1990. The veins intersect mainly Cambrian and Brioverian shaly-silts, where they occupy tension cracks and shear fractures related to two faulting episodes of major South-Armorican brittle structures (Marcoux et al. 1984). The major north-south to northeast-southwest oriented veins result from a dextral displacement. The paragenesis is complex (Table 3), dominated by stibnite set in a gangue of chalcedony quartz. A late lead-bearing paragenesis exists in some veins.

The Quimper-Cap Sizun district is the third most extensive antimony district of the Armorican Massif, especially since the discovery (in the last decade) of numerous occurrences and the mining of the small Ty Gardien deposit near Quimper. It is clearly related spatially to the Hercynian granites and the major South-Armorican shear zones (Chauris et al. 1977). The mineralized veins occur in tension cracks caused by the South Armorican dextral shear movement (Fouquet 1980) and intersect migmatites, micaschists, granites, and trondhjemite massifs. The hydrothermal alteration was accompanied by neoformation of montmorillonite-illite. The paragenesis of the veins derives from three mineralized episodes (Fouquet 1980) variably expressed according to the veins: the first was complex with dominant arsenopyrite; the second was antimoniferous, responsible for the deposition of berthierite and massive stibnite; and the third consisted mainly of a lead-zinc supply that allowed formation of Pb/Sb sulfosalts. The Quimper-Cap Sizun district is a unique example of superimposed lead and antimony paragenesis in the Armorican Massif. However, the second introduction of lead cannot be considered as the limited and distal expression of the extensive Armorican lead-zinc veins.

Also to be mentioned are the uncommon Bellligné-La Rouxière occurrences, characterized by the association of cinnabar with stibnite, in the Silurian schists north of the Ancenis basin. The spatial distribution of the Sb and Sb(Hg) deposits fits fairly well with a perigranitic zoning observable at the scale of the Armorican Massif; most deposits are located in the east where Hercynian batholiths do not crop out (Chauris and Guigues 1969).

The deposits are distributed in a north-northeasterly oriented strip, about 60 km wide, which is displaced in the South-Armorican fault by sinistral movements. This transverse disposition on major Armorican structural axes suggests that the source of antimony is pre-Hercynian, possibly lying in the Cambrian-Tremadocian volcanites (Marcoux et al. 1984). The antimony veins were therefore preferentially formed in this original preconcentration zone of antimony as a result of Hercynian magmatic events and Late Hercynian hydrothermalism. The Hercynian granites redistributed the scattered antimony, redepositing it farther away, and this accounts for the obvious antithesis of antimony and tin-uranium deposits.

The erosion level in the granites is different from east to west and explains the asymmetric distribution of antimony in the Armorican Massif. The plunge of granites in the east of the Massif preserved the Sb aureoles with major distal (Rochetrejoux) and upper veins (La Lucette). The exposure of the granites in the west of the Armorican Massif resulted in most of the Sb deposits being eroded, with only the Quimper-Cap Sizun deposits being preserved (Chauris and Guigues 1969). The major shear faults thus appear to be the major metallotect – most of the antimony veins are located in their proximity, generally in extension and conjugate fractures generated by the fault movements (Chauris et al. 1977).

The precise age of this hydrothermal event is not known. The spatial relationship between the Sb veins and the Stephanian basins led Fouquet (1980) to assign an Upper Stephanian to Permian age to the Quimper mineralization. The antimony mineralization pre-dates the N10–20 ° W fractures, which displaced the major South-Armorican shear faults, and consequently the extensive lead-zinc veins.

Table 4. Mineralogy of the lead-zinc and lead-antimony veins (in order of decreasing abundance) (Pierrot et al. 1973, 1975, 1980; Fouquet 1980; Cantinolle 1984; Pillard et al. 1985)

	Pb-Zn vein deposits									Pb-Sb-Cu-Zn vein deposits	
Deposit	La Touche	Pontpéan	Huelgoat Poullaouen	Plélauff	Trémuson	Châte-laudren	Baud	La Telhaie	Carnoët	La Chapelle-Saint-Melaine	Plogoff
Main Minerals	Sphalerite, galena, marcasite, pyrite	Sphalerite, galena, pyrite, marcasite	Galena, sphalerite, chalcopyrite, pyrite, marcasite	Sphalerite, galena, marcasite, pyrite	Sphalerite, galena, tetrahedrite, marcasite	Galena, sphalerite, hematite	Galena, sphalerite, pyrite	Marcasite, pyrite, sphalerite, galena	Sphalerite, chalcopyrite, tetrahedrite, galena	Sphalerite, galena, boulangerite, pyrite, arsenopyrite, meneghinite	Sphalerite, galena, bournonite, semseyite, boulangerite, chalcopyrite
Accessory minerals	Chalcopyrite, linneite, pyrrhotite	Wurtzite, pyrrhotite, bravoite, proustite, rammelsbergite, tetrahedrite, chalcopyrite, nickeline	Native silver, bournonite, bravoite, cobaltite, tetrahedrite, linneite, millerite, stannite, ullmannite, cassiterite	Arsenopyrite, greenockite, stannite, chalcopyrite, ullmannite, magnetite	Bournonite, chalcopyrite, bornite, arsenopyrite, pyrite, pearceite, pyrargyrite, gudmundite	Chalcopyrite, tetrahedrite, bournonite, diaphorite, arsenopyrite, pyrite, polybasite, pyrrhotite	Chalcopyrite, cubanite, marcasite, pyrargyrite, polybasite, andorite	Bravoite, chalcopyrite, pyrrhotite, roquesite, ullmannite	Pyrite, marcasite, bournonite, bravoite, native silver	Chalcopyrite, cubanite, enargite, gold, pyrrhotite, gersdorffite, tennantite	Tetrahedrite, marcasite, pyrite
Gangue	Quartz, (barite)	Quartz, calcite	Quartz	Quartz	Quartz, siderite	Quartz, siderite	Quartz	Quartz, (barite), (siderite)	Quartz	Quartz	Quartz
Production	1000 t Pb, 1000 t Zn	250000 t Pb, 230 t Ag	> 50000 t Pb + Zn	Unknown	× 1000 t Pb	Unknown (× 100 t Pb)	60 t Pb	–	Unknown	Unknown	–

7.4 Lead-Zinc Veins

Lead-zinc deposits are numerous in the Armorican Massif and most are spatially related to the major faults (Chauris and Guigues 1969; Chauris 1977) with the mineralization being located in transverse fractures rather than in the main faults. The deposits are asymmetrically distributed (Fig. 5). The major deposits (Pontpéan, Huelgoat-Poullaouen, Trémuson, Châtelaudren, La Touche, and Plélauff) are localized in a relatively narrow, 30- to 50-km-wide strip which is spatially related to the Medio-Armorican shear fault; only the small deposits of Baud, Plumelin, Locrio, Crossac, and La Telhaie mark the South-Armorican shear faults. The majority of the veins occupy northerly trending fractures, which intersect all the geological formations. These fractures were developed mainly during extensional episodes with dextral displacement and they post-date the induration of the Hercynian basement. They affect the entire Breton Peninsula and enabled the emplacement of fissure magmatism (microgranite, lamprophyrite and dolerite) (Bellon et al. 1988). The very few veins located in linear N80 ° W fractures (La Chapelle-Saint-Melaine, Plogoff) have original characteristics, such as a mineralogy with abundant Sb and Cu minerals (Table 4), and very probably derive from an older metallogenesis pre-dating the antimoniferous veins (Fouquet 1980; Marcoux 1980).

The mineralization belongs to the classical "BGPC" formation with dominant sphalerite, pyrite, galena and seldom chalcopyrite. The accessory minerals of Ag, Sn, Sb and Cu are common in a few veins (Huelgoat) and better illustrate the complex geochemistry of the mineralization (Pierrot et al. 1973; Chauris et al. 1976) (Table 4).

The lead isotope geochemistry results (Marcoux 1987) show that lead and most of the metallic elements of the veins come from lithological formations of the investigated districts (Fig. 3). The lead isotope compositions of the veins are fairly consistent: $^{206}Pb/^{204}Pb$ ranges from 18.01 to 18.74 with a concentration between 18.20 and 18.50. They clearly reflect the various geological environments of the mineralization and their characteristics. The granite massifs are the main source of lead for the veins enclosed in or located close to the Hercynian (Plélauff, Huelgoat and Trémuson) or Cadomian (Bois-Neuf) massifs. The isotopic differences between the veins of a same district generally result from mixing of varying proportions of lead derived from the main lithological units.

The physical and chemical emplacement conditions were only studied in La Telhaie vein which underwent argillic hydrothermal alteration (Cantinolle 1984). The alteration paragenesis of dominant interstratified illite-smectite with illite and kaolinite indicates mesothermal conditions (150–180 °C), which are consistent with the nature of the sulfide paragenesis. The same type of argillic alteration with interstratified illite-smectite, kaolinite and montmorillonite had already been identified at Plélauff by Ziserman (in: Sarcia et al. 1965). At La Touche and Pontpéan, the parageneses and textures (fibrous, colloform sphalerite and fibroradiate marcasite and pyrite) suggest the same depositional conditions as those of La Telhaie.

The lead-zinc veins are very likely Triassic in age (210–230 Ma) and thus would be contemporaneous with the extensive fluorite-barite (galena) veins of the Massif Central (210 ± 20 Ma) (Jébrak 1985; Bril, pers. commun.) which also formed during a major extension episode. These assertions are based on K/Ar datings of three illite samples (4.0 to 5.6% K) from La Telhaie vein which give a mean value of 222 ± 10 Ma corresponding really to the formation of illites synchronous with Pb-mineralization, and not to a rejuvenation (Cantinolle 1984). A rejuvenated alteration paragenesis is known in the vein, but has different mineralogical features, notably with high proportions of kaolinite. This age is supported by dating of eruptive dykes in western Brittany, related to major Late Hercynian or pre-Atlantic extension phases, which developed from the Permian to the Lias (274–190 Ma) (Bellon et al. 1988).

8 Post-Hercynian Mineralization

Barite occurrences (in places with fluorite, pyrite, galena, and sphalerite) are known in the Mesozoic formations which are transgressive on the peneplaned Hercynian basement along the eastern margin of the Armorican Massif, particularly in the Cotentin, near Alençon, in Vendée and in the Poitou strait (Melle deposit) (Fig. 5). The occurrences are located in Triassic, Liassic, and Dogger arkose sandstones and dolomite limestones (Lougnon et al. 1974) and consist of fairly stratiform impregnations or veins, which may be intensely silicified. The age and origin of the mineralization are both difficult to define. In Normandy, stratigraphic observations indicate at least two distinct periods, (pre-Hettangian and post-Bathonian), and intermediate periods very likely exist. The sedimentary hypothesis suggests that the barium (and lead) may have resulted from leaching of granite sands and may have concentrated in shoal basins. According to the "thermal spring" hypothesis, post-Hercynian fracturing of the peneplaned Hercynian basement during extension movements related to the first opening phase of the Atlantic Ocean may have reactivated old faults – a phenomenon accompanied by magmatism (Bellon et al. 1988) and deposition of mineralization in favorable layers.

9 Conclusions

9.1 The Armorican Massif: a Metallogenic Province

The Armorican Massif is a true metallogenic province characterized by the concentration of the same elements during several episodes and in different structural and metallogenic settings. Iron, lead-zinc, and cassiterite mineralization exemplifies this assertion. The iron concentrations are related to volcanosedimentary processes, marine sedimentation or alteration processes, and occurred during various periods: Upper Brioverian (the Rouez massive sulfide deposit), Ordovician (the oolitic deposits of Anjou and Normandy), Silurian (the Haie-Claire polymetallic deposit), Devonian (La Porte-aux-Moines volcanosedimentary pyrite deposit, the Crozon and Gouarec sedimentary deposits) and Carboniferous (jasper with volcanosedimentary magnetite of the Châteaulin basin); they also developed during the Mesozoic (the Bercons Cenomanian sedimentary occurrences), the Tertiary (the Rougé lateritic coatings) and the Quaternary (black magnetite-bearing beach sands). The lead-zinc concentrations occur in various forms: (1) hydrothermal (or volcanosedimentary) deposits of the Upper Brioverian (Rouez), Silurian (Saint-Georges-sur-Loire), Devonian (La Porte-aux-Moines and Bodennec); (2) classical Late Hercynian vein formations (Huelgoat), and (3) stratiform deposits in the Mesozoic transgressive cover rocks (Melle). Cassiterite is observed in fragments of detrital quartz of the Brioverian sandstones and in the Rouez massive sulfide deposit; it occurs with Late Cadomian granites, but mainly with the various Hercynian granites which were emplaced over several tens of millions of years. With the destruction of the primary deposits, cassiterite occurs as detrital grains in the Triassic, Cenomanian, and Pliocene, and especially in the Quaternary fluviatile alluvium and on rare Recent beaches.

In the Armorican Massif, the spatial distribution of the various mineral concentrations appears to indicate two metallogenic polarities. Variation in the parageneses and distribution of the cassiterite-wol-

Table 5. Chronology of the Armorican mineralizations

Age		Main geological units	Type of ore deposit	Major ore deposits
Cenozoic	0	Erosion-sedimentation	Stanniferous alluvial deposits	St Renan, Piriac
	160			
		Shelf carbonaceous sedimentation	Ba-Pb occurrences (210–160 Ma)	Melle
	210			
Mesozoic		Basement fracturing (beginning of the Atlantic Ocean	Pb-Zn-Ag veins (230–210 Ma) Sb veins	Pontpéan, Huelgoat La Lucette, Rochetrejoux
	250	extension and associated magmatism)	U deposits (290–270 Ma)	La Commanderie, L'Ecarpière
			Pb-Sb-Cu-Zn veins	La Chapelle-Saint-Melaine
	290		Au deposits	La Bellière, Beslé
			Kaolin deposits	Ploëmeur, Berrien
		Hercynian magmatism	Sn-W pneumatolytic deposits	St Renan, Abbaretz, La Villeder
			Andalusite hornfels	Guerphalès
	345			
Paleozoic	395	Oolitic epicontinental sedimentation	Iron deposits	Diélette
	410 (?)	Intracratonic rift	Polymetallic sulfide deposits	Bodennec, La-Porte-aux-Moines
	430	Final Mancellian magmatism	Sn-W pneumatolytic deposits	Montbelleux
	435	Back-arc basin	Polymetallic sulfide deposits	La Haie-Claire
	470	Pelitic sedimentation with phosphate genesis	Monazite schists	Grand Fougeray
	480	Oolitic epicontinental sedimentation	Iron deposits	Segré, Soumont Acqueville, Bailleul
		Arenaceous sedimentation of the epicontinental Arenigian	Littoral Zr-Ti deposits	
	570			
		Cadomian orogenesis	Polymetallic sulfide deposits	Rouez
Brioverian	650		Sulfide disseminations	Trégan

framite mineralization associated with the three main Hercynian granite belts have been recorded. Tin is associated mainly with the granites of the ilmenite series (leucogranites) and is not found in the red granites (magnetite series). A similar polarity also exists in the distribution of the uranium mineralization, with a clear decrease in the size of the occurrences from south to north. The distribution of the lead-zinc occurrences is particularly remarkable: the major deposits are pratically all located in central Armorica, in an east-west-oriented strip, whatever their age and depositional conditions; in South Armorica, these occurrences are numerous, of varied types and economically insignificant. Such distributions must reflect deep metallogenic factors which may be related to an evolution connected with the plate tectonics of the area (source of generating magmas, linear faults); advances in research should confirm this. Late erosional and plunging processes may have obscured the delineation of the metalliferous belts, but their contours can still be distinguished.

9.2 Metallotects and Sources of Elements

The various types of mineralization in the Armorican Massif suggest that the mineralizing processes developed since the Cadomian were diverse (Table 5).

Volcanism and the associated hydrothermal processes were responsible for the emplacement of the Brioverian (Trédrez), Silurian (Saint-Georges-sur-Loire) and Devonian (Châteaulin Basin) polymetallic sulfide mineralization. These processes reveal an increase in the geothermal gradient related to intracrustal extensions which took place during an intracratonic rifting phase (Châteaulin basin) or during the formation of an island-arc basin (Saint-Georges-sur-Loire synclinorium). This genesis may be applied to the Rouez massive sulfide deposit, although no volcanic event has been identified so far in the area. The mineralized fluids belonged to major hydrothermal circuits which leached the entire crustal segment, and all the lithological units of this segment constitute the source of the metals. The lead isotope composition of

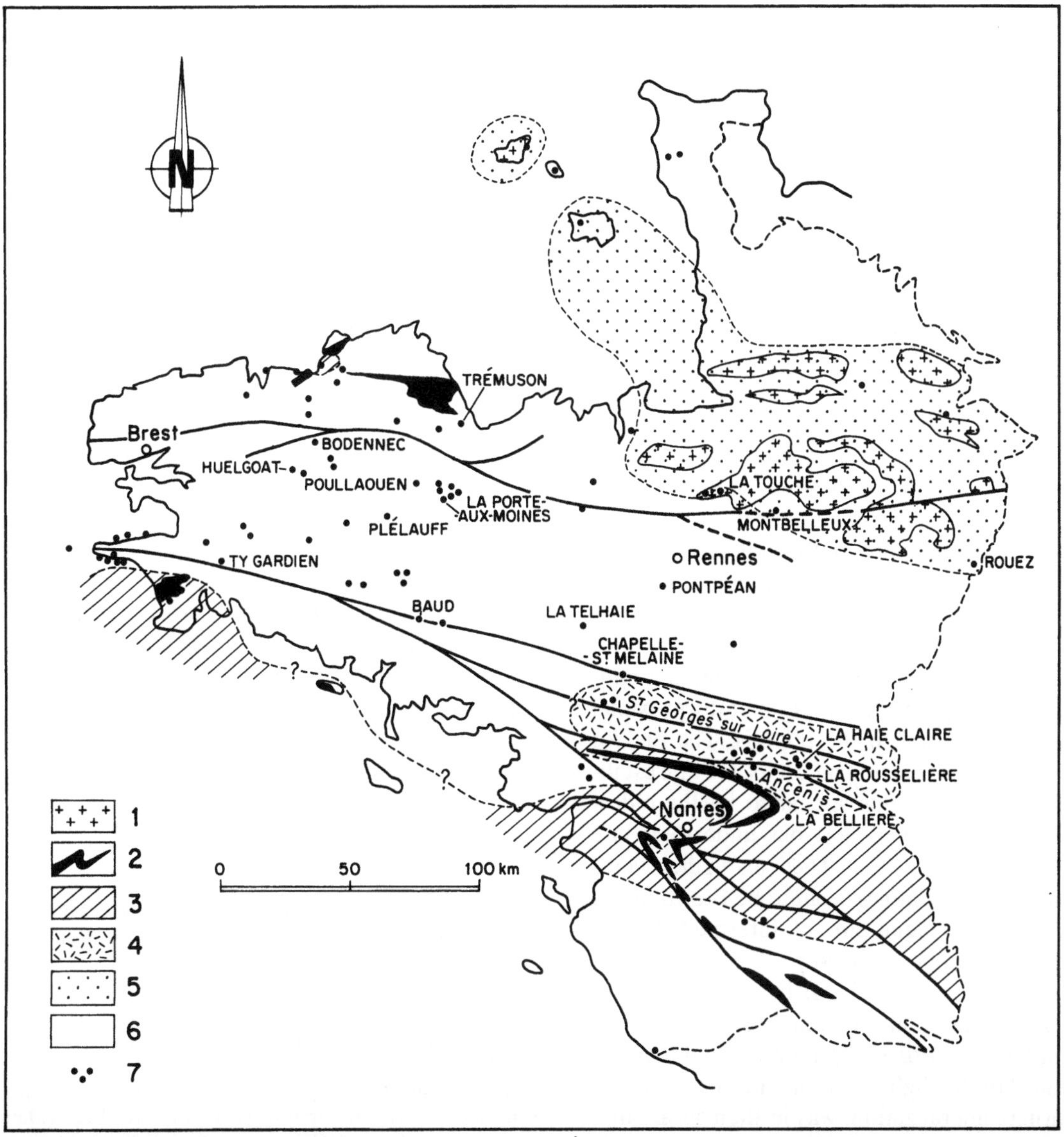

Fig. 6. Origin of the lead in the mineralization. *1* Cadomian granites of the Mancellian batholith; *2* mantle-derived rocks; *3* domain with minor mantle-lead; *4* domain of the lead of the uranium-depleted continental crust (rocks of the granulitic facies); *5* Cadomian lead domain ($^{206}Pb/^{204}Pb < 17.90$); *6* domain of the Hercynian continental crust (Hercynian granites, Paleozoic sediments and metasediments); *7* (Pb, Sb, Au, Zn-Cu-Pb, etc.) mineralization subjected to isotope analysis

the orebody gives an average image of the source environment: granite rocks and slightly metamorphosed sediments (Bodennec and La Porte-aux-Moines) or granulitic rocks (La Haie-Claire).

The Cadomian and Hercynian magmatism is the second most important metallotect of the Armorican mineralization. The intensely eroded Mancellian batholith only shows mineralization related to the satellite massifs that were emplaced at greater depth (taking into account the strong erosion of the Mancellian batholith), such as the Montbelleux pneumatolytic Sn-W veins. The more complex and diversified Hercynian magmatism is directly responsible for the extensive pneumatolytic Sn, W, mineralization with tin and tungsten localized in the upper greisenized portions of the granite massifs or along their margins (Charoy 1979; Chauris 1981). The metallogenic part played by the granites is also shown by the presence of economically significant andalusite concentrations in the contact aureole of some massifs (Rostrenen). The late magmatic hydrothermal alteration produced major kaolin deposits at the expense of Hercynian granitic plutons (Ploëmeur and Berrien). Although they are located in the major South Armorican shear fractures, the gold deposits may be subcontemporaneous with, or slightly post-date, the emplacement of the last Hercynian granites (~ 300 Ma); acting as thermal "propellers", the plutons may have remobilized the gold concentrations disseminated in the Brioverian. Uranium is common in the Hercynian leucogranites (Pontivy, Guérande, and Mortagne) and in monzogranites (Saint-Renan and Plouaret). Mineable pitchblende concentrations formed during major Late Hercynian fracturing episodes.

The major shear faults are also one of the main metallotects of the Armorican vein-type mineralization. As early as the Permian, major extension periods related to the opening of the Atlantic Ocean produced N10–20 ° W fracture systems. Associated hydrothermal circulations deposited uranium, antimony, and lead-zinc mineralization in these fractures. The extensive deposits are localized near major Armorican shear faults, thereby emphasizing their important role as drains of the mineralizing fluids. The spatial relationship between the deposits and granites differs according to the substance considered: uranium is intra- or perigranitic, lead is generally proximal and antimony is distal to the plutons. The precise chronology of these mineralizations is not well established. The uranium veins appear to be the oldest (290–270 Ma by U/Pb dating) but underwent late remobilization processes. The lead-zinc veins (230–210 Ma, K/Ar dating) seem to be the youngest and the antimony veins are of intermediate age (structural arguments, see above).

The source of the metals is partially known. Uranium is mainly granitic in origin and the lead isotope geochemistry (Marcoux 1987) showed that the source of the lead of the lead-zinc deposits, and also of all the lead-bearing mineralization, was in their local environment. Furthermore, the lead-zinc veins are not genetically related to volcanosedimentary deposits. Lead is then the source marker of all the various types of lead-bearing mineralization. At the regional scale, the extent of the various geological source environments can be determined by their distinct isotopic characteristics (Marcoux 1987) and is consistent with the geodynamic setting (Fig. 6). The various types of southern mineralization commonly contain some mantle lead ($^{207}Pb/^{204}Pb$ lower than 15.60 in the Hercynian mineralization). This characteristic indicates that mantle rocks are more abundant in this domain than suggested by outcrops, and that they underwent infra-Hercynian oceanic subduction and contributed to the formation of various types of mineralization. The mineralization of the Ancenis and Saint-Georges-sur-Loire basins displays original unradiogenic and retarded lead signatures which are detectable in the Silurian polymetallic deposits and in the Late Hercynian Sn-W deposits. This specificity demonstrates the presence of a granulitic metamorphic basement that occurs in a complex structural position only in this domain. The isotopic compositions of most of the other mineralization are consistent with a mainly Paleozoic, continental-crust environment. The influence of the Mancellian batholith does not extend beyond its current extension; its role as the source of the mineralization is evidenced by the fact that the Mancellian mineralization shows the lowest isotopic ratios in the Armorican Massif ($^{206}Pb/^{204}Pb$ ranges from 17.50 to 17.90) (Fig. 3).

Exogenetic processes also played an important part in the Armorican metallogenesis. As early as the Ordovician, the marine sedimentation abundantly accumulated rutile and zircon sands or europium-bearing monazite in the shales along the coast. During the same period, the chemical shelf sedimentation deposited iron-bearing layers, a process which was resumed during the Devonian. Erosion gradually dismantled the primary cassiterite deposits to produce cassiterite alluvium, which was mined at Saint-Renan.

References

Allon A (1980) Premières observations sur les indices polymétalliques de la Rousselière (Loire-Atlantique). Chron Rech Min 455: 72–92

Aye F (1983) Contrôles géologiques des gîtes stratiformes de Pb, Zn, Cu, Ag de la bordure du bassin de Châteaulin (Massif Armoricain, France). Mém Bur Rech Géol Min (Orléans) 120, 264 pp

Bellon H, Chauris L, Hallegouet B, Thonon P (1988) Magmatisme fissural permien à triasique dans le Pays de Léon (Massif armoricain, France). C R Acad Sci Paris II 307: 2049–2054

Bernard-Griffiths J, Peucat JJ, Sheppard S, Vidal P (1985) Petrogenesis of Hercynian leucogranites from the southern Armorican Massif: contribution of REE and isotopic (Sr, Nd, Pb and O) geochemical data to the study of source rocks characteristics and age. Earth Planet Sci Lett 74: 235–250

Bernazeaud J (1981) Introduction à la connaissance actuelle du gisement sulfuré massif de Rouez: sa découverte, son avenir. Chron Rech Min 458: 5–11

Bonnemaison M, Marcoux E (1987) Les zones de cisaillement aurifères du socle hercynien français. Chron Rech Min 488: 29–42

Bril H (1975) La minéralisation à cuivre, plomb, zinc de Bodennec en Bolazec (Finistère) dans son cadre géologique. Thèse 3è cycle, Paris

Bur Rech Géol Min (1960–1964) Carte des gîtes minéraux de la France à 1/320 000. Feuilles Brest, Rennes et Nantes

Bur Rech Géol Min (1979) Carte des gîtes minéraux de la France à 1/500 000. Feuilles, Nantes et Rouen

Callier L (1989) Le porphyre à molybdène d'âge hercynien de Beauvain en Basse-Normandie. Chron Rech Min 496: 3–23

Cantinolle P (1984) Altération hydrothermale et minéralisa-

tion liées à un filon plombo-zincifère de basse température: le gîte de la Telhaie (Morbihan). Thèse 3è cycle, Poitiers
Cathelineau M (1981 a) Les gisements d'uranium liés spatialement aux leucogranites sud-armoricains et à leur encaissant métamorphique. Sci Terre, Nancy, Mémoires 42, 375 pp
Cathelineau M (1981 b) Les gisements uranifères de la presqu'île guérandaise (Sud-Bretagne). Approche structurale et métallogénique. Mineral Deposita 16: 227–240
Cathelineau M, Nieva D (1985) A chlorite solid-solution geothermometer. The Los Azufres (Mexico) geothermal system. Contrib Mineral Petrol 91: 235–244
Charoy B (1975) Ploëmeur kaolin deposit (Brittany), an example of hydrothermal alteration. Petrology 1.4: 253–266
Charoy B (1979) Définition et importance des phénomènes deutériques et des fluides associés dans les granites. Conséquences métallogéniques. Sci Terre, Nancy Mémoires 37: 364 pp
Chauris L (1977) Linéaments, granites et gîtes plombo-zincifères en Armorique. C R Somm Soc Géol Fr 2: 85–88
Chauris L (1980) Plate tectonics and ore deposits in Western Europe: the example of the Armorican Massif (France) In: Ridge JD (ed) Proc of Fifth Quad IAGOD Symp Snowbird, Utah, USA. Stuttgart Schweizerhart'sche Verlagsbuchhandlung 1978, pp 401–413
Chauris L (1981) Une province stannifère: le Massif Armoricain (France). Chron Rech Min 462: 5–42
Chauris L (1984) Accidents linéamentaires et minéralisations uranifères: l'exemple de la ceinture batholitique hercynienne médioarmoricaine (France). Bull Soc Géol Fr 7. 26: 1375–1380
Chauris L (1988) Processus post-magmatiques à Berrien dans le massif granitique polyphasé de Huelgoat (Finistère, France). Bull Soc Linn Normandie 110–111: 101–122
Chauris L (1989) Les exploitations minières dans le Massif armoricain. Déclin ou progrès? Norois 141: 5–32
Chauris L, Guigues J (1969) Gîtes minéraux de la France, vol I. Massif armoricain. Mém Bur Rech Géol Min (Orléans) 74: 96 pp
Chauris L, Jonin M (1978) Introduction à la métallogénie cadomienne (Protérozoïque supérieur) dans un massif hercynien: le domaine normano-breton en Armorique (France). Bull Soc Linn Normandie 106: 35–48
Chauris L, Laforêt C, Picot P, Pierrot R (1976) Individualité paragénétique de la formation filonienne BGPC en Bretagne occidentale. Bull Soc Fr Minér Cristall 99: 287–293
Chauris L, Fouquet Y, Moëlo Y (1977) Les occurrences antimonifères de l'île de Sein et du Cap Sizun (Finistère). Leur place dans la métallogénie armoricaine. Bull Soc Fr Minér Cristall 100: 208–213
Chauris L, Houlgatte E, Laforêt C, Picot P (1985) Un district antimono-aurifère à gangue quartzo-carbonatée: Le Semnon (Ille-et-Vilaine, Massif armoricain, France). Hercynica 1–2: 111–119
Chauris L, Lulzac Y, Cotten J (1989) Une lame de granite albitique tardi-cadomienne: le gisement stanno-wolframifère de Montbelleux (Massif armoricain, France). Chron Rech Min 496: 25–39
Cumming GL, Richards JR (1975) Ore lead isotope ratios in a continuously changing Earth. Earth Planet Sci Lett 28: 155–171
Darnley AG (1982) The relationship between uranium distribution and some major crustal features in Canada. Mineral Mag 44: 425–436
Donnot M, Guigues J, Lulzac Y, Magnien A, Parfenoff A, Picot P (1973) Un nouveau type de gisement d'Europium: la monazite grise à europium en nodules dans les schistes paléozoïques de Bretagne. Mineral Deposita 8: 7–18
Faure P (1978) Les grès à rutile et zircon du Massif armoricain. Thèse Doct Ing E. N. S. M. Paris
Fehn V, Cathles LM, Holland HD (1978) Hydrothermal convection and uranium deposits in abnormally radioactive plutons. Econ Geol 73: 1556–1566
Fouquet Y (1980) Le district antimonifère du Cap Sizun-Quimper. Thèse 3è cycle, Univ. Clermont-Ferrand, 191 pp
Geffroy J, Sarcia J, Chervet J (1960) Les minerais uranifères français et leurs gisements. Presses universitaires de France. Paris I, 324 pp
Germain C, Kervella F, Le Bail F (1964) Gisements et indices uranifères de Bretagne. In "Les minerais uranifères français". PUF Paris 3: 209–275
Gerstner A, Baras L, Pinaud C, Tayeb G (1962) Monographie des gisements vendéens. In "Les minerais uranifères français". PUF Paris. 2: 293–399
Guigues J (1978) Histoire de la découverte des gisements Pb-Zn-Cu-Ag armoricains. Méthodologie de la prospection. Chron Rech Min 445: 11–32
Guigues J, Devismes P (1969) La prospection minière à la batée dans le Massif armoricain. Mém Bur Rech Géol Min (Orléans) 71, 172 pp
Holliger P, Cathelineau M (1986) Le chronomètre U-Pb en milieu uranifère: application aux gisements hydrothermaux d'uranium liés spatialement au batholite de Mortagne (Vendée, France). Chron Rech Min 485: 33–43
Icart JC, Safa P (1981) Rouez: environnement géologique et minéralisation. Chron Rech Min 458: 12–32
Jébrak M (1985) Contribution à l'histoire naturelle des filons (F, Ba) du domaine varisque. Doc Bur Rech Géol Min 99: 1–473
Kosztolanyi C (1971) Géochronologie des gisements uranifères français par la méthode uranium-plomb. Thèse, Univ Nancy I, 279 pp
La Roche H de, Stussi JM, Chauris L (1980) Les granites à deux micas hercyniens français. Essais de cartographie et de corrélations géochimiques appuyés sur une banque de données. Implications pétrologiques et métallogéniques. Sci Terre. Nancy Mémoires XXIV 1, pp 5–121
Le Fur Y (1978) Les minéralisations volcano-sédimentaires de Trémuson (Côtes-du-Nord). Chron Rech Min 445: 70–76
Leroy J, Holliger P (1984) Mineralogical, chemical and isotopic (U-Pb method) studies of Hercynian uraniferous mineralizations (Margnac and Fanay mines, Limousin, France). Chem Geol 45: 121–134
Lougnon JM, Duthou JL, Lasserre M (1974) Les gisements plombozincifères du seuil du Poitou et de sa bordure limousine. Bull Bur Rech Géol Min 2. 5: 453–476
Lulzac Y (1968) Les filons à wolfram et or natif de la région de Locarn-Duault (Côtes-du-Nord). Bur Rech Géol Min, 31 pp (Unpublished report)
Marcoux E (1980) Le district de Pontivy. Sa place dans la métallogénie plombo-zincifère du Massif armoricain (France). Thèse 3è cycle, Univ. de Clermont-Ferrand, 252 pp
Marcoux E (1987) Isotopes du plomb et paragenèses métalliques, traceurs de l'histoire des gîtes minéraux. Doc Bur Rech Géol Min (Orléans) 117, 289 pp
Marcoux E, Serment R, Allon A (1984) Les gîtes d'antimoine de Vendée (Massif armoricain, France). Historique des recherches de synthèse métallogénique. Chron Rech Min 476: 3–30
Mirabail H, Terrée P (1986) Le gisement de tungstène de Coat-an-Noz (Côte-du-Nord). SNEA (P), 7 pp (unpublished report)
Moussu R (1963) Le gisement d'étain de Saint-Renan. Géologie et prospection. Ann Mines IX: 555–572

Nicolas J (1956) Contribution à l'étude géologique et minéralogique de quelques gisements de kaolins bretons. Thèse, Univ. Paris, 254 pp

Oh MS, Marcoux E, Calvez JY, Touray JC 1989 Anomalous unradiogenic lead from stratabound Zn-Pb-Cu mineralizations in the Saint-Georges-sur-Loire synclinorium (Armorican Massif, France). Econ Geol 8: 1436–1443

Pierrot R, Chauris L, Laforêt C (1973) Inventaire minéralogique de la France 3, vol. Finistère Bur Rech Géol Min, Orléans 118 pp

Pierrot R, Chauris L, Laforêt C (1975) Inventaire minéralogique de la France 5, vol. Côtes-du-Nord. Bur Rech Géol Min, Orléans, 220 pp

Pierrot R, Chauris L, Laforêt C (1980) Inventaire minéralogique de la France 9, vol. Morbihan. Bur Rech Géol Min, Orléans, 316 pp

Pillard F, Chauris L, Laforêt C (1985) Inventaire minéralogique de la France 13, vol. Ille-et-Vilaine, Bur Rech Géol Min, Orléans, 148 pp

Renard JP (1971) Etude pétrographique et géochimique des granites du district uranifère de Vendée. Liaison entre l'évolution minéralogique et le comportement de l'uranium. Conséquences pour la prospection. Sc Terre. Nancy Mémoires 30, 214 pp

Ruhlmann F, Gillet Y, Lissilour J (1980) Une occurrence uranifère à sélénium et bismuth dans la ceinture métamorphique du leucogranite de Mortagne-sur-Sèvre (Vendée). Bull Minéral 103: 245–249

Sarcia JA, Moussu R, Sakovitch W, Munck F, Bertraneu J, Guigues J, Ziserman A (1965) Le gîte plombo-zincifère de Plélauff (Côte-du-Nord). Bull Bur Rech Géol Min 1: 1–39

Stacey JS, Kramers JD (1975) Approximation of terrestrial lead isotope evolution by a two-stage model. Earth Planet Sci Lett 6: 369–376

Part III The Massif Central

Introduction

D. S. Santallier

The Massif Central is one of the two main segments in the French Paleozoic Orogen. It occupies some 80000 km^2 in the central part of France, with an average height above sea level of 800 m. It is mainly composed of high grade metamorphic rocks and widespread granitic plutons. The possible existence of a Precambrian basement in the Massif Central is still debated. There is no evidence for a basal Cambrian unconformity. The metasedimentary formations apparently represent both lower Paleozoic and upper Precambrian (Eo-Cambrian). A few radiometric ages on orthogneisses provided Eo-Cambrian ages (Lévêque 1985) of 600 ± 10 Ma (U-Pb on zircons), but there is no clear evidence of Precambrian plutons intruding well established tectono-metamorphic Cadomian formations (Pin 1989). The presently known age constraints on the Paleozoic orogeny are the following: the youngest strata involved are the uppermost Silurian (Guillot and Doubinger 1971; Guillot and Lefèvre 1975). The oldest post-tectonic granitoids are dated at 350–360 Ma (Berthier et al. 1979; Bernard-Griffiths et al. 1985). The oldest post-tectonic deposits are Stephanian and Permian molassic deposits.

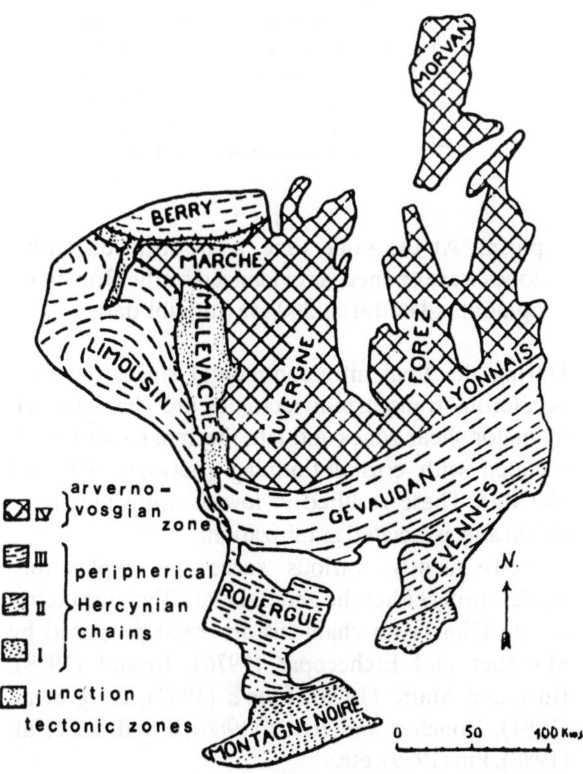

Fig. 1. Orogenic zones in the French Massif Central (After Jung 1954)

Historically, a "static" interpretation of the orogen was favored until 1970–75, despite the pioneering works of Demay (1942). In 1936, Jung and Roques considered that, in presumably autochtonous successions, the metamorphism grade increased with depth. They thus proposed distinguishing five successive zones which, from top to bottom, were called upper micaschist zone, lower micaschist zone, upper gneiss zone, lower gneiss zone and migmatite zone. A few years later, Jung (1954), Roques (1971) and Chenevoy and Ravier (1971) (Figs. 1, 2, and 3) proposed successively slightly different schemes for the French Massif Central. They distinguished:

A high grade metamorphic Precambrian basement they respectively called the Arverno-Vosgian, Lemovico-Arverne, and Arverne zone. (The Arverni and Lemovices were Gallic tribes established in the central and western Massif Central 2000 years ago.)

A surrounding high grade metamorphic Paleozoic chain was called the Peripherical Hercynian chain or Peripherical Schist zone by Jung and Roques. This was subdivided by Chenevoy and Ravier into an inner Ruthéno-Limousin zone (the Ruteni was another Gallic tribe living in the Rouergue area) and an outer Epizonal Schist zone.

After that, and till recently, geologists favored the German zonation of Kossmat (1927), and distinguished (Fig. 4) an external Rheno-Hercynian zone, a middle Saxo-Thuringian zone, and an internal high grade metamorphic Moldanubian zone, which was originally supposed to represent a Precambrian basement.

J. D. Keppie (Ed.): Pre-Mesozoic Geology
in France and Related Areas

Fig. 2. Structure of the pre-Stephanian basement in the French Massif Central. (After Roques 1971). *Shaded areas* crystalline schists in the Lemovico-Arverne zone (main metamorphism middle Briovenian); lined areas: peripherical schists zone (main metamorphism and deformation Caledonian); *cross-hatched areas* peripherical crystalline massifs; *dots in the northeastern part* Devonian and Dinantian formations unconformably overlying the Lemovico-Arverne zone (Hercynian metamorphism in the Lyonnais area; *dots in the southern part* pre-Stephanian Paleozoic formations in the southern Montagne Noire zone (main deformation Hercynian); *in white* granites and post-Dinantian cover

It has now been known for several years (Santallier 1983) that the tectono-metamorphic history of the French Paleozoic orogen is polyphase. Structural, petrological, and geophysical data collected during the last 12 years have shown that it displays all the characteristics of a collision belt (Burg and Matte 1978; Autran and Cogné 1980; Kornprobst et al. 1980; Matte 1986; Robinson et al. 1988; Pin 1989) with the following classical stages (Fig. 6):

1. A main distensive episode of Cambro-Ordovician age (530–480 Ma) characterized by widespread emplacement of basic magmas.
2. A high pressure metamorphism induced by a subduction process during the Silurian (450–410 Ma).
3. Collisional tangential tectonics which produced the present tectonic pile during Devonian and lower Carboniferous (400–350 Ma). These are associated with a medium pressure type polyphase metamorphism. At the same time, some local distensive zones were opened, as evidenced by the Brévennes ophiolites dated at ca. 360 Ma (unpubl. data).

During the final intra-continental stage, the Carboniferous transcurrent tectonics are characterized by major shear zones and late thermal evolution associated with crustal thinning (between 350 and 300 Ma). These resulted in low-pressure metamorphism and associated granitization.

Consequently, various successive explanatory models using either the general Plate Tectonic theory or the Himalayan chain history were proposed by Mattauer and Etchecopar (1976), Briand (1978), Burg and Matte (1978), Floc'h (1983), Burg et al. (1984), Bouchez and Jover (1986), Friedrich et al. (1988), Pin (1989), etc.

The presently accepted nappe pile, as proposed by Ledru et al. (1989), is composed of five major com-

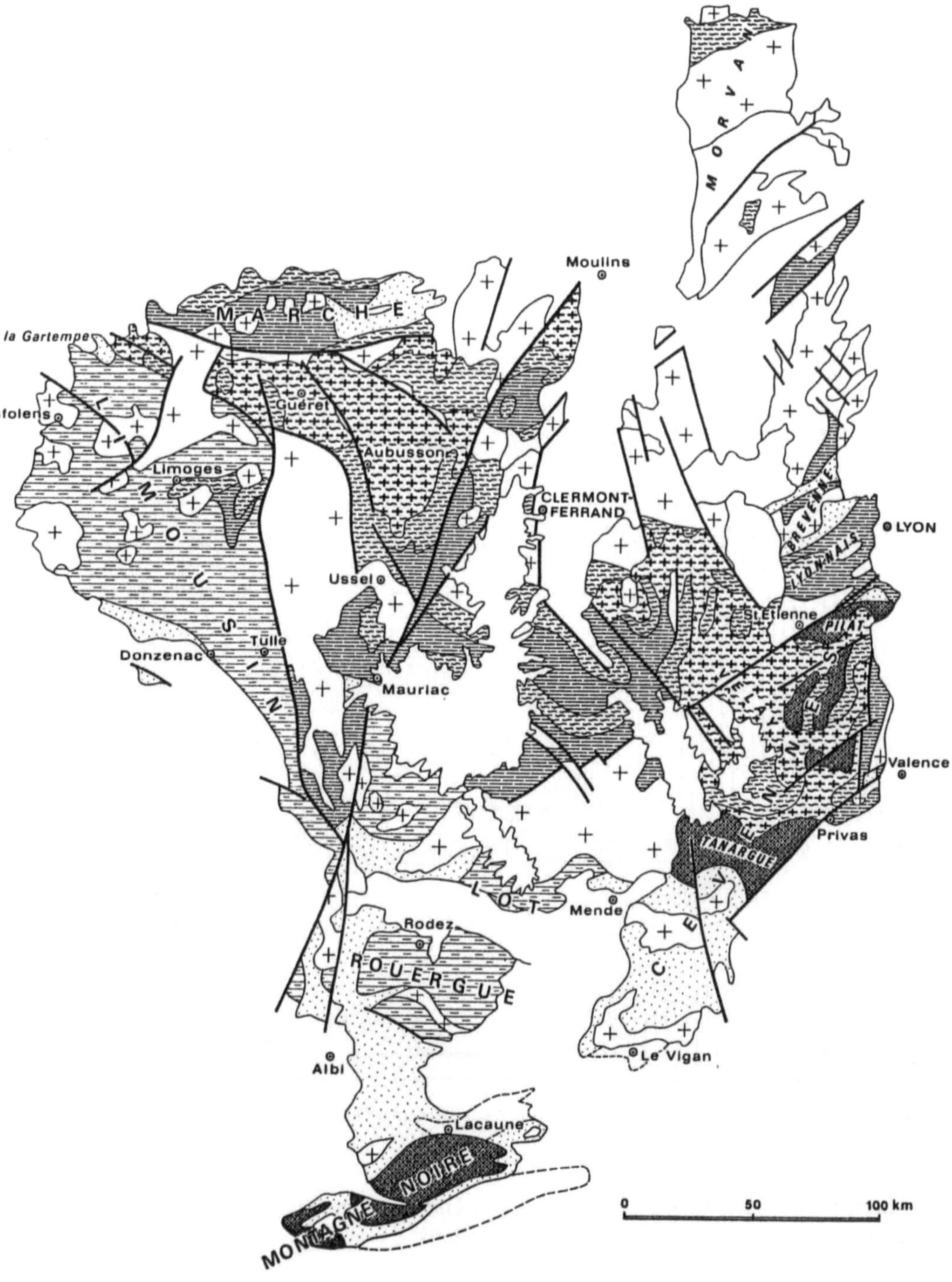

Fig. 3. Main tectono-metamorphic subdivisions in the French Massif Central (after Chenevoy and Ravier 1971). [symbol] Epizonal schists. *Meso- and catazonal Medium Pressure series:* [symbol] Rutheno-Limousin zones: quartzo-feldspathic gneisses and micaschists. Arverne zone: [symbol] pelitic micaschists and gneisses: [symbol] cordierite-bearing Aubusson migmatites; [symbol] Guéret and Velay cordierite-bearing granites; *Meso- and catazonal Low Pressure series:* [symbol] quartzo-feldspathic gneisses and micaschists; [+] other granites

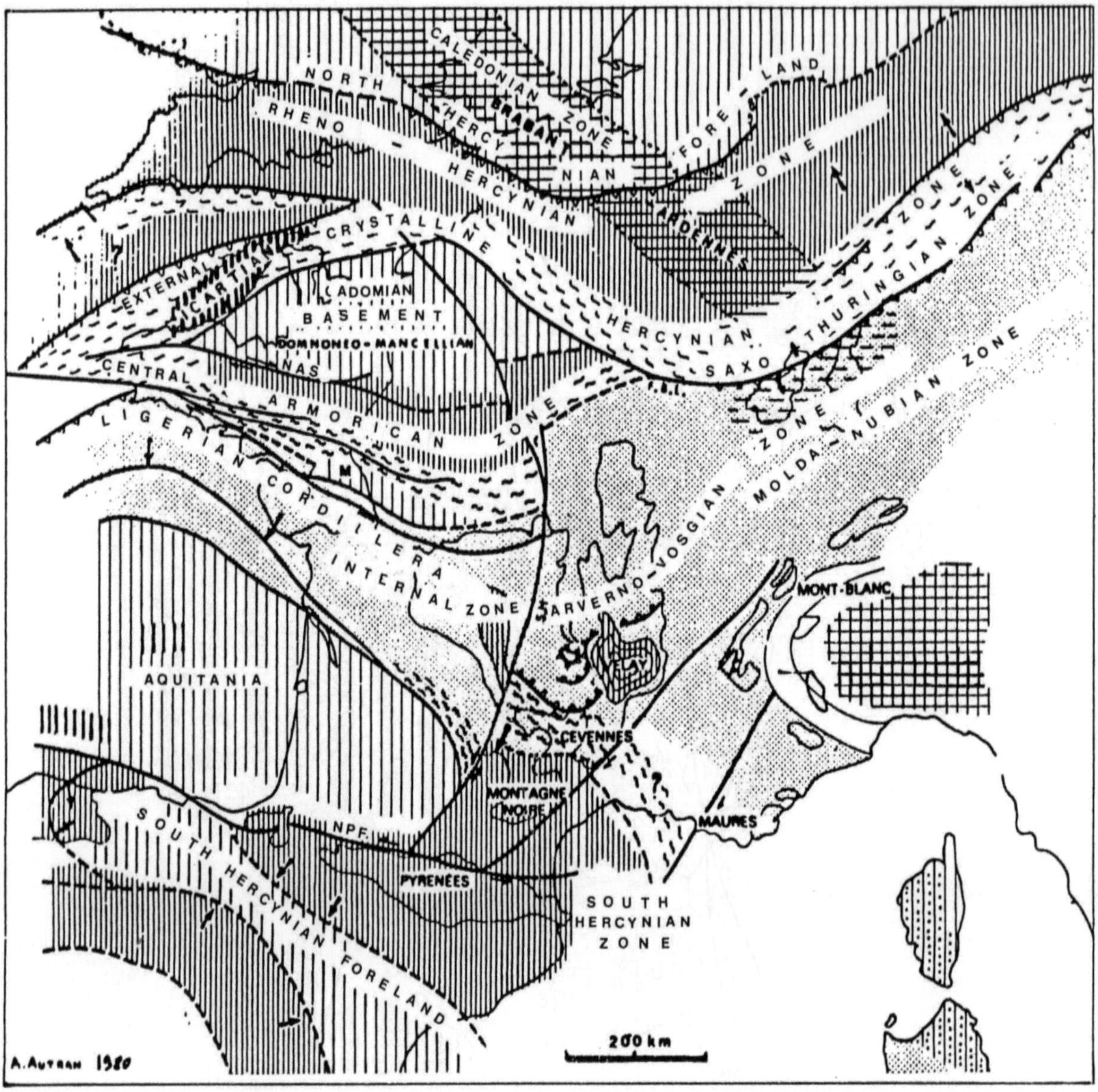

Fig. 4. Main structural subdivisions in the French Massif Central. (After Kossmat 1927; Autran and Cogné 1980)

plexes which, from top to bottom, are respectively called (Figs. 5 and 6):

1. The upper allochthon: epizonal units (Genis, Thiviers-Payzac, and Brévenne units).
2. The middle allochthon (upper gneissic unit of Floc'h 1983, Floc'h et al. 1984, and Ledru et al. 1989). This is characterized by widespread high pressure relicts.
3. The medium to high grade lower allochthon (lower gneissic unit of Floc'h 1983. Floc'h et al. 1984 and Ledru et al. 1989).
4. The mostly epizonal relatively autochtonous complex.
5. The paleontologically dated southern Paleozoic units.

The tectonic contacts between these five tectonic units are thrust planes (between units 1, 2, 3, and 4) or shear zones (between units 1 and 2 and units 4 and 5) of different ages. The thrusts are younger than the high pressure metamorphism: they are penecontemporaneous with the Devonian or Lower Carboniferous medium pressure metamorphism. The shear zones are younger (between 350 and 320 Ma). The widespread granitic magmatism is related to crustal anatexis with sometimes moderate mantle contributions; it lasted for about 60 Ma between 360 and

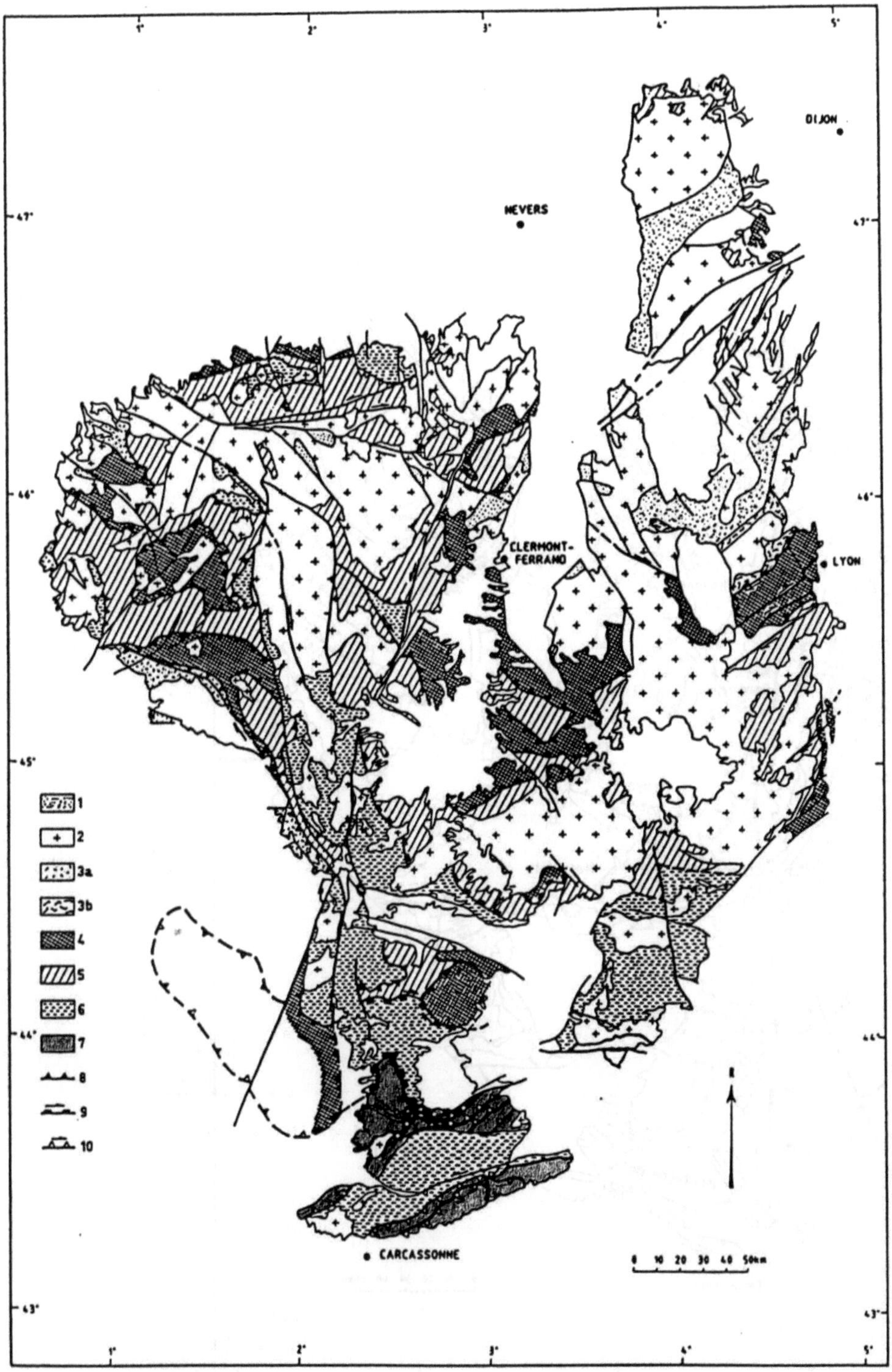

Fig. 5. Sketch map of the main thrust nappes of the French Massif Central. (Ledru et al. 1989).
1 Visean deposits; 2 Undifferentiated granitic bodies; 3a Thiviers-Payzac and Génis units; 3b Brévenne unit; 4 Upper gneissic unit; 5 Lower gneissic unit; 6 Relative autochton unit; 7 Paleozoic southern units; 8 Thrusts nappes (370–390 Ma); 9 Thrusts and transcurrent zones (350–320 Ma); 10 Thrusts and transcurrent zones (320 +/ – 10 Ma)

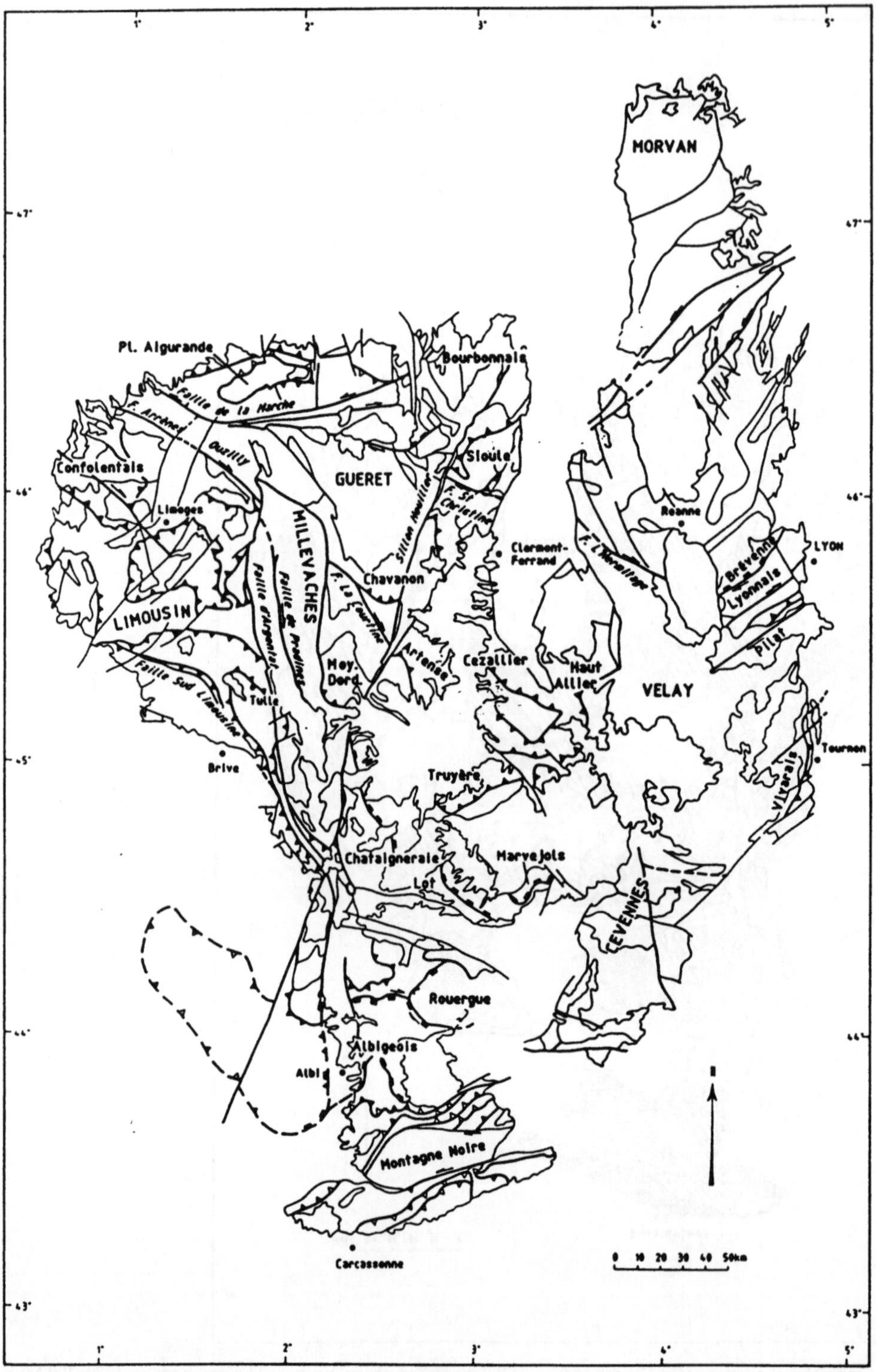

Fig. 6. Geographic location of the main geologic regions

280 Ma. The older granites (pre-320 Ma) were emplaced in thickened crust the latter, whereas, are linked to crustal thinning at least in the Velay area (Mallavieille et al. 1989). Ninety five percent of them range from grano-diorites to two-mica leucogranites.

The following five chapters of Part III are devoted to the stratigraphic and lithologic constraints, the magmatic, metamorphic, and deformation processes, and to the metallogenic characteristics of the orogen.

Chapter III. 1 is subdivided into two parts. The first part (P. Ledru et al.) provides lithological descriptions of the metamorphic formations in the internal zone. The second part (R. Feist and H. Echtler et al.) is devoted to the paleontologically dated formations involved in the orogen and to the post-tectonic deposits in the southern external Montagne Noire area.

In Chapter III. 2, R. Feist, H. Echtler et al. analyze the structural and stratigraphic evolution in the southern external part (Montagne Noire) of the French Massif Central (Fig. 6). The area studied is composed of a high grade metamorphic core overthrust on its northern boundary by northern internal metamorphic tectonic units and on its southern flank by nonmetamorphic Paleozoic nappes (Fig. 5). The whole pile was later affected by a doming phase. The Montagne Noire dome and the northern metamorphic units underwent the same polyphase tectono-metamorphic evolution that affected the whole internal zone of the orogen during Devonian and Lower Carboniferous. The southern nappes were tectonically emplaced during the Upper Carboniferous. They exhibit pre- and syn-orogenic sediments preserved in the Massif Central domain. R. Feist et al. show how the evolving sedimentologic characteristics are closely related to the northern development of the orogen (see sedimentary logs in Chap. III. 2). During the Lower Paleozoic (pre-orogenic) time, in the Montagne Noire southern flank area, the Lower Cambrian is represented by platform sediments, the Upper Cambrian is missing, and the Ordovician detritus is the youngest pre-Devonian deposits. On the contrary, on the northern flank, Cambrian detritus linked with bimodal or basic volcanics indicate an early continental proto-rifting with northward deepening into basinal conditions on the northern margin of Gondwana. The Devonian to Lower Carboniferous deposits (including Upper Visean flysch) started with an angular unconformity. This follows Silurian block tilting and erosion related to the Lower Paleozoic crustal thinning. The Devono-Carboniferous rocks are represented by epicontinental shelf carbonate deposits contemporaneous with the main tectono-metamorphic events in the northern internal domain. Their northward progressively proximal character is good evidence for a southward – migrating orogenic front and progressive uplift in the internal zone. The Lower Paleozoic sedimentologic polarity was thus inverted during the Carboniferous. It appears that the post-orogenic molassic deposits are older (Middle to Late Devonian) in the northern part of the French Massif Central than in the southwestern part (Brive basin), where they are Stephanian and Permian.

In the internal zones (Ledru et al., Chap. III. 1), the Lower Paleozoic and/or Eo-Cambrian formations were affected by a generally high grade polyphase metamorphism. Only in the epizonal upper allochthon and in part of the relatively autochthonus units are the original sedimentary or magmatic characteristics easily recognizable. In the high grade lower and middle allochthons, the meta-sedimentary formations are mostly meta-graywackes mixed with volcanic tuffs. These are associated with a few acid meta-volcanics, widespread metabasites (including a few dismembered ophiolites) and orthogneisses. The relatively autochthonous micaschists and gneisses are mostly derived from graywackes and pelites. They represent epicontinental marginal deposits of Eo-Cambrian and Lower Paleozoic age deposited on a presumably Cadomian basement. They are devoid of any volumetrically important meta-magmatic intercalations. The lower allochthon formations are also mostly metapelites and metagraywackes of Eo-Cambrian and Lower Paleozoic age. They are associated with metabasites and with Cambrian and Ordovician metagranitoids either corresponding to the Late Cadomian plutonic activity or resulting from the crustal thinning episode. The lower allochthon formations are presumably representative of an old continental rise. The middle allochthon high grade metamorphic rocks comprise meta-graywacke gneisses associated with widespread metabasites. These are good evidence for a more distal and partly oceanic environment. High Pressure relicts such as ca. 435-Ma-old eclogites, whiteschists, and granulites are widespread, supporting the existence of Silurian subduction. The epizonal upper allochthonous units are more varied. The western Thiviers-Payzac unit (Fig. 5) contains only calc-alkaline meta-magmatic products (rhyo-dacitic tuffs with minor bimodal suites, cross-cut by calc-alkaline doleritic dykes and overstepped by ignimbritic tuffs). The age of these island-arc-type formations is completely unknown. In the overlying Génis unit, Ordovician and Silurian, paleontologically dated, meta-pelitic sediments are associated with, a presumably lower Paleozoic, dismembered ophiolite. On the other hand, in the eastern Brévenne unit, bimodal magmatism (including an ophiolitic complex) (Figs. 5 and 6) is good evidence for a Devonian crustal thinning episode in an external part of the orogen. This is contemporaneous with the major tectono-metamorphic culmination in the internal zones.

The geochemical characteristics of the various acid and basic pre-orogenic magmatic products are explained in Chapter III. 5 by Briand, Girardeau, Ohnenstetter, Stussi et al. More or less dismembered ophiolitic associations have been easily identified in Limousin and Brevenne areas (Fig. 6). The other metabasite protoliths are not directly recognizable because they were completely deformed and metamorphosed in high grade conditions. Trace-elements and REE studies established that they belonged to various magmatic series (N-MORBs, T-MORBs, continental tholelites, island-arc basalts, etc.) emplaced in various geodynamic sites such as intraplate continental rifts, marginal basins, and island arcs. This fits well the epicontinental or marginal character of the associated meta-sediments. The various orthogneisses in the middle and lower allochthons are Cambrian, Ordovician, and a few of them even Silurian. The oldest are mainly calc-alkaline; they were also mainly related to the lower Paleozoic distensive episode. The few Silurian metagranitoids are presumably related to the subduction episode; they have a different geochemical signature that implies the involvment of a thinned continental lithosphere.

As outlined by Ledru et al. (Chap. III. 3) and by Santallier et al. (Chap. III. 4) the main tectono-metamorphic events in the French Massif Central occurred during Devonian and lower Carboniferous. There is very little structural evidence for an early Silurian deformation phase associated to the high pressure metamorphism. Except in the Albigeois area (Fig. 6), this high pressure metamorphism, which is related to the subduction stage, generated high temperature eclogites. The P/T conditions of crystallization (ca. 15–20 kbar and 650–800 °C) are not compatible with the subduction of a well-developed cold oceanic crust, but it fits better either the subduction of a still hot crust (in a narrow oceanic domain or an oceanic ridge), or a very slow uplift of the buried rocks. The low temperature eclogites associated with white schists in the western Albigeois region are similar to some of the South-Armorican high pressure rocks. This supports the presence of a true Lower Paleozoic oceanic suture in the western part of France. The main overthrusting corresponding to the collision are Middle to Upper Devonian and Lower Carboniferous. It post-dates the high pressure metamorphism. The amount of displacement is unknown and the direction of thrusting is still debated. It seems that a Devonian westward or northwestward transport is better supported in Limousin but there is better evidence for a Lower Carboniferous southward vergence in the southern part of the French Massif Central. During the Lower Carboniferous, the structural evolution is characterized by mainly dextral wrenching and thrusting. The major tectono-metamorphic event resulted in the nappes pile described above. The thrusts were syn-metamorphic and occurred in medium pressure epizonal, or catazonal conditions, depending upon structural position. It appears that the stacking occurred under regularly decreasing pressure and ended with a 375–390 Ma-old migmatization. This indicated the main collision tectonics took place during the uplift of the chain. During the Upper Carboniferous, the internal zone is characterized mainly by sinistral wrenching and limited thrusting like those in the southern external Montagne Noire area. At the same time, several thermal anomalies in the thickened crust induced thermal domes characterized by low pressure metamorphism and widespread granitization. At least locally, the thermal gradient was so high that crustal thinning phenomena presumably also occurred as in the Velay dome (Fig. 6).

The post-orogenic magmatism described by Ploquin et al. in Chapter III. 5 mainly comprises granitic plutons emplaced between roughly 360 and 280 Ma with two major periods of activity: 360/325 and 320/280 Ma. From the chemical and mineralogical points of view, three different granitoids suites are distinguished. (1) The calc-alkaline suite comprises granodiorites, biotite-bearing granites, and leucogranites. It is mainly of crustal origin with some minor mantle contributions. (2) The subalkaline suite is composed of hornblende-bearing monzogranites, granites and leucogranites; (3) The aluminopotassic suite comprises biotite- and cordierite- or sillimanite-bearing grano-diorites or granites, biotite-bearing granites and two-mica leucogranites. These three suites have different associated mineralization. There is no discernable time-related evolution of the chemical compositions, only a geographical zonation with three different domains. (1) A central region extending from northwestern Limousin to southeastern Velay (Fig. 6) is characterized by widespread aluminopotassic granites. (2) A northeastern region exhibits both aluminopotassic and subalkaline granitoids with some volcanics of similar compositions. (3) The southern region extends from the eastern Cévennes to the southwestern Montagne Noire containing all the three granitic suites.

The ore deposits in the Massif Central (as outlined by Bril et al. in Chap. III. 6) are related to the tectonic and magmatic features. The authors distinguish a pre-collision metallogenesis (Zn, Cu, Au, and W), which is directly related to the pre-collision acid magmatism. These are mainly massive sulphides or strata-bound and sedimentary exhalative deposits. The authors emphasize there are none of the classical ophiolite- or island arc-related mineralization (no Cyprus-type massive sulphides or porphyry deposits) in the French Massif Central. The lack of collision-related mineralization is also noted, but there are many ore deposits related to the hyper-collision (intra-continental) stage. These are related either to deformation (Au, As, Sb) or to post-tectonic granites (W, F, U, and rare metals such as Li, Sn, Ta, Be, Nb, and Sn).

References

Autran A, Cogné J (1980) La zone interne de l'orogène varisque dans l'ouest de la France et sa place dans le développement de la chaîne hercynienne. 26th I. G. C. Mém Bur Rech Géol Min 108: 90–111

Bernard-Griffiths J, Gebauer D, Grunenfelder M, Piboule M (1985) The tonalite belt of Limousin (French Central Massif): U-Pb zircon ages and geotectonic implications. Bull Soc Geol Fr (8) I 4: 523–529

Berthier F, Duthou JL, Roques M (1979) Datations géochronologiques Rb/Sr sur roche totale du granite de Guéret, âge fini-Dévonien de mise en place de l'un de ses faciès types. Bull Bur Rech Geol Min (2) I 2: 59–72

Bouchez JL, Jover D (1986) Le Massif Central: un chevauchement de type himalayen vers l'ouest-nord-ouest. C R Acad Sci Paris 302. II: 675–680

Briand B (1978) Métamorphisme inverse et chevauchement de type "himalayen" dans la vallée du Lot (Massif Central français). C R Acad Sci Paris 286 D: 729–731

Burg JP, Matte P (1978) A cross section through the french Massif Central and the scope of its variscan geodynamic evolution. Z Dt. Geol Ges 129: 429–460

Burg JP, Leyreloup A, Marchand J, Matte P (1984) Inverted metamorphic zonation and large scale thrusting in the Variscan belt: an example in the French Massif Central. In: Hutton DHW, Sanderson DJ (eds) Variscan tectonics of the north Atlantic Region. Blackwell, London, pp 47–61

Chenevoy M, Ravier J (1971) Caractères généraux des métamorphismes du Massif Central français. In: Symposium Jean Jung, Géologie, Géomorphologie et Structure profonde du Massif Central français. Plein Air Service, Clermont-Ferrand, pp 109–132

Demay A (1942) Microtectonique et tectonique profonde. Mém Serv Carte Géol France, Paris, 247 pp

Floc'h JP (1983) La série métamorphique du Limousin central: une traverse de la branche ligérienne de l'orogène varisque, de l'Aquitaine à la zone broyée d'Argentat (Massif Central français). State thesis, Limoges University, 445 pp

Floc'h JP, Santallier D, Henry B (1984) La série limousine: réflexion axée sur une synthèse géologique régionale et son utilisation pour la prospection de l'or. Chron Rech Min 474: 5–10

Friedrich M, Marignac C, Floc'h JP (1988) Sur l'existence de trois chevauchements ductiles "himalayens" successifs à vergence NW en Limousin. C R Acad Sci Paris 306, II: 663–669

Guillot PL, Doubinger J (1971) Découverte d'acritarches dans les schistes supérieurs de Génis (Dordogne). C R Acad Sci Paris 272: 2763–2764

Guillot PL, Lefèvre J (1975) Découverte de conodontes dans le calcaire à entroques de Génis en Dordogne (série métamorphique du Bas-Limousin. C R Acad Sci Paris 280: 1529–1530

Jung J (1954) Prolèmes géologiques dans les vieux terrains du Massif Central français. Ann Hébert Haug VIII: 245–258

Jung J, Roques M (1936) Les zones d'isométamorphisme dans les terrains cristallophylliens du Massif Central français. Rev Sci Nat Auvergne II 2: 179–191

Kornprobst J, Bard JP, Briand B, Cantagrel JM, Guitard G, Kienast JR, Lasnier B, Le Corre C, Santallier D (1980) Le métamorphisme en France. 26ème. C. G. I., Paris, C7, "Géologie de la France". Mém Bur Rech Géol Min 107: 161–189

Kossmat F (1927) Gliederung des Varistischen Gebirgbaues. Abh Sächs Geol Landesamts 1, 39 pp

Ledru P, Lardeaux JM, Santallier D, Autran A, Quenardel JM, Floc'h JP, Lerouge G, Maillet N, Marchand J, Ploquin A (1989) Où sont les nappes dans le Massif Central français? Bull Soc Géol Fr (8) V 3: 605–618

Levêque MH (1985) Mise en évidence d'un socle Précambrien dans le Massif Central français: l'orthogneiss des Palanges (Aveyron). C R Acad Sci Paris 300 II: 277–282

Malavieille J, Guilhot P, Costa S, Lardeaux JM, Gardien V (1990) Collapse of a thickened Variscan crust in the French Massif Central: Mont Pilat extensional shear zone and Saint-Etienne Upper Carboniferous basin. Tectonophysics 177: 139–149

Mattauer M, Etchecopar A (1976) Arguments en faveur de chevauchements de type himalayen dans la chaîne hercynienne du Massif Central français. In: Ecologie et géologie de l'Himalaya. CNRS Int Coll Paris 268, pp 261–267

Matte P (1986) La chaîne varisque parmi les chaînes paléozoïques péri-atlantiques, modèle d'évolution et position des grands blocs continentaux au Permo-Carbonifère. Bull Soc Géol Fr (8) II, 1: 9–24

Pin C (1989) Essai sur la chronologie et l'évolution géodynamique de la chaîne herynienne d'Europe. State thesis, Clermont-Ferrand University. 470 pp

Robinson P, Tracy RJ, Santallier DS, Andreasson PG, Gili Barguchi JI (1988) Scandian-Acadian-Caledonian sensu stricto metamorphism in the age range 430–360 Ma. In: Harris AL, Fettes DJ (eds) The Caledonian-Appalachian orogen. Geol Soc Spec Publ 38: 453–467

Roques M (1971) Structure géologique du Massif Central. In: Symposium Jean Jung, Géologie, géomorphologie et structure profonde du Massif Central Français. Plein Air Service, Clermont-Ferrand, pp 17–32

Santallier DS (1983) Main metamorphic features of the Paleozoic Orogen in France. In: Schenk PE (ed) Regional trends in the Appalachian-Caledonian-Hercynian-Mauritanide Orogen. Reidel, Dordrecht, pp 263–274

Lithostratigraphy of Variscan Terranes in the French Massif Central: A Basis for Paleogeographical Reconstruction

P. LEDRU, A. AUTRAN, and D. SANTALLIER

1 Definition of the Main Geological Units: Lithological Content of the Allochthonous and Autochthonous Units

As a rule, it is quite difficult to correlate lithotectonic units in different regions that have complex and polyphase tectonic history. In fact, if the succession resulted from the stacking of nappes during collisional tectonics, it is very unlikely that such units would consist of the same rock types, with an equivalent lithologic content and with the same deformational history.

In the case of the French Massif Central (Fig. 1, Ledru et al. 1989), it can be seen that metamorphism and structures are related to the main crustal thrust zones and to large, late, anatectic domes, all of which cross-cut the initial lithology. The main geological units may be defined by considering: (1) the present geometrical position; (2) the original site of deposition; (3) the age of the units; (4) the metamorphic evolution. Lithological correlations will only be meaningful if they consider the tectonic history of the different geological units, which has to be common at least in part.

1.1 The Upper Epizonal Units (Upper Allochthon): the Thiviers-Payzac, Génis, and Brevenne Units

The Thiviers-Payzac and Génis units occur in the southwestern part of the Limousin and occupy an upper allochthonous position in relation to the highly metamorphic units of the Limousin (Roubichou 1980; Guillot 1981). However, the major dextral shear zone at the contact of these units with the upper and lower gneiss units (Ledru and Autran 1987) has obliterated most primary relationships, and the exact position of the Thiviers-Payzac and Génis units is still under debate.

The ophiolitic Génis unit overlies the Thiviers-Payzac unit; it is a portion of oceanic crust that appears to be overturned (Guillot 1981). Although shales and limestones located at the base of this unit are of Late Ordovician-Pridolian age, the age of this ophiolite is still unknown. However, detailed mapping has led Myashita (unpubl. data) to the conclusion that this ophiolite is an Early Devonian olistostrome, which would date the first obduction at around 400 Ma.

The Brevenne unit occurs on the eastern margin of the Massif Central and corresponds to an extensional rift filled by MORB-like basalts, gabbros, and some soda rhyolite and trondhjemite. It was obducted during the Early Visean (340 Ma) over the Early Devonian Lyonnais nappes and occupies an upper allochthonous position.

1.2 Upper Gneiss Unit (Middle Allochthon): Dismembered Ophiolitic Crust and Peri-Arc Basin Deposits, Early Paleozoic Age

This unit occupies the highest geometrical position and contains the characteristic so-called leptynite-amphibolite complexes, which are composed of quartzo-feldspathic gneisses, amphibolites and amphibolite gneisses. The former gneisses are derived from granites, microgranites, acid volcanics, graywackes, and acid tuffs (Pin and Lancelot 1982), and the amphibolite were ultrabasic and basic plutons (gabbro-diorite), and volcanic or tuffaceous rocks. Radiometric data indicate Late Cambrian to Early Ordovician ages (510–470 Ma) for the emplacement of the protoliths (Pin 1979; Duthou et al. 1981; Pin and Lancelot 1982).

Fig 1. Geographic location of the main geologic regions and ▶ sketch map of the main thrust nappes of the French Massif Central. *1* Visean sedimentary deposit/Visean granite; *2* undifferentiated granites; *3* Upper epizonal units; *3a* Payzac and Génis units; *3b* Brevenne unit; *4* Upper Gneiss unit (middle allochthon); *5* Lower Gneiss unit (lower aloochthon); *6* relative autochthonous complexes; *7* Paleozoic southern units; *8* Montagne Noire metamorphic core complex; *9* thrust nappes (390–360 Ma); *10* thrust and transcurrent zones (350 Ma); *11* neo-Variscan thrust zones (320 Ma)

J. D. Keppie (Ed.): Pre-Mesozoic Geology
in France and Related Areas

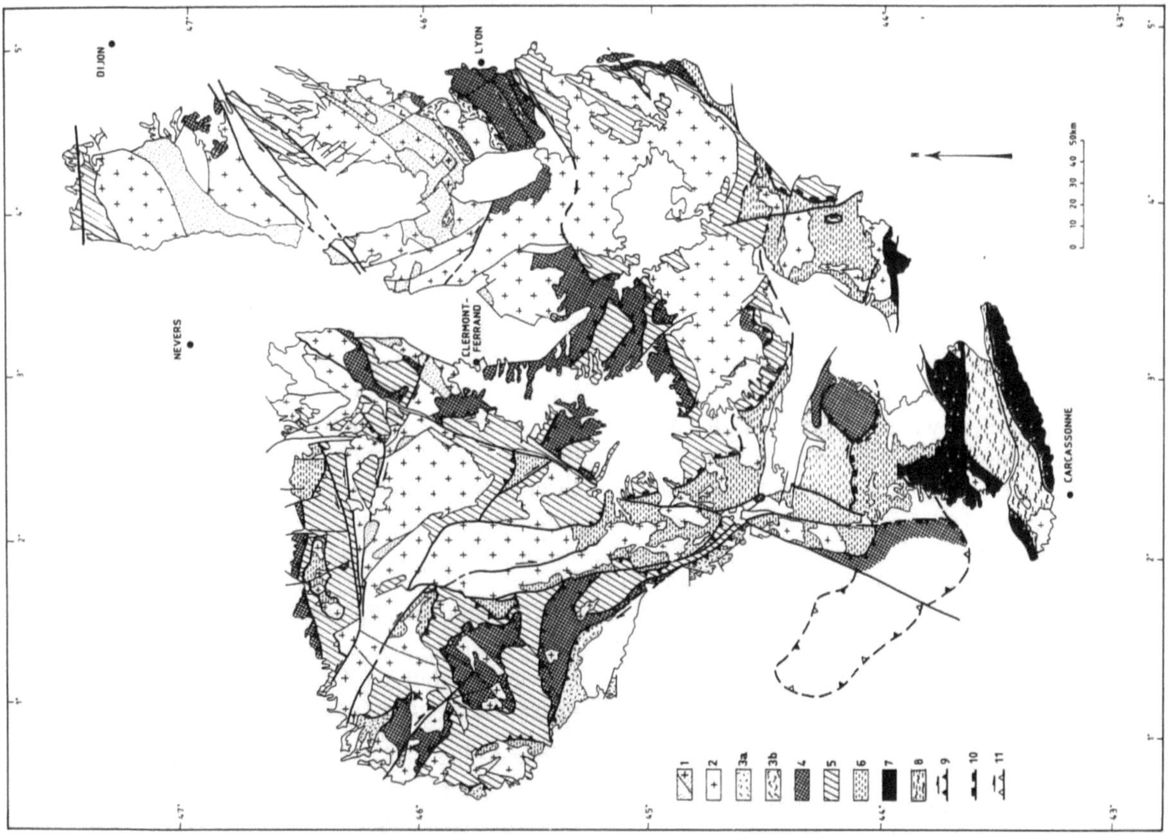
DIJON
LYON
NEVERS
CLERMONT-FERRAND
CARCASSONNE
1
2
3a
3b
4
5
6
7
8
9
10
11
0 10 20 30 40 50km

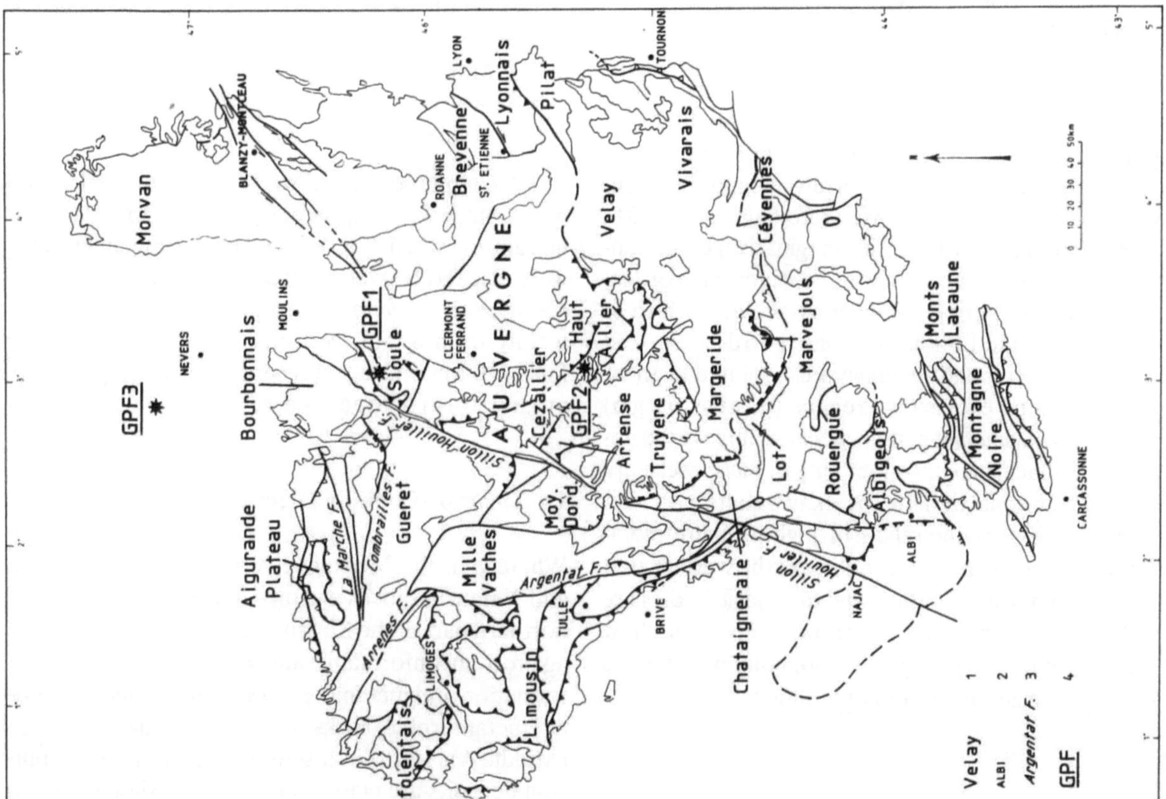
GPF3
NEVERS
Morvan
BLANZY-MONTCEAU
MOULINS
Bourbonnais
Aigurande Plateau
La Marche F.
Combrailles F.
Gueret
Arrenes F.
Confolentais
LIMOGES
Limousin
TULLE
Mille Vaches
Argentat F.
BRIVE
Chataigneraie
Sillon Houiller
NAJAC
ALBI
Albigeois
Rouergue
Lot
Marvejols
Margeride
Truyere
Artense
Moy. Dord.
Cezallier
GPF2
Haut Allier
GPF1
Sioule
CLERMONT FERRAND
AUVERGNE
ROANNE
Brevenne
ST. ETIENNE
Lyonnais
LYON
Pilat
Velay
Vivarais
TOURNON
Cevennes
Monts Lacaune
Montagne Noire
CARCASSONNE
0 10 20 30 40 50km
Velay 1
ALBI 2
Argentat F. 3
GPF 4

The geochemistry of these complexes is of tholeitic to calc-alkaline affinity (Bodinier et al. 1986, 1988; Briand et al. 1988). The banded texture is in most cases of tectonic origin and these complexes may be considered as metamorphosed, mainly bimodal, volcano-plutonic complexes. High-pressure relicts (eclogite and granulite) occur in boudins of basic or acid composition and are dated between 450 and 415 Ma (Pin 1979; Pin and Lancelot 1982).

The base of the upper allochthon is usually outlined by dismembered ultrabasic and basic bodies, whose ophiolitic origin was demonstrated in several districts and in particular in the Limousin (Girardeau et al. 1986). However, where recognizable, the matrix of this unit consists of volcanogenic graywacke without clear oceanic sedimentary rocks, which is more indicative of a back-arc basin than of an open ocean. Finally, in some areas, such as the Rouergue and the Limousin (Bernard-Griffiths et al. 1985; Pin and Piboule 1988), calc-alkaline laccoliths of gabbro-diorite, dated between 380 and 360 Ma, were emplaced after the high-pressure stage towards the end of the tectonic evolution.

1.3 Lower Gneiss Unit (Lower Allochthon): Continental Rise on a Cadomian Crust with Magmatism Related to Early Ordovician Crustal Thinning

This allochthonous unit is essentially composed of former sedimentary rocks, such as pelite and graywacke. Its age is assumed to be Late Proterozoic, as it is cross-cut by a major porphyritic granite intrusion and minor quartz-diorite dated around 540 Ma (Pin and Lancelot 1978). This magmatism could represent a late plutonic activity of the Cadomian orogeny. A second period of igneous activity, which is known from the Limousin, is dated Ordovician and probably corresponds to magmatism related to an intra-continental extensional regime (passive margin) than to a volcanic-arc system, because it lacks the andesitic to dacitic composition and all basic rocks are either extensional tholeites or alkali basalts.

This unit contains relicts of high-pressure metamorphism (eclogite and granulite) in the western part of the Massif Central. Basic and ultrabasic bodies are rather rare except in the Limousin, where some show eclogite relicts (Sauviat). Even so, none of them has been recognized as being of ophiolitic origin.

1.4 Presumed Autochthonous Complexes

These complexes are thought to be autochthonous as they occupy the lowest lithostratigraphic position and do not show any sign of high-pressure metamorphism. Essentially composed of meta-pelites and graywackes (micaschist and paragneiss), they correspond to platform and basin deposits of Brioverian to Early Paleozoic age (Cambro-Ordovician) in an epicontinental domain. There is no indication of a rift zone or an oceanic domain, which prior to Variscan time could have separated this zone and the "southern Paleozoic nappe complexes" from the "continental rise" crust of the Lower Allochthon. In fact, in Ordovician times these three units may have formed three paleogeographical zones of the southern continental plate.

1.5 The Southern Paleozoic Nappe Complexes

Based on lithological criteria, three main districts have to be differentiated:

1. The Albigeois-Cevennes, with two pelitic and volcanic basin sequences of Early Cambrian and Ordovician age, separated during the Late Cambrian by more detrital platform sandstones.
2. The Monts de Lacaune, composed essentially of pelite and limestones.
3. The Montagne Noire, composed of detrital Cambro-Ordovician rocks, of platform-type deposits of Devonian age and Visean flysch.

These complexes extend south from the front of the main nappes and apparently formed the southern margin on which the upper and lower allochthons were stacked between 380 and 340 Ma. The present configuration of these nappe complexes resulted mainly from a later collision of Carboniferous age (340–310 Ma). This southward-directed migration is progressive in time and space.

1.6 The Post-Collision Deposits

Whereas in the Montagne Noire the Upper Visean and Namurian rocks are pre- to syn-orogenic, in the northern part of the Massif Central rocks of the same age rest unconformably, and without penetrative deformation and metamorphism, over the deeply eroded metamorphic rocks involved in the Devonian (Middle Variscan) collision event. These post-collision deposits are clastic sediments of Middle Visean age and a huge Late Visean volcanic pile. During the

Middle Visean, the marine deposits extended over a large area, from Gueret in the west to the Morvan in the east. As they are well preserved, it can be deduced that post-collisional thinning of the Devonian crust had, by this time, ended in the northern part of the Massif Central. In the Morvan, Upper Devonian to Lower Carboniferous unconformable sedimentary and volcanic deposits show an approximately NNE-SSW-oriented configuration, in which a well-defined calc-alkaline belt in the western Morvan, developed during Famennian to Tournaisian times, is superimposed over a Frasnian marine limestone platform.

2 Lithological Content of the Main Geological Districts

The high-grade metamorphic units that were involved in the Variscan collision belt, have preserved variable relict traces of their depositional sites by means of their pre-metamorphic lithology and geochemistry (Limousin, Auvergne, Lyonnais). Their ages have to be determined by radiometric methods. However, available data are as yet insufficient to permit clear stratigraphical correlations. For the same reasons, all attempts at paleogeographical reconstruction must remain tentative, the more so in view of the allochthonous nature of these units. At present, the only reliable key to the reconstruction of Variscan tectonics and depositional environments is formed by the observed litho-tectonic stacking of thrust sheets.

On the other hand, the low-grade and nonmetamorphic units found in the Albigeois, Montagne Noire, Morvan, and southwest Limousin enable a reasonable stratigraphic and paleogeographic reconstruction.

The comparison between these units, from the internal zone (northern and central Massif Central) to the external zone (southern Massif Central), is thus very instructive.

2.1 **Limousin** (Fig. 2)

The stacking of the metamorphic series of the Limousin is clearly visible in the major antiforms and synforms (Floc'h 1983)

In the core of the antiforms occur micaschists and paragneisses with rare basic-volcanic levels (presumed autochthonous complexes of Millevaches, Sussac, St Goussaud, and Dronne). This unit is overlain by the lower gneiss unit (Lower Allochthon), which is composed of several granitic complexes, including cordierite, two-mica, biotite-aluminum granites, dated around 540 Ma in the southern Limousin, and subalkaline granites (Aubazine granite) in the northern part. The Rb/Sr isochron ages are generally 20 to 40 Ma younger than U/Pb ages on zircons of the same orthogneisses (Pin 1981; Lafon 1986). The subalkaline intrusions may represent Late Cadomian orogenic magmatism, similar to that of central Brittany and Normandy, which is of calc-alkaline affinity and dated between 500 and 480 Ma. This second plutonism may represent either an intra-plate extensional setting (as suggested by the tholeitic extensional affinity of the associated gabbro and its ferrous biotite and hornblende) or a volcanic-arc series. A succession composed of alternating leptynite and amphibolite of clear volcanic origin, is present at the top of this unit and are dated at 499 Ma (Calvez unpublished data). Some ultrabasic bodies occur near the base of the unit.

The synforms are occupied by plagioclase-rich gneisses, including abundant basic layers of MORB affinity with eclogite relicts, belonging to the upper gneiss unit, or Middle Allochthon. A banded succession composed of quartzo-feldspathic gneiss and amphibolites ("leptynite-amphibolite complex"), represents primary volcanic assemblages or rhythmic tuffaceous deposits. Amphibole gabbro-diorite without eclogite and of calc-alkaline affinity, seems to have been emplaced during the last stage of metamorphic evolution (Piboule 1979). Dismembered ophiolite bodies are located at the contact between the two gneiss units (Girardeau et al. 1986; Maillet 1987). In the southwestern part of the Limousin, the Thiviers-Payzac and Génis units occur in outcrop, representing upper epizonal units, or the Upper Allochthon.

2.2 **Aigurande Plateau** (Fig. 3)

Overthrust nappe units (Quenardel et al. 1984; Lerouge 1987; Rolin 1987) in the northwestern-most Massif Central, the Aigurande plateau, are well documented from mapping, even though exposures are commonly poor. Four tectono-lithological units have been defined. The region, as a whole, is a large antiform, the core of which is occupied by the lowermost Fougères unit. A Late-Brioverian age (650 Ma), based upon lead-isotope dating of galena from mineralization associated with amphibolite gneiss, was proposed for the supracrustal rocks of this unit (Ahmadzadeh and Petitpierre 1984).

The two overlying units (Eguzon and Gargilesse/Dun-le-Palestel) have been compared to the lower and upper gneiss units of the Limousin (Floc'h

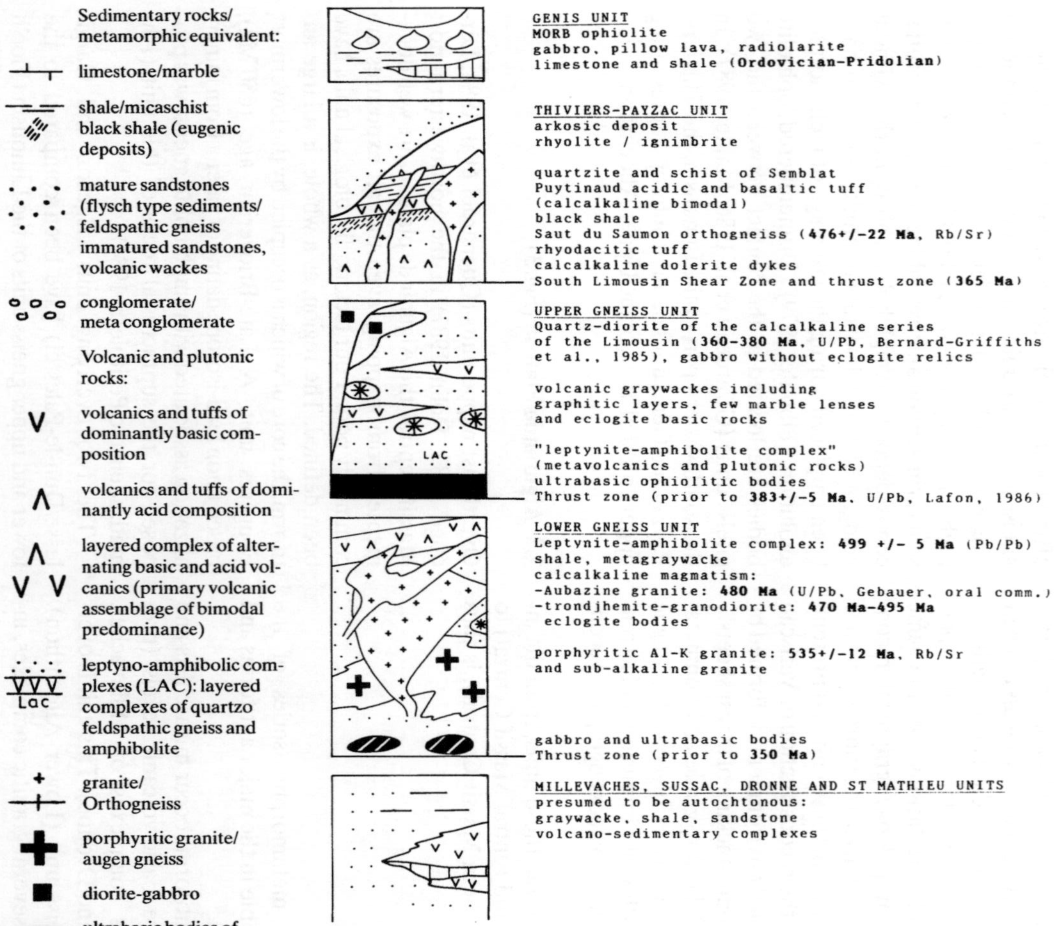

Fig. 2. Limousin

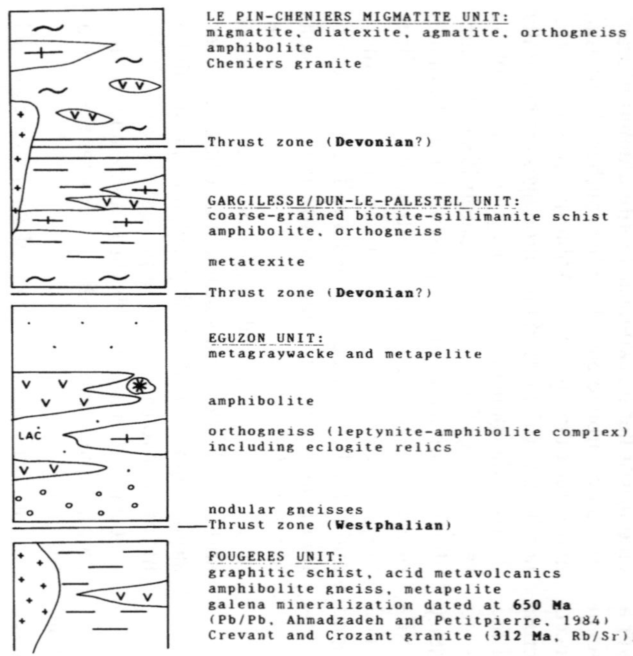

Fig. 3. Aigurande Plateau

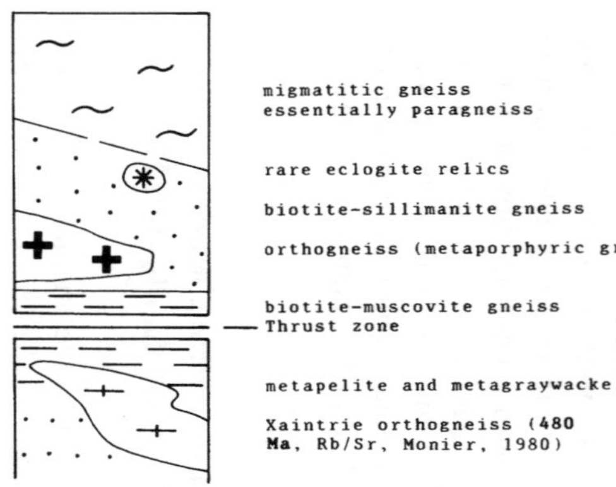

Fig. 4. Millevaches/Moyenne Dordogne

et al. 1983). They are mainly composed of metagraywacke and metapelite associated with orthogneiss and the leptynite-amphibolite complex. In the lowermost part of the Gargilesse/Dun-le-Palestel unit, quartzo-feldspathic metatexite appears.

The uppermost unit of the sequence, the Le Pin-Chéniers migmatite unit, is composed of mainly sedimentary-derived migmatites with rare interbedded orthogneissic and amphibolitic rocks. The age of tectonic emplacement of the upper nappe units has been estimated to be Middle Devonian (Quenardel and Rolin 1984). The latest thrusting of the nappes over the Fougères unit was dated Westphalian by whole-rock Rb-Sr method (Rolin et al. 1982).

2.3 Millevaches, Moyenne Dordogne Region (Fig. 4)

These two regions are situated between the Argentat and the Sillon Houiller faults. They are mainly underlain by micaschists with rare orthogneiss, dated at 480 Ma (Rb/Sr, whole rock, Monier 1980), over which a paragneiss unit was thrust, which is anatectic in its upper part (Tempier 1985; Feix 1988; Ledru 1988). The contact between the two units is the geometrical equivalent of the contact between the lower gneiss unit and the presumed autochthon in the Limousin. A few high-pressure relicts are known, but are still poorly documented in the lower allochthon.

2.4 Auvergne-Lot (Fig. 5)

The Variscan nappes were first described in this region by Demay (1942); Briand (1978), Burg and Matte (1978), Briand and Piboule (1979), Burg et al. (1984), and Pin (1980) further refined the knowledge of these rocks. The gneiss unit that occupies the highest position, consists of paragneiss (metagraywacke) and of a leptynite-amphibolite complex, including metarhyolite dated at 478 ± 6 Ma (Pin 1979; Pin and Lancelot 1982) and metagabbro dated at 484 ± 7 Ma (Pin 1979). This unit contains abundant and varied eclogite relics in boudins.

A main thrust zone is located at the contact between this unit and the lower gneiss unit (lower allochthon), which is composed of paragneiss cross-cut by granitic orthogneiss dated between 500 and 550 Ma, and quartz diorite dated at 540 ± 15 Ma (Pin and Lancelot 1978). This unit shows no trace of high-pressure metamorphism. It rests upon a presumed autochthonous succession, composed essentially of metapelite of Brioverian to Early Paleozoic age, and of lower metamorphic grade than the overlying units, which occur as outcrops in the Lot river. This micaschist unit is very similar to that of Middle Dordogne area.

2.5 Sioule-Echassières (Fig. 6)

The upper gneiss unit is composed of a leptynite-amphibolite complex at its base, overlain by feldspathic gneiss and migmatite. High-pressure metamorphic relicts are present (Gentilhomme 1975; Ravier and Chenevoy 1979).

This unit, correlated with the Middle Allochthon, rests upon a micaschist unit that occupies the core of the antiform. However, the original Devonian stack of thrust sheets was further affected by intense Carboniferous thrusting that has been dated around 340 Ma by syn-tectonic intrusions, and migmatitic gneiss is again found under the micaschists.

2.6 Morvan (Fig. 7)

The basement of the Morvan is essentially composed of anatectic gneiss, including high-pressure metamorphism relics (Godard 1990). The knowledge of this high grade terrane is still poor, although its position is fundamental for the understanding of the Variscan belt: the overstepping sequences start with reefal Givetian and Frasnian limestone, interlayered in marine clastic sediments. Famennian to Tournaisian shales are associated with calc-alkaline volcanic rocks in the western part. Carboniferous sequences start in the Tournaisian and end with a major acid-volcanic sequence of calc-alkaline or crustal anatectic affinity, during the Late Visean.

2.7 Lyonnais (Fig. 8)

The Lyonnais region shows the stacking of three allochthonous units, with from bottom to top:

The Lower Allochthon or Mont Pilat unit, consisting of paragneiss (mostly metapelite) and orthogneiss (porphyritic granite dated at 540 Ma).

The Middle Allochthon, comprising a leptynite-amphibolite complex whose protolith is in part dated at 504 Ma (Dufour et al. 1983), including peridotite and eclogite bodies, overlain by para- and orthogneisses.

The Upper Allochthon or Brevenne unit. This MORB rift formation was thrust during the Early

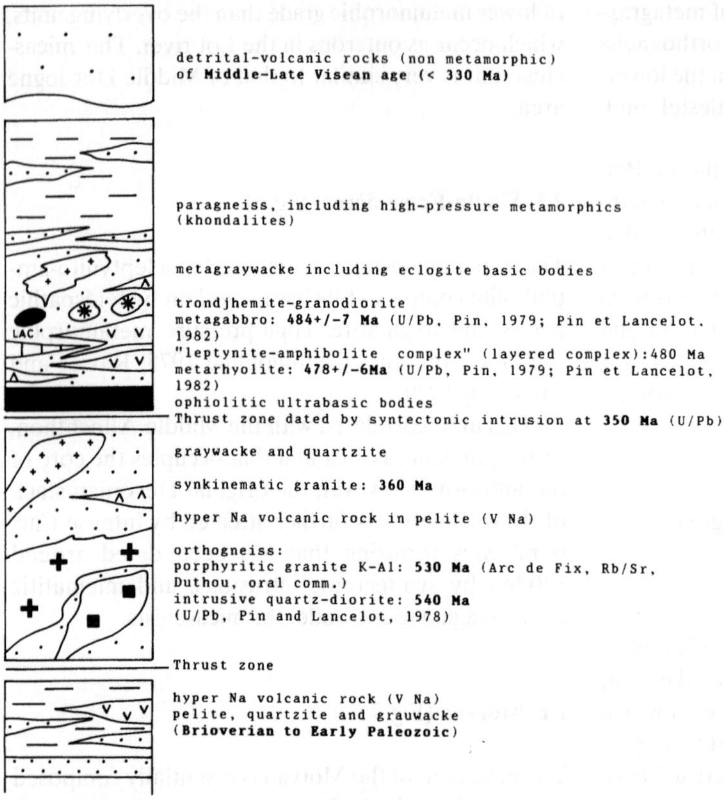

Fig. 5. Auvergne/Lot

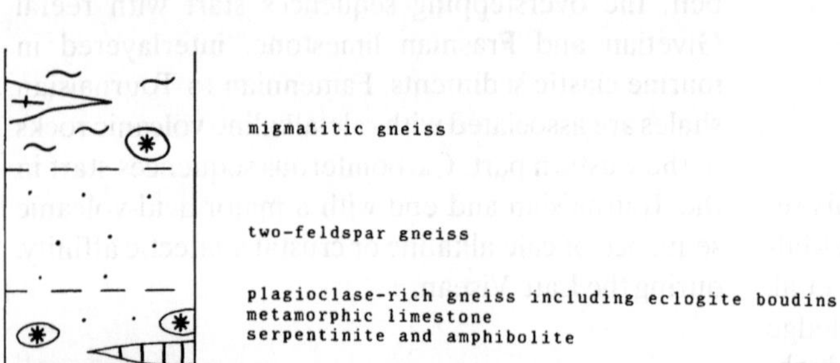

Fig. 6. Sioule-Echassières

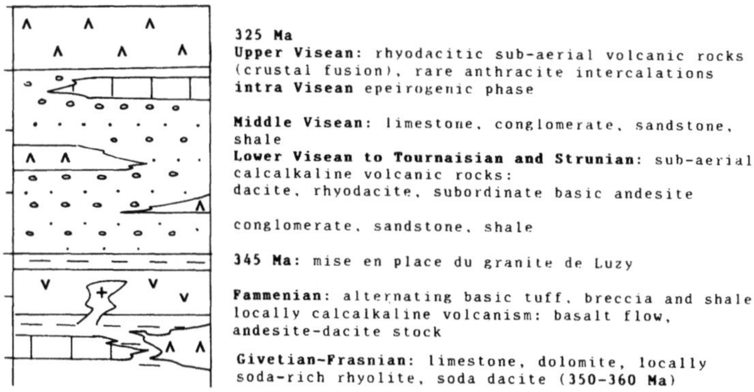

Fig. 7. Morvan/Montagne Bourbonnaise

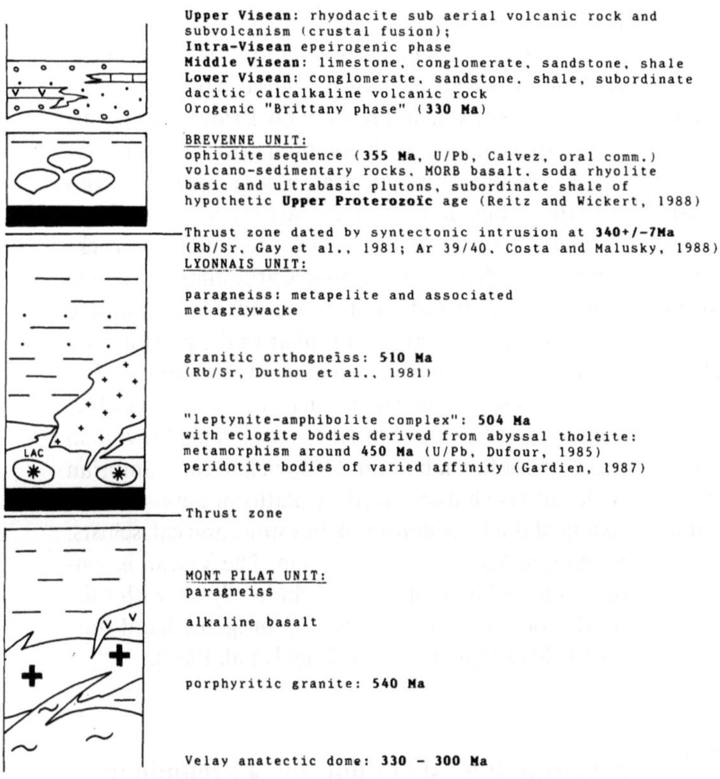

Fig. 8. Brevenne-Lyonnais

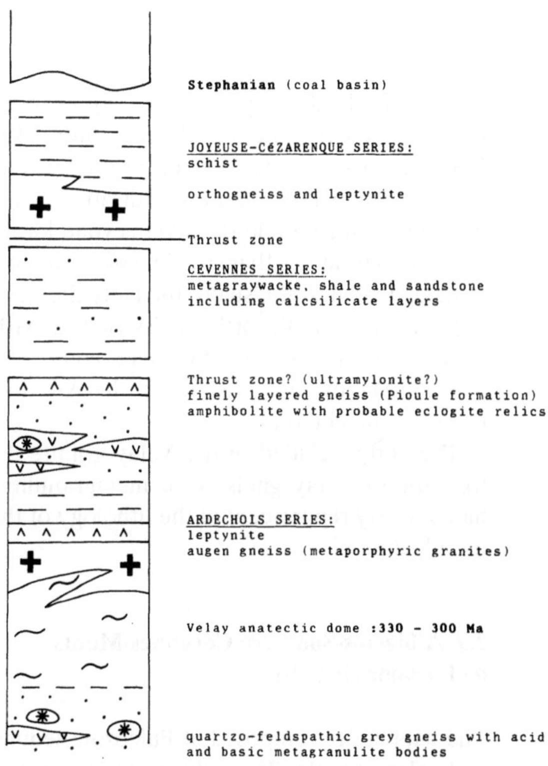

Fig. 9. Norther Cevennes(Ardèche/Southern Velay

Carboniferous over the underlying stack of nappes, which preceded the deposition of Visean conglomerates.

Unpublished data (Calvez, pers. commun.) give an age of 355 Ma for the emplacement of trondhjemmite. Reitz and Wickert (1988) have recently identified Upper Proterozoic microfossils in low grade intercalated phyllites. This can be interpreted as a tectonic imbrication of the Proterozoic basement during the Carboniferous thrust tectonics. However, according to the descriptions, these microfossils are no so characteristic and may be also of Paleozoic age (J.J. Chauvel, pers. comm.).

2.8 Northern Cevennes (Fig. 9)

The tectonic units in this area are partly overprinted by a strong migmatization developed between about 320 and 300 Ma, as is shown by the emplacement of the late-anatectic Velay granite. Nevertheless, it is possible to recognize two main units separated by a thrust contact:

1. At the top, the Joyeuse Cézarenque series, consisting of low-grade orthogneiss and leptynite, with a "cover" of biotite-chlorite schist; this unit is probably related to the lower allochthon.
2. At the bottom, a complex of low- to high-grade metasediments (micaschist and gneiss) of the "Cevennes series", and the even deeper metagranitic rocks (augen gneiss and leptynite) of the "Ardéchois series", Marignac et al. 1980), the Cevennes and Ardéchois series being interlayered due to an early tectonic event (Macaudière et al. 1987).

A level of calc silicate gneiss and the presence of amphibolite lenses (initially alkali basalts, Weisbrod 1970) characterize the Cevennes series. The amphibolite contains relict garnets, indicative of an early medium- to high-grade Barrovian event; locally they show ocellar fabrics that could be interpreted as meta-eclogite features. Usually, the leptynites are at the apparent top of the orthogneiss of the Ardéchois series. The age of the orthogneiss could be about 530–550 Ma, by comparison with the Arc de Fix, in the lower allochthon.

Presently included in the Velay granite dome, a formation of gray gneiss with metagranulitic remnants is very reminiscent of the lithology of the middle allochthon.

2.9 Albigeois-Southern Cévennes-Monts de Lacaune (Fig. 10)

The nappes are composed of Paleozoic sedimentary and volcanic rocks. The only recognized remnant of basement is a slice of orthogneiss, dated at 717 Ma, that was caught up in the thrust tectonics. The oldest sedimentary rocks are thick detrital series attributed to the Lower Cambrian K1 (in part Infracambrian?). They have the chemical signature of sodic quartzitic arkoses and may have derived from the erosion of the Cadomian mountains. A first acid volcanic episode occurred at the top of this detrital series in the eastern part of the Monts de Lacaune and probably also in the St Sernin nappe.

During the Early Cambrian K2, carbonate sedimentation and associated bimodal volcanism were well developed in the eastern part of the Monts de Lacaune, indicating the presence of an unstable platform margin with an extensional basin to the north. In the Albigeois, the carbonates are absent while the volcanism is well developed. The middle Cambrian is represented in the Monts de Lacaune by the Paradoxides schists. The Lower Ordovician O_1–O_2 rocks show to what extent pelitic sedimentation became widespread. All sections known show an hiatus that represents the Middle Ordovician. The youngest formations of the nappes are of Silurian age.

2.10 Montagne Noire (Fig. 10)

This region is composed of a highly metamorphic axial zone, that lies in a late antiform beneath a stack of thrust sheets consisting of Paleozoic sedimentary rocks (including eclogite relics and showing kyanite-zone metamorphism along its southern border).

The lower axial unit is probably Brioverian to Early Paleozoic in age, with common orthogneiss occurring in a sedimentary pile that shows great similarities with the Albigeois or the Central Pyrenees.

The nappe complex, whose sedimentary rocks range in age from Cambrian to Carboniferous, shows a very different sedimentary record: Cambrian and Ordovician rocks are very similar to the platform facies of the lowest northern nappes in the Monts de Lacaune. These are then overlain by the first Lower Devonian strata of the Massif Central. This Devonian rests unconformably on Ordovician or Cambrian rocks and is characterized by platform deposits consisting of thick sequences of limestone and calcschists, ranging in age up to Tournaisian. The Visean is synorogenic sedimentation. It includes flysch with olistoliths, derived the approaching orogenic front during the Mid-Upper Visean (Engel et al. 1980).

3 Chronological Outline and a Preliminary Paleogeographical Reconstruction

The lithological data clearly indicate two domains in the high grade metamorphic units that make up the Variscan terranes of central and southern France. They are:

1. The lower gneiss unit (lower allochthon) and the presumed autochthon.
2. The upper gneiss unit, or middle allochthon.

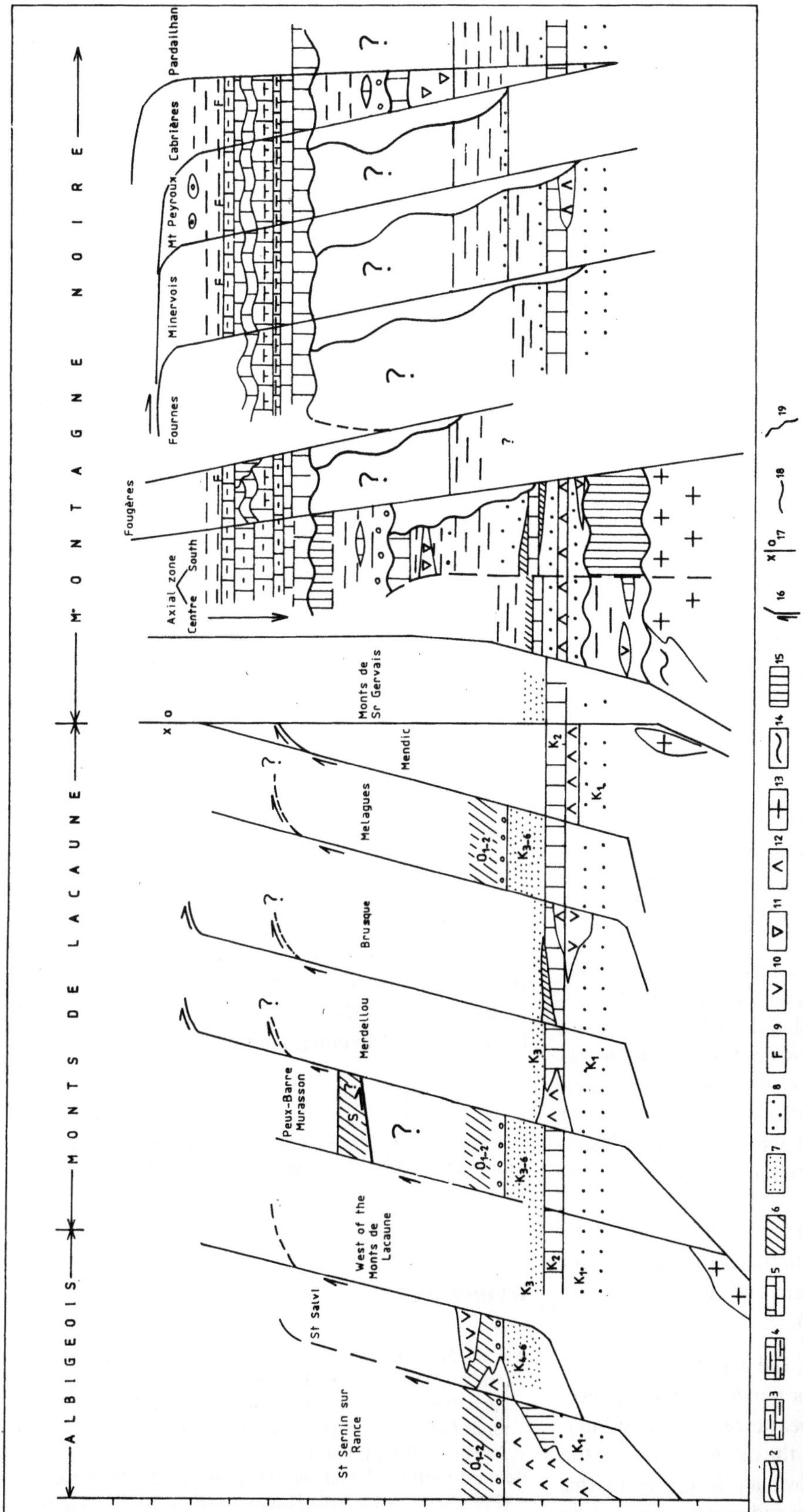

Fig 10. Albigeois/monts de Lacaune/Montagne Noire. *1* Olistolith; *2* nodular limestone; *3* calc schist; *4* reef limestone; *5* platform limestone; *6* black shale; *7* pelite; *8* sandstone; *9* flysch-type sequence; *10* basic volcanism; *11* intermediate volcanism; *12* acid volcanism; *13* granite; *14* migmatite; *15* thrust zone; *16* strike-slip fault; *17* unconformity; *18* hiatus; *19* erosion. *S* Silurian; O_{1-2} Lower Ordovician; K_{1-6} Cambrian

The first domain (1), which now occurs in a lower geometrical position, consists of sedimentary rocks of mostly pelitic to graywacke origin, which overlie or are intruded by Cadomian porphyritic granite intrusions. These rocks can be considered as the basement of the Variscan mountains in what today is the Massif Central, and their age ranges from Late Proterozoic (Brioverian) to Early Paleozoic.

These terranes can be considered to occupy a para-autochthonous position in the presumed autochthonous complexes (Millevaches, Albigeois, Cevennes), but they are allochthonous in the lower gneiss unit and in the Paleozoic complexes further south, where their transformation into thrust sheets thus post-dated the first collision. These three complexes can be considered as three paleogeographic units that are allochthonous to varying extents, but deriving from the same southerly continental plate.

At the base of the Ordovician, a major extensional regime was recorded in different ways:

1. Oceanization was marked by the formation of new crust with oceanic affinity, which today is preserved as ophiolite wedges in displaced suture zones that can be observed in the upper gneiss unit (the middle allochthon).
2. Major magmatic activity occurred on the continental rise, which witnessed of an intra-continental extensional regime with a strong thermal anomaly. Certain authors, such as Matte (1986) and Girardeau et al. (1986) attribute this magmatism to a volcanic arc, but this hypothesis is as yet not well documented.
3. In the Albigeois, the Late Cambrian sandy sediments and the black shales deposited in the Ordovician basin that were associated with tholeites of an extensional character, show that intra-continental extension was reactivated after the quiet of the uniform platform sedimentation that was prevalent during the Late Cambrian, which in itself followed upon the much more varied paleogeography of the Early Cambrian.
4. The deposits now found in the Montagne Noire apparently did not record these events, but rather show an extensional regime during the Cambrian and the Lower Ordovician.

Towards 440 to 450 Ma ago, the general mechanism reversed, resulting in a major period of convergence. This was marked by the appearance of the first high-pressure parageneses and by the disappearance of the intrusions that marked Ordovician. To the south, this period is not recorded as no sediments or volcanics subsist, being probably eroded prior Early Devonian times.

From 400 Ma onwards, during the middle of the Variscan events, the thrust sheets were emplaced, indicating continental collision. It was during this period that a clearly calc-alkaline magmatism became dominant, and was probably related to the intense tectonic activity.

During this period a southern platform existed, which can be deduced from the continuity of the Devonian sedimentary deposits found in the thrust sheets of the southern Montagne Noire. There, the base of the Devonian is marked by a return to platform sediments, following upon the erosion period that existed since the Ordovician. The southern area thus recorded platform sedimentation with a basin to the south, while at the same time major deformation took place farther north.

The post-tectonic deposits found in the north of the Massif Central, which overlie highly metamorphic units, are of Middle of Late Visean age. This indicates an upper limit of about 340 Ma for the main erosion event that affected the upper and lower gneiss units. In any case, the Morvan area shows proof of a marine transgression from the Frasnian onwards; furthermore, data from the Brevenne area show that an extensional regime existed around 360 Ma ago, before returning to a compressional regime by 340 Ma. These remarks indicate that the nappe emplacement clearly was diachronous over the Massif Central.

Finally, a last tectonic-collision event took place in what is today the axial zone of the Montagne Noire, between 330 and 300 Ma. This evolution corresponds to a southerly migration of areas of deep crustal shortening. It was during this period that the platform with continuous Devonian deposition, which was earlier situated north of the axial zone, disappeared beneath the Albigeois-Cévennes nappes.

Acknowledgments. The authors thank B. Briand, J. Delfour, M. Demange, T. L. Duthou, R. Feist, T. L. Feybesse, J. P. Floc'h, J. M. Lardeaux, C. Marignac., and J. M. Quenardel for their constructive remarks on the regional geology of the French Massif Central.

References

Ahmadzadeh H, Petitpierre E (1984) Age briovérien supérieur par la méthode au plomb sur galène d'un niveau de gneiss à amphibole de la série métamorphique du Plateau d'Aigurande (unité de Fougères, Massif Central français). C R Acad Sci Paris (D) 98: 491–494

Bernard-Griffiths J, Gebauer D, Grunenfelder M, Piboule M (1985) The tonalite belt of Limousin (French Central Massif): U-Pb zircon age and geotectonic implications. Bull Soc Geol Fr. I 4: 523–529

Bodinier J L, Giraud A, Dupuy C, Leyerloup A, Dostal J (1986)

Caracterisation géochimique des métabasites associées à la suture hercynienne, Massif central français et Chamrousse (Alpes). Bull Soc Géol Fr. II 1: 115–123

Bodinier J. L, Burg J. P, Leyreloup A, Vidal H (1988) Reliques d'un bassin d'arrière arc subducté, puis obducté dans la région de Marvejols (Massif Central). Bull Soc Géol Fr IV 1: 21–33

Briand B (1978) Métamorphisme inverse et chevauchement de type "himalayen" dans la série de la vallée du Lot (Massif Central français). C R Acad Sci Paris 286: 729–731

Briand B, Piboule M (1979) Les métabasites de la série de Marvejols (Massif central): témoins d'un magmatisme tholéiitique d'arrière-arc cambro-ordovicien? Bull Bur Rech Géol Min 2 I 2: 131–171

Briand B, Piboule M, Bouchardon J L (1988) Diversité géochimique des métabasites des groupes leptyno-amphiboliques du Rouergue et de Marvejols (Massif Central). Origine et implication. Bull Soc Géol Fr (8) IV 3: 489–498

Burg J P, Matte P (1978) A cross-section through the French Massif central and the scope of its Variscan geodynamic evolution Z dtsch Geol Ges 109: 429–460

Burg J P, Leyreloup A, Marchand J, Matte Ph (1984) Inverted metamorphic zonation and large-scale thrusting in the Variscan belt: an example in the French Massif Central In: Hutton D H W Sanderson D J (Eds) Variscan tectonics of the North-Atlantic region. Spec Publ Geol Soc London 14: 47–61

Costa S, Maluski H (1988) Datations par la méthode ^{39}Ar-^{40}Ar de matériel magmatique et métamorphique paléozoïque provenant du forage de Couy-Sancerre (Cher, France). Programme G. P. F. C R Acad Sci Paris II 306: 351–356

Demay A (1942) Microtectonique et tectonique profonde. Mem Serv Carte Géol France vol 1. Paris, 247 pp

Dufour E (1985) Granulite faciès metamorphism and retrogressive evolution of the monts du Lyonnais metabasites (Massif Central, France). Lithos 18: 97–113

Dufour E, Piboule M, Duthou J L (1983) Les granulites des Monts du Lyonnais (Massif Central français). Evolution métamorphique et premiers résultats radiométriques Rb/Sr. Terra Cognita 3, 2–3: 197

Duthou J L,. Piboule M, Gay M, Dufour E (1981) Datations radio-métriques Rb-Sr sur les orthogranulites des Monts du Lyonnais (Massif Central français). C R Acad Sci Paris II 292: 749–752

Engel W, Feist R, Franke W (1980) Le Carbonifère anté-Stéphanien de la Montagne Noire: rapports entre mise en place des nappes et sédimentation. Bull Bur Rech Géol Min 2, I: 341–389

Feix I (1988) Etude géologique dans le Sud Millevaches: lithologie, géochimie, métamorphisme et structure des séries métamorphiques situées au Sud de la vallée de la Dordogne. Place dans la Massif Central occidental. These 3ème cycle, Univ Orléans, 535 pp

Floc'h J P (1983) Le socle métamorphique du Limousin Central: une traverse de la branche ligérienne de l'orogène varisque de l'Aquitaine à la zone broyée d'Argentat (Massif Central français). Thèse d'Etat, Univ. Limoges, 445 pp

Floc'h J P, Quenardel J M, Rolin P, Santallier D (1989) Une géotraverse dans le Massif Central occidental (France). lère partie: les unités lithostructurales. In: Le Maroc et l'orogène paléozoïque, Symposium, Rabat 1983, Projet 27 du P. I. C. G. Orogène calédonien Note Mem Serv Geol Maroc, Rabat, 335, pp 367–376

Gardien V (1987) Les laminations croûte-manteau dans les monts du Lyonnais (M. C. F.): Approches microstructurale, thermodynamique et cinétique. Mém D E A, Grenoble, 75 pp

Gay M., Peterlongo J M, Caen-Vachette M (1981) Age radiométrique des granites allongés et en feuillets minces syntectoniques dans les Monts du Lyonnais (Massif Central français). C R Acad Sci Paris II 293: 993–996

Gentilhomme P (1975) Leptynites et quartzites de Breuilly-Cesset. Existence de roches granulitiques rétromorphosées en Bourbonnais. Rev Sci Bourbonnais, Moulins, pp 82–101

Girardeau J, Dubuisson G, Mercier J C (1986) Cinématique de mise en place des ophiolites et nappes cristallophylliennes du Limousin, Quest du Massif Central français. Bull Soc Géol Fr (8) II 5: 849–860

Godard G (1990) Découverte d'éclogites, de péridotites à spinelle et d'amphibolites à anorthite, spinelle et corindon dans le Morvan. C R Acad Sci II, 310: 227–232

Grolier J (1971) La tectonique du socle hercynien dans le Massif Central. In: Symposium J. Jung, Géologie, Géomorphologie et structure profonde du Massif Central français. Plein Air Service, Clermont Ferrand, pp 215–248

Guillot P L (1981) La série métamorphique du Bas-Limousin: de la vallée de l'Isle à la vallée de la Corrèze, le socle en bordure du Bassin aquitain. Thèse d'Etat, Univ Orléans, 369 pp

Lafon J M (1986) Géochronologie U-Pb appliquée à deux segments du Massif Central français. Le Rouergue oriental et le Limousin central. Thèse d'Etat, Univ. Montpellier, 152 pp

Ledru P (1988) Le chevauchement crustal de Moyenne Dordogne. Bull Soc Géol Fr (8) IV 3: 471–478

Ledru P, Autran A (1987) L'édification de la chaine varisque dans le Limousin. Rôle de la faille d'Argentat à la limite Limousin-Millevaches. In: Bur Rech Geol Min (ed) Programme Géologie profonde de la France. Doc Bur Rech Géol Min Orléns, 140, pp 51–81

Ledru P, Lardeaux J M, Santallier D, Autran A, Quenardel J M, Lerouge G, Floc'h J P, Maillet N, Marchand J, Ploquin A (1989) Où sont les nappes du Massif Central Français? Bull Soc Géol Fr 8, V, 3: 605–618

Lerouge G (1987) Tectogenèse comparée de deux segments de la chaîne hercynienne: Le Massif Central français septentrional et le Sud du Massif armoricain. Thèse Sci Univ Paris-Sud, Orsay, et Mémoire Géodiffusion 2 Paris (1988) 363 pp

Macaudière J, Marignac C, Weisbrod A (1987) Grandes nappes synschisteuses collisionnelles dans la catazone hercynienne des Cévennes médianes (Massif Central, France). C R Acad Sci Paris II, 304: 19: 1195–1199

Maillet N (1987) Dualité d'origine des massifs ultrabasiques limousins. Implications géodynamiques. Thèse 3ème cycle, Univ. Lyon, 149 pp

Marignac C, Leroy J, Macaudiere J, Pichavant M, Weisbrod A (1980) Evolution tectonometamorphique d'un segment de l'orogène hercynien: les Cévennes médianes, Massif Central français. C R Acad Sci Paris, D, 291: 605–608

Matte P (1986) Tectonics and plate tectonics model for the variscan belt of Europe. Tectonophysics 126: 329–374

Monier G (1980) Pétrologie des granitoïdes du Sud-Millevaches (M. C. F.) Minéralogie, géochimie, géochronologie. Thèse 3ème cycle, Univ Clermont, 288 pp

Piboule M (1979) L'origine des amphibolites: approche géochimique et mathématique. Application aux amphibolites du Massif central français. Thèse d'Etat, Univ Lyon, 1003 pp

Pin C (1979) Géochronologie U-Pb et microtectonique des séries métamorphiques antéstéphaniennes de l'Aubrac et de la région de Marvejols (Massif Central). Thèse 3ème cycle, Univ Montpellier, 205 pp

Pin C (1980) Données microstructurales sur les terrains métamorphiques de la série du Lot. Bull Univ Rech Géol Min I, 4: 293–314

Pin C (1981) Old inherited zircons in two synkinematic variscan granitoïds: the "granite du Pinet" and "the orthogneiss de Marvejols" (Southern Massif Central). Neues Jahrb Miner Abh 142: 27–48
Pin Ch, Lancelot J R (1978) Un exemple de magmatisme cambrien dans le Massif Central: les métadiorites quartziques intrusives dans les séries du Lot. Bull Soc Géol Fr (7) XX 2: 203–208
Pin C, Lancelot J R (1982) U-Pb dating of an early paleozoïc bimodal magmatism in the French Massif Central and of its further metamorphic evolution. Contrib Mineral Petrol 79: 1–12
Pin C, Piboule M (1988) Age Dévonien supérieur de la série calcoalcaline de la ceinture basique du Lévezou (Rouergue). Un exemple de complexe leptyno-amphibolique composite. – Bull Soc Géol Fr (8) IV 2: 261–265
Quenardel J M, Rolin P (1984) Paleozoic evolution of the Plateau d'Aigurande (N-W. Massif Central, France). In: Hutton D, Sanderson D (eds) Variscan tectonics of the North Atlantic region. Geol Soc London Spec Publ 14 pp 63–77
Quenardel J M, Lerouge G, Rolin P, Schmitt P, Bouvier P (1984) Grands traits géologiques du Massif Central nord occidental (France). 109e Cong Nat Soc Sav, Dijon, Sciences 1: 203–214
Ravier J, Chenevoy M (1979) Présence de formations granulitiques jalonnant un linéament crustal dans la série cristallophyllienne de la Sioule (Massif Central français). C R Acad Sci Paris 288: 1703–1706
Reitz E, Wickert F (1988) Upper Proterozoïc microfossils in low grade phyllites of the Brevenne Unit, NE Massif Central (France). C R Acad Sci Paris 307 II: 1717–1721
Rolin P (1987) Tectogenèse comparée de deux segments de la chaîne hercynienne: le Massif Central français septentrional et la Meseta marocaine septentrionale. Thèse Doctorat ès Sciences, Univ de Paris Sud, Orsay, 484 pp
Rolin P, Duthou J L, Quenardel J M (1982) Datation Rb/Sr des leucogranites de Crozant et d'Orsennes. Conséquences sur l'âge de la dernière phase tectonique tangentielle sur le Plateau d'Aigurande. C R Acad Sci Paris II 294: 799–802
Roubichou P (1980) Les formations métamorphiques de la région de Leyme près de St-Céré (lot). Doc Bur Rech Géol Min (Orléans) 18, 161 pp
Tempier P (1985) Nouvelles données structurales sur le socle oriental du Cézallier. 110è Congr Nat Soc Sav, Montpellier, Bibliothèque Nationale Paris, fasc 1, pp 33–42
Weisbrod A (1970) Découverte de gneiss alcalins à eckermanite et augite aegyrinique dans la catazone ardéchoise. C R Somm Soc Géol Fr: 31

Biostratigraphy and Dynamics of the Nonmetamorphic Sedimentary Record

Coordinator: R. Feist
Contributors: H. Echtler, R. Feist, J. Galtier, B. Mouthier

1 Pre-Collisional and Syn-Tectonic Evolution

1.1 Internal Zones

(B. Mouthier)

Low grade to nonmetamorphic sediments overstepping exhumed deeper crustal Variscan formations represent important markers for the dynamic evolution of the internal zone in the Massif Central. They build up major units in the northeastern part (Fig. 1) with (f) the Morvan and Beaujolais units; and adjacent to the south (g) the Brevenne unit. Here we describe essentially the lithologic and biostratigraphic characteristics of these sequences; for further geodynamic implications see Chapter III.4 Structure.

1. The Morvan and Beaujolais Units. The Morvan and Beaujolais Paleozoic units involved in the Variscan orogeny reflect two successive dynamic events: firstly, pretectonic crustal extension, which was marked progressively by an increase of magmatic activity and effects of low-grade metamorphism. This was contemporaneous with the southern thrusting and crustal thickening during the Devonian and Early Carboniferous. Secondly, the synorogenic Visean period is characterized by volcano-sedimentary deposits.

The oldest known unmetamorphosed rocks of the northeastern Massif Central constitute the Devonian Somme Group, which is restricted to small outcrops in the western Morvan (Fig. 1). The succession is as follows:

- bioclastic carbonates with crinoid packstones yielding corals *(Heliophyllum, Alveolites)* and brachiopods *(Stringocephalus burtini)* of Late mid-Devonian age (Julien 1896; Michel-Levy 1908; Guffroy 1964);
- peloid and bioclastic wackestones alternating with laminated siltstones, changing gradually to dolomicrites with rhythmic sedimentation of sand layers of storm origin. Local occurrences of syn-sedimentary breccias with bioclasts and volcanic fragments may result from gravitational slides on submarine slopes. An early Upper Frasnian age of the succession is indicated by conodonts, ostracods and foraminiferids (Lys et al. 1961; Delfour and Gigot 1985);
- laminated siltstones, shale with minor carbonate layers changing gradually into conglomeratic fluvial sequences with quartzite and chert pebbles, and volcanic intercalations (Robelin 1981). Age according to trilobites, conodonts, and microfloras: Uppermost Famennian (Guffroy et al. 1966; Coulon et al. 1980; Delfour and Gigot 1985).

Volcanic intercalations of partly reworked pyroclastic tuffs and lavas are more frequent in the lower part of the Famennian than in its upper part. Chemistry indicates a continuous nonbimodal calc-alkaline suite ranging from basalts and andesites to quartz-keratophyres (soda dacites). The calc-alkaline origin of tholeittic affinity indicates an ensialic domain related to crustal extension (Bebien 1970; Delfour et al. 1984).

The Dinantian Culm facies of the Haut Morvan Group consists of volcano-sedimentary sequences with conglomerates, sandstones, siltstones, and shales yielding limestone lenses as well as intermediate to basic volcanics with rhyodacitic tuffs and andesite-dacite to andesite-labradorite lavas. Age according to micofaunas and -floras: Lower to Middle Visean (Mamet 1968a); until now no unambiguous Tournaisian age has been obtained. In the Beaujolais area an analoguous volcano-sedimentary succession rests unconformably on metamorphic basement. Brachiopod data from the Goujet limestone quarry indicate a Lower Visean age (Julien 1896; Michel-Levy 1908; Lacroix 1980). Middle Visean fossiliferous carbonates unconformably supersede the basal volcano-sedimentary complex or the basement (Bertaux and Rubiello 1981).

The volcanics consist of dacitic ash flows, which are intruded by stocks and sills of contemporaneous alkali granite, secondary potassified K-microtrondhjemites, gabbros, and andesites. The evolution of the

J. D. Keppie (Ed.): Pre-Mesozoic Geology in France and Related Areas

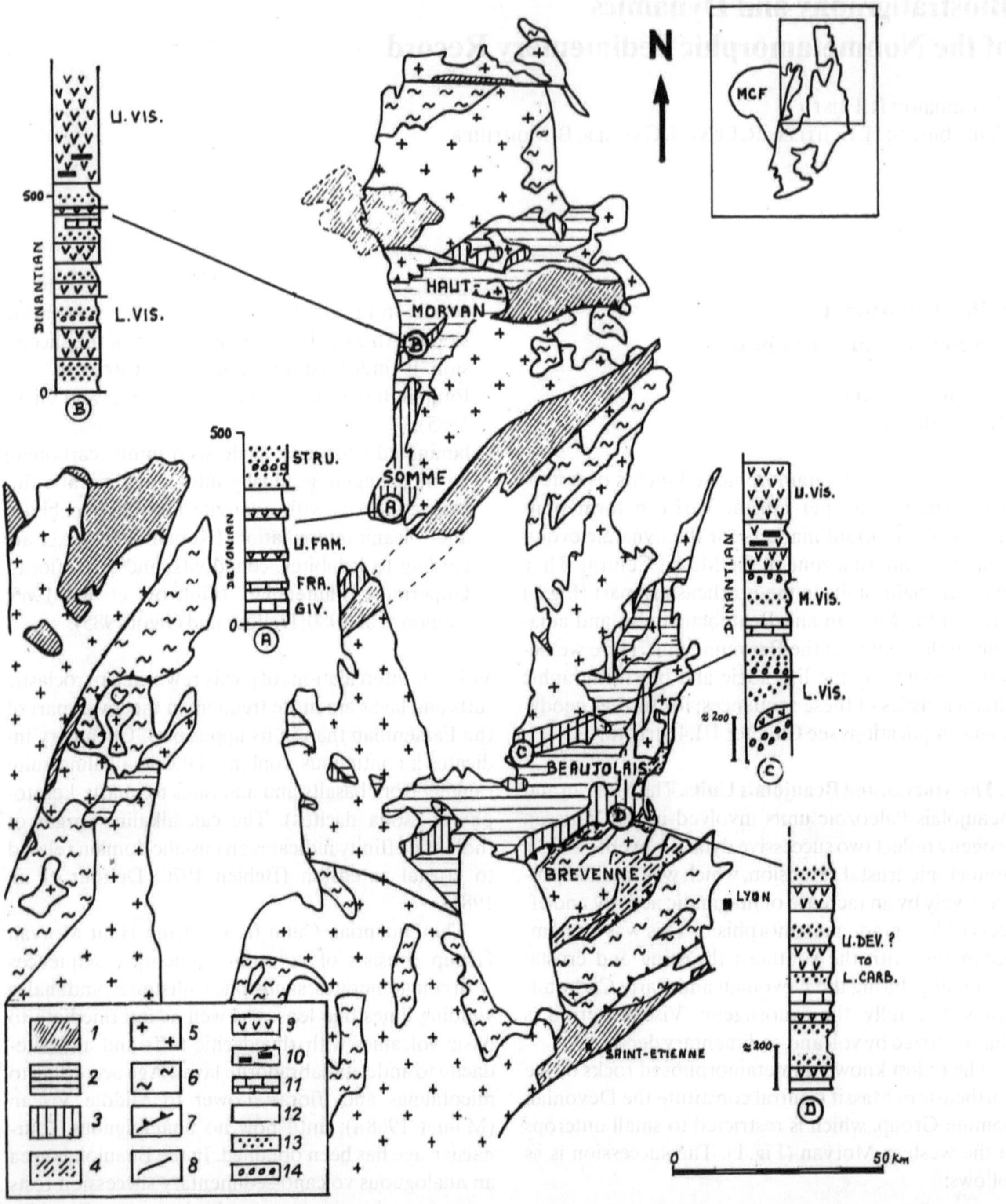

Fig. 1. Low grade and nonmetamorphic Paleozoic deposits of the northeastern Massif Central. Geological map and schematic columnar sections. *A* Somme Group; *B* Haut Morvan Group; *C* Beaujolais unit; *D* Brevenne unit. *1* Stephanian-Permian continental deposits; *2* Upper Visean Tufs anthracifères pyroclastics; *3* volcano-sedimentary Lower to Middle Visean deposits; *4* low to medium grade Paleozoic of Brevenne metamorphics; *5* Visean to Namurian synorogenic granites; *6* metamorphics; *7* main thrust folds; *8* wrench faults; *9* volcanics; *10* anthracite; *11* carbonates; *12* siltstones and pelitic shales; *13* sandstones; *14* conglomerates. Data from: Laville (1986); Delfour et al. (1984); Berthaux and Rubiello (1981); Peterlongo (1960); Wickert (1985); Sider and Ohnenstetter (1986); Coulon et al. (1980)

Beaujolais and Brevenne magmatism is interpreted either within a back-arc or an ensialic marginal basin environment during Early Paleozoic to Early Carboniferous subduction (Piboule et al. 1982; Sider and Ohnenstetter 1986; Wickert 1988).

The Upper Visean Tufs anthracifères deposits fill volcano-tectonic graben structures (Leistel and Gagny 1984). They consist of volcano-sedimentary successions with pyroclastics of calc-alkaline-potassic affinity (Bebien and Gagny 1980). In the lower part, arkoses and conglomerates containing fossiliferous anthracitic levels with silicified wood and miospores are intercalated (Galtier 1970; Scott et al. 1984). Various sedimentary environments from shallow marine with brachiopods (Chichery 1938) through paralic lagoonal and continental are represented (Rex 1986). In the upper part of the succession, acid, garnet-bearing volcanics comprise ignimbrites, breccias of garnet-bearing tuffs of rhyodacitic to rhyolitic composition; these may indicate anatectic melting of continental crust (Bertaux 1982; Leistel and Gagny 1984).

Upper Devonian to Lower Carboniferous sedimentary and volcanic rocks of the entire area are very similar to successions in the southern Vosges and Black Forest. Paralic lagoons and a continental environment for the Upper Visean show progressive crustal thickening and shoaling of the depositional environment (Wickert 1988). In Late Visean to Early Namurian times regional deformation resulted in mylonitic zones and strike-slip faulting, thereby providing channels for the spread of monzonitic magmas along the Morvan-Beaujolais basins.

2. The Brevenne Unit. The metamorphic Brevenne unit lies adjacent to the south of the Beaujolais area juxtaposed by tectonic imbrication. It is the result of Late Visean to Namurian N-S to NW-SE-directed N-vergent thrusting (Wickert 1988). The rock succession consists of metasediments and a metavolcanic pile of spilitic-keratophyric character considered to be an ensialic ophiolite complex (Sider and Ohnenstetter 1986). Similar lithologies in the Beaujolais unit are striking.

The Brevenne unit experienced a low to medium grade metamorphism increasing from NW to SE with a penetrative foliation. Secondary crenulation cleavage and fold vergence indicate tectonic transport to the SE. The southern contact with the Lyonnais metamorphic basement is interpreted as a pre-Visean to possibly pre-Devonian dextral normal fault (Wickert 1988). Other interpretations suggest a Carboniferous dextral ductile thrust-shear system for the emplacement of the allochthonous Brevenne nappes (Feybesse et al. 1988).

Being unconformably overlain by the Upper Visean Tufs anthracifères (Jung and Raguin 1935), the age of the Brevennes series is still debated. Classically attributed to the Upper Devonian-Lower Carboniferous on lithological and chemical analogies (Peterlongo 1960; Piboule et al. 1982; Sider and Ohnenstetter 1986), older pre-middle Devonian ages had been proposed for the structural and sedimentary setting (Wickert 1986). Finally, Late Proterozoic spheroidal microfossils attributed to *Protosphaeridium* have been recently reported from low grade phyllites from the Brevenne series (Reitz and Wickert 1988).

Equivalent sedimentary successions as in the northeastern Massif Central are also characteristic in the Vosges and Black Forest further to the NE, which present other segments of the internal Variscan belt of central Europe. This area of marine and volcanoclastic sedimentation during Devonian and Lower Carboniferous times characterizes a very specific paleogeographic domain. The sediments are overstepping essentially migmatitic rocks which correspond to exhumed and deeply eroded (Eo-Variscan?) deposits. This setting is coherent with an important extensional event in the most internal zone of the belt before a period of instability during late Visean and subsequent shortening and slight crustal stacking during latest Variscan compressional phases.

1.2 External Zones
(R. Feist and H. Echtler)

1.2.1 Zonation and Tectono-Metamorphic Setting of the Southern Massif Central

The Variscan structures of the Southern Massif Central comprising the main Cevennes (east), Albigeois (west), and Montagne Noire (south) areas (Fig. 2) represent the southern external margin of the French Variscan chain (Guérangé-Lozes and Burg 1990). Paleozoic sediments constitute the main component. Towards the north, they underlie the southward-thrust internal crystalline units (Burg and Matte 1978; Matte 1983; Burg et al. 1984); they are bordered to the south by the synorogenic foreland flysch (Engel et al. 1982; Engel 1984). Following tectonometamorphic and lithological criteria (see Ledru et al., Part III.2), the internal segmentation of the belt represents three major zones which are from north to south (Fig. 2):

- the unfossiliferous Northern Albigeois and Cevennes, folded and schistose epimetamorphic sediments of assumed Early Paleozoic age (Marig-

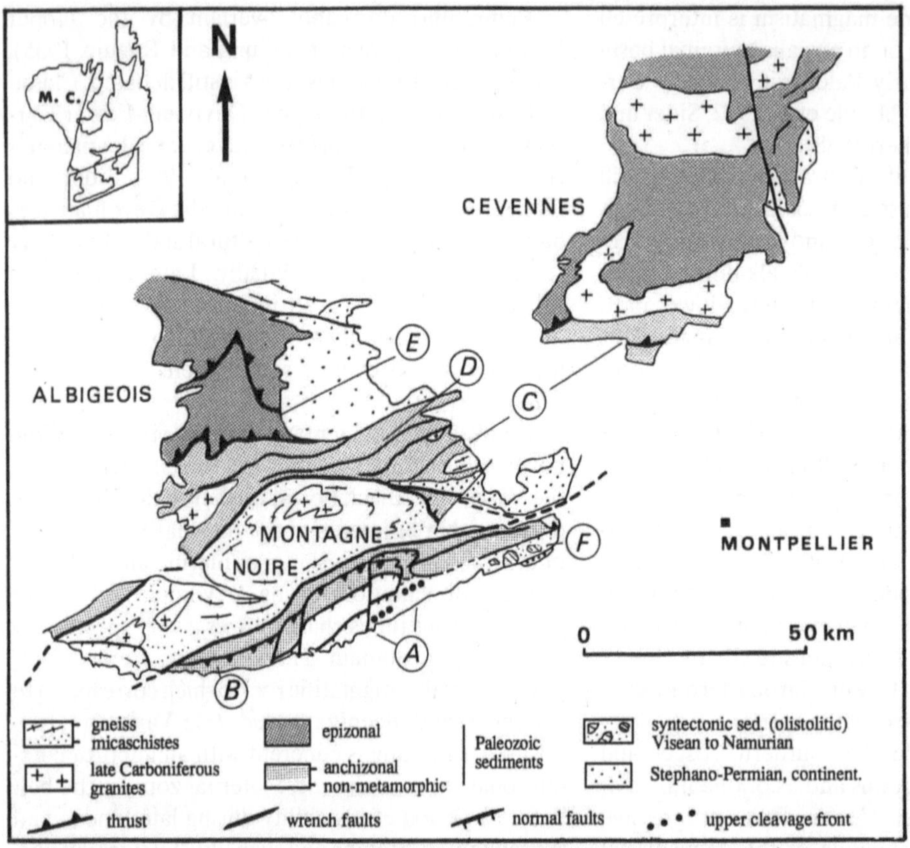

Fig. 2. Geological and structural map of the Southern Massif Central. *A, B, C, D, E, F* Emplacement of compared stratigraphic logs of Fig. 3

nac et al. 1980; Guérangé-Lozes 1987). This unit overthrusts to the south;

- the southern Albigeois ("Monts de Lacaune, Northern Montagne Noire" auct.) and Southern Cevennes units characterized by schistose and folded fossiliferous Lower Paleozoic sediments which are weakly (anchi-zonal) metamorphosed (Thoral 1935; Donnot and Guérangé 1978; Guérangé-Lozes and Guérangé 1984). They are considered to constitute the more external structure of the northern metamorphic area (Guérangé-Lozes 1987);
- the Montagne Noire units, separated from the Albigeois units by Late Variscan wrench faults (Arthaud and Matte 1977).

This complex geological structure (Fig. 2) is characterized by an ENE-WSW-elongated antiformal metamorphic core (Axial Zone) of gneissic, migmatitic, and anatectic rocks (Schuiling 1960; Ellenberger 1967; Bogdanoff 1973; Bard and Rambeloson 1973, 1978; Bard and Loveyit 1978; Demange 1975, 1982; Beaud 1985). To the south, an adjacent pile of large scale recumbent folds (Arthaud et al. 1966; Arthaud 1970; Echtler 1990) affect very low grade or even nonmetamorphic fossiliferous Paleozoic rocks (Cambrian to Lower Carboniferous, Bergeron 1889; Thoral 1935; Gèze 1949). To the southeast a wild-flysch basin with olistolithic blocks (Cabrières klippes) constitutes the most external part of the Massif Central Variscan belt (Engel et al. 1982; Engel 1984; Franke and Engel 1986).

1.2.2 *Stratigraphic Evolution*

Fossiliferous Paleozoic strata involved in the Variscan tectogenesis of the Southern Massif Central are the unequivocal witnesses of an orogenic evolution in time and space. The detailed litho- and biostratigraphical data that have been obtained from the southern, slightly metamorphosed and nonmetamorphic units permit tracing the geodynamic evolution of the former northern Gondwana margin before its collision with Laurobaltica. In relation to the Variscan orogenic evolution, the sedimentary succession of the southern external belt is subdivided into three cycles with individual depositional histories and dynamic implications (Fig. 3):

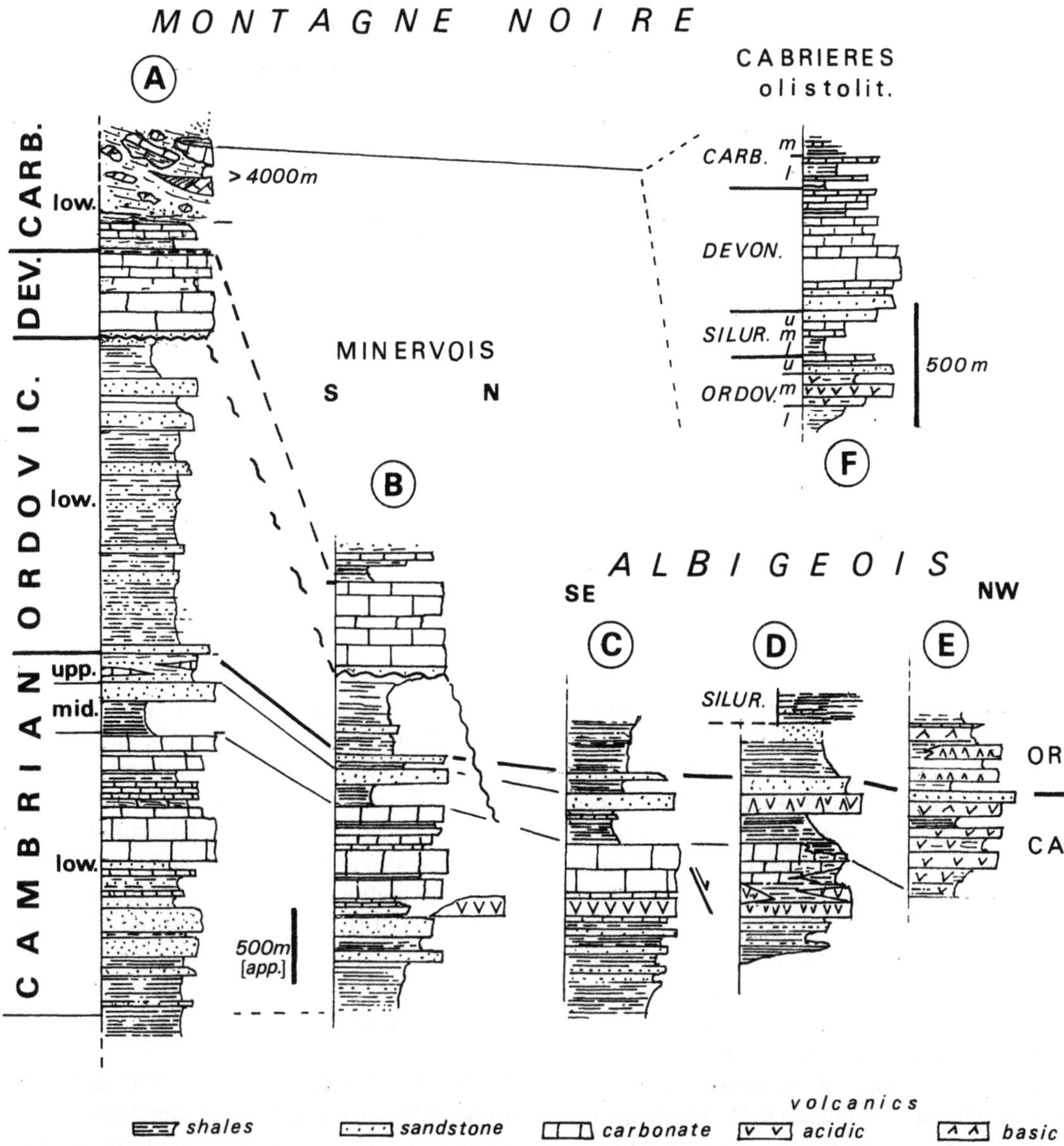

Fig. 3. Schematic columnar sections of Paleozoic sequences from the Southern Massif Central. Vertical scale is approximate. Emplacement see Fig. 2. Data from (Thoral (1935); Gèze (1949); Donnot et Guérangé (1978); Engel et al. (1982); Prian (1980); Feist and Courtessole (1984, 1985); Lépine et al. (1988); Courjault-Radé (1988a); Guérangé-Lozes (1987); Alsac et al. (1987)

- The pre-collision succession (Cambrian to Silurian) constitutes the main body of strata throughout the area.
- The pre-tectonic Variscan deposits (Devonian to Lower Carboniferous) are best exposed in the southwestern and southeastern parts of the Montagne Noire.
- The syn-tectonic sediments (uppermost Lower Carboniferous) which are found from the extreme southeast of the Montagne Noire (Cabrières area) and further southwards and southwestwards under the post-Paleozoic cover of the Languedoc.

Pre-Collision Succession (Cambrian to Silurian). The oldest Paleozoic deposits unconformably overlie medium grade metasediments of probable Brioverian age (Fournier-Vinas and Debat 1970) which constitute a thick detrital succession of fine conglomerates and sandstones. These originated from the erosion of a Cadomian-Panafrican basement. The lowermost Cambrian sandstone are developed throughout the area without notable lateral changes. In the highest part of the succession, where limestone interbeds occur in increasing numbers, the first macrofossils appear: hyoliths, protolenid trilobities, and small archeocyathids of late Early Cambrian age (Debrenne and Courjault-Radé 1986).

In the central Southern Montagne Noire the overlying later Lower Cambrian deposits are characterized, as elsewhere in the Mediterranean area (Sardinia, Morocco), by the establishment of a shallow neritic carbonate platform (Archaeocyathus Limestone) which clearly deepens towards the North; there the carbonates are progressively replaced by anoxic black shales (Courjault-Radé 1988a, b).

In this southern area there are no volcanics, but rhyolitic lavas and tuffs of alkaline to tholeittic affinities (blavierites) are intercalated in the southwestern Montagne Noire (Lépine et al. 1988), southeastern Albigeois (eastern Lacaune area) (Donnot and Guérangé 1978), and southern Cevennes (Prian 1980) at the top of the early Lower Cambrian detrital sequences. Concomitant extensional tectonics, sedimentary instability, the northwestward increase and younging of volcanic activity and a lesser proportion of carbonates indicate slope environments and early continental proto-rifting.

To the NW of this subsident unstable margin, the adjacent northern Albigeois and Cevennes areas display deep basinal conditions with black shales, basic volcanism, tuffs, and pillow lavas deposited during the entire Cambrian and Early Ordovician. These may indicate the effect of oceanic extension and rifting in the north at that time (Alsac et al. 1987).

At the commencement of the Middle Cambrian the southern carbonate platform was progressively drowned and clayey siltstones were deposited. The famous trilobite and echinoderm faunas of these *Paradoxides* shales allow a detailed biostratigraphical correlation within an individualized Mediterranean subprovince of the Acado-Baltic faunal province (Courtessole 1973). A tendency towards paleogeographical endemism and the subsequent separation from the Acado-Baltic province is indicated by the appearance of Asian faunal elements associated with the last *Paradoxides* (Feist 1986).

The highest Cambrian deposits below the slightly unconformable basal Ordovician consist of alternating quartzitic sandstones and silty slates which, in the southernmost areas, include sporadic limestone lenses with a typical peri-Gondwana early Upper Cambrian trilobite association (Feist and Courtessole 1984). As no late Upper Cambrian has been established so far, a gap corresponding to a period of nonsedimentation and erosion may be inferred prior to the ubiquitously uniform reappearance of fossiliferous red slates and quartzites of the basal Tremadoc (Sdzuy 1958; Martin 1972).

The Lower Ordovician consists of a rather thick and monotonous detrital succession of alternating quartzitic sandstones and slates which had been deposited on a quickly subsiding shallow platform. Rich assemblages of trilobites, echinoderm, and molluscs, occurring both in dark slates and in quartzitic nodules, are found throughout the Tremadocian. A characteristic feature of the quartzitic sandstone interbeds is the occurrence of abundant sole marks, mainly trilobite trails (Cruziana sandstone).

The presence of pelagic cyclopygid trilobites with giant eyes may indicate the contiguity of an oceanic basin situated further northwards (South-Armorican ocean?, Henry 1989). Whereas fine-grained sediments predominate in the north (Fournier-Vinas and Donnot 1977), shallower conditions prevailed in the south where a lower Arenig *Lingula* sandstone has been deposited near the shoreline under storm conditions (Eschard in Courtessole et al. 1985). Graptolites which appear first of all at the base of the Arenig become abundant in the highly fossiliferous higher Lower Arenigian shales (Dean 1966). These constitute the youngest pre-Devonian deposits of the Montagne Noire (the Cabrières Klippes excepted).

However, in the south Albigeois area, which contains fossiliferous Lower Ordovician strata (Thoral 1935), an isolated slab has been found of anoxic pelagic carbonates and marls with molluscs and conodonts of undoubted Silurian age (Chaubet 1937; Centène 1977). This occurrence can only be compared with Silurian preserved in the form of olistoliths within the syntectonic sediments which crop out in the southeastern Montagne Noire (Cabrières klippes).

The dynamic evolution of the Early Paleozoic of the Southern Massif Central is characterized by stable epicontinental conditions on the Gondwana margin extending southeastwards, in contrast to more unstable conditions in the northwest where the effects of subsidence and rifting led to the opening of the Mid-European Ocean.

Pre-Tectonic Variscan Deposits (Devonian to Lower Carboniferous). Devonian to Dinantian deposits are found exclusively in the Southern Montagne Noire. They unconformably overlie Lower Ordovician to Lower Cambrian successions which exhibit no sign of pre-Devonian compressional deformation. The angular unconformity is interpreted as the result of Late Silurian fracturing and tilting of blocks and consequent erosion, the reason being that large olistoliths with complete Upper Ordovician and Silurian successions exist in the Cabrières Wildflysch (Fig. 3 and below).

The Devonian and Lower Carboniferous strata display a nearly continuous carbonate succession without apparent stratigraphic gaps, as is demonstrated by detailed biostratigraphic investigations which are mainly based on conodonts (Feist 1985). This succession constitutes a major positive cycle illustrating the paleogeographical evolution of an epicontinental shelf. The following sedimentary environments can be recognized from the base upwards:

- a post-"Caledonian", early Lochkovian transgression on an eroded substrate, with near-shore deltaic detrital facies; followed by dolomites deposited under supra- to intertidal pre-evaporitic and lagoonal conditions; and cherty marls and limestones with abundant neritic biota characterizing a shallow platform environment (Feist and Schönlaub 1974; Feist and Groos-Uffenorde 1979);
- transition from neritic to pelagic conditions with still-water build-ups and their biodetrital cover at the end of the Early Devonian (Feist et al. 1985); open pelagic platform conditions with a differentiation into submarine rises with slowly deposited oxygenated styliolinid- and cephalopod-bearing calcilutites, seamound slopes with brecciated and nodular limestones and shallow euxinic depressions with shales, cherts and ferro-manganese-oxide deposits during the Middle and early Upper Devonian (Tucker 1974; Feist and Klapper 1985; Feist et al. 1985);
- distal pelagic outer platform "griotte" facies (nodular cephalopod-bearing calcilutites) during the latest Devonian and earliest Carboniferous comprising the Devonian/Carboniferous boundary Hangenberg-Schiefer horizon and the middle to upper Tournaisian black radiolarian chert (Boyer et al. 1968; Engel et al. 1982; Feist and Flajs 1987; Galtier et al. 1988);
- shelf slope and slope bottom deposits characterized by light-colored cherts with intercalated allodapic biodetrital limestones ranging in age from early Visean to the commencement of late Visean (Vachard 1977; Engel et al. 1978).

The palinspastic situation shows a rather similar facies development throuthout the whole area (Cabrières Klippes excepted); the Lower and Middle Devonian successions, however, are commonly affected by late diagenetic dolomitization obscuring the primary features. In contrast, Visean limestone turbidites show unequivocal evidence of transport from a shallow to a deeper area from north to south.

Coincident with an early orogenic collisional event in the northern Massif Central during the early Devonian, a flexural basin development on a continental crust (Raymond 1987) adjacent to uplift at the southward migrating orogenic front. This is indicated by the increasing proximality of epicontinental deposits from south to north.

Syn-Tectonic Sedimentation (Highest Lower Carboniferous). The existence of Carboniferous flysch is a characteristic feature of the most external part of the "Variscan" Massif Central. The thickest and most complete flysch succession in outcrop is located in the southeastern Montagne Noire (Fig. 1). It extends further southwards under the post-orogenic cover.

Starting with a matrix-rich debris flow in the basal Upper Visean, the well over 4000 m thick succession contains a wide-ranging assortment of sediments. The internal organization suggests an increase of dynamism upwards and the concomitant approach of the source area (Engel et al. 1982). The succession is as follows:

- channelized turbidity current and overbank deposits;
- conglomeratic graywackes and mid-fan turbidites;
- density current and debris-flow deposits (pebbly mudstones);
- slumps with exotic blocks of contemporaneous shallow water deposits and slabs of channel deposits;
- a chaotic set of pre-flysch exotic blocks (ranging from Lower Ordovician to Visean) with sizes from 1 m^3 to more than 1 km^3 (i.e., the Cabrières wildflysch).

Goniatite index fossils have been discovered only in the lower part of the flysch succession, and indicate a latest Visean Go gamma age (Kullmann in Engel et al. 1982). Higher up, the siliciclastic matrix of the flysch is barren of stratigraphically significant fossils. As far as the evidence goes, the youngest exotic blocks contain a floral assemblage of early Namurian age. Gravitational syntectonic transport lasted, in this part of the Variscan belt, until at least the commencement of the Late Carboniferous (Pennsylvanian) (Feist and Galtier 1985).

The Cabrières Klippes provide precious information on the stratigraphical content (Fig. 2F) of the source area from which the olistoliths originated. Compared to the contemporaneous nappe sequences, there are significant differences in the stratigraphic content and environmental development. Upon fossiliferous Lower Ordovician slates with nodules characterised by cone-in-cone structures (Thoral 1946; Bérard 1986), Mid- and Upper Ordovician deposits have been laid down. These comprise trachytic volcanic deposits, pyroclastic tuffs and volcaniclastic sediments (Gonord et al. 1964). They are overlain by fossiliferous Caradocian sandstones and Lower Ashgillian cystoid and bryozoan-bearing limestones (Dreyfuss 1948). Although widespread in the Mediterranean area, from the Carnic Alps and Sardinia to the eastern Pyrenees, this late Ordovician facies development is lacking in the Montagne Noire nappe succession.

Another outstanding characteristic is the occurrence of a complete and highly fossiliferous Silurian succession with black anoxic graptolitic shales in its lower part and pelagic dark limestones and marls with graptolites and shelly faunas in the upper part (Chaubet 1937). However, at the base of the uppermost Silurian (Pridoli) stage, obvious shallowing occurred, the euxinic complexes being replaced by coarse-grained, sandy carbonates with neritic faunas (Feist and Schönlaub 1974; De Bock 1982): this shallowing is accompanied by a major expansion of the marine sedimentary area, a development which is in advance of that occurring in the nappe succession. In consequence, the stated emersion and consequent removal of material as is shown by stratigraphic gaps in the pre-Devonian nappe succession probably started no earlier than the basal Pridolian (Late Silurian). This interval is contemporaneous to the earliest Variscan collisional events and movements of uplift in the internal Massif Central (Pin 1989).

Devonian-Early Carboniferous times are characterized, as they are in the nappe successions, by the development of a carbonate platform. However, the depositional environment of the klippe material appears to have been shallower and was situated nearer to the shoreline than the contemporaneous nappe succession. This is obvious from the following characteristics of the klippe succession:

- development of shallow neritic, cherty limestones with sessile benthos dominated by reef builders such as tabulate and stromatoporoid colonies during the late Early and Middle Devonian. In particular, true reefal limestones constituted by Amphipora boundstones with *Stringocephalus* developed in places (Frech 1887; Vachard 1974; Feist and Groos-Uffenorde 1979);
- a clearly subsequent change from neritic to pelagic conditions no earlier than in the late Middle Devonian;
- restricted anoxic environment under occasional influence of a nearby neritic area, as is suggested by intercalated tempestites with brachiopod-, trilobite- and echinoderm-coquinas (Feist and Klapper 1985);
- shallowing-upward Famennian "griotte", changing to laminated plant-bearing siltstones and detrital oolitic sequences across the Devonian-Carboniferous boundary (Flajs and Feist 1988); whereas carbon-rich detrital marls and plant-bearing shales with only a few radiolarian cherts developed in places during the Tournaisian (Galtier et al. 1988). Emerging land areas existed contemporaneously, since stratigraphic gaps developed prior to the conformable terminal Tournaisian transgression (Maurel 1966; Flajs and Feist 1988);
- various biogenic platform limestones ("Productus Limestone") with neritic faunas (mainly foraminifera and algae, corals, brachiopods and echinoderms), besides shallow-water sandstones and marls which developed during the entire Visean (Böhm 1935; Gèze 1949; Mamet 1968b; Vachard 1977). At the same time, gravitational movements took place in the nappe depositional area. The base of the limestone turbidites, which developed in the nappes during early and mid-Visean times, can be traced by conodont biostratigraphy into the time-equivalent slope and platform facies, the relicts of which are now preserved in the form of displaced blocks within the Cabrières wildflysch.

The syn-tectonic sedimentation of the olistolith material exhibits the same directional movement of tectonic and gravitational transport (Engel 1984; Franke and Engel 1986). Consequently, the source area of the Cabrières klippes has to be found further to the north than the site from where the nappes originated (Engel et al. 1978, 1982).

1.2.3 Conclusion

The Early Paleozoic history of the Massif Central is related to the opening of the Mid-European (Rheic + South-Armorican) Ocean which separated Baltica from Gondwana as early as in the late Mid-Cambrian. The subsequent closure of this ocean and the collision between Gondwana and Lauro-Baltica during the Devonian resulted in the Variscan orogen (see Part III.3 Ledru et al.). South of the internal collisional zones and their southward facing crystalline nappes, the weakly or nonmetamorphic external zone of the orogen is constituted by the epicontinental deposits on the northern Gondwana margin of early Paleozoic times. Precise biostratigraphic data and the sedimentary dynamics of this zone allow us to trace the palinspastic and paleogeographical evolution; they reflect tectogenetic events of the more internal metamorphic zones of the belt.

Accordingly, a polyphase evolutionary history of the sedimentation in the southern Massif Central can be distinguished in pre- and syn-tectonic times.

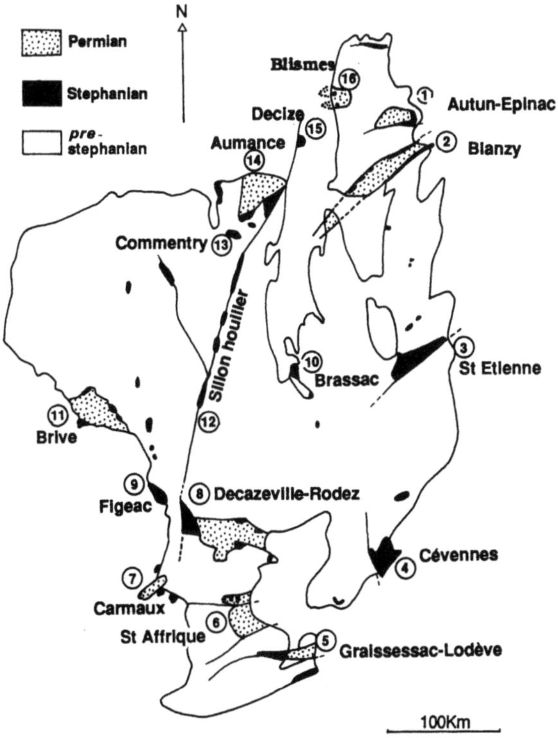

Fig. 4. Geographical distribution of the main Stephanian-Autunian basins in the Massif Central

During the first phase, which started in the early Cambrian, increasing deepening of the depositional area from south to north resulted in the opening of the Mid-European oceanic basin. Extensional dynamics accompanied by volcanic activity in subsident and deep basinal environments which are characterized by more distal deposits in the north contrast with stable shallower platform conditions without volcanics in the south. This tendency continued from Cambrian to Early Ordovician times. After the eustatic changes due to the Late Ordovician glacial period on the Gondwana Continent, an overall uniformity of more stable sedimentary conditions resulted in the Silurian, although a progressive shallowing is obvious by the increase of carbonate production at higher levels.

The second phase, beginning with the late Upper Silurian transgression, is characterized by the inversion of the former polarity. Early events, contemporaneous with the first orogenic stages in the northern Massif Central, relate to the tilting of blocks and their consequent erosion on the adjacent Gondwana margin. Convergence in the north resulted in the formation of a southward deepening epicontinental flexural basin which developed during Devonian through Early Carboniferous times, as is indicated by the southward increase in depth and increasing distance from shore of the platform deposits.

The dynamics of sedimentation during the third phase are controlled by the approach of the orogenic front in mid-base Visean times. The climax of Variscan tectogenesis during the late Visean and the early Namurian, on the southern edge of the Massif Central, results in the emplacement of recumbent fold nappes. The concomitant syn-tectonic sedimentary transport with gravitational flows of exotic blocks such as the Cagrières klippes were emplaced with the same north (internal) to south (external) polarity.

2 The Late- to Post-Orogenic Continental Sedimentation

(H. Echtler, J. Galtier, and B. Mouthier)

2.1 Structural Setting and Basin Formation

Late Paleozoic sedimentation in the Massif Central is characterized by the formation of contintental basins in isolated intramontane troughs. From Stephanian to Permian times coarse coal-bearing sequences were deposited unconformably upon exhumed crystalline basement (Courel 1987). These basins document advanced stages of the intense crustal uplift and extension process that dominated the late-orogenic evolution subsequent to major Variscan crustal shortening (Lorenz and Nicholls 1976, 1984). Their structural setting throughout the Variscan belt (Fig. 1) is linked to large-scale tectono-metamorphic discontinuities such as late Hercynian wrench-fault zones (Arthaud and Matte 1975, 1977). Basin formation is controlled by active fault zones: either sinistral (the N-S-oriented Sillon Houiller, Fig. 1) (Gélard et al. 1986; Bonijoly and Castaing 1987); or dextral on E-W- to NE-SW-oriented wrench-faults in most cases (Arthaud and Matte 1977; Bonijoly and Castaing 1987).

Based on the microtectonic analysis in brittle structures, Bonijoly and Castaing (1987) related the strike-slip tectonics and associated basin subsidence to continuous compression and varying stress patterns during Late Carboniferous times which preceded a general N-S-oriented extension during the Permian. In contrast, Echtler et al. (1988), Echtler and Malavieille (1990) and Malavieille et al. (1990) used the evidence from two examples to propose that the Stephanian basins of the Massif Central resulted from generalized late orogenic crustal extension and thinning subsequent to orogenic crustal thickening. The tectonic features of the Graissessac and Saint-Etienne basins (Fig. 4) show that basin subsidence along faults was linked to the uplift of metamorphic terranes and tectonic denudation. During the early late

Carboniferous, widespread HT/LP metamorphism was associated with intense migmatization and the intrusion of anatectic granites. This metamorphism overprints deeper crustal structures (Pin 1989).

Ductile low-angle detachment and transcurrent wrench faulting constituted major extensional shear zones related to crustal thinning that took place essentially during the Westphalian. N-NE-trending elongate Stephanian basins formed parallel to such tectonic and metamorphic discontinuities, when deformation under progressive retrograde metamorphic conditions became shallow and brittle during continuous uplift and erosion. Coarse continental sediments were deposited along active, normal or strike-slip fault zones above the previously eroded hanging walls. Asymmetric synclines show progressive unconformities and thinned layers along the fault-bounded margins of the basins. Tectonic fabrics of synsedimentary and brittle deformation are generally coaxial to extensional features of ductile deformation in the extended substratum. This structural setting on the edge of high-grade and plutonic metamorphic complexes attests to basin formation which is controlled by low-angle normal faults and related to the collapse of the thick Variscan crust. The progressive superposition of brittle faulting on pre-existing ductile shear zones which are also observed in the Blanzy-Montceau basin (Castaing 1982; Vallé et al. 1988) and the Autun basin (Rolin 1987) as well as the geometry of deformation emphasise a general extensional evolution during the Late Paleozoic.

Permian clastics, mainly unconformably overlying Stephanian strata, were deposited in large asymmetric half-grabens controlled by steeply dipping, brittle, normal and strike-slip faults. This late general N-S extension (Bonnion 1983) marks the end of late orogenic crustal extension.

One of the most notable features of the late Paleozoic sedimentary history in the Massif Central is the essentially synchronous development of continental basins throughout the extensive Variscan belt (Fig. 5). Apart from some basins where sedimentation started during the early Stephanian or as an exception during latest Westphalian (Grangeon et al. 1968), the oldest deposits are generally of late Stephanian (B) age. This pattern is consistent with a relatively instantaneous extensional evolution immediately following the long period of Variscan orogenic shortening. Consequently, Stephano-Permian continental basins are important markers for the timing and mechanisms of the late- to post-orogenic tectonic evolution in the Massif Central.

2.2 Lithology and Stratigraphy

Stephano-Permian basin fills of the Massif Central show rather homogeneous coal-bearing detrital sedimentation in fluvio-lacustrine environments. The coarse detritus resulted from the erosion of the Variscan chain. Consequently, the sediments contain a wide range of lithics, comprising material of metamorphic, magmatic, and sedimentary origin of various ages and Late Carboniferous plutons (Courel 1987); this demonstrates the importance of uplift and erosion during the Late Paleozoic.

In the following, the deposits of the Blanzy-Montceau and Autun-Epinac basins (Fig. 4) are taken as representatives of the outstanding stratigraphic succession of continental sedimentation in the Massif Central. In addition, the Blismes basin in northern Morvan (Fig. 4) is briefly described as it is characterized by exceptionally important volcanic and magmatic activity.

1. The Blanzy-Montceau Basin. Stephanian deposits in the Blanzy-Montceau basin are estimated to be more than 1200 m in thickness (Feys and Greber 1958). The sedimentary environment is characterized by an irregularly subsiding lacustrine basin, which was episodically filled with torrentially transported coarse detritus (Vallé et al. 1988). Four lithological units can be distinguished from the base upwards:

- basal conglomerate (up to 400 m thick) containing crystalline and metamorphic meter-scale blocks and pebbles and, occasionally, Devono-Dinantian sediments;
- up to 600 m coal measures which consist of arkosic sandstones, siltstones, and fossiliferous dark shales with 7–10 coal seams (individual coal beds reach 35 m in thickness in some cases). Numerous cineritic horizons are intercalated;
- 250 m of arkosic sand- and siltstones with minor coal beds;
- 200 m of polygenic conglomerates changing laterally to sand- and siltstones without coal layers.

A Stephanian B-C age is attributed to these units and the occurrence of a Stephanian D is questioned according to macrofloral studies (Langiaux 1984).

The overlying Autunian succession which has a tansitional contact with the underlying top Stephanian is composed of more than 1000 m of alternating arkosic sandstones, conglomerates, and bituminous shales.

In addition to plant impressions present in shales, fish remains, tetrapod and invertebrate tracks occur at Montceau (Langiaux 1984). Furthermore, nodules

	WESTPHALIAN	STEPHANIAN				AUTUNIAN
		A	B	C	D	
AUTUN-EPINAC			+	?	+	+
BLANZY-MONCEAU			+	+	+	+
ST-ETIENNE		+	+	+	+	+?
CEVENNES	?	+	+			
GRAISSESSAC-LODEVE		?	+			+
ST-AFFRIQUE			+			+
CARMAUX		+	+			+
DECAZEVILLE-RODEZ		+?	+	+		+
ST PERDOUX-FIGEAC		+	+			+
BRASSAC			+			
BRIVE			+	+	+?	+
SILLON HOUILLER			+			
COMMENTRY			+	+		+
AUMANCE			+			+
DECIZE	+		+	+?		+?
BLISMES			+		+	+

Fig. 5. Stratigraphical distribution of the main Stephanian deposits in the Massif Central and relationship to the Westphalian and Autunian. (Doubinger and Vetter 1985; Vetter 1986; Courel 1987). N.B. New names and stratotypes have been proposed and adopted by the Subcommission on Carboniferous Stratigraphy to replace Westphalian A, B, C, and Stephanian A stages (Carboniferous Newsletter no. 7, 1989, pp 7–8)

containing exceptionally well preserved plants and animals (mainly arthropods, especially insects and vertebrates) have been recently discovered in the second unit (Pacaud and Sotty 1979; Heyler and Poplin 1988).

The stratigraphy of the Stephanian coal basins of the Massif Central has long been based on the subdivisions established in the former stratotypic basin of Saint-Etienne from the succession of macrofloras (Gand'Eury 1877). The use of miospore zonation has been developed more recently and, in combination with the study of cinerites, allows precise stratigraphic correlations (Bouroz and Doubinger 1978).

The early Stephanian (A) flora is characterized by the occurrence and subsequent extinction of a few species which are not known in younger strata. In contrast, the widespread plant assemblages of late Stephanian (B–C) age represent the classic Stephanian flora; about 30 very common species are found in all basins (Doubinger and Vetter 1985). The uppermost Stephanian shows a reduction in the number of species and corresponds to the Miospore zone NBM (Clayton et al. 1977).

Where detrital deposits are lying in apparent conformity on Stephanian C, which is the case in most of the easterly basins of the Massif Central (Vetter 1986), the Stephanian-Autunian transition is not clear and paleobotanical data do not always concur with those of sequential analysis or X-ray logging. For example, the exclusive use of first appearances of Callipteris and Walchia to characterize the Autunian has been questioned, since these plants are long known to occur in undisputed Stephanian deposits. It is established, however, that Stephanian and Autunian floral associations were coeval for a rather long period of time (Broutin et al. 1986). In other basins the Autunian is lying in disconformity on Stephanian B (Fig. 5).

2. The Autun Basin. The Autun basin corresponds to an E-W orientated half-graben overlying the Devonian-Lower Carboniferous Haut Morvan unit to the north (Marteau 1983). To the south, it is separated from the Luzy batholith by a major fault of Autunian age (Carrat 1969; Gélard 1978); the fault defines a wide and complex ductile shear zone (Rolin 1987). It covers and overlaps the underlying Stephanian B Epinac basin which shares the NE-SW orientation with Blanzy and Saint-Etienne (Fig. 4). Both superimposed basins exhibit the same orientation of progressive subsidence: basin depth increases from E to W during Stephanian through Autunian times.

Sediments consist mainly of fluviatile and lacustrine deposits with associated fossiliferous bituminous shales (Autunian stratotype: Munier-Chalmas and De Lapparent 1893). Four units can be distinguished from the lower to the upper parts:

- 250 m of coal measures with bituminous shales intercalated in a rhythmic argillaceous sandstone and local conglomeratic beds;
- 400 m of similar lithology but with fewer coal-bearing intervals. Bituminous shales contain an abundant and well preserved fish fauna.

Both units are attributed to the Lower Autunian on macrofloral contents (Doubinger 1956).

- The Upper Autunian begins with 250 m bituminous argilites which are overlain by
- the 250-m-thick Boghead Group characterized by argillaceous and carbonaceous laminites associated with a well-known algal bed. The uppermost Autunian is composed of hematised arkose.

No Saxonian deposits are recognized in this basin (Falke and Feys 1965) but are known from other basins that are mainly situated in the southwestern Massif Central, e.g., the Saint-Affrique basin (3000 m), Rodez and Lodève (1500 m). There, the sediments consist of alternating coarse detritus, argilites, and siltstones with rare trace fossils (Vetter 1971). Red beds, already present in the underlying Autunian lacustrine facies, become predominant in the Saxonian. Disappearance of coal deposits and of certain plant assemblages at the Autunian-Saxonian transition may result from a major climatic change (Odin et al. 1986; Courel 1987).

As far as the Stephanian-Autunian transition in the Autun basin is concerned, there has been much debate regarding the use of palynostratigraphy (Doubinger and Elsass 1979; Farjanel et al. 1985) and the interpretation of occurrences of Stephanian-like assemblages including silicified plants within the Autunian succession (Galtier 1980, Broutin et al. 1986). It has been suggested that the Autunian should be characterized by a distinct type of sedimentation related to extensive tectonics (Farjanel et al. 1985).

3. The Blismes Basin. The Blismes caldera provides a good example of Late Paleozoic volcanism on the northernmost edge of the belt. This E-W orientated elliptical caldera shows two outer grabens separated by a NE-SW orientated central horst (Guérin and Monnier 1984). The volcaniclastica formations, 200 to 800 m thick, incorporate coal-bearing sandstone-siltstone sequences. The numerous pyroclastic bands are composed of tuffs, lahars, interbedded reworked volcanics, and minor ash-flows, yielding an age of 291 ± 7 Ma (fission track in zircon by Carpena et al. 1984). The two major volcano-sedimentary sequences of the basin are of late Stephanian (B–C) and late Autunian age respectively. A thick Saxonian ignimbritic unit (265 ± 5 Ma, Carpena et al. 1984) covers the caldera infilling. The volcanics are quartz-latites with a subalkaline-potassic S-type composition (Monnier and Stussi 1984).

The last phase of Paleozoic sedimentation in the Massif Central is characterized by the filling of intramontane troughs in Stephano-Permian times. Clastics derive from the erosion of the Variscan belt during peneplanation and uplift processes. Volcanic activity and sedimentation in asymmetric half-grabens account to a general extensional evolution at the end of the Variscan orogeny.

References

Alsac C, Cabanis B, Guérangé-Lozes J, Béziat D (1987) Caractères magmatiques du volcanisme basique ordovicien de la nappe de Saint-Salvi-de-Carcavès dans l'Albigeois cristallin (Tarn-Aveyron, France). C R Acad Sci 305: 1199–1205

Arthaud F (1970) Etude tectonique et microtectonique comparée de deux domaines hercyniens: les nappes de la Montagne Noire (France) et l'anticlinorium de l'Iglesiente (Sardaigne). Thesis, University Montpellier, 175 pp

Arthaud F, Matte Ph (1975) Les décrochements tardi-hercyniens du sud-ouest de l'Europe. Geometrie et essai de reconstitution des conditions de la déformation. Tectonophysics 25: 139–171

Arthaud F, Matte Ph (1977) Late Paleozoic strike-slip faulting in Southern Europe and Northern Africa: result of right lateral shear zone between the Appalachians and the Urals. Geol Soc Am Bull 88: 1305–1320

Arthaud F, Mattauer M, Proust F (1966) La structure et la microtectonique des nappes hercyniennes de la Montagne Noire. Coll Etages tectoniques, Neuchâtel, pp 231–243

Autran A, Cogné J (1980) La zone interne de l'orogène varisque dans l'Ouest de la France et sa place dans le développement de la chaîne hercynienne. Mém Bur Rech Geol Min 108: 90–111

Bard J-P, Loueyit J (1978) Sur l'origine des gneiss oeillés de l'Espinouse dans la zone axiale de la Montagne Noire (Massif Central): consequences tectoniques. Bull Soc Geol Fr 20: 751–772

Bard J-P, Rambeloson R (1973) Metamorphisme plurifacial et sens de variation du degré géothermique durant la tectogenèse polyphasée hercynienne dans la partie orientale de la zone axiale de la Montagne Noire (Massif du Caroux, sud du Massif Central Français). Bull Soc Geol Fr 15: 579–586

Beaud F (1985) Etude structurale de la Zone Axiale orientale de la Montagne Noire (Sud du Massif Central Français). Détermination des mécanismes de déformation. Relation avec les nappes du Versant Sud. Thesis, University Montpellier, 191 pp

Bebien J (1970) Etude pétrographique et géochimique des formations volcaniques dévono-dinantiennes de l'extrémité sud-ouest du faisceau synclinal du Morvan (Massif Central français). Thèse 3e cycle, Nancy, 110 pp

Beben J, Gagny C (1980) Volcanites du Précambrien au Crétacé et leur signification géostructurale. Mém Bur Rech Geol Min, Orléans 107: 99–135

Becq-Giraudon J-F (1973) Etude géologique du bassin houiller de Graissessac (Hérault). Bull Bur Rech Geol Min, Orléans 3: 151–163

Bérard P (1986) Trilobites de l'Ordovicien inférieur des Monts de Cabrières. Mém Cent Etud Rech Géol Hydrol Montpellier 24, 220 pp

Bergeron J (1889) Etude géologique du massif ancien situé au sud du Plateau Central. Ann Soc Géol Paris 22, 362 pp

Berthaux J (1982) Origine métamorphique des grenats des volcanites acides d'âge Viséen supérieur dans le nord-est du Massif Central français. Bull Minéral 105: 212–222

Berthaux J, Rubiello MF (1981) Cartographie de la feuille géologique de Roanne au 1/50000. Pétrologie, géochimie et pétrologie structurale du magmatisme Viséen supérieur. Essai de synthèse régionale. Thèses de 3e cycle, Nancy

Böhm R (1935) Etudes sur les faunes du Dévonien supérieur et du Carbonifère de la Montagne Noire. Thesis, University Montpellier, 203 pp

Bogdanoff S (1973) Contribution à l'étude géologique de l'extrémité orientale de la zone axiale granito-gneissique de la Montagne Noire (Monts de l'Espinouse, Hérault). Bull Bur Rech Geol Min (Orléans) 1: 28

Bonijoly D, Castaing C (1987) Ouverture et évolution structurale de quelques bassins houillers de directions orthogonales, dans le Massif Central français. Ann Soc Géol Nord 106: 189–200

Bonnion S (1983) Structuration du bassin houiller de l'Aumance (Allier). Analyse structurale des dépots de charbon et de stériles, géophysique. Thesis, University Dijon, 303 pp

Bouroz A, Doubinger J (1978) La position stratigraphique du Faisceau de la Talaudière et du Poudingue mosaïque dans le Carbonifère de Saint-Etienne. Rev Ind Min 60: 485–492

Boyer F, Krylatov S, Lefèvre J, Stoppel D (1968) Le Dévonien supérieur et la limite dévono-carbonifère en Montagne Noire (France). Lithostratigraphie – biostratigraphie (conodontes). Bull Centr Rech S N P A Pau 2: 5–33

Broutin J, Doubinger J, Langiaux J, Primey D (1986) Conséquences de la coexistence de flores à caractères stéphaniens et autuniens dans les bassins limniques d'Europe occidentale. Mém Soc Géol Fr 149: 15–25
Burg JP, Matte Ph (1978) A cross-section through the French Massif Central and the scope of its Variscan geodynamic evolution. Z Dtsch Geol Ges 129: 429–460
Burg JP, Leyreloup A, Marchand J, Matte Ph (1984) Inverted metamorphic zonation and large-scale thrusting in the Variscan belt: an example in the French Massif Central. In: Hutton D, Sanderson D (eds) Variscan tectonics of the North Atlantic region. Blackwell, Oxford, pp 47–61
Capéra JC, Courtessole R (1974) Contribution à la paléontologie et à la stratigraphie du Cambrien moyen et du Tremadoc inférieur des monts de Lacaune. Bull Soc Hist Nat Toulouse 110: 93–101
Carpéna J, Doubinger J, Guérin R, Juteau J, Monnier M (1984) Le volcanisme acide de l'Ouest-Morvan dans son cadre géologique: caractérisation géochimique, structurale et chronologique de mise en place. Bull Soc Géol Fr 5: 839–859
Carrat HG (1969) Le Morvan cristallin. Thesis, University Nancy
Castaing C (1982) Gîtologie prévisionnelle des charbons dans le Nord-Est du Massif Central. Rapport Bur Rech Geol Min (Orléans) 82 SGN 326 GEO
Centène A (1977) Nouvelles données biostratigraphiques (conodontes) sur le Silurien des Monts de Lacaune (Montagne Noire, France). 5e Reunich Annuelles des Sci de la Terre, Rennes, Soc Géol Fr, p 140
Chaubet M-C (1937) Contribution à l'étude géologique du Gothlandien du versant méridional de la Montagne Noire. Trav Lab Géol, University Montpellier, 223 pp
Chichery M (1938) Stratigraphie des formations dinantiennes du bassin de l'Ardoisière. C R Acad Sci 206: 263
Clayton G, Coquel R, Doubinger J, Gueinn K, Loboziak S, Owens B, Streel M (1977) Carboniferous miospores of Western Europe: illustration and zonation. Meded Rijks Geol Dienst 29: 1–71
Costa S, Maluski H (1988) 39Ar-40Ar dating for terranes boundaries definition. The example of the French Massif Central. IGCP 233 Int Conf, Montpellier (Abstr)
Coulon M, Corsin P, Doubinger J (1980) Données nouvelles sur la stratigraphie du Dévono-dinantien du Morvan. Bull Soc Hist Nat et Amis Mus Autun 95: 3–12
Courel L (1987) Les bassins houillers du Paléozoique supérieur continental du Massif Central français; dynamique sédimentaire. Ann Soc Géol Nord 106: 229–237
Courjault-Radé P (1988 a) Proposition d'un profil paléogéographique au Cambrien du versant sud (nappes de Pardailhan et du Minervois) de la Montagne Noire (France). C R Acad Sci 307: 1917–1923
Courjault-Radé P (1988 b) Analyse sédimentologique de la formation de l'Orbiel ("alternances gréso-calcaires" auct., Cambrien inférieur). Evolution tectono-sédimentaire et climatique (versant sud de la Montagne Noire, Massif Central, France). Bull Soc Géol Fr 6: 1003–1013
Courtessole R (1973) Le Cambrien moyen de la Montagne Noire (biostratigraphie). Publ Lab Géol Toulouse, 238 pp
Courtessole R, Pillet J, Vizcaïno D, Eschard R (1985) Etude biostratigraphique et sédimentologique des formations arenacées de l'Arenigien du Saint Chinianais oriental (Hérault), versant sud de la Montagne Noire (France méridionale). Mém Soc Étud Sci Aude, Carcassonne, 99 pp
Dean WT (1966) The Lower Ordovician stratigraphy and trilobites of the Landeyran Valley and the neighbouring district of the Montagne Noire, south-western France. Bull Br Mus Geol 12: 247–353
De Book F (1982) Présence de Chitinozoaires dans le passage siluro-dévonien de la Montagne Noire sud-orientale (Hérault, France). Géobios 15: 845–871
Debrenne F, Courjault-Radé P (1986) Découverte de faunules d'Archéocyathes dans l'Est des monts de Lacaune, flanc nord de la Montagne Noire. Implications biostratigraphiques. Bull Soc Géol Fr 2: 285–292
Delfour J, Gigot Cl (1985) Données stratigraphiques apportées par l'étude de conodontes du Dévono-Dinantien de la région de Boubon-Lancy (Saône et Loire). Geol Fr Bur Rech Geol Min 4: 331–340
Delfour J, Isnard P, Lecuyer E, Lemiere B, Lhote F, Moine B, Piboule M, Picot P, Ploquin A, Tegyey M (1984) Etude du gite de pyrite de Chizenil (Saône et Loire) et de son environment volcano-sédimentaire dévonien et dinantien. Doc Bur Rech Geol Min (Orléans) 73, 142 pp
Demange M (1975) Style pennique de la zone axiale de la Montagne Noire entre Saint-Pons et Murat-sur-Vébre (Massif Central). Bull Bur Rech Geol Min 2: 91–139
Demange M (1982) Etude géologique de Massif de l'Agout, Montagne Noire, France. Thesis, University Paris, 647 pp
Donnot M, Guérangé B (1978) Le synclinorium de Brusque; implications stratigraphiques et structurales dans les Monts est de Lacaune (Tarn, Aveyron, Hérault). Bull Bur Rech Geol Min 4: 333–363
Doubinger J (1956) Contritubtion à l'étude des flores autuno-stéphaniennes. Mém Soc Géol Fr (Paris) 75, 180 pp
Doubinger J, Elsass F (1979) Le bassin permo-carbonifère d'Autun: nouvelles données stratigraphiques et palynologiques. Bull Soc Hist Nat Autun 91: 9–24
Doubinger J, Vetter P (1985) Le Stéphanien en France. C R X. Congr Strat Geol Carboniferous I. Inst Geologico y Minero de España, Madrid, pp 45–54
Dreyfuss M (1948) Contribution à l'étude géologique et paléontologique de l'Ordovician supérieur de la Montagne Noire. Mém Soc Géol Fr (Paris) 58, 63 pp
Echtler H (1990) Geometry and kinematics of recumbent folding and low-angle detachment in the Pardailhan nappe (Montagne Noire, Southern French Massif Central). In: Matte Ph (ed) Terranes in the variscan Belt of Europe and Circum-Atlantic Palaeozoic Orogens. Teconophysics 177: 109–123
Echtler H, Malavielle J (1989) Basement uplift, extensional tectonics and collapse basin emplacement in the Montagne Noire. EUG V, Terra Abstracts, Strasbourg, p 255 (Abstr)
Echtler H, Malavieille J (1990) Extensional tectonics, basement uplift and Stephano-Permian collapse basin in a late Variscan metamorphic core complex (Montagne Noire, Southern Massif Central). In: Matte Ph (ed) Terranes in the variscan Belt of Europe and Circum-Atlantic Palaeozoic Orogens. Teconophysics 177: pp 125–138
Echtler H, Malavieille J, Matte Ph, Brunel M (1988) Kinematics of late orogenic uplift, extension and wrenching in the Montagne Noire (southern French Massif Central). IGCP 233 Int Conf Montpellier. Univ Montpellier (Abstr)
Ellenberger F (1967) Replis de micaschistes et tectonique d'infrastructure au sein du massif gneissique de Caroux (zone axiale de la Montagne Noire). C R Soc Géol Fr (Paris), pp 227–228
Engel W (1984) Migration of folding and flysch sedimentation on the southern flank of the Variscan Belt (Montagne Noire, Mouthoumet Massif, Pyrenees). Z Dtsch Geol Ges 135: 279–292
Engel W, Feist R, Franke W (1978) Synorogenic gravitational transport in the Carboniferous of the Montagne Noire (S France). Z Dtsch Geol Ges 129: 461–472
Engel W, Feist R, Franke W (1982) Le Carbonifère an-

testéphanien de la Montagne Noire: rapports entre mise en place des nappes et sédimentation. Bull Bur Rech Geol Min 1980–81, 4: 341–389

Falke H, Feys R (1965) Il n'y a pas de véritable grès rouges ni de Saxonien dans le Permien du bassin d'Autun. Bull Serv Carte Géol Fr 278/61: 101–106

Farjanel G, Feys R, Marteau P (1985) Le Stéphanien et l'Autunien du bassin d'Autun (France). Nouvelles données stratigraphiques. C R X. Int Congr Carboniferous Madrid Proc 4: 487–492

Feist R (1985) Devonian stratigraphy of the southeastern Montagne Noire (France). Cour Forschungsinst Senckenb 75: 331–352

Feist R (1986) Late Cambrian trilobite biogeography of the western mediterranean area: relationship between Europe and Gondwana. Int Conf Iberian Terranes, IGCP 23 Conf Abstr, Univ of Oriedo, p 50

Feist R, Courtessole R (1984) Découverte de Cambrien supérieur à trilobites de type est-asiatique dans la Montagne Noire (France méridionale). C R Acad Sci 298: 177–182

Feist R, Flajs G (1987) La limite Dévonien-Carbonifère dans la Montagne Noire (France). Biostratigraphie et environnement. C R Acad Sci 305: 1537–1544

Feist R, Galtier J (1985) Découverte de flores d'âge Namurien probable dans le flysch à olistolites de Cabrières Hérault). Implication sur la durée de la sédimentation synorogénique dans la Montagne Noire (France méridionale). C R Acad Sci 300: 207–212

Feist R, Groos-Uffenorde H (1979) Die "calcaires à polypiers siliceux" und ihre Ostracoden-Faunen (Oberes Unter-Devon, Montagne Noire, S-Frankreich). Senckenb. Lethaea 60: 83–187

Feist R, Klapper G (1985) Stratigraphy and conodonts in pelagic sequences across the Middle-Upper Devonian boundary, Montagne Noire, France. Palaeontographica 188: 1–18

Feist R, Schönlaub H-P (1974) Zur Silur-/Devon-Grenze in der östlichen Montagne Noire Südfrankreichs. Neues Jahrb Geol Paläontol 4: 200–219

Feist R, Schönlaub H-P, Bultynck P (1985) Faciès et biostratigraphie (conodontes) du passage Dévonien inférieur-moyen dans la Montagne Noire (France). Hercynica 2: 81–97

Feybesse JL, Lardeaux JM, Johan V, Tegyey M, Dufour E, Lemiere B, Delfour J (1988) La série de la Brevenne (Massif Central français): une unité dévonienne chariée sur le complexe métamorphique des Monts du Lyonais à la fin de la collision varisque. C R Acad Sci Paris 307 (2): 991–996

Feys R, Greber Ch (1958) Le bassin houiller de Blanzy et du Creusot. Rev Ind Minéral 40: 490–530

Flajs G, Feist R (1988) Index conodonts, trilobites and environment of the Devonian-Carboniferous Boundary beds at La Serre (Montagne Noire, France). Cour Forschungsinst Senckenb 100: 53–107

Fournier-Vinas Ch, Debat P, (1970) Présence de micro-organismes dans les terrains métamorphiques précambriens (schistes X) de l'Ouest de la Montagne Noire. Bull Soc Géol Fr 12: 351–355

Fournier-Vinas Ch, Donnot M (1977) Présence d'Acritarches dans l'Ordovicien inférieur et la Cambrien des monts de l'est de Lacaune (Aveyron). Versant nord de la Montagne Noire. Bull Bur Rech Geol Min 2: 121–129

Franke W, Engel W (1986) Synorogenic sedimentation in the Variscan belt of Europe. Bull Soc Géol Fr 1: 25–33

Frech F (1887) Die Paläozoischen Bildungen von Cabrières (Languedoc). Z Dtsch Geol Ges 39: 360–487

Galtier J (1970) Recherches sur les végétaux à structure conservée du Carbonifère inférieur français. Paleobiol contin I (4), Montpellier, 221 pp

Galtier J (1980) Les végétaux silicifiés du Permien d'Autun: analyse quantitative et interprétation possible. Bull Soc Hist Nat Autun 95: 35–39

Galtier J, Meyer-Berthaud B, Rowe NP (1988) Tournaisian plants from the "Lydiennes" Formation of the Montagne Noire (France). Cour Forschungsinst Senckenb 100: 109–117

Gélard JP (1978) La fracturation de la Bourgogne méridionale: essai d'interprétation et implications tectoniques régionales. Rev Géol Dyn Géogr Phys 20: 5–28

Gélard JP, Castaing C, Bonijoly D, Grolier J (1986) Structure et dynamique de quelques bassins houillers limniques du Massif Central. Mém Soc Géol Fr 149: 57–72

Gèze B (1949) Etudes géologiques de la Montagne Noire et des Cévennes méridionales. Mém Soc Géol Fr (Paris) 62, 215 pp

Gonord H, Ragot J-P, Saugy L (1964) Observations lithostratigraphiques nouvelles sur la série de base (Ordovicien inférieur) des nappes de Cabrières, région de Gabian-Glauzy (Montagne Noire, Hérault). Bull Soc Géol Fr 6: 419–427

Grand'Eury C (1877) Flore carbonifère du département de la Loire et du centre de la France. Mém Acad Sci II: 511–517

Grangeon M, Feys R, Greber Ch, Lefavrais-Raymond A (1968) Géologie profonde de la région de Decize (Nièvre). Essai de synthèse d'après les sondages profonds. Bull Bur Rech Geol Min 4: 43–108

Guérangé-Lozes J (1987) Les nappes varisques de l'Albigeois cristallin. Lithostratigraphie, volcanisme et déformations. Doc Bur Rech Geol Min (Orléans) 135, 259 pp

Guérangé-Lozes J, Burg J-P (1990) Les nappes varisques du sud-ouestdu Massif Central (cartes géologiques et structurales 1/250 000, Montpellier et Aurillac). Géol France 3–4: 71–106

Guérangé-Lozes J, Guérangé B (1984) Les nappes de charriage de l'Albigeois cristallin (Tarn et Aveyron, France). C R Acad Sci 299: 471–474

Guérin R, Monnier M (1984) Relation entre histoires structurale et volcanisme dans le complexe volcanique acide stéphano-permien du Morvan occidental (Blismes-Montreuillon, Massif Central français). C R Acad Sci 299: 37–42

Guffroy J (1964) Sur L'existence du Givétien dans le Morvan. C R Acad Sci Paris 258: 2619–2620

Guffroy J, Pillet J (1966) Découverte d'un nouveau gisement de trilobites dans le Morvan. C R Soc Geol Fr 6: 236–237

Guillot PL, Doubinger J (1971) Découverte d'acritarches dans les schistes supérieurs de Génis (Dordogne). C R Acad Sci Paris 272: 2763–2764

Guillot PL, Lefevre J (1975) Découverte de conodontes dans le calcaire à entroques de Génis en Dordogne (série métamorphique du Bas-Limousin). C R Acad Sci Paris 280: 1529–1530

Henry J-L (1989) Paléoenvironnements et dynamique de faunes de trilobites dans l'Ordovicien (Llanvirn supérieur – Caradoc basal) du Masif Armoricain (France). Palaeogeogr Palaeoclimatol Palaeoecol 73: 139–153

Heyler D, Poplin CM (1988) The fossils of Montceau-les-Mines. Sci Am 256: 104–110

Julien A (1896) Le terrain carbonifère marin de la France Centrale. Masson, Paris, 304 pp

Jung J, Raguin E (1935) Discordance du Viséen sur le socle cristallophyllien entre Balbigny, Méronde, Violay (Loire). C R Soc Géol Fr 16: 247–249

Lacroix P (1980) Etude déologique du groupe de Violay (Rhône). Contribution à l'étude des granites sodiques et des formations éruptives et volcano-sédimentaires associées. Thesis Lyon, 139 pp

Langiaux J (1984) Flores et Faunes des formations supérieures du Stéphanien de Blanzy-Montceau. Stratigraphie et Paléoé-

cologie. Rev La Physiophile, Soc Et Sci Nat H ist Montceau 100, 270 pp
Laville P (1986) La série carbonatée dans la région de Gilly-sur-Loire (SW du Morvan-Frasnien-Famennien). D E A Thesis, University Dijon, 23 pp
Leistel JM, Gagny Cl (1984) Mise en évidence d'une fosse volcano-tectonique au Viséen supérieur dans le Nord-Est du Massif Central Français. Rev Géol Dyn Géogr Phys 25: 17–29
Lépine J, Ben Ayad MA, Béziat D (1988) Mise en évidence de métabasites alcalines Cambrien inférieur dans le district aurifère de Salsigne (Versant Sud-Ouest de la Montagne Noire). C R Acad Sci 306: 1087–1092
Lorenz V, Nicholls IA (1976) The Permocarboniferous basin and range province of Europe. An application of plate tectonics. – In: Falke H (ed) The continental Permian in Cetral, West and South Europe. NATO Adv Study Inst Ser C, Math Phys Sci 22: 313–342
Lorenz V, Nicholls IA (1984) Plate and intraplate processes of Hercynian Europe during the Late Paleozoic. Tectonophysics 107: 25–56
Lys M, Serre B, Mauvier A, Grekopf N (1961) Contribution à la connaissance des microfaunes du Plaéozoique. Etudes microplaéontologiques (conodontes, ostracodes) dans le Dévonien du Morvan. Rev Inst Fr Petrol 16 (5): 530–567
Malavieille J, Guilhot P, Costa S, Lardeaux JM, Gardien V (1990) Collapse of a thickened Variscan crust in the French Massif Central: Mont Pilat extensional shear zone and St. Etienne Upper Carboniferous basin. In: Matte Ph (ed) Terranes in the variscan Belt of Europe and Circum-Atlantic Palaeozoic Orogens. Tectonphysics 177: 139–149
Mamet B (1968a) Sur quelques microfaciès carbonifères du Morvan et du Forez. Bull Bur Rech Geol Min (Orléans) 1 (2): 57–62
Mamet B (1968b) Sur les microfaciès calcaires du Viséen de la Montagne Noire (France). Rev Micropal 3: 121–136
Marignac C, Leroy J, Macaudière J, Pichavant M, Weisbrod A (1980) Evolution tecto-métamorphique d'un segment de l'orogène hercynien: les Cévennes médianes, Massif Central français. C R Acad Sci 291: 605–608
Marteau P (1983) Le bassin permo-carbonifère d'Autun. Stratigraphie, sédimentologie et aspects structuraux. Doc Bur Rech Geol Min (Orléans) 64, 198 pp
Martin F (1972) Les Acritarches de l'Ordovicien inférieur de la Montagne Noire (Hérault, France). Bull Inst R Sci Nat Belg 10: 1–61
Matte Ph (1983) Two geotraverses across the Ibero-Armorican Variscan Arc of Western Europe. Profiles and Orogenic Belts. Geodyn Ser 10: 53–81
Matte Ph (1986) Tectonics and plate tectonics model for the Variscan Belt of Europe. Tectonophysics 126: 329–374.
Maurel M (1966) Etudes géologiques sur le Dévonien et le Carbonifère inférieur du versant méridional de la Montagne Noire. Thesis, University Montpellier, 194 pp
Michel-Levy A (1908) Les terrains primaires du Morvan et de la Loire. Bull Serv Carte Géol Fr 120: 193–489
Monnier M, Stussi JM (1984) Le bassin de Montreuillon (Nord Morvan): éléments d'une association volcano-plutonique subalcaline hercynienne. Données géochimiques. C R Acad Sci 298: 601–604
Munier-Chalmas E, De Lapparent A (1893) Note sur la nomenclature des terrains sédimentaires. Bull Soc Géol Fr 21: 451
Odin B, Doubinger J, Conrad G (1986) Attribution des formations détritiques rouges du Permien du Sud de la France au Thuringien d'après l'étude du bassin de Lodève: implications géologiques, paléontologiques et paléoclimatiques. C R Acad Sci 302: 1015–1020
Pacaud JG, Sotty AD (1979) Inventaire de la collection Sotty. Bull Soc Hist Nat Autun 90, 36 pp
Peterlongo JM (1960) Les terrains cristallins des Monts du Lyonnais (Massif Central, France). Ann Fac Sci Clermont-Ferrand, 4, 187 pp
Piboule M, Briand B, Beurrier M (1982) Géochimie de quelques granites albitiques dévoniens de l'est du Massif Central français. Neues Jahrb Miner Abh 143, 3: 279–308
Pin C (1989) Essai sur la chronologie et l'évolution géodynamique de la Chaine Hercynienne d'Europe. Thesis, University Clermont-Ferrand, 470 pp
Prian JP (1980) Les prophyrites cambriennes du versant septentrional de la Montagne Noire, au Sud du bassin permien de Camarés (Aveyron). Thesis, University Paris, 407 pp
Raymond D (1987) Le Dévonien et le Carbonifère inférieur du sud-ouest de la France (Pyrénées, Massif de Mouthoumet, Montagne Noire): sédimentation dans un bassin flexural en bordure sud de la chaîne de collision varisque. Geol Rundsch 76: 795–803
Reitz E, Wickert F (1988) Existence de microfossiles d'âge Protérozoique supérieur dans des phyllites épizonales de l'unité de la Brévenne, NE du Massif Central (France). C R Acad Sci Paris 307, II: 1717–1721
Rex GM (1986) The preservation and palecology of the Lower Carboniferous silicified plant deposits at Esnost, near Autun, France. Géobios 19, 6: 773–800
Robelin C (1981) Le Dévono-Dinantien dans la région de Bourbon-Lancy (SW Morvan-Massif Central français). Enchainement des phénomènes sédimentaires et magmatiques. Thesis 3e cycle, University Dijon, 159 pp
Rolin P (1987) Tectogénèse comparée de deux segments de la chaine hercynienne: le Massif Central français septentrional et la Meseta marocaine septentrionale. Thesis, University Orsay, 484 pp
Schuiling RD (1960) Le dôme gneissique de l'Agout (Tarn-Hérault). Mém Soc Géol Fr (Paris) 39, 59 pp
Scott AC, Galtier J, Clayton G (1984) Distribution of anatomically preserved floras in the Lower Carboniferous in Western Europe. Trans R Soc, Edingburgh 75: 311–340
Sdzuy K (1958) Fossilien aus dem Tremadoc der Montagne Noire. Senckenb Lethaea 39: 255–288
Sider H, Ohnenstetter M (1986) Field and petrological evidences for the development of an ensialic marginal basin related to the Hercynian orogeny in the Northeastern Massif Central. Geol Rundsch 75, 2: 421–443
Thoral M (1935) Contribution à l'étude géoloique des monts de Lacaune et des terrains cambriens et ordoviciens de la Montagne Noire. Bull Serv Carte Géol Fr (Orléans) 318 pp
Thoral M (1946) Cycles géologiques et formations nodulifères de la Montagne Noire. Nouv Arch Muséum Hist Nat Lyon 1, 103 pp
Tucker ME (1974) Sedimentology of Plaeozoic limestones: the Devonian Griotte (Southern France) and Cephalopodenkalk (Germany). Oxford Spec Publ Int Ass Sediment 1: 71–92
Vachard D (1974) Contribution à l'étude stratigraphique et micropaléontologique (Algues et Foraminifères) du Dévonien-Carbonifère inférieur de la partie orientale du versant méridional de la Montagne Noire (Hérault, France). Thesis, University Paris VI, 408 pp
Vachard D (1977) Etude stratigraphique et micropaléontologique (Algues et Foraminifères) du Viséen de la Montagne Noire (Hérault, France). Mém Inst Géol Univ Louvain 29: 111–195
Vallé B, Courel L, Gélard JP (1988) Les marqueurs de la tectonique synsédimentaire et syndiagénétique dans le bassin

stéphanien à régime cisaillant de Blanzy-Montceau (Massif Central, France). Bull Soc Géol Fr 4: 529–540

Vetter P (1971) Le Carbonifère supérieur et le Permien du Massif Central. Sympos Jung, Plein-Air Service, Clermont-Ferrand, pp 169–213

Vetter P (1986) Les formations limniques du Carbonifère supérieur et le l'Autunien en France. Mém Soc Géol Fr 149: 7–14

Vignard G (1976) Mise en évidence du Viséen dans la région de Saint-Pons. Conséquences sur l'interprétation tectonique des séries réputées autochtones de l'unité du Saint-Ponais (versant sud de la Montagne Noire, France). C R Acad Sci 283: 607–610

Wickert F (1985) Geologische Untersuchungen im NE-Teil des Massif Central (Frankreich) unter besonderer Berücksichtigung der Metamorphose und Tektonik der Brévenne-Serie. Thesis, University Freiburg 123 pp

Wickert F (1986) Ein Beitrag zur Metamorphose-Geschichte der Brévenne-Serie, nordöstliches Massif Central (Frankreich). Neues Jahrb Geol Pal Mh 10: 626–640

Wickert F (1988) Palaeozoic evolution of continental crust in the Beaujolais-Lyonnais-area, northeastern part of the Massif Central, France. Geol Rundsch 77, 2; 67–482

Structure

P. LEDRU, S. COSTA, and H. ECHTLER

1 Introduction

The Variscan orogeny resulted from the continental convergence and collision between Gondwana and Laurasia. In the French Massif Central, convergence and crustal shortening began as early as 450 Ma with a phase of lithospheric subduction (Burg and Matte 1978; Autran and Cogne 1980; Bard et al. 1980; Brun and Burg 1982; Autran 1984; Matte 1986) and ended around 320 Ma. This was followed by extensional processes that affected the previously thickened crust.

The climax of collisional tectonics occurred during Late Devonian and Early Carboniferous times, with overthrusting of the main deep crustal nappes in the internal zone. The observed structures and microstructures show the deformation of the geological units during their tectonic passage from the site of deposition to the present location.

The collisional evolution of the Variscan orogeny involved different tectonic units in the main thrusts at various periods and with varying deformational intensity, and evidence of polyphase structures, non-coaxial deformation and metamorphism implies that the creation of the belt was a progressive process. This evolution is recognizable in both the chronology and geometrical position of the units, and shows that the tectonic style evolved from thrusting to major transcurrent faulting which gave rise to transtensional and transpressional areas in juxtaposed parts of the orogenic belt (for instance the Marches Combrailles, Sillon Houiller fault zones; Fig. 1). Major regional antiforms are controlled by the intrusion of large granitic domes (such as Millevaches, Velay, and the axial zone of the Montagne Noire) associated with extensional tectonics during Late Carboniferous and Permian times. The formation of detrital and coal-bearing basins marks this late orogenic evolution.

The definition of the nappes (Fig. 2) takes into account the lithological content, their age and site of emplacement, and their structural and metamorphic characteristics. Several maps have been published of the Massif Central (Burg and Matte 1978; Autran et al. 1980; Autran 1984; Matte 1986) and recently Ledru et al. (1989) defined the following major nappe complexes:

1. The Upper Epizonal Unit (the upper allochthon) located in restricted peripheral areas in the northeast and southwest of the Massif Central. This unit (comprising the Thiviers Payzac, Génis and Brevenne nappes) is considered to have overthrust other tectonic units under epizonal metamorphic conditions.
2. The Upper Gneiss Unit (the middle allochthon) which has a long and complex tectonic history reflected by numerous eclogite and granulite relicts in the so-called leptynite-amphibolite complex (Burg and Matte 1978). It mainly represents peri-arc basin deposits of Early Paleozoic age (Pin and Lancelot 1982), including locally dismembered ophiolitic suites (Dubuisson et al. 1989).
3. The Lower Gneiss Unit (the lower allochthon), which contains mainly continental-rise metasedimentary rocks of Late Upper Proterozoic to Early Paleozoic age, and some products of Early Ordovician magmatism.

These first three units constitute the so-called internal zone of the Massif Central. The main difference between this concepts and earlier nappe concepts (Burg et al. 1984; Matte 1986), is the distinction of the lower allochthon which had been considered mainly as an autochthonous unit.

4. The southern Paleozoic nappe complexes, which apparently formed the southern margin that was overthrust by the first three units. They constitute the external zone of the Massif Central and were involved in the thickening processes during the Carboniferous, but the metamorphism was not as severe as in the internal zone. The sedimentary record (Feist et al. Chap. III.3 this Vol.) shows a southward progression of the orogenic front.

The related tectonic evolution of the three allochthons of the internal zone started during the Silurian and continued during the Devonian (Pin and Peucat 1986). It is marked by the thrusting and a polyphase tectonometa-

J. D. Keppie (Ed.): Pre-Mesozoic Geology
in France and Related Areas

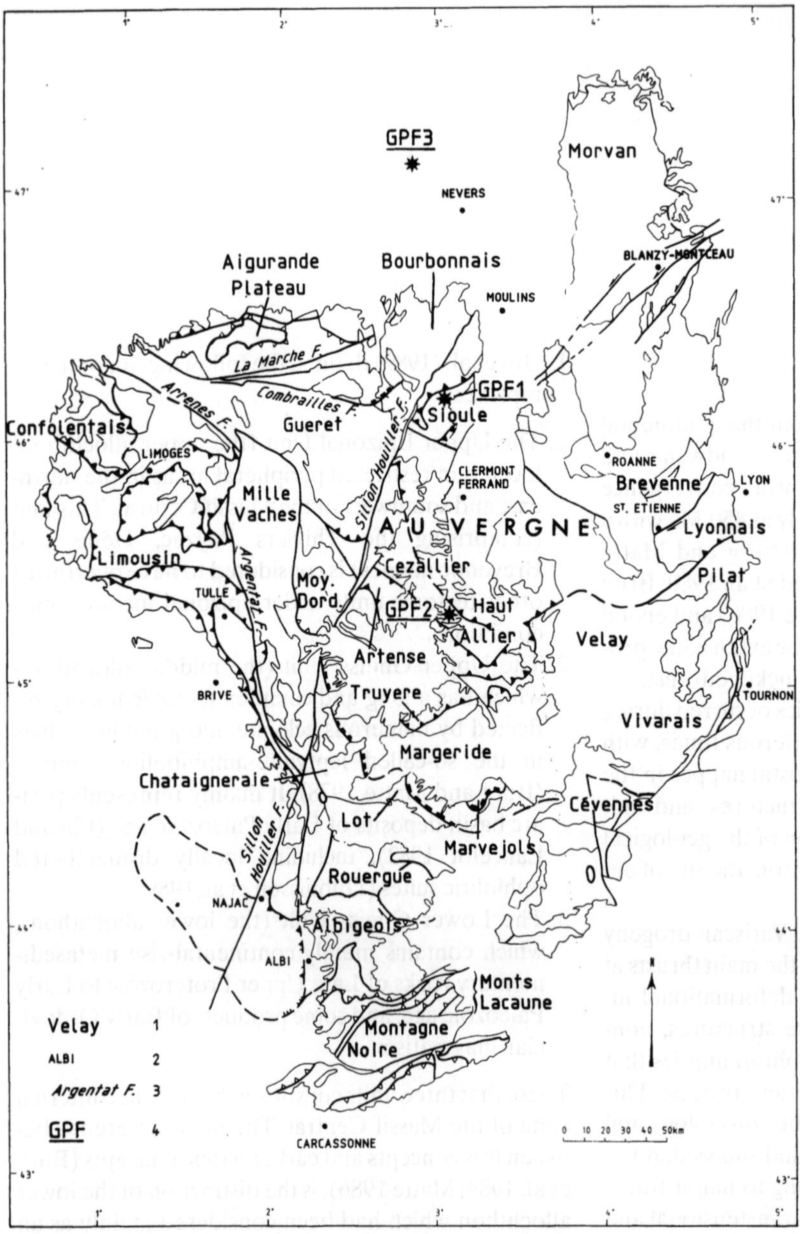

Fig. 1. Location map of the main regions of the French Massif Central. *1* Region; *2* town; *3* fault; *4* deep drill core program site (*1* Echassières granite; *2* Cézallier; *3* magnetic anomaly of the Paris basin)

morphic evolution. Where present, the Late Paleozoic sequences were not involved in the tectonics. The units of the internal zone rest upon presumed para-autochthonous complexes, which mainly comprise micaschists of Late Proterozoic to Early Paleozoic age without any relicts of high-pressure metamorphism. These para-autochthonous complexes appear in tectonic antiform windows in the northwestern part of the Massif Central (Millevaches, Aigurande plateau) and underlie the southern areas (Albigeois, Cévennes).

In the southern Paleozoic nappe complexes of the external zone of the Massif Central, the sedimentary sequences overstepping the exhumed orogenic sequences become younger from north to south: Late Devonian reef limestones unconformable on migmatites of unknown age in the Morvan block; Late Visean sediments unconformable in the Limousin and Auvergne; Stephanian deposits constituting an upper age bracket in the southern Montagne Noire region.

The structural setting presented here follows the proposed general succession and position within the internal and external zones; the structures are described in relation to: (1) the successive chronological periods (Eo-, Medio-, and Neo-Variscan), and (2) the lithotectonic units involved (nappes and rock types), their structural characteristics (tangential, transcur-

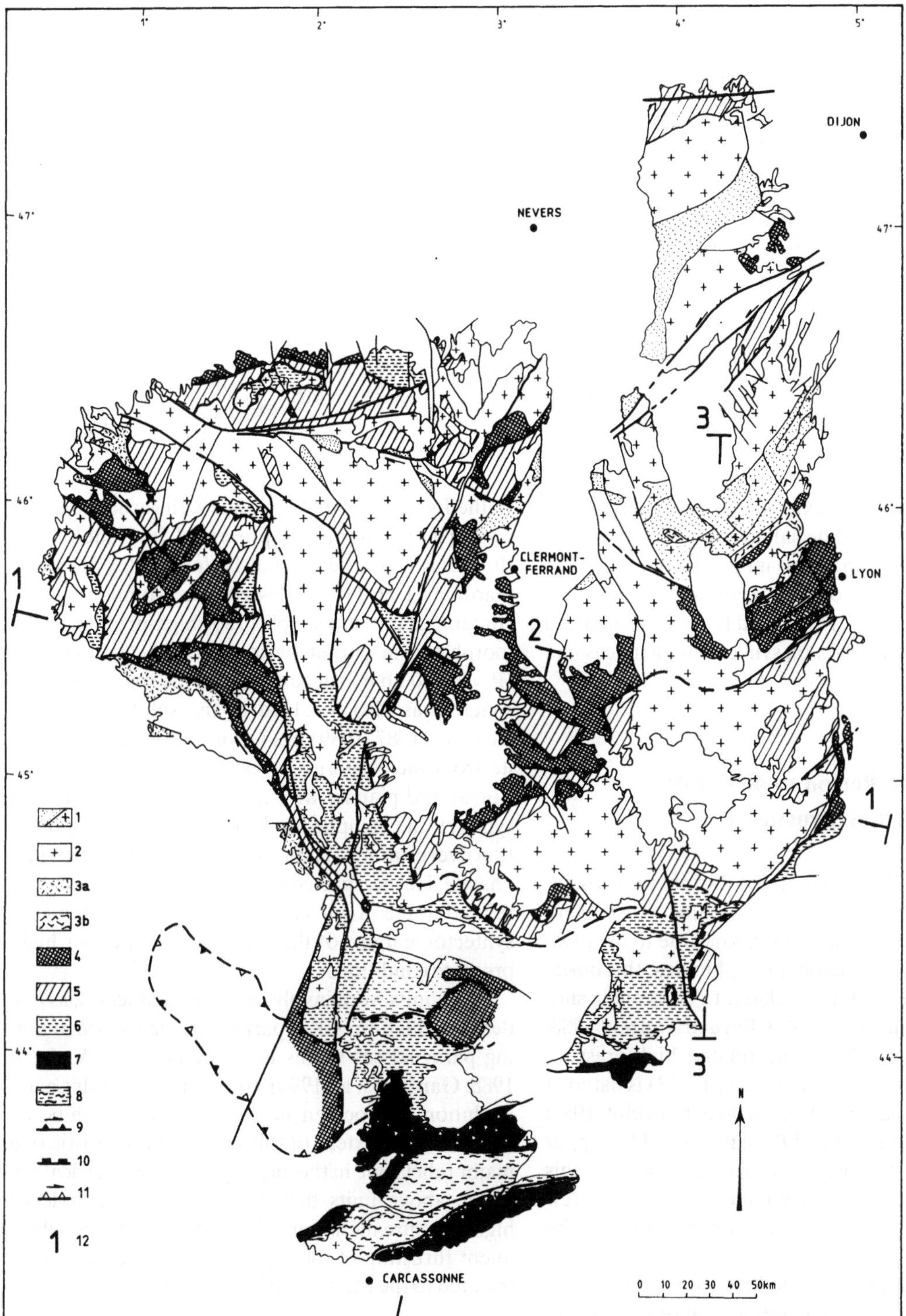

Fig. 2. The main nappes and lithotectonic units of the French Massif Central: *a* map and *b* cross-section. *1* Visean sedimentary deposit/Visean granite; *2* undifferentiated granite; *3* Upper epizonal units; *3a* Thiviers Payzac and Genis units; *3b* Brevenne unit; *4* Upper Gneiss Unit; *5* Lower Gneiss Unit; *6* relatively autochthonous complexes; *7* Paleozoic southern units; *8* Montagne Noire metamorphic core complex; *9* thrust nappes (390–360 Ma); *10* thrust and transcurrent zones (350 Ma); *11* Neo-Variscan thrust zone (320 Ma); *12* location of the cross-sections

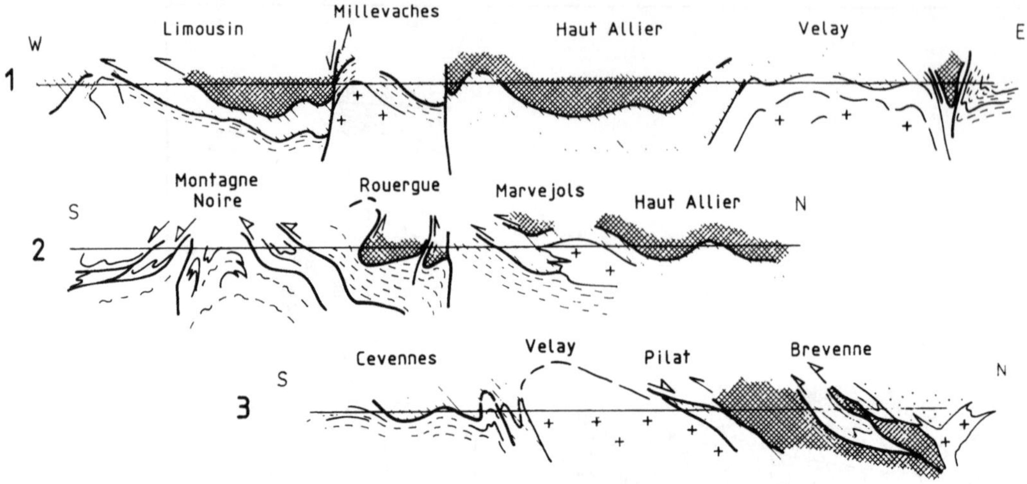

Fig. 2.b

rent, gravitational, and extensional) and their associated metamorphism. Special attention is paid to the age of the tectonic phases as inferred from overstepping sequences, syn- or post-tectonic intrusions, radiometric data (U-Pb, Rb-Sr, and K-Ar methods) and from new $^{39}Ar/^{40}Ar$ data on syn-kinematic micas and amphiboles.

2 Eo-Variscan Period (> 400 Ma): Relict Subduction Structures

During the Eo-Variscan period, lithospheric slabs, including both continental and oceanic crust, were buried to depths greater than 50 km. Due to the presence of relicts of dismembered ophiolites, this burial is currently interpreted as related to an oceanic subduction (Burg and Matte 1978; Pin and Vielzeuf 1988; Dubuisson et al. 1989). The related high-pressure metamorphism (Santallier et al. Part IV.3) is dated at between 430 and 415 Ma (Pin and Lancelot 1982; Ducrot et al. 1983; Pin and Peucat 1986). The Upper Gneiss Unit best shows the structures related to this episode; the Lower Gneiss Unit displays only few high-pressure relicts, located exclusively in the Limousin.

The eclogites are restricted to scattered outcrops covering less than a few hundred square meters and occur essentially within basic lenses. Preserved structures associated with this high-pressure metamorphism are very rare. Retrograde metamorphism is neither static (kelyphitic coronas) nor is it synchronous with the granulitic or Barrovian metamorphic foliation. In the Haut-Allier, however, Burg (1977) described isoclinal folds older than a static recrystallization under granulite facies metamorphism. In the Lyonnais, rare lineations with garnet, feldspar, and orthopyroxene are preserved in granulitic orthogneisses (Lardeaux and Dufour 1987). Recently, Gardien et al. (1988) described mylonitic textures, contemporaneous with prograde high-pressure metamorphism, in garnet peridotite from the Lyonnais (P = 20–30 kbar) suggesting that the mantellic lithosphere was also involved in the subduction. Bouchardon (1987, 1989) has reported a mylonitic eclogite from the Rouergue area, showing an L–S fabric (linear and planar) with quartz ribbon textures and syn-tectonic omphacite-kyanite-zoisite crystallization overprinting a static growth of chromiferous omphacite. He interpreted the mylonitic texture as being related to the major shearing that brought about the syntectonic uplift of these eclogites still under high-pressure conditions.

If all units were involved in the subduction, as evidence of high-pressure paragenesis in the surrounding paragneisses seems to imply (Pin and Vielzeuf 1988; Gardien et al. 1990), the main signs of this early evolution are located in the Peptynite-amphibolite complex. The extent of this unit is characteristic of a major relict zone in the nappe pile. Its geometrical position, above units that generally did not undergo high-pressure metamorphism, attests to major subsequent thrusting of the Upper Gneiss Unit and is attributed to the Medio-Variscan period.

3 Medio-Variscan Period (400–340 Ma): Main Nappe Tectonics

The main intracontinental deformation and collision tectonics took place in the internal zone of the Massif Central during the Medio-Variscan period, between

380 and 350 Ma according to Matte (1986). Matte (1986) also suggests a diachronism in nappe emplacement, showing that the thrusting is older in the Limousin than in the Marvejols area; this diachronism has been supported recently by Ledru et al. (1989), who show variations in kinematics and cooling periods between different regions. Most of the planar and linear penetrative structures and ductile deformation of the nappes are assigned to this period, with the syntectonic metamorphism being of amphibolite-facies grade (P = 5–9 kbar; T = 400–800 °C).

3.1 The Internal Zone

3.1.1 Structures and Microstructures

The major tectonic features related to the Medio-Variscan period are low-angle thrusts characterized by an intense mylonitic deformation, such as in the Haut-Allier (Burg 1977), Marvejols (Faure et al. 1979) and Rouergue (Burg et al. 1989). Continuing displacement along these faults is indicated by retrograde mineral assemblages which commonly mark the boundaries between the allochthonous and autochthonous units, as in Moyenne Dordogne (Ledru 1988). One of the main boundaries is the Peptynite-amphibolite complex mapped at the base of the Upper Gneiss Unit. This marks a major thrust zone in which the intensity of deformation is very high, with dispersed boudins of subcontinental lherzolite metamorphosed to garnet grade in the Lyonnais, Guéret and Limousin, and sheath folds and L tectonites locally present in the Haut Allier (Burg 1977) and Limousin (Girardeau et al. 1986); a minimum displacement of 150 km along the main thrust is suggested by Burg and Matte (1978). One of the most southerly extensions of this nappe is the La Bessenoit klippe, where Burg and Matte (1977) have defined lithological, structural, and metamorphic arguments for large-scale thrusting of crystalline units.

Mylonitic zones are not constant throughout the belt, however, and delineation of the thrust zones calls for the aid of other criteria. Lithological discontinuities, for example, are meaningful: ophiolitic sequences in the Limousin (Girardeau et al. 1986) or serpentinized harzburgites in the Haut Allier (Burg and Matte 1978) overlying continental-rise sediments. Inverse zonation of metamorphism can also be an argument for the presence of a thrust, as the prograde evolution in the parautochthon may be related to the emplacement of an allochthon (Briand 1978; Burg and Matte 1978; Burg et al. 1989). Also, abnormal age superposition indicates the allochthonous character of the Lower Gneiss Unit in the Lot region, where Late Proterozoic sedimentary rocks over Early Paleozoic rocks (Briand and Gay 1978).

The progressive character of the deformation contemporaneous with the Barrovian facies metamorphism is shown by the following features at all scales: microfolds with synmetamorphic minerals, such as kyanite or sillimanite, crystallizing both in the limbs and the axial plane; S-C planes (Berthé et al. 1979); asymmetric pressure shadows; and helicitic inclusions in minerals such as plagioclase and garnet. The dominant planar structure resulting from this progressive deformation appears as a regional composite foliation generally referred to as S_{1-2} (Fig. 3).

Major isoclinal recumbent folding had been presumed, until recently, in several regions in order to explain the inverted metamorphism; this inverted zonation is now explained by the stacking of nappes (Demay 1942; Burg and Matte 1978). However, folds do occur at different scales: isoclinal mesoscopic folds with a general southward vergence are described beneath the thrust zone of the Upper Gneiss Unit in the Haut Allier (Burg and Matte 1978; Burg et al. 1984); major recumbent folds of kilometric amplitude are described by Macaudière et al. (1987) in the para-autochthonous complex in the Cévennes. These folds involve metapelites and orthogneisses and are related to thrust tectonics.

The dominant linear structure is a stretching lineation observed on the S_{1-2} foliation plane; its orientation varies (Fig. 4): E-W to N120 °E in the western part (Limousin, Floc'h 1983) and in the Sioule (Feybesse and Tegyey 1987); N120 °E to N-S in the Haut Allier (Burg and Matte 1978); N60 °E in the Aigurande plateau (Quenardel and Rolin 1984), the Lyonnais (Lardeaux and Dufour 1987), the Artense and the Cévennes (Lacassin and Van den Driessche 1982). The stretching lineation may be parallel to the displacement vector in simple shear zones, or be controlled by the geometry of folds in pure shear zones (Burg et al. 1987); an ambiguity that has important implications for the kinematics of nappe emplacement. Mattauer and Echecopar (1977) have suggested that the stretching lineation of the Massif Central represents a transport direction, as has been inferred for the Himalayas and the Scandinavian Caledonides. Burg et al. (1984), however, describe an intense mineral lineation trending N100 °–120 °E, parallel to fold axes and perpendicular to the apparent vergence. Moreover, in some areas, the lineation has two orientations on the composite foliation plane without clear evidence of polyphased deformation, the mineral assemblages being the same; such is the case in the Cezallier [the GPF (Geologie Profonde de la France)

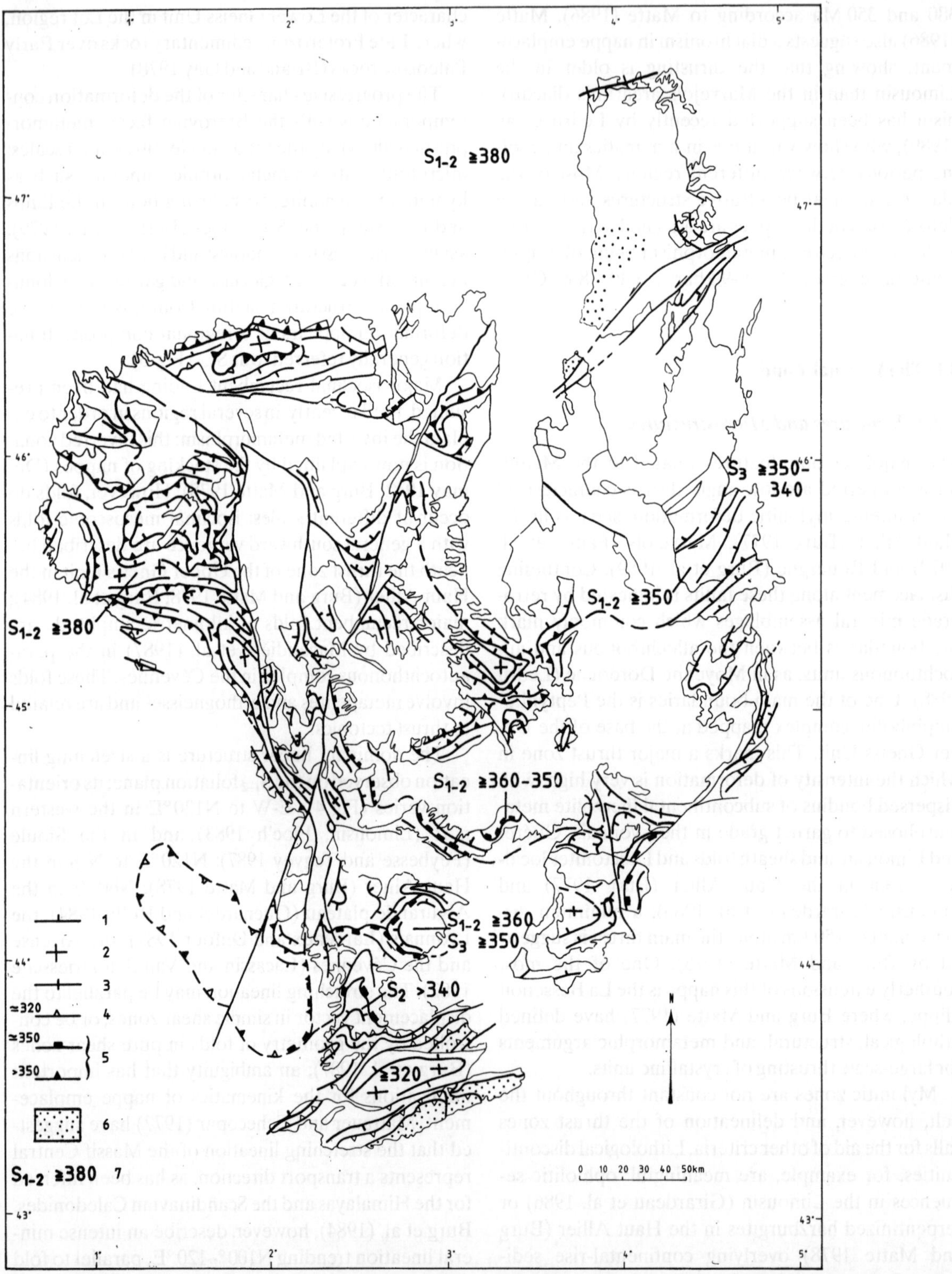

Fig.3. Trajectories of the foliation and schistosity planes. $^{39}Ar/^{40}Ar$ ages of syntectonic minerals. *1* trajectories of the foliation; *2* horizontal foliation; *3* vertical foliation; *4* Neo-Variscan thrust zone; *5* Medio-Variscan thrust zone; *6* Pre-Carboniferous units without penetrative schistosity; *7* minimum age of the foliation (S_{1-2} or S_3) in Ma

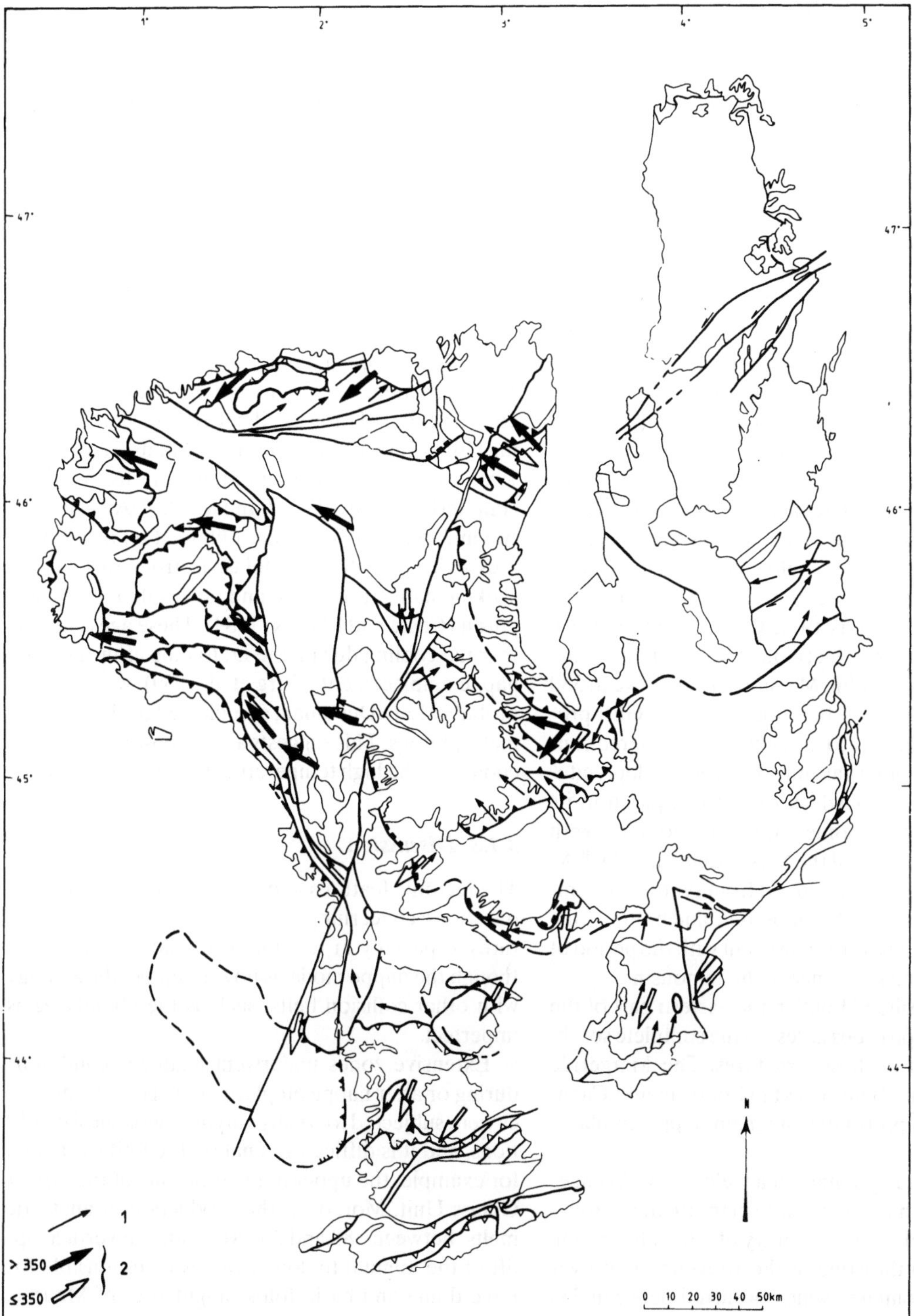

Fig. 4. Stretching lineation and kinematic indication. *1* Stretching lineation; *2* kinematic indications during Medio-Variscan (> 350 Ma) and Neo-Variscan (< 350 Ma) periods

deep-borehole program; Marquer et al. 1987], where the N60 °E and N120 °E stretching directions are synchronous and are related to a complex mechanism of stretching during pure shear deformation.

3.1.2 Chronology

In the Limousin, the contact between the Upper and Lower Gneiss units and the foliation are cut by anatectic melts dated at 382 ± 5 Ma (U/Pb, Lafon 1986) and 375 ± 6 Ma (Rb/Sr, Duthou 1977). These melts are associated with the culmination of the Barrovian metamorphism, after the emplacement of the crystalline nappes. Late-kinematic diorite to tonalite bodies were emplaced between 380 and 360 Ma (Bernard-Griffiths et al. 1985) and numerous granites dated at 360–350 Ma (such as Guéret; Duthou 1977) cut across the structures. These dates are corroborated by K/Ar data on amphibole, muscovite, and biotite, which indicate postmetamorphic cooling below 500–300 °C of the basement between 360 and 350 Ma (Cantagrel 1973; Bernard-Griffiths 1975; Pin and Peucat 1986). This suggests intense thermotectonic activity in the northwestern part of the Massif Central between 380 and 350 Ma. The deep borehole at Couy (GPF program), located farther to the north in the covered continuation of the internal zone, has cored a leptynite-amphibolite complex which can be compared in its lithology and tectonometamorphic evolution with other such complexes of the Massif Central. $^{39}Ar/^{40}Ar$ ages of about 380 Ma, obtained from syntectonic amphibolite facies minerals (amphibole and biotite; Costa and Maluski 1988), are consistent with the period of nappe emplacement defined in the Limousin.

East of the Sillon Houiller, the eastern part of the Massif Central provides access to the complete southward extension of these structures. The Margeride pluton, dated at about 330 Ma (Couturie and Caen-Vachette 1979) post-dates the main nappe emplacement.

The Monts du Lyonnais area yields $^{39}Ar/^{40}Ar$ ages of 350 Ma on syntectonic biotite in the main S_{1-2} foliation (Costa et al. 1989). Intensity of late deformation (D_3) related to thrusting of the Brevenne unit over the Lyonnais complex, which occurred between 350 and 340 Ma, could have partially reset these minerals. Nevertheless, this date of 350 Ma provides a minimum value for the age of the main deformation as it occurred under high-temperature metamorphic conditions.

In the Marvejols area, the age of the deformation spans 20 Ma, between 360 and 340 Ma: the $^{39}Ar/^{40}Ar$ method yields ages of 360 Ma on within-foliation biotites and 340 Ma on post-foliation muscovite (Costa 1989), both within the mylonitic contact zone between the Upper and Lower Gneiss units. The U/Pb method on zircons in the contact zone gives a lower intercept date of 346 ± 8 Ma for synkinematic granites (Pin 1981) and an age of 344 ± 12 Ma on a syn- to late-tectonic pegmatite vein in the Upper Gneiss Unit (Pin and Lancelot 1982). Finally, a time span of 350–340 Ma is indicated by rocks of the Lower Gneiss Unit (Costa 1989). Here, a late metamorphic imprint (T > 500 °C), associated with the D_3 folding of the S_{1-2} foliation (Briand and Gay 1978; Pin 1980), may explain these younger ages.

These ages are in the same range as those from the Rouergue area, where syntectonic intrusions are dated at 350–360 Ma for the Pinet granite (Pin 1981, using U/Pb; Burg and Teyssier 1983; Maluski and Monié 1988, using $^{39}Ar/^{40}Ar$,) and at 360 Ma for a gabbro (Pin and Piboule 1988). Similar $^{39}Ar/^{40}Ar$ ages (between 355 and 350 Ma) are obtained on amphibole, muscovite, biotite, and feldspar from the various synkinematic intrusions and associated metamorphism (Costa and Maluski 1989). These ages date the thermal cooling during the D_3 folding after the main nappe emplacement (Burg et al. 1989), which is thus probably Late Devonian, as indicated by the synkinematic plutons. This suggests a rapid postmetamorphic cooling of the high temperature metamorphic zone.

3.1.3 Discussion

The tectonic features and the ages measured lead to questions concerning the significance of the structures, especially in relation to the kinematics. Even though the nappe model is now accepted, the analogy with other collision belts, such as the Himalayas, is imperfect.

Extensive zones underwent anatectic conditions during or after nappe emplacement. Thus the gneisses were subjected to widely varying rheological conditions, and it is difficult to analyze the finite strain in, for example, the upper migmatitic part of the Upper Gneiss Unit. Moreover, the production of anatectic melts, between 380 and 360 Ma, induced vertical uplift of the migmatite domes that was responsible for large dome and basin folds (migmatites of Meuzac, Guéret, Neuvic, etc.).

The strain ellipsoid generally falls either within the field of plane-strain deformation (k = 1) (Burg and Matte 1978; Girardeau et al. 1986) or in the field of flattening; constrictional domains (k > 1) are rare. Linear fabrics (L > S) are restricted in some areas, such as at the base of the Lower and Upper Gneiss units in the Limousin (Girardeau et al. 1986; Ledru and Autran 1987). Planar fabrics are dominant in many regions, such as the Haut Allier (Burg et al. 1984), Cézallier, Artense, or Lyonnais (Lardeaux and Dufour 1987). In such cases, where the deformation is not close to plane-strain, the direction of transport cannot be immediately inferred from the stretching lineation (Burg et al. 1987). Moreover, the direction of transport deduced from kinematic criteria in plane-strain zones is difficult to correlate from one region to another. In the Limousin, where tectonic evolution took place before 380 Ma, most nappe movement was westward to northwestward (Floc'h 1983; Girardeau et al. 1986; Burg et al. 1987; Ledru and Autran 1987), although early southward displacements are mentioned by Girardeau et al. (1986) and Burg et al. (1987). Unlike those in the Limousin, the major thrusts in the Haut-Allier and Marvejols areas (dated prior to the 360–340 Ma interval which constitutes a minimum age) were towards the south according to Burg et al. (1984) and Matte (1986). Finally, in many areas, no kinematic criteria are available for any reconstruction of the overall direction of transport of the nappes. In such conditions the rooting of the thrusts remains unclear.

The paleogeographical positions of the overstep sequences hampers investigation of the root zone and of the duration of thrust-tectonic activity. Givetian to Frasnian marine platform limestones in the Morvan rest on migmatites of unknown age, providing northern and eastern limits for the rooting of nappes emplaced after 380 Ma. The Middle to Late Visean deposits are weakly deformed without any penetrative schistosity, which excludes any possibility that the stacking of nappes continued after the Late Visean (around 340 Ma): thickening of the crust throughout the Limousin and Auvergne took place before then. On other grounds, a gravimetric high oriented NW-SE from Moulins to St Etienne suggests a possible rooting zone for the earlier Devonian nappes under the northeastern part of Massif Central (Lefort et al., Chap. I.2 this Vol.)

3.2 The External Zone of the Belt

A distinction is possible between the internal and external zones of the orogenic belt. The external zone is implied in the orogenic evolution from the Early Carboniferous and did not undergo such intense strain and metamorphic conditions as the internal zone. Thus, the boundary should be located at the southern edge of the terranes affected by the Medio-Variscan period. The foliation S1–2 of the upper nappe of the Albigeois area (Saint Sernin nappe; Gueranges Lozes and Guerange 1984) is of Devonian age as it is affected by the same D_3 folding dated at about 350 Ma in the Rouergue. The D_3 folds are absent to the south and it is difficult to assert the equivalence of the early schistosities from a geometrical reconstruction alone. Moreover, $^{39}Ar/^{40}Ar$ ages from epizonal syntectonic muscovite of the lower Albigeois nappe fall between 340 and 330 Ma (Costa, unpubl.) and no indication exists of earlier ages.

Finally, the Devonian platform sequence of the Montagne Noire external nappes provides no evidence of any tectonic activity during Medio-Variscan times. This implies that while the nappes were emplaced in the internal zone, shallow-water marine limestones were deposited on a stable continental rise in the external zone. Whatever the late tectonic event, this paleogeographic constraint implies that until the Early Carboniferous, the internal and external zones were largely separated. It is considered, therefore, that on the structural maps (Fig. 3) the southern Paleozoic nappes of the Albigeois represents the southernmost extension of the Medio-Variscan nappes.

4 Neo-Variscan Period (350–290 Ma)

After the major stacking related to Medio-Variscan nappe tectonics, the orogenic front migrated southward during the Carboniferous. Thus, the tectonic evolution of the internal zone was represented mainly by deformation along deep-seated structures, essentially transcurrent shear zones, prior to the deposition of the overstepping Visean sequences. This period was also characterized by intense plutonism. The evolution of the external zone, especially the emplacement of the nappes, is reflected in certain sedimentary features in the Carboniferous formations. After a long period of crustal shortening during the Early and Middle Carboniferous, orogenic uplift and extension began during the Westphalian in the north and slightly later in the south.

4.1 Intracontinental Evolution of the Internal Zone

4.1.1 Combined Thrusting and Wrenching at the Contact Between the Lyonnais and Brevenne Units

The overstepping sequences of the Medio-Variscan period indicate that stacking of the nappes was completed before Middle Visean times. The tectonics of the Lyonnais-Brevenne contact zone may well illustrate the transition to the Neo-Variscan era in the northern internal zone.

During this period (D_3), the major S_{1-2} and sheared structures of the Lyonnais complex were intensely refolded under retrograde metamorphic conditions, contemporaneous with the regional development of dextral ductile wrench faults and the emplacement of synkinematic granites (Lardeaux and Dufour 1987). A well-supported age of 340 Ma can be inferred for these deformations through the $^{39}Ar/^{40}Ar$ dating of the granites and associated synmetamorphic deformation within the high-grade rocks of the Lyonnais unit (Costa et al. 1989).

To the north, the Brevenne unit underwent a different evolution from that described above. This Upper Epizonal Unit consists of a typical spilite-keratophyre series attributed to an ensialic marginal basin (Piboule et al. 1982; Pin 1989). The age is debatable: Late Proterozoic to Early Paleozoic (Reitz and Wickert 1988), Silurian-Devonian (Sider et al. 1986) or Late Devonian (350–360 Ma, $^{207}Pb/^{206}Pb$, Calvez, unbublished); also the age determined from microfossils is poorly constrained as the sphaeromorphes are not considered as characteristic of any particular period within the Late Proterozoic-Paleozoic sedimentary successions of the Armorican Massif (Chauvel, written comm.; Mansuy 1983). The Brevenne unit is juxtaposed with the Lyonnais complex along a dextral N60°E-striking wrench fault (Lardeaux and Dufour 1987; Wickert 1988) with a major reverse component that was responsible for the thrusting (Feybesse et al. 1988). The activity of this major wrench fault occurred in the period 350–340 Ma, as can be inferred from the dating of the synkinematic granite underlying the contact zones: 339 ± 8 Ma by the Rb/Sr method (Gay et al. 1981); 349 ± 3 and 346 ± 3 Ma by $^{39}Ar/^{40}Ar$ dating on muscovite and biotite (Costa et al. 1989).

4.1.2 Major Strike-Slip Faults

Major strike-slip faults, several hundred kilometers long, deformed and displaced blocks during the Late-Variscan period (Fig. 5). The grade of the associated metamorphism is generally greenschist facies (biotite-muscovite-plagioclase-chlorite-green amphibole) and the geometrical characteristics are typical of transcurrent-tectonic zones (vertical strike-slip faults, association of S-C-C' planes, plane-strain to constrictional deformation with subhorizontal stretching lineation, intense gradient of deformation, constant asymmetric criteria such as pressure shadows, folds, etc. (Berthé et al. 1979, Gapais et al. 1987).

The main period of activity of these strike-slip faults is defined by the age of emplacement of the granites (Fig. 6) affected or not by the movement (Lerouge 1988) and shows two main phases: 355–325 Ma with dextral displacement along faults striking E-W to NNW-SSE and clearly overstepped in the northern Massif Central by Middle-Late Visean deposits, and 325–300 Ma with essentially sinistral displacements. These dextral and sinistral faults are cut by granites dated at 300–290 Ma.

The Sillon Houiller ("coal-bearing trough"), which crosses the Massif Central along a NNE-SSW axis, shows a complex evolution. Early ductile sinistral displacement, estimated at about 70 km, probably occurred around 310 Ma ago during the Westphalian (Grolier 1971; Feybesse 1981; Belin 1983). Then, according to Bonijoly and Castaing (1984), the coal basins were deposited along brittle faults in transtensional zones (pull-apart basins, for example) during Stephanian compression. Certain granites (Montmarault in the north, Glenat in the south) were emplaced around 300 Ma, after the ductile shear deformation (Grolier et al. 1984).

Reconstructing the Massif Central before the sinistral displacement of the Sillon Houiller fault, the following deformation zones would have been present from north to south (Fig. 5):

1. The E-W Marche fault (Choukroune et al. 1983; Lespinasse and Pecher 1984; Lerouge and Quenardel 1985, 1988), which was an active sinistral shear zone during the Namurian-Westphalian stages (325–310 Ma). This fault, according to Lerouge and Quenardel (1985), had an Early to Middle Carboniferous dextral movement that was recorded only in granites emplaced between 356 Ma and 340 Ma (Guéret type). To the south of this fault, Lerouge and Quenardel (1985) describe another dextral shear zone, the Combrailles fault, which changes to the south into a succession of lateral overthrusts with an apparent vergence toward the southeast (Lerouge 1988).
2. The inner part of the internal zone is cut by a series of dextral NW-SE faults which overlap from west to east: these are the Oradour, Arrènes-Ouzilly (Lespinasse et al. 1986), Pradines (Mezure 1981), La Courtine (Martin 1980), and L'Hermitage (Barbarin and Belin 1982; Gapais and Barbarin 1986) faults.
Several granitic intrusions were deformed by these shear zones, during or after their emplacement. Thus the Saint Sylvestre granite, dated at 325 Ma (Duthou 1977, using Rb/Sr; Leroy and Holiger 1984, using U/Pb), is considered to be contemporaneous with the last movements along the Arrènes-Ouzilly fault (Lespinasse et al. 1986), and the Egle-

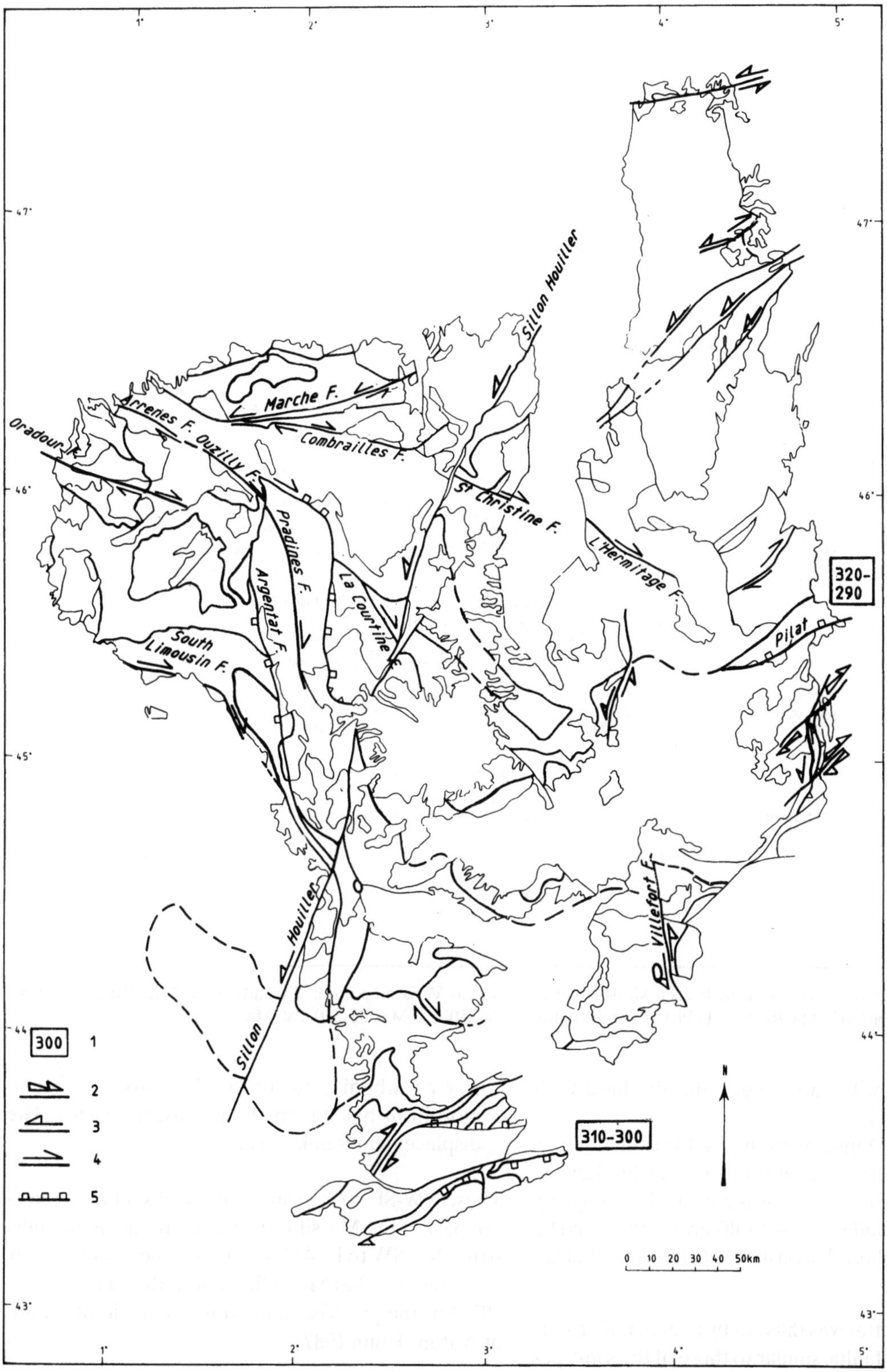

Fig. 5. Transcurrent and extensionnal tectonics (Neo-Variscan period). *1* $^{39}Ar/^{40}Ar$ ages of extensionnal tectonics from synkinematic minerals; *2* late dextral strike slip fault (300 Ma); *3* sinistral strike slip fault (320–300 Ma); *4* dextral strike slip fault (350–330 Ma); *5* normal fault

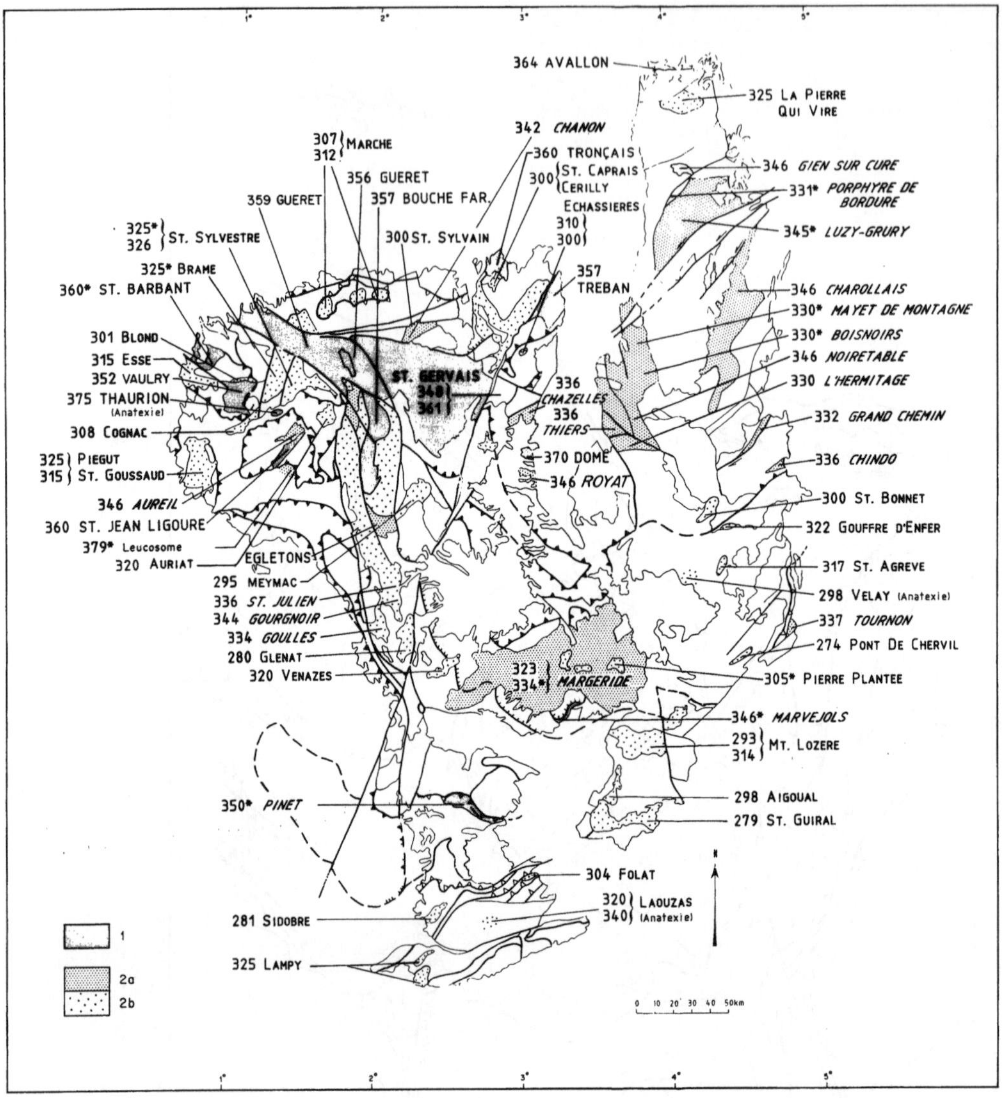

Fig. 6. The main granitic massifs of the French Massif Central and their absolute ages (in Ma, Rb/Sr, *: U/Pb). *1* granite of the Medio-Variscan period; *2* granite of the Neo-Variscan period; *2a* 350–330 Ma; *2b* 330–280 Ma

tons granite with those along the Pradines fault (Mezure 1981).

Grolier (1971) interprets the La Courtine fault as being the western equivalent of the St Gervais-l'Hermitage fault, which was cut, at a late stage, by the Sillon Houiller. These faults are overstepped by unmetamorphosed deposits of Middle-Late Visean age.

The Massif Central was thus cut by a set of major, early, transcurrent faults, similar to those of the same period observed in the Armorican Massif (Lerouge and Quenardel 1988).

3. The dextral N110-150°E South Limousin fault (Ledru and Autran 1987), located within the Upper Epizonal Unit which was probably active during Early Carboniferous times. The associated metamorphism belongs to the greenschist facies. The displacement is unknown.

These NW-SE strike-slip faults are absent in the eastern part of the Massif Central, where the main faults strike NE-SW to E-W; for example, the sinistral shear zone dated at 340 Ma in the South Morvan (Barrère 1971) or the pre-Visean dextral E-W mylonite south of Autun (Rolin 1987).

4.1.3 Granitic Intrusion and Deformation

The granitic intrusions constitute large areas of the internal zone. Some are syn or late tectonic relative to the Medio-Variscan period of nappe emplacement,

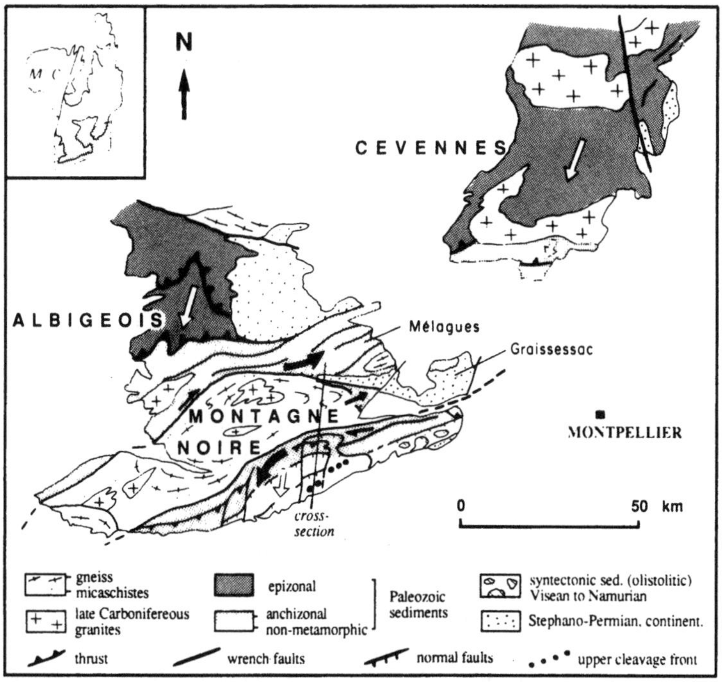

Fig. 7. Geological and structural scheme map of the southern Massif Central. *Open arrows* dispacement vector of early major south-southwestward thrusting (Lower Carboniferous); *black arrows* extensional movements associated with the uplift of the Montagne Noire metamorphic core complex

but most have ages between 350 and 290 Ma, the oldest being located mainly in the western part of the Massif Central (Fig. 6). Their modes of emplacement are good markers of the successive tectonic environments of the Neo-Variscan period.

The intrusions vary from very large, thick, homogeneous batholiths such as the Margeride, through composite batholiths incorporating anatectic zones such as Millevaches and Velay, to leucogranitic laccoliths a few kilometers thick as in the Limousin (Petrequin 1979). Except for tonalites and hornblende granodioritic plutons in the Limousin and Auvergne, most of the granite is dominantly of crustal anatectic origin with a small content of mantle-derived material (Pin 1989). The plutons generally show an internal fabric typically controlled by the shape of the intrusion and by the regional tectonic setting. The internal structures are planar and linear mineral fabrics, foliations, and cataclastic zones; some of the planar structures have been refolded by late folding episodes. Deformation associated with the emplacement of the plutons can be high in the vicinity of the plutons, but more generally consists of low-amplitude folds (marginal synclines) in the host rock. Contact metamorphism is difficult to observe as the earlier grade of regional metamorphism is generally in the amphibolite facies.

Macroscopic fabric and magnetic susceptibility anisotropy fabric in the western part of the Massif Central were studied on a regional scale in the large plutonic bodies of Guéret (360–340 Ma) and Millevaches (350–330 Ma) by Bouchez and Jover (1986) and in the Limousin leucogranites (325 Ma) by Mollier and Lespinasse (1985). One major result was the discovery of a quite constant viscous elongation in the granites oriented around N120 °E. This persistence of structural direction through time is striking as this direction is parallel to the stretching lineation of the Medio-Variscan period in the Limousin gneisses. This has led several authors to envisage continuous westerly thrusting from the time of the Medio-Variscan nappes to the intrusion of the leucogranite dated at 330 Ma. However, the kinematics of the emplacement of these granites are in doubt, as it has never been demonstrated that the lineation in the granites is parallel to any transport direction during a plane-strain deformation.

4.2 Major Thrusts and Nappe Tectonics in the External Zone

The external zone of the Massif Central was involved in the collision belt during the Carboniferous (Figs. 7 and 8). The succession of the two major structural and metamorphic events (Gèze 1949) during the Neo-Variscan period followed the same pattern as the evolution of the internal zone during the Medio-Variscan period: crustal thickening, transcurrent faulting, and post-collision crustal collapse.

The first major event corresponded to early thrusting related to southward directed orogenic convergence. The thrusting produced folding and flat to

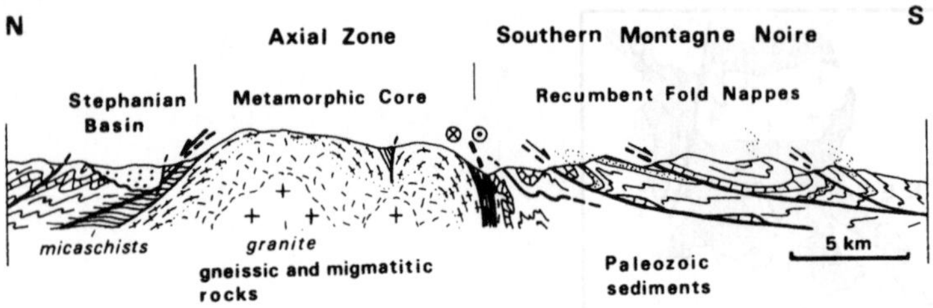

Fig. 8. Geological and structural cross-section through the Montagne Noire

slightly north dipping synmetamorphic foliation in the north of the Massif Central (Albigeois, Cevennes) and more slaty cleavage in the nappes of the southern edge of the massif. An axial zone, formed by the Montagne Noire crystalline complex, roughly separates these two domains.

The southern edge (Fig. 7) and the Ecailles de Melagues just north of the axial zone of the Montagne Noire, show no trace of cleavage (Brunel 1974; Donnot and Guerange 1978). Southwestward shearing parallel to prominent stretching lineation is transverse to the axes of south-facing recumbent folds, which are the predominant feature of shortening in the southern Cevennes and northern Montagne Noire (Arthaud et al. 1966; Arthaud 1970; Echtler 1990).

In the axial zone of the Montagne Noire, the foliation is composite, related to isoclinal folds (Ellenberger 1967; Bogdanoff 1973) and large penninic-type folds and thrusts (Demange 1975; Bard and Loueyit 1978). The thickening of the crust, due to stacking of the units, is shown by few relicts of high-pressure metamorphism (Demange and Leger 1980) and by an intermediate pressure metamorphism with staurolite-garnet assemblages (Bard and Rambeloson 1973) and staurolite-kyanite-garnet assemblages (Bouchardon et al. 1979; Demange and Jamet 1985). These thrust-related structures were then tightened and refolded by late compressional tectonics during which the thickening continued.

Radiometric $^{39}Ar/^{40}Ar$ data on syntectonic muscovite in the northern Albigeois crystalline nappes yield essentially Early Carboniferous ages (345 ± 3 and 337 ± 3 Ma; Costa, unpubl.). These ages were obtained on muscovite which marked N-S stretching L_2 lineations on S_2 planes in metamorphic quartzites and low-grade Cambrian-Ordovician schists. As metamorphic conditions did not exceed 400 °C, the radiometric data obtained can be considered as an indicator of the date of this tectonometamorphic imprint. In the Montagne Noire crystalline complex, 320 Ma (Hamet and Allègre 1976, using Rb/Sr; Maluski et al. 1990, using $^{39}Ar/^{40}Ar$) is the youngest age for the high-temperature metamorphism and for the emplacement of plutons and related metamorphism. It can thus be assumed as the younger limit to the end of the orogenic compressional regime and of the anatexis process.

The age of south-verging folding and tangential shortening in the slightly metamorphosed Montagne Noire nappes is constrained by biostratigraphic data (Engel et al. 1981) to post mid-Late Visean (latest Early Carboniferous). This slight southward younging is consistent with the migration of the early convergent tectonics toward the external orogenic zones (Engel 1984). Partly interfolded synorogenic Late Visean (Feist and Galthier 1985) clastics and refolded thrust planes indicate the progressive and continuous nature of this major shortening event.

4.3 The Late-Orogenic Extension Period

4.3.1 The Internal Zone

The main effects of the extension in the internal zone are related to the migmatitic domes of the Millevaches and Velay areas (Fig. 1).

The Argentat fault is a major normal fault located at the western flank of the Millevaches dome; its vertical displacement is estimated to be more than 3 km (Ledru and Autran 1987). Faulting probably occurred during Westphalian times, if one takes the 330–320 Ma age of the granites affected in a solid state by this tectonic episode (Duthou et al. 1984).

The Mont Pilat unit (Lower Gneiss Unit) is located on the north-eastern edge of the Velay granitic dome whose late anatectic melts are dated around 300 Ma (Caen Vachette et al. 1982, using Rb/Sr). Large-scale extensional tectonics, associated with a low-pressure high-temperature (LP-HT) metamorphism, is superimposed on the previous compressional events (Malavieille et al. 1990). Regional N-S lineations and shear criteria indicate a northward displacement of the hanging wall during the extension. Normal faulting continued at shallower depths and the Stephanian continental basin of St Etienne was deposited on the hanging wall of a normal shear zone during the last stages of this post-orogenic extension (Malavieille et al. 1990).

The syntectonic granite of Gouffre d'Enfer was dated at 322 ± 9 Ma (Caen-Vachette et al. 1984). Minerals that crystallized during the extensional process and associated LP-HT metamorphism yielded $^{39}Ar/^{40}Ar$ ages between 313 ± 3 Ma (in the most ductile structures) and 291 ± 3 Ma (in a synkinematic granite intruding the basal contact of the Early Stephanian coal deposits, Costa, unpubl. data).

4.3.2 The External Zone

Regionally, the second major event is dominated by the late tectonometamorphic evolution of the Montagne Noire crystalline complex (Figs. 7 and 8). The uplift of the migmatitic and anatectic core (Beaud 1985) is associated with HT/LP conditions in polymetamorphic Proterozoic and Early Paleozoic gneisses and sediments (Ellenberger 1967; Bard and Rambeloson 1973; Bogdanoff 1973; Demange 1975, 1982). Metamorphism and uplift were associated with and succeeded by dextral extensional wrenching along the northern and southern flanks of the domal antiform (Echtler et al. 1988; Echtler and Malavieille 1990). Ductile transcurrent shearing along syn- to late-metamorphic mylonites is parallel to the ENE-WSW elongation of the dome and the pervasive subhorizontal or low-angle stretching lineation. Very low grade to non-metamorphosed Paleozoic sediments, juxtaposed tectonically upon the high- to medium-grade gneisses and micaschists, have been detached in a dextral/normal sense relative to the uplifted core (Figs. 7 and 8). This west-southwestward shearing has overprinted the basal thrusts and lower parts of the southern fold nappes. The upright axial planar cleavage and the axes of en echelon folds trend subparallel to the mylonitic wrench zones; this intense ductile deformation overprints earlier overturned sedimentary successions (Vignard 1976; Byung-Joo Lee et al. 1988). Southwestward to southward movements on low-angle, southward-dipping brittle/ductile normal fault zones controlled the final emplacement of the upper recumbent fold units (Echtler 1990). Extensional wrench and detachment tectonics throughout the massif post-date and overprint the earlier tangential and compressional structures.

Syntectonic mineral assemblages of the mylonitic contacts yield essentially Late Westphalian to Early Stephanian $^{39}Ar/^{40}Ar$ ages (Maluski et al. 1990). Thus, synkinematic minerals (amphibole, muscovite, biotite) of the major dextral wrench fault zones yield ages that cluster around 308 ± 3 Ma and muscovites from low grade metamorphic sediments that were involved in detachment zones to the northern and southern boundaries give ages of 297 ± 3 Ma.

In agreement with such ages, the unconformably overlying Stephanian terrestrial sediments show only slight anchi-metamorphism (Becq Giraudon 1973). Structural evidence of a detachment-controlled development of the Graissessac basin, northeast of the metamorphic core (Fig. 7), also exists, thus indicating a continuous extensional evolution during latest Carboniferous times (Echtler et al. 1988; Echtler and Malavieille 1990).

5 Structural Setting of the Late- to Post-Orogenic Continental Basin Formation

Late Paleozoic sedimentation in the Massif Central is characterized by the formation of isolated intramontane troughs as basins of deposition. From Stephanian to Permian times, coarse, coal-bearing sequences were deposited unconformably upon exhumed crystalline basement (Courel 1987). These basins document advanced stages of the intense crustal uplift and extensional process that dominated the late-orogenic evolution subsequent to the major Variscan crustal shortening (Lorenz and Nicholls 1976, 1984). Their structural setting throughout the Variscan belt (Fig. 1) is linked to large-scale tectonometamorphic discontinuities such as Late Hercynian wrench fault zones (Arthaud and Matte 1975, 1977). Basin formation was controlled by active fault zones – either sinistral such as the N-S-oriented Sillon Houiller (Gelard et al. 1986; Bonijoly and Castaing 1984) or dextral generally on E-W to NE-SW-oriented wrench faults (Arthaud and Matte 1977).

Based on the microtectonic analysis of brittle structures, Bonijoly and Castaing (1987) related the strike-slip tectonics and associated basin subsidence to continuous compression and varying stress patterns during the Late Carboniferous, preceding a general N-S extension during the Permian. In contrast, Echtler et al. (1988), Echtler and Malavieille (1990) and Malavieille et al. (1990) used the evidence from two examples to propose that the Stephanian basins of the Massif Central resulted from generalized late-orogenic crustal extension and thinning subsequent to orogenic crustal thickening. The tectonic features of the Graissessac and St Etienne basins show that basin subsidence along faults was linked to the uplift of metamorphic terranes and tectonic denudation.

Ductile low-angle detachment and transcurrent wrench faulting constituted major extensional shear zones related to crustal thinning that took place essentially during the Westphalian. Elongate east-northeast-trending Stephanian basins formed parallel to such tectonic and metamorphic discontinuities, when deformation under progressive retrograde metamorphic conditions became shallow and brittle during continuous uplift and erosion. Coarse continental sediments were deposited along active, normal or strike-slip fault zones. Asymmetric synclines show progressive unconformities and thinned layers along the fault-bounded margins of the basins. Tectonic fabrics of syn-sedimentary and brittle deformation are generally coaxial to extensional features of ductile deformation in the extended substratum. This structural setting on the edge of high-grade and plutonic metamorphic complexes attests to basin formation controlled by low-angle normal faults and related to the collapse of the thick Variscan crust. The progressive superposition of brittle faults on pre-existing ductile shear zones which are also observed in the Blanzy-Montceau basin (Fig. 1; Castaing 1982; Valle et al. 1988) as well as the geometry of the deformation indicates a general extensional evolution during the Late Paleozoic.

Permian clastics, mainly unconformably overlying Stephanian strata, were deposited in large asymmetric half-grabens controlled by steeply dipping, brittle, normal and strike-slip faults. This late general N-S extension (Bonnion et al. 1983) marks the end of late-orogenic crustal extension.

One of the most notable features of the Late Paleozoic sedimentary history in the Massif Central is the essentially synchronous development of continental basins throughout the extensive Variscan belt. Apart from some basins where sedimentation started during the Early Stephanian or as an exception during latest Westphalian (Grangeon et al. 1968), the oldest deposits are generally of Late Stephanian (B) age. This pattern is consistent with a relatively "instantaneous" extensional evolution immediately following the long period of Variscan orogenic shortening. Consequently, Stephano-Permian continental basins are important markers for the timing and mechanisms of the late- to post-orogenic evolution in the Massif Central.

Acknowledgments. The authors are grateful for helpful reviews and suggestions made by A. Autran, J. M. Lardeaux and J. L. Feybesse. They also thank Sir Patrick Skipwith for his translation of this contribution and Anne-Lise Chassignol for the drawing.

References

Autran A (1984) Chevauchements synmétamorphes varisques (Limousin), note de synthèse. In: Bur Rech Géol Min (ed) Programme Géologie profonde de la France. Doc Bur Rech Géol Min, Orléans, 81–83, pp V–XXI

Autran A, Cogné J (1980) La zone interne de l'orogénèse varisque dans l'Ouest de la France et sa place dans le développement de la chaine hercynienne. – In: Cogné J, Slansky M (eds) Géologie de l'Europe du Précambrien aux bassins sédimentaires post-hercyniens, 26ème Cong Géol Int Coll C6, Paris 1980. Ann Soc Géol Nord, Lille, XCIX, pp 90–111

Autran A, Breton JP, Chantraine J, Chiron JC, Gros Y, Roger P (1980) Introduction à la carte tectonique de la France. Mém Bur Rech Géol Min (Orléans) 110, 52 pp

Arthaud F (1970) Etude tectonique et microtectonique comparée de deux domaines hercyniens: les nappes de la Montagne Noire (France) et de l'anticlinorium de l'Iglesiente (Sardaigne). Publ Univ Sci Tech Languedoc, Univ. Montpellier, 175 pp

Arthaud F, Matte P (1975) Les décrochements tardi-hercyniens du Sud-Ouest de l'Europe. Géométrie et essai de reconstitution des conditions de la déformation. Tectonophysics 25: 139–171

Arthaud F, Matte P (1977) Late Paleozoic strike slip faulting in southern Europe and northern Africa: result of right lateral shear zone between the Appalachians and the Urals. Geol Soc Am Bull 88: 1305–1320

Arthaud F, Mattauer M, Proust F (1966) La structure et la microstructure des nappes hercyniennes de la Montagne Noire. Ann Univ Neuchâtel, Coll "Etages tectoniques", pp 231–243

Barbarin B, Belin JM (1982) Mise en évidence du cisaillement ductile hercynien "St Gervais–l'Hermitage" (Massif Central français). C R Acad Sci Paris II 294: 1377–1380

Bard JP, Loueyit J (1978) Sur l'origine des gneiss oeillés de l'Espinouze, zone axiale de la Montagne Noire. Bull Soc Géol Fr (7) 5: 751–772

Bard JP, Rambeloson R (1973) Métamorphisme plurifacial et sens de variation du degré géothermique durant la tectogenèse polyphasée hercynienne dans la partie orientale de la zone axiale de la Montagne Noire (Massif du Caroux, Sud du Massif Central français). Bull Soc Géol Fr 7 XV 5–6: 579–586

Bard JP, Burg JP, Matte P, Ribeiro A (1980) La chaîne hercynienne d'Europe Occidentale en termes de tectonique des plaques. In: Cogné J, Slansky M (eds) Géologie de l'Europe du Précambrien aux bassins sédimentaires post hercyniens, 26ème Cong Géol Int Coll C6, Paris 1980. Ann Soc Géol Nord, Lille, XCIX, pp 233–246

Barrere J (1971) Notice et carte géologique à 1/50 000. Feuille de St Bonnet de Joux et Montceau les Mines. Bur Rech Géol Min, Orléans

Beaud F (1985) Etude structurale de la Zone axiale orientale de la Montagne Noire (Sud du Massif Central français). Détermination des mécanismes de déformation. Relation avec les nappes du Versant Sud. Thèse de 3ème cycle, Univ. Montpellier, 191 pp

Becq Giraudon JF (1973) Etude géologique du bassin houillier de Graissessac (Hérault). Bull Bur Rech Géol Min 3: 151–163

Belin JM (1983) Le Massif de St Gervais d'Auvergne (Massif Central français). Evolution d'un ensemble granitique composite. Thèse 3ème cycle. Univ Clermont Ferrand, 151 pp

Bernard-Griffiths J (1975) Essai sur la signification des âges au strontium dans une série métamorphique. Le Bas-Limousin (Massif Central français). Ann Fac Sci Clermont 55, 243 pp

Bernard-Griffiths J, Gebauer D, Grunenfelder M, Piboule M (1985) The tonalite belt of Limousin (French Central Massif): U-Pb zircon age and geotectonic implications. Bull Soc Géol I, 4: 523–529

Berthé D, Choukroune P, Jegouzo P (1979) Orthogneiss, mylonite and non-coaxial deformation of granites: the example of the South Armorican Shear Zone. J Struct Geol 1: 31–42

Bogdanoff S (1973) Contribution à l'étude géologique de l'extrémité orientale de la zone axiale granito-gneissique de la Montagne Noire (Monts de l'Espinouse, Hérault). Bull Bur Rech Géol Min 2: 1–28

Bonijoly D, Castaing C (1984) Fracturation et genèse des bassins stéphaniens du Massif Central français en régime compressif. Ann Soc Géol Nord, Lille, CIII: 187–193

Bonnion S, Courel L, Gelard JP, Paquette Y (1983) L'organisation des dépôts de charbon et de stérile dans le bassin de l'Aumance (Allier): tectonique sédimentaire et syndiagénétique. Mém Geol Univ Dijon 8: 87–97

Bouchardon JL (1987) Evaluation des conditions de Pression-Température de "top" métamorphique des éclogites du Lévezou (Rouergue). C R Acad Sci Paris II 305: 271–275

Bouchardon JL (1989) Témoins de cisaillements syn-éclogitisation: les éclogites blastomylonitiques à clastes omphacitiques du Rouergue. C R Acad Sci Paris II 309: 69–72

Bouchardon JL, Dechomets R, Demange M (1979) A propos du disthène en roche dans les micaschistes et les gneiss du synclinal de Rossis et du flanc sud, zone axiale de la Montagne Noire (Massif Central français). C R Acad Sci Paris D 288: 1067–1071

Bouchez JL, Jover O (1986) Le Massif Central français: un chevauchement de type himalayen vers l'Ouest-Nord-Ouest. C R Acad Sci Paris II 302: 675–680

Briand B (1978) Métamorphisme inverse et chevauchement de type "himalayen" dans la série de la vallée du Lot (Massif Central français). C R Acad Sci Paris II, 286: 729–731

Briand B, Gay M (1978) La série inverse de Saint-Geniez-d'Olt: évolution métamorphique et structurale. Bull Bur Rech Géol Min (2) I 3: 167–186

Brun JP, Burg JP (1982) Combined thrusting and wrenching in the Ibero-Armorican arc: a corner effect during continental collision. Earth Planet Sci Lett 61: 319–332

Brunel M (1974) La phase hercynienne majeure dans le versant nord de la Montagne Noire (Ecaille de Lacaune). C R Acad Sci Paris II, 278: 991–994

Burg JP (1977) Tectonique et microtectonique des séries cristallophylliennes du Haut Allier et de la vallée de la Truyère. Thèse 3 ème cycle, Univ Montpellier, 79 pp

Burg JP, Matte P (1977) La klippe de la Bessenoits (Decazeville, Aveyron): un nouvel argument en faveur de l'existence d'un chevauchement majeur vers le Sud dans le Massif Central. C R Somm Soc Géol Fr 6: 325–329

Burg JP, Matte P (1978) A cross-section through the French Massif Central and the scope of its Variscan geodynamic evolution. Z Dtsch Geol Ges 109: 429–460

Burg JP, Teyssier C (1983) Contribution à l'étude tectonique et microtectonique des séries cristallophylliennes du Rouergue oriental: la déformation des laccolithes syntectoniques, type Pinet. Géol Fr 2, 1, 1–2: 3–30

Burg JP, Leyreloup A, Marchand J, Matte Ph (1984) Inverted metamorphic zonation and large-scale thrusting in the Variscan belt: an example in the French Massif Central. In: Hutton DHW, Sanderson DJ (eds) Variscan tectonics of the North-Atlantic region. Spec Publ Geol Soc London 14: 47–61

Burg JP, Bale P, Brun JP, Girardeau J (1987) Stretching lineation and transport direction in the Ibero-Armorican arc during the Siluro-Devonian collision. Geodin Acta 1, 1: 71–87

Burg JP, Delor CP, Leyreloup AF, Romney F (1989) Inverted metamorphic zonation and Variscan thrust tectonics in the Rouergue area (Massif Central, France): P-T-t record from mineral to regional scale. In: Daly JS, Cliff RA, Yardley BWD (eds) Evolution of metamorphic Belts. Geol Soc Lond Spec Publ 439: 423–439

Caen Vachette M, Couturie JP, Didier J (1982) Age radiométrique des granites anatectiques et tardimigmatitique du Velay (Massif Central français). C R Acad Sci Paris II, 294: 135–138

Caen Vachette M, Gay M, Peterlongo JM, Pitiot P, Vitel G (1984) Age radiométrique du granite syntectonique du gouffre d'Enfer et du métamorphisme hercynien dans la série de basse pression du Pilat (Massif Central Français). C R Acad Sci Paris II, 299: 1201–1204

Cantagrel JM (1973) Signification des âges à l'argon déterminés sur amphiboles dans les socles métamorphiques anciens: application au Massif central français et à l'Aleksod, Sahara algérien. Ann Sci Univ Clermont, 50

Castaing C (1982) Gitologie prévisionnelle des charbons dans le Nord Est du Massif Central. Rapport Bur Rech Géol Min 82 SGN 326 GEO

Choukroune P, Gapais D, Matte P (1983) Tectonique hercynienne et déformation cisaillante: la faille ductile de la Marche (Massif Central français). C R Acad Sci Paris II, 296: 859–862

Costa S (1989) Age radiométrique 39Ar/40Ar du métamorphisme des séries du Lot et du charriage du groupe leptyno-amphibolique de Marvejols (MCF). C R Acad Sci Paris II, 309: 561–567

Costa S, Maluski H (1988) Datations par la méthode ^{39}Ar–^{40}Ar de matériel magmatique et métamorphique paléozoïque provenant du forage de Couy-Sancerre (Cher, France). Programme GPF. C R Acad Sci Paris II 306: 351–356

Costa S, Maluski H (1989) 39Ar/40Ar dating of Variscan tectonometamorphic events: the example of the French Massif Central. IGCP 233 Conference, Athens-Georgia, July 1989 (Abstr)

Costa S,Maluski H, Lardeaux JM (1989) 40Ar/39Ar dating in the Monts du Lyonnais area (French Massif Central): an uplifted complex in a collisional context. TERRA, vol 1, 1, EUG, Strasbourg, March 1989, p 265 (Abstr)

Courel L (1987) Les bassins houillers du Paléozoique supérieur continental du Massif Central français: dynamique sédimentaire. Ann Soc Géol Nord 106: 229–237

Couturie JP, Caen-Vachette M (1979) Age Namurien d'un laccolite granitique différencié par gravité: le granite de la Margeride (Massif Central français). C R Acad Sci Paris II, 289: 449–452

Demange M (1975) Style pennique de la zone axiale de la Montagne Noire. Bull Bur Rech Géol Min 2 °S, 2: 91–139

Demange M (1982) Etude géologique du massif de l'Agout, Montagne Noire, France. Thèse d'état, Univ. Paris VI, 1050 pp

Demange M, Jamet P (1985) Le stade majeur du métamorphisme est de type moyenne pression sur le flanc sud de la montagne Noire. C R Acad Sci Paris II, 301 9: 603–606

Demange M, Leger M (1980) Présence d'éclogites en Cabardes: un nouveau témoin de métamorphisme de haute à moyenne pression sur le flanc sud de la zone axiale de la Montagne Noire (Massif Central français). C R Acad Sci Paris II, 567: 2410–2412

Demay A (1942) Microtectonique et tectonique profonde. Mem Serv Carte Géol Fr, Paris, 1 vol, 247 pp

Donnot M, Guerange B (1978) Le synclinorium de Brusque: implications stratigraphiques et structurales dans les Monts de l'Est de Lacaune (Tarn, Aveyron, Hérault). Bull Bur Rech Géol Min 4: 333–363

Dubuisson G, Mercier JCC, Girardeau J, Frison JY (1989) Evidence for a lost ocean in Variscan terranes of the western Massif Central, France. Nature 337, 23: 729–732

Ducrot J, Lancelot JR, Marchand J (1983) Datation U-Pb sur zircons de l'éclogite de la Borie (Haut-Allier, France) et conséquences sur l'évolution anté-hercynienne de l'Europe Occidentale. Earth Planet Sci Lett 18: 97:113

Duthou JL (1977) Chronologie Rb-Sr et géochimie des granitoïdes d'un segment de la chaîne varisque. Relations avec le

métamorphisme: le Nord-Limousin. Ann Fac Sci Clermont, 63, fasc. 30, 290 pp
Duthou JL, Cantagrel JM, Didier J, Vialette Y (1984) Paleozoic granitoids from the French Massif Central: age and origin studied by the 87Rb/87Sr system. Phys Earth Planet Int Amsterdam 35: 131–144
Echtler H (1990) Geometry and kinematics of recumbent folding and low angle detachment in the Pardailhan nappe (Montagne Noire, Southern French Massif Central). Tectonophysics 177: 109–123
Echtler H, Malavieille J (1990) Extensional tectonics, basement uplift and Stephano-Permian collapse basin in a late Hercynian Metamorphic core complex from the French Massif Central (Montagne Noire). Tectonophysics 177: 125–138
Echtler H, Malavieille J, Matte P, Brunel M (1988) Kinematics of Late orogenic uplift, extension and wrenching in the Montagne Noire (Southern French Massif Central). IGCP Project n°233, Intern Conf Montpellier, p 21 (Abstr)
Ellenberger F (1967) Replis de micaschistes et tectonique d'infrastructure au sein du massif granitique du Caroux (zone axiale de la Montagne Noire). Comptes Rendus de la Société Géologique de France, pp 227–228
Engel W (1984) Migration of folding and flysch sedimentation on the southern flank in the Variscan belt (Montagne Noire, Mouthoumet massif, Pyrenes). Ret Dtsch Geol Ges 135: 279–292
Engel W, Feist R, Franke W (1981) Le Carbonifère anté-Stéphanien de la Montagne Noire: rapports entre mise en place des nappes et sédimentation. Bull Bur Rech Géol Min 2, I: 341–389
Faure M, Pin C, Mailhe D (1979) Les roches mylonitiques associées au charriage du groupe leptyno-amphibolique sur les schistes du Lot dans la région de Marvejols (Lozère, MCF). C R Acad Sci Paris, D: 288: 1267–1270
Feist F, Galthier J (1985) Découverte de flancs d'age namurien probable dans le flysch à olistolites de Cabrières (Hérault). Implication sur la durée de la sédimentation synorogénique dans la Montagne Noire. C R Acad Sci Paris II, 4, 300: 207–212
Feybesse JL (1981) Tectonique et microtectonique de la région de Laroquebrou (Cantal, Massif Central français). Rôle de la déformation ductile et évolution du Sillon Houiller. Thèse 3ème cycle, Univ. Clermont-Ferrand, 182 pp
Feybesse JL, Tegyey M (1987) Evolution tectonométamorphique dévonienne et carbonifère de la série de la Sioule. Géol Fr 2–3: 33–41
Feybesse JL, Lardeaux JM, Johan V, Tegyey M, Dufour E, Lemiere B, Delfour J (1988) La série de la Brévenne (Massif Central Français): une unité dévonienne charriée sur le complexe métamorphique des Monts du Lyonnais à la fin de la collision varisque. C R Acad Sci Paris II 307: 991–996
Floc'h JP (1983) Le socle métamorphique du Limousin Central: une traverse de la branche ligérienne de l'orogène varisque de l'Aquitaine à la zone broyée d'Argentat (Massif Central français): Thèse d'Etat, Univ Limoges, 445 pp
Gapais D, Barbarin B (1986) Quartz fabric transition in a cooling syntectonic granite (Hermitage Massif, France). Tectonophysics 125: 357–370
Gapais D, Bale P, Choukroune P, Cobbold PR, Mahjoub Y, Marquer D (1987) Bulk kinematics from shear zone patterns: some field examples. J Struct Geol 9, 5/6: 635–646
Gardien V, Lardeaux JM, Misseri M (1988) Les péridotites des Monts du Lyonnais (M.C.F.): témoins privilégiés d'une subduction de lithosphère paléozoique. C R Acad Sci Paris II, 307: 1967–1972
Gardien V, Tegyey M, Lardeaux JM, Misseri M, Dufour E (1990) Crustal-mantle relationships in the french Variscan chain: the example of the Southern Monts du Lyonnais unit (eastern french Massif Central). Metamorph Geol 8: 477–492
Gay M, Peterlongo JM, Caen-Vachette M (1981) Age radiométrique des granites allongés et en feuillets minces syntectoniques dans les Monts du Lyonnais (Massif Central français). C R Acad Sci Paris II 293: 993–996
Gelard JP, Castaing C, Bonijoly D, Grolier J (1986) Structure et dynamique de quelques bassins houillers limniques du Massif Central. Mem Soc Géol Fr 149: 57–72
Geze B (1949) Etude géologique de la Montagne Noire et des Cévennes méridionales. Mem Soc Géol Fr, Paris XXIX, 62, 215 pp
Girardeau J, Dubuisson G, Mercier JC (1986) Cinématique de mise en place des ophiolites et nappes cristallophylliennes du Limousin, Ouest du Massif Central français. Bull Soc Géol Fr (8) II 5: 849–860
Grangeon M, Feys R, Greber CH, Lefavrais Raymond A (1968) Geologie profonde de la région de Decize (Nièvre). Essai de synthèse d'après les sondages profonds. Bull Bur Rech Géol Min 4: 43–108
Grolier J (1971) La tectonique du socle hercynien dans le Massif Central français. In: Géologie et structure profonde du Massif central de France. Plein Air Service, Clermont Ferrand, 610 pp
Grolier J, Herz N, Duthou JL, Mergoil J, Monier G, Riss J, de la Roche H, Roques M, Sossa M, Stussi JM, Turland M, Vachette M (1984) Idées actuelles (mars 1984) sur la géologie, la géochimie et la géochronologie du complexe granitique de Montmarault (Allier, Puy de Dôme). Doc Bur Rech Géol Min 81-2: 65–80
Guerange Lozes J, Guerange B (1984) Les nappes de charriage de l'Albigeois cristallin (Tarn et Aveyron, France). C R Acad Sci Paris II 299: 471–474
Hamet J, Allègre CJ (1976) Hercynian orogeny in the Montagne Noire (France). Application of 87Rb-87Sr systematics. Geol Soc Am Bull 87: 1429–1442
Lacassin R, van den Driessche J (1982) Analyse de la déformation hercynienne majeure des gneiss de la Cézarenque (Cévennes, Massif Central). C R Acad Sci Paris II 295: 1027–1030.
Lafon JM (1986) Géochronologie U-Pb appliquée à deux segments du Massif Central français. Le Rouergue oriental et le Limousin central. Thèse d'Etat, Univ. Montpellier, 152 pp
Lardeaux JM, Dufour E (1987) Champs de déformation superposés dans la chaîne varisque. Exemple de la zone nord des Monts du Lyonnais (Massif Central français). C R Acad Sci Paris II 305: 61–64
Ledru P (1988) Le chevauchement crustal de Moyenne Dordogne. Bull Soc Géol Fr (8) IV 3: 471–478
Ledru P, Autran A (1987) L'édification de la chaine varisque dans le Limousin. Rôle de la faille d'Argentat à la limite Limousin-Millevaches. In: Bur Rech Géol Min (ed) Programme Géologie profonde de la France. Doc Bur Rech Géol Min, Orléans, 140, pp 51–81
Ledru P, Lardeaux JM, Santallier D, Autran A, Quenardel JM, Floc'h JP, Lerouge G, Maillet N, Marchand J, Ploquin A (1989) Où sont les nappes du Massif Central Français? Bull Soc Géol Fr 8 V 3: 605–618
Lee BJ, Faure M, Cluzel D, Cadet JP (1988) Mise en évidence d'un cisaillement ductile d'Ouest en Est dans les nappes du versant Sud de la Montagne Noire (S du Massif Central). C R Acad Sci Paris II, 306: 455–462
Lerouge G (1988) Tectogenèse comparée de deux segments de la chaîne hercynienne: Le Massif Central français septentrional et le Sud du Massif armoricain. Thèse Sci Univ Paris-Sud, Orsay, et Mémoire Géodiffusion, 2 (1988), Paris, 363 pp
Lerouge G, Quenardel JM (1985) Chronologie des évènements tectoniques dans le Nord-Ouest du Massif Central français et

le Sud du bassin de Paris du Carbonifère inférieur au Plio-quaternaire. C R Acad Sci Paris 301 II 9: 621–626

Lerouge G, Quenardel JM (1988) Les zones de cisaillement carbonifères dans les plutons vendéens et leur prolongement dans le NO du Massif Central. Bull Soc Géol Fr (8) IV 5: 813–838

Leroy J, Holiger P (1984) Mineralogical, chemical and isotopic (U-Pb method) studies of hercynian uraniferous mineralizations (Margnac and Fanay mines, Limousin, France). Chem Geol 45: 121–134

Lespinasse M, Pecher A (1984) Déformation ductile et rupturelle au voisinage de la faille de la Marche. C R Acad Sci Paris 298: 17–20

Lespinasse M, Mollier B, Delair J, Bladier Y (1986) Structuration tangentielle et chevauchements carbonifères dans les leucogranites du NW du Massif Central français: l'exemple des failles de Bussière-Madeleine et D'Arrènes-Ouzilly. C R Acad Sci Paris II 303: 1575–1580

Lorenz V, Nicholls IA (1976) Plate and intraplate processes of Hercynian Europe during the Late Paleozoic. Tectonophysics 107: 25–56

Lorenz V, Nicholls IA (1984) The Permo-Carboniferous basin and range province of Europe. An application of plate tectonics. In: Falke H (ed) The continental Permian in Central, West and South Europe. Reidel, Dordrecht-Holland, pp 313–342

Macaudiere J, Marignac C, Weisbrod A (1987) Grandes nappes synschisteuses collisionnelles dans la catazone hercynienne des Cévennes médianes (Massif Central, France). C R Acad Sci Paris II 304 19: 1195–1199

Malavieille J, Guihot P, Costa S, Lardeaux JM, Gardien V (1990) Collapse of the thickened Variscan crust in the French Massif Central: Mont Pilat extensional shear zone and St Etienne upper Carboniferous basin. Tectonophysics 177: 139–149

Maluski H, Monie P (1988) 40Ar/39Ar laser probe multi-dating inside single biotites of a Variscan orthogneiss (Pinet, Massif Central, France). Chem Geol (Isotope Geoscience section) 73: 245–263

Maluski H, Costa S, Echtler H (1990) Late Variscan tectonic evolution by thinning of earlier thickened crust: 39Ar-40Ar study of the Montagne Noire, Southern Massif Central, France. Lithos 26, 314: 287–304

Mansuy C (1983) Les microsphères du Protérozoique supérieur du Massif Armoricain (Briovérien): affinités biologiques et utilisation stratigraphique. Thèse, Univ. Rennes, 107 pp

Marquer D, Feybesse JL, Gapais D, Bale P, Lecorre C, Maggion R, Charvet J, Faure M, Burg JP (1987) Dispersion des linéations d'étirement sous un chevauchement varisque majeur: le cas de l'orthogneiss de St Alyre. Sondage GPF de Chassolle. Géol Fr 4: 33–43

Martin D (1980) Etude structurale, pétrologique et gîtologique de la vallée du Chavanon (Massif Central français). Thèse 3ème cycle, Univ. Clermont-Ferrand, 132 pp

Mattauer M, Echecopar A (1977) Arguments en faveur de chevauchements de type himalayen dans la chaîne hercynienne du Massif Central français. In: Coll Intern CNRS, Ecologie et Géologie de l'Himalaya. Paris, 1976, CNRS 268, pp 261–267

Matte P (1986) Tectonics and plate tectonics model for the Variscan belt of Europe. Tectonophysics 126: 329–374

Mezure JF (1981) Orientation préférentielle des mégacristaux de feldspath potassique dans les granites: application à l'étude structurale des granites d'Egletons et de Meymac (Corrèze, Massif Central français). Bull Soc Géol Fr (7) XXIII 6: 641–649

Mollier B, Lespinasse M (1985) Déformation magmatique et plastique en limite nord du granite de St Sylvestre (Nord-Ouest du Massif Central français): la faille d'Arrènes Ouzilly. C R Acad Sci Paris II 300: 681–686

Petrequin M (1979) Etude gravimétrique du massif de Margeride et de sa bordure orientale. Thèse 3éme cycle, Univ Montpellier, 130 pp

Piboule M, Briand B, Beurrier M (1982) Géochimie de quelques granites albitiques dévoniens de l'Est du Massif Central (France). Neues Jahrb Miner Abh 143 3: 279–308

Pin C (1980) Données microstructurales sur les terrains métamorphiques de la série du Lot. Bull Bur Rech Géol Min I 4: 293–314

Pin C (1981) Old inherited zircons in two synkinematic variscan granitoïds: the "granite du Pinet" and "the orthogneiss de Marvejols" (Southern Massif Central). Neues Jahrb Miner Abh 142: 27–48

Pin C (1989) Essai sur la chronologie et l'évolution géodynamique de la chaîne hercynienne d'Europe. Thèse, Clermont Ferrand II, 470 pp

Pin C, Lancelot JR (1982) U-Pb dating of an early paleozoïc bimodal magmatism in the french Massif Central and of its further metamorphic evolution. Contrib Mineral Petrol 79: 1–12

Pin C, Peucat JJ (1986) Ages des épisodes de métamorphisme paléozoïque dans le Massif Central et le Massif Armoricain. Bull Soc Géol Fr (8) II 3: 461–469

Pin C, Piboule M (1988) Age Dévonien supérieur de la série calco-alcaline de la ceinture basique du Lévezou. Bull Soc Géol Fr 2: 261–266

Pin C, Vielzeuf D (1988) Les granulites de haute-pression d'Europe moyenne témoins d'une subduction éo-hercynienne. Implications sur l'origine des groupes leptyno-amphibolitiques. Bull Soc Géol Fr (8) IV 1: 13–20

Quenardel JM, Rolin P (1984) Paleozoic evolution of the Plateau d'Aigurande (N-W Massif Central, France). In: Hutton D, Sanderson D (ed) Variscan tectonics of the North Atlantic region. Geol Soc London Spec pub 14 pp, 63–77

Reitz E, Wickert F (1988) Upper Proterozoïc microfossils in low-grade phyllites of the Brevenne Unit, NE Massif Central (France). C R Acad Sci Paris II 307: 1717–1721

Rolin P (1987) Tectogenèse comparée de deux segments de la chaîne hercynienne: le Massif Central français septentrional et la Meseta marocaine septentrionale. Thèse Doctorat ès Sciences, Univ de Paris Sud, Orsay, 484 pp

Sider H, Leistel JM, Gagny C (1986) Réflexion sur l'évolution géodynamique de la chaine hercynienne au Paléozoique dans le Nord Est du Massif Central français. Bull Soc Géol Fr 8 II 4: 637–644

Valle B, Courel L, Gelard JP (1988) Les marqueurs de la tectonique synsédimentaire et syndiagenetique dans le bassin Stéphanien à régime cisaillant de Blanzy-Montceau (Massif Central, France). Bull Soc Géol Fr 4: 529–540

Vignard G (1976) Mise en évidence du Viséen dans la région de St Pons. Conséquences sur l'interprétation tectonique des séties réputées autochtones de l'unité du St Ponais (versant Sud de la Montagne Noire). C R Acad Sci Paris D, 283: 607

Wickert F (1988) Paleozoic evolution of continental crust in the Beaujolais-Lyonnais area, northeastern part of the Massif Central, France Geol Rundsch 77/2: 467–482

Metamorphism

D. S. Santallier, J. M. Lardeaux, J. Marchand, and Ch. Marignac

1 Introduction

Structural, petrological, and geophysical data collected during the last decade have shown that the Variscan nappes of the French Massif Central are the result of subduction and collision tectonics which evolved during ca. 150 Ma (Cogné 1976; Burg and Matte 1978; Autran and Cogné 1980; Bard et al. 1980; Kornprobst et al. 1980; Matte 1983, 1986; Pin and Peucat 1986; Robinson et al. 1988; Ledru et al. 1989). Thus, the tectono-metamorphic history of the French Paleozoic orogen is polyphase (Santallier 1983b). In such a context, single estimates of the pressure and temperature (P-T) conditions that prevailed during regional metamorphism provide little information on the orogenic history of a nappe belt (England and Richardson 1977; Thompson and England 1984; Davy and Gillet 1987). A dynamic approach of metamorphism is required to define the pressure-temperature-time (P-T-t) paths of each tectono-metamorphic unit stacked in the orogenic zone (see, for example, the high-pressure and low-temperature metamorphism in the western Alps, Caby et al. 1978; Lardeaux et al. 1982; Rubie 1984; Gillet et al. 1986). The purpose of this chapter is to discuss the P-T-t evolution of the main nappes of the French Massif Central, in order to delineate the major factors that control the development of regional metamorphism in the internal part of the Variscan chain.

The structural features observed in the Massif Central indicate that the microstructures of the Variscan metamorphic re-equilibration display a range of morphologies. These are related to heterogeneities in the finite strain distribution. The older metamorphic assemblages are found, as relicts, within undeformed (or slightly deformed) volumes (1–10 m^3) of rocks in which corona-type reactions are well developed, allowing the superimposed metamorphic stages to be defined. These small, generally lenticular, volumes of rocks, which are weakly affected by Variscan foliations, are embodied in larger masses of polydeformed tectonites which are pervasively reequilibrated at the granular scale. The latter metamorphic equilibrium is achieved during the major deformation event. The metamorphic parageneses can only be dated relatively when the structural relationships can be observed in a single outcrop. Unfortunately, this is generally not the case, so the P-T-t paths must be reconstructed using a combination of several types of information from both undeformed and deformed samples collected in different but equivalent locations.

2 The Nappe Pile (Figs. 1 and 2)

A "static" interpretation of the orogen responsible for the Massif Central had been favored for 20 years, despite the pioneering works of Demay (1942). Many authors produced sketches of successive allochthonous tectonic events of the French Massif Central (Mattauer and Etchecopar 1976; Briand 1978; Burg and Matte 1978; Floc'h 1983; Burg et al. 1984; Bouchez and Jover 1986; Friedrich et al. 1988). The five geometrically stacked units, recently defined by Ledru et al. (1989) (Fig. 2), are described here. The tectono-metamorphic evolution of each unit is presented in order to discuss its geodynamic significance. The units are described in order from top to bottom.

2.1 The Upper Allochthon Epizonal Units

These comprise two different epizonal series which are thrust over the middle allochthon units. They represent a minor volume of the tectonic pile, and are exposed in the western (Limousin) and northeastern (Brévenne) Massif Central. The Limousin unit is composed of more or less reworked rhyodacitic tuffs (associated with calc-alkaline dolerites, Cabanis et al. 1983), meta-ignimbrites, and Ordovician and Silurian marine sediments (Guillot and Doubinger 1971; Guillot and Lefevre 1975). Some of the Brévenne schists have been attributed recently to the upper Precambrian (Reitz and Wickert 1988) although this is still

J. D. Keppie (Ed.): Pre-Mesozoic Geology
in France and Related Areas

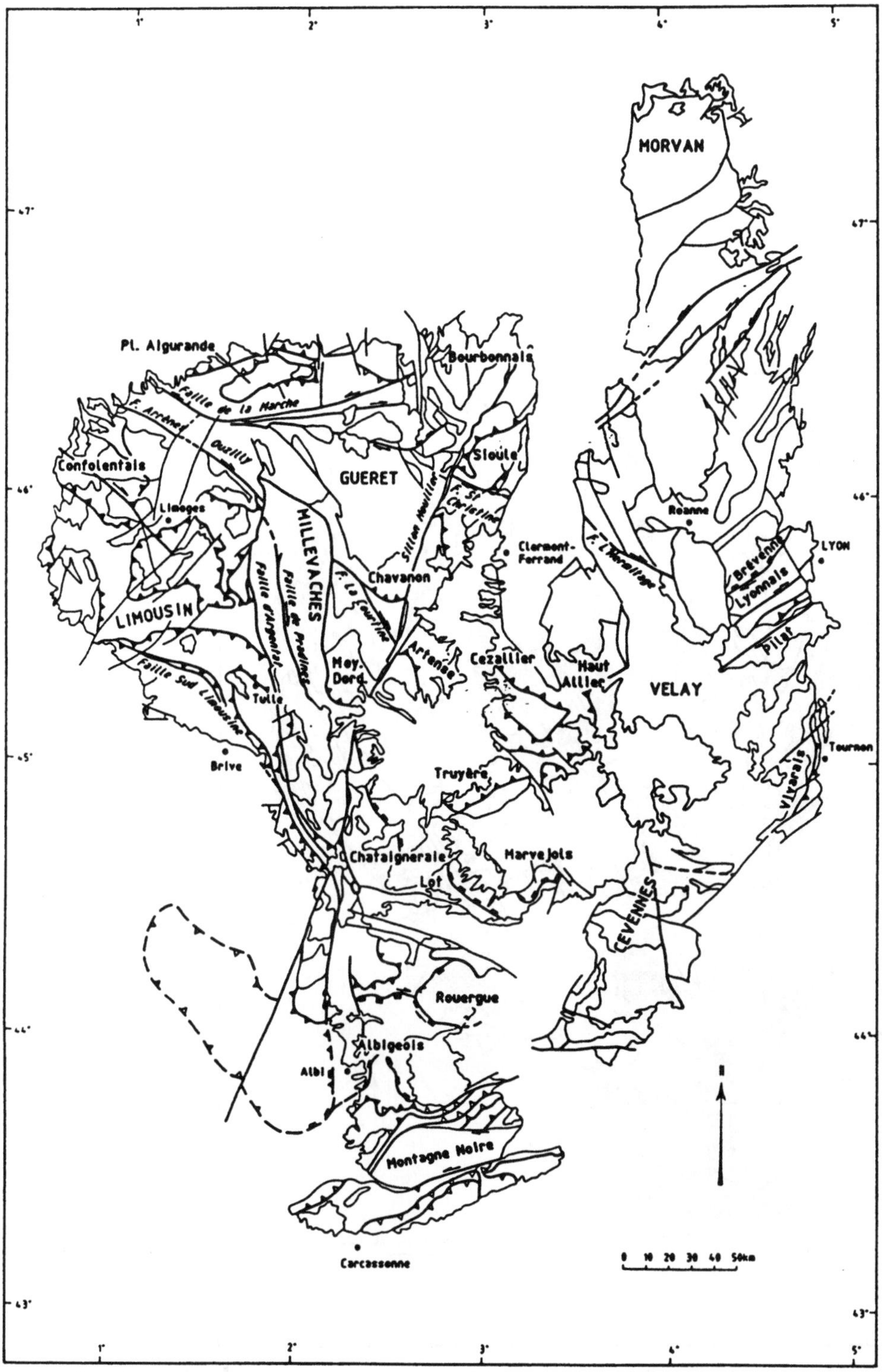

Fig. 1. The nappes pile in the French Massif Central. Toponymic map. (Ledru et al. 1989)

under debate. Dismembered ophiolites have been described in both areas (Durand 1981; Guillot 1981; Piboule et al. 1982). The units have been affected by a Devonian (Limousin) or lower Carboniferous (Brévenne) mainly epizonal metamorphism and are devoid of any high pressure relicts. The tectonic contacts with neighboring units are strike-slip ductile faults with a more or less pronounced thrusting component (Feybesse et al. 1988; Ledru and Autran 1988). These tectonic contacts are well dated only in

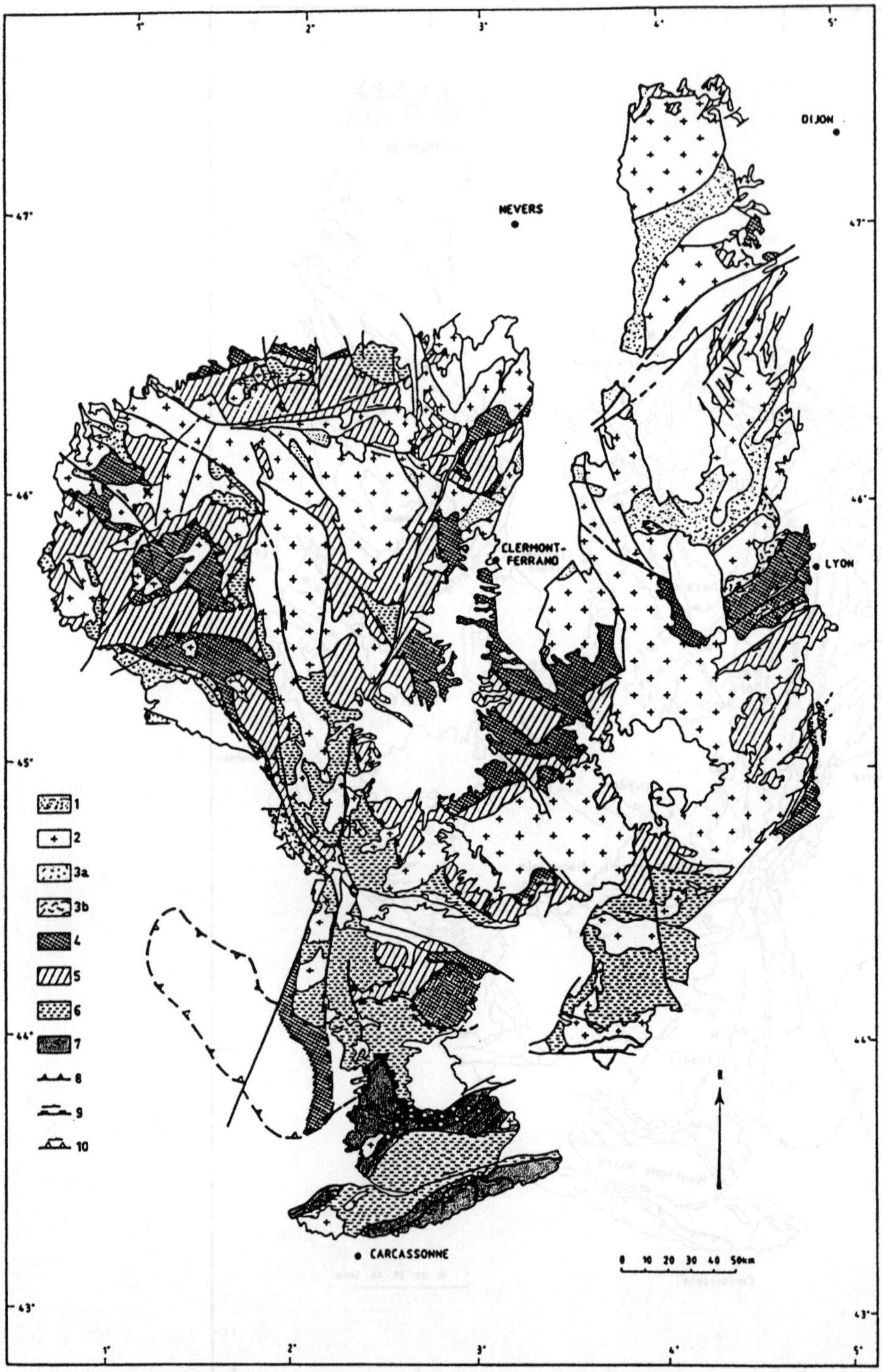

Fig. 2. The nappes pile in the French Massif Central. Geological sketch map. (Ledru et al. 1989). *1* mesozoic cover; *2* post-tectonic granitoids; *3* upper allochthon; *3a* Thiviers-Payzac and Génis Units; *3b* Brévenne unit; *4* middle allochthon; *5* lower allochthon; *6* presumed autochthon; *7* southern Paleozoic units; *8* thrust nappes (390 ± 370 Ma); *9* thrusts and transcurrent zones (350 ± 320 Ma); *10* thrusts and transcurrent zones (320 ± 10 Ma)

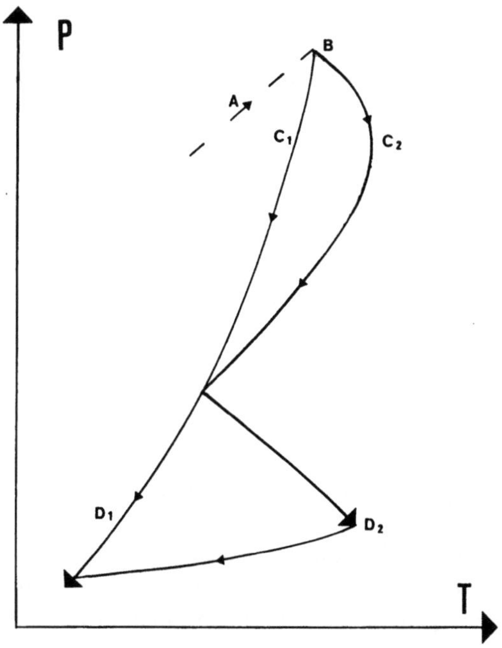

Fig. 3. The different kinds of P-T paths described in the French Masif Central (see text for discussion)

the Brévenne area, where their age is 332 ± 10 Ma (Rb-Sr whole rock) (Gay et al. 1981) or between 340 and 350 Ma (39Ar–40Ar Bt) (Costa and Maluski 1988).

2.2 The Middle Allochthon Units

These are composed of more or less evolved graywacke sediments containing a great amount of orthogneissic and/or basic intercalations, particularly widespread in the leptynitic-amphibolitic complexes (LAC) (Forestier et al. 1973; Santallier et al. 1988). All lithologies commonly exhibit high-pressure and generally high-temperature relicts such as acidic and basic granulites, high-pressure amphibolites, granulitic orthogneisses, and eclogites (Marchand 1974; Lasnier 1977; Pin and Vielzeuf 1988). The relicts are particularly well exposed in the northern Massif Central, from the western Limousin to the eastern Lyonnais areas and in the southern Lot and Rouergue areas (Nicollet and Leyreloup 1978; Kornprobst et al. 1980; Santallier 1981; Dufour et al. 1985; Delor et al. 1986; Bouchardon 1987; Mercier et al. 1989). These units have been affected by widespread migmatization of Devonian and Carboniferous age, and are intruded by large granitic plutons (Bernard-Griffiths et al. 1977; Duthou et al. 1984). The prograde relicts are scarce but they allow distinction of two prograde A courses, HT and LT (high temperature and low temperature), prior to the metamorphic climax (Fig. 5). The LT course is restricted to the small Najac area (Delor et al. 1986, 1987). In the middle allochthon, crustal and mantelic rocks are tectonically juxtaposed, and both exhibit prograde and retrograde courses (Maillet 1987; Gardien et al. 1988). Their tectonic coupling occurred relatively late in the tectonometamorphic history, during the main deformation of the crustal rocks.

2.3 The Para- and Ortho-Gneissic Medium- to High-Grade Lower Allochthon Units

These are devoid of any high pressure relicts in the eastern areas (Lyonnais, Haut-Allier) but exhibit some in situ eclogitic relicts (Santallier and Floc'h 1979) in the west (Limousin, Plateau d'Aigurande). Although the metamorphic grade is highly variable, the entire lower allochthon has apparently suffered Devonian (ca. 400–360 Ma) tectonometamorphic evolution (well dated only in Limousin: Bernard-Griffiths et al. 1977) under medium pressure conditions. The thrust plane between the middle and lower allochthons is generally marked by ductile mylonites (amphibolite facies conditions) and sporadic ophiolitic remnants (Girardeau et al. 1986; Maillet 1987) which are synmetamorphic with the Devonian metamorphism.

2.4 The Presumed Autochthonous Units

These are composed mainly of micaschists and locally orthogneisses, somewhat similar to the lower allochthon. The autochthon apparently represents an old stable continental margin on which the nappes were later emplaced during Devonian times with the development of ductile mylonites in low amphibolite or greenschist facies conditions. The metamorphic imprint is highly variable. In the southwestern areas (Albigeois and Lot), there are only epizonal chlorite and biotite bearing assemblages of Devonian age (380 ± 10 Ma, $^{39}Ar/^{40}Ar$ Ms in Guérangé-Lozes (1987); see also Briand and Gay (1978), Guérangé-Lozes and Alsac (1986)). A later LP-HT thermal event is well developed in the Millevaches (Ledru 1988) and Cévennes (Weisbrod 1970; Marignac et al. 1980; Weisbrod et al. 1980; Toteu 1981) areas, giving rise to medium- to high-grade andalusite-sillimanite- and cordierite-bearing assemblages.

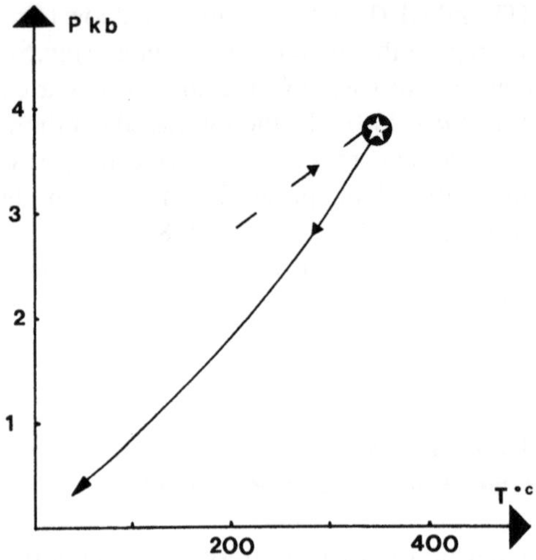

Fig. 4. The P-T path of the upper allochthon epizonal units. The *star* indicates the P-T conditions of the main deformation event of this unit

2.5 The Southern Paleozoic Units

These are composed of paleontologically dated schists and limestones, weakly or nonmetamorphosed, corresponding to margin sediments (Guérangé–Lozes 1987).

3 Tectono-Metamorphic Evolution in the Different Nappes

From the observation of the mineralogical assemblages and their evolution and from thermobarometrical calculations, it must be emphasized that only a restricted number of shapes of the PT curves have been recognized in the Massif Central. These several possible PTt paths are presented in Fig. 3. All of these curves are in accord with the theoretical tectono-thermal considerations of England and Richardson (1977) and England and Thompson (1984). The relict parageneses are distinguished as either prograde (A) or representative of the Maximum Pressure event (B). The most widespread typomorphic (C) and retromorphic (D) assemblages of the decompressional course are also distinguished. Stage C can be subdivided into two different paths: C1 corresponds to a rapid uplift (with a less pronounced thermal effect: Davy and Gillet 1987) and C2 corresponds to a slower uplift with more thermal re-equilibration (Thompson and England 1984). Stage D is also divided into D1 (low temperature) and D2 (high temperature) trajectories corresponding to a possible heat supply related to magmatism and/or crustal thinning at the end of the orogenic history.

Tables 1 to 4 display the lithologies, nature, and ages of the protoliths, mineral assemblages and P-T estimates for the different formations. The absolute ages of the stages A, B, C and D and the related timing of deformation may be different in the different units. Consequently, for each unit, one or two detailed regional example(s) will be described. In some cases, metamorphic maps with isograds have been drawn. They were used as a basis for the definition of the nappes and the construction of the P-T curves. Readers interested in these maps, as well as in facies analyses and mineral chemistry, will find them in the referenced papers.

3.1 The Upper Allochthon Epizonal Units

In the Upper Allochthon Epizonal Units (Fig. 4 and Table 1) only two episodes are identified, with reference to our general diagram (Fig. 3). In both areas, B is the P and T maximum; it is Devonian in the Limousin area and lower Carboniferous in the Brévenne area. The later path corresponds to a decrease of both P and T, which gives rise to very low grade metamorphic mineral assemblages (C-D stage). Both areas are devoid of any high pressure relicts.

3.1.1 The South-Limousin Example
(D. S. Santallier and J. P. Floc'h)

The western Massif Central provides a good example of the upper allochthon tectono-metamorphic evolution. The upper allochthon unit is subdivided (Guillot 1981) into two subunits. The lower Thiviers-Payzac subunit is composed of calc-alkaline formations: more or less completely reworked rhyodacitic tuffs and interbedded graywackes, bimodal volcaniclastic rocks and cross-cutting doleritic dykes all overlain by ignimbritic welded tufs. The upper Génis subunit comprises Ordovician and Silurian schists and a dismembered ophiolite of unknown age. The whole upper subunit and most of the lower one are affected by an epizonal, medium pressure Devonian metamorphism. Characteristically, sericite and chlorite developed as index minerals in the acidic and pelitic formations, and chlorite, actinolite, epidote, and albite characterize the basic ones. The metamorphic assemblages developed on an S1 plane, which is subsequently affected by an S2 strain-slip cleavage. The estimated "phase 1" P-T conditions are 3–3.5 kbar and 350 °C. The tectonic contact between the upper and middle allochthons is a dextral shear zone which generated a vertical foliation (Ledru and Autran 1988).

Table 1. Upper allochthon

Lithology	Age and nature of the protoliths	A	B	C–D
I Limousin				
Metagabbros, metadolerites, meta-basalts, metatuffs	Unknown age Abyssal. th. and calc-alkaline dolerites		Chl. + Act. + Ab. +/– Ep.	Pump. + Calc. + Chl.
Carbonates	Limestones and recifal dolomites		Calc. and Dol.	
Chlorite and serici-te-schists	Ordovician, Silu-rian, and Siluro-Devonian		Chl.-Ser.	
"Porphyroides"	Ignimbrites Trémadoc?		Ser.	
"Thiviers sandstones"	Rhyo-dacitic tuffs		Chl. +/– Bt. +/– Gt.	
P-T Estimates			350–400 °C/3–3.5 kbar	
II Brévenne				
Amphibolites	Th. basalts and gabbros estimated Devonian		Act. (or Hbl.) + Bt. +/– Chl. + Ep.	Calc. + Sph. + Chl. + Pump. + Ab.
Ab-bearing Q-feldsp. gneisses	Plagiogranites + keratophyres Estimated age: 370 Ma		Bt + Ms + Spess. +/– Plagio. + Chl. + Ep.	Chl. + Ab.
Schists	Pèlites + quartzites		Ms + Bt + Chl.	Ab. + Chl.
Serpentinites	Peridotites		Serp.	
P-T estimates			350–400 °C and 3–4 kbar	< 350 °C 2 kbar

Abbreviations

Ab = albite	Dol = dolomite	Mag = magnetite	Saph = sapphirine
Act = actinolite	Ep = epidote	Micr = microcline	Scap = scapolite
Alk = alkaline	Fo = forsterite	Ms = muscovite	Ser = sericite
Amph = amphibole	Gl = glaucophane	Ol = olivine	Serp = serpentine
An = anorthite	Gr = graphite	Olig = oligoclase	Sill = sillimanite
And = andesine	Gross = grossular	Omph = omphacite	Sp = spinel
Andal = andalusite	Gt = garnet	Opx = orthopyroxene	Spess = spessartine
Bt = biotite	Hbl = hornblende	Pal = Paleozoic	Sph = sphene
CA = calc-alkaline	HT = high temperature	Parag = paragonite	St = staurolite
Calc = calcite	Hum = humite	Parg = pargasite	Ta = talc
Chl = chlorite	Hy = hypersthene	PC = Precambrian	Th = tholeiite
Chtd = chloritoid	Ilm = ilmenite	Perth = perthite	Tr = tremolite
Co = corindon	Kb = kilobar	Ph = phengite	Zo = zoisite
Cord = cordierite	KF = K-feldspar	Phl = phlogopite	Zn = zinc
Cpx = clinopyroxene	Korn = kornerupine	Plagio = plagioclase	
Cumm = cummingtonite	Ky = kyanite	Prehn = prehnite	
Di = diopside	LT = low temperature	Py = pyrope	

The shear zone itself is not dated, but it is penecontemporaneous with a syntectonic granite of estimated lower Carboniferous age, and a neighboring shear zone in the Thiviers-Payzac subunit is dated at 355 ± 5 Ma (Rb-Sr whole rock) (Bernard-Griffiths et al. 1977). Close to the shearing process and induced by it, first biotite and then garnet developed in a zone a few hundred meters wide in the rhyodacitic metatuffs and interbedded metagraywackes (Santallier et al. 1978). In the basic rocks, hornblende and plagioclase are found in the same narrow zone. Thus the shear zone evolved in temperature conditions higher than the previous epizonal ones and it induced retrograde assemblages in the neighboring middle allochthon gneisses. The corresponding estimated P-T conditions increase up to 4 kbar and 450 °C. This localized metamorphic evolution results from the tectonic coupling of the LT upper allochthon and higher

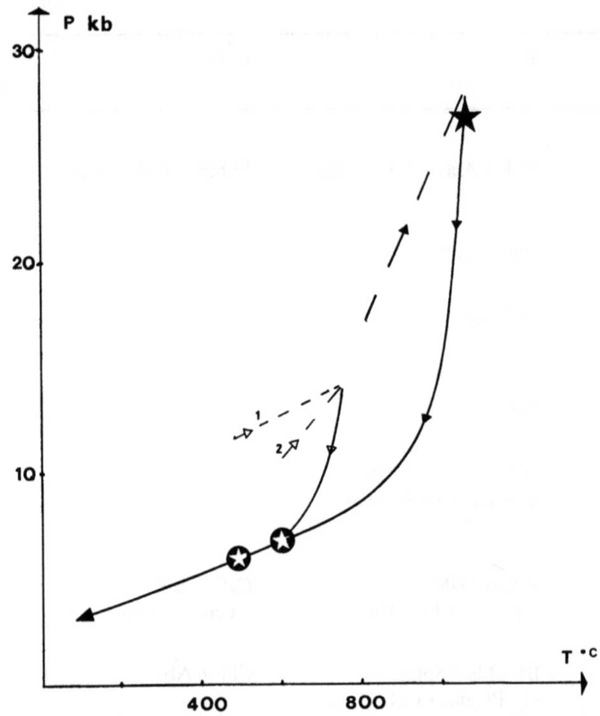

Fig. 5. The middle allochthon P-T path. *Black star* main deformation episode in mantle assemblages; *open stars* main deformation events in crustal assemblages; *1* low-temperature prograde course (crustal rocks); *2* high-temperature prograde course (crustal rocks)

T middle allochthon. A comparable evolution occurred at the Lyonnais and Brévenne tectonic contact in eastern Massif Central.

3.2 The High Pressure Middle Allochthon Nappes

The High-Pressure Middle Allochthon Nappes (Fig. 5 and Table 2) are almost exclusively found high in the stacked pile. These nappes exhibit the most complete story of metamorphism because they show a great amount of relicts of the different metamorphic stages. Lithologically, they are highly variable in the different locations, which is more consistent with the nappe complex concept than with a single nappe scheme.

3.2.1 The Monts du Lyonnais Example

(J. M. Lardeaux and V. Gardien)

In the Eastern Massif Central, the Monts du Lyonnais crystalline complex consists of granulite and amphibolite facies rocks, the most abundant of which are garnet-biotite- and sillimanite-bearing gneisses (Peterlongo 1958). These provide the most complete PT evolution for the middle allochthon in spite of local overprinting by late anatexis. Several generations of structures are recognized, including a widespread N20–60 °E-trending foliation plane. On this plane, a well-developed mineral lineation is defined by micas, quartz, and sillimanite. The structural features have been affected by two deformational phases, P2 and P3 (Lardeaux and Dufour 1987). Local S2/S1 transposition is observed. P3 folds are contemporaneous with a dextral strike-slip system in which a mylonitized syntectonic granite is dated at 332 ± 10 Ma (Rb-Sr whole rock) (Gay et al. 1981) and between 340 and 350 Ma ($^{39}Ar/^{40}Ar$ biotite) (Costa and Maluski 1988; Costa et al. 1989). A combined structural, microstructural, and mineralogical study shows that P1 and P2 structures developed in high amphibolite facies conditions (6–8 kbar and 650–700 °C), whereas P3 occurred under lower pressure conditions (ca. 5 kbar). Under such conditions, lenses or layers of granulites, eclogites and garnet peridotites preserved earlier assemblages. The data from the different lithologies have been used to establish Fig. 6, in which the most prominent feature is a regular trend of decreasing pressure and temperature observed in all crustal rocks through the eclogite, HP granulite, LP granullite and amphibolite facies (Dufour 1985; Dufour et al. 1985; Lardeaux et al. 1989). The prograde path is observed only in the mantle rocks (Gardien et al. 1988). The HP metamorphism has been dated at about 440–400 Ma (Duthou et al. 1981; Dufour 1985; Pin and Peucat 1986). The authors suggest that the HP event records a subduction stage, the retromorphic evolution reflecting the uplift of the middle allochthon unit in a collisional context. The main deformational phases P1 and P2 also occurred during the collision event. Biotite, muscovite and amphibole pertaining to the S1–2 foliation plane have been dated between 380 and 360 Ma (^{39}Ar–^{40}Ar micas and amphiboles) (Costa et al. 1989). The synmetamorphic P_3 strike-slip system occurred at the end of the continental collision; the corresponding intracontinental deformation in this unit is 350–340 Ma old.

In at least two cases, rather different evolution courses have been described. For example, in Limousin, the P-T evolution is much simpler. After the initial HP eclogitic event, the rocks were directly retrogressed under either hydrous or anhydrous amphibolite facies conditions (Santallier 1983) (Table 2 and Fig. 5). In the Najac area, a few LT pre-eclogitic relicts are known (400–500 °C) and the eclogitization itself is achieved at relatively HT (650–700 °C).

Table 2. Middle allochthon

Lithology	Age and nature of the protoliths	A	B	C	D
Amphibolitized eclogites	Th. gabbros, basalts, dolerites 480–530 Ma	LT: Gl HT: Zo. + Hbl. + Q	Omph. + Gtl + Ru. + Q +/– Ky. +/– Zo. +/– Ph. +/– Amph. +/– Co.	(1) Di. + Olig. + Ilm. +/– An. +/– Co. +/– Sp. +/– St. +/– Gt2 + Hbl. +/– And. +/– Bt. or (2) Hbl. + And. +/– Na Cpx	Bt. + Act. + Chl. + Olig. + Sph. +/– Prehn.
Granulites:					
Basic	Th. Alk. and CA. basalts, gabbros		Cpx + Plagio. + Gt. + Ru. +/– Parg. +/– Opx +/– Ky.	Opx + Plagio. + Parg. or Hbl. + Ilm. +/– Bt.	Bt. + Hbl. Cumm. + Plagio. + Sph. +/– Chl.
Acid	Lower Paleozoic Granites, rhyolites Age: 500 Ma or unknown	St. +/– Gtl	Gt2 + Perth. KF +/– Ky. +/– Opx. +/– Saph. + Korn. + Ru. +/– Gr.	KF + Sill. + Bt. + Co. + Sp. + Cord. + Ilm. +/– Cumm.	Ms. + Bt. + Sph. +/– Cord. +/– Cumm. +/– Chl.
Paragneisses, micaschistes, and quartzites	Graywackes and pelites ± Sandstones Unknown age	Chtd. ?	HT: Gt. + Ky. + KF + Ru. LT: Ph. + Gt. + Chl. +/– Zn-St. +/– Ky. + Ru.	Bt. + Sill. + Gt. + KF +/– St. +/– Ilm. +/– Sp.	Bt. + Ms. +/– Chl. or KF + Cord. +/– And.
Marbles	Limestones		Calc. +/– Di. +/– Gross. +/– Zo. +/– Scap. +/– Fo. +/– Hum. +/– Phl. +/– Ru. +/– Sp.	Calc. +/– Hbl. +/– Ep. +/– Plagio.	Chl. + Serp. + Calc. + Sph.
Amphibolites, striped amphibolites	Volcano-sedimentary tuffs, unknown age		Di. + Zo. + Scap. +/– Gt. +/– Hbl.+ Ilm. +/– Calc.	Hbl. + And. + Ep. + Sph.	Chl. + Sph. + Ep.
Ordinary amphibolites	Th. and CA gabbros, dolerites, basalts, Devonian		Hbl. + And. + Sph.		
Mg amphibolites	Cumulates (wherlites, gabbros anorthosites, pyroxenites) unknown age		(1) Cpx + Opx + Sp. +/– Gt. +/– Parg (2) Parg. + Mg.Chl +/– Zo. +/– Ky. +/– Co. +/– Ged. +/– Hy +/– Sp. +/– Gt. +/– Ph. +/– Parag	Hbl. + Plagio. + Chl + Ilm. +/– Sph. Hbl. + Plagio. +/– Chl. + Ilm.	Chl. + Sph. +/– Tr.
Serpentinites	Peridotites (lherzolites, dunites, hartzburgites) +/– related pyroxenites, unknown age	Sp. + Ol. +/– Opx +/– Cpx	Gt. + Ol. + Opx + Cpx	Sp. + Opx. + Ol. +/– Cpx +/– Amph. +/– Mg-Chl.	Chl. + Tr. +/– Ta. + Serp. +/– Prehn. + Mag.

1: Anhydrous assemblages
2: Hydrous assemblages

	A	B	C	D
P-T estimates: crustal assemblages	HT= LT=P 10 kbar 400–500 °C	12–22 kbar 600–800 °C	6–10 kbar 600–750 °C	3–5 kbar 400–700 °C
P-T estimates: mantle assemblages	< 20 kbar < 750–850 °C	25–30 kbar 900–1200 °C	10–20 kbar 750–850 °C	5 kbar 500 °C

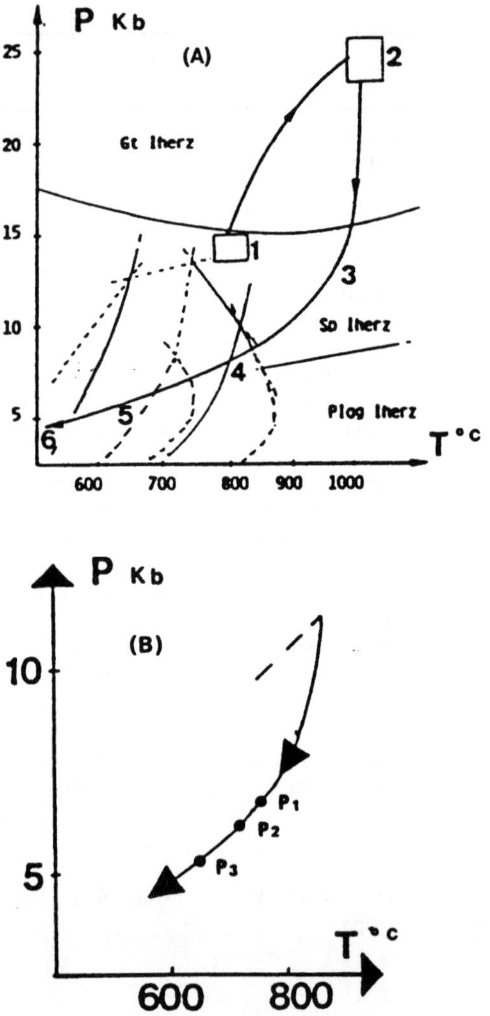

Fig. 6 A, B. Detailed P-T paths for different lithologies in the Lyonnais area. **A** Peridotitic assemblages. (After Gardien et al. 1988): *1* to *6* represents the successive mineralogical stages. *1* Spinel-lherzolite stage; *2* garnet-lherzolite stage; *3* spinel-lherzolite stage; *4* amphibole-lherzolite stage; *5* chlorit-amphibole-lherzolite stage; *6* serpentinization event. Reaction curves after Jenkins (1981) and Gasparik (1987). **B** Granulitic assemblages. (After Dufour 1985; Lardeaux et al. 1989). *P1* first deformation event; *P2* second deformation event; *P3* third deformation event

3.3 The Lower Allochthon Units

The Lower Allochthon Units (Fig. 7 and Table 3) exhibit highly variable lithology and metamorphic evolution.

In the western areas (Limousin, Aigurande Plateau) (Fig. 7 a), eclogites and other high pressure relicts have been described (Coffrant 1974); Santallier and Floc'h 1979; Floc'h et al. 1984; Quenardel and Rolin 1984; Guillot and Lacour 1988). The latest metamorphic stage corresponds either to a classical retromorphic course (muscovite- and chlorite-bearing assemblages) or to HT and LP assemblages related to thermal doming (Floc'h 1979, 1983). In the central areas (Fig. 7 b), HP relicts are unknown. An inverted metamorphic succession has been described (Briand and Gay 1978; Burg et al. 1984) in the Rouergue, Lot, Marvejols, and Haut-Allier areas, related to the emplacement of the upper middle allochthon nappe. These evolution trends are only clearly visible far from the latest intrusive granites which developed localized overprinting thermal contact aureoles (Delor et al. 1984). In the eastern areas (Fig. 7 c), HP assemblages still have not been studied in detail, although a polyphase metamorphism has been recognized (Chenevoy and Ravier 1971, 1972). The end of the tectono-metamorphic evolution is characterized by contrasting PT paths: decreasing P and T giving rise to greenschist assemblages, or increasing T/decreasing P related to a large scale thermal anomaly in the Velay dome (Weisbrod 1970; Chenevoy and Ravier 1972; Marignac et al. 1980; Weisbrod et al. 1980; Toteu 1981; Gardien et al. 1989).

The greatest contrast between examples is provided by the Limousin and Marvejols areas.

3.3.1 The Limousin Example (Fig. 7 a) (D. S. Santallier and J. P. Floc'h)

The HP-HT assemblages are observed in several different lithologies (Floc'h 1983; Santallier 1983 a; Maillet 1987). Phengitic white mica and rutile are widespread in most micaceous gneisses; mesoperthitic K-feldspar, kyanite, and Mg-rich garnet assemblages are found in a few acidic granulite lenses and a few eclogitic lenses are also present. In the dismembered ophiolites located between the middle and lower allochthons (Girardeau et al. 1986; Maillet 1987), gedrite, kyanite, and corundum-bearing amphibolites developed at the expense of feldspar-rich cumulates. Spinel, orthopyroxene, clinopyroxene, and sometimes garnet are found in ultramafics or as coronas in meta-olivine-rich gabbros. The estimated PT conditions are 750 °C and 20 kbar in the eclogites, 700–800 °C and 10–16 kbar in the granulites and HP amphibolites, 750 °C and 12 kbar in the coronitic rocks. Such PT values are good evidence for an early subduction stage, probably between 400 and 440 Ma (Pin and Peucat 1986). The medium pressure event is contemporaneous with an S1 + S2 plane. The two corresponding successive assemblages crystallized under slightly decreasing pressure and constant or slightly increasing temperature. Kyanite ± staurolite ± sillimanitel developed in S1 structures under 7 kb and 600 °C. Sillimanite2 ± K-feldspar crystallized in

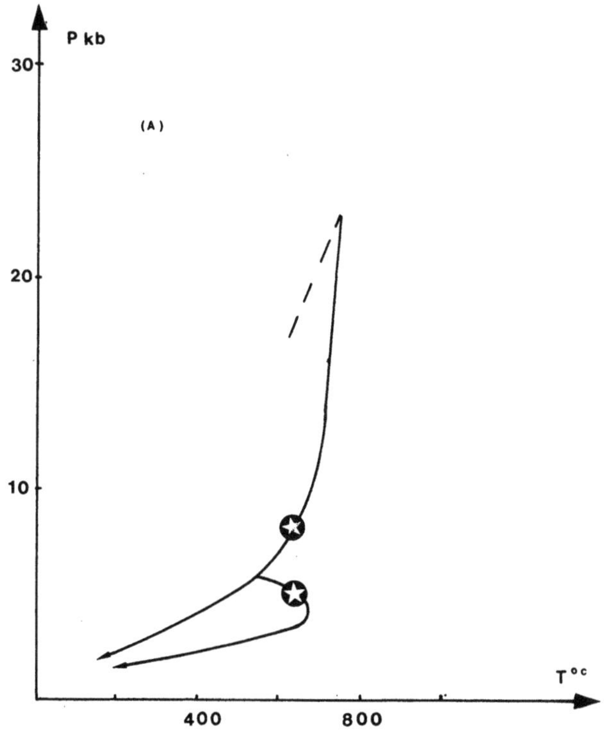

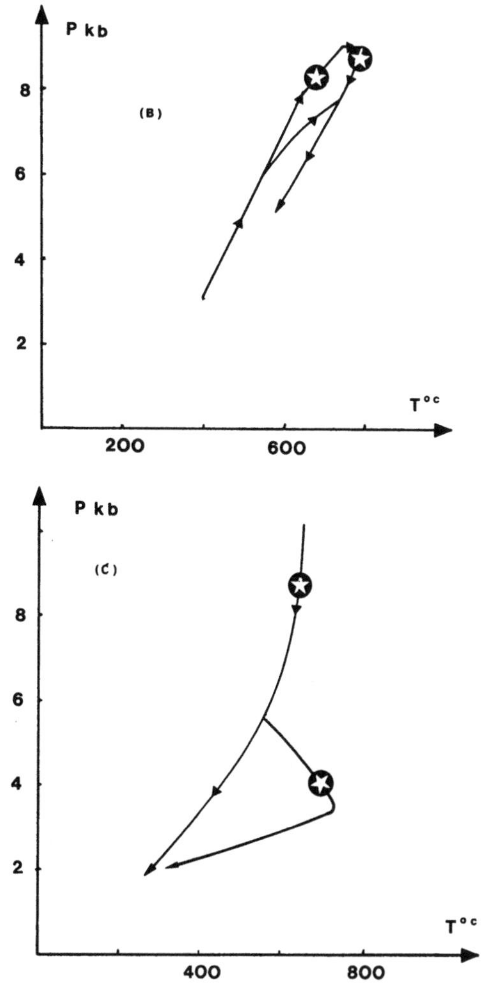

Fig. 7 A–C. The lower allochthon contrasted P-T paths (see text for discussion). **A** Western areas. *Open stars* main deformation events. **B** Central areas. The evolution enhances an inverted medium-pressure metamorphism. *Open stars* main deformation episodes. **C** Eastern areas. *Open stars* main generations of structures

S2 under 4.5 kbar and 750 °C. Late S2 anatexis occurred under 4.5 kbar and 650 °C (Floc'h 1983). The anatexis is dated at 383 ± 5 Ma (Rb-Sr whole rock) and 375 ± 6 Ma (U-Pb Zr) (Duthou 1977; Lafon 1986). The two main deformation events, which occurred during the collision process, are 400–380 Ma old, and apparently slightly older than in the eastern Massif Central area (Costa and Maluski 1988).

3.3.2 The Marvejols Example (Fig. 7 b) (B. Briand)

In the Lot valley, west of Marvejols, the lower allochthon exhibits a prograde metamorphic zonation, in striking contrast to that of the overthrust middle allochthon. It is characterized by a polyphase inverted metamorphism. The inverted metamorphic succession was first interpreted as representing the reverse limb of a large recumbent fold (Boule 1899–1900: Roques 1941; Demay 1946, 1948; Lapadu-Hargues 1947; Peyretti 1971; Poulain 1972). Kornprobst and Poulain (1972) proposed a different interpretation, advocating the thrusting of an upper mesozonal unit on a lower epizonal one. Northwards and from bottom to top, toward the trust zone, Briand and Gay (1978), Pin (1980), and Burg et al. (1984) described in the underlying lower allochthon series, the successive occurrence of biotite, garnet, staurolite, kyanite, and (close to the thrusting zone) sillimanite + muscovite. These index minerals, indicative of the metamorphic climax, are linked to the regional D2 deformation event and S2 associated structures and are syn- to post-D2. Later, post-D2 and syn-D3 andalusite and cordierite developed locally in the staurolite and kyanite zones, indicating an increase of the thermal gradient during the late metamorphic stages. The thrusting event is contemporaneous with the regional D2 deformation and the inverted metamorphic zonation, corresponds to the metamorphic climax. It is syn- to post-thrusting. The late D3 and local S3 structures which post-date the metamorphic climax and the corresponding isograds generally exhibit an up-

Table 3. Lower allochthon

Lithology	Age and nature of the protoliths	A	B	C	D
Western French Massif Central (Limousin, Aigurande Plateau)					
Amphibolitized eclogites	Th. basalts and gabbros 500 Ma		Omph. + Gt. + Q. + Ru. +/– Ky.	(1) Di. + Olig. + Ilm. +/– An. +/– Co +/– Sp. or (2) Hbl. + And.	Hbl. + Plagio. Chl.
Acid granulites	Granites pro parte, Unknown age		KF + Ky. + Gt. + Plagio. + Ru.	KF + Ms + Bt + Plagio. +/– Ilm?	?
Micaschists and paragneisses	Pelites and more or less evolved greywackes Probably PC and/or lower Pal.		Ph 3.25 + Ru. +/– St. +/– Ky. + Gt1	Ms + Sill. + Gt2 + Olig. +/– KF	Ms. + Bt. +/– Chl. or Bt. + Sill. + KF +/– Cord.
Striped amphibolites	Th. basic tuffs, unknown age		Di. +/– Zo. +/– Hbl. +/– Ru.	Hbl. + And. + Ep. + Sph. +/– St. +/– Bt.	Ep. + Chl. + Sph.
Orthogneisses	Granites and tonalites, between 475 and 540 Ma		B and or C Bt. + Ms. +/– Gtl +/– Hbl.		Ms. + Chl. or KF + Bt. +/– Gt2
Central French Massif Central (Ht-Allier, Marvejols, Moyenne Dordogne, E. Rouergue, Sioule, Northern Millevaches)					
Paragneisses and micaschists	Greywackes and pelites, unknown age	Bt 1 + Ms +/– Gt1	B and or C (inverted metamorphism advocated) a) far from the thrust plane Bt2 + Gt2 +/– Ms. + Sill.1 +/– KF +/– St. b) midway from the thrust plane Ms. + Bt2 + Gt2 + St. +/– Ky. c) close to the thrust plane Bt2 +/– Cord. + Sill.2 +/– Ms. +/– KF +/– Ky. +/– Gt2		Bt. + Ms. +/– Chl.
Orthogneisses	Granites and leuco-granites 406 Ma and 540 Ma		B and or C Bt. +/– Gt. +/– Sill. +/– Ms.		Ms. + Chl.
Amphibolites	Gabbros		B and or C Plagio. + Hbl. + Sph. +/– Bt.		
Eastern French Massif Central (Pilat, Vivarais, Cévennes médianes)					
Micaschists and paragneisses	Pelites and greywackes, unknown age		St. + Ky. + Gt 1 + Ph.? + Ru.	Ms + Bt + Olig. + Ilm.	Ms. + Chl. + Ab. or KF + Sill. + Bt. + Gt2 + Cord. + And.
Amphibolites	Alk. basalts pro parte, unknown age		Di. + Zo. + Hbl.	Ep. + Hbl. + And. +/– Bt.	Chl. + Sph. + Ab. + Ep. + Calc.
Orthogneisses	CA granites ca. 500 Ma		B and or C Bt. + Ms. +/– Gt1 +/– Sill.		Ms. + Chl. or Sill. + KF +/– And. + Bt. +/– Gt2 +/– Ms.

right attitude and are of too local significance to be responsible for an eventual large scale post-metamorphic inversion of the isograds. Consequently, Briand and Gay (1978), Pin (1980), and Burg et al. (1984) considered the inverted isograd pattern as a primary feature, similar to the one described in the Himalayan chain by Pecher (1978). The thrust has been successively dated at 346 ± 8 Ma (Pin, 1981), 344 ± 13 Ma (Pin and Lancelot 1982) (U-Pb Zr) and 360 Ma (Costa and Maluski 1988) (^{39}Ar–^{40}Ar Bt).

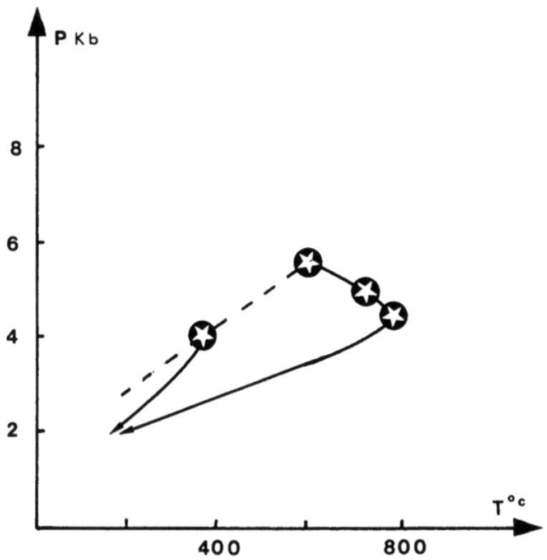

Fig. 8. The two kinds of P-T curves described for the relative autochthon. *Open stars* main deformation events

3.4 The Presumed Autochthon Units

The Presumed Autochthon Units (Fig. 8 and Table 4) are not so well defined: the lithologic content is quite variable. In the southern and southeastern areas, they apparently comprise mostly epizonal units (e.g., Albigeois, Lot, southern Cévennes). In the western and southeastern parts (e.g., Limousin, Millevaches Plateau, and northern Cévennes) domal thermal effects are well developed, giving rise to andalusite and cordierite and/or sillimanite-bearing HT assemblages and anatexis.

3.4.1 The Lot and Western Cévennes Examples (B. Briand)

The Lot series is composed mostly of epizonal metapelites with thin quartzitic and sometimes sodic volcaniclastic intercalations. The Devonian metamorphism was of medium pressure (MP) (Costa and Maluski 1988) and there is associated polyphase deformation (Pin 1980). The index minerals (Chl + Ab + Ser ± Bt 1 ± Gt) mostly developed on the S1 + S2 regional schistosity, and locally, Bt 2 + Ms 2 developed on S3 (Briand and Gay 1978). Eastwards, in the western Cévennes area, an equivalent epizonal MP series (comprising metapelites with a few graphite-bearing quartzites, amphibolites and metarhyolites) is overprinted by the upper Carboniferous thermal re-equilibration. As previously described [under (3.3) above] in the lower allochthon unit, the thermal dome developed a medium grade LP metamorphism characterized by Bt 2 + Cord + Andal assemblages. Cordierite and andalusite are penecontemporaneous with P3 folds; they overprint the earlier mylonitic S1 + S2 Barrovian schistosity.

3.4.2 The Cévennes Médianes Example (Fig. 9) (J. M. Montel, Ch. Marignac, P. Barbey, J. Macaudière, P. Pichavant, and A. Weisbrod)

In the eastern area, the Cévennes médianes also records a polymetamorphic history. Two main episodes are identified which completely overprint a probable early granulitic M1 stage. The first episode (M2), also almost completely overprinted, was of MP-MT type, and the second (M3) LP-HT. The late LP-HT M3 metamorphism is characterized by a well-defined isograd pattern broadly enveloping the Velay dome, except when they are deformed in the vicinity of granitic bodies.

The characteristic M3 mineral associations (Table 3) are: Q + Chl + Ms + Bt + pyrophyllite in low grade metashales and Cord, Gt + Cord, Cord + Andal, St + Andal or Gt + Andal in medium-grade metashales. The low- to medium-grade transition is expressed by reaction (1): Chl + Ms $\rightarrow$ Bt + Cord and the prograde medium grade path corresponds to reaction (2): Cord + Ms $\rightarrow$ Andal + Bt. The Sill + Ms association is only represented in a zone a few hundred meters wide. In high grade metashales, after the Ms + Q breakdown into Sill + KF, reaction (3): Sill + Bt + Q $\rightarrow$ Cord + KF + water is observed. Typically, reaction (3) coincides with the beginning of anatexis. In the more internal zones, reaction (3′) Sill + Bt + Q $\rightarrow$ Gt + KF + water is possible. The occurrence of hercynitic spinel is also significant.

There are two major deformational phases (D2 and D3) overprinting an earlier S1 structure. D2 is responsible for the development of a regional S2 schistosity in all lithologies, of rare P2 similar folds (up to 10 m) and of recumbent folds (many kilometers in size) (Macaudière et al. 1987). D3 is responsible for east-west P3 folds which verge northward. In the medium-grade areas, cordierite and andalusite develop between D2 and D3, andalusite grew during D3 and sillimanite is pre- and syn-D3. P3 folds are cross-cut by a 315 ± 5-Ma-old granite (Miahle 1980); this should also be the age of the M3 stage (Fig. 9). In the high grade area, P3 folds represent the main deformation event and a S3 schistosity is common. Sillimanite is pre- and syn-D3 and anatexis starts with S3 giving rise to a conspicuous banding parallel to S3 in the migmatites. The HT assemblages are involved in a late M4 melting event leading to the Velay granitic dome emplacement. The static development of nodules (Cord + Q cores and KF + Q rims), widespread in large parts of the anatectic area, is penecontemporaneous with M4.

Table 4. Presumed autochthonous

Lithology	Age and nature of the protoliths	A	B	C	D
Micaschists	Pelites and graywackes, unknown age		Various assemblages (1) Gt. + St. + Ky. + Plagio. + Bt. + Ms. or (2) Chl. + Ms. +/– Bt or Ky.	(1) Sill. + Ms.	(1) Sill. +/– Cord. +/– KF +/– Bt. or Ms + Chl
				C and/or D (2) Ms. + Chl.	
Striped amphibolites	Gabbros, pyroxenites, and basic tuffs 600 Ma pro parte		Gross. + Hbl. +/– Cpx. + Plagio.	Hbl. + Ilm. + Plagio.	Chl. + Sph.
Graphite-bearing quartzites	Sandstones		Gr. + Ms.		
Orthogneisses and porphyrites	CA granites and rhyolites 488–360 Ma or unknown age		Ab. + Ms.		
	Soda-rhyolites, unknown age		Ab + Ms.		

Assemblages (1) are in Limousin and Millevaches Plateau
Assemblages (2) are in Lot, southern Cévennes, and Albigeois areas

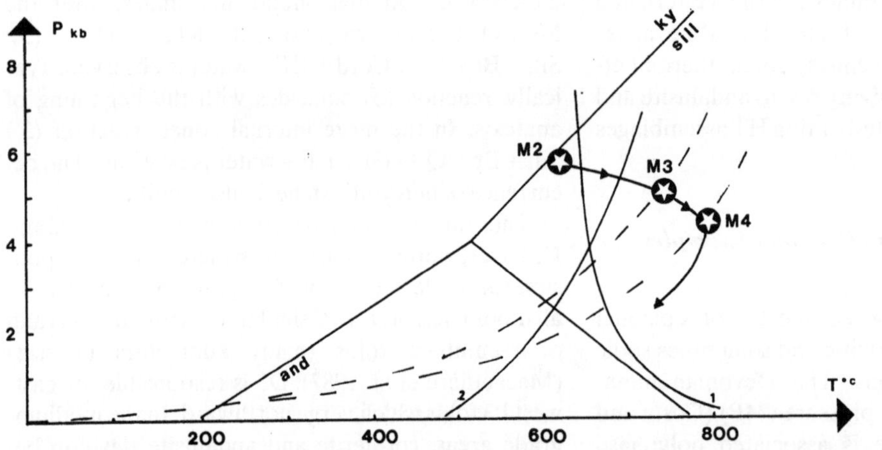

Fig. 9. The relative autochthon P-T paths exemplified by the Cévennes médianes area. *Open stars* main deformation events (M2, M3, M4 refer to the successive metamorphic stages). *M2* Barrovian metamorphism (relict); *M3* HT-LP metamorphism (320–315 Ma); *M4* Velay anatexis (300 Ma); *1* wet haplogranite solidus; *2* maximal muscovite stability

The M4 leucocratic migmatites are locally dated at 298 ± 8 Ma (Rb-Sr whole rock) (Caen-Vachette et al. 1982). Both migmatites and granites were eroded before the early Stephanian, attesting a high rate of denudation at the end of the Hercynian cycle.

Relying on assemblages such as Gt + Plagio, Gt + Bt, Gt + Cord (sillimanite being present) and Sp + Q + Cord, calculation of PT conditions indicated a pressure decrease and a temperature increase (Fig. 9), i.e., an increasing thermal flow in the area. The heat source provided by the thermal relaxation after the collision is not sufficient to explain the observed pattern. An additional heat supply may be found in a pervasive basic magmatism (Pin and Vielzeuf 1983).

Relicts of an earlier MP-MT metamorphism are clearly syn-S2. They are thus attributed to a low grade M2 metamorphism. In low- and medium-grade schists, garnets are often syn-S2, sometimes later, in which case they have a syn-S2 core and an M3 rim. In high-grade amphibolites, peculiar ocellar facies result from the transformation during M3 of earlier syn-S2 garnets into Hbl + Plagio assemblages. The M2 metamorphism was thus locally medium-grade.

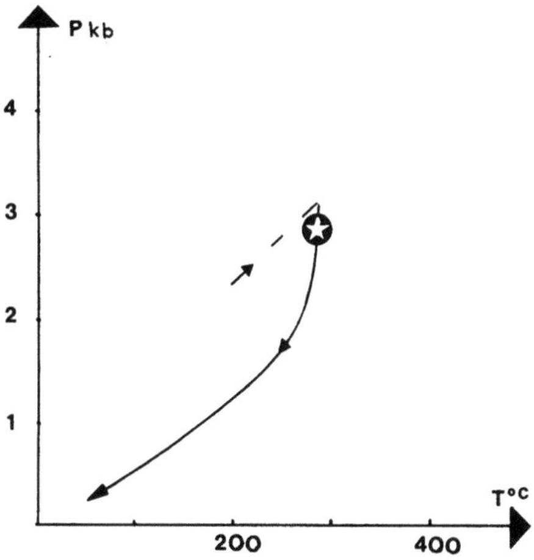

Fig. 10. Metamorphic evolution of the southern Paleozoic units. *Open stars* main deformation event

Table 5. Southern Paleozoic units

Lithology	Age and nature of the protoliths	A	B	C–D
Micaschists	Lower Cambrian		Bt.	Ser.+Bt.
Limestones			± Bt.	Bt. + Ser.
Schists and sandstones	Lower Cambrian			

Finally, it is noteworthy that bodies of acidic and basic relict granulites (meters to tens of meters in diameter) occur in some migmatized quartzo-feldspathic gneisses. In acidic granulites, the earlier assemblage was anhydrous (devoid of biotite) with kyanite (now retrogressed into prismatic sillimanite) and very Mg-rich garnet. Py-rich-garnets are also found in the basic granulites.

3.5 The Southern Paleozoic Units

The Southern Paleozoic Units (Fig. 10 and Table 5) are a non- or very weakly metamorphosed series. The recrystallizations are inhomogeneous and mainly located in thrust planes.

4 Discussion: Metamorphic Constraints for the Variscan Geodynamics

In conclusion to the study of metamorphism in the French Massif Central, the following points must be emphasized.

1. All metamorphic events are Paleozoic (Pin and Peucat 1986). It is no longer necessary to distinguish Caledonian and Variscan metamorphic cycles in France. The polyphase continuous orogen building lasted for ca. 150 Ma from roughly 450 to 300 Ma ago. The magnitude is comparable with what is observed in recent orogenic belts, such as the Alpine belt.

2. High pressure relicts are particularly abundant in two of the allochthonous units. The calculated pressures (although under-estimated) are generally 20 ± 5 kbar. Such values imply a subduction-type process. The corresponding temperature values are fairly high (750 ± 50 °C), apparently too high for a classical oceanic subduction. However, the relict HP parageneses now observed result from recrystallization under HT conditions and consequently are not necessarily representative of the initial HP thermal environment. Generally speaking, an HP-HT metamorphism can be expected in two situations:

- when a young, and therefore still hot, oceanic crust (e.g., corresponding to a narrow oceanic domain) or an oceanic ridge is subducted (Sisson and Hollister 1988). In such conditions one must not expect the development of contemporaneous island arc magmatics (the lack of such magmatics in the French Massif Central supports this situation); and
- when the subduction process stops very suddenly, and is followed by a slow uplift of the buried series, thus allowing a complete thermal re-exquilibration.

3. The nappe stacking is a long diachronic process which ended late in the tectono-metamorphic history. It post-dates the HP metamorphism and is sometimes late compared to the MP-MT re-equilibration. The stacking occurred under decreasing pressure and indicates that the main collision tectonics took place during the uplift of the chain.

4. Except for the upper allochthon, the more metamorphosed units are higher in the tectonic pile. How can this fact be interpreted? (a) Is it a post-metamorphic inversion as in the La Chataigneraie area (Bogdanoff et al. 1989) and should the Scandinavian Caledonides model be referred to? (b) Considering the existence of a primary inverted metamorphic series in Marvejols area, must the metamorphic evolution of the French Massif Central be compared with the Himalayan metamorphism (Burg et al. 1984)? (c) Were both mechanisms active?

5. The end of the tectono-metamorphic history is, in some places, characterized by remarkably high temperatures. Indeed, the existence of LP granulitic xenoliths in tertiary basalts in the southeastern Massif Central (Leyreloup 1973), supports a thermal environment

which was probably similar to that of the Pyrénées. To explain such a thermal anomaly, different processes can be invoked: heat supply by infra-crustal basic magmatism (Pin and Vielzeuf 1983), and/or post-thickening crustal thinning accommodated by low angle detachment faults subsequent to the collapse of the chain (Gardien et al. 1989; Malavieille et al. 1990).

6. In conclusion, the French Massif Central orogenic history involves (a) an early subduction episode; (b) a subsequent collisional tectonic culmination with widespread nappe piling and (c) a late collapse of the thickened crust. The authors suggest the names Eo-Paleozoic, Medio-Paleozoic, and Neo-Paleozoic to describe the three main thermo-mechanical episodes, and to emphasize that the Paleozoic orogen continued from the Silurian to the Carboniferous.

Acknowledgements. The authors are indebted to J. L. Duthou and C. Pin, who improved the geochronological background. Detailed comments and suggestions by A. Autran and P. Ledru are greatly appreciated. We gratefully acknowledge J. Duncan Keppie, R. P. Raeside, and two anonymous reviewers for critical reviews of the manuscript and also clarifying the English translation.

References

Autran A, Cogne J (1980) La zone interne de l'orogène varisque dans l'Ouest de la France et sa place dans le développement de la chaîne hercynienne: 26th IGC Paris, C6, Mém Bur Rech Géol Min 108, pp 90–111

Bard JP, Burg JP, Matte P, Ribeiro A (1980) La chaîne hercynienne d'Europe occidentale en termes de tectonique des plaques. 26th IGC, C6, Mém Bur Rech Géol Min 108, pp 233–246

Bernard-Griffiths J, Cantagrel JM, Duthou JL (1977) Radiometric evidence for an Acadian tectono-metamorphic event in western Massif Central français. Contrib Mineral Petrol 61: 199–212

Bogdanoff S, Cirodde JL, Donnot M (1989) The nappes of La Chataigneraie, southwest Massif Central, France. Tectonophysics 157: 69–79

Bouchardon JL (1987) Evaluation des conditions de Presion-Température du "top" métamorphique des éclogites du Lévezou (Rouergue, Massif Central français). CR Acad Sci Paris 305: 271–275

Bouchez JL, Jover O (1986) Le Massif Central: un chevauchement de type himalayen vers l'Ouest-Nord-Ouest. CR Acad Sci Paris 286 D: 729–731

Boule M (1899–1900) Géologie des environs d'Aurillac. Bull Serv Carte Géol Fr 11 (3): 253–262

Briand B (1978) Métamorphisme inverse et chevauchement de type "himalayen" dans la série de la vallée du Lot (Massif Central français). CR Acad Sci Paris 286: 729–731

Briand B, Gay M (1978) La série inverse de Saint-Geniez-d'Olt: évolution métamorphique et structurale. Bull Bur Rech Géol Min 2 I 3: 167–186

Burg JP, Matte P (1978) A cross-section through the French Massif Central and age of its Variscan geodynamic evolution. Z Dtsch Geol Ges 129: 429–460

Burg JP, Leyreloup A, Marchand J, Matte P (1984) Inverted metamorphic zonation and large scale thrusting in the Variscan belt: an example in the French Massif Central. In: Hutton DHW, Sanderson DJ (eds) Variscan tectonics of the north Atlantic region. Blackwell, London, pp 47–61

Cabanis B, Guillot PL, Santallier D, Jaffrezic H, Meyer P, Treuil M (1983) Apport des eléments en traces à l'étude géochimique des métabasites du Bas-Limousin. Bull Soc Géol Fr (7) XXV 4: 563–574

Caby R, Kienast JR, Saliot P (1978) Structure, métamorphisme et modèle d'évolution tectonique des Alpes occidentales. Rev Géol dyn Géogr phys 20, 4: 39–43

Caen-Vachette M, Couturie JP, Didier J (1982) Ages radiométriques des granites anatectiques et tardimigmatitiques du Velay (M.C.F.). CR Acad Sci Paris 294: 135–138

Chenevoy M, Ravier J (1971) Caractères généraux des métamorphismes du Massif Central français. Plein Air Service, Clermont-Ferrand, pp 17–32

Chenevoy M, Ravier J (1972) Les grands ensembles cristallophylliens et granitiques du Massif Central français. 14th Int Geol Congr Montréal, 2, pp 123–134

Coffrant D (1974) Les éclogites et les roches basiques et ultrabasiques associées du massif de Sauviat-sur-Vige (Massif Central français). Bull Soc Fr Minéral Cristallogr 97: 70–78

Cogne J (1976) La chaîne hercynienne ouest-européenne correspond-t-elle à un orogène par collision? Propositions pour une interprétation géodynamique globale. In: Ecologie et Géologie de l'Himalaya. Int Coll CNRS Paris 268, pp 111–129

Costa S, Maluski H (1988) 39 Ar–40 Ar dating for terranes boundaries definition. The example of the French Massif Central. Oral Comm Int Conference "Terranes in the Variscan belt of France and western Europe", Montpellier

Costa S, Lardeaux JM, Maluski H (1989) 39 Ar–40 Ar dating in the Monts du Lyonnais area (French Massif Central): an uplifted complex in a collisional context. Terra Cognita 1, 1: 265 (Abstr)

Davy P, Gillet P (1987) The stacking of thrust slices in collision zones and its thermal consequences. Tectonics 5, 6: 913–929

Delor C, Burg JP, Leyreloup AF (1984) Staurolite producing reactions and geothermobarometry of a high pressure thermal aureole in the French Massif Central. J Metam Geol 2, 1: 55–72

Delor C, Leyreloup A, Bodinier JL, Burg JP (1986) Découverte d'éclogites à glaucophane dans la klippe de Najac (Massif Central, France): nouveaux témoins océaniques d'un stade haute pression dans la chaîne de collision varisque. CR Acad Sci Paris 302 II 11: 739–744

Delor C, Burg JP, Guiraud M, Leyreloup A (1987) Les métapélites à phengite – chloritoide – grenat – staurotide – disthène de la klippe de Najac-Carmaux: nouveaux marqueurs d'un métamorphisme de haute pression varisque en Rouergue occidental. CR Acad Sci Paris 305 II: 589–595

Demay A (1942) Microtectonique et tectonique profonde. Mém Serv Carte Géol Fr Paris, 247 pp

Demay A (1946) Sur la nappe anté-stéphanienne de la Margeride dans la région médiane du Massif Central. CR Acad Sci Paris 222: 1119–1121

Demay A (1948) Tectonique anté-stéphanienne du Massif Central. Mém Explic Carte Géol Fr Paris, 259 pp

Dufour E (1985) Granulite facies metamorphism and retrogressive evolution of the Monts du Lyonnais metabasites (Massif Central, France). Lithos 18: 97–113

Dufour E, Lardeaux JM, Coffrant D (1985) Eclogites et granulites dans les Monts du Lyonnais: une évolution métamorphique plurifaciale éohercynienne. CR Acad Sci Paris II 300: 141–144

Durand C (1981) Etude pétrographique et géochimique des spilites kératophyres et roches associées de la série dévono-dinantienne de la Brévenne (Massif Central français). 3rd cycle Thesis, Lyon 1 University

Duthou JL (1977) Chronologie Rb-Sr et géochimie des granitoides d'un segment de la chaîne varisque. Relations avec le métamorphisme: le Nord-Limousin. Ann Sci Univ Clermont-Ferrand 63, 294 pp

Duthou JL, Piboule M, Gay M, Dufour E (1981) Datations radiométriques Rb-Sr sur les orthogranulites des Monts du Lyonnais (Massif Central français). CR Acad Sci Paris 292 II: 749–752

Duthou JL, Cantagrel JM, Didier J, Vialette A (1984) Paleozoic granitoids from the French Massif Central; age and origin studied 87Rb/87Sr system. Phys Earth Planet Int 35: 131–144

England PC, Richardson SW (1977) The influence of erosion upon the mineral facies of rocks from different metamorphic environments. J Geol Soc (Lond) 134: 201–213

England PC, Thompson AB (1984) Pressure-temperature-time paths of regional metamorphism. I. Heat transfer during the evolution of regions of thickened continental crust. J Petrol 125, 4: 894–928

Feybesse JL, Lardeaux JM, Johan V, Tegyey M, Dufour E, Lemière B, Delfour J (1988) La série de la Brévenne (Massif Central français): une unité dévonienne charriée sur le complexe métamorphique des Monts du Lyonnais à la fin de la collision varisque. CR Acad Sci Paris 307 II: 991–997

Floc'h JP (1979) Le métamorphisme et la mise en place des granites du dôme de Saint-Mathieu (Limousin occidental). Bull Bur Rech Géol Min I 2: 89–107

Floc'h JP (1983) La série métamorphique du Limousin central: une traverse de la branche ligérienne de l'orogenèse varisque, de l'Aquitaine à la zone broyée d'Argentat (Massif Central français). State Thesis, Limoges University, 445 pp

Floc'h JP, Santallier D, Henry B (1984) La série limousine: réflexion axée sur une synthèse géologique régionale et son utilisation pour la prospection de l'or. Chron Rech Min 474: 5–10

Forestier FH, Lasnier B, Leyreloup A, Marchand J (1973) Vues nouvelles sur la catazone dans le Massif Central français et le Massif Armoricain de l'affleurement au Moho. Bull Soc Géol Fr XV: 562–568

Friedrich M, Marignac C, Floc'h JP (1988) Sur l'existence de trois chevauchements ductiles "himalayens" successifs à vergence NW en Limousin. CR Acad Sci Paris 306 II: 663–669

Gardien V, Lardeaux JM, Misseri M (1988) Les péridotites des Monts du Lyonnais (M.C.F.): témoins privilégiés d'une subduction de lithosphère paléozoïque. CR Acad Sci Paris 307 II: 1967–1972

Gardien V, Lardeaux JM, Malavieille J (1989) P-T path related to uplift during extensional shearing: an example from the Pilat unit (Eastern French Massif Central). Terra Cognita 1, 1: 260 (Abstr)

Gasparik T (1987) Orthopyroxene thermobarometry in simple and complex systems. Contrib Mineral Petrol 96: 357–370

Gay M, Peterlongo JM, Caen-Vachette M (1981) Age radiométrique des granites allongés et en feuillets minces syntectoniques dans les Monts du Lyonnais (Massif Central français). CR Acad Sci Paris 293: 993–996

Gillet P, Choukroune P, Ballèvre M, Davy P (1986) Thickening history of the western Alps. Earth Plante Sci Lett 78: 44–52

Girardeau J, Dubuisson G, Mercier JCC (1986) Cinématique de mise en place des ophiolites et nappes cristallophylliennes du Limousin, Ouest du Massif Central. Bull Soc Géol Fr (8) II 5: 849–860

Guérangé-Lozes J (1987) Les nappes varisques de l'Albigeois cristallin. Lithostratigraphie, volcanisme et déformations. Doc Bur Rech Géol Min 135: 259 pp

Guérangé-Lozes J, Alsac C (1986) Les nappes varisques de l'Albigeois cristallin. Lithostratigraphie, volcanisme et déformations. Géol France 3: 309–337

Guillot PL (1981) La série métamorphique du Bas-Limousin: de la vallée de l'Isle à la vallée de la Corrèze. Le socle en bordure du Bassin Aquitain. State Thesis, Orléans University, 3 vol

Guillot PL, Doubinger J (1971) Découverte d'acritarches dans les schistes supérieurs de Génis (Dordogne). CR Acad Sci Paris 272: 2763–2764

Guillot PL, Lacour A (1988) Présence d'éclogites au sein de la formation leptyno-amphibolique "quercynoise" (Lot, Massif Central, France). CR Acad Sci Paris 307 II: 1977–1980

Guillot PL, Lefevre J (1975) Découverte de conondontes dans les calcaires à entroques de Génis en Dordogne (série métamorphique du Bas-Limousin). CR Acad Sci Paris 280: 1529–1530

Jenkins DM (1981) Experimental phase relations of hydrous peridotites modelled in the system H_2O-CaO-MgO-Al_2O_3-SiO_2. Contrib Mineral Petrol 77: 166–176

Kornprobst J, Poulain D (1972) Relations structurales entre les "micaschistes supérieurs" et les "micaschistes inférieurs" dans la région de St-Geniez-d'Olt (Aveyron). CR Acad Sci Paris 274: 3511–3514

Kornprobst J, Bard JP, Briand B, Cantagrel JM, Guitard G, Kienast JR, Lasnier B, le Corre C, Santallier D (1980) Le métamorphisme en France. 26th I.G.C. Paris, C7, Mém Bur Rech Géol Min 107, 161–189

Lafon JM (1986) Géochronologie U-Pb appliquée à deux segments du Massif Central français: le Rouergue oriental et le Limousin central. Thesis, Montpellier University, 152 pp

Lapadu-Hargues P (1947) Les massifs de la Margeride et du Mont Lozère et leurs bordures. Bull Serv Carte Géol Fr 46, 222: 379–532

Lardeaux JM, Dufour E (1987) Champs de déformation superposés dans la chaîne varisque. Exemple de la zone nord des Monts du Lyonnais (Massif Central français). CR Acad Sci Paris II 305: 61–64

Lardeaux JM, Gosso G, Kienast JR, Lombardo B (1982) Relations entre le métamorphisme et la déformation dans la zone de Sésia-Lanzo (Alpes occidentales) et le problème de l'éclogitisation de la croûte continentale. Bull Soc Géol Fr (7) XXIV 4: 793–800

Lardeaux JM, Reynard B, Dufour E (1989) Granulites à kornérupine et décompression postorogénique des Monts du Lyonnais (M.C.F.). CR Acad Sci Paris 308 II: 1443–1449

Lasnier B (1977) Persistance d'une série granulitique au coeur du Massif Central français (Haut-Allier). Les termes basiques, ultrabasiques et carbonatés. State Thesis, Nantes University, 351 pp

Ledru P (1988) Le chevauchement crustal de Moyenne Dordogne. Bull Soc Géol Fr (8) IV 3: 471–478

Ledru P, Autran A (1988) L'édification de la chaîne varisque dans le Limousin. Rôle de la faille d'Argentat à la limite Limousin-Millevaches. G.P.F. 3 Program. Doc Bur Rech Géol Min (Orléans) 140, pp 51–81

Ledru P, Lardeaux JM, Santallier D, Autran A, Quenardel JM, Floc'h JP, Lerouge G, Maillet N, Marchand J, Ploquin A (1989) Où sont les nappes dans le Massif Central français? Bull Soc Géol Fr 8 V 3: 605–618

Leyreloup A (1973) Le socle profond en Velay d'après les enclaves remontées par les volcans néogènes. Son thermométamorphisme et sa lithologie: granites et série charnockitique (Massif Central français). 3rd Cycle Thesis, Nantes University, 356 pp

Macaudiere J, Marignac C, Weisbrod A (1987) Grandes nappes synschisteuses collisionnelles dans la catazone hercynienne des Cévennes médianes (Massif Central, France). CR Acad Sci Paris 304: 1195–1199

Maillet N (1987) Dualité d'origine des massifs ultrabasiques limousins. Implications géodynamiques. Thesis, Lyon University 149 pp

Malavielle J, Guihot P, Costa S, Lardeaux JM, Gardien V (1990) Collapse of the thickened Variscan crust in the French Massif Central: Mont Pilat extensional shear and Saint-Etienne upper Carboniferous basin. Tectonophysics 177: 139–149

Marchand J (1974) Persistence d'une série granulitique au coeur du Massif Central français (Haut-Allier). Les termes acides. 3rd Cycle Thesis, Nantes University, 207 pp

Marignac C, Leroy J, Macaudiere J, Pichavant M, Weisbrod A (1980) Evolution tectonométamorphique d'un segment de l'orogène hercynien: les Cévennes médianes, Massif Central français. CR Acad Sci Paris 291: 605–608

Mattauer M, Etchecopar A (1976) Arguments en faveur de chevauchements de type himalayen dans la chaîne hercynienne du Massif Central français. In: Ecologie et Géologie de l'Himalaya. Int. Coll. CNRS Paris 268, pp 261–267

Matte P (1983) Two geotraverses across the Ibero-Armorican Variscan arc of western Europe. In: Rast N, Delany FM (eds) Profiles of Orogenic Belts. Gedyn. Ser., Washington, 10, AG4, pp 53–81

Matte P (1986) Tectonics and plate tectonics model for the Variscan belt of Europe. Tectonophysics 126: 329–374

Mercier L, Johan V, Lardeaux JM, Ledru P (1989) Découverte d'éclogites dans l'Artense (M.C.F.): implications pour la définition des nappes à l'est du Sillon Houiller. CR Acad Sci Paris 308 II: 315–320

Miahle J (1980) Le massif granitique de la Borne (Cévennes). Thèse 3ème Cycle, Clermont-Ferrand II University

Nicollet C, Leyreloup A (1978) Pétrologie des niveaux trondjémitiques de haute pression associés aux éclogites et amphibolites des complexes leptynoamphiboliques du Massif Central français. Can J Earth Sci 15, 5: 696–707

Pêcher A (1978) Déformations et métamorphisme associés à une zone de cisaillement. Exemple du grand chevauchement central himalayen (M.C.T.), transversale des Annapurnas et du Manaslu, Népal. State Thesis, Grenoble University, 354 pp

Peterlongo JP (1958) Terrains cristallins des Monts du Lyonnais (Massif Central français). Ann Fac Sci Clermont, 4, 187 pp

Peyretti G (1971) Etude géologique des formations à l'Ouest de Marvejols (Lozère). 3rd Cycle Thesis, Lyon University, 133 pp

Piboule M, Briand B, Beurrier M (1982) Géochimie de quelques granites albitiques dévoniens de l'est du Massif Central français. Neus Jahrb Mineral Abh 143, 3: 279–308

Pin C (1980) Données microstructurales sur les terrains métamorphiques de la série du Lot. Bull Bur Rech Géol Min 1 (4): 293–313

Pin C (1981) Old inherited zircons in two synkinematic Variscan granitoids: the "granite du Pinet" and the "orthogneiss de Marvejols" (southern french Massif Central). Neues Jahrb Mineral Abh 142, 1: 27–48

Pin C, Lancelot J (1982) U-Pb Dating of an Early Paleozoic bimodal magmatism in the French Massif Central and of its further metamorphic evolution. Contrib Mineral Petrol 79: 1–12

Pin C, Peucat JJ (1986) Ages des épisodes de métamorphisme paléozoiques dans le Massif Central et le Massif Armoricain. Bull Soc Géol Fr (8) II 3: 461–469

Pin C, Vielzeuf D (1983) Granulites and related rocks in Variscan median Europe: a dualistic interpretation. Tectonophysics 93: 47–74

Pin C, Vielzeuf D (1988) Les granulites de haute-pression d'Europe moyenne témoins d'une subduction éohercynienne. Implications sur l'origine des groupes leptyno-amphibolitiques. Bull Soc Géol Fr (8) IV 1: 13–20

Poulain D (1972) Les micaschistes des environs de St-Geniez-d'Olt (Aveyron). 3rd Cycle Thesis, Paris University, 78 pp

Quenardel JM, Rolin P (1984) Palaeozoic evolution of the Plateau d'Aigurande (NW Massif Central, France). In: Hutton DHW, Sanderson DJ (eds) Variscan tectonics of the north Atlantic region. Geol Soc London spec publ 14 pp 63–70

Reitz E, Wickert F (1988) Upper Proterozoic microfossils in low-grade phyllites of the Brévenne Unit, NE Massif Central (France). CR Acad Sci Paris 307 II: 1717–1721

Robinson P, Tracy RJ, Santallier DS, Andreasson PG, Gil Ibarguchi JI (1988) Scandian-Acadian Caledonian sensu stricto metamorphism in the age range 430–360 Ma. In: Harris AL, Fettes DJ (eds) The Caledonian – Appalachian orogen. Geol Soc Spec Publ 38: 453–467

Roques M (1941) Les schistes cristallins de la partie Sud-Ouest du Massif Central français. State Thesis, Clermont-Ferrand University, 527 pp

Rubie DC (1984) A thermal tectonic model for H.P. metamorphism and deformation in the Sésia-Lanzo zone, Western Alps. J Geol 92: 21–36

Santallier D (1981) Les roches basiques dans la série métamorphique de Bas Limousin, Massif Central (France). State Thesis, Orléans University, 340 pp

Santallier DS (1983 a) Les éclogites du Bas-Limousin, Massif Central français. Comportement des clinopyroxènes et des plagioclases antérieurement à l'amphibolitisation. Bull Minéral 106: 691–707

Santallier D (1983 b) Main metamorphic features of the Paleozoic orogen in France. In: Schenk P (ed) Regional trends in the geology of the Appalachian – Caledonian – Hercynian – Mauritanide orogen. Reidel, Dordrecht, pp 263–274

Santallier D, Floc'h JP (1979) Les éclogites de la Faurie (Bas-Limousin). Feuille d'Uzerche à 1.50.000. Bull Bur Rech Géol Min 2: 109–119

Santallier D, Floc'h JP, Guillot PL (1978) Quelques aspects du métamorphisme dévonien en Bas-Limousin. Massif Central, France. Bull Soc Fr Minéral Cristallogr 101: 77–88

Santallier D, Briand B, Menot RP, Piboule M (1988) Les complexes leptyno-amphiboliques (C.L.A.): revue critique et suggestions pour un meilleur usage de ce terme. Bull Soc Géol Fr (8) IV 1: 3–13

Sisson VB, Hollister LS (1988) Low-pressure facies series metamorphism in an accretionary sedimentary prism, southern Alaska. Geology 16: 358–361

Thompson AB, England PC (1984) Pressure-temperature-time paths of regional metamorphism. II Their influence and interpretation using mineral assemblages in metamorphic events. J Petrol 25: 928–954

Toteu F (1981) Utilisation des analyses chimico-minéralogiques et microstructurales dans la reconstitution des évènements tectono-métamorphiques des formations polycycliques. Exemple du Vivarais cristallin (Ardèche). 3rd Cycle Thesis, Nancy, 207 pp

Weisbrod A (1970) Pétrologie du socle métamorphique des Cévennes médianes. Reconstitution sédimentologique et approche thermodynamique du métamorphisme. State Thesis, Nancy University, 4 vol

Weisbrod A, Pichavant M, Marignac C, Macaudiere J, Leroy J (1980) Relations structurales et chronologiques entre le magmatisme basique, les granitisations et l'évolution tectono-métamorphique tardi-hercynienne dans les Cévennes médianes. CR Acad Sci Paris 291: 665–668

Igneous Activity. Caledono-Hercynian Magmatism in the French Massif Central

Coordinator: A. Ploquin
Contributors: B. Briand, G. Dubuisson, J. Girardeau, J. C. Mercier, M. Ohnenstetter, A. Ploquin, D. Santallier, and J. M. Stussi

1 Introduction

A. Ploquin and D. Santallier

This chapter presents Caledono-Hercynian magmatism under two main headings:

- magmatism, Sections 2–5
- granitoid magmatism, Section 6

As explained in the Introduction to Part III, the Massif Central is now interpreted as a collision chain with classical stages that can be sketched as follows:

- Cambro-Ordovician distension(s) in a complicated crust, with production of basic rocks in the series and of a more or less advanced, or more or less aborted oceanization.
- Silurian convergence and closing stage with high pressure metamorphism.
- Devonian collisional stage, or Acadian Phase, with nappes tectonic and medium pressure metamorphism, locally with production of migmatites which may affect some Cambro-Silurian granitoid gneisses.
- During the Upper Devonian, Carboniferous, and Early Permian, huge masses of granites invaded the whole Massif Central. Some ensialic opening may occur.
- The post-orogenic stage is described in Chap. III 2

Basic magmatism does not occur as large intrusions but is instead represented by intrafolial metamorphic units, or by some bodies of ultrabasic and basic rocks which have been interpreted to be ophiolite remnants. Permo-Carboniferous volcanic rocks occur in the northeast.

The descriptions of (meta)volcanics are included within the descriptions of the supracrustal series (Ledru and Mouthier Chap. III 3). In Section 2 Briand et al. give a synthesis of metabasite units intercalated within gneissic units. These rocks may be of Cambro-Ordovician age or older. Some metabasites belong to the so-called Leptynitic-Amphibolitic Complexes (LAC), see below; their chemical affinities are variable (Santallier et al. 1988).

Regarding the ophiolite remnants, two inferred hypotheses prevail. One corresponds to a period of Devonian distension, as in the Beaujolais and Brévenne area (Ohnenstetter, Sect. 5), while the second, older, is more problematic. In Limousin many ultrabasic and basic bodies occur, mainly near the limits between Lower and Middle Allochthon Units but occasionally between Lower and Upper Allochthon Units; are all these bodies remnants of one dismembered ophiolite older complex? Does it represent a large lost ocean? This is the opinion clearly expressed by Girardeau et al. (Sect. 3); however, some geologists do not agree, and a discussion by Santallier is added (Sect. 4). This controversy raises the question of the existence and position of a major suture inside Limousin, e.g., see in Duthou et al. (1989, pp. 671–4, 676, 687). Recent geophysical data draw attention to the Saintes graben Lefort et al. Chap. III 2; if this line in fact represents the major Siluro-Devonian suture, the problem will be different!

The second half of this chapter is concerned with the granitoids and felsic orthogneisses. This section presents conclusions from a large compilation of data on the French Massif Central, first started during IGCP 154 and 27, and an attempted synthesis by means of chemical typology. The granitoid massifs are widespread and mainly of Carboniferous age. Some orthogneisses cover large areas, as, for example, the arcs in Limousin, SE Auvergne, Ardèche, and in Montagne Noire, and they are also wide spread within some gneissic units. During the course of this compilation the Leptyno-Amphibolites Complexes (LAC) was not well defined, but was used in different ways and with local definitions. The use of this term has been widened but we do not have the information to order our results on orthogneisses according to the new concepts of LAC. In the analyzed samples, chemical characters reveal two types of "leptynites" according to Na/K variations. A recent review was published by Santallier et al. (1988).

The French use of the term Leptynitic-Amphibolitic Complexes (LAC) was created by Forestier in 1961. Originally. It was purely descriptive and applied

J. D. Keppie (Ed.): Pre-Mesozoic Geology in France and Related Areas

to bimodal acid-basic associations in high grade metamorphic series of the French Massif Central, commonly associated with minor amounts of carbonates and garnet peridotites. Since then, specific aspects have been emphasized by successive authors:

1. The associated basic and ortho- or para-derived acid rocks must contain relict high pressure assemblages such as eclogites and/or granulites (Lasnier et al. 1973; Marchand 1974; Burg 1977; Lasnier 1977; Nicollet 1978; Godard 1981; Dufour et al. 1985).
2. The LAC are a bimodal magmatic series affected by at least one metamorphic event of variable intensity. Initially, they were mostly tholeiites and are generally interpreted as indicators of extensional processes (Montigny and Allègre 1974; Nicollet and Leyreloup 1978; Briand and Piboule 1979; Le Métour and Bernard-Griffiths 1979; Piboule 1979; Bernard-Griffiths and Jahn 1981; Piboule et al. 1982, 1983; Lancelot et al. 1984; Bernard-Griffiths and Cornichet 1985; Piboule and Briand 1985; Bodinier et al. 1986; Seyler 1986; Thiéblemont et al. 1987; Bodinier et al. 1988; Briand et al. 1988.
3. The LAC are typical of orogenic mobile zones. They comprise terrigeneous rocks such as graywackes, associated with the basic magmatic products of the associated extensional event (Carme 1975 a, b; Burg et al. 1984; Dubuisson et al. 1986).
4. The LAC are litho-tectonic formations emphasizing major ductile deep-seated shear zones in continental environment (Burg and Matte 1978; Autran and Peterlongo 1980; Marchand 1981; Bouchez and Jover 1986).

Consequently, the LAC are no longer precisely defined (Santallier et al. 1988), and any attempt to use the term based upon a single criterion can lead to erroneous geological interpretations.

References

Autran A, Peterlongo JM (1980) Introduction à la géologie du Massif central. In: C. Lorenz (ed) Géologie des pays européens. 26e CGI Paris Dunod, Paris, pp 9–121

Bernard-Griffiths J, Cornichet J (1985) Origin of eclogites from south Brittany, France. A Sm-Nd isotopic and REE study. Chem Geol 52, 2: 185–201

Bernard-Griffiths J, Jahn BM (1981) REE geochemistry of eclogites and associated rocks from Sauviat-sur-Vige, Massif central français. Lithos 14: 263–274

Bodinier JL, Burg JP, Leyreloup A, Vidal H (1988) Reliques d'un bassin d'arrière-arc subducté, puis obducté dans la région de Marvejols (Massif Central). Bull Soc Géol Fr (8) IV 1: 21–33

Bodinier JL, Giraud A, Dupuy C, Leyreloup A, Dostal J (1986) Caractérisation géochimique des métabasites associées à la suture méridionale hercynienne: Massif central français et Chamrousse (Alpes). Bull Soc Géol Fr (8) II 1: 115–123

Bouchez JL, Jover O (1986) Le Massif central: un chevauchement de type himalayen vers l'Ouest-Nord-Ouest. C R Acad Sci Paris 302 II 9: 675–680

Briand B, Piboule M (1979) Les métabasites de la série de Marvejols (Massif central): témoins d'un magmatisme tholéiitique d'arrière-arc cambro-ordovicien? Bull Bur Rech Geol Min 2 I 2: 131–171

Briand B, Piboule M, Bouchardon JL (1988) Diversité géochimique des métabasites des groupes leptyno-amphiboliques du Rouergue et de Marvejols (Massif Central). Origine et implications. Bull Soc Géol Fr (8) IV 3: 489–498

Burg JP (1977) Tectonique et microtectonique des séries cristallophylliennes du Haut-Allier et de la vallée de la Truyère. These 3e cycle, Montpellier, 106 pp

Burg JP, Matte P (1978) A cross-section through the French Massif Central and the scope of its Variscan geodynamic evolution. Z dtsch Geol Ges 129: 429–460

Burg JP, Leyreloup A, Marchand J, Matte P (1984) Inverted metamorphic zonation and large scale thrusting in the variscan belt: an example in the French Massif central. In: Hutton DHV, Sanderson DJ (eds) Variscan tectonics of the North Atlantic region. Geol Soc London, sp. publ., 14 47–61

Carme F (1975 a) Essai d'interprétation unitaire des corps de roches basiques et ultrabasiques et de leur contexte gneissique dans la partie moyenne du Massif central et dans les massifs cristallins externes alpins. C R Acad Sci Paris 281 D: 351–354

Carme F (1975 b) Caractères chimiques de deux ensembles magmatiques pré-orogéniques du socle anté-houiller des massifs cristallins externes alpins et des monts du Lyonnais. C R Acad Sci Paris 281 D: 89–92

Dubuisson G, Galdeano A, Girardeau J, Mercier JCC (1986) La synforme de Saint-Germain-les-Belles en Limousin central. Etude magmatique et relations ophiolites – "groupe leptyno-amphibolique". Séance spéc. Soc géol France: Les complexes leptyno-amphiboliques, Rennes

Dufour E, Lardeaux JM, Coffrant D (1985) Eclogites et granulites dans les monts du Lyonnais: une évolution métamorphique plurifaciale éohercynienne. C R Acad Sci Paris 300 II 4: 141–144

Duthou JL, Floc'h JP (1989) Evolution tectonométamorphique du Massif Central. Réunion extraordinaire de la Société Géologique de France. Bull Soc Geol Fr Paris (8) V 4: 667–692

Forestier FH (1961) Métamorphisme hercynien et anté-hercynien dans le bassin du Haut-Allier (Massif central français). Thèse d'Etat Clermont-Ferrand, 2 vol 300 pp

Godard G (1981) Lambeaux probables d'une croûte océanique subductée: les éclogites de Vendée (Massif armoricain, France). Thèse 3e cycle, Nantes, 153 pp

Lancelot JR, Allegret A, Ducrot J, Gallibert F, Lafon JM, Lévêque MH, Respaut JP (1984) Datation des evênements anté-hercyniens en Europe méridionale à partir de la géochronologie U-Pb sur zircons. Séance spéc. Soc. géol. France: Structure profonde de la croûte hercynienne d'Europe, Montpellier

Lasnier B (1977) Découverte d'une série granulitique au cœur du Massif central français (Haut-Allier). Les termes basiques, ultrabasiques et carbonatés. Thèse d'Etat, Nantes, 333 pp

Lasnier B, Leyreloup A, Marchand J (1973) Découverte d'un granite "charnockitique" au sein de "gneiss œillés". Perspectives nouvelles sur l'origine de certaines leptynites du Massif

armoricain méridional, France. Contrib Mineral Petrol 41: 131–144
Le Métour J, Bernard-Griffiths J (1979) Age (limite Ordovicien-Silurien) de mise en place du massif hypovolcanique de Thouars (Massif vendéen); implications géologiques. Géol Fr 1, 4: 365–371
Marchand J (1974) Persistance d'une série granulitique au cœur du Massif central français, Haut-Allier. Les termes acides. Thèse 3e cycle, Nantes, 207 pp
Marchand J (1981) Ecaillage d'un "mélange tectonique" profond: le complexe cristallophyllien de Champtoceaux (Bretagne méridionale). C R Acad Sci Paris 293 II: 223–226
Montigny R, Allègre C J (1974) A la recherche des océans perdus: les éclogites de Vendée, témoins métamorphisés d'une ancienne croûte océanique. C R Acad Sci Paris 279 D: 543–545
Nicollet C (1978) Etude pétrologique, géochimique et structurale des terrains cristallins anté-permiens du versant sud du Lévezou (Massif central français). Thèse 3e cycle, Montpellier
Nicollet C, Leyreloup A (1978) Pétrologie des niveaux trondhjemitiques de haute pression associés aux éclogites et amphibolites des complexes leptyno-amphiboliques du Massif central français. Can J Earth Sci 15, 5: 696–707
Piboule M (1979) L'origine des amphibolites: approche géochimique et mathématique. Application aux amphibolites du Massif central français. Thèse d'Etat, Lyon, 2 vol, 1003 pp
Piboule M, Briand B (1985) Geochemistry of eclogites and associated rocks of the southeastern area of the French Massif Central: origin of the protoliths. Chem Geol 50, 1–3: 189–199
Piboule M, Briand B, Beurrier M (1982) Géochimie de quelques granites albitiques dévoniens de l'Est du Massif central (France). Neues Jahrb Miner Abh 143, 3: 279–308
Piboule M, Beurrier M, Briand B, Lacroix P (1983) Les trondhjémites de Chindo et de Saint-Vérand et le magmatisme kératophyrique associé. Pétrologie et cadre géostructural de ce magmatisme dévono-dinantien. Géol (2) I 1–2: 55–72
Pin C (1989) Essai sur la chronologie et l'évolution geodynamique de la chaîne herognienne d'Europe. Thésis, Clermont)
Santallier D, Briand B, Ménot R P, Piboule M (1988) Les complexes leptyno-amphiboliques (C. L. A.): revue critique et suggestions pour un meilleur emploi de ce terme. Bull Soc Géol Fr (8) IV 1: 3–12
Santallier D, Floc'h J P (1978) Les roches métamorphiques basiques et ultrabasiques de la feuille de Thiviers à 1/50 000. Bull Bur Rech Geol Min 1, 2: 125–147
Seyler M (1986) Petrology and genesis of Hercynian Alkaline Orthogneisses from Provence, France. J Petrol 27, 5: 1229–1251
Thiéblemont D, Cabanis B, Le Métour J (1987) Etude géochimique d'un magmatisme de distension intracontinentale: la série bimodale ordovicienne du Choletais (massif vendéen). Géol Fr 1: 65–76

2 Basic and Ultrabasic Magmatism: Metabasites and Palaeozoic Openings in the French Massif Central

2.1 Highly Transformed Metabasites

B. Briand and D. Santallier

2.1.1 Introduction

The original nature of the highly transformed basic rocks is generally not easily determined by direct petrological observation. Among them, there are three groups:

1. Variably retrogressed eclogites, high pressure granulites, and coronitic meta-gabbros have been affected by the Silurian high pressure metamorphic event. They are widespread in Limousin, eastern and western Rouergue, Lot, Haut-Allier, and Lyonnais areas (Fig. 1), in the middle allochthon unit (Fig. 2). They are apparently mainly of Cambrian and Ordovician age.
2. Metagabbros and metadolerites with relict magmatic textures. These have been affected only by the Medium Pressure Devonian metamorphism. They are exposed particularly well in Limousin and eastern Rouergue, in the middle allochthon or lower allochthon high grade metamorphic units, as well as in the epizonal upper allochthon Thiviers-Payzac unit in Limousin. They are of either Lower Paleozoic or Devonian age.
3. Ordinary amphibolites or meta-tuffs are widespread in the middle and lower allochthons units as well as in the presumed autochthon. They could be of either Lower Paleozoic or Upper Proterozoic age.

Even for the second group, the nature of the protoliths is best determined using major and trace element geochemistry, including REE, despite the possible influence of crustal contamination or element mobility during metamorphism.

2.1.2 Middle Allochthon Units

In the Middle Allochthon Units two populations can be distinguished: widespread Cambro-Orodovician and Devonian.
Most of the Cambrian and Ordovician metabasites of the middle allochthon are interlayered in the LAC. We must emphasize that the protoliths were originally heterogeneous. Cabanis et al. (1983), Giraud et al. (1984), Piboule and Briand (1985), Bodinier et al. (1986), and Briand et al. (1988) record N-MORB types characterized by flat or slightly LREE-depleted patterns for the Lot (Fig. 1) granulites (Fig. 3 A and Fig. 4, type I), and Haut-Allier common amphibolites

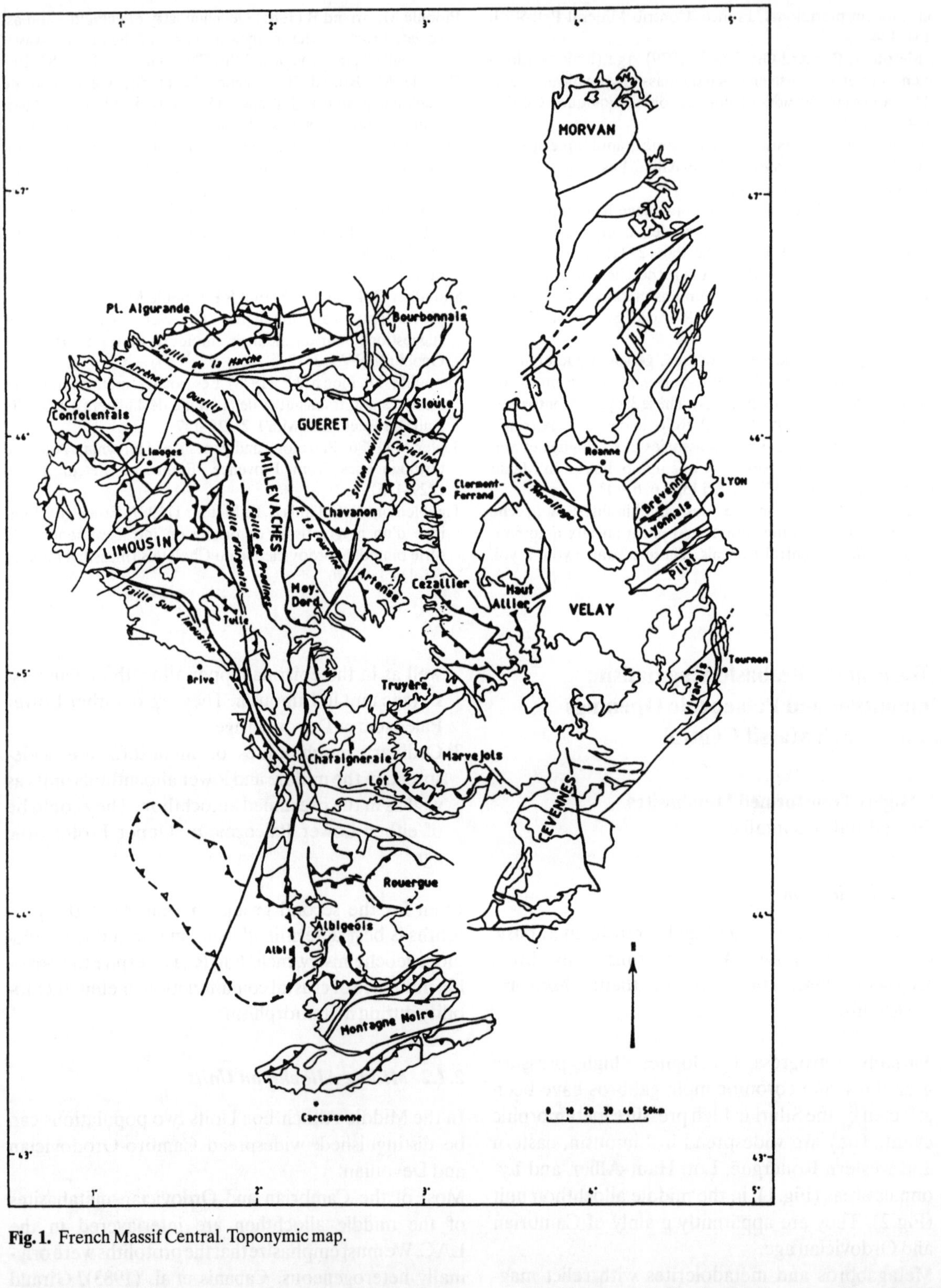

Fig. 1. French Massif Central. Toponymic map.

(Fig. 5). In contrast, slightly LREE-enriched patterns are found for the Limousin eclogites and mesozonal meta-dolerites (Fig. 6, curves 3 and 4). All these authors concluded that the protoliths were emplaced during an oceanization process of undetermined importance.

Outside the LAC, the Najac granulites (Fig. 7) have characteristics similar to N-MORB magmas.

In a few cases, more precise studies provided better information on the original settings in the Lot and Rouergue areas. In the Lot granulites, Bodinier

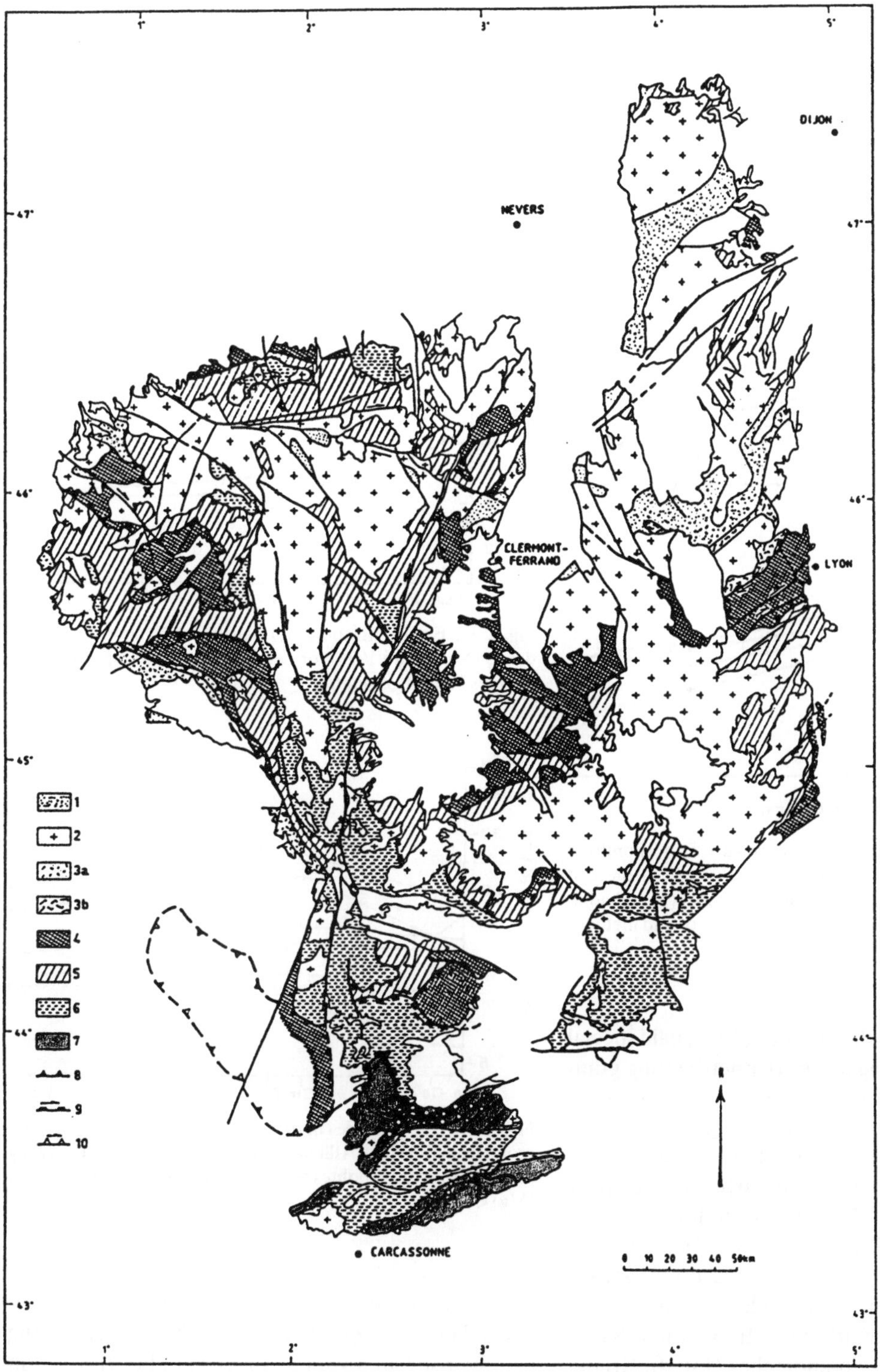

Fig. 2. French Massif Central. The nappes pile. *1* Mesozoic cover; *2* Post-tectonic granitoids; *3* Upper Allochthon – *3a* Thiviers-Payzac and Génis Units, *3b* Brévenne Unit; *4* Middle Allochthon; *5* Lower Allochthon; *6* Presumed Autochthon; *7* Southern Paleozoic Units; *8* Thrust nappes (390–370 Ma); *9* Thrusts and transcurrent zones (350–320 Ma); *10* Thrusts and transcurrent zones (320 ± 10 Ma).

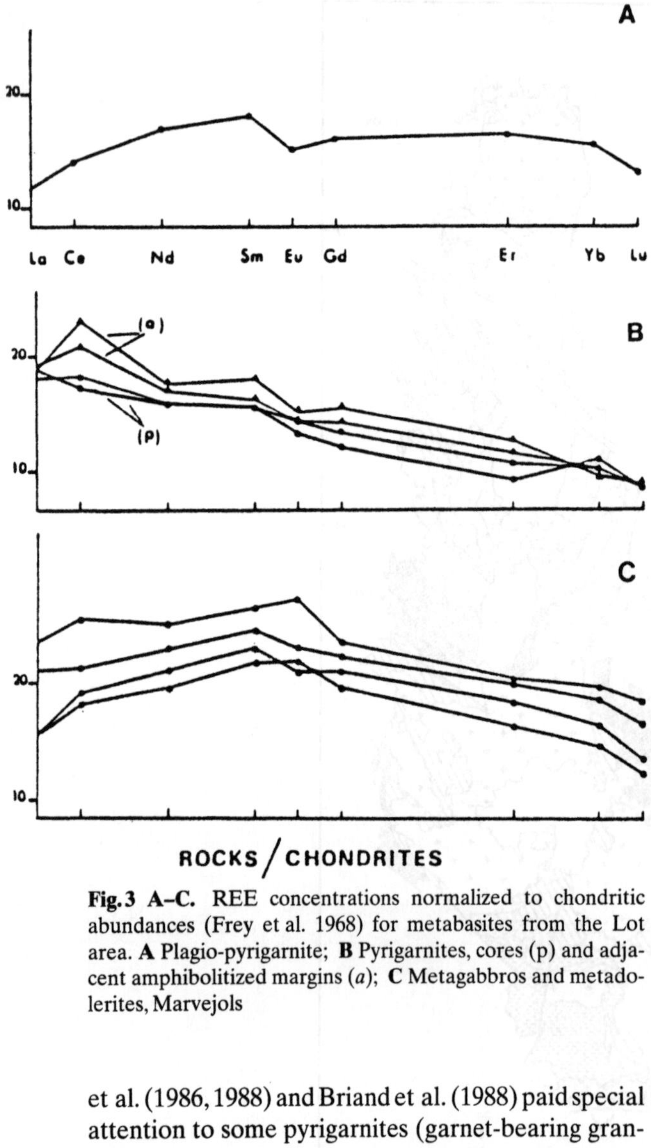

Fig. 3 A–C. REE concentrations normalized to chondritic abundances (Frey et al. 1968) for metabasites from the Lot area. **A** Plagio-pyrigarnite; **B** Pyrigarnites, cores (p) and adjacent amphibolitized margins (*a*); **C** Metagabbros and metadolerites, Marvejols

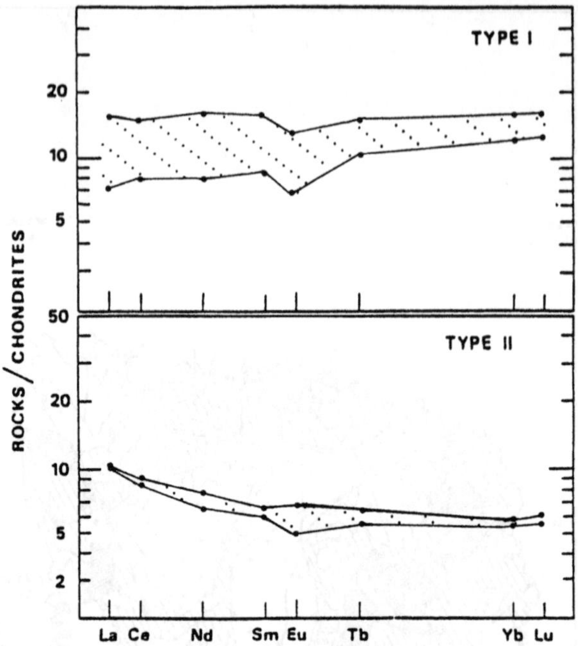

Fig. 4. REE concentrations normalized to the chondrite abundance (Nakamura 1974) for two tectonic lenses representative of types I and II granulites

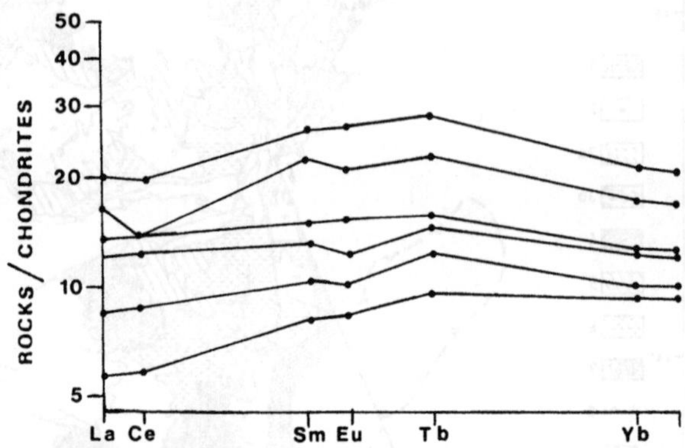

Fig. 5. REE abundances in rocks of the leptynitic-amphibolitic group from the Haut-Allier area, normalized to chondrite (Frey et al. 1968). Amphibolites without high grade relicts and with (Mg) < 0.7

et al. (1986, 1988) and Briand et al. (1988) paid special attention to some pyrigarnites (garnet-bearing granulites). All these authors noted that these rocks are characterized by their low HREE contents and slight LREE enrichment (Fig. 4, type II and Fig. 3B, curve P). Bodinier et al. (1988) particularly emphasized the similarity of Ti/V ratios with those of island arcs lavas, the low Ti, P and V contents (Ti/V< 20) and the importance of some specific ratios, e.g. V/Sc. All these characteristics allowed Bodinier et al. to suggest possible boninitic affinities. In addition, based upon strong negative Nb anomalies (Fig. 8), Briand et al. (1988) emphasized that the Lot granulites have strong affinites with subduction zone magmatism. Occurring in close association with the previous granulites, are 484 ± 7 Ma old (U/Pb zr: Pin and Lancelot 1982) coronitic meta-grabbos which exhibit convex-shaped REE patterns (Fig. 3C). Such patterns have been interpreted by Saunders (1983), in his study of magmatism in the Gulf of California region, as being characteristic of a discrete "subcontinental" influence. This fits well with the Lot basic rocks being closely associated with 478 ± 6 Ma old (U/Pb zr: Pin and Lancelot 1982) high-silica rhyolites generated by crustal melting. All of these geochemical characteristics, in addition to the lithologic association with meta-graywackes and orthogneisses, led Bodinier et al. (1986, 1988) and Briand et al. (1988) to suggest that the basic rocks were emplaced in a marginal basin site, probably in an ensialic environment. The same conclusion (the basic rocks were em-

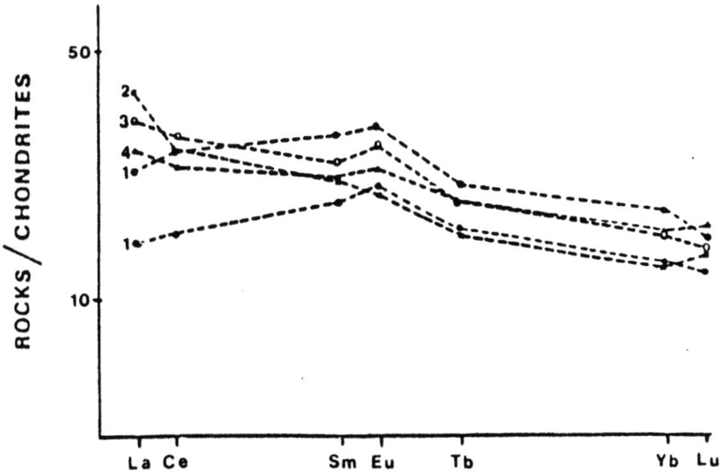

Fig. 6. Diagram of distribution of lanthanides. *Curve 1* Genis greenstones; *curve 2* Puytinaud tuffs; *curve 3* eclogites; *curve 4* Seilhac-type dolerites

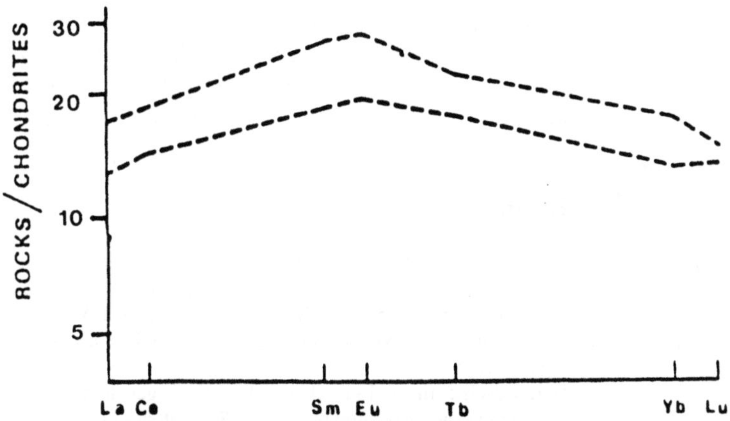

Fig. 7. REE concentrations normalized to chondrite (Frey et al. 1968) for the Najac non-cumulative rocks

placed in back-arc basins) has also been proposed, based upon lithologic associations in other parts of the Middle Allochthon. In Haut-Allier, the metabasites are closely associated with metapelites carbonates, and meta-graywackes. In Limousin, they are in close association with continental meta-granitoids and meta-graywackes. The same associations have been described also in eastern Rouergue and eastern Vivarais (Piboule and Briand 1985). In the Monts du Lyonnais, Dufour (1985) described tholeiitic intraplate metabasites in close association with crustal-derived rhyolites, suggesting an intracontinental origin.

The Devonian Generation. It has recently been demonstrated in Limousin, the southern Plateau d'Aigurande and eastern Rouergue that the middle and lower allochthons contain Devonian syn-tectonic metagabbros, the nature of which is highly variable. For example, in eastern Rouergue, a metagabbro of calç-alkaline composition (Piboule 1979) was dated at 367 ± 10 Ma (U/Pb zr) by Pin and Piboule (1988). In contrast, in the Limousin area, apparently contemporaneous gabbros have an oceanic tholeiitic composition (Cabanis and Santallier unpublished data on trace elements).

2.1.3 Lower Allochthon and Presumed Autochthonous Units

The Lower Allochthon and Presumed Autochthonous Units contain metabasites, some of which have high pressure relicts. For example, in the lower allochthon, there are eclogites, a 496 + 25/-19 Ma old (U/Pb zr and monazite: Gebauer et al. 1981) coronitic gabbro in Limousin, high pressure amphibolites, etc. These are, at least partly, of alkaline affinity, based upon LREE-enriched patterns and immobile elements (Cabanis et al. 1983). In western Vivarais, Cévennes médianes, Lyonnais, and Limousin, Chenevoy et al. (1986) and Bernard-Griffiths and Jahn (1981) advocated an intraplate setting (continental in Vivarais and Cévennes médianes, continental or oceanic in Limousin). The fact that they are, in both areas, closely associated with metagranitoids supports an intracontinental setting.

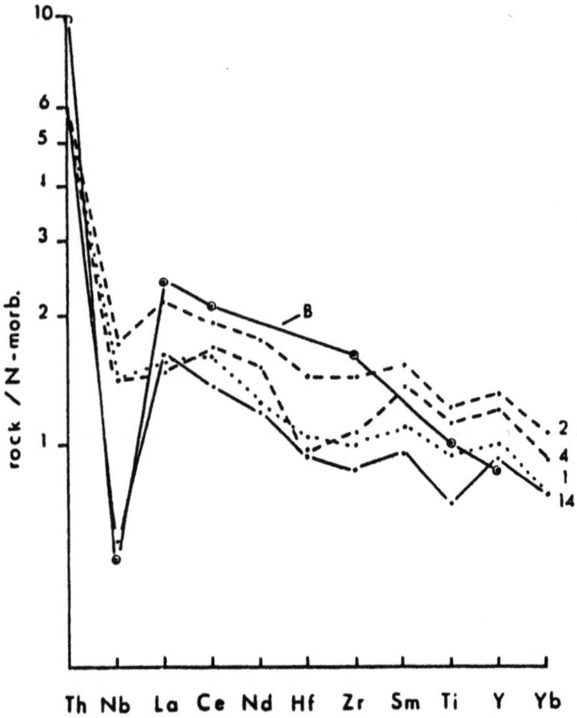

Fig. 8. Normalized multi-elements plots of some metabasites from Lot (L) and Rouergue (R) areas (N-type MORB values are from Pearce 1983). *Light dashed line* meta-gabbro (1 L); *heavy dashed line* meta-dolerites (2, 4 L); *continuous lines* meta-gabbro (14 R) and *(B):* basalt from ensialic back-arc basin (Bransfield Strait) in Weaver et al. 1979

In the presumed autochthon, the only well-documented occurences of volcanic intercalations are the Albigeois (Guérangé-Lozes 1987) rhyolitic tuffs of lower Cambrian to lower Ordovicin age, considered to have been produced by crustal melting during an episode of crustal extension.

2.1.4 Epizonal Upper Allochthon Units

The Epizonal Upper Allochthon Units also contain metabasites. In Limousin, the Genis ophiolite has an oceanic tholeiitic composition characterized by Th/Ta < 1 and La/Ta = 22 (Cabanis et al. 1983) (Fig. 6, curve 1). The underlying Thiviers-Payzac unit is composed of rhyodacitic and bimodal meta-tuffs cross-cut by dolerite dykes. All these are calc-alkaline (Cabanis et al. 1983) and are similar to convergent zones magmatism as supported by their Th/Ta = 6 (Fig. 6, curve 2).

2.1.5 Southern Paleozoic Units

In the Southern Paleozoic Units North of the Montagne Noire dome, Alsac et al. (1987), Guérangé-Lozes (1987) and Marini (1988) described low-grade metabasic lava flows, sills and volcanoclastics with some felsic layers. All these units are of Tremadocian and Arenigian age. Based mainly upon stable incompatible trace elements and REE studies, these authors have concluded that the basic rocks are continental tholeiites (Alsac et al. 1987), or "initial rifting tholeiites" or "transitional to tholeiitic basalts related to continental break-up or abortive rifting" (Marini 1988). These authors concluded that the basic rocks were emplaced in an extensional environment, upon a thinned continental crust. Such a conclusion concurs with the association of terrigeneous metasediments and acidic magmas produced by crustal melting. Additionally, in the Paleozoic nappes of the southern limb of the Montagne Noire dome, basic metavolcanics of alkaline nature have also been described (Lepine et al. 1988). These are associated with debris flows and turbidites, which suggests that they were also emplaced in an extensional geotectonic setting.

References

Alsac C, Cabanis B, Guérangé-Lozes J, Beziat D (1987) Caractéristiques magmatiques du volcanisme basique Ordovicien de la nappe de Saint-Salvi-de-Carcavès dans l'Albigeois cristallin (Tarn-Aveyron, France). CR Acad Sci Paris 305 II: 1199–1205

Bernard-Griffiths J, Jahn BM (1981) REE geochemistry of eclogites and associated rocks from Sauviat-sur-Vige (Massif Central, France). Lithos 14: 263–274

Bodinier JL, Giraud A, Dupuy C, Leyreloup A, Dostal J (1986) Caractérisation géochimique des métabasites associées à la suture méridionale hercynienne: Massif Central français et Chamrousse (Alpes). Bull Soc Géol Fr (8) II 1: 115–123

Bodinier JL, Burg JP, Leyreloup A, Vidal H (1988) Reliques d'un bassin d'arrière-arc subducté, puis obducté dans la région de Marvejols (Massif Central). Bull Soc Géol Fr (8) IV 1: 21–33

Briand B, Piboule M, Bouchardon JL (1988) Diversité géochimique des métabasites des groupes leptyno-amphiboliques du Rouergue et de Marvejols (Massif Central). Origine et implications. Bull Soc Géol Fr (8) IV 3: 489–498

Cabanis B, Guillot PL, Santallier D, Jaffrezic H, Meyer G, Treuil M (1983) Apport des éléments traces à l'étude géochimique des métabasites du Bas-Limousin. Bull Soc Géol Fr (7) XXV 4: 563–575

Chenevoy M, Gay M, Lochon P (1986) Métabasaltes alcalins et méta-leucogranites de la série métamorphique du Vivarais occidental (Massif Central français): marqueurs successifs de processus paléozoïques de distension et de collision. CR Acad Sci Paris 303 II: 725–730

Dufour E (1985) Granulite facies metamorphism and retrogressive evolution of the Monts du Lyonnais metabasites (Massif Central, France). Lithos 18: 97–113

Frey FA, Haskin MA, Poetz JA, Haskin LA (1968) Rare earth abundances in some basic rocks. J Geophys Res 73 (18): 6085–6098

Gebauer D, Bernard-Griffiths J, Grünenfelder M (1981) U-Pb zircon and monazite dating of a Mafic – Ultramafic Complex

and its Country rocks. Example: Sauviat-sur-Vige, French Central Massif. Contrib Mineral Petrol 76: 292–300
Giraud A, Marchand J, Dupuy C, Dostal J (1984) Geochemistry of leptyno-amphibolitic complex from Haut-Allier (French Massif Central). Lithos 17: 203–214
Guérangé-Lozes J (1987) Les nappes varisques de l'Albigeois cristallin. Lithostratigraphie, volcanisme et déformations. Doc Bur Rech Geol Min (Orléans) 135, 259 pp
Lépine J, Ben Ayad M A, Béziat D (1988) Mise en évidence de métabasites alcalines Cambrien inférieur dans le district aurifères de Salsigne (versant Sud-Ouest de la Montagne Noire). C R Acad Sci Paris 306 II: 1087–1092
Marini F (1988) "Phase" sarde et distension ordovicienne du domaine sud-varisque, effets de point chaud? Une hypothèse fondée sur les données nouvelles du volcanisme albigeois. C R Acad Sci Paris 306 II: 443–450
Nakamura N (1974) Determination of REE, Ba, Fe, Mg, Na and K in carbonaceous and ordinary chondrites. Geochim Cosmochim Acta 38: 757–775
Pearce J A (1983) Role of the sub-continental lithosphere in magma genesis at active continental margins. In: Hawkesworth C J, Norry M J (eds) Continental basalts and mantle xenoliths. Shiva Publ, Nantwich, pp 230–249
Piboule M (1979) L'origine des amphibolites: Approche géochimique et mathématique. Application aux amphibolites du Massif Central français. State thesis, Lyon University, 1001 pp
Piboule M, Briand B (1985) Geochemistry of eclogites and associated rocks of the southeastern area of the French Massif Central: origin of the protoliths. Chem Geol 50: 189–199
Pin C (1989) Essai sur la chronologie et l'évolution géodynamique de la chaine herynienne d'Europe. State Thesis, Clermont-Ferrand University, 470 pp
Pin C, Lancelot J (1982) U-Pb dating of an Early Paleozoic bimodal magmatism in the French Massif Central and of its further metamorphic evolution. Contrib Mineral Petrol 79: 1–12
Pin C, Piboule M (1988) Age dévonien supérieur de la série calco-alcaline de la ceinture basique du Lévezou (Rouergue). Un exemple de complexe leptyno-amphibolique composite. Bull Soc Géol Fr (8) IV 2: 261–265
Saunders A D (1983) Geochemistry of basalts recovered from the gulf of California during Leg 65 of the Deep Sea Drilling Project LXV. In: Lewis B T R, Robinson P et al. (eds) Init. Rep. of the Deep Sea Drilling Project LXV. Washington U S Gov Print Off pp 591–621
Weaver B L, Saunders A D, Pankurst R J, Tarney J (1979) A geochemical study of magmatism associated with the initial stages of back-arc spreading: the Quaternary volcanism of Bransfield Strait, from South Shetland Islands. Contrib Mineral Petrol 68: 151–169

3 Early Paleozoic Mantle Formations, Western Massif Central, a Discussion

3.1 The Limousin Ophiolite Complexes: Evidence for Oceanic Lithosphere

(J. Girardeau, G. Dubuisson, and J.-C. Mercier)

3.1.1 Introduction

According to regional tectonic studies, the French Variscides partly result from the Late Silurian-Early Devonian collision of at least two continental blocks separated by oceanic domains (Burg and Matte 1978; Matte and Burg 1981; Burg 1984; see review in Matte 1986). Until recently, the only geological evidence for a Cambro-Ordovician oceanic lithosphere was limited to some lenses of retrogressed eclogites and amphibolites showing N-MORB-type REE patterns (Montigny and Allègre 1974; Briand and Piboule 1979; Cabanis et al. 1983; Giraud et al. 1984; Piboule and Briand 1985; Bodinier et al. 1986, 1988; Gardien et al. 1988; Godard 1988). Flattened lenses of mafic and ultramafic rocks, interbedded within the leptyno-amphibolitic complexes and intensely deformed, were commonly interpreted as cumulates formed in aborted ensialic extension zones (Santallier 1981; Maillet et al. 1984; Piboule and Briand 1985; Maillet 1987; Santallier et al. 1988), with the exception of some peridotite boudins from the Mont du Lyonnais Gardien et al. 1988). In the western part of the French Massif Central (Limousin), however, such rocks often represent moderately dismembered ophiolitic complexes, as is clearly shown by the original mineral assemblages, textures, and relationships still locally preserved (Mercier et al. 1982, 1984; Dubuisson et al. 1983, 1989; Dubuisson 1988; Girardeau 1986; Girardeau et al. 1986). Such ophiolitic massifs are well exposed in the St. Germain-Les-Belles area, where they constitute a 10-km-long, 5-km-wide subcontinuous belt and as isolated massifs within 50 km SW of this locality (Fig. 9). The ophiolitic massifs generally form large boudins (several tens of kilometers long by a few kilometers wide) enclosed within sialic rocks comprising micaschists, leptynites, and minor gneisses. All these massifs may belong to a specific litho-tectonic unit, the intermediate allochthon of Girardeau et al. (1986) and Dubuisson et al. (1988), at the base of the upper gneissic unit defined by Floc'h, (1983) and Floc'h et al. (1984). This unit, which suffered very intense high-temperature shear deformations (Girardeau et al. 1986), has been interpreted as a paleo-suture thrust zone (Dubuisson et al. 1983, 1984, 1988). The purpose of this section is to present the main characteristics of the ophiolite.

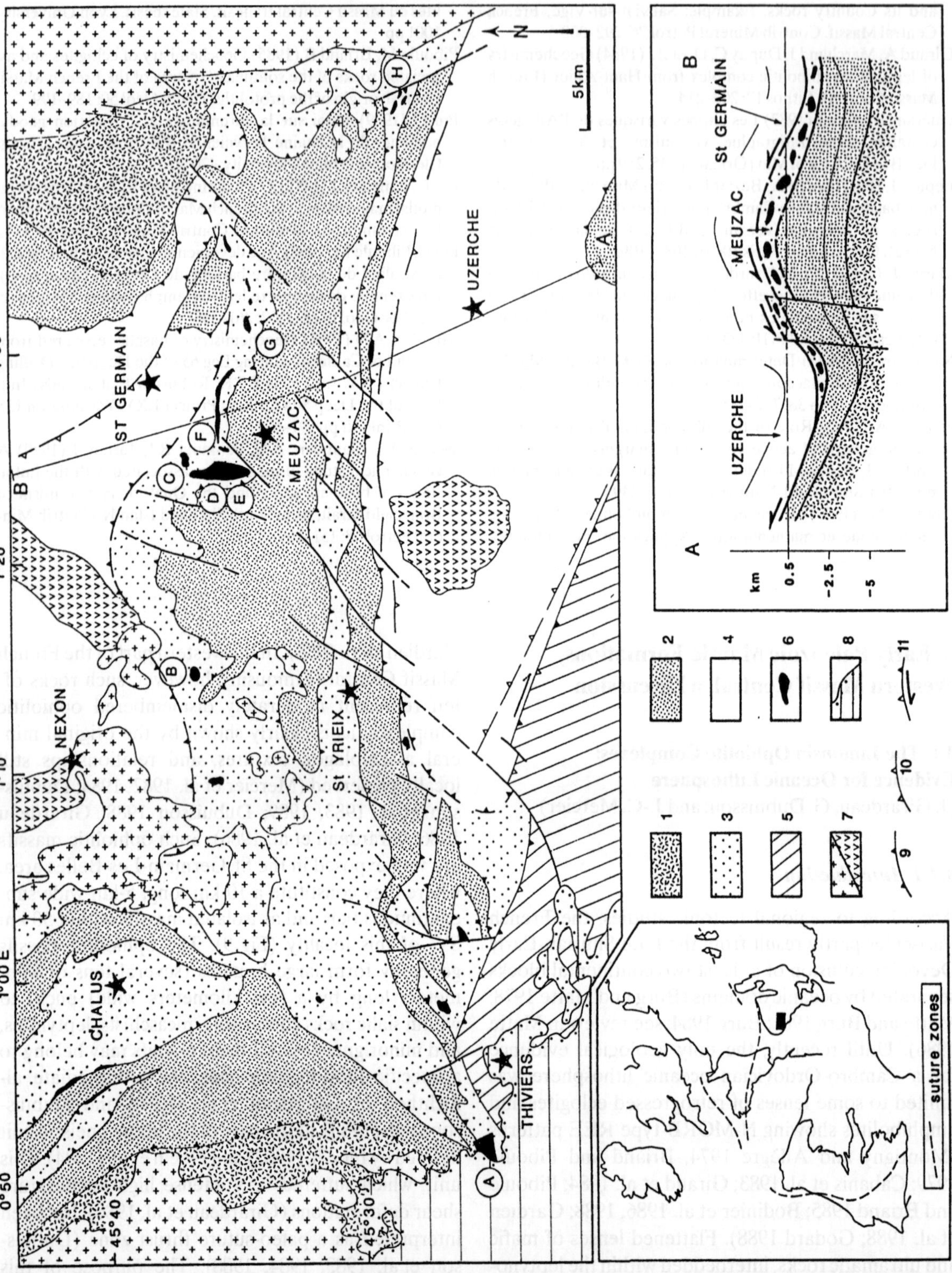

Fig. 9. Ophiolite location, geologic sketch map and schematic cross-section of the Limousin, western French Massif Central. (After Girardeau et al. 1986; Dubuisson et al. 1988). The nomenclature used to describe the litho-tectonic series is that defined by Girardeau et al. (1986) and Dubuisson et al. (1988). It is not exactly the one used in this book, the upper and intermediate allochthon of the figure corresponding, generally, to the middle one. *1* Presumed autochthon; *2* lower/basal allochthon; *3* intermediate allochton; *4* upper allochthon; *5* Thiviers-Payzac and Génis units; *6* ophiolite; *7* granite/diorite; *8* Aquitaine sedimentary basin; *9* early thrust; *10* late ductile thrust; *11* late thrust. *A* La Queue d'Ane; *B* La Roche l'Abeille; *C* Flacquecorne; *D–E* La Flotte *(D)* – Le Cluzeau *(E)* massif; *F* La Plagne; *G* La Porcherie; *H* Le Lonzac

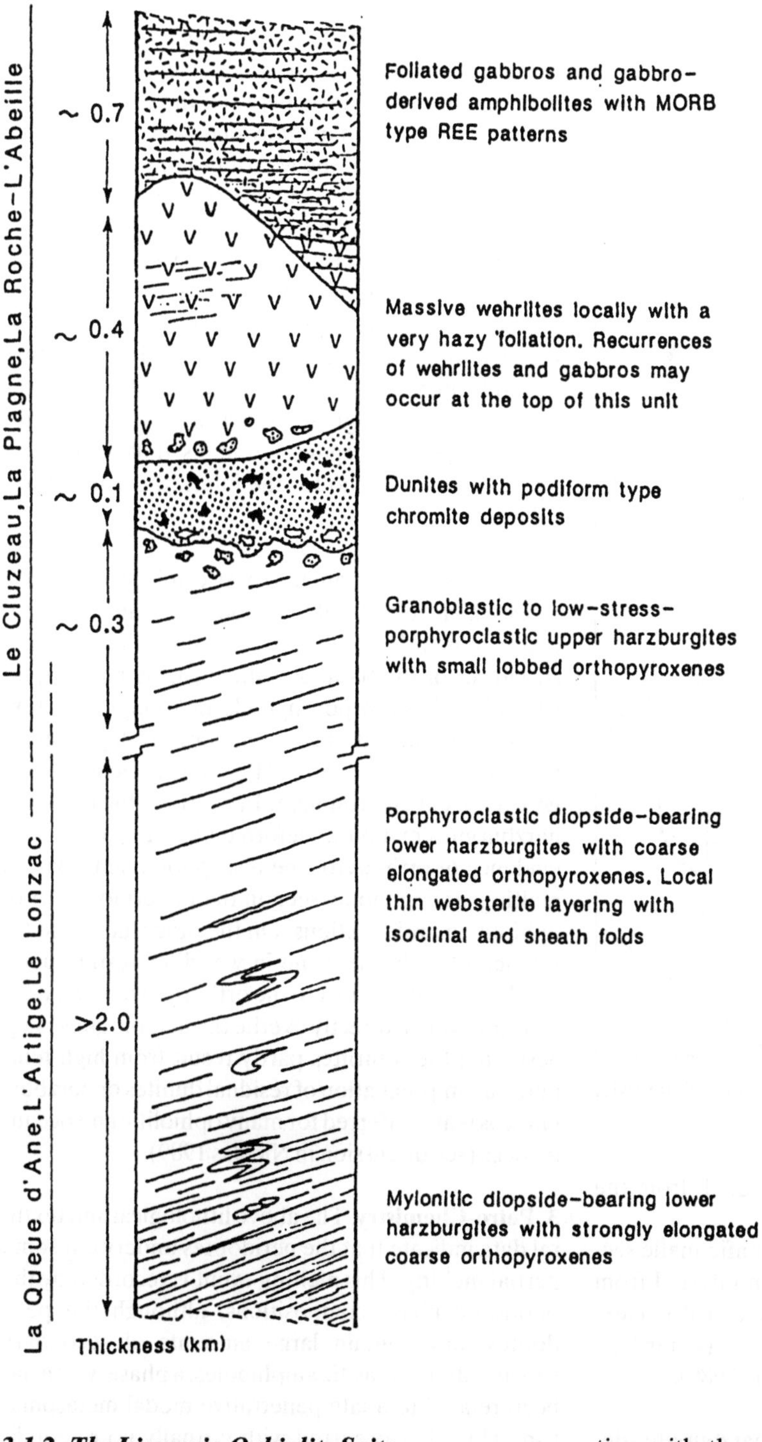

Fig. 10. Schematic composite section of the Limousin ophiolite. Dubuisson et al. 1989)

3.1.2 The Limousin Ophiolite Suite

Figure 10 shows a schematic, composite section of the Limousin ophiolite which takes into account the data obtained on all the mafic and ultramafic massifs present in the intermediate allochthon unit. This section comprises foliated gabbros, wehrlites, dunites, depleted harzburgites, and diopside-bearing harzburgites. Intrusives have not been found in close association with these rocks. Poorly preserved metamorphosed lavas with MORB compositions (Cabanis et al. 1983) occur some 50 km to the south, in the Génis Unit (Guillot and Lefèvre 1975), the uppermost unit of the Limousin series (Floc'h 1983; Floc'h et al. 1984). Although most of the massifs are strongly tectonized and contain only parts of the above sequence, some of them (the La Flotte-Le Cluzeau massif, for instance; Fig. 11), display virtually complete sections

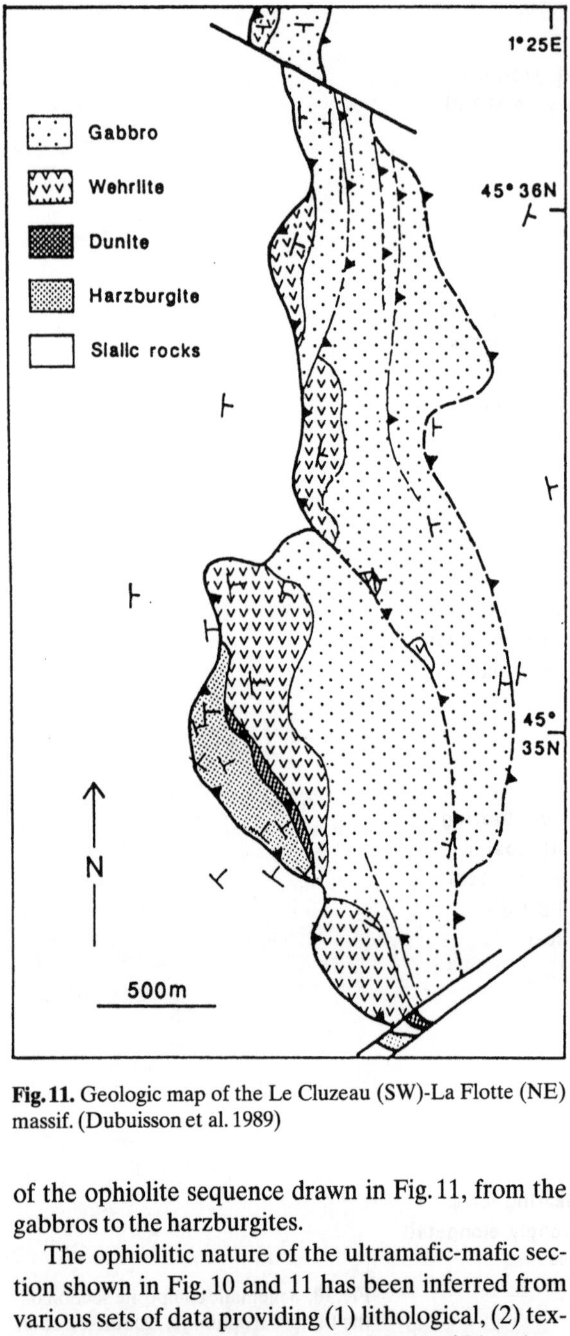

Fig. 11. Geologic map of the Le Cluzeau (SW)-La Flotte (NE) massif. (Dubuisson et al. 1989)

of the ophiolite sequence drawn in Fig. 11, from the gabbros to the harzburgites.

The ophiolitic nature of the ultramafic-mafic section shown in Fig. 10 and 11 has been inferred from various sets of data providing (1) lithological, (2) textural, and (3) petro-chemical evidence (Girardeau 1986; Dubuisson 1988; Dubuisson et al. 1989).

1. Lithology. The diopside-bearing harzburgite, depleted harzburgite, dunite with/without podiform chromite, wehrlite, and gabbro section can be directly compared with typical ophiolitic sections thought to have formed in true oceanic environments (see review in Nicolas 1989). Unfortunately, the original thickness of the sequence cannot be reconstructed, most units having been thinned during nappe emplacement. This makes it impossible to know whether the present lack of cumulate gabbros, sheeted dikes, and lavas is a primary feature of the ophiolite suite or results from tectonic processes. Nevertheless, primary magmatic contacts between the different units have been preserved in many places, particularly the gabbro-wehrlite and the wehrlite-dunite or wehrlite-depleted harzburgite contacts in the La Flotte-Le Cluzeau massif. On the other hand, the Le Lonzac massifs exhibit primary contacts between dunite and depleted harzburgite and diopsite-bearing harzburgite. All of these primary contacts appear progressive and irregular. Figure 11 clearly shows that the wehrlites define small pockets, probably representing discrete magmatic zones in between the gabbros and the harzburgites or dunites, thus defining an irregular paleo-Moho for the inferred oceanic lithosphere.

2. Textures. The ultramafic rocks display various textures which reflect deformation and magmatic processes of mantle origin. This is best displayed by the structures of the basal diopside-bearing harzburgite which locally show porphyroclastic to mylonitic textures acquired at high temperature-high stress ($>1000\,°C$, $\sigma_1 - \sigma_3 \simeq 500$ MPa), quite incompatible with a crustal environment. In contrast, the depleted harzburgites exhibit undeformed textures with clear evidence of orthopyroxene resorption due to partial melting, as commonly seen in the uppermost part of ophiolite mantle sections. On the other hand, the total lack of orthopyroxene in wehrlites resting upon the dunites (in contrast to the orthopyroxene-bearing wehrlites which directly overlie the harzburgites) suggests that the wehrlites partly result from high-temperature impregnation of residual dunites or harzburgites, as is also inferred for many ophiolitic and oceanic rocks (see discussion in Nicolas 1989).

3. Petro-Chemistry. The above lithological and textural data indicate that the peridotites underwent some partial melting. This melting event is recorded by the peridotite phase compositions, although the peridotites may contain large amounts of secondary subeuhedral pargasitic amphiboles, a phase which has been related to a late penetrative modal metasomatism which has occurred with virtually no re-equilibration of the primary phases (Dubuisson et al. 1989), This melting event is particularly well manifested by olivine and spinel compositional variations: Mg#[1] for olivine varies from 0.890 to 0.915 from the diopside-bearing harzburgites to the dunites and Al#[2] for spinel from 0.91 to 0.72. Spinel from the wehrlites is Cr-rich (Al# = 0.92), as expected for residual rocks. However, their olivine is richer in Fe (Mg# = 0.857) than for the

[1] Mg# means $\dfrac{\text{Mg} + \text{Fe}}{\text{Mg}}$

[2] Al# means $\dfrac{\text{Al} + \text{Cr}}{\text{Mg}}$

harzburgites, a feature ascribed to chemical re-equilibration during peridotite impregnation (Dubuisson et al. 1989). On the other hand, the foliated gabbros have REE concentrations and patterns with a clear depletion in light REE relative to chondrites, $(Ce/Sm)_N < 1$) similar to those of the ophiolitic upper gabbros derived from the crystallization of MORB tholeiites.

3.1.3 Conclusion

All the lithological, textural, and petro-chemical arguments developed above point to an oceanic origin for the Limousin mafic and ultramafic suite, by comparison with ophiolite studies (Nicolas 1989). This ophiolite section is, however, very incomplete, as no layered cumulate gabbros, dolerites, or basalts have been identified. It is also highly reduced in thickness and dismembered, two features we attribute to shearing during its emplacement onto the continent during the Variscan orogeny. This makes it quite impossible to determine in which environment this ophiolite formed. However, the residual character of its mantle part, as indicated by the spinel peridotites and the MORB nature of its basaltic crustal part, both suggest that this ophiolite formed at an oceanic ridge, true oceanic domain or marginal basin, and not in a continental environment.

References

Briand B, Piboule M (1979) Les métabasites de la série de Marvejols (Massif Central): témoins d'un magmatisme tholéitique d'arrière-arc cambro-ordovicien? Bull Bur Rech Geol Min 2 I 2: 131–171

Burg J-P (1984) Tectogenèse comparée de deux segments de chaîne de collision: le sud Tibet (suture du Zangbo) et la chaine hercynienne en Europe (suture du Massif Central). Thèse d'Etat, Montpellier, 360 pp

Burg J-P, Matte (1978) A cross-section through the french Massif Central and the scopes of its variscan geodynamic evolution. P Z Dtsch Geol Ges 129: 429–460

Bodinier J L, Giraud A, Dupuy C, Leyreloup A, Dostal J (1986) Caractérisation géochimique des métabasites associées à la suture hercynienne du Massif Central Français. Bull Soc Géol Fr 8 II: 115–123

Bodinier J-L, Burg J-P, Leyreloup A, Vidal H (1988) Reliques d'un bassin d'arrière arc subducté puis obducté dans la région de Marvejols (Massif Central). Bull Soc Géol Fr 8 IV 1: 21–33

Cabanis B, Guillot P-L, Santallier D, Laffrezic H, Meyer G, Treuil M (1983) Apport des éléments traces à l'étude géochimique des métabasiques du Limousin Bull Soc Geol Fr 7 XXV: 563–574

Dubuisson G (1988) Etude d'un segment d'orogène ancien et de marqueurs ophiolitiques: approche tectonique géophysique et pétrologique. Thèse de Doctorat, Univ Paris 7 323 pp

Dubuisson G, Mercier J-C C, Girardeau J (1983) Geodynamic significance of the Bas Limousin mafic and ultramafic complexes. Terra Cognita 3: 156

Dubuisson G, Hirn A, Girardeau J, Mercier J-C C, Veinante J-L (1984) Evidence sismologique de l'existence de multiples nappes en Limousin. Doc Bur Rech Geol Min 95-3: 34–58

Dubuisson G, Hirn A, Girardeau J, Mercier J-C C, Veinante J-L (1988) Multiple variscan nappes in Limousin, France. Geophysical constraints to the geological model and geodynamic significance. Tectonophysics 147: 19–31

Dubuisson G, Mercier J-C C, Girardeau J, Frison J-Y (1989) Evidence for a lost ocean in Variscan terranes of the western Massif Central, France. Nature 337: 729–732

Floc'h J-P (1983) Le socle métamorphique du Limousin central: une traverse de la branche Ligérienne de l'orogène varisque de l'Aquitaine à la zone broyée d'Argentat. Thèse d'Etat univ., Limoges. 2 vol, 590 pp

Floc'h J-P, Santallier D, Henri B (1984) La série limousine: réflection axée sur une synthèse géologique régionale et son utilisation pour la prospection minière (or). Chron Rech Min 474: 5–10

Gardien V, Lardeaux J-M, Misseri M (1988) Les péridotites du Mt. Du Lyonnais (MCF): témoins privilégiés d'une subduction de lithosphère paléozoique. C R Acad Sci Paris T 307, 2; 1967–1972

Girardeau J (1986) Ophiolites et chaînes de collision: exemples des chaînes Himalaya-Tibet et Hercynienne. Etude pétrostructurale des ophiolites et de leur encaissant. Thèse d'Etat, Univ. P7-IPGP, 404 pp

Girardeau J, Dubuisson G, Mercier J-C C (1986) Cinématique de mise en place des ophiolites et nappes cristallophylliennes du Limousin. Bull Soc Geol Fr 8 II: 849–860

Giraud A, Marchand J, Dupuy C, Dostal J (1984) Geochemistry of leptyno-amphibolitic complexes from Haut-Allier (French Massif Central). Lithos 17: 203–214

Godard C (1988) Petrology of some eclogites in the Hercynides: the eclogites of southern massif Armoricain, France. In: Smith D C (ed) Ecologites and eclogite facies rocks, chap 8. Elsevier, Amsterdam, pp 451–519

Guillot P L, Lefèvre J (1975) Découverte de conodontes dans le calcaire à entroques de Génis en Dordogne (série métamorphique du Bas-Limousin). C R Acad Sci D 280: 1529–1530

Maillet N (1987) Dualité d'origine des massifs ultrabasiques limousins. Implications géodynamiques. Thèse Doctorat, Université de Lyon, 149 pp

Maillet D, Piboule M, Santallier D (1984) Diversité d'origine des ultrabasites dans la série métamorphique du Limousin. Doc Bur Rech Geol Min 81-3: 1–24

Matte P (1986) Tectonics and plate tectonic model for the variscan belt of Europe. Tectonophysics 126: 329–374

Matte Ph, Burg J-P (1981) Sutures, thrust and nappes in the variscan arc of western Europe: plate tectonic implications. In: Mc Clay K R, Price N J (eds) Thrust and nappe tectonics. Geol Soc London Spec pub 9, pp 353–358

Mercier J-C C, Pozzo Di Borgo M, Frison J-Y, Girardeau J (1982) Les associations basiques et ultrabasiques du Limousin: restes d'un complexe ophiolitique démembré. 9ème RAST, Paris, pp 430

Mercier J-C C, Girardeau J, Prinzhofer A, Dubuisson G (1984) Les complexes ophiolitiques du Limousin: structure, pétrologie et géochimie. Doc Bur Rech Geol Min 95-3: 35–48

Montigny R, Allègre C J (1974) A la recherche des océans perdus: les éclogites de Vendée témoins métamorphisés d'une ancienne croûte océanique. C R Acad Sci Paris 279 D: 543–545

Nicolas A (1989) Structure of ophiolites and dynamics of oceanic lithosphere. Kluwer Academic, Amsterdam 367 pp

Piboule M, Briand B (1985) Geochemistry of eclogites and associated rocks of the southeastern area of the French Massif Central: origin of the protoliths. Chem Geol 50: 189–199
Santallier D (1981) Les roches basiques dans la série métamorphique du Bas-Limousin, Massif Central, France. Thèse d'Etat Univ. Orléans
Santallier D, Briand B, Ménot R-P, Piboule M (1988) Les complexes leptyno-amphiboliques (C.L.A.): revue critique et suggestions pour un meilleurs emploi de ce terme. Bull Soc Géol Fr 8 IV 1: 3–12

4 A Reply: Another Way of Considering the Ophiolite Question

D. Santallier

Concerning the Limousin ophiolitic complexes (Girardeau et al. section 3), one must keep in mind a few controversial issues. The schematic ophiolitic section proposed (Fig. 10) by Girardeau et al. incorporates data from several different occurrences (the so-called La Queue d'Ane, l'Artige, etc. massifs). However, according to Maillet (1987), and based upon geochemistry (comparison with Ophra Data Bank) and P/T estimates, all these massifs are not genetically associated.

The La Flotte massif can reasonably be considered as representing all Limousin layered massifs. What is the regional significance of these massifs? The question has been debated for several years. Briefly summarizing, the successive interpretations are:

- The layered bodies represent small laccoliths differentiated within the metamorphic series (Chenevoy et al. 1963),
- They represent remnants of an extensional episode of moderate importance in a thinned continental crust (Piboule 1979; Santallier 1981).
- They represent remnants of a non metamorphosed ophiolite suite of major importance (Mercier et al. 1982).

As a result of new field and laboratory studies, authors have agreed on the following point: the layered massifs are part of a (?) dismembered ophiolitic complex(es), the main argument being the existence of mantle tectonites located at the base of the section.

Nevertheless, there still remain some debated points:

1. Have the ophiolites been affected by high pressure metamorphism? Girardeau et al. (1986) answered this point negatively because piling of the nappes and the tectonic emplacement of the ophiolites series occurred directly in medium pressure metamorphic conditions.

Santallier (1981), Maillet et al. (1984), and Maillet (1987) answered this question positively, as they argued that high pressure assemblages are clearly visible in several layered massifs, either as Cpx + Opx ± Sp coronas (La Flotte, Sarrazac, La Roche-Noire, la Reboulie massifs, for example), or as varied high pressure amphibolites containing:

Pargasitic Hbl + Gedrite + Ky + Co + Ru (La Flotte),
Pargasitic Hbl + Mg chlorite + Ky + Co (Sarrazac),
Parg Hbl + Mg Chl + Ky + Zo + Co + Ru ± Paragonite ± Margarite (la Roche Noire),
Parg Hbl + Mg Chl + Gedrite + Ky + Co and Parg Hbl + Mg Chl + Opx + Sp (la Boissonie),
Parg Hbl + Mg Chl + Opx + Cpx + Mg Gt + Sp (la Geneste), etc.

It is noteworthy that gedrite is known only in the northern massifs, which implies some differences between the various P–T paths (Fig. 12) for the ophiolitic remnants. This is also supported by the occurrence of pyrope-rich garnet in only one body. Significantly, in some massifs (Sarrazac, la Boissonie, and la Geneste), the high pressure assemblages are clearly associated with tectonic structures such as foliations and fold hinges overprinted by garnets, gedrites, and Opx. These data do not conform well with the interpretation of the present thrust plane between the two gneissic units as representing a direct transposition of the initial oceanic suture as proposed by Dubuisson et al. (1988).

2. Do all the Limousin ultrabasic bodies represent parts of one and the same ophiolitic pile, as advocated by Dubuisson et al. (1987a and b)? There are two types of ultrabasic bodies in the Limousin series: the layered massifs and the purely ultramafic ones. Two of these are of peculiar interest because they are well preserved. The first (la Rebière) is a spinel and two-pyroxenes peridotite interpreted as a subcontinental lherzolite (Maillet 1987), or as a harzburgite modified by mantle metasomatism (Mercier et al. 1985). The other (l'Artige) is a banded garnet websterite/garnet lherzolite, also interpreted as a subcontinental peridotite (Maillet et al. 1984), or as an eclogitized spinel peridotite by Dubuisson et al. (1987a and b). According to the latter authors, this is the only Limousin harzburgite in which a high pressure event is expressed. The debated points include the low-Cr content in the spinel (Fig. 13), the high Na-content in the Cpx (Fig. 14), the primary character of the garnet and the whole rock compositions.

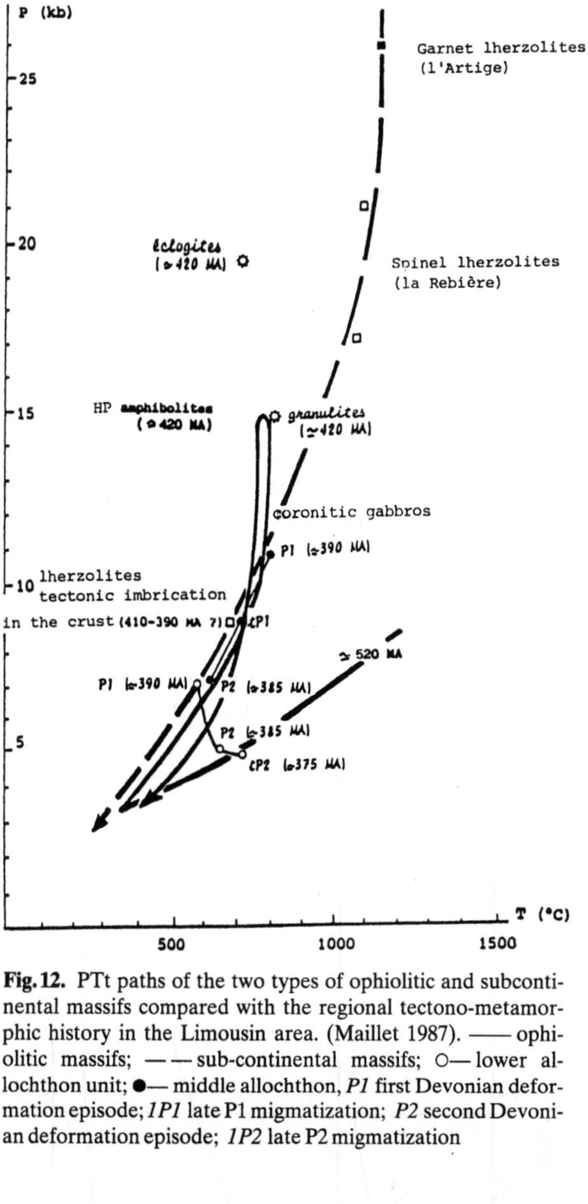

Fig. 12. PTt paths of the two types of ophiolitic and subcontinental massifs compared with the regional tectono-metamorphic history in the Limousin area. (Maillet 1987). —— ophiolitic massifs; — — sub-continental massifs; ○— lower allochthon unit; ●— middle allochthon, *P1* first Devonian deformation episode; *1P1* late P1 migmatization; *P2* second Devonian deformation episode; *1P2* late P2 migmatization

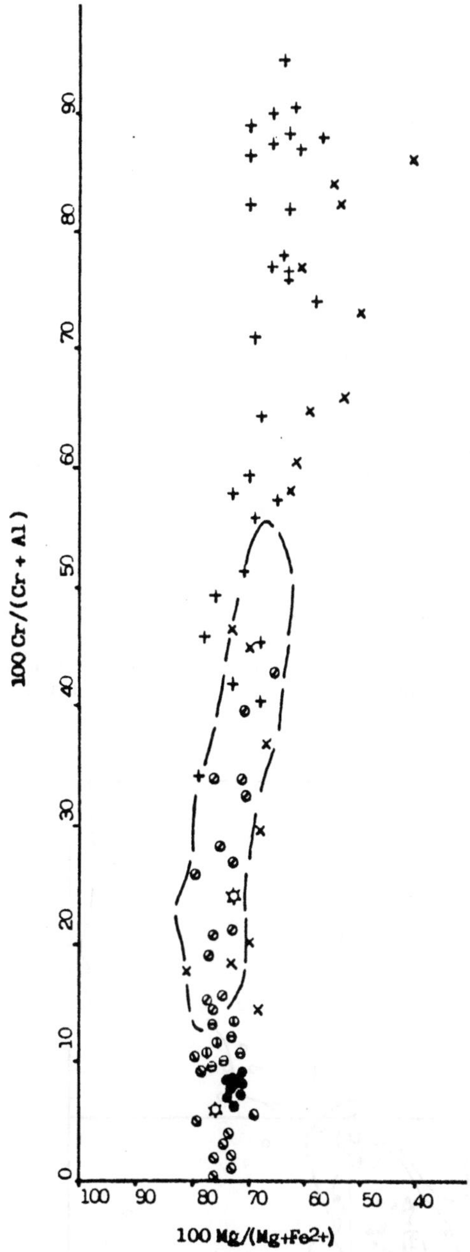

Fig. 13. La Rebière spinels in the 100 Cr/Cr + Al) versus 100 Mg/(Mg + Fe^{2+}) diagram after Dick and Bullen (1984). A comparison with spinels in mantle peridotites. (After Maillet 1987): ● la Rebière. Reference points for: *Subcontinental peridotites:* ⌀ Rhonda (Spain); ✡ Beni-Bouchera (Morocco); ⦸ Lherz (France); ⊖ Balmuccia (Italy); ⏀ Baldissero (Italy). *Ophiolitic massifs:* + New Caledonia); × Bay of Island (Newfoundland). All analyses in Dick and Bullen (1984), except for Lherz (in Conquéré 1977) and la Rebière (in Maillet 1987)

– Following Dubuisson et al. (1987b), (1) the spinel compositions in the La Flotte massif are similar to those in Newfoundland ophiolites, (2) the high Na content in the Cpx is good evidence of a high temperature (!) mantle metasomatism and (3) the garnets in the l'Artige peridotite represent high pressure secondary ones because they contain spinel inclusions (but following Piboule pers. commun. the spinels also contain garnet inclusions, both minerals being more or less in equilibrium).

– Following Maillet (1987), (1) the low Cr content in spinel fits what is observed in the subcontinental peridotites, and (2) the high Na content of the Cpx (up to 2.21 % Na_2O for 0.25 % Na_2O in the corresponding whole rock analysis) proves high pressure conditions still unknown in the case of mantle metasomatism (Boudier, pers. commun.). In addition, the P-T conditions of crystallization as determined by Maillet (1100 °C and 20 Kbar for the la Rebière massif and 1150 °C and 26 kbar for l'Artige) are inconsistent with a suboceanic mantle origin.

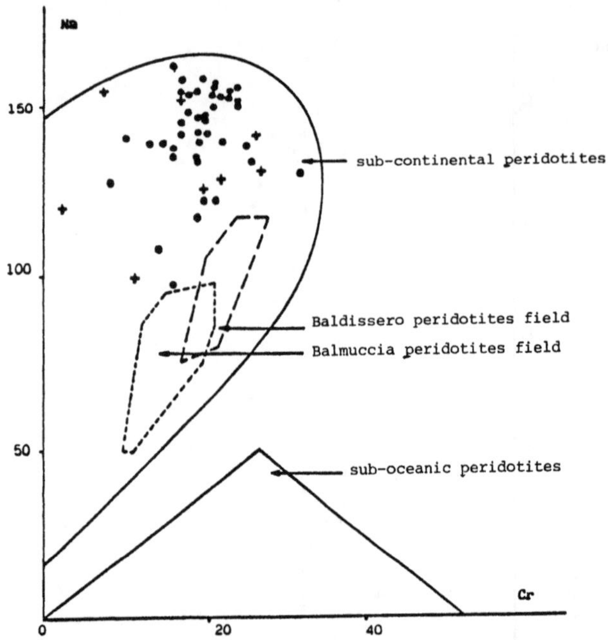

Fig. 14. La Rebière and l'Artige massifs. Clinopyroxenes in the Na-Cr diagram of Kornprobst et al. (1981). (After Maillet 1987) ● la Rebière; + l'Artige

These values were obtained using the following methods:

Geothermometry based upon (1) two-pyroxene equilibria (Mori (1977), Mercier (1980), Sachtleben and Seck (1981)), (2) olivine-spinel equilibrium (Fabriès (1979, 1984). Roeder et al. (1979) and Sack (1982)), (3) Si content in spinel (Berger and Vannier 1984) and (4) olivine-Opx equilibrium (Berger and Vannier 1978).

Geobarometry was based upon Al solubility in Opx (Lane and Ganguly (1980) and Gasparik (1984)).

Still following Maillet (1987), the la Flotte and la Rebière massifs cannot be genetically linked because, when compared to some classical ophiolite suites in the world (Canyon Mountain, Newfoundland, western Alps, Corsica, Elba, Yugoslavia, Turkey, Cyprus, Syria, Greece, Iran, Oman) they do not conform to chemically plausible trends. The comparison has been established with the OPHRA ophiolite bank (CRPG, Nancy), using the FPY 1/FDE 1 diagram (Fig. 15) of de la Roche (1977), itself derived from the mineralogical tetrahedron of Yoder and Tilley (1962). The two values FPY 1 = 1/5 (4A–5E) and FDE

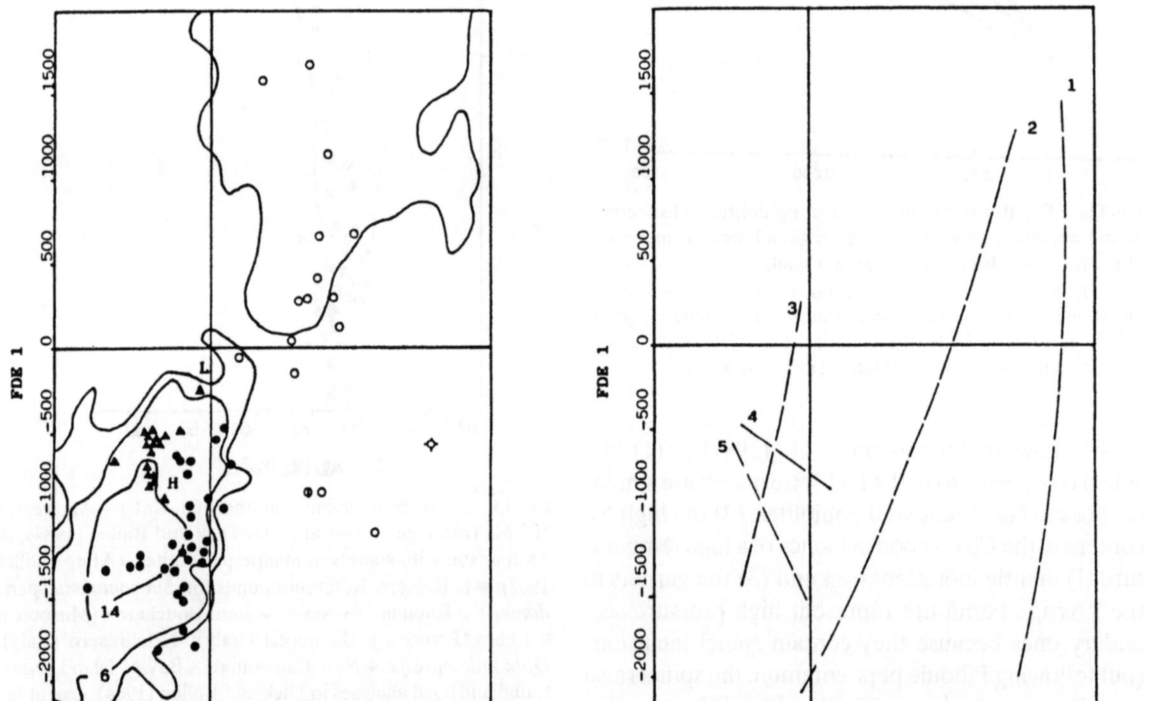

Fig. 15. La Flotte and la Rebière massifs. *FPY 1-FDE 1 diagram.* (After Maillet 1987). ▲ la Rebière massif. La Flotte massif: ● s. l. metaperidotites; ○ s. l. meta-gabbros; Φ amphibolite with a basaltic composition; ✧ meta-anorthosite; ⑥ iso-density curve for peridotites and gabbros in non-mediterranean ophiolites. **L** Lherzolites field center, based upon the OPHRA data bank. **H** Harzburgites field center, based upon the OPHRA data bank. *Dashed lines* different possible differentiation trends: *1, 2,* and *3* classical ophiolitic trends in the world; *4* and *5* only possible trends between the la Flotte and la Rebière massifs if they belonged to the same ophiolitic suite

1 = 5D-2/3E depend upon the four multicationic parameters F (Forsterite), E (Enstatite), D (diopside) and A (anorthite-albite):
$F = -3\,Si + 3/2\,Al + 3/4\,Fe + 3\,Mg + 3\,Ca + 15/2\,Na + 11\,K$
$E = 4\,Si - Al - 2\,Mg - 6\,Ca - 11\,Na - 11\,K$
$D = 4\,Ti - 2\,Al - Fe + 4\,Ca + 2\,Na - 4\,K$
$A = 5\,Ti + 5/2\,Al + 5/4\,Fe + 5/2\,Na + 5\,K$.
In the diagram, the five visible trends are: 1, 2, and 3 = classical ophiolitic trends in the world: 4 and 5 represent the only possible trends in la Flotte-la Rebière if the two massifs belonged to the same ophiolitic suite. Such trends are at present unknown among currently recognized ophiolites, thus these two massifs cannot belong to the same suite. Consequently, Maillet concludes that the la Flotte massif was part of an ophiolitic suite, since the two purely lherzolitic massifs of la Rebière and l'Artige were of subcontinental origin.

3. The existence of dismembered ophiolites in the Limousin series is good evidence for an oceanization process of probable Lower Paleozoic age (it is at least pre-Devonian and, although not precisely dated in Limousin, it can reasonably be considered as equivalent to the Cambrian and Ordovician basic rocks emplaced everywhere in the French Paleozoic orogen). Was there a large ocean or a narrow extensional zone such as a back-arc basin? A large east-west oceanic zone has been advocated (Matte 1983) through the Armorican Massif and north of the Massif Central. Such a root zone for the Limousin ophiolite (Dubuisson et al. 1988) would imply a southward nappe displacement which does not fit the observations of Floc'h (1983), Bouchez and Jover (1986), and Ledru and Autran (1987), who argued for a westward or northwestward displacement. However, more recently, there has been a consensus (Dubuisson et al. 1987b, Maillet 1987) for a back-arc basin site. However, Dubuisson et al. argue for fast spreading because of the absence of plagioclase in the spinel harzburgites, while Maillet argues for slow spreading by (1) comparison with the Macedonian ophiolites and (2) because the presence of anorthosites fits better with small magmatic chambers (Boudier, pers. commun.). In addition, a preliminary Sm-Nd study reveals crustal contamination (Pin, pers. commun.), which apparently precludes the hypothesis of a large oceanic zone (Paquette 1987).

Unfortunately, the presence of widespread basic rocks of basaltic composition in the surrounding area cannot help solve the problem because it is not known to which one of the recognized magmatic series (Cabanis et al. 1983) (N-MORB, transitional tholeiites, or calc-alkaline series) the layered massifs could be linked. In addition, one must not forget that the existence of another ophiolitic remnant in the southwestern epizonal Génis nappe supports the possible existence of more than one extensional zone in the Limousin domain.

4. The last debated point is: do the la Flotte and neighboring massifs indicate a primary oceanic suture between the two gneissic units, thus signifying that the two nappes originally belonged to two different continental blocks (Dubuisson et al. 1988)? or do they merely outline a late intra-crustal shear-zone? In the latter case there is no argument for the primary location of the oceanic suture. And as (1) there are several belts of layered bodies at different levels in the Limousin series, (2) the ophiolitic sequence suffered a high pressure tectono-metamorphic event prior to the nappe piling, the second possibility cannot be ruled out at first sight.

References

Berger E, Vannier M (1978) Un géothermomètre reposant sur le partage du nickel et du magnésium entre olivine et orthopyroxène: application à l'étude des péridotites. CR Acad Sci Paris 286 D: 733–736

Berger E, Vannier M (1984) Approche expérimentale de la géothermométrie spinelle-olivine: température d'équilibre et teneur en silicium du spinelle. Bull Liaison Soc Fr Mineral 107 (3): 43

Bouchez JL, Jover D (1986) Le Massif Central: un chevauchement de type himalayen vers l'ouest-nord-ouest. CR Acad Sci Paris 302 II: 675–680

Chenevoy M, Forestier FH, Gay M (1963) L'étude des complexes ultrabasiques différenciés du Haut-Limousin (Feuille de Limoges à 1/80 000). Bull Carte Géol Fr 273 (54): 89–116

Cabanis B, Guillot PL, Santallier D, Jaffrezic H, Meyer G, Treuil M (1983) Apports des éléments traces à l'étude géochimique des métabasites du Bas-Limousin. Bull Soc Géol Fr (7), XXV, 4: 563–574

Conquéré F (1977) Pétrologie des pyroxénites litées dans les complexes ultramafiques de l'Ariège (France) et autres gisements de lherzolites à spinelle. I – Comparaison minéralogique et chimique, évolution des conditions d'équilibre des pyroxénites. Bull Soc Fr Min Crist 100: 42–80

Dick HJB, Bullen T (1984) Chromian spinel as petrogenetic indicator in abyssal and alpine type peridotites and spacially associated lavas. Contrib Mineral Petrol 86: 54–76

Dubuisson G, Galdeano A, Pozzi JP (1987a) Etude magnétique et géologique de la synforme de Saint-Germain-lès-Belles: implications tectoniques: présence d'une paléo-suture varisque charriée. Doc Bur Rech Min 140: 1–20

Dubuisson G, Hirn A, Girardeau J, Mercier JCC, Veinante JL (1988) Multiple Variscan nappes in Limousin, western Massif Central, France: geophysical constraints to the geological model and geodynamic implications. Tectonophysics 147: 19–31

Dubuisson G, Mercier JC, Girardeau J (1987b) Lithologie, textures et pétrologie des ophiolites du Limousin: reliques d'un paléo-océan varisque. Doc Bur Rech Min 140: 21–50

Duthou JL (1977) Chronologie Rb-Sr et géochimie des granitoïdes d'un segment de la chaine varisque, relations avec le métamorphisme: le nord Limousin (Massif Central français). Ann Sci Clermont-Ferrand, 63, 30, 294 pp

Fabries J (1979) Spinel-olivine geothermometry in peridotites from ultramafic complexes. Contrib Mineral Petrol 69 (4): 329–336

Fabries J (1984) Utilisation des échanges Fe/Mg en géothermométrie. Application aux roches mafiques et ultramafiques. In: Lagache M (ed) Thermométrie et barométrie géologiques. Soc Fr Mineral Cristallogr, Paris

Floc'h J P (1983) La série métamorphique du Limousin central: une traverse de la branche ligérlenne de l'orogène varisque, de l'Aquitaine à la zone broyée d'Argentat (Massif Central français). State Thesis, Limoges University, 590 pp

Frey F A, Haskin M A, Poetz J A, Haskin L A (1968) Rare earth abundances in some basic rocks. J Geophys Res 73, 18: 6085–6098

Gasparik T (1984) Two pyroxenes thermobarometry with new experimental data in the system CaO-MgO-Al_2O_3–SiO_2. Contrib Mineral Petrol 87: 87–97

Girardeau J, Dubuisson G, Mercier J C C (1986) Cinématique de mise en place des ophiolites et nappes cristallophylliennes du Limousin, Ouest du Massif Central français. Bull Soc Géol Fr (8) II 5: 849–860

Kornprobst J, Ohnenstetter D, Ohnenstetter M (1981) Na and Cr contents in Cpx from peridotites: a possible discriminant between "sub-continental" and "sub-oceanic" mantle. Earth Planet Sci Lett 53: 241–254

Lafon J M (1986) Géochronologie U-Pb appliquée à deux segments du Massif Central français: le Rouergue oriental et le Limousin central. Thesis, Montpellier University, 152 pp

Lane D L, Ganguly J (1980) Al_2O_3 solubility in orthopyroxene in the system MgO-Al_2O_3-SiO_2: a reevaluation and mantle geotherms. J Geophys Res 85: 6963–6972

La Roche H de (1977) Un problème de pétrologie quantitative: dans un tétraèdre, projection sur un plan quelconque à partir d'un point quelconque. Exemple d'application à la chimie des roches basiques et ultrabasiques. Sci Terre (Sér Info Géol) 12: 5–20

Ledru P, Autran A (1987) L'édification de la chaine varisque dans le Limousin. Rôle de la faille d'Argentat à la limite Limousin-Millevaches. Colloque G.P.F. 3, Paris, thème 3 "Chevauchements varisques en Limousin". Doc Bur Rech Géol Min 140: 51–81

Maillet N (1987) Dualité d'origine des massifs ultrabasiques limousins. Implications géodynamiques. Thesis, Lyon University, 149 pp

Maillet N, Piboule M, Santallier D (1984) Diversité d'origine des ultrabasiques dans la série métamorphique du Limousin. Doc Bur Rech Min 81 (3): 1–24

Matte P (1983) Two geotraverses across the Ibero-Armorican variscan arc of western Europe. In: Rast N, Delany F M (eds) Profiles of Orogenic Belts. Geodyn Ser 10: 53–81

Mercier J C C, Pozzo di Borgo M, Frison J Y, Girardeau J (1982) Les associations basiques et ultrabasiques du Bas-Limousin: restes d'un complexe ophiolitique démembré d'une frai cheur remarquable. 9th Earth Sci Ann Meet, Paris, Abstr vol 430

Mercier J C C (1980) Single pyroxene thermobarometry. Tectonophysics 70: 1–37

Mercier J C C, Girardeau J, Prinzhofer A, Dubuisson G (1985) Les complexes ophiolitiques du Limousin: structure, pétrologie et géochimie. Doc Bur Rech Géol Min 95 (3): 35–48

Mori T (1977) Geothermometry of spinel lherzolites. Contrib Mineral Petrol 59 (3): 261–279

Nakamura N (1974) Determination of REE, Ba, Fe, Mg, Na and K in carbonaceous and ordinary chondrites. Geochim Cosmochim Acta 38: 757–775

Paquette J L (1987) Comportement des systèmes isotopiques U-Pb et Sm-Nd dans le métamorphisme éclogitique. Chaine hercynienne et chaine alpine. Mém Doc Centre Arm Etudes Struc Socles, Rennes, 14, 190 pp

Pearce J A (1983) Role of the sub-continental lithosphere in magma genesis at active continental margins. In: Hawksworth C J, Norry M (eds) Continental basalts and mantle xenoliths. Shiva Publ, Nantwich pp 230–249

Piboule M (1979) L'origine des amphibolites: approche géochimique et mathématique. Application aux amphibolites du Massif Central français. State Thesis, Lyon University, 1003 pp

Roeder P L, Campbell I H, Jamieson H E (1979) A re-evaluation of the olivine-spinel geothermometer. Contrib Mineral Petrol 68 (3): 325–334

Sachtleben T, Seck H A (1981) Chemical control of Al solubility in clinopyroxene and its implications on pyroxenes thermometry. Contrib Mineral Petrol 78 (2): 157–165

Sack R O (1982) Spinels as petrogenetic indicators: activity-composition relations at low pressures. Contrib Mineral Petrol 79 (2): 169–186

Santallier D (1981) Les roches basiques dans la série métamorphique du Bas-Limousin (Massif Central, France). State Thesis, Orléans University, 470 pp

Weaver B L, Saunders A D, Pankhurst R J, Tarney J (1979) A geomchemical study of magmatism associated with the initial stages of back-arc spreading: the Quaternary volvanism of Bransfield Strait, from South Shetland Islands. Contrib Mineral Petrol 68: 151–169

Yoder H S, Tilley C E (1962) Origin of basaltic magmas: an experimental study of natural and synthetic rocks systems. J Petrol 3: 342–532

5 Devonian Ophiolites, Eastern Massif Central

(M. Ohnenstetter)

5.1 The Ensialic Marginal Basin of Devonian Age

The Variscan chain includes several Devonian ophiolitic complexes (Ohnenstetter and Rossi 1987), such as the Lizard (Styles, and Kirby 1980; Le Gall and Poncet 1983; Davies, 1984) and the Beaujolais-Brévenne massifs (Sider and Ohnenstetter 1986), which indicate limited oceanization. In the Brévenne massif, the age proposed is based on that of the acid differentiates (Piboule et al. 1983) which are coeval with certain basic volcanics. Matte (1986) proposed that ophiolites of Devonian age belong to a young Paleozoic suture located north of the 500–480 Ma-old suture zone.

In the Beaujolais area, Sider and Ohnenstetter (1986) defined six tectonolithologic units, in tectonic

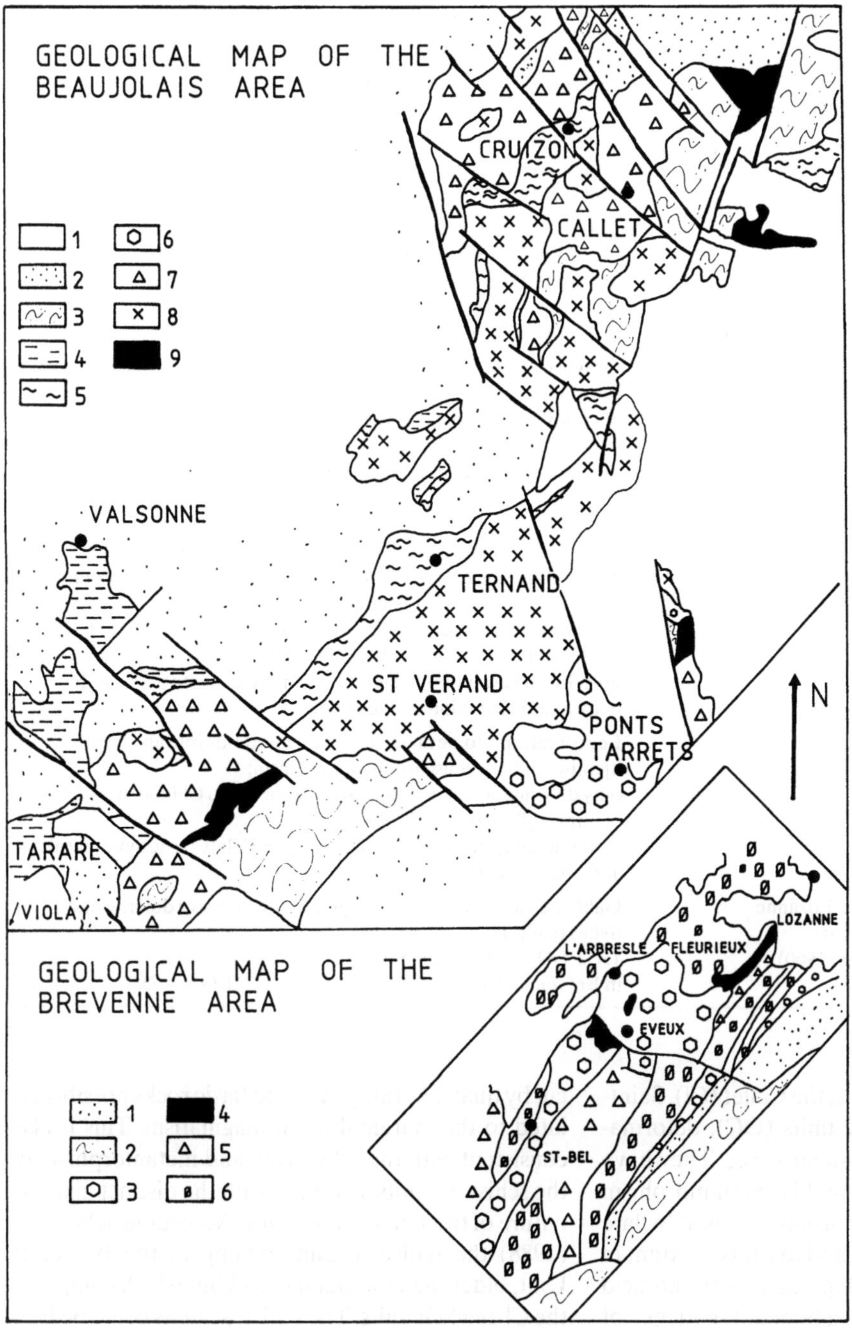

Fig. 16. Geological maps of the Monts du Beaujolais and of the Brévenne area, modified after Bardin (1971) and Sider and Ohnenstetter (1986). The location of the map areas is indicated in Fig. 1. *Beaujolais area: 1* post-Paleozoic formations; *2* Namuro-Westphalian granites and Middle to Upper Visean formations; *3* gneissic shield; *4* Valsonne unit; *5* Ternand-Cruizon unit; *6* Ponts Tarrets unit, Alaskan-type complex, arkoses, and graywackes; *7* Callet unit; *8* St. Vérand unit; *9* Ophiolitic complex. *Brévenne area: 1* Visean syntectonic granites: *2* gneissic shield; *3* L'Arbresle metagraywackes; *4* Fleurieux ultrabasic rocks; *5* and *6* Rocher formation: *5* quartz-keratophyres with minor spilites; *6* spilites with minor quartz-keratophyres

contact with each other (Fig. 16), forming an elongated NE-trending zone 20 km long and 3 to 6 km wide. Visean rocks unconformably overlie these units (Sider et al. 1986). Table 1 gives the main lithological and geochemical features of these units, as well as the possible correlation with the Brevenne formations.

In the Beaujolais area, three units have been affected by regional dynamothermal metamorphism under low epidote-amphibolite to greenschist facies conditions, the Valsonne (V), Ternand-Cruizon (TC) and Ponts Tarrets (PT) units. All of these units are characterized by abundant sedimentary deposits, commonly graywackes, locally associated with transitional (suite 1) to tholeiitic rocks (Lacroix 1980). A mildly alkaline character is indicated for the TC unit by high TiO_2, and Nb and LREE enrichment (Fig. 17). Serpentinite, phlogopite dunite, wehrlite, hornblendite, syenite, and dolerite in the PT unit represent the dismembered part of an alkaline basic-ultrabasic complex (suite 5).

Table 1. Composition and characteristics of the Beaujolais units with the possible correspondence with the Brévenne formations

Beaujolais units	Brévenne formations	Composition and characteristics of the Beaujolais units
Units affected by regional dynamothermal metamorphism		
Valsonne (V)	Violay group (Peterlongo 1960)	Volcanosedimentary units; Graywackes ≥ 200 m; tholeiitic volcanic rocks (Lacroix 1980); $1.48 < TiO_2 < 3.2\%$
Ternand-Cruizon (TC)		Orthoamphibolitic schist; paraschist (metapsammite; quartzite; calcschist; marble); greenschist facies metamorphism (high temperature); transitional basalt ($TiO_2 > 2.7\%$); positive Nb anomaly; $(La/Yb)_n = 3.25$ (Ohnenstetter and Sider 1988)
Ponts Tarrets (PT)	L'Arbresle metagraywackes (St-Antoine schist; Fleurieux serpentinite) (Peterlongo 1960)	Graywackes; arkoses (Lacroix 1980); clasts of serpentinite, phlogopite dunite, wehrlite, hornblendite, dolerite of an Alaskan-type complex derived from an alkaline magma (suite 5); $6.43 < (La/Yb)_n < 23.79$; N 120 to N 140 schistosity; greenschist to low epidote amphibolite facies
Units mainly affected by a static greenschist metamorphism		
Callet (Cl)	Rocher formation (Peterlongo 1970; Bardin 1971; Piboule et al. 1982)	Bimodal volcanic sequence intercalated with sediments; pelitic schist and calcareous lens; sodi-potassic quartz keratophyre (suite 2; 200–300 m; $K_2O/Na_2O < 1$); increasing proportion of basic rocks towards the top and to the east; (suite 3 = low TiO_2 suite); $TiO_2 = 1.2\%$; doleritic dykes: suite 4 = TiO_2-rich suite; TiO_2 up to 3.2%; only deformation in mylonitic shear zone
Saint-Verand (StV)		Leucocratic plutonic massif (suite 2): quartz-diorite and trondhjemite; static greenschist metamorphism; only deformation in mylonitic shear zone. Also Chindo massif, SW of Tarare (Beurrier et al. 1980); trondhjemite; inclusion of dolerite (suite 3); late basic dyke: basalt to ferrobasalt (suite 4)
Ophiolitic Complex (OC)	L'Arbresle-Lozanne (pillow-lavas) (Peterlongo 1960)	Gabbro to ferrogabbro and plagiogranite; massive dolerite and dykes (suite 4); $1.4 < TiO_2 < 2.9\%$; $(La/Yb)_n \approx 2$; in Brévenne: $1.46 < (La/Yb)_n < 3.2$ for $1.35 < TiO_2 < 1.75\%$

In the other Beaujolais units, the Callet (Cl), Saint-Vérand (StV), and ophiolitic units (OC), deformation is restricted to mylonitic shear zones. They have been affected by static greenschist metamorphism and locally by thermal metamorphism caused by late granitoid intrusions. In the Cl and to a lesser extent in the StV unit, close relationships exist between acid and basic rocks. The Cl unit is a bimodal sequence of lava flows with keratophyre to quartz-keratophyre and basic lavas, the proportion of which increases towards the top of the sequence and to the east. The StV unit comprises a large intrusion of trondhjemite and subordinate quartz-diorite locally with inclusions of dolerite. Several dismembered gabbro, ferrogabbro, and dolerite intrusions, cut by dykes, may belong to an incomplete ophiolitic complex, in which peridotite is lacking.

Chronological relationships indicate that units related to the marginal basin were younger than the schistose units (V, TC, PT). The PT and TC units are cut by quartz keratophyre and basic rocks possibly related to the marginal basin magmatism. This is also consistent with the schistosity and metamorphism of the schistose units as well as with the distinct alkaline nature of their magmatic suites. According to Wickert (1988) the schistose units belong to the Brévenne Unit, older than the Beaujolais Unit which comprises the Cl and StV units. The schistose units could be Early Paleozoic, or even older as some formations of the Brévenne Unit are dated as Late Proterozoic (Reitz and Wickert 1988).

5.2 Geochemistry of the Ensialic Marginal Basin

Three magmatic suites are recognized in the Cl, StV, and OC units. Quartz-keratophyre lavas of the Cl unit and diorite to trondhjemite of the StV unit are cogenetic and define an acid plutonic-volcanic ridge (suite 2). Rare earth patterns are quite similar in plutonic

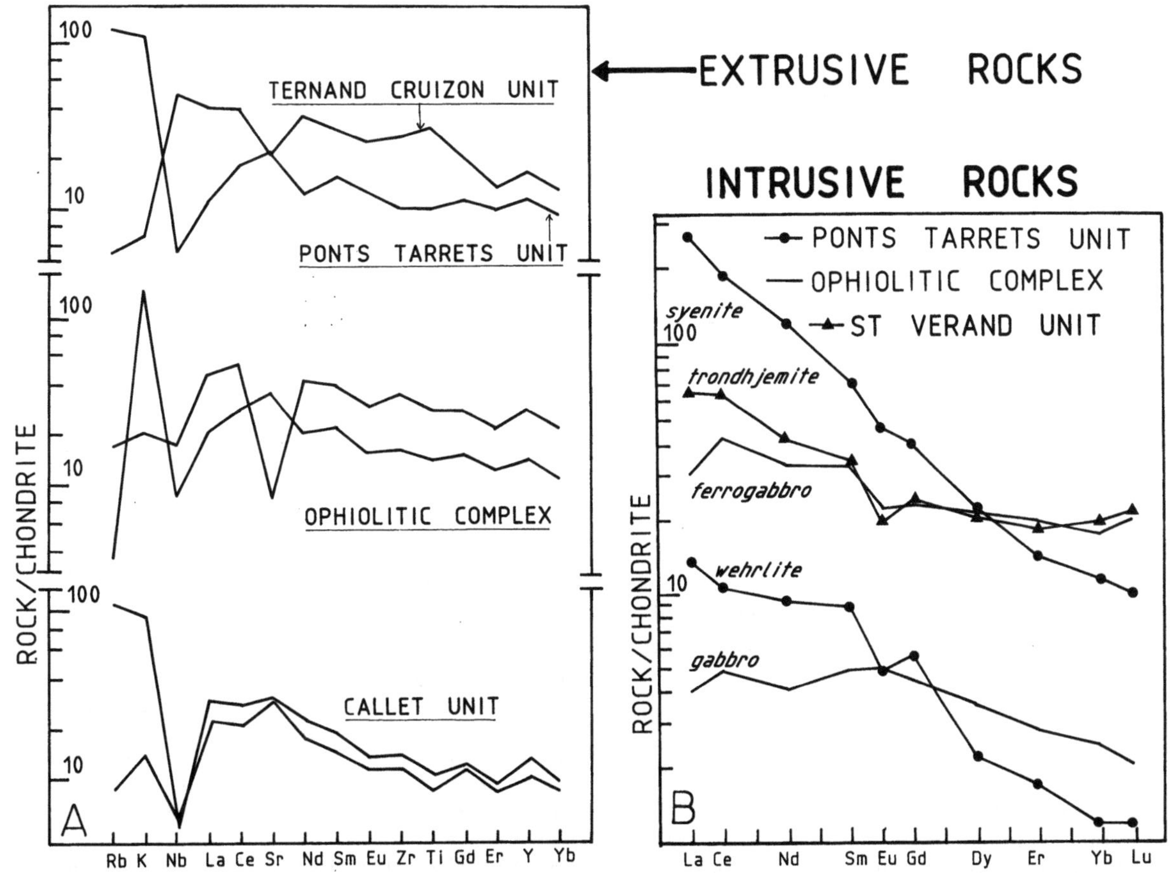

Fig. 17 A, B. Trace element patterns of the Beaujolais magmatic suites. **A** Diagram of Thompson (1982) and Thompson et al. (1982) for extrusive rocks. Elements are ordered according to their incompatible characters. **B** Rare earth distribution in intrusive rocks

and effusive rocks (Fig. 17 B), being characterized by a high $(La/Sm)_n$ ratio (1.89 to 2.3) as in acid calc-alkaline intrusions (Ohnenstetter and Sider 1988). Petrogenetic considerations as well as the huge volume of acid rocks related to suite 2 argue against derivation from a MORB-type ophiolitic magma having only a limited volume of acid differentiates. The parental magma of suite 2 may derive from an anatectic melt of acid composition (69–71 % SiO_2) which has suffered fractional crystallization leading to the evolved sodi-potassic quartz keratophyre. The REE pattern of suite 2 is close to that of the Chilean trondhjemite-granophyre plutons of continental origin which are associated with ophiolites.

The low-TiO_2 tholeiitic suite (suite 3) of calc-alkaline affinity comprises most of the basic lavas from the Cl unit and the included dolerite from the StV unit. Major and trace element distributions are similar to those of immature island-arc magmas. In particular, this suite shows low contents of REE and incompatible elements and a pronounced Ti, Zr, and Nb negative anomaly compared to MORB lavas. Its geochemical characteristics are also consistent with the low-Ti clinopyroxene belonging to the subalkaline and orogenic fields of the Leterrier et al. (1982) discriminant diagram.

Massive dolerites and dykes in the OC and late dykes in the Cl and StV units belong to a TiO_2-rich magmatic suite with MORB major-element characteristics. However, the LREE enrichment ($La/Yb_n \approx 2$), as in some enriched MORB lavas, and the Nb negative anomaly, as in some marginal basin lavas (Saunders and Tarney 1979), testify to the existence of a geochemical influence comparable to that occurring in a back-arc environment. On the basis of major and trace element variation, the ophiolitic cumulates could be cogenetic with the TiO_2-rich lavas and dykes. The Brévenne lavas and cumulates show the same kind of geochemical variation and could be related to a similar geotectonic environment (Pin et al. 1982).

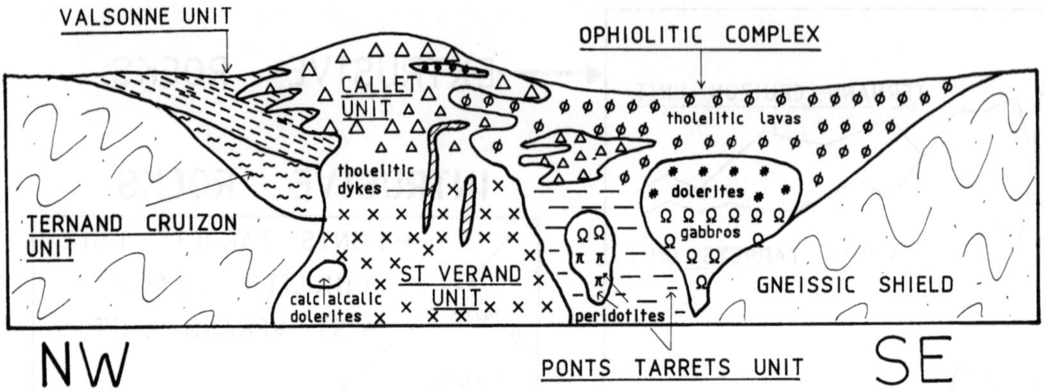

Fig. 18. Composite NW-SE cross-section showing the position of the acid ridge with respect to the ophiolitic basin and other volcano-sedimentary units

5.3 Evolution of the Ensialic Marginal Basin

The distribution of the three magmatic suites could be related to the zonation of an arc-back-arc system with significant participation of remobilized continental crust.

A NW-SE section (Fig. 18) through the Beaujolais area indicates that the acid ridge lies west of the marginal basin, with the StV unit as a root zone. The low-TiO_2 suite was most often emplaced at the same time as that of the acid ridge, but before the TiO_2-rich suite. If the OC is cogenetic with the TiO_2-rich lavas and dykes of the Cl and StV units, the marginal basin could have opened after the formation of the acid ridge, above a subduction zone dipping to the east. This is consistent with the less pronounced influence of the subduction zone for the TiO_2-rich suite than for the low-TiO_2 suite, which is closer to the western arc. These events are closely similar to those interpreted in the Mesozoic continental back-arc basin in Chile (Bruhn et al. 1978; Saunders et al. 1979). This type of environment can explain some of the peculiarities of the ensialic ophiolites: remobilization of continental crust, scarcity of ultrabasic rocks, influence of a subduction zone component with the Nb negative anomaly and the LREE enrichment which decreases with time during the widening of the basin. In all the basic remnants of the Beaujolais area, the persistence of high LREE contents, which is related to the influence of a subduction zone component, may indicate limited back-arc extension. The existence of a narrow basin agrees with the general plate configuration, indicating major oceanic closure at this time.

References

Bardin D (1971) Les amas pyriteux de Sain-Bel (Rhône) liés au groupe spilites-kératophyres de la Brévenne. Bull Bur Rech Geol Min 2, II: 17–41

Beurrier M, Piboule M, Durand C (1980) Gisement, pétrographie et géochimie des plagiogranites dévono-dinantiens des Monts de Tarare et du Beaujolais méridional (Massif Central français). C R Acad Sci Paris D 290: 147–150

Bruhn R L, Stern C, De Wit M J (1978) Field and geochemical data bearing on the development of a Mesozoic volcano-tectonic rift zone and back-arc in southernmost South America. Earth Planet Sci Lett 41: 32–46

Davies G R (1984) Isotopic evolution of the Lizard complex. J geol Soc London 141: 3–14

Lacroix P (1980) Géologie du groupe de Violay (Rhône): contribution à l'étude des granites sodiques et des formations éruptives et volcanosédimentaires associées. Thèse 3ème cycle, Univ. Lyon, 169 pp

Le Gall B, Poncet J (1983) Le front du Lizard (Cornouaille anglaise): un contact tectonique majeur jalonné par des écailles de calcaires paléozoïques. C R Acad Sci Paris D 296: 1683–1688

Leterrier J, Maury R C, Thonon P, Girard D, Marchal M (1981) Clinopyroxene composition as a method of identification of the magmatic affinities of paleo-volcanic series. Earth Planet Sci Lett 59: 139–154

Matte Ph (1986) Tectonics and plate tectonic model for the Variscan belt of Europe. Tectonophysics 126: 329–374

Ohnenstetter M, Rossi Ph (1987) Signification des complexes basiques-ultrabasiques associés aux granitoïdes dans l'évolution de la chaîne varisque du Sud de l'Europe. Bull Soc Géol Fr 8 III. 801–810

Ohnenstetter M, Sider H (1988) Contraintes géochimiques apportées par le magmatisme sur le développement du bassin marginal ensialique du Beaujolais au Dévonien. Bull Soc Géol Fr 8 IV: 499–510

Peterlongo J M (1960) Les terrains cristallins des Monts du Lyonnais (Massif Central français). Ann Fac Sci Univ Clermont-Ferrand 4: 1–187

Peterlongo J M (1970) Pillow lavas à bordure variolitique et matrice basique dans la série métamorphique de la Brévenne. Bull Soc Géol Fr (7) 12: 190–194

Piboule M, Briand B, Beurrier M (1982) Géochimie de quelques granites albitiques dévoniens de l'Est du Massif Central français. Neues Jahrb Miner Abh 143: 279–308

Piboule M, Beurrier M, Briand B, Lacroix P (1983) Les trondhjémites de Chindo et de St Vérand (Massif Central français) et le magmatisme kératophyrique associé. Bull Bur Rech Geol Min 2 1, 1–2: 55–72

Pin C, Dupuy C, Peterlongo J M (1982) Répartition des Terres Rares dans les roches volcaniqes basiques dévono-dinantiennes du Nord-Est du Massif Central. Bull Soc Géol Fr 24: 669–676

Reitz E, Wickert F (1988) Upper Proterozoic microfossils in low-grade phyllites of the Brévenne Unit, NE Massif Central (France). C R Acad Sci Paris II 307: 1717–1721

Saunders A D, Tarney J (1979) The geochemistry of basalts from a back-arc spreading centre in the East Scotia sea. Geochim CosmoChim Acta 43: 555–572

Saunders A D, Tarney J, Stern C R, Dalziel I W D (1979) Geochemistry of Mesozoic marginal basin floor igneous rocks from southern Chile. Geol Soc Am Bull 90: 237–258

Sider H, Ohnenstetter M (1986) Field and petrological evidence for the development of an ensialic marginal basin related to the Hercynian orogeny in the Massif Central, France. Geol Rundsch 75: 421–443

Sider H, Leistel J M, Gagny C (1986) Réflexion sur l'évolution géodynamique de la chaîne hercynienne au Paléozoïque dans le Nord-Est du Massif Central français. Bull Soc Géol Fr 8: 637–644

Styles M T, Kirby G A (1980) New investigations of the Lizard complex, Cornwall, England and a discussion of an ophiolite model. In "Ophiolites. Proceedings International Ophiolite Symposium Cyprus 1979" A Panayiotou ed., Ministry of Agriculture and Natural Ressources, Geological Survey Department: 517–526

Thompson R N (1982) Magmatism of the British Tertiary volcanic province. Scott J Geol 18: 49–107

Thompson R N, Dickin A P, Gibson I L, Morrison M A (1982) Element fingerprints of isotopic contamination of Hebridean Paleocene mantle derived magmas by Archean sial. Contrib Mineral Petrol 79: 159–168

Wickert F (1988) Paleozoic evolution of continental crust in the Beaujolais-Lyonnais area, northeastern part of the Massif Central, France. Geol Rundsch 77: 467–482

6 Felsic Plutonism and Volcanism in the Massif Central

A. Ploquin and J. M. Stussi

6.1 Introduction

The geodynamic evolution of the Massif Central was accompanied by an important magmatic activity ranging in age from early Lower Paleozoic to the Stephanian and Permian. The plutonic rocks occur throughout the Massif Central and are associated with the various lithotectonic units. They are more prominent than the volcanic rocks, which have a more restricted distribution.

The plutons consist dominantly of felsic rocks; intermediate and mafic phases are sparse and of limited volumetric development and are observed as separate intrusions or as inclusions in the felsic rocks. The plutonic rocks can be subdivided into two groups. The oldest or Pre-Variscan group consists of orthogneisses (including rocks of possibly volcanic origin) and amphibolites and represents about 15 to 20% of the outcropping metamorphics. These rocks range from 410 to 550 Ma in age. No radiometric ages older than ca. 550 Ma have been observed, hence the existence of an outcropping Precambrian basement in the Massif Central is unlikely (Duthou et al. 1984). The younger or Variscan group is composed mainly of granitoids with minor intermediate and mafic phases. These rocks represent about 50% of the outcropping areas and occur throughout the Massif Central (Fig. 19). Radiometric ages suggest continuous magmatic activity from 280 to 370 Ma. It is separated from the Pre-Variscan group by an approximately 40 Ma gap in igneous activity between 370 and 410 Ma. This time-span corresponds to the Devonian folding and climax of Barrovian metamorphism and anatexis (Duthou et al. 1984; Duthou and Floc'h 1989), all of which is related to the collisional stage of the oceanic subduction – continent-continent collision – continent subduction tectonics which characterizes the French Variscan (Matte 1986; Friedrich et al. 1988).

The present chapter summarizes the principal characteristics of the Pre-Variscan and Variscan igneous rocks with special emphasis on the Variscan granitoids. Some implications resulting from their chemical features and pattern of distribution are discussed.

6.2 Granitoid Characteristics

6.2.1 *Petrography*

The Pre-Variscan igneous rocks consist of orthogneisses with compositional variation from amphibolites, leptynites, and trondhjemites to granodiorites and granites. The range of composition of the Variscan plutonic rocks varies from minor gabbros to numerous granitoids and two-mica leucogranites. The modal compositions of the granitoid rocks plot mainly in the granodiorite and monzogranite fields of the I. U. G. S. plutonic rock classification (Streckeisen 1976) (Fig. 20). Igneous inclusions have compositions varying from diorites and tonalites to granodiorites, monzodiorites, and syenites (Fig. 20).

The textures of the granitoid rocks range from porphyritic through coarse- and medium-grained to fine-

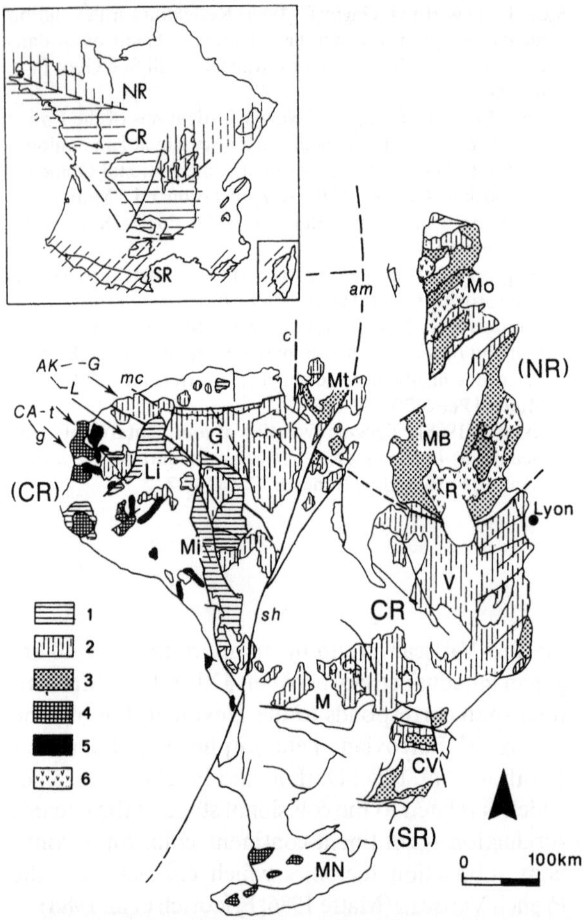

Fig. 19. Regional distribution of the granitoid types in the Massif Central. *1* Aluminopotassic granitoids (Limousin type); *2* aluminopotassic granitoids (Guéret type); *3* subalkaline granitoids; *4* calc-alkaline granitoids; *5* mafic and intermediate calc-alkaline rocks; *6* Dinantian to Permian volcanics and volcano-sedimentary rocks. Abbreviations: *(NR)* northern region; *(CR)* central region; *(SR)* southern region; *CV* Cévennes; *G* Guéret; *M* Margeride; *MB* Montagne Bourbonnaise; *Li* Limousin; *Mi* Millevaches; *MN* Montagne Noire; *Mo* Morvan; *Mt* Montmarault; *R* Roanne; *V* Velay; *AK* aluminopotassic Guéret type *(G)* and Limousin *(L)* type granitoid belts; *CA* calc-alkaline granitoid *(g)* and tonalitic *(t)* belts; *am* magnetic anomaly of the Paris basin; *c* Cher fault; *mc* Marches-Combrailles fault; *sh* Sillon Houiller

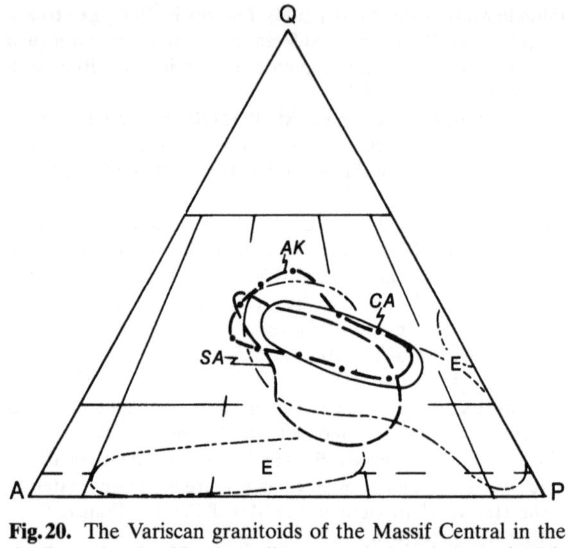

Fig. 20. The Variscan granitoids of the Massif Central in the QAP rock classification diagram of Streckeisen (1976). *AK* Aluminopotassic granitoids; *CA* calc-alkaline granitoids; *SA* subalkaline granitoids; *E* enclaves. (Didier 1973)

Table 2. Selected compositions of tholeïtic plagiogranites from the Massif Central. (Data source. Piboule et al. 1982)

N°	8	2	5	3	6
SiO_2	66.05	68.52	72.03	73.86	76.43
Al_2O_3	15.85	14.76	13.78	13.53	12.45
FeOt	4.07	4.02	2.71	2.07	2.14
MnO	0.06	0.06		0.02	0.06
MgO	2.57	1.48	1.36	0.67	0.53
CaO	0.69	1.76	0.56	0.56	0.60
Na_2O	6.50	5.36	5.83	5.37	5.81
K_2O	0.82	1.24	1.08	1.30	1.23
TiO_2	0.52	0.24	0.29	0.23	0.18
P_2O_5	0.20	0.14	0.06	0.06	0.02
Total	97.33	97.58	97.70	97.67	99.45

grained varieties. In an igneous complex, the fine-grained varieties generally represent the youngest intrusions and may be marked by metallogenic potential (cf. Cuney and Autran 1987; Friedrich et al. 1987). A frequent magmatic foliation is manifested by the planar and linear orientation of feldspar megacrysts and/or biotite. This foliation can be observed in the various textural and petrographic varieties of the granitoids (Mezure 1981; Mollier and Lespinasse 1985; Jover and Bouchez 1986; Peiffer 1986). Magmatic layering characterized by biotite-rich bands has also been described (Couturié 1977). Augen-gneiss textures are observed in the syn-metamorphic deformed Pre-Variscan orthogneisses and also in the Variscan granitoids deformed by late- to post-magmatic ductile shear zones (Mezure 1981; Lespinasse 1984; Mollier and Lespinasse 1985; Rolin 1987; Lerouge 1988). Subsolidus fluid-rock interaction, manifested by the presence of discrete late-magmatic deformation, is the origin of albitization, muscovitization, and tourmalinization processes. They are observed mainly in the two-mica leucogranites (Turpin 1984; cf. Friedrich et al. 1987).

Volcanic rocks consist of lavas, ignimbrites (Alsac et al. 1979), breccias, and tuffs. Their composition varies from minor andesites to dacites and volumetrically more developped rhyodacites and rhyolites of Upper Carboniferous to Stephanian and Permian age.

Table 3. Summary of the petrographical, mineralogical and isotopic characteristics of the Massif Central granitoid varieties. (Data sources Duthou et al. (1984); Sheppard (1986)

	Aluminopotassic		Subalkaline	Calc-alkaline	Tholeitic
	AK-Guéret	AK-Limousin			
Petrographic association	gd, g, lg, 2mlg	g, lg, 2mlg	mg, g, lg qm, m, md, mgb	gd, g, lg to, qd, d, gb	plg, trj
Inclusions	Metasediment Restites	Metasediment Restites	Igneous (m, md)	Igneous (to, d)	
Characteristic minerals	cord, gt Al-bi, mu, and, tour, lep, il, ur, mo	sil, gt, Al-bi, mu, lep, il, ur, mo	(cpx), hb, Mg-bi, all, ti, il, mt, th, ur	hb, Fe-bi, (all, ti), il	Fe-bi
Mineralization	SiK: U SiNa: W(Sn), Be, Li, Nb-Ta W skarn	SiK: U SiNa: Sn(W) Be, Li, Nb-Ta	Cu, Mo, Cu-Mo Cu-Sn, Mo-W U W skarn	W skarn	Sulfides (Fe)
Age (Ma)	Upper Proteroz. Cambrian 357–285	370–290	345–285	375–275	Middle-Upper Devonian
initial Rb/Sr ratio	0.705–0.714	0.705–0.718	0.706–0.717	0.706–0.713	
δ 18O	+9.0 to +13.5	+9.0 to +13.5	+8.1 to +9.5	+7.4 to +10.4	
ε NdT	−2.3 to −7.9			−4.0 to −9.6	

g: granites; gd: granodiorites; lg: leucogranites; 2mlg: two mica leucogranites; to: tonalites; d: diorites; mg: monzogranites; qm: quartz monzonites; m: monzonites; md: monzodiorites; mgb: monzogabbros; qd: quartz diorites; plg: plagiogranites; trj: trondhjemites.

Table 4. Selected and mean (S1, S5) compositions of calc-alkaline granitoids from the Massif Central. (Dutreuil 1978; Isnard 1985)

	4	29	S1	49	S5	16
SiO_2	63.9	67.06	69.34	70.63	71.67	73.52
Al_2O3	15.9	15.44	15.06	14.90	14.55	13.64
Fe_2O3t	5.04	3.34	3.23	2.58	2.20	1.78
MnO	0.10	0.08	0.07	0.08	0.06	0.07
MgO	2.12	2.02	0.92	0.75	0.62	0.54
CaO	4.00	3.00	2.63	2.38	1.94	0.59
Na_2O	3.40	3.70	3.25	4.25	3.33	3.56
K_2O	2.81	3.20	3.76	2.39	4.09	4.50
TiO_2	0.72	0.41	0.40	0.32	0.25	0.12
L.O.I.	1.17	1.67	1.01	0.66	0.92	1.18
Total	99.16	99.92	99.67	98.94	99.63	99.50
Ba	434	1013		1202		336
Sr	291	605		452		89

6.2.2 Granitoid Associations

The Variscan granitoids have traditionally been divided in two principal groups (Didier and Lameyre 1969): (1) the granodiorite-monzogranite group and (2) the two-mica leucogranite group studied by Lameyre (1966). Orsini (1979) discriminates three associations among the granodiorites and monzogranites: (1) the calc-alkaline association with tonalites, granodiorites, and related monzogranites and granites, (2) the metaluminous potassic subalkaline association including monzodiorites, quartz monzonites, monzogranites, syenogranites, and related granites, (3) the peraluminous potassic subalkaline association with cordierite- and muscovite-bearing granitoids. This classification is comparable to the chemical classification of granitoids proposed by La Roche (1980) and La Roche et al. (1980) as well as by Debon and Le Fort (1988).

The classification of La Roche et al. (1980) individualizes three principal associations among the orogenic granitoids: (1) the calc-alkaline association (i.e., granodioritic), (2) the subalkaline association (i.e., monzonitic), and (3) the aluminopotassic association. This classification is completed with the tholeïtic associations (plagiogranites and trondhjemites) and the anorogenic alkaline-peralkaline associations (Bonin 1982). The tholeïtic rocks have limited occurrence in the NE Massif Central (Piboule et al. 1982), while the alkaline-peralkaline granitoids are notably absent. In the present chapter, the La Roche et al. (1980) chemical classification has been used. The principal petrographical and mineralogical characteristics of the various granitoids are summarized in Table 2. The mean chemical composi-

Table 5. Selected compositions of subalkaline granitoids from the Massif Central (Sossa et al. (1984); CRPG Geochemical Data Bank)

	M36	M43	2330	25	12	42	63	54
SiO_2	60.12	63.47	65.70	67.10	69.90	71.60	74.90	76.50
Al_2O_3	17.17	16.88	15.40	14.90	13.05	14.10	13.20	11.85
Fe_2O3t	6.60	4.70	3.62	3.09	2.95	1.87	1.29	0.89
MnO	0.08	0.03	0.06	0.05	0.07	0.06	0.11	0.03
MgO	2.71	1.79	2.10	1.80	1.77	0.84	0.45	0.20
CaO	3.65	3.37	2.70	2.00	0.88	0.79	0.85	0.05
Na_2O	3.93	3.79	3.50	3.25	3.30	3.31	3.25	2.96
K_2O	3.14	3.91	5.00	5.30	5.12	5.38	5.30	5.28
TiO_2	1.24	0.82	0.50	0.50	0.34	0.20	0.10	0.05
L.O.I.	0.24	0.99	0.85	1.50	1.55	1.28	0.22	1.27
Total	98.88	99.75	99.43	99.49	98.93	99.43	99.67	99.08
Ba	887	1380	660	625	1000	720	200	160
Sr	596	842	370	320	400	280	110	100
Rb	179	173	200	255			295	320
Th					37	55		51
U					10	9		9

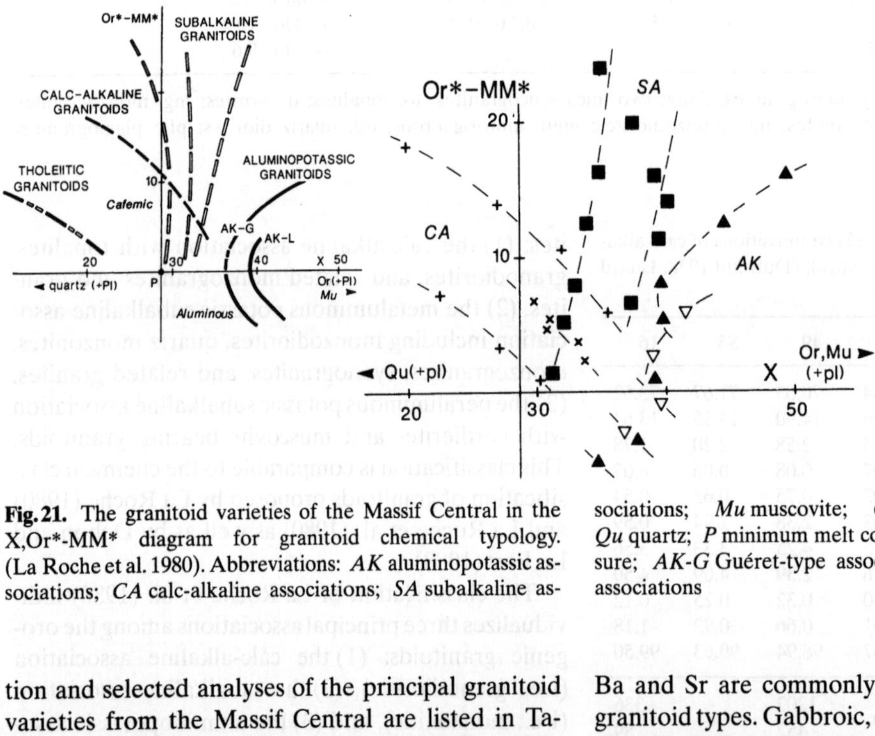

Fig. 21. The granitoid varieties of the Massif Central in the X,Or*-MM* diagram for granitoid chemical typology. (La Roche et al. 1980). Abbreviations: *AK* aluminopotassic associations; *CA* calc-alkaline associations; *SA* subalkaline associations; *Mu* muscovite; *Or* orthoclase; *pl* plagioclase; *Qu* quartz; *P* minimum melt composition for 5 kb water pressure; *AK-G* Guéret-type associations; *AK-L* Limousin-type associations

tion and selected analyses of the principal granitoid varieties from the Massif Central are listed in Tables 3, 4, and 5. In Fig. 21, these compositions are plotted in the X,Or*-MM* diagram of granitoid chemical typology (La Roche et al. 1980), while in Fig. 22 the data are plotted in the AFM and alkalies vs. silica diagrams.

The calc-alkaline granitoids consist of granodiorites, granites, and associated leucogranites (less than 7% mafic minerals). They exhibit a compositional range from 66 to 74% SiO_2; K_2O/Na_2O ratios vary from 0.56 to 1.29 and low peraluminous compositions are observed only in the most differentiated phases; Ba and Sr are commonly higher than in the other granitoid types. Gabbroic, dioritic, and tonalitic rocks of calc-alkaline nature with compositions ranging from 44 to 63% SiO_2 (Table 7) occur as inclusions or as small, discrete intrusions in the Boussac area (N Massif Central). In the SW Limousin area, they define a 250-km-long NW-SE lineament composed of isolated and scattered plutons (Peiffer 1986) and are intruded diachronously by calc-alkaline granitoids (Floc'h 1979) with which they do not show an obvious genetic relation.

The subalkaline granitoids are represented by quartz monzonites, monzogranites, granites, and as-

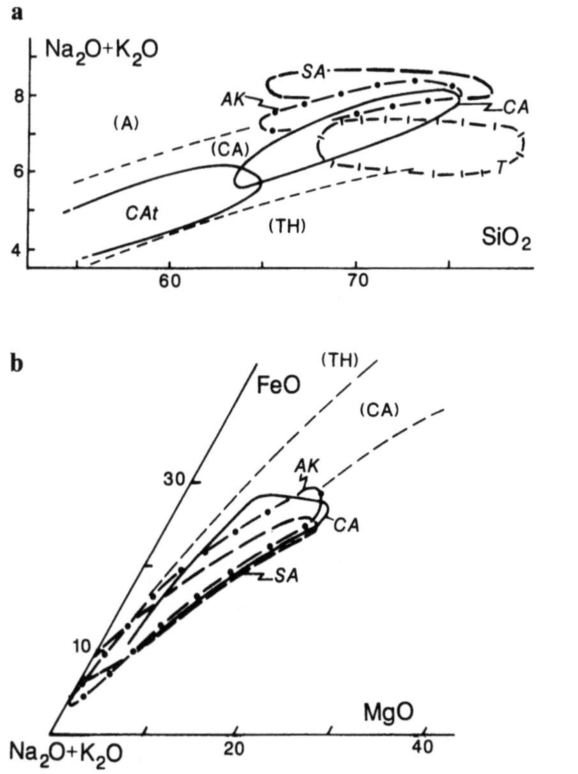

Fig. 22 a, b. The Variscan granitoids of the Massif Central in the total alkalies vs. SiO_2 **(a)** and AFM variation diagram **(b)**. *TH* tholeïtic field; other abbreviations see Fig. 21

sociated leucogranites. Although their range of SiO_2 (63 to 77%) is comparable to that of the calc-alkaline granitoids, total K_2O + Na_2O (Fig. 22) and K_2O/Na_2O ratios (0.68 to 1.65) are higher than in these rocks. In addition, (1) for an equivalent color index the subalkaline granitoids are more cafemic, (2) a slight peraluminous composition is observed in the only most evolved members, (3) the Fe/Fe + Mg ratio usually is lower than in the calc-alkaline granitoids, and (4) the Ba and Sr are commonly high, while the U, Th, and REE abundance (presence of titanite, allanite, Th-rich uraninite) is higher compared with the calc-alkaline granitoids. Spatially associated but less evolved phases (inclusions, low volumetric intrusions) have compositions which vary from monzodiorites to quartz monzonites (Table 5) (i.e., Montmarault massif: Sossa et al. 1984). A subalkaline magma composition is also observed for most of the volcanics of the Massif Central (i.e., the Stephano-Permian caldeira of Blismes-Montreuillon in the Morvan area: Monnier and Stussi 1984; the Upper Visean volcanics of the Roanne area: Leistel and Gagny 1984).

The aluminopotassic granitoids are characterized by the presence of aluminous minerals (cordierite, sillimanite, andalusite, muscovite). They can be classified into two associations. (1) The Guéret-type association consists of cordierite-bearing granodiorites (granodiorite field of the I. U. G. S. QAP rock classification; Fig. 2), granites and two-mica leucogranites. The SiO_2 contents range from 63 to 75%. (2) The Limousin-type association consists of sillimanite-bearing granites and leucogranites and associated two-mica leucogranites. The SiO_2 contents range vary from 69 to 74%, while the K_2O content is higher and the Na_2O abundance of the granites lower than in the Guéret-type granites (Table 6). In both associations, the atomic proportion of alumina is in excess of the combined total of (K + Na +2Ca). For equivalent MgO contents, the composition of the biotite is more aluminous (Al_2O_3 > 18%) than in the calc-alkaline (16 to 18%) and subalkaline (13–16%) granitoids (Nachit et al. 1985). The granites and especially the two-mica leucogranites of these two associations are commonly enriched in lithophile elements (Rb, Li, Cs, B, Be, F, Sn, W, U). Their U abundance is higher and the REE contents lower than in the subalkaline granitoids.

This classification has also been applied to the Pre-Variscan orthogneisses. It generally confirms their calc-alkaline or tholeïtic nature as defined using different approaches. However, in some cases the definition of their precise magma composition may present some difficulties because of the modification of their chemistry during the orthogneissification processes. Na, Ca, and subsidary K can be removed resulting in an apparent increase of the Al-(K + Na +2Ca) balance, thus modifying the original chemistry, i.e., calc-alkaline and subalkaline granitoids may present an aluminopotassic nature of secondary origin (Bourguignon 1988; Ploquin et al. 1988) (Fig. 23). An alkaline nature has been ascribed to some orthogneisses; however, this has not been confirmed by recent studies (e.g., Nages and Mendic massifs in Montagne Noire).

6.2.3 *Size and Form*

The form of the Pre-Variscan orthogneisses generally has been modified by Devonian metamorphic and deformational events. Most of the orthogneisses actually exhibit sheet- or lense-like forms conformable to the foliation of their surrounding metamorphics. The absence of augen-gneiss textures and of unconformable mappable contacts can hinder the determination of their plutonic or volcanic origin.

The outcrop areas of the Variscan plutonic bodies range from more than 5000 km^2 (Velay, Guéret, Margeride) to less than 1 km^2 (Beauvoir Sn-Li-Nb-Ta mineralized intrusion). The smaller bodies (< 200 km^2) consist of single, frequently composite, isolated intrusions (Crevant, Sidobre). The larger bodies

Table 6. Mean composition of the aluminopotassic granitoids of the Massif Central. C. R. P. G.-Nancy Geochemical data bank

	Aluminopotassic Granitoids										
	Guéret association							Limousin association			
	G 1	G 2	G 3	G 4	G 5	G 6	G 7	L 3	L 4	L 5	L 6
SiO_2	63.26	65.31	68.79	69.34	71.69	73.48	74.72	70.17	71.78	72.74	73.87
Al_2O_3	16.35	16.35	15.33	15.46	14.83	14.66	14.43	15.08	14.56	14.57	14.44
Fe_2O_3t	5.62	4.38	2.99	2.58	1.87	1.17	0.77	2.56	1.95	1.66	1.42
MnO	0.09	0.07	0.05	0.05	0.05	0.03	0.04	0.03	0.03	0.02	0.03
MgO	2.49	1.84	1.29	1.09	0.74	0.27	0.20	0.72	0.49	0.27	0.17
CaO	2.74	2.48	1.60	1.43	0.70	0.43	0.41	1.04	0.77	0.44	0.46
Na_2O	2.82	3.21	3.27	3.37	3.28	3.39	3.91	2.88	2.99	2.95	3.31
K_2O	3.98	4.24	4.52	4.53	4.66	4.68	4.10	5.74	5.40	5.31	4.75
TiO_2	0.71	0.65	0.43	0.38	0.26	0.14	0.11	0.38	0.26	0.20	0.14
L. O. I.	1.80	1.37	1.25	1.34	1.43	1.41	1.14	1.39	1.29	1.42	1.31
Total	99.86	99.90	99.52	99.57	99.51	99.66	99.83	99.99	99.52	99.58	99.90
n	3	7	13	9	76	46	26	9	9	19	10
Ba	955	565	488	418	362	239	38	343	285	258	189
Sr	326	190	168	180	94	77	19	80	73	67	52
Rb	250	207	239	245	309	435	520	396	394	409	333
Cs				24	30	52	100	< 20	30–70	< 20	100
Li	92	100	116	145	193	271	406	100	220	180	360
Sn			15	33	58	83	138	8	15–30	8	23
W			0.6	2.4	4.7	10	7.7				
Be		2.6	4.3	3.5	6	6	< 10	4	13	14	27
Ta			1.4	2.4	2.2	4.5	10				
U		5.1	7.3	5.6	8.4	7.8	1.2	13.2	18.8	12.2	13.9
Th		20.6	17.9	17.5	11.0	10.0	0.7	45.1	43.1	20.0	10.5

n: number of analyses. G1 to G7 and L3 to L6: mean compositions determined from statistical distribution in Fig. 3.

Table 7. Selected compositions of intermediate and mafic calc-alkaline rocks from the southwestern (tonalite belt: Peiffer 1986) and northern (Boussac area: Bouvier 1985) Massif Central

	18	8	6	10	6	18	BJ8	BJ9	BJ24	BJ1
SiO_2	43.45	50.30	51.16	55.01	59.92	64.41	44.22	51.42	54.6	63.48
Al_2O_3	13.11	18.30	17.87	18.22	17.20	15.44	21.68	18.65	16.62	16.77
Fe_2O3t	12.88	10.55	8.63	7.85	7.28	4.56	8.35	9.52	9.59	3.43
MnO	0.15	0.14	0.18	0.11	0.13	0.10	0.23	0.19	0.16	0.19
MgO	13.03	4.15	4.21	3.95	2.56	2.11	5.73	4.85	3.88	1.25
CaO	10.59	8.00	7.75	6.59	6.16	3.91	13.32	7.12	6.07	2.91
Na_2O	1.73	3.00	3.89	3.51	3.51	3.61	1.96	3.17	2.93	4.55
K_2O	0.92	1.45	2.50	1.57	1.64	2.50	0.17	1.91	2.08	3.42
TiO_2	2.22	1.35	1.10	1.02	1.07	0.60	1.04	0.85	1.19	1.43
P_2O_5	0.00	0.31	0.46	0.21	0.00	0.07	1.23	0.14	0.45	0.19
L. O. I.	1.79	1.20	1.38	1.22	0.77	1.18	1.37	1.46	1.31	1.39
Total	99.87	98.75	99.13	99.26	100.34	98.50	99.30	98.46	98.88	98.96
Ba	252		1723	603	441	902	63	361	762	3632
Sr	117		968	415	442	436	560	396	366	815
Rb			72	40		74	2	63	62	105
Ni	94		28	76	32	21				
Cr	42		38	39	27	24				
V	438		236	203	182	97				

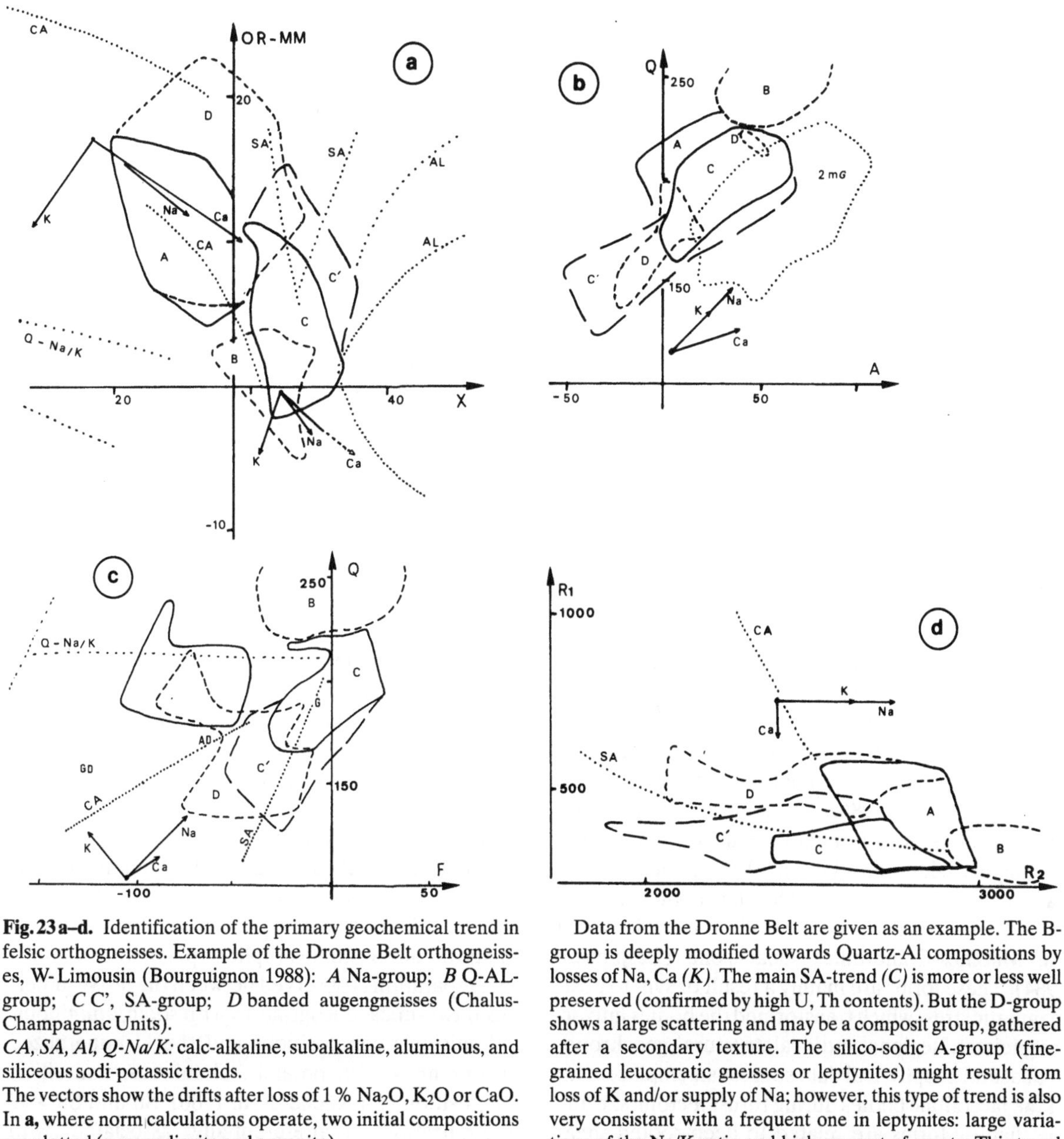

Fig. 23a–d. Identification of the primary geochemical trend in felsic orthogneisses. Example of the Dronne Belt orthogneisses, W-Limousin (Bourguignon 1988): *A* Na-group; *B* Q-AL-group; *C C'*, SA-group; *D* banded augengneisses (Chalus-Champagnac Units).
CA, SA, Al, Q-Na/K: calc-alkaline, subalkaline, aluminous, and siliceous sodi-potassic trends.
The vectors show the drifts after loss of 1 % Na_2O, K_2O or CaO. In **a,** where norm calculations operate, two initial compositions are plotted (a granodiorite and a granite).
Parameters: OR-MM versus X, (cf. Fig. 21)
Q = Si/3 – (Na + K + 2Ca/3); A = Al – (Na + K + 2Ca);
F = K – (Na + Ca); R1 = 6Ca + 2Mg + Al;
R2 = 4Si – 11(Na + K) – 2(Fe + Ti).
Comments: In the used typology (cf. Fig. 21), the drift, induced by losses of Na, Ca, and/or K, is not easy to control: e. g., **a,** the value and the orientation are not the same according to the initial composition and to the ratio of losses. Thus, a set of diagrams was used to check the Al-balance and drifts from inferred trends in comparison with the positions and trends of kindred data: e. g., it is not possible to shift a CA-trend into a coherent SA- or Th-trend, but trends become more or less dim, and the A, and R2 values increase. The problem is more difficult if losses and supplies occurred together: drifts towards neo-minerals may be clues. About 1500 analyses of orthogneisses from the Massif Central were compiled and computed.

Data from the Dronne Belt are given as an example. The B-group is deeply modified towards Quartz-Al compositions by losses of Na, Ca *(K)*. The main SA-trend *(C)* is more or less well preserved (confirmed by high U, Th contents). But the D-group shows a large scattering and may be a composit group, gathered after a secondary texture. The silico-sodic A-group (fine-grained leucocratic gneisses or leptynites) might result from loss of K and/or supply of Na; however, this type of trend is also very consistant with a frequent one in leptynites: large variations of the Na/K ratio and high amount of quartz. This trend may be associated with a quartz/Na-plagioclase one (the relative weight of these two trends varies from one occurrence to another, cf. **c** and **a**). Felsic gneisses from bimodal association (CA or Th series), plagiogranites and quartz-keratophyres may show these trends; e. g., Albussac and Vergonzac Units (S-Limousin), a part of the Meuzac Belt, "ordovician" leptynites in the Thaurion Belt (Limousin), in the Levezou Complex (Rouergue), in the Marvejol Complex (Lot), etc. Some felsic gneisses from the Leptyno-Amphibolites Complex belong also to this type.

The absence of correlation between A values and L. O. I., and the development of devonian barrovian minerals on Na-Ca-(K) depleted gneisses reveal a Prevariscan age for this depletion and confirm the secondary nature of the Al-character; these depletions need the destructions of Na-, Ca-, K-carrier minerals, i. e., a green schist metamorphic event.

(Velay, Millevaches) are igneous complexes composed of multiple phases with various petrographic and textural characteristics. In some complexes, magmatic activity extended over 40 Ma (Millevaches: Monier 1980; Duthou et al. 1984; Razafimahefa 1987; Morvan: Monnier and Stussi 1984; Carrat 1987; Rolin 1987).

There is no clear correlation between the size and the form of the plutonic bodies. Diapiric forms characterize both single intrusions (Meymac: Mezure 1981; Crevant: Lemaire et al. 1986) and igneous complexes. The Velay complex is interpreted as a large anatectic diapir dome composed of a number of intrusions of varying composition (peraluminous cordierite-bearing granites and leucogranites, two-mica leucogranites and anatexites; spatially associated non-cogenetic vaugnerites) (Montel and Weisbrod 1986). A half-funnel form has been ascribed to the syn-kinematic Egletons massif (Mezure 1981), while the Millevaches massif is interpreted to be a diapir-window like igneous complex with marginal ductile mylonitic to phyllonitic texture (Lameyre 1982). Petrographic fabrics (magmatic foliation and lineation), borehole indications and geophysical data suggest a laccolith form with a vertical petrographical zonation of the Guéret, Saint-Sylvestre, Margeride, and Northern Millevaches complexes (Couturié 1977; Mollier and Lespinasse 1985; Jover and Bouchez 1986; Lameyre et al. 1988). Similar forms are observed in minor intrusions (Mayet-de-Montagne: Didier and Peyrel 1980; Aigoual massif: Sabourdy 1975). Vertical sheet-like bodies are observed in the North Massif Central along the length of the Marches-Combrailles shear zone (Lespinasse 1984). These forms are mainly a result of syn- to post-magmatic stress which transformed the bodies into orthogneisses. Regional and local structural, geochronological and geophysical data suggest an emplacement of the laccolithic bodies during thrust tectonics, while the diapiric forms are related to deep-rooted shear zones (Mezure 1981; Jover and Bouchez 1986; Rolin 1987; Lerouge 1988). Laccolithic and diapiric forms are independent of granitoid composition. Most of the granitoid bodies are truncated by late-Variscan and Alpine strike-slip faults, fractures, and block-faults. These structures, together with the deep erosional level, prevent the precise determination of the dimensions and shapes of the granitoid bodies.

6.2.4 Contacts

Pre-Variscan orthogneisses commonly exhibit conformable and sharp to "gradational" contacts with their surroundings. Locally, the gradational contacts have been ascribed to deformed granitoid bedrock overlain by an arkosic cover. At present, structural and geochemical data suggest the development of mylonitic textures in the orthogneisses. The Variscan plutons have either conformable or discordant and sharp contacts with the surroundings; both types of contacts can be observed around the same pluton. Autochthonous granitoids are sparse and of limited exposures in high-grade and anatectic environments.

Allochthonous granitoids intrude high- to medium-grade (central part of the Massif Central) and low-grade metamorphics (S Massif Central). In the NE Massif Central, they cut Upper Visean volcano-sedimentary series with very low-grade metamorphism and weak folding. The intrusions generally superimpose metamorphic aureoles (sillimanite, andalusite, cordierite, muscovite) in their surrounding metamorphics and granitoids. The original intrusion contacts have frequently been modified by post-granitic fracturing.

6.2.5 Associated Mineralization

Numerous economic deposits (U) and occurrences (Sn, W, Li, Be, Nb, Ta, Cu) are observed inside and in the proximity of the Variscan granitoids. The local and regional distribution of mineralization exhibits a close relationship to the composition of their host rocks (Stussi 1989). Disseminated, vein-type and stockwork (U, Sn-W, and related elements) mineralization is intimately associated with aluminopotassic two-mica leucogranites and their altered (greisens, stockscheiders, or contact pegmatites, episyenites) or unaltered differentiates (aplites, pegmatites). A distinct separation is observed between the Sn-W mineralized two-mica leucogranites with sodic, aluminous, and siliceous compositions and the U mineralized leucogranites with potassic, aluminous, and silicic composition (La Roche et al. 1980). Sn and Sn > W mineralization is preferentially associated with Limousin-type leucogranites, while W and W > Sn mineralization is observed mainly with Guéret-type granites and related outcropping or hidden two-mica leucogranites (Burnol et al. 1980). The subalkaline granitoid-related mineralization is typically represented along the French Variscan by Cu and Mo occurrences (Stussi 1989). In the Massif Central, this type of mineralization is limited to localized complex Cu-Sn occurrences. A number of U deposits and occurrences are also observed in close proximity to evolved subalkaline granites and leucogranites (Bois-Noirs, Morvan) which ascribe to the subalkaline magmatism a U potential. The influence of late two-mica leucogranites has been advocated for the formation of this mineralization (Cuney 1978).

Table 8. Rb/Sr chronology of the Prevariscan and Variscan granitoid varieties of the Massif Central

Age Ma.	Orogenic events	Stratigraphy	Granite associations AK-G	AK-L	CA	SA	TH	A-PA	Tectonics
230									Extension
250	Saalian	Permian							
280									
		Stephanian							
290	Asturian								
		Westphalian							Shearing
310	Erzgebirgian								
		Namurian							
325	Sudetan								to
		Visean							
340									
		Tournaisian							
345	Bretonian (Acadian)								Thrusting
		Devonian							
380	Erian								
395	Ardennian	Silurian							Collision
435	Taconian	Ordovician							Oceanic subduction
500	Sardinian								Extension
		Cambrian							
570	Cadomian								
		Proterozoic							

Isotopic data sources Michard-Vitrac et al. (1980), Bonin (1982), Duthou et al. (1984), Bernard-Griffiths et al. (1985). A-PA: alkaline-peralkaline granitoids. Other abbreviations see Fig. 3

6.3 Chronology of Emplacement

Radiometric determinations (Rb/Sr; U/Pb) are summarized in Table 8 and Fig. 24 together with the magma chemistry of the dated igneous rocks. The Pre-Variscan orthogneisses have been subdivided into three chronological groups (Duthou et al. 1984). The Cambrian group (496 to 550 Ma) consists of augengneiss, leptynites, and some mafic rocks. They are observed in the Limousin (Thaurion leucocratic orthogneisses and leptynites: 532 ± 24 to 521 ± 12 Ma; Tulle orthogneisses: 525 ± 12 Ma; in the Montagne Noire: Plaisance orthogneisses are dated by U/Pb systematics at 552 Ma, while the Somail-Agout orthogneisses may be possibly older). The composition of these orthogneisses is mainly calc-alkaline. The Ordovician group is represented by the trondhjemites of Levezou (486 Ma), by the Saint-Yrieix gneisses (468 ± 8 Ma) of subalkaline chemistry and by minor aluminopotassic rocks (Châteauneuf belt: 483 ± 22). Some granite orthogneisses from the Thaurion belt exhibit 462 Ma ages, thus demonstrating that within an individual lithotectonic unit different members can be characterized by varying ages and magma composition. In addition, the Thaurion belt contains also Middle Devonian diatexites dated at 375 ± 6 Ma. The tholeïtic to calc-alkaline leptyno-amphibolitic group (CLA; cf. Santallier et al. 1988) belongs to this chronological group. The augengneiss of Celoux (406 ± 20 Ma) and the trondhjemites of Marvejols (415 Ma) characterize the Silurian calc-alkaline to tholeïtic magmatism. No obvious magmatic gap can be observed between these groups.

The Rb/Sr age determinations of the Variscan granitoids (Duthou et al. 1984) show that the magmatic activity has been continuous throughout the Massif Central from the Middle-Upper Devonian to the Stephanian. Most of the granitoids (including

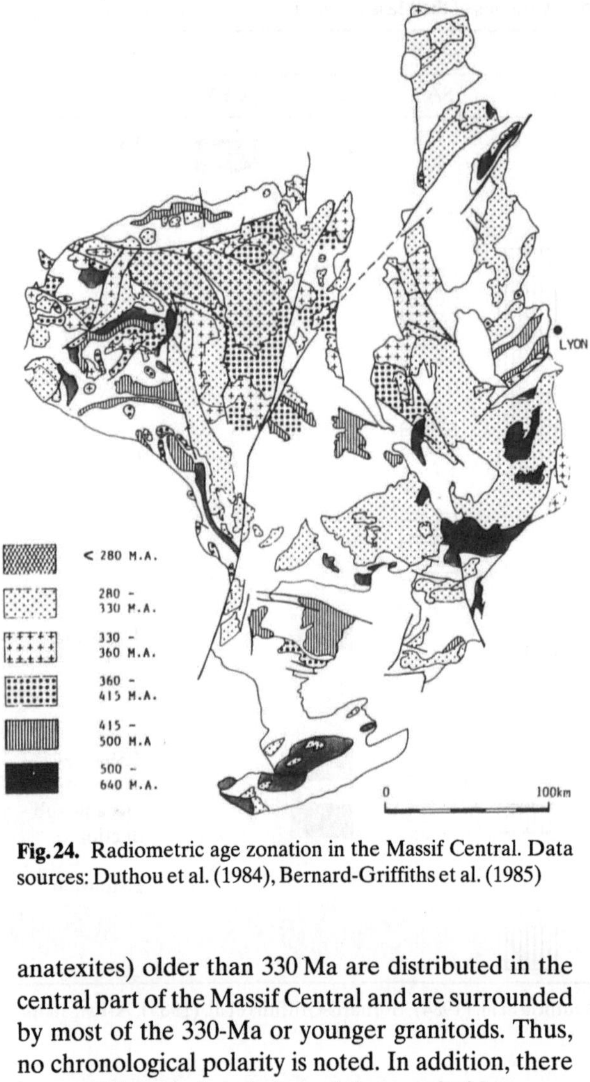

Fig. 24. Radiometric age zonation in the Massif Central. Data sources: Duthou et al. (1984), Bernard-Griffiths et al. (1985)

anatexites) older than 330 Ma are distributed in the central part of the Massif Central and are surrounded by most of the 330-Ma or younger granitoids. Thus, no chronological polarity is noted. In addition, there is no evidence of evolution of the granitoid magma chemistry with age because from the Upper Devonian to the Stephanian granitoids of differing magma composition were emplaced in the same plutonic body as well as in different areas. The only exception to this pattern concerns the calc-alkaline to tholeïtic rocks which characterize the Middle-Upper Devonian magma activity of the SW Limousin (calc-alkaline tonalites and quartz diorites with subordinate tholeïtic phases) and of the NE Massif Central (tholeïtic plagiogranites and related keratophyres) (Piboule et al. 1982; Bernard-Griffiths et al. 1985; Pfeiffer 1986). The subcontemporaneous emplacement of these rocks with the aluminopotassic Guéret (356 Ma), Vaulry (357 Ma) and Bouchefarolle (352 Ma) massifs (cf. Duthou et al. 1984) is noted. Similar interspersed emplacement of coeval or subcontemporaneous granitoids of either calc-alkaline (SW Limousin) or subalkaline (Morvan, Montagne Bourbonnaise, Cévennes) and aluminopotassic granitoids intruding each other characterize the Upper Visean to the Westphalian and Stephanian times.

As in the Massif Armoricain (Vidal 1980), the evolution of the initial strontium ratio with time does not indicate the presence in the Massif Central of an outcropping crust older than 600 Ma. Only isolated U/Pb data on 1000- to 2000-Ma-old zircons from some metasedimentary rocks and orthogneisses indicate the probable contribution of an old crustal component.

6.4 Regional Pattern of Granitoid Distribution

Because of their low abundance and scattered distribution across the Massif Central, no obvious magmatic zonation can be observed among the Pre-Variscan orthogneisses. Most of the granitoid-derived orthogneisses are associated with the Lower Allochthonous Unit. From the Western Limousin towards the Lot area (SE Massif Central) through the Auvergne, the calc-alkaline orthogneisses are more prominent in the southern than in the northern part of this region, while the subalkaline orthogneisses are more prominent in the NW part of Limousin. The most common but also frequently debated magmatic activity is represented by the bimodal calc-alkaline or tholeïtic derived leptyno-amphibolitic complex which is generally associated with the Middle Allochthonous Unit (cf. Santallier et al. 1988).

The regional distribution of the granitoid associations is shown in Fig. 1. Three principal regions can be defined. (1) The Central Region extends from the NW Limousin towards the SE Velay complex. It is mainly composed of diachronous aluminopotassic granitoids which are distributed in three provinces.(1 a) A NE province consisting of Guéret-type granitoids (Guéret, Margeride, Velay massifs, and adjacent bodies) which are associated with medium- to high-grade lithotectonic units. It is characterized by U deposits and by W and W > Sn mineralization. (1 b) A SW province composed dominantly of Limousin-type granitoids with some subordinate Guéret-type intrusions (Aureil, axial zone of Millevaches). The surroundings consist of medium- to high-grade metamorphics and orthogneisses. The mineralization is represented by the most important U deposits and by Sn > W, Li, and Be occurrences. Some Au mineralization is observed in these two provinces. However, their precise relationships with the Variscan granitoid varieties are not clearly known. (1 c) The SW Limousin calc-alkaline province is composed of granodiorites and granites (315 to 325 Ma) which transect the diachronous (355 to 375 Ma) calc-alkaline or tholeïtic tonalite belt (Pfeiffer 1986). It also intrudes coeval to subcontemporaneous aluminopotassic granites and leucogranites (Saint-Mathieu-Piégut massif) (cf. Floc'h 1979). Some minor localized and scattered subalkaline intrusions with monzodioritic to granitic composition cut these provinces and belts.

The NE Region extends northeasterly from the Guéret-type province (N Auvergne, Montagne Bourbonnaise, Morvan, Charolais, Beaujolais). It is characterized by the alternating emplacement between 345 and 280 Ma of aluminopotassic and subalkaline granitoids (Cuney 1978; Carrat 1987; Rolin 1987). Calc-alkaline phases are known only as minor inclusions. Coeval or diachronous volcanics (Upper Visean "tufs anthracifères" associated with some andesites overlying and/or intruded by subalkaline plutonic and hypabyssal phases) and hypabyssal phases with the same range of composition and magma chemistry are associated with the subalkaline granitoids. They occur in association with volcano-tectonic troughs and calderas (Leistel and Gagny 1984; Monnier and Stussi 1984). Typical mineralization is represented by U deposits and occurrences, and by the only Cu-porphyry-type mineralization observed in the French Variscan (Icart et al. 1980). Minor Cu-Sn and W-skarn mineralization is locally observed (Gagny et al. 1983).

The S Region includes the Cévennes and the Montagne Noire. It consists of localized bodies composed of subalkaline (Aigoual, Mont-Lozère) or calc-alkaline (Sidobre, Folat) granitoids with some minor, spatially associated aluminopotassic phases. They cut medium- to low-grade Lower Paleozoic metasedimentary series. Granitoid related mineralization is sparse and of limited interest (W-skarn: Sidobre, Borne massif; W-Sn veins: Saint-Cierges) (Bril and Ramboz 1982), except for the W-Sn mineralization of Montredon-Labessonié, which is related possibly to a concealed granitoid pluton (Beziat et al. 1980).

6.5 Discussion

6.5.1 Pattern of Magma and Granitoid Distribution

The Pre-Variscan, predominantly calc-alkaline magmatism contrasts markedly with the varieties of granitoids which characterize the overall Massif Central throughout the Variscan times (280 to 370 Ma). The observed pattern of magma types (Fig. 18) exhibits more a zonation than a magmatic polarity across the Massif Central. The alternating and interspersed emplacement of granitoids of differing magma chemistry is specific to the SW Limousin province, and the NE and the SE Massif Central. It contrasts with the very prominent aluminopotassic magmatism of the Central Region. This pattern of magma types and of magma distribution is enhanced by the close relationships with the associated mineralization. It implies the involvement of various source materials in a vertically heterogeneous crust. In addition, the observed pattern of granitoid distribution and the transversal widespreading of the igneous activity across the Massif Central, cannot be interpreted in terms of a single subduction model as previously proposed using less data (Autran and Lameyre 1980; Pupin 1985).

6.5.2 Granitoid Source Materials

The relatively low strontium initial ratio of some orthogneisses (0.703 to 0.711), suggests an origin by melting of a Rb-depleted granulitic lower crust rather than a mantle derivation. This crust, which is not exposed but may be represented by granulitic inclusions found in the recent basaltic volcanics (Leyreloup 1974), must have formed during the late Proterozoic (Duthou et al. 1984). The Variscan granitoids usually are characterized by high strontium initial ratios (0.705 to 0.718) and have formed by the anatexis of a late Proterozoic crust. Strontium initial ratios increase with age, therefore suggesting the increasing involvement of the crust with progressively younger materials (cf. Duthou et al. 1984). Restitic aluminous inclusions (biotite, sillimanite, cordierite, spinel, corundum) and isotope constraints ($\delta^{18}O$; ε Nd) argue for an origin of the aluminopotassic granitoids by the melting of peraluminous metasedimentary rocks or of peraluminous orthogneisses (Didier and Lameyre 1969; Duthou et al. 1984; Montel et al. 1986 a, b; Sheppard 1986). Thermo-barometric constraints give evidence for the formation of the Velay anatectic granitoids at temperatures in the range of 790 to 830° C for $P = 4.8$ to 5.3 kbar (Montel et al. 1986 a). These PT constraints can be compared to the experimental temperature of crystallization ($T < 600°$ C for 1 kbar) obtained on the F-, Li-, P-rich and specialized (Sn, Nb) lepidolite-bearing leucogranite of Beauvoir (Pichavant et al. 1987). According to their aluminous metasedimentary origin, the aluminopotassic granitoids may be compared to the S-type granitoids (White and Chappell 1983). However, differences with this granitoid type are observed concerning their higher peraluminous composition which increases with decreasing color index, the alkali abundance (with $Na_2O > 2.8\%$), higher K/Na ratios, and the more extensive development of two-mica leucogranites. Conversely, as shown by Didier et al. (1982), the origin of some aluminopotassic granitoids by the anatexis of peraluminous orthogneisses suggests an I-type origin. These peraluminous orthogneisses include those which have inherited their peraluminous composition from the orthogneissification processes prior to the Devonian Barrovian metamorphism and anatexis (Bourguignon 1988; Ploquin et al. 1988). Geochemical modeling has shown an origin for the Saint-Sylvestre peraluminous leucogranites, which is consistent with the melting of the Arc de la Dronne peraluminous orthogneisses of mainly primitive subalkaline nature (Turpin et al. 1988).

The calc-alkaline and subalkaline granitoids exhibit comparable strontium initial ratios (0.705 to 712) (Table 2) which are close to those of the aluminopotassic granitoids. These ratios indicate that their origin cannot be of pure mantle derivation and involves a notable crustal contribution (Duthou et al. 1984). According to Sheppard (1986), the calc-alkaline granitoids have formed by the anatexis of a mixed series of rocks including basalts, shales and graywackes, whereas the subalkaline granitoids are derived from the melting of long-lived crustal materials composed of amphibolites, graywackes and granulite series. A mantle contribution, shown either by the mafic inclusions or by the spatially associated mafic and intermediate phases, is not precluded but it must have been of small proportions (Albarède et al. 1980). Michard-Vitrac et al. (1980) propose a crustal origin of the various granitoids for the Pyrenean Maladetta massif by the melting at different depths of a vertically zoned crust. Compared to the S-type-I-type granitoid classification of White and Chappel (1983), the observed geochemical characteristics and isotopic signatures would infer for the calc-alkaline and subalkaline granitoids a more probable mixed I- and S-type origin rather than a typical I-type nature. Thus they would be more comparable to the granitoids of mixed S-type – I-type origin described by Kerrich and Beckinsale (1985) in Southeast Asia.

Geophysical data provide evidence for a crust beneath the Massif Central composed of a 20-km-thick migmatitic upper crust (including granites, anatectic granites and anatexites underlain by HP granulites), underlain by a lower crust composed W of the Sillon Houiller of granulitic series of metamorphic origin (upper layer) and of charnockites (lower layer) (Bayer and Hirn 1987). The crust is thinner east of the Sillon Houiller (Auvergne) where the lower crust is composed of either HP metasedimentary rocks or granulite-charnockite mélange. This crust model ascribes to the aluminopotassic granitoids a typical upper crustal origin, while the calc-alkaline and subalkaline granitoids may result from the melting of various parts of the lower crust, possibly including granitoids resulting from the contamination or hybridization between upper crustal and lower crustal materials. The more important development of the subalkaline granitoids in the NE Massif Central can be related to the thinner upper crust which allows the uprising of more deeply generated magmas.

6.5.3 Geodynamic Constraints

The collisional model currently proposed for the geodynamic evolution of the Massif Central (Matte 1986) still suffers some limitations because of the uncertainties concerning the structural relations among the lithotectonic units and their more or less allochthonous origin. In addition, the range of nappe displacement and of the crustal shortening which occurred during the post-collisional thrusting tectonics remains unknown. These uncertainties also hinder the significance and the geodynamic interpretation of magma zonation presently observed across the Massif Central as well as across all the French Variscan.

The significance of the magma varieties and of the pattern of granitoid distribution can be constrained by some factors. (1) Three types of magma characterize the post-collisional magmatic activity of the Massif Central (Stussi and La Roche 1984). In the SW-Limousin and in the NE region, aluminopotassic and either calc-alkaline or subalkaline granitoids are observed in the same area or plutonic body, while the regional pattern of granitoids and related mineralization is not consistent with Andean-type subduction-related magmatic activity and polarity (cf. Mitchell and Garson 1981; Pitcher 1983). (2) The notable development of mainly crustal-derived calc-alkaline and subalkaline granitoids with only minor and generally diachronous intermediate to mafic rocks is emphasized, as already shown by Pitcher (1983). Their diachronous to coeval occurrence with crustal-derived aluminopotassic granitoids can be compared only with the magma pattern observed along subduction-related S-type line and elongated back-arc magmatic zones and thrust belts (Mitchell and Garson 1981; Kontak et al. 1986). In these regions, the narrow and elongated distribution of plutonic rocks differs highly from the widespreading granitoid distribution observed in the Massif Central. In addition, within these Cordilleran environments the frequent lower peraluminous compositions of the two-mica leucogranites compared to the Variscan two-mica leucogranites must be noted (Asmerom et al. 1988). (3) The crustal isotopic signatures of the granitoid varieties involved extensive ensialic magmatic activity in the Massif Central which developed over more than 300 km across this orogeny. This magmatism implies continuous heat production between 270 to 370 Ma, whose origin has been related to post-collisional crustal thickening (up to 60 km), to radioactive decay, and possibly to frictional heating (Strong and Hanmer 1981), although the latter kind of heat production would be of limited importance (Brun and Cobbold 1980). The influence of a hot mantle (Zwart and Dornsiepen 1978) is shown by the localized mafic to intermediate intrusions. (4) The magmatic activity of the French Variscan shows that calc-alkaline and subalkaline granitoids with major and trace element composition comparable to that of Cordilleran granitoids, but with distinct isotopic signature, have formed in a tectonic environment which may be different from subduction zones. These remarks show that a too restricted interpretation of granitoid magma nature can be misleading in the sense of strict regional geotectonic settings. (5) The magmatic types and distribution pattern observed in the Massif Central is enhanced by the quite similar pattern observed all along the French Variscan (Stussi and La Roche 1984; Stussi 1989). The Central Region of the Massif Central with its highly prominent aluminopotassic magmatism can be extended towards the S Brittany leucogranite belt (Lerouge 1988; Lerouge and Quenardel 1988), while the NE Region can be related, although the structural relations among these regions are not yet well known, with the NW part of the Massif Armoricain and with the Vosges (cf. Fig. 18, inset). It can also be compared with the pattern of granitoids observed in the Variscan External crystalline massifs of (1) the Alps and (2) the Pyrenees-Montagne Noire region. This last region is the only one along which calc-alkaline magmatism is more developed than the other magmatic types.

6.6 Conclusion

The igneous activity of the Massif Central develops almost continuously from the Cambrian to the Permian. The Lower Paleozoic magmatism is related to post-Cadomian extensional tectonics and to the Ordovician to Silurian oceanic subduction stage. This stage generates mainly calc-alkaline to tholeïtic rocks which have been orthogneissified during Variscan times.

The relatively homogeneous nature of the Pre-Variscan igneous activity contrasts with the Variscan magma varieties generated between 270 and 370 Ma in response to post-collisional crustal thickening and intra-crustal thrusting and shearing tectonics. The generally interspersed emplacement of the granitoid types and the absence of an evident magmatic polarity emphasizes the ensialic origin of the magmatic activity. Some problems with respect to magma generation are still debated, i. e., the widespread transversal development of the magma activity across the Massif Central and the correlative source of the heat which must have been involved for crustal anatexis more than 80 Ma after the closure of the presumed southern ocean.

The granitoid magma pattern observed in the Massif Central, as well as along the other French Variscan segments, shows that at present it cannot be easily interpreted in terms of terranes: a similar magmatic pattern characterizes the observed lithotectonic units among which structural relations are not yet definitely known. Although a comparative petrographical and geochemical study would be necessary in order to establish the similarities and differences of the magmatic features, the observed pattern of granitoid varieties can be partly compared with that described in other chains of the Circum-Atlantic Paleozoic orogen (Strong 1982; Pitcher 1983; Don Hermes and Murray 1988). Thus, the varying granitoid types observed on both sides of the Atlantic Ocean seems to characterize specifically the Circum-Atlantic Paleozoic orogen, emphasizing its distinct geodynamic and magmatic evolution with regard to the Circum-Pacific orogenies.

References

Albarède F, Dupuis C, Taylor HR (Jr) (1980) $^{18}O/^{16}O$ evidence for non-cogenetic magmas associated in a 300-Ma-old concentric pluton at Ploumanac'h (Brittany, France). J Geol Soc Lond 137: 641–647

Alsac C, Bertaux J, Rubiello MF (1979) Examen de quelques phénomènes volcanologiques dans les paléo-ignimbrites du Viséen supérieur de la région de Roanne (Loire). Bull Bur Rech Géol Min I, 2: 73–87

Asmerom Y, Ikramuddin M, Kinart K (1988) Geochemistry of Late Cretaceous granitoids from northeastern Washington: implication for genesis of two-mica Cordilleran granites. Geology 16: 431–435

Autran A, Lameyre J (1980) Les granitoïdes de France. In: Autran A, Dercourt J (eds) Evolutions géologiques de la France. Mém Bur Rech Géol Min 107 and 26th I. G. C. Paris, 1980, Colloquium C7, pp 87–91

Bayer R, Hirn A (1987) Données géophysiques sur la structure profonde de la croûte hercynienne dans l'arc-ibéro-armoricain et le Massif Central français. Bull Soc Géol Fr (8) III 3: 561–574

Bernard-Griffiths J, Gebauer D, Grünenfelder M, Piboule M (1985) The tonalite belt of Limousin (French Massif Central): U-Pb zircon age and geotectonic implications. Bull Soc Géol Fr (8) I 4: 523–529

Beziat P, Prouhet JP, Tollon F (1980) Le district de Montredon-Labessonié (Tarn). W, Sn, F. 26th I. G. C. Paris, 1980, Gisements français, E7, 44 pp

Bonin B (1982) Les granites des complexes annulaires. Bur Rech Géol Min, Orléans, Manuels et méthodes, 4, 184 pp

Bourguignon A (1988) Origine des formations paradérivées et orthodérivées acides du Limousin central. Une source possible pour les leucogranites uranifères. Thèse, Univ. Lyon-I, 208 pp

Bouvier P (1985) Les gabbro-diorites du domaine de Boussac. Pétrographie, minéralogie, géochimie. Relations avec le magmatisme acide du domaine de la Marche-Guéret. Thèse Spécialité 3° Cycle, Univ. Paris-Sud, Orsay, 262 pp

Bril H, Ramboz C (1982) Les concentrations stanno-wolframifères du district de Brioude-Massiac (Cantal) et du Sud du Massif Central: analyse comparée de la minéralogie et des phases fluides associées. C R Acad Sci Paris 294: 387–390

Brun JP, Cobbold PR (1980) Strain heating and thermal softening in continental shear zones: a review. J Struct Geol 2, 1–2: 149–158

Burnol L, Péronne Y, Vaucorbeil H de (1980) La coupole cachée de leucogranites de Neuf-Jours (Corrèze) et les minéralisations en tungstène associées. Chron Rech Min 455: 93–116

Carrat HG (1987) Les origines magmatiques du massif orogénique de Luzy (Morvan) dans son cadre structural. Bull Soc Hist Nat Autun 122: 7–34

Couturié JP (1977) Le massif granitique de la Margeride (Massif Central français). Ann Sci Univ. Clermont, Géol Minéral 62, 319 pp

Cuney M (1978) Geologic environment, mineralogy and fluid inclusions of the Bois Noirs-Limouzat uranium veins, Forez France. Econ Geol 73: 1567–1610

Cuney M, Autran A (1987) Le forage scientifique d'Echassières (Allier). Une clé pour la compréhension des mécanismes magmatiques et hydrothermaux associés aux granites à métaux rares. Bur Rech Géol Min, Orléans, Géologie de la France, 2–3, 350 pp

Debon F, Le Fort P (1988) A cationic classification of common plutonic rocks and their magmatic associations: principles, method, application. Bull Minéral Fr 111: 493–510

Didier J (1973) Granites and their enclaves. Elsevier, Amsterdam, 393 pp

Didier J, Lameyre J (1969) Les granites du Massif Central français. Etude comparée des leucogranites et granodiorites. Contrib Mineral Petrol 24: 219–238

Didier J, Peyrel JY (1980) Structure laccolitique et âge carbonifère terminal du granite du Mayet-de-Montagne (Massif Central français). C R Acad Sci Paris 291: 797–800

Didier J, Duthou JL, Lameyre J (1982) Mantle and crust granites: genetical classification of orogenic granites and the nature of their enclaves. J Volc Geoth Res 14: 125–132

Don Hermes O, Murray DP (1988) Middle Devonian to Permian plutonism and volcanism in the N American Appalachians. In: Harris AL, Fettes DJ (eds) The Caledonian-Appalachian Orogen. Geol Soc Spec Pap 38, London, pp 559–573

Duthou JL, Floc'h JP (1989) Evolution tectonométamorphique du Massif Central. Réunion extraordinaire de la Société Géologique de France. Bull Soc Géol Fr (8) V 4: 667–692

Duthou JL, Cantagrel JM, Didier J, Vialette Y (1984) Paleozoic granites from the French Massif Central age and origin studied by the 87Rb-87Sr system. Phys Earth Planet Inter 35: 131–144

Dutreuil JP (1978) Les granites de l'Ouest du Limousin, leur pétrologie, leur altération, leurs sols. Thèse Etat, Univ. Limoges, 519 pp (unpublished)

Floc'h JP (1979) Le métamorphisme et la mise en place des granites du dôme de Saint-Mathieu (Limousin occidental). Bull Bur Rech Géol Min I 2: 89–107

Friedrich M, Cuney M, Poty B (1987) Uranium geochemistry in peraluminous leucogranites. Uranium 3: 353–385

Friedrich M, Marignac C, Floc'h JP (1988) Sur l'existence de trois chevauchements ductiles "himalayens" successifs à vergence NW en Limousin. C R Acad Sci Paris 306: 663–669

Gagny C, Leistel JM, Sider H (1983) Le skarn à scheelite de Cherves: un exemple d'indice contrôlé par la tectonique tangentielle sudète dans le Beaujolais (NE du Massif Central français). Chron Rech Min 471: 23–30

Gauthier JC (1974) Evolution granitique, développement des granites à deux micas et géochimie des alcalins dans la Marche Orientale (Massif Central français). 2° partie: géochimie des alcalins rares. Sci Terre Nancy, XIX 2: 119–151

Icart JC, Lacharpagne JC, Sider H, Walgenwitz F (1980) Altération et minéralisation de type "porphyry" à Sibert (Rhône). Chron Rech Min Paris 455: 7–35

Jover O, Bouchez JL (1986) Mise en place syntectonique des granitoïdes de l'Ouest du Massif Central français. C R Acad Sci Paris 303, 10: 969–974

Kerrich R, Beckinsale RD (1985) $^{18}O/^{16}O$-$^{87}Sr/^{86}Sr$ evidence for the origin of granitoids in the tin belt of Southeast Asia. In: Taylor RP, Strong DF (eds) Granite-related mineral deposits. Geology, petrogenesis and tectonic setting. CIM Conference on granite-related mineral deposits, Halifax, Canada, pp 156–161

Kontak DJ, Clark AH, Farrar E, Pearce T, Strong DF, Baadsgaard H (1986) Petrogenesis of a Neogene shoshonitic suite, Cerro Moromorom, Puno, Southeastern Pern. Can Miner 24: 117–135

Lameyre J (1966) Leucogranites et muscovitisation dans le Massif Central français. Ann Fac Sciences Univ. Clermont-Ferrand, Géol Mineral, 29, fasc. 12, 264 pp

Lameyre J (1982) Contribution à la géologie du Limousin: arguments pour des fenêtres ouvertes dans un grand charriage par des diapirs leucogranitiques. C R Acad Sci Paris 294: 1237–1240

Lameyre J, Durandau A, Laurent O, Sagon JP, Vauchelle L, Duzelier D, Juteau J, Leriche A, Vuillemenot P, Pocachard J (1988) Démonstration, par sondage, de la présence de gneiss d'Aubusson sous les granites du batholite de Guéret (Massif Central français) et de la nature tectonique du contact. C R Acad Sci Paris 307: 2077–2083

La Roche H de (1980) Granites' chemistry through multicationic diagrams. Sci Terre Nancy, Série Informatique Géol 13: 67–88

La Roche H de, Stussi JM, Chauris L (1980) Les granites à deux micas hercyniens français. Essai de cartographie et de corrélations géochimiques appuyées sur une banque de données. Implications pétrologiques et métallogéniques. Sci Terre Nancy XXIX 1: 5–121 and 26th I. G. C. Paris, Proc G25, Pétrographie, minéralogie, banques de données pétrographiques, Guide book excursion 150C, pp 5–121

Leistel JM, Gagny C (1984) Mise en évidence d'une fosse volcano-tectonique au Viséen supérieur dans le Nord-Est du Massif Central français. Rev Géol Dyn Géogr Phys 26,1: 19–31

Lemaire D, Debeglia N, Quenardel JM (1986) Apport de la gravimétrie dans l'élaboration de modèles de mise en place des ensembles magmatiques du Nord du Massif Central français. I. Le massif granitique de Crevant. C R Acad Sci Paris 303: 799–804

Lemaire D, Quenardel JM, Debeglia N (1987) Apport de la gravimétrie dans l'élaboration de modèles de mise en place des ensembles magmatiques du Nord du Massif Central français II: Les massifs granitiques de la Marche Orientale. C R Acad Sci Paris 305: 1073–1079

Lerouge G (1988) Tectogenèse comparée de deux segments de la chaîne hercynienne: le Massif Central français septentrional et Sud du Massif Armoricain. Thèse Etat, Université Paris Sud, Orsay

Lerouge G, Quenardel JM (1988) Les zones de cisaillement carbonifères dans les plutons vendéens et leurs prolongations dans le Nord-Ouest du Massif Central français. Bull Soc Géol Fr (8) IV 5: 831–838

Lespinasse M (1984) Contexte structural des gisements d'uranium de la Marche occidentale. Fracturation, circulations fluides, propagation de l'épisyénitisation. Géol. et Géochimie Uranium, Nancy, Mém. 8, 196 pp

Lespinasse M, Mollier B, Delair J, Bladier Y (1986) Structuration tangentielle et chevauchements carbonifères dans les leucogranites du NW du Massif Central français: l'exemple des failles de Bussières-Madeleine et d'Arrènes-Ouzilly. C R Acad Sci Paris 303: 1575–1580

Leyreloup A (1974) Les inclusions catazonales remontées par les éruptions néogènes de France: nature de la croûte inférieure. Contrib Mineral Petrol 46: 17–27

Matte P (1986) Tectonics and plate tectonics model for the Variscan belt of Europe. Tectonophysics 126: 329–374

Mezure JF (1981) Orientation préférentielle de mégacristaux de feldspath potassique dans les granites: application à l'étude structurale des granites d'Egletons et de Meymac (Corrèze, Massif Central français). Bull Soc Géol Fr (7) XXIII 6: 641–650

Michard-Vitrac A, Albarède F, Dupuis C, Taylor HR (Jr) (1980) The genesis of Variscan (Hercynian) plutonic rocks: inferences from Sr, Pb and O studies on the Maladetta igneous complex, Central Pyrénées (Spain). Contrib Mineral Petrol 72: 57–72

Mitchell AHG, Garson MS (1981) Mineral deposits and global tectonic settings. Academic Press, Geol Ser London 405 pp

Mollier B, Lespinasse M (1985) Déformation magmatique et plastique en limite Nord du granite de Saint-Sylvestre (Nord-Ouest du Massif Central français): la faille d'Arrènes-Ouzilly. C R Acad Sci Paris 300: 681–686

Monier G (1980) Pétrologie des granitoïdes du Sud Millevaches (Massif Central français). Thèse Spécialité 3° Cycle, Univ. Clermont II, 333 pp

Monnier M, Stussi JM (1984) Le bassin de Montreuillon (Nord Morvan): élément d'une association volcano-plutonique subalcaline hercynienne. Données géochimiques. C R Acad Sci Paris 298: 601–604

Montel JM, Weisbrod A (1986) Characteristics and evolution of "vaugneritic magmas": an analytical and experimental approach, on the example of the Cévennes Médianes (French Massif Central). Bull Minéral Fr 109: 575–587

Montel JM, Weber C, Pichavant M (1986a) Biotite-sillimanite-spinel assemblages in high grade metamorphic rocks: occurrences, chemographic analysis and thermobarometric interest. Bull Minéral Fr 109: 555–573

Montel JM, Weber C, Barbey P, Pichavant M (1986b) Thermobarométrie du domaine anatectique du Velay (Massif Central, France) et conditions de genèse des granites tardimigmatitiques. C R Acad Sci Paris 302: 647–652

Nachit H, Razafimahefa N, Stussi JM, Carron JP (1985) Composition chimique des biotites et typologie magmatique des granitoïdes. C R Acad Sci Paris 301: 813–818

Orsini JB (1979) Existence de trois associations magmatiques dans les granitoïdes post-viséen moyen (groupe des granodiorites-monzogranites) de la chaîne varisque française. C R Acad Sci Paris 289: 981–984

Peiffer MT (1986) La signification de la ligne tonalitique du Limousin. Son implication dans la structuration varisque du Massif Central français. C R Acad Sci Paris 303: 305–310

Piboule M, Beurrier M, Briand B (1982) Géochimie de quelques granites albitiques dévoniens de l'Est du Massif Central français. Neues Jahrb Mineral Abh 143, 3: 279–308

Pichavant M, Boher M, Stenger JF, Aissa M, Charoy B (1987) Relations de phase des granites de Beauvoir à 1 et 3 kbar en conditions de saturation en H_2O. Géologie de la France, Bur Rech Géol Min 2–3: 77–86

Pitcher WS (1983) Granite type and tectonic environment. In: Hsu K (ed) Mountain building processes. Academic Press, New York, pp 19–40

Ploquin A, Stussi JM, Bourguignon A, Cuney M (1988) French Paleozoic granitoid mapping: problems on orthogneisses typology. Int Conf IGCP Proj 233, Terranes in the Circum-Atlantic Paleozoic orogens, Montpellier, Univ. Montpellier (Abstr)

Pupin JP (1985) Magmatic zoning of the Hercynian granitoids in France based on zircon typology. Schweiz Mineral Petro Mitt 65: 29–65

Razafimahefa N (1987) Granites et leucogranites alumineux du Nord-Est du Millevaches (Massif Central français): pétrologie, géochimie et synthèse cartographique. Thèse, Univ. Nancy, 164 pp (unpublished)

Rolin P (1987) Tectogenèse comparée de deux segments de la chaîne hercynienne: le Massif Central français septentrional et la Meseta marocaine septentrionale. Thesis Dr.-ès-Sci. Nat., Université Paris-Sud, Orsay, 350 pp (unpublished)

Sabourdy G (1975) Apport de la géochimie à la connaissance de la pétrogenèse des granitoïdes des Cévennes méridionales, Massif Central français. Ann Sci Univ. Clermont-Ferrand, 52, 278 pp

Santallier D, Briand B, Ménot RP, Piboule M (1988) Les complexes leptyno-amphiboliques (C. L. A.): revue critique et suggestions pour un meilleur emploi de ce terme. Bull Soc Géol Fr (8) IV 1: 3–12

Sheppard SMF (1986) Isotopic case studies of magmatism in Africa, Eurasia and oceanic islands. In: Valley JW, Taylor HP (Jr), O'Neil JR (eds) Review in mineralogy, 16. Stable isotopes in high temperature geological processes. Min Soc America, pp 319–371

Sossa M, La Roche H de, Stussi JM, Grolier J (1984) Le massif granitique de Montmarault (Massif Central français) dans son domaine centre-oriental. Fondements pétrographiques et géochimiques d'une hypothèse de bi-magmatisme avec contamination. Géol. Profonde de la France, thème 2, Socle et anomalie magnétique du Bassin de Paris. Bur Rech Géol Min Doc 81–2, Bur Rech Géol Min, ed, Orléans

Streckeisen A (1976) To each rock its proper name. Earth Sci Rev 12: 1–33

Strong DF (1982) Granitoid rocks and associated mineral deposits of Eastern Canada and Western Europe. In: Strangway DW (ed) The continental crust and its mineral deposits. Geol Assoc Canada, Spec. Paper 20, pp 741–769

Strong DF, Hanmer SK (1981) The leucogranites of Southern Brittany: origin by faulting, frictional heating, fluid flux and fractional melting. Can Miner 19: 163–176

Stussi JM (1989) Granitoid chemistry and associated mineralization in the French Variscan. Econ Geol 84: 1363–1381

Stussi JM, La Roche H de (1984) Le magmatisme orogénique granitique de la chaîne varisque française. Typologie chimique et répartition spatiale. C R Acad Sci Paris 298: 43–48

Turpin G (1984) Altérations hydrothermales et caractérisation isotopique (O-H-C) des minéraux et fluides dans le massif uranifère de Saint-Sylvestre. Extension à d'autres gisements intragranitiques d'uranium français. Géol Géochim Uranium, Mém. Nancy, 6, 290 pp

Turpin G, Friedrich M, Cuney M, Bouchez JL, Aubertin M

(1988) Metaigneous origin of Hercynian peraluminous granites in the French Massif Central: implications for crustal history reconstructions. Contrib Mineral Petrol 104: 163–172
Vidal P (1980) L'évolution polyorogénique du Massif Armoricain: apport de la géochronologie et de la géochimie isotopique du strontium. Soc Géol Minéral Bretagne, Rennes, Mém. 21, 162 pp
White AJR, Chappell BW (1983) Granitoid types and their distribution in the Lachlan fold belt, Southeastern Australia. In: Roddick JC (ed) Circum-Pacific plutonic terranes. Geol Soc Am Mem. 159, pp 21–34
Zwart HJ, Dornsiepen UF (1978) The tectonic framework of Central and Western Europe. Geol Mijnbouw 57: 627–654

Metallogenesis of the French Massif Central: Time-Space Relationships Between Ore Deposition and Tectono-Magmatic Events

H. Bril, Ch. Marignac, M. Cathelineau, F. Tollon, M. Cuney, and M. C. Boiron

1 Introduction

The French Massif Central (FMC) is an old mining area. As early as in the Gallo-Roman times, it was mined for Au, Ag, Pb, and Sn; but the great mining epoch was in the 18th–19th centuries, with Sb as the main product. After a decline at the onset of the 20th century, exploration and mining were developed again after the Second World War, mainly for U, but also for Ba, Zn, W, and F. The FMC is nowadays one of the major uranium-producing districts in the world and the main gold producer of Western Europe. It is also mined for F (containing, in the Morvan, 6% of the world reserves), Ba, Zn, Ge, kaolin, and coal. Moreover, it contains important subeconomic deposits of W, Sn, Li, and Ta, as well as minor deposits of Sn, Be, Sb, Pb, Zn, Cu and Mo. The main deposits are located in Fig. 1 and listed in Table 1.

This chapter will review the present state of our knowledge on the relationships between ore deposition and the tectono-magmatic evolution of the FMC. Ore deposits will be presented within a broad frame of tectonomagmatic history, in order to take account of the fact that ore genesis is in many cases a multistage process, strongly dependent on geodynamic events (Cathelineau et al. 1989b). For this purpose, we will follow the distinction between pre-collision, collision, hyper-collision, and post-collision stages (see Santallier et al., Chap. III. 4 this Vol.).

The tectonostratigraphic pile as evidenced by Ledru et al. (1989) will be used systematically (see Fig. 1); the relevance of the terrane concept to the understanding of metallogenesis in the FMC will be discussed at the end of the paper.

2 Metallogenetic Development of the French Massif Central (FMC)

2.1 Pre-Collision Metallogenesis

All known deposits of the pre-collision stage are strata-bound, mostly of the sedimentary exhalative ("sedex") type.

Despite the importance of pre-Variscan acid and/or basic magmatism (Ploquin and Stussi, Chap. III. 6 this volume) true volcanic-sedimentary deposits are very scarce, apart from very small pyrite bodies in the "Génis porphyroides" (ignimbrite formation of supposed Ordovician age, in the Upper Allochthon of Limousin) (Bouladon 1986). The small exhalative volcanic-sedimentary sulfide deposits from the Beaujolais and Morvan were formed in distensive environments at the collision stage (see below).

It must be emphasized that neither Cyprus-type massive sulfide deposits are presently known in the whole FMC, nor are pre-Variscan arc-related ones, such as porphyry deposits.

On the contrary, exhalative sedimentary (sedex) occurrences are not rare; but, except for the Pb-Zn La Loubatière ore body, Montagne Noire (Oksengorn 1958), and possibly the Salsigne gold deposit (see below) they are uneconomic. However, they may have been significant as protore for the main ore deposition (Les Malines, Saint-Salvy, see below). They are mostly found within the relative autochthon or the southern Paleozoic units.

1. Zn-Cu-(Pb-Sn). Very small pyrite-rich bodies found within micaschists in Limousin (Sussac) and Marche (Montmarçon), are interpreted as metamorphosed exhalative sedimentary deposits (Bouladon 1986), although, at Sussac, there is also a quartz-cassiterite stockwork apparently related to a nearby small leucogranite cupola.

2. Pb-Zn. Stratabound lenses and beds within Lower Cambrian rocks (Aubague et al. 1977; Courjault-Ra-

J. D. Keppie (Ed.): Pre-Mesozoic Geology in France and Related Areas

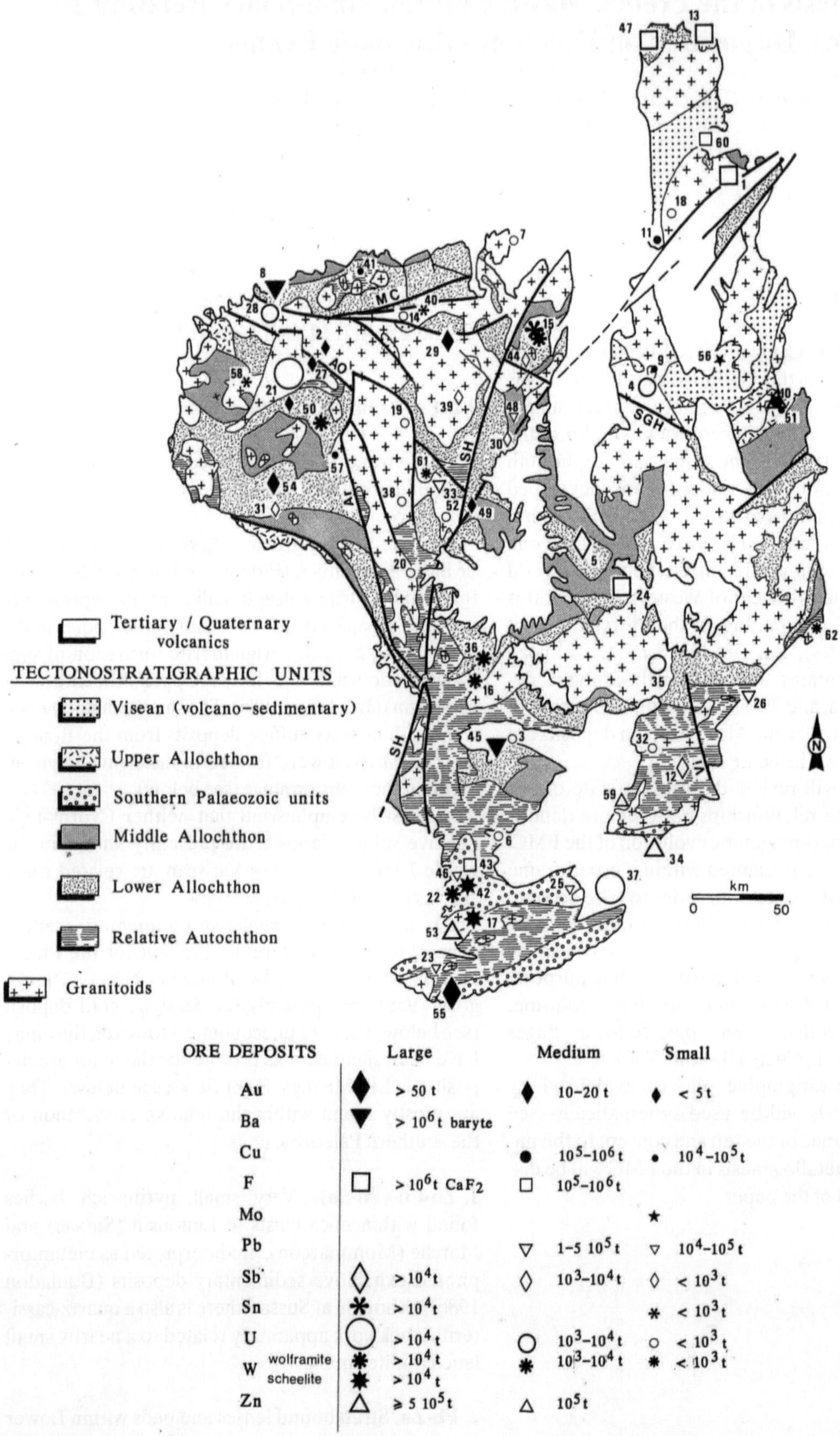
Tertiary / Quaternary volcanics
TECTONOSTRATIGRAPHIC UNITS
Visean (volcano-sedimentary)
Upper Allochthon
Southern Palaeozoic units
Middle Allochthon
Lower Allochthon
Relative Autochthon
Granitoids
km
0
50
N
ORE DEPOSITS
Large
Medium
Small
Au
> 50 t
10–20 t
< 5 t
Ba
> 10⁶ t baryte
Cu
10⁵–10⁶ t
10⁴–10⁵ t
F
> 10⁶ t CaF₂
10⁵–10⁶ t
Mo
Pb
1-5 10⁵ t
10⁴–10⁵ t
Sb
> 10⁴ t
10³–10⁴ t
< 10³ t
Sn
> 10⁴ t
< 10³ t
U
> 10⁴ t
10³–10⁴ t
< 10³ t
W wolframite
> 10⁴ t
10³–10⁴ t
< 10³ t
scheelite
> 10⁴ t
Zn
≥ 5 10⁵ t
10⁵ t

dé 1989) are demonstrably earlier than the Variscan tectonometamorphic events, on a structural basis (Les Malines, Lacerda and Bernard 1984; Saint-Salvy, Barbanson and Tollon 1980) as well as on lead isotope grounds (Brevart et al. 1982).

At *Saint-Salvy* (Montagne Noire), sandy levels enriched in sphalerite, siderite and pyrite are contained in an horizon of Lower Cambrian black shales. These levels are mineralized on several km^2. Their Zn content is consistently high (1000–2000 ppm), locally up to 7 wt % Zn; Ge content is also high (25 to 80 ppm); Ge is only in part contained in sphalerite (Barbanson and Geldron 1983).

This occurrence is the more prominent of a set of similar very small sedex (?) deposits in the Lower Cambrian Série noire on the northern side of the Montagne Noire (Lescuyer and Giot 1988).

At *Les Malines,* lenses of pyrite ± sphalerite + quartz (10 to 100 m × 1 to 7 m) are intercalated into Lower Cambrian dolomitic rocks; they are interpreted as syngenetic (Lacerda and Bernard 1984; Chantal and Leblanc 1986) on evidences of deformed pyrite; however, the syngenetic origin of sphalerite has recently been doubted (Macquar et al. 1988).

3. Au. At Salsigne (Montagne Noire) gold is contained in strata-bound sulfide deposits and in ore veins (Reynolds 1965); numerous genetic hypotheses have been proposed (see Lépine et al. 1988). There are two series of stratabound sulfide bodies:

◀ **Fig. 1.** Location of main ore deposits of the FMC. Tectonostratigraphic units redrawn from Ledru et al. (1989). Main faults: *AO* Arrènes-Ouzilly; *Ar* Argentat; *MC* Marche-Combrailles; *SGH* Saint-Gervais-L'Hermitage; *SH* Sillon Houiller; *Vi* Villefort.
Numbers (*in bold:* mineralized district. See also Table 1): *Antimony: 5* Brioude-Massiac; *12* Collet-de-Dèze; *30* Le Colombier; *31* Les Biards; *39* Mérinchal; *44* Nades; *49* Pontvieux. *Barite: 8* Chaillac; *45* Pessens. *Copper: 9* Charrier; *10* Chessy; *11* Chizeuil; *41* Montmarçon; *51* Sain-Bel; *56* Sibert; *57* Sussac. *Fluorite: 1* Antully; *13* Courcelles; *24* Langeac; *43* Montroc; *47* Pierre Perthuis; *60* Voltenne; *Gold: 2* Bénévent (Northern Limousin); *27* Laurière (Northern Limousin); *29* Le Châtelet-Villeranges (Marche-Combrailles); *49* Pontvieux; *54* Saint-Yrieix (Southern Limousin); *55* Salsigne. *Lead: 23* La Loubatière; *25* La Rabasse; *26* Largentière; *33* Les Farges; *46* Peyrebrune; *48* Pontgibaud. *Molybdenum: 56* Sibert. *Tin: 15* Echassières; *40* Montebras; *58* Vaulry. *Tungsten: 15* Echassières; *16* Enguialès; *17* Fumade; *22* Lacoste; *36* Leucamp; *42* Montredon; *50* Puy-les-Vignes; *61* Neuf-Jours showing; *62* Vivarais showings. *Uranium: 3* Bertholène; *4* Bois Noirs-Limouzat; *6* Brousse-Broquies; *7* Cérilly; *14* Coussa; *18* Grury; *19* Hyverneresse; *20* La Besse; *21* La Crouzille; *28* Le Bernardan; *32* Les Bondons; *35* Les Pierres Plantées; *37* Lodève; *38* Longeron; *52* Saint-Pierre-du-Cantal. *Zinc: 25* La Rabasse; *34* Les Malines; *53* Saint-Salvy; *59* Villemagne.

a) A set of layers (traditional mine, 90 t Au mined since the beginning of the 20th century; Fig. 2) contained in the Lower Cambrian formations of the Salsigne-Fournes nappe (mainly sandy-carbonate layers, associated with volcanogenic sediments of alkaline basalt affinity, indicative of a distensive environment) (Lépine et al. 1988 b). From bottom to top are successively encountered: the rich sandstones (alternation of millimetric arsenopyrite and quartz-sericite laminae) (200–1000 ppm Au); the arsenopyrite + pyrite ± chalcopyrite layers (8–75 ppm Au); the pyrite + pyrrhotite layers (0–25 ppm Au). As all sulfides have suffered plastic deformation in the regional schistosity, these bodies are interpreted as sedex deposit, and gold deposition is considered syngenetic as well (Lépine et al. 1988 b; Lépine 1989). Lead isotope studies (Javoy et al. 1983) indicate a Lower Paleozoic age for the lead in these strata-bound sulfides.
b) At the top of an undated Schistes X unit, overlain by the Salsigne Fournes nappe (Fig. 2), is found a volcano-sedimentary formation with rocks ranging from turbidites to spilite and quartz-keratophyre lavas. Within this formation, massive sulfide bodies with arsenopyrite-pyrite and pyrrhotite are spatially associated with Ba-rich green siltstone. They contain an average gold content of 5 to 10 ppm over 1 km^2. They have been interpreted as sedex deposits (Bonnemaison et al. 1986), but another interpretation has recently been proposed (Lépine 1989): it would be an epigenetic mineralization of unknown age, affected by the Variscan folding events.

Gold deposition is supposed to have taken place concurrently with the formation of the sulfide deposits (Bonnemaison et al. 1986; Lépine et al. 1988 b; Lépine 1989). However, taking into account the importance of late Variscan (re)mobilization events for gold (vein deposits, see below), the reality of an early gold appearance at Salsigne is debatable, and further work is needed.

Other types of sedimentary ores are rare, contrary to the Armorican Massif (see Chauris and Marcoux, Chap. II. 2 this Vol.).

An arsenopyrite-scheelite-bearing metaquartzite horizon with anomalous Sn content (up to 1500 ppm Sn, as detrital cassiterite, and 2500 ppm W) was described in low grade schists from the Eastern Vivarais (relative autochthon ?) and interpreted as a Sn-W paleoplacer (Samama 1971). There are Sn and W showings around the Saint-Cierge granite in the same area (Ramboz 1980; Sandras et al. 1988); the W-Sn-bearing metaquartzite is a possible source rock for these showings.

Table 1. Main characteristics of the major ore deposits in the French Massif Central (see Fig. 1)

Stage	Type of deposit	Examples	Metals	Age	Size	Comments
Pre-collision	1. sedimentary exhalative	Sussac (57)	Zn-Cu-(Pb-Sn)	Unknown	Uneconomic showing	
		Montmarçon (41)	Pb-Zn-(Cu)	Unknown	Uneconomic showing	
		Les Malines (34)	Zn	Early Cambrian	Uneconomic	Possible protore
		Saint-Salvy (53)	Zn-Ge	Early Cambrian	1000–2000 ppm Zn/n km^2	Protore
		Salsigne (55)	Au?-As	Early Cambrian	90 t Au mined	Early Au not fully demonstrated
		La Loubatière (23)	Pb-(Zn)	Early Cambrian	100000 t mined	Closed
	2. sedimentary	Vivarais (62)	W-Sn	Unknown	showing (up to 2500 ppm W)	Metamorphosed paleoplacer
Collision	1. volcanic-sedimentary	Charrier (9)	Cu-Sn	Late Devonian (?)	760 t Sn mined	Acid volcanism
		Chessy (10)	Zn-Cu	Late Devonian (?)	500000 t Zn, 100000 t Cu	idem
		Chizeuil (11)	Cu	Famennian	5 Mt pyrite mined	idem
		Sain-Bel (51)	Cu	Late Devonian (?)	60000 t Cu mined	idem
Hyper-collision	1. metamorphic	Montredon (42)	W (scheelite)	Carboniferous	12–15000 t WO3	Two deposits: Hom-Haut, Auriole
	2. granite related					
	a) quartz-wolframite stockworks	Leucamp (36)	W	Carboniferous	8000 t WO3	Closed; AK granite related
		Enguialès (16)	W	Carboniferous	1500 t WO3 mined	idem
		Echassières (15)	W	Carboniferous	10000 t WO3	Closed; indirect evidence for the AK relation
		Lacoste (22)	W	Carboniferous	10000 t WO3	Assumed AK relation
	b) rare metals Li-F granites	Echassières (15)	Sn-Ta-Li	310 Ma (Rb-Sr)	150000 t Sn	Low-grade (0.10 wt % Sn) prospect
		Montebras (40)	Sn-(Be)	Carboniferous	300 t Sn mined	
	c) skarns	Fumade (17)	W (scheelite)	280 Ma (Rb-Sr)	12000 t WO3	
	d) vein-type	Saint-Salvy (53)	Zn-Ge	Unknown	450000 t Zn, 480 t Ge	
	e) substitution	La Rabasse (25)	Pb-Zn	Unknown	46000 t Pb and 46000 t Zn mined	
	f) breccia pipes	Puy-les-Vignes (50)	W (wolframite)	Unknown	5400 t WO3 mined	Closed; concealed granite (unknown)
	g) porphyry-type	Sibert (56)	Cu-Mo	Post-Visean	Very low grade prospect	
	3. vein-type					
	a) gold veins	Northern Limousin (2, 27)	Au	Carboniferous	Minor deposits	Bénévent (2), Laurière (27), and Ambazac districts
		Chatelet-Villeranges (29)	Au	315 ± Ma (K-Ar)	11 t mined (26 g/t)	Le Châtelet deposit and the Villeranges basin prospects
		Saint-Yrieix (54)	Au	Carboniferous	≥ 15 t	Cheni (8 t, closed); Cros-Gallet, Lauriéras deposits
		Salsigne (55)	Au-Ag-As-(Bi)	unknown	90 t Au mined (reserves about 20 t)	Late Hercynian Au (re?)mobilization; As: 10 % world production
	b) Sb-As veins	Brioude-Massiac (5)	Sb-(Au)	Carboniferous (?)	39000 t Sb	Closed; Ouche (9250 t), La Bessade (8500 t) veins
		Collet-de-Dèze (12)	Sb-(Au)	Unknown	4600 t Sb	Closed; La Felgerette vein (2500 t); Au showings (Alteyrac veins)
Permian distension	1. vein deposits	Bois Noirs-Limouzat (4)	U	Permian (?)	7000 t U	Closed
		La Besse (20)	U	Permian (?)	600 t U	
		Hyverneresse (19)	U	Permian (?)	500 t U	Mostly in episyenites
		La Crouzille (21)	U	276 Ma (U-Pb)	40000 t U	Veins and episyenites
		Le Bernardan (28)	U	Permian (?)	8000 t U	Episyenites (Le Bernardan, 6000 t); some veins
		Pontgibaud (48)	Pb-(Ag)	Permian (?)	50000 t Pb, 100 t Ag mined	Closed
	2. strata-bound	Cérilly (7)	U	Autunian	3600 t U	
Post-Variscan Atlantic/Tethys opening	1. vein deposits	Les Pierres Plantées (35)	U	188 ± 12 Ma (U-Pb)	4000 t U	Remobilization of early ore
		Grury (18)	U	270/175 Ma (U-Pb)	1500 t U	Remobilization of early ore
		Les Farges (33)	Pb	Liassic	120000 t Pb (5.7 wt % Pb)	
		Peyrebrune (46)	Pb-Zn	Unknown	64000 t Pb–35000 t Zn mined	Closed
		Villemagne-Trèves (59)	Zn-Pb-Ag	Liassic (?)	> 200000 t Zn	Trèves: strata-bound
		Langeac (24)	F	Liassic	1 Mt CaF2	
		Montroc (43)	F	Liassic	1 Mt CaF2	Montroc (0.7 Mt), Le Burc (0.3 Mt) veins
		Voltenne (60)	F	Liassic	0.5 Mt CaF2	Voltenne (closed), Maine veins
	2. unconformity-related	Bertholène (3)	U	173 ± 9 Ma (U-Pb)	1000 t U	
		Brousse-Broquiès (6)	U	194/160 Ma (K-Ar)		
		Les Malines (34)	Zn-Pb-Ag	Triassic to Bathonian	700000 t Zn mined (6 wt % Zn)	Mainly flooded karsts; minor strata-bound lenses
	3. strata-bound	Lodève (37)	U	173/110 Ma (U-Pb)	20000 t U	
		Largentière (26)	Pb-Zn	Triassic	450000 t Pb	Closed
		Antully (1)	F	Liassic (?)	0.9 Mt CaF2	
		Courcelles (13)	F-Ba	Liassic (?)	1 Mt CaF2; 0.3 Mt BaSO4	
		Pierre-Perthuis (47)	F	Liassic (?)	1.8 Mt CaF2	Pierre-Perthuis (1.3 Mt); Pontaubert (0.5 Mt)
		Chaillac (8)	Ba-F	Unknown	3 Mt BaSO4	Le Rossignol vein (0.5 Mt CaF2) at the bottom of Ba ore-body
Tertiary	1. Strata-bound	Pessens (45)	Ba	Liassic	1.5 Mt BaSO4	
		Saint-Pierre-du-Cantal (52)	U	Oligocene (?)	1100 t U	

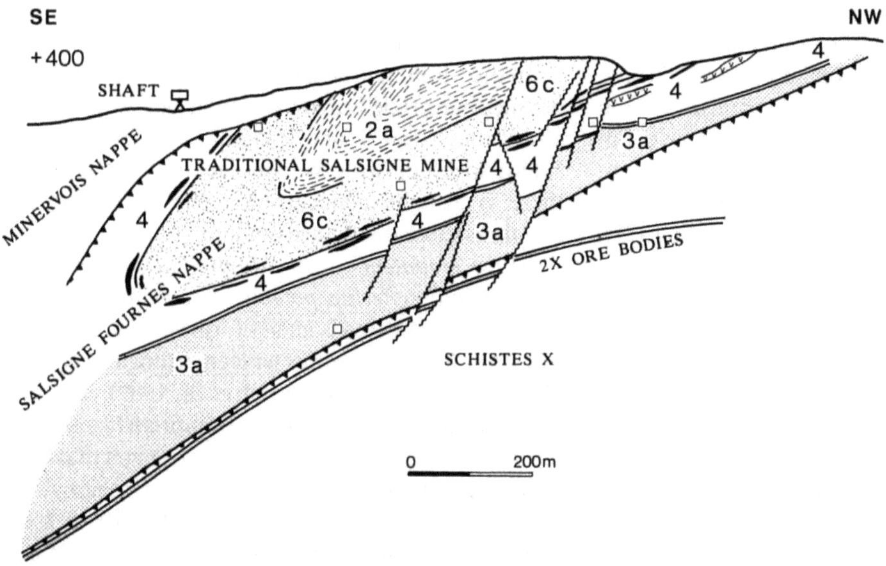

Fig. 2. NW-SE schematic cross-section of Salsigne gold mine. (After Bonnemaison et al. 1986). Early Cambrian: *2a* shale; *6c* sandstone; *4* limestone-dolostone-shale-sandstone and basic volcanic rocks with strata-bound orebodies (schematic black lenses). Devonian: *3a* limestones

2.2 The Lack of Collision-Related Metallogenesis

Although correlated with major generation of alumino-potassic (AK) granite magmas (the Gueret type, at ca. 355 Ma; and the Limousin type at ca. 325 Ma; see Briand et al., Chap. III. 6 this Vol.), the collision events leading to nappe emplacement during the Devonian (and possibly the early Carboniferous; Sider et al. 1986; Friedrich et al. 1988) are not accompanied by any significant ore deposition. This is possibly a consequence of the deep level of intrusion of the collision-related water-rich granites which was not compatible with the development of significant peri-batholithic hydrothermal circulation. It is interesting to notice that in the Himalayas, which are a possible structural model for the FMC at the collision stage of its development, the collision-related leucogranites of the Manaslu type are practically devoid of ore deposits, although they contain high U, Sn, and W values (Cuney et al. 1984b; France-Lanord and Le Fort 1988). In the same way, the leucogranites of the Limousin type, which are demonstrated on geochemical and isotopic grounds to result from the melting of orthogneisses from the lower allochthon (or from the relative autochthon) Bourguignon 1988; Turpin et al. 1989), have a relatively high U and Sn potential (Friedrich 1984; Friedrich et al. 1987) but were not apparently associated with ore deposition at this stage of the tectonic development in the FMC.

During collision, Late Devonian calk-alkaline granitoids were also emplaced (the so-called Limousin tonalite line, Didier and Lameyre 1971); they are sometimes considered as resulting of the activity of a subduction zone (Pfeiffer 1986), for the existence of which there is not, however, other evidence. There are no associated mineralizations.

In some of the upper allochthon units distensive environments are recorded during the Late Devonian-Early Carboniferous, i.e., at the transition between the collision and hyper-collision stages in the FMC. Bi-modal volcanic series (spilite-keratophyres) and ophiolites are typical (Sider and Ohnenstetter 1986). To such environments pertain a series of small massive sulfide deposits, most of supposed Late Devonian age, in the upper allochthon (Beaujolais and Brévenne units, NE FMC) (cf. Fig. 1):

The Cu *Chizeuil* prospect (Lécuyer 1982; Lemière 1983; Lemière et al. 1986) is the most noticeable of a series of small pyrite-(chalcopyrite-sphalerite-galena) deposits interspersed in calc-alkaline acid volcanics of Famennian age, in a shallow intracontinental marine basin setting. Up to 5 Mt pyrite were mined at Chizeuil. A prominent pyrite-barite stockwork, with minor Cu minerals, is characteristic of an exhalative origin. There is an important sericitic-argillic alteration halo in the surrounding lavas. The deposit was metamorphosed in the thermal aureola of the Carboniferous Luzy granite.

The old Cu-Sn *Charrier* deposit (Picot 1965), with 760 t Sn mined, included a series of small lenses in chloritites, with cassiterite and bornite-chalcopyrite, cassiterite being earlier. It was characterized by the presence of an indium-rich sphalerite. The age of the enclosing volcanic-sedimentary sequence is poorly constrained, either Late Devonian or Visean; these rocks suffered the thermal metamorphism of the nearby Bois-Noirs granite (330 Ma, Binon and Pin 1989). The relationships of the deposit with this granite are not known.

In the old Cu-(Zn) mine at *Sain-Bel* (Bardin 1971), a series of pyritic lenses associated with quartz-keratophyre rocks were the host for a minor chalcopyrite and sphalerite mineralization (60000 t Cu mined).

The Cu-Zn *Chessy* deposit was recently reassessed and was proved to contain 5 Mt of an ore with average contents of 2.5 wt.% Cu, 10 wt.% Zn and 15 wt.% Ba. The mineralization occurs in two acid effusive volcanic units (mainly submarine volcanic flows) characterized by dacitic to rhyolitic composition (Vuagnat 1984; Lacomme et al. 1987). Ore bodies are stratiform, and display two parts: a central core of banded sulfides (pyrite-sphalerite) interlayered with the volcanic flows, and a distal barite halo; the main ore body is rooted in quartz-pyrite-white mica stockworks. Similar stockworks occur at several volcanic levels, demonstrating a long-lived hydrothermal activity in the Chessy setting (Lacomme et al. 1987).

These Cu-(Zn) deposits, associated with acid volcanics, are not related to the neighboring ophiolites.

2.3 Hyper-Collision Metallogenesis

Much of the ore deposition in the FMC occurred in relation to the progressive development of hyper-collision.

2.3.1 Tectonometamorphic Events and Ore Deposition

In correlation with the development of low P/high T metamorphic areas in the FMC since ca. 320 Ma (Pin and Peucat 1986; Santallier et al., this III.4 volume), various kinds of W mobilization and concentration took place.

a) *Formation of numerous scheelite showings* in association with calc-silicate gneissic layers (skarnoids) (see, for instance, Fig. 4).

Such occurrences, essentially located in the Montagne Noire and the Cévennes, were at first thought to represent syngenetic W preconcentrations (Routhier et al. 1973; Boyer and Routhier 1974). As demonstrated in the FMC (Gibert et al. 1988) and in other areas (Derré et al. 1982; Cheilletz 1988), they are rather the effect of synmetamorphic hydrothermal mobilization of the tungsten present at a low concentration level in the metamorphic pile and its trapping in Ca-rich layers of the metamorphosed series.

In the Montagne Noire, the skarnoids, formed under medium grade Barrovian conditions, were mineralized under low T and P conditions (Gibert et al. 1988); together with W enrichment (as scheelite), there is also Sn concentration (up to 2000 ppm), mainly as substituted Sn in epidote.

b) *Development of the Montredon subeconomic strata-bound scheelite potential ore bodies* (Fig. 1). The Montredon area is an antiform structure formed by Lower Cambrian series with an orthogneissic core (330 Ma metagranite) (Béziat et al. 1980). The deposits occur in a specific Lower Cambrian black shale setting. For this reason, it was at first thought that they resulted from the remobilization of a regional W synsedimentary enrichment (Béziat et al. 1980). Further studies demonstrated, however, that W content in the black shales is usually lower than 4 ppm, except for local enrichment (W-bearing Ti-oxides) related to strata-bound tourmalinization (Caleffi et al. 1988).

Tourmalinization is a regional process in the Montredon area, generally related to late quartz-wolframite lodes (see below), but also to an earlier low P/high T tectono-metamorphic event. Thus, an exhalative-sedimentary origin for the W-rich tourmalinites may be ruled out.

Scheelite occurs in K-feldspar + fluorite + quartz lenses (low P synmetamorphic pegmatites) developed along the regional schistosity during the low P/high T metamorphic event (Caleffi et al. 1988) and deformed by some folds (Béziat et al. 1986). Scheelite and fluorite, together with sulfides, are also disseminated along the schistosity in the surrounding rocks.

At *Hom-Haut*, small lenses (6 × 30 cm) are found within pyroxene-bearing mica schists (20 m with 0.2 wt.% WO_3 on ca. 1 km^2); this is reminiscent of the skarnoid occurrences. Metamorphic conditions are evaluated at ca. 450° C, 1–2 kbar (Caleffi et al. 1988).

At *Auriole*, the pegmatite lenses (K-feldspar rich) are more prominent; they are interlayered in the metamorphic foliation of black schists along a definite level (10 m at 0.35 wt.% WO_3).

Scheelite-bearing lenses are cross-cut by quartz-(wolframite bearing) veinlets related to a swarm of quartz-wolframite lodes at lacoste (10000 t WO_3 at 0.6 wt.% WO_3; Guion et al. 1985) (Fig. 1); these are probably related to a concealed granite under the Montredon area (Guérangé-Lozes and Alsac 1986), and are possible representative of the AK granite related W deposits (see below).

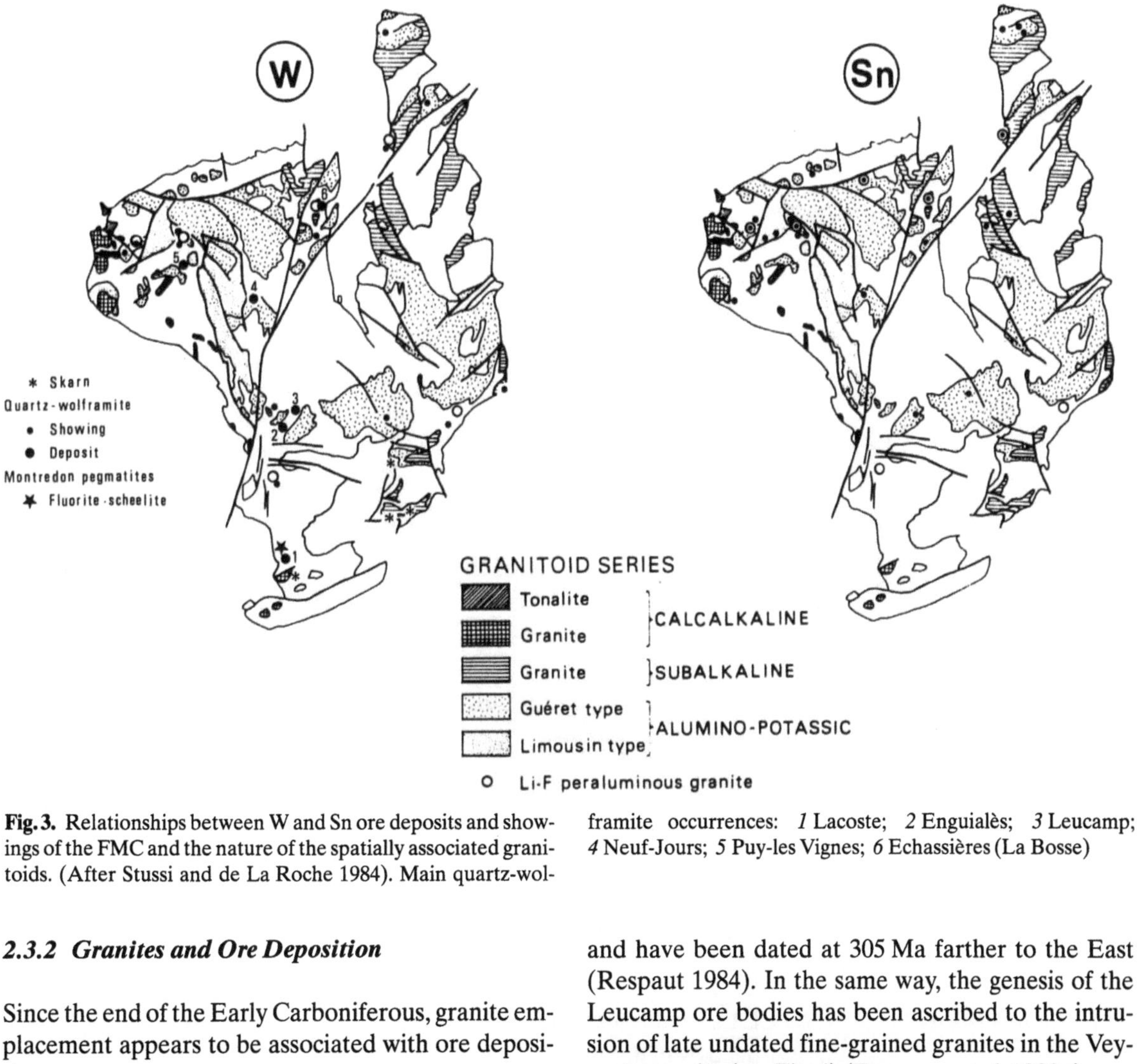

Fig. 3. Relationships between W and Sn ore deposits and showings of the FMC and the nature of the spatially associated granitoids. (After Stussi and de La Roche 1984). Main quartz-wolframite occurrences: *1* Lacoste; *2* Enguialès; *3* Leucamp; *4* Neuf-Jours; *5* Puy-les Vignes; *6* Echassières (La Bosse)

2.3.2 Granites and Ore Deposition

Since the end of the Early Carboniferous, granite emplacement appears to be associated with ore deposition.

a) *Alumino-potassic granites (Margeride type).* Many of the W showings and most of the quartz-wolframite deposits in the FMC are spatially associated (Fig. 3) with alumino-potassic (AK) granites of the Margeride association (see Ploquin and Stussi, this Vol.), most of which are Late Visean (320–335 Ma, see references in Duthou et al. 1984; Respaut 1984). With a Westphalian Rb/Sr age of 295 ± 10 Ma, the Meymac granite is the youngest one. The temporal relationship between the W deposits and those AK granites is, however, debatable.

For instance, at Meymac (Neuf Jours prospect) Burnol et al. 1980), greisens and the W mineralization are associated with a leucogranite cupola intrusive in the main Meymac granite (Raimbault 1984). It has also been proposed that the Enguialès deposit could be related to an (unknown) leucogranite stock (Derré 1982) rather than to the Margeride granite, in which such leucogranites are indeed intrusive (see Fig. 4) and have been dated at 305 Ma farther to the East (Respaut 1984). In the same way, the genesis of the Leucamp ore bodies has been ascribed to the intrusion of late undated fine-grained granites in the Veynazès granite (see Fig. 4) (Demange et al. 1988 a).

Additionally, an associated granite is unknown at Lacoste, Montredon (see above) and Echassières; however, indirect arguments exist at Echassières to relate the W-bearing quartz stockwork to a concealed AK-type granite (Cuney and Autran 1987 b; see below).

These quartz-wolframite deposits and showing amount to ca. 30 000 t WO_3. Most are peri-batholithic with a micaschist wall-rock, but some are of the greisen, intra-granitic type (e.g., Neuf Jours, Raimbault 1984). Compositions of W oxides vary from true wolframites to ferberites, inside a deposit and from one deposit to another; hubnerites are rare (there is a small deposit at Echassières, Aubert 1969). At Echassières, Nb-rich ferberites are found; they exhibit a significant vertical zoning (Aïssa et al. 1987 a).

The Leucamp and Enguiales deposits are good representatives. They are spatially associated with the Margeride batholith and the Veynazès granite

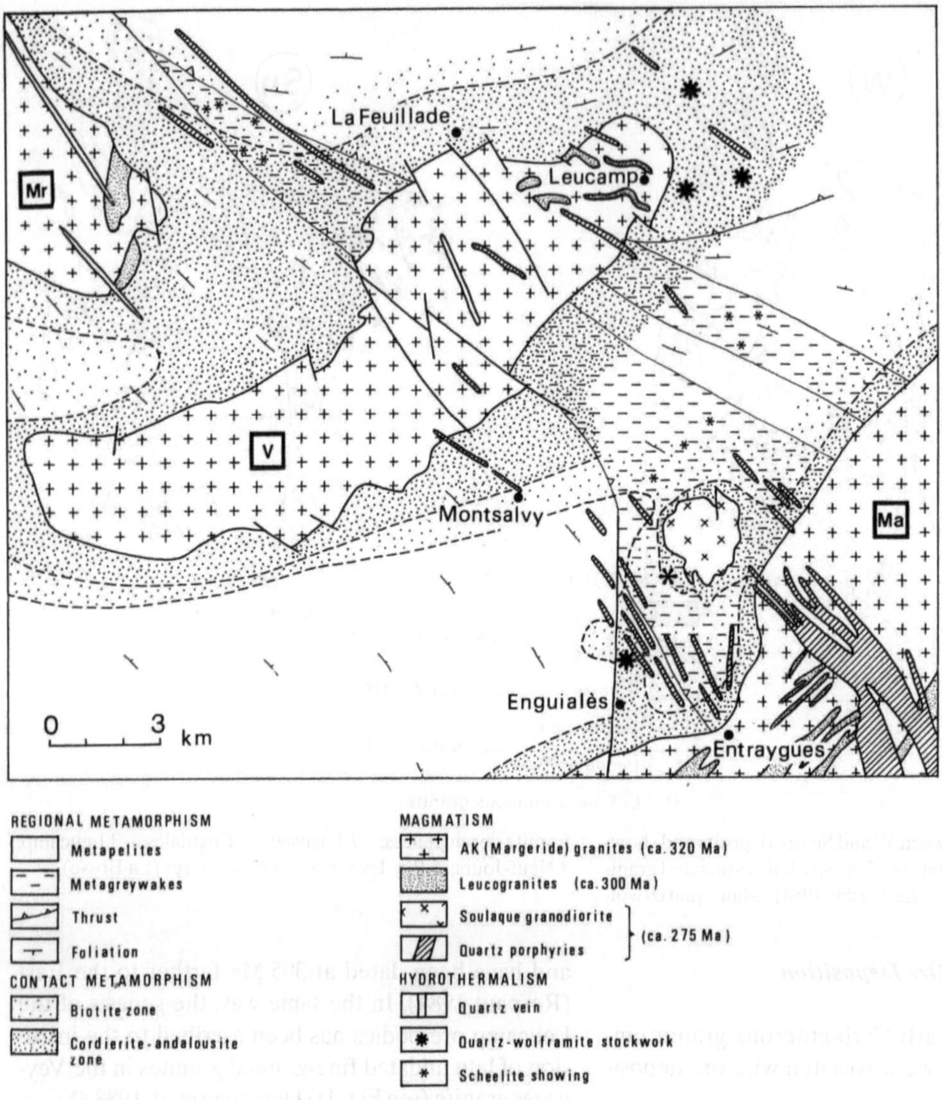

Fig. 4. Regional setting of the Enguiales and Leucamp quartz-wolframite stockworks. (After maps by Joubert 1978; Derré 1982; Nicolas 1985; Demange et al. 1988b; Bogdanoff et al. 1989). Granites, batholites, and stocks: *Mr* Marcolès; *Ma* Margeride; *V* Veynazès

stock, both dated at ca. 330 Ma (Fig. 4). They are hosted by low grade metapelites and metagraywackes, with intercalated metacinerites (Na-leptynites) Bogdanoff et al. 1989). The deposits are located in andalusite-cordierite contact metamorphic aureole above flat-lying part of the granite roof (Leucamp), or a tongue of the main mass (Enguialès) (Fig. 4). The quartz-wolframite-apatite-(anatase-muscovite-arsenopyrite) veins follow a complex set of late fold-related tectonic joints (some possibly related to granite emplacement) (Bonne and Moreau 1973; Derré 1982; Nicolas 1985; Demange et al. 1988b). Wall-rock tourmalinization is extensive and predates the opeing of the veins. The simultaneous opening of all joint directions indicate a hydrothermal overpressure during ore deposition; this behavior could be related to the CO_2-rich nature of early fluids (see below) (Ramboz 1986). Wolframite is deposited either during quartz filling or in later microcracks.

At Enguiales, post-wolframite tourmalinization is also known, but predates a late sulfide deposition event related to microgranite dikes (see below). This late tourmalinization is associated with a reheating event (Ramboz 1980), possibly related to the 300 Ma leucogranite emplacement (Respaut 1984) in the Margeride batholith (Fig. 4). This is an argument against the Derré (1982) proposal of a genetic link between ore deposition at Enguiales and the emplacement of leucogranites rather than with Margeride-type granites.

Fluid inclusion studies of quartz-wolframite deposits and showings in the FMC (Amossé 1978; Ramboz 1980; Ramboz et al. 1985; Aïssa et al. 1987b; Dubessy et al. 1987; Guillot et al. 1988; Noyé and Weisbrod 1988) evidence a common pattern of hydrothermal circulation in all these granite-related occurrences (Weisbrod 1988).

Early circulation of low density C-H-O-N fluids occurred at high temperatures (ca. 500° C) under a lithostatic pressure of ca. 1 kbar. Dilution and cooling by low saline aqueous fluids followed and induced W deposition at ca. 375° C.

The early fluid compositions were controlled by graphite-water equilibrium which fixed a low fO_2 apparently favorable to W transport. Such fluids are probably of metamorphic derivation.

A pro parte metamorphic origin for fluids involved in the genesis of the Neuf-Jours wolframites is also evidenced by Raimbault (1985) on a REE study basis.

Source of tungsten is usually sought in more or less specialized granite magmas (e.g., Raimbault et al. 1987; Rossi et al. 1988). However, geochemical studies have not evidenced a strong W enrichment in the AK granites related to quartz-wolframite deposits in the FMC (Boissavy-Vinau 1979; Boissavy-Vinau et al. 1979, 1980; Raimbault et al. 1987). Indeed, the very concept of a granite derivation for these mainly peribatholitic deposits is questionable (Marignac and Weisbrod 1986).

A brannerite concentration at Les Bondons (cf. Fig. 1) is related to the contact metamorphism of leucogranite dikes, intrusive in the SE border of the Mt Lozere subalkaline granite (Cathelineau 1983a). However, the deposit was repeatedly remobilized in successively hydrothermal (Cathelineau 1983a) and supergene (Eulry and Vargas 1979) events.

b) *Peraluminous Li-F magmatism.* Peraluminous high-level granites with Li-mica and topaz, as well as volcanites of the "topaz-rhyolite" type, are known from the Westphalian (on a radiometric basis) to the Lower Stephanian (on a stratigraphic basis) locally throughout the FMC. They are important both from uranium and rare metals metallogenesis.

Uranium. There are some minor showings directly linked with similar high level cupolas (Badia and Fuchs 1981, 1987; Cuney and Brouand 1987). Far more relevant, however, are the late small intrusions of medium- to fine-grained peraluminous granites within the Saint Sylvestre (Limousin) leucogranite. They display a significant primary uranium enrichment as uraninite, which is of great importance for the genesis of the economic uranium mineralization in the area.

The Saint Sylvestre leucogranite displays a high U content (10 to 30 ppm). Further uranium enrichment occurs along late magmatic shear zones controlling the emplacement of successive injections of medium- to fine-grained granite rich in uraninite (Friedrich et al. 1987). These phenomena lead to the formation of strongly favorable source rocks, in which uranium is mainly located in uraninite, easily leachable by oxidizing hydrothermal solutions. The late magmatic shear zones are grouped in spatially well defined areas. They were preferentially reworked at the brittle stage: in fact, overlapping between these uranium-enriched magmatic shear zones and the later brittle fracture zones represents one of the main factor controlling the efficiency of the uranium remobilization and concentration (see below).

Rare metals. The Li-mica and topaz bearing peraluminous granites are responsible for most of the Sn mineralization in the FMC (Fig. 3). These are the Echassières, Montebras (Aubert 1969), Blond (Burnol 1974) and Chavence (Mourey 1985) cupolas. They are also relevant for W (numerous quartz-wolframite showings), Li and Ta (Montebras, Echassières: Aubert 1969) and Be (small uneconomic pegmatite deposits) (Burnol 1974).

At Echassières, a deep drill-hole of the GPF program has recently given the opportunity of studying the characteristics of such "specialized" granites in great detail (see references in: Cuney and Autran 1987a) (Fig. 5). The albite-lepidolite-topaz granite is locally known as the Beauvoir granite. It is dated (Rb/Sr) at ca. 310 Ma.

The Beauvoir granite was intrusive as several sheets (successively, B3, B2, and B1), progressively enriched in Li, F, Sn, Ta, W, Be, Nb, U, etc. The rare metals enrichment is correlated with a decreasing shift in $\delta^{18}O$, which may be indicative of an early fluid-rock interaction. Although the total Sn, Li, Ta and W contents are very high, the values are too low for the Beauvoir granite to be presently an economic target (for instance, ca. 150000 t Sn at ca. 0.10 wt. %).

Quite surprisingly, there is no vein-type rare metals mineralization genetically associated with the Beauvoir granite (except for the very minor Chaillats quartz-cassiterite-sulfide vein, Fig. 5a): this is at strong variance with, for instance, the Erzgebirge Younger Intrusives. A first reason is that Sn fractionation in the fluids generated from the Li-F magma crystallization was lowered by different factors such as the F content, a fO_2 increase of the magma towards the roof, or the low solidus temperature of the magma. Besides, post-magmatic remobilization was of limited extent, probably because of a rather deep level of intrusion for the Beauvoir granite, which is ca. 2.7 km (lithostatic pressure of ca. 700 bar, estimated from fluid inclusion study), to be compared with a depth of

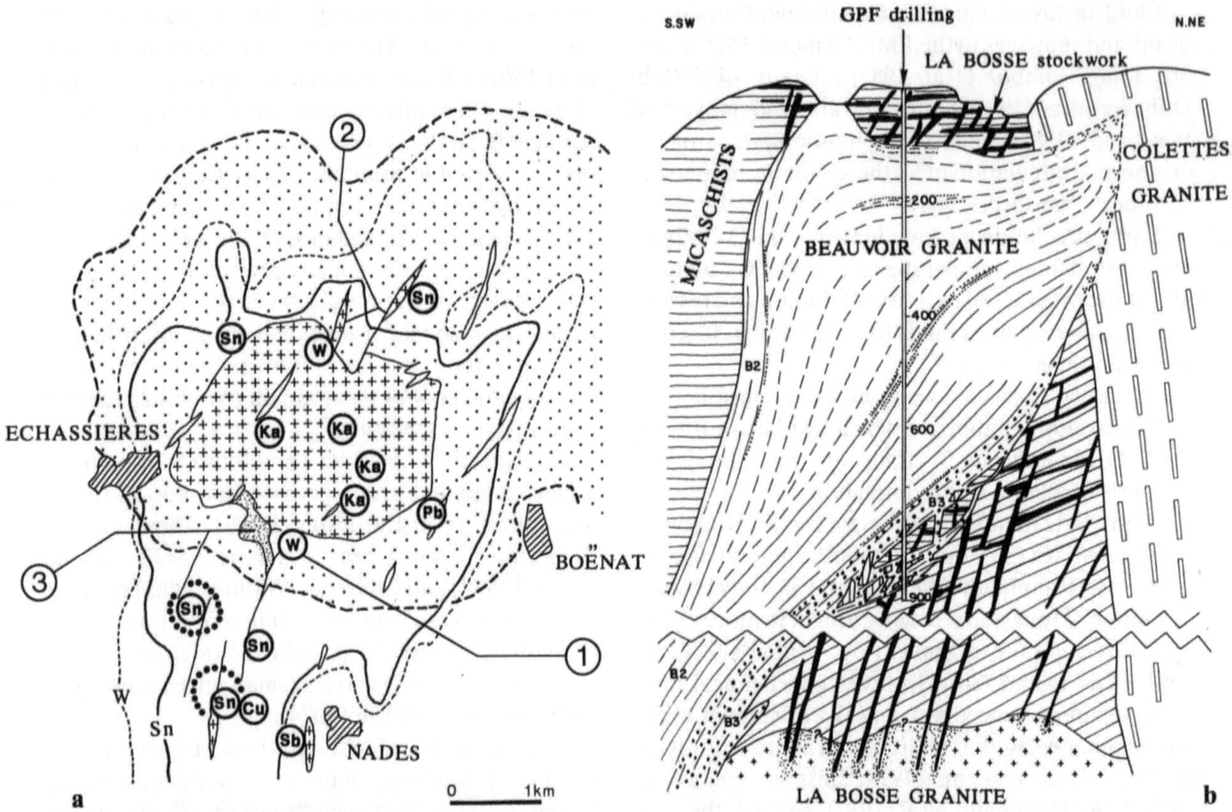

Fig. 5 a, b. The Beauvoir Li-F peraluminous granite and its context. (After Cuney and Autran 1987b). **a** Geological map. *Crosses* Colette granite; *fine dots* Beauvoir granite; *stippled line* and *lightly dotted area* extension of the negative gravimetric anomaly related to the concealed part of the granite complex; *dotted and solid lines* respectively, W and Sn anomalic geochemical haloes. *Ka* kaolin; *Pb* galena vein; *Sb* stibnite vein; *Sn* cassiterite vein; *W* quartz-wolframite veins. Deposits: *1* La Bosse ferberite stockwork; *2* Le Mazet hübnerite veins; *3* Beauvoir cassiterite-lepidolite-microlite disseminated mineralization. **b** Interpretative cross-section according to the deep GPF drilling

emplacement of ca. 1.5 km for the Erzgebirge granites (Durisova 1988).

However, the Beauvoir granite cross-cuts an earlier quartz-wolframite stockwork, related to the concealed (unknown) La Bosse granite (Fig. 5b); comparing the characteristics of the geochemical haloes associated to the stockwork emplacement with the geochemical haloes around outcropping granites in the vicinity of Echassières, it has been deduced that the La Bosse granite is of the AK type (Margeride subtype) (Cuney and Autran 1987b).

There is a probable relationship between Li-F magma generation and deep-seated thermal anomalies, apparently linked with the activity of transcurrent faults (Gagny et al. 1984; Friedrich et al. 1987).

c) *Meta-aluminous granites.* True calk-alkaline type granites are rare in the FMC (Ploquin and Stussi, this volume). They are late in the history, and do not appear clearly related to any subduction-type process. Of significance from a metallogenetic point of view are:

The Montagne Noire calk-alkaline plutons. Intrusion of the Sidobre granite (ca. 290 Ma) was associated with the formation of the recently discovered *Fumade W-Skarn* deposit (Safa et al. 1987; Couilloud 1988). The skarn, with 1.2 Mt at 1 wt. % WO_3, is typical of the reduced W-skarn association (Einaudi et al. 1981).

The skarns are developed within calcareous-pelitic rocks of the Lower Cambrian, at the contact with a differentiated satellite stock of the Sidobre granite. Skarns are mainly stratiform. The richest ore is in prograde skarn formed of F-bearing idocrase and an andradite free garnet, with minor quartz and a few sulfides (mainly pyrite). Metamorphic conditions are estimated at ca. 500–550° C, 1–2 kbar, $XCO_2 < 0.1$. The retrograde hydrothermal stage is poorly expressed; it is not associated with a significant W deposition.

The Sidobre granite played also a part in the genesis of the *Saint-Salvy Zn* deposit (Foglierini et al. 1980b; Béziat et al. 1982; Barbanson and Geldron 1983; Munoz et al. 1988).

The Saint-Salvy vein deposit (Fig. 1) is located along a major shear zone associated with the emplacement of the Sidobre granite. Hydrothermal alteration was first phengitic, at 250–350° C, then with quartz dissolution and albite crystallization, at ca. 250° C. The formation of the economic ore vein (450 000 t Zn, 480 t Ge; Ge is contained in sphalerite: 1600 ppm), with five stages of sphalerite deposition, took place thereafter, from 200° C down to 100° C. Ore deposition is thought to result from the hydrothermal remobilization of the early stratifrom deposit (see above), induced by the residual thermal anomaly associated with the emplacement of the Sidobre granite. It has been proved, on a geological basis, that ore deposition was completed before the end of the Autunian.

The Folat microgranite-rhyolite complex (290 Ma) has been related to the formation of the old *La Rabasse Pb-Zn* deposit (Lescuyer and Giot 1988) (Fig. 1).

The La Rabasse deposit was located at the contact between Lower Cambrian dolomites and a rhyolite sill of the Folat complex. The deposit resulted from the replacement of dolomites by massive sulfides. There was three stages of ore deposition: (1) pyrrhotite-Sb-rich arsenopyrite-pyrite; (2) Fe-sphalerite (with stannite inclusions); (3) galena-sulphosalts. It is significant that a similar succession is observed in the nearby Bournac Sb-As vein, which is clearly related to the emplacement of the Folat complex, at temperatures decreasing from 380 to 140° C (Munoz and Shepherd 1987).

In the Loire synclinorium (North-East FMC), the *Sibert microgranites* of unknown age, but cross-cutting Upper Visean ignimbrites, are associated with the only known Cu-Mo porphyry-type deposit in the FMC (Icart et al. 1980; Beaufort and Meunier 1983) (Fig. 1).

In a volume of more than 1 km^3 (1 km × 2 km × 200–300 m), strong fracturing allowed development of a classic porphyry pattern of zoned alterations. The grades, however, are very low: 250 pp Cu (200–500); 55 ppm Mo (up to 200).

In the Loire synclinorium (North-East FMC), a small scheelite skarn showing is related to a Namuro-Westphalian *subalkaline granite* (Gagny et al. 1983).

d) *The formation of the Puy-les-Vignes tourmaline breccia pipe* (Weppe 1951), with quartz-wolframite, is probably related to the emplacement of a concealed granite, as all breccia-pipes of this type (Sillitoe 1985). A Late Carboniferous age may be assumed: indeed, the Puy-les-Vignes pipe cross-cuts two dikes, one leucogranite and one microsyenite (Weppe 1951), which are probably not older than ca. 300 Ma, by comparison with the nearby Saint-Sylvestre granite.

2.3.3 Brittle Fracturing and Ore Deposition

During the Carboniferous, there was a great development of shear-zones at various scales in the FMC (see Ledru et al. Chap. III. 4 this Vol.). These shear-zones were usually long-lived and experienced deformation from ductile (or brittle/ductile) to more and more brittle conditions (usually, from compressive to extensive regimes), leading to the formation of complex fracture zones. This behaviour is related both to the progressive uplift of the FMC and the changes in stress field orientation during the Late Carboniferous (Santallier et al. Chap. III. 5 this Vol.). Repeated magmatic activity throughout the FMC during the Late Carboniferous (Ploquin and Stussi, Chap. III. 6 this Vol.) testifies for the maintenance of high heat flows. Therefore, conditions were quite often created for the space and time association of brittle deformation zones and thermal anomalies, leading to the formation of a wide range of vein-type hydrothermal deposits, however not necessarily directly related to granite emplacement; granite-related deposits have been presented in the preceding section.

a) *Au-As-(Sb) deposits.* There are numerous gold deposits and showings in the FMC (Crouzet et al. 1979). They are mostly of the quartz vein-type, and associated with long-lived deep crustal structures (so-called gold-bearing shear zones, Bonnemaison 1986), characterized by successive events from ductile to brittle deformation (Bouchot et al. 1988; Bouchot 1989). In Northwestern FMC, some mafic to ultramafic rocks (amphibolites, serpentinites) are locally associated to the deposits: along the Bourneix shear zone (Roques 1969), and as wall-rocks of the Chatelet veins (Zappettini 1983). These rocks are sometimes listwaenitized and may concentrate gold, but do not represent significant volumes of the deposits.

The internationally competitive Salsigne district (see above), is not of the vein type although there is a small vein component in the Cambrian layers, and the Schistes "X" are mostly mineralized in quartz rich structures.

Most of the economic vein-type gold deposits are located in the Western part of the FMC (Fig. 1; see also Fig. 6): Limousin, with the Saint-Yrieix district (Cros-Gallet, Le Bourneix: Ahmadzadeh et al. 1984; Hubert 1986; Bouchot et al. 1988; Calli et al. 1988), and smaller districts such as the Bénévent-L'Abbaye (Nenert 1989) and Ambazac-Laurière ones; Marche-Combrailles, with the Le Châtelet-Villeranges district (Zappettini 1983; Zappettini et al. 1983; Boiron 1987; Boiron and Cathelineau 1988; Boiron et al. 1989b). Some minor deposits and showings are known along the Sillon Houiller (Pontvieux-Labes-

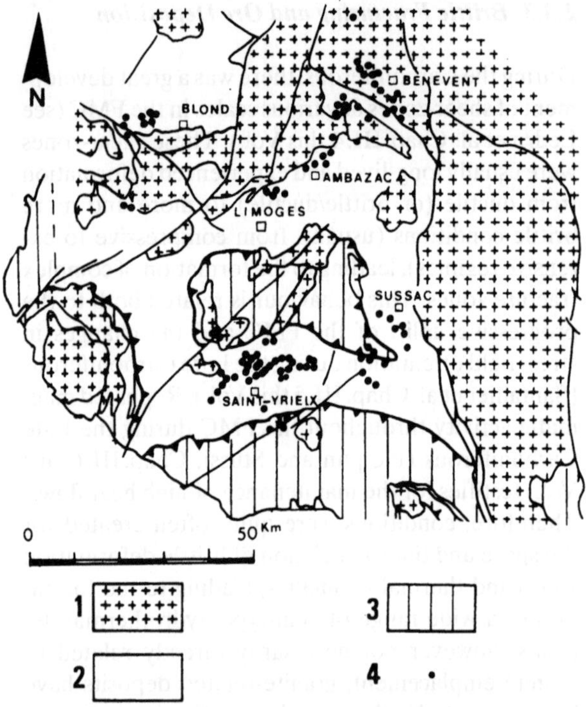

Fig. 6. Gold deposits repartition in the Limousin and adjacent areas in the FMC. (After Floc'h et al. 1984). *1* Granites; *2* lower allochthon; *3* middle allochthon; *4* gold occurrences. A severe restriction of gold occurrences to formations of the lower allochthon unit as well as their relationship with regional lineaments are obvious

sette: Marcoux et al. 1985), in the Haut-Allier (Sb district of Brioude-Massiac: Périchaud 1980; Sandras 1988) and in the Cévennes (Sb district of the Collet-de-Dèze: Ahmadzadeh et al. 1988).

These veins share common characteristics, such as: their setting in long-lived shear zones, reflected in successive quartz habits (Hubert 1986; Bouchot et al. 1988); their limited paragenesis, dominated by quartz with minor sulfides (essentially arsenopyrite and pyrite); their dominant association of native gold particles together with late sulfides and sulphosalts (Bi, Pb, Sb minerals), at the exception of the Villeranges-Châtelet occurrences; or the typical occurrence of K-mica alteration zones associated to gold deposition (Boiron and Cathelineau 1988; Proust 1988). Gold is found either as native gold (or electrum) or, rather frequently, as invisible gold in substitution in sulfides, mainly arsenopyrite (Marion et al. 1988; Cathelineau et al. 1989 a; Marcoux et al. 1989), at a highly variable average content (Cathelineau et al. 1989 a) which varies from 1 to 50 ppm at Bourneix, Salsigne, and from 100 to 2000 ppm in the Chatelet-Villeranges district.

Bonnemaison and Marcoux (1987) proposed a unified model of the gold-bearing shear zone: firstly concentrated as invisible gold in sulfides deposited at an early stage of the activity of the shear-zone (pyrrhotite, then ubiquitous arsenopyrite), gold would be progressively reconcentrated and expressed as free gold during the following (usually extensive) stages of activity. However, on the basis of (1) the fluid evolution characteristics, as recorded by fluid inclusions; (2) the sequence of mineral assemblages and (3) the data on gold content and distribution in minerals; it is possible to distinguish two kinds of vein-type gold deposits (Touray 1987; Boiron et al. 1988):

1. Early high T/high conditions are recorded in *deep-seated veins:* ca. 350–400° C and 1–2 kb, the dominant fluids being metamorphic, belonging to the CO_2-CH_4-H_2S-N_2 (± H_2O-NaCl) system. A (sudden ?) drop of temperature and especially pressure, which became hydrostatic (320° C and 0.6 kb at Cros-Gallet; Hubert 1986), was correlated with sulfide and free gold deposition; CO_2-rich fluids are observed together with meteoric aqueous fluids of low salinity. However, these veins display also late native gold (electrum) particles deposited together with sulfide (galena)-sulfosalts assemblage from aqueous fluids at lower temperatures (< 200° C). These veins are mostly found within the metamorphic series of the Lower Allochthon (Fig. 6), in the vicinity of small late cupolas and dikes of peraluminous granites, in the Limousin. The late gold-bearing quartz veins in Montagne Noire (Salsigne district) display the same range of conditions (Boiron et al. 1989 a) but in a distinct environment (see above).
2. There are veins in which low P/low T conditions are recorded during their entire evolution *("epithermal" veins).* These are mainly veins associated with late hydrothermal circulations in the vicinity of the Marche-Combrailles shear zone (Fig. 1) especially in the Villeranges basin (Boiron 1987) and in quartz comb veins from Le Châtelet (Boiron et al. 1988). Almost purely aqueous fluids of low salinity (1–4 wt. % NaCl), migrating at relatively low T and P (150–250° C, hydrostatic pressure) throughout the basement and the Visean volcano-sedimentary cover, are responsible for the deposition of gold-rich arsenopyrite. It thus appears that "invisible gold" deposition was not here associated with the early stage of activity of shear-zone.

The ages of ore-forming events are not well constrained. K-Ar dating on a late illite-smectite assemblage from the Cros-Gallet deposit gives a 305–287 Ma range (Bonhomme et al. 1987); 305 ± 9 Ma K-AR age were also obtained in Limousin deposits (Baubron 1982, unpubl.: in Marcoux et al. 1988). At Villeranges,

K-micas yield K-Ar ages of 315 ± 7 Ma (Boiron 1987). In the Bourneix area, Au-As veinlets cross-cut feldspathic episyenites; such episyenites are known to have been formed at 301 ± 3 Ma in other Limousin areas – see (d) below. Gold-bearing quartz pebbles are found in the Early Stephanian conglomerates at the bottom of Cévennes Coal Measures (Ahmadzadeh et al. 1988). The gold paragenesis is very similar to that in the quartz veins of the Cévennes. As a whole, a Late Carboniferous age may be confidently assumed.

The source of gold is usually sought in ultramafic to mafic rocks (amphibolites) intercalated in the metamorphic series (according to the general theory of Boyle, 1979) (Floc'h et al. 1984).

There is little direct evidence (Zappettini et al. 1983): apart from listwaenites (which would rather indicate late hydrothermal enrichment: Buisson and Leblanc 1985), the only possible indication of primary gold enrichment in amphibolites is the presence of early sulfides (with associated gold) in an unaltered amphibolite, in the vicinity of the Cros-Gallet district (Calli et al. 1988).

Lead isotope studies yield evidence for a mantellic source for at least part of the gold in the Marche (Marcoux and Bonnemaison 1988), thus strengthening the case for an origin in the ultramafic rocks. However, in most deposits, lead isotopes evidence leaching from a great variety of source rocks.

b) *Sb-As deposits.* There are more than 200 Sb-bearing veins in the FMC (Périchaud 1980), but 80% of the 48000 t Sb yield (1850–1931) were produced in the Brioude-Massiac district. Other significant districts are the Collet de Dèze district (4600 t Sb), located in the Mt Lozère micaschists, and the south and northeastern part of the Limousin area (Emberger and Périchaud 1979) (Fig. 1).

There are two paragenetic types, Sb-As and Pb-Sb (see below). The Sb-As veins were the only mineable ones. Arsenopyrite, pyrite, sphalerite, berthierite and stibnite are usually the only minerals found in the veins. They are locally associated with minor gold, at Les Biards (Saint Yrieix gold district, Limousin) and La Petite Faye (Bénévent gold district, Creuse).

In the Brioude-Massiac district (Haut-Allier, see Fig. 1), arsenopyrite (Sb-bearing; Bril 1985) is the first mineral to crystallize. Mineralizing fluids were moderately saline CO_2-bearing aqueous fluids circulating at 300–400° C (Bril 1982), inducing a quartz 2M1 muscovite alteration assemblage in gneisses. A second hydrothermal stage controls a new superimposed alteration with quartz – 2M1 phengite ± carbonate and the deposition of the main Sb ore at 250–300° C (Bril and Beaufort 1989). The Sb-As veins are found in metamorphic series (Haut Allier gneisses, Cévennes schists), but often in close spatial relationship with andalusite-bearing leucogranite and/or microgranite dike swarms, frequently of Permian age (Bellon and Gibert 1981; Bril 1983); this is the case for Brioude-Massiac (Sandras 1988), Mérinchal (Creuse) and Nades (Allier). In the Brioude-Massiac district, dating of sericites (Bril et al. 1991) from vein selvages yielded Permian ages (260 Ma). However, the characteristics of the Sb ore deposition display numerous similarities with the Au-As-(Sb) mineralization previously described, and at least some of these Sb deposits could be older than Permian (possibly in relation with small leucogranite cupolas): indeed, the Sb-As veins of the small Pontvieux district (Sillon Houiller, Fig. 1) yield K-Ar ages from 315 to ca. 300 Ma (Marcoux et al. 1988). Anomalous Sb contents (30 ppm) have been reported in the Brioude-Massiac gneisses as well as in the Cévennes epischists (Roger 1972); Sb is mainly contained in ilmenite (and also in rutile and hematite). The source of antimony has therefore been sought in the hydrothermal leaching of the country rocks (Roger 1972). A derivation of the lead in the veins from the surrounding metamorphic rocks has been evidenced on isotopic grounds at Brioude-Massiac (Marcoux and Bril 1986).

c) *W-Sn-Bi veins.* In the Limousin gold district (Marcoux and Bonnemaison 1988), as well as the Brioude-Massiac antimony district (Bril 1983) and the Pontgibaud lead district (Marcoux and Picot 1985), are found rare As-Sn (as cassiterite)-W (as wolframite, or rarely scheelite: Cros-Gallet)-Bi noneconomic veins, which predate the main metallogenic events.

At Pontgibaud, such mineralizations have been related to the Gelles granite (319 ± 12 Ma, Mezure and Negroni 1983); however, they have been dated at 299 ± 9 Ma (K-Ar, Baubron, unpubl.: in Marcoux and Picot 1985); and the Pb in them comes from the metamorphic rocks and not from the Gelles granite (Marcoux and Picot 1985).

In the Brioude-Massiac district, W-, Sn- or Au-bearing veins were formed at the same high temperatures, from the same kind of metamorphic-derived C-H-O-N fluids, as the Sb-As veins (Bril 1982). It is possible, however, that the pressure conditions were different.

d) *Episyenites.* Quartz leaching processes leading to bodies of porous feldspathic rocks (the so-called episyenites) from initial magmatic rocks (often, peraluminous granites) are common in the FMC (Leroy 1978; Cathelineau 1986). Alteration proceeds along veins and pipes in definite microcrack networks (Pecher et al. 1985; Lespinasse and Pecher 1986) and is characterized by the complete hydrothermal leach-

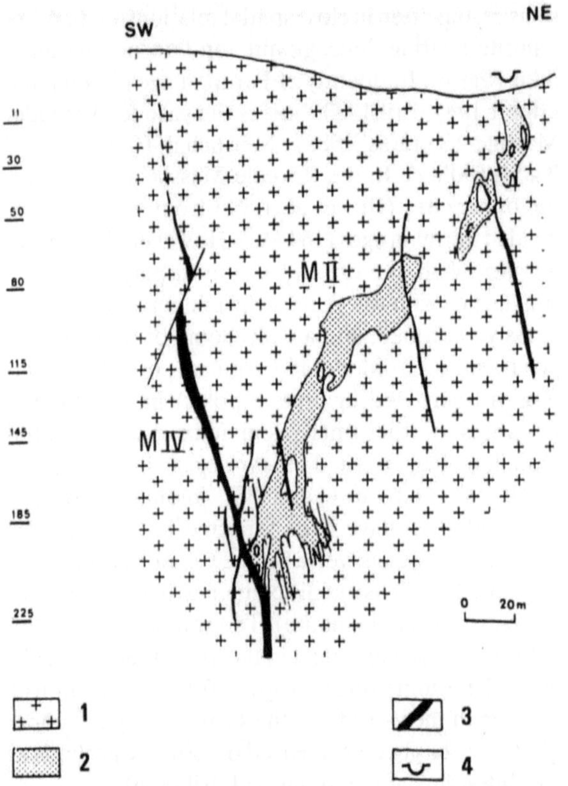

Fig. 7. The episyenites as passive traps for uranium deposition. Cross-section of a mineralized episyenitic pipe at La Crouzille. (After Leroy 1978). *1* Saint-Sylvestre granite; *2* Mineralized episyenite; *3* U veins; *4* open pit

ing of magmatic quartz during a fluid mixing event (Cathelineau 1986). These processes are not directly significant from a metallogenic point of view, with the exception of a few occurrences (U-Th-REE showing at Les Bombes, close to Les Bondons; Eulry and Vargas 1979; Cathelineau 1987a). However, the episyenitic bodies were, in some instances, porous traps for later U deposition (see below).

In the Limousin area, thermal anomalies associated with the emplacement of microgranite and lamprophyre dikes have been related to the episyenite alteration of peraluminous granites, at around 301 ± 3 Ma (Turpin 1985), i.e., ca. 25 Ma after the Saint-Sylvestre main granite crystallization. In Margeride, a similar time gap is observed between granite emplacement and episyenitic alteration (Respaut 1984).

2.4 End of the Variscan Cycle: Permian Distension and Related Metallogenesis

Although the Permian distension marks a major change in the geodynamic evolution of the FMC, no such break is revealed in metallogenesis. On the contrary, as active fracturing and thermal anomalies (in relation with the continued magmatic activity throughout the FMC) were present during Permian times as well as in the earlier period, metallogenesis persisted. In fact, it is during the Permian that the major U concentrations were formed in the FMC. It is often difficult, in a single locality, to distinguish between pre-Permian and Permian events: this is particularly relevant for Sb deposits (see above). In the same way, in some locatities, later stages of ore deposition occurred under the same kind of P-T conditions as during the Permian: this is particularly relevant for U deposits (see below).

2.4.1 Uranium

Most U deposits are spatially associated with AK granites, most deposits being found in the Limousin sub-type leucogranites.

a) *In the Limousin area,* permanence of thermal anomalies allowed development of deep-seated geothermal systems responsible for the formation of the large La Crouzille district (Leroy 1978) (Figs. 7, 8), at ca. 276 Ma (Leroy and Holliger 1984).

In this district, the interaction of meteoric oxidizing fluids, able to leach preconcentrated U within the host Saint-Sylvestre leucogranite (see above), and of "connate" reducing fluids (originating in a surincombent Permian basin ?) is deduced from fluid inclusion and stable isotope studies (Turpin 1984; Dubessy et al. 1987). Their mixing in favorable sites, such as fault zones and porous episyenite bodies, is thought to be responsible for U deposition (Fig. 8).

b) *Other localities.* Numerous other important deposits have been encountered in AK granites of Millevaches (Hyverneresse, Martin 1981), Marche (Le Bernardan: Michel 1983; Leroy 1984; Lespinasse 1984) and Margeride (Les Pierres Plantées, Cathelineau 1987b) (cf. Fig. 1). The exhausted Bois Noirs-Limouzat deposit (Cuney 1978) is the only major one to have been located within a sub-alkaline granite; however, the U source was probably a concealed AK granite (Limousin subtype) at depth (Cuney et al. 1984a).

The Permian period represents a major stage for uranium mobility and concentration on an European scale (Poty et al. 1986). However, U ores actually mined in most deposits result from strong remobilization of the Permian ores during later hydrothermal stages (see below).

In the Permian basins themselves, probable synsedimentary U deposition is recorded for instance at the Cérilly deposit (Landais and Conan 1980).

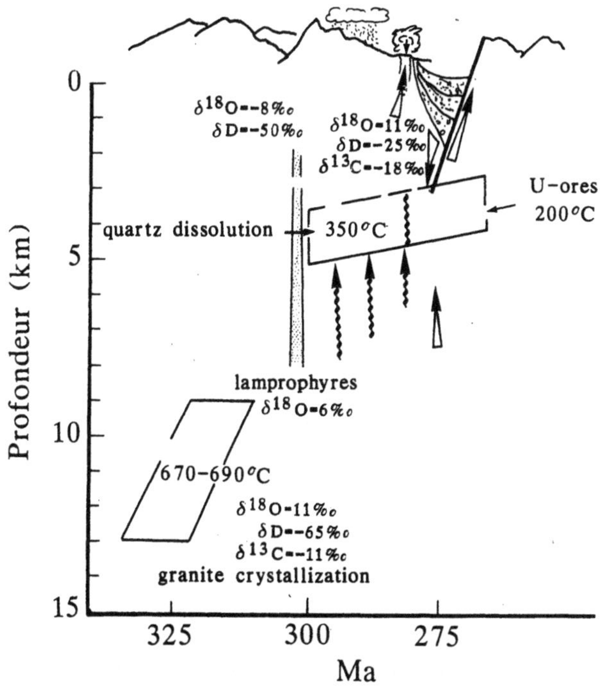

Fig. 8. Space-time relationships of the magmatic and hydrothermal events in the Saint-Sylvestre granite. (After Turpin 1984). Explanation in the text

2.4.2 Sb Deposits (Pb-Sb Veins)

Characterized by the abundance and variety of lead sulfosalts (Moëlo 1982) related to a late and intense overprinting by a Pb-remobilization (Kosakevitch and Moëlo 1982; Marcoux 1987), these noneconomic veins were formed, for the most part, at significantly lower temperatures (100–180° C) than the Sb-As veins (see above), from rather saline, purely aqueous fluids (Bril 1982, 1983). They evidently post-date the Sb-As veins, as demonstrated by cross-cutting relationships and, locally, by the successive emplacement of the two parageneses in the same vein. Permian ages (see above) could be related to these events; but Liasic remobilizations (as evidenced for F-Ba veins: Baubron et al. 1980; Jebrak 1985) are also possible (Marcoux and Bonnemaison 1988).

2.4.3 Base Metal Sulfide Deposition

At least some of the base metal vein deposits found in the FMC, as for instance the Pontgibaud lead district (Fig. 1), were probably formed at this stage.

The main Pb-Ag veins at Pontgibaud have been deposited at ca. 200° C; Pb was mainly leached from the nearby Gelles granite (Marcoux and Picot 1985).

Reworking of some earlier quartz-wolframite stockworks (see above) also occurred during the Permian.

In the Leucamp-Enguiales district, microgranite and microgranodiorite dikes (Fig. 4) of Permian age (Bellon and Gibert 1981) are associated with a low-temperature sulfide mineralization (arsenopyrite, pyrrhotite, bismuthinite, chalcopyrite, sphalerite) overprinting the quartz-wolframite ore bodies (Derré 1982; Demange et al. 1988 b).

2.5 Post-Variscan Metallogenesis

Extension tectonic phases related to the Tethys and North Atlantic rifting from Early Triassic (235 Ma) to Middle Jurassic (170 Ma), and to the Alpine and Pyrenean orogenies from Upper Cretaceous to Miocene (8 Ma), are recorded in the FMC metallogenesis (Bouladon and de Grasciansky 1985). Most of the related mineralizing events are restricted to the eastern and southeastern borders (Morvan, Forez and Cévennes), reflecting the dominance of the Tethysian events; but great fracture zones such as the Sillon Houiller were also reactivated.

Deposition took place either in the basement rocks or in the sedimentary cover; in the latter case, it is often possible to demonstrate that the concentrated elements (U, Pb-Zn, Ba, F) were (at least in part) leached from the basement rocks.

This is in particular the case for: (1) the stratabound F deposits of Antully, Courcelles, and Pierre Perthuis in Morvan (Soulé de Lafont and Lhégu 1980); (2) the Pb-Zn sulfides in the flooded paleokarst of les Malines (Cévennes) (Foglierini et al. 1980 a) or in the red-bed deposit of Largentière (Cévennes) (Samama 1968); and (3) the Tertiary U deposits of Saint-Pierre-du-Cantal (Carré 1979) and Coutras (20000 t U) (Ballery et al. 1985); the latter is found outside the FMC, but is thought to concentrate leached U from the FMC (Bouladon and de Graciansky 1985).

However, for a given deposit, it is often difficult to discriminate between the effects of the *in situ* remobilization of early concentrations and of the new metal supply during the long-lived multi-stage story of ore deposition.

This is particularly relevant for uranium deposits, in which numerous ore deposition events are usually revealed by U-Pb dating (e.g., Lancelot et al. 1984; Respaut 1984; Vella and Lancelot 1986; Leveque et al. 1988; Lancelot and Vella 1989). Mineralogical studies demonstrate a strong reworking of the early U ores into new mineral assemblages [1: coffinite in clay (smectite) alteration zones, 2: newly formed pitchblende in oxidation-reduction fronts] at low temperatures (below 150° C) (Cathelineau 1983b). This is the case also in some strata-bound Pb-Zn deposits. At Les Malines, for instance, Middle and Upper Jurassic mineralization stages related to brine migration from the adjoining Subalpine Basin overprint the earlier karstic stage of ore deposition (Charef and Sheppard 1988; Ramboz and Charef 1988).

Many ages recorded either in U deposits (either hydrothermal veins, like Bertholène and Les Pierres Plantées, or strata-bound deposits in Permian basins, such as Lodève), or in Pb-Zn-Ba-F deposits, range from 170 to 200 Ma and may be attributed to a major period of tectonic movements and anomalous heat flow associated with the onset of the Atlantik-Tethys opening. In particular, a widespread event of formation of F-Ba and Pb-Ba veins is recorded at ca. 190 Ma in the whole FMC (Jebrak 1985; Marcoux and Jebrak 1987).

Subsequent crustal thinning associated with the Tethys opening process resulted in abnormal heat flows which favored convection of hydrothermal and diagenetic fluids on the eastern border of the FMC: brine migration from the Subalpine Basin towards the Cévennes "margin" during the whole Jurassic is responsible for formation, often in presence of organic matter, of Mississippi Valley type Pb-Zn deposits (e.g., Disnar et al. 1986; Disnar and Sureau 1990).

Late geotectonic stages are more rarely witnessed in metallic deposits from FMC. During these stages, the Tethys was nearly closed, and the complete opening of the Gascogne Gulf, together with the first distensive movements related to opening of the North Atlantic Ocean, seem to have had little thermic and tectonic influence in Brittany and in Massif Central, except in the southern part of FMC.

By contrast, opening of the Algero-Provençal Basin during the Upper Tertiary triggered the FMC uplift; the subsequent outcropping and surficial leaching of U-rich leucogranites was then responsible for sedimentary U deposition (Coutras and Saint-Pierre-du-Cantal deposits, see above) (Bouladon and de Grasciansky 1985).

3 Discussion

3.1 Main Characteristics of FMC Metallogenesis

1. The dominant types of mineralization in the FMC pertain to the broad class of hydrothermal deposits and were formed at the end of the geodynamic evolution (essentially between ca. 330 and ca. 270 Ma, i.e. during the Late Carboniferous and Permian), expanding from the hyper-collision stage to the post-collision one without major break. The lack of collision-related ore deposition is outstanding, whereas the early (pre-collision) mineralizations are usually minor showings or possible protores for further reconcentration [Zn: Saint Salvy, Les Malines (?), for instance].

This major characteristic is readily explained by four factors:

a) The repeated generation of peraluminous granitic magmas of crustal derivation, enriched in such elements as U, Sn, and W, practically since the beginning of the hyper-collision stage.

b) The continuous uplifting of the nappe system since the end of the collision stage, as evidenced by the reconstructed P-T-t paths (Santallier et al., this Vol.).

c) The development of strong anomalous heat flows linked to crustal thinning (Santallier et al., this Vol.), beginning during the hypercollision stage and following during the Permian and proto-Tethysian stages of distension (i.e., at least for ca. 50 Ma).
Factors (2) and (3) explain the progressive development of geothermal fields involving more and more surficial waters, from granite-centered systems to fault-related ones.

d) Progressive development of brittle fracture zones owing a) to the uplifting of the chain and b) to the continent-scale faulting associated with the end of geodynamic evolution (pp. 305). The existence of extensive faulting in the vicinity of major drains is indeed essential for the formation of ore deposits implying metal leaching from great rock volumes: this is particularly relevant in the FMC for Au and Sb, and, to a lesser extent, for U.

2. There is an overprinting by late metallogenic events, yielding late and heterochronous F-Ba-Zn-Pb-U mineralizations in the whole FMC, but especially in its southern and eastern parts. Repeated hydrothermal activity from Permian to Mesozoïc times is recorded, with decreasing temperature events. Although there is an apparent slackening of activity during the Late Permian-Early Triassic period, it is difficult to evidence a significant break between the end of the Variscan metallogenesis and the late events related to distinct geodynamic processes.

3.2 Metallogenic Constraints for FMC Evolution

1. *At the Pre-Collision Stage,* the following facts are relevant for a geodynamic reconstruction:

- absence of true oceanic-related ore deposition: neither Cyprus-type massive sulfide deposits, nor podiform chromite ore bodies;
- absence of arc-related ore bodies: the fact that no pre-collision porphyry-type mineralizations are known emphasizes the scarcity of true early calc-alkaline magmatism in the FMC;
- scarcity of massive sulfide deposits for which a back-arc setting is possible;
- exhalative sedimentary nature of most of the early mineralizations and protores, implying anomalous heat flows in the corresponding (usually continental) settings.

All these facts argue against the existence of large ocean(s) at the early stage of the history of the FMC; they point, however, towards the existence of small basins with either thinned continental crust or true oceanic curst.

2. *For the Collision Stage,* the absence of porphyry-type ore bodies again emphasizes the scarcity of true calc-alkaline magmatism in the whole FMC belt. In particular, there is no mineralization associated to the small plutons of the so-called Devonian tonalite line in Limousin (Didier and Lameyre 1971; Stussi and de La Roche 1984; Pfeiffer 1986; Briand et al. III.5 this Vol.). The only porphyry-type deposit known in the FMC is the Sibert Cu-Mo showing, related to Late Carboniferous magmatic activity; it cannot be ascribed to the activity of a subduction zone.

On the contrary, the few Late Devonian volcanic-exhalative massive sulfide deposits could be indicative of back-arc settings (e.g., Autran and Lameyre 1980; Sider and Ohnenstetter 1986). Minor element distribution in the main sulfides of these deposits is consistent with a Kuroko-type setting (Lardeau 1988).

This apparent contradiction (no evidence for a subduction zone, evidence for subduction-related distensive settings) may be solved by considering the Late Devonian-Early Carboniferous basins, with bimodal (spilite-keratophyre) volcanism and related mineralization, as pertaining to the broad class of "pull apart" basins. They could be related to the early activity of the transcurrent ductile faults which characterize the hyper-collision stage of the FMC; such early faults are known, for instance, the Saint-Gervais-L'Hermitage pre-Late Visean ductile fault (Barbarin and Belin 1982) (see Fig. 1).

3.3 Terranes and Metallogenesis

It is quite evident that some kind of mineralizations are consistently associated with some tectonostratigraphic units, thus indicating a relationship which can help for the indentification of these units as terranes (see Piqué et al., this Vol.).

The obvious restriction of massive sulfide deposits to the *upper allochthon units* is a strong argument for the identification of this tectonostratigraphic set as a terrane.

The concentration of Sn, W, and U deposits in and around granite bodies intrusive into the *lower allochthon* or the *relative autochthon* is another striking fact. For the Limousin (La Crouzille) U district, such a relationship is clearly related to the derivation of the host granites from orthogneisses pertaining to the considered tectonostratigraphic units; moreover, it has been shown that many of these orthogneisses (metagranites of subalkaline to possibly peraluminous affinity) are themselves enriched in Sn, W, and U (Bourguignon 1988; Turpin et al. 1989; Cuney et al. 1989).

The restriction of Au deposits in the Western FMC to the sole *lower allochthon* (Floc'h et al. 1984) (Fig. 6) is a third fact of possible significance for the terrane appraisal. It has been suggested that such a definite location was primary controlled by the presence of leachable gold in the ultramafic rocks (presently, amphibolites and serpentinites) included in the metamorphic series (Floc'h et al. 1984).

Last, the absence of significant mineralization within the *middle allochthon* unit is an interesting negative character, this being the only tectonostratigraphic unit apparently devoid of metallogenetic potential throughout the FMC.

4 Conclusions

The French Massif Central (FMC), the main metallogenic area in France, a major mining region on a world scale for uranium, contains numerous economic deposits of U, Au, W, Sb, F, Ba, Pb-Zn.

1. It is a striking fact that only a few, small, mainly base metals strata-bound deposits were formed during the pre-collision stage of the development of the FMC; and that no granite-related deposits were formed at the collision stage, although many peraluminous Sn-U-rich granites were emplaced at this time.

Most ore deposits are of the hydrothermal type and developed in relation with anomalous heat flows after the thickening of the Variscan crust, during the hyper-collision stage and the Permian distension (ca. 330–270 Ma). Very significant are: the granite-related quartz-wolframite stockworks, mainly formed between 330 and 300 Ma; the gold deposition in long-lived deep crustal structures, during the Late Carboniferous; the formation of large U vein deposits in geothermal systems during the Permian.

After the end of the Variscan cycle, a second period of anomalous heat flows (200–170 Ma) resulted from crustal thinning related to the Atlantic-Tethys openings. Significant Pb-Zn-F-Ba deposition occurred at that time, concurrently with a reworking of previously formed mineralizations (U in particular).

2. The metallogenic features yield some constraints for a reconstruction of the geodynamic evolution of the FMC.

The absence of true oceanic-related deposits, together with the absence of subduction-related (porphyry-type) deposits, argue against the existence of large ocean(s) prior to the Variscan collision.

The existence of volcanic-exhalative massive sulfide deposits in distensive settings at the very time of the end of collision/beginning of hyper-collision may be related to the formation of pull-apart basins due to the activity of early transcurrent faults.

3. Metallogenetic characteristics may be used for an interpretation of the tectonostratigraphic units in terms of terranes.

The Lower Allochthon and the Relative Autochthon units host most of the U, Au, Sb, Sn, and W hydrothermal ore deposits. This fact could be related to the enrichments in Sn and U observed in the Early Paleozoic acid protoliths of the Variscan orthogneisses. Such a specificity supports the interpretation of these two units as one or two similar terranes of continental basement derivation, as proposed elsewhere (see Piqué et al., this Vol.). From another point of view, these units are good representative of the "metallogenic province" concept devised by Routhier (1980).

Acknowledgments. Comments by M. Zentilli and two anonymous reviewers helped to improve the text and the English manuscript.

References

Ahmadzadeh H, Auriol M, Calli M, Vaucorbeil H de, Foglierini F, Gelas M, Olivie C, Picot P, Tollon F (1984) Le gisement aurifère de Cros-Gallet, Le Bourneix, district de Saint-Yrieix (Haute-Vienne). Chron Rech Min 474: 11–32

Ahmadzadeh H, Picot P, Legendre O, Marcoux E (1988) Caractérisation des minéralisations aurifères des Cévennes (Massif Central, France) dans leur environment géologique. In: Johan Z, Ohnenstetter D (eds) Gisements métallifères dans leur contexte géologique. Doc Bur Rech Géol Min, Orléans 158, vol. 2, pp 527–563

Aïssa M, Marignac Ch, Weisbrod A (1987 a) Le stockwerk à ferbérite d'Echassières: évolution spatiale et temporelle; cristallochimie des ferbérites. In: Cuney M, Autran A (eds) Echassières: le forage scientifique. Une clé pour la compréhension des mécanismes magmatiques et hydrothermaux associés aux granites à métaux rares. Géol France 2–3: 311–333

Aïssa M, Weisbrod A, Marignac Ch (1987 b) Caractéristiques chimiques et thermodynamiques des circulations hydrothermales du site d'Echassières. In: Cuney M, Autran A (eds) Echassières: le forage scientifique. Une clé pour la compréhension des mécanismes magmatiques et hydrothermaux associés aux granites à métaux rares. Géol France 2–3: 335–350

Amossé J (1978) Variation in wolframite composition according to temperature at Borralha, Portugal, and Enguialès, France. Econ Geol 73: 1170–1175

Aubague M, Orgeval J-J, Soulie M (1977) Les gîtes minéraux de la terminaison méridionale du Massif Central et de sa bordure languedocienne (essai de synthèse). Bull Bur Rech Géol Min (2) II: 139–181

Aubert G (1969) Les coupoles granitiques de Montebras et d'Echassières (Massif Central français) et la genèse de leurs minéralisations en étain, tungstène, lithium et béryllium. Mém Bur Rech Géol Min 46: pp 1–345

Autran A, Lameyre J (1980) Conclusions générales. In: Lameyre J and Autran A, Les granitoïdes de France. In: Géologie de la France. Mém Bur Rech Géol Min 107: 87–99

Badia D, Fuchs Y (1981) Permo-carboniferous volcanism in the Massif Central (France) and its relations to ore deposits. In: Schneider HJ (ed) Mineral deposits of the Alps and of the Alpine Epoch in Europe, S. G. A. Spec. Publ. 3. Springer, Berlin Heidelberg New York, pp 275–290

Badia D, Fuchs Y (1987) The geodynamic-magmatic evolution of the European plate during Late Carboniferous and Lower Permian and its significance for the formation of volcanic hosted uranium deposits. Monogr Ser Mineral Depos 27: 59–69

Ballery JL, Cariou L, Le Goff P, Obelianne A, Thirion G (1985) Les recherches d'uranium de la COGEMA dans le nord de l'Aquitaine et leurs résultats. Bull Soc Géol Fr (8) I: 305–308

Barbanson L, Geldron A (1983) Distribution du germanium, de l'argent et du cadmium entre les schistes et les minéralisations stratiformes et filoniennes à blende-sidérite de la région de Saint-Salvy (Tarn). Chron Rech Min 470: 33–42

Barbanson L, Tollon F (1980) Les minéralisations zincifères stratiformes associées au gisement filonien de Saint-Salvy (Tarn). C R Acad Sci Paris 291 D: 9–12

Barbarin B, Belin JM (1982) Mise en évidence du cisaillement ductile hercynien "Saint-Gervais-L'Hermitage" (Massif Central français). C R Acad Sci Paris 294 II: 1377–1380

Bardin D (1971) Les amas pyriteux de Sain-Bel (Rhône) liés au groupe spilite-kératophyre de la Brévenne. Bull Bur Rech Géol Min (2) II 6:17–41

Baubron JC, Jebrak M, Joannes C, Lhegu J, Touray JC, Ziserman A (1980) Nouvelles données K/Ar sur des filons de quartz et fluorine du Massif Central français. C R Acad Sci Paris 290 D: 951–953

Beaufort D, Meunier A (1983) A petrographic study of phyllic alteration superimposed on potassic alteration. The Sibert porphyry deposit (Rhône, France). Econ Geol 78: 1514–1527

Bellon H, Gibert J-P (1981) Analyses radiométriques K-Ar des associations plutoniques et hypovolcaniques de la région Nord-Margeride (Massif Central français). Bull Soc Géol Fr (7) XXIII: 429–437

Béziat P, Prouhet J-P, Tollon F (1980) Le district de Montredon-Labessonnié (Tarn): W, Sn, F. Int Geol Cong., 26th, Paris, Gisements français, E7, pp 1–44

Béziat P, Michaud JG, Tollon F (1982) Le gisement filonien de Noailhac-Saint-Salvy (Tarn). Relations entre les corps minéralisés et la géologie de l'encaissant. Eléments de métallogénie régionale. In: Symp. gîtes filoniens Pb-Zn-F-Ba de basse température du domaine varisque d'Europe et d'Afrique du Nord. Bull Bur Rech Géol Min, Orléans (2) II: 249–252

Béziat P, Boudrie M, Guion JL, Tollon F (1986) Le district à W (Sn-F) de Montredon-Labessonnié (Tarn). Présence de concentrations stratiformes et filoniennes. In: Abstr European Meet "Tungsten Deposits", Toulouse, 12–14 Mai 1986, pp 65–66

Binon M, Pin C (1989) Geochronologie Rb/Sr et U/Pb des granites du Mayet de Montagne et des Bois Noirs, Montagne Bourbonnaise. Bull Soc Géol Fr (8) V: 695–704

Bogdanoff S, Cirodde J-L, Donnot M (1989) The nappes of La Chataigneraie, southwest Massif Central, France. Tectonophysics 157: 69–79

Boiron MC (1987) Minéralisations à Au, As, Sb, altérations hydrothermales et fluides associés dans le bassin de Villeranges (Combrailles, Massif Central français). Géol Géochim Uranium Mém, Nancy, 15, pp 1–310

Boiron MC, Cathelineau M (1988) Les altérations hydrothermales et les minéralisations associées des tufs viséens du bassin de Villeranges (Massif Central français). Données géothermométriques déduites de l'étude des inclusions fluides. In: Johan Z, Ohnenstetter D (eds) Gisements métallifères dans leur contexte géologique. Doc Bur Rech Géol Min 158, 2, pp 651–665

Boiron MC, Cathelineau M, Dubessy J, Bastoul AM (1988) Contrasted behaviour of Au and U in granites at the hydrothermal stage: the role of fO2 and pH. Proceedings of the SIMP meeting, Verbania, Italia, Granites and their surroundings, 1987, Rend Soc Ital Min Petrol 43: 485–498

Boiron MC, Cathelineau M, Dubessy J, Kibonzi B, Tollon F, Normand M (1989) Physical and chemical conditions of gold deposition in quartz veins from the Montagne Noire Schistes X (Salsigne gold district, France). Terra Abstr Gold 89 in Europe Toulouse 3/2: 28

Boiron MC, Cathelineau M, Trescases JJ (1989b) Conditions of gold-bearing arsenopyrite crystallization in the Villeranges basin, Marche-Combrailles shear zone, France. A mineralogical and fluid inclusion study. Econ Geol (Special monography on the Hercynian deposits in France) 84: 1340–1362

Boissavy-Vinau M (1979) Processus géochimiques de concentrations liées à l'évolution des magmas granitiques. Application aux filons à Sn-W du Massif Central et du Nord Portugal. 3d Cycle Thesis, Univ. Paris 6 (unpublished)

Boissavy-Vinau M, Derré C, Jaffrezic H, Roger G, Treuil M (1979) L'étain et le tungstène dans l'évolution granitique: le secteur N de Viseu (Portugal). Comparaison avec le Massif Central français. Publ Mus Laboratorio Mineral Geol Fac Cienc Porto (4) 91: 65–74

Boissavy-Vinau M, Derré C, Jaffrezic H, Roger G, Treuil M (1980) La cristallisation fractionnée: un processus efficace de concentration de l'étain dans les magmas granitoïdes. Int Geol Congr 26th Paris III: 911 (Abstr)

Bonhomme MG, Baubron JC, Jebrak M (1987) Minéralogie, géochimie, terres rares et âge K-Ar des argiles associées aux minéralisations filoniennes. Chem Geol 65: 321–339

Bonne A, Moreau J (1973) Etude structurale du gisement d'Enguialès (Massif Central français). Mineral Deposita 8: 57–63

Bonnemaison M (1986) Les "filons de quartz aurifère": un cas particulier de shear-zone aurifère. Chron Rech Min 482: 55–66

Bonnemaison M, Marcoux E (1987) Les zones de cisaillement aurifères du socle hercynien français. Chron Rech Min 488: 29–42

Bonnemaison M, Crouzet J, Thiercelin F, Tollon F (1986) Controls on exhalative gold deposits hosted by volcanoclastic sediments in the Schistes X, Salsigne gold district, Montagne Noire, Southern France. Proc gold '86 symp Toronto 1986: 457–469

Bouchot V (1989) Contexte géologique et structural des minéralisations aurifères du district de Saint Yrieix (Limousin, Massif Central français). Thesis, Orléans Univ. (unpublished)

Bouchot V, Gros Y, Calli M, Auriol M, Bonnemaison M, Charvet J, Tollon F (1988) La mine d'or de Cros-Gallet, Le Bourneix (Massif Central français): chronologie relative et rôles respectifs des déformations dans la mise en place de la minéralisation. In: Johan Z, Ohnenstetter D (eds) Gisements métallifères dans leur contexte géologique. Doc Bur Rech Géol Min 158, vol 2, pp 565–590

Bouladon J (1986) Les gisements de plomb-zinc-argent. In: Bril H, Watelet P (eds) Les richesses du sous-sol en Auvergne et Limousin. Aurillac City Editions, pp 190–198

Bouladon J, de Graciansky PC (1985) Les minéralisations dites de couverture (plomb, zinc, cuivre, uranium, barytine, fluorine, ...) du Trias au Pliocène, en France. Chron Rech Min 480: 17–33

Bourguignon A (1988) Origine des formations paradérivées et orthodérivées acides du Limousin central. Une source possible pour les leucogranites uranifères. Thesis, Univ. Lyon, 208 pp (unpublished)

Boyer F, Routhier P (1974) Extension régionale de couches à scheelite dans la couverture métamorphique de la zone axiale en Montagne Noire (Hérault). C R Acad Sci Paris 279: 1829–1832

Boyle RW (1979) The geochemistry of gold and its deposits (together with a chapter on geochemical prospecting for the element). Geol Surv. Can Bull 280: 1–584

Brevart O, Dupré B, Allègre CJ (1982) Metallogenic provinces and the remobilization process studied by lead-isotopes: Lead-zinc ore deposits from the Southern Massif Central, France. Econ Geol 77: 564–575

Bril H (1982) Fluid inclusion study in the W, Au, As, Sb and Pb-Zn district of Brioude-Massiac (Massif Central français). Tschermaks Mineral Petrol Mitt 30: 1–16

Bril H (1983) Etude métallogénique des minéralisations à antimoine et associées du district de Brioude-Massiac (Massif Central français). Conditions géochimiques de dépôt; implications génétiques. Ann Sci Univ Clermont-Ferrand II 77: 1–340

Bril H (1985) Conditions de stabilité des sulfures dans les filons de haute température du district de Brioude-Massiac (Massif Central français). Bull Minéral 108: 161–171

Bril H, Beaufort D (1989) Hydrothermal alteration and fluid circulation related to W, Au, and Sb vein mineralizations, Haut Allier, Massif Central, France. Econ Geol 84: 2237–2251

Bril H, Bonhomme MG, Marcoux E, Baubron JC (1991) Ages K/Ar des minéralisations de Brioude-Massiac (W-Au-As-Sb; Pb-Zn), Pontgibaud (Pb-Ag; Sn), et Labessette (As-Pb-Sb-Au): Place de les districts dans l'évolution géotectonique du Massif Central français. Mineral Depos 26: 189–198

Buisson G, Leblanc M (1985) Gold in carbonatized, ultramafic rocks (listwaenites) from ophiolite complexes. Ecòn Geol 80: 2028–2029

Burnol L (1974) Géochimie du béryllium et types de concentrations dans les leucogranites du Massif Central français. Relations entre les caractéristiques géochimiques des granitoïdes et les gisements endogènes de type départ acide (Be, Sn, Li) ou de remaniement tardif (U, F, Pb et Zn). Mém Bur Rech Géol Min 85: 1–168

Burnol L, Peronne Y, Vaucorbeil H de (1980) La coupole cachée de Neuf-Jours (Corrèze) et les minéralisations en tungstène associées. Chron Rech Min 455: 93–116

Caleffi D, Moine B, Béziat P, Tollon F (1988) Etude géochimique et minéralogique de la "série noire" du Cambrien inférieur de la Montagne Noire: origine des minéralisations stratiformes à scheelite de l'Hom-Haut (district de Montredon). In: Johan Z, Ohnenstetter D (eds) Gisements métallifères dans leur contexte géologique. Doc Bur Rech Géol Min 158 (1): 195–217

Calli M, Bonnemaison M, Auriol M, Tollon F (1988) Etude lithogéochimique (éléments majeurs et traces) de la minéralisation aurifère du gisement de Cros-Gallet, Le Bourneix (Haute-Vienne, district de Saint Yrieix, Massif Central français): Implications pour la prospection de l'or en Limousin. In: Johan Z, Ohnenstetter D (eds) Gisements métallifères dans leur contexte géologique. Doc Bur Rech Géol Min 158 (2): 591–607

Carré J-L (1979) Les minéralisations uranifères des dépôts oligocènes de Saint-Pierre (Cantal, France) dans leur cadre géologique régional et local. 3d Cycle Thesis, INPL, Nancy (unpublished)

Cathelineau M (1983 a) Le gisement d'uranium des Bondons: un exemple typique de métallogénèse polyphasée. C R Acad Sci Paris 296 II: 985–988

Cathelineau M (1983 b) Les minéraux phylliteux dans les gisements hydrothermaux d'uranium. II Distribution et évolution cristallochimique des illites, interstratifiés, smectites et chlorites. Bull Minéral 103: 553–569

Cathelineau M (1986) The hydrothermal alkali metasomatism effects on granitic rocks: quartz dissolution and related subsolidus changes. J Petrola 27: 945–965

Cathelineau M (1987 a) U-Th-REE mobility during albitization and quartz dissolution in granitoïds: Evidence from South East french Massif Central. Bull Mineral 110: 249–259

Cathelineau M (1987 b) Les interactions entre fluides et roches: thermométrie et modélisation. Exemples d'un système géothermique actif "Los Azufrès" et d'altérations fossiles dans la chaîne varisque. Doct. Etat Thesis, INPL, Nancy, 503 pp (unpublished)

Cathelineau M, Boiron MC, Holliger P, Marion P, Denis M (1989 a) Gold-rich arsenopyrite: crystal chemistry, gold location and state, physical chemical conditions of crystallization. Econ Geol Monogr 6, The geology of gold deposits: the perspectives in 1988: 328–340

Cathelineau M, Boiron MC, Holliger P, Poty B (1990) Metallogenesis of the French part of the Variscan orogen. Part II: Time-space relationships between U, Au, Sn-W ore deposition and geodynamic events: mineralogical and U-Pb data. Tectonophysics 177: 59–79

Chantal C, Leblanc M (1986) Interprétation métallogénique des amas sulfurés du Cambrien des Malines (France): dépôts synsédimentaires sur un talus continental. C R Acad Sci Paris 320 II: 579–582

Charef A, Sheppard SMF (1988) Les Malines Cambrian shale-hosted Pb-Zn deposit, France: Thermometric and isotopoic (H, O) evidence for pulsating hydrothermal mineralization. Mineral Deposita 23: 86–95

Cheilletz A (1988) Stratiform tungsten deposits: a review. With implications for the Yxsjörberg-Sandudden deposits of Sweden. Geol Mijnbouw 67: 293–311

Couilloud D (1988) Etude pétrographique, minéralogique et géochimique des skarns à tungstène de Fumade (Tarn, Montagne Noire). 3d Cycle, Univ. Toulouse (unpublished)

Courjault-Radé P (1989) Environnements sédimentaires et contextes géodynamiques des minéralisations polymétalliques de la province asturo-sarde, Europe méridionale. Comparaison Montagne Noire-Sardaigne Sud-Occidentale. Thesis, Univ. Toulouse, 250 pp (unpublished)

Crouzet J, Recoing M, Tollon F (1979) Les gisements aurifères du Massif Central français. Chron Rech Min 452: 5–38

Cuney M (1978) Geologic environment, mineralogy, and fluid inclusions of the Bois Noirs-Limouzat uranium vein, Forez France. Econ Geol 73: 1567–1610

Cuney M, Autran A (eds) (1987 a) Echassières: le forage scientifique. Une clé pour la compréhension des mécanismes magmatiques et hydrothermaux associés aux granites à métaux rares. Géol France 2–3: 350

Cuney M, Autran A (1987 b) Objectifs généraux du projet GPF Echassières nº 1 et résultats essentiels acquis par le forage de 900 m sur le granite albitique à topaze-lépidolite de Beauvoir. In: Cuney M, Autran A (eds) (1987) Echassières: le forage scientifique. Une clé pour la compréhension des mécanismes magmatiques et hydrothermaux associés aux granites à métaux rares. Géol France 2–3: 7–24

Cuney M, Brouand M (1987) Minéralogie et géochimie de U-Th dans le granite de Beauvoir et les micaschistes environnants. In: Cuney M, Autran A (eds) (1987) Echassières: le forage scientifique. Une clé pour la compréhension des mécanismes magmatiques et hydrothermaux associés aux granites à métaux rares. Géol France 2–3: 247–259

Cuney M, Friedrich M, Poty B (1984 a) Pétrologie et géochimie des radioéléments dans les granites et dans les leucogranites en particulier. Conséquences génétiques et applications à la prospection. In: Valorisation des ressources du sous-sol. Doc Bur Rech Géol Min 79: 91–125

Cuney M, Le Fort P, Wang ZX (1984 b) Uranium and thorium geochemistry and mineralogy in the Manaslu leucogranite (Nepal, Himalaya). In: Geology of granites and their metallogenetic relations. University Press, Beijing, pp 853–873

Cuney M, Friedrich M, Blumenfeld Ph, Bourguignon A, Vigneresse J-L, Boiron M-C, Poty B (1990) Metallogenesis in the French part of the Variscan orogen. Part I: U preconcentration in pre-Variscan and Variscan formations. A comparison with Sn, W and Au. Tectonophysics 177: 39–57

Demange M, Nicolas VA, Soler P, Giouse H (1988 a) Le batholite composite du Veynazès (Cantal, France): pétrographie, géologie et géochimie. Relations avec le gisement tungstifère de Leucamp. Bull Soc Géol Fr (8) IV: 839–850

Demange M, Nicolas VA, Soler P, Giouse H (1988 b) Le gisement tungstifère de Leucamp (Cantal, France). Contrôles géologiques et minéralisations. Bull Soc Géol Fr (8) IV: 559–570

Derré C (1982) La province à Sn-W Ouest-Européenne. Histoire de divers types de gisement du Massif Central, des Pyrénées et du Portugal. Distribution des gisements. Doct. Etat Thesis, P. and M. Curie University (Paris 6), 2 vol (unpublished)

Derré C, Lécolle M, Roger G (1982) Les quartzites à silicates calciques et scheelite: Préconcentrations familières ou pièges pour un tungstène étranger lié à l'hydrothermalisme périgranitique? Exemple du Nord-Est Transmontain (Portugal). Mineral Deposita 17: 363–385

Didier J, Lameyre J (1971) Les roches granitiques du Massif Central. In: Jung J (ed) Symp. Plein-Air Service, Clermont-Ferrand, pp 133–155

Disnar J-R, Gauthier B, Carter JS (1986) Utilisation des hydrocarbures gazeux en prospection régionale des minéralisations plombo-zincifères de couverture: application au gîte de Trèves (Gard, France). Chron Rech Min 482: 67–75

Disnar J-R, Sureau JF (1990) Organic matter in ore genesis: Progress and perspectives. Org Geochem 16: 577–599

Dubessy J, Ramboz C, Nguyen Trung C, Cathelineau M, Charoy B, Cuney M, Leroy J, Poty B, Weibrod A (1987) Physical and chemical control (fO2, T, pH) of the opposite behaviour of U and Sn-W as exemplified by hydrothermal deposits in France and Great Britain, and solubility data. Bull Minéral 110: 261–281

Durisova J (1988) Diversity of fluids in the formation of ore assemblages in the Bohemian Massif (Czechoslovakia). Bull Minéral 111: 477–492

Duthou J-L, Cantagrel J-M, Didier J, Vialette Y (1984) Palaeozoic granitoids of the French Massif Central: age and origin studied by 87Rb–87Sr system. Phys Earth Planet Int 35: 131–144

Einaudi MT, Meinert LD, Newberry RJ (1981) Skarn deposits. Econ Geol 75th Anniv Vol: 317–391

Emberger A, Périchaud J-J (1979) Carte des ressources minérales du Massif Central à l'échelle du 1/1 000 000ème. Bur Rech Géol Min, Paris, 105 pp

Eulry M, Vargas JM (1979) Les altérations météoriques anté-liasiques des Cévennes Méridionales. Leur rôle dans la métallogénèse des concentrations uranifères. 3d Cycle Thesis, INPL, (Nancy), 388 pp (unpublished)

Floc'h J-P, Santallier D, Henry B (1984) La série limousine: réflexion axée sur une synthèse géologique régionale et son utilisation pour la prospection de l'or. Chron Rech Min 474: 5–10

Foglierini F, Bernard A, Verraes G (1980a) Le gisement des Malines (Gard): Zn, Pb. Int Geol Cong., 26th, Paris, Gisements français, E5, pp 1–56

Foglierini F, Béziat P, Tollon F, Chabod JC (1980b) Le gisement filonien de Noailhac-Saint-Salvy (Tarn): Zn (Ag, Ge, Pb, Cd). Int Geol Cong., 26th, Paris, Gisements français, E6, pp 1–43

France-Lanord C, Le Fort P (1988) Crustal melting and granite genesis during the himalayan collision orogenesis. Trans R Soc Edinb Earth Sci 79: 183–195

Friedrich M (1984) Le complexe granitique hyperalumineux de St-Sylvestre, nord-ouest du Massif Central français. Géol Géochim Uranium Mém, Nancy, 5: 1–361

Friedrich M, Cuney M, Poty B (1987) Uranium geochemistry in peraluminous leucogranites. Uranium 3: 353–385

Friedrich M, Marignac Ch, Floc'h J-P (1988) Sur l'existence de trois chevauchements ductiles "himalayens" successifs à vergence NW en Limousin. C R Acad Sci Paris 306 II: 663–669

Gagny Cl, Leistel J-M, Sider H (1983) Le skarn à scheelite de Cherves: un exemple d'indice contrôlé par la tectonique tangentielle sudète dans le Beaujolais (NE du Massif Central français). Chron Rech Min 471: 23–30

Gagny Cl, Courrioux G, Cuénin O, Jacquot Th, Leistel J-M (1984) Contrôle structural de la métallogénèse dans la région d'Echassières par une zone de cisaillement régionale orientée à N 60°E (Massif Central français). C R Acad Sci Paris 298 II: 459–462

Gibert F, Moine M, Béziat P (1988) Les gneiss à silicates calciques minéralisés en W-Sn dans la série des schistes X, à la terminaison orientale de la Montagne Noire. Résultats préliminaires. In: Johan Z, Ohnenstetter D (eds) Gisements métallifères dans leur contexte géologique. Doc Bur Rech Géol Min 158: 219–244

Guérangé-Lozes J, Alsac C (1986) Les nappes varisques de l'Albigeois cristallin. Lithostratigraphie, volcanisme et déformation. Géol Fr 3: 309–337

Guillot C, Moine B, Beny C, Tollon F, Touray JC (1988) Repartition and origin of N2 in the fluid inclusions of the dome of Montredon, Montagne Noire, France. Chem Geol 70: 163

Guion JL, Béziat P, Tollon F (1985) Données nouvelles sur les paragenèses du champ filonien à wolframite de Lacoste, district de Montredon-Labessonié (Tarn). Bull Minéral 108: 391–402

Hubert P (1986) Textures et inclusions fluides des quartz aurifères. Application au gîte de Cros Gallet (Haute Vienne, France) et au prospect de Sanoukou (District de Kinieba, Mali). Doc Bur Rech Géol Min 114;: 1–350

Icart JC, Lacharpagne JC, Sider H, Walgenwitz F (1980) Altération et minéralisation de type porphyry à Sibert (Rhône). Chron Rech Min 455: 421–443

Javoy M, Brevart O, Tollon F (1983) L'origine des minerais filoniens et des fluides minéralisateurs de Salsigne d'après les données de la géochimie isotopique. In: Proc Coll ATP Géochimie-Métallogénie du CNRS, Bonas, pp 196–204

Jebrak M (1985) Contribution à l'histoire naturelle des filons (F, Ba) du domaine varisque. Do Bur Rech Géol Min 99: 1–473

Joubert M (1978) Etude pétrographique, structurale et métallogénique de la Châtaigneraie (secteur de Veynazès, Cantal), Massif Central français. 3d Cycle Thesis, Ecole Nationale Supérieure des Mines de Paris (unpublished)

Kosakévitch A, Moëlo Y (1982) Le plomb dans les gisements d'antimoine du Maroc central. Chron Rech Min 467: 5–32

Lacerda H, Bernard AJ (1984) Existence de minéralisations plombo-zincifères syngénétiques du substratum Cambrien du district des Malines (Gard, France). Mineral Deposita 19: 152–157

Lacomme A, Milesi JP, Gros Y (1987) L'amas sulfuré à Cu-Zn de Chessy (Rhône), évolution historique des recherches, état actuel des connaissances sur le gisement et son environment. Chron Rech Min 489: 3–20

Lancelot J, Vella V (1989) Datation U-Pb liasique de la pechblende de Rabejac. Mise en évidence d'une préconcentration uranifère permienne dans le bassin de Lodève (Hérault). Bull Soc Géol Fr (8) V: 309–316

Lancelot JR, De Saint André B, La Boisse H de (1984) Systématique U/Pb et évolution du gisement d'uranium de Lodève (France). Mineral Deposita 19: 44–53

Landais P, Conan J (1980) Relation uranium-matière organique dans deux bassins permiens français: Lodève (Hérault) et Cérilly-Bourbon L'Archambault (Allier). Bull Cent Rech Explor Prod Elf-Aquitaine 4: 709–757

Lardeau M (1988) Etude des principaux sulfures du minerai de Chessy-les-Mines (Rhône) à l'aide de la microsonde électronique; implications génétiques. In: Johan Z, Ohnenstetter D (eds) Gisements métallifères dans leur contexte géologique. Doc Bur Rech Géol Min 158 (1): 307–324

Lecuyer E (1982) Etude du gîte de pyrite de Chizeuil (Saône et Loire) et de son environnement géologique. 3rd Cycle Thesis, Univ. Paris IV (unpublished)

Ledru P, Lardeaux J-M, Santallier D, Autran A, Quenardel J-M, Floc'h J-P, Lerouge G, Maillet N, Marchand J, Ploquin A (1989) Qù sont les nappes dans le Massif Central français? Bull Soc Géol Fr (8) V: 605–618

Lemière B (1983) L'amas sulfuré de Chizeuil (Saône et Loire). Doc Bur Rech Géol Min 54: 1–144

Lemière B, Delfour J, Moine B, Piboule M, Ploquin A, Isnard P, Tegyey M (1986) Acidic model for the formation of aluminous haloes around sulphide deposits at Chizeuil (Morvan). Mineral Deposita 21: 147–155

Lépine J (1989) Le gisement sulfo-arsénié aurifère cambrien inférieur de la mine traditionnelle de Salsigne (Montagne Noire). 3d Cycle Thesis, Univ. Toulouse (unpublished)

Lépine J, Ben Ayad MA, Béziat D (1988 a) Mise en évidence de métabasites alcalines Cambrien inférieur dans le district aurifère de Salsigne (versant Sud de la Montagne Noire). C R Acad Sci Paris 306 II: 1087–1092

Lépine J, Talayssat M, Crouzet J, Tollon F (1988 b) Les minéralisations sulfo-arséniées aurifères cambriennes de la mine traditionnelle de Salsigne (Aude, France). In: Johan Z, Ohnenstetter D (eds) Gisements métallifères dans leur contexte géologique. Doc Bur Rech Géol Min 158 2: pp 627–650

Leroy J (1978) The Margnac and Fanay uranium deposits of the La Crouzille district (Western Massif Central, France): geologic and fluid inclusion studies. Econ Geol 73: 1611–1634

Leroy J (1984) Episyénitisation dans le gisement d'uranium du Bernardan (Marche). Comparaison avec des gisements similaires du nord-ouest du Massif Central français. Mineral Deposita 19: 26–35

Leroy J, Holliger P (1984) Mineralogical, chemical and isotopic (U-Pb method) studies of hercynian uraniferous mineralizations (Margnac and Fanay mines, Limousin, France). Chem Geol 45: 121–134

Lescuyer J-L, Giot D (1988) Les amas de substitution tardi-hercyniens de La Rabasse (Monts de l'Orb, Hérault): un type de gisement de Pb-Zn peu connu. Chron Rech Min 490: 3–16

Lespinasse M (1984) Contexte structural des gisements d'uranium de la Marche occidentale: fracturation, circulations fluides, propagation de l'épisyénitisation. Géol Géochim Uranium Mém, Nancy, 8, pp 1–196

Lespinasse M, Pecher A (1986) Microfracturing and regional stress field: a study of the preferred orientations of fluid inclusion planes in a granite from the Massif Central, France. J Struct Geol 8: 169–180

Levèque MH, Lancelot JR, George E (1988) The Bertholène uranium deposit. Mineralogical characteristics and U/Pb dating on the primary U mineralization and its subsequent remobilization: Consequences upon the evolution of the U deposits of the Massif Central, France. Chem Geol 69: 147–163

Macquar JC, Marini D, Rouvier H (1988) Métallogénie post-triasique dans le district plombo-zincifère des Malines (Gard). In: Johan Z, Ohnenstetter D (eds) Gisements métallifères dans leur contexte géologique. Doc Bur Rech Géol Min 158 (2): 773–802

Marcoux E (1987) Isotopes du plomb et paragenèses métalliques, traceurs de l'histoire des gîtes minéraux. Doc Bur Rech Géol Min 117: 1–289

Marcoux E, Bonnemaison M (1988) La géochimie isotopique du plomb et la prospection de l'or en France. In: Johan Z, Ohnenstetter D (eds) Gisements métallifères dans leur contexte géologique. Doc Bur Rech Géol Min 158 (2): 489–508

Marcoux E, Bril H (1986) Héritage et source des métaux d'après la géochimie isotopique du plomb. Exemple des minéralisations filoniennes (Sb, Pb, Ba, F) du Haut-Allier (Massif Central, France). Mineral Deposita 21: 35–43

Marcoux E, Jebrak M (1987) Approche géochimique de l'origine et de la durée des dépôts hydrothermaux dans le district d'Ussel (Massif Central Français). C R Acad Sci Paris 305 II: 377–381

Marcoux E, Picot P (1985) Les minéralisations de Pontgibaud (Puy-de-Dôme): une approche par la géochimie isotopique du blomb et les paragenèses. Chron Rech Min 481: 27–38

Marcoux E, Picot P, Moëlo Y (1985) Evolution paragenétique des minéralisations sulfurées aurifères de Pontvieux-Labessette (Puy-de-Dôme). Apport des études isotopiques du plomb. Bull Minéral 108: 465–481

Marcoux E, Moëlo Y, Picot P, Baubron J-C (1988) Evolution minéralogique et isotopique (Pb) du filon sulferé des Borderies (Massif central français). Implications métallogéniques. Mineral Deposita 23: 58–70

Marcoux E, Bonnemaison M, Braux Ch, Johan Z (1989) Distribution of Au, Sb, As and Fe in gold-bearing arsenopyrite from Le Châtelet and Villeranges deposits (Creuse, french Massif Central). C R Acad Sci Paris 308 II: 293–300

Marignac Ch, Weisbrod A (1986) L'origine des gisements stanno-wolframifères: une approche critique. In: Abstr European Meet Tungsten Deposits, Toulouse, 12–14 Mai 1986, pp 24–25

Marion P, Wagner FE, Jumas JC, Regnard JR (1988) Apports de la spectroscopie Mössbauer à la connaissance de l'état de liaison de l'or dans les sulfures. In: Johan Z, Ohnenstetter D (eds) Gisements métallifères dans leur contexte géologique. Doc Bur Rech Géol Min 158 (2): 509–525

Martin S (1981) Episyénitisation et minéralisation dans le leucogranite d'Hyverneresse (Nord Millevaches, M. C. F.). 3d Cycle Thesis, Univ. Clermont II, 222 pp (unpublished)

Mezure J-F, Negroni J-M (1983) Relations structurales, pétrographiques et géochimiques de deux intrusions magmatiques à potentialité métallogénique: les granites de Gelles et de Meymac (Massif Central français). Bull Soc Géol Fr (7) XXV: 71–82

Michel J-J (1983) Episyénites et concentrations uranifères associées dans le Massif de Saint-Sulpice-Les-Feuilles (Massif Central, France). 3d Cycle Thesis, INPL, Nancy (unpublished)

Moëlo Y (1982) Contribution à l'étude des conditions naturelles de formation des sulfures complexes d'antimoine et de plomb (sulfosels de Pb-Sb), et signification métallogénique. Doc Bur Rech Géol Min 55: 1–624

Mourey Y (1985) Le leucogranite à topaze de Chavence. Un nouvel exemple de massif à Sn, W, Li dans le Nord du Massif Central Français. C R Acad Sci Paris 300 II: 951–954

Munoz M, Shepherd TJ (1987) Fluid inclusion study of the Bournac polymetallic (Sb-As-Pb-Zn-Fe-Cu) vein deposit (Montagne Noire, France). Mineral Deposita 22: 11–17

Munoz M, Kim YD, Verraes G, Tollon F (1988) Hydrothermal alteration parageneses associated with the zinc vein deposit of Noailhac-Saint Salvy (Tarn, France). Conditions of mineralization deposition. Proc. 7th IAGOD Symp. Schweizerbart'sche Verlag-Buchhandlung, Stuttgart, pp 171–178

Nenert S (1989) Les minéralisations aurifères des districts de Bénévent-L'Abbaye et Laurière dans leur contexte géologique et métallogénique. Limousin, Massif Central français. Thesis, Univ. Limoges, 268 pp (unpublished)

Nicolas VA (1985) Le gisement wolframifère de Leucamp dans son environnement géologique (Châtaigneraie du Cantal, France). 3d Cycle Thesis, Ecole Nationale Supérieure des Mines de Paris (unpublished)

Noyé F, Weisbrod A (1988) Les indices stanno-wolframifères de Saint-Mélany (Cévennes Médianes, Ardèche, Massif Central français). In: Johan Z, Ohnenstetter D (eds) Gisements métallifères dans leur contexte géologique. Doc Bur Rech Géol Min 158 (1): 175–194

Oksengorn S (1958) Métallogénie du gisement de La Loubatière. Bull Soc Géol Fr (7) VIII: 281–287

Pecher A, Lespinasse M, Leroy J (1985) Relations between fluid inclusion trails and regional stress field: a tool for fluid chronology – An example of an intragranitic uranium ore deposit (nortwest Massif Central, France). Lithos 18: 229–237

Pfeiffer MT (1986) La signification de la ligne tonalitique du Limousin. Son implication dans la structuration varisque du Massif Central français. C R Acad Sci Paris 303 II: 305–310

Périchaud JJ (1980) L'antimoine, ses minerais et ses gisements. Synthèse gîtologique sur les gisements du Massif Central français. Chron Rech Min 456: 5–64

Picot P (1965) Le gîte de cuivre, étain et fer de Charrier (Allier). In: Colloque sur la géologie et la minéralogie du Massif Central. Publ Fac Sci Clermont-Ferrand, fasc. 2, pp 120–131

Pin Ch, Peucat JJ (1986) Ages des épisodes de métamorphisme paléozoïque dans le Massif Central et le Massif Armoricain. Bull Soc Géol Fr (8) II: 461–469

Poty B, Leroy J, Cathelineau M, Cuney M, Friedrich M, Lespinasse M, Turpin L (1986) Uranium deposits spatially related to granites in the french part of the Hercynian orogen. In: Vein type Uranium deposits, IAEA, Vienna, pp 215–246

Proust D (1988) Altérations hydrothermales du gisement aurifère de Cros-Gallet, Le Bourneix, district de Saint-Yrieix (Haute-Vienne). In: Johan Z, Ohnenstetter D (eds) Gisements métallifères dans leur contexte géologique. Doc Bur Rech Géol Min 158, vol 2: 609–626

Raimbault L (1984) Géologie, pétrographie et géochimie des granites et minéralisations associées de la région de Meymac (Haute Corrèze, France). Ing.-Doct. Thesis, Ecole Nationale Supérieure des Mines de Paris (unpublished)

Raimbault L (1985) Utilisation des spectres de terres rares des minéraux hydrothermaux (apatite, fluorine, scheelite, wolframite) pour la caractérisation des fluides minéralisateurs et l'identification des magmas sources et des processus évolutifs. Bull Minéral 108: 737–744

Raimbault L, Meyer G, Treuil M (1987) Comportements différenciés de W, Sn, U, Ta, Nb dans quelques complexes granitiques du Massif Central français. Bull Minéral 110: 591–601

Ramboz C (1980) Géochimie et étude des phases fluides de gisements et indices d'étain-tungstène du Sud du Massif Central (France). 3d Cycle Thesis, INPL, Nancy (unpublished)

Ramboz C (1986) Paramètres physiques et facteurs de contrôle des circulations hydrothermales convectives impliquées dans la formation des concentrations en tungstène du Sud du Massif Central français. In: Abstr European Meet Tungsten Deposits, Toulouse, 12–14 Mai 1986, pp 22–23

Ramboz C, Charef A (1988) Temperature, pressure, burial history, and paleohydrology of the Les Malines Pb-Zn deposit: reconstruction from aqueous inclusions in barite. Econ Geol 83: 784–800

Ramboz C, Schnapper D, Dubessy J (1985) Reconstitution from fluid inclusion studies of the P-V-T-X evolution of a carbonaceous hydrothermal event in the wolframite-vein of Serrecourte (MCF, France). Geochim Cosmochim Acta 49: 205–219

Respaut JP (1984) Géochronologie et géochimie isotopique U-Pb de la minéralisation uranifère des Pierres Plantées (Lozère) et de son encaissant: le massif granitique de la Margeride. 3d Cycle Thesis, U. S. T. L. Montpellier, 122 pp (unpublished)

Reynolds DG (1965) Geology and mineralization of the Salsigne gold mine, France. Econ Geol 60: 772–791

Roger G (1972) Un type de minéralisations épigénétiques familières: les filons à antimoine du Massif Central français. Hypothèse de la sécrétion latérale. Mineral Deposita 7: 360–382

Roques M (1969) Notice explicative, carte géologique 1/80000, feuille Périgueux, 2nd edn. Bur Rech Géol Min, Orléans

Rossi Ph, Cocherie A, Meyer G, Fouillac AM, Autran A (1988) Metallogenic W and Sn granites: genesis and main distinguishing features. In: Boissonnas J, Ohmenetto P (eds) Mineral deposits within the European Community. Springer, Berlin Heidelberg New York

Routhier P (1980) Où sont les métaux pour l'avenir? Les provinces métalliques. Essai de métallogénie globale. Mém Bur Rech Géol Min 105: 1–409

Routhier P, Brouder P, Fleischer P, Macquar J-C, Pavillon J-M, Roger G, Rouvier M (1973) Some major concepts of metallogeny. Mineral Deposita 8: 237–258

Safa P, Couilloud D, Tessier B, Moine B (1987) Le gisement de tungstène de Fumade (Tarn, France): Gîtologie, caractères pétrographiques et géochimiques des skarns et des granites. Chron Rech Min 487: 39–51

Samama J-C (1968) Contrôle et modèle génétique de minéralisations en galène de type "Red-Beds". Mineral Deposita 3: 261–271

Samama J-C (1971) Paléoplacers à tungstène-étain dans les séries métamorphiques du Vivarais oriental. C R Acad Sci Paris 272: 516–518

Sandras A (1988) Les structures auro-antimonifères du district de Brioude Massiac (Massif Central français): Gitologie et metallogénie des concentrations aurifères. Thesis, Nancy I Univ., 216 pp (unpublished)

Sandras A, Gagny Cl, Moyroud B (1988) Contribution de la pétrologie structurale à la connaissance du prospect à tungstène des Chênelettes, granite de Saint-Cierge (Ardèche). Chron Rech Min 490: 25–34

Sider H, Ohnenstetter M (1986) Field and petrological evidence for the development of an ensialic marginal basin related to the Hercynian orogeny in the Massif Central, France. Geol Rundsch 75: 421–443

Sider H, Leistel JM, Gagny C (1986) Réflexions sur l'évolution géodynamique de la chaîne hercynienne au Paléozoïque dans le Nord-Est du Massif Central français. Bull Soc Géol Fr (8) II: 637–644

Sillitoe R (1985) Ore related breccias in volcano-plutonic areas. Econ Geol 80: 1467–1514

Soulé de Lafont D, Lhégu J (1980) Les gisements stratiformes de fluorine du Morvan. Int Geol Cong., 26th, Paris, Gisements français, E2

Stussi J-M, La Roche H de (1984) Le magmatisme orogénique granitique de la chaîne varisque française. Typologie chimique et répartition spatiale. C R Acad Sci Paris 298 II: 43–48

Touray J-C (1987) Transport et dépôt de l'or dans les fluides de la croûte continentale, l'apport des inclusions fluides. Chron Rech Min 488: 43–53

Turpin L (1984) Altérations hydrothermales et caractérisation isotopique (O-H-C) des minéraux et des fluides dans le massif uranifère de Saint-Sylvestre – extension à d'autres gisements intragranitiques d'uranium français. Géol Géochim Uranium Mém, Nancy, 6, pp 1–190

Turpin L (1985) Rb-Sr dating of hydrothermal alteration in Saint-Sylvestre leucogranite (French Massif Central). EUG III. Terra Cognita 5: 280

Turpin L, Cuney M, Friedrich M, Bouchez J-L, Aubertin M (1990) Meta-igneous origin of Hercynian peraluminous granites in N-W. French Massif Central: implications for crustal history reconstructions. Contrib Mineral Petrol 104: 163–172

Vella V, Lancelot JR (1986). Evolution pluriphasée du gîte uranifère de Lodève. In: 11ème RST, Paris. Abstr Soc Géol France, p 180

Vuagnat Ph (1984) L'amas sulfuré de Chessy (Beaujolais méridional). Etude cartographique, pétrographique et géochimique des formations dévono-dinantiennes de son enveloppe sédimentaire. 3d Cycle Thesis, Univ. Lyon, 137 pp (unpublished)

Weisbrod A (1988) Caractères généraux des phases fluides dans les indices et gisements de tungstène et d'étain. In: Johan Z,

Ohnenstetter D (eds) Gisements métallifères dans leur contexte géologique. Doc Bur Rech Géol Min 158 (1): 9–14
Weppe M (1951) Contribution à l'étude des gîtes de tungstène français: Puy-les-Vignes (Haute-Vienne), la Châtaigneraie (Cantal). Société d'Impressions typographiques, Nancy, 210 pp
Zappettini EO (1983) Le gisement d'or du Châtelet (Creuse, Massif Central français). Cadre géologique régional. Etude métallogénique. 3d Cycle Thesis, Univ. Limoges, 151 pp (unpublished)
Zappettini EO, Picot P, Sabourdy G (1983) Nouvelles données sur le gisement aurifère du Châtelet (Massif Central français), liaison génétique probable entre l'or et les roches ultrabasiques. C R Acad Sci Paris 297: 351–354

Part IV Correlative Areas

The Aquitaine Basin

F. PARIS and G. LE POCHAT

1 Introduction

The Aquitaine basin occupies about 50000 km^2 on the mainland. It is bounded to the east by the French Massif Central and to the south by the Pyrénées. It extends seawards on the South Armorican margin as far as the Bay of Biscay, where oceanic crust occurs. The Paleozoic basement rocks of the Aquitaine are only known from boreholes and consist mainly of terrigenous deposits ranging from Late Cambrian up to Permian. Subordinate calcareous sections occur in the Devonian and possibly in the Early Carboniferous.

Some features of the Mesozoic and Cenozoic history of the Aquitaine basin are summarized here as they have direct influence on the distribution of Paleozoic subsurface material intersected by exploratory oil wells. The major feature is the Celt-Aquitaine flexure (Winnock 1971), which subdivides the basin into two regions of different post-Paleozoic evolution. To the north is a fairly stable area covered by a relatively thin Mesozoic and Cenozoic succession (usually less than 3000m). Southward, several episodes of rifting and crustal thinning resulted in an acceleration of the subsidence rate. These events, related to continental extension (opening of the Atlantic Ocean and subsequently to the opening of the Bay of Biscay) occurred respectively during the Trias-Early Lias, the Late Jurassic-Late Aptian, and the Albian (Bur Rech Geol Min et al. 1974; Brunet 1984; Montadert 1984; Mullan 1984). The last event, corresponding to a crustal flexuring of the southern part of the basin, resulted from its loading with northvergent Pyrenean thrusts and molasse deposits during Late Cretaceous and Eocene times (see references in Brunet 1984). South of the Celt-Aquitaine flexure the basement is very deep (e.g., 6000 m beneath the Parentis basin and up to 10000 m beneath the Adour basin; Fig. 1) and most of the available data dealing with the Paleozoic basement of the Aquitaine basin are concentrated in its northern part (Fig. 1). Some additional information is provided by wells located on the northern border of the Pyrénées. However, as this Paleozoic material is fairly similar to that of the neighboring outcrops, these data will not be considered further.

2 Stratigraphy and Paleontology

The knowledge of the Paleozoic succession of the Aquitaine basement is mainly based on data from about 40 wells. Despite recent improvements (Paris et al. 1987), it seems premature to establish a single synthetic stratigraphic column for the whole Paleozoic of this area, as the available data remain sparse and fragmentary. Moreover, the subsurface material comes from three distinct terranes (Fig. 2), namely the Western, the Median, and the Eastern units (Paris et al. 1988). These terranes may have a different paleogeographic origin and thus, in order to prevent any stratigraphic miscorrelation, separate columns will be described in ascending chronological order for each unit.

2.1 Paleozoic Succession in the Western unit (Figs. 3, 4)

Uppermost Cambrian (?) and Early Ordovician faunas have been recorded in wells BSM.1, Non.1 and Mc.101. Other wells (e.g., Lc.101, Len.1, LTn.1 bis, Lo.1, and LPg.1) are barren, but they are tentatively referred to the Early Ordovician on the basis of lithologic similarities.

So far, the oldest paleontologically dated strata in the Aquitaine basin belong to the Sainte Marthe Formation (at least 200m thick, as its base is unknown). This formation includes gray-green to dark mudstones and siltstones with thin interbedded sandstones and quartzites. The main characteristic of this formation is the occurrence of calcareous and reddish dolomitic horizons near the bottom of well BSM.1 (Bugnicourt et al. 1959). Trilobite remains and Late Cambrian (?)-Early Tremadoc acritarchs are report-

J. D. Keppie (Ed.): Pre-Mesozoic Geology in France and Related Areas

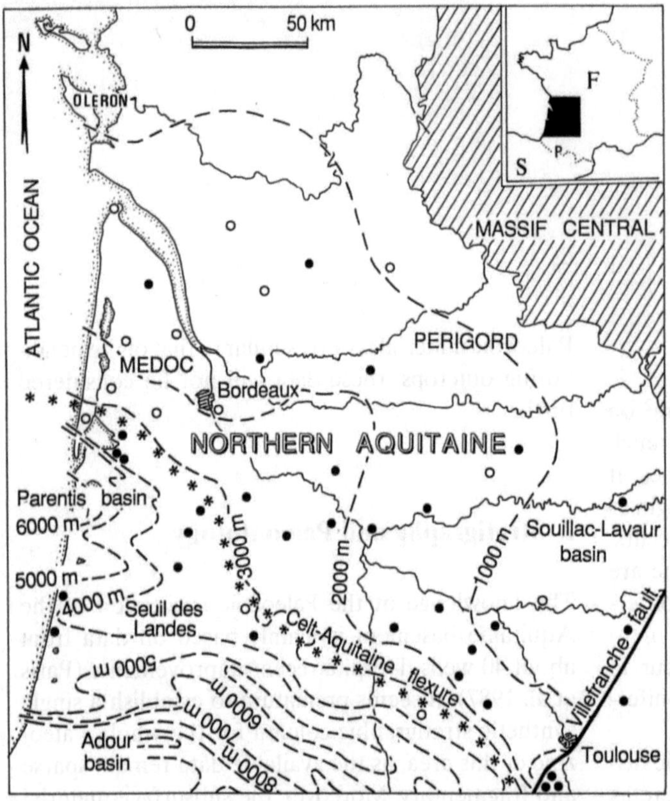

Fig. 1. Index map showing the isobaths of the basement roof and the location of the main geologic features of the Aquitaine basin. The position of the wells which have reached the basement is indicated by *black circles* (age assignment based on paleontologic or lithologic evidence) or by *open circles*. (age assignment not documented) (isobath curves from Bur Rech Geol Min et al. 1974) (*F* France; *S* Spain; *P* Pyrénées)

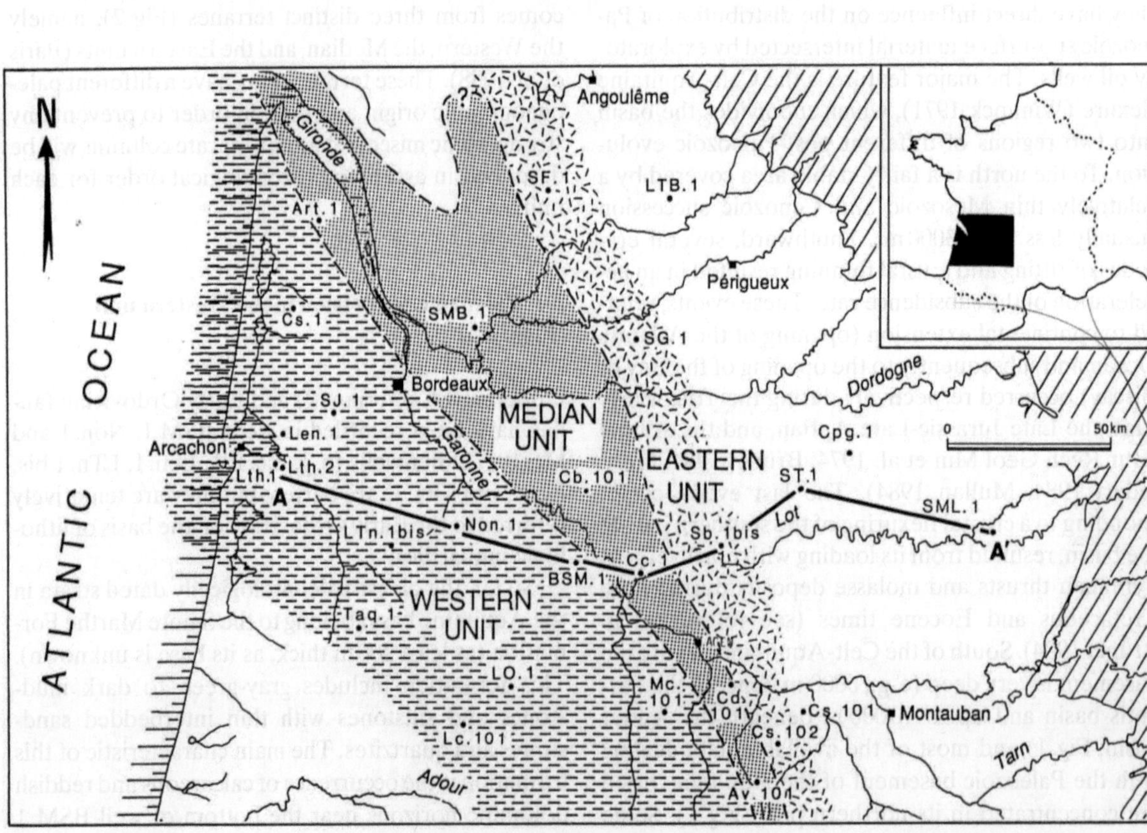

Fig. 2. Simplified map showing the distribution of the three units identified in the Paleozoic basement of the Aquitaine basin and location of the main oil exploration wells penetrating this basement. (Paris et al. 1988). *A–A'* cross-section of Fig. 6

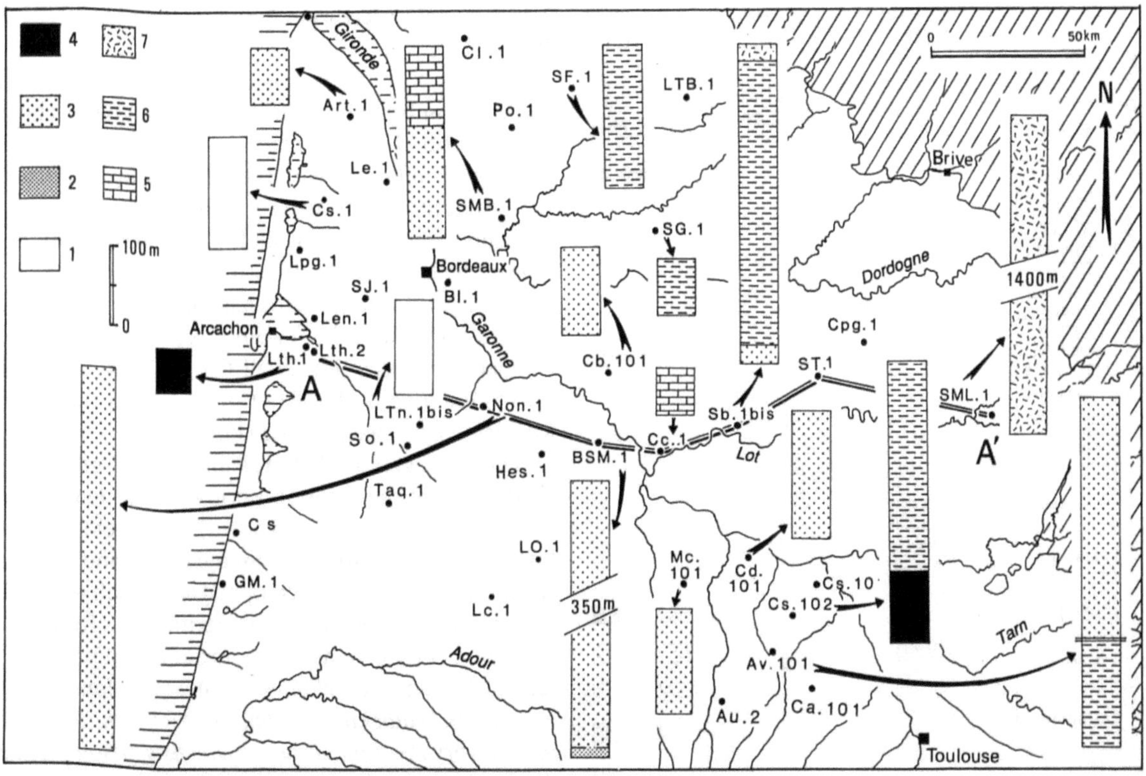

Fig. 3. Thickness and age assignment of Paleozoic strata intersected by oil exploration wells in northern Aquitaine. *1* unspecified age; *2* Cambrian; *3* Ordovician; *4* Silurian; *5* Devonian; *6* Early Carboniferous; *7* Stephano-Permian; *A–A'* location of the cross-section of Fig. 6

ed from these strata (Baudelot and Fournier-Vinas 1984). The strata assigned to the Cambrian on lithologic grounds (i. e., occurrence of dolomites) by Mullan (1984), specifically the dolomites of wells Cd. 101 and LTh. 1, are respectively of Middle Ordovician and Silurian age as demonstrated on faunal evidence (Paris et al. 1987)

The overlying Bouglon Formation is a thick quartzitic succession (about 250 m) with thin argillaceous interbeds yielding rich Tremadocian acritarch assemblages (Baudelot and Fournier-Vinas 1984). This formation is completely penetrated by well BSM. 1 and partly intersected in well Non. 1. The quartzitic material of well Lo. 1 probably belongs to the Bouglon Formation.

The succeeding Moncrabeau Formation (about 250 m thick) corresponds to somewhat deeper marine deposits represented by black mudstones and siltstones, with minor sandy intercalations. A Middle Arenig fauna, including trilobites *(Taihungshania* cf. *shuilandeyranensis),* graptolites *(Azygograptus suecicus, Tetragraptus* sp.), bivalves, acritarchs, and chitinozoans *(Lagenochitina esthonica, Eremochitina brevis,* and *Tanuchitina achabae),* is reported from wells Non. 1, Mc. 101 and BSM. 1 (Paris et al. 1987).

The youngest Ordovician strata so far paleontologically dated in this Western unit correspond to the Noaillan Formation. Strongly bioturbated alternating shales and sandstones occur in its lower part, while quartzitic beds prevail toward the top. The upper part of the Noaillan Formation is still unknown and thus its thickness may exceed 300 m. The poorly preserved chitinozoan assemblage *(Lagenochitina obeligis, Jenkinochitina vulgaris)* recovered from well Non. 1 suggests a late Arenig-lowermost Llanvirn age, at least for the middle part of the formation (Paris et al. 1987).

Suitable material to illustrate the Middle and Late Ordovician succession of the Western unit is still lacking. Moreover, Variscan cleavage and shearing have altered the original lithology of the Paleozoic deposits drilled in several wells northwest of Bordeaux (e. g., Len. 1, SJ. 1, and Cs. 1). Thus their precise stratigraphic assignment, in spite of some lithologic similarities with Early Ordovician formations, cannot be specified.

In the Western unit, fossiliferous Silurian strata of the Le Teich Formation are tectonically disturbed (recumbent folds, schistosity, shearing). The prevailing lithology, as seen in well LTh. 1, corresponds to

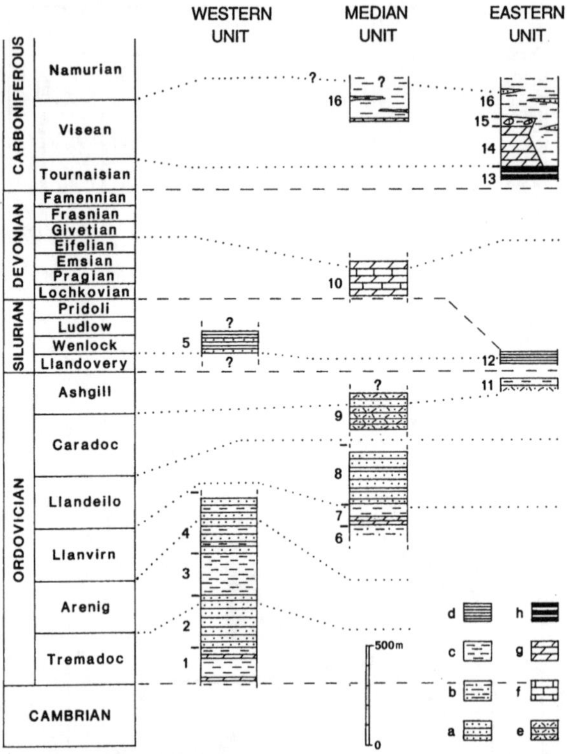

Fig. 4. Age assignment and correlations between the pre-Stephanian formations of the northern Aquitaine basement. (Paris et al. 1988). Name of the formations: *1* Saint Marthe; *2* Bouglon; *3* Moncrabeau; *4* Noaillan; *5* Le Teich; *6* Caubon; *7* Caudecoste; *8* Avensac; *9* Saint Martin du Bois; *10* Clairac; *11* Soubirous; *12* black shales; *13* Castelsarrasin; *14* dolostones of Soubirous; *15* diamictites of Soubirous; *16* Saint Géry. Lithology: *a* quartz arenites; *b* siltstones; *c* mudstones; *d* black shales; *e* volcaniclastic rocks; *f* limestones; *g* dolostones; *h* cherts

banded black shales and gray millimetric silty levels. The occurrence of low grade metamorphic minerals (e. g., talc, margarite) has to be stressed. Due to the recumbent folds, the thickness of the formation cannot be specified (a few tens of meters?). Extremely poorly preserved monograptids supporting a Silurian age have been recorded near the bottom of well LTh. 1 (Paris et al. 1987). In all likelihood, the Le Teich Formation is represented in the neighboring well LTh. 2 and probably eastwards.

No definite assignment either to Devonian or to Early Carboniferous strata can be documented in the Western unit. However, the dolomitic succession reported by the operator in well LTh. 1 and possibly in well Cts. 1 might represent part of the Devonian. The youngest Paleozoic deposits recorded in this terrane are probably the well-sorted sandstones and the associated reddish mudstones assigned to the Late Carboniferous-Early Permian by the operator in well SGM. 1 (south of the Parentis basin).

2.2 Paleozoic Succession in the Median Unit (Fig. 3, 4)

Neither Cambrian nor Early Ordovician faunas have been identified in this terrane. Nevertheless, based on lithologic similarities with the Sainte Marthe Formation, the unfossiliferous green shales, siltstones and micaceous sandstones drilled in well Art. 1 may belong to the Late Cambrian or to the Tremadoc.

The fossiliferous Ordovician succession includes: the Caubon Formation (wells Cb. 101, Cd. 101), the Caudecoste Formation (well Cd. 101), the Avensac Formation (wells Av. 101, Cd. 101), and the Saint Martin du Boiș (well SMB. 1) Formation.

The gray-green siltstones and fine argillaceous sandstones of the Caubon Formation (at least 90 m thick, as its base is still unknown) have yielded trilobite remains, graptolites (Diplograptids and diverse siculae), inarticulate brachiopods and a rich and well preserved paleoplankton of acritarchs (Rauscher 1973) and chitinozoans (Taugourdeau 1961). The latter, including *Halochitina retracta, Belonechitina communis, Kalochitina inflata, Laufeldochitina stentor* and very few *Linochitina pissotensis,* belong to the the *L. pissotensis* Biozone (Paris 1989) and therefore indicate a lower Llandeilo age for the whole formation (Paris et al. 1987; Paris unpublished).

The gray dolomites and overlying dark shales with sandy interbeds intersected in well Cd. 101 are provisionally regrouped into the Caudecoste Formation (less than 100 m thick). The shales contain poorly preserved and nondiagnostic acritarchs (Baudelot and Fournier-Vinas 1984) and chitinozoans (Paris 1984) tentatively assigned to the Early Llandeilo (Paris et al. 1987).

The Avensac Formation corresponds to well-bedded, gray, fine-grained micaceous sandstones with black argillaceous or silty interbeds. The top of this formation is still unknown, thus its thickness may exceed the 280 m drilled in well Av. 101. No macrofossils have been recorded in these shallow marine deposits. The age assignment is provided by abundant and well-preserved early Llandeilian chitinozoans associated with acritarchs (Rauscher 1973) and a few graptolite remains. Among the most diagnostic chitinozoans are *Linochitina pissotensis, Pogonochitina simplex, Conochitina pellucida, Belonechitina turris,* and *Hercochitina* sp. (Taugourdeau 1961; Paris et al. 1987).

The Late Ordovician deposits of the Median unit are represented by the tuffites and wackes of the Saint Martin du Bois Formation. Volcanic clasts (e. g., glass fragments, embayed quartz, feldspars) are variously distributed in a terrigenous and calcareous matrix (Pelhate and Holtzapffel in Paris et al. 1987). This formation, at least 160 m thick (the base was not drilled in well SMB. 1), contains moderately abundant bioclasts (bryozoans and corals remains) and a few organic microfossils such as scolecodonts, graptolite tissues, acritarchs, and chitinozoans *(Spinachitina* sp. aff. *bulmani, Belonechitina* e.g. *robusta, Lagenochitina baltica* ?). The chitinozoans, in spite of a poor preservation, are indicative of a Late Caradoc-Early Ashgill age. However, they may be reworked in younger deposits.

Silurian deposits have not been identified in the subsurface of the Median unit. The absence of Silurian black shales between Late Ordovician and Early Devonian strata is more likely related to thrusting processes (decollement horizon) than to a depositional gap.

Devonian strata are represented by the limestones and dolomites of the Clairac Formation (more than 150 m thick). The lower part is drilled in well SMB. 1. It consists of argillaceous limestones with locally abundant quartz grains and a few reworked volcanic clasts. Ostracodes, brachiopods *(Howellella* e.g. *mercuri, Atrypa* pl. sp.), bryozoans, crinoids as well as a few acritarchs, scolecodonts, miospores, and chitinozoans *(Eisenachitina bohemica, Margachitina catenaria)* occur in the lower part of the formation. They indicate an early Lochkovian age. Higher up in the sequence explored in well SMB. 1, Pragian and probably Emsian chitinozoans are reported (Paris et al. 1987). The rest of the Clairac Formation is represented in well Cc. 1 by shelf packstones and wackestones with trilobites (proetids), numerous ostracodes, brachiopods *(Atrypa* sp., *Schizophoria* sp.), bryozoans, crinoids, and an abundant coral fauna *(Calceola sandalina,* stromatopores) of Late Eifelian-Early Givetian age (Bugnicourt et al. 1959). A well-preserved paleoplankton of tasmanitids, acritarchs, scolecodonts, miospores, and chitinozoans *(Ancyrochitina cornigera, Gotlandochitina gilbertsoni)* supports this stratigraphic assignment.

Late Devonian strata have not been identified in the Median unit. On the other hand, the dark shales, siltstones, and a reddish sandstone bed which occur in the lower part of well Av. 101, beneath the Middle Ordovician Avensac Formation, are tentatively referred to the Early Carboniferous Saint Géry Formation. This age assignment is supported by the record of tracheids (with complex pits) and dark miospores near the bottom of well Av. 101. These fossils, which are incompatible with an Ordovician age (Paris 1984), are associated with a few acritarchs (Rauscher 1973) regarded now as reworked (Paris et al. 1987). Consequently, in well Av. 101, the Avensac Formation overlaps much younger strata (see below).

2.3 Paleozoic Succession in the Eastern Unit (Fig. 3, 4)

Five wells (SF. 1, SG. 1, Sb. 1 bis, Cs. 102, and Ca. 101) have explored fossiliferous Paleozoic sections (Fig. 5). The oldest dated strata are the siltstones and mudstones of the Soubirous Formation (about 25 m thick). Heterometric quartz grains, scattered within an argillaceous matrix, could be reminiscent of the late Ordovician glaciomarine deposits largely distributed over the northern Gondwana margin (Robarded and Doré 1988). The acritarchs and chitinozoans (e.g., *Armoricochitina nigerica, Calpichitina lenticularis)* of Late Ashgill age recorded in the Soubirous Formation (Paris et al. 1987) support such a topmost Ordovician glaciomarine origin.

The dark pyroclastic horizon drilled near the bottom of well Sb. 1 bis, beneath the Soubirous Formation, is unfossiliferous and cannot be dated.

The Silurian is represented in well Cs. 102 by graptolite bearing black shales (= Ampélites de Castelsarrasin, a more or less time-equivalent of the Le Tech Formation in the Western unit). These anoxic deposits have a high organic carbon content (up to 10%) and are a few tens of meters thick. The oldest available samples indicate a Late Llandovery age (probably the *crispus* graptolite zone). This black shale formation extends up into the Pridoli and possibly into the early Lochkovian (based on the occurrence of spiny mazuelloids with complex siliceous and organic walls).

In well Cs. 102, the Silurian black shales are overlain by the dark shales, grey cherty mudrocks, and black cherts of the Castelsarrasin Formation. These strata are folded, so that their real thickness may not exceed 50 m. A few miospores and tracheids with complex pits, as well as numerous conodonts (in the upper cherts), suggest an Early Carboniferous age (Tournaisian?) for the whole formation (Paris et al. 1987). The cherty deposits recorded in well Ca. 101 are probably of the same age.

In well Cs. 102 the Devonian is absent. In another borehole (Sb. 1 bis), about 150 m of brecciated dolomites with shaly levels were tentatively assigned by the operator to the Devonian (occurrence of crinoids and cephalopod orthocones?). However, no definite paleontological data are available and these dolomites could be assigned to the Early Carboniferous. They are overlain by the Soubirous diamictites bearing angular blocks of fossiliferous (Fig.5) Early Devonian limestones (trilobites, ostracodes, dacryoconarids), fragments of cherts, shales, and sandstones included in a dark argillaceous matrix. Such chaotic deposits are similar to Early Carboniferous

	FORMATIONS	Acritarchs	Chitinozoans	Scolecodonts	Spores	Conodonts	Bryozoans	Brachiopods	Bivalves	Orthocones	Crinoids	Corals	Ostracodes	Trilobites	Tentaculites	Graptolites	Plant remains
PER.	sandstones,shales...				●												●
CARBONIF.	St Gery	○			●			●									●
	Diamictites		○					○			○		○	○	○		
	Dolomites									●	●						
	Castelsarrasin				●	●											●
DEV.	Clairac	●	●	●	●		●	●			●	●	●	●	●		
SIL.	Le Teich / Castelsarrasin	●														●	
ORDOVICIAN	Soubirous	●	●													●	
	St Martin du Bois	●	●	●			●										
	Avensac	●	●	●												●	
	Caudecoste	●	●														
	Caubon	●	●	●				●						●		●	
	Noaillan	●	●											●		●	
	Moncrabeau	●	●						●					●		●	
	Bouglon	●	●														
CAM.	Ste Marthe	●	●											●			

Fig. 5. Occurrence of the main macro- and microfossil groups in the Paleozoic formations of the northern Aquitaine basement. *Open circles* indicate reworked fossils

olistolite-bearing strata described by Engel et al. (1982) in southern France.

The Early Carboniferous Saint Géry Formation corresponds to dark shales and siltstones with quartz grains of various origin, feldspars, fragments of cherts, pyrite (see Pelhate and Holtzapffel in Paris et al. 1987). These relatively immature deposits yield a few dark miospores and abundant tracheids and plant debris. Poorly preserved plant fossils (Sphenopterids) reported from wells SG. 1 and SF. 1 suggest a Dinantian to Namurian age for this material (Bugnicourt et al. 1959). The Saint Géry Formation occurs along a NW-SE line, the so-called Carboniferous graben of previous authors. It rests unconformably on strata of various Paleozoic age and supports the existence of Late Devonian events.

2.4 The Eastern Aquitaine Basement

Eastward to the previous terrane is a more metamorphosed zone explored by well LTB. 1 and possibly by well ST. 1. This unfossiliferous material of unknown age (possibly Proterozoic) bounds the post-orogenic graben of Souillac-Lavaur which is similar to the neighboring Brive basin. This Souillac-Lavaur basin is filled up by thick (up to 3000 m) Stephano-Permian lacustrine and fluvial deposits (wells Cpg. 1, SML. 1, Gr. 1, and Lv. 101). The succession consists of conglomerates, sandstones, red shales with locally dolomitic nodules (well SML. 1), gray bituminous shales (well Lv. 101), and a few coal-bearing horizons. In addition to plant remains, these post-orogenic Stephano-Permian strata yield miospores and disaccate pollen.

3 Structure

Until recently, the Paleozoic basement of the Aquitaine basin was believed to have been slightly affected or unaffected by Variscan events because of its external position with respect to the main Variscan fold belt (Autran and Cogné 1980; Debelmas 1986; Matte 1986). New investigations, however, have demonstrated, at least locally, the occurrence of recumbent folds, low-angle overthrusts and shear planes, low dipping schistosities, and later crenulation cleavages (Paris et al. 1987, 1988). Moreover, the structure of the Aquitaine crust has partly been investigated by deep seismic profiles, i. e., the Bay of Biscay (Pinet et al. 1987) and the Pyrenees (ECORS Pyrénées team, 1988) ECORS profiles. Despite these improvements, the structure of the Aquitaine basement remains unclear because of the scarcity of accurate subsurface structural data. It would be premature at this time, therefore, to propose even a very

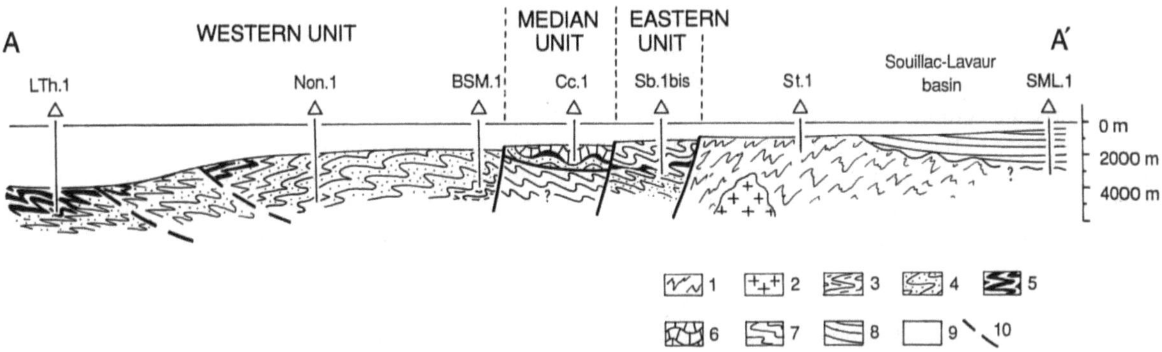

Fig. 6. Preliminary E-W cross-section through the Paleozoic basement of the Aquitaine basin, using the normal fault boundary interpretation of the Median and the Eastern units. For location, see Figs. 2 and 3. *1* Basement of unknown age; *2* granitoid; *3* Cambrian; *4* Ordovician; *5* Silurian; *6* Devonian; *7* Early Carboniferous; *8* Stephano-Permian; *9* Mesozoic and Cenozoic cover; *10* thrusts and normal faults

schematic geological map of the basement. The available stratigraphic and structural data are too scanty to provide reliable information on orientations of the main deformations. Moreover, the azimuths of both bedding and cleavage planes are unknown. For such reasons the basement of the Aquitaine basin is provisionally subdivided into three "units" of relatively different tectonometamorphic evolution (Fig. 2).

3.1 Faults and Thrusts

Major faults, including the Garonne and the Villefranche faults, have long been recognized in the Aquitaine basement. Others faults are documented both from subsurface data on Paleozoic rocks (Le Pochat 1984; Paris et al. 1987) and from exploration late Mesozoic and Cenozoic rocks. Three NW-SE-oriented faults bound respectively the Western, the Median, and the Eastern units (Fig. 6). Their nature is still debated: they may be either normal or thrust faults. Some evidence related to post-Paleozoic reactivations supports a normal fault type for these stuctures. In this case, since they are parallel to the present margin of the south Armorican shelf, these NW-SE faults may be related to a very early extensional episode (Late Carboniferous-Permian) preceding the main rifting phase culminating in the opening of the Bay of Biscay. Similar extensional events seem to control the initiation and loading of the Stephano-Permian basins (e.g., the Souillac-Lavaur basin).

Another group of faults with orientations from E-W to N 50° E, have been recognized based on geophysical data and post-Paleozoic reactivation evidence (Le Pochat 1984).

The most striking feature of the Aquitaine basement is the occurrence of thrusts. The best-documented case is in well Av. 101, where a Middle Ordovician sequence overlies younger rocks referred to the Early Carboniferous Saint Géry Formation. Evidence of thrusting is provided by the existence of a gently dipping brecciated zone separating the Ordovician Avensac Formation from the underlying Carboniferous St Géry Formation. In addition, numerous slickensides are present in the Carboniferous of well Av. 101 and in the Ordovician of wells Cb. 101 and SMB. 1. The whole Median unit may be regarded as allochtonous (Paris et al. 1987, 1988). The setting of this exotic terrane is, however, still unsatisfactorily explained and several alternative hypothesis can be proposed:

1. The Median unit corresponds to a large scale nappe, similar to those occurring, for instance, in the Cantabrian Mountains (NW Spain) during Late Carboniferous. The roots of such a nappe could be situated eastwards such that the material of the Median unit would represent part of the missing Paleozoic cover of the Perigord basement. In this case, the thrusting occurred during Late Carboniferous times and predated the NW-SE normal fault system which cut off the nappe. The subsequent graben would have preserved the material of the Median unit from post-orogenic erosion. Such a hypothesis seems to be consistent with the deep structure of the northern Aquitaine basement as revealed by the Bay of Biscay ECORS profile (Pinet et al. 1987).

2. A second alternative involves thrusting such that the NW-SE faults bounding the Median terrane are interpreted as thrust planes rather than high angle faults. The terrane would correspond to an unmetamorphosed slice caught between stacked, low grade metamorphic slices, with a south vergence.

3. A third hypothesis is based on early gravity-slide processes similar to those reported in the Early Car-

boniferous rocks of the Montagne Noire (Engel et al. 1982; Franke and Engel 1986). Apart from the synsedimentary setting of the exotic material, the rest of the Variscan tectonic evolution can be regarded as similar to those of the first hypothesis. This involved differential fault movement between the three units (NW-SE late Variscan faults) followed by deeper post-Triassic erosion of the Western and Eastern units.

Other thrusts probably occur in the western part of the para-autochtonous Western unit as suggested by the brecciated to sheared zones visible in core samples from wells Cs. 1 and Sj. 1 located in the Medoc (Fig. 3). They are likely related to deep overthrusts recorded in the northern part of the Bay of Biscay deep seismic profile (Pinet et al. 1987).

3.2 Cleavages and Folds

Dips of bedding and slaty cleavage are fairly well documented from available cores and operator drill-logs (Karnay et al. 1984). In some wells (e.g., BSM. 1, Non. 1, Cb. 101, Cc. 1, Av. 101) the dips remain rather constant within formations. In other cases, the bedding plane dip is highly variable (e.g., wells Cs. 102, Cd. 101, SMB. 1). However, no obvious stratigraphic repetitions can be documented even in these boreholes. In most cases, long wave folds are present and locally, as in well LTh. 1, there are decimeter-scale recumbent folds (Paris et al. 1987).

In the Western Unit, the small angular relationship between the slaty cleavage (dips up to 50°) and the bedding planes (30° to 45°) in well Non. 1 support the existence of overturned folds. Their vergence cannot be specified as the azimuths of both the bedding and the cleavage are unknown. Nevertheless, based on the occurrence of crustal slices, including Paleozoic sediments, overthrust to the south during the Variscan orogeny (Pinet et al. 1987), a south vergence is likely for these folds (Fig. 6). The peculiar distribution of the Paleozoic strata, with the oldest rocks eastward (Cambro-Tremadoc in well BSM. 1) and the youngest ones westward (Silurian in well LTh. 1), suggests a broad anticline disposition for the Paleozoic material of the Western Unit. The wide geographic extent of the Ordovician relative to its thickness may result from second order folds and possibly from low angle internal thrusts (Fig. 6).

The dips of Paleozoic strata of the Median unit are variable and range from 10–15° (well Av. 101) up to 60–65° (well Cb. 101). Thus this sequence is probably folded, despite the lack of a true schistosity. Because the directions of bedding are unknown, the orientation of these possible weak folds cannot be specified. A second alternative exists, however: the variation in dips may be related to the tilting of fault-bounded blocks.

The Eastern unit was classically regarded as a graben filled by Carboniferous (Bur Rech Geol Min et al. 1974; Durand-Delga et al. 1980). The sequence recorded in the southern wells (Cs. 102 and Sb. 1 bis) does not support this interpretation, as only a few hundreds of meters of Early Carboniferous deposits overlie Ordovician or Silurian strata (Fig. 3). Moreover, the northern boreholes (SG. 1 and SF. 1) stopped without completely penetrating the Carboniferous sequence so that the actual thickness of these sediments is unknown (Fig. 3). The only obvious tectonic feature is the folding both of the Carboniferous and older strata (dips ranging from 10 to 90 in well Cs. 102). Pre-Carboniferous tectonic events and unconformities are likely (i. e., absence of Devonian strata in well Cs. 102). Late Devonian folding or cleavage are not recorded despite the interpretation that the neighboring South Armorican Ocean closed during Devonian time (Cogné 1977).

3.3 The South Armorican Suture

So far, no oceanic material has been intersected in boreholes penetrating the Aquitaine basement. Consequently, it is almost impossible to accurately locate the South Armorican Suture based upon the subsurface well data. Gravity and magnetic survey data indicate that the remnants of the South Armorican Ocean may be located along a line running from Oleron Island southward to the north of Toulouse. The suture is most likely cut off by the sinistral Villefrance fault and its possible prolongation would be situated north of the Montagne Noire. The obvious paleogeographic relationship between the Paleozoic sequences of the Montagne Noire and of the Aquitaine basement (Fig. 7) would support this conclusion. On paleobiogeographic grounds, the Paleozoic faunas and lithological successions of the Aquitaine basement appear different from those of the Medio-North Armorican Domain (Paris and Robardet 1985). The intermediate Ancenis and Vendée Paleozoic material, in contrast, seems more similar to the Aquitaine basement. Thus the suture may be located north of the Aquitaine Paleozoic basement (Fig. 7).

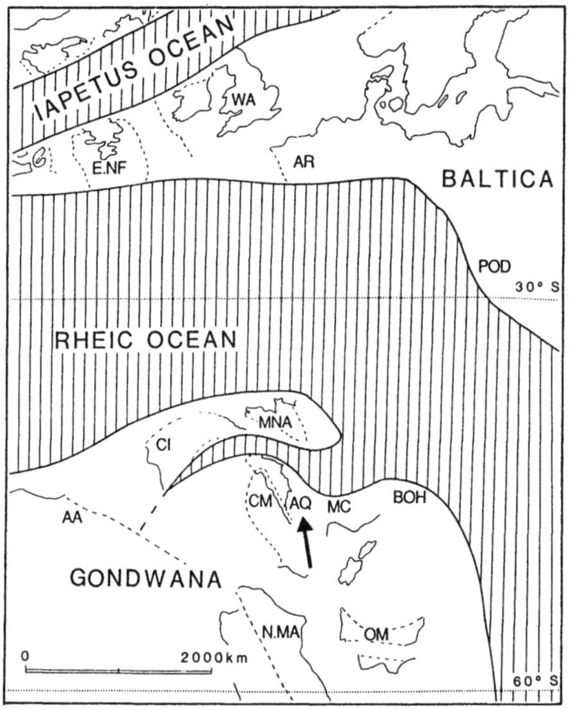

Fig. 7. Paleogeographic location of the Aquitaine basement during the Silurian. (Paris and Robardet 1990). *E.NF* East Newfoundland; *WA* Wales; *AR* Ardennes; *POD* Podolia: *MNA* Medio-North Armorican Domain; *CI* Central Iberian Domain; *AA* Anti-Atlas; *CM* Cantabrian Mountains; *AQ* Aquitaine basement; *MC* French Massif Central; *BOH* Bohemia; *N.MA* North Maghreb; *OS* Ossa Morena; *hatched area* corresponds to oceanic crust

4 Metamorphism

The metamorphic evolution recognized in the Paleozoic rocks of the northern Aquitaine basement appears to be related to the Variscan Orogeny. In the northern Aquitaine basin subsequent shallow burial beneath Mesozoic and Cenozoic cover apparently had no significant effect, as indicated by the variable metamorphic grade observed in Paleozoic sediments at similar depths in neighboring wells (e.g., Cc. 1 and BSM. 1). In areas with a high subsidence rate during post-Paleozoic times (e.g., under the Parentis and Adour basins), the thermal evolution of Paleozoic rocks cannot be studied but may result from both Variscan and Late Mesozoic events. A specific metamorphism related to the South Armorican Orogeny (= Ligerian Orogeny, sensu Cogné 1977) has not been documented to date.

In the Median terrane the Ordovician and Devonian deposits are not metamorphosed, based upon the very poor crystallinity of the illite and on a low thermal alteration of the organic matter present in the sediments. In well Cc. 1 and Cb. 101 the transparent to pale yellow color of the acritarchs clearly indicates a very low maturation, within the oil window.

Metamorphism in the Western and Eastern units ranges from the anchizone up to the lower epizone. The crystallinity of illite is fairly similar in all rocks older than Late Carboniferous in both terranes. Thus no Late Devonian metamorphic event can be firmly identified from the available material. The maturation of the organic matter, as expressed by the common occurrence of dark, highly coalified acritarchs, chitinozoans, and spores, is much higher in these two units than in the Median terrane. It is worth noting that metamorphism increases northwestward in the Western unit. Greenschist facies are present under the Medoc area (northwest Bordeaux) where minerals of low epizone grade are reported (e.g., talc and margarite in well LTh. 1). The "ancient basement" of the Medoc area, assigned to the Proterozoic by previous authors (Bur Rech Geol Min et al. 1974; Durand-Delga et al. 1980), corresponds in fact to a slightly more metamorphic Paleozoic section (Paris et al. 1988).

The Permian Souillac-Lavaur basin is bounded to the east by a metamorphic terrane (e.g., well LTb. 1 and possibly ST. 1) of unknown age. This terrane may represent the southern end of the Essarts complex. So far, neither high pressure nor high temperature metamorphic rocks have been reported from the Aquitaine basement.

5 Igneous Activity

Gravity anomalies (e.g., the Libourne anomaly) suggest that granitoid bodies are present in the Aquitaine basement (Bur Rech Geol Min et al. 1974; Durand-Delga et al. 1980). However, firm evidence of such granites is known only in well Cs. 101 (shallow granite covered by Eocene strata). The boreholes drilled in the other gravity anomalies have not intersected granite or contactmetamorphosed Paleozoic rocks. Thus a sedimentary origin of the gravity anomalies appears to be more likely (Marchal 1986; Mullan 1984; Paris et al. 1987).

Igneous activity is also expressed by the volcaniclastic rocks included in the Paleozoic sequence, particularly in the Ordovician of the Median unit where the main volcaniclastic deposits are represented by the Saint Martin du Bois Formation. Subordinate occurrences are present as pyroclastic rocks underlying the Late Ordovician Soubirous Formation in the Eastern unit. The present lack of geochemical data precludes assessment of the nature, setting, and geodynamic significance of this Late Ordovician volcanic event.

6 Economic Geology

The Paleozoic strata of the Aquitaine basement are too deeply buried (Fig. 1) to allow conventional mining. Consequently, the mineral occurrences reported from several formations are not considered herein. The hydrocarbon potential of the Paleozoic rocks, in contrast, is more attractive even though exploratory wells have only intersected noneconomic oil and gas occurrences. The most promising areas are the Median unit and the Stephano-Permian basins. The existence of low grade metamorphism in the Western and Eastern units (at least in the northern part) clearly indicates overmature rocks and limited potential, although thermally mature rocks with dry gas may occur southward in the less anchimetamorphic material.

The present maturation of the organic matter in the Median unit indicates that the Ordovician deposits in well Av. 101 and Cb. 101 and the Devonian material of wells Cc. 1 and SMB. 1 are situated in the oil zone. Analysis of the available material indicates that the total organic carbon ratio content is fairly low (Mullan 1984; IFP, unpubl. data). Nevertheless, values up to 20% are known in the Silurian black shales of the Mediterranean Province (Dabard and Paris 1986). These Silurian black shales have not been drilled in the Median terrane (gap or decollement ?) although they occur in a neighboring well (Cs. 102) of the Eastern unit. Due to the poor porosity of the Ordovician sandstones and of the Devonian dolomites and limestones, no good reservoirs are known in the Early Paleozoic sequence. The overlying Triassic rocks, however, provide more attractive reservoir units including coarse sandstones and fractured dolomites that are sealed by salt and anhydrite (see Mullan 1984). As a matter of fact, small hydrocarbon occurrences, most likely generated during Mesozoic burying of the Paleozoic deposits, are reported from the base of the Triassic sequence in some wells (e.g., Cc. 1)

Hydrocarbon potential may occur in Stephanian-Permian post-orogenic deposits filling up half grabens located east of the main Paleozoic basement (e.g., the Souillac-Lavaur basin; cf. Figs. 1 and 6). Although coarse-grained strata and reddish shales (e.g., well SMB. 1) are unattractive as hydrocarbon sources, very bituminous lacustrine shales (e.g., well Lav. 101) could be a source rock for economic oil concentrations.

7 Conclusions

In spite of the recent advances discussed above, the detailed stratigraphy and Variscan tectonometamorphic evolution of the Paleozoic basement of the Aquitaine basin remain virtually unknown. Nevertheless, some important basic conclusions can already be drawn:

1. Based on lithologic (e.g., thick terrigenous Tremadoc-Arenig sequence; Middle-Upper Ordovician volcaniclastic rocks; Devonian carbonates; Early Carboniferous cherts) and on faunal evidence (e.g., occurrence of *Taihungshania* in the Arenig; rich Devonian coral fauna, etc.), the Paleozoic deposits of the Aquitaine basement and those of the Montagne Noire belong to the same paleogeographic entity. This domain also includes the Pyrénées (Mirouse and Bourrouilh 1984) and in all likelihood the Paleozoic basement of the Ebre basin and the Cantabrian Mountains (Paris and Robardet 1990). The paleogeographic position of the Vendée is more questionable owing to the scarcity of faunal record and to its tectonically disturbed succession. Some peculiar features suggest a closer affinity with the "Cantabro-Ebro-Aquitaine" Domain rather than with the Medio-North Armorican Domain. Thus, the main paleogeographic limit (i.e., the South Armorican Suture) would be located northeast of the Aquitaine basement (Fig. 7).

2. The Paleozoic material of the Aquitaine basement obviously was affected by Variscan (Late Carboniferous) tectonometamorphic events whose effects increased northwestward (e.g., occurrence of overturned or recumbent folds, slaty cleavage, thrusts). All the pre-Stephanian strata were affected by these Variscan events.

3. The effects of an Early-Middle Devonian continental collision between the Cantabro-Ebro-Aquitaine block and the Armorican block have not been clearly identified. One possible feature related to this event would be a pre-Tournaisian hiatus, although the absence of Late Devonian strata could have other origins (e.g., decollement).

4. The ancient basement of the Medoc corresponds in fact to Paleozoic material of early epizonal metamorphic grade.

5. Nonmetamorphic Paleozoic sediments with organic maturation levels indicative of the oil zone are preserved in the Central unit and may represent potential hydrocarbon sources.

Acknowledgments. The authors thank F. Doré, R. Boehme, and an anonymous reviewer for critically reviewing the manuscript.

References

Autran A, Cogné J (1980) La zone interne de l'orogène varisque dans l'Ouest de la France et sa place dans le développement de la chaine hercynienne. Ann Soc Geol Nord 99, 1: 90–111

Baudelot S, Fournier-Vinas C (1984) Bassins paléozoiques cachés sous l'Aquitaine: étude micropaléontologique des carottes prélevées à l'occasion de forages pétroliers. Doc Bur Rech Geol Min 81-7: 1–11

Brunet MF (1984) Subsidence history of the Aquitaine basin determined from subsidence curves. Geol Mag 5: 421–428

Bugnicourt D, Dalbiez F, Dufaure P (1959) Contribution à l'étude du Paléozoïque nord-aquitain. Bull Soc Géol Fr 7 1: 583–587

Bur Rech Geol Min, Elf-Re, Esso-Rep. SNEA (1974) Géologie du Bassin d'Aquitaine. Bur Rech Geol Min, Paris, 21 pl

Cogné J (1977) La chaîne hercynienne correspond-elle à un orogène par collision? Proposition pour une interprétation géodynamique globale. Colloq Int CNRS Géol Himalaya 268: 111–129

Dabard MP, Paris F (1986) Palaeontological and geochemical characteristics of Silurian black shale formations from the Central Brittany domain of the Armorican massif (Northwest France). Chem Geol 55: 17–29

Debelmas J (1986) L'héritage hercynien à l'origine des grands bassins sédimentaires français. Bull Cent Rech Explor Prod Elf-Aquitaine 10 1: 151–161

Durand-Delga M (coord) (1980) La chaîne des Pyrenées et son avant-pays aquitain-languedocien. In: Itinéraires géologiques: Aquitaine, Languedoc, Pyrénées. Bull Cent Rech Explor Prod Elf-Aquitaine Mem 3, pp 1–77

ECORS Pyrénées team (1988) The ECORS deep reflection seismic survey across the Pyrénées. Nature 331: 508–511

Engel W, Feist R, Franke W (1982) Le Carbonifère anté-Stéphanien de la Montagne Noire: rapports entre mise en place des nappes et sédimentation. Bull Bur Rech Geol Min 2, 1: 341–389

Franke W, Engel W (1986) Synorogenic sedimentation in the Variscan Belt of Europe. Bull Soc Géol Fr (8) 2 1: 25–33

Karnay G, Le Pochat G, Platel JP (1984) Bassins paléozoïques cachés sous l'Aquitaine. Etude des pendages dans le Paléozoïque des forages pétroliers et comparaison avec la structuration de la couverture mésozoïque. Doc Bur Rech Geol Min 81-7: 19–28

Le Pochat G (1984) Bassins paléozoïques cachés sous l'Aquitaine. Note de synthèse. Doc Bur Rech Geol Min 81-7: 5–9

Marchal F (1986) Apport de la gravimétrie à la connaissance des structures profondes sous le Bassin d'Aquitaine; extension à la chaîe hercynienne. Thèse de 3eme cycle, Montpellier University

Matte P (1986) La chaîne varisque parmi les chaînes paléozoïques péri-atlantiques, modèle d'évolution et position des grands blocs continentaux au Permo-Carbonifère. Bull Soc Geol Fr (8) 2 1: 9–24

Mirouse R, Bourrouilh R (1984) Le Paléozoïque supérieur des bordures du Bassin d'Aquitaine: analyse de données bibliographiques inédites. Doc Bur Rech Geol Min 81-7: 29–39

Montadert L (1984) La marge continentale du Nord du Golfe de Gascogne et le Bassin d'Aquitaine. In: Boillot HG (coord) Les marges continentales actuelles et fossiles autour de la France. Masson, Paris, pp 83–154

Mullan HS (1984) Deep gas potential of the Aquitaine basin, France. Am Assoc Pet Geol 68 12: 1857–1869

Paris F (1984) Bassins paléozoïques cachés sous l'Aquitaine: biostratigraphie par les Chitinozoaires, Ostracodes et Tentaculites. Doc Bur Rech Geol Min 87-1: 13–17

Paris F (1990) The Ordovician chitinozoan biozones of the Northern Gondwana Domain. Rev Palaeobot Palynol 66: 181–209

Paris F, Robardet M (1985) Evaluation des affinités entre le Paléozoïque caché sous l'Aquitaine et les formations armoricaines contemporaines. Doc Bur Rech Geol Min 95-7: 11–36

Paris F, Robardet M (1990) Early Paleozoic palaeogeography of the Variscan regions. Tectonophysics 177: 193–213

Paris F, Le Pochat G, Pelhate A, Holtzapffel T (1987) Synthèse des connaissances sur le Paléozoïque nord-aquitain. Doc Bur Rech Geol Min 144: 1–57

Paris F, Le Pochat G, Pelhate A (1988) Le socle paléozoïque nord-aquitain: caractéristiques principales et implications géodynamiques. C R Acad Sci Paris 306 II: 597–603

Pinet B, Montadert L, Curnelle R, Cazes M, Marillier F, Rolet J, Tomassino A, Galdeano A, Patriat P, Brunet MF, Olivet JL, Schaming M, Lefort JP, Arrieta A, Riaza C (1987) Crustal thinning on the Aquitaine shelf, Bay of Biscay, from deep seismic data. Nature 325: 513–516

Rauscher R (1973) Recherches micropaléontologiques et stratigraphiques dans l'Ordovicien et le Silurien de France. Etude des Acritarches, des Chitinozoaires et des spores. Sciences géologiques Strasbourg Mem 38: 1–224

Robardet M, Doré F (1988) The Late Ordovician diamictic formations from southwestern Europe: north Gondwana glaciomarine deposits. Palaeogeogr Palaeoclimatol Palaeocol 66: 19–31

Taugourdeau P (1961) Chitinozoaires du Silurien d'Aquitaine. Rev Micropal 4 3: 135–154

Winnock E (1971) Géologie succincte du bassin d'Aquitaine (Contribution à l'histoire du Golfe de Gascogne). In: Histoire structurale du Golfe de Gascogne. Technip, Paris, pp IV. 1.1-IV. 1.30

The Vosges Massif

A. Piqué, P. Fluck, J.-L. Schneider, and H. Whitechurch

1 Introduction

The Vosges Massif has been known since the last century as a part of the European Variscan (= Hercynian) chain. Like other European massifs, it is composed of two parts. The Northern Vosges consist of weakly metamorphosed and unmetamorphosed rocks, representing the whole Paleozoic sequence, deformed after Visean times. The Northern Vosges are related to the Saxo-Thuringian zone of Central Europe (Kossmat 1927). The second part, the Central and Southern Vosges, lies south of the Lalaye-Lubine zone (Fig. 1) a roughly east-west major tectonic boundary. The domain consist of crystalline rocks, either highly metamorphosed or granitic (Central Vosges) and their Upper Devonian – Dinantian cover (Southern Vosges). This second part is referred to as the Moldanubian zone. The crystalline rocks of the Central Vosges, like those of other Moldanubian massifs in Czechoslovakia, Germany, and in the French Massif Central, have generally been regarded as an old Precambrian core. Isotopic dating of the mother rocks have demonstrated that they correspond to Lower Paleozoic sequences contemporaneous with (although with very different facies) the lower part of the Paleozoic sequence in the Saxo-Thuringian zone.

The present chapter is an attempt to summarize in English the main aspects of the Vosges geology. It is drawn largely from a recent synthesis, in French (Fluck et al. 1987), that contains a comprehensive bibliography.

2 Stratigraphy and Paleontology

Lower Paleozoic strata, paleontologically dated, are found only in the Northern Vosges. The presence of contemporaneous rocks, metamorphosed presently, is, however, suspected in the central part of the massif. Upper Devonian and Dinantian strata occur in both the Northern and Southern Vosges.

J. D. Keppie (Ed.): Pre-Mesozoic Geology
in France and Related Areas

2.1 Northern Vosges

The Northern Vosges sequence comprises:

1. Villé Schists and Slates. These are greenish fine-grained pelitic rocks, with intercalations of sandy layers and "porphyroids", the latter derived from felsic tuffs. They have been affected by a low grade metamorphism, whose intensity is maximum at the southern rim of the Villé schists, along the Lalaye-Lubine shear zone. The Villé schists and slates are attributed traditionally to the uppermost Proterozoic but some Spongiae (Doubinger and von Eller 1963; Ross 1964) suggest a Cambrian or even Ordovician age.

2. Steige Slates. These purple and greenish slates are commonly fine-grained, with subordinate millimetric sandy beds, and show effects of a very low grade metamorphism. They have been dated as Middle Ordovician to Silurian using chitinozoons (Doubinger 1963). The slates are overthrust by the Villé schists.

3. Champ-du-Feu and Saint-Nabor Volcaniclastic Rocks. These are dark tuffs, and volcanic breccias with conglomeratic lenses alternating with mafic and intermediate effusive rocks that are at least 150 m thick. They occur to the north of the Steige slates from which they are separated by plutonic massifs. They seem to be overlain by the Middle to Upper Devonian lower levels of the Bruche sequence. The Schirmeck and Rabodeau volcanic series are spilites and keratophyres, of dominantly tholeiitic character (Rizki and Baroz 1988). These rocks were traditionary attributed to Mid-Devonian (Givetian) volcanic activity, but their close magmatic affinities with Dinantian magmatic rocks suggest a possible Visean age.

4. Bruche Sequence. Formerly described as a continuous Middle Devonian to Dinantian series (Blanalt 1969), this sequence comprises, in ascending order: graywackes of the Russ forest; gray argillites with *Asteroxylon hostimensis;* Russ conglomerate, 200 m

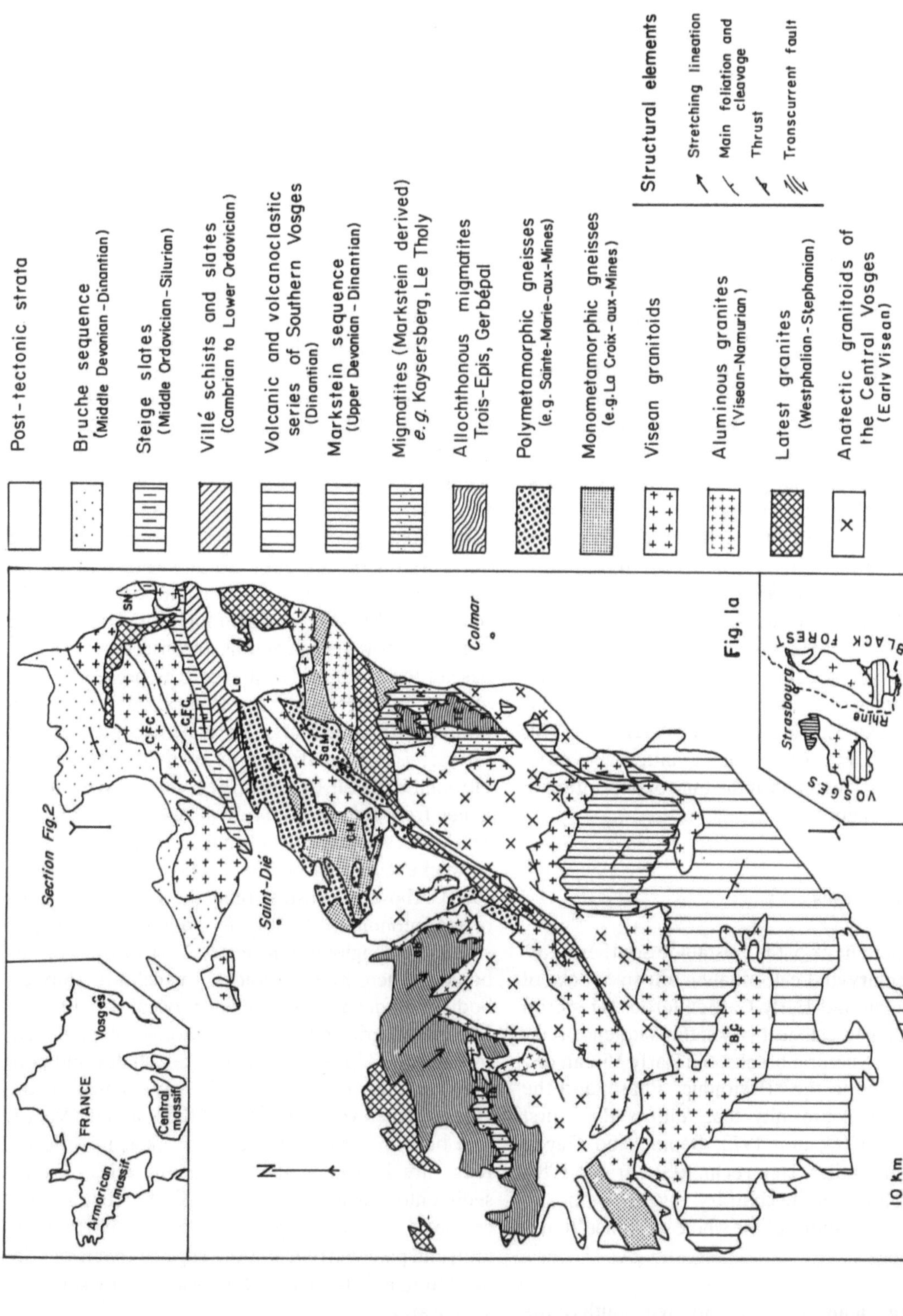

Fig. 1a, b. Simplified map of the Vosges massif. **a** Map. **b** Legend. *BC* Ballons Granitic Complex; *CFC* Champ-du-Feu granitic Complex; *CM* La Croix-aux-Mines; *Gb* Gerbépal; *Ky* Kaysersberg; *La* Lalaye; *Lu* Lubine; *SaM* Sainte-Marie-aux-Mines; *SN* Saint Nabor; *TE* Trois-Epis; *Th* Le Tholy

thick, with well-rounded pebbles and cobbles of various lithologies, granitoids, dolerites, keratophyres, arkoses, and numerous lenses of a reefal limestone (Russ "marble"), itself well dated by *Stringocephalus burtini* (Givetian); graywackes and argillites, with isolated blocks of limestone and conglomeratic lenses; black shales, phthanites, graywackes and arkoses (Hersbach quarries), more than 100 m thick and dated Frasnian by bivalves, plants, and goniatites *(Crickites sahlgrundensis);* Netzenbach and Petite-Wisches breccias, graywackes, and slates (ca. 300 m); and Bergerie graywackes and shales, in turbiditic sequences, several hundred meters thick, dated Early Visean (Spongiae, *Posidonomya becheri)*. The whole sequence has been folded and is unconformably overlain by the Nideck red beds and felsic volcanics of Permian age.

Recent studies suggest that the Bruche sequence might be more complicated and younger than previously thought. The graywackes and argillites that rest upon the Russ conglomerate are interpreted as an olistostrome where blocks of various sizes of the Russ conglomerate are dispersed. This limestone is Givetian in age, whereas the olistostrome matrix and the autochthonous, undisturbed beds, have yielded Famennian chitinozoons (Rais 1988). Thus the Frasnian strata of the Hersbach quarries, which rest upon this olistostrome, constitute an allochthonous unit. This hypothesis is corroborated by the presence, within the olistostrome, of isolated blocks of graywackes and phthanites that are very similar to the facies found in the Hersbach quarries. Another interesting, but as yet unpublished observation, concerns the red granitic pebbles and cobbles of the Russ conglomerate. A recent U-Pb dating, performed by Dr. W. Dörr (Giessen University, Federal Republic of Germany) has furnished a 493-Ma age, similar to those found in other granitic pebbles of some Mid-Paleozoic conglomerates of the Saxo-Thuringian zone of Germany. This indicates that Lower Ordovician granites were cropping out during Devonian at the time of deposition of the Bruche sequence.

2.2 Southern Vosges

At the southern flank of the crystalline Southern Vosges, sedimentary and calc-alkaline effusive rocks of Late Devonian and Early Carboniferous ages crop out. According to classical views, this sequence is divided by two structural breaks, the Early Visean disconformity, due to the "Bretonnic" Phase, and the Late Visean unconformity that follows the "First Sudetic" Phase (Coulon 1973; Bébien and Gagny 1978). Recent studies (Hahn et al. 1981; Grimm 1983; Schneider, 1990), on the contrary, suggest a more or less complete continuity of the whole sequence, that is described below.

1. Upper Devonian. In the Belfort area, south of the Vosges massif, calcareous beds rest upon mafic and felsic lavas and Middle and Late Devonian limestone (Morre-Biot et al. 1972). These strata are time-equivalent with the Treh shales, at the base of the Markstein unit (see below). These reddish and greenish shales are now attributed to Upper Devonian on the basis of their Conodonts (Maass and Stoppel 1982), and they may be the northern basinal equivalent of the Belfort limestones.

2. Tournaisian. These rocks have not been dated paleontologically but resedimented trilobites and Foraminiferae have been found at the base of Upper Visean strata (Hahn et al. 1981). The Tournaisian may be represented by the lower part of the Markstein unit.

3. Visean. The Visean is represented by two structural and lithostratigraphic units separated by a thrust plane, the "Ligne des klippes". The "klippen" that outline the thrust are blocks and cobbles, randomly dispersed within an underformed matrix of graywacke. The petrographic nature of these clastic elements is highly variable and includes: gneisses similar to those of the Central Vosges Sainte-Marie-aux-Mines area; leptynites; tholeiitic gabbros and foliated harzburgites of ophiolitic affinity (Pin and Carme 1988). Initially considered as mylonitic breccia at the base of the Markstein allochthon (Jung 1928), these rocks are now recognized as an olistostrome at the base of the Markstein sequence. The presence of the various petrographic facies, especially the peridotites, within the Markstein sequence, is very interesting; it is explained either by an in situ erosion of an ultramafic massif (Maass 1988), or a gravity nappe issued from a northern suture (Fluck et al. 1987). The Markstein sequence, which may have been structurally thickened, is more than 5000 m thick (Grimm 1983). Above the olistostrome, the sequence consists of monotonous turbiditic rhythmites with slumped strata and conglomeratic layers. Very distinctive carbonate concretions ("ovoides") which are elliptical bodies, of decimetric diameter, occur locally. The Markstein series remains paleontologically undated, overlies the Upper Devonian Treh shales and contains pebbles of volcanic rocks very similar to the Upper Visean volcanic series of the Southern Vosges (see below). Apart from these pebbles, no volcanic rocks are known in the Markstein series. Several sedimentological arguments, e.g., the regional decrease of the coarse input from south to north in the unit, and paleocurrent analysis suggest that it was deposited in a distal part of the basin, fed by southern source areas.

South of the "Ligne des klippes" thrust, the other Southern Vosges sequences are of a volcanic and volcaniclastic character. The Lower Visean strata are mainly turbiditic and alternate with bimodal volcanic rocks, spilites, and keratophyres. Higher in the sequence, indication of resedimentation processes com-

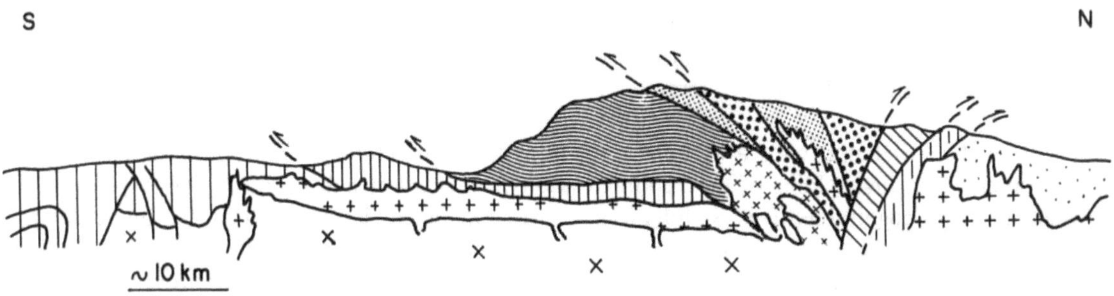

Fig. 2 Schematic section (located on Fig. 1) of the Vosges massif, during Namurian times (same symbols as in Fig. 1)

monly occur. These chaotic rocks have been compared to the Markstein "klippen" (Maass 1988), although they do not contain any ultramafic fragments and are unrelated to any structural contact. The upper levels of the Visean sequence show evidence of explosive volcanic activity, with pyroclastic breccias associated with ignimbrites and rhyolites. Upper Visean rocks conformably overlie the Lower Visean strata and volcanic rocks. Schneider (1990) documents the progressive change from marine sediments, often resedimented, through aerial deposits, with sandstones, conglomerates, and thin coal beds, to felsic volcanic rocks, andesites, latites, rhyodacites, and rhyolites.

3 Structure and Metamorphism

3.1 Central Vosges

1. General Structure (Figs. 1, 2). In the Central Vosges, the Sainte-Marie-aux-Mines area was studied in great detail by Fluck (1980), who distinguished two units.

The upper unit (groupe varié) consists of gneisses and migmatites with granulites and eclogites as preserved relicts. In the granulitic gneisses, some intercalations of leptynites (garnet gneisses), skarns, amphibolites, eclogites, and peridotites are present. At its base, this unit comprises a thick horizon of amphibolites. The whole unit is considered as the metamorphic equivalent of a thick graywacke sequence (granulitic gneisses) that overlies basalts and andesites (amphibolites) and its base is truncated by a thrust plane. In the rest of the Central Vosges, the La-Croix-aux-Mines gneisses and the Gerbépal and Trois-Epis migmatitic units correspond to the Sainte-Marie-aux-Mines upper unit.

The lower unit (groupe monotone) consist of biotite and sillimanite gneisses. Outside the Sainte-Marie-aux-Mines area, it corresponds to the Urbeis gneisses.

2. Metamorphic and Structural Evolution. Hameurt (1967) and Fluck (1980) recognized two metamorphic episodes in the Sainte-Marie-aux-Mines upper unit and the related units. The first episode developed garnet, kyanite, and other minerals corresponding to granulitic conditions and was attributed to an Early Ordovician orogenic phase. The second metamorphism developed low-pressure-high-temperature parageneses and an incipient anatexis. It has been dated at 386 ± 15 Ma (Bonhomme and Fluck 1981) by whole-rock Rb-Sr method. Only the second episode was known in the lower unit which was therefore thought to have been deposited during the time interval between the two metamorphisms.

Recent work of Rey (1988) provides a new approach to the interpretation of the metamorphic and structural evolution of the Sainte-Marie-aux-Mines area. The gneisses of the upper unit show a major S_1 foliation, initially flat-lying, determined by the disposition of quartz, plagioclase (An 30), K-feldspar, garnet and sillimanite. A N30E lineation is indicated by the preferential alignment of sillimanite, garnet and biotite. This S_1 foliation, contemporaneous with the development of the sillimanite, post-dates the granulitic paragenesis. No structure that could be related to the early HP metamorphism has been found. On the other hand, the S_1 foliation was itself folded by several deformational episodes.

In conclusion, the tectonometamorphic evolution of the upper gneisses records the following.

- A granulitic metamorphism, developed under pressures from 12.5 to 20 kb and a temperature higher than 750 °C (Rey 1988). Contemporaneous structures have not been preserved.
- Decrease of the intensity of metamorphism (pressure, and later temperature) initiated a second, retrograde, metamorphic paragenesis, during which

the S_1 foliation was created. Subsequent folding and deformation episodes occurred during the last stages of the temperature and pressure decrease.

The gneisses of the lower unit have a S_1 foliation which corresponds to a first folding of the bedding plane which is still recognizable. This S_1 foliation therefore represents the first deformation event in this unit and was developed during the crystallization of sillimanite and the beginning of the regional anatexis. The P-T conditions were about 600–650 °C and 8–13 kbar. This event was contemporaneous with the S_1 development in the upper unit and it was followed by the same subsequent deformations, accompanied by a general decrease of pressure and temperature. Both upper and lower units show a parallel P-T metamorphic evolution but their P-T paths are not strictly identical. The metamorphic conditions determined for the upper unit imply metamorphism at very deep levels, whereas the lower unit was located in a different, higher, structural level (Rey 1988). Their tectonic superposition occurred later, during the last stages of their metamorphic evolution, since their P-T paths coincide only in the chlorite stability field. The last greenschist metamorphic episode was contemporaneous with extensive planar fabrics, themselves probably linked to the ascent of Carboniferous plutonic bodies.

A comparable evolution is recorded elsewhere in the Central Vosges, especially in the Gerbépal and Trois-Epis migmatites studied by Hameurt (1967), Fluck (1980), and Fluck and Whitechurch (1983) respectively. Their frequent association with granulitic relicts and garnet peridotites suggests that they are derived from older granulites. Both series of migmatites are allochthonous. The Gerbépal migmatites have been thrust southeasterly upon the gneissic series of the Central Vosges, still cropping out in the Tholy tectonic window (Blumenfeld 1981) (Fig. 1). The Trois-Epis migmatites are a tectonic klippe resting upon the autochthonous Kaysersberg migmatites, which are lithologically comparable to the Tholy window rocks. The Kaysersberg and Trois-Epis areas allow discussion of the nature and age of the autochthonous unit and the age and kinematics of the thrust as well. Geochemical study of the Kaysersberg migmatites demonstrated that they are derived from turbiditic sequences, mainly of graywacke. They include lenticular bodies, centimeters to decimeters long, made up of calc-silicates and pyroxenes. These bodies correspond to the carbonate concretions that are found in the Markstein graywacke, and Fluck (1976) suggested that the Kaysersberg migmatites are the metamorphic and migmatic equivalent of the Markstein unit, from which they are separated by about 10 km. This hypothesis is strongly supported by the existence, within this interval, of an outcrop where the Markstein sequence, still recognizable from the sedimentological and petrographic points of view, is affected by a low grade metamorphism. The thrust movement is indicated by the anatectic lineation within the overthrusting Trois-Epis migmatites and is roughly southward. The root of the allochthonous units is therefore likely to the north, probably in the La Croix-aux-Mines upper unit. This suggests that the thrust of the Gerbépal and Trois-Epis migmatites upon the Markstein metasedimentary sequences was connected to, and probably contemporaneous with, the thrusts (described above) within the Sainte-Marie-aux-Mines units.

At the same time, or slightly later, ductile and cataclastic motion occurred along steeply dipping fault planes like the north-northeast-trending Sainte-Marie-aux-Mines fault, which acted as a left-lateral strike-slip fault (Fluck 1980).

3.2 Southern Vosges

In the Markstein unit, the regional metamorphism is weak and very low grade. The folds have subhorizontal axes, trend north or northeast, and their axial plane is marked by a discrete cleavage that is less pervasive in the southern part of the unit. Local northeast-trending left-lateral faults cross-cut the primary structures (Fluck 1980).

The Markstein unit is thrust southward upon the Southern Vosges Late Devonian and Dinantian volcanic and sedimentary rocks. The "Ligne des klippes" unit, at the base of the Markstein unit, outlines the tectonic contact. The thrust shows local evidence of right-lateral motion. South of the thrust, the Southern Vosges display a complex pattern of thrusts and folds. The folds are east to southeast-trending and commonly plunge steeply. The deformation of the Markstein and the Southern Vosges units post-dated Late Visean time and is commonly referred to as the Sudetic Phase. Recent mapping and observations from Maass et al. (1980) and Schneider (1990) suggest that the intra-Visean ("Sudetic 1") Phase theory earlier postulated by Coulon (1973) must be abandoned.

3.3 Northern Vosges

The contact between Central and Northern Vosges is the Lalaye-Lubine zone, which is a complex tectonic lineament trending east-northeast. This area, several kilometers wide, consists of gneisses and schists. Some of the more strongly metamorphosed and deformed rocks are similar to the Villé schists and slates. The deformation of the Lalaye-Lubine zone (Bouyalaoui 1992) is exhibited by a fundamental polyphased foliation, trending roughly east-northeast, which is generally refolded by isoclinal folds and affected by thrusts. The structural vergence is to the north-northwest and a right lateral component of the deformation is visible. Of particular interest are some gneissic slices (ecailles du Climont) considered fragments of the basement of the Villé series which have been thrust, in metamorphic conditions, upon the Villé schists and slates.

The intensity of both regional metamorphism and deformation decreases from the southern rim of the Villé schists and slates to the northern part of the Steige slates and, beyond, in the Bruche valley. The Villé schists and slates were affected by two sets of coaxial folds which may have developed in a single progressive deformation. The Steige slates ordinarily show one single folding, accompanied by an axial-plane cleavage. The cleavage is poorly marked in the Bruche Valley sedimentary rocks. Metamorphism reached medium grade in the southern Villé area, low grade in the rest of the Villé zone, very low grade in the Steige area, and is weakly perceptible in the Bruche strata. All over the Northern Vosges, the general structural vergence indicated by the facing direction of the folds and the thrusts (e. g., the thrust of the Villé rocks upon the Steige slates) is to the north. The exact timing of the synmetamorphic deformation, formerly determined from Rb/Sr analyses by Clauer and Bonhomme (1970) and Prévot (1969), for the Villé-Steige and Bruche rocks respectively, is being reviewed (Bouyalaoui, work in progress). It was obviously post-Late Viséan (= Sudetic?) in the Bruche area, but it could be slightly older the Lalaye-Lubine area.

4 Granitoids

Granitoids are widespread in the Vosges massif and have a wide range of petrographic composition and structural locations.

In the Northern Vosges, the oldest granitoids are derived from dioritic and granodioritic magmas. Their emplacement was followed by the emplacement of batholiths of U-rich, monzogranitic composition. All of the batholiths have a calc-alkaline affinity (de la Roche and von Eller 1969) and yield ages scattered between 330 and 320 Ma (Edel et al. 1986). The most recent plutonic intrusion is a leucogranite, characterized by subvolcanic textures, that passes laterally to Permian peralkaline rhyolites (Hahn-Weinheimer et al. 1971).

In the Southern Vosges, likewise, the oldest plutons are the gabbros, diorites, and monzodiorites that form the outer rim of the Ballons complex. Their affinity is tholeiitic (André and Bebien 1983 a) to shoshonitic (Pagel and Leterrier 1980), they are comagmatic with the basalts of the volcanic series of Southern Vosges. The plutons result from the Viséan crustal extension (André and Bébien 1983 a). According to Fluck (1984), the magma produced during this extension could have been emplaced in the Ballons massif. After crustal contamination, it would have produced the subalkaline and potassic monzogranites that occupy the centre of the Ballons complex and the Crêtes granite that was emplaced as a large laccolith between the basement and its Dinantian cover. The nappe motion of the "Sudetic" Phase was accompanied by the anatexis of both the allochthonous and the autochthonous units, and produced granitic bodies from the partial melting of the basement (Hameurt 1967; Fluck 1980) or the cover. The aluminous monzogranites are characterized by large K-feldspar crystals and were emplaced preferentially within the thrust plane. The youngest Variscan granites (Fluck 1980; Schneider 1983) are leucogranites of Westphalian or Stephanian age.

5 Correlations with the Black Forest

Although a careful comparison of the two Variscan massifs, Vosges and Black Forest, has not been completed, several restricted observations (Fluck et al. 1987; Eisbacher and Luschen 1988) have shed light on many similarities, briefly summarized below. The similarities are strong enough to allow the postulation that these two massifs belonged to the same Variscan segment before the Cenozoic Rhine rifting.

The Lalaye-Lubine shear zone's eastern equivalent is the Baden-Baden lineament, which is concealed beneath the Permian cover. North of the Baden-Baden line, small outcrops from south to north are related to the southern Villé schists, the Steige slates, and the Hohwald granodiorite.

South of the Baden-Baden line, the central and southern Black Forest massif has a greater ratio of metamorphic rocks, gneisses, and schists to granitoids than the Vosges massif. Rocks similar to those of the Vosges massif are recognized in the Black Forest. These include: kinzigites, similar to the Sainte-Marie-

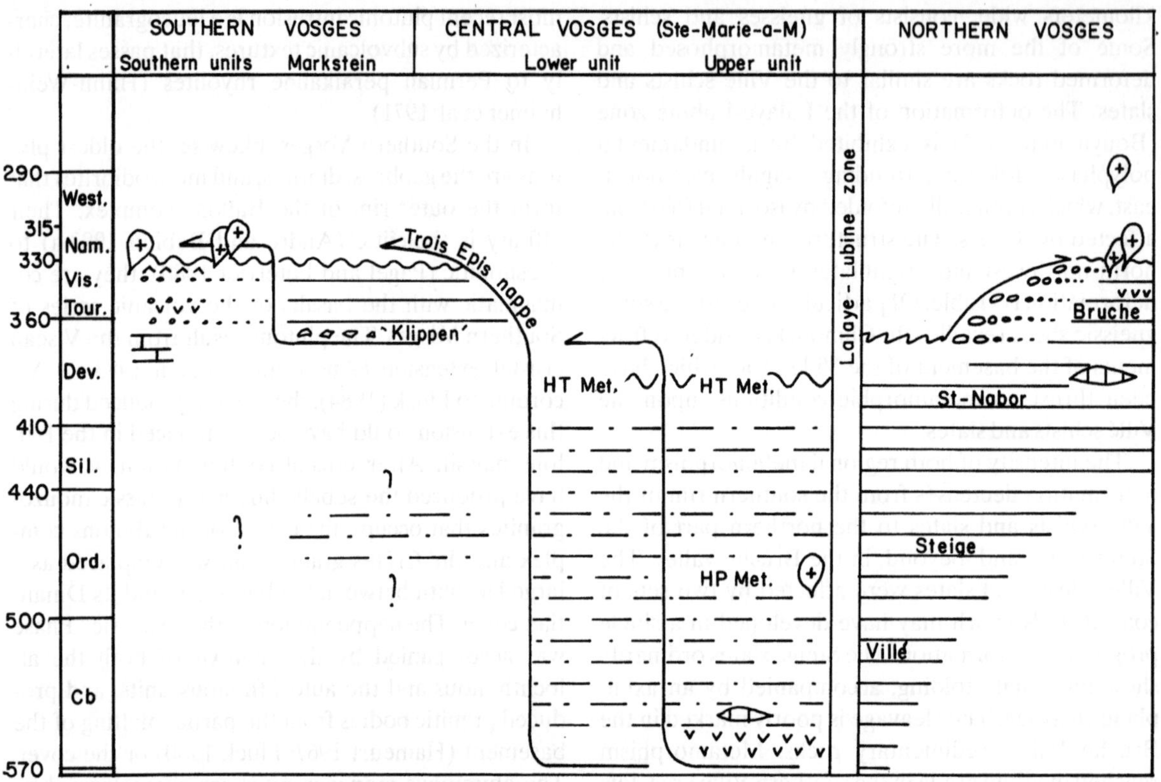

Fig. 3 Time-space diagram of the Paleozoic evolution of the Vosges massif

aux-Mines garnet schists and gneisses; granulites, which have a HP metamorphic age around 480 Ma (Schleicher and Kramm 1986), relicts of ultramafic rocks later migmatized; leptynites and amphibolites which are interpreted as an old spilite keratophyre complex (the Spiesshorn series, a gneiss with leptynitic and amphibolitic intercalations, is comparable with the Val d'Ajol gneisses, which in turn corresponds to the Sainte-Marie-aux-Mines series); graywackes, conglomerates, and volcaniclastic sequences of the Badenweiler-Lenzkirch zone, which are similar to the Southern Vosges series, Markstein excepted, and many granitoids which are comparable to the Vosges granitoïds (e. g., Bärhalde and Brezouard, or Forbach-Seebach and Dambach granitoids).

6 Discussion and Conclusion

An unambiguous interpretation of the tectonic development of the Vosges massif is not possible with the current knowledge of its geological evolution. Any future attempt at such an interpretation should include the stages outlined below (Fig. 3).

The Early Paleozoic crustal extension is recorded in the highly metamorphosed series of the Central Vosges. The Sainte-Marie-aux-Mines "groupe varié", for instance, with its turbiditic sequences interstratified with bimodal magmatic rocks and cross-cut by Ordovician granitoids (like those preserved as cobbles in the Devonian Russ conglomerates) or metamorphosed into orthogneisses, is characteristic of crustal thinning. It has many similarities with the Leptyno-amphibolitic Complexes of other Moldanubian massifs in France (Santallier et al. 1988). Whether or not this thinned crust evolved toward the development of a passive margin and a true oceanic rift is not known, since no direct nor indirect evidence for this (e.g., an ophiolitic complex) has been preserved. Whether or not the ophiolitic clasts of the Markstein "Klippen" originated north of their present location, they suggest that the crustal thinning reached at least the stage of the denudation of the mantle, if not a true oceanic rift.

The Cambro-Ordovician age of the crustal thinning coincides, in the Black Forest, with the age of the HP granulitic metamorphism, determined by U-Pb method around 480–490 Ma (Schleicher and Kramm 1986). This implies that this metamorphic episode, which developed at the deepest levels of the crust, was contemporaneous with the ascent of generally bimodal magma, through the crust and with the sedimentation of the Cambro-Ordovician turbiditic sequences. Weber (1984), among others, has furnished a tectonic model for this pattern. During Early Paleozoic rifting, the thinning of the crust was expressed, in its upper levels, by block-faulting contemporaneous with more or less continuous sedimentation and the ascent of felsic and mafic magmas. The brittle extensive structures passed through ductile shear zones along which a stretching regime produced synmetamorphic fabrics. At the base of the crust, the metamorphic conditions were those of the granulitic and eclogitic facies. Meanwhile, the prograde metamorphism related to the crustal stretching and the contemporaneous dewatering allowed the production of alkaline and calcalkaline granitoids that became orthogneisses later in the course of the Variscan orogeny. Other models, on the contrary, are based upon the existence of a compressive regime, already active during Cambro-Ordovician time (Santallier 1983; Matte 1986).

The Northern Vosges, north of the Lalaye-Lubine shear zone, show no evidence of this crustal thinning. The Northern Vosges were characterized from Cambrian to at least Silurian times by progressively finer deposits free of any magmatic activity, except some tuffaceous beds in the Villé schists.

The onset of the orogenic compression certainly began at different times in the Northern and Southern Vosges. In the Central Vosges, if further dating confirm the Early Ordovician age of the HP metamorphism, this metamorphism could be explained by crustal thinning. An alternative explanation would be required if dating indicates a younger age (e. g., Silurian or Early Devonian) for the metamorphism. In the latter case, the granulitic and eclogitic episode could have resulted from crustal thickening of the Central Vosges crust (Rey 1988). In any case, the subsequent HT metamorphism, tentatively dated as Late Devonian, developed during the uplift of the gneissic series of the Central Vosges and during a succession of deformation episodes that ended with the thrust of the "groupe varié" upon the "groupe monotone".

The emplacement of the gravity klippen unit, at the base of the Markstein sequence preceded the deposition of the Markstein sequence itself. The Markstein sequence and other Southern Vosges sedimentary and volcaniclastic complexes were deposited during the Late Devonian over the gneissic series, then uplifted and exposed to aerial erosion. The persistance of the tectonic compression (Fig. 2) all over the Vosges massif resulted in the opening of Late Devonian-Dinantian sedimentary basins, probably controlled by transcurrent faults, in the Northern Vosges (Bruche) and the Southern Vosges (Markstein and southernmost basin) on both sides of the Lalaye-Lubine "suture". Magmas were emplaced within these basins (Champ-du-Feu and Ballons complexes) and in some compressive zones (i. e., Crêtes granite). The ascent of these granitoid complexes and their comagmatic effusive series (Schirmeck and Southern Vosges series, respectively) was contemporaneous with a clastic, often turbiditic, sedimentation.

The continuation of the crustal shortening in the Northern Vosges was expressed by the deformation of the Paleozoic column, which is more severely folded and metamorphosed in its southern part than in the Bruche valley. The tectonic vergence (Wickert and Eisbacher 1988) developed here is the result of the northward vergence of the folds and the northward synmetamorphic thrust of the Lalaye-Lubine zone upon the Villé schists and slates. This thrust, still poorly dated, may be older than the final folding of the Bruche sequence. In the Southern Vosges, the structural vergence is to the south or the southeast (in present coordinates: large scale rotations are demonstrated by Edel, in Fluck et al. 1987). The shortening was then expressed by the emplacement of the allochthonous units of the Central and Southern Vosges after Late Visean times. Many successively emplaced granitoids marked the progressive consolidation of the orogenic belt.

The only oceanic material which has been found in the Vosges massif is the peridotites and gabbros presently as blocks in the Klippe unit of the Markstein series. Their age is obviously pre-Visean, but it is not otherwise determined. If they are Siluro-Ordovician, they are contemporaneous with the ophiolites of the Southern Armorican massif and Northwestern Central massif (Rolet et al.; Santellier et al., Chap. III. 5 this Vol.), and of those of the Belledonne massif in the Western Alps (Menot et al., Chap. IV. 6 this Vol.). They could therefore indicate either a roughly east-west-trending suture, from Southern Brittany to the Lalaye-Lubine zone, as proposed earlier by Matte (1986), or another suture, trending northwards from the Western Alps.

References

Andre F, Bebien J (1983 a) Minéralogie et pétrologie des cumulats gabbros-dioritiques situés en bordure septentrionale du massif des Ballons (Vosges méridionales, France): cristallisation d'un magma basique en domaine "orogénique" intercontinental. Bull Minéral 106: 341–351

Andre F, Bebien J (1983 b) Déchirures continentales et plutonisme: étude pétrologique et structurale comparée de l'association ignée de Guévguéli (Macédoine grecque) et des formations plutoniques basiques situées en bordure nord du Massif des Ballons (Vosges, France). Bull Soc Géol Fr (7) 25: 291–299

Bebien J, Gagny Cl (1978) Le plutonisme viséen des Vosges méridionales: un nouvel exemple de combinaison magmatique entre roches tholéiitiques et calco-alcalines. CR Acad Sci Paris 286 D: 1045–1048

Blanalt JG (1969) Contribution à l'étude du conglomérat givétien de Russ (Vallée de la Bruche, Vosges). Essai de paléogéographie. Thèse de 3 e cycle, Strasbourg, 72 pp

Blumenfeld P (1981) Etude d'une zone de cisaillement ductile avec anatexie: de la genèse de l'orthogneissification des granites syncinématiques de Vologne-Straiture (Vosges occidentales). DEA rep Univ of Nancy I, 91 pp (unpublished)

Bonhomme M, Fluck P (1981) Nouvelles données isotopiques Rb-Sr obtenues sur les granulites des Vosges. Age protérozoïque terminal de la série volcanique calco-alcaline et âge acadien du métamorphisme régional. C R Acad Sci Paris 293 D II: 771–774

Bouyalaoui J (1992) Schistes de Steige, phyllades de Villé et mylonites du Climont (Vosges du Nord); le développement de la zone cisaillée de Lalaye-Lubine. Thesis, Univ Bretagne occidentale, 207 pp

Clauer N, Bonhomme M (1970) Datations rubidium-strontium dans les schistes de Steige et la série de Villé (Vosges). Bull Serv Carte Géol Als Lorr 23, 3–4: 191–208

Coulon M (1973) La discordance du Viséen supérieur dans les Vosges méridionales. Réun Ann Sci Terre, Paris, p 143

Doubinger J (1963) Chitinozoaires ordoviciens et siluriens des schistes de Steige dans les Vosges. Bull Serv Carte Géol Als Lorr 16: 125–136

Doubinger J, von Eller JP (1963) Présence de spongiaires dans les schistes précambriens métamorphiques des Vosges. Bull Serv Carte Géol Als Lorr 16, 3: 111–123

Edel JB, Montigny R, Royer JY, Thuizat R, Trolard F (1986) Paleomagnetic investigations and K-Ar dating on the variscan plutonic massif of the Champ du feu and its volcanic-sedimentary environment, northern Vosges, France. Tectonophysics 122: 165–185

Eisbacher B, Luschen G (1988) Major tectonic boundaries in the hercynian Schwarzwald-Vosges region of central Europe from surface geology and deep reflection seismics. In: Matte P (ed) Terranes in the Variscan belt of France and Western Europe, meeting. IGCP, Montpellier, p 233 (Abstr)

Fluck P (1976) Contribution géochimique à l'étude du problème des migmatites du Kaysersberg (Vosges). Sci Géol Bull 29: 93–119

Fluck P (1980) Métamorphisme et magmatisme dans les Vosges moyennes d'Alsace. Contribution à l'histoire de la chaîne varisque. Sci Géol Mém, 62, 248 pp

Fluck P (1984) Le socle anté-dévonier et la couverture dévonodinantien dans le bloc Vosges moyennes – Vosges du Sud. Schweiz Mineral Petrogr Mitt: 292–296

Fluck P, Whitechurch H (1983) Relations socle-couverture dans les Vosges moyennes lorraines. EUG Strasbourg, Terra Cognita 3, 2–3: 197–198

Fluck P, Edel JB, Gagny C, Montigny R, Pique A, Schneider JL, Whitechurch H (1987) Le socle vosgien, un segment de la chaîne varisque d'Europe. Doc Bur Rech Geol Min 146, 97 pp

Grimm B (1983) Petrographische Untersuchungen in Geröllen des Markstein-Bereiches (Paläozoikum, Südvogesen). Dissert Doct, Freiburg i. Br., 72 pp

Hahn G, Hahn R, Maass R (1981) Trilobiten aus dem Unter-Karbon der S-Vogesen. Oberrhein Geol Abh 30: 1–26

Hahn-Weinheimer P, Propach G, Raschka H (1971) Zur Genese des Kagenfels Granits. Bull Serv Carte Géol Als Lor 24: 5–56

Hameurt J (1967) Les terrains cristallins et cristallophylliens du versant occidental des Vosges moyennes. Mém Serv Carte Géol Als Lorr 26, 402 pp

Jung J (1928) Contribution à la géologie des Vosges hercyniennes d'Alsace. Mém Serv Carte Géol Als Lorr 2, 481 pp

Kossmat F (1927) Gliederung des varistischen Gebirgsbaues. Abh Sächs Geol Landesamt 1: 1–39

Maass R (1988) Die Südvogesen in variszischer Zeit. Neues Jahrb Geol Palaontol Monatsh 10: 611–638

Maas R, Stoppel D (1982) Nachweis von Oberdevon bei Markstein (Bl. Munster, Südvogesen). Z Dtsch Geol Ges 133: 403–408

Maass R, Grimm B, Vogt Ch, Weinsziehr R, Wickert F (1980) Das Variszikum der südlichen Teile von Schwarzwald und Vogesen. Nachr Dtsch Geol Ges 23, Vortrags-Kurzreferate, pp 11–13

Matte P (1986) Le chaîne varisque parmi les chaînes paléozoïques périatlantiques, modèle d'évolution et position des grands blocs continentaux au Permo-Carbonifère. Bull Soc Géol Fr 8, 2: 9–24

Morre-Biot N, Denner C, Monnot J (1972) Contribution à l'étude des roches vertes du massif de Chagey (Haute-Saône). Ann Sci Univ Besançon 17: 47–65

Pagel M, Leterrier J (1980) The subalkaline potassic magmatism of thje Ballons massif (Southern Vosges, France). Shoshonitic affinity. Lithos 13: 1–10

Pin C, Carme F (1988) Ecailles de matériaux d'origine océanique dans le charriage herecynien de la "ligne des Klippes", Vosges méridionales (France). CR Acad Sci Paris 306 II: 217–222

Prevot L (1969) Application de la méthode rubidium-strontium à l'étude de l'âge radiométrique de quelques dépôts devonodinantiens du massif de la Bruche (Vosges du Nord). Thèse 3 e cycle, Strasbourg

Rais N (1988) Sédimentologie et géodynamique d'un bassin syn-orogénique: "La série de la Bruche" (Vosges du Nord, Dévonien sup.-Carbonifère inf.). DEA, Univ L Pasteur, Strasbourg, 31 pp (unpublished)

Rey P (1988) Etude pétrologique et structurale des gneiss des Vosges moyennes d'Alsace: Un segment de croûte sous-charrié puis tectoniquement uplifté? DEA, Université de Montpellier, USTL, 95 pp

Rizki A, Baroz F (1988) Le volcanisme tholéiitique du massif de Schirmeck (Vosges septentrionales, France), témoin d'une zone de convergence de plaques au Paléozoïque supérieur. C R Acad Sci Paris 307 II: 511–516

Roche de la H, Eller JP von (1969) Caractères et tendances géochimiques des bandes granodioritiques et granitiques formant le Massif du Champ du Feu (Vosges cristallines du Nord). Bull Serv Carte Géol Als Lorr 22: 199–233

Ross PH (1964) Fossilfunde in den Steiger und Weiler Schiefern (Vogesen). Nachr Akad Wiss Göttingen II, 3: 37–43

Santallier D (1983) Main metamorphic features of the Paleozoic orogen in France. In: Schenk P (ed) Regional trends in the Geology of the Appelachiron-Caledonian-Hercynian-Mauritanide Orogen. Reidel, Dordrecht, pp 263–274

Santallier D, Briand B, Menot RP, Piboule M (1988) Les complexes leptyno-amphiboliques (CLA): revue critique et suggestions pour un meilleur emploi de ce terme. Bull Soc Géol Fr (8) IV: 3–12

Schleicher H, Kramm U (1986) Altersbestimmungen nach der Uran/Blei Methode an mittel- bis hochdruckmetamorphen Gesteinen des Schwarzwaldes. 76. Jahrestagg. Geol Ver Giessen, p 69 (Abstr)

Schneider C (1983) Aspects structuraux liés aux leucogranites du Bolstein et du Brézouard (Vosges moyennes). European Union of Geosciences Meeting, Strasbourg. Terra Cognita 3: 260

Schneider JL (1990) Enregistrement de la dynamique varisque dans les bassins volcano-sédimentaires dévono-dinantiens: exemple des Vosges du Sud (zone moldanubienne). Thesis Univ L Pasteur, Strasbourg, 222 pp

Weber K (1984) Variscan events: Early Paleozoic continental rift metamorphism and late Paleozoic crustal shortening. In: DHW Hutton, Sanderson HJ (eds) Variscan tectonics of the North Atlantic region. Blackwell, Oxford, pp 3-22

Wickert F, Eisbacher G (1988) Two-sided Variscan thrust tectonic in the Vosges Mountains, northeastern France. Geodin Acta 2: 101–120

Crystalline Provence: Structure and Variscan Evolution

G. Crevola and J.-P. Pupin

1 Introduction

Crystalline Provence, located in southeastern France and located along the Mediterranean Sea, constitutes one of the basement inliers which belong to the southern branch of the internal zone of the western European Variscan fold belt (Arthaud and Matte 1974; Autran and Cogné 1980; Matte 1986 a, b). Its relationships with the French Massif Central, the External Crystalline Massifs of the Alps, and the Axial Zone of the Pyrenees are not yet well understood. However, numerous data indicate that together with the Corso-Sardinian block it formed the continuous "Thyrrenian block" in Variscan times. This block was subsequently disrupted as a result of the Tertiary opening of the north-Balearic basin (Vellutini 1977; Orsini et al. 1980).

Crystalline Provence stretches for nearly 110 km from Toulon in the SW to Cannes in the NE. It is composed of three main geographical and geological units (Fig. 1):

1. The Maures Massif, to which the small massifs of the Toulon area and the Hyeres islands are attached.
2. The Tanneron Massif, which constitutes the northern continuation of the Maures Massif.
3. The Permian basin which surrounds the Maures Massif and, together with the Esterel Permian volcanic massif, separates the Maures Massif from the Tanneron Massif.

The Maures and Tanneron Massifs consist of Variscan metamorphic formations intruded by Carboniferous granitoids. These massifs, in which major structures display a submeridian trend, expose an 80-km transect. The western half of the transect exhibits epi- to mesozonal formations and the eastern half is primarily composed of migmatitic formations.

The Maures and Tanneron Massifs, Permian basin, and Esterel Massif are cut by N-S and E-W sets of extensional faults which have been active at different points during the region's evolution. The N-S faulting led to the development of northerly-trending Stephanian coal basins (Plan-de-la-Tour in the Maures Massif, Pennafort and Reyran in the Tanneron Massif). The E-W faulting active throughout the Permian formed grabens and controlled their sedimentary and volcanic evolution. On the eastern and northern edges of Crystalline Provence, detrital basal Triassic formations rest unconformably on deeply eroded metamorphic basement and Permian formations.

2 Stratigraphy and Paleontology

2.1 Major Lithologic Sequences

2.1.1 Introduction

The first lithostratigraphic and structural schemes for the Provencal basement were developed in the 1950s and 1960s by Bordet (1951, 1957, 1961, 1966) and Gueirard (1957). The works of Bordet, on which are based the regional geological maps (seven 1:50000 maps published between 1966 and 1976) highlight two fundamental points:

1. The continuity between the Maures and Tanneron Massifs as evidenced by similarities between their post-metamorphic features; consequently, the continuity of their metamorphic formations appears probable.
2. In the Maures and Tanneron Massifs the metamorphic series appears continuous from West to East. Isoclinal folding is responsible for repeating the lithologic formations. Among them, an amphibolite formation is regarded as an important stratigraphic reference.

In the 1970s and early 1980s, detailed field studies (Maluski 1968; Orsini 1968; Giraud et al. 1975; Seyler 1975; Le Marrec 1976; Crevola 1977; Olives Baños 1979; de Groulard 1982; Seyler 1983) resulted in many local modifications of Bordet's scheme. New attempts were also made to erect improved general lithostrati-

J. D. Keppie (Ed.): Pre-Mesozoic Geology
in France and Related Areas

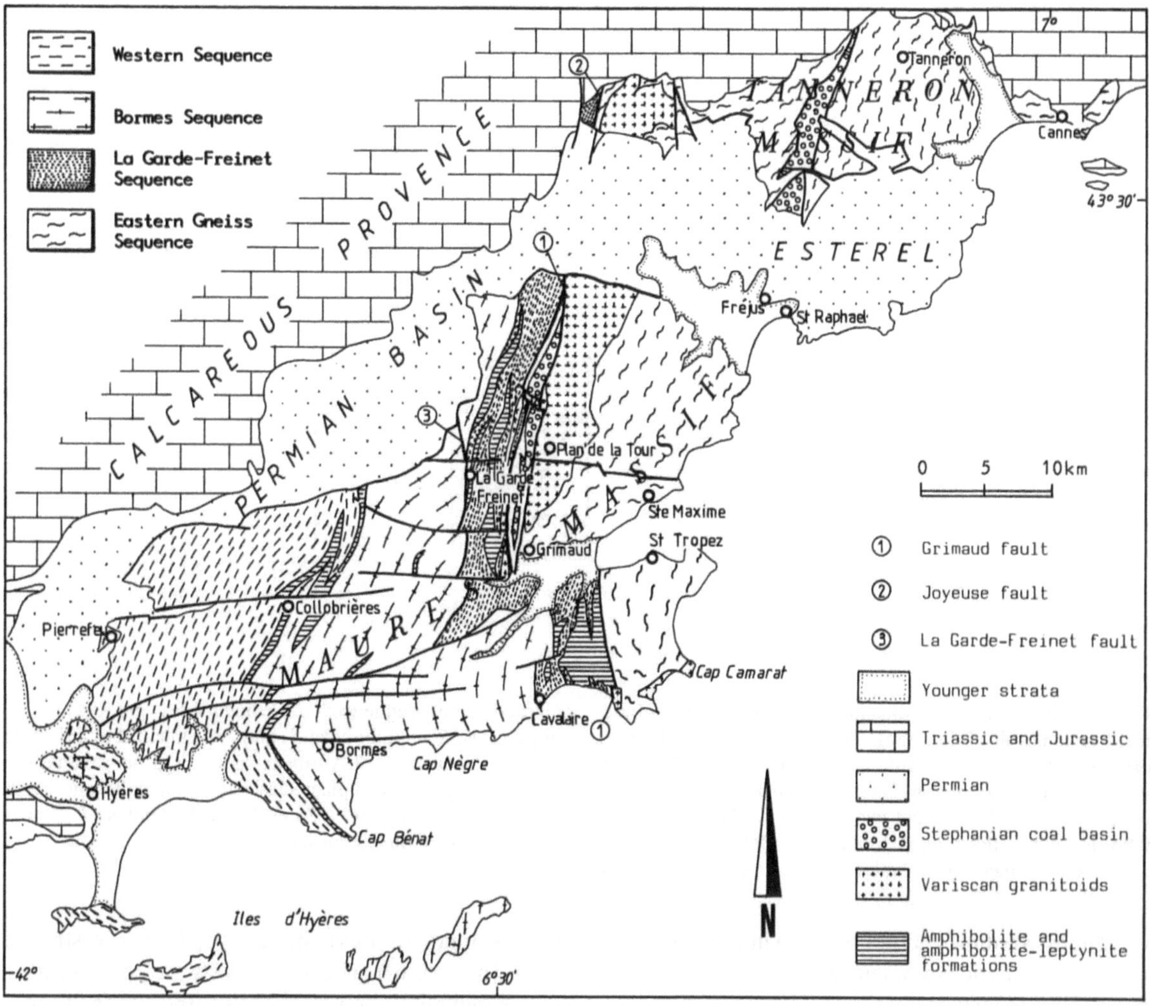

Fig. 1. General sketch map of crystalline Provence

graphies (Caruba and Turco 1976; Seyler and Boucarut 1978; Seyler 1983).

However, in recent years, the continuity of the metamorphic series has been questioned due to the recognition of the importance of the Grimaud-Joyeuse fault, which separates two sequences regarded as structurally contrasting (Caruba 1983; Vauchez and Bufalo 1985, 1988).

In the lithostratigraphical scheme adopted in this chapter, four major lithologic and/or lithotectonic sequences can be distinguished from west to east (Figs. 1, 2, and 4) on the basis of lithology, degree of metamorphism, and tectonic discontinuities (Tempier 1978; Crevola 1985). The first and second sequences are restricted to the Maures Massif, whereas the third and fourth ones extend over the Maures and Tanneron Massifs. The significance of major discontinuities will be discussed later.

2.1.2 The 4.2Sequence (WS)

This essentially epizonal sequence, 15 to 18 km wide, (Bordet 1966; Maluski 1968; Seyler 1975; Olives Baños 1979; de Groulard 1982) extends over the westernmost part of the Maures Massif (Figs. 1, 2). Its base is located at the structural top of the Bormes orthogneisses. The WS consists predominantly of thick and monotonous phyllite and quartzphyllite formations with frequent flysch-like facies. Some particular layers are intercalated in this series: (1) massive quartzites which form conspicuous ridges in the landscape, (2) amphibolites and leptynites (Collobrières formation), and (3) locally (Mont Fenouillet) rare lenses of crinoidal limestone and a fossiliferous black schist layer of middle Silurian age.

Thick Cambrian limestone formations characteristic of other Variscan regions (Montagne Noire,

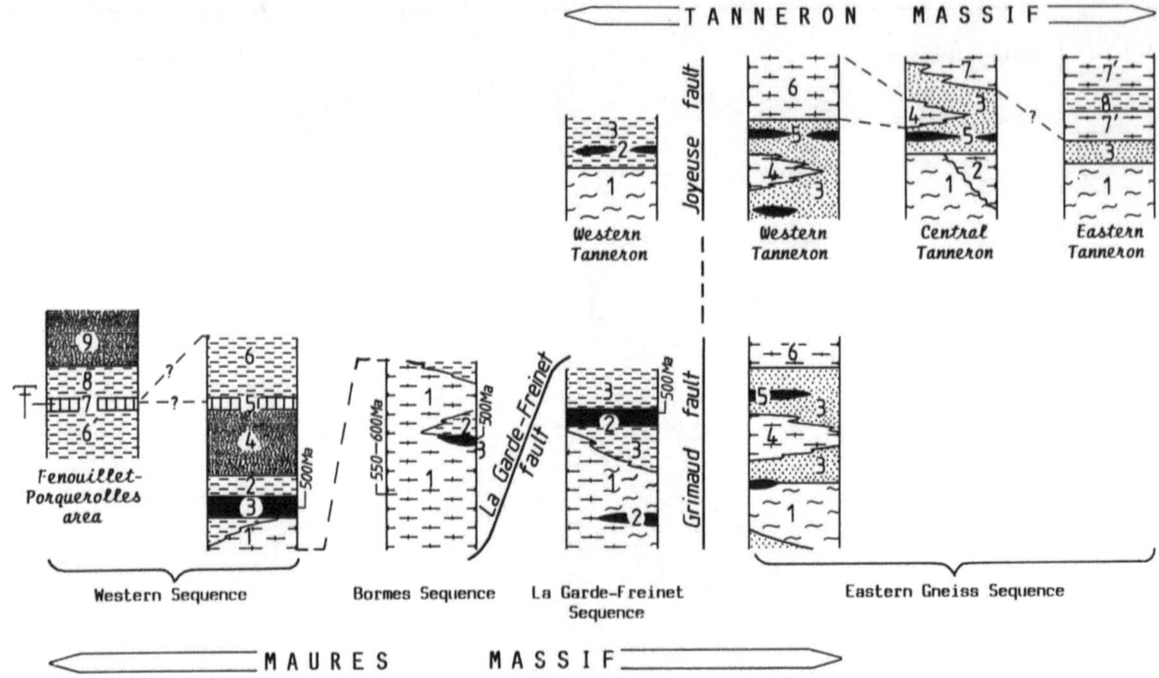

Fig. 2. Tentative general lithostratigraphy of Provencal basement metamorphic formations. Western sequence: *1* Bormes orthogneisses; *2* mica schists; *3* Collobrières amphibolites; *4* Loli flysch-like formation; *5* Temple quartzite; *6* and *6'* Maurettes phyllites; *7* Fenouillet quartzite; *8* Les Salins phyllites; *9* Porquerolles flysch-like formation. Bormes sequence: *1* Bormes orthogneisses; *2* mica schists; *3* amphibolites. La Garde-Freinet sequence: *1* Orthogneisses grading up to migmatites; *2* amphibolites; *3* sillimanite mica schists. Eastern gneisses sequence: *1* Auribeau and Peygros migmatites; *2, 4, 6,* orthogneisses; *3* plagioclasic and sillimanite micaceous gneisses; *5* amphibolites; *7* Les Adrets leptynites; *7'* Cannes leptynitic migmatites; *8* mica schists

Cévennes, Sardinia) do not outcrop in this sequence, and formations of Devonian age have not been recognized in its upper part.

2.1.3 The Bormes Sequence (BS)

The BS, which is overlain by the WS, (Gueirard 1957; Maluski 1968; Seyler 1975; Crevola 1985) is 10 to 20 km wide and has an apparent thickness of about 10 km (Figs. 1, 2). A narrow band of BS occurs along the Grimaud fault separated from the main outcrop area.

The BS consists mainly of augen to flaser gneisses (Bormes gneisses) intimately associated with micaceous gneisses or garnet-staurolite-kyanite mica schists (the micaschistes à minéraux). These two formations are locally (Vaucaude) intercalated with thin leptynite and amphibolite layers.

The Bormes gneisses, previously divided into ortho- and paragneisses (Maluski 1968; Seyler and Boucarut 1978; Seyler 1983), are now considered totally orthogneisses (Gueirard 1982; Crevola 1985). They originated through intense blastomylonitization of previous aluminous porphyritic monzogranites of which some remnants (Barral, Aire du Lac, Route des Crêtes metagranites) are exposed over restricted areas (m^2 to km^2). The mica schists are considered as metasediments originally constituting either the cover (Seyler 1983) or the country rock (Caruba, 1983) of the older granites. However, they may also originate through blastomylonitization with mass transfer of the older granites (ortho-mica schist, Crevola 1985).

2.1.4 The La Garde-Freinet Sequence (LGFS)

The LGFS is structurally overlain by the BS, the contact being normal or marked by faults of La Garde-Freinet fault system (Figs. 1, 2). This sequence is slightly more metamorphic than the BS, as kyanite is replaced by sillimanite and anatexis is reached.

In the Maures Massif (Gueirard 1957; Bordet 1957; Seyler 1975; Crevola 1985), the LGFS is 3 to 6 km wide and is bounded to the east by the Grimaud fault. Four main lithologic types can be distinguished: (1) prominent aluminous orthogneisses, (2) sillimanite mica schists, (3) cordierite migmatites, and (4) amphibolites associated with leptynites, subordinate metagabbros and serpentinites (the leptynitic-amphibolitic complex). The aluminous orthogneisses (La Garde-Freinet orthogneisses, Crevola 1985) contain huge masses of metagranite (the Moure-Jaumet metagranite: Seyler 1975; Crevola 1985) which outcrops over about 15 km from north to south. The orthogneisses and mica schists of this sequence are distinct from those of the BS: nonporphyritic orthogneisses vs. porphyritic orthogneisses, sillimanite mica schists vs. kyanite mica schists. The establishment of a lithostratigraphy is problematic in this sequence, which contains many orthogneisses grading up to mica schists and migmatites.

In the Tanneron Massif the LGFS is restricted to a narrow band extending to the west of the Joyeuse fault which is the northern continuation of the Grimaud fault (Bordet 1957, 1961; Crevola 1994). It consists of sillimanite mica schists and cordierite migmatites with subordinate amphibolite layers. These formations are in the same structural position as their equivalents of the Forêt des Arcs at the northern edge of the Maures Massif.

2.1.5 The Eastern Gneisses Sequence (EGS)

This sequence, extending to the east of the Grimaud-Joyeuse fault for nearly 40 km, consists mainly of migmatitic gneisses (Bordet 1961; Orsini 1968; Giraud et al. 1975; Le Marrec 1976; Crevola 1977, 1994). Its continuity is not affected by the La Moure fault which forms the eastern limit of the Reyran coal basin. A tentative lithostratigraphy can be established in the Tanneron and northeastern Maures Massifs in which the folded structures are fairly well-defined (Crevola 1994).

In the Tanneron Massif (Fig. 2), the structurally lower part of the series is made up of fairly homogeneous, banded and augen migmatites, which outcrop in the Tanneron and Auribeau areas (Auribeau migmatites, Crevola 1977). Of possibly granitic origin, they are themselves cross-cut by granitic orthogneisses (Tanneron orthogneisses). They are overlain by a complex of plagioclasic and micaceous sillimanite gneisses of sedimentary origin, with subordinate marble and calc-silicate gneiss lenses. This complex contains huge orthogneiss masses (Rioutard and Bois de Bagnols orthogneisses, Adrets leptynites) as early premetamorphic intrusions and/or tectonic repetitions. The Mandelieu and Cannes leptynitic migmatites, which are exposed at the eastern edge of the Tanneron Massif over tens of kilometers, are structurally situated under the lower migmatites. Nevertheless, their lithological analogies with the Adrets leptynites and their staurolite-kyanite-sillimanite mica schists suggest that they may be regarded as part of the upper complex which has been tectonically repeated. Mafic formations consisting of lensoidal masses of amphibolite with eclogitic relicts are intercalated at various levels of the series.

In the northeastern Maures Massif (Fig. 2), banded and augen migmatites similar to Auribeau migmatites outcrop near the Plan-de-la-Tour granite (Peygros migmatites, Le Marrec 1976), being intercalated near the base of the series. They are overlain by a thick sillimanite micaceous gneiss formation in which several huge masses of orthogneiss are intercalated (Fournel and Les Issambres orthogneisses). As in the Tanneron Massif, lensoidal masses of amphibolite with eclogitic relicts occur at various levels of the series. The upper part of this lithostratigraphic column is quite similar to the western Tanneron Massif, but banded migmatites are missing in the Tanneron Massif due to present-day different levels of erosion (Crevola 1994).

2.2 The Metamorphic Series: General View

2.2.1 Age of the Metamorphic Series: Paleontological and Radiometric Data

The original age of the metamorphic series from the Provencal basement can be estimated based on scarce paleontological and radiometric data which were acquired only in the western and central part of the Maures Massif. In the WS, the occurrence within the Mont Fenouillet quartzite of a thin black schist layer which yielded graptolites of Middle Silurian age (Upper Llandovery to Lower Tarannonian) allows dating of the uppermost part of the WS (Schoeller 1938; Gueirard et al. 1970). The Bormes orthogneisses were formed from ancient Cadomian granites which were emplaced at about 550-600 Ma (Rb-Sr dating[1]: Maluski, 1971; ^{39}Ar–^{40}Ar dating: Maluski and Gueirard 1978). Leptynites, which belong to the bimodal metamagmatic formations of the WS, BS, and LGFS in the Maures Massif, were formed from felsic volcanics emplaced at about 500 Ma (U-Pb zircon dating: Lancelot et al. 1992).

Thus, these data, at least for the WS and BS, enable the establishment of a loose stratigraphical framework. These two sequences, taken together, appear to comprise ancient Cadomian granites and Lower Paleozoic formations. The age of the formations of the LGFS and EGS is still unresolved due to the lack of reliable radiometric data.

[1] All Rb-Sr ages are calculated using the constant $\lambda_{^{87}Rb} = 1.42 \times 10^{-11}$/year.

2.2.2 Occurrence of a Basement-to-Cover Relationship

The occurrence, prior to metamorphism and deformation of a pre-paleozoic basement intruded by Cadomian granites and a Paleozoic cover, suggests the potential of local cartographic unconformities (Crevola 1977; Seyler and Boucarut 1978; Bourrouilh et al. 1980; Seyler 1983). This scheme, though criticized (Bard and Caruba 1983), might apply to the two more westerly sequences (WS and BS) for which a suitable geochronological framework exists (Seyler 1983). However, the age and significance of the mica schists intimately associated with the Bormes orthogneisses remain poorly known. In the LGFS and the EGS, the occurrence of several orthogneiss masses at various levels of the series and the scarcity of geochronological markers hampers the correlation with western formations and does not enable the definition of a basement-to-cover relationship.

2.2.3 Importance of Granite-Derived Gneisses

The granitic origin of numerous masses of augen to flaser gneisses and augen migmatites has been progressively recognized over the two last decades, in three (BS, LGFS, and EGS) of the four lithologic sequences (Maluski 1968; Seyler 1975; Crevola 1977, 1994; Gueirard 1982). Because they appear at various levels in different sequences, it is not clear that they were emplaced contemporaneously. Some of these orthogneissic formations (Bormes, Moure-Jaumet, Bois de Bagnols, Fournel, and Tanneron orthogneisses) contain slightly deformed metagranitic remnants of various sizes (m^2 to km^2) displaying mafic microgranular enclaves. Petrographical (magmatic cordierite and sometimes garnet), zircon typology, and geochemical features indicate that they belong to an aluminous granitic suite of crustal origin.

3 Structure

3.1 Succession of Phases of Deformation

Three main phases of deformation have been recognized in various areas of the Provencal basement (Arthaud and Matte 1966; Maluski 1968; Seyler 1975; Le Marrec 1976; Crevola 1977). The two first are synmetamorphic and characterized by isoclinal, subhomoaxial F1 and F2 folds with S1 and S2 axial-planar foliations. The F1 folds are scarce and small (cm to m). In contrast, the F2 folds are more frequent and developed on all scales (cm to km). These two phases are clearly distinguished in the WS by the superposition of their folds and foliations (Arthaud and Matte 1966; Maluski 1968). In the more deformed and metamorphosed sequences to the east (BS, LGFS, and EGS), only F2 folds are clearly visible. Some sheath folds (cm to m) related to F1 or F2 folding have also been locally described (BS: Goldberg 1983; LGFS: Seyler 1983; EGS: Vauchez and Bufalo 1988). They trend E-W to NW-SE in the BS and LGFS and N-S to N 30 in the EGS.

A pervasive blastomylonitization or mylonitization is related to these two first phases of deformation and affects the whole metamorphic pile of the BS, LGFS and EGS (Le Marrec 1976; Crevola 1977; Seyler 1983; Vauchez and Bufalo 1988). As a result, the S1 foliation is progressively transposed into a composite S1-S2 foliation with blastomylonitic or mylonitic character in orthogneisses, banded migmatites, late-migmatitic granites, and bimodal metavolcanics. Stretching lineations are well developed in the orthogneisses as well as in the late-migmatitic granites. They are regarded as mainly related either to the first phase of deformation (Vauchez and Bufalo 1985, 1988) or to the second phase (Le Marrec 1976; Crevola 1977; Seyler 1983; Caruba 1983). In the EGS, they exhibit a submeridian trend, being broadly parallel to regional trends, in particular to axes of large F2 folds. To the west of the Grimaud-Joyeuse fault (i.e., in the WS, BS, and LGFS) the pattern of the stretching lineations is poorly known. E-W to N 130-trending stretching lineations have been locally described (Goldberg 1983; Seyler 1983; Vauchez and Bufalo 1988) as well as N-S-trending stretching lineations (Crevola 1994) and relationships between the two sets of lineations are unknown (Fig. 3).

The third phase (F3), developed on all scales, is post-metamorphic and clearly post-dates the extensive blatomylonitic or mylonitic episode (Crevola 1977). It is characterized at outcrop scale, in the eastern Maures and Tanneron Massifs, by concentric F3 folds which grade to chevron folds or kink-bands in the western Maures Massif.

3.2 Large-Scale Folds

In the Maures and Tanneron Massifs, the general trend of cartographic bands, fold axes, and foliation is broadly N-S. However, in some areas it is diverted to an E-W direction. These trends are not modified by late E-W and N-S trending major faults.

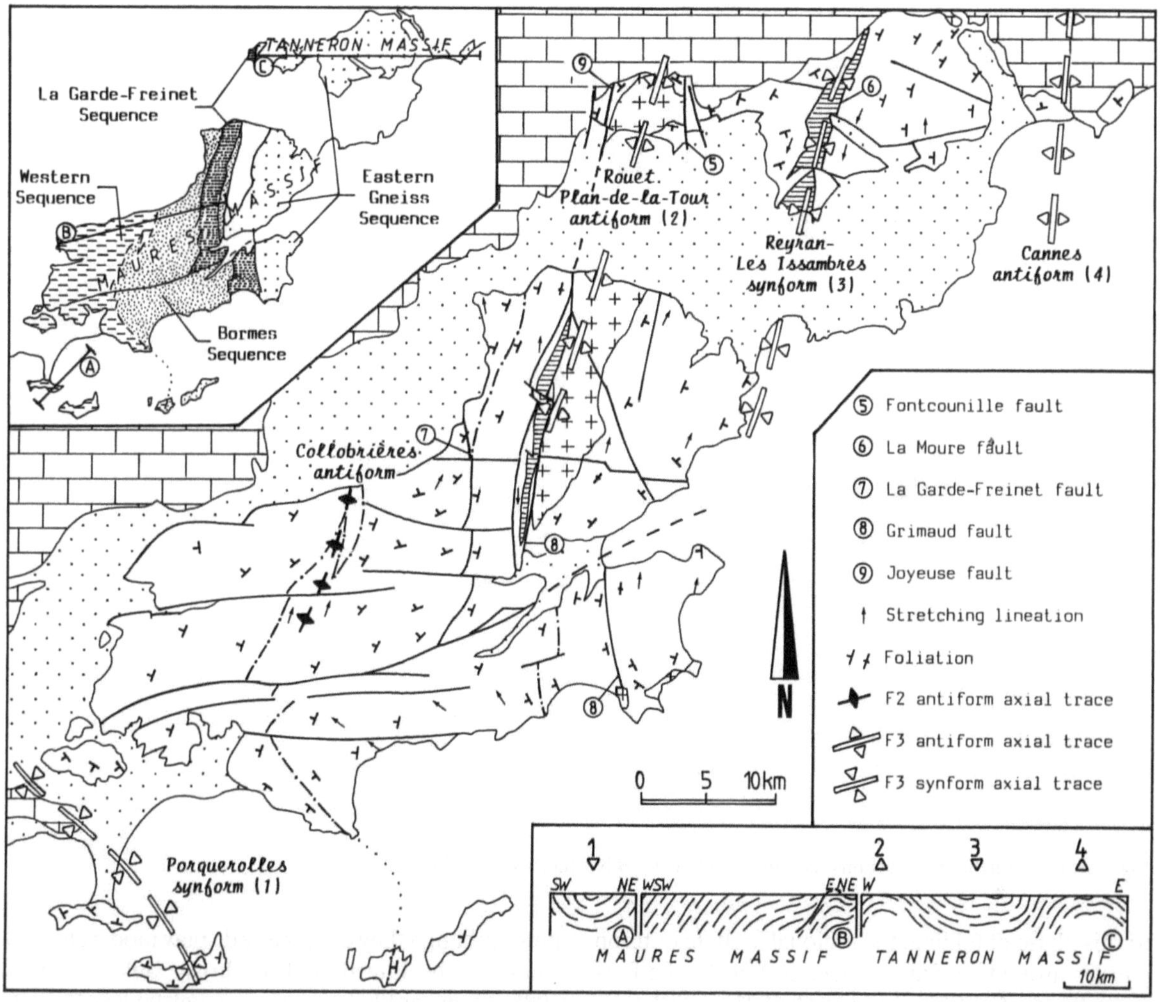

Fig. 3. Synthetic map of the Provencal basement and synthetic cross-section showing major F3 structures shown by foliation attitudes

To date, regional mapping has not identified any large recumbent fold related to the first phase (F1), as suggested by microstructural markers found in the WS (Maluski 1968). In fact, the only conspicuous map-scale structures are F2 and F3 folds, which affect the metamorphic banding and S1 foliation. By their superposition these large-scale structures can account for the overall observed geometry.

The large-scale F2 structures are hectometric to kilometric subisoclinal folds with thinned limbs and thickened hinges, displaying a fairly good axial continuity. By their close spacing, they define an "isoclinal style" (Bordet 1957; Gueirard 1957). However, only a few of these large F2 folds have been thoroughly mapped in suitable areas such as the "Chaîne de la Sauvette" (between Collobrières and La Garde-Freinet in the Maures Massif), and the central part of the Tanneron Massif. Among these large F2 folds the "Collobrières anticline" and the "Sauvette syncline" (Bordet 1957; de Groulard 1982), with an amplitude nearing 10 km, appear as major structures of the Provencal basement.

The third phase plays a major part, long unrecognized, in the structure of the Provencal basement (Crevola 1977). The large-scale F3 structures are hectometric to decakilometric open or box folds, with NS trending axes (Figs. 3, 4). They constitute an overall "dome and basin" structure. This kind of structure is shown by large arcuate outcrop patterns and by alternating areas of N-S-trending steep foliation and E-W-trending flat-lying foliation. F3 major folds are well developed in the Tanneron and the northeastern Maures Massifs. Three large decakilometric structures can be distinguished from east to west: the Cannes antiform, the Reyran-Les Issambres synform, the Rouet/Plan-de-la-Tour antiform (Figs. 3, 4). In the rest of the Maures Massif this pattern is less clear. However, various map-scale structures can be related

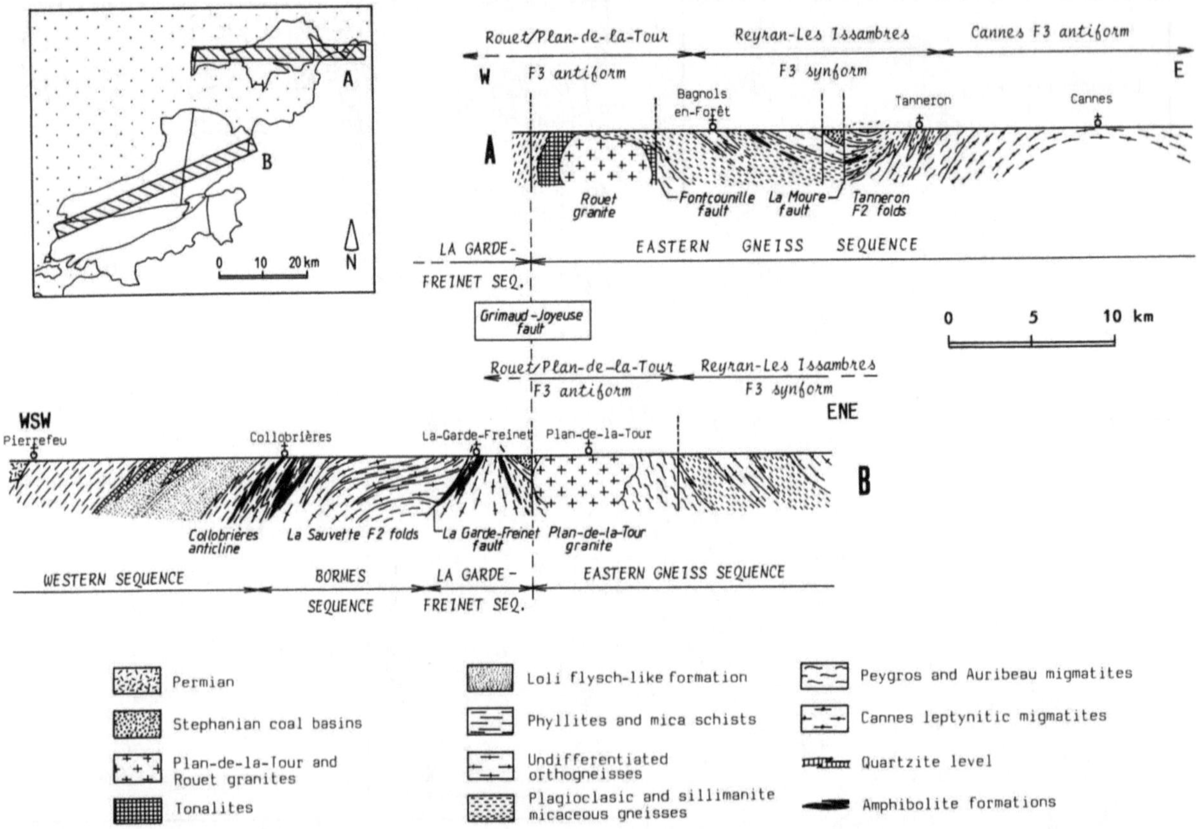

Fig. 4. A, B. Synthetic cross-sections. **A** Tanneron Massif. **B** Northern Maures Massif.

to this phase of folding: the regional arcuate outcrop pattern around the Saint-Tropez peninsula, the Porquerolles synform (Bronner et al. 1971) and the so-called La Garde-Freinet virgation, which corresponds to a giant bend in the western limb of the Rouet/Plan-de-la-Tour antiform.

3.3 Major Faults

The metamorphic formations are cut by sets of E-W- and N-S-trending faults. The E-W faults, of Permian age, may have significant vertical displacement (1 or 2 km). In the Maures Massif, some of them also have sinistral strike-slip motion, with increasing lateral displacement toward the south of the massif (i. e., the Saint-Tropez Gulf fault, whose main movement is probably Permian). In contrast, the N-S faults may have been active earlier in the structural evolution. Two groups of N-S faults are of particular importance.

The La Garde-Freinet fault (Figs. 1, 3, 4) and the faults which bound to the west the narrow band of the BS stretching along the Grimaud fault are steeply dipping, although they can locally display moderate dips. They can be interpreted as thrust faults (Crevola 1985) or more likely as extensional faults (Crevola 1994). They appear as late metamorphic, as they induced local low-grade retromorphic crystallization, and possibly predated the third phase of folding.

The Grimaud-Joyeuse fault (Figs. 1, 3, 4) belongs to a system of N-S mylonitic faults closely related to the development of the coal basins and the emplacement of the Plan-de-la-Tour and Rouet granitic line. Unlike the Fontcounille and the La Moure faults, which in the Tanneron Massif do not affect the continuity of the EGS, the Grimaud-Joyeuse fault separates two blocks regarded as lithologically and structurally contrasting. Owing to its structural position and to repeated episodes of mylonitization it has been widely studied.

The identity of the Grimaud fault, in the Maures Massif, and the Joyeuse fault, in the Tanneron Massif, has been established by Bordet (1951, 1961) and subsequently confirmed by geophysical data (Debeglia, 1977) and recent mapping (Crevola 1994). In the Maures Massif, classical studies (Demay 1927; Gueirard 1957; Bordet 1951, 1957, 1961; Triat 1964) have shown that cataclastic rocks developed along the fault were locally intruded by the Plan-de-la-Tour granite and were found in the Stephanian conglomerate, demonstrating a pre-Stephanian movement. Subsequently modern studies have shown that the Grimaud fault may correspond to a synmetamorphic and/or post-tectonic strike-slip fault of major importance (Caruba 1983; Vauchez and Bufalo 1985, 1988). Based on detailed microstructural studies, Vauchez and Bufalo have indentified three stages of evolution:

- a sinistral ductile strike-slip movement under epi- to mesozonal metamorphic conditions contemporaneous with F2 folding. This stage corresponds to the mylonitization of a 2 km wide zone of the EGS adjacent to the fault, particularly developed in late-migmatitic granites;
- a moderate cataclastic reactivation of the fault with a dextral sense of shear which predated the emplacement of the Plan-de-la-Tour granite;
- a late post-Stephanian and pre-Triassic brittle movement.

3.4 Problems with the Structural Interpretation of the Provencal Basement

3.4.1 Tectonic Significance of the Leptynitic-Amphibolitic Complex of the LGFS

This complex, which consists of bimodal metavolcanics associated with subordinate mafic metaplutonics, has undergone an early high pressure metamorphism and a subsequent blastomylonitization under lower grade metamorphic conditions. The geochemical features of its rock types (Sect. 5.1, below) suggest a pre-Silurian extensional magmatism related to a phase of crustal thinning (Bard and Caruba 1981; Caruba 1983; Seyler 1986).

However, the tectonic setting of this complex remains a debated question. Based on general considerations about the allochthonous setting of many leptynitic-amphibolitic complexes within the western European Variscan fold belt, Bard and Caruba (1981) and Caruba (1983) have regarded this complex a allochtonous. They have suggested that it may represent the sole of a nappe emplaced by westward ductile thrusting over the Bormes gneisses, during the first phase of deformation. Other authors (Seyler and Crevola 1982; Seyler 1983; Crevola 1994) have considered this complex to be autochtonous as no cartographic evidence has been found for a major overthrust, and as a pervasive blastomylonitization affected all the metamorphic pile of the central and eastern Maures Massif, being not restricted to the narrow zone of the complex.

3.4.2 Significance of the Grimaud-Joyeuse Fault

The Grimaud-Joyeuse fault has been regarded either as a post-tectonic normal fault (Gueirard 1957; Bordet 1951, 1957, 1961; Triat 1964; Seyler and Crevola 1982; Seyler 1983) or as a strike-slip fault of major importance (Caruba 1983; Vauchez and Bufalo 1985, 1988). The significance of the Grimaud-Joyeuse fault appears as a key problem for the understanding of the structure of Crystalline Provence and even of the southern part of the European Variscan fold belt. The most recent contributions to this problem are based on different kinds of arguments and develop contrasting points of view.

For Caruba (1983), in the Maures Massif the two blocks separated by the Grimaud fault exhibit contrasting lithologies, lithostratigraphies, and structures (opposite dips and plastic flow directions). Moreover, general considerations about late-Variscan wrench tectonics suggest that the Grimaud fault may correspond to a late-Variscan strike-slip fault with a dextral strike-slip movement. For Vauchez and Bufalo (1985, 1988) microstructural markers indicate distinct plastic flow directions for the two blocks and repeated transcurrent displacement along the Grimaud fault even as early as the second phase of deformation.

For Seyler and Crevola (1982), Seyler (1983) and Crevola (1994), the two blocks are similar in lithology and metamorphic grade and are both involved in F3 megastructures. Indeed recent mapping (Crevola 1994) has shown common foliation attitudes when the two blocks are in contact without the interposition of the Plan-de-la-Tour granite, i.e., in three main areas along the Grimaud-Joyeuse fault: western Tanneron, north of Grimaud, and Saint-Tropez Peninsula. These data appear consistent only with a minor strike-slip movement.

3.4.3 Structural Vergences

1. Structural Vergences Inferred from Stretching Lineations. Stretching lineations are well developed in the BS, LGFS, and EGS. In the Maures Massif, Vauchez and Bufalo (1985, 1988) have regarded these lineations as tectonic transport markers of the first phase of deformation indicating an overall ductile thrusting towards the north in the eastern Maures Massif and towards the west in the central Maures Massif.

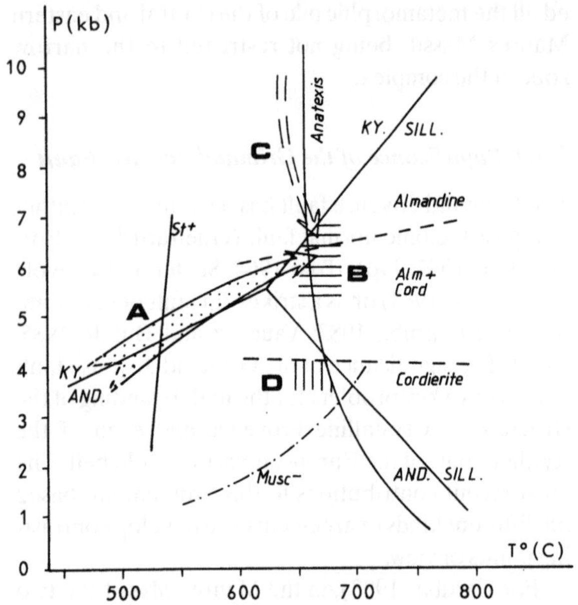

Fig. 5. Temperature-pressure conditions of crystalline Provence metamorphism. *A* Metamorphic path of western, Bormes, and La-Garde-Freinet sequences (Seyler 1975; Caruba 1983; Goldberg 1983); *B* field of eastern gneisses sequence metamorphism (Le Marrec 1976; Crevola 1977); *C* hypothetic retromorphic path of eclogitic rocks (after Maquil 1976); *D* field for late-migmatitic granites (after Maquil 1976) and retrograde assemblages (after Seyler 1975)

2. Structural Vergences Inferred from Fold Trends. A general westerly fold vergence has frequently been proposed, due to local microstructural observations of F1 microfolds in the WS (Arthaud and Matte 1966; Maluski 1968). In fact, early major F1 folds coeval with the development of S1 foliation have not been recognized. On the other hand, the large F2 folds display an apparent easterly vergence that has been observed in some areas: Chaîne de la Sauvette in the Maures Massif and central part of the Tanneron Massif (Crevola 1977, 1985, 1994). It should be stressed that, in the EGS where the pattern of stretching lineations is clear, the vergence inferred from fold trends appears orthogonal to the one inferred from stretching lineations.

4 Metamorphism

4.1 Tectonometamorphic Evolution (Fig. 5)

Numerous studies carried out over the two last decades in various areas of Maures and Tanneron Massifs have resulted in the establishment of a coherent tectonometamorphic scheme (Maluski 1968; Seyler 1975; Le Marrec 1976; Maquil 1976; Crevola 1977; Caruba 1983; Goldberg 1983). Three successive metamorphic stages with decreasing pressure conditions may be distinguished. The second stage which is the most extensive and the best known can be correlated with the two first deformation phases (F1 and F2) recognized.

The metamorphic evolution began with a high pressure stage of unknown age, which predated the two first phases of deformation. This high pressure event is particularly visible in relict meta-igneous rocks found in two metamorphic sequences (GFS and EGS). Metagabbros from amphibolite formations of the LGFS display a coronitic stage of granulite facies grade (Lasnier 1970). Amphibolite formations of the EGS contain widespread eclogitic relict lenses displaying various stages of amphibolitization (Le Marrec 1976; Maquil 1976; Crevola 1977; Caruba 1983). Among these eclogitic rocks a kyanite-sapphirine-hypersthene eclogite, which displays a retromorphic stage of granulite facies grade (P: 8–10 kbar, T: 750–850 °C), has been found (Les Cavalières eclogite: Le Marrec 1976; Bard and Caruba 1982). In the BS, LGFS, and EGS, metagranites showing coronas of garnet around magmatic biotite and pseudomorphs of garnet and sillimanite after magmatic cordierite might also record this high pressure event (Crevola 1977; Maluski and Gueirard 1978).

The main metamorphic event corresponds to the second stage of intermediate pressure type. Accompanying the first two isoclinal phases of deformation, it grades from epizone in the west up to catazone in the east. The metamorphic suite displays a complete Barrovian succession with a characteristic staurolite-kyanite zone, well-exposed in the BS (Gueirard 1957; Seyler 1975; Caruba 1983; Goldberg 1983). Locally andalusite, which occurs alone or associated with kyanite, appears as a typomorphic mineral (Seyler 1975; Goldberg 1983). Anatexis is reached just to the west of the Grimaud-Joyeuse fault in the sillimanite-muscovite zone and affects the entire EGS. Anatexis is partial, mainly developed in quartzo-feldspathic lithologies, particularly in orthogneisses, and rarely produces homogeneous migmatites. In the epizone and mesozone, the crystallization-deformation relationships show that the metamorphic mineral growth began before the first deformation phase and continued until the end of the second deformation phase, when the metamorphic peak was reached. In the anatectic domain the development of migmatitic banding was broadly contemporaneous with the F1 tectonic phase (Le Marrec 1976; Crevola 1977). Pegmatites or biotite-muscovite-(cordierite) leucogranites were emplaced between the two first tectonic phases, as dikes and stocks cross-cutting the metamorphic foliation. Migmatitic gneisses, late-migmatitic granites and pegmatites were reworked during the F2 tectonic phase sometimes with sillimanite development. However, locally in the EGS, thin leucocratic mobilizates were emplaced along shear planes affecting small F2 folds (Le Marrec 1976).

The third metamorphic stage is ill-defined and not clearly related to the second or the third deformation phase. It is characterized by a widespread but slight retrogression of earlier metamorphic minerals. Locally, cordierite or andalusite crystallized during this stage (Seyler 1975). Retrograde assemblages are also associated with some major faults: the La Garde-Freinet thrust fault and the Grimaud fault. In two Tanneron areas numerous dikes and veinlets of tourmaline granite were emplaced along a predominantly E-W direction, cross-cutting concentric folds of the F3 phase.

4.2 Age and Significance of Tectonometamorphic Events

The chronology of the tectonometamorphic evolution is poorly constrained due to scarce paleontological and radiometric data. The basic framework is as follows:

1. The epizonal formations of the BS, which contain the Mont Fenouillet fossiliferous layer of Silurian age, exhibit the same sequence of structures as more highly metamorphic formations. The lacustrine coal deposits of Westphalian D to Stephanian B age represent the first sedimentary formations overlying a deeply eroded metamorphic basement.
2. The post-tectonic Plan-de-la-Tour granite was emplaced at about 313 Ma (Rb-Sr dating: Maluski 1972) and the post-tectonic Camarat granite at about 300 Ma (Rb-Sr dating: Roubault et al. 1970a; Amenzou 1988).
3. The age of the metamorphism itself can be only broadly estimated due to paucity of radiometric dates. The metamorphism of the Barral metagranite took place between 340 and 310 Ma (^{39}Ar–^{40}Ar dating: Maluski and Gueirard 1978), like that of some gneisses of the EGS (about 336 Ma, Rb-Sr dating: Roubault et al. 1970a). However, some gneisses from the Tanneron Massif yield older ages, at about 400 Ma (Rb-Sr dating: Roubault et al. 1970a).

Finally, owing to the scarcity of radiometric data, it is difficult to know precisely whether the tectonometamorphic evolution is Devonian to Carboniferous or confined solely to the Carboniferous. However, no radiometric evidence has been found for the existence of remnants of an older metamorphic basement displaying Cadomian or Precambrian metamorphism and deformation.

The tectonometamorphic evolution of the Provencal basement is similar to that of other regions belonging to the southern branch of the internal zone of the western European Variscan fold belt, e.g., the French Massif Central (Santallier 1983; Ledru et al. 1989). As in these regions, the tectonometamorphic evolution should be interpreted as the result of collision tectonics during Devonian and/or lower to middle Carboniferous: first, high pressure stage related to subduction of continental crust, then continent-continent collision with Barrovian metamorphism, anatexis, and major folding, finally crustal reactivation and late granite emplacement.

5 Igneous Activity

5.1 Lower Paleozoic Mafic and Felsic Magmatism

Amphibolite or amphibolite-leptynite formations, occur within the four lithologic sequences. In recent years detailed petrographical and geochemical studies have been carried out in order to define their original magmatic affinities and evolutions and their original tectonic settings (Ricci and Sabatini 1978; Seyler and Boucarut 1979; Caruba 1983; Seyler 1986). Owing to their different locations, these formations exhibit varied original magmatic features:

1. The Collobrières formation, part of the WS, corresponds to a moderately alkaline differentiated suite in which basaltic and acid products (trachyte, rhyolite) have a co-magmatic origin (Caruba 1983; Seyler 1986).
2. The so-called leptynitic-amphibolitic complex of LGFS (Forêt des Arcs and Gassin formations) represents bimodal magmatism. It consisted originally of tholeiitic basic rocks (basalts, gabbros) with continental affinities associated with alkali rhyolites, which originated either by differentiation or crustal anatexis (Seyler 1986).
3. The amphibolitic formations of the EGS were derived from a tholeiitic protolith of oceanic or continental affinities (Caruba 1983; Seyler 1986).

All these original rock types can account for an intracontinental magmatism related to the development of extensional basins or even of an incipient oceanic rifting. In the western and central Maures Massif, this event took place at the Cambrian-Ordovician boundary, as shown by recent radiometric datings of some felsic occurrences (Lancelot et al. 1992). Such an event is now recognized throughout the West European Variscan fold belt (Ricci and Sabatini 1978; Matte 1986b).

5.2 Variscan Granitoids

(J.P. Pupin)

The Variscan granitoids appear as numerous small intrusions mainly confined to a N-S-trending belt in the Maures and Tanneron Massifs (Fig. 6).

5.2.1 Typology

Two main groups can be distinguished from a genetic standpoint, using classical data (field mapping, petrography, geochemistry) along with typologic studies of zircons (Pupin 1980, 1985; Chèvremont et al. 1988). The granitoids are of either crustal or mantle-crustal origin.

The first group of entirely or mainly crustal origin includes:

1. Late-migmatitic syn-tectonic granites and pegmatites (Saint-Pons-les-Mures and Moulin Blanc deformed leucogranites, the latter with cordierite-andalusite; Ramatuelle biotite granite) or late anatectic dikes of tourmaline Grime-type leucogranite from the Tanneron Massif.
2. Aluminous cordierite-aluminium silicate granites (some of which contain mafic micrograngular enclaves) resulting from melting at various crustal levels (Rouet and Plan-de-la-Tour monzogranites, Gigaro granodiorite, Plan-de-la-Tour and Camarat micromonzogranites, various dikes of biotite-cordierite or two-mica-cordierite-andalusite leucogranites, dikes of granophyric microgranite from the Tanneron Massif).

The second group is calc-alkaline, resulting from a mantle-crust hybridization. It includes varied biotite and biotite-amphibole granitoids (Fontcounille quartz-gabbro; Prignonet, Reverdi and Bagarède tonalites; Figarets granodiorite; L'Hermitan, Colle Dure, La Mente monzogranites; Vignon and Minuti microgranite dikes) and the Camarat biotite-muscovite granite.

5.2.2 Age of Emplacement and Structure

Structural relationships and radiometric data (Roubault et al. 1970a) indicate multiple episodes of emplacement.

Apart from the late-migmatitic granites, the oldest Variscan intrusions have a tonalite mean composition (Boucarut 1963; Orsini 1968; Pupin 1976; Amenzou 1988) similar to those of older intrusions in the western French Massif Central. They are unfoliated or slightly foliated, small intrusions of assumed pre-upper Visean age (Rb-Sr dating: Roubault et al. 1970a) emplaced along a N-S-trending "tonalitic line" which extends from the Tanneron Massif to the Saint-Tropez Peninsula. The Figaret granodiorite and the Hermitan-type monzogranite, which have varying degrees of deformation, exhibit petrographical and mineralogical characteristics indicating genetic links with the tonalitic formations (Amenzou 1988).

At about 315–335 Ma, in upper Visean-Namurian times (Rb-Sr dating: Roubault et al. 1970 a; Maluski 1972), a 45-km-long belt of cordierite porphyritic granites was emplaced (Rouet: Boucarut 1963; Plan-de-la-Tour: Triat 1964; Serment 1965; Serment and Triat 1967; Pupin 1976; Amenzou 1988; Gigaro: Amenzou and Pupin 1986; Amenzou 1988) along the Grimaud-Joyeuse fault. This linear intrusion crosscuts the "Provencal tonalitic line" in the Tanneron Massif. It is interrupted to the north by the Permian trough and offset to the south by the Saint-Tropez Gulf fault. As a general rule, the outcrops are, from north to south, increasingly more basic and have less cordierite. The main bodies are cross-cut by dikes of cordierite-biotite leucogranites, dikes, and stocks of cordierite-andalusite submiarolitic leucogranite (Plan-de-la-Tour: Triat 1964; Pupin 1976; Amenzou 1988) and granophyre dikes (Rouet: Boucarut 1963), all of essentially anatectic origin (Pupin 1976; Amenzou 1988). In the eastern Tanneron Massif, K-feldspar megacryst-bearing microgranites, which have been emplaced as dikes, belong to this aluminous group (Crevola 1977).

During the Stephanian (Rb-Sr dating: Roubault et al. 1970 a; Amenzou 1988) the magmatism was mainly marked by the Camarat granitic intrusion dated at 300 Ma, where detailed mapping has been recently carried out (Amenzou and Pupin 1986). Zircon typology indicates the calc-alkaline affinities of this two-mica granite (Pupin 1976; Amenzou 1988), suggesting that this southerly Provencal area may represent a transition zone between the Provencal plutonism and the wholly calc-alkaline Corsican orogenic plutonism developed to the SSE (Pupin 1985, 1988).

Anatectic magmatism resumed in the upper Stephanian. During this episode a subvolcanic stock and microgranular to felsitic dikes were emplaced in the Plan-de-la-Tour coal basin (Bordet 1951; Pupin and Turco 1973). Ash layers were deposited in the Reyran coal basin (Basso 1985; Begassat 1985) and

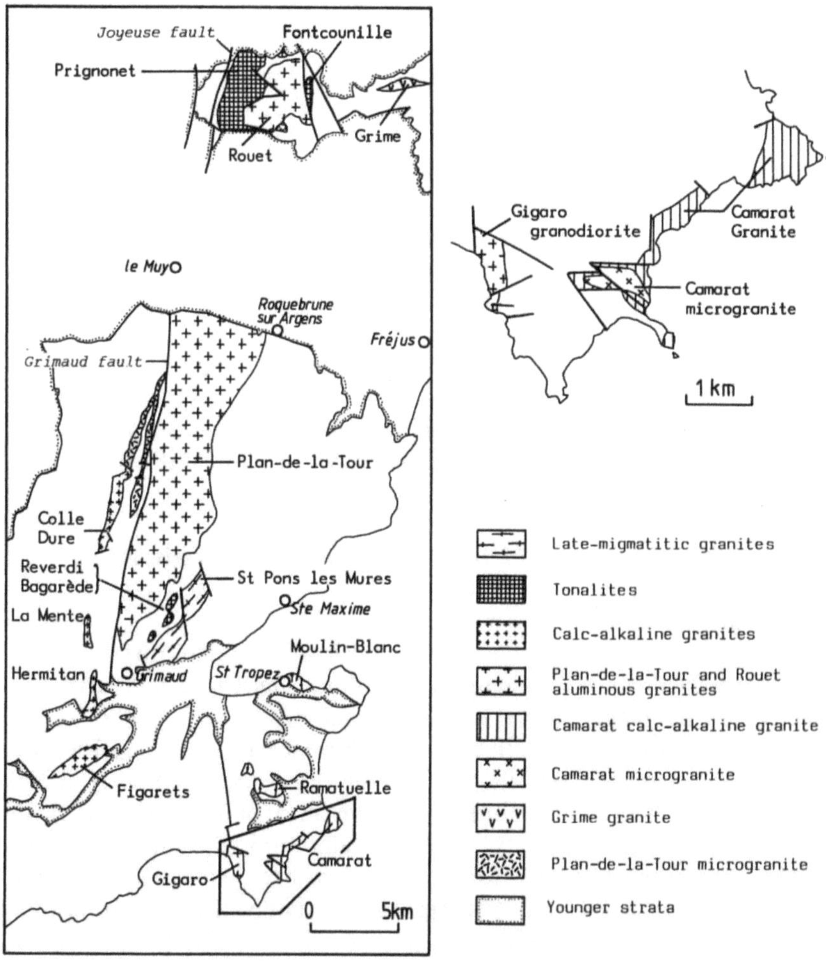

Fig. 6. The granitoid belt of crystalline Provence

the Camarat cordierite-andalusite-dumortierite microgranite was emplaced in older Camarat two-mica granite (Amenzou and Pupin 1986; Amenzou 1988).

The orogenic magmatism (the first Permian magmatic cycle of calc-alkaline type; Vellutini 1977) appears to have continued until the Autunian, as revealed by volcanic markers which occur within the detrital fill of Permian basins. In particular, calc-alkaline rhyolite pebbles derived from a more southerly volcanic source are widespread in Permian deposits of southern Esterel and Toulon areas (Pupin 1987).

5.2.3 Geodynamic Significance

In agreement with current plate tectonic models proposed for the West European Variscan orogenesis (Bard et al. 1980), the late emplacement of granitoids of either hybrid calc-alkaline or crustal anatectic type appears to be the result of a subduction-collision coupling (Pupin 1985; Amenzou 1988). The major suture would be situated to the west of Crystalline Provence. It may be the extension of the proposed suture located to the west of the Massif Central (Lameyre and Autran 1980). The waning stages of this long-lived magmatic event would be related to the slow resorption of thermal anomalies generated during the subduction-collision stage.

On the other hand, the comparison between Provencal and Corsican granitoids based on zircon typology reveals a magmatic complimentarity between the two domains (Amenzou 1988; Pupin 1988): crustal anatectic and calc-alkaline granitoids in Provence, calc-alkaline and K-calc-alkaline granitoids in Corsica. Thus, Provence and Corsica, when placed in their Variscan time positions, display an overall magmatic zoning: from crustal anatectic granites to the west, to increasingly hybrid calc-alkaline granitoids towards the east (Amenzou 1988; Pupin 1988).

5.3 Middle and Upper Permian Alkaline Magmatism

The Saalian phase at the Autunian-Saxonian boundary resulted in a tilting of previous Autunian formations of the small Avellan graben (Esterel Massif), deposition of widespread sedimentary breccias and local outpourings of alkali-basalts. These eruptions marked the onset of extensive alkaline volcanism (the second Permian magmatic cycle of alkaline type; Vellutini 1977) in downthrown E-W elongated basins (Toutin 1980; Baudemont 1985), particularly in the Esterel massif at the eastern edge of crystalline Provence (Bordet 1951; Boucarut 1971). The volcanic products are felsic (ignimbritic rhyolites, flow-banded rhyolites), mafic (basalts, hawaites, mugearites) and intermediate (trachytes). The volcanic activity occurred between 278 and 228 Ma (K-Ar dating: Roubault et al. 1970b) with a peak in activity at about 270 Ma, corresponding to the A7 ignimbritic flow. This volcanism was marked by three main stages (Boucarut 1971): initial outpouring of several voluminous ignimbritic rhyolite flows along E-W-trending faults which form the northern border of the Permian basin; secondary emplacement of small rhyolitic domes accompanying or post-dating caldera collapses (Vellutini et al. 1976; Gondolo 1989); and ultimate emplacement of basaltic or weakly differentiated basaltic flows and sills.

Thus in Provence as well as in Corsica, after the waning of orogenic magmatism, the second alkaline magmatism occurred in an extensional regime related to an incipient rifting already heralding the Alpine orogenesis (Vellutini 1977; Bonin 1987; Toutin-Morin 1987).

6 Metallogeny

In crystalline Provence, aside from a few isolated mineralizations related to the metamorphic evolution, the main metallogenic epoch was Permian to Triassic in age. In particular, it should be stressed that the well-developed Variscan granitoids are not accompanied by any mineralization.

6.1 Metallogeny Related to Metamorphic Evolution

Two mineralizations are clearly related to metamorphic evolution:

1. In the Tanneron Massif near les Adrets de l'Estérel, scattered scheelite mineralizations with economic concentrations (5 to 6% wt.% WO_2 at La Favière Mine) occur within calc-silicate gneiss lenses intercalated in a plagioclasic gneiss formation. A "peri-anatectic" model has been proposed to explain these mineralizations: tungsten-bearing anatectic fluids induced chemical changes as they crossed calc-silicate gneiss lenses (Sonnet et al. 1985).
2. Near the boundary between WS and BS the so-called Collobriérite (i.e., an iron ore consisting of magnetite, almandine garnet, and fayalite) occurs as thin lensoidal bodies within garnet-staurolite mica schists. This mineralization is classically regarded as a metamorphosed sedimentary iron ore which was intercalated in a sedimentary pelitic formation (Gueirard 1957; de Groulard 1982).

6.2 Metallogeny Related to Post-Variscan Evolution

Two types of mineralization are related to this stage of development of fault-bounded basins with detrital infill and bimodal volcanism:

1. In crystalline Provence, numerous Qz-Pb-Zn-F Ba vein-type deposits occur near major E-W or N-S faults either in the basement or in the Permian cover (Solety 1965; Vervialle 1975; Rostand and Turco 1982). Most of them have been extensively mined until recent times. They can be divided in two groups distinguished by their principal metal content, morphology, and relative age: a Qz-Pb-Zn group occurring in the western Maures Massif, and an F-Ba group occurring in the eastern Maures, Tanneron, and Esterel Massifs. The first group could be of end-Stephanian age while the second could be of Permian to lower Triassic age (Baudemont 1985).

Three kinds of metallogenic hypothesis have been successively proposed to explain these mineralizations (review in Begassat 1985).

- A magmatic hypothesis: mineralization is related to Permian late volcanic fluids.
- A "per descensum" hypothesis: mineralization originates through filling of previous fractures by surface brines.
- A paleothermal hypothesis: fluids are related to an extensive geothermal event during Permian and lower Triassic times which accompanied the break up of the Variscan basement.

2. Several small uranium deposits occur in the Permian trough, either as stratiform lenses hosted in sandstones or as veins cross-cutting the volcanics or the metamorphic basement. Among them, the Le Capellan stratiform deposit, near Vidauban, is the

only one to reach an economic size. Up to now, only scarce information about these deposits was available.

References

Amenzou M (1988) Les granitoïdes hercyniens du Massif des Maures (Var, France). Etude géologique et minéralogique. Implications génétiques. Thesis, Nice University, 280 pp

Amenzou M, Pupin JP (1986) Le granite de Camarat, un ensemble polygénique: zone clé entre la Provence et la Corse. C R Acad Sci Paris 303 II 8: 697–700

Arthaud F, Matte P (1966) Contribution à l'étude des tectoniques supersosées dans la chaîne hercynienne: étude microtectonique des séries métamorphiques du Massif des Maures (Var). C R Acad Sci Paris 262: 436–439

Arthaud F, Matte P (1974) Synthèse provisoire sur l'évolution tectonique et les raccords entre les segments hercyniens péribaléares. In: La Chaîne varisque d'Europe Moyenne et Occidentale. Colloq Int CNRS, Rennes, 243 pp 497–513

Autran A, Cogné J (1980) La zone interne de l'orogène varisque dans l'Ouest de la France et sa place dans le développement de la chaîne hercynienne. Colloque C6 Géologie de l'Europe 26ème C G I, Mém Bur Rech Géol Min, Orléans, 108: 90–111

Bard JP, Caruba C (1981) Les séries leptyno-amphiboliques à éclogites relictuelles et serpentinites des Maures, marqueurs d'une paléosuture varisque affectant une croûte amincie. C R Acad Sci Paris 292 II: 611–614

Bard JP, Caruba C (1982) Texture et minéralogie d'une éclogite à disthène-sapphirine-hypersthène-quartz en inclusion dans les gneiss migmatitiques des Cavalières, Massif de Sainte-Maxime (Maures orientales, France). C R Acad Sci Paris 294 II: 103–106

Bard JP, Caruba C (1983) Mise au point sur l'application du concept de socle-couverture dans le massif des Maures-Tanneron: une nouvelle interprétation des orthogneiss oeillés dans ce segment varisque du Sud de la France. C R Acad Sci Paris 296 II: 1341–1344

Bard JP, Burg JP, Matte P, Ribeiro A (1980) La chaîne hercynienne d'Europe occidentale en termes de tectonique de plaques. Colloque C6 Géologie de l'Europe 26ème C G I, Mém Bur Rech Géol Min, Orléans, 108: 233–246

Basso AM (1985) Le Carbonifère de Basse Provence (SE de la France), évolution sédimentaire et structurale. Thesis, Aix-Marseille University, 319 pp

Baudemont D (1985) Relations tectoniques socle-couverture en Provence orientale. Evolution tectonosédimentaire permienne du Bassin du Luc. Thesis, Strasbourg University, 204 pp

Begassat P (1985) Les bassins stéphaniens des Maures et du Tanneron: Pétrologie, Géochimie du volcanisme, métallogénie. Thesis, Pierre et Marie Curie University, Paris, 163 pp

Bonin B (1987) Réflexions à propos de la répartition des granitoïdes dans les massifs cristallins externes des Alpes françaises. Géol Alp 63: 137–149

Bordet P (1951) Etude géologique et pétrographique de l'Estérel. Mém Serv Carte Géol Fr 28: 207 pp

Bordet P (1957) Géologie de la partie centrale des Maures. Bull Serv Carte Géol Fr 54: 89–101

Bordet P (1961) Sur la géologie du massif de Tanneron (Var et Alpes Maritimes). C R Acad Sci Paris 252 D: 913–915

Bordet P (1966) Sur la structure du NW du Massif des Maures. C R Acad Sci Paris 262 D: 2677–2680

Bordet P (1966–1976) Cartes géologiques à 1/50000 Collobrières, Draguignan, Fayence, Fréjus-Cannes, Cannes-Grasse, Hyères. Bur Rech Géol Min, Orléans

Boucarut M (1963) Etude tectonique du Tanneron occidental (Var). DES, Marseille University, 93 pp

Boucarut M (1971) Etude volcanique et géologique du Massif de l'Esterel. Thesis, Nice University, 469 pp

Bourrouilh R, Cocozza T, Demange M, Durand-Delga M, Gueirard S, Guitard G, Julivert M, Martinez FJ, Massa D, Mirouse R, Orsini JB (1980) Essai sur l'évolution paléogéographique, structurale et métamorphique du Paléozoîque du Sud de la France et de l'Ouest de la Méditerranée. Mém Bur Rech Géol Min, Orléans, 108: 159–188

Bronner G, Lécorché JP, Orsini JB (1971) Un pli conique kilométrique: l'île de Porquerolles, fragment méridional du Massif hercynien des Maures (Var, France). C R Acad Sci Paris 272 D: 20–23

Caruba C (1983) Nouvelles données pétrographiques, minéralogiques et géochimiques sur le massif métamorphique hercynien des Maures (Var, France). Comparaison avec les segments varisques voisins et essai d'interprétations géotectoniques. Thesis, Nice University, 265 pp

Caruba C, Turco G (1976) Etude lithostratigraphique du massif du Fenouillet: corrélations lithostratigraphiques à l'échelle des Maures (Var, France). C R Acad Sci Paris 283 D: 1473–1476

Chevremont P, Gagny C, La Roche H de, Pupin JP, Rossi P (1988) Représentation cartographique du plutonisme acide et du plutonisme intermédiaire à basique associé. Doc Bur Rech Géol Min, Orléans, 152, 61 pp

Crevola G (1977) Etude pétrographique et structurale de la partie orientale du massif du Tanneron (Provence cristalline). Thesis, Nice University, 355 pp

Crevola G (1985) Les orthomicaschistes du Massif des Maures (France): produits ultimes du tectonométamorphisme métasomatique d'anciens granites. C R Acad Sci Paris 301 II: 99–104

Crevola G (1994) Carte géologique Fréjus-Cannes à 1/50000: terrains métamorphiques. Bur Rech Géol Min, Orléans (to be published)

Debeglia N (1977) Interprétation des données aéromagnétiques disponibles sur les cartes à 1/250000 de Nice et Gap. Bull Bur Rech Géol Min, Orléans, (2) II 4: 213–229

Demay A (1927) La zone mylonitique de Grimaud et la tectonique du Massif des Maures. Bull Soc Géol Fr (4) 28: 279–336

Giraud JD, Boucarut M, Caruba C, Turco G (1975) Données nouvelles concernant la géologie de la terminaison orientale du massif des Maures, zone nord massif de Sainte-Maxime. Bull Suisse Minéral Pétrogr 55: 461–469

Goldberg JM (1983) Etude pétrologique et structurale des séries pélitiques du Cap Benat à Saint-Tropez (Massif des Maures). D E A, Montpellier University, 60 pp

Gondolo A (1989) Le volcanisme alcalin permo-triasique du SE de la France. Scandola et Sénino (Corse). Estérel et Bas-Argens (Provence). Thesis, Nice University, 280 pp

Groulard P de (1982) Etude pétrographique et géochimique d'une série métamorphique à concentrations en fer: le groupe de Collobrières, massif des Maures (Var). Thesis, Lyon University, 167 pp

Gueirard S (1957) Description pétrographique et zonéographique des schistes cristallins des Maures (Var, France). Thesis, Marseille University, 171 pp

Gueirard S (1982) A propos de l'origine des gneiss de Bormes (Massif des Maures, Var, France). C R Acad Sci Paris 295: II 1125–1128

Gueirard S, Waterlot G, Gherzi A, Samat M (1970) Sur l'âge llandovérien supérieur à taronnien inférieur des schistes à graptolites du Fenouillet, massif des Maures (Var). Bull Soc Géol Fr (7) 12, 2: 195–199

Lameyre J, Autran A (1980) Les granitoïdes de France. Colloque C7, 26ème C G I. Mem Bur Rech Géol Min, Orléans, 107: 51–97

Lancelot JR, Ducrot J, Pupin JP, Seyler M (1992) U-Pb chronology of Prevariscan ortholeptynites from the Massif des Maures (S-E France). Earth Planet Sci Lett (to be published)

Lasnier B (1970) Le métamorphisme régional des gabbros d'après la littérature internationale. Etude préliminaire des gabbros coronitiques du massif armoricain et du massif des Maures. Thesis, Nantes University, 297 pp

Ledru P, Lardeaux JM, Santallier D, Autran A, Quénardel JM, Floc'h JP, Lerouge G, Maillet N, Marchand J, Ploquin A (1989) Où sont les nappes dans le Massif Central francais? Bull Soc Géol Fr (8) V 3: 605–618

Le Marrec A (1976) Reconnaissance pétrographique et structurale des formations cristallophylliennes catazonales du massif de Sainte-Maxime (quart NE du massif varisque des Maures, Var, France). Thesis, Marseille University, 125 pp

Maluski H (1968) Etude tectonique, microtectonique et géochronologique de la partie méridionale du massif des Maures (Var, France). Thesis, Montpellier University, 87 pp

Maluski H (1971) Etude au Rb/Sr des minéraux des gneiss de Bormes (Maures, Var, France). C R Acad Sci Paris 273 D: 1470–1473

Maluski H (1972) Etude au Rb/Sr du massif granitique de Plan de la Tour (Maures). C R Acad Sci Paris 274 D: 520–523

Maluski H, Gueirard S (1978) Mise en évidence par la méthode au $^{39}Ar/^{40}Ar$ de l'âge à 580 Ma du granite de Barral (Massif des Maures, Var, France). C R Acad Sci Paris 287 D: 195–198

Maquil R (1976) Contribution à l'étude pétrographique et structurale de la région S-E du massif des Maures (Var, France). Ann Soc Géol Belg 99: 601–613

Matte P (1986 a) La chaîne varisque parmi les chaînes paléozoïques péri-Atlantiques, modèle d'évolution et position des grands blocs continentaux au Permo-Carbonifère. Bull Soc Géol Fr 8: 9–24

Matte P (1986 b) Tectonics and plate tectonics model for the Variscan belt of Europe. Tectonophysics 126: 329–374

Olives Baños J (1979) Géologie des terrains paléozoïques dans le Sud-Quest du massif des Maures (Var, France). Thesis, Marseille University, 122 pp

Orsini JP (1968) Etude pétrographique et structurale du massif du Tanneron (Var) (partie occidentale et centrale). Thesis, Grenoble University, 109 pp

Orsini JP, Coulon C, Cocozza T (1980) La dérive cénozoïque de la Corse et de la Sardaigne. Géol Alp 56: 169–202

Pupin JP (1976) Signification des caractères morphologiques du zircon commun des roches en pétrologie. Base de la méthode typologique. Applications. Thesis, Nice University, 394 pp

Pupin JP (1980) Zircon and granite petrology. Contrib Mineral Petrol 33: 207–220

Pupin JP (1985) Magmatic zoning of hercynian granitoids in France based on zircon typology. Schweiz Min Petrol Mitt 65: 29–56

Pupin JP (1987) Origine des grès de la »dépression permienne« (Var, France) par la typologie des zircons. Conséquences Paléogéographiques. Géologie Alpine, Mém. h. s., 13: 81–90

Pupin JP (1988) Granites as indicators in paleogeodynamics. Rend Soc Ital Mineral Petrol 43, 2: 237–262

Pupin JP, Turco G (1973) Le microgranite de Plan de la Tour (Maures): Relations avec le granite de cette localité et détermination de son âge C R Acad Sci Paris 276 D: 3261–3264

Ricci CA, Sabatini G (1978) Petrogenetic affinity and geodynamic significance of metabasic rocks from Sardinia, Corsica and Provence. Neues Jahrb Mineral Mh 2,1: 28–38

Rostand P, Turco G (1982) Essai de synthèse sur la distribution des filons fluo-barytiques du district provençal (Maures, Tanneron, Estérel). Bull Bur Rech Géol Min, Orléans (2), 2: 179–185

Roubault M, Bordet P, Leutwein F, Sonet J, Zimmermann JL (1970 a) Ages absolus des formations cristallophylliennes des massifs des Maures et du Tanneron. C R Acad Sci Paris 271 D: 1067–1070

Roubault M, Bordet P, Leutwein F, Sonet J, Zimmerman J (1970 b) Ages absolus des formations volcaniques du Tanneron - Estérel (Var). C R Acad Sci Paris 271 D: 1157–1160

Santallier D (1983) Main metamorphic features of the Paleozoic orogen in France. In: Schrenk P (ed) Regional trends in the geology of the Appalachian-Caledonian-Hercynian-Mauritanide orogen. Reidel, Dordrecht, pp 263–274

Schoeller H (1938) Sur la présence de Graptolites dans les schistes métamorphiques du Massif des Maures. C R Somm Soc Géol Fr Paris 8: 147–148

Serment R (1965) Etude structurale du granite de Plan-de-la-Tour. D E S, Marseille University

Serment R, Triat JM (1967) Etude tectonique du granite de Plan-de-la-Tour. Bull Soc Géol Fr 7: 662–669

Seyler M (1975) Pétrologie et lithostratigraphie des formations cristallophylliennes dans la chaîne de la Sauvette (Maures, Var, France). Thesis, Nice University, 175 pp

Seyler M (1983) Géologie de la partie centrale des Maures. Conséquences sur l'évolution géodynamique du massif. Géol Fr, Orléans (2) I: 81–96

Seyler M (1986) Magmatologie des séries volcaniques métamorphiques. L'exemple des métavolcanites cambro-ordoviciennes, en particulier alcalines, du socle provencal (France). Thesis, Lyon University and Doc Lab Géol Lyon, 371 pp

Seyler M, Boucarut M (1978) Données nouvelles sur la lithostratigraphie du massif des Maures le long d'une transversale Réal Martin - Grimaud. Bull Bur Rech Géol Min, Géologie de la France, Orléans, (2) 1: 3–18

Seyler M, Boucarut M (1979) Existence d'un paléovolcanisme alcalin et transitionnel dans le massif hercynien des Maures (Var). Etude géochimique des amphibolites et résultats préliminaires concernant la chimie des leptynites. Bull Soc Géol Fr (7) 21, 1: 11–20

Seyler M, Crevola G (1982) Mise au point sur la structure et l'évolution géodynamique de la partie centrale du massif des Maures. C R Acad Sci Paris 295 II: 243–246

Solety P (1965) Les gisements de fluorine de la Provence cristalline (Maures, Estérel, Tanneron). Mém Bur Rech Géol Min, Orléans, 39, 264 pp

Sonnet P, Verkaeren J, Crevola G (1985) Scheelite-bearing calcsilicate gneisses in the Provence cristalline basement (Var, France). Bull Minéral, Paris, 108: 377–390

Tempier C (1978) Les événements calédoniens dans les massifs varisques du Sud-Est de la France, Corse et Sardaigne. P I C G 27, Geol Surv Can Pap 78–13: 177–181

Toutin N (1980) Le Permien continental de la Provence orientale (France). Thesis, Nice University, 2 t, 594 pp

Toutin-Morin N (1987) Les bassins permiens provencaux, témoins de l'orogénèse hercynienne et de l'ouverture de la Téthys dans le Sud-Est de la France. Ann Soc Géol Nord CVI: 183–187

Triat JM (1964) Etude géologique du granite de Plan-de-la-Tour, au Nord des départementales 44 et 72. D E S, Marseille University, 81 pp

Vauchez A, Bufalo M (1985) La limite Maures occidentales-Maures orientales (Var, France): un décrochement ductile sénestre majeur entre deux provinces structurales très contrastées. C R Acad Sci Paris 301 II 14: 1059–1062

Vauchez A, Bufalo M (1988) Charriage crustal, anatexie et décrochements ductiles dans les Maures orientales (Var, France) au cours de l'orogenèse varisque. Geol Rundsch 77/1: 45–62

Vellutini P (1977) Le magmatisme permien de la Corse du Nord-Quest. Son extension en Méditerranée occidentale. Thesis, Marseille University, 276 pp

Vellutini P, Vincent PM, Boucarut M, Brisset F, Bruneton P, Cocheme JJ (1976) Découverte d'une structure en »cauldron« dans le massif volcanique de l'Estérel. C R Acad Sci Paris 282 D: 33–36

Vervialle JP (1975) Les minéralisations filoniennes à barytine, fluorine et plomb-zinc du socle hercynien des Maures orientales, de l'Estérel et du Tanneron (Var, France). Thesis, Orléans University, 96 pp

Pyrenees

C. Majeste-Menjoulas and P. Debat

1 Introduction

From the Basque Country to the Roussillon area the Pyrenees consist of an approximately rectilinear intracontinental chain of mountains stretching from west to east and framing the North-Pyrenean zone, the Axial zone or Variscan High Chain, and the South-Pyrenean zone (Fig. 1).

The Variscan High Chain occupies the median part between Cap Cerbère and Pic d'Anie, thus forming the present orographic axis. The Alpine tectonic axis is situated more to the north at the level of the North-Pyrenean front. To the east the High Chain is gradually covered by the Ampurdan and the Roussillon basins. To the west from pic d'Anie it disappears under the Mesozoic cover and reappears as a discontinuous outcrop at the level of the Basque Massifs (Igounce, Mendibelza, Aldudes-Quinto Real, Ursuya-Cinco villas, Rhune-Haya). The upper Paleozoic outcrops broadly in the West Pyrenees and the lower Paleozoic in the East and Central Pyrenees.

The metamorphic rocks (gneiss and Precambrian mica schist) and the Paleozoic deposits were folded, sometimes metamorphosed, and intruded by plutons of granitic rock during the Variscan orogeny. The Alpine orogeny was more or less distinct in these ancient formations and the southern border of the axial structure was involved in major tangential movements towards the south (southern border of the Basque Massifs, the nappe of Gavarnie, the plunging tip of Nogueras); this is also the case for some internal areas such as the Cinq Monts-Gentiane thrust (Eaux Chaudes) in the Ossau valley (Fig. 1).

North of the Variscan High Chain, the North-Pyrenean zone is composed of Mesozoic terranes with locally a substratum of Precambrian or Paleozoic rocks. The substratum, identical to that of the Variscan High Chain, outcrops in elliptical North-Pyrenean Hercynian massifs with long axes oriented WNW-ESE, among which the most important are: Agly, Salvezines, Bessède, Saint Barthélémy, Trois Seigneurs,

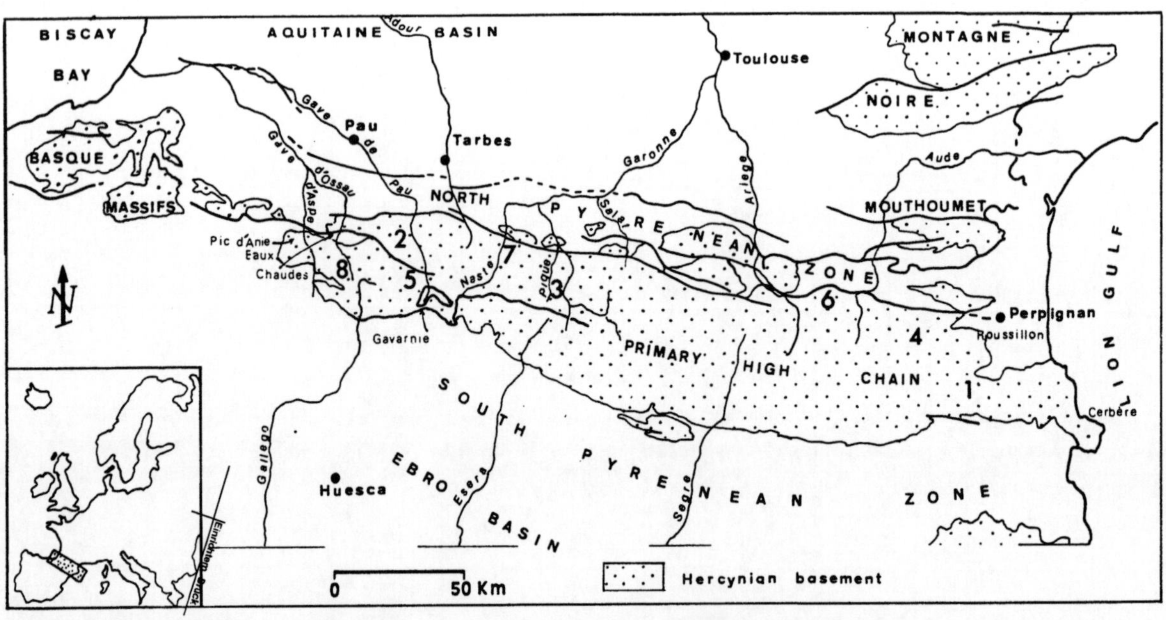

Fig. 1. General setting of the Pyrenees. *1, 2, 3, 4, 5, 6, 7, 8* Locations of the stratigraphic columns of Fig. 3

J. D. Keppie (Ed.): Pre-Mesozoic Geology
in France and Related Areas

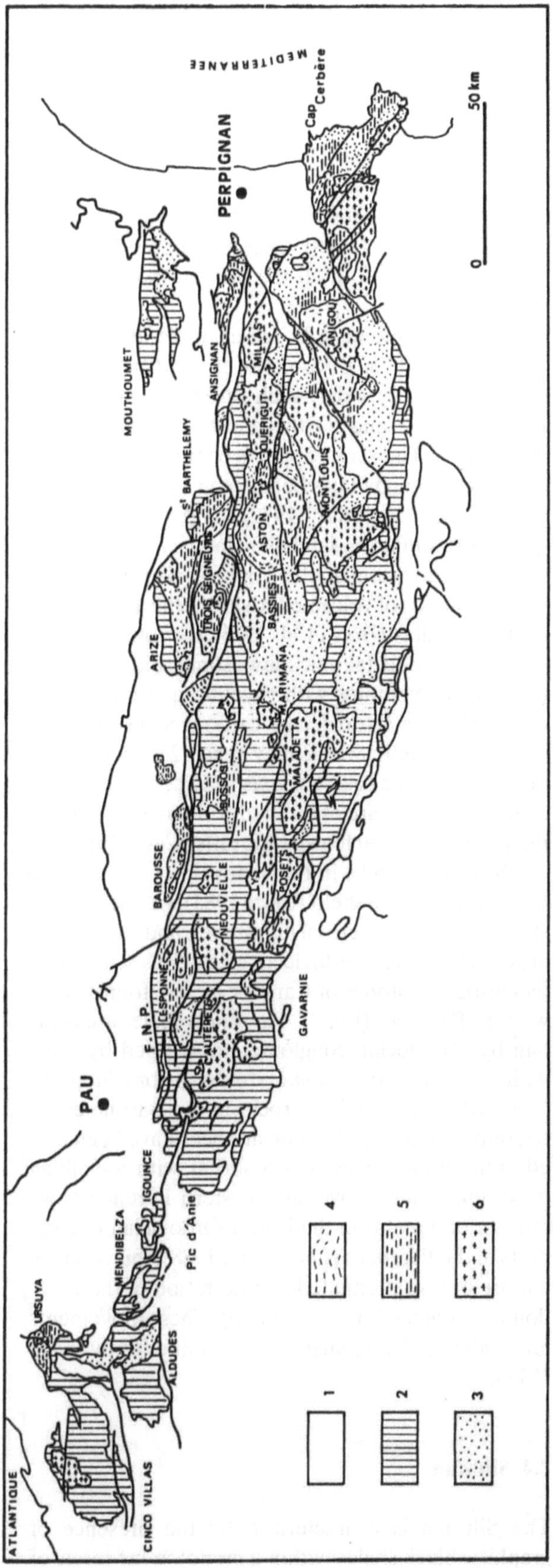

Fig. 2. The Pyrenees. *1* Mesozoic and Tertiary; *2* Silurian, Devonian, Carboniferous; *3* Cambro-Ordovician; *4* orthogneiss and paragneiss (Cadomian basement ?); *5* metamorphism; *6* granitoids

Castillon, Arize, Milhas, Barousse, and Bagnères-de-Bigorre (Fig. 2).

To the south, the South-Pyrenean zone comprises terranes dating from the Permo-Triassic to the Tertiary. There are no ancient massifs as in the North-Pyrenean zone. Paleozoic rocks of the South-Pyrenean zone (on its northern border) belong to the wedge thrust towards the south during the Alpine orogeny. Hence, between the Iberian and the European cratons, the Pyrenees form a polyorogenic chain where the Paleozoic formations were vigorously deformed during the Variscan orogeny, then strongly reshaped during the Alpine orogeny.

2 Lithostratigraphic Formations

Two main lithostratigraphic units can be distinguished within successive formations of the ancient massifs of the Pyrenees (High Chain and North-Pyrenean zone): a Cadomian basement and an unconformable Paleozoic cover.

2.1 Precambrian

Being wholly metamorphic (orthogneiss and paragneiss), Precambrian rocks have been dated only in the eastern Pyrenees (Canigou). Elsewhere only relative ages have been suggested, based either on facies analogies, stratigraphic unconformities (Ursuya, Aston, Canigou, Albères), or petrographic criteria.

The Precambrian of the Ursuya Basque Massif (Boissonnas et al. 1974) is composed of orthogneiss (ancient gabbros and norites), garnet-bearing leptynites and paragneiss with biotite (ancient rhythmic deposits of pelites and arkosic sandstones). Metamorphism of these mafic orthogneisses reaches the catazone (granulite facies, subfacies with orthose sillimanite), and the paragneisses the beginning of ultrametamorphism.

The Cadomian basement outcrops much more broadly in the central and eastern part of the chain (Aston, Hospitalet, Mont Louis, Canigou, Roc de France, Albères in the Variscan High Chain; Castillon, Saint Barthélémy, Agly in the North-Pyrenean massifs). In this region it is composed (Autran et al. 1966; Guitard 1970) of augen orthogneisses due to the transformation of more or less porphyritic calc-alkaline granites with a rapakivi texture on the one hand and aluminous and feldspathic paragneiss resulting from the metamorphism of pelites and graywackes on the other hand.

The age of the G2 orthogneiss of Canigou is roughly 580 million years (Vitrac-Michard and Allègre 1975). These are Cadomian granitoids belonging to the orogeny that marks the end of the Precambrian. Thus this basement must have been composed of massifs of porphyritic granitic rock with enclaves of metasediments (pelites and graywackes).

2.2 Cambro-Ordovician

Formations lying between the lower fossiliferous levels (Caradoc, Ashgillian) and the orthogneiss attributed to the Precambrian are assigned to the Cambro-Ordovician S. 1.

Various heterogeneous metamorphic formations (gneiss, mica schist, meta-arkoses, metapelites, and marbles), which lie unconformably on the Precambrian basement, correspond to the earliest subdivisions of the Cambrian period. In the Basque Massif of Ursuya Baygoura, they are conformably covered by Ordovician quartzites.

In the eastern Pyrenees, Cambro-Ordovician formations consist of the Canaveille series at the bottom and the Jujols series (Lower Ordovician) at the top, according to Cavet (1957). Laumonier and Guitard (1986) and Laumonier (1988) subdivide the group of Canaveilles with varied lithology (arkosic sandstone pelites, sandstone pelites, volcanic or volcano-sedimentary rocks) into three formations from bottom to top: (1) the metamorphic Canaveilles formation (amphibolite facies) with limestone, dolomite, orthoamphibolites and granular gneiss (Vallespir); (2) Cabrils formation with limestone, carbonate layers, and at the top volcanics (rhyolites in Albères); (3) the Evol formation with rhyolites (microconglomerates, pelites, and alternating sandstones and pelites; Fig. 3). In 1988, Laumonier assigned the whole Canaveilles group to the lower Cambrian, the Evol formation being equivalent to the Marcory sandstone (Montagne Noire).

The detrital series of Jujols overlies the Canaveilles group. It is coarser to the west and more pelitic to the east. It consists of pelites, sandstone, banded quartzitic phyllite (Jujols slates), a few quartzites and mircoconglomerates, limestones (Aspre), and carbonaceous granular slates (Conflent). Laumonier (1988) suggests a Mid- or Upper Cambrian age without excluding the possibility of a Tremadocian age. In fact, in the Mouthoumet the uppermost part of the series, comparable to the Jujols, was dated as Tremadocian by acritarchs (Cocchio 1982).

2.3 Ordovician

Due to the lack of fossils, except in the upper formations, Laumonier (1988) suggests a hiatus during the Lower Ordovician in all the ancient massifs of the Pyrenees.

In the Basque massifs, for example at Baygoura (Heddebaut 1973; Boissonnas et al. 1974), massive quartzites and then slaty shales are overlain by psammitic quartzites and "Orthis plicata" shale (Caradoc, Ashgillian). The series appears complete in this section. The Arenigian has been recognized (slates with graptolites; Degardin 1979) and the trilobite quartzites contain a Llandeilian fauna.

The Ordovician of the High Chain shows local manifestations of volcanic activity. In the area of Pierrefitte Nestalas (Bois et al. 1976) the lower (Lower Ordovician) detrital unit shows some acid volcanism (microconglomerates with rhyolitic quartz) and the upper (Upper Ordovician) carbonaceous volcano-sedimentary unit, which may reach 250 m in thickness, shows some basic volcanism (Fig. 3). The Aspres Ashgillian slates are associated with lava flows (porphyr), breccias, and feldspathic tuffs (Cavet 1957).

Ordovician rocks in the central part of the High Chain formed on a shelf(André 1985) with deposits of shales, quartzites, detrital limestones, and conglomerates in the Lower Ordovician calcareous schists and bioclastic limestones of Caradocage are found in the western Pyrenees (Fig. 3). The Jujols shales are overlain by Caradocian conglomerates capped by shale with Ashgillian bryozoans, cystoids, and brachiopods.

Apart from Arenigian rocks in the Basque country, only Ashgillian-Caradocian rocks have been dated. If the Jujols shales are Cambrian, with Ashgillian rocks immediately overlain (eastern Pyrenees), we must infer a hiatus in the Lower Ordovician in large parts of the Pyrenees (Laumonier 1988); this could be a more or less attenuated manifestation of the Caledonian orogeny (Taconian phase). These movements are evidenced by angular unconformities (Santanach 1972).

2.4 Silurian

The Silurian is characterized by the presence of graphitic black shales within a monotonous series of fine detrital deposits. The detrital sedimentation of

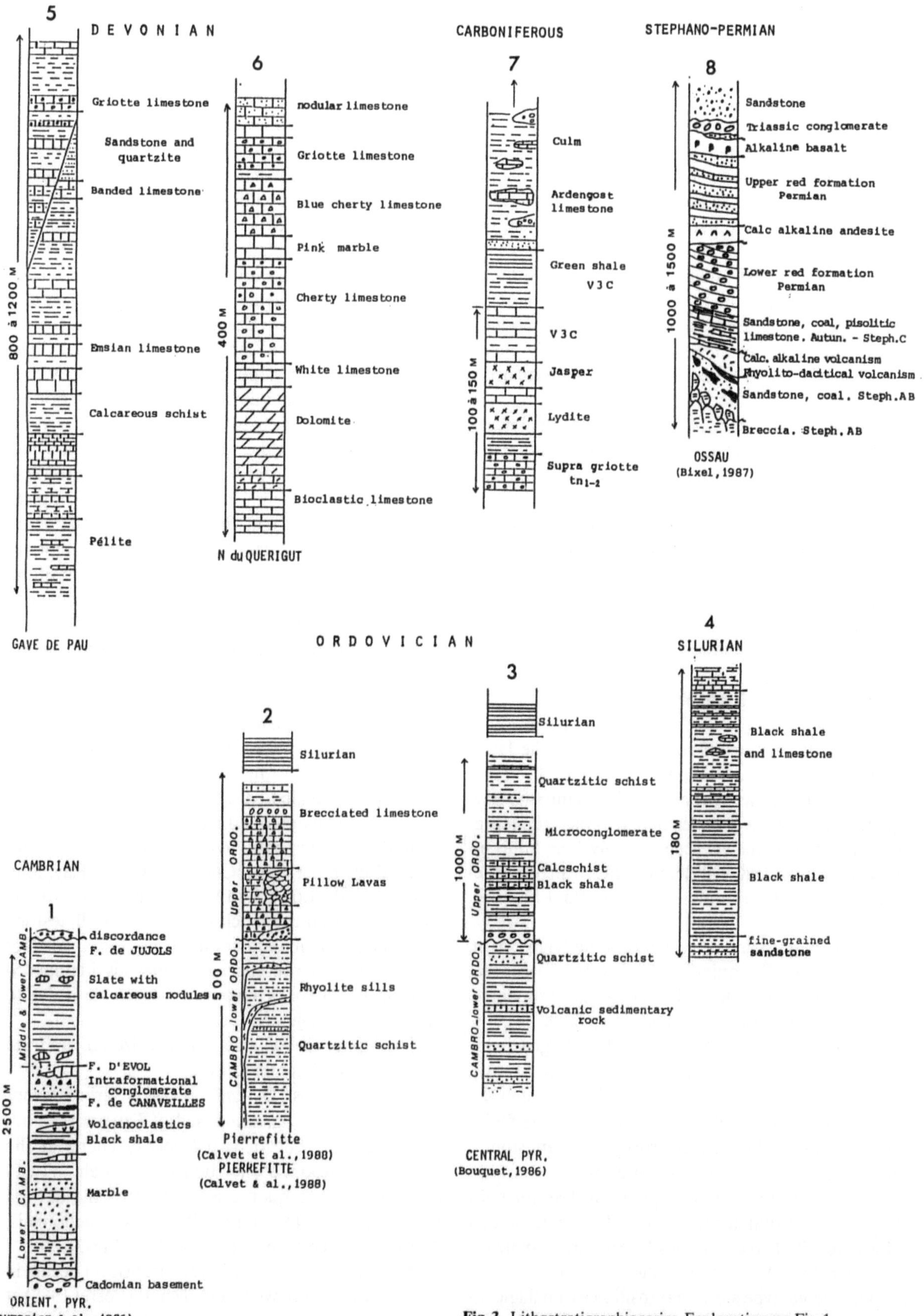

Fig. 3. Lithostratigraphic series. For location see Fig. 1

the latest Ordovician continued in the Silurian with fine-grained sediments; the transition is indicated by an increase in the organic content (planktonic organisms). Despite the occurrence of a few décollement levels, the series seems to be complete from the Llandoverian to the Lludoverian (Degardin 1987).

In the Aldudes massif (Basque Country), blue limestones with *Orthoceratidae* appear on top of bituminous black shale. In the eastern part of the High Chain limestones are even more abundant: calcareous nodules and black limestones of the Wenlockian (Central Pyrenees), multicoloured shales with calcareous nodules of the Early Llandoverian, and lilac shales with conodont bearing calcareous nodules from the Lludlovian in the Eastern Pyrenees. Quarzites and sandstones in the Upper Silurian occur in the central part of the High Chain (Fig. 3).

These thin deposits (200 m maximum thickness), rich in organic matter, result from sedimentation in a quiet marine environment, hence confined and subjected to reducing conditions. During the Variscan and Alpine orogenesis, the Silurian sediments played an important role as décollement level, reacting as a lubricant and injected in the surrounding sediments (Majesté-Menjoulas 1979). The structural relationship to the underlying Ordovician rocks is complex.

2.5 Devonian and Carboniferous

The regional pattern of sedimentation in the latest Silurian, namely detrital in the western and calcareous-pelitic in the eastern part, continued into the Early Devonian. More complex in detail, the sedimentation announces the Variscan orogeny. The Silurian-Devonian boundary is ill-defined, due to an often tectonic contact and a more or less important hiatus in the basal units.

In the western and central part of the High Chain, from the basal Devonian to the Givetian, a general trend occurs from detrital deposits to carbonates. Three types of sequences can be recognized:

The first type (northern border of the High Chain) occurs in the northwestern facies (Aspe and Ossau), and in the northern facies (Bodin 1987) between the Pique and the Salat valleys. In the northwestern facies detrital rocks (pelites with microrhythms, sandstones, graywackes) pass to fossiliferous formations (crinoids, bryozoans, brachiopods). In the northern facies, fine detrital deposits with pelagic macrofauna alternate with two calcareous layers, one of which is Praguian, the other Emsian in age.

The second type shows a transition to a pelagic facies, occurring in the median axis of the High Chain. The series passes vertically from pelite to Emsian limestone with alternating silty and calcareous layers (Fig. 3). The detrital Upper Emsian sediments and the Eifelian blue slates with Catherville trilobites are overlain by Givetian shelve limestones. In the median area, Emsian limestones pass through detrital layers into banded Frasnian limestones, the latter representing a transitional phase to flysch-like detrital deposits.

The third type (southwestern area) shows fossiliferous fine-grained deposits intercalated with carbonate levels indicating faunal reef-like characteristics in the Eifelian. The Devonian series are often incomplete here, underlying an uppermost Devonian and/or a transgressive Carboniferous sequence.

In the western and central part of the High Chain clastics have been deposited along the WNW-ESE axis of the mountain chain, indicating an important phase in the geological history of the Pyrenees at the transition from Middle to Upper Devonian. These detrital rocks (series of Sia and Las Bordas), 500 to 700 m thick, appear in the Upper Givetian to basal Frasnian. Some of them (Sia) resemble the characteristics of turbiditic series of deep water environments, others (Lariste) seem to be intermediate transitional to calcareous clastics, corresponding to shallower detrital deposits. East of the Salat valley, the Devonian becomes more and more carbonaceous, as the terrigenous deposits decrease from west to east.

Two areas take on individual characteristics. In the la Fayolle unit, the lower unit of Conflent (Raymond and Weyant 1982), the Devonian starts with limestones, often decalcified. The Givetian is pelitic and in the Frasnian layers of sedimentary breccia are intercalated with limestones. In a more eastern area (unit of Ourtiset, upper unit of Conflent, Aspres; Raymond et al. 1982), the Lower and Middle Devonian is totally carbonate, from the bioclastic Gedinnian limestone to the Famennian nodular limestone (Fig. 3, north of Querigut). The presence of numerous juxtaposed heterogeneous rocks in complex structures is due to late Variscan tangential thrusting.

The Upper Devonian is characterized by uniform Famennian nodular limestone.

The Uppermost Devonian-Lower Carboniferous is, broadly speaking, composed of Famennian nodular limestone, one or two levels of cherts at the Devonian-Carboniferous boundary, then Visean limestone. This sequence is overlain by the detrital Culm series. In the central Pyrenees (Luchonnais, Aure valley) the sequence is complete (Fig. 3).

The Devonian-Carboniferous transition occurs in limestone between two levels of chert (Perret 1988) at the bottom of the Tournaisian. At the Visean-Namurian boundary detrital units give way to carbonate formations which are marine deposits accumulated in ocean basins. In the SW of the western Pyrenees, the complete series shows Namurian limestones with laminites resulting from mechanisms of bottom slope sedimentation.

In the southwestern some Infracarboniferous or Lower Devonian units are missing (Mirouse 1966; Perret 1988). Cherts are rare and the transgressive base of the Carboniferous may rest on various Devonian formations. The reworking of Frasnian and Famennian conodonts, in the lowermost units of the Carboniferous, may reveal a temporary emergence of Devonian deposits (Perret 1977). In the western Pyrenees, the series show one or two layers of cherts but do not contain any laminite. In the Basque Massif, two areas take on individual characteristics: a western area where the Devonian and the Carboniferous are transgressive on older formations, and an eastern area where there is stratigraphic continuity from the Ordovician to the Namurian.

2.6 Carboniferous

Deposition of detrital deposits (Culm) by turbidity currents, which announced the Variscan orogeny, give way to the biochemical sedimentation in the Dinantian. The detrital sedimentation begins in a diachronous manner, earlier in the east than in the west: in the Visean V3c in the autochthon of the Mouthoumet Massif (Bessière et al. 1980) and in the Catalan Pyrenees (Boersma 1973); in the Namurian-A-Namurian-B transition in the Central and Western Pyrenees (Perret 1979; Delvolvé 1981); in the Westphalian in the Pays Basque (Mirouse et al. 1983). A gradation of flysch-like deposits to the south and the SW postulated by Engel (1984). The top of these series is not wellknown and the Stephanian deposits unconformably overlie the Westphalian deposits. The flysch deposits are composed of shale and turbiditic sandstone contains conglomerates and exotic blocks. The deposits are the result of gravitational flow in a deep marine environment (bottom slope or slope). Channels and carbonate slope facies are abundant within the turbidite deposits (Delvolvé 1987). The flow direction of the turbidity currents was to the southwest (Delvolvé 1987).

The Carboniferous platform limestone was reworked into pebbles or, in some cases, into fragmented sedimentary klippes (e.g., Ardengost limestone; Perret 1971). The age of the base of the carbonates decreases from the northeast towards the southwest. Thus, the deep marine deposits accumulated in a basin following an east-west axis. Sedimentation was from the northeast and migrated toward the southwest (based on the age of the deposits, reworking of internal flysch, and their sedimentary substratum).

2.7 Stephano-Permian

The molasse deposits, unconformable on the Variscan structures, outcrop discontinuously. Permo-Triassic red sandstones (Fig. 3) generally conformably overlie the Stephanian deposits, but locally Permian deep channels may cut the formations. The Permo-Triassic sandstones may in places unconformably overlie the folded Paleozoic rocks.

These formations correspond to lateral sedimentary accumulations of glacial and alluvial fans and to longitudinal fluviatile deposits (Lucas 1985). The Stephano-Autunian basins formation was controlled by submeridian fractures (NS steep-sided troughs). The red deposits, which are kilometers thick (Lucas 1985) are intercalated with deposits of tectonic slides and talus accumulation, deposits in altiplano mud flow, deposits of currents (sides of basins), and deposits in distal zones of oblate fans. The deposits are more distal on the southern slope than on the northern one. To the west of Lerida the Stephanian half-grabens were oriented NW–SE and the primary half-grabens, NNE–SSW and ENE–SSW. To the east, during the Stephanian, they were NE–SW and during the Permian E–W and NW–SE.

The Permian formed in multikilometric arid and subarid intramontane basins in which post-Variscan fracturing was still active in the form of volcanism.

3 Variscan Tectonics

3.1 The Pyrenees, a Polyorogenic Chain

The rocks of the Pyrenees have undergone several orogenies, mainly Variscan and Alpine and, to a lesser degree, Cadomian and Caledonian. A stratigraphic unconformity separates the Cadomian basement from the lower Paleozoic (Autran et al. 1966; Boissonnas et al. 1974), but current knowledge does not allow an accurate definition of the characteristics of a Precambrian tectonic event.

Sedimentation was continuous from the Ordovician to the Silurian and from the Silurian to the Devonian. Yet Caradocian sedimentological variations (conglomerates giving way to sandstones) or volcanic units (Caradocian volcano-sedimentary unit unconformably overlying the black slates of the Middle Ordovician; Calvet et al. 1988) have been interpreted as more or less discrete manifestations of Caledonian events. During the Lower Devonian period, distinct conditions of sedimentation along an E-W axis suggest the existence of pre-existing (Caledonian) faults, if not active ones.

If only the last deformation of the Pyrenean Paleozoic massifs is Alpine, then most of its structure is Variscan. The ductile reaction of the Paleozoic, throughout Alpine deformation, has been clearly shown for definite sectors (Eaux Chaudes, Gavarnie nappe), and that the Variscan strike faults operated in stages of deformation (Majesté-Menjoulas 1979). In such places, the Paleozoic is directly involved in large Alpine structures. Elsewhere, it is often impossible to differentiate the structures which resulted from the different orogenies.

3.2 Regional Tectonics – Phases of Deformation

The Variscan structures of the Pyrenees have been modeled by a succession of deformations which occurred in a diachronous manner earlier in the east than in the west, between the middle Westphalian and the lower or middle Stephanian. The most recent marine Carboniferous formations (Westphalian C-Kachirian) in the Pyrenees are located in the Basque country and were resedimented in the Culm (Delvolvé et al. 1987) on which the first post-Hercynian deposits rest unconformably.

During the paroxysmal tectonics, large folds trending east-west were formed. WNW–ESE upright faults, which were slightly oblique in comparison to the previous folds, then divided the Variscan structure into WNE–ESE blocks.

The presence of interference figures, sometimes on a cartographic scale, indicates the existence of other tectonic phases before and after the major tectonism. Several phases of ductile or brittle deformation have been recognized. In the central and western Pyrenees where mainly the upper units of the Paleozoic series outcrop, some authors (Zwart 1968, 1979; Soula 1971; Muller and Roger 1977; Majesté-Menjoulas 1979) have counted a maximum of four phases.

In the eastern Pyrenees, where the metamorphic basement outcrops, seven phases have been described (Guitard 1970; Laumonier and Guitard 1978; Guitard et al. 1984; Laumonier et al. 1984), but they are difficult to correlate with the phases in the more westerly sectors. The different phases took place during a very short period and generally represent continuous deformation.

1. In the central and western parts of the High Chain, an early tectonic phase corresponds to drastic changes in thickness with multikilometric vertical movements (Aure Valley; Muller et al. 1977), non-schistose folds, isoclinal folds with décollements on the deeper levels, and large submeridian hectometric recumbent folds.

The major tectonic phase (phase 2) associated with the main foliation is responsible for the large structures. It is characterized by the presence of schistosity with a relatively constant strike of 110°, throughout the chain. The schistosity surrounds the granitoid plutons, suggesting to some authors a syn-phase 2 pluton emplacement. Pluton emplacement was completed during the late phases. The peak Variscan metamorphism is also connected with this phase.

Phase 2 structures were deformed by submeridian folds and two schistosities are more or less apparent in the different sectors. Strike-slip faults separate the Variscan structure into blocks. In the southern part of the High Chain, overlapping of units occurs at the end of the Variscan cycle: the overlap of Erta in the Nogueras (Muller and Roger 1977) and overlaps in the sectors of Freser and Ter (Munoz et al. 1983). These overlaps are unconformably overlain by the Permo-Triassic deposits.

2. In the eastern Pyrenees, during the F1/F3 early phases (Laumonier and Guitard 1978), large structures were formed. The first two were not schistose, the third was synchronous with the regional schistosity S3, axial planar to large recumbent folds. Connected with the third paroxysmal phase are the large tangential structures interpreted either as basement thrusts sheets (Guitard 1970) or as duplexes in metamorphic material (Soula et al. 1986a, b). Five late phases follow: F4 with large 110° folds overturned to the north; F5 with large horizontal recumbent 75° folds verging toward the SSE; F6 a phase of tilting shaped like a large upright 110° fold; F7 with large upright folds oriented north-south, and F8 with small 140° folds, but hardly conspicuous.

A partition of the High Chain into domes and basins results from the superimposition of phases F5 and F6. Phase 7 is related to the emplacement of the Mont Louis granite (Guitard et al. 1984; Laumonier et al. 1984).

3.3 Variscan Tangential Tectonics

In various Pyrenean sectors, a tangential early phase, with originally subhorizontal faults which were subsequently deformed by the Variscan and Alpine phases, has been identified from the existence of mixed Devonian facies. The southward vergence of this early deformation has been established; yet the extent of the transport and the structural scheme for the entire High Chain are still uncertain.

The sectors where this tectonic setting has been established are relatively numerous. In the upper Gallego and in the upper Aure Valley (Muller et al. 1977), increases in thickness of series with fault boundaries subparallel to the stratification indicate kilometric displacements. In the upper Gave de Pau Valley, fault contacts separate mixed Devonian facies. The Bachebirou-Chinipro, Soulan Peyrefitte fault separates the Devonian southern facies from the Devonian intermediate facies (Bodin 1987, cf. Sect. 1). This fault is cross-cut by the Neouvielle granites (Majesté-Menjoulas 1982). Early Hercynian nappes with mixed Devonian material exist in the Ariège area (Central Pyrenees) (Bodin and Ledru 1986). From the dome of Bosost to Marimanya (Losantos et al. 1986), the fault contact between the Moredo-Salat series and the Ordovician of the Isil series is interpreted as a great Variscan thrust verging south (Moret and Weyant 1986). In the upper Pays de Sault, Devonian series are allochthonous and overlie more recent or ancient terranes (Raymond 1980). The fault contact, present at the core of the Devono-Carboniferous complex, is exposed in the Querigut Massif.

Between Cerdagne and Llobregat, late Hercynian nappes and gravity flow deposits of material from the north (Llac 1979) were deposited on the Devono-Carboniferous complex after the folding phases. Munoz et al. (1983) indicate a shearing tectonic phase (overthrust) in the infra-Silurian in the Freser area. The hypothesis of tangential tectonics is thus verified in the whole Variscan High Chain and when the faults are cross-cut by granitoid plutons their connection with Variscan tectonic phases is unquestionable.

4 Metamorphism

Paleozoic formations of the Pyrenees show large structural-metamorphic units which correspond to more or less complete domes centered on migmatite (Trois Seigneurs and Gavarnie massifs), orthogneiss (Canigou and Aston-Hospitalet massifs), the most metamorphosed rocks (Bosost dome), granulitic gneisses (Agly, Saint-Barthélémy, and Castillon massifs), and granitoid rock (Néouvielle and Quérigut massifs). The general type of dome will be described, using the Trois Seigneurs massif as an example, before briefly discussing a few characteristics of the metamorphic rocks around orthogneiss and granitoid rocks and a few controversial viewpoints about the granulitic series and the relationships between metamorphism and deformation.

4.1 Metamorphic Evolution of the Trois Seigneurs Massif

The North-Pyrenean Trois Seigneurs Massif (Ariège) is bounded to the south, west, and east by Mesozoic sediments and to the north by the Hercynian Arize massif (Fig. 4). It is a large structural-metamorphic half-dome centered on migmatite and gneiss and cut by granodioritic and granitic plutons. It is one of the best examples for defining the characteristics of Hercynian metamorphism (Mercier 1988; Mercier et al. 1988) because: (1) the metamorphic and anatectic series is nearly complete, from shale containing chlorite to partial melt and gneiss, a roughly 15-km-thick section of crust can be analyzed; (2) the relations between regional metamorphism and metamorphism associated with granitoids are completely exposed; (3) the Alpine tectonism does not disturb the Variscan structure which may be regarded as complete.

The metamorphic evolution (Mercier 1988) is outlined by the succession of isograds oriented east-west and often subparallel to the stratification but sometimes clearly cross-cutting. In the western part of the massif the biotite and andalusite isograds cut the Ordovician-Silurian contact. On the border of the granitoids, the regional isograds deviate and follow the outlines of the plutons.

A north-south section of the massif enables definition of a sequence of metamorphosed zones:

a) the chlorite zone;
b) the biotite zone;
c) the andalusite zone, associated with cordierite in the lower zones;
d) the sillimanite zone, in which andalusite coexists with sillimanite;
e) the prismatic sillimanite zone, in which andalusite has disappeared;
f) the intermediate gneiss zone along the discontinuous border of migmatite;
g) a zone characterized by the appearance of K feldspar;
h) a zone where phenomena of partial fusion become very important;
i) the Lapège gneiss zone, which might be equivalent to the granulitic gneiss of Saint-Barthélémy.

Temperature and pressure conditions are defined: (1) for the beginning of the partial melting from the

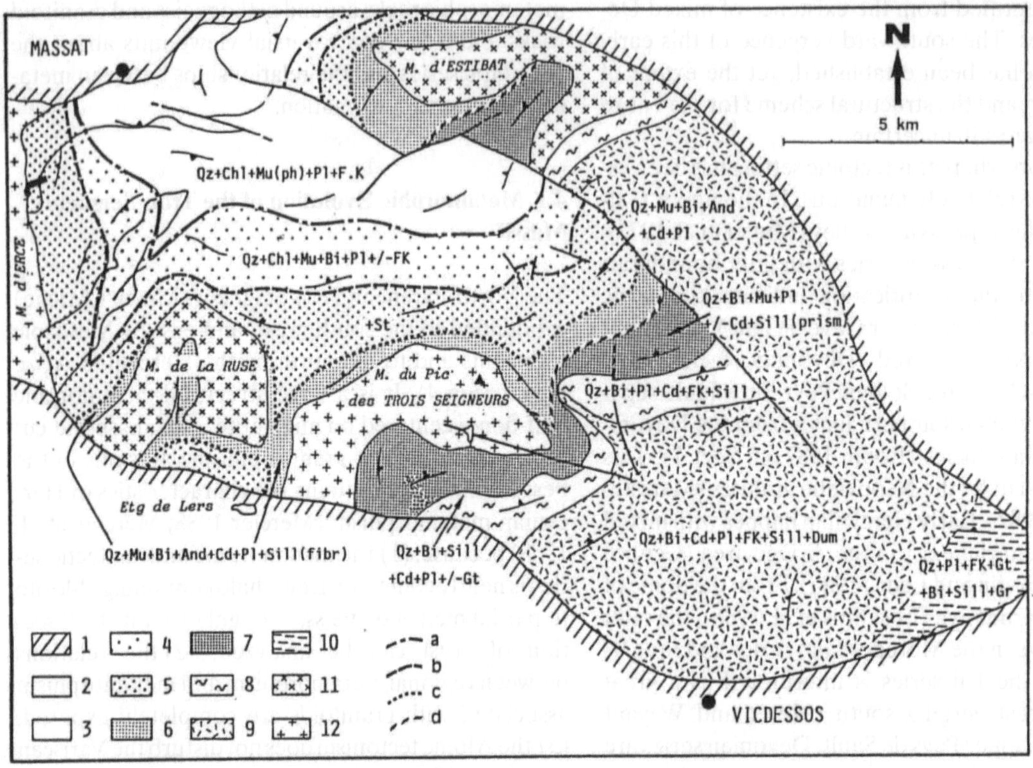

Fig. 4. Trois-Seigneurs Massif (Mercier 1988). *1* Host rocks; *2* Ordovician; *3* Silurian; *4* biotite zone; *5* andalusite zone; *6* fibrous sillimanite + andalusite zone; *7* prismatic sillimanite zone; *8* intermediate gneiss zone; *9* migmatite; *10* Lapege gneiss; *11* biotite and muscovite granite; *12* granodiorite. *a* biotite isograd; *b* andalusite isograd; *c* fibrous sillimanite isograd; *d* prismatic sillimanite isograd

association biotite-garnet and garnet-cordierite (650 to 670 °C) and the assemblage cordierite-garnet-sillimanite-quartz (3 kbar); (2) for the bottom of gneiss series (Lapège gneisses) from the same assemblages (730 °C for 4 to 5.5 kbar).

The evolution of the thermodynamic conditions: (1) determined on a classical P-T diagram from the reactions equilibrium observed and (2) compared with the experimental data and on a T-depth diagram from the cartographic thickness of the different zones, enables a determination of an average linear geothermal gradient of the order of 60 °C/km. In contrast, around the granitoid plutons where the isograds are compressed, the same analysis leads to the recognition of higher than average linear gradients of the order of 100 to 120 °C/km.

The study of other complex metamorphic domes (e.g., Gavarnie, Lesponne-Chiroulet, and Bosost) displays the general case of the two types of gradients corresponding to two different metamorphic evolutions which are the result of a single metamorphic phenomenon:

1. A metamorphism with a medium gradient (50 to 70 °C/km) characteristic of the large regional domes (Trois Seigneurs Massif, Mercier 1988; Gavarnie Massif, Dahmani 1987; Chiroulet Massif, Pouget 1984; but also Bosost, Pouget 1989; Aston; Verhoef et al. 1984; Hospitalet, van den Eeckhout 1986; Canigou).

These anatectic and metamorphic transformations required the addition of a major heat source to a major water supply (sea water, according to Wickham and Taylor 1985). The geological phenomenon responsible for such a thermal anomaly can be associated with the emplacement of basic and ultrabasic rocks which are abundant in many migmatitic massifs (Gavarnie Massif; Debat 1965; Driouch 1987; Arize and Aston massifs; Soula et al. 1986 a, b). The general character of anatexis in the lower part of the series shows a good thermal homogenization and enables the assumption of a transfer of heat by convection and conduction (Vielzeuf 1984; Wickham 1987).

2. A metamorphic event with a gradient of 100 to 120 °C/km was centered around granodioritic plutons at the core of the regional domes or in their immediate vicinity. These gradients result from a supplementary heat supply to the general thermal budget through the plutons, the transfer of heat was mainly

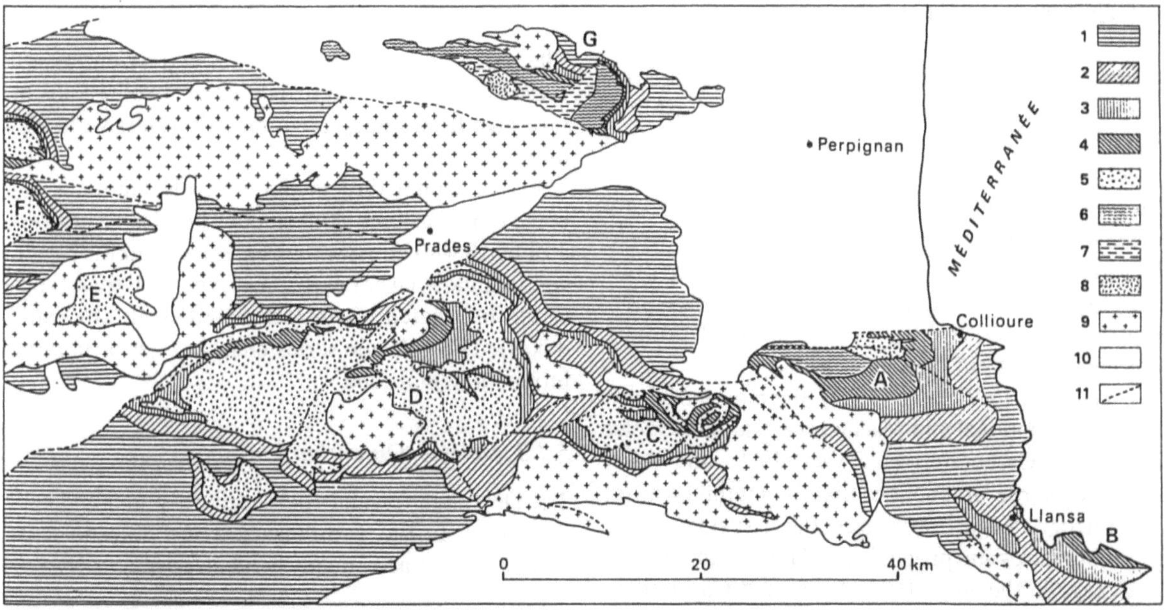

Fig. 5. Map of the Hercynian metamorphic zones in the Eastern Pyrenees (from Autran, Fonteilles, Guitard in Bard et al. 1980). Paleozoic cover: *1* chlorite zone; *2* biotite zone; *3* cordierite and andalusite zone; *4* sillimanite zone (sillimanite + muscovite, sillimanite + K feldspar); in the Canigou massif the sillimanite zone rims the deep granite of Canigou. Zones 1 to 4 are defined in the pelitic sequence. Basement: 2 augen gneisses derived from cadomian granitoids; *6* and *7* paragneisses; *8* charnockite of Ansignan; *9* Hercynian granitoids emplaced into epizonal and mesozonal series; *10* post-Hercynian formations; *11* faults. *A* Albères; *B* Llansa and Cap Creus country; *C* Roc de France; *D* Canigou; *E* Mont-Louis; *F* Aston and Hospitalet; *G* Agly

by advection and favored by the over-abundant fluid around the intrusions.

The complex evolution of the thermal flux occurred during the development of the general dome-shaped structures and is interpreted in a polydiapiric geodynamic context characterized by the quasi-simultaneous emplacement of crystalline units with various dimensions throughout the orogeny (Soula 1982; Soula et al. 1986a, b; Pouget 1987). Other geodynamic models have been proposed to explain the characteristics of Hercynian metamorphism (Guitard 1965; Vielzeuf 1984; Wickham and Oxburgh 1985; Van den Eeckhout and Zwart 1988).

4.2 Metamorphism in the Massifs with an Orthogneiss Core

In the eastern (Fig. 5) part of the chain, there are large metamorphic units with orthogneiss cores (e.g., Roc de France, Canigou, Aston, Hospitalet). The orthogneiss and/or paragneiss correspond, according to Guitard (1965), to a pre-Hercynian basement 600–520 Ma old (date from Canigou; Vitrac-Michard and Allègre 1975). Around the gneissic core, there are a series of subconcentric metamorphic zones, in order of decreasing distance from the core, bounded by biotite, andalusite and sillimanite + potassic feldspar isograds with migmatite in contact with the gneissic core (Aston). The isograds are extremely close and sometimes follow the gneiss-mica schist series contact. This indicates that the basement had some control over or effect on the metamorphism of the overlying series and corresponds to the "basement effect" of Guitard (1965). Soula (1982) and Soula et al. (1986a, b) state that geometrical relations are due to the intrusive character of the gneiss in comparison to mica schist.

4.3 Metamorphism Around the Upper Granodioritic Massifs

Biotite, andalusite-cordierite, and sillimanite isograds, in order of decreasing distance, surround the granitoid rocks emplaced in the Upper Paleozoic formations (e.g., Quérigut, Maladetta, Bassiès, Néouvielle, Cauterets) and indicate a thermal metamorphism.

4.4 Granulitic Metamorphism

Granulites appear exclusively in some North-Pyrenean massifs (e.g., Agly, Saint-Barthélémy, Castillon, Ursuya). They correspond to ancient sedimentary or volcano-sedimentary series with basic intercalations (amphibolite, gabbro and ultrabasic rocks) which are often lenticular. These formations have been affected by a granulitic metamorphism of low to medium pressure (4–7 kbar). The limit of the granulitic metamorphism is marked by the crystallization of primary cordierite (Vielzeuf 1984).

According to Vielzeuf (1984), the metamorphism was progradational and reached the granulitic facies during the Carboniferous. Zwart (1979) and Roux (1977) disagree, and believe that the metamorphism was retrograde with two stages: the first stage in the granulite facies and probably pre-Hercynian; the second stage in the amphibolite facies and Carboniferous. This granulitic metamorphism has been dated in the Agly massif at 300–310 Ma (Postaire 1983).

4.5 Relations Between Metamorphism and Deformation

Relations between metamorphism and deformation are highly controversial. Guitard (1965) believes that metamorphism followed the development of nappe structures. Zwart (1979, 1986) and other Dutch workers believe metamorphism was later, as the degree of transformation is higher. Soula (1982) and Soula et al. (1986a, b) state that metamorphism and major deformation were two quasi-simultaneous phenomena.

5 Igneous Activity

5.1 Volcanics

5.1.1 Cambrian Volcanics

In the eastern part of the High Chain, the Cambrian unconformably overlies the Cadomian basement and contains units of volcanic origin, either acidic (rhyolite) or basic (tholeiitic), associated with detrital sediments (e.g., sills, lava flows, pyroclastic rocks, volcano-sedimentary units).

5.1.2 Ordovician Volcanic Rocks

The base of the Ordovician series in the Central and West Pyrenees has intercalations of rhyolite (sills or rare lava flows) and conglomeratic layers with rhyolitic quartz.

In the Upper Ordovician, the volcanic rocks become more basic with intermediate andesite and keratophyres with effusive rock types (e.g., albitic basalt, spilite, pillow-lavas) and some explosive rock types (e.g., brecciated tuffs associated with andesitic dykes). The volcanic series characterizes a thinning intracontinental zone.

5.1.3 Devono-Dinantian Volcanism

The Devonian volcanic rocks include spilites and keratophyres. In the Tournaisian, a few volcano-sedimentary layers accompany lydite deposits.

5.1.4 Permo-Carboniferous Volcanism

Stephano-Permian volcanism is mainly exhibited by the Pic du Midi d'Ossau formations (Atlantic Pyrenees) which have five successive volcanic phases. Five phases are represented in the eastern area (from Cadi to the Mediterranean Sea) and three in the western area (to the west of Lerida; Bixel 1987; Fig. 3, Ossau).

a) Stephanian volcanism has calc-alkaline characteristics (calc-alkaline basic magma from the mantle – a basaltic or tholeiitic source – and calc-alkaline acidic magma from the crust, anatectic fusion of metamorphics). With the exception of Ossau, there are no intermediate rocks. These magmas took shape in an area of large disconnected uneven terrains where basins were filled with volcanics and detrital deposits (Biscel 1987).
b) Terminal Autunian-Stephanian volcanism appears only in the West Pyrenees. The magma was calc-alkaline, acidic, and resulted from the assimilation of potassic siliceous materials by andesites.
c) The character of the volcanism of the Autunian located in the East Pyrenees, in Catalonia (Cadi, Camprodon), and in the Corbières (Baixas) is alkaline rhyolite (east of Cadi). Volcanism in extrusive ignimbritic domes is comparable to rhyolitic volcanism in the Esterel and in Corsica, and to Corsican alkaline plutonism.
d) Autuno-Saxonian volcanism is indicated by the presence of acidic and basic andesite at the boundary of the lower and upper red series. Aphyric or slightly porphyritic materials form large dykes (50 to 100 m thick).

e) Basaltic and doleritic dykes mark the volcanism of the terminal Permian along the whole length of the chain (Rhune in the Basque Country, Labastide of Serou and Sainte-Colombe in Ariège). The volcanism ceased during the transition from the Permian to the Triassic.

5.2 Granitoids

5.2.1 Cadomian Metagranites – Metagranites of the Lower Cambrian

These granitoids were transformed by Variscan anatexis and metamorphism and became orthogneiss massifs (e.g., Aston, Canigou, Albères). Autran (1980) distinguishes two types of granitoids.

The first orthogneisses are Cadomian, and are composed of ancient calc-alkaline granite with biotite, commonly prophyritic with rapakivi feldspars. They are associated with granodiorites which contain biotite and hornblende, and with leucocratic granite (Aston). Evidence for these rocks being basement includes dating (Vitrac-Michard and Allègre 1975) and the identification of potassic perthitic feldspar in some levels of the Canaveilles group.

The second group of granitoids, granular leucocratic gneiss with biotite, is associated with leptynite and ferrohastingsite (e.g., Casemi gneiss in the Canigou massif; Riberoles gneiss in the Agly massif). These granitoids form laccoliths in the Cambrian and have features analogous to the Montagne Noire Cambrian metagranites which tend to be alkaline.

5.2.2 Variscan Synorogenic Granites

All the large Pyrenean granitoid massifs belong to the family of Variscan synorogenic granites. Their formation is highly controversial, and is in the opinion of some geologists more or less contemporaneous with the major Variscan deformation, whilst, in the opinion of others, subsequent to it.

1. Classification of Pyrenean Granitoids. The nomenclature of Autran et al. (1970) will be adopted here, where granitoids altered by regional anatexis are distinguished from granitoids unaltered by anatexis.

The granitoids unaltered by regional anatexis include the upper and the intermediate massifs. The upper massifs situated in the epizone are large calc-alkaline plutonic complexes (with hornblende granodiorite, monzonitic granite, prophyritic granite with phenocrysts of orthoclase and microcline). The basic enclaves are numerous and dioritic. In general, the basic unit occurs in the periphery and acidic units in the central part of the upper massifs. These include the large complex massifs (Quérigut, Cauterets) and the massifs with two petrographic natures (Eaux Chaudes, Néouvielle, Marimanya, Maladetta).

The intermediate granitoids emplaced in the mesozone are calc-alkaline massifs with automorphic muscovite and are devoid of amphibole. The plagioclases are more acidic and the orthoclase and microline phenocrysts are as rare as the basic enclaves. Garnet and cordierite are found locally. The tops of these massifs are homogeneous but their heterogeneous deeper parts are migmatitic. Their periphery shows migmatites with pegmatitic intrusions. Examples of these granitoids include the deep granite of Canigou, Ax Massif in the Aston gneiss dome, the outcrops in Hospitalet and Roc de France, and the centre of the Mont Louis Massif.

The granitoids linked to regional anatexis are located in the Precambrian basement or at the base of the Paleozoic. Because of the lack of absolute dating, they are difficult to distinguish from the Cadomian metagranites. They are small and their upper limit corresponds to the cordierite isograde. Granitoids located in the Paleozoic series have sometimes large randomly scattered potassic feldspar crystals (cordierite granite in the Albères and the Gavarnie massifs). Granitoids located in the Precambrian basement form the lower massifs and appear like sills within the surrounding rocks (e.g., the Ansignan granitoid and Cassagne granite in the Agly massif). The charnockite of Ansignan (Guitard 1960; Autran et al. 1970; Fonteilles 1976; Andrieux 1982) is located in the hypersthene zone and forms thick lenses.

This magmatism may have been synchronous with granulitic metamorphism.

In the lower part of the Paleozoic cover, above the isograd with anatexis, granite and pegmatite veins have been injected (e.g., western part of the Albères, anatectic border of Agly, Saint-Barthélémy Massif).

2. Age of Pyrenean Granitoids. Age dating of the Pyrenean granitoids based on their geometrical relations with the surrounding formations remains uncertain. Hence, several hypotheses have been inferred. Absolute dating should be more reliable and eliminate ambiguity.

The oldest rock may be the precursor of a gneiss from Aston dated at 450 to 470 Ma using Rb-Sr methods. A series of orthogneisses aged 435 Ma (Siluro-Ordovician boundary) are next in the sequence. The age of the large granitoid plutons in the Pyrenees ranges from 300 to 277 Ma, the Stephano-Autunian boundary. The Montnegre pluton (Catalonia) located

Period	Age	Granitoid	Reference
AUTUNIAN	268 ± 10 Ma	COSTABONNE	Vitrac-Michard & Allègre, 1975
AUTUNIAN	268 Ma	NEOUVIELLE AXIAL ZONE	Alibert & al., 1988
AUTUNIAN	269 ± 4 Ma	MONTNEGRE Rb-Sr	Enrique & Debon, 1987
AUTUNIAN	275 ± 12 Ma	QUERIGUT-MILLAS-MONTLOUIS	Vitrac-Michard & Allègre, 1975
AUTUNIAN	277 ± 7 Ma	MALADETTA Rb-Sr	Vitrac-Michard *et al.* 1980
AUTUNIAN	280 Ma	MALADETTA Rb-Sr	Palau & al., 1988
AUTUNIAN	280 - 274 Ma	MONTNEGRE K-Ar	Enrique & Debon, 1987
STEPHANIAN	280 Ma	LYS-CAILLAOUAS (Relict porphyroblast)	Majoor, 1988
STEPHANIAN	289 Ma	NEOUVIELLE Rb-Sr	Alibert & al., 1988
STEPHANIAN	290 ± 270 Ma	AX-CANIGOU (large massifs)	Vitrac-Michard & Allègre, 1975
STEPHANIAN	292 ± 269 Ma	LYS-CAILLAOUAS Rb-Sr, K-Ar	Majoor, 1988
WESTPHALIAN	290 ± 300 Ma	AX, ASTON (muscovite granite) BASSIES (granodiorite) Rb-Sr (whole rock)	Majoor, 1988
WESTPHALIAN	300 ± 20 Ma	NEOUVIELLE Rb-Sr	Alibert & al., 1988
NAMURIAN	320 ± 15 Ma	AX-CANIGOU (small massifs) Rb-Sr	Vitrac-Michard & Allègre, 1975
NAMURIAN	339 ± 23 Ma	MARIMANYA Rb-Sr	Palau & al., 1988
ORDO. SILUR. DEVON	435 Ma	ASTON orthogneiss	Majoor, 1988
ORDO. SILUR. DEVON	450 ± 470 Ma	Précursor ASTON gneiss Rb-Sr	Majoor, 1988

Time scale (Ma): 260, 270, 280, 290, 300, 310, 320, 345, 435, 450. Stephanian subdivisions: A, B, C; Westphalian subdivisions: A, B, C, D.

Fig. 6. Absolute ages of some Pyrenean granitoids

farther southeast could be more recent (269 ± 4 Ma) (Fig. 6). Thus, the granitoids of the Pyrenees are younger than the major deformation. The younger folded series under the Westphalian D-Stephanian unconformity are the Westphalian A series (Delvolvé 1987).

Emplacement and Origin of Granitoid Plutons in the Pyrenees. The age of the emplacement of the magmas, compared to that of the deformations of the Pyrenees, remains a controversial topic.

According to Soula (1982), Soula et al. (1986 a, b), and Pouget et al. (1988), granodioritic magmas were emplaced diapirically and developed a dome-like structure during the major Hercynian deformation (D 2). Then, during D 3, a retrograde metamorphism accompanied the cooling of the pluton. The leucogranites crystallized later, thus creating satellite domes.

According to Marre (1973), magmas were emplaced after the compressive phases, during the relaxation of pressure, within previously existing structures (Marre 1973). Thus there was a conformity between the internal fabrics of the pluton and certain structures of the surrounding rocks. Most dating (Fig. 6) gives the granitoids an age ranging from 300 to 277 Ma, Westphalian C to Stephanian C, and supports this second hypothesis.

The origin of the magmas has been much debated. Whereas the ultrabasic and basic rocks are interpreted as evidence of magmas of mantle origin and basal crust (Autran et al. 1970; Leterrier 1972; Marre 1973; Vitrac-Michard et al. 1980; Ben Othman et al. 1984; Wickham 1987), the origin of the granitoids is attributed to two different types of granitoids derived from mantle magmas by fractional crystallization and contamination. For Autran et al. (1988) different magmatic units of a pluton coming from different crustal protoliths result from stages of crustal fusion (Maladetta, Quérigut, Mont Louis) or from one single crustal source more or less hybridized, with mantle gabbros (Néouvielle, Saint Laurens de Cerdans). For Zwart (1979), Vitrac-Michard et al. (1980), Ben Othman et al. (1984) and Wickham (1987), granitoids result from crustal fusion. The upper massifs (Quérigut, Maladetta) came from magmas from a depth of about 16 km (Ben Othman et al. 1984; Wickham 1987), whereas the granites with muscovite resulted from the crystallization of shallower hydrated magmas formed at about 12 km depth according to Pouget (1984), Dahmani (1987), Wickham (1987), and Mercier (1988).

Finally, the charnockite magmas (Ansignan granite) are produced by a more or less homogeneous mixture between a noritic member and acid end generated by the partial or total fusion of the deep granulitic crust.

References

Alibert Ch, Debon F, Ternet Y (1988) Le pluton à structure concentrique du Néouvielle (H.P.): typologie chimique, âge, genèse. CR Acad Sci Paris 306 II: 49–54

Andre JP (1985) Contribution à l'étude des terrains paléozoïques de la vallée du Lys Zone Axiale des Pyrénées. Bull Soc Géol Fr (8) 1, 4: 515–521

Andrieux P (1982) La charnockite d'Ansignan. Mise en place et évolution paragénétique. Introduction à l'étude des équilibres grenat-orthopyroxène. Thèse 3e cycle, Clermont-Ferrand

Autran A (1980) L'évolution du Protérozoïque aux distensions post hercyniennes. Colloque C 7 "Géologie de la France", 26e CGI, Extrait Bur Rech Geol Min 107, pp 8–17

Autran A, Fonteilles M, Guitard G (1966) Discordance du Paléozoïque inférieur métamorphique sur un socle gneissique antéhercynien dans le massif des Albères. CR Acad Sci Paris 263 D: 317–320

Autran A, Fonteilles M, Guitard G (1970) Relations entre les intrusions de granitoïdes, l'anatexie et le métamorphisme régional considérés principalement du point de vue du rôle de l'eau; cas de la chaîne hercynienne des Pyrénées orientales. Bull Soc Géol Fr (7) XII 4: 673–731

Autran A, Cocherie A, Rossi Ph (1988) Zonalité du plutonisme varisque sur une transversale du segment méridional NS de l'orogène à travers les blocs actuellement disjoints Pyrénées Sardaigne Corse. Symposium on the geology of the Pyrenees and Betics, University of Barcelona, 11–15 April, 1988, p 53

Bard JP, Briand B, Cantagrel JM, Guitard G, Kienast JR, Kornprobst J, Lasnier B, Le Corre C, Santallier D (1980) Le métamorphisme en France. In: Autran A, Dercourt J (eds) Evolutions géologiques de la France, éd. par 26ème cong. géol inter., colloque C 7. Mém Bur Rég Géol Min Fr 107

Ben Othman D, Fourcade S, Allegre CJ (1984) Recycling processes in granite-granodiorite complex genesis: the Querigut case studied by Rd-Sr isotope systematics. Earth Planet Sci Lett 69: 290–300

Bessiere G, Mirouse R, Perret MF (1980) Découverte de faunes de la limite Viséen-Namurien sous le Culm carbonifère du Mouthoumet (Aude). CR Acad Sci Paris 291 II: 521–524

Bixel F (1987) Le volcanisme stephano-permien des Pyrénées. Pétrographie – Minéralogie – Géochemie. Cuad Geol Iberica II: 41–55

Bodin J (1987) Le Dévonien inférieur et moyen des Pyrénées ariégeoises et centrales. Biostratigraphie, séries hétéropiques et mise en évidence de nappes hercyniennes précoces. Thèse Université, Toulouse-III

Bodin J, Ledru P (1986) Nappes hercyniennes précoces à matériel dévonien hétéropique dans les Pyrénées arégeoises. CR Acad Sci Paris 302, 15: 969–974

Boersma KT (1973) Devonian and Lower Carboniferous conodont biostratigraphic Spanish Central Pyrenees. Leidse Geol Meded 49: 303–377

Bois JP, Pouit G, Gros Y, Guerange B, Picot P (1976) Les minéralisations de Zn (pb) dans l'anticlinorium de Pierrefitte: un exemple de gisements hydrothermaux et sédimentaires associés au volcanisme dans le Paléozoïque des Pyrénées Centrales. Bull Bur Rech Geol Min 2, 6: 543–567

Boissonnas J, Le Pochat G, Thibault Cl, Bernatsky M (1974) Carte géologique au 1/50 000 Iholdy. Bur Rech Geol Min, Orléans

Bouquet C (1986) Coupe stratigraphique des Pyrénées Centrales. In: Synthèse des Pyrénées. Bur Rech Geol Min, IGME, Orléans

Calvet P, Lapierre H, Charvet J (1988) Diversité du volcanisme ordovicien dans la région de Pierrefitte (Hautes Pyrénées): rhyolites calco-alcalines et basaltes alcalins. CR Acad Sci Paris 307 II: 805–812

Cavet P (1957) Le Paléozoïque de la Zone Axiale des Pyrénées orientales françaises entre le Roussilon et l'Andorre (étude stratigraphique et paléontologique). Bull Serv Carte Géol Fr 254

Cocchio AM (1982) Données nouvelles sur les Acritarches du Trémadoc et de l'Arénig dans le massif de Mouthoumet (Corbières, France). Rev Micropal 25, 1: 26–59

Dahmani A (1987) Les formations métamorphiques et anatectiques de Gavarnie, Héas, Barroude et Plan de Larri (Pyrénées), Thèse 3e cycle, University of Toulouse-III

Debat P (1965) Les formations métamorphiques des vallées de Gavarnie et d'Héas (Hautes Pyrénées). Bull Soc Hist Nat Toulouse 100: 137–173

Degardin JM (1979) Découverte du genre *Phyllograptus* (Graptolites) dans l'Ordovicien des Pyrénées Atlantiques: conséquences stratigraphiques. Geobios 12, 2: 321–329

Degardin JM (1987) Le Silurien des Pyrénées: biostratigraphie et paléogéographie. Thesi sci University of Lille

Delvolve JJ (1981) Arguments en faveur de l'âge namurien du Culm des Pyrénées centrales françaises. CR Acad Sci Paris 293 II: 219–222

Delvolve JJ (1987) Un bassin synorogénique varisque: le Culm des Pyrénées centro-occidentales. Thèse Sci, Toulouse-III

Delvolve JJ, Perret MF, Vachard D (1987) Découverte du Kachirien (Moscovien inf.) à fusulines et algues dans le massif de Cinco Villas (Pyrénées basques, Espagne). Geobios 20, 4: 541–548

Driouch Y (1987) Les formations plutoniques basiques métamorphisées du complexe de Gavarnie, Héas, Barroude et Plan de Larri. Thèse 3e cycle, University of Toulouse-III

Engel W (1984) Migration of folding and flysch sedimentation on the southern flank of the Variscan belt (Montagne Noire, Mouthoumet massif, Pyrenees). Z dtsch Geol Ges Hannover 135: 279–292

Enrique P, Debon F (1987) Le pluton permien de Montnègre chaînes côtières catalanes (Espagne): étude isotopique Rb-Sr et comparaison avec les granites hercyniens des Pyrénées Sardaigne Corse. CR Acad Sci Paris 305 II: 1157–1162

Fonteilles M (1976) Essai d'interprétation des compositions chimiques des roches d'origine métamorphique et magmatique du massif hercynien de l'Agly. Thèse Sci, University of Paris VI

Guitard G (1960) Sur la présence et l'âge d'un granite à hypersthène d'affinité charnockitique dans le massif de l'Agly (P. O.). CR Acad Sci Paris 251: 2554–2555

Guitard G (1965) Phases de plissement dans les terrains métamorphiques de la Zone Axiale Pyrénéenne du Canigou durant l'orogenèse hercynienne. CR Acad Sci Paris 265: 1357–1360

Guitard G (1970) Le métamorphisme hercynien mésozonal et les gneiss oeillés du massif du Canigou (Pyrénées orientales). Mém Bur Rech Geol Min 63

Guitard G, Geyssant J, Laumonier B (1984) Les plissements hercyniens tardifs dans le Paléozoïques inférieur du versant nord du Canigou: analyse géométrique et chronologie des phases superposées. Relations avec le granite du Mont Louis et le métamorphisme régional. Géologie de la France, Bul Rech Geol Min 4: 95–125

Heddebaut Cl (1973) Etudes géologiques dans les massifs Paléozoïques basques. Thèse Sci, University of Lille

Laumonier B (1988) Sur l'âge cambrien des groupes de Canaveilles et de Jujols (Pyrénées catalanes, Espagne). Symposium on the Geology of the Pyrenees and Betics, University of Barcelona, 11–15 april 1988, p 51

Laumonier B, Guitard C (1978) Contribution à l'étude de la tectonique superposée hercynienne des Pyrénées orientales: le problème des plissements précoces dans le Paléozoïque inférieur épizonal du synclinal de Villefranche de Conflent. Rev Géogr Phys Géol Dyn Fr 2, 20, 2: 177–212

Laumonier B, Guitard G (1986) Le Paléozoïque inférieur de la moitié orientale de la Zone Axiale des Pyrénées. Essai de synthèse. CR Acad Sci Paris 302 II: 473–478

Laumonier B, Geyssant J, Guitard G (1984) Les plissements hercyniens tardifs dans le Paléozoïque inférieur du versant nord du Canigou: essai de restitution rétrotectonique et cartographique des mégastructures. Géologie de la France, Bur Rech Geol Min 4: 127–145

Leterier J (1972) Etude pétrographique et géochimique du massif de Quérigut. Thèse Sci, University of Nancy

Llac F (1979) Les nappes tardi-hercyniennes entre Cerdagne et Llobregat (versant sud des Pyrénées catalanes). Bull Soc Géol Fr (7) XXI 4: 467–473

Losantos M, Palau J, Sanz J (1986) Considérations about Hercynian thrusting in the Marimanya massif (Central Pyrenees). Tectonophysics 129: 71–79

Lucas Cl (1985) Le grès rouge du versant nord des Pyrénées. Essai sur la géodynamique des dépôts continentaux du Permien au Trias. Thèse Sci, University of Toulouse III

Majeste-Menjoulas Cl (1979) Evolution alpine d'un segment de chaînes varisque: Nappe de Gavarnie – Chevauchement Cinq-Monts Gentiane. Thèse Sci, University of Toulouse III

Majeste-Menjoulas Cl (1982) L'unité Paléozoïque de Bachebirou Chinipro, témoin d'une tectonique tangentielle varisque dans les Pyrénées Centrales. CR Acad Sci Paris 294 II: 145–150

Majoor FJM (1988) A geochronological study of the Axial Zone of the Central Pyrenees, with emphasis on Variscan events and Alpine resetting. Verhandeling NR 6-ZWO Laboratorium voor isotopen-geologie, Amsterdam, 117 pp

Marre J (1973) Le complexe éruptif de Quérigut. Pétrologie, structurologie, cinématique de mise en place. Thèse Sci, University of Toulouse III

Mercier A (1988) Illustration du métamorphisme hercynien dans les Pyrénées: le massif nord pyrénéen des Trois Seigneurs. Modalités, implications géodynamiques. Thèse, University of Toulouse III

Mercier A, Dahmani A, Debat P, Driouch Y, Pouget P, Roux L, Soula JC (1988) Définition et signification des différents types de gradients dans le métamorphisme hercynien des Pyrénées centrales et occidentales. CR Acad Sci Paris 306 II: 1179–1185

Mirouse R (1966) Recherches géologiques dans la partie occidentale de la zone primaire axiale des Pyrénées. Mém Carte Géol Fr, 441 pp

Mirouse R, Barrouquere G, Bessiere G, Delvolve JJ, Perret MF (1983) Amorce de la sédimentation synorogénique dans les Pyrénées varisques. Données chronologiques; implications paléogéographiques. Geol Rundsch 72, 1: 253–287

Moret JF, Weyant M (1986) Datation de l'Emsien-Dévonien moyen des calcaires de Campaüs et des schistes d'Escala alta équivalents occidentaux de la série de Salau (Zone Axiale pyrénéenne, Haute Noguera Pallaresa, province de Lerida, Espagne); conséquences structurales. CR Acad Sci Paris 302 II 6: 353–356

Muller G, Roger Ph (1977) L'évolution structurales des Pyrénées (domaine central et occidental). Le segment hercynien, la chaîne de fond alpine. Gèol Alp 53, 2: 149–151

Munoz JA, Sabat F, Santanach P (1983) Cisaillements alpins et hercyniens dans le versant méridional de la zone axiale des Pyrénées entre de Freser et le Ter (Pyrénées catalanes). CR Acad Sci Paris 296 II: 1453–1458

Palau J, Bickle M, Chapman HJ, Vaquer R (1988) Geological and geochemical constraints on the Marimanya pluton. Symposium on the geology of the Pyrenees and Betics, University of Barcelona, 11-15 april 1988, p 55

Perret MF (1971) Les calcaires carbonifères d'Ardengost (Hautes Pyrénées). Thèse 3e cycle, University of Toulouse III

Perret MF (1977) Données récentes de la micropaléontologie dans l'étude du Carbonifère marin des Pyrénées. Ann Soc Géol Nord XCVII: 77–85

Perret MF (1979) Le passage du Mississipien au Pensylvanien dans les bassins marins pyrénées, coupures biostratigraphiques, corrélations. Ann Soc Géol Nord XCIX: 443–448

Perret MF (1988) Le passage du Dévonien au Carbonifère dans les Pyrénées. Zonation par Conodontes. Com Forschungsinst Senckenb Frankfurt 100: 39–52

Pouget P (1984) Géologie de la région de Lesponne (Hautes Pyrénées). Mise en place syncinématique de dômes cristallins hercyniens. Evolution structurale au cours de l'orogenèse alpine. Thèse 3e cycle, University of Toulouse III

Pouget P (1987) Le massif migmatitique du Chiroulet (Hautes Pyrénées, France): un diapir hercynien syncinématique. Conséquences tectono-métamorphiques. CR Acad Sci Paris 305 II: 1545–1549

Pouget P (1989) Evolution géodynamique hercynienne des Pyrénées centrales. Thèse, University of Toulouse III

Pouget P, Lamouroux Ch, Debat P (1988) Typologie et mode de mise en place des granitoïdes dans les Pyrénées hercynienne. 12e Réun Ann Sci Terre, Lille, p 11

Raymond D (1980) Dévouverte d'une unité allochtone varisque dans le haut pays de Sault (Zone Axiale pyrénéenne, confins de l'Aude et de l'Ariège). CR Somm Soc Géol Fr 6: 250–252

Raymond D, Weyant M (1982) Individualisation de deux séries hétéropiques au sein du Dévonien et du Carbonifère inférieur de la Zone Axiale pyrénéenne entre les vallées de l'Aude et de l'Ariège (France). Bull Soc Géol Fr (7) XXIV 2: 265–274

Roux L (1977) L'évolution des roches de faciès granulite et le problème des ultramafitites dans le massif de Castillona (Ariège). Thèse Sci, University of Toulouse III

Soula JC (1971) Evolution des structures hercyniennes de l'Arize métamorphique (Pyrénées ariégeoises). Rev Géogr Phys Géol Dyn 2 XIII: 233–244

Soula JC (1982) Characteristics and mode of emplacement of gneiss domes and plutonic domes in central eastern Pyrenees. J Struct Geol 4: 313–342

Soula JC, Debat P, Deramond J, Pouget P (1986a) A dynamic model of the structural evolution of the Hercynian Pyrenees. Tectonophysics 129: 29–51

Soula JC, Debat P, Deramond J, Guchereau JY, Lamouroux Ch, Pouget P, Roux L (1986b) Evolution structurale des ensembles métamorphiques, des gneiss et des granitoïdes dans les Pyrénées centrales. Bull Soc Géol Fr (8) II 1: 79–93

Van den Eeckhout B (1986) A case study of a mantled gneiss antiform, the Hospitalet massif, Pyrenees (Andorra, France). Geologica Ultraiectina, 45, 196 pp

Van den Eeckhout B, Zwart HJ (1988) Hercynian crustal-scale extensional shear zone in the Pyrenees. Geology 16: 135–138

Verhoef PNW, Vissers RLM, Zwart HJ (1984) A new interpretation of the structural and metamorphic history of the western Aston massif (Central Pyrenees, France). Geol Mijnbouw 63: 399–410

Vielzeuf D (1984) Relations de phases dans le faciès granulite et implications géodynamiques. L'exemple des granulites des Pyrénées. Thèse Sci, Clermont-Ferrand

Vitrac-Michard A, Allegre Cl (1985) 238U-236Pb, 235U-207Pb systematics on Pyrenean basement. Contrib Mineral Petrol 51, 3: 205–212

Vitrac-Michard A, Albarede F, Dupuis C, Taylor HP (1980) The genesis of Variscan (Hercynian) plutonic rocks: inferences from Sr Pb and O studies on the Maladetta igneous complex (Central Pyrenees, Spain). Contrib Mineral Petrol 72: 57–72

Wickham SM (1987) Crustal anatexis and granite petrogenesis during low-pressure regional metamorphism: The Trois Seigneurs massif, Pyrenees, France. J Petrol 28 I: 127–169

Wickham SM, Oxburgh ER (1985) Continental rifts as a setting for regional metamorphism. Nature 318: 330–333

Wickham SM, Taylor HP jr (1985) Stable isotopic evidence for large-scale seawater infiltration in a regional metamorphic terrane: Trois Seigneurs Massif, Pyrenees, France. Contrib Mineral Petrol 91: 122–137

Zwart HJ (1968) The Paleozoic crystalline rocks on the Pyrenees in their structural setting. Kristalini Kom 6: 125–140

Zwart HJ (1979) The geology of the Central Pyrenees. Leidse Geol Meded 50: 1–74

Zwart HJ (1986) The Variscan geology of the Pyrenees. Tectonophysics 129: 9–27

Variscan Basement of the Western Alps: the External Crystalline Massifs

R.-P. MÉNOT, J. F. VON RAUMER, S. BOGDANOFF, and G. VIVIER

1 Introduction

In the western part of the Alps, the basement relicts occur in the Helvetic (Alpine external zone or Dauphiné zone) and Penninic (Briançon and Piemont zones) domains, which correspond respectively to the External and Internal Crystalline Massifs (ECMs and ICMs).

In the Penninic realm of the western Alps, the pre-Permian evolution is difficult to recognize because of the severe Alpine metamorphism and tectonics. Thus the paleogeographic and paleotectonic situation of the basement relicts (ICMs) is still largely ambiguous (von Raumer 1988). Consequently, this chapter deals only with the ECMs: e.g., from north to south, the massifs of Aar-Gotthard, Aiguilles Rouges, Mont Blanc, Belledonne, Grandes Rousses, Haut Dauphiné, Argentera (Fig. 1).

The ECMs appear as dome-like structures of crystalline basement, each considered to be subautochtonous and to override its unconformable sedimentary cover towards west and northwest (Bellière 1988). The cover sequences, ranging from Late Carboniferous to Permian and Triassic in age, consist of continental deposits (Debelmas and Kerkhove 1980). They are associated with volcanics and pyroclastics of silicic and intermediate composition and having calcalkaline affinities (Banzet et al. 1985; Schenker 1986; Mercolli and Oberhänsli 1988; Oberhänsli et al. 1988). Such sequences express late-Variscan rifting, pull-apart and strike-slip tectonics related to the evolution of intramontane basins (Ménard and Molnar 1988).

The Alpine overprint leads to anchizonal to epizonal recrystallizations (Monié and Maluski 1983; Apprahamian 1988) disturbing the isotopic Rb/Sr signature of the Variscan parageneses (Demeulemeester et al. 1986). Moreover, a rejuvenation of the Variscan faults is indicated by the occurrence of Late Paleozoic to Liassic tectonic slices.

2 Paleontology and Lithology

Paleontological data are scarce. Visean palynological relics (Bellière and Streel 1980) have been observed in metapelites of the southwestern edge of the Aiguilles Rouges. These can be compared with the probable Late Visean microfaunas of the Taillefer series (southernmost part of the Belledonne) (Gibergy 1968), whilst Cambrian to Ordovician microfossils (Giorgi et al. 1979) have been discovered within the black and green schists of the Huez formation (Grandes Rousses). Such an Early Paleozoic age for epizonal sediments, which are lithologically similar to the Visean schists of the Aiguilles Rouges and Taillefer, highlights the weakness of correlations based only upon facies comparison.

The majority of the Paleozoic sediments have suffered greenschist and amphibolite facies metamorphism. As a consequence, a comparative lithostratigraphy of the Late Proterozoic and Early Paleozoic formations is difficult to define.

Despite the few radiometric data, three main groups may be tentatively distinguished within the pre Late Carboniferous basement. These are respectively related to the Late Proterozoic-Early Paleozoic, to the Devonian and to the Early Carboniferous periods (Fig. 1).

1. The Older Group (Late Proterozoic-Early Paleozoic) includes three different types of formations:

a) gneissic and amphibolitic formations, which have undergone a long-lived orogenic evolution (Silurian to Visean) under deep crustal conditions. Such conditions are evidenced by the wide extent of migmatization and the occurrence of relics of granulitic and/or eclogitic parageneses. These formations could contain some relics of an older basement [Aar-Gotthard, Haut Dauphiné, Aig. Rouges (?)] which is as yet poorly defined.

In the Aig. Rouges, Belledonne (domain NE), and Argentera massifs, the gneissic and amphibolitic formations could represent a taphrogenic evolution from Cambrian to Ordovician that leads to a general

J. D. Keppie (Ed.): Pre-Mesozoic Geology in France and Related Areas

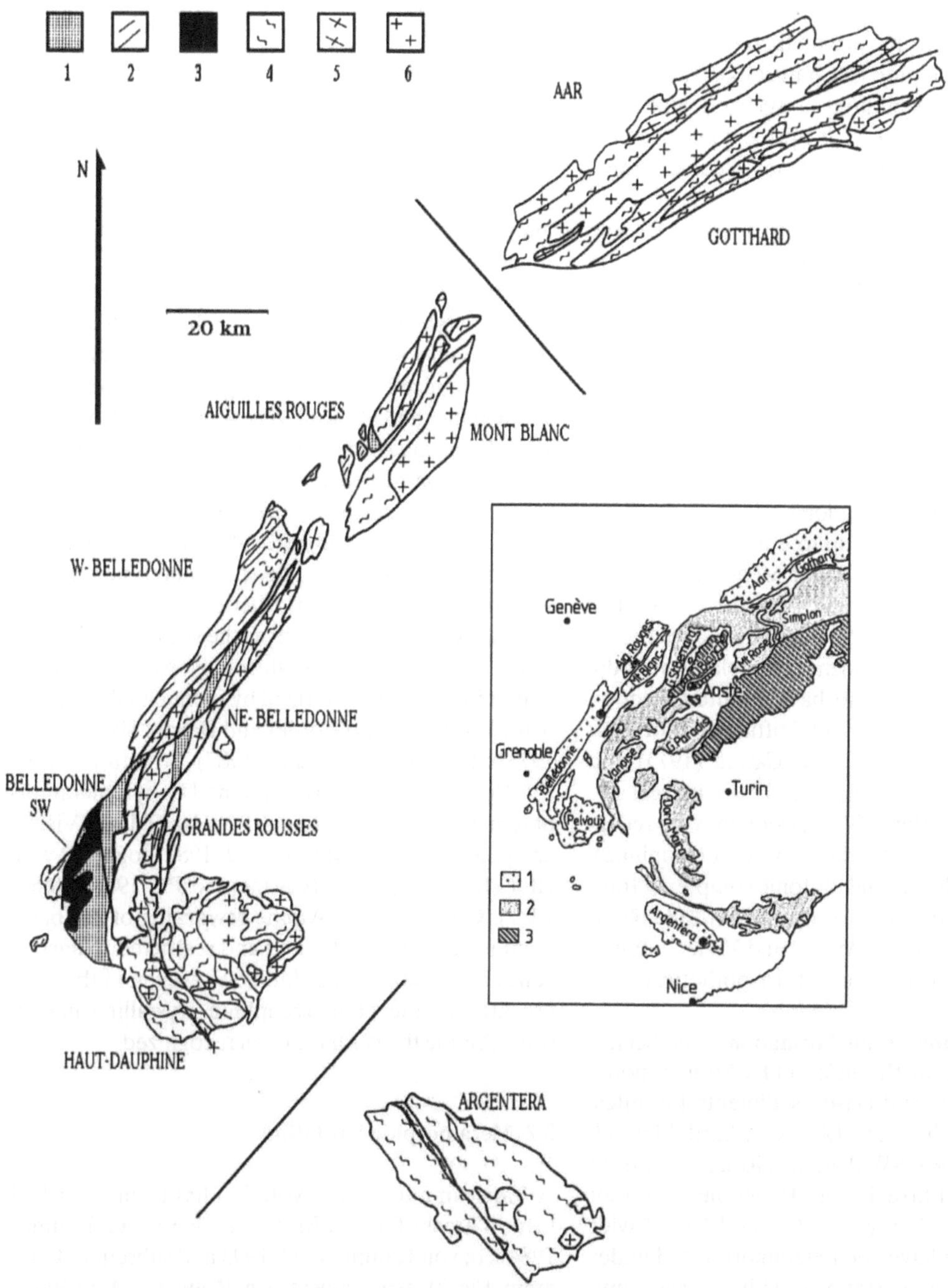

Fig. 1. Lithological sketch map of the ECMs. *1* Devonian to Early Carboniferous formations; *2* metapelitic formations, Late Proterozoic (?) to Early Paleozoic; *3* 496-Ma-old ophiolite of Chamrouse; *4* gneissic and amphibolitic basement, Late Proterozoic to Early Paleozoic. *5* orthogneisses, Early Paleozoic, and Late Devonian; *6* granite, Carboniferous. *Inset:* Location of the ECMs and of the pre-Alpine basement within the Alpine chain: *1* External crystalline massifs (ECMs); *2* Penninic domain; *3* Austro-Alpine and South-Alpine domains

Ordovician rifting (Paquette et al. 1989). Such a conclusion is supported by an important mafic and/or bimodal volcanism (Liégeois and Duchesne 1981; von Raumer 1987; Vivier et al. 1987; Paquette et al. 1989; Ménot and Paguette 1993). In the same way, the Aar-Gotthard ultramafics are considered as ophiolitic remnants (Oberhänsli et al. 1988; Pfeifer et al. 1993). However, some amphibolites of the Gotthard, perhaps older, are representative of volcanic arc magmas (Gebauer et al. 1988).

b) Early Paleozoic formations, of mainly mafic igneous nature, which have suffered less severe (no

HP episode) and younger (Late Devonian to Dinantian) metamorphic events than the formations of (1 a). The ophiolite of Chamrousse is the most spectacular representative of such terrains. This complex being overthrust, in upside down position, upon Devonian-Dinantian series (see 2, below) (Carme 1973b; Ménot et al. 1987).

c) Metapelitic formations, arising from a flysch series [St Gervais, Série satinée, Sarenne (?), Valetta (?)]. Their age of deposition is badly constrained but is probably of Precambrian or Early Paleozoic (Carme 1970a; Siméon 1979; Bogdanoff 1980). The orogenic evolution is related to supra-crustal conditions.

2. The Devonian Group has only been described and dated in the south-western area of the Belledonne massif (Rioupéroux-Livet formations). Early Carboniferous recrystallization grade up to epizonal and mesozonal conditions (Carme 1973a; Ménot 1986, 1987; Ménot et al. 1987; Ménot and Paquette 1993). The trondhjemitic intrusives 352 ± 56 and 365 ± 17 Ma; Ménot et al. 1984, 1988b) and the time-related bimodal volcanism are considered to have resulted from an aborted mechanism of continental rifting (Ménot 1987; Menot and Paquette 1993), but Carme (1975) and Carme and Pin (1987) argue for an active margin setting. A possible southwards extension may be recognized within the Chaillol series (SW Haut Dauphiné). The Lac Blanc volcanic and plutonic complex (Grandes Rousses) has also been compared with the Rioupéroux-Livet formations. (Ménot and Vivier, unpubl. observ.) but such a correlation is still hypothetical.

3. The Early Carboniferous Formations (see Sect. 2, above) outcrop within the different ECMs as bimodal volcanic sequences and clastic sediments (arenites, pelites, and conglomerates) (Aar-Gotthard: Mercolli and Oberhänsli 1988; SW Aiguilles Rouges: Laurent 1968; Bellière and Streel 1980; Belledonne: Bordet and Bordet 1963; Gibergy 1968; Carme 1970a; Vivier et al. 1987). These have been metamorphosed under epizonal conditions, sometimes with a strong mylonitic fabric (NE Belledonne). Such formations could represent syn-(to late-) orogenic basins of pull-apart (?) type.

3 Tectonics and Metamorphism

The basement displays deformation patterns and a metamorphic evolution which can be explained only by a sequence of Paleozoic events (von Raumer 1988).

3.1 Structures

The common present day structures of the ECMs are characterized by vertical to steeply dipping foliation planes. They are related to large-scale upright folds, with horizontal axes (von Raumer 1984a, b; Bogdanoff 1986) and/or with noncylindrical shear folds associated with regional strike-slip zones (NE Belledonne: Carme 1970b; Vivier et al. 1987).

The late Variscan tectonic events occurred under epizonal "shear-zone" metamorphic conditions and overprinted former nappes and thrusts, which have been recognized in the Aig. Rouges (von Raumer 1984; von Raumer and Ménot 1989), Argentera (Bogdanoff 1980, 1986) and Aar (Oberhänsli et al. 1988) massifs.

The central and eastern parts of the Haut Dauphiné seem to be structurally more complicated, with both gently dipping and subvertical foliations within the prebatholith basement. The space and time relationships of these two kinds of structures are unknown.

Finally, some areas, such as the southwestern edge of Belledonne and the westernmost part of the Haut Dauphiné, are characterized by monoclinal or broadly folded structures (Bordet and Bordet 1963; Carme 1973b; Le Fort 1973; Ménot 1987). Structures of the SW Belledonne area are explained by syn-metamorphic nappe piling (Carme 1973a, b) of Late Visean age (Carme 1973b; Ménot et al. 1987) together with later shear megafolds (Carme and Pin 1987; Ménot 1987). Generally, the Alpine reworking of the basement is nevertheless limited and restricted to narrow zones reactivating older faults. However, in the eastern area of the Haut Dauphiné crystalline massif, some Alpine thrusts have been recognized.

3.2 Metamorphic Evolution

A Variscan multistage evolution has been described in most areas of the ECMs (see review in von Raumer 1984a, b; von Raumer et al. 1993 and subsequent diagram: Fig. 2), especially in Aig. Rouges – Mont Blanc (Bellière 1958, 1980; von Raumer 1976, 1983, 1987) Belledonne (Carme 1973a; Carme and Pin 1987; Ménot 1987; Vivier et al. 1987) and Argentera massifs (Bogdanoff 1980, 1986):

1. Early HP and HT event of Late Silurian to Early Devonian age (U/Pb: 425–395 Ma: Paquette 1987; Paquette et al. 1989), supported by eclogitic and granulitic relics: Aig. Rouges (780 °C/11 kbar: Liégeois and Duchesne 1981), Argentera (710–760 °C/12–14 kbar and 710–760 °C/9–11 kbar: Latouche and Bogdanoff 1987). Recent petrological observations lead to the conclusion that not only the eclogites but

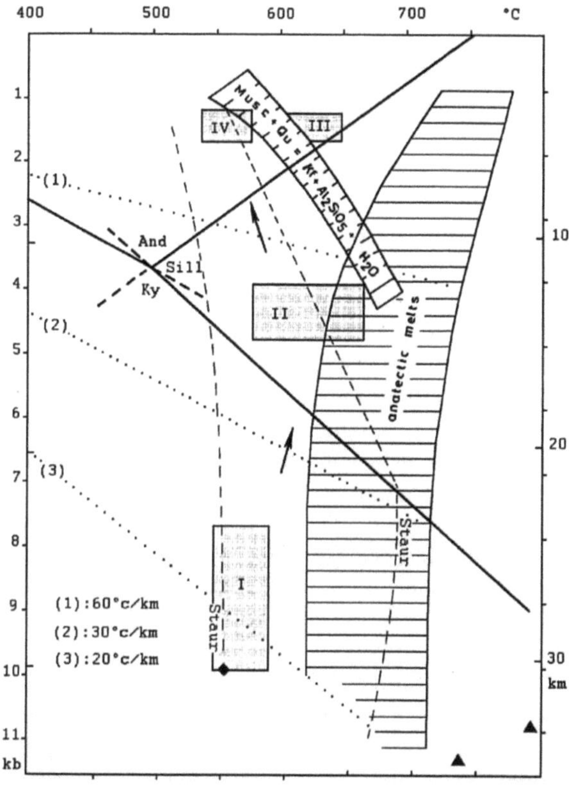

Fig. 2. Variscan metamorphic evolution of the crystalline basement of ECMs. (after von Raumer 1984a). Al_2SiO_5 triple point (Holdaway 1971); anatectic areas (Thompson 1982); staurolite limits (Richardson 1968; Pigage and Greenwood 1982). *Triangles.* PT conditions for eclogites and granulites (Aiguilles Rouges: Liégeois and Duchesne 1981; Argentera: Latouche and Bogdanoff 1987); *square* pressure determination (garnet-plagioclase) from metapelites (Aiguilles Rouges: von Raumer 1983). *I* Late Silurian–Early Devonian HP stage; *II* to *IV.* Devonian to Carboniferous Barrovian evolution and subsequent formation of cordierite-bearing granitoids (biotite + sillimanite; cordierite + sillimanite + K-feldspar; K-feldspar + andalusite; muscovite + andalusite)

also major parts of the basement have reached lower crustal conditions at this stage (von Raumer 1988).

2. Subsequent evolution is expressed by a P-T path where early kyanite (500–550 °C/10 kbar: von Raumer 1983) is replaced by sillimanite-bearing assemblages. Mineral assemblages with cordierite-potassic feldspar anatectic first melts and even incongruent cordierite-bearing melts appear at the thermal peak (650–700 °C/4–6 kbar: von Raumer 1984b, 1987: Bogdanoff 1986). A late growth of andalusite is locally recorded (Aig. Rouges: von Raumer 1984b). Such a P-T-t path underlines a general changes towards higher geothermal gradients. Amphibolite facies conditions are Devonian in age (Belledonne NE: 373 Ma, Demeulemeester 1982; Argentera: 375–350 Ma, Monié and Maluski 1983).

The metamorphic recrystallizations were accompanied by structural transformations which, in general sequence, are the expression of an early nappe tectonics followed by strike-slip shearing. The latter is partly contemporaneous with intrusions of Middle Carboniferous granites (Carme 1970b; Vivier et al. 1987; Joye 1989).

3. Epizonal retrogression related to strike-slip shearing during Early to Mid-Carboniferous times (Carme 1970b; Demeulemeester 1982; Bogdanoff 1986; Vivier et al. 1987).

The interpretation of the sequence of events leads to a paleotectonic model involving subduction-obduction, collision, and hypercollision processes (nappe stacking, crustal thickening, uplift and strike-slip shearing), from Late Silurian to Mid-Carboniferous times. Such an evolution can be compared with that of the axial zone of the Variscan belt (innermost crystalline nappes: Matte 1986) (von Raumer 1984a; von Raumer and Ménot 1989; Bogdanoff 1980, 1986; Vivier et al. 1987; von Raumer and Neubauer 1993, Schulz and von Raumer 1993).

However, some parts of the ECMs display a rather different tectonic and metamorphic evolution (the southwestern edge and the western part of Belledonne) or a more complicated history (polycyclic formations: Aar-Gotthard, Haut Dauphiné, Argentera).

The southwestern area of Belledonne, around the lower Romanche valley, shows the effects of a major tectonic and metamorphic event of Dinantian age (post 365 Ma and pre- 324 Ma: Ménot et al. 1987) which resulted in syn- to late-metamorphic nappe piling (epi-to mesozonal conditions) (Carme 1973a, b; Ménot 1986; Carme and Pin 1987; Ménot et al. 1987). This Early Carboniferous event implies an increasing PT evolution for most of the tectonic units (496-Ma-old ophiolite of Chamrousse: Pin and Carme 1987; Ménot et al. 1988a; Devonian plutonic and volcanic complex of Rioupéroux-Livet: Ménot et al. 1984, 1988b; Visean volcanic and sedimentary series of Taillefer). The nappe pile also includes an amphibolitic and gneissic unit (Allemont formation), the evolution of which is comparable to the polyphase evolution described in the major parts of the ECMs (see above). The SW domain of Belledonne was involved in the orogenic belt only during the Late Devonian–Early Carboniferous times (Ménot 1987). It probably represents an outer zone of the Variscan belt (Ménot 1987, 1988).

On the western border of Belledonne, a metapelitic formation, the "série satinée" (see 1c) has undergone a polymetamorphic evolution, with successive events in the upper amphibolite and greenschist conditions. This latest metamorphism is related to Early to Mid-Carboniferous shearing (Carme 1970 b; Gasquet et al. 1981). No relics of high grade assemblages have been recognized.

The oldest metamorphic events have been recorded in the Aar-Gotthard, Haut Dauphiné, and Argentera massifs:

Following Arnold (1970) and Grauert and Arnold (1968), some formations from Gotthard suffered successive recrystallizations under amphibolite facies, with subsequent anatexis, and then granulite facies conditions (470 Ma: Gebauer et al. 1988) before being intruded by Caledonian granitoids.

In the central Haut Dauphiné and the southeastern Argentera massifs, two successive anatectic stages have been observed (Le Fort and Pécher 1971; von Raumer 1984 a; Bogdanoff 1986). The second event post-dates relics of staurolite-kyanite-garnet gneisses and of sillimanite-cordierite-bearing orthogneisses (Le Fort 1973; Bogdanoff 1986). The two anatectic stages are commonly considered as Cadomian (or earlier) and Variscan (Devonian to Carboniferous). Such a scheme leads to a Precambrian – Early Paleozoic age for the orthogneisses.

From the above, we conclude that most parts of the differents ECMs exhibit a similar lithological, tectonic, and metamorphic evolution. An older basement with its cover (Proterozoic to Early Paleozoic) suffered a complex metamorphic evolution during Late Silurian to Early Carboniferous times, being characteristic for the central part of the Variscan orogen. Particular areas of Belledonne (W and SW) show a rather distinct tectonic and metamorphic history related to their supracrustal or marginal location with respect to Variscan orogenic zonation. The different parts of the Belledonne massif (W, SW, and NE) have been tectonically juxtaposed along major late orogenic strike-slip faults (Ménot 1987, 1988; Vivier et al. 1987).

4 Igneous Activity

4.1 Volcanic and Subvolcanic Events

Products of volcanic and subvolcanic events consist of basaltic and bimodal series, mainly representative of crustal distension stages at different times (Ménot and Paquette 1993).

Proterozoic and Early Paleozoic metavolcanics: Related amphibolites are included in the gneissic and amphibolitic basement of the ECMs. Several types can be distinguished according to the origin and the age of the protoliths, (and thus their geodynamic significance) and to the nature of the subsequent tectonic and metamorphic evolution:

The oldest amphibolites (protolith age: ca. 870 Ma) are known from the Gotthard massif and could result from an active margin magmatism. The HP parageneses are related to an Early Ordovician event (468 Ma) (Gebauer et al. 1988).

The most common type is defined by retrogressive eclogites and mafic granulites. They testify to an ensialic rifting (Aig. Rouges, Belledonne, Argentera), as shown by the bimodal igneous activity, and the chemistry of the eclogite protoliths, N and T-MORBS with a slight crustal component (Liégeois and Duchesne 1981; Paquette et al. 1989; von Raumer et al. 1990) or to oceanic relics (Aar-Gotthard: Oberhänsli et al. 1988). Dates give Ordovician ages (475–450 Ma) for the magmatic event and Silurian ages (425–395 Ma) for the high grade recrystallizations (Paquette et al. 1989).

The 496-Ma-old meta-ophiolite of Chamrousse (Pin and Carme 1987; Ménot et al. 1988 a) differs from the above described amphibolites by the spectacular preservation of its primary magmatic characters and by the lack of any high pressure metamorphic event (Guillot et al. 1992). Its chemical evolution, through a multistage accretion (Bodinier et al. 1981) expresses the transition from an attenuated continental setting to a true oceanic domain (Ménot 1987; Ménot et al. 1988 a). A subduction zone influence is argued by Pin and Carme (1987).

Middle Paleozoic metavolcanics. They have only been recognized in the southwestern part of Belledonne (i.e., the Rioupéroux-Livet formations) and consist of leptynitic and micaceous gneisses, acid and basic metavolcanics, and trondhjemites. They are considered a single plutonic and volcanic complex, now dismembered into five main units (Ménot 1986). The crystallization age of the trondhjemitic bodies has been dated as 352 ± 53 and 365 ± 17 Ma (Ménot et al. 1984, 1988 b). This igneous event could be representative of an aborted continental rift (Ménot 1987; Ménot and Paquette 1993) or of an active margin (Carme and Pin 1987).

Late Paleozoic metavolcanics: Only the syn- to late-orogenic volcanics (i.e., Early Carboniferous) are taken into account. Spilites, keratophyres, and tuffs are associated with sedimentary facies, arenites, pelites, and conglomerates. Visean microfaunas have been recognized in the southern edge of Aig. Rouges (Bellière and Streel 1980) and in the SW area of Belledonne

(Gibergy 1968). Chemical characters are ambiguous with tholeiitic and calc-alkaline affinities (Carme and Pin 1987; Ménot 1987; Mercolli and Oberhänsli 1988).

4.2 Plutonic Events

Intermediate to silicic intrusives can be related to two different igneous stages: an early event, which is badly defined, of Late Proterozoic–Early Paleozoic age and a second event during the Late Devonian–Early Carboniferous.

Proterozoic–Lower Paleozoic Plutonism. Some orthogneisses from Aar-Gotthard (Arnold 1970; Mercolli and Oberhänsli 1988), Aig. Rouges (von Raumer 1987), Haut Dauphiné (Le Fort and Pécher 1971) and Argentera (Bogdanoff 1980, 1986), sometimes bearing high grade parageneses (sillimanite), are considered as part of a pre-Devonian basement. Recently, comparable orthogneisses from the Mont-Blanc area yielded an Ordovician age (460 Ma; U-Pb zircon, Bussy and von Raumer, 1993).

Late Devonian to Mid-Carboniferous Plutonism. The earliest plutonic event related to the Variscan orogenic cycle is also represented by a series of orthogneisses. In the Belledonne chain, they are spatially and genetically associated with the Devonian anatectic stage (Vivier et al. 1987). They mainly arise from heterogenous biotitic and amphibolitic porphyritic granitoids, diorites and leucogabbros. Locally, acid-basic igneous associations (magma mingling) are preserved. Geochemical patterns indicate both a mantle contribution and a crustal origin by partial melting of metagraywackes (Vivier in Bogdanoff et al. 1990). In the other ECMs, where the late orogenic shearing is less efficient, cordierite-bearing anatexites and granites (von Raumer 1984a, b, 1987; Bonin et al. 1993) represent time equivalents of the Devonian orthogneisses.

These earliest products of Variscan plutonism are followed by discordant granitoids of Carboniferous age (cordierite-bearing granites, and subsequent calc-alkaline and Mg-rich subalkaline granites). Their time-related chemical evolution is characterized by an increasing mantle source component and a corresponding decrease of the crustal contribution (Aumaitre et al. 1985; Vivier in Bogdanoff et al. 1990). These syn- to late-kinematic intrusions are partly controlled by the Early to Mid-Carboniferous strike slip tectonics (Carme 1970b; Vivier et al. 1987; Bellière 1988; Joye 1989). They display magmatic anisotropies (phenocryst and enclaves orientation) and shear zones. The latest intrusions (Mont Blanc: 315 Ma, Bussy et al. 1989) are post-kinematic and devoid of internal magmatic fabrics. The Carboniferous plutonism of the ECMs is similar to that of the innermost parts of the Variscan belt (French Massif Central, Corsica, etc.) (Orsini 1979; Ménot and Orsini 1990).

5 Metallogeny

Although numerous ore concentrations occur within the ECMs (Fig. 3), few are, or have been, of economical importance. They have undergone considerable alteration during the Late-Variscan and Alpine overprinting, giving rise to a multitude of secondary products.

Most of the occurrences are found in the neighborhood of the basement-cover boundary, and some extend and cross-cut the Triassic dolomitic limestones. Such field relationships explain the peculiar distribution of the concentrations along the borders of the crystalline massifs or within intra-basement "pinched" Mesozoic sedimentary "basins" (Beaufortin, Rocheray, "syncline" of Morges, Argentera). Most of these ore deposits are of hydrothermal origin, with siliceous and sometimes carbonate guangues. The main parageneses of arsenopyrite-gold and galena-sphalerite-chalcopyrite-pyrite are reported. Ankerite and siderite are common whilst barite and fluorite are more subordinate. The sketch map (Fig. 3) points out the essential elements for each site.

Gold, together with mispickel or galena, is observed along the western (external) border of the ECMs: Salanfe (Aig. Rouges), St Paul, and Tours sur Isère (N-Belledonne), Le Perailler (S-Belledonne), La Gardette (Haut Dauphiné), Duminière (W-Argentera).

Silver, nickel, and cobalt have been worked in the Chalanches site (S-Belledonne) where 10 tons of silver have been extracted from a stockwerk in amphibolitic gneisses. Associated minerals are galena, blende, cinabar, antimony and bismuth.

Barite concentrations, together with blende-pyrite-galena-chalcopyrite mineralization, are frequently enclosed within dikes that intrude the metagraywackes of the basement. Fluorite is often associated with barite but can also be strictly related to the neighbourhood of granites (Mont Blanc, Rocheray, NE-Belledonne).

Also genetically related to the Late Variscan granites, scattered occurrences of molybdenite and wolframite/scheelite are found in the late leucocratic differentiates, aplitic and pegmatitic veins.

A scheelite-skarn observed in the Aiguilles-Rouges, and bound to metavolcanics of a possible Late Proterozoic age, was formed during a Late

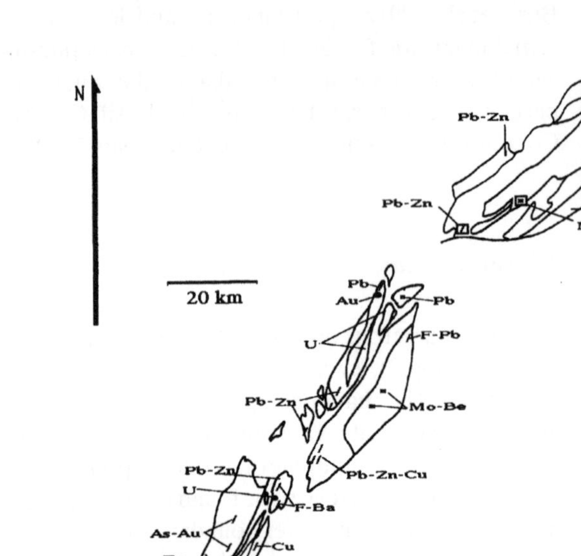

Fig. 3. Location of the main ore concentrations, after the French 1/500000 map of ore deposits (simplified), Bureau Recherches géologiques et Minières (BRGM) edit., Orléans France: Concentrations of economic importance □; Field settings: Dikes /; masses or pockets ●; stockworks ★

Variscan high temperature metamorphic event (Chiaradia 1993). Such mineralizations are not restricted to the ECMs and can be followed up from Europe to Northern Africa.

Uranium-rich minerals are known from the southern border of the Aar massif and from the Vallorcine (Aig. Rouges) and Lauzière (NE-Belledonne) granites. The whole-rock U content of the latter sites is low (5–9 ppm and 5–15 ppm respectively). Alteration of the primary minerals give rise to a multitude of secondary minerals that have been reworked and concentrated, probably during the Late Variscan retrograde metamorphic stages (Negga 1984). The tectonic control of the concentration processes is obvious.

Finally, only a podiform-type mineralization can be strictly related to Paleozoic mafic igneous formations: chromite-rich layers of several meters extension are associated with magnetite-bearing serpentinites of the Chamrousse ophiolitic complex. The low Cr/Fe ratio of chromites precluded any working.

6 Final Considerations

The preceding very short characteristics of the External Massifs in the context of the French Pre-Mesozoic basement areas cannot but represent an incomplete view on the general palaeotectonic situation of the ECMs during the evolution since the Late Proterozoic. Referring to our recent discussion (von Raumer et al. 1993) they, in fact, represent the long and complex geological evolution beginning with early rifting in an older basement during the Late Proterozoic – Cambrian, followed by the general crustal shortening since the Ordovician. Based on the interpretation of the pre-Mesozoic basement in the entire Alps (von Raumer and Neubauer 1993), the ECMs, since the Ordovician, are part of the active continental margin and testify the whole evolution in the complex suture zone of asymmetric collision between Gondwana and Laurasia, producing the supercontinent Pangea at the Late Carboniferous.

Acknowledgments. We thank A. Michard, F. Neubauer (Graz, Austria), and anonymous reviewers for constructive remarks, and R. Oliver (I. L. L., Grenoble) for improving the English version.

References

Apprahamian J (1988) Cartographie du métamorphisme faible à très faible dans les Alpes françaises externes par l'utilisation de la cristallinité de l'illite. Geodyn Acta 2, 1: 25–32

Arnold A (1970) On the history of the Gotthard Massif (Central Alps, Switzerland). Eclogae Geol Helv 63: 29–30

Aumaitre R, Giraud P, Poncerry E, Vivier G (1985) Les granitoïdes sub-alcalins potassiques de la Lauzière, massif cristallin externe de Belledonne. Géol Alp 61: 17–32

Banzet G, Lapierre H, Le Fort P, Pécher A (1985) Le volcanisme carbonifère supérieur du massif des Grandes Rousses (zone dauphinoise – Alpes externes françaises): un magmatisme à affinités shoshonitiques lié à la fracturation crustale tardi-varisque. Géol Alp 61: 33–60

Bellière J (1958) Contribution à l'étude pétrogénétique des schistes cristallins du massif des Aiguilles Rouges (Hte Savoie). Mém Soc Géol Belg 81, 204 pp

Bellière J (1980) Massif du Mont Blanc et des Aiguilles Rouges. Structures et pétrologie du socle. Géol Alp 56: 237–249

Bellière J (1988) On the age of the mylonite within the Mont Blanc massif. Geodyn Acta 2, 1: 13–16

Bellière J, Streel M (1980) Roches d'âge viséen supérieur dans le massif des Aiguilles Rouges (Hte Savoie). CR Acad Sci Paris 290 D: 1341–1343

Bodinier JL, Dupuy C, Dostal J, Carme F (1981) Geochemistry of ophiolites from the Chamrousse complex (Belledonne massif, Alps). Contrib Mineral Petrol 78: 379–388

Bogdanoff S (1980) Analyse structurale dans la partie occidentale de l'Argentera-Mercantour (Alpes Maritimes). Thesis, Orsay Univ., France, pp 1–316 (unpublished)

Bogdanoff S (1986) Evolution de la partie occidentale du massif cristallin externe de l'Argentera. Place dans l'arc alpin. Bull Bur Rech Geol Min, sér géol France, 4: 433–453

Bogdanoff S, Ménot RP, Vivier G (1991) Les massifs cristallins externes des Alpes françaises. In: Piqué A (ed) Les massifs cristallins anciens. Sciences Géologiques, Strasbourg 44: 3–4 pp 237–285

Bonin B, Brändelin P, Bussy F, Desmons J, Eggenberger U, Finger F, Graf K, Marro Ch, Mercolli I, Oberhänsli R, Ploquin A, von Quadt A, von Raumer JF, Schaltegger U, Steyrer HP, Visona D, Vivier G (1993) Late Variscan magmatic evolution of the Alpine basement. In J. von Raumer and F. Neubauer (Eds) The pre-Mesozoic Geology in the Alps. Springer, Heidelberg, pp. 171–201

Bordet P, Bordet C (1963) Belledonne – Gdes Rousses et Aiguilles Rouges: quelques données nouvelles sur leurs rapports structuraux. In: Livre à la mém. Prof. Fallot. Mém Soc Géol Fr: 311–316

Bussy F, Schaltegger U, Marro Ch (1989) The age of the Mont Blanc granite (Western Alps): a heterogeneous isotopic system dated by Rb-Sr whole rock determinations on its microgranular enclaves. Schweiz Mineral Petrogr Mitt 69: 3–13

Bussy F and von Raumer JF (1993): U-Pb dating of Palaeozoic event in the Mont-Blanc crystalline massif, Western Alps. Terra Nova Vol. 5, suppl. 1, 1993, 382–383

Carme F (1970 a) Age briovérien probable de la majeure partie des séries supposées dévonodinantiennes et existence d'un cycle orogénique anté-hercynien, sans doute cadomien, dans chaîne de Belledonne (Alpes francaises). CR Acad Sci Paris 271 D: 631–633

Carme F (1970 b) Caractères fondamentaux de l'architecture d'ensemble des chaîne de Belledonne et des Aiguilles Rouges. Rôle des décrochements. CR Acad Sci Paris 271 D: 2232–2235

Carme F (1973 a) Précisions sur le métamorphisme majeur des schistes cristallins de la chaîne de Belledonne: essai de zonéographie et distribution verticale des zones au niveau de la coupe de Romanche. CR Acad Sci Paris 277 D: 2133–2136

Carme F (1973 b) Précisions nouvelles sur l'ampleur de la tectonique tangentielle hercynienne dans la chaîne de Belledonne: CR Acad Sci Paris 277 D: 2309–2312

Carme F (1975) Caractères chimiques de deux ensembles de roches magmatiques préorogéniques du socle anté-houiller des Massifs Cristallins Externes alpins et des Monts du Lyonnais. CR Acad Sci Paris 281 D: 89–92

Carme F, Pin C (1987) Vue d'ensemble sur le magmatisme préorogénique et évolution métamorphique et tectonique varisque dans le sud de la chaîne de Belledonne (massifs cristallins externes, Alpes françaises). CR Acad Sci Paris 304 II 19: 1177–1180

Chiaradia M (1993): The Scheelite-skarn of Salanfe (Valais, Switzerland). Schweiz Min Petr Mitt 73, 41–51

Debelmas J, Kerckhove Cl (1980) Géologie des Alpes franco-italiennes. Géol Alp 56: 21–58, et Géologie des pays européens, Dunod, pp 227–314

Demeulemeester P (1982) Contribution à l'étude radio métrique à l'argon et au strontium des massifs cristallins externes (Alpes françaises). Thesis, Grenoble I Univ., France, pp 1–227 (unpublished)

Demeulemeester P, Roques M, Giraud P, Vivier G, Bonhomme M (1986) Influence du métamorphisme alpin sur les âges istopiques Rb/Sr et k/ar des biotites dans les massifs cristallins externes (Alpes françaises). Géol Alp 62: 31–44

Favre D (1979) Les gneiss oeillés d'Emosson (Aiguilles Rouges): une définition. Thesis, Fribourg Univ., Swiss (unpublished)

Gasquet D, Giraud P, Ploquin A, Vivier G (1981) Géochimie des mylonites relations entre les Rameaux Interne et externe du massif de Belledonne (Alpes françaises). CR Acad Sci Paris 92 II: 607–610

Gebauer D, Quadt A, Compston W, Williams IS, Grünenfelder M (1988) Archean zircons in retrograded Caledonian eclogite of the Gotthard massif (Central Alps, Switzerland) Schweiz Mineral Petrogr Mitt 68: 485–490

Gibergy P (1968) Découverte de "grès à trous" renfermant des débris d'organismes dans les schistes noirs du Valbonnais (série cristallophylienne des massifs cristallins des Alpes françaises). CR Acad Sci Paris 267 D: 1251–1254

Giorgi L, Giraud P, Vachard D (1979) Sur la présence de micro-organismes d'âge cambrien dans les schistes cristallins du versant occidental du massif cristallin externes des Gdes Rousses (Alpes occidentales). CR Acad Sci Paris 288 D: 1079–1082

Grauert, Arnold A (1968) Deutung diskordanter Zirkonalter der Silvretta-Decke und des Gotthardmassivs (schweizer Alps). Contrib Mineral Petrol 20: 34–56

Guillot S, Ménot RP, Lardeaux JM (1992) Tectonique intraocéanique distensive dans l'ophiolite paléozoïque de Chamrousse (Alpes occidentales). Bull Soc Géol 163 (3): 229–240

Holdaway MJ (1971) Stability of andalusite and the aluminium silicate phase diagram. Amer J Sci 271: 97–131

Joye JB (1989) Une zone de cisaillement du Carbonifère inférieur dans les Aiguilles Rouges. Thesis, Fribourg Univ., Swiss (unpublished)

Latouche L, Bogdanoff S (1987) Evolution précoce du massif de l'Argentera: apport des éclogites et des granulites. Géol Alp 63: 151–164

Laurent R (1968) Etude géologique et pétrographique de l'extrémité méridionale du massif des Aiguilles Rouges (Htes Savoie). Arch Sci Genève 20, 2: 223–254

Le Fort P (1973) Géologie du Ht Dauphiné cristallin (Alpes françaises). Etude pétrologique et structurale de la partie occidentale. Sci Terre, Nancy 25, pp 1–373

Le Fort P, Pécher A (1971) Présentation d'un schéma structural du Ht. Dauphiné cristallin. CR Acad Sci Paris 273 D: 3–5

Liégeois J-P, Duchesne J-C (1981) The lac Cornu retrogressed eclogites (Aiguilles Rouges massif, western Alps, France): evidence of crustal origin and metasomatic alteration. Lithos 14: 35–48

Matte Ph (1986) Tectonics and plate tectonics model for the Variscan belt of Europe. Tectonophysics 126: 329–374

Menard G, Molnar P (1988) Collapse of an hercynian tibetan plateau into a late Paleozoic European basin and range province. Nature 334: 235–237

Ménot R-P (1986) Les formations plutono-volcaniques dévoniennes de Rioupéroux-Livet (massifs cristallins externes des Alpes françaises): nouvelles définitions lithostratigraphique et pétrographique. Schweiz Mineral Petrogr Mitt 66: 229–258

Ménot R-P (1987) Magmatismes paléozoiques et structuration carbonifère du massif de Belledonne (Alpes françaises). Contraintes nouvelles pour les schémas d'évolution de la chaîne varisque ouest-européene. Mém et Doc. Centre Armoricain d'Etudes Struct. des Socles, Rennes, 1988, 21, pp 1–465

Ménot R-P (1988) An overview of the geology of the Belledonne massif (External crystalline massifs of western Alps). Schweiz Minéral Petrogr Mitt 68, 3: 531–542

Ménot R-P, Orsini JB (1990) Evolution du socle anté-stéphanien de Corse: évènements magmatiques et métamorphiques. Schweiz Mineral Petrogr Mitt 70: 33–53

Ménot R-P, Peucat JJ, Scarenzi D, Piboule M (1984) A continental (?) volcanic arc of devono-carboniferous age in the French alpine external massifs: The Rioupéroux and Livet formations (Romanche valley, Belledonne massif). Meet. "Ophiolites through time", Nancy. Ofioliti 10, 2–3: 528 (Abstr)

Ménot R-P, Bonhomme MG, Vivier G (1987) Structuration tectono-métamorphique carbonifère dans le massif de Belledonne (Alpes occidentales françaises). Apport de la géochronologie K/Ar des amphiboles. Schweiz Mineral Petrogr Mitt 67, 3: 273–284

Ménot R-P, Paquette JL (1993) Geodynamic significance of basic and bimodal magmatic events in the external domain. In: J von Raumer and F Neubauer (Eds) The pre-Mesozoic Geology of the Alps. Springer, Heidelberg, New York 1993 pp. 241–254

Ménot R-P, Peucat JJ, Scarenzi D, Piboule M (1988 a) 496 My age of plagiogranites in the Chamrousse ophiolite complex (external crystalline massifs in the French Alps): evidence of a Lower Paleozoic oceanization. Earth Planet Sci Lett 88: 82–92

Ménot R-P, Peucat JJ, Paquette JL (1988 b) Les formations leptyno-amphiboliques paléozoïques et les Complexes Leptyno-amphiboliques: les corrélations hasardeuses. Exemple des massifs cristallins externes alpins. Bull Soc Géol Fr (8) IV 6: 917–926

Mercolli I, Oberhänsli R (1988) Variscan tectonic evolution in the Central Alps: a working hypothesis. Schweiz Mineral Petrogr Mitt 68, 3: 491–500

Monié P, Maluski H (1983) Données géochronologiques 39 Ar–40 Ar sur le socle anté-permien du massif de l'Argentera-Mercantour (Alpes maritimes, France). Bull Soc Géol Fr (7) XXV 2: 247–257

Negga HS (1984) Comportement de l'uranium lors des métamorphismes tardi-hercynien et alpin dans les massifs des Aiguilles Rouges et de Belledonne (Vallorcine et Lauzière), Alpes occidentales. Thesis, CREGU, Nancy Univ., France, pp 1–400 (unpublished)

Paquette JL (1987) Comportement des systèmes isotopiques U-Pb et Sm-Nd dans le métamorphisme éclogitique. Chaîne hercynienne et chaîne alpine. Centre Armoricain d'Etudes Struct. Socles, Rennes, 14, pp 1–189

Paquette JL, Ménot RP, Peucat JJ (1989) Sm-Nd and U-Pb zircon study of eclogites from the Alpine External Massifs (Western Alps): evidence for crustal contamination. Earth Planet Sci Lett 96, 1–2: 181–198

Pfeifer HR, Biino G, Ménot RP, Stille P (1993) Ultramafic rocks in the pre-Mesozoic basement of the Central and External Western Alps. In: J von Raumer and F Neubauer (Eds) The pre-Mesozoic Geology in the Alps. Springer, Heidelberg, pp 199–143

Pigage LC, Greenwood HJ (1982) Internally consistent estimates of pressure and temperature: the staurolite problem. Am J Sci: 942–969

Pin C, Carme F (1987) A Sm-Nd isotopic study of 500 Ma old oceanic crust in the Variscan belt of western Europe: the Chamrousse ophiolite complex, western Alps (France). Contrib Mineral Petrol 96: 406–413

Oberhänsli R, Shenker F, Mercolli I (1988) Indications of Variscan nappe tectonics in the Aar massif. Schweiz Mineral Petrogr Mitt 68, 3: 509–520

Orsini JB (1979) Existence de trois associations magmatiques dans les granitoïdes post-viséen moyen (groupe des granodiorites-monzogranites) de la chaîne varisque française. CR Acad Sci Paris 289 D: 981–984

Raumer JF von (1976) Variszicum in den Zentral- und Westalpen. Nova Acta Leopold 224 (45): 119–144

Raumer JF von (1983) Die Metapelite von Emosson (Aiguilles Rouges Massiv) als Beispiel spätkaledonisch-frühvarizischer Metamorphose im Altkristallin des helvetischen Bereichs. Schweiz Mineral Petrogr Mitt 63: 421–455

Raumer JF von (1984 a) The External Massifs, relics of Variscan basement in the Alps. Geol Rundsch 73, 1: 1–31

Raumer JF von (1984 b) The Hercynian basement in the Helvetic Realm (Western and Central Alps). In: Dal Piaz, GV (ed) Alpi Occidentali. Mem Soc Geol Ital 29: 57–69

Raumer JF von (1987) Les massifs du Mont Blanc et des Aiguilles Rouges témoins de la formation de croûte varisque dans les Alpes occidentales. Géol Alp 63: 7–24

Raumer JF von (1988) Caledonian-Variscan structures in the Alps. Schweiz Mineral Petrogr Mitt 68, 3

Raumer JF von, Ménot R-P (1989) Evolution paléozoique du socle ouest-alpin et place des Massifs Cristallins Externes dans l'orogène varisque. CR Acad Sci Paris 309 II: 397–402

Raumer JF von, Galetti G, Pfeifer HR, Oberhänsli R (1990) Amphibolites from Lac d'Emosson, Aiguilles Rouges, Switzerland: tholeiitic basalts of a Paleozoic continental rift zone. Schweiz Mineral Petrogr Mitt 70: 419–435

Raumer JF von, Ménot RP, Abrecht J, Biino G (1993): The Pre-Alpine evolution of the External Massifs. In: J von Raumer and F Neubauer (Eds) The pre-Mesozoic Geology in the Alps. Springer, Heidelberg, pp 221–239

Raumer JF von, Neubauer F (1993) Late Precambrian and Palaeozoic Evolution of the Alpine Basement – an overview. In: J von Raumer and F Neubauer (Eds) The pre-Mesozoic Geology of the Alps. Springer, Heidelberg, New York 1993 pp 641–663

Richardson SW (1968) Saturolite stability in a part of the system Fe-Al-Si-O-H-. J Petrol 9: 467–488

Schenker F (1986) Saurer Magmatismus und Beckendbildungen im Aarmassiv unter kompressiver Tektonik. Dissertation, Bern Univ., Swiss

Schulz B. and von Raumer J. (1993) Syndeformational Uplift of Variscan High-pressure Rocks (Col de Bérard, Aiguilles Rouges Massif, Western Alps). Z dt geol Ges 144, 104–120

Siméon Y (1979) Etude pétrologique, géochimique et structurale des terrains cristallins de Belledonne entre l'Arc et l'Isère (Alpes françaises). Thesis, Grenoble Univ., France, pp 1–224 (unpublished)

Thompson AB (1982) Dehydration melting of pelitic rocks and the generation of H_2O-undersaturated granite liquids. Am J Sci 282: 1567–1595

Vivier G, Ménot R-P, Giraud P (1987) Magmatismes et structuration orogénique paléozoiques de la chaîne de Belledonne (Massifs cristallins externes). Le domaine nord-oriental. Géol Alp 63: 25–53

Corsica and Sardinia in the Variscan Chain

J. M. Lardeaux, R. P. Menot, J. B. Orsini, Ph. Rossi, G. Naud, and G. Libourel

1 Introduction: The Main Structures of the Corsica-Sardinian Segment

(J. M. Lardeaux, R. P. Menot, and J. B. Orsini)

Corsica and Sardinia are two islands which separate the Western Mediterranean Sea into the Provençal Basin in the West and the Tyrrhenian Sea in the East (Fig. 1). Paleomagnetic, structural, and lithological similarities indicate that Corsica and Sardinia can be correlated with the Provence basement in southeastern France (Arthaud and Matte 1966; Westphal et al. 1973; Cocozza 1974). The islands have rotated away from the European continent to their present position and have had a common geological history up to the early Tertiary (Westphal et al. 1976; Orsini et al. 1980; Edel et al. 1981). In this introduction, we briefly summarize the main structural features of Corsica and Sardinia. The following sections are respectively devoted to stratigraphic and lithologic features, magmatic, metamorphic, and metallogenic activities.

1.1 Crustal-Scale Structures

1. Corsica has been classically divided into two tectonic domains (Figs. 1, 2):

- A western autochthonous domain consisting mainly of Variscan granites and Permian magmatic rocks which intruded mostly medium – to high-grade Variscan metamorphic series.
- A northeastern allochthonous Alpine domain comprising the Jurassic and Cretaceous ophiolitic

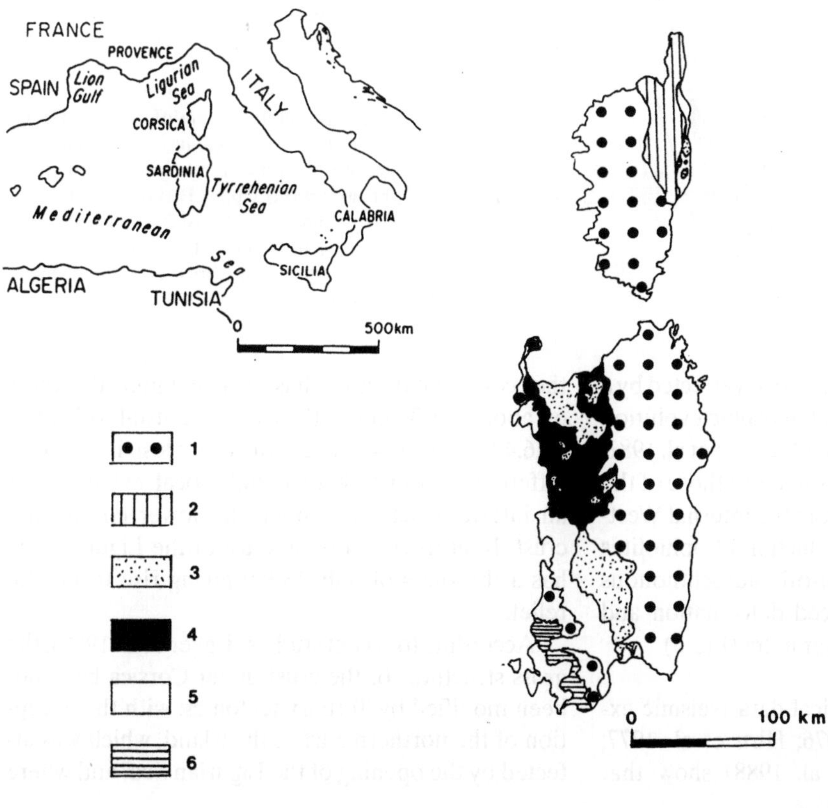

Fig. 1. Location map and main structural domains of Corsica and Sardinia. *1* Paleozoic domain (western autochthon); *2* Alpine domain of Corsica (northeastern allochthon); *3* Miocene deposits; *4* volcanic deposits (Quaternary to Oligocene); *5* Quaternary sediments; *6* Cambrian

J. D. Keppie (Ed.): Pre-Mesozoic Geology in France and Related Areas

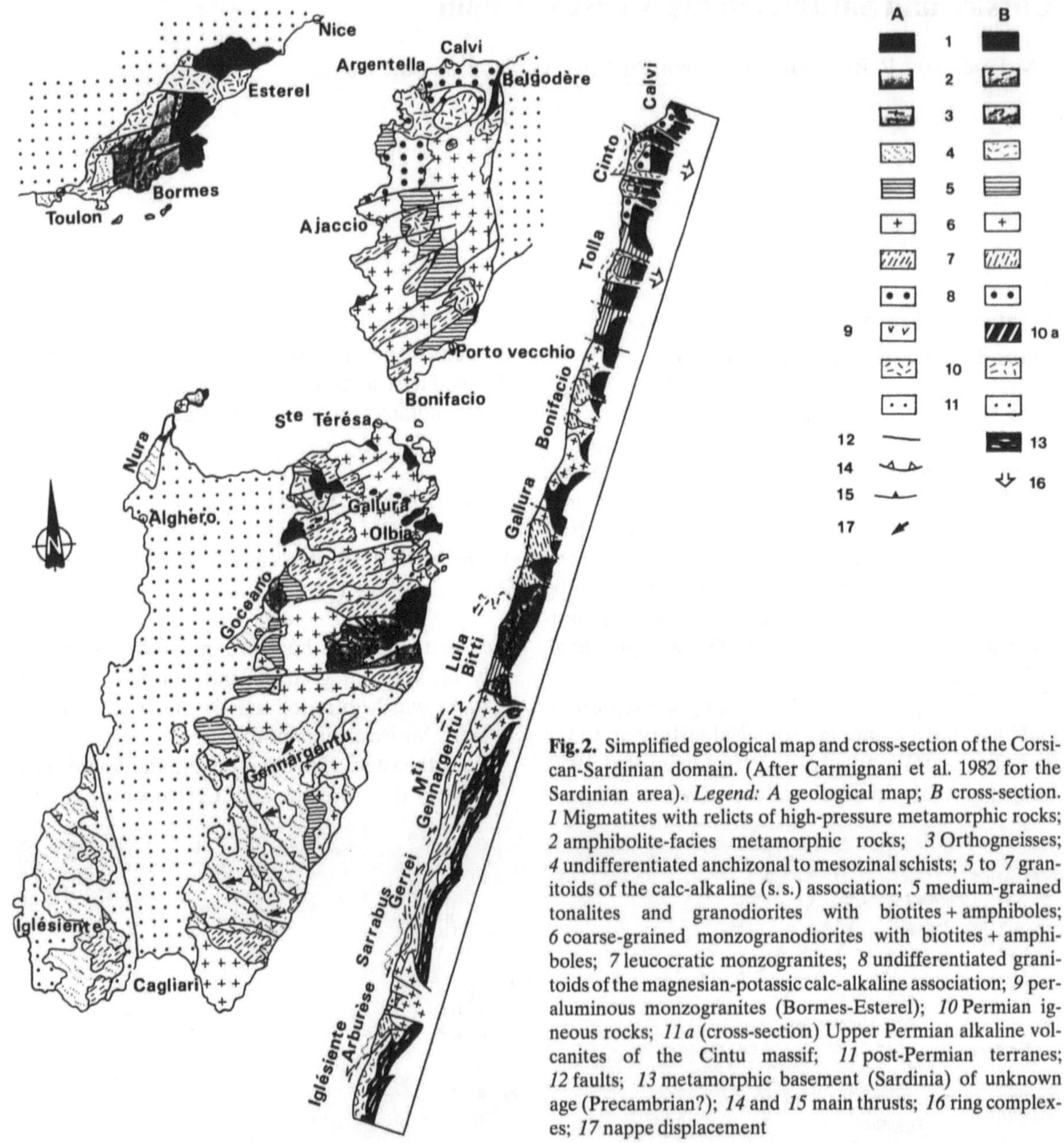

Fig. 2. Simplified geological map and cross-section of the Corsican-Sardinian domain. (After Carmignani et al. 1982 for the Sardinian area). *Legend: A* geological map; *B* cross-section. *1* Migmatites with relicts of high-pressure metamorphic rocks; *2* amphibolite-facies metamorphic rocks; *3* Orthogneisses; *4* undifferentiated anchizonal to mesozinal schists; *5* to *7* granitoids of the calc-alkaline (s. s.) association; *5* medium-grained tonalites and granodiorites with biotites + amphiboles; *6* coarse-grained monzogranodiorites with biotites + amphiboles; *7* leucocratic monzogranites; *8* undifferentiated granitoids of the magnesian-potassic calc-alkaline association; *9* peraluminous monzogranites (Bormes-Esterel); *10* Permian igneous rocks; *11 a* (cross-section) Upper Permian alkaline volcanites of the Cintu massif; *11* post-Permian terranes; *12* faults; *13* metamorphic basement (Sardinia) of unknown age (Precambrian?); *14* and *15* main thrusts; *16* ring complexes; *17* nappe displacement

Schistes Lustrés (s. l.) which have been affected by a polyphase Alpine tectono-metamorphic evolution (Caron 1977; Caron et al. 1981; Mattauer et al. 1981; Harris 1985). These units are similar to those of the well-known Piemont Zone from the internal Western Alps. The Cretaceous obduction of ophiolites and their sedimentary cover onto the autochthonous terranes have locally produced deformation and the gneissification of Variscan granites (Fig. 3).

At a crustal scale, geophysical data (seismic experiments, Hirn and Sapin 1976; Hirn et al. 1977; Morelli et al 1977; Egger et al. 1988) show that the continental crust has a maximum thickness of about 30–33 km with average crustal velocities of 6.4 km/s (Fig. 4). The Corsican crust is not well differentiated and there is only local evidence for an interface between upper and lower continental crust. In northern Corsica, crust of the Ligurian Sea has a thickness of only 18 km giving rise to mantle relief.

According to recent studies (Egger et al. 1988), the gross structures of the crust under Corsica have not been modified by Tertiary tectonics, with the exception of the northern part of the island, which was affected by the opening of the Ligurian Sea, and where

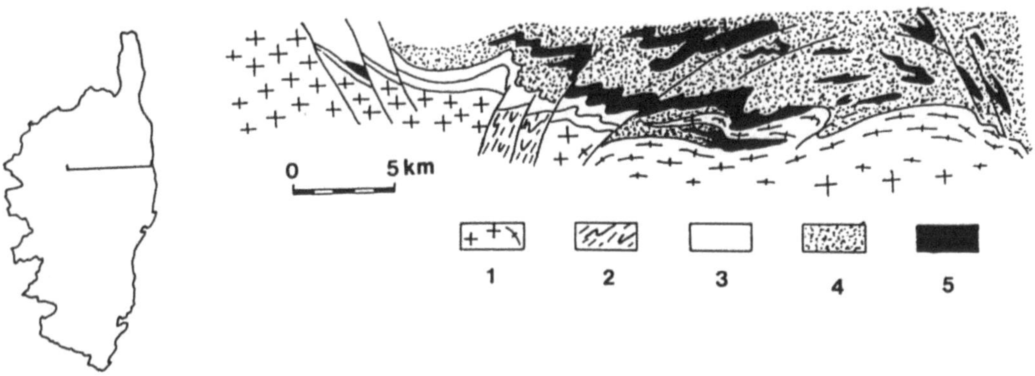

Fig. 3. Geometrical relationships between Paleozoic and Alpine domains in Corsica. (After Caron et al. 1981). *1* Variscan granites; *2* pre-Carboniferous basement; *3* para-autochtonous units; *4* schistes lustrés nappe; *5* ophiolitic rocks

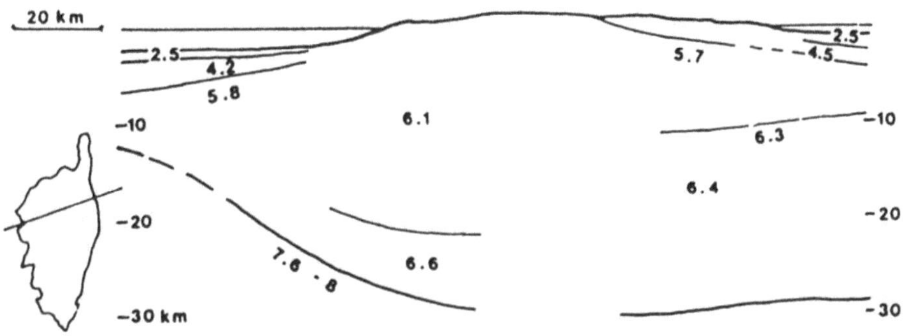

a

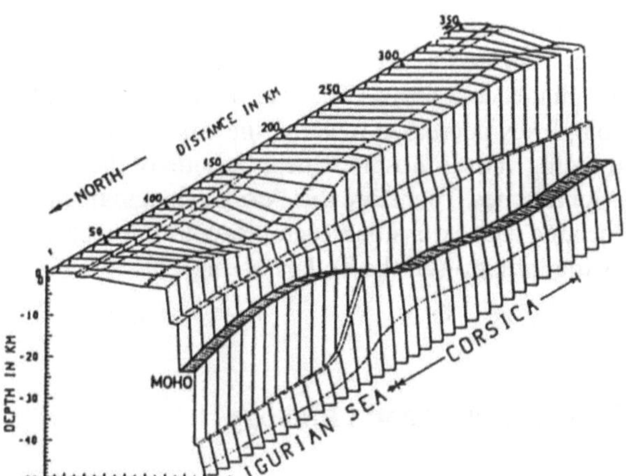

b

Fig. 4a, b. Crustal scale structures in Corsica. **a** After Hirn and Sapin (1976). **b** After Egger et al. (1988)

the heat flow value is maximum (about 90 mW/m^2 for an average of 76 mW/m^2, Lucazeau and Mailhé 1986).

According to Bayer et al. (1973), positive gravimetric and magnetic anomalies are related to gabbros or Permian volcanites in the Western Corsica, while in Alpine Corsica they can be correlated to the main ophiolitic masses.

2. Sardinia can also be divided into contrasting tectonic domains (Fig. 2). The principal is the Variscan in the eastern part of the island. The western part consists mainly of Oligocene to Quaternary volcanic rocks, Miocene deposits and Quaternary sediments emplaced within two structures: the Oligocene-Miocene Sardinia rift striking north-south and the northwest-southeast Pliocene-Quaternary Campidano graben. It is in the latter area that the higher heat flow values have been found in Sardinia. Eastern Sardinia thus offers one of the most complete Paleozoic successions in Italy, unaffected by the Alpine orogeny (Fig. 2, Carmignani et al. 1982).

Following geophysical investigations (Egger et al. 1988), a maximum crustal thickness of 34 km is proposed, with average crustal velocities of 6.3 km/s, for Sardinia. The depth of the Moho reaches a maximum of 34 km in the center of the island. The crust-mantle boundary is located at 27 km depth in the north, at 30 km in the east, at 23 km in the south. In the Sardinian channel the depth of the Moho is thought to be at 20 km. As it was the case in Corsica, the Sardinian crust is relatively undifferentiated.

1.2 Variscan Tectonics

1. Corsica. As shown in Figs. 2 and 3, the metamorphic basement in Corsica occurs as dismembered, rotated, and thus isolated panels within large domains of Variscan granites. Therefore in this island, the Variscan tectonics cannot be studied because of the lack of geometrical constraints. However, some present day field observations indicate that:

- The finite strain pattern is quite complex as the result of polyphase deformation. Indeed, a generally well-developed foliation plane is reactivated by at least two generations of structures.
- The main deformation of the Variscan crust occurred during the uplift of metamorphic rocks, thus under retrogressive P-T paths (mainly under amphibolite facies conditions). The main tectonics post-dates the high pressure metamorphism recognized in many areas.
- Variscan tectono-metamorphic events are accompanied and/or followed by intrusion of widespread syntectonic plutonic masses.

2. Sardinia. According to the literature (Arthaud and Matte 1977; Arthaud and Sauniac 1981; Carmignani et al. 1981, 1982), three main tectonic zones can be recognized in the Sardinian Variscan Orogen (Fig. 2).
a) A southwestern zone, restricted to the southwest of the Campidano graben, including the Iglesiente and Sulcis regions. This region is characterized by "dome and basin" structures (Poll and Zwart 1964; Arthaud 1970; Cocozza et al. 1974) resulting from the superposition of N-S folds upon E-W structures. More precisely, in the Iglesiente and northern Sulcis areas, the N-S folds are associated with the development of fracture and/or strain slip vertical cleavage (Arthaud 1970). On the contrary, in the southern Sulcis, a subhorizontal cleavage is associated with recumbent isoclinal folds. This southwestern zone can therefore be regarded as an external orogenic domain (foreland) with slight shortening and metamorphism.
b) A central zone, characterized by syn-metamorphic tangential tectonics and well-developed overthrusts. This zone runs across the island from Nurra to Sarrabus (Carmignani et al. 1981). A polyphase tectonic evolution is responsible for the refolding of the nappe structures (first generation of structures, Carmignani et al. 1979). A third generation of structures is seen as open folds and strike-slip fractures. In this domain the metamorphic grade does not exceed the greenschist facies conditions.
c) A northeastern zone affected by a strong tectono-metamorphic reequilibration and a widespread granitization. The metamorphic grade seems to increase from the southwest to the northeast, as shown by the transition from amphibolite-facies rocks to well-developed migmatites. However, such a metamorphic zonation is questionable because the northeastern zone is characterized by the superimposition of several tectono-metamorphic episodes and the contrasted metamorphic facies can also suggest the tectonic juxtaposition of two different units. It is worth pointing out that relicts of high-pressure metamorphic rocks (eclogites and/or granulites, Miller et al. 1976; Ghezzo et al. 1979, 1982) occur in the migmatites.

In Sardinia, an external domain (southwestern zone) and a granitized root zone (northeastern zone) are separated by an intermediate tectono-metamorphic zone allowing study of both external and internal orogenic domains of the Variscan chain. Geological, petrological, and structural constraints indicate that the main features of the Sardinian basement have been produced within a large-scale ensialic shear zone during Variscan times (Vai and Cocozza 1986). This underthrusting event was followed by the uplift of deep-seated nappes accompanied by the intrusion of widespread granitoids at around 310-300 Ma (Ferrara et al. 1978; Carmignani et al. 1981, 1982).

1.3 Permian Tectonics

In Corsica, after development of structures related to the Variscan orogeny, a generalized extensional regime formed well-developed grabens and annular magmatic complexes. The size of these magmatic complexes decreases towards the east, indicating a thinning of the lithosphere near the sites of the future Alpine zones. Therefore this Permian magmatism associated with extensional tectonics could be regarded as the initiation of the Tethys ocean.

2 Stratigraphy and Lithology

(G. Naud)

2.1 Formations Involved in the Variscan Tectonics

2.1.1 Sardinia

1. External Zone (Southwestern Domain). In this domain a formation (Bithia formation, Junker and Schneider 1979) mainly composed of meta-sandstones, meta-arkoses, meta-conglomerates, and meta-limestones is considered to unconformably overlie polymetamorphic micaschists and orthogneisses (of Precambrian age?, Cocozza et al. 1972; Naud 1979). However, recent studies (Sassi in prep.) indicate that it is a tectonic contact. This weakly metamorphosed formation is considered to be lower Cambrian, as it is overlain by paleontologically dated units: the Nebida formation with meta-sandstones and the Gonnesa formation composed of meta-dolomites are lower Cambrian; the Cabitza formation, which consists of meta-argillites, is middle to upper Cambrian and lower Ordovician (Tremadocian-Arenigian, Carmignani et al. 1986; Vai and Cocozza 1986). These are overlapped by a thick layer of polygenic meta-conglomerate (Barca et al. 1981). The latter is followed by upper Ordovician meta-sandstones, meta-silts and meta-argillites with scarce intercalations of meta-encrinitic limestones and meta-volcanics (Cocozza and Leone 1977). The Silurian is characterized by black graptolite meta-argillites and lenses of orthocerid limestones and the Devonian by meta-argillites and meta-crinoidal limestones. The Carboniferous might be represented by meta-sandstones but there are no fossils present.

2. Central Zone. Even with distinct local variations, two main units can be recognized (Naud and Tempier 1977; Naud 1982; Carmignani et al. 1986):

a) The lower unit mainly composed of meta-conglomerates, meta-sandstones, meta-silts, and meta-argillites of Middle Cambrian to Lower Ordovician age (Arenigian) (Barca et al. 1981; Naud and Pittau Demelia 1985). Locally, acidic volcanics can be found (San Vito-Solanas sandstones).

b) The upper unit is an Upper Ordovician to Lower Carboniferous unit which unconformably overlies the lower unit (Calvino 1961; Naud 1981). At its base, a meta-conglomerate is overlain by a thick complex of meta-rhyolites or meta-rhyodacites (Memmi et al. 1982). The latter is followed by meta-conglomerates, meta-arkoses, meta-sandstones and green meta-silts (Caradocian), the upper part (Ashgillian) of which has meta-limestone intercalations. In the Gerrei area, meta-basalts and meta-gabbros were recognized in this a formation. The Silurian is characterized, on the one hand by meta-silts, black graptolite meta-argillites, meta-lydites, and scarce meta-basic rocks, and on the other hand by levels of meta-orthocerid limestones. The Devonian consists of meta-tentaculitid limestones. The Carboniferous is known in only one region (Corona Mizziu and Villasalto areas) and is represented by conodont limestones of Tournaisian age (Olivieri 1969).

3. Internal Zone (Northeastern domain). Despite the high metamorphic and tectonic grade, the formations of the internal zone include meta-rhyolites, meta-basic rocks, marbles, calcschists, graphitic phyllites, and micaschists, and are similar to the terranes known in central Sardinia from the lower Ordovician to the lower Carboniferous.

2.1.2 Corsica

Near Galeria (northern Corsica), sandstones and quartzites, with minor lenses of conglomerates of Ordovician age (Tremadocian-Arenigian), probably lie unconformably upon the gneisses of the Argentella Complex (Durand-Delga 1978; Nardi et al. 1979). This series is followed by upper Ordovician detrital sediments and sandstones, shales, and Silurian lydites. Foraminiferous limestones of Upper Devonian and Lower Carboniferous age have been found, but with no clear geometrical relation to the previous formations. A mixed sandstone and pelitic formation (Culm facies), including lenses of conglomerates rich in quartz pebbles and quartzites (Baudelot et al. 1976), completes the Paleozoic series. However, recent studies (Gonnord and Ménot 1993), have indicated the presence of tectonic contacts within the Paleozoic sedimentary series of the Argentella area. The pelitic Silurian formation could also be considered as an allochthonous unit and therefore cannot be used as a chronological reference for the underlying formations.

2.2 Post-Tectonic Formations

Continental to lacustrine clastic deposits (including hard coal levels), of Carboniferous age (Stephanian-Westphalian?) are known within two tectonic basins near San Giorgio (southwestern Sardinia) and Osani (Corsica). The lower Permian consists of clastic deposits, often red-colored, associated with volcanics (rhyolites and andesites) and distributed in many small basins in Sardinia (Sulcis, Nurra, Gerrei, Barbagia, Gallura) and Corsica (Fontana et al. 1982).

3 Igneous Activity

(J. B. Orsini and Ph. Rossi)

The Corsican-Sardinian batholith is a composite domain in which three main magmatic associations, emplaced over more than 100 Ma, can be distinguished as follows.

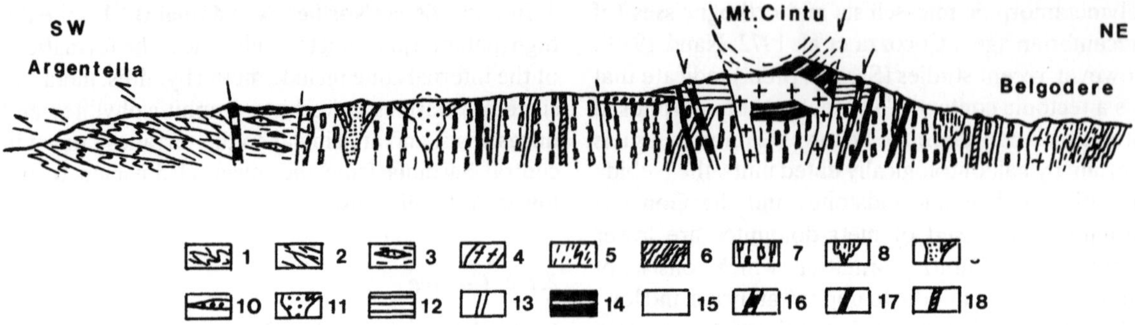

Fig. 5. Cross-section of the Paleozoic basement in northwestern Corsica (Calvi area). *1* Argentella metamorphic unit; *2* Silurian schists; *3* Upper Devonian-Lower Carboniferous sediments; *4* orthogneisses; *5* migmatitic gneisses; *6* leptyno-amphibolitic complex; *7* megacrysts-bearing granitoids of the magnesian-potassic calc-alkaline association; *8* magnesian-potassic leucocratic granite; *9* late leucocratic microgranite; *10* Stephanian sediments; *11* Lower Permian monzogranodiorites with subvolcanic characters; *12* Lower Permian calc-alkaline volcanics; *13* Lower Permian microdiorite and microgranitic dykes; *14* Upper Permian alkaline volcanics of the Cintu Complex; *15* alkaline and hyperalkaline granites involved within Upper Permian ring complexes; *16* alkaline ring dykes; *17* alkaline microgranite and rhyolite sills; *18* dolerites

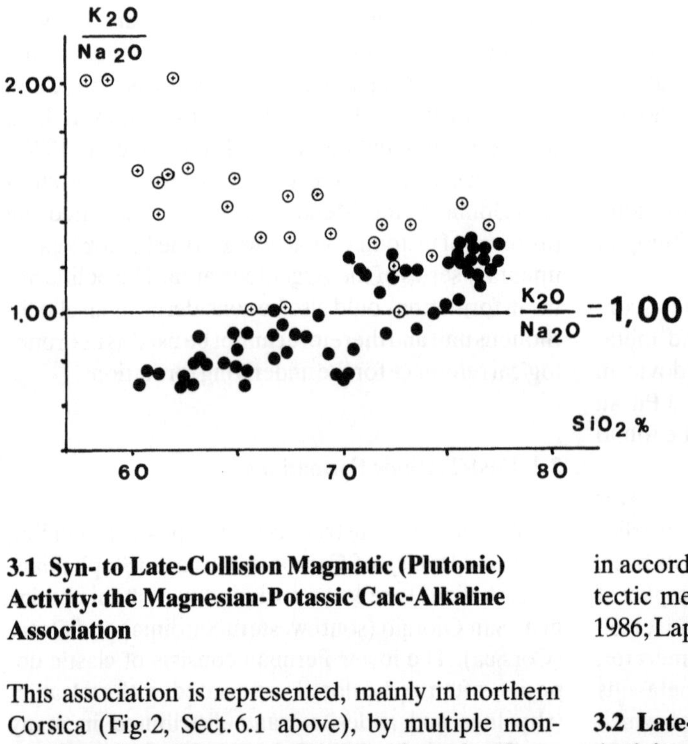

Fig. 6. Geochemical characters of the two main syn- to post-collision magmatic associations. *Open circles* magnesian-potassic calc-alkaline association; *black dots* calc-alkaline (s. s.) association

3.1 Syn- to Late-Collision Magmatic (Plutonic) Activity: the Magnesian-Potassic Calc-Alkaline Association

This association is represented, mainly in northern Corsica (Fig. 2, Sect. 6.1 above), by multiple monzonitic and syenitic-monzonitic intrusions closely associated to basic rocks. This juxtaposition is regarded as incompleted magma mixing (Orsini 1976, 1980; Rossi 1986). Geochemical data are consistent with the generation of the granites by partial melting of a granulitic crust (Cocherie 1984). The related basic rocks are derived from the hybridation of mantle-derived basaltic magma. Moreover, mineralogical data suggest that the granites have been produced under high-temperature and low-P_{H_2O} conditions (Rossi 1986). These early intrusions, emplaced around 350 (+36, −18) Ma (Rossi et al. 1988), show well-developed internal magmatic structures (mostly meridian) in accordance with the regional structures of the anatectic metamorphic basement (Fig. 5; Laporte et al. 1986; Laporte 1987).

3.2 Late- to Post-Collision Magmatic (Plutonic) Activity: the Calc-Alkaline (s. s.) Association

This magmatic association consists, over the whole Corsica-Sardinian segment, of multiple and coalescent intrusions (Marre and Rossi 1980) which evolve from tonalites and granodiorites to more acid compositions (Orsini 1979, 1980; Ghezzo and Orsini 1982). Basic rocks occur as enclaves, septa, or small massifs synchronous or related in comparison with granitic rocks. This calc-alkaline (s. s.) association differs from the previous one in modal compositions (Fig. 7), mineralogical and geochemical characters (Fig. 6) and also the nature of the basic rocks (Orsini 1976, 1980).

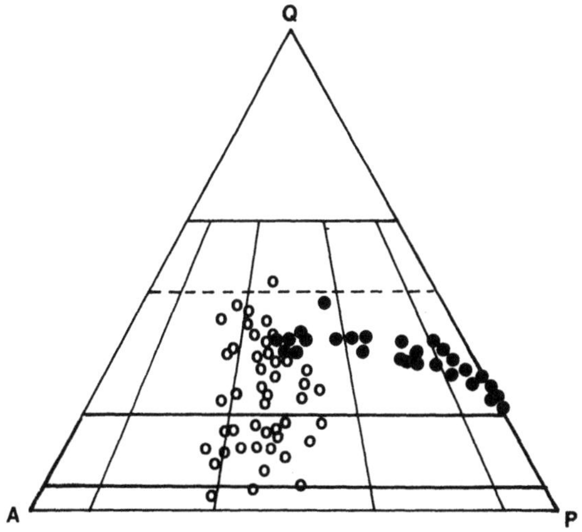

Fig. 7. Contrasting modal compositions of Corsican granites in the Streckeisen (1974) diagram. *Open circles* magnesian-potassic calc-alkaline association; *black dots* calc-alkaline (s. s.) association

The calc-alkaline granitic rocks have been generated, under lower temperature and higher oxidation conditions than the magnesian-potassic ones (Rossi 1986), around 313±9 Ma in Corsica (Cocherie 1984) and 302±5 in Sardinia (Del Moro et al. 1975). The plutons enhance a NW-SE-trending magmatic structure, indicating their emplacement within a deformed metamorphic basement at the end of its uplift (Legall et al. 1985; Orsini and Fernandez 1987). Geochemical investigations (Cocherie 1984) indicate a crustal origin, under hydrous conditions, for monzonitic rocks, while the basic rocks result from fractional crystallization of aluminous basalts, themselves originated by partial melting of spinel-lherzolite. Intermediate rocks are regarded as the result of acid-basic magma mixing (Cocherie 1984; Cocirta et al. 1989).

3.3 Extensional Magmatism (Subvolcanic to Volcanic Activity)

3.3.1 The 300-280-Ma Period

In the Corsican Sardinian segment, hololeucocratic biotite-muscovite-garnet (almandine-spessartine)-bearing granites were emplaced at around 300-280 Ma (Del Moro et al. 1975; Maluski 1977; Cocherie 1984). These granites are related to the fractional crystallization of a monzogranitic magma produced by crustal partial melting under hydrous conditions (Cocherie 1984). The subvolcanic character of the intrusions (fine-grained rocks, margins, dyke networks, microlithic facies), and their structural arrangement as NE-SW elongated ellipses indicate an emplacement at a shallow structural level. This magmatic event testifies to the presence of a radical modification, on the scale of the Corsican Sardinian block, of a tectonic regime which can be regarded as a precursor to the Permian distension.

3.3.2 The Autunian Period

An Autunian calc-alkaline magmatic event (Vellutini 1977; Fontana et al. 1982) has generated dyke swarms and volcanic rocks, mainly ignimbritic rhyolites, andesites, and scarce dacites. These volcanics are only locally preserved within collapsed zones mainly located near the center of alkaline ring complexes. The dykes, mainly oriented NE-SW, cross-cut all the previous intrusions. In northwestern Corsica, two kinds of dykes can be distinguished (Fumey-Humbert et al. 1986; Fumey-Humbert 1989): peraluminous microgranites and a calc-alkaline sequence involving microdiorites, microtonalites, and microgranites. The chemical and mineralogical affinities between the volcanic rocks and the dykes indicate that the latter could have acted as magma supply pipes for fissure volcanism during crustal extension.

3.3.3 The Upper Permian Period

This period is characterized by the occurrence of alkaline magmatism, related to extension of the lithosphere. This is represented, mainly in Corsica, by annular or linear (NE-SW trending) volcano-plutonic structures (ring-like complexes). Differences in the erosion levels expose the lower parts (Tolla example, Bonin 1983), the intermediate parts (Cintu example, Vellutini 1977) or the upper parts (Scandola-Seninu example, Brisset and Cocheme 1976) of the ring complexes to be observed. Both aluminous and hypo-aluminous sequences have been recognized (Vellutini 1977; Bonin 1980). The Scandola-Seninu basic volcanics define a transitional volcanism (Vellutini 1977), while the Ota and Pelosa gabbros have been related both to the alkaline series (Bonin 1980; Platevoet 1985) or the calc-alkaline magmatic suites (Cocherie 1984; Cocherie et al. 1984). With regard to the origin of this alkaline magmatism, both mantle (Bonin 1980) and crustal (Rossi and Cocherie 1987) sources have been envisaged.

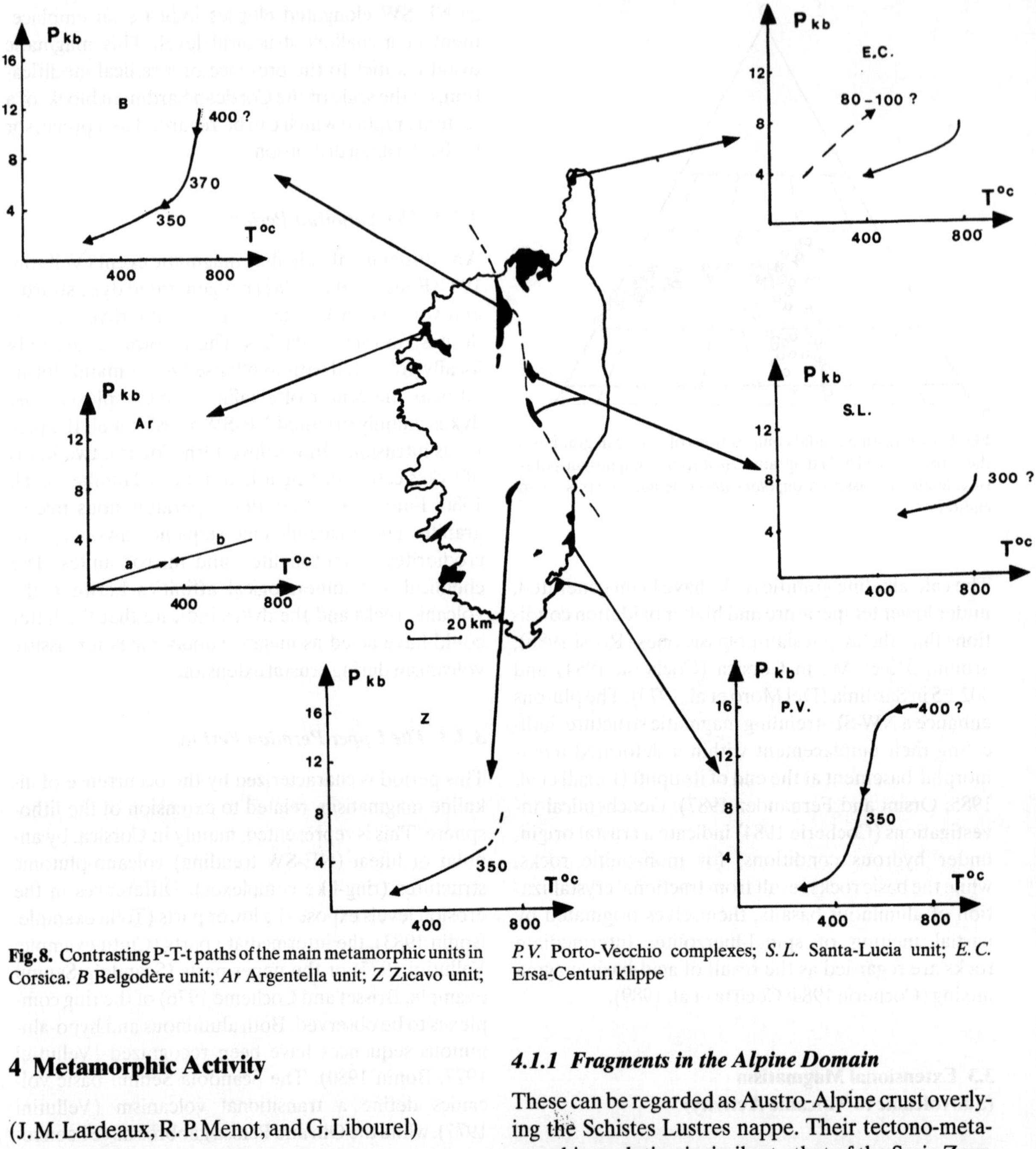

Fig. 8. Contrasting P-T-t paths of the main metamorphic units in Corsica. *B* Belgodère unit; *Ar* Argentella unit; *Z* Zicavo unit; *P.V.* Porto-Vecchio complexes; *S.L.* Santa-Lucia unit; *E.C.* Ersa-Centuri klippe

4 Metamorphic Activity

(J.M. Lardeaux, R.P. Menot, and G. Libourel)

4.1 Paleozoic Metamorphism in Corsica

The Corsican pre-Carboniferous basement occurs as dismembered outcrops within both eastern and western Corsican domains (Fig. 8). These fragments are located in the Alpine domain (Ersa-Centuri klippe), in the Variscan zone (Belgodere, Argentella, Zicavo, Porto Vecchio) and finally in the basement rocks at the Alpine front (Fig. 8, Santa Lucia unit).

4.1.1 Fragments in the Alpine Domain

These can be regarded as Austro-Alpine crust overlying the Schistes Lustres nappe. Their tectono-metamorphic evolution is similar to that of the Sesia Zone (Italian Western Alps; Compagnoni et al. 1977; Lardeaux et al. 1982; Rubie 1984) involving a late Paleozoic retrogression and a subsequent Alpine prograde evolution (Harris 1984).

4.1.2 Fragments in the Variscan Domain

As illustrated in Fig. 8, the different basement rocks display distinct and diachronic tectonometamorphic evolution. Evidence of eclogitic and HP granulitic parageneses of Mid-Paleozoic (?) age are known in

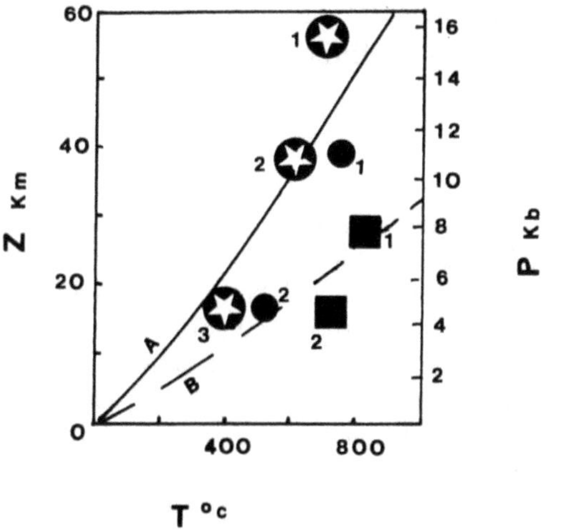

Fig. 9. The thermal setting of Paleozoic metamorphism in Corsica. *A* Geotherm of a stable crust; *B* maximally relaxed geotherm. (After England and Thompson 1984). *Stars* Successive metamorphic stages of the Porto-Vecchio unit; *black dots* metamorphic evolution of the Belgodère-Zicavo units; *black squares* metamorphic evolution of the Santa-Lucia unit

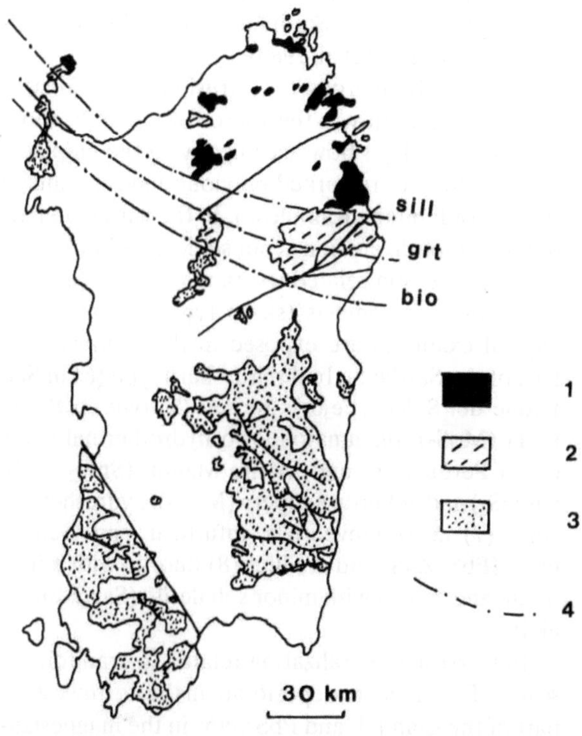

Fig. 10. Metamorphic zonation in Sardinia. (After Carmignani et al. 1981, 1982; Ghezzo and Orsini 1982). *1* Migmatites including relict high-pressure assemblages; *2* amphibolite facies rocks; *3* anchi- to epizonal rocks; *4* boundaries between the above metamorphic zones

the Belgodere and Porto-Vecchio areas (Libourel 1985; Palagi et al. 1985). Subsequent events, with a migmatitic and anatectic stage, are controlled by isobaric cooling followed by decompression (Porto-Vecchio) or by isothermal decompression followed by a decrease of both pressure and temperature (Belgodere). The retrograde path of the Zicavo fragment, involving the generation of cordierite-bearing melts, is comparable to that of Belgodere but, as yet, no earlier HP assemblages have been described (Vezat 1986, 1988).

In the Argentella region, a more complex situation has been deciphered, with a nappe pile (Gonnord and Ménot 1993) in which two different units are tectonically juxtaposed. They show different metamorphic evolution. A prograde recrystallization under epizonal conditions is given by the Silurian schists of the Argentella unit. On the other hand, the staurolite-andalusite gneisses of the Argentella basement are retrogressed under epizonal conditions.

From the thermal point of view (Fig. 9), the whole metamorphic history resides in the thermal setting of a subduction-collision regime. Indeed, the temperature of the HP crystallization is less than those expected for a geotherm in a stable crust and during uplift the retrograde temperatures never exceed the maximum value of the relaxed geotherm even with low erosion rates.

4.1.3 Fragments at the Alpine Front

From available petrologic data (Libourel 1988), the Santa Lucia unit is characterized by a low pressure granulitic episode of probably Lower to Mid-Carboniferous age. This metamorphic evolution cannot be explained in a collision context (Fig. 9) and reflects extensional processes at the end of the orogenic evolution of the Variscan chain.

In conclusion the diversity of the P-T-t evolutions suggests that the pre-Carboniferous basement of Corsica results from the tectonic accretion of various segments of the Hercynian orogenic belt which are representative of strikingly different tectonic settings. Such a scheme does not involve some basement pieces (Agriates, Roches Brunes) which are, as yet, not well-known.

4.2 Paleozoic Metamorphism in Sardinia

In Sardinia a progressive metamorphic zonation, with a southwards decreasing gradient, is commonly proposed (Fig. 10; Carmignani et al. 1979, 1982). Following this point of view, such a metamorphic evolution is

Fig. 11. Main mineralization in the Corsica-Sardinian block. Numbers *(1* to *13)* refer to the text

reported to reflect the intracontinental subduction of a crustal segment during Variscan times (Carmignani et al. 1981). However, such a simple history seems to be contradictory with the occurrence of well-developed nappes at different crustal levels and with the polyphase character of the tectono-metamorphic evolution (Miller et al. 1976; Ghezzo et al. 1979; Franceschelli et al. 1982). Furthermore, the thickness of each metamorphic zone is too large and cannot be simply related to any reasonable value of thermal gradient.

Recent investigations (Cappelli 1991; Carmignani et al. 1991; Ricci 1991) indicate that after thickening the Sardinian metamorphic units followed contrasted P-T-t paths. In all cases, the basement exhibits metamorphic evidence for continental thickening and uplift (i.e., high pressure stage followed by a decompressional evolution). The post-thickening exhumation of the different metamorphic zones was probably controlled by contrasted processes like isostasy, tectonic onroofing, and crustal thinning.

5 Paleozoic Metallogeny of the Corsica-Sardinian Block

(Ph. Rossi)

In the lower Cambrian, in the Iglesias area of southwestern Sardinia (1, in Fig. 11) concentrations of sphalerite-pyrite-galena and barite of stratabound type are related to a carbonate platform. Deposition was related to submarine volcanic activity, but the present shape of the deposits may result from reconcentration by karstic phenomena or magmatic activity (Boni 1982).

In the Silurian of central eastern Sardinia, the establishment of "black shale" reducing environments associated with submarine volcanic events, led to the deposition of polymetallic ores; mixed sulfide complexes in Correboi and Funtana Raminosa (2); antimony and scheelite deposits in South Ballao and Villasalto (3); Argentiferous veins of Sarrabus (4) have Ag, ZnS, PbS and CaF2. Oolitic iron deposits (5) in Nurra (northwestern Sardinia) should also be mentioned (Garbarino et al. 1980).

In the Carboniferous, the intrusion of the Corsica-Sardinia batholith was the source of mineralization. In places, the heat flow accompanying the emplacement of granites reworked previous deposits (mixed sulfides of Funtana Raminosa). In the Sardinian part of the batholith, erosion is not so deep as in Corsica, so mineralization related to granitic apexes (greisen types) or with contacts (skarn types) is preserved. Typical examples are exposed in the southwestern part of the Sardinian batholith: skarn type (6) in San Leone del Sulcis (Fe_3O_4); greisen (7) type in Perda Lada (MoS_2), pneumatolitic to hydrothermal veins (7) in Perdu Cara and Monte Mannu (SnO_2, ZnS, $CuFeS_2$) and at Perdu e Pipera (MoS_2); hydrothermal veins (7) in Montevecchio, Infurtosu, and Gennamare (PbS; ZnS) and at Silius (8) fluorite with minor barite and galena with minor sphalerite (Salvadori et al. 1986).

In Corsica, mineralization related to granitoids is scarce. The main ore deposits are in the northwestern part of the island. F and PbS veins in the magnesian-potassic calc-alkaline granitoids have been recognized at Monticello (8). In Argentella (10), south of Calvi, veins of Ag-bearing galena cross-cutting magnesian-potassic calc-alkaline leucocratic granite were mined in the 1890 s. A strong whole-rock geochemical anomaly in the surrounding leucocratic granites is due to very fine disseminated sulfides (argento-jarosite) deposited during a cataclastic episode. At Prunelli (11), East of Porto, Cu, Fe, Zn, Mo, and Sn mineralization is located in calc-alkaline granitoids and associated with widely exposed silicic and potassic metasomatic processes which may be related to Permian volcanic-plutonic activity. Large leucocratic calc-alkaline monzogranite intrusions marked by Be halos in stream sediments in places show Sn and/or W anomalies. Cassiterite-bearing quartz veins associated with greisens are known in places near Pianottoli (12) and at Punta Telica (north of Porto Vecchio). Alluvial concentrations of REE minerals in the NE of Porto area (13) result from the erosion of the hyperalkaline granites of Evisa.

Acknowledgments. This chapter has benefited from discussions with E. Egal, M. I. Spalla, and H. L. M. Van Roermund and helpful reviews of F. P. Sassi, C. B. Vai, and an anonymous referee. We are grateful to C. Uberty for typing the manuscript.

References

Arthaud F (1970) Etude tectonique et microtectonique comparée de deux domaines hercyniens: les nappes de la Montagne Noire (France) et l'anticlinorium de l'Iglesiente (Sardaigne). Thèse Doc Sci, Montpellier, USTL Publ Sér Geol Struct, 1, 175 pp

Arthaud F, Matte P (1966) Contribution à l'étude des tectoniques superposées dans la chaîne hercynienne: étude microtectonique des séries métamorphiques du Massif des Maures (Var). CR Acad Sci Paris 262; 436–439

Arthaud F, Matte P (1977) Synthèse provisoire sur l'évolution tectonique et les raccords entre les segments hercyniens situés autour du bassin nord-Baléare (Sud de la France, Espagne, bloc corso-sarde). Coll Int CNRS, Rennes, 243, pp 497–513

Arthaud F, Sauniac S (1981) Une coupe synthétique à travers la chaîne varisque de Sardaigne. Commentaires sur l'évolution tectono-métamorphique. Bull Soc Géol Fr (7) XXIII 5: 535–539

Barca S, Del Rio M, Pittau Demelia P (1981) Acritarchs in the "Arenarie di San Vito" of south-east Sardinia: stratigraphical and geological implications. Bull Soc Geol Ital 100: 369–375

Baudelot S, Doubinger J, Durand-Delga M, Vellutini P (1976) Caractéristiques et âges de cinq cycles paléozoïques du Nord-Ouest de la Corse. Bull Soc Géol Fr (7) XVIII 5: 1221–1228

Bayer R, Le Mouel JL, Le Pichon X (1973) Magmatic anomaly pattern in the western Mediterranean. Earth Planet Sci Lett 19: 168–176

Boni M (1982) Paleocarsismo e mineralizzazioni nella Sardegna Sud-Occidentale. Guide Geologiche Regionali. Soc Geol Ital, Cagliari, pp 83–85

Bonin B (1980) Les complexes alcalins acides anorogéniques: l'exemple de la Corse. Thèse Doc Sci, Paris VI, 779 pp

Bonin B (1983) Permian events in Corsica: vanishing Variscan orogenesis or new geodynamical history. IGCP 5, Newsletter 5: 217–219

Brisset F, Cocheme JJ (1976) Etude géologique des presqu'îles de Scandola et Seninu (Corse). Thèse 3è Cycle, Univ. Aix-Marseille III, 64 pp

Calvino F (1961) Lineamenti strutturali del Sarrabus-Gerrei (Sardegna sud-orientale). Bull Serv Geol Ital 81: 489–556

Cappelli B (1991) LP-HT metamorphic core complexes in the Nappe Zone of the Hercynian chain in Sardinia. (Abstr.) In: Cappelli B, Liotta D (eds) Geologia del basamento Italiano. Accademia dei Fisiocritici, Siena, p 118 (Abstr)

Carmignani L, Franceschelli M, Pertusati PC, Ricci CA (1979) Evoluzione tettonicometamorfica del basamento della Nurra (Sardegna NW). Mem Soc Geol Ital 20: 57–84

Carmignani L, Cocozza T, Minzoni N, Pertusati PC (1981) Structural and paleogeographic lineaments of the Variscan Cycle in Sardinia. In: Zwart, Dornsiepen (eds) Geologie en Mijnbown 60: 171–181

Carmignani L, Cocozza T, Ghezzo C, Pertusati PC, Ricci CA (1982) Lineamenti del basamento Sardo. Guida alla Geologia del Paleozoico Sardo. Guide Geologiche Regionali. Soc Geol Ital, Cagliari, pp 11–23

Carmignani L, Cocozza T, Gandin A, Pertusati PC (1986) The geology of Iglesiente. In: Carmignani L, Cocozza T, Ghezzo C, Pertusati PC, Ricci CA (eds) guidebook to the excursion on the Paleozoic basement of Sardinia. IGCP, Project n° 5, Final Meeting, Sardinia. IGCP, Pisa, pp 31–41

Carmignani L, Barca S, Cappelli B, Di Pisa A, Gattiglio M, Oggiano G, Pertusati PC (1991) A tentative geodynamic model for the hercynian basement of Sardinia. In: Cappelli B, Liotta D (eds) Geologia del basamento Italiano. Accademia dei Fisiocritici, Siena, pp 23–28

Caron JM (1977) Lithostratigraphie et tectonique des Schistes lustrés dans les Alpes cottiennes septentrionales et en Corse orientale. Sci Géol Strasbourg, Mém. 48, 326 pp

Caron JM, Kienast JR, Triboulet C (1981) High pressure-low temperature metamorphism and polyphase deformation at San Andrea di Cotone (Eastern Corsica, France). Tectonophysics 78: 419–453

Cocherie A (1984) Interaction manteau-croûte: son rôle dans la genèse d'associations plutoniques calco-alcalines, contraintes géochimiques (éléments en traces et isotopes du strontium et de l'oxygène). Thèse Doc Sci, Rennes I, 245 pp

Cocherie A, Rossi Ph, Le Bel L (1984) The Variscan calc-alkaline plutonism of Western Corsica: Mineralogy and major and traces element geochemistry. Phys Earth Planet Interiors 35: 145–178

Cocirta C, Orsini JB, Coulon C (1989) Un exemple de mélange de magmas en contexte plutonique: les enclaves des tonalites-granodiorites du massif de Bormes (Sardaigne Septentrionale). Can J Earth Sci 26: 1264–1281

Cocozza T (1974) Structural pattern of Sardinia. In: Structural pattern of Italy. Quaderni de la ricerca scientifica 90. CNR, Roma, pp 607–642

Cocozza T (1979) The Cambrian of Sardinia. Mem Soc Geol Ital 20: 163–187

Cocozza T, Leone F (1977) Sintesi della successione stratigrafica paleozoica della Sardegna sud-occidentale. Escursione in Sardegna 1977: risultati e commenti (a.c. di G.B. VAI). GLP 2: 15–23 (Suppl)

Cocozza T, Conti L, Lombardi G, Traversa G (1972) Le Précambrien en Sardaigne. 24e Int Geological Congress, Canadian Geological Society, Montreal, p 7

Cocozza T, Jacobacci A, Nardi R, Salvadori I (1974) Schema stratigrafico-strutturale del massicio Sardo-Corso e minerogenesi della Sardegna. Mem Soc Geol Ital 13: 85–186

Compagnoni R, Dal Piaz GV, Hunziker JC, Gosso G, Lombardo B, Williams PF (1977) The Sesia Lanzo zone, a slice of continental crust with high pressure-low temperature assemblages in the Western Italian Alps. Rend Soc Ital Mineral Petrol 33: 335–374

Del Moro A, Di Simplicio P, Ghezzo C, Guasparri G, Rita F, Sabatini G (1975) Radiometric data and intrusive sequence in the Sardinian Batholith. Neues Jahrb Miner Abh 126 (1): 28–44

Durand-Delga M (1978) Corse. Guides géologiques régionaux. Masson, Paris, 208 pp

Edel JB, Montigny R, Thuizat R (1981) Late Paleozoic rotations of Corsica and Sardinia. New evidence from paleomagnetic and K-Ar studies. Tectonophysics 79: 201–223

Egger A, Demartin M, Ansorge J, Banda E, Maistrella M (1988) The gross structure of the crust under Corsica and Sardinia. Tectonophysics 150: 363–389

England PC, Thompson AB (1984) Pressure-Temperature-Time paths of regional metamorphism. I. Transfer during the evolution of regions of thickened continental crust. J Petrol 25 (4): 894–928

Ferrara G, Ricci CA, Rita F (1978) Isotopic ages and tectono-metamorphic history of the metamorphic basement of north-eastern Sardinia. Contrib Mineral Petrol 68: 99–106

Fontana D, Gelmini R, Lombardi G (1982) Le successioni sedimentarie e vulcaniche carbonifere e permo-triassiche della Sardegna. In: Soc Geol Ital (ed) Guida alla Geologia del Paleozoico Sardo. 1 Cent Soc Geol Ital, Cagliari, pp 183–192

Franceschelli M, Memmi I, Ricci CA (1982) Zoneografia metamorfica della Sardegna settentrionale. In: Guida alla Geologia del Paleozoico Sardo. Guide Geologiche Regionali. Soc Geol Ital, Cagliari, pp 137–149

Fumey-Humbert F (1989) Le magmatisme hypovolcanique tardi hercynien. L'exemple du réseau filonien de Balagne (Nord-Ouest de la Corse). Etude pétrographique, minéralogique et géochimique. Thèse Doc Sci, Univ. St Etienne, 237 pp

Fumey-Humbert F, Menot RP, Orsini JB (1986) Mise en évidence d'un magmatisme filonien original, d'âge Permien inférieur en Balagne (Corse du Nord-Ouest). CR Acad Sci Paris II 303: 1105–1110

Garbarino C, Maccioni L, Minzoni N, Padalino G, Tocco S, Violo M (1980) The strata-bound copper-lend-zinc – mineralization of the Sardinian microplate – Italy. Symp European Copper Deposits, Belgrade, pp 251–261

Ghezzo C, Orsini JB (1982) Lineamenti strutturali e composizionali del batolite ercinico Sardo-Corso in Sardegna. In: Guida alla Geologia del Paleozoico Sardo. Guide Geologiche Regionali. Soc Geol Ital, Cagliari, pp 165–181

Ghezzo C, Memmi I, Ricci CA (1979) Un evento granulitico nel basamento metamorfico della Sardegna nord-orientale. Mem Soc Geol Ital 20: 23–38

Ghezzo C, Memmi I, Ricci CA (1982) Le granuliti e le eclogiti della Sardegna nord-orientale. In: Guida alla Geologia del Paleozoico Sardo. Guide Geologiche Regionali. Soc Geol Ital, Cagliari, pp 151–156

Gonnord H, Ménot RP (1993) Sur la structure du socle antébatholique de l'Argentella (Corse Septentrionale): allochtonie probable de certains termes et implications. Bull Soc Géol Fr (in press)

Harris LB (1984) Déformations et déplacements dans la Chaîne Alpine: l'exemple des Schistes Lustrés du Cap Corse. Thèse Doc Sci, Rennes, 307 pp

Harris LB (1985) Direction changes in thrusting of the Schistes Lustrés in Alpine Corsica. Tectonophysics 120: 37–56

Hirn A, Sapin M (1976) La croûte terrestre sous la Corse: données sismiques. Bull Soc Géol Fr 7: 1195–1199

Hirn A, Steinmetz L, Sapin M (1977) A long range seismic profile in the Western Mediterranean basin: structure of the Upper Mantle. Ann Géophys 33: 373–384

Junker B, Schneider HH (1979) L'Infracambriano della Sardegna sud-occidentale. Mem Soc Geol Ital 20: 462

Laporte D (1987) Un exemple d'intrusion syntectonique: l'intrusion d'Ile-Rousse, Corse du Nord-Ouest. Etude pétrographique, minéralogique et géochimique. Analyse structurale. Thèse Doc Sci, St Etienne University, 422 pp

Laporte D, Fumey-Humbert F, Michon G, Orsini JB, Fernandez A (1986) Sur le caractère syntectonique des granitoïdes calco-alcalins magnésiopotassiques de Balagne (Corse du Nord-Ouest). CR Acad Sci Paris 303: 1321–1326

Lardeaux JM, Gosso G, Kienast JR, Lombardo B (1982) Relations entre le métamorphisme et la déformation dans la zone Sesia Lanzo (Alpes Occidentales) et le problème de l'éclogitisation de la croûte continentale. Bull Soc Géol Fr 26: 793–800

Legall B, Orsini JB, Fernandez A, Ghezzo C (1985) Signification tectonique des structures internes du massif granitique de Tempio (Sardaigne septentrionale). CR Acad Sci Paris 300 II: 1013–1018

Libourel G (1985) Le complexe de Santa Lucia di Mercurio (Corse). Ultramafites mantelliques, intrusion basique stratifiée, paragneiss granulitiques. Un équivalent possible des complexes de la zone d'Ivrée. Thèse Doc Sci, Toulouse, 461 pp

Libourel G (1988) Le complexe de Santa Lucia di Mercurio (Corse): un nouveau jalon de la base de la croûte varisque en Méditerranée Occidentale. CR Acad Sci Paris 307 II: 1067–1073

Lucazeau F, Mailhé D (1986) Heat flow, heat production and fission track data from the Hercynian basement around the Provençal Basin (Western Mediterranean). Tectonophysics 128: 335–356

Maluski H (1977) Application de la méthode 39 Ar/40 Ar aux minéraux des roches cristallines perturbées par les évènements thermiques et tectoniques en Corse. Thèse Doc Sci, Montpellier

Marre J, Rossi P (1980) Extension et antériorité du plutonisme subalcalin potassique dans l'ensemble des granitoïdes varisques de Corse. CR Acad Sci Paris 291: 1015–1018

Mattauer M, Faure M, Malavieille J (1981) Transverse lineation and large-scale structures related to Alpine obduction in Corsica. J Struct Geol 3 (4): 401–409

Memmi I, Barca S, Carmignani L, Cocozza T, Franceschelli M, Gattiglio M, Ghezzo C, Minzoni N, Naud G, Pertusati PC, Ricci C (1982) Il magmatismo pre-ercinico della Sardegna. In: Guida alla Geologia del Paleozoico Sardo. Guide Geologiche Regionali. Soc Geol Ital, Cagliari, pp 157–164

Miller L, Sassi FP, Armari G (1976) On the occurrence of altered eclogite rocks in north-eastern Sardinia and their implications. Neues Jahrb Geol Paläont Mh 1976, 11: 683–689

Morelli C, Giese P, Carrozzo MT, Colombi B, Guerra I, Hirn A, Letz H, Nicolich R, Prodehl C, Reichert C, Rower P, Sapin M, Scarascia S, Wigger P (1977) Crustal and upper mantle structure of the northern Apennines, the Ligurian Sea and Corsica, derived from seismic and gravimetric data. Boll Geofis Teor Appl 75/76: 199–260

Nardi R, Puccinelli A, Verani M (1979) Carta geologica della Balagne "sedimentaria" (Corsica). Alla scala 1:25000 e note illustrative. Bull Soc Geol Ital 97: 11–30

Naud G (1979) Tentative de synthèse sur l'évolution géodynamique de la Sardaigne anté-permienne. Mem Soc Geol Ital 20: 85–96

Naud G (1981) Confirmation de l'existence de la discordance angulaire anté-ordovicienne dans le Sarrabus (Sardaigne sud-orientale): conséquences géodynamiques. CR Acad Sci Paris II 292: 1153–1156

Naud G (1982) Un exemple de structuration des formations paléozoïques de Sardaigne orientale: les unités chevauchantes de la Basse Vallée du Flamendosa (Sardaigne, Italie). 9e Réun. Ann. Sci. Terre, Soc Géol Fr, Paris, p 469

Naud G, Pittau Demelia P (1985) Première découverte d'Acritarches du Cambrien moyen à supérieur basal et du Trémadocien-Arénigien dans la basse vallée du Flumendosa: mise en évidence d'un nouveau témoin de la phase sarde en Sardaigne orientale. In: Evol. strati., tett., metam. e magm. del Paleoz. italiano. Accademia Dei Fisiocritici, Siena, pp 39–40

Naud G, Tempier C (1977) Schéma stratigraphique et tectonique des formations paléozoïques de Sardaigne sud-orientale. 5 e Réun. Ann. Sci. Terre, Soc Géol Fr, Rennes, p 348

Olivieri R (1969) Conodonti e zonatura del Devoniano superiore e riconoscimento di Carbonifero inferiore nei Calcari di Corona Mizziu (Gerrei-Sardegna). Bull Soc Geol Ital 8: 63–152

Orsini JB (1976) Les granitoïdes hercyniens corsosardes: mise en évidence de deux associations magmatiques. Bull Soc Géol Fr (7) 18: 1203–1206

Orsini JB (1979) Existence de trois associations magmatiques dans les granitoïdes post-viséen moyen (groupe des granodiorites-monzogranites) de la chaîne varisque française. CR Acad Sci Paris 289: 981–984

Orsini JB (1980) Le batholite corso-sarde: un exemple de batholite hercynien (structure, composition, organisation d'ensemble). Sa place dans la chaîne varisque de l'Europe moyenne. Thèse Doc Sci, Aix-Marseille III, 390 pp

Orsini JB, Fernandez A (1987) Signification de la discordance structurale entre fluidalité magmatique et zonalité pétrographique dans les intrusions de granitoïdes: l'exemple de l'intrusion de Budduso (Sardaigne septentrionale). CR Acad Sci Paris 304 II: 993–996

Orsini JB, Coulon C, Cocozza T (1980) Dérive cénozoïque de la Corse et de la Sardaigne et ses marqueurs géologiques. Geol Mijnbouw 59: 385–396

Palagi P, Laporte D, Lardeaux JM, Menot RP, Orsini JB (1985) Identification d'un complexe leptyno-amphibolique au sein des "gneiss de Belgodère" (Corse occidentale). CR Acad Sci Paris 301 II: 1047–1052

Platevoet J (1985) Etude pétrologique d'une association acide-basique dans le complexe annulaire anorogénique de Porto (Corse). Thèse 3 ème cycle, Univ. Paris VI, 182 pp

Poll JJK, Zwart HJ (1964) On the tectonics of the Sulcis area, S. Sardinia. Geol Mijnbouw 43 (4): 144–146

Rossi P (1986) Organisation et genèse d'un grand batholite orogénique: le batholite calco-alcalin de la Corse. Thèse Doc Sci, Toulouse, Doc Bur Rech Geol Min (Orléans) 107, 292 pp

Rossi P, Cocherie A (1987) Comportement des éléments en trace dans les granites lors de la transition calco-alcalin/alcalin dans le batholite permo-carbonifère de Corse. Principaux résultats scientifiques et techniques pour 1986. Doc Bur Rech Geol Min, Orleans, pp 90–91

Rossi P, Calvez JY, Cocherie A (1988) Age varisque précoce du plutonisme magnésio-potassique en Corse occidentale: conséquences géodynamiques. CR Acad Sci Paris 307, 2: 1541–1547

Rubie DC (1984) A thermal-tectonic model for high-pressure metamorphism and deformation in the Sesia zone, Western Alps. J Geol 92: 21–36

Salvadori I, Uras I, Zuffardi P (1986) The paleozoic metallogenesis of Sardinia in relation to its geological evolution. In: Carmignani L, Cocozza T, Ghezzo C, Pertusati PC, Ricci CA (eds) Guide-Book to the Excursion on the paleozoic basement of Sardinia. Newsletter, Spec Issue IGCP Project N° 5

Streckeisen A (1974) Classification and nomenclature of plutonic rocks. Geol Rundsch 68: 773–786

Vai CB, Cocozza T (1986) Tentative schematic zonation of the hercynian chain in Italy. Bull Soc Géol Fr (8) 2: 95–114

Vellutini P (1977) Le magmatisme permien de la Corse du Nord-Ouest; son extension en Méditerranée Occidentale. Thèse Doc Sci, Aix-Marseille III, 317 pp

Vezat R (1986) Le batholite calco-alcalin de Corse; les formations métamorphiques calédono-varisques de Zicavo; la mise en place du batholite calco-alcalin. Thèse Doc Sci, Toulouse University, 366 pp

Vezat R (1988) Les formations métamorphiques de Zicavo (Corse centrale) et leur signification dans la cadre de l'orogenèse varisque en Méditerranée. CR Acad Sci Paris II 306: 725–729

Westphal M, Bardon C, Bossert A, Hamzeh R (1973) A computer fit of Corsica and Sardinia against Southern France. Earth Planet Sci Lett 18: 137–140

Westphal M, Orsini JB, Vellutini P (1976) Le micro-continent corso-sarde, sa position initiale: données paléomagnétiques et raccords géologiques. Tectonophysics 30: 141–157

Part V

The French Paleozoic Terranes

A. Piqué, S. Bogdanoff, J.-M. Quénardel, J. Rolet, and D. Santallier

In this Part we review the stratigraphic, magmatic, metamorphic, structural, and geophysical constraints on the evolution of the entire Paleozoic belt of France in order to build a tectonic model for the belt based on the terrane concept. The areas reviewed in this synthesis are shown in their present position and orientation in Fig. 1. However, their spatial and mutual relationships were certainly different during the Paleozoic period, before the thrust and transcurrent motions that occurred throughout the orogenic and late orogenic development of the belt.

1 Main Constraints

1.1 Stratigraphic, Sedimentological, and Paleontological Data

1.1.1 Platforms

These successions are characterized by shallow-water depositional environments indicated by orthoquartzites, sandstones, and reef limestones, with frequent unconformities and benthic faunas.

1. Cambrian to Silurian. During Cambrian to Silurian times, a northern platform (relative to present geographic coordinates) extended throughout the northern and central Armorican zone and probably continued eastward beneath the Paris Basin. A southern platform, represented by the Montagne Noire, possibly extended westwards under the Aquitaine basin (Paris, Chap. II.B.9, this Vol.). In both areas, all facies are those of an epicontinental shelf. The disconformities are traditionally assigned to a late Cambrian "Sardic" or a post-Arenigian pre-Devonian "Caledonian" phase. This event is generally thought to represent a compressive event. An alternative hypothesis, however, favors extensional deformation accompanied by tilting of parts of the epicontinental platform (Ballard et al. 1986).

2. Devonian. In the southern Montagne Noire, Early Paleozoic tilting is marked by a disconformity between Lower Devonian strata and Cambrian or Ordovician sequences in the westernmost units. The nature of the sedimentation during Late Ordovician and Silurian time is consequently uncertain, sediments of this age having been found only in Lower Carboniferous olistostromes in the southeasternmost nappe unit of the Montagne Noire. In the northern platform, from the Armorican Massif to the Vosges, and in the Génis area, within the Limousin, shallow water environments persisted into the Devonian, as indicated by rare mid-Devonian reef limestones. In the central Armorican massif, the terrigenous component increased during the Devonian, probably derived from elevating adjacent areas. In the Ardenne, an Early Paleozoic outer shelf and slope succession (see below) became an epicontinental shelf south of the Old Red Sandstone continent after the Caledonian orogeny. There, a northward transgressive sea deposited a thick Lower Devonian clastic sequence, lying disconformably upon Caledonian structures (Meilliez 1989) and followed by silty layers and reef limestones.

3. Early Carboniferous. During the Early Carboniferous, shallow water conditions persisted in most parts of the Ardenne which, at that time, was a wide carbonate platform separated by marginal oolitic deposits from southern orogenic domains that are inferred from comparisons with Germany (Franke and Engel 1986). In the Armorican and Vosges massifs, synorogenic basins, terrestrial in Western Brittany, developed within the former shelf, in which chaotic and turbiditic sediments were deposited. In the southern Montagne Noire, the starved basin that was established during the Late Devonian began to fill with turbiditic Culm deposits during Late Visean that were probably derived from the North (Franke and Engel 1986).

4. Late Carboniferous. From Namurian to Westphalian time, the Lower Carboniferous carbonate platform of the central and northern Ardenne be-

J. D. Keppie (Ed.): Pre-Mesozoic Geology in France and Related Areas

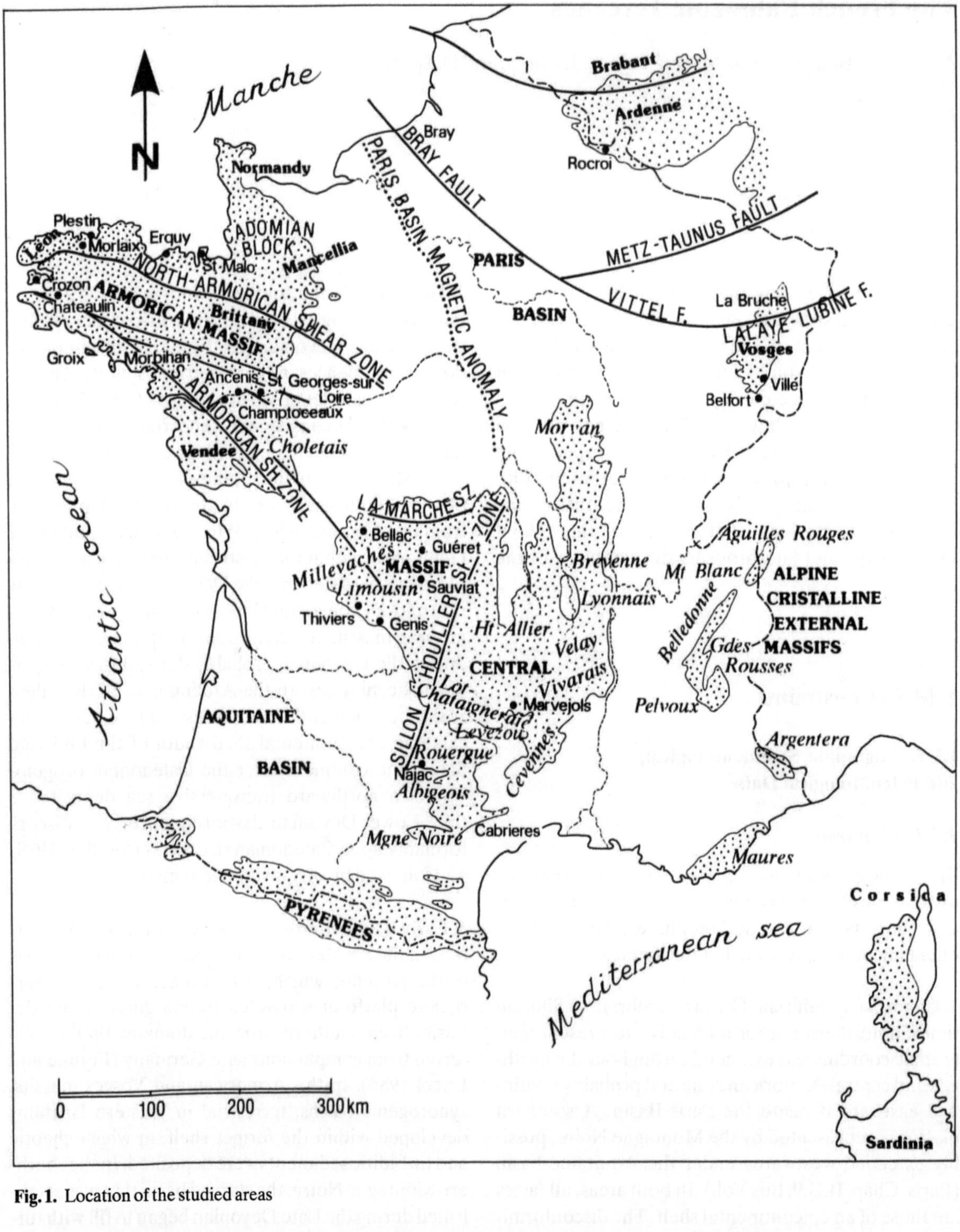

Fig. 1. Location of the studied areas

came a foreland basin where paralic sediments were deposited. The sedimentary deposits recorded in the Montagne Noire are characterized by increasingly proximal sequences. These sequences are broadly contemporaneous with the limnic series of the central zones of the orogen (see below).

1.1.2 Outer Platforms and Slopes

These successions are characterized by neritic facies. Limestones are generally absent and the arenites and pelitic rocks that characterize the sequences are often intercalated with quartzitic horizons and volcaniclastic rocks. The subsequent structural and metamorphic evolution of these successions usually prevents determination of their initial thickness and destroys their fossil record.

1. Cambrian to Silurian. In the southern Ardenne (Rocroi and Stavelot massifs), the Cambro-Ordovician sequence is clastic. Initially it comprised clays, silts and sandstones with thick quartzitic beds. Similar facies containing volcaniclastic sequences are encountered north of the Montagne Noire platform, in the Albigeois (Guérangé-Lozes and Alsac 1986), and in the southern and middle Cévennes. The Champtoceaux (South Armorican massif) and the Najac (southwestern Massif Central) sequences are now highly metamorphosed white schists, but initially comprised sediments that were dominated by siliceous and aluminous minerals. In the External Crystalline Massifs (ECM) of southeastern France, similar sequences of presumed Cambro-Ordovician age are present in the western part of the Maures massif (Gueirard et al. 1979) and in the Grandes Rousses massif (Giorgi et al. 1979), where dark schists are Early Paleozoic in age.

2. Devonian. In none of the above areas have sedimentary or metasedimentary rocks been described that could have resulted from the continuation of the same environmental conditions after Silurian times. Devonian deposits have either been subsequently removed or the areas no longer received sediments after the Silurian. The Ardenne domain is a notable exception where a Cambro-Ordovician deep marine region became a Devonian shelf after the Caledonian deformation and uplift.

1.1.3 Intracontinental Rift Sequences and Ocean-Related Complexes

Strongly affected by an intense and polyphase metamorphic evolution, these successions are characterized by clastic sedimentary facies, that could represent graywackes or more mature clastic rocks, mainly graywackes. The graywackes series show alternating fine and coarse materials and are possibly remnants of turbiditic sequences. The successions also contain minor amounts of limestones and important intervals of magmatic rocks. Since very few fossils have been preserved, these sequences are mainly dated radiometrically on the extrusive and intrusive magmatic series which also give valuable indications for the tectonic environment in which they have been emplaced. The magmatic data discussed below show that the metasedimentary sequences are spatially associated with either intracontinental rifts, or thinned passive margins or magmatic arcs.

During Early Paleozoic times, these usually high-grade meta-sedimentary sequences were deposited in the southern Armorican Massif (mainly in Morbihan and Vendée), in various parts of the Central Massif north of the Albigeois-Cévennes area and in the central Vosges. Other massifs such as the Alpine External Crystalline Massifs, the eastern Maures, and Corsica show similar sequences.

Following the Early-Middle Devonian, these areas either became terrestrial or were covered by a shallow sea that resulted in the deposition of unconformable strata such as mid-Devonian reef limestones at Ville Dé in Vendée and the Upper Devonian limestones at Belfort, south of the Vosges massif. These are locally followed (e.g., in the southern Vosges, Morvan, Aiguilles Rouges (ECM), etc.) by Lower Carboniferous clastic and volcaniclastic sequences and, finally, by Upper Carboniferous limnic beds.

1.1.4 Conclusion

1. During the Cambrian and Ordovician, present day France was characterized by various deposits and faunas that allow us to distinguish several tectonic domains, described here in present coordinates.

a) A northern platform, probably extending from the northern and central Armorican massif to the northern Vosges, was underlain by continental crust exposed in the remnants of the Cadomian orogen in the Northern Armorican massif. The southern Ardenne (Rocroi and Stavelot massifs) was the northern margin of this platform and was deformed and uplifted prior to the earliest Devonian. The southern limit of the platforms is the South Armorican Shear Zone (Jegouzo (see p. 195ff.)). It is concealed beneath the Paris Basin and it reappears in the Vosges massif along the Lalaye-Lubine fault (Piqué et al., p. 416 ff.). Its eastern extension within the Rheno-Hercynian and Saxo-Thuringian zones of Germany can be advocated for Cambrian and Ordovician times, since the Mid-German Crystalline Rise that separated the Saxo-Thuringian zone from the Rheno-Hercynian zone developed later, between 400–420 Ma and 370 Ma (Franke 1989 a).

Beyond its sedimentological facies, a characteristic feature of the northern platform is the presence of uppermost Ordovician pebbly mudstones that have been noticed from Normandy to Thuringia (Robardet and Doré 1988). These rocks are probably of glaciomarine origin and their occurrence in various areas is a strong argument for these areas to have been located at the same latitude during latest Ordovician times.

b) A central domain comprising the southern Armorican massif, the main part of the Massif Central and the central and southern Vosges. This domain is highly metamorphosed and better characterized by magmatic and geochemical than by sedimentological data. Meta-sedimentary rocks of Early Paleozoic age are indicative of relatively deep basins established upon either thin continental crust (rifts or continental margin) or on oceanic crust.
c) A southern platform represented in the Montagne Noire and probably extending beneath the Aquitaine basin. Its northern margin, defined by outer shelf or slope deposits, lies in the Albigeois, in the southern Cévennes and, probably, northward in the Lot and Châtaigneraie areas (Briand and Gay 1978; Bogdanoff et al. 1984 a, b).
d) The western margin of another oceanic domain extends discontinuously (but perhaps initially continuously) within the Alpine External Crystalline Massifs and into the Maures-Corsica areas. Its present trend is roughly perpendicular to that of the central domain.

2. During Middle to Late Paleozoic Times the central domain either experienced terrestrial conditions or was covered by a shallow Devonian sea and finally by Carboniferous limnic deposits. However, Ardenne and Southern Montagne Noire, parts of the northern and southern platform, underwent strong subsidence during the Late Devonian and Early Carboniferous and became foreland basins.

1.2 Magmatic Data

1.2.1 Early Paleozoic: Cambrian to Silurian

1. In the Northern Platform and in its Southern Margin (i. e., the northern and central Armorican massif, the northern Vosges, and the Ardenne), Early Paleozoic magmatism is minor but tectonically significant. Mid-Cambrian calc-alkaline volcanism in the westernmost part of the Armorican Massif is generally considered to represent late Cadomian magmas. Most of the lavas and plutonic rocks of Ordovician and Silurian age in the Armorican massif, (i.e., at Erquy, Crozon, etc.), are tholeiitic or alkaline and characteristic of intraplate extension (Cabanis 1986). However, contemporary magmatic rocks differ on either sides of the northern platform. On its northern side, the Siluro-Devonian tholeiitic diabases of southern Ardenne, perhaps recording a back-arc site, are genetically related to the Caledonian orogeny of the Brabant Massif (Meilliez 1989). On the southern side, the Silurian magmatic suite of Saint-Georges-sur-Loire (Southeastern Armorican Massif) is interpreted as an arc and fore-arc association (Thieblemont et al. 1987, 1988).

In the southern platform (Montagne Noire), Mid-Ordovician lavas occur only in the Cabrières unit where Mid-Upper-Ordovician strata are preserved and in the Mouthoumet massif, between the Montagne Noire and the Pyrenees. In the Albigeois and Cévennes, which is considered to represent the northern margin of the Lower Paleozoic platform, magmatic activity is important and grows in volume northwards. It is expressed as the volcanic rocks and volcaniclastic sequences. The Lower Ordovician tholeiites of the Albigeois suggest crustal thinning (Marini 1988).

2. The Central Domain is characterized by a large volume of Early Paleozoic magmatic rocks. Most of these rocks have been subsequently affected by a polyphase metamorphic evolution, including HP-HT conditions. These mafic or bimodal magmatic associations include basalts, gabbros and ultramafic rocks (mainly garnet-peridotites) grouped in the leptynitic-amphibolitic complexes (LAC) (Santallier et al. 1988). Protolith ages are between 530 and 450 Ma.

Several tectonic settings have been recognized in the central domain magmatic series.
a) Oceanic crust. An ophiolite suite has been petrographically identified in the Limousin (Girardeau et al. 1986) and represents a remnant of oceanic crust of unknown dimensions. In central Limousin, these ophiolites have suffered HP metamorphism (Maillet 1987). In southwestern Limousin, however, another highly dismembered ophiolitic suite (the Génis series) has been affected by only a low grade metamorphism. Its structural position is not yet understood.
b) Magmatic arc and back-arc basins. Island arc and back-arc basin volcanics have been identified in Southern Armorican Massif [Saint-Martin-des-Noyers, Saint Georges sur Loire (Thieblemont et al. 1987, 1988)] and include transitional basalts, MORB, arc-tholeiites and andesites. Calc-alkaline associations in the western Massif Central (Bellac, Thiviers-Payzac) (Piboule 1979) may have a similar significance. Roughly contemporary magmatic series indicative of back-arc settings also characterize several areas of the Massif Central (Marvejols, Rouergue, Vivarais and Limousin) (Bodinier and Burg 1980; Cabanis et al. 1983; Briand et al. 1988).
c) Intracontinental rifts or extensional settings. These occur in Choletais, Sauviat (Thieblemont et al. 1987), and Lyonnais.

3. The Eastern Domain (Alpine External Crystalline Massifs, Maures and Corsica) shows strong evidence of Early Paleozoic magmatic activity with the intraplate volcanism of Maures (Seyler 1986) and Corsica, indicative of crustal thinning. This resulted locally in the opening of an ocean of unknown width, now recorded in the Chamrousse and Belledonne ophiolites (Pin and Carme 1987; Ménot 1988).

1.2.2 Late Paleozoic: Devonian to Permo-Carboniferous

In the northern platform evidence of the crustal extension is represented in the Devonian by the volcanic rocks of the Northern Armorican Massif (Cabanis 1986) (Morlaix, Plestin, Bolazec) and by the Bruche magmatic series in the northern Vosges. These rocks are contemporaneous with a rifting episode that culminated during Early and Middle Devonian times. Elsewhere, (e.g., Cornwall, U.K.) this extension resulted in the creation of a true oceanic domain, represented by the Lizard ophiolite. This oceanic domain has an exact equivalent in the Giessen-Harz nappes of Germany which have MOR-type basalts of Devonian age, overlain by condensed pelagics and Frasnian flysches (Franke 1989b, and pers. commun., 1990). Geophysical features (Bois et al., Lefort et al., Chaps. I.1, I.2, this Vol.) do not afford clear evidence for this suture beneath the Paris Basin, that could be located along the Metz-Taunus fault (Franke et al. 1989; Lefort et al., Chap. I.2, this Vol.). Some undated rocks of oceanic affinity (the so-called "klippes") that are found as olistoliths in the southern Vosges are thought to have a northern origin, since they are found within a northern-derived allochthonous synsedimentary unit (Piqué et al., Chap. IV.2, this Vol.). The magnetic anomaly of the Paris Basin has not yet been reached by any borehole. However, one of the hypotheses that have been proposed for this anomaly suggests it corresponds to mafic or ultramafic rocks emplaced during the Devonian (Autran et al. 1986).

During latest Devonian and Early Carboniferous times, the opening of the Chateaulin pull-apart basin in western Brittany (Rolet 1984) and the Bruche basin in northern Vosges was accompanied by the ascent of intracontinental tholeiitic lavas that preceeded the emplacement of various granitoids rocks.

In the central zone, following the HT metamorphism (see below) and the emplacement of synmetamorphic gabbros dated around 367 Ma in the Rouergue (Pin and Piboule 1988), anatectic, syntectonic melts were produced and crystallized during the late Devonian in the southern Armorican massif (e.g., Morbihan) through the Massif Central (e.g., Limousin) and into External Crystalline Massifs. They were followed by an important Carboniferous phase of post-tectonic granitoid emplacement with various compositions and ages. Very different magmatic activity, often mafic or bimodal, affects restricted areas of the central domain. It is represented by the magmatic and volcaniclastic series of the southern Vosges, Eastern Massif Central, Morvan and Brévenne (Ohnenstetter, this volume) and is related to local extensional regimes also reflected by the development of fault-induced sedimentary basins.

1.3 Metamorphic Data

1.3.1 Central and Eastern Domains

The oldest metamorphic event recognized in the Paleozoic belt is the Cambrian granulitic episode of the central German Black Forest. This metamorphic phase also affected the Vosges massif and other areas in the Paleozoic belt of Germany, and has been attributed there to an extensional regime during which the Leptynitic-Amphibolitic Complexes (LAC) were produced (Weber 1984; Piqué et al., Chap. IV.2, this Vol.). A second event of HP metamorphism affects scattered areas at the Ordovician-Silurian boundary and during the Silurian. HP-LT paragenesis characterize the Ile de Groix and Bois-de-Céné blueschists (Peucat 1984), the Champtoceaux white schists (Ballèvre et al. 1987) and the Najac glaucophane-bearing eclogites (Delor et al. 1986). They are attributed to some form of subduction processes. Other HP paragenesis here indicative of HT conditions, are found in many eclogites and granulites of the Massif Central and the Vendée (e.g., Les Essarts), in the allochthonous units of Lyonnais and Haut-Allier etc., but not in the parautochthonous units (see below). Most of the HP parageneses were later retrograded under MP conditions that are generally Devonian in age. During the Carboniferous, LP-HT metamorphism locally developed thermal domes where volumes of anatectic melt were important as in the Velay and in the axial zone of the Montagne Noire (Santallier et al., Chap. III.5, this Vol.).

In general, this polyphase metamorphic evolution suggests that the central domain, from the southern Armorican Massif to the western Massif Central, has been affected by Late Ordovician-Silurian subduction of oceanic crust and its associated continental margins. The higher temperatures yielded in eastern parts of the central domain suggest that the underlying crust in this region was hot oceanic or thinned continental. The subsequent Devonian metamorphic evolution records a collisional thickening of the crust and its progressive uplift.

The Eastern domain, especially the Alpine External Crystalline Massifs, shows a comparable evolution, from HP to LP conditions. This evolution ended with a HT-LP post-kinematic cordierite anatexis comparable to that of the Velay.

1.3.2 Northern Domain

The oldest metamorphic event in the northern domain is Late Devonian in age and associated in the western Armorican Massif with the Bretonic phase (Rolet, Chap. II.B.10, this Vol.). It was accompanied by partial melting (Martinez and Rolet 1988). In the central part of this massif, the major synorogenic metamorphism was low-grade and occurred at the end of Early Carboniferous. Similar ages have been found along the Bray fault in NW France (Matte et al. 1986), in the southern Ardenne area (Rocroi massif, Piqué et al. 1984) and in the northern Vosges. This metamorphic event accompanied regional deformation assigned in central Europe to the Sudetic Phase. Northwards, the metamorphism becomes progressively younger and its intensity decreases. The northern part of the Ardenne is affected by a post-Early Westphalian diagenetic or at most a very low-grade metamorphic event that is virtually coeval with the Asturian deformation.

1.3.3 Southern Domain

In the southern domain, the gneissic axial zone of the Montagne Noire and its envelopes were affected by a HP to MP metamorphism of debated age, followed by a Westphalian post-kinematic LP anatexis that is Stephanian in the Velay (Santallier et al., this volume) and in the External Alpine crystalline massifs. Elsewhere, Late Carboniferous metamorphism is low-grade and more generally only very low-grade. The Paleozoic rocks of the Pyrénées that share many aspects of the evolution of the Montagne Noire, were affected at the end of Carboniferous by a low-grade to high-grade LP metamorphism.

1.4 Structural Data

1.4.1 Central Domain

Structures contemporaneous with the high grade metamorphism of the central domain are not sufficiently well preserved to indicate unambiguously the geometry and kinematics of the Late Ordovician-Silurian deformation. However, observations in the Groix and Vendée areas suggest a roughly west-east to northwest-southeast displacement direction (Quinquis and Choukroune 1981).

Subsequent deformation is better constrained, especially in the Massif Central. Regionally, it corresponds to the emplacement of large-scale thrusts. These post-date the high-grade metamorphic episode because the Leptynitic Amphibolitic Complexes (LAC) that record this event are thrust upon units which show only rare relics or no relics at all of high-grade metamorphism. An often flat-lying polyphase foliation and a variably trending lineation (N-S, E-W or NW-SE) are associated with these southwards directed thrusts (Ledru et al. 1989; Quenardel and Lerouge 1990). They are coeval with the MP metamorphism and their age, although probably slightly variable, is Devonian. The main result of this deeply rooted thrusting event was the thickening of the crust. Subsequent crustal uplift, following the end of the Devonian, was marked by local extension and the development of extensional fabrics (Malavieille et al. 1990). Deformation of Late Devonian-Early Carboniferous sequences in the southern Vosges, in various areas of the Massif Central (Morvan, Brévenne etc.), and in the Vendée is thought to represent a transpressional episode.

1.4.2 Northern Domain

Except for the Late Devonian Bretonic phase (Rolet et al. 1986 a), deformation of the Northern domain is Carboniferous in age and broadly youngs towards the north. In the Armorican massif, the Cadomian block was affected only along its margins by a penetrative Paleozoic deformation. In Brittany, the synmetamorphic deformation developed in a dextral transcurrent regime, becoming increasingly pervasive towards its southern limit: the South Armorican Shear Zone (Jegouzo 1980). During Carboniferous times, this shear zone evolved from ductile to brittle conditions. In the northern Vosges, Early Carboniferous synmetamorphic deformation is characterized by a northward vergence sometimes combined with a dextral component. The Ardenne no longer belonged to the same block as the northern Vosges at that time (Early Carboniferous). It was a part of the Rheno-Hercynian zone, where dating of the associated metamorphism demonstrates the progressive northward younging of the synmetamorphic deformation, prior to the regional thrusting indicated by geophysical data (see below).

1.4.3 Southern Domain

In the Albigeois and Northern Montagne Noire, polyphase deformation developed during several main episodes of Late Devonian and Early Carboniferous folding and thrusting, the latest showing a southward vergence (Guérangé-Lozes 1987). In the Southern Montagne Noire, the synmetamorphic anchi- to epizonal structures are younger than those of the Albigeois and developed during mid-Carboniferous. The allochthonous MP-MT units situated beyond the northern and southern flanks of the gneissic Axial Zone are consequently thought to show a Late Devonian to mid-Carboniferous southward vergence. This could also be the case for the Axial Zone, although the importance of more or less ductile dextral motions along east-west-trending shear zones should not be underestimated (Bogdanoff et al. 1982, 1984 a, Echtler 1989).

1.5 Geophysical Data

1.5.1 Paleomagnetism

Although many paleomagnetic studies have been carried out on the Paleozoic rocks, few results are, to date, reliable enough to afford constraints on the tectonic evolution of the Paleozoic orogen. However, two broad generalities seem well established:

- the progressive separation of a microplate ("Armorica") from Gondwana, since Late Ordovician times, and its drift towards low latitudes (Perroud et al. 1984).
- important rotations, especially well documented in the Vosges massif, during the Late Paleozoic (Edel 1982).

1.5.2 Magnetism and Gravimetry

Magnetic and gravity investigations show major anomalies in the basement beneath the Paris basin: the Bray-Vittel lineament and the Paris Basin Magnetic Anomaly (PBMA) that could represent the trace of an undated (Upper Devonian?) rift. Other features are discussed by Lefort et al. (Chap. I.2, this Vol.).

1.5.3 Seismic Data

Seismic investigations are still in progress. To date, ECORS profiles beneath the Paris Basin suggest that the entire Ardenne massif is allochthonous (Bois et al., Chap. I.1, this Vol.). Thrusting of the Cadomian block is more contentious. However, some discontinuities, interpreted as thrust planes, are very important and could affect the lower crust.

2. Elements for a Synthetic Approach

2.1 Plate Tectonic Evolution of the French Paleozoic Orogen

2.1.1 Early Paleozoic

By the end of the Cadomian orogeny, most of present day France was established on continental crust, except for the internal zones which were already affected by latest Precambrian or Cambrian rifting. Evidence for sialic basement is afforded by exposures of pre-Paleozoic rocks in and around the "Cadomian block" (Fig. 1) where they record the effects of the Cadomian orogeny. Elsewhere, in the central Armorican Massif and in the Montagne Noire, only indirect evidence of continental crust occurs in the form of clastic facies of latest Precambrian and earliest Cambrian age and their shallow marine depositional setting. These areas formed a shelf (the northern and southern platforms) that was part of a continental plate. The first marine deposits represented (the "Calcaires à Archeocyathus" and overlying clays and siltstones) are very similar in their facies and associated fauna to the contemporary sequences of Morocco. Similarly, the Lower Ordovician "Grès armoricains" of Brittany and Normandy, a very characteristic and well-known arenaceous formation, is found practically unchanged in central Spain and corresponds to the Lower Ordovician clastic sediments of Morocco. The storm deposits and frequent bioturbation that characterize this formation suggest that it was deposited upon a shallow platform. Facies similarities between contemporary sequences of the Armorican Massif, the Montagne Noire and central Spain and Morocco, and the presence within the Cambrian and Lower Ordovician strata of fossils of Gondwanan affinity, suggest that all of these regions were connected at that time to Gondwana.

Paleomagnetic results (Perroud et al. 1984) and paleoclimatic observations (Robardet and Doré 1988) suggest that the northernmost part of the Gondwana plate separated from the rest of Gondwana by progressive rifting and drifted northwards, faster than the Gondwana plate itself. The resulting Gondwana-derived microplate, which has been called the Armorican microplate since it included the central and northern Armorican Massif, also included

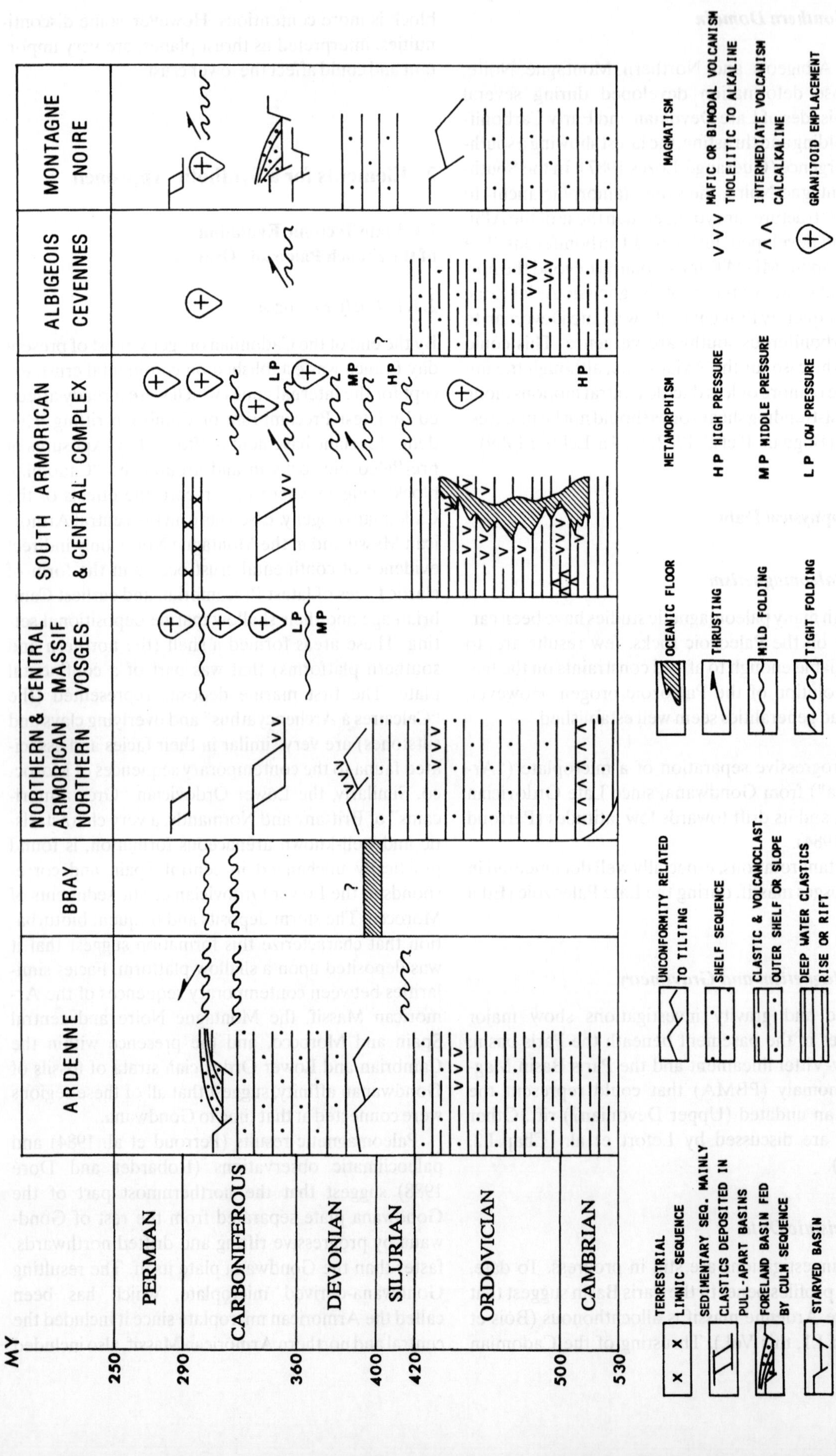

Fig. 2. Diagrammatic sketch of the French Paleozoic evolution

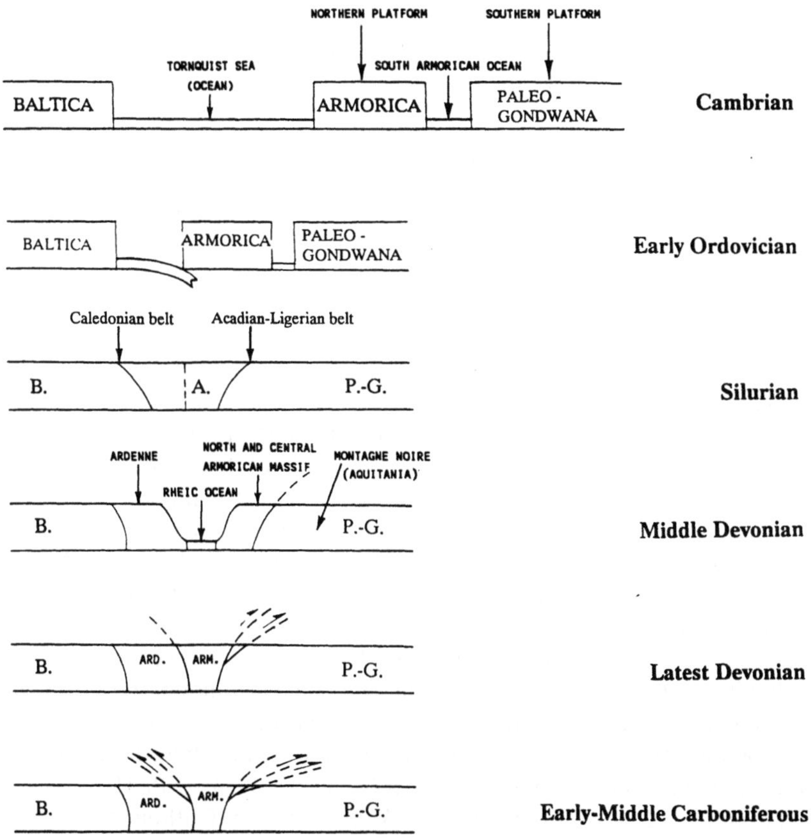

Fig. 3. Schematic plate tectonic model for the evolution of France during Paleozoic times

the other parts of France that preserved portions of the northern platform. This rifting may have started during the Late Precambrian and occurred throughout Cambrian times (Figs. 2, 3).

Early Paleozoic crustal extension is also recorded well within the developing platforms in the form of block tilting and within-plate magmatic activity. Facies comparisons between the Montagne Noire and the Albigeois suggest that the Montagne Noire platform, which was probably connected at that time to Paleo-Gondwana, was continued to the north by a continental margin in the present Albigeois area. If so, the central zone of the Central Massif could represent the continental rise and the oceanic domain that developed north of the Montagne Noire platform and the Albigeois margin. Late Paleozoic motions have obscured the original relationships between the northern and southern platforms. Nevertheless, it is clear that the central domain, (i. e., southern Brittany, the main part of the Massif Central and probably central and southern Vosges), was characterized during the Cambrian and Early Ordovician by strong crustal extension. However, within this complex area, it is not yet possible to distinguish between intracontinental rifts, passive margins, magmatic arcs and ocean floor, although each of these tectonic elements is likely to be represented. In fact, rather than forming a well-defined plate boundary, this domain was most likely composed of unequally thinned continental crust and perhaps several oceanic areas of unknown dimensions and importance. The absence of true ophiolitic complexes within eastern parts of this central domain, and the fact that the subsequent structural and metamorphic evolution indicates the presence of thinned continental crust in this region, suggest that the true oceanic rift zone was located in southern Brittany and western Central Massif, along a northwest-southeast trend (in present coordinates).

During this time, the southern platform (Montagne Noire, Aquitaine basin basement, and central Pyrénées) is thought to have remained part of the Paleo-Gondwana plate. An oceanic rift (or a continuation of the Massif Central basin?) is locally documented within the External Crystalline Massifs, where an ophiolitic sequence of Early Paleozoic age has been described (Carme 1975; Menot et al., Chap. IV.5, this Vol.). This rifting was coeval with and is an extension of the crustal thinning that affected the Maures massif (Bard and Caruba 1981).

North of the Armorican plate, a poorly known domain separated the Armorica and Baltica plates. This domain may have been a branch of the Tornquist sea which formed a presumed extension of the Iapetus ocean from Denmark to Poland (Ziegler 1984).

2.1.2 *Ordovician – Devonian*

The northernmost part of France and Belgium was affected by an orogeny usually known as the Caledonian, that developed during Late Ordovician and Silurian times. It is considered to be the result of the continental collision between Baltica and Armorica. At the same time, the South Armorican and the External Alps oceans started closing, later inducing Devonian continental collision.

This tectonic inversion is dated by blueschist metamorphism recorded in the western part of the central domain. There, oceanic lithosphere was subducted and a collisional event developed during Silurian and throughout Devonian times. Some structural arguments would call for an oblique sinistral collision along the suture located south of Brittany. During this event, remnants of oceanic lithosphere (ophiolites and continental rise deposits) were structurally amalgamated to form the Leptynitic Amphibolitic Complex (LAC) and subsequently thrust upon the former slope and outer shelf deposits. The ophiolitic and LAC sequences were affected by a HP eclogitic to granulitic metamorphism, that developed during crustal thickening. The deeply rooted thrusts are contemporaneous with MP metamorphism, dated as Devonian. Scarce structural data yielded north-south-directed thrusting whereas others show a SSW-NNE, E-W or even S-N vergence. Rotation of the shortening direction during Devonian times is thus probable, although it is not yet corroborated by any paleomagnetic data.

During the same period, the northern platform was submitted to crustal extension that resulted in the opening of a rift during the Early Middle Devonian. This oceanic rift separated the northern platform into the Rheno-Hercynian Zone to the north and the Saxo-Thuringian Zone to the south. As shown above, there is no direct evidence in France of this rift whose trace is hidden beneath the Mesozoic cover of the Paris Basin. Its existence is, however, inferred from the development of a passive margin at that time in the Southern Ardenne (Meilliez 1989), and from the existence of oceanic rocks in southwestern England and central Germany (Franke 1989b; Franke et al. 1989). On the basis of some geophysical arguments (Lefort et al., this volume), its trace in France could be found along the Metz fault (Fig. 1). In the absence of precise age control, it is not yet possible to define accurately the latest Devonian and Carboniferous deformation (including continental collisional structures) that followed the closure of the oceanic domain along the Alpine External Crystalline massifs, Maures and Corsica.

2.1.3 *Early Carboniferous*

During the Early Carboniferous, the closure of the Rheic ocean is likely to have been followed by the initiation of the crustal thrusting of the Cadomian block upon the Ardenne (Bois et al., this volume) and by the syn-metamorphic Late Visean deformation in the southern Ardenne. South of the Rheic suture, pull-apart basins such as like the Chateaulin and Laval basins, were opened by dextral transcurrent motion along the southern limit of the Cadomian block (Houlgatte et al. 1988). A dextral motion is also evidenced along the Bray fault (Matte et al. 1986). Farther south, the Late Devonian-Early Carboniferous basins that developed in the central domain, (i.e., the southern Vosges, Morvan, etc.) may also be pull-apart basins controlled by the same rejuvenation of old fractures. At the same time, in other parts of the central domain (west and northwest of Central Massif) the emplacement of dioritic to gabbrodioritic intrusions (Bouvier et Quenardel 1984; Peiffer 1985) is followed by calc-alkaline granites (Duthou et al. 1984). The starved basin of the Montagne Noire that had developed as a foreland basin at the front of the thickening central domain during the Late Devonian began to be filled by north-derived Culm turbidites.

2.1.4 *Mid-Carboniferous*

The Mid-Carboniferous was characterized by the emergence of much of the belt: the central domain and wide areas on both sides of it; to the south, the Albigeois-Cévennes and northern Montagne Noire; to the north, the central and northern Armorican massif and northern Vosges on the one hand, and southern Ardenne on the other hand, separated by deep crustal structures, namely the Bray-Vittel and Metz faults. Seaways were confined to two elongate troughs on both margins of the emerging belt, in the northern Ardenne and in the southern Montagne Noire and Pyrenees. These troughs filled rapidly during the Namurian and Early Westphalian with coal-bearing paralic sequences that are more or less coeval with the limnic sequences deposited in small basins within the central domain. The final closure of these residual foreland basins occurred in response to north-south directed folding and thrusting that is contemporaneous with the Asturian phase in Spain. Recent seismic profiles in the Ardenne and Paris basin (ECORS program, Bois et al., this volume) show that the thrusts which accompanied this Late Carboniferous deformation are more important than was previously believed. They affect the entire crust and define a similar crustal structure to that developed in the Himalaya range by ongoing continental collision.

2.1.5 Late Carboniferous

A Late Carboniferous rifting event, indicated by calc-alkaline intrusions in the Pyrénées, may have occurred in response to the motion of continental-scale faults. Elsewhere, a roughly north-south trend of Late Carboniferous shortening is revealed by the northern vergence in the north and southern vergence in the south of associated thrusts and by the reactivation of conjugate sets of transcurrent faults. During Late Carboniferous, the South Armorican Shear Zone shows successive sinistral and dextral displacements as the northern border fault of the Millevaches granitic massif, while the La Marche-Combrailles and Sillon Houiller shear zones only record sinistral movements (Quenardel et al. 1984; Lerouge and Quenardel 1985, 1988; Lerouge 1987). Dextral and sinistral faults were also partly active during the Stephanian within the "Late-Hercynian fault system" of Arthaud and Matte (1977).

Extensional structures of Permian age are recorded in the central domain, where they may reflect late-orogenic uplift. They have also been detected seismically beneath the Paris Basin.

2.2 Plate Tectonic Correlations Between the French Paleozoic Orogen and the Central European Variscan Belt

A simple approach to test regional correlations at the beginning of Paleozoic times is to determine which of the areas concerned, in Western Europe, northern Africa, and eastern North America were parts of, or close to, the stable Paleo-Gondwana plate. The various terranes which compose the Avalon lithostratigraphic zone (Williams 1978) of the eastern Appalachians are of Paleo-Gondwanan origin (Rast and Skehan 1983; Gromet 1989) and were probably parts of its northwestern margin. Consequently, they were close to the zones of France which share the same provenance, and all formed part of the same Armorica microplate. This is also the case for the northern zones of Central Europe which contain Gondwana-related facies such as Late Ordovician tillites (Robardet and Doré 1988).

Synthesis of the Variscan belt of Western Europe has long been based on the pattern proposed by Kossmat (1927), who subdivided the belt into several zones on the basis of their stratigraphic, sedimentological, and structural evolution. From north to south they include Rheno-Hercynian, Saxo-Thuringian and Moldanubian zones (Fig. 4). The main characteristics of each zone, as defined by Kossmat and refined by later workers (e.g., Behr et al. 1984; Franke 1989b), are very briefly summarized here. The Rheno-Hercynian zone of Germany is characterized by thick Devonian detrital sequences that were deposited upon a subsiding shelf and fed from the northern Caledonian continent. Locally, reef building and volcanic activity occurred within the shelf. At least, from the Lower Frasnian south-derived turbidites were deposited. By the end of the Namurian, this flysch had evolved progressively into paralic deposits, molasse, and coal sequences. The zone was affected by important north vergent thrusts one of which emplaced the allochthonous Giessen-Harz nappe which comprises sedimentary sequences and volcanic series of oceanic affinities.

In contrast to the Rheno-Hercynian zone, the Saxo-Thuringian zone is characterized by a Cambrian through mid-Ordovician neritic sequence which includes a bimodal volcanic suite, overlain by Late Ordovician to Middle Frasnian pelagic sequence represented mainly by shales. The floor of this pelagic basin was locally broken by ridges along which alkaline and tholeiitic lavas were erupted during the Frasnian. The top of sequence comprises hemi-pelagic carbonates and shales. Subsequent Frasnian through Upper Visean turbidites came from the southern Nappes (e.g., the Münchberg nappe, Behr et al. 1982) that derived from the southern boundary of the Saxo-Thuringian zone and were emplaced during the latest Visean.

In the Moldanubian zone, high-grade meta-sedimentary and meta-igneous sequences are broadly similar to the Leptynitic-Amphibolitic Complexes (LAC) of France. The timing of deposition and intrusion is bracketed between Latest Precambrian and Early to Middle Ordovician, whereas deformation and metamorphism occurred during the Late Ordovician and Silurian. According to recent summaries (e.g., Franke 1989b) this Moldanubian zone is divided in Germany and Czechoslovakia into several units, or microplates, among which is the Tepla-Barrandian unit. This is characterized by the presence of the Cadomian basement, a Lower and Middle Paleozoic meritic sedimentary sequence and a weak synmetamorphic deformation, except along its northern and southern limits, which are strongly deformed and thrust upon the Saxo-Thuringian and Moldanubian s. s. zones respectively.

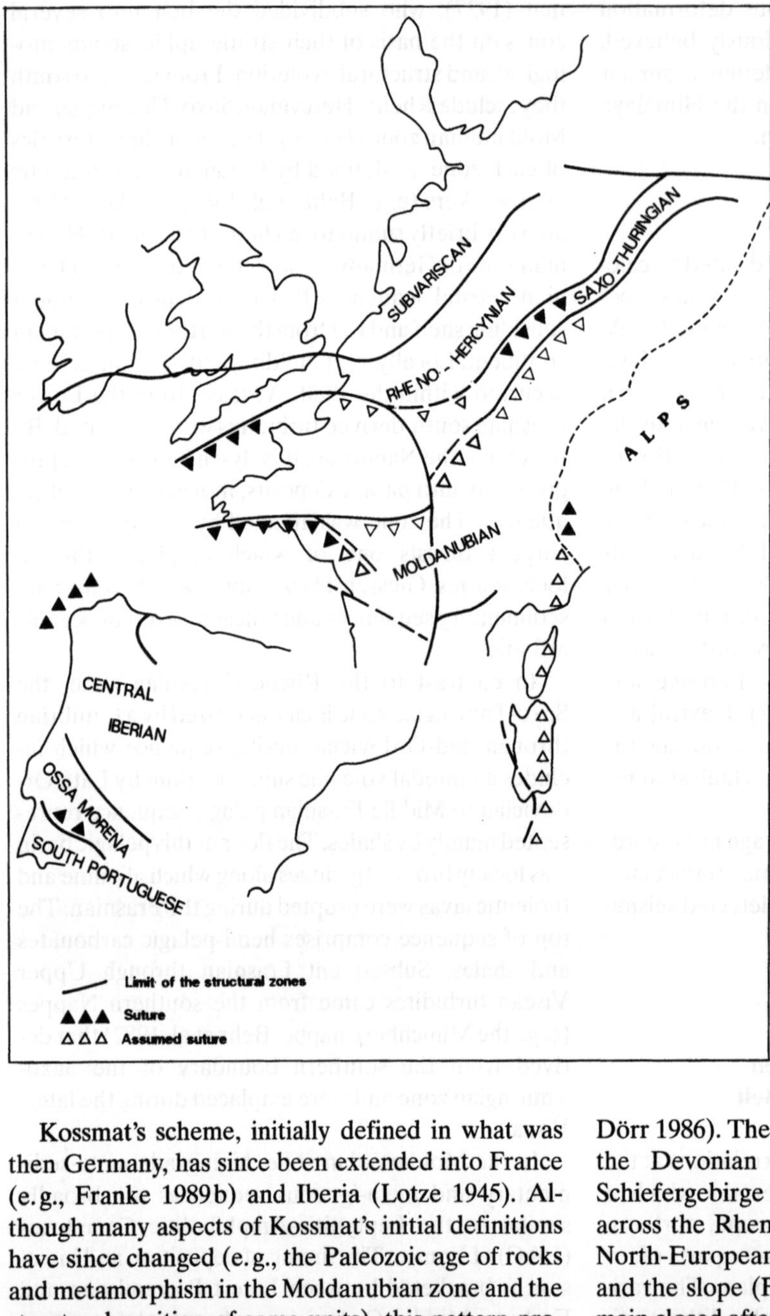

Fig. 4. Lithostructural zones of Western Europe

Kossmat's scheme, initially defined in what was then Germany, has since been extended into France (e.g., Franke 1989b) and Iberia (Lotze 1945). Although many aspects of Kossmat's initial definitions have since changed (e.g., the Paleozoic age of rocks and metamorphism in the Moldanubian zone and the structural position of some units), this pattern provides a useful framework in which to test models of paleo-plate tectonics.

In Germany, the Rheno-Hercynian zone comprises the Rheinisches Schiefergebirge and the Harz Mountains. It is separated from the Saxo-Thuringian zone to the south by a metamorphic lineament, the Mid-German Crystalline Rise, which is well defined in Germany but disappears westwards beneath the Mesozoic sedimentary cover of the Paris Basin. Recent studies of the Rheno-Hercynian zone north of this crystalline rise have shown that it contains rocks of oceanic affinity (Wedepohl et al. 1983; Grösser and Dörr 1986). The opening of this ocean is recorded in the Devonian paleogeography of the German Schiefergebirge which preserves from north to south across the Rheno-Hercynian zone, the shore of the North-European continent, the inner and outer shelf, and the slope (Franke et al. 1978). The oceanic domain closed after the end of Devonian. Although the ocean is poorly documented magmatically, its existence is nevertheless beyond doubt. It corresponds to the Rheic ocean whose suture is documented south of the Lizard zone (Le Gall and Darboux 1986) and is suspected beneath the Paris Basin. The Rheic ocean was roughly contemporaneous with another ocean or oceanic rift that developed between the Ossa Morena and South Portuguese zones of Iberia, the existence of which is documented by the oceanic nature of the Pulo de Lobo complex (Bard 1977; Munha et al. 1989).

During the Devonian the Rheic ocean in Germany separated the Rheno-Hercynian zone to the north from the Saxo-Thuringian zone to the south. The Ardenne represents the westernmost extension of the Schiefergebirge and, despite the absence of a widespread Devonian volcanism comparable with the Lahn-Dill magmatic activity in the Schiefergebirge (Wedepohl et al. 1983), shows a parallel Devonian and Carboniferous evolution. This involves the development of a passive margin to the south during the Devonian, followed by the initiation and the northward development of a deformation associated with the closure of the Rheic ocean.

South of this ocean, the German Saxo-Thuringian zone is generally thought to extend into the northern Vosges, the hidden Paleozoic basement beneath the Paris Basin, and into the northern and central Armorican massif. However, neither the Cadomian block (northern Armorican massif) nor the central Armorican massif show an evolution comparable to that of the Saxo-Thuringian zone of Germany. The shelf environment characteristic of the Armorican Ordovician is totally different from the deep facies characteristic of the Ordovician shales of the Saxo-Thuringian zone. For this reason, the Saxo-Thuringian zone is thought to terminate westwards under the Paris Basin along the Bray fault (Franke 1989b; Lefort et al., this volume). However, according to Edel (1990), the northern extension of the Vosges massif, hidden beneath the northwestern Paris basin, presents the same gravimetric signature as the Cadomian block and the Manche area. Therefore, rather than the Saxo-Thuringian zone, this Kraichgau-Saverne-Sarrebourg structure, and in extension in the Northern Vosges, could represent a distinct unit, comparable to the Cadomian block and the Tepla-Barrandian unit. On the other hand, the correlations are far better between the central Armorican massif and the Central Iberian zone where many stratigraphic sequences are essentially identical (Paris and Robardet 1977). It is clear that Kossmat's pattern, which is perhaps still reliable in Germany, should at least be used with caution in the deciphering of the Paleozoic plate tectonics of France.

The internal zones of France compare well with the "Moldanubian s. s." units of the Bohemian massif (Franke 1989b) which are composed of high-grade metamorphic rocks where the distinction between a Proterozoic basement and a Lower Paleozoic cover is not possible and, for some of them, have suffered Devonian tectonothermal events. The Leptynitic-Amphibolitic Complexes correspond to a complex domain where metamorphic rocks of oceanic affinity are structurally amalgamated with remnants of ancient passive margins, areas of thinned continental crust, and magmatic arcs. The Moldanubian zone, extending westward through the central and southern Vosges, the Central massif, the external Alpine crystalline massifs, and the southern Armorican massif, corresponds to a complex zone where, at our present stage of knowledge, several sutures could be tentatively trace (Fig. 4). The South Armorican "Acadian" suture, southeasterly trending from the southern Armorican massif to the northern part of Aquitaine Basin, an east-west-trending suture (?) north of the Massif Central (Lefort et al., this volume), and a south-southwest-trending suture, from the Alpine ECM to Corsica and, further, eastern Marocco (Piqué and Michard 1989). We do not know presently if these sutures are connected each to the others. In particular, the continuous trend of the South Armorican Suture from west to Southern Vosges and Germany is but a schematic representation of a probably more complicated scheme.

2.3 French Paleozoic Terranes?

The French Paleozoic orogen has been intensively studied since the last century and many of its exposures, stratotypes, and characteristic fossils assemblages are now classical in the international geological literature. New geochemical and geophysical techniques, and recent advances in structural geology and metamorphic petrology, have given us a rough idea of its geodynamic evolution. At the scale at which most orogenic belts, (e.g., the Appalachians and Caledonides), are studied, the French Paleozoic orogen is well mapped and interpreted. However, the detailed scale of investigations suggests very important complexities in regions of widespread Mesozoic cover and areas where the overprint of the Alpine orogeny is apparent. Knowledge of this belt is therefore sufficient to allow us to trace its evolution and examine it in terms of regional Paleozoic terranes. However, more precise analysis will lead to a more precise definition of these terranes and probable identification of small scale terranes, particularly in the central and eastern domains.

During the Early Paleozoic (Cambrian to Silurian) four terranes are identified at regional scale (Fig. 5):

1. Armorica. This is the northern platform and, in France, comprises the northern and central Armorican massif previously affected by the Cadomian orogeny, the northern basement of the Paris basin, the northern Vosges, and the Ardenne. Westwards, it extends into southern England and Ireland, and the Avalon zone of the northern Appalachians. Estwards, it corresponds to the classical Saxo-Thuringian and

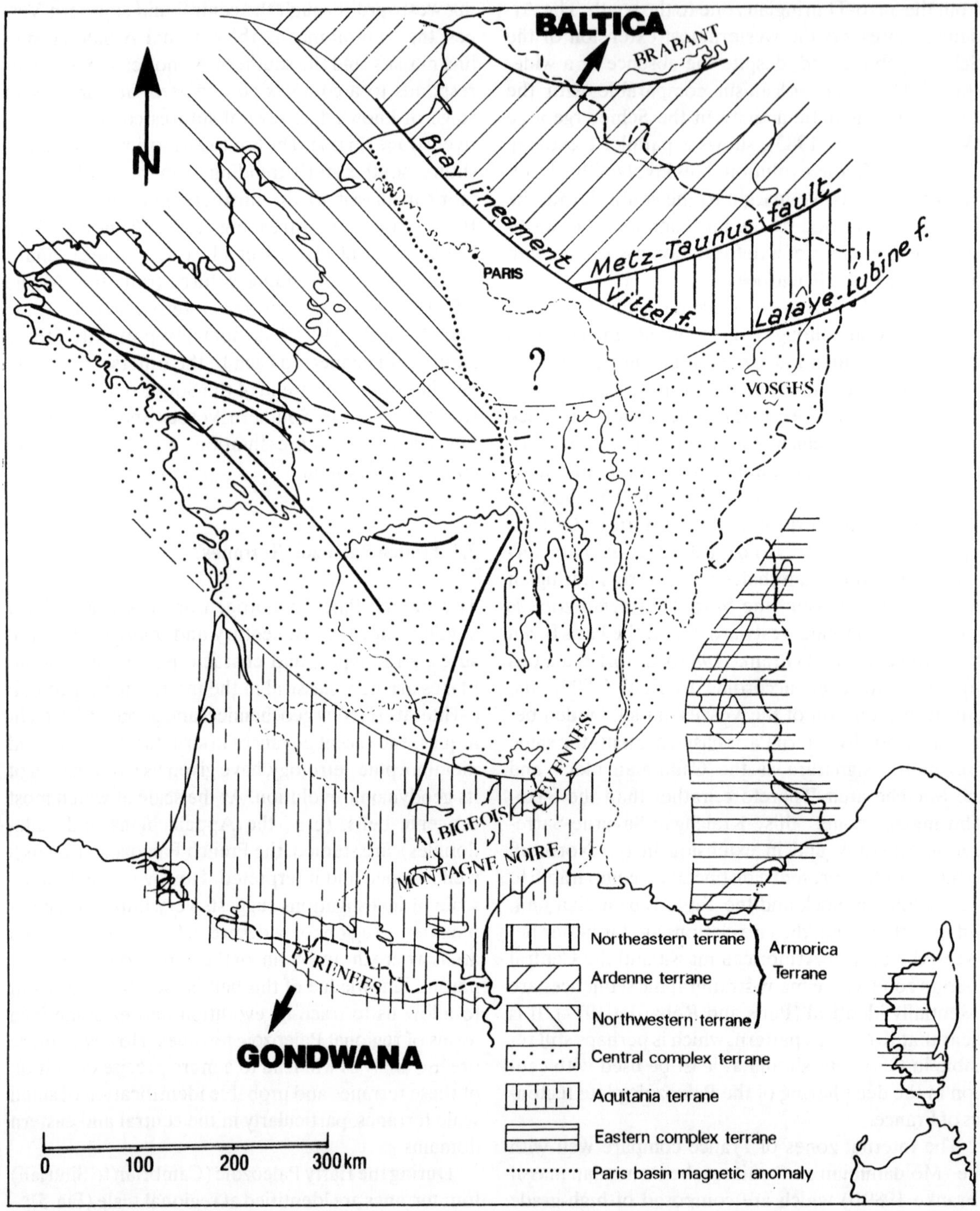

Fig. 5. French Paleozoic terranes

Rheno-Hercynian zones of Central Europe. It was separated from Baltica to the north by the Tornquist sea, a branch of the Iapetus ocean. To the south, it was connected to Paleo-Gondwana during Early Paleozoic times. It eventually collided with Baltica and, by the end of Silurian, Armorica and Baltica constitued a composite terrane across which Lower Devonian overstep sequences were deposited.

2. Aquitania. This is the southern platform and, in France, comprises the basement of the Aquitaine basin, the Pyrénées, the Montagne Noire and on its

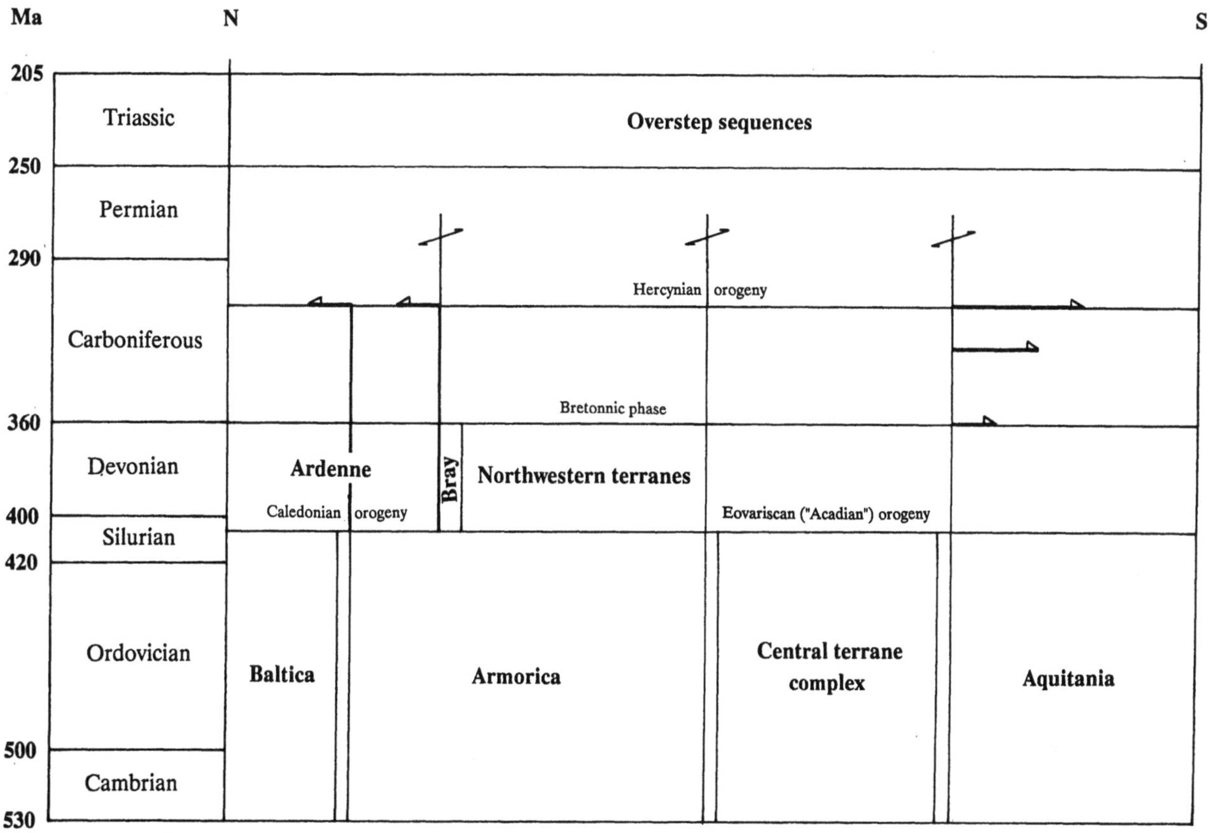

Fig. 6. Diagramatic sketch of the terrane evolution

northern margin, Albigeois and southern Cévennes. To the southwest, it comprises several zones of the Iberian peninsula. To the east, its extension was probably interrupted by the Corsica-Maures-Alpine ECM zone. Aquitania was connected to Gondwana during the Early and perhaps during all the Paleozoic. To the north it was separated from Armorica by the Complex Central Terrane.

3. Complex Central Terrane. This area which separates Armorica from Aquitania is here provisionally considered as a single terrane although further studies will probably identify several terranes within it. It includes the southern Armorican massif, most of the Massif Central, and probably the southern Vosges. Eastwards, it extends into the Moldanubian zone of Central Europe. To the west, it could be followed into the central Iberian Zone. It comprises distinct terranes as yet incompletely described and mapped. These include ophiolitic complexes, magmatic arcs, and remnant of continental platforms, that were structurally imbricated from the Silurian to Late Carboniferous.

4. Eastern Terrane. Like the Central terrane, with which it has numerous features in common, this is a complex mosaic of blocks indicative of various tectonic settings. The nature of the Paleozoic orogeny, and Alpine overprint in this area are not yet fully understood so any effort to delineate terranes would be premature.

During Late Silurian-Early Devonian time, continental convergence between Aquitania, Armorica and Baltica was completed and the entire French Paleozoic belt became a composite superterrane. Within this superterrane, continuation of lithospheric convergence as a result of intracontinental subduction processes caused large scale deformations, particularly thrusting, that isolated allochthonous slices of various sizes, each of which could be interpreted as an individual terrane. More important, however, is the crustal extension within the former Armorica terrane that was already active since Late Ordovician.

During the Early and Middle Devonian, the developing Rheic ocean divided the Armorica terrane into three terranes, that correspond respectively to the Rheno-Hercynian and Saxo-Thuringian zones of Central Europe, and the Northwestern Terrane, centered upon the northern and central Armorican massif. The nature, origin, and geodynamic significance of the Paris Basin Magnetic Anomaly (PBMA), is still unknown. However, the Ardenne, Northeastern, and Northwestern Terranes maintained their identity after the closure of the Rheic ocean, since they were involved in the Carboniferous deformation. At this time, the Northwestern and Northeastern terranes (and the Complex Central terrane?) were thrust upon the Ardenne terrane which, in turn, was thrust upon the Brabant. The Permo-Triassic red beds and the Jurassic strata overstep these terranes, as they also do for part of the southern and eastern terranes.

Acknowledgments. We thank Drs. P. Gell and D. Nance for rewriting part of the manuscript as for scientific improvement of the first draft of the paper. We are indebted to Dr. J. B. Edel for many helpful discussions. Dr. W. Franke is greatly acknowledged for his constructive criticism. Dr. D. Lemaire and G. Roche are thanked for their tireless help.

References

Arthaud F, Matte P (1977) Late Paleozoic strike-slip faulting in Southern Europe and northern Africa. Result of a right-lateral shear-zone between the Appalachians and the Urals. Geol Soc Am Bull 88: 1305–1320

Autran A, Castaing C, Debeglia N, Guillen A, Weber C (1986) Nouvelles contraintes géophysiques et géodynamiques pour l'interprétation de l'anomalie magnétique du bassin de Paris: hypothèse d'un rift paléozoïque refermé au Carbonifère. Bull Soc Géol Fr (8) 2: 125–141

Ballard J-L, Brun J-P, Durand J (1986) La discordance Briovérien-Paléozoïque inférieur en Bretagne centrale: signature d'une étape de distension ordovicienne. CR Acad Sci Paris II 303: 1327–1332

Ballèvre M, Kienast J-R, Paquette JL (1987) Le métamorphisme éclogitique dans la nappe hercynienne de Champtoceaux (Massif Armoricain) CR Acad Sci Paris II 305: 127–131

Bard J-P (1977) Signification tectonique des métatholéiites d'affinité abyssale de la ceinture métamorphique de basse pression d'Aracena (Huelva, Espagne). Bull Soc Géol Fr 19, 2: 385–393

Bard JP, Caruba C (1981) Les séries leptyno-amphiboliques à éclogites relictuelles et serpentinites des Maures, marqueurs d'une paléosuture varisque affectant une croûte amincie. CR Acad Sci Paris II 292: 611–614

Behr H-J, Engel W, Franke W (1982) Variscan wildflysh and nappe tectonics in the Saxothuringian zone (northeast Bavaria, West Germany). Am J Sci 282: 1438–1470

Behr H-J, Engel W, Franke W, Viese P, Weber K (1984) The Variscan belt in central Europe. Main structures, geodynamic implications, open questions. Tectonophysics 109: 15–40

Bodinier J-L, Burg J-P (1980) Evolution métamorphique et tectonique des séries cristallophylliennes du Rouergue occidental: mise en évidence d'un chevauchement dans la région de Najac (Aveyron). Bull Bur Rech Geol Min I 4: 315–339

Bogdanoff S, Donnot M, Quenardel J-M (1982) Carte géologique de France au 1:50000 Bédarieux. Bur Rech Geol Min, Orléans

Bogdanoff S, Donnot M, Ellenberger F (1984a) Notice de la carte géologique de France au 1:50000 Bédarieux. Bur Rech Geol Min, Orléans, 105 pp

Bogdanoff S, Cirodde JL, Donnot M (1984b) The nappes of La Chataigneraie, south-west Massif Central France. Tectonophysics 157: 69–79

Bouchardon J-L, Santallier D, Briand B, Ménot RP, Piboule M (1989) Eclogites in the French Paleozoic orogen: geodynamic significance. Tectonophysics 169: 317–332

Bouvier P, Quenardel J-M (1984) Principales caractéristiques des intrusions dévono-carbonifères du Massif Central nord occidental (France). 109ème Congr Nat Soc Sav, Dijon, 1, pp 227–237

Briand B, Gay M (1978) La série inverse de Saint Geniès d'Olt: évolution métamorphique et structurale. Bull Bur Rech Geol Min (2) I 3: 167–186

Briand B, Piboule M, Bouchardon J-L (1988) Diversité géochimique des métabasites des groupes leptyno amphiboliques du Rouergue et de Marvejols (Massif Central). Origine et implication. Bull Soc Géol Fr 8 IV 2: 489–498

Cabanis B (1986) Identification des séries magmatiques dans les socles métamorphiques, sur la base de critères géologiques pétrographiques et géochimiques. Thèse Doc Sci, Université Pierre et Marie Curie, Paris 6, 682 pp (unpublished)

Cabanis B, Guillot P-L, Santallier D, Jaffrezic H, Meyer G, Treuil M (1983) Apport des éléments traces à l'étude géochimique des métabasites du Bas-Limousin. Bull Soc Géol Fr 7, 25, 4: 563–574

Carme F (1975) Le massif basique de Chamrousse et de Tabor, une variante originale du complexe alpinotype et un jalon majeur d'une ceinture basique briovérienne. CR Acad Sci Paris II 281: 223–226

Delor C, Leyreloup A, Bodinier J-L, Burg J-P (1986) Découverte d'éclogites à glaucophane dans la klippe de Najac (Massif Central, France), nouveaux témoins océaniques d'un stade de haute pression dans la chaîne de collision varisque. CR Acad Sci Paris II 302: 739–744

Duthou J-L, Cantagrel J-M, Didier J, Vialette Y (1984) Paleozoic granitoids from the French Massif Central: age and origin studied by Rb/Sr system. Phys Earth Planet Interiors 35: 131–144

Echtler H (1989) Les nappes du versant sud dans l'évolution tectonique varisque de la Montagne Noire (Sud du Massif Central): de l'épaississement crustal à l'extension tardi varisque. Thèse Doc Sci, Université des Sciences et Techniques du Languedoc, Montpellier, 228 pp (unpublished)

Edel J-B (1982) Le socle varisque de l'Europe moyenne. Apports du magnétisme et de la gravimétrie. Sci Géol Bull 35: 207–224

Edel J-B (1987) Paleomagnetic evolution of the Massif Central (France) during the Carboniferous. Earth Planet Sci Lett 82: 180–192

Edel J-B (1990) The Variscan basement between the Paris basin and the Bohemian massif deduced from geophysical data. In: Weber K (ed) Abstr I. G. C. P. 233 meeting, Göttingen

Edel J-B, Coulon M, Hernot M-P (1984) Mise en évidence par le paléomagnétisme d'une importante rotation antihoraire des Vosges méridionales entre le Viséen terminal et le Westphalien supérieur. Tectonophysics 106: 239–257

Engel W, Franke W (1983) Flysch-sedimentation: its relations to tectonism in the European Variscides. In: Martin H, Eder FW (eds) Intracontinental Fold Belts. Springer, Berlin Heidelberg New York, pp 289–321

Engel W, Franke W, Grote C, Weber K, Ahrendt H, Eder FW (1983) Nappe tectonics in the Southeastern part of the Rheinisches Schiefergebirge. In: Martin H, Eder FW (eds) Intracontinental Fold Belts. Springer, Berlin Heidelberg New York, pp 267–287

Fluck P, Edel JB, Gagny C, Montigny R, Piqué A, Schneider JL,

Whitechurch H (1987) Le socle vosgien, segment de la chaine varisque d'Europe. Etat des connaissances, propositions d'axes de recherches. Doc Bur Rech Geol Min (Orléans) 146, 102 pp

Franke W (1989 a) Variscan plate tectonics in Central Europe – current ideas and open questions. Tectonophysics 169: 221–228

Franke W (1989 b) Tectonostratigraphic units in the Variscan belt of Central Europe. Geol Soc Am Sp paper 230: 67–90

Franke W, Engel W (1986) Synorogenic sedimentation in the Variscan belt of Europe. Bull Soc Géol Fr 8, 2: 25–33

Franke W, Eder W, Engel W, Langenstrassen F (1978) Main aspects of geosynclinal sedimentation in the Rhenohercynian zone. Z dtsch Geol Ges 129: 201–216

Franke W, Floyd PA, Holver M, Leveridge B (1989) The Rhenohercynian ocean revisited. EUG Abstract., Strasbourg, Terra cognita 365

Frisch W, Neubauer F (1989) Pre-Alpine terranes and tectonic zoning in the eastern Alps. Geol Soc Am Sp paper 230: 91–100

Gil Ibarguchi JI (1989) Age, tectonothermal evolution and palinspastic derivation of exotic terranes in northwest Iberia. In: Dallmeyer R (ed) I. G. C. P. 233 meeting, Tectonostratigraphic expression of terranes accretion in the circum-Atlantic Paleozoic orogen, Athens, July 1989. Abstract vol, pp 35–40

Giorgi L, Giraud P, Vachard D (1979) Sur la présence de microorganismes d'âge cambrien dans les schistes cristallins du versant occidental du massif cristallin externe des Grandes Rousses (Alpes occidentales). CR Acad Sci Paris 288 II: 1079–1082

Girardeau J, Dubuisson G, Mercier J-C (1986) Cinématique de mise en place des ophiolites et nappes cristallophylliennes du Limousin, Ouest du Massif Central français. Bull Soc Géol Fr 8, 2: 849–860

Gromet P (1989) Avalonian terranes and late Paleozoic tectonism in southeastern New-England; constraints and problems. Geol Soc Am Sp paper 230: 193–211

Grösser J, Dörr W (1986) MOR-Typ-Basalte im östlichen Rheinischen Schiefergebirge. Neues Jahrb Min Pal Monatsh 12: 705–712

Gueirard S, Waterlot G, Gherzi A, Samat M (1979) Sur l'âge llandoverien supérieur à tarannonien inférieur des schistes à graptolites du Fenouillet. Massif des Maures (Var). Bull Soc Géol Fr 8, XII: 195–199

Guérangé-Lozès J (1987) Les nappes varisques de l'Albigeois cristallin. Lithostratigraphie, volcanisme et déformations. Thèse Doc Sci, Univ. P. Sabatier, Toulouse, Doc Bur Rech Geol Min, Orléans, 135, 259 pp

Guérangé-Lozès J, Alsac C (1986) Les nappes varisques de l'Albigeois cristallin. Lithostratigraphie, volcanisme et déformation. Bull Bur Rech Geol Min, Géologie de la France 3: 309–337

Guillot P-L (1981) La série métamorphique du Bas-Limousin: de la vallée de l'Isle à la vallée de la Couille. Le socle en bordure du Bassin Aquitain. Thèse Doc Sci, Université d'Orléans, 557 pp (unpublished)

Houlgatte E, Le Herisse A, Pelhate A, Rolet J (1988) Evolution géodynamique du bassin carbonifère de Laval. Bull Bur Rech Geol Min Géologie de la France 1: 27–46

Jegouzo P (1980) The South Armorican Shear Zone. J Struct Geol 2, 1–2: 39–47

Julivert M, Martinez E, Ribeiro A (1980) The Iberian segment of the European Hercynian foldbelt. Mém Bur Rech Geol Min 108: 132–158

Kossmat F (1927) Gliederung des varistischen Gebirgsbaues. Abh Sächs Geol Landesamt, 1, 39 pp

Ledru P, Lardeaux J-M, Santallier D, Autran A, Quenardel J-M, Floc'h J-P, Lerouge G, Maillet N, Marchand J, Ploquin A (1989) Où sont les nappes dans le Massif Central français. Bull Soc Géol Fr 8, 3: 605–618

Le Gall B, Darboux JR (1986) Variscan strain pattern in the Paleozoic series at the Lizard front, Southwest England. Tectonics 5, 4: 589–606

Lerouge G (1984) Contribution à l'étude de la fracturation du Nord-Ouest du Massif Central et Sud du Bassin de Paris (France). Thèse Doc 3 ème cycle, Univ. Orléans, 2 vol, 394 pp (unpublished)

Lerouge G (1987) Tectogenèse comparée du deux segments de la chaîne hercynienne: le Massif Central français septentrional et le Sud du Massif armoricain. Thèse Doc Sci, Univ. Paris-Sud, Orsay, et Mém Géodiffusion, Paris, 1988, 2, 363 pp, 130 fig

Lerouge G, Quenardel J-M (1985) Chronologie des événements tectoniques dans le Nord-Ouest du Massif Central français et le Sud du bassin de Paris du Carbonifère inférieur au Plio-quaternaire. CR Acad Sci Paris 301 II 9: 621–626

Lerouge G, Quenardel J-M (1988) Les zones de cisaillement carbonifères dans les plutons vendéens et leurs prolongations dans le Nord-Ouest du Massif Central français. Bull Soc Géol Fr 8, 4, 5: 831–838

Lotze F (1945) Zur Gliederung der Varisziden der Iberischen Meseta. Geotekton Forsch 6: 78–82

Maillet N (1987) Dualité d'origine des massifs ultrabasiques limousins. Implications géodynamiques. Thèse Doc Sci, Univ. Lyon, 149 pp (unpublished)

Malavielle J, Guillot P, Costa S, Lardeaux J-M, Gardien V (1990) Collapse of the thickened Variscan crust in the French Massif Central: Mont Pilat extensionnal shear and Saint-Etienne upper Carboniferous basin. Tectonophysics 177: 139–149

Marini F (1988) "Phase" sarde et distension ordovicienne du domaine sudvarisque, effets de point chaud? Une hypothèse fondée sur les données nouvelles du volcanisme albigeois. CR Acad Sci Paris (II) 306: 443–450

Martinez F, Rolet J (1988) Late paleozoic metamorphism in the Northwestern Iberian peninsula, Brittany and related areas in SW Europe. In: Harris AL, Fettes DJ (eds) The Caledonian-Appalachian Orogen. Geol Soc Spec Publ 38: 611–620

Matte P (1986) La chaîne varisque parmi les chaînes paléozoïques periatlantiques, modèle d'évolution et position des grands blocs continentaux au Permo-Carbonifère. Bull Soc Géol Fr 8, 2: 9–24

Matte P, Respaut J-P, Maluski H, Lancelot J, Brunel M (1986) La faille NW-SE du Pays de Bray: un décrochement ductile dextre hercynien; déformation à 320 Ma d'un granite à 570 Ma dans le sondage Pays de Bray 201. Bull Soc Géol Fr 8, 11: 55–69

Meilliez F (1989) Importance de l'événement calédonien dans l'allochtone ardennais; essai sur une cinématique paléozoïque de l'Ardenne dans la chaîne varisque. Thèse Doc Sci, Université du Maine, Le Mans (inédit)

Ménot R-P (1988) The geology of the Belledonne massif: an overview (External Crystalline Massif of Eastern Alps). Schweiz Mineral Petrogr Mitt 68: 531–542

Munha J, Oliveira JT, Quesada C, Fonseca P, Castro P (1989) Acreted terranes in Southern Iberia: The Beja-acebuches ophiolite and related oceanic sequences. In: Dallmeyer R (ed) I. G. C. P. 233 meeting, Tectonostratigraphic expression of terranes accretion in the circum-Atlantic Paleozoic orogens, Athens, July 1989. Abstract vol, pp 67–69

Paris F, Robardet M (1977) Paléogéographie et relations ibéroarmoricaines au Paléozoïque anté-Carbonifère. Bull Soc Géol Fr 19: 1121–1126

Peiffer M-T (1984) La ligne tonalitique du Limousin. 10ème Réun. Ann Sci Terre, Bordeaux. Soc Geol France, Paris

Peiffer M-T (1985) La ligne tonalitique du Limousin. Sa composition, sa signification son implication dans la structuration varisque du Limousin. Thèse Doc Sci, Univ. Limoges, 585 pp (unpublished)

Perroud H, Van der Voo R, Bonhommet N (1984) Paleozoic evolution of the Armorica plate on the basis of paleomagnetic data. Geology 12: 579–582

Peucat J-J (1984) Rb/Sr and U/Pb dating of the blueschists of the Ile de Groix (France). In Blueschists and eclogites. Geol Soc Am Mem 164: 229–238

Piboule M (1979) L'origine des amphibolites. Approche géochimique et mathématique. Application aux amphibolites du Massif Central français. Thèse Doc Sci, Univ. Lyon, 1003 pp (unpublished)

Pin C, Carme F (1987) Vue d'ensemble sur le magmatisme préorogénique et l'évolution tectonique et métamorphique dans le sud de la chaîne de Belledonne (Massifs Cristallins Externes, Alpes françaises). CR Acad Sci Paris II 304: 1177–1181

Pin C, Piboule M (1988) Age dévonien supérieur de la série calcoalcaline de la ceinture basique du Lévezou (Rouergue). Un exemple de complexe leptyno-amphibolique composite. Bull Soc Géol Fr 8, 4: 261–266

Piqué A, Michard A (1989) Moroccan Hercynides: a synopsis. The Paleozoic sedimentary and tectonic evolution of the northern margin of West Africa. Am J Sci 289: 286–330

Piqué A, Huon S, Clauer N (1984) La schistosité hercynienne et le métamorphisme associé dans la vallée de la Meuse, entre Charleville-Mézières et Namur (Ardennes franco-belges). Bull Soc Belge Géol 93: 55–70

Quenardel J-M, Lerouge G (1990) Structural evolution of the French Massif Central. Abstract I. G. C. P. 233 meeting, Göttingen

Quenardel J-M, Lerouge G, Bouvier P, Guezou J-C, Langaney C, Lemaire D, Rolin P, Schmitt P (1984) Nature et évolution géologique du Massif Central septentrional à l'Ouest de l'Anomalie Magnétique du Bassin de Paris. G.P.F. 1, Thème 2, Doc Bur Rech Geol Min (Orléans) 81-2: 131–141

Quesada C (1989) Late-proterozoic terranes within the basement of the paleozoic iberian terrane. In: Dallmeyer R (ed) I. G. C. P. 233 meeting. Tectonostratigraphic expression of terranes accretion in the circum-atlantic paleozoic orogens, Athens, July 1989. Abstract vol, pp 83–86

Quinquis H, Choukroune P (1981) Les schistes bleus de l'Ile de Groix dans la chaîne hercynienne: implications cinématiques. Bull Soc Géol Fr (7) 23: 409–418

Rast N, Skehan J (1983) Evolution of the Avalonian plate. Tectonophysics 100: 257–286

Robardet M, Doré F (1988) The Late Ordovician diamictic formations from southwestern Europe: North Gondwana glaciomarine deposits. Paleogeogr Paleoclimatol Paleoecol 66: 19–31

Rolet J (1984) Grabens losangiques (pull-apart) en régime de décrochement. Le rôle des coulissements hercyniens dans l'individualisation des bassins carbonifères du Massif Armoricain. Ann Soc Géol Nord C III: 209–220

Rolet J, Le Gall B, Darboux J-R, Thonon P, Gravelle M (1986 a) L'évolution géodynamique dévono-carbonifère de l'extrémité occidentale de la chaîne hercynienne d'Europe sur le transect Armorique-Cornwall. Bull Soc Géol Fr 8 II 1: 43–54

Rolet J, Plusquellec Y, Babin C, Deunff J (1986 b) Famennian regression and Strunian grabens in the Armorican massif. A key-area: western Brittany. Ann Soc Géol Belg 109: 197–203

Santallier D, Briand B, Ménot R-P, Piboule M (1988) Les complexes leptyno-amphiboliques (C. L. A.): revue critique et suggestions pour un meilleur usage de ce terme. Bull Soc Géol Fr 8, 4: 3–12

Seyler M (1986) Petrology and genesis of Hercynian alkaline orthogneiss from Provence. France. J Petrol 27: 1229–1251

Thiéblemont D, Cabanis B, Wyns R, Treuil M (1987) Etude géochimique de la formation amphibolique de Saint-Martin des Noyers (complexe cristallophyllien des Essarts, Vendée). Mise en évidence d'un paléo-arc insulaire dans la partie vendéenne de la zone interne de l'orogène varisque. Bull Soc Géol Fr 8, 3: 159–166

Thiéblemont D, Triboulet C, Godard G (1988) Mineralogy petrology and P. T. path of Ca-Na amphibole assemblages. Saint-Martin des Noyers formation, Vendée, France. J Metamorphic Geol 6: 697–715

Vai GB, Cocozza T (1986) Tentative schematic zonation of the Hercynian chain in Italy. Bull Soc Géol Fr 8, 2: 95–114

Weber K (1984) Variscan events: early Paleozoic continental rift metamorphism and late Paleozoic crustal shortening. In: Hutton DH, Sanderson DJ (eds) Variscan tectonics of the North Atlantic region. Blackwell, Oxford, pp 3–22

Wedepohl KH, Meyer KD, Mueckeg KDJ (1983) Chemical composition and genetic relations of metavolcanic rocks from the Rheno-Hercynian belt of northwest Germany. In: Martin H, Eder FW (eds) Intracontinental fold belts. Springer, Berlin Heidelberg New York, pp 231–257

Williams H (1978) Tectonic lithofacies map of the Appalachian orogen. St. John's Memorial University of Newfoundland, Map I, Scale 1/1 000 000

Ziegler PA (1984) Caledonian and Hercynian crustal consolidation of western and central Europe. A working hypothesis. Geol Mijnbouw 63: 93–108

Subject Index

Springer-Verlag and the Environment

We at Springer-Verlag firmly believe that an international science publisher has a special obligation to the environment, and our corporate policies consistently reflect this conviction.

We also expect our business partners – paper mills, printers, packaging manufacturers, etc. – to commit themselves to using environmentally friendly materials and production processes.

The paper in this book is made from low- or no-chlorine pulp and is acid free, in conformance with international standards for paper permanency.

MIX
Papier aus verantwortungsvollen Quellen
Paper from responsible sources
FSC® C105338
FSC
www.fsc.org

If you have any concerns about our products,
you can contact us on
ProductSafety@springernature.com

In case Publisher is established outside the EU,
the EU authorized representative is:
Springer Nature Customer Service Center GmbH
Europaplatz 3, 69115 Heidelberg, Germany

Printed by Libri Plureos GmbH
in Hamburg, Germany